# Get the Most Out of
# MyStatLab® for Business Statistics

MyStatLab, Pearson's online homework, tutorial, and assessment program, creates personalized experiences for students and provides powerful tools for instructors. With a wealth of tested and proven resources, each course can be tailored to fit your specific needs. Talk to your Pearson representative about ways to integrate MyStatLab into your course for the best results.

## Learning in any Environment

- Because classroom formats and student needs continually change and evolve, MyStatLab has built-in flexibility to accommodate various course designs and formats.

- With a new, streamlined, mobile-friendly design, students and instructors can access courses from most mobile devices to work on exercises and review completed assignments.

# Career Readiness

Preparing students for careers is a priority. Now with the Open in Excel functionality in homework problems, students can open data sets directly in Excel and gain experience with the tools they will use in their careers.

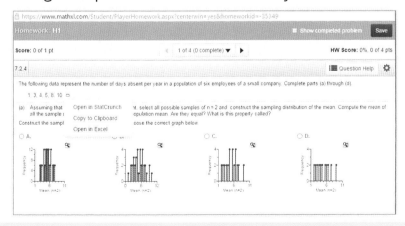

**Visit www.mystatlab.com and click Get Trained to make sure you're getting the most out of MyStatLab.**

## StatCrunch

Integrated directly into MyStatLab, StatCrunch is powerful, web-based statistical software that allows students to quickly and easily analyze data sets from their text and exercises.

In addition, MyStatLab includes access to www.StatCrunch.com, the full web-based program where users can access tens of thousands of shared data sets, create and conduct online surveys, interact with applets and simulations, and perform complex analyses using the powerful statistical software.

# Bring Statistics to Life

Use the StatCrunch Twitter app to analyze trending topics in real time. Easily load tweets using screen names or keywords to construct an interactive word wall.

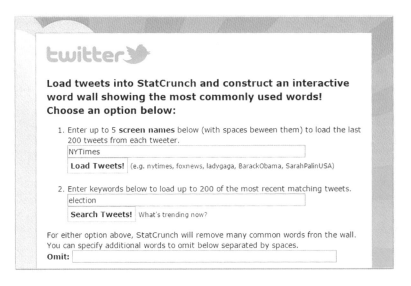

# BUSINESS
## Statistics
A Decision-Making Approach

TENTH EDITION

# BUSINESS
## Statistics

### A Decision-Making Approach

**David F. Groebner**

Boise State University, Professor Emeritus of Production Management

**Patrick W. Shannon**

Boise State University, Professor Emeritus of Supply Chain Management

**Phillip C. Fry**

Boise State University, Professor of Supply Chain Management

 Pearson

**Director, Portfolio Management:** Deirdre Lynch
**Portfolio Management Assistants:** Justin Billing and Jennifer Snyder
**Content Producer:** Kathleen A. Manley
**Managing Producer:** Karen Wernholm
**Media Producer:** Jean Choe
**Manager, Courseware QA:** Mary Durnwald
**Manager, Content Development:** Robert Carroll
**Product Marketing Manager:** Kaylee Carlson
**Product Marketing Assistant:** Jennifer Myers
**Senior Author Support/Technology Specialist:** Joe Vetere
**Cover and Text Design, Production Coordination, Composition:** Cenveo® Publisher Services
**Illustrations:** Laurel Chiapetta and George Nichols
**Cover Image:** Arthimedes/Shutterstock

**Library of Congress Cataloging-in-Publication Data**
Names: Groebner, David F., author. | Shannon, Patrick W., author. | Fry, Phillip C., author.
Title: Business statistics : a decision-making approach / David F. Groebner,
 Boise State University, Professor Emeritus of Production Management,
 Patrick W. Shannon, Boise State University, Professor Emeritus of Supply Chain
 Management, Phillip C. Fry, Boise State University, Professor of Supply Chain Management.
Description: Tenth edition. | Boston : Pearson, 2016. | Revised edition of Business
 statistics, 2014.
Identifiers: LCCN 2016016744 | ISBN 9780134496498 (hardcover) | ISBN
 0134496493 (hardcover)
Subjects: LCSH: Commercial statistics. | Statistical decision.
Classification: LCC HF1017 .G73 2016 | DDC 519.5–dc23
LC record available at https://lccn.loc.gov/2016016744

16 2019

 Pearson

ISBN-10:    0-13-449649-3
ISBN-13:    978-0-13-449649-8

To Jane and my family, who survived the process one more time.

DAVID F. GROEBNER

To Kathy, my wife and best friend; to our children, Jackie and Jason.

PATRICK W. SHANNON

To my wonderful family: Susan, Alex, Allie, Candace, and Courtney.

PHILLIP C. FRY

# About the Authors

**David F. Groebner, PhD,** is Professor Emeritus of Production Management in the College of Business and Economics at Boise State University. He has bachelor's and master's degrees in engineering and a PhD in business administration. After working as an engineer, he has taught statistics and related subjects for 27 years. In addition to writing textbooks and academic papers, he has worked extensively with both small and large organizations, including Hewlett-Packard, Boise Cascade, Albertson's, and Ore-Ida. He has also consulted for numerous government agencies, including Boise City and the U.S. Air Force.

**Patrick W. Shannon, PhD,** is Professor Emeritus of Supply Chain Operations Management in the College of Business and Economics at Boise State University. He has taught graduate and undergraduate courses in business statistics, quality management and lean operations and supply chain management. Dr. Shannon has lectured and consulted in the statistical analysis and lean/quality management areas for more than 30 years. Among his consulting clients are Boise Cascade Corporation, Hewlett-Packard, PowerBar, Inc., Potlatch Corporation, Woodgrain Millwork, Inc., J.R. Simplot Company, Zilog Corporation, and numerous other public- and private-sector organizations. Professor Shannon has co-authored several university-level textbooks and has published numerous articles in such journals as *Business Horizons, Interfaces, Journal of Simulation, Journal of Production and Inventory Control, Quality Progress,* and *Journal of Marketing Research.* He obtained BS and MS degrees from the University of Montana and a PhD in statistics and quantitative methods from the University of Oregon.

**Phillip C. Fry, PhD,** is a professor of Supply Chain Management in the College of Business and Economics at Boise State University, where he has taught since 1988. Phil received his BA. and MBA degrees from the University of Arkansas and his MS and PhD degrees from Louisiana State University. His teaching and research interests are in the areas of business statistics, supply chain management, and quantitative business modeling. In addition to his academic responsibilities, Phil has consulted with and provided training to small and large organizations, including Boise Cascade Corporation, Hewlett-Packard Corporation, the J.R. Simplot Company, United Water of Idaho, Woodgrain Millwork, Inc., Boise City, and Intermountain Gas Company.

# Brief Contents

# Contents

## CHAPTER 16  Analyzing and Forecasting Time-Series Data  *636*

## CHAPTER 17  Introduction to Nonparametric Statistics  *687*

## CHAPTER 18  Introducing Business Analytics  *718*

# Preface

In today's workplace, students can have an immediate competitive edge over both new graduates and experienced employees if they know how to apply statistical analysis skills to real-world decision-making problems.

Our intent in writing *Business Statistics: A Decision-Making Approach* is to provide an introductory business statistics text for students who do not necessarily have an extensive mathematics background but who need to understand how statistical tools and techniques are applied in business decision making.

This text differs from its competitors in three key ways:

1. Use of a direct approach with concepts and techniques consistently presented in a systematic and ordered way.

2. Presentation of the content at a level that makes it accessible to students of all levels of mathematical maturity. The text features clear, step-by-step explanations that make learning business statistics straightforward.

3. Engaging examples, drawn from our years of experience as authors, educators, and consultants, to show the relevance of the statistical techniques in realistic business decision situations.

Regardless of how accessible or engaging a textbook is, we recognize that many students do not read the chapters from front to back. Instead, they use the text "backward." That is, they go to the assigned exercises and try them, and if they get stuck, they turn to the text to look for examples to help them. Thus, this text features clearly marked, step-by-step examples that students can follow. Each detailed example is linked to a section exercise, which students can use to build specific skills needed to work exercises in the section.

Each chapter begins with a clear set of specific chapter outcomes. The examples and practice exercises are designed to reinforce the objectives and lead students toward the desired outcomes. The exercises are ordered from easy to more difficult and are divided into categories: Conceptual, Skill Development, Business Applications, and Computer Software Exercises.

This text places on data and how data are obtained. Many business statistics texts assume that data have already been collected. We have decided to underscore a more modern theme: Data are the starting point. We believe that effective decision making relies on a good understanding of the different types of data and the different data collection options that exist. To highlight our theme, we begin a discussion of data and data collection methods in Chapter 1 before any discussion of data analysis is presented. In Chapters 2 and 3, where the important descriptive statistical techniques are introduced, we tie these statistical techniques to the type and level of data for which they are best suited.

We are keenly aware of how computer software is revolutionizing the field of business statistics. Therefore, this textbook purposefully integrates Microsoft Excel throughout as a data-analysis tool to reinforce taught statistical concepts and to give students a resource that they can use in both their academic and professional careers.

## New to This Edition

- **Textual Examples:** Many new business examples throughout the text provide step-by-step details, enabling students to follow solution techniques easily. These examples are provided in addition to the vast array of business applications to give students a real-world, competitive edge. Featured companies in these new examples include Dove Shampoo and Soap, the Frito-Lay Company, Goodyear Tire Company, Lockheed Martin Corporation, the National Federation of Independent Business, Oakland Raiders NFL Football, Southwest Airlines, and Whole Foods Grocery.
- **More Excel Focus:** This edition features Excel 2016 with Excel 2016 screen captures used extensively throughout the text to illustrate how this highly regarded software is used as an aid to statistical analysis.
- **New Excel Features:** This edition introduces students to new features in Excel 2016, including Statistic Chart, which provides for the quick construction of histograms and box and whisker plots. Also, Excel has a new Data feature—Forecasting Sheet—for time-series forecasting, which is applied throughout this edition's forecasting chapter. Also new to this edition is the inclusion of the XLSTAT Excel add-in that offers many additional statistical tools.
- **New Business Applications:** Numerous new business applications have been included in this edition to provide students current examples showing how the statistical techniques introduced in this text are actually used by real companies. The new applications covering all business areas from accounting to finance to supply chain management, involve companies, products, and decision-making scenarios that are familiar to students. These applications help students understand the relevance of statistics and are motivational.
- **New Topic Coverage:** A new chapter, Introducing Business Analytics, is now a part of the textbook. This chapter introduces students to basic business intelligence and business analytics concepts and tools. Students are shown how they can use Microsoft's Power BI tool to analyze large data sets. The topics covered include loading data files into Power BI, establishing links between large data files, creating new variables and measures, and creating dashboards and reports using the Power BI tool.
- **New Exercises and Data Files:** New exercises have been included throughout the text, and other exercises have been revised and updated. Many new data files have been added to correspond to the new Computer Software Exercises, and other data files have been updated with current data.

- **Excel 2016 Tutorials:** Brand new Excel 2016 tutorials guide students in a step-by-step fashion on how to use Excel to perform the statistical analyses introduced throughout the text.
- **Improved Notation:** The notation associated with population and sample proportions has been revised and improved to be consistent with the general approach taken by most faculty who teach the course.
- **New Test Manual:** A new test manual has been prepared with well-thought-out test questions that correspond directly to this new edition.
- **MyStatLab:** The latest version of this proven student learning tool provides text-specific online homework and assessment opportunities and offers a wide set of course materials, featuring free-response exercises that are algorithmically generated for unlimited practice and mastery. Students can also use a variety of online tools to independently improve their understanding and performance in the course. Instructors can use MyStatLab's homework and test manager to select and assign their own online exercises and can import TestGen tests for added flexibility.

## Key Pedagogical Features

- **Business Applications:** One of the strengths of the previous editions of this textbook has been the emphasis on business applications and decision making. This feature is expanded even more in the tenth edition. Many new applications are included, and all applications are highlighted in the text with special icons, making them easier for students to locate as they use the text.
- **Quick Prep Links:** Each chapter begins with a list that provides several ways to get ready for the topics discussed in the chapter.
- **Chapter Outcomes:** At the beginning of each chapter, outcomes, which identify what is to be gained from completing the chapter, are linked to the corresponding main headings. Throughout the text, the chapter outcomes are recalled at the appropriate main headings to remind students of the objectives.
- **Clearly Identified Excel Functions:** Text boxes located in the left-hand margin next to chapter examples provide the Excel function that students can use to complete a specific test or calculation.
- **Step-by-Step Approach:** This edition provides continued and improved emphasis on providing concise, step-by-step details to reinforce chapter material.
  - **How to Do It** lists are provided throughout each chapter to summarize major techniques and reinforce fundamental concepts.
  - **Textual Examples** throughout the text provide step-by-step details, enabling students to follow solution techniques

easily. Students can then apply the methodology from each example to solve other problems. These examples are provided in addition to the vast array of business applications to give students a real-world, competitive edge.

- **Real-World Application:** The chapters and cases feature real companies, actual applications, and rich data sets, allowing the authors to concentrate their efforts on addressing how students apply this statistical knowledge to the decision-making process.
  - **Chapter Cases**—Cases provided in nearly every chapter are designed to give students the opportunity to apply statistical tools. Each case challenges students to define a problem, determine the appropriate tool to use, apply it, and then write a summary report.
- **Special Review Sections:** For Chapters 1 to 3 and Chapters 8 to 12, special review sections provide a summary and review of the key issues and statistical techniques. Highly effective flow diagrams help students sort out which statistical technique is appropriate to use in a given problem or exercise. These flow diagrams serve as a mini-decision support system that takes the emphasis off memorization and encourages students to seek a higher level of understanding and learning. Integrative questions and exercises ask students to demonstrate their comprehension of the topics covered in these sections.
- **Problems and Exercises:** This edition includes an extensive revision of exercise sections, featuring more than 250 new problems. The exercise sets are broken down into three categories for ease of use and assignment purposes:
  1. **Skill Development**—These problems help students build and expand upon statistical methods learned in the chapter.
  2. **Business Applications**—These problems involve realistic situations in which students apply decision-making techniques.
  3. **Computer Software Exercises**—In addition to the problems that may be worked out manually, many problems have associated data files and can be solved using Excel or other statistical software.
- **Computer Integration:** The text seamlessly integrates computer applications with textual examples and figures, always focusing on interpreting the output. The goal is for students to be able to know which tools to use, how to apply the tools, and how to analyze their results for making decisions.
  - **Microsoft Excel 2016** integration instructs students in how to use the Excel 2016 user interface for statistical applications.
  - **XLSTAT** is the Pearson Education add-in for Microsoft Excel that facilitates using Excel as a statistical analysis tool. XLSTAT is used to perform analyses that would otherwise be impossible, or too cumbersome, to perform using Excel alone.

# Resources for Success

## Student Resources

### Student Solutions Manual

The Student Solutions Manual contains worked-out solutions to odd-numbered problems in the text. It displays the detailed process that students should use to work through the problems. The manual also provides interpretation of the answers and serves as a valuable learning tool for students. Students can purchase this solutions manual by visiting www.mypearsonstore.com and searching for ISBN: 0-13-450646-4.

## MyStatLab™ Online Course (access code required)

MyStatLab from Pearson is the world's leading online resource for teaching and learning statistics, integrating interactive homework, assessment, and media in a flexible, easy-to-use format. MyStatLab is a course management system that helps individual students succeed.

- MyStatLab can be implemented successfully in any environment—lab-based, traditional, fully online, or hybrid—and demonstrates the quantifiable difference that integrated usage has on student retention, subsequent success, and overall achievement.

- MyStatLab's comprehensive gradebook automatically tracks students' results on tests, quizzes, homework, and in the study plan. Instructors can use the gradebook to provide positive feedback or intervene if students have trouble. Gradebook data can be easily exported to a variety of spreadsheet programs, such as Microsoft® Excel®.

MyStatLab provides **engaging experiences** that personalize, stimulate, and measure learning for each student. In addition to the resources below, each course includes a full interactive online version of the accompanying textbook.

- **Personalized Learning:** Not every student learns the same way or at the same rate. Personalized homework and the companion study plan allow your students to work more efficiently, spending time where they really need to.

- **Tutorial Exercises with Multimedia Learning Aids:** The homework and practice exercises in MyStatLab align with the exercises in the textbook, and most regenerate algorithmically to give students unlimited opportunity for practice and mastery. Exercises offer immediate helpful feedback, guided solutions, sample problems, animations, videos, statistical software tutorial videos, and eText clips for extra help at point of use.

- **Learning Catalytics™:** MyStatLab now provides Learning Catalytics—an interactive student response tool that uses students' smartphones, tablets, or laptops to engage them in more sophisticated tasks and thinking.

- **Videos** tie statistics to the real world.

  - **StatTalk Videos:** Fun-loving statistician Andrew Vickers takes to the streets of Brooklyn, NY, to demonstrate important statistical concepts through interesting stories and real-life events. This series of 24 fun and engaging videos will help students actually understand statistical concepts. Available with an instructor's user guide and assessment questions.

  - **Business Insight Videos** Ten engaging videos show managers at top companies using statistics in their everyday work. Assignable questions encourage discussion.

- **Additional Question Libraries:** In addition to algorithmically regenerated questions that are aligned with your textbook, MyStatLab courses come with two additional question libraries:

  - **450 exercises** in **Getting Ready for Statistics** cover the developmental math topics students need for the course. These can be assigned as a prerequisite to other assignments, if desired.

  - **Nearly 1,000 exercises** in the **Conceptual Question Library** require students to apply their statistical understanding.

- **StatCrunch™:** MyStatLab integrates the web-based statistical software StatCrunch within the online assessment platform so that students can

# Resources for Success

easily analyze data sets from exercises and the text. In addition, MyStatLab includes access to www.statcrunch.com, a vibrant online community where users can access tens of thousands of shared data sets, create and conduct online surveys, perform complex analyses using the powerful statistical software, and generate compelling reports.

- **Statistical Software, Support, and Integration:** We make it easy to copy our data sets, from both the eText and the MyStatLab questions, into software such as StatCrunch, Minitab®, Excel®, and more. Students have access to a variety of support tools—Technology Tutorial Videos, Technology Study Cards, and Technology Manuals for select titles—to learn how to effectively use statistical software.

## MyStatLab Accessibility

- MyStatLab is compatible with the JAWS screen reader, and enables multiple choice, fill-in-the-blank, and free-response problem types to be read and interacted with via keyboard controls and math notation input. MyStatLab also works with screen enlargers, including ZoomText, MAGic®, and SuperNova. And all MyStatLab videos accompanying texts with copyright 2009 and later have closed captioning.

- More information on this functionality is available at http://mystatlab.com/accessibility.

And, MyStatLab comes from an **experienced partner** with educational expertise and an eye on the future.

- Knowing that you are using a Pearson product means knowing that you are using quality content. That means our eTexts are accurate and our assessment tools work. It means we are committed to making MyStatLab as accessible as possible.

- Whether you are just getting started with MyStatLab or have a question along the way, we're here to help you learn about our technologies and how to incorporate them into your course.

To learn more about how MyStatLab combines proven learning applications with powerful assessment, visit www.mystatlab.com or contact your Pearson representative.

### Student Online Resources

Valuable online resources for both students and professors can be downloaded from www.pearsonhighered.com/groebner; these include the following:

- **Online Chapter—Introduction to Decision Analysis:** This chapter discusses the analytic methods used to deal with the wide variety of decision situations a student might encounter.

- **Online Chapter—Introduction to Quality and Statistical Process Control:** This chapter discusses the tools and techniques today's managers use to monitor and assess process quality.

- **Data Files:** The text provides an extensive number of data files for examples, cases, and exercises. These files are also located at MyStatLab.

- **Excel Simulations:** Several interactive simulations illustrate key statistical topics and allow students to do "what if" scenarios. These simulations are also located at MyStatLab.

## Instructor Resources

**Instructor Resource Center:** The Instructor Resource Center contains the electronic files for the complete Instructor's Solutions Manual, the Test Item File, and Lecture PowerPoint presentations (www.pearsonhighered.com/groebner).

- **Register, Redeem, Login:** At www.pearsonhighered.com/irc, instructors can access a variety of print, media, and presentation resources that are available with this text in downloadable, digital format.

- **Need help?** Our dedicated technical support team is ready to assist instructors with questions about the media supplements that accompany this text. Visit http://247pearsoned.com/ for answers to frequently asked questions and toll-free user-support phone numbers.

## www.mystatlab.com

# Resources for Success

## Instructor's Solutions Manual

The Instructor's Solutions Manual, created by the authors and accuracy checked by Paul Lorczak, contains worked-out solutions to all the problems and cases in the text.

## Lecture PowerPoint Presentations

A PowerPoint presentation is available for each chapter. The PowerPoint slides provide instructors with individual lecture outlines to accompany the text. The slides include many of the figures and tables from the text. Instructors can use these lecture notes as is or can easily modify the notes to reflect specific presentation needs.

## Test Item File

The Test Item File contains a variety of true/false, multiple choice, and short-answer questions for each chapter.

## TestGen®

TestGen® (www.pearsoned.com/testgen) enables instructors to build, edit, print, and administer tests using a computerized bank of questions developed to cover all the objectives of the text. TestGen is algorithmically based, allowing instructors to create multiple but equivalent versions of the same question or test with the click of a button. Instructors can also modify test bank questions or add new questions.

The software and test bank are available for download from Pearson's Instructor Resource Center.

# MathXL® for Statistics

# Online Course (access code required)

MathXL® is the homework and assessment engine that runs MyStatLab. (MyStatLab is MathXL plus a learning management system.)

## With MathXL for Statistics, instructors can

- Create, edit, and assign online homework and tests using algorithmically generated exercises correlated at the objective level to the textbook.
- Create and assign their own online exercises and import TestGen® tests for added flexibility.
- Maintain records of all student work, tracked in MathXL's online gradebook.

## With MathXL for Statistics, students can

- Take chapter tests in MathXL and receive personalized study plans and/or personalized homework assignments based on their test results.
- Use the study plan and/or the homework to link directly to tutorial exercises for the objectives they need to study.
- Access supplemental animations and video clips directly from selected exercises.
- Copy our data sets, both from the eText and the MyStatLab questions, into software like StatCrunch™, Minitab, Excel, and more.

MathXL for Statistics is available to qualified adopters. For more information, visit our website at www.mathxl.com, or contact your Pearson representative.

**www.mystatlab.com**

## Acknowledgments

Publishing this tenth edition of *Business Statistics: A Decision-Making Approach* has been a team effort involving the contributions of many people. At the risk of overlooking someone, we express our sincere appreciation to the many key contributors. Throughout the two years we have worked on this revision, many of our colleagues from colleges and universities around the country have taken time from their busy schedules to provide valuable input and suggestions for improvement. We would like to thank the following people:

Rob Anson, *Boise State University*

Paul Asunda, *Purdue University*

James Baldone, *Virginia College*

Al Batten, *University of Colorado – Colorado Springs*

Dave Berggren, *College of Western Idaho*

Robert Curtis, *South University*

Joan Donohue, *University of South Carolina*

Mark Gius, *Quinnipiac University*

Johnny Ho, *Columbus State University*

Vivian Jones, *Bethune-Cookman University*

Agnieszka Kwapisz, *Montana State University*

Joseph Mason, *Rutgers University – New Brunswick*

Constance McLaren, *Indiana State University*

Susan McLoughlin, *Union County College*

Jason Morales, *Microsoft Corporation*

Stefan Ruediger, *Arizona State University*

A special thanks to Professor Rob Anson of Boise State University, who provided useful comments and insights for Chapter 18, Introducing Business Analytics. His expertise in this area was invaluable.

Thanks, too, to Paul Lorczak, who error checked the manuscript and the solutions to every exercise. This is a very time-consuming but extremely important role, and we greatly appreciate his efforts.

Finally, we wish to give our utmost thanks and appreciation to the Pearson publishing team that has assisted us in every way possible to make this tenth edition a reality. Jean Choe oversaw all the media products that accompany this text. Mary Sanger of Cenveo expertly facilitated the project in every way imaginable and, in her role as production project manager, guided the development of the book from its initial design all the way through to printing. And finally, we wish to give the highest thanks possible to Deirdre Lynch, the Editor in Chief, who has provided valuable guidance, motivation, and leadership from beginning to end on this project. It has been a great pleasure to work with Deirdre and her team at Pearson.

—David F. Groebner

—Patrick W. Shannon

—Phillip C. Fry

# 1 The Where, Why, and How of Data Collection

## WHY YOU NEED TO KNOW

A transformation is taking place in many organizations involving how managers are using data to help improve their decision making. Because of the recent advances in software and database systems, managers are able to analyze data in more depth than ever before. Disciplines called **business analytics/business intelligence** and **data mining** are among the fastest-growing career areas. Data mining or knowledge discovery is an interdisciplinary field involving primarily computer science and statistics. While many data mining statistical techniques are beyond the scope of this text, most are based on topics covered in this course.

**Data Mining**
The application of statistical techniques and algorithms to the analysis of large data sets.

**Business Analytics/Business Intelligence**
The application of tools and technologies for gathering, storing, retrieving, and analyzing data that businesses collect and use.

---

### Quick Prep

**Locate** a recent copy of a business periodical, such as *The Economist, Fortune*, or *Bloomberg Businessweek*, and take note of the graphs, charts, and tables that are used in the articles and advertisements.

**Recall** any recent experiences you have had in which you were asked to complete a written survey or respond to a telephone survey.

**Make sure** that you have access to Excel software. Open Excel and familiarize yourself with the software.

Chapter 18 provides an overview of business analytics and introduces you to Microsoft analytics software called Microsoft Power BI. People working in this field are referred to as "data scientists." Doing an Internet search on data mining will yield a large number of sites that describe the field.

In today's workplace, you can have an immediate competitive edge over other new employees, and even those with more experience, by applying statistical analysis skills to real-world decision making. The purpose of this text is to assist in your learning and to complement your instructor's efforts in conveying how to apply a variety of important statistical procedures.

Cell phone companies such as Apple, Samsung, and LG maintain databases with information on production, quality, customer satisfaction, and much more. Amazon collects data on customers' online purchases and uses the data to suggest additional items the customer may be interested in purchasing. Walmart collects and manages massive amounts of data related to the operation of its stores throughout the world. Its highly sophisticated database systems contain sales data, detailed customer data, employee satisfaction data, and much more. Governmental agencies amass extensive data on such things as unemployment, interest rates, incomes, and education. However, access to data is not limited to large companies. The relatively low cost of computer hard drives with massive data storage capacities makes it possible for small firms and even individuals to store vast amounts of data on desktop computers. But without some way to transform the data into useful information, the data these companies have gathered are of little value.

Transforming data into information is where business statistics comes in—the statistical procedures introduced in this text are those that are used to help transform data into information. This text focuses on the practical application of statistics; we do not develop the theory you would find in a mathematical statistics course. Will you need to use math in this course? Yes, but mainly the concepts covered in your college algebra course.

Statistics does have its own terminology. You will need to learn various terms that have special statistical meaning. You will also learn certain dos and don'ts related to statistics. But most importantly, you will learn specific methods to effectively convert data into information. Don't try to memorize the concepts; rather, go to the next level of learning called *understanding*. Once you understand the underlying concepts, you will be able to *think statistically*.

Because data are the starting point for any statistical analysis, Chapter 1 is devoted to discussing various aspects of data, from how to collect data to the different types of data that you will be analyzing. You need to gain an understanding of the where, why, and how of data and data collection, because the remaining chapters deal with the techniques for transforming data into useful information.

---

### 1.1    What Is Business Statistics?

**Business Statistics**
A collection of procedures and techniques that are used to convert data into meaningful information in a business environment.

Articles in your local newspaper and on the Internet, news stories on television, and national publications such as *The Wall Street Journal* and *Fortune* discuss stock prices, crime rates, government-agency budgets, and company sales and profit figures. These values are statistics, but they are just a small part of the discipline called **business statistics**, which provides a wide variety of methods to assist in data analysis and decision making.

Business statistics can be segmented into two general categories. The first category involves the procedures and techniques designed to *describe data*, such as charts, graphs, and numerical measures. The second category includes tools and techniques that help decision makers *draw inferences* from a set of data. Inferential procedures include estimation and hypothesis testing. A brief discussion of these techniques follows.

## Descriptive Statistics

**BUSINESS APPLICATION**　Describing Data

**Independent Textbook Publishing, Inc.**　Independent Textbook Publishing, Inc. publishes 15 college-level texts in the business and social sciences areas. Figure 1.1 shows an Excel spreadsheet containing data for each of these 15 textbooks. Each column in the spreadsheet corresponds to a different factor for which data were collected. Each row corresponds to a different textbook. Many statistical procedures might help the owners describe these textbook data, including descriptive techniques such as *charts*, *graphs*, and *numerical measures*.

**FIGURE 1.1　Excel 2016 Spreadsheet of Independent Textbook Publishing, Inc.**

> **Excel 2016 Instructions:**
> 1. Open File: **Independent Textbook.xlsx.**

| | A | B | C | D | E | F | G |
|---|---|---|---|---|---|---|---|
| 1 | Book | Units Sold | Pages | Competing Books | Advertising Budget | Age of Author | Market Classification |
| 2 | 1 | 15000 | 176 | 5 | 25000 | 49 | Social Sciences |
| 3 | 2 | 140000 | 296 | 10 | 83000 | 57 | Business |
| 4 | 3 | 75000 | 483 | 7 | 40000 | 29 | Business |
| 5 | 4 | 100000 | 811 | 14 | 29000 | 37 | Social Sciences |
| 6 | 5 | 26000 | 302 | 9 | 52000 | 35 | Business |
| 7 | 6 | 33000 | 411 | 15 | 33000 | 43 | Business |
| 8 | 7 | 59000 | 333 | 7 | 19000 | 51 | Social Sciences |
| 9 | 8 | 103000 | 602 | 4 | 37000 | 62 | Business |
| 10 | 9 | 88000 | 504 | 12 | 51000 | 33 | Social Sciences |
| 11 | 10 | 10000 | 204 | 3 | 30000 | 50 | Business |
| 12 | 11 | 9000 | 376 | 4 | 19000 | 26 | Business |
| 13 | 12 | 77000 | 600 | 7 | 41000 | 40 | Business |
| 14 | 13 | 59000 | 400 | 3 | 26000 | 44 | Social Sciences |
| 15 | 14 | 183000 | 597 | 8 | 51000 | 59 | Business |
| 16 | 15 | 16000 | 126 | 1 | 27000 | 38 | Social Sciences |

**Charts and Graphs**　Chapter 2 will discuss many different charts and graphs—such as the one shown in Figure 1.2, called a *histogram*. This graph displays the shape and spread of the distribution of number of copies sold. The *bar chart* shown in Figure 1.3 shows the total number of textbooks sold broken down by the two markets, business and social sciences.

Bar charts and histograms are only two of the techniques that can be used to graphically analyze the data for the textbook publisher. In Chapter 2, you will learn more about these and other techniques.

**FIGURE 1.2　Histogram Showing the Copies Sold Distribution**

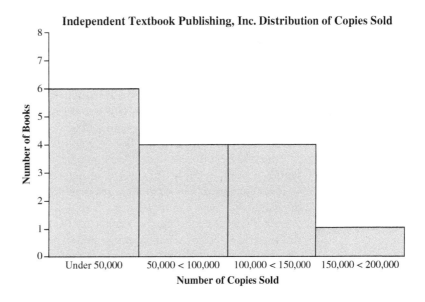

**FIGURE 1.3 Bar Chart Showing Copies Sold by Sales Category**

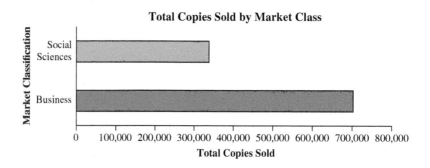

In addition to preparing appropriate graphs, you will compute a variety of numerical measures. Chapter 3 introduces the most important measures that are used along with graphs, charts, and tables to describe data.

## Inferential Procedures

Advertisers pay for television ads based on the audience level, so knowing how many viewers watch a particular program is important; millions of dollars are at stake. Clearly, the networks don't check with everyone in the country to see if they watch a particular program. Instead, they pay a fee to the Nielsen company (www.nielsen.com/), which uses **statistical inference procedures** to *estimate* the number of viewers who watch a particular television program.

**Statistical Inference Procedures**
Procedures that allow a decision maker to reach a conclusion about a set of data based on a subset of that data.

There are two primary categories of statistical inference procedures: *estimation* and *hypothesis testing*. These procedures are closely related but serve very different purposes.

**Estimation**   In situations in which we would like to know about all the data in a large data set but it is impractical to work with all the data, decision makers can use techniques to estimate what the larger data set looks like. These techniques arrive at estimates by looking closely at a subset of the larger data set.

For example, energy-boosting drinks such as Red Bull, Rockstar, Monster, and Full Throttle have become very popular among college students and young professionals. But how do the companies that make these products determine whether they will sell enough to warrant the product introduction? A typical approach is to do market research by introducing the product into one or more test markets. People in the targeted age, income, and educational categories (*target market*) are asked to sample the product and indicate the likelihood that they would purchase the product. The percentage of people who say that they will buy forms the basis for an *estimate* of the true percentage of *all* people in the target market who will buy. If that estimate is high enough, the company will introduce the product.

In Chapter 8, we will discuss the estimating techniques that companies use in new product development and many other applications.

**Hypothesis Testing**   Media advertising is full of product claims. For example, we might hear that "Goodyear tires will last at least 60,000 miles" or that "more doctors recommend Bayer Aspirin than any other brand." Other claims might include statements like "General Electric light bulbs last longer than any other brand" or "customers prefer McDonald's over Burger King." Are these just idle boasts, or are they based on actual data? Probably some of both! However, consumer research organizations such as Consumers Union, publisher of *Consumer Reports*, regularly test these types of claims. For example, in the hamburger case, *Consumer Reports* might select a sample of customers who would be asked to blind taste test Burger King's and McDonald's hamburgers, under the hypothesis that there is no difference in customer preferences between the two restaurants. If the sample data show a substantial difference in preferences, then the hypothesis of no difference would be rejected. If only a slight difference in preferences was detected, then *Consumer Reports* could not reject the hypothesis. Chapters 9 and 10 introduce basic hypothesis-testing techniques that are used to test claims about products and services using information taken from samples.

## 1.1 EXERCISES

### Skill Development

**1-1.** For the following situation, indicate whether the statistical application is primarily descriptive or inferential.

"The manager of Anna's Fabric Shop has collected data for 10 years on the quantity of each type of dress fabric that has been sold at the store. She is interested in making a presentation that will illustrate these data effectively."

**1-2.** Consider the following graph that appeared in a company annual report. What type of graph is this? Explain.

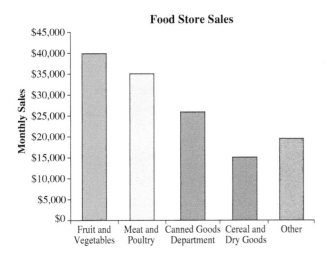

**Food Store Sales**

**1-3.** Review Figures 1.2 and 1.3 and discuss any differences you see between the histogram and the bar chart.

**1-4.** Think of yourself as working for an advertising firm. Provide an example of how hypothesis testing can be used to evaluate a product claim.

### Business Applications

**1-5.** Describe how statistics could be used by a business to determine if the dishwasher parts it produces last longer than a competitor's brand.

**1-6.** Locate a business periodical such as *Fortune* or *Forbes* or a business newspaper such as *The Wall Street Journal*. Find three examples of the use of a graph to display data. For each graph,
   a. Give the name, date, and page number of the periodical in which the graph appeared.
   b. Describe the main point made by the graph.
   c. Analyze the effectiveness of the graphs.

**1-7.** The human resources manager of an automotive supply store has collected the following data showing the number of employees in each of five categories by the number of days missed due to illness or injury during the past year.

| Missed Days | 0–2 days | 3–5 days | 6–8 days | 8–10 days |
|---|---|---|---|---|
| Employees | 159 | 67 | 32 | 10 |

Construct the appropriate chart for these data. Be sure to use labels and to add a title to your chart.

**1-8.** Suppose *Fortune* would like to determine the average age and income of its subscribers. How could statistics be of use in determining these values?

**1-9.** Locate an example from a business periodical or newspaper in which estimation has been used.
   a. What specifically was estimated?
   b. What conclusion was reached using the estimation?
   c. Describe how the data were extracted and how they were used to produce the estimation.
   d. Keeping in mind the goal of the estimation, discuss whether you believe that the estimation was successful and why.
   e. Describe what inferences were drawn as a result of the estimation.

**1-10.** Locate one of the online job websites and pick several job listings. For each job type, discuss one or more situations in which statistical analyses would be used. Base your answer on research (Internet, business periodicals, personal interviews, etc.). Indicate whether the situations you are describing involve descriptive statistics or inferential statistics or a combination of both.

---

### 1.2   Procedures for Collecting Data

We have defined business statistics as a set of procedures that analysts use to transform data into information. Before you learn how to use statistical procedures, it is important that you become familiar with different types of data collection methods.

**OUTCOME 1**   **Primary Data Collection Methods**

Many methods and procedures are available for collecting data. The following are considered some of the most useful and frequently used data collection methods:

- Experiments
- Telephone surveys
- Written questionnaires and online surveys
- Direct observation and personal interviews

---

**BUSINESS APPLICATION**    Experiments

**Food Processing**   A company often must conduct a specific experiment or set of experiments to get the data managers need to make informed decisions. For example, Con-Agra Foods, Inc., McCain Foods from Canada, and the J. R. Simplot Company are the primary suppliers of french fries to McDonald's in North America. These companies have testing facilities where they conduct experiments on their potato manufacturing processes. McDonald's has strict standards on the quality of the french fries it buys. One important attribute is the color of the fries after cooking. They should be uniformly "golden brown"— not too light or too dark.

French fries are made from potatoes that are peeled, sliced into strips, blanched, partially cooked, and then freeze-dried—not a simple process. Because potatoes differ in many ways (such as sugar content and moisture), blanching time, cooking temperature, and other factors vary from batch to batch.

Company employees start their **experiments** by grouping the raw potatoes into batches with similar characteristics. They run some of the potatoes through the line with blanch time and temperature settings at specific levels defined by an **experimental design**. After measuring one or more output variables for that run, employees change the settings and run another batch, again measuring the output variables.

Figure 1.4 shows a typical data collection form. The output variable (for example, percentage of fries without dark spots) for each combination of potato category, blanch time, and temperature is recorded in the appropriate cell in the table. Chapter 12 introduces the fundamental concepts related to experimental design and analysis.

**Experiment**
A process that produces a single outcome whose result cannot be predicted with certainty.

**Experimental Design**
A plan for performing an experiment in which the variable of interest is defined. One or more factors are identified to be manipulated, changed, or observed so that the impact (or influence) on the variable of interest can be measured or observed.

**FIGURE 1.4 Data Layout for the French Fry Experiment**

| Blanch Time | Blanch Temperature | Potato Category 1 | 2 | 3 | 4 |
|---|---|---|---|---|---|
| 10 minutes | 100 110 120 | | | | |
| 15 minutes | 100 110 120 | | | | |
| 20 minutes | 100 110 120 | | | | |
| 25 minutes | 100 110 120 | | | | |

---

**BUSINESS APPLICATION**    Telephone Surveys

**Public Issues**   Chances are that you have been on the receiving end of a telephone call that begins something like: "Hello. My name is Mary Jane and I represent the XYZ organization. I am conducting a survey on . . ." Political groups use telephone surveys to poll people about candidates and issues. Marketing research companies use phone surveys to learn likes and dislikes of potential customers.

Telephone surveys are a relatively inexpensive and efficient data collection procedure. Of course, some people will refuse to respond to a survey, others are not home when the calls come, and some people do not have home phones—they only have a cell phone—or cannot be reached by phone for one reason or another. Figure 1.5 shows the major steps in conducting a telephone survey. This example survey was run a number of years ago by a Seattle television station to determine public support for using tax dollars to build a new football stadium for the National Football League's Seattle Seahawks. The survey was aimed at property tax payers only.

Because most people will not stay on the line very long, the phone survey must be short—usually one to three minutes. The questions are generally what are called

**FIGURE 1.5 Major Steps for a Telephone Survey**

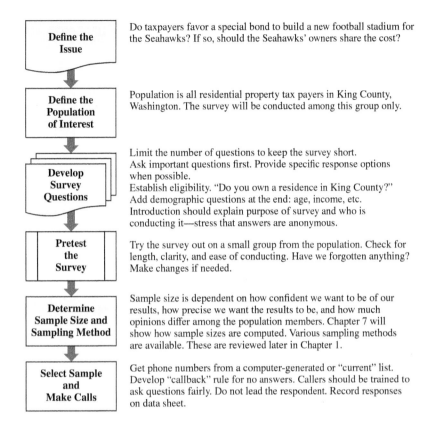

**Define the Issue** — Do taxpayers favor a special bond to build a new football stadium for the Seahawks? If so, should the Seahawks' owners share the cost?

**Define the Population of Interest** — Population is all residential property tax payers in King County, Washington. The survey will be conducted among this group only.

**Develop Survey Questions** — Limit the number of questions to keep the survey short. Ask important questions first. Provide specific response options when possible. Establish eligibility. "Do you own a residence in King County?" Add demographic questions at the end: age, income, etc. Introduction should explain purpose of survey and who is conducting it—stress that answers are anonymous.

**Pretest the Survey** — Try the survey out on a small group from the population. Check for length, clarity, and ease of conducting. Have we forgotten anything? Make changes if needed.

**Determine Sample Size and Sampling Method** — Sample size is dependent on how confident we want to be of our results, how precise we want the results to be, and how much opinions differ among the population members. Chapter 7 will show how sample sizes are computed. Various sampling methods are available. These are reviewed later in Chapter 1.

**Select Sample and Make Calls** — Get phone numbers from a computer-generated or "current" list. Develop "callback" rule for no answers. Callers should be trained to ask questions fairly. Do not lead the respondent. Record responses on data sheet.

**Closed-End Questions**
Questions that require the respondent to select from a short list of defined choices.

**Demographic Questions**
Questions relating to the respondents' characteristics, backgrounds, and attributes.

**closed-end questions**. For example, a closed-end question might be, "To which political party do you belong? Republican? Democrat? Or other?"

The survey instrument should have a short statement at the beginning explaining the purpose of the survey and reassuring the respondent that his or her responses will remain confidential. The initial section of the survey should contain questions relating to the central issue of the survey. The last part of the survey should contain **demographic questions** (such as gender, income level, education level) that will allow researchers to break down the responses and look deeper into the survey results.

A researcher must also consider the survey budget. For example, if you have $3,000 to spend on calls and each call costs $10 to make, you obviously are limited to making 300 calls. However, keep in mind that 300 calls may not result in 300 usable responses.

The phone survey should be conducted in a short time period. Typically, the prime calling time for a voter survey is between 7:00 P.M. and 9:00 P.M. However, some people are not home in the evening and will be excluded from the survey unless there is a plan for conducting callbacks.

Telephone surveys are becoming more problematic as more and more households drop their landlines in favor of cell phones, which makes it difficult to reach prospective survey responders. Additionally, many people refuse to answer if the caller ID is not a number they recognize.

**Written Questionnaires and Surveys** The most frequently used method to collect opinions and factual data from people is a written questionnaire. In some instances, the questionnaires are mailed to the respondents. In others, they are administered directly to the potential respondents. Written questionnaires are generally the least expensive means of collecting survey data. If they are mailed, the major costs include postage to and from the respondents, questionnaire development and printing costs, and data analysis. Online surveys are being used more frequently for written surveys now that software packages such as Survey Monkey are readily available. This technology eliminates postage costs and makes it

**FIGURE 1.6 Written Survey Steps**

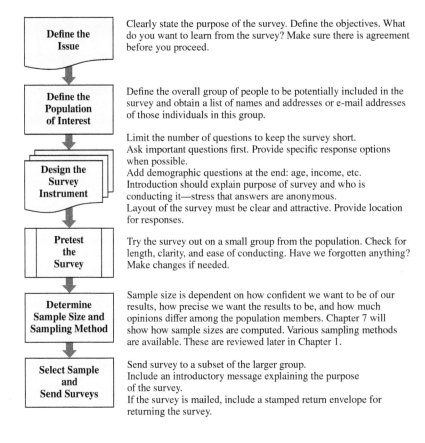

| **Define the Issue** | Clearly state the purpose of the survey. Define the objectives. What do you want to learn from the survey? Make sure there is agreement before you proceed. |
| **Define the Population of Interest** | Define the overall group of people to be potentially included in the survey and obtain a list of names and addresses or e-mail addresses of those individuals in this group. |
| **Design the Survey Instrument** | Limit the number of questions to keep the survey short. Ask important questions first. Provide specific response options when possible. Add demographic questions at the end: age, income, etc. Introduction should explain purpose of survey and who is conducting it—stress that answers are anonymous. Layout of the survey must be clear and attractive. Provide location for responses. |
| **Pretest the Survey** | Try the survey out on a small group from the population. Check for length, clarity, and ease of conducting. Have we forgotten anything? Make changes if needed. |
| **Determine Sample Size and Sampling Method** | Sample size is dependent on how confident we want to be of our results, how precise we want the results to be, and how much opinions differ among the population members. Chapter 7 will show how sample sizes are computed. Various sampling methods are available. These are reviewed later in Chapter 1. |
| **Select Sample and Send Surveys** | Send survey to a subset of the larger group. Include an introductory message explaining the purpose of the survey. If the survey is mailed, include a stamped return envelope for returning the survey. |

easier to format the data for statistical analysis. Figure 1.6 shows the major steps in conducting a written survey. Note how written surveys are similar to telephone surveys; however, written surveys can be slightly more involved and, therefore, take more time to complete than those used for a telephone survey. You still must be careful to construct a questionnaire that can be easily completed without requiring too much time.

A written survey can contain both closed-end and **open-end questions**. Open-end questions provide the respondent with greater flexibility in answering a question; however, the responses can be difficult to analyze. Note that telephone surveys can use open-end questions, too. However, the caller may have to transcribe a potentially long response, and there is risk that the interviewees' comments may be misinterpreted.

**Open-End Questions**
Questions that allow respondents the freedom to respond with any value, words, or statements of their own choosing.

Written surveys also should be formatted to make it easy for the respondent to provide accurate and reliable data. This means that proper space must be provided for the responses, and the directions must be clear about how the survey is to be completed. A written survey needs to be pleasing to the eye. How it looks will affect the response rate, so it must look professional.

You also must decide whether to manually enter or scan the data gathered from your written survey. The approach you take will affect the survey design. If you are administering a large number of surveys, scanning is preferred. It cuts down on data entry errors and speeds up the data gathering process. However, you may be limited in the form of responses that are possible if you use scanning.

If the survey is administered directly to the desired respondents, you can expect a high response rate. For example, you probably have been on the receiving end of a written survey many times in your college career, when you were asked to fill out a course evaluation form right in the classroom. In this case, most students will complete the form. On the other hand, if a survey is administered through the mail or online, you can expect a low response rate—typically 5% to 10% for mailed surveys. Although there are mixed findings about online survey response rates, some authors suggest that online response rates tend to be lower than rates for mailed surveys. (See A. Bryman, *Social Research Methods*, Fifth Edition, Oxford University Press, 2015.) Therefore, if you want 200 responses, you might need to distribute as many as 4,000 questionnaires.

Overall, written surveys can be a low-cost, effective means of collecting data if you can overcome the problems of low response. Be careful to pretest the survey and spend extra time on the format and look of the survey instrument.

Developing a good written questionnaire or telephone survey instrument is a major challenge. Among the potential problems are the following:

- Leading questions

    Example: "Do you agree with most other reasonably minded people that the city should spend more money on neighborhood parks?"

    Issue: In this case, the phrase "Do you agree" may suggest that you should agree. Also, since the question suggests that "most reasonably minded people" already agree, the respondent might be compelled to agree so that he or she can also be considered "reasonably minded."

    Improvement: "In your opinion, should the city increase spending on neighborhood parks?"

    Example: "To what extent would you support paying a small increase in your property taxes if it would allow poor and disadvantaged children to have food and shelter?"

    Issue: The question is ripe with emotional feeling and may imply that if you don't support additional taxes, you don't care about poor children.

    Improvement: "Should property taxes be increased to provide additional funding for social services?"

- Poorly worded questions

    Example: "How much money do you make at your current job?"

    Issue: The responses are likely to be inconsistent. When answering, does the respondent state the answer as an hourly figure or as a weekly or monthly total? Also, many people refuse to answer questions regarding their income.

    Improvement: "Which of the following categories best reflects your weekly income from your current job?
    _____Under $500    _____$500–$1,000
    _____Over $1,000"

    Example: "After trying the new product, please provide a rating from 1 to 10 to indicate how you like its taste and freshness."

    Issue: First, is a low number or a high number on the rating scale considered a positive response? Second, the respondent is being asked to rate two factors, taste and freshness, in a single rating. What if the product is fresh but does not taste good?

    Improvement: "After trying the new product, please rate its taste on a 1 to 10 scale with 1 being best. Also rate the product's freshness using the same 1 to 10 scale.
    _____Taste    _____Freshness"

The way a question is worded can influence the responses. Consider an example that occurred in 2008 that resulted from the sub-prime mortgage crisis and bursting of the real estate bubble. The bubble occurred because home prices were driven up due to increased demand by individuals who were lured into buying homes they could not afford. Many financial organizations used low initial interest rates and little or no credit screening to attract customers who later found they could not make the monthly payments. As a result, many buyers defaulted on their loans and the banks were left with abandoned homes and no way of collecting the money they had loaned out. Three surveys were conducted on the same basic issue. The following questions were asked:

"Do you approve or disapprove of the steps the Federal Reserve and Treasury Department have taken to try to deal with the current situation involving the stock market and major financial institutions?" (Dan Balz and Jon Cohen, "Economic fears give Obama clear lead over McCain in poll," www.washingtonpost.com, Sep. 24, 2008) 44% Approve—42% Disapprove—14% Unsure

"Do you think the government should use taxpayers' dollars to rescue ailing private financial firms whose collapse could have adverse effects on the economy and market, or is it not the government's responsibility to bail out private companies with taxpayer dollars?" (Doyle McManus, "Americans reluctant to bail out Wall Street," *Los Angeles Times*/Bloomberg Poll, Sep. 24, 2008) 31% Use Tax Payers' Dollars—55% Not Government's Responsibility—14% Unsure

"As you may know, the government is potentially investing billions to try and keep financial institutions and markets secure. Do you think this is the right thing or the wrong thing for the government to be doing?" (PewResearchCenter, www.people-press.org, Sep. 23, 2008) 57% Right Thing—30% Wrong Thing—13% Unsure

Note the responses to each of these questions. The way the question is worded can affect the responses.

**Direct Observation and Personal Interviews**   *Direct observation* is another procedure that is often used to collect data. As implied by the name, this technique requires researchers to actually observe the data collection process and then record the data based on what takes place in the process.

Possibly the most basic way to gather data on human behavior is to watch people. If you are trying to decide whether a new method of displaying your product at the supermarket will be more pleasing to customers, change a few displays and watch customers' reactions. If, as a member of a state's transportation department, you want to determine how well motorists are complying with the state's seat belt laws, place observers at key spots throughout the state to monitor people's seat belt habits. A movie producer, seeking information on whether a new movie will be a success, holds a preview showing and observes the reactions and comments of the movie patrons as they exit the screening. The major constraints when collecting observations are the amount of time and money required. For observations to be effective, trained observers must be used, which increases the cost. Personal observation is also time-consuming. Finally, personal perception is subjective. There is no guarantee that different observers will see a situation in the same way, much less report it the same way.

*Personal interviews* are often used to gather data from people. Interviews can be either **structured** or **unstructured**, depending on the objectives, and they can utilize either open-end or closed-end questions.

Regardless of the procedure used for data collection, care must be taken that the data collected are accurate and reliable and that they are the right data for the purpose at hand.

**Structured Interviews**
Interviews in which the questions are scripted.

**Unstructured Interviews**
Interviews that begin with one or more broadly stated questions, with further questions being based on the responses.

## Other Data Collection Methods

Data collection methods that take advantage of new technologies are becoming more prevalent all the time. For example, many people believe that Walmart is one of the best companies in the world at collecting and using data about the buying habits of its customers. Most of the data are collected automatically as checkout clerks scan the UPC bar codes on the products customers purchase. Not only are Walmart's inventory records automatically updated, but information about the buying habits of customers is also recorded. This allows Walmart to use *analytics* and *data mining* to drill deep into the data to help with its decision making about many things, including how to organize its stores to increase sales. For instance, Walmart apparently decided to locate beer and disposable diapers close together when it discovered that many male customers also purchase beer when they go to the store for diapers.

Bar code scanning is used in many different data collection applications. In a DRAM (dynamic random-access memory) wafer fabrication plant, batches of silicon wafers have bar codes. As the batch travels through the plant's workstations, its progress and quality are tracked through the data that are automatically obtained by scanning.

Every time you use your credit card, data are automatically collected by the retailer and the bank. Computer information systems are developed to store the data and to provide decision makers with procedures to access the data. For example, a number of years ago Target executives wanted to try marketing to pregnant women in their second trimester, which is when most expectant mothers begin buying products like prenatal vitamins and maternity clothing. (See Charles Duhigg, "How companies learn your secrets," *The New York Times Magazine*, Feb. 16, 2012.) If Target could attract women to buy these products, then once the baby was born, the women would be likely to buy many other products as well. But Target needed a way to know when a woman was in her second trimester. Analysts observed that women on their baby registry tended to buy certain products in larger amounts early on in their pregnancy and other products later in the pregnancy. They found that pregnant women also tended to purchase certain types of products such as washcloths closer to their delivery date. By applying statistical analytics to the data they collect on their customers, Target was able to identify about 25 products that, when analyzed together, allowed them to assign each shopper a "pregnancy prediction" score. They could also estimate her due date to within a small window, so Target could send coupons timed to very specific stages of her pregnancy.

In many instances, your data collection method will require you to use *physical measurement*. For example, the Andersen Window Company has quality analysts physically measure the width and height of its windows to assure that they meet customer specifications, and a state Department of Weights and Measures physically tests meat and produce scales to determine that customers are being properly charged for their purchases.

## Data Collection Issues

**Data Accuracy**    When you need data to make a decision, we suggest that you first see if appropriate data have already been collected, because it is usually faster and less expensive to use existing data than to collect data yourself. However, before you rely on data that were collected by someone else for another purpose, you need to check out the source to make sure that the data were collected and recorded properly.

Such organizations as Bloomberg, Value Line, and Fortune have built their reputations on providing quality data. Although data errors are occasionally encountered, they are few and far between. You really need to be concerned with data that come from sources with which you are not familiar. This is an issue for many sources on the World Wide Web. Any organization or any individual can post data to the web. Just because the data are there doesn't mean they are accurate. Be careful.

**Bias**
An effect that alters a statistical result by systematically distorting it; different from a random error, which may distort on any one occasion but balances out on the average.

**Interviewer Bias**    There are other general issues associated with data collection. One of these is the potential for **bias** in the data collection. There are many types of bias. For example, in a personal interview, the interviewer can interject bias (either accidentally or on purpose) by the way she asks the questions, by the tone of her voice, or by the way she looks at the subject being interviewed. We recently allowed ourselves to be interviewed at a trade show. The interviewer began by telling us that he would only get credit for the interview if we answered all of the questions. Next, he asked us to indicate our satisfaction with a particular display. He wasn't satisfied with our less-than-enthusiastic rating and kept asking us if we really meant what we said. He even asked us if we would consider upgrading our rating! How reliable do you think these data will be?

**Nonresponse Bias**    Another type of bias that can be interjected into a survey data collection process is called *nonresponse bias*. We stated earlier that mail surveys suffer from a high percentage of unreturned surveys. Phone calls don't always get through, people refuse to answer, or e-mail surveys are deleted. Subjects of personal interviews may refuse to be interviewed. There is a potential problem with nonresponse. Those who respond may provide data that are quite different from the data that would be supplied by those who choose not to respond. If you aren't careful, the responses may be heavily weighted by people who feel strongly one way or another on an issue.

**Selection Bias**    Bias can be interjected through the way subjects are selected for data collection. This is referred to as *selection bias*. A study on the virtues of increasing the student athletic fee at your university might not be best served by collecting data from students attending a football game. Sometimes, the problem is more subtle. If we do a telephone survey during the evening hours, we will miss all of the people who work nights. Do they share the same views, incomes, education levels, and so on as people who work days? If not, the data are biased.

Written and phone surveys and personal interviews can also yield flawed data if the interviewees *lie* in response to questions. For example, people commonly give inaccurate data about such sensitive matters as income. Lying is also an increasing problem with *exit polls* in which voters are asked who they voted for immediately after casting their vote. Sometimes, the data errors are not due to lies. The respondents may not know or have accurate information to provide the correct answer.

**Observer Bias**   Data collection through personal observation is also subject to problems. People tend to view the same event or item differently. This is referred to as *observer bias*. One area in which this can easily occur is in safety check programs in companies. An important part of behavioral-based safety programs is the safety observation. Trained data collectors periodically conduct a safety observation on a worker to determine what, if any, unsafe acts might be taking place. We have seen situations in which two observers will conduct an observation on the same worker at the same time, yet record different safety data. This is especially true in areas in which judgment is required on the part of the observer, such as the distance a worker is from an exposed gear mechanism. People judge distance differently.

**Measurement Error**   A few years ago, the authors worked as consultants to a wood window manufacturer. The company was having a quality problem with one of its saws. A study was developed to measure the width of boards that had been cut by the saw. Two people were trained to use digital calipers and record the data. This caliper is a U-shaped tool that measures distance (in inches) to three decimal places. The caliper was placed around the board and squeezed tightly against the sides. The width was indicated on the display. Each person measured 500 boards during an 8-hour day. When the data were analyzed, it looked like the widths were coming from two different saws; one set showed considerably narrower widths than the other. Upon investigation, we learned that the person with the narrower width measurements was pressing on the calipers much more firmly. The soft wood reacted to the pressure and gave narrower readings. Fortunately, we had separated the data from the two data collectors. Had they been merged, the measurement error might have gone undetected.

**Internal Validity**

A characteristic of an experiment in which data are collected in such a way as to eliminate the effects of variables within the experimental environment that are not of interest to the researcher.

**Internal Validity**   When data are collected through experimentation, you need to make sure that proper controls have been put in place. For instance, suppose a drug company such as Pfizer is conducting tests on a drug that it hopes will reduce cholesterol. One group of test participants is given the new drug, while a second group (a control group) is given a placebo. Suppose that after several months, the group using the drug saw significant cholesterol reduction. For the results to have **internal validity**, the drug company would have had to make sure the two groups were controlled for the many other factors that might affect cholesterol, such as smoking, diet, weight, gender, race, and exercise habits. Issues of internal validity are generally addressed by randomly assigning subjects to the test and control groups. However, if the extraneous factors are not controlled, there could be no assurance that the drug was the factor influencing reduced cholesterol. For data to have internal validity, the extraneous factors must be controlled.

**External Validity**

A characteristic of an experiment whose results can be generalized beyond the test environment so that the outcomes can be replicated when the experiment is repeated.

**External Validity**   Even if experiments are internally valid, you will always need to be concerned that the results can be generalized beyond the test environment. For example, if the cholesterol drug test had been performed in Europe, would the same basic results occur for people in North America, South America, or elsewhere? For that matter, the drug company would also be interested in knowing whether the results could be replicated if other subjects are used in a similar experiment. If the results of an experiment can be replicated for groups different from the original population, then there is evidence the results of the experiment have **external validity**.

An extensive discussion of how to measure the magnitude of bias and how to reduce bias and other data collection problems is beyond the scope of this text. However, you should be aware that data may be biased or otherwise flawed. Always pose questions about the potential for bias and determine what steps have been taken to reduce its effect.

## 1.2 EXERCISES

### Skill Development

**1-11.** If a pet store wishes to determine the level of customer satisfaction with its services, would it be appropriate to conduct an experiment? Explain.

**1-12.** Define what is meant by a leading question. Provide an example.

**1-13.** Briefly explain what is meant by an experiment and an experimental design.

**1-14.** Suppose a survey is conducted using a telephone survey method. The survey is conducted from 9 A.M. to 11 A.M. on Tuesday. Indicate what potential problems the data collectors might encounter.

**1-15.** For each of the following situations, indicate what type of data collection method you would recommend and discuss why you have made that recommendation:
   a. collecting data on the percentage of bike riders who wear helmets
   b. collecting data on the price of regular unleaded gasoline at gas stations in your state
   c. collecting data on customer satisfaction with the service provided by a major U.S. airline

**1-16.** Assume you have received a class assignment to determine the attitude of students in your school toward the school's registration process. What are the validity issues you should be concerned with?

### Business Applications

**1-17.** According to a December 2004 report in *AgResearch Magazine* titled "Putting out the fire" issued by the U.S. Department of Agriculture (USDA), the agency estimates that about 40% of all people who live in urban areas infested with imported fire ants are stung by the ants each year. What data collection method do you think was used to collect these data? Explain your answer.

**1-18.** Suppose you are asked to survey students at your university to determine if they are satisfied with the food service choices on campus. What types of biases must you guard against in collecting your data?

**1-19.** Briefly describe how new technologies can assist businesses in their data collection efforts.

**1-20.** Assume you have used an online service such as Orbitz or Travelocity to make an airline reservation. The following day, you receive an e-mail containing a questionnaire asking you to rate the quality of the experience. Discuss both the advantages and disadvantages of using this form of questionnaire delivery.

**1-21.** In your capacity as assistant sales manager for a large office products retailer, you have been assigned the task of interviewing purchasing managers for medium and large companies in the San Francisco Bay area. The objective of the interview is to determine the office product buying plans of the company in the coming year. Develop a personal interview form that asks both issue-related questions and demographic questions.

**1-22.** The regional manager for Macy's is experimenting with two new end-of-aisle displays of the same product. An end-of-aisle display is a common method retail stores use to promote new products. You have been hired to determine which is more effective. Two measures you have decided to track are which display causes the higher percentage of people to stop and, for those who stop, which causes people to view the display for a longer time. Discuss how you would gather such data.

**1-23.** In your position as general manager for United Fitness Center, you have been asked to survey the customers of your location to determine whether they want to convert the racquetball courts to an aerobic exercise space. The plan calls for a written survey to be handed out to customers when they arrive at the fitness center. Your task is to develop a short questionnaire with at least three "issue" questions and at least three demographic questions. You also need to provide the finished layout design for the questionnaire.

**1-24.** As part of a consulting project for a local ABC television affiliate, a survey was conducted with 744 respondents. Of those responding, 32% indicated that they prefer to watch local news on this station. How might this survey have been conducted, and what type of bias could occur if that data collection method was used?

---

**1.3**    **Populations, Samples, and Sampling Techniques**

**OUTCOME 2**

### Populations and Samples

**Population**
The set of all objects or individuals of interest or the measurements obtained from all objects or individuals of interest.

**Sample**
A subset of the population.

Two of the most important terms in statistics are **population** and **sample**.

The list of all objects or individuals in the population is referred to as the *frame*. Each object or individual in the frame is known as a sampling unit. The choice of the frame depends on what objects or individuals you wish to study and on the availability of the list of these objects or individuals. Once the frame is defined, it forms the list of sampling units. The next example illustrates this concept.

Populations and Samples

**U.S. Bank**   We can use U.S. Bank to illustrate the difference between a population and a sample. U.S. Bank is very concerned about the time customers spend waiting in the drive-up teller line. At a particular U.S. Bank, on a given day, 347 cars arrived at the drive-up.

A population includes measurements made on all the items of interest to the data gatherer. In our example, the U.S. Bank manager would define the population as the waiting time for all 347 cars. The list of these cars, possibly by license number, forms the frame. If she examines the entire population, she is taking a **census**. But suppose 347 cars are too many to track. The U.S. Bank manager could instead select a subset of these cars, called a *sample*. The manager could use the sample results to make statements about the population. For example, she might calculate the average waiting time for the sample of cars and then use that to conclude what the average waiting time is for the population. How this is done will be discussed in later chapters.

There are trade-offs between taking a census and taking a sample. Usually the main trade-off is whether the information gathered in a census is worth the extra cost. In organizations in which data are stored on computer files, the additional time and effort of taking a census may not be substantial. However, if there are many accounts that must be manually checked, a census may be impractical.

Another consideration is that the measurement error in census data may be greater than in sample data. A person obtaining data from fewer sources tends to be more complete and thorough in both gathering and tabulating the data. As a result, with a sample there are likely to be fewer human errors.

**Census**
An enumeration of the entire set of measurements taken from the whole population.

**Parameters and Statistics**   Descriptive numerical measures, such as an average or a proportion, that are computed from an entire population are called *parameters*. Corresponding measures for a sample are called *statistics*. Suppose, in the previous example, the U.S. Bank manager timed every car that arrived at the drive-up teller on a particular day and calculated the average. This population average waiting time would be a parameter. However, if she selected a sample of cars from the population, the average waiting time for the sampled cars would be a statistic. These concepts are more fully discussed in Chapters 3 and 7.

## Sampling Techniques

Once a manager decides to gather information by sampling, he or she can use a sampling technique that falls into one of two categories: **statistical** or **nonstatistical**.

Decision makers commonly use both nonstatistical and statistical sampling techniques. Regardless of which technique is used, the decision maker has the same objective—to obtain a sample that is a close representative of the population. There are some advantages to using a statistical sampling technique, as we will discuss many times throughout this text. However, in many cases, nonstatistical sampling represents the only feasible way to sample, as illustrated in the following example.

**Statistical Sampling Techniques**
Those sampling methods that use selection techniques based on chance selection.

**Nonstatistical Sampling Techniques**
Those methods of selecting samples using convenience, judgment, or other nonchance processes.

Nonstatistical Sampling

**Sun-Citrus Orchards**   Sun-Citrus Orchards owns and operates a large fruit orchard and fruit-packing plant in Florida. During harvest time in the orange grove, pickers load 20-pound sacks with oranges, which are then transported to the packing plant. At the packing plant, the oranges are graded and boxed for shipping nationally and internationally. Because of the volume of oranges involved, it is impossible to assign a quality grade to each individual orange. Instead, as the sacks move up the conveyor into the packing plant, a quality manager periodically selects an orange sack, grades the individual oranges in the sack as to size, color, and so forth, and then assigns an overall quality grade to the entire shipment from which the sample was selected.

**Convenience Sampling**
A sampling technique that selects the items from the population based on accessibility and ease of selection.

Because of the volume of oranges, the quality manager at Sun-Citrus uses a nonstatistical sampling method called **convenience sampling**. In doing so, the quality manager is willing to assume that orange quality (size, color, etc.) is evenly spread throughout the many sacks of oranges in the shipment. That is, the oranges in the sacks selected are of the same quality as those that were not inspected.

There are other nonstatistical sampling methods, such as *judgment sampling* and *ratio sampling*, that are not discussed here. Instead, the most frequently used statistical sampling techniques will now be discussed.

**Statistical Sampling**    *Statistical sampling* methods (also called *probability sampling*) allow every item in the population to have a known or calculable chance of being included in the sample. The fundamental statistical sample is called a *simple random sample*. Other types of statistical sampling discussed in this text include *stratified random sampling*, *systematic sampling*, and *cluster sampling*.

OUTCOME **3**

**BUSINESS APPLICATION**    Simple Random Sampling

**Cable ONE**    A salesperson at Cable ONE wishes to estimate the percentage of people in a local subdivision who have satellite television service (such as DIRECTV). The result would indicate the extent to which the satellite industry has made inroads into Cable ONE's market. The population of interest consists of all families living in the subdivision.

For this example, we simplify the situation by saying that there are only five families in the subdivision: James, Sanchez, Lui, White, and Fitzpatrick. We will let $N$ represent the population size and $n$ the sample size. From the five families ($N = 5$), we select three ($n = 3$) for the sample. Ten possible samples of size 3 could be selected:

{James, Sanchez, Lui}          {James, Sanchez, White}          {James, Sanchez, Fitzpatrick}

{James, Lui, White}          {James, Lui, Fitzpatrick}          {James, White, Fitzpatrick}

{Sanchez, Lui, White}          {Sanchez, Lui, Fitzpatrick}          {Sanchez, White, Fitzpatrick}

{Lui, White, Fitzpatrick}

Note that no family is selected more than once in a given sample. This method is called *sampling without replacement* and is the most commonly used method. If the families could be selected more than once, the method would be called *sampling with replacement*.

**Simple random sampling** is the method most people think of when they think of random sampling. In a correctly performed simple random sample, each possible sample would have an equal chance of being selected. For the Cable ONE example, a simplified way of selecting a simple random sample would be to put each sample of three names on a piece of paper in a bowl and then blindly reach in and select one piece of paper. However, this method would be difficult if the number of possible samples were large. For example, if $N = 50$ and a sample of size $n = 10$ is to be selected, there are more than 10 billion possible samples. Try finding a bowl big enough to hold those!

Simple random samples can be obtained in a variety of ways. We present two examples to illustrate how simple random samples are selected in practice.

**Simple Random Sampling**
A method of selecting items from a population such that every possible sample of a specified size has an equal chance of being selected.

**BUSINESS APPLICATION**    Random Numbers

**State Social Services**    Suppose the state director for a Midwestern state's social services system is considering changing the timing on food stamp distribution from once a month to once every two weeks. Before making any decisions, he wants to survey a sample of 100 citizens who are on food stamps in a particular county from the 300 total food stamp recipients in that county. He first assigns recipients a number (001 to 300). He can then use the random number function in Excel to determine which recipients to include in the sample. Figure 1.7 shows the results when Excel chooses 10 random numbers. The first recipient sampled is number 214, followed by 47, and so forth. The important thing to remember is that assigning each recipient a number and then randomly selecting a sample from those numbers gives each possible sample an equal chance of being selected.

Excel Tutorial

**Excel 2016 Instructions**

1. On the **Data** tab, click **Data Analysis.***
2. Select **Random Number Generation** option.
3. Set the **Number of Random Numbers** to 10.
4. Select **Uniform** as the distribution.
5. Define range as between 1 and 300.
6. Indicate that the results are to go in cell A1.
7. Click **OK**.

**FIGURE 1.7** **Excel 2016 Output of Random Numbers for State Social Services Example**

To convert the output to random numbers with zero decimal places, select the data in column A and on the Home tab in the Number group click the Decrease Decimal button several times times to display zero decimal places.

**Random Numbers Table**   If you don't have access to computer software such as Excel, you can determine the items in the population to be sampled by using the *random numbers table* in Appendix A. Begin by selecting a starting point in the random numbers table (row and digit). Suppose we use row 5, digit 8 as the starting point. Go down 5 rows and over 8 digits. Verify that the digit in this location is 1. If we ignore the blanks between columns that are there only to make the table more readable, the first three-digit number is 149. Recipient number 149 is the first one selected in the sample. Each subsequent random number is obtained from the random numbers in the next row down. For instance, the second number is 127. The procedure continues selecting numbers from top to bottom in each subsequent column. Numbers exceeding 300 and duplicate numbers are skipped. When you have found enough numbers for the desired sample size, the process is completed. Food-stamp recipients whose numbers are chosen are then surveyed.

---

**BUSINESS APPLICATION**    **Stratified Random Sampling**

**Federal Reserve Bank**   Sometimes, the sample size required to obtain a needed level of information from a simple random sampling may be greater than our budget permits. At other times, it may take more time to collect than is available. **Stratified random sampling** is an alternative method that has the potential to provide the desired information with a smaller sample size. The following example illustrates how stratified sampling is performed.

**Stratified Random Sampling**
A statistical sampling method in which the population is divided into subgroups called *strata* so that each population item belongs to only one stratum. The objective is to form strata such that the population values of interest within each stratum are as much alike as possible. Sample items are selected from each stratum using the simple random sampling method.

Each year, the Federal Reserve Board asks its staff to estimate the total cash holdings of U.S. financial institutions as of July 1. The staff must base the estimate on a sample. Note that not all financial institutions (banks, credit unions, and the like) are the same size. A majority are small, some are medium sized, and only a few are large. However, the few large institutions have a substantial percentage of the total cash on hand. To make sure that a simple random sample includes an appropriate number of small, medium, and large institutions, the sample size might have to be quite large.

As an alternative to the simple random sample, the Federal Reserve staff could divide the institutions into three groups called *strata*: small, medium, and large. Staff members could then select a simple random sample of institutions from each stratum and estimate the total cash on hand for all institutions from this combined sample. Figure 1.8 shows the stratified random sampling concept. Note that the combined sample size ($n_1 + n_2 + n_3$) is the sum of the simple random samples taken from each stratum.

The key behind stratified sampling is to develop a stratum for each characteristic of interest (such as cash on hand) that has items that are quite *homogeneous*. In this example, the size of the financial institution may be a good factor to use in stratifying. Here the combined

---

* If Data Analysis does not appear as an Analysis tool, you can download it using the following steps:

• PC users: On File tab, select Options > Add-Ins. Under Manage, select Excel Add-Ins with down arrow. Click Go and check the boxes next to Analysis ToolPak and Analysis ToolPak VBA. Click OK.
• MAC users: Go to www.support.microsoft.com and search "How to find and install Data Analysis ToolPak or Solver for Excel for Mac."

**FIGURE 1.8 Stratified Sampling Example**

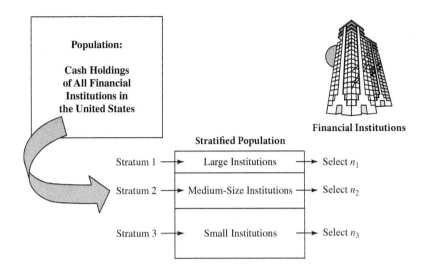

**FIGURE 1.8 Stratified Sampling Example**

sample size ($n_1 + n_2 + n_3$) will be less than the sample size that would have been required if no stratification had occurred. Because sample size is directly related to cost (in both time and money), a stratified sample can be more cost effective than a simple random sample.

Multiple layers of stratification can further reduce the overall sample size. For example, the Federal Reserve might break the three strata in Figure 1.8 into *substrata* based on type of institution: state bank, interstate bank, credit union, and so on.

Most large-scale market research studies use stratified random sampling. The well-known political polls, such as the Gallup and Harris polls, use this technique also. For instance, the Gallup poll typically samples between 1,800 and 2,500 people nationwide to estimate how more than 60 million people will vote in a presidential election. We encourage you to go to the Gallup and Harris websites for more information about their polling methods and many examples of the types of surveying they do (www.gallup.com and www.theharrispoll.com).

**BUSINESS APPLICATION**    ## Systematic Random Sampling

**State University Associated Students**    Student council officers at a mid-sized state university in the Northeast decided to survey fellow students on the legality of carrying firearms on campus. To determine the opinion of its 20,000 students, the council sent a questionnaire to a sample of 500 students. Although they could have used a simple random sample, they chose an alternative method called **systematic random sampling**.

**Systematic Random Sampling**
A statistical sampling technique that involves selecting every *k*th item in the population after a randomly selected starting point between 1 and *k*. The value of *k* is determined as the ratio of the population size over the desired sample size.

The university's systematic random sampling plan called for sending the questionnaire to every 40th student ($20,000/500 = 40$) from an alphabetic list of all students. The process could begin by using Excel to generate a single random number in the range 1 to 40. Suppose this value was 25. The 25th student in the alphabetic list would be selected. After that, every 40th student would be selected (25, 65, 105, 145, . . .) until there were 500 students selected.

Systematic sampling is frequently used in business applications. Use it as an alternative to simple random sampling only when you can assume the population is randomly ordered with respect to the measurement being addressed in the survey. In this case, students' views on firearms on campus are likely unrelated to the spelling of their last name.

**BUSINESS APPLICATION**    ## Cluster Sampling

**Oakland Raiders Football Team**    The Oakland Raiders of the National Football League plays its home games at O.co (formerly Overstock.com) Coliseum in Oakland, California. Despite its struggles to win in recent years, the team has a passionate fan base. Consider a situation in which an outside marketing group was retained by the Raiders to interview season ticket holders about the potential for changing how season ticket pricing is structured. The seating chart at the O.co Coliseum is available on the Oakland Raiders website (www.raiders.com).

The marketing firm plans to interview season ticket holders just prior to home games during the current season. One sampling technique is to select a simple random sample of size $n$ from the population of all season ticket holders. Unfortunately, this technique would likely require that interviewer(s) go to each section in the stadium to do the interviews. This would prove to be an expensive and time-consuming process. A systematic or stratified sampling procedure also would probably require visiting each section in the stadium. The geographical spread of those being interviewed in this case causes problems.

A sampling technique that overcomes the geographical spread problem is **cluster sampling**. The stadium sections would be the clusters. Ideally, each cluster would have the same characteristics as the population as a whole.

After the clusters have been defined, a sample of $m$ clusters is selected at random from the list of possible clusters. The number of clusters to select depends on various factors, including our survey budget. Suppose the marketing firm randomly selects eight clusters:

$$104 - 142 - 147 - 218 - 228 - 235 - 307 - 327$$

These seating sections are the *primary clusters*. Next, the marketing company can either survey all the ticketholders in each cluster or select a simple random sample of ticketholders from each cluster, depending on time and budget considerations.

**Cluster Sampling**

A method by which the population is divided into groups, or clusters, that are each intended to be mini-populations. A simple random sample of $m$ clusters is selected. The items chosen from a cluster can be selected using any probability sampling technique.

---

## 1.3 EXERCISES

### Skill Development

**1-25.** Indicate which sampling method would most likely be used in each of the following situations:
   a. an interview conducted with mayors of a sample of cities in Florida
   b. a poll of voters regarding a referendum calling for a national value-added tax
   c. a survey of customers entering a shopping mall in Minneapolis

**1-26.** A company has 18,000 employees. The file containing the names is ordered by employee number from 1 to 18,000. If a sample of 100 employees is to be selected from the 18,000 using systematic random sampling, within what range of employee numbers will the first employee selected come from?

**1-27.** Describe the difference between a statistic and a parameter.

**1-28.** Why is convenience sampling considered to be a nonstatistical sampling method?

**1-29.** Describe how systematic random sampling could be used to select a random sample of 1,000 customers who have a certificate of deposit at a commercial bank. Assume that the bank has 25,000 customers who own a certificate of deposit.

**1-30.** Explain why a census does not necessarily have to involve a population of people. Use an example to illustrate.

**1-31.** If the manager at First City Bank surveys a sample of 100 customers to determine how many miles they live from the bank, is the mean travel distance for this sample a parameter or a statistic? Explain.

**1-32.** Explain the difference between stratified random sampling and cluster sampling.

**1-33.** Use Excel to generate five random numbers between 1 and 900.

### Business Applications

**1-34.** Based on The College Board's Annual Survey of Colleges in 2014–2015, the average tuition increase was 3.7% at private colleges and 2.9% at public universities (trends.collegeboard.org). Are these percentages statistics or parameters? Explain.

**1-35.** During the 2016 presidential campaign, an NBC News/*Wall Street Journal* poll conducted between December, 16, 2015, and January 13, 2016, showed that Marco Rubio would beat Hillary Clinton in the general election 47% to 44.5% (www.realclearpolitics.com). Are these values statistics or parameters? Explain.

**1-36.** Give the name of the kind of sampling that was most likely used in each of the following cases:
   a. a *Wall Street Journal* poll of 2,000 people to determine the president's approval rating
   b. a poll taken of each of the General Motors (GM) dealerships in Ohio in December to determine an estimate of the average number of Chevrolets not yet sold by GM dealerships in the United States
   c. a quality-assurance procedure within a Frito-Lay manufacturing plant that tests every 1,000th bag of Fritos Corn Chips produced to make sure the bag is sealed properly
   d. a sampling technique in which a random sample from each of the tax brackets is obtained by the Internal Revenue Service to audit tax returns

**1-37.** Your manager has given you an Excel file that contains the names of the company's 500 employees and has asked you to sample 50 employees from the list. You decide to take your sample as follows. First, you assign a random number to each employee using Excel's random number function **Rand**(). Because the random number is volatile (it recalculates itself whenever you modify the file), you freeze the random numbers using the Copy—Paste Special—Values feature. You then sort by the random numbers in ascending order. Finally, you take the first 50 sorted employees as your sample. Does this approach constitute a statistical or a nonstatistical sample?

## Computer Applications

**1-38.** Sysco Foods has a very large warehouse in which the food is stored until it is pulled from the shelves to be delivered to the customers. The warehouse has 64 storage racks numbered 1−64. Each rack is three shelves high, labeled A, B, and C, and each shelf is divided into 80 sections, numbered 1−80. Products are located by rack number, shelf letter, and section number. For example, breakfast cereal is located at 4-A-52 (rack 4, shelf A, section 52).

Each week, employees perform an inventory for a sample of products. Certain products are selected and counted. The *actual count* is compared to the *book count* (the quantity in the records that should be in stock). To simplify things, assume that the company has selected breakfast cereals to inventory. Also for simplicity's sake, suppose the cereals occupy racks 1 through 5.

a. Assume that you plan to use simple random sampling to select the sample. Use Excel to determine the sections on each of the five racks to be sampled.

b. Assume that you wish to use cluster random sampling to select the sample. Discuss the steps you would take to carry out the sampling.

c. In this case, why might cluster sampling be preferred over simple random sampling? Discuss.

**1-39.** To evaluate the quality of its airline service, airline managers want to randomly sample passengers traveling between two large airports. On a particular day, all 578 passengers flying between the two airports were given an identification number between 001 and 578. Use Excel to generate a list of 40 flyer identification numbers so that those identified can be surveyed.

**1-40.** The National Park Service has started charging a user fee to park at selected trailheads and cross-country ski lots. Some users object to this fee, claiming they already pay taxes for these areas. The agency has decided to randomly question selected users at fee areas in Colorado to assess the level of concern.

a. Define the population of interest.

b. Assume a sample of 250 is required. Describe the technique you would use to select a sample from the population. Which sampling technique did you suggest?

c. Assume the population of users is 4,000. Use Excel to generate a list of users to be selected for the sample.

**1-41.** Mount Hillsdale Hospital has 4,000 patient files listed alphabetically in its computer system. The office manager wants to survey a statistical sample of these patients to determine how satisfied they were with service provided by the hospital. She plans to use a telephone survey of 100 patients.

a. Describe how you would attach identification numbers to the patient files; for example, how many digits (and which digits) would you use to indicate the first patient file?

b. Describe how you would obtain the first random number to begin a simple random sample method.

c. How many random digits would you need for each random number you selected?

d. Use Excel to generate the list of patients to be surveyed.

1.4

## Data Types and Data Measurement Levels

Chapters 2 and 3 will introduce a variety of techniques for describing data and transforming the data into information. As you will see in those chapters, the statistical techniques deal with different types of data. The level of measurement may vary greatly from application to application. In general, there are four types of data: *quantitative, qualitative, time-series*, and *cross-sectional*. A discussion of each follows.

**OUTCOME 4**

## Quantitative and Qualitative Data

**Quantitative Data**
Measurements whose values are inherently numerical.

In some cases, data values are best expressed in purely numerical, or **quantitative**, terms, such as in dollars, pounds, inches, or percentages. As an example, a cell phone provider might collect data on the number of outgoing calls placed during a month by its customers. In another case, a sports bar could collect data on the number of pitchers of beer sold weekly.

In other situations, the observation may signify only the category to which an item belongs. Categorical data are referred to as **qualitative** data.

**Qualitative Data**
Data whose measurement scale is inherently categorical.

For example, a bank might conduct a study of its outstanding real estate loans and keep track of the marital status of the loan customer—*single, married, divorced*, or *other.* The same study also might examine the credit status of the customer—*excellent, good, fair*, or *poor.* Still another part of the study might ask the customers to rate the service by the bank on a 1 to 5 scale with 1 = very poor, 2 = poor, 3 = neutral, 4 = good, and 5 = very good. Note, although the customers are asked to record a number (1 to 5) to indicate the service quality, the data would still be considered qualitative because the numbers are just codes for the categories.

## Time-Series Data and Cross-Sectional Data

**Time-Series Data**
A set of consecutive data values observed at successive points in time.

**Cross-Sectional Data**
A set of data values observed at a fixed point in time.

Data may also be classified as being either **time-series** or **cross-sectional**.

The data collected by the bank about its loan customers would be cross-sectional because the data from each customer relate to a fixed point in time. In another case, if we sampled 100 stocks from the stock market and determined the closing stock price on March 15, the data would be considered cross-sectional because all measurements correspond to one point in time.

On the other hand, Ford Motor Company tracks the sales of its F-150 pickup trucks on a monthly basis. Data values observed at intervals over time are referred to as time-series data. If we determined the closing stock price for a particular stock on a daily basis for a year, the stock prices would be time-series data.

## Data Measurement Levels

Data can also be identified by their *level of measurement*. This is important because the higher the data level, the more sophisticated the analysis that can be performed. This will be clear when you study the material in the remaining chapters of this text.

We shall discuss and give examples of four levels of data measurements: *nominal, ordinal, interval,* and *ratio*. Figure 1.9 illustrates the hierarchy among these data levels, with nominal data being the lowest level.

**Nominal Data**    *Nominal data* are the lowest form of data, yet you will encounter this type of data many times. Assigning codes to categories generates nominal data. For example, a survey question that asks for marital status provides the following responses:

    1. Married     2. Single     3. Divorced     4. Other

For each person, a code of 1, 2, 3, or 4 would be recorded. These codes are nominal data. Note that the values of the code numbers have no specific meaning, because the order of the categories is arbitrary. We might have shown it this way:

    1. Single     2. Divorced     3. Married     4. Other

With nominal data, we also have complete control over what codes are used. For example, we could have used

    88. Single     11. Divorced     33. Married     55. Other

All that matters is that you know which code stands for which category. Recognize also that the codes need not be numeric. We might use

    S = Single     D = Divorced     M = Married     O = Other

**FIGURE 1.9 Data-Level Hierarchy**

**Ordinal Data**    *Ordinal* or *rank data* are one notch above nominal data on the measurement hierarchy. At this level, the data elements can be rank-ordered on the basis of some relationship among them, with the assigned values indicating this order. For example, a typical market research technique is to offer potential customers the chance to use two unidentified brands of a product. The customers are then asked to indicate which brand they prefer. The brand eventually offered to the general public depends on how often it was the preferred test brand. The fact that an ordering of items took place makes this an ordinal measure.

Bank loan applicants are asked to indicate the category corresponding to their household incomes:

|  |  |  |
|---|---|---|
| _____Under $20,000 | _____$20,000 to $40,000 | _____over $40,000 |
| (1) | (2) | (3) |

The codes 1, 2, and 3 refer to the particular income categories, with higher codes assigned to higher incomes.

Ordinal measurement allows decision makers to equate two or more observations or to rank-order the observations. In contrast, nominal data can be compared only for equality. You cannot order nominal measurements. Thus, a primary difference between ordinal and nominal data is that ordinal data can have both an equality ($=$) and a greater than ($>$) or a less than ($<$) relationship, whereas nominal data can have only an equality ($=$) relationship.

**Interval Data**    If the distance between two data items can be measured on some scale and the data have ordinal properties ($>$, $<$, or $=$), the data are said to be *interval data*. The best example of interval data is the temperature scale. Both the Fahrenheit and Celsius temperature scales have ordinal properties of $>$ or $<$ and $=$. In addition, the distances between equally spaced points are preserved. For example, $32°F > 30°F$, and $80°C > 78°C$. The difference between $32°F$ and $30°F$ is the same as the difference between $80°F$ and $78°F$, two degrees in each case. Thus, interval data allow us to precisely measure the difference between any two values. With ordinal data this is not possible, because all we can say is that one value is larger than another.

**Ratio Data**    Data that have all the characteristics of interval data but also have a true zero point (at which zero means "none") are called *ratio data*. Ratio measurement is the highest level of measurement.

Packagers of frozen foods encounter ratio measures when they pack their products by weight. Weight, whether measured in pounds or grams, is a ratio measurement because it has a unique zero point—zero meaning no weight. Many other types of data encountered in business environments involve ratio measurements—for example, distance, money, and time.

As was mentioned earlier, a major reason for categorizing data by level and type is that the methods you can use to analyze the data are partially dependent on the level and type of data you have available.

| EXAMPLE 1-1 | **Categorizing Data** |

Figure 1.10 shows a portion of the data in the file named **Colleges and Universities**. Each column corresponds to a different variable for which data were collected from various colleges and universities. Before doing any statistical analyses with these data, we need to determine the type and level for each of the factors. Limiting the effort to only those factors that are shown in Figure 1.10, we use the following steps:

STEP 1    **Identify each factor in the data set.**
The 14 factors (or variables) in the data set shown in Figure 1.10 are

| College Name | State | Public (1) Private (2) | Math SAT | Verbal SAT | # appli. rec'd. | # appli. acc'd | # new stud. enrolled | # FT under- grad | # PT under- grad | in-state tuition | out-of- state tuition | % fac. w/PHD | stud./ fac. ratio |
|---|---|---|---|---|---|---|---|---|---|---|---|---|---|

**FIGURE 1.10 Data for U.S. Colleges and Universities**
(*Source:* mathforum.org/workshops)

| | A | B | C | D | E | F | G | H | I | J | K | L | M | N |
|---|---|---|---|---|---|---|---|---|---|---|---|---|---|---|
| | | | Public (1) Private (2) | Math SAT | Verbal SAT | # appli. rec'd | # appl. Acc'd | # new stud. enrolled | # FT u-grad | # PT u-grad | in-state tuition | out-of-state tuition | % fac. w/PHD | stud./fac. ratio |
| 1 | College Name | State | | | | | | | | | | | | |
| 2 | Adelphi University | NY | 2 | 500 | 457 | 2186 | 1924 | 512 | 2683 | 1227 | 12280 | 12280 | 29 | 12.2 |
| 3 | Alaska Pacific University | AK | 2 | 490 | 482 | 193 | 146 | 55 | 249 | 869 | 7560 | 7560 | 76 | 11.9 |
| 4 | Albertson College | ID | 2 | 524 | 487 | 587 | 479 | 158 | 678 | 41 | 13500 | 13500 | 67 | 9.4 |
| 5 | Albertus Magnus College | CT | 2 | 460 | 430 | 353 | 340 | 103 | 416 | 230 | 13290 | 13290 | 90 | 11.5 |
| 6 | Albright College | PA | 2 | 537 | 486 | 1038 | 839 | 227 | 973 | 306 | 15595 | 15595 | 79 | 11.3 |
| 7 | Alderson-Broaddus College | WV | 2 | 443 | 394 | 582 | 498 | 172 | 799 | 78 | 10468 | 10468 | 40 | 11.5 |
| 8 | Allentown Coll. of St. Francis de Sales | PA | 2 | 515 | 470 | 1179 | 780 | 290 | 1130 | 638 | 9690 | 9690 | 60 | 13.3 |
| 9 | Alvernia College | PA | 2 | 410 | 430 | 707 | 614 | 168 | 801 | 464 | 8374 | 8374 | 42 | 10.8 |
| 10 | Ambassador University | TX | 2 | 516 | 493 | 397 | 244 | 223 | 1083 | 58 | 2700 | 2700 | 34 | 17.2 |

(Excel 2016)

STEP 2 **Determine whether the data are time-series or cross-sectional.**
Because each row represents a different college or university and the data are for the same year, the data are cross-sectional.

STEP 3 **Determine which factors are quantitative data and which are qualitative data.**
Qualitative data are codes or numerical values that represent categories. Quantitative data are those that are purely numerical. In this case, the data for the following factors are qualitative:

College Name
State
Code for Public or Private College or University

Data for the following factors are considered quantitative:

| | | |
|---|---|---|
| Math SAT | Verbal SAT | # new stud. enrolled |
| # appl. rec'd. | # appl. accepted | in-state tuition |
| # PT undergrad | # FT undergrad | out-of-state tuition |
| | | % fac. w/PHD |
| | | stud./fac. ratio |

STEP 4 **Determine the level of data measurement for each factor.**
The four levels of data are nominal, ordinal, interval, and ratio. This data set has only nominal- and ratio-level data. The three nominal-level factors are

College Name
State
Code for Public or Private College or University

The others are all ratio-level data.

# 1.4 EXERCISES

## Skill Development

**1-42.** For each of the following, indicate whether the data are cross-sectional or time-series:
a. quarterly unemployment rates
b. unemployment rates by state
c. monthly sales
d. employment satisfaction data for a company

**1-43.** What is the difference between qualitative and quantitative data?

**1-44.** For each of the following variables, indicate the level of data measurement:
a. product rating {1 = excellent, 2 = good, 3 = fair, 4 = poor, 5 = very poor}
b. home ownership {own, rent, other}

c. college grade point average
d. marital status {single, married, divorced, other}

**1-45.** What is the difference between ordinal and nominal data?

**1-46.** *Consumer Reports*, in its rating of cars, indicates repair history with circles. The circles are either white, black, or half and half. To which level of data does this correspond? Discuss.

## Business Applications

**1-47.** Verizon Wireless has a support center customers can call to get questions answered about their cell phone accounts. The manager in charge of the support center has recently conducted a study in which she surveyed

2,300 customers. The customers who called the support center were transferred to a third party, who asked the customers a series of questions.

a. Indicate whether the data generated from this study will be cross-sectional or time-series. Explain why.

b. One of the questions asked customers was approximately how many minutes they had been on hold waiting to get through to a support person. What level of data measurement is obtained from this question? Explain.

c. Another question asked the customer to rate the service on a scale of 1–7, with 1 being the worst possible service and 7 being the best possible service. What level of data measurement is achieved from this question? Will the data be quantitative or qualitative? Explain.

1-48. Suppose the following information is found in a petroleum company's annual report to shareholders. Indicate the level of data measurement for each:

a. a list of countries where the company has exploration and production activity

b. the number of operating oil rigs

c. dollars spent during the year for oil exploration

d. capital expenditures measured in dollars

e. a list of its peer companies

1-49. The following information on 15 different real estate investment trusts (REITs) has been collected. Identify whether the data collected are cross-sectional or time-series:

a. income distribution by region in 2016

b. per-share funds from operations (FFO) for the years 2012–2016

c. number of residential apartment buildings owned on December 31, 2015

d. overall percentage of leased space for the 137 properties in service as of December 31, 2015

e. annual dividends paid to shareholders in 2010–2015

1-50. As part of an economics research study, an analyst has accessed data compiled by the U.S. Bureau of Labor Statistics. The data are in a file named **BLS County Data** (*source:* www.ers.usda.gov). Consider the data in columns A-L, and indicate what level of data is represented by the variables in each column.

1.5

## A Brief Introduction to Data Mining

OUTCOME 5

## Data Mining—Finding the Important, Hidden Relationships in Data

What food products have an increased demand during hurricanes? How do you win baseball games without star players? Is my best friend the one to help me find a job? What color car is least likely to be a "lemon"? These and other interesting questions can and have been answered using data mining. Data mining consists of applying sophisticated statistical techniques and algorithms to the analysis of big data (i.e., the wealth of new data that organizations collect in many and varied forms). Through the application of data mining, decisions can now be made on the basis of statistical analysis rather than on only managerial intuition and experience. The statistical techniques introduced in this text provide the basis for the more sophisticated statistical tools that are used by data mining analysts.

Walmart, the nation's largest retailer, uses data mining to help it tailor product selection based on the sales, demographic, and weather information it collects. While Walmart managers might not be surprised that the demand for flashlights, batteries, and bottled water increased with hurricane warnings, they were surprised to find that there was also an increase in the demand for strawberry Pop-Tarts before hurricanes hit. This knowledge allowed Walmart to increase the availability of Pop-Tarts at selected stores affected by the hurricane alerts. The McKinsey Global Institute estimates that the full application of data mining to retailing could result in a potential increase in operating margins by as much as 60% (*source:* McKinsey Global Institute, "Big data: The next frontier for innovation, competition, and productivity" by James Manyika, Michael Chui, Brad Brown, Jacques Bughin, Richard Dobbs, Charles Roxburgh, and Angela Hung Byers, May 2011).

Data are everywhere, and businesses are collecting more each day. Accounting and sales data are now captured and streamed instantly when transactions occur. Digital sensors in industrial equipment and automobiles can record and report data on vibration, temperature, physical location, and the chemical composition of the surrounding air. But data are now more than numbers. Much of the data being collected today consists of words from Internet search engines such as Google and from pictures from social media postings on such platforms as Facebook. Together with the traditional numbers that make up quantitative data, the availability of new unstructured, qualitative data has led to a data explosion. IDC, a technology research firm, estimates that the amount of data available is growing at a rate of 50% a year. All these data—referred to as big data—have created a need not only for highly skilled data scientists who can mine and analyze them but also for managers who can make decisions using them. McKinsey Global Institute, a consultancy firm, believes that big data offer an opportunity for organizations to create competitive advantages for themselves if they can understand and use the information to its full potential. McKinsey reports that the use of big data "will become a key basis of competition and growth for individual firms." This will create a need for highly trained data scientists and managers who can use data to support their decision making. Unfortunately, McKinsey predicts a shortage in the United States of 140,000 to 190,000 people with deep analytical skills as well as 1.5 million managers and analysts with the know-how needed to use big data to make meaningful and effective decisions (*source:* McKinsey Global Institute, "Big data: The next frontier for innovation, competition, and productivity" by James Manyika et al., May 2011). The statistical tools you will learn in this course will provide you with a good first step toward preparing yourself for a career in data mining and business analytics. Chapter 18 provides an overview of business analytics and introduces you to Microsoft analytics software called Microsoft Power BI.

# 1 Overview

## Summary

- **Descriptive statistics** includes visual tools such as charts and graphs and numerical measures. The role of descriptive statistics is to describe data and help transform data into usable information.
- **Inferential statistics** allows decision makers to draw conclusions about a large body of data by examining a smaller subset of those data. We describe two areas of inference: estimation and hypothesis testing.

**OUTCOME 1** Know the key data collection methods.

- Before we can analyze data using business statistics techniques, we must collect the data. We review these data collection methods: **experiments, telephone surveys, written questionnaires, direct observation**, and **personal interviews** (see Figure 1.11).
- Be aware of data collection problems such as **interviewer bias, nonresponse bias, selection bias, observer bias**, and **measurement error**.
- We define the concepts of **internal validity** and **external validity**.

**OUTCOME 2** Know the difference between a population and a sample.

**OUTCOME 3** Understand the similarities and differences between different sampling methods.

- We define the important concepts of **population** and **sample** and provide examples of each (see Figure 1.12). Because many statistical applications involve samples, we emphasize how to select samples.
- Two main sampling categories are **nonstatistical sampling** and **statistical sampling**. We focus on statistical sampling and discuss four statistical sampling methods: **simple random sampling, stratified random sampling, cluster sampling**, and **systematic random sampling** (see Figure 1.12)

**FIGURE 1.11 Data Collection Techniques**

| Data Collection Method | Advantages | Disadvantages |
| --- | --- | --- |
| Experiments | Provide controls<br>Preplanned objectives | Costly<br>Time-consuming<br>Requires planning |
| Telephone Surveys | Timely<br>Relatively inexpensive | Poor reputation<br>Limited scope and length |
| Written Surveys (Direct, Mailed, and Online) | Inexpensive<br>Can expand length<br>Can use open-end questions | Low response rate<br>Requires exceptional clarity |
| Direct Observation<br>Personal Interview | Expands analysis opportunities<br>No respondent bias | Potential observer bias<br>Costly |

**FIGURE 1.12 Sampling Techniques**

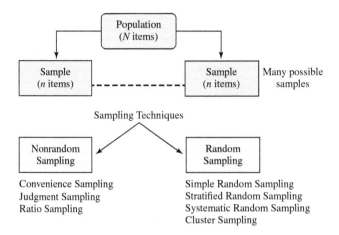

**1.4**    ## Data Types and Data Measurement Levels (pg. 19–23)

**OUTCOME 4** Understand how to categorize data by type and level of measurement.

- Knowing the type of data you have is important because the data type influences the kind of statistical procedures you can use.
- Data can be classified as either **quantitative** or **qualitative**. Data can also be **cross-sectional** or **time-series** (see Figure 1.13).
- Another way to classify data is by the level of measurement. There are four levels from lowest to highest: **nominal, ordinal, interval**, and **ratio**.

**FIGURE 1.13 Data Classification**

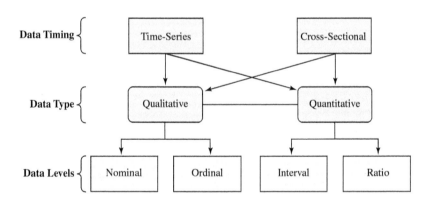

**1.5**    ## A Brief Introduction to Data Mining (pg. 23–24)

**OUTCOME 5** Become familiar with the concept of data mining and some of its applications.

- Because electronic data storage is so inexpensive, organizations are collecting and storing greater volumes of data than ever before. As a result, a relatively new field of study called data mining has emerged.
- **Data mining** involves the art and science of delving into the data to identify patterns and conclusions that are not immediately evident in the data.

## Key Terms

**Bias** pg. 11
**Business analytics/business intelligence** pg. 1
**Business statistics** pg. 2
**Census** pg. 14
**Closed-end questions** pg. 7
**Cluster sampling** pg. 18
**Convenience sampling** pg. 15
**Cross-sectional data** pg. 20
**Data mining** pg. 1
**Demographic questions** pg. 7

**Experiment** pg. 6
**Experimental design** pg. 6
**External validity** pg. 12
**Internal validity** pg. 12
**Nonstatistical sampling
 techniques** pg. 14
**Open-end questions** pg. 8
**Population** pg. 13
**Qualitative data** pg. 19
**Quantitative data** pg. 19

**Sample** pg. 13
**Simple random sampling** pg. 15
**Statistical inference procedures** pg. 4
**Statistical sampling techniques** pg. 14
**Stratified random sampling** pg. 16
**Structured interviews** pg. 10
**Systematic random sampling** pg. 17
**Time-series data** pg. 20
**Unstructured interviews** pg. 10

## Chapter Exercises

### Conceptual Questions

**1-51.** Several organizations publish the results of presidential approval polls. Movements in these polls are seen as an indication of how the general public views presidential performance. Comment on these polls within the context of what was covered in this chapter.

**1-52.** With what level of data is a bar chart most appropriately used?

**1-53.** With what level of data is a histogram most appropriately used?

**1-54.** Two people see the same movie; one says it was average and the other says it was exceptional. What level of data are they using in these ratings? Discuss how the same movie could receive different reviews.

**1-55.** The University of Michigan publishes a monthly measure of consumer confidence. This is taken as a possible indicator of future economic performance. Comment on this process within the context of what was covered in this chapter.

### Business Applications

**1-56.** In a business publication such as *The Wall Street Journal* or *Bloomberg Businessweek*, find a graph or chart representing time-series data. Discuss how the data were gathered and the purpose of the graph or chart.

**1-57.** In a business publication such as *The Wall Street Journal* or *Bloomberg Businessweek*, find a graph or chart representing cross-sectional data. Discuss how the data were gathered and the purpose of the graph or chart.

**1-58.** *The Oregonian* newspaper has asked readers to e-mail and respond to the question, "Do you believe police officers are using too much force in routine traffic stops?"

a. Would the results of this survey be considered a random sample?

b. What type of bias might be associated with a data collection system such as this? Discuss how you might reduce this bias potential.

**1-59.** The makers of Mama's Home-Made Salsa are concerned about the quality of their product. The particular trait of concern is the thickness of the salsa in each jar.

a. Discuss a plan by which the managers might determine the percentage of jars of salsa believed to have an unacceptable thickness by potential purchasers. Define (1) the sampling procedure to be used, (2) the randomization method to be used to select the sample, and (3) the measurement to be obtained.

b. Explain why it would or wouldn't be feasible (or, perhaps, possible) to take a census to address this issue.

**1-60.** A maker of energy drinks is considering abandoning can containers and going exclusively to bottles because the sales manager believes customers prefer drinking from bottles. However, the vice president in charge of marketing is not convinced the sales manager is correct.

a. Indicate the data collection method you would use.

b. Indicate what procedures you would follow to apply this technique in this setting.

c. State which level of data measurement applies to the data you would collect. Justify your answer.

d. Are the data qualitative or quantitative? Explain.

**1-61.** You want to sample 40 NBA players. Your friend who is a big NBA fan has offered to give you a list of his 40 favorite players.

a. Discuss why your friend's list is not a random sample and is biased.

b. Develop a sampling plan for selecting a simple random sample of 40 NBA players.

**1-62.** You want to survey students at your university concerning the hours of operation for the Student Fitness Center. You want to sample 300 students, but you need your sample to contain students from each class (freshmen, sophomore, junior, senior, and graduate). What sampling design would you recommend for selecting your sample?

# 2 Graphs, Charts, and Tables— Describing Your Data

## WHY YOU NEED TO KNOW

We live in an age in which presentations and reports are expected to include high-quality graphs and charts that effectively transform data into information. Although the written word is still vital, words become even more powerful when coupled with an effective visual illustration of data. The adage that a picture is worth a thousand words is particularly relevant in business decision making. We are constantly bombarded with visual images and stimuli. Much of our time is spent watching television, playing video games, or working at a computer. These technologies are advancing rapidly, making the images sharper and more attractive to our eyes. Flat-panel, high-definition televisions and high-resolution monitors represent significant improvements over the original technologies they replaced. However, this phenomenon is not limited to video technology, but has also become an important part of the way businesses communicate with customers, employees, suppliers, and other constituents.

When you graduate, you will find yourself on both ends of the data analysis spectrum. On the one hand, regardless of what you end up doing for a career, you will almost certainly be involved in preparing reports and making presentations that require using visual descriptive statistical tools presented in this chapter. You will be on the "do it" end of the data analysis process. Thus, you need to know how to use these statistical tools.

On the other hand, you will also find yourself reading reports or listening to presentations that others have made. In many instances, you will be required to make important decisions or to reach conclusions based on the information in those reports or presentations. Thus, you will be on the "use it" end of the data analysis process. You need to be knowledgeable about these tools to effectively screen and critique the work that others do for you.

Charts and graphs are not just tools used internally by businesses. Business periodicals such as *Fortune* and *Bloomberg Businessweek* use graphs and charts extensively in articles to help readers better understand key concepts. Many advertisements will even use graphs and charts effectively to convey their messages. Virtually every issue of *The Wall Street Journal* contains different graphs, charts, or tables that display data in an informative way.

Throughout your business career, you will be both a producer and a consumer of the descriptive statistical techniques known as graphs, charts, and tables. You will create a competitive advantage for yourself throughout your career if you obtain a solid understanding of the techniques introduced in Chapter 2. This chapter introduces some of the most frequently used tools and techniques for describing data with graphs, charts, and tables. Although this analysis can be done manually, we will provide output from Excel software showing that software can be used to perform the analysis easily, quickly, and with a finished quality that once required a graphic artist.

## 2.1 | Frequency Distributions and Histograms

Today, businesses collect massive amounts of data they hope will be useful for making decisions. Every time a customer makes a purchase at a store like Macy's or the Gap, data from that transaction are updated to the store's database. Major retail stores like Walmart capture the number of different product categories included in each "market basket" of items purchased. Table 2.1 shows these data for all customer transactions for one morning at one Walmart store in Dallas. A total of 450 customers made purchases on the morning in question. The first value in Table 2.1 is a 4, which indicates that the customer's purchase included four different product categories (for example food, sporting goods, photography supplies, and dry goods).

Although the data in Table 2.1 are easy to capture with the technology of today's cash registers, in this form, the data provide little or no information that managers could use to determine the buying habits of their customers. However, these data can be converted into useful information through descriptive statistical analysis.

**OUTCOME 1**

### Frequency Distributions

**Frequency Distribution**
A summary of a set of data that displays the number of observations in each of the distribution's distinct categories or classes.

A more effective way to display the Dallas Walmart data would be to construct a **frequency distribution**.

The product data in Table 2.1 take on only a few possible values (1, 2, 3, . . . , 11). The minimum number of product categories is 1 and the maximum number of categories in these data is 11. These data are called **discrete data**.

**Discrete Data**
Data that can take on a countable number of possible values.

When you encounter discrete data, where the variable of interest can take on only a reasonably small number of possible values, a frequency distribution is constructed by counting the number of times each possible value occurs in the data set. We organize these counts into the *frequency distribution table* shown in Table 2.2. From this frequency distribution we are able to see how the data values are spread over the different number of possible product

**TABLE 2.1 Product Categories per Customer at the Dallas Walmart**

| | | | | | | | | | | | | | | |
|---|---|---|---|---|---|---|---|---|---|---|---|---|---|---|
| 4 | 2 | 5 | 8 | 8 | 10 | 1 | 4 | 8 | 3 | 4 | 1 | 1 | 3 | 4 |
| 1 | 4 | 4 | 5 | 4 | 4 | 4 | 9 | 5 | 4 | 4 | 10 | 7 | 11 | 4 |
| 10 | 2 | 6 | 7 | 10 | 5 | 4 | 6 | 4 | 6 | 2 | 3 | 2 | 4 | 5 |
| 5 | 4 | 11 | 1 | 4 | 1 | 9 | 2 | 4 | 6 | 6 | 7 | 6 | 2 | 3 |
| 6 | 5 | 3 | 4 | 5 | 6 | 5 | 3 | 10 | 6 | 5 | 7 | 7 | 4 | 3 |
| 8 | 2 | 2 | 6 | 5 | 11 | 9 | 9 | 5 | 5 | 6 | 5 | 3 | 1 | 7 |
| 6 | 6 | 5 | 3 | 8 | 4 | 3 | 3 | 4 | 4 | 4 | 7 | 6 | 4 | 9 |
| 1 | 6 | 5 | 5 | 4 | 4 | 7 | 5 | 6 | 6 | 9 | 5 | 6 | 10 | 4 |
| 7 | 5 | 8 | 4 | 4 | 7 | 4 | 6 | 6 | 4 | 4 | 2 | 10 | 4 | 5 |
| 4 | 11 | 8 | 7 | 9 | 5 | 6 | 4 | 2 | 8 | 4 | 2 | 6 | 6 | 6 |
| 6 | 4 | 6 | 5 | 7 | 1 | 6 | 9 | 1 | 5 | 9 | 10 | 5 | 5 | 10 |
| 5 | 4 | 7 | 5 | 7 | 6 | 9 | 5 | 3 | 2 | 1 | 5 | 5 | 5 | 5 |
| 5 | 9 | 5 | 3 | 2 | 5 | 7 | 2 | 4 | 6 | 4 | 4 | 4 | 4 | 4 |
| 6 | 5 | 8 | 5 | 5 | 5 | 5 | 5 | 2 | 5 | 5 | 6 | 4 | 6 | 5 |
| 5 | 7 | 10 | 2 | 2 | 6 | 8 | 3 | 1 | 3 | 5 | 6 | 3 | 3 | 6 |
| 5 | 4 | 5 | 3 | 3 | 7 | 9 | 4 | 4 | 5 | 10 | 6 | 10 | 5 | 9 |
| 4 | 3 | 8 | 7 | 1 | 8 | 4 | 3 | 1 | 3 | 6 | 7 | 5 | 5 | 5 |
| 4 | 7 | 4 | 11 | 6 | 6 | 3 | 7 | 9 | 4 | 4 | 2 | 9 | 7 | 5 |
| 1 | 6 | 6 | 8 | 3 | 8 | 4 | 4 | 1 | 9 | 3 | 9 | 3 | 4 | 2 |
| 9 | 5 | 5 | 7 | 10 | 5 | 3 | 4 | 7 | 7 | 6 | 2 | 2 | 4 | 4 |
| 4 | 7 | 3 | 5 | 4 | 9 | 2 | 3 | 4 | 3 | 2 | 1 | 6 | 4 | 6 |
| 1 | 8 | 1 | 4 | 3 | 5 | 5 | 10 | 4 | 4 | 4 | 6 | 9 | 2 | 7 |
| 9 | 4 | 5 | 3 | 6 | 5 | 5 | 3 | 4 | 6 | 5 | 7 | 3 | 6 | 8 |
| 3 | 6 | 1 | 5 | 7 | 7 | 5 | 4 | 6 | 6 | 6 | 3 | 6 | 9 | 5 |
| 4 | 5 | 10 | 1 | 5 | 5 | 7 | 8 | 9 | 1 | 6 | 5 | 6 | 6 | 4 |
| 10 | 6 | 5 | 5 | 5 | 1 | 6 | 5 | 6 | 4 | 7 | 9 | 10 | 2 | 6 |
| 4 | 4 | 6 | 11 | 9 | 5 | 4 | 4 | 3 | 5 | 4 | 6 | 2 | 6 | 7 |
| 3 | 5 | 6 | 7 | 4 | 5 | 4 | 6 | 9 | 4 | 3 | 3 | 6 | 9 | 4 |
| 3 | 7 | 5 | 6 | 11 | 4 | 4 | 8 | 4 | 2 | 8 | 2 | 4 | 2 | 3 |
| 6 | 5 | 1 | 10 | 5 | 9 | 5 | 4 | 5 | 1 | 4 | 9 | 5 | 4 | 4 |

**TABLE 2.2 Dallas Walmart Product Categories Frequency Distribution**

| Number of Product Categories | Frequency |
|---|---|
| 1 | 25 |
| 2 | 29 |
| 3 | 42 |
| 4 | 92 |
| 5 | 86 |
| 6 | 68 |
| 7 | 35 |
| 8 | 19 |
| 9 | 29 |
| 10 | 18 |
| 11 | 7 |
| | Total = 450 |

categories. For instance, you can see that the most frequently occurring number of product categories in a customer's "market basket" is 4, which occurred 92 times. You can also see that the three most common numbers of product categories are 4, 5, and 6. Only a very few times do customers purchase 10 or 11 product categories in their trip to the store.

**TABLE 2.3** Frequency Distributions of Years of College Education

| Philadelphia | | Knoxville | |
|---|---|---|---|
| Years of College | Frequency | Years of College | Frequency |
| 0 | 35 | 0 | 187 |
| 1 | 21 | 1 | 62 |
| 2 | 24 | 2 | 34 |
| 3 | 22 | 3 | 19 |
| 4 | 31 | 4 | 14 |
| 5 | 13 | 5 | 7 |
| 6 | 6 | 6 | 3 |
| 7 | 5 | 7 | 4 |
| 8 | 3 | 8 | 0 |
| | Total = 160 | | Total = 330 |

**Relative Frequency**
The proportion of total observations that are in a given category. Relative frequency is computed by dividing the frequency in a category by the total number of observations. The relative frequencies can be converted to percentages by multiplying by 100.

Consider another example in which a consulting firm surveyed random samples of residents in two cities, Philadelphia and Knoxville. The firm is investigating the labor markets in these two communities for a client that is thinking of relocating its corporate offices to one of the two locations. Education level of the workforce in the two cities is a key factor in making the relocation decision. The consulting firm surveyed 160 randomly selected adults in Philadelphia and 330 adults in Knoxville and recorded the number of years of college attended. The responses ranged from zero to eight years. Table 2.3 shows the frequency distributions for the two cities.

Suppose now we wished to compare the distribution for years of college for Philadelphia and Knoxville. How do the two cities' distributions compare? Do you see any difficulties in making this comparison? Because the surveys contained different numbers of people, it is difficult to compare the frequency distributions directly. When the number of total observations differs, comparisons are easier to make if **relative frequencies** are computed. Equation 2.1 is used to compute the relative frequencies.

Table 2.4 shows the relative frequencies for each city's distribution. This makes a comparison of the two much easier. We see that Knoxville has relatively more people without any college (56.7%) or with one year of college (18.8%) than Philadelphia (21.9% and 13.1%). At all other levels of education, Philadelphia has relatively more people than Knoxville.

**TABLE 2.4** Relative Frequency Distributions of Years of College

| Years of College | Philadelphia | | Knoxville | |
|---|---|---|---|---|
| | Frequency | Relative Frequency | Frequency | Relative Frequency |
| 0 | 35 | 35/160 = 0.219 | 187 | 187/330 = 0.567 |
| 1 | 21 | 21/160 = 0.131 | 62 | 62/330 = 0.188 |
| 2 | 24 | 24/160 = 0.150 | 34 | 34/330 = 0.103 |
| 3 | 22 | 22/160 = 0.138 | 19 | 19/330 = 0.058 |
| 4 | 31 | 31/160 = 0.194 | 14 | 14/330 = 0.042 |
| 5 | 13 | 13/160 = 0.081 | 7 | 7/330 = 0.021 |
| 6 | 6 | 6/160 = 0.038 | 3 | 3/330 = 0.009 |
| 7 | 5 | 5/160 = 0.031 | 4 | 4/330 = 0.012 |
| 8 | 3 | 3/160 = 0.019 | 0 | 0/330 = 0.000 |
| Total | 160 | | 330 | |

**Relative Frequency**

$$\text{Relative frequency} = \frac{f_i}{n} \qquad\qquad (2.1)$$

where:

$f_i$ = Frequency of the $i$th value of the discrete variable

$n = \sum\limits_{i=1}^{k} f_i$ = Total number of observations

$k$ = The number of different values for the discrete variable

**TABLE 2.5** TV Source
Frequency Distribution

| TV Source | Frequency |
|-----------|-----------|
| DISH | 80 |
| DIRECTV | 90 |
| Cable | 20 |
| Other | 10 |
| | Total = 200 |

The frequency distributions shown in Table 2.2 and Table 2.3 were developed from *quantitative data*. That is, the variable of interest was numerical (number of product categories or number of years of college). However, a frequency distribution can also be developed when the data are *qualitative data*, or nonnumerical data. For instance, if a survey asked homeowners how they get their TV signal, the possible responses in this region are:

DISH    DIRECTV    Cable    Other

Table 2.5 to the left shows the frequency distribution from a survey of 200 homeowners.

---

**EXAMPLE 2-1**    **Frequency and Relative Frequency Distributions**

**Real Estate Transactions**    In late 2008, the United States experienced a major economic decline thought to be due in part to the sub-prime loans that many lending institutions made during the previous few years. When the housing bubble burst, many institutions experienced severe problems. As a result, lenders became much more conservative in granting home loans, which in turn made buying and selling homes more challenging. To demonstrate the magnitude of the problem in Kansas City, a survey of 16 real estate agencies was conducted to collect data on the number of real estate transactions closed in December 2008. The following data were observed:

| | | | |
|---|---|---|---|
| 3 | 0 | 0 | 1 |
| 1 | 2 | 2 | 0 |
| 0 | 2 | 1 | 0 |
| 2 | 1 | 4 | 2 |

**HOW TO DO IT** (Example 2-1)

**Developing Frequency and Relative Frequency Distributions for Discrete Data**

**1.** List all possible values of the variable. If the variable is ordinal level or higher, order the possible values from low to high.

**2.** Count the number of occurrences at each value of the variable and place this value in a column labeled "Frequency."

To develop a relative frequency distribution, do the following:

**3.** Use Equation 2.1 and divide each frequency count by the total number of observations and place in a column headed "Relative Frequency."

The real estate analysts can use the following steps to construct a frequency distribution and a relative frequency distribution for the number of real estate transactions.

**STEP 1**    **List the possible values.**
The possible values for the discrete variable, listed in order, are 0, 1, 2, 3, 4.

**STEP 2**    **Count the number of occurrences at each value.**
The frequency distribution follows:

| Transactions | Frequency | Relative Frequency |
|:---:|:---:|:---:|
| 0 | 5 | 5/16 = 0.3125 |
| 1 | 4 | 4/16 = 0.2500 |
| 2 | 5 | 5/16 = 0.3125 |
| 3 | 1 | 1/16 = 0.0625 |
| 4 | 1 | 1/16 = 0.0625 |
| | Total = 16 | 1.0000 |

**STEP 3**    **Determine the relative frequencies.**
The relative frequencies are determined by dividing each frequency by 16, as shown in the right-hand column above. Thus, just over 31% of those responding reported no transactions during December 2008.

**TRY EXERCISE 2-1** (pg. 46)

**EXAMPLE 2-2** | **Frequency Distribution for Qualitative Data**

**Automobile Accidents** State Farm Insurance recently surveyed a sample of the records for 15 policy holders to determine the make of the vehicle driven by the eldest member in the household. The following data reflect the results for 15 of the respondents:

| | | | | |
|---|---|---|---|---|
| Ford | Dodge | Toyota | Ford | Buick |
| Chevy | Toyota | Nissan | Ford | Chevy |
| Ford | Toyota | Chevy | BMW | Honda |

The frequency distribution for this qualitative variable is found as follows:

STEP 1 **List the possible values.**
For these sample data, the possible values for the variable are BMW, Buick, Chevy, Dodge, Ford, Honda, Nissan, Toyota.

STEP 2 **Count the number of occurrences at each value.**
The frequency distribution is

| Car Company | Frequency |
|---|---|
| BMW | 1 |
| Buick | 1 |
| Chevy | 3 |
| Dodge | 1 |
| Ford | 4 |
| Honda | 1 |
| Nissan | 1 |
| Toyota | 3 |
| Total = | 15 |

**TRY EXERCISE 2-7** (pg. 47)

---

**BUSINESS APPLICATION** | **Frequency Distributions**

**Excel Tutorial**

**Athletic Shoe Survey** In recent years, a status symbol for many students has been the brand and style of athletic shoes they wear. Companies such as Nike and Adidas compete for the top position in the sport shoe market. A survey was conducted in which 100 college students at a southern state school were asked a number of questions, including how many pairs of Nike shoes they currently own. The data are in a file called **SportsShoes**.

The variable *Number of Nike* is a discrete quantitative variable. Figure 2.1 shows the frequency distribution (output from Excel) for the number of Nike shoes owned by those surveyed. The frequency distribution shows that, although a few people own more than six pairs of Nike shoes, most of those surveyed own two or fewer pairs.

---

**OUTCOME 1** | ## Grouped Data Frequency Distributions

In the previous examples, the variable of interest was a discrete variable and the number of possible values for the variable was limited to only a few. However, there are many instances in which the variable of interest will be either **continuous** (e.g., weight, time, length) or discrete and will have many possible outcomes (e.g., age, income, stock prices), yet you want to describe the variable using a frequency distribution.

**Continuous Data**
Data whose possible values are uncountable and that may assume any value in an interval.

**Excel 2016 Instructions**

1. Open File: **SportsShoes.xlsx**.
2. Enter the Possible Values for the Variable; i.e., 0, 1, 2, 3, 4, etc.
3. Select the cells to contain the Frequency values.
4. Select the **Formulas** tab.
5. Click on the $f_x$ button.
6. Select the **Statistical-FREQUENCY** function.
7. Enter the range of data and the bin range (the cells containing the possible number of shoes).
8. Press **Ctrl-Shift-Enter** to determine the frequency values.

**FIGURE 2.1 Excel 2016 Output—Nike Shoes Frequency Distribution**

| | A | B | C | D | E | F | G | H | I | J | K | L | M |
|---|---|---|---|---|---|---|---|---|---|---|---|---|---|
| 1 | Age | Annual Income | Pairs per Year | Pairs Now | Number Nike | Number Adidas | Brand Preference | Comfort | Advertising | Appearance | Price | Level of Activity | Gender |
| 89 | 34 | 18,500 | 1 | 6 | 3 | 0 | 1 | 1 | 2 | 1 | 1 | 2 | 2 |
| 90 | 65 | 27,500 | 1 | 2 | 1 | 0 | 1 | 1 | 2 | 1 | 1 | 1 | 2 |
| 91 | 29 | 21,000 | 5 | 1 | 0 | 1 | 2 | 1 | 2 | 2 | 2 | 2 | 2 |
| 92 | 56 | 63,000 | 2 | 7 | 4 | 2 | 1 | 1 | 1 | 1 | 1 | 1 | 1 |
| 93 | 24 | 13,000 | 2 | 3 | 2 | 1 | 2 | 2 | 2 | 1 | 1 | 2 | 1 |
| 94 | 72 | 80,000 | 2 | 5 | 1 | 4 | 2 | 2 | 2 | 1 | 1 | 1 | 2 |
| 95 | 25 | 23,000 | 1 | 6 | 6 | 0 | 1 | 1 | 2 | 2 | 1 | 3 | 2 |
| 96 | 21 | 32,000 | 3 | 2 | 1 | 0 | 1 | 1 | 2 | 1 | 2 | 1 | 1 |
| 97 | 56 | 51,000 | 2 | 3 | 0 | 1 | 2 | 1 | 1 | 1 | 1 | 1 | 2 |
| 98 | 27 | 26,500 | 6 | 3 | 2 | 0 | 1 | 1 | 2 | 2 | 1 | 3 | 2 |
| 99 | 19 | 11,000 | 1 | 2 | 1 | 1 | 2 | 1 | 2 | 1 | 2 | 1 | 1 |
| 100 | 26 | 32,000 | 1 | 2 | 1 | 0 | 1 | 1 | 2 | 1 | 2 | 2 | 2 |
| 101 | 51 | 90,000 | 2 | 5 | 4 | 1 | 1 | 1 | 2 | 1 | 1 | 2 | 1 |

| Number of Nikes | Freq. |
|---|---|
| 0 | 27 |
| 1 | 27 |
| 2 | 20 |
| 3 | 9 |
| 4 | 6 |
| 5 | 4 |
| 6 | 4 |
| 7 | 1 |
| 8 | 2 |

**BUSINESS APPLICATION**    ## Grouped Data Frequency Distributions

**Video Streaming**   Video streaming services such as Netflix, Cloudload, Amazon Prime, and Hulu have grown in popularity in recent years. A distribution manager for Netflix collected data for a random sample of 230 Netflix customers and recorded the number of movies they streamed during the previous month. Table 2.6 shows the results. These data are discrete, quantitative data. The values range from 0 to 30.

**TABLE 2.6 Number of Streamed Movies**

| | | | | | | | | | |
|---|---|---|---|---|---|---|---|---|---|
| 9 | 4 | 13 | 10 | 5 | 10 | 13 | 14 | 10 | 19 |
| 0 | 10 | 16 | 9 | 11 | 14 | 8 | 15 | 7 | 15 |
| 10 | 11 | 9 | 7 | 6 | 12 | 12 | 14 | 15 | 16 |
| 15 | 14 | 10 | 13 | 9 | 12 | 12 | 10 | 10 | 11 |
| 15 | 14 | 9 | 19 | 3 | 9 | 16 | 19 | 15 | 9 |
| 4 | 2 | 4 | 5 | 6 | 2 | 3 | 4 | 7 | 5 |
| 6 | 2 | 2 | 0 | 0 | 8 | 3 | 4 | 3 | 2 |
| 2 | 5 | 2 | 5 | 2 | 2 | 6 | 2 | 5 | 6 |
| 5 | 2 | 7 | 3 | 5 | 1 | 6 | 4 | 3 | 6 |
| 3 | 7 | 7 | 1 | 6 | 2 | 7 | 1 | 3 | 2 |
| 4 | 0 | 2 | 2 | 4 | 6 | 2 | 5 | 3 | 7 |
| 4 | 16 | 9 | 10 | 11 | 7 | 10 | 9 | 10 | 11 |
| 11 | 12 | 9 | 8 | 9 | 7 | 9 | 17 | 8 | 13 |
| 14 | 13 | 10 | 6 | 12 | 5 | 14 | 7 | 13 | 12 |
| 9 | 6 | 10 | 15 | 7 | 7 | 9 | 9 | 13 | 10 |
| 9 | 3 | 17 | 5 | 11 | 9 | 6 | 9 | 15 | 8 |
| 11 | 13 | 4 | 16 | 13 | 9 | 11 | 5 | 12 | 13 |
| 0 | 3 | 3 | 3 | 2 | 1 | 4 | 0 | 2 | 0 |
| 3 | 7 | 1 | 5 | 2 | 2 | 3 | 2 | 1 | 3 |
| 2 | 3 | 3 | 3 | 0 | 3 | 3 | 3 | 1 | 1 |
| 13 | 24 | 24 | 17 | 17 | 15 | 25 | 20 | 15 | 20 |
| 21 | 23 | 25 | 17 | 13 | 22 | 18 | 17 | 30 | 21 |
| 18 | 21 | 17 | 16 | 25 | 14 | 15 | 24 | 21 | 15 |

The manager is interested in transforming these data into useful information by constructing a frequency distribution. Table 2.7 shows one approach in which the possible values for the number of movies streamed are listed from 0 to 30. Although this frequency distribution is a step forward in transforming the data into information, because of the large number of possible values, the 230 observations are spread over a large range, making analysis difficult. In this case, the manager might consider forming a *grouped data frequency distribution* by organizing the possible number of streamed movies into discrete categories or *classes*.

To begin constructing a grouped frequency distribution, sort the quantitative data from low to high. The sorted data are called a *data array*. Now, define the classes for the variable of interest. Care needs to be taken when constructing these classes to ensure each data point is put into one, and only one, possible class. Therefore, the classes should meet four criteria.

First, they must be **mutually exclusive**.
Second, they must be **all-inclusive**.
Third, if at all possible, they should be of **equal width**.
Fourth, if possible, classes should not be empty.

**Mutually Exclusive Classes**
Classes that do not overlap, so that a data value can be placed in only one class.

**All-Inclusive Classes**
A set of classes that contains all the possible data values.

**Equal-Width Classes**
The distance between the lowest possible value and the highest possible value in each class is equal for all classes.

**TABLE 2.7** Frequency Distribution of Streamed Movies

| Streamed Movies | Frequency |
| --- | --- |
| 0 | 8 |
| 1 | 8 |
| 2 | 22 |
| 3 | 22 |
| 4 | 11 |
| 5 | 13 |
| 6 | 12 |
| 7 | 14 |
| 8 | 5 |
| 9 | 19 |
| 10 | 14 |
| 11 | 9 |
| 12 | 8 |
| 13 | 12 |
| 14 | 8 |
| 15 | 12 |
| 16 | 6 |
| 17 | 7 |
| 18 | 2 |
| 19 | 3 |
| 20 | 2 |
| 21 | 4 |
| 22 | 1 |
| 23 | 1 |
| 24 | 3 |
| 25 | 3 |
| 26 | 0 |
| 27 | 0 |
| 28 | 0 |
| 29 | 0 |
| 30 | 1 |
| | Total = 230 |

Equal-width classes make analyzing and interpreting the frequency distribution easier. However, there are some instances in which the presence of extreme high or low values makes it necessary to have an open-ended class. For example, annual family incomes in the United States are mostly between $15,000 and $200,000. However, there are some families with much higher family incomes. To best accommodate these high incomes, you might consider having the highest income class be "over $200,000" or "$200,000 and over" as a catch-all for the high-income families.

Empty classes are those for which there are no data values. If this occurs, it may be because you have set up classes that are too narrow.

**Steps for Grouping Data into Classes**    There are four steps for grouping data, such as those in Table 2.6, into classes.

STEP 1    **Determine the number of groups or classes to use.**
Although there is no absolute right or wrong number of classes, one rule of thumb is to have *between 5 and 20 classes*. Another guideline for helping you determine how many classes to use is the $2^k \geq n$ rule, where $k =$ the number of classes and is defined to be the smallest integer so that $2^k \geq n$, where $n$ is the number of data values. For example, for $n = 230$, the $2^k \geq n$ rule would suggest $k = 8$ classes ($2^8 = 256 \geq 230$ while $2^7 = 128 < 230$). This latter method was chosen for our example. Our preliminary step, as specified previously, is to produce a frequency distribution from the data array as in Table 2.7. This will enhance our ability to envision the data structure and the classes.

Remember, these are only guidelines for the number of classes. There is no specific right or wrong number. In general, use fewer classes for smaller data sets; use more classes for larger data sets. However, using too few classes tends to condense data too much, and information can be lost. Using too many classes spreads out the data so much that little advantage is gained over the original raw data.

STEP 2    **Establish the class width.**
The minimum **class width** is determined by Equation 2.2.

**Class Width**
The distance between the lowest possible value and the highest possible value for a frequency class.

$$W = \frac{\text{Largest value} - \text{Smallest value}}{\text{Number of classes}} \qquad (2.2)$$

For the streamed movies data using eight classes, we get

$$W = \frac{\text{Largest value} - \text{Smallest value}}{\text{Number of classes}} = \frac{30 - 0}{8} = 3.75$$

This means we could construct eight classes that are each 3.75 units wide to provide mutually exclusive and all-inclusive classes. However, because our purpose is to make the data more understandable, we suggest that you *round up to a more convenient class width*, such as 4.0. If you do round the class width, always round up.

STEP 3    **Determine the class boundaries for each class.**

**Class Boundaries**
The upper and lower values of each class.

The **class boundaries** determine the lowest possible value and the highest possible value for each class. In the streamed movies example, if we start the first class at 0, we get the class boundaries shown in the first column of the following table. Notice the classes have been formed to be *mutually exclusive* and *all-inclusive*.

STEP 4    **Determine the class frequency for each class.**
The count for each class is known as a class frequency. As an example, the number of observations in the first class is 60.

| Streamed Movies | Frequency |
|---|---|
| 0–3 | 60 |
| 4–7 | 50 |
| 8–11 | 47 |
| 12–15 | 40 |
| 16–19 | 18 |
| 20–23 | 8 |
| 24–27 | 6 |
| 28–31 | 1 |
| | Total = 230 |

**Cumulative Frequency Distribution**
A summary of a set of data that displays the number of observations with values less than or equal to the upper limit of each of its classes.

**Cumulative Relative Frequency Distribution**
A summary of a set of data that displays the proportion of observations with values less than or equal to the upper limit of each of its classes.

Another step we can take to help analyze the streamed movies data is to construct a relative frequency distribution, a **cumulative frequency distribution**, and a **cumulative relative frequency distribution**.

| Streamed Movies | Frequency | Relative Frequency | Cumulative Frequency | Cumulative Relative Frequency |
|---|---|---|---|---|
| 0–3 | 60 | 0.261 | 60 | 0.261 |
| 4–7 | 50 | 0.217 | 110 | 0.478 |
| 8–11 | 47 | 0.204 | 157 | 0.683 |
| 12–15 | 40 | 0.174 | 197 | 0.857 |
| 16–19 | 18 | 0.078 | 215 | 0.935 |
| 20–23 | 8 | 0.035 | 223 | 0.970 |
| 24–27 | 6 | 0.026 | 229 | 0.996 |
| 28–31 | 1 | 0.004 | 230 | 1.000 |
| | Total = 230 | | | |

The cumulative frequency distribution is shown in the "Cumulative Frequency" column. We can then form the cumulative relative frequency distribution as shown in the "Cumulative Relative Frequency" column. The cumulative relative frequency distribution indicates, as an example, that 85.7% of the sample have streamed fewer than 16 movies.

**HOW TO DO IT** (Example 2-3)

**Developing Frequency Distributions for Continuous Variables**

**1.** Determine the desired number of classes or groups. One rule of thumb is to use 5 to 20 classes. The $2^k \geq n$ rule can also be used.

**2.** Determine the minimum class width using

$$\frac{\text{Largest value} - \text{Smallest value}}{\text{Number of classes}}$$

Round the class width up to a more convenient value.

**3.** Define the class boundaries, making sure that the classes that are formed are *mutually exclusive* and *all-inclusive*. Ideally, the classes should have equal widths and should all contain at least one observation.

**4.** Determine the class frequency for each class.

**EXAMPLE 2-3** **Frequency Distribution for Continuous Variables**

**Emergency Response Communication Links** One of the major efforts of the U.S. Department of Homeland Security has been to improve the communication between emergency responders, like the police and fire departments. The communications have been hampered by problems involving linking divergent radio and computer systems, as well as communication protocols. While most cities have recognized the problem and made efforts to solve it, Homeland Security has funded practice exercises in 72 cities of different sizes throughout the United States. The data below are the times, in seconds, it took to link the systems.

| | | | | | |
|---|---|---|---|---|---|
| 35 | 339 | 650 | 864 | 1,025 | 1,261 |
| 38 | 340 | 655 | 883 | 1,028 | 1,280 |
| 48 | 395 | 669 | 883 | 1,036 | 1,290 |
| 53 | 457 | 703 | 890 | 1,044 | 1,312 |
| 70 | 478 | 730 | 934 | 1,087 | 1,341 |
| 99 | 501 | 763 | 951 | 1,091 | 1,355 |
| 138 | 521 | 788 | 969 | 1,126 | 1,357 |
| 164 | 556 | 789 | 985 | 1,176 | 1,360 |
| 220 | 583 | 789 | 993 | 1,199 | 1,414 |
| 265 | 595 | 802 | 997 | 1,199 | 1,436 |
| 272 | 596 | 822 | 999 | 1,237 | 1,479 |
| 312 | 604 | 851 | 1,018 | 1,242 | 1,492 |

Homeland Security wishes to construct a frequency distribution showing the times until the communication systems are linked. The frequency distribution is determined as follows:

STEP 1 **Group the data into classes.**

The number of classes is arbitrary, but typically will be between 5 and 20, depending on the volume of data. In this example, we have $n = 72$ data items. A common method of determining the number of classes is to use the $2^k \geq n$ guideline. We get $k = 7$ classes, since $2^7 = 128 \geq 72$ and $2^6 = 64 < 72$.

STEP 2 **Determine the class width.**

$$W = \frac{\text{Largest value} - \text{Smallest value}}{\text{Number of classes}} = \frac{1{,}492 - 35}{7} = 208.1429 \Rightarrow 225$$

Note that we have rounded the class width up from the minimum required value of 208.1429 to the more convenient value of 225.

STEP 3 **Define the class boundaries.**

| | | |
|---|---|---|
| 0 | and under | 225 |
| 225 | and under | 450 |
| 450 | and under | 675 |
| 675 | and under | 900 |
| 900 | and under | 1,125 |
| 1,125 | and under | 1,350 |
| 1,350 | and under | 1,575 |

These classes are mutually exclusive, all-inclusive, and have equal widths.

STEP 4 **Determine the class frequency for each class.**

| Time to Link Systems (in seconds) | Frequency |
|---|---|
| 0 and under 225 | 9 |
| 225 and under 450 | 6 |
| 450 and under 675 | 12 |
| 675 and under 900 | 13 |
| 900 and under 1,125 | 14 |
| 1,125 and under 1,350 | 11 |
| 1,350 and under 1,575 | 7 |

This frequency distribution shows that most cities took between 450 and 1,350 seconds (7.5 and 22.5 minutes) to link their communications systems.

**TRY EXERCISE 2-5** (pg. 46)

OUTCOME 2

# Histograms

Although frequency distributions are useful for analyzing large sets of data, they are in table format and may not be as visually informative as a graph. If a frequency distribution has been developed from a quantitative variable, a **frequency histogram** can be constructed directly from the frequency distribution. In many cases, the histogram offers a superior format for transforming the data into useful information. (*Note:* Histograms cannot be constructed from a frequency distribution in which the variable of interest is qualitative. However, a similar graph, called a *bar chart*, discussed later in this chapter, is used when qualitative data are involved.)

A histogram shows three general types of information:

**Frequency Histogram**
A graph of a frequency distribution with the horizontal axis showing the classes, the vertical axis showing the frequency count, and (for equal class widths) the rectangles having a height equal to the frequency in each class.

1. It provides a visual indication of where the approximate center of the data is. Look for the center point along the horizontal axes in the histograms in Figure 2.2. Even though the shapes of the histograms are the same, there is a clear difference in where the data are centered.
2. We can gain an understanding of the degree of spread (or variation) in the data. The more the data cluster around the center, the smaller the variation in the data. If the data

FIGURE 2.2 **Histograms Showing Different Centers**

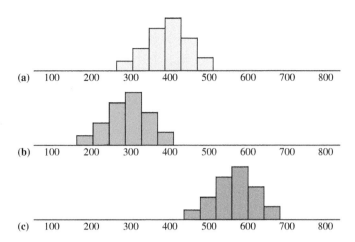

are spread out from the center, the data exhibit greater variation. The examples in Figure 2.3 all have the same center but are different in terms of spread.

3. We can observe the shape of the distribution. Is it reasonably flat, is it weighted to one side or the other, is it balanced around the center, or is it bell-shaped?

---

**BUSINESS APPLICATION**     ## Constructing Histograms

**Excel Tutorial**

**Capital Credit Union** Even for applications with small amounts of data, such as the streamed movies example, constructing grouped data frequency distributions and histograms is a time-consuming process. Decision makers may hesitate to try different numbers of classes and different class limits because of the effort involved and because the "best" presentation of the data may be missed.

Fortunately, current data analysis and statistical software, including Microsoft Excel 2016, makes the construction of histograms quick and easy. Using such software frees decision makers from the more tedious and time-consuming tasks associated with computing frequencies and class widths so that they can devote more time to data analysis and interpretation.

Let's illustrate the use of Excel 2016's Statistic Chart capability by considering Capital Credit Union (CCU) in Mobile, Alabama. CCU recently began issuing a new credit card. Managers at CCU are interested in how customers use the card, so they selected a random sample of 300 customers. Data on the current credit card balance (rounded to the nearest dollar) and the gender of the 300 sampled cardholders are contained in the file named **Capital.** To construct a histogram using Excel 2016, the analyst selects the data and inserts a Statistic Chart as indicated in the instructions in Figure 2.4. Excel automatically calculates the number of classes (bins) using a bin width algorithm. The resulting chart is displayed on the same sheet with the data (see Figure 2.4), but it can be moved to another sheet if desired. Excel's histogram formatting employs standard notation for indicating whether a value is included in a class, with brackets [], or not included in a class, with parentheses (). For example, the class (249.00, 399.00] does not include the value 249.00 but does include the value 399.00.

Although Excel 2016 automatically selects the number of classes and the class width for the histogram, analysts can adjust these two values if they desire. Selecting the horizontal axis will open a Format Axis dialog box on the right of the Excel screen. Axis options is one of the choices available and grants the decision maker the ability to select a different bin (class) width or a different number of bins (classes). Changes to these values will result in a new histogram of the data. Analysts now have the ability to quickly see the effects of different numbers of classes or class widths on the histogram without having to spend time updating frequency tables. Excel's Design Option on the ribbon allows for alternative displays of the histogram by changing the background color or adding frequency counts to the histogram bars.

The resulting histogram in Figure 2.4 shows that the data are centered in the class ($699, $849]. The customers vary considerably in their credit card balances, but the distribution is quite symmetric and somewhat bell-shaped. CCU managers must decide whether the usage rate for the credit card is sufficient to warrant the cost of maintaining the credit card accounts.

**FIGURE 2.3 Histograms—Same Center, Different Spread**

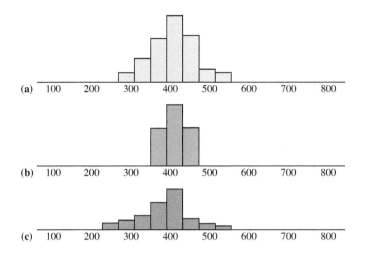

**FIGURE 2.4 Excel 2016 Histogram of Credit Card Balances**

**Excel 2016 Instructions**

1. Open file: **Capital.xlsx**.
2. Select the **Credit Card Account Balance** column.
3. On the **Insert** tab, in the **Charts** group, click the **Insert Statistic Chart** arrow and click **Histogram**.
4. Select the numbers on the Horizontal Category Axis, right mouse click and select **Format Axis**. Change the Number Category to Currency with 0 Decimal places.
5. Add the Chart Title to the Histogram.

| | A | B | C |
|---|---|---|---|
| 1 | Customer Number | Credit Card Account Balance | Gender 1 = Male 2 = Female |
| 2 | 51130 | $ 1,018.00 | 1 |
| 3 | 61413 | $ 1,021.00 | 1 |
| 4 | 54713 | $ 1,081.00 | 1 |
| 5 | 79247 | $ 300.00 | 1 |
| 6 | 94899 | $ 769.00 | 2 |
| 7 | 48035 | $ 486.00 | 1 |
| 8 | 81473 | $ 716.00 | 1 |
| 9 | 47715 | $ 1,013.00 | 1 |
| 10 | 77905 | $ 440.00 | 1 |
| 11 | 87443 | $ 1,246.00 | 2 |
| 12 | 67666 | $ 1,254.00 | 1 |
| 13 | 51470 | $ 191.00 | 1 |
| 14 | 79898 | $ 797.00 | 1 |
| 15 | 98615 | $ 1,127.00 | 1 |
| 16 | 54278 | $ 646.00 | 1 |
| 17 | 58800 | $ 1,212.00 | 2 |
| 18 | 73680 | $ 1,036.00 | 1 |
| 19 | 75732 | $ 477.00 | 1 |
| 20 | 48393 | $ 591.00 | 1 |
| 21 | 49325 | $ 886.00 | 1 |
| 22 | 87544 | $ 1,060.00 | 1 |
| 23 | 62798 | $ 671.00 | 2 |

---

**EXAMPLE 2-4    Frequency Histograms**

**NBA Player Weights**    The weights of 505 NBA players during the 2013 season are in the file **NBA Player Weight**. To develop the frequency histogram of player weight using Excel 2016, perform the following steps:

**STEP 1    Open the data file.**
Open the file **NBA Player Weight** and select the column (B2:B506) containing the 505 players' weights.

**STEP 2    Select desired chart.**
On the Insert tab in the Charts group, select the *Insert Statistic Chart* option arrow and then click Histogram.

**STEP 3    Apply the chart tool.**
With the chart still selected, on the Design tab, in the Charts Layouts group, click the Quick Layout arrow and click Layout 2. This will place the frequency at the top of each bar. Use the Design tool to also format the histogram by adding titles and axis labels.

The formatted histogram, developed with the default classes, is shown here:

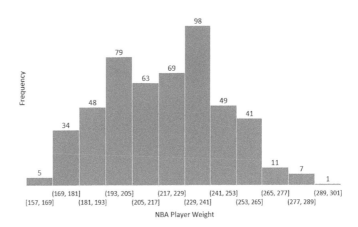

STEP **4**   **Format the histogram.**

Make changes in the number of classes or the class width, if desired.

Select the histogram to activate the Format Data Series feature. Under Series option, use the drop-down arrow to select Horizontal Category Axis. Under Axis options, make desired changes.

Below is a modified histogram of NBA player weights with the bin width (class width) set to 20.

With this modified histogram, NBA player weights appear to be symmetrically distributed about a center in the range of just above 217 pounds to 237 pounds: (217, 237].

**TRY EXERCISE 2-17** (pg. 49)

## Relative Frequency Histograms and Ogives

Histograms can also be used to display relative frequency distributions and cumulative relative frequency distributions. A relative frequency histogram is formed in the same manner as a frequency histogram, but the vertical axis displays relative frequencies, rather than frequencies. The cumulative relative frequency is presented using a graph called an **ogive**. Example 2-5 illustrates each of these graphical tools.

**EXAMPLE 2-5** Relative Frequency Histograms and Ogives

**NBA Player Weights** (*continued*) Example 2-4 introduced an analysis of the weights of 505 NBA players during the 2013 season. To graph the relative frequencies with the ogive, perform the following steps:

STEP **1** **Create the frequency distribution.**
Using the last histogram in Example 2-4, create the following frequency distribution:

| NBA Player Weight | Frequency |
|---|---|
| [157, 177] | 22 |
| (177, 197] | 90 |
| (197, 217] | 117 |
| (217, 237] | 135 |
| (237, 257] | 97 |
| (257, 277] | 36 |
| (277, 297] | 8 |

STEP **2** **Create relative and cumulative relative frequencies.**
Relative frequencies are calculated by dividing each frequency by the sum of the frequencies (505). Cumulative frequencies are found by the running sum of the relative frequencies.

| NBA Player Weight | Frequency | Relative Frequency | Cumulative Relative Frequency |
|---|---|---|---|
| [157, 177] | 22 | 0.0436 | 0.0436 |
| (177, 197] | 90 | 0.1782 | 0.2218 |
| (197, 217] | 117 | 0.2317 | 0.4535 |
| (217, 237] | 135 | 0.2673 | 0.7208 |
| (237, 257] | 97 | 0.1921 | 0.9129 |
| (257, 277] | 36 | 0.0713 | 0.9842 |
| (277, 297] | 8 | 0.0158 | 1.0000 |
| | 505 | 1.0000 | |

STEP **3** **Construct the relative frequency histogram.**
Copy the relative frequency table into Excel to create the relative frequency histogram. Place the quantitative variable on the horizontal axis and the relative frequencies on the vertical axis. The vertical bars are drawn to heights corresponding to the relative frequencies of the classes.

STEP **4** **Format the relative frequency histogram.**
Right mouse click on the bars and use the Format Data Series options to set the gap width to zero. Use the Fill and Line option to add solid black lines to the bars. Add an appropriate title.

NBA Player Weight Relative Frequency Histogram

(Excel 2016)

Note that the relative frequency histogram has exactly the same shape as the frequency histogram. However, the vertical axis has a different scale.

STEP 5 **Construct the ogive.**

a. Right mouse click Chart, click Select Data, and then add a Series. In the Edit Series Dialog box under Series Value, highlight the cumulative relative frequency column (exclude the label) and then click OK. It will show both charts at this point as two bar charts.

b. Right mouse click Chart and select Change Series Chart Type. Under Combo, change Series 2 to a Line chart and select Secondary Axis. Click OK.

NBA Player Weight Relative Frequency Histogram and Ogive

(Excel 2016)

**TRY EXERCISE 2-15** (pg. 48)

OUTCOME **3**    ## Joint Frequency Distributions

Frequency distributions are effective tools for describing data. Thus far, we have discussed how to develop grouped and ungrouped frequency distributions for one variable at a time. For instance, in the Capital Credit Union example, we were interested in customer credit card balances for all customers. We constructed a frequency distribution and histogram for that variable. However, often we need to examine data that are characterized by more than one variable. This may involve constructing a *joint frequency distribution* for two variables. Joint frequency distributions can be constructed for qualitative or quantitative variables.

**EXAMPLE 2-6** Joint Frequency Distribution

**Miami City Parking** Parking is typically an issue in large cities like Miami, Florida. Problems seem to occur for customers and employees both in locating a parking spot and in being able to quickly exit a lot at busy times. The parking manager for Miami City Parking has received complaints about the time required to exit garages in the downtown sector. To start analyzing the situation, she has collected a small sample of data from 12 customers showing the type of payment (cash or charge) and the garage number (Garage Number 1, 2, or 3). One possibility is that using credit card payments increases exit times at the parking lots. The manager wishes to develop a joint frequency distribution to better understand the paying habits of those using her garages. To do this, she can use the following steps:

**HOW TO DO IT** (Example 2-6)

**Constructing Joint Frequency Distributions**

**1.** Obtain a set of data consisting of paired responses for two variables. The responses can be qualitative or quantitative. If the responses are quantitative, they can be discrete or continuous.

**2.** Construct a table with *r* rows and *c* columns, in which the number of rows represents the number of categories (or numeric classes) of one variable and the number of columns corresponds to the number of categories (or numeric classes) of the second variable.

**3.** Count the number of joint occurrences at each row level and each column level for all combinations of row and column values, and place these frequencies in the appropriate cells.

**4.** Compute the row and column totals, which are called the *marginal frequencies*.

**STEP 1** **Obtain the data.**
The paired data for the two variables for a sample of 12 customers are obtained.

| Customer | Payment Method | Parking Garage |
|----------|----------------|----------------|
| 1 | Charge | 2 |
| 2 | Charge | 1 |
| 3 | Cash | 2 |
| 4 | Charge | 2 |
| 5 | Charge | 1 |
| 6 | Cash | 1 |
| 7 | Cash | 3 |
| 8 | Charge | 1 |
| 9 | Charge | 3 |
| 10 | Cash | 2 |
| 11 | Cash | 1 |
| 12 | Charge | 1 |

**STEP 2** **Construct the rows and columns of the joint frequency table.**
The row variable will be the payment method, and two rows will be used, corresponding to the two payment methods. The column variable is parking garage number, and it will have three levels, because the data for this variable contain only the values 1, 2, and 3. (Note that if a variable is continuous, classes should be formed using the methods discussed in Example 2-3.)

| | Parking Garage | | |
|--------|---|---|---|
| | 1 | 2 | 3 |
| Charge | | | |
| Cash | | | |

**STEP 3** **Count the number of joint occurrences at each row level and each column level for all combinations of row and column values, and place these frequencies in the appropriate cells.**

| | Parking Garage | | | |
|--------|---|---|---|-------|
| | 1 | 2 | 3 | Total |
| Charge | 4 | 2 | 1 | 7 |
| Cash | 2 | 2 | 1 | 5 |
| Total | 6 | 4 | 2 | 12 |

**STEP 4** **Calculate the row and column totals (see Step 3).**
The manager can now see that for this sample, most people charged their parking fee (seven people), and Garage number 1 was used by most people in the sample used (six people). Likewise, four people used Garage number 1 and charged their parking fee.

**TRY EXERCISE 2-11** (pg. 47)

### Joint Frequency Distribution

Excel Tutorial

**Capital Credit Union** (*continued*)   Recall that the Capital Credit Union discussed earlier was interested in evaluating the success of its new credit card. Figure 2.4 showed the frequency distribution and histogram for a sample of customer credit card balances. Although this information is useful, the managers would like to know more. Specifically, what does the credit card balance distribution look like for male versus female cardholders?

One way to approach this is to sort the data by the gender variable and develop frequency distributions and histograms for males and females separately. You could then make a visual comparison of the two to determine what, if any, difference exists between males and females. However, an alternative approach is to jointly analyze the two variables: gender and credit card balance.

Excel provides a means for analyzing two variables jointly. In Figure 2.4, we constructed the frequency distribution for the 300 credit card balances using ten classes. The class width was set at $150. Figure 2.5 shows a table that is called a *joint frequency distribution*. This type of table is also called a *cross-tabulation* table.[1]

### FIGURE 2.5  Excel 2016 Joint Frequency Distribution for Capital Credit Union

**Excel 2016 Instructions**

1. Open file: **Capital.xlsx**.
2. Place cursor anywhere in the data.
3. On the **Insert** tab, click on **PivotTable** and click **OK**.
4. Drag **Credit Card Account Balance** to **Rows**.
5. Right click in **Credit Card Account Balance** numbers and click **Group**.
6. Change **Start** to 90. Change **End** to 1589. Change **By** to 150.
7. Drag **Gender** to **Columns**.
8. Drag **Credit Card Account Balance** to **Values**.
9. Change **Summarize Values** to **Count**.
10. On the **Design** tab in the **Layout Group**, click the **Report Layout** arrow and select **Show in Tabular Form**.

| Count of Credit Card Account Balance | Gender 1 = Male 2 = Fema... | | |
|---|---|---|---|
| Credit Card Account Balance | 1 | 2 | Grand Total |
| 90-239 | 11 | 2 | 13 |
| 240-389 | 16 | 3 | 19 |
| 390-539 | 33 | 9 | 42 |
| 540-689 | 45 | 16 | 61 |
| 690-839 | 35 | 12 | 47 |
| 840-989 | 41 | 9 | 50 |
| 990-1139 | 28 | 8 | 36 |
| 1140-1289 | 14 | 6 | 20 |
| 1290-1439 | 8 | 2 | 10 |
| 1440-1589 | 1 | 1 | 2 |
| Grand Total | 232 | 68 | 300 |

The Capital Credit Union managers can use a joint frequency table to analyze the credit card balances for males versus females. For instance, for the 42 customers with balances of $390 to $539, Figure 2.5 shows that 33 were males and 9 were females. Previously, we discussed the concept of relative frequency (proportions, which Excel converts to percentages) as a useful tool for making comparisons between two data sets. In this example, comparisons between males and females would be easier if the frequencies were converted to proportions (or percentages). The result is the *joint relative frequency table* shown in Figure 2.6. Notice that the percentages in each cell are percentages of the total 300 people in the survey. For example, the $540-to-$689 class had 20.33% (61) of the 300 customers. The male customers with balances in the $540-to-$689 range constituted 15% (45) of the 300 customers, whereas females with that balance level made up 5.33% (16) of all 300 customers. On the surface, this result seems to indicate a big difference between males and females at this credit balance level.

Suppose we really wanted to focus on the male versus female issue and control for the fact that there are far more male customers than female. We could compute the percentages differently. Rather than using a base of 300 (the entire sample size), we might instead be interested in the percentages of the males who have balances at each level, and the same measure for females.[2]

There are many options for transferring data into useful information. Thus far, we have introduced frequency distributions, joint frequency tables, and histograms. In the next section, we discuss one of the most useful graphical tools: the bar chart.

---

[1] In Excel, the joint frequency distribution is developed using a tool called Pivot tables.
[2] Such distributions are known as *marginal distributions*.

**FIGURE 2.6 Excel 2016 Joint Relative Frequencies for Capital Credit Union**

**Excel 2016 Instructions**

1. Place cursor in the **Gender** numbers of the PivotTable.
2. Right click and select **Value Field Settings**.
3. On the **Show values as** tab, click on the down arrow and select **% of Grand Total**.
4. Click **OK**.

| B8 | | | $f_x$ | 15% | | |
|---|---|---|---|---|---|---|
| | A | B | C | D | E | F |
| 1 | | | | | | |
| 2 | | | | | | |
| 3 | Count of Credit Card Account Balance | Gender 1 = Male 2 = Fema | | | | |
| 4 | Credit Card Account Balance | 1 | 2 | Grand Total | | |
| 5 | 90-239 | 3.67% | 0.67% | 4.33% | | |
| 6 | 240-389 | 5.33% | 1.00% | 6.33% | | |
| 7 | 390-539 | 11.00% | 3.00% | 14.00% | | |
| 8 | 540-689 | 15.00% | 5.33% | 20.33% | | |
| 9 | 690-839 | 11.67% | 4.00% | 15.67% | | |
| 10 | 840-989 | 13.67% | 3.00% | 16.67% | | |
| 11 | 990-1139 | 9.33% | 2.67% | 12.00% | | |
| 12 | 1140-1289 | 4.67% | 2.00% | 6.67% | | |
| 13 | 1290-1439 | 2.67% | 0.67% | 3.33% | | |
| 14 | 1440-1589 | 0.33% | 0.33% | 0.67% | | |
| 15 | Grand Total | 77.33% | 22.67% | 100.00% | | |

In Figure 2.6, we have used the Value Field Settings in the Excel PivotTable to represent the data as percentages.

# 2.1 EXERCISES

## Skill Development

**2-1.** Given the following data, develop a frequency distribution:

| | | | | |
|---|---|---|---|---|
| 5 | 3 | 2 | 6 | 6 |
| 7 | 3 | 3 | 6 | 7 |
| 7 | 9 | 7 | 5 | 3 |
| 12 | 6 | 10 | 7 | 2 |
| 6 | 8 | 0 | 7 | 4 |

**2-2.** Assuming you have data for a variable with 2,000 values, using the $2^k \geq n$ guideline, what is the smallest number of groups that should be used in developing a grouped data frequency distribution?

**2-3.** A study is being conducted in which a variable of interest has 1,000 observations. The minimum value in the data set is 300 points and the maximum is 2,900 points.
   a. Use the $2^k \geq n$ guideline to determine the minimum number of classes to use in developing a grouped data frequency distribution.
   b. Based on the answer to part a, determine the class width that should be used (round up to the nearest 100 points).

**2-4.** Produce the relative frequency distribution from a sample size of 50 that gave rise to the following ogive:

**Ogive**

**2-5.** You have the following data:

| | | | | |
|---|---|---|---|---|
| 8 | 6 | 11 | 14 | 10 |
| 11 | 9 | 7 | 2 | 8 |
| 9 | 5 | 5 | 5 | 12 |
| 7 | 8 | 4 | 17 | 8 |
| 12 | 7 | 8 | 8 | 7 |
| 10 | 8 | 6 | 9 | 9 |
| 11 | 16 | 2 | 7 | 4 |
| 8 | 4 | 4 | 5 | 5 |
| 9 | 9 | 6 | 6 | 7 |
| 7 | 9 | 5 | 4 | 5 |
| 14 | 2 | 9 | 0 | 6 |
| 1 | 1 | 12 | 11 | 4 |

   a. Construct a frequency distribution for these data. Use the $2^k \geq n$ guideline to determine the number of classes to use.
   b. Develop a relative frequency distribution using the classes you constructed in part a.
   c. Develop a cumulative frequency distribution and a cumulative relative frequency distribution using the classes you constructed in part a.
   d. Develop a histogram based on the frequency distribution you constructed in part a.

**2-6.** The following cumulative relative frequency distribution summarizes data obtained in a study of the ending overages (in dollars) for the cash register balance at a business:
   a. Determine the proportion of the days in which there were no shortages.
   b. Determine the proportion of the days the cash register was less than $20 off.
   c. Determine the proportion of the days the cash register was less than $40 over or at most $20 short.

| Class | Frequency | Relative Frequency | Cumulative Relative Frequency |
|---|---|---|---|
| −60.00−<−40.00 | 2 | 0.04 | 0.04 |
| −40.00−<−20.00 | 2 | 0.04 | 0.08 |
| −20.00−<−00.00 | 8 | 0.16 | 0.24 |
| 00.00−< 20.00 | 16 | 0.32 | 0.56 |
| 20.00−< 40.00 | 20 | 0.40 | 0.96 |
| 40.00−< 60.00 | 2 | 0.04 | 1.00 |

**2-7.** You are given the following data:

| | | | | | |
|---|---|---|---|---|---|
| 6 | 10 | 6 | 4 | 9 | 5 |
| 5 | 5 | 5 | 7 | 6 | 2 |
| 5 | 5 | 5 | 4 | 5 | 7 |
| 6 | 7 | 8 | 6 | 8 | 4 |
| 7 | 5 | 5 | 5 | 5 | 7 |
| 8 | 7 | 6 | 7 | 5 | 4 |
| 6 | 4 | 4 | 7 | 4 | 6 |
| 6 | 7 | 8 | 6 | 7 | 6 |
| 7 | 8 | 5 | 6 | 5 | 7 |
| 3 | 6 | 4 | 7 | 4 | 4 |

a. Construct a frequency distribution for these data.
b. Based on the frequency distribution, develop a histogram.
c. Construct a relative frequency distribution.
d. Develop a relative frequency histogram.
e. Compare the two histograms. Why do they look alike?

**2-8.** Using the data from Exercise 2-7:
  a. Construct a grouped data relative frequency distribution of the data. Use the $2^k \geq n$ guideline to determine the number of classes.
  b. Construct a cumulative frequency distribution of the data.
  c. Construct a relative frequency histogram.
  d. Construct an ogive.

## Business Applications

**2-9.** A Burger King manager selected a random sample of sales receipts. She was interested in the number of line items on the receipts. For instance, if a customer ordered two 1/4-pound hamburgers, one side of fries, and two soft drinks, the number of line items would be five. The following data were observed:

| | | | | | | | | | |
|---|---|---|---|---|---|---|---|---|---|
| 7 | 5 | 7 | 6 | 5 | 5 | 4 | 8 | 6 | 5 |
| 8 | 7 | 6 | 5 | 6 | 2 | 9 | 4 | 4 | 5 |
| 8 | 4 | 9 | 6 | 6 | 5 | 8 | 9 | 9 | 1 |
| 6 | 5 | 10 | 6 | 7 | 6 | 5 | 5 | 5 | 6 |
| 8 | 7 | 6 | 8 | 6 | 6 | 9 | 6 | 12 | 7 |

| | | | | | | | | | |
|---|---|---|---|---|---|---|---|---|---|
| 5 | 6 | 7 | 11 | 4 | 4 | 3 | 4 | 1 | 4 |
| 11 | 2 | 5 | 5 | 8 | 2 | 3 | 4 | 9 | 6 |
| 6 | 5 | 8 | 6 | 3 | 6 | 4 | 5 | 8 | 10 |

a. Develop a frequency distribution for these data. Discuss briefly what the frequency distribution tells you about these sample data.
b. Based on the results in part a, construct a frequency histogram for these sample data.

**2-10.** In a survey conducted by a brokerage company, investors were asked to rate how knowledgeable they felt they were as investors. Both online and traditional investors were included in the survey. The survey resulted in the following data:

Of the online investors, 8%, 55%, and 37% responded they were "savvy," "experienced," and "novice," respectively.

Of the traditional investors, the percentages were 4%, 29%, and 67%, respectively.

Of the 600 investors surveyed, 200 were traditional investors.

a. Use the information to construct a joint frequency distribution.
b. Use the information to construct a joint relative frequency distribution.
c. Determine the proportion of investors who were both online investors and rated themselves experienced.
d. Calculate the proportion of investors who were online investors.

**2-11.** The sales manager for a TV station affiliate in a southern Florida city recently surveyed 20 advertisers and asked each one to rate the service of the station on the following scale:

| Very Good | Good | Fair | Poor | Very Poor |
|---|---|---|---|---|
| 1 | 2 | 3 | 4 | 5 |

He also tracked the general time slot when the advertiser's commercials were shown on the station. The following codes were used:
1 = morning     2 = afternoon
3 = evening     4 = various times
The following sample data were observed:

| Rating | Time Slot | Rating | Time Slot |
|---|---|---|---|
| 2 | 1 | 4 | 3 |
| 1 | 1 | 2 | 2 |
| 3 | 3 | 3 | 3 |
| 2 | 1 | 3 | 3 |
| 1 | 1 | 2 | 1 |
| 4 | 4 | 1 | 1 |
| 2 | 2 | 1 | 1 |
| 1 | 1 | 5 | 3 |
| 2 | 1 | 2 | 4 |
| 2 | 2 | 3 | 4 |

a. Construct separate relative frequency distributions for each of the two variables.
b. Construct a joint frequency distribution for these two variables.
c. Construct a joint relative frequency distribution for these two variables. Write a short paragraph describing what the data imply.

**2-12.** A St. Louis–based shipping company recently selected a random sample of 49 airplane weight slips for crates shipped from an automobile parts supplier. The weights, measured in pounds, for the sampled crates are as follows:

| | | | | | | |
|----|----|----|-----|----|----|----|
| 89 | 83 | 97 | 101 | 86 | 89 | 86 |
| 91 | 84 | 89 | 87  | 93 | 86 | 90 |
| 86 | 92 | 92 | 88  | 88 | 92 | 86 |
| 93 | 80 | 93 | 77  | 98 | 94 | 95 |
| 94 | 88 | 95 | 87  | 99 | 98 | 90 |
| 91 | 87 | 89 | 89  | 96 | 88 | 94 |
| 95 | 79 | 94 | 86  | 92 | 94 | 85 |

a. Create a data array of the weights.
b. Develop a frequency distribution using five classes having equal widths.
c. Develop a histogram from the frequency distribution you created in part b.
d. Develop a relative frequency and a cumulative relative frequency distribution for the weights using the same five classes created in part b. What percent of the sampled crates have weights greater than 96 pounds?

**2-13.** The bubble in U.S. housing prices burst in 2008, causing sales of houses to decline in almost every part of the country. Many homes were foreclosed because the owners could not make the payments. Below is a sample of 100 residential properties and the total balance on the mortgage at the time of foreclosure.

| | | | | |
|----------|----------|----------|----------|----------|
| $172,229 | $211,021 | $159,205 | $247,697 | $247,469 |
| $176,736 | $240,815 | $195,056 | $315,097 | $257,150 |
| $129,779 | $207,451 | $165,225 | $178,970 | $319,101 |
| $ 87,429 | $219,808 | $242,761 | $277,389 | $213,803 |
| $153,468 | $205,696 | $210,447 | $179,029 | $241,331 |
| $117,808 | $188,909 | $376,644 | $185,523 | $168,145 |
| $158,094 | $135,461 | $131,457 | $263,232 | $256,262 |
| $240,034 | $289,973 | $302,341 | $178,684 | $226,998 |
| $176,440 | $268,106 | $181,507 | $118,752 | $251,009 |
| $196,457 | $195,249 | $195,986 | $201,680 | $233,182 |
| $271,552 | $123,262 | $212,411 | $246,462 | $177,673 |
| $103,699 | $252,375 | $192,335 | $265,992 | $232,247 |
| $320,004 | $213,020 | $192,546 | $295,660 | $211,876 |
| $265,787 | $207,443 | $203,043 | $133,014 | $289,645 |
| $251,560 | $302,054 | $185,381 | $284,345 | $184,869 |
| $237,485 | $282,506 | $278,783 | $335,920 | $199,630 |
| $248,272 | $232,234 | $188,833 | $168,905 | $357,612 |
| $241,894 | $186,956 | $114,601 | $301,728 | $251,865 |
| $207,040 | $221,614 | $318,154 | $156,611 | $219,730 |
| $201,473 | $174,840 | $196,622 | $263,686 | $159,029 |

a. Using the $2^k \geq n$ guideline, what is the minimum number of classes that should be used to display these data in a grouped data frequency distribution?
b. Referring to part a, what should the class width be, assuming you round the width up to the nearest $1,000?
c. Referring to parts a and b, develop a grouped data frequency distribution for these mortgage balance data.
d. Based on your answer to part c, construct and interpret a frequency histogram for the mortgage balance data.

**2-14.** A national executive placement company provides compensation information on more than 170 benchmark positions, including human resource managers at Fortune 500 companies. It recently reported that salaries ranged from $127,735 to $209,981 (before bonuses). Suppose the following data represent a sample of the annual salaries for 25 HR directors. Assume that data are in thousands of dollars.

| | | | | | | | |
|-------|-------|-------|-------|-------|-------|-------|-------|
| 173.1 | 171.2 | 141.9 | 112.6 | 211.1 | 156.5 | 145.4 | 134.0 |
| 192.0 | 185.8 | 168.3 | 131.0 | 214.4 | 155.2 | 164.9 | 123.9 |
| 161.9 | 162.7 | 178.8 | 161.3 | 182.0 | 165.8 | 213.1 | 177.4 |
| 159.3 | | | | | | | |

a. Using 11 classes, construct a cumulative frequency distribution.
b. Determine the proportion of HR director salaries that are at least $175,000.
c. Determine the proportion of HR director salaries that are less than $205,000 and at least $135,000.

**2-15.** One effect of the great recession was to lower the interest rate on fixed-rate mortgages. A sample of 30-year fixed-rate mortgage rates taken from financial institutions in the Pacific Northwest resulted in the following:

| | | | | |
|------|------|------|------|------|
| 3.79 | 4.03 | 3.92 | 3.87 | 3.86 |
| 3.93 | 3.87 | 3.69 | 3.99 | 3.88 |
| 3.91 | 3.81 | 3.85 | 3.81 | 3.65 |
| 4.15 | 3.98 | 3.82 | 4.08 | 3.84 |
| 3.95 | 4.03 | 3.96 | 3.69 | 3.97 |
| 4.08 | 3.86 | 3.82 | 3.83 | 4.19 |
| 3.49 | 3.49 | 3.75 | 3.46 | 3.92 |
| 3.67 | 3.63 | 3.81 | 3.86 | 4.15 |
| 3.85 | 4.01 | 3.93 | 4.02 | 4.09 |

a. Using Excel, construct a histogram with eight classes.
b. Determine the proportion of 30-year fixed mortgage rates that are at least 3.76%.
c. Produce an ogive for the data.

## Computer Software Exercises

**2-16.** The 21st annual J.D. Power and Associates' customer-satisfaction survey, the Automotive Performance, Execution and Layout (APEAL) Study[SM], measures owners' satisfaction with the design, content, layout, and performance of their new vehicles. A file titled **APEAL** contains the satisfaction ratings for 2015 for each make of car.
   a. Using Excel, construct a histogram that has class widths of 15 for the APEAL ratings.
   b. The industry average APEAL rating was 790 for 2008. What do the 2015 data suggest in terms of the relative satisfaction with the 2015 models?

**2-17.** The Franklin Tire Company is interested in demonstrating the durability of its steel-belted radial tires. To do this, the managers have decided to put four tires on 100 different sport utility vehicles and drive them throughout Alaska. The data collected indicate the number of miles (rounded to the nearest 1,000 miles) that each of the SUVs traveled before one of the tires on the vehicle did not meet minimum federal standards for tread thickness. The data file is called **Franklin**.
   a. Construct a histogram using eight classes.
   b. The marketing department wishes to know the tread life of at least 50% of the tires, the 10% that had the longest tread life, and the longest tread life of these tires. Provide this information to the marketing department. Also provide any other significant items that point out the desirability of this line of steel-belted tires.
   c. Construct a histogram using 12 classes. Compare your results with those in parts a and b. Which distribution gives better information about the desirability of this line of steel-belted tires? Discuss.

**2-18.** The California Golf Association recently conducted a survey of its members. Among other questions, the members were asked to indicate the number of 18-hole rounds that they played last year. Data for a sample of 294 members are provided in the data file called **Golf Survey**.
   a. Using the $2^k \geq n$ guideline, what is the minimum number of classes that should be used to display these data in a grouped data frequency distribution?
   b. Referring to part a, what should the class width be, assuming you round the width up to the nearest integer?
   c. Construct and interpret a frequency histogram for the data.

**2-19.** A marketing research firm was hired to conduct a study aimed at determining the number of alcoholic drinks individuals consumed during a week's time in the Seattle, Washington, area. The firm sampled 100 adults and asked each to record the number of *alcoholic drinks* consumed during the third week of October. The file called **Alcoholic Drinks** contains the data for the 84 people who completed the study. Construct a histogram that shows the distribution for the number of drinks consumed.

**2-20.** The Orlando International Airport is busy throughout the year. Among the variety of data collected by the Greater Orlando Aviation Authority is the number of passengers by airline. The file **Orlando Airport** contains passenger data for December 2011. Suppose the airport manager is interested in analyzing the column labeled "Total" for these data.
   a. Using the $2^k \geq n$ guideline, what is the minimum number of classes that should be used to display the data in the "Total" column in a grouped data frequency distribution?
   b. Referring to part a, what should the class width be, assuming you round the width up to the nearest 1,000 passengers?
   c. Construct and interpret a frequency histogram for the data.

**2-21.** The manager of AJ's Fitness Center, a full-service health exercise club, recently conducted a survey of 1,214 members. The objective of the survey was to determine the satisfaction level of his club's customers. In addition, the survey asked for several demographic factors such as age and gender. The data from the survey are in a file called **AJFitness**.
   a. One of the key variables is "Overall Service Satisfaction." This variable is measured on an ordinal scale as follows:
   5 = very satisfied 4 = satisfied 3 = neutral
   2 = dissatisfied 1 = very dissatisfied
   Develop a frequency distribution for this variable and discuss the results.
   b. Develop a joint relative frequency distribution for the variables "Overall Service Satisfaction" and "Typical Visits per Week." Discuss the results.

**2-22.** The file **Danish Coffee** contains data on individual coffee consumption (in kg) for 144 randomly selected Danish coffee drinkers.
   a. Construct a histogram of the coffee consumption data. Briefly comment on what the histogram reveals concerning the data.
   b. Develop a relative frequency distribution and a cumulative relative frequency distribution of the coffee data. What percentage of the coffee drinkers sampled consume 8.0 kg or more annually?

| 2.2 | Bar Charts, Pie Charts, and Stem and Leaf Diagrams |

**OUTCOME 4**

## Bar Charts

Section 2.1 introduced some of the basic tools for describing numerical variables, both discrete and continuous, when the data are in their raw form. However, in many instances, you will be working with categorical data or data that have already been summarized to some extent. In these cases, an effective presentation tool is often a **bar chart**.

**Bar Chart**

A graphical representation of a categorical data set in which a rectangle or bar is drawn over each category or class. The length or height of each bar represents the frequency or percentage of observations or some other measure associated with the category. The bars may be vertical or horizontal. The bars may all be the same color or they may be different colors depicting different categories. Additionally, multiple variables can be graphed on the same bar chart.

| BUSINESS APPLICATION | Developing Bar Charts |

**New Car Sales**   The automobile industry is a significant part of the U.S economy. When car sales are up, the economy is up, and vice versa. Table 2.8 displays data showing the total number of cars sold in the United States in 2015 by the ten largest automobile companies in the world. Although the table format is informative, a graphical presentation is often desirable. Because the car sales data are given by car company, a bar chart would work well in this instance. The bars on a bar chart can be vertical (called a *column bar chart*) or horizontal (called a *horizontal bar chart*). Figure 2.7 is an example of a column bar chart. The height of the bars corresponds to the number of cars sold by each company. This gives you an idea of the sales advantage held by General Motors and Ford in 2015.

One strength of the bar chart is its capability of displaying multiple variables on the same chart. For instance, a bar chart can conveniently compare new car sales data for 2015 and sales for the previous year. Figure 2.8 is a horizontal bar chart that does just that. Notice that all companies posted gains between 2014 and 2015, with Fiat Chrysler highest at 7.22%.

**TABLE 2.8** 2015 New Car Sales for the Top Ten Automobile Companies (United States)

| Car Company | 2015 Sales |
|---|---|
| Daimler | 380,461 |
| BMW | 405,377 |
| Volkswagen | 606,840 |
| Hyundai Kia | 1,387,528 |
| Nissan | 1,484,918 |
| Honda | 1,586,551 |
| Fiat Chrysler | 2,257,728 |
| Toyota | 2,499,313 |
| Ford | 2,603,082 |
| General Motors | 3,082,366 |

*Source:* www.goodcarbadcar.net

**FIGURE 2.7** **Bar Chart Showing 2015 New Car Sales**
(*Source:* www.goodcarbadcar.net)

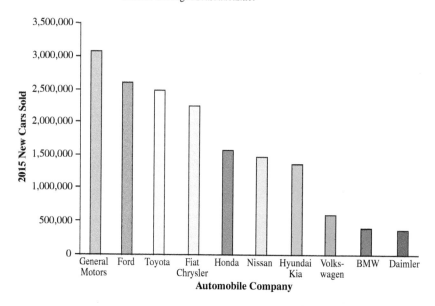

People sometimes confuse histograms and bar charts. Although there are some similarities, they are two very different graphical tools. Histograms are used to represent a frequency distribution associated with a single quantitative (ratio- or interval-level) variable. Refer to the histograms in Section 2.1. In every case, the variable on the horizontal axis was numerical, with values moving from low to high. The vertical axis shows the frequency count, or relative frequency, for each numerical value or range of values. There are no gaps between the histogram bars. On the other hand, bar charts are used when one or more variables of interest are categorical, as in this case in which the category is "car company."

**FIGURE 2.8 Bar Chart Comparing 2014 and 2015 New Cars Sold**

(*Source:* www.goodcarbadcar.net)

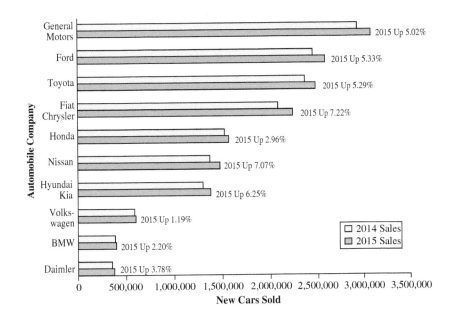

---

**HOW TO DO IT** (Example 2-7)

**Constructing Bar Charts**

**1.** Define the categories for the variable of interest.

**2.** For each category, determine the appropriate measure or value.

**3.** For a column bar chart, locate the categories on the horizontal axis. The vertical axis is set to a scale corresponding to the values in the categories. For a horizontal bar chart, place the categories on the vertical axis and set the scale of the horizontal axis in accordance with the values in the categories. Then construct bars, either vertical or horizontal, for each category such that the length or height corresponds to the value for the category.

**EXAMPLE 2-7**     **Bar Charts**

**Price/Earnings Ratios**     The S&P 500, or the Standard & Poor's 500, is an American stock market index based on the market capitalizations of 500 large companies that have common stock listed on the NYSE or NASDAQ. A key measure of importance to investors is the price earnings (PE) ratio. The PE ratio is computed by dividing a stock's price by the earnings of the company. Companies with high PE ratios may be overpriced, while companies with low PE ratios may be underpriced. Each year on January 1, Standard & Poor's computes the combined PE ratio for all 500 companies in the S&P 500 Index. We are interested in developing a bar chart that displays the S&P 500 PE ratios for the years 2010 through 2016.

STEP 1     **Define the categories.**

Data for the PE ratios for the seven years are shown as follows:

| Year | S&P 500 PE Ratio |
|------|------------------|
| 2016 | 21.21 |
| 2015 | 20.02 |
| 2014 | 18.15 |
| 2013 | 17.03 |
| 2012 | 14.87 |
| 2011 | 16.3 |
| 2010 | 20.07 |

The category to be displayed is the year.

STEP 2     **Determine the appropriate measure to be displayed.**

The measure of interest is the PE ratio.

STEP 3   **Develop the bar chart.**

A column bar chart is developed by placing the seven years on the horizontal axis and constructing bars whose heights correspond to the values of the S&P PE ratio. Each year is assigned a different-colored bar. The resulting bar chart is

STEP 4   **Interpret the results.**

The bar chart shows that the S&P PE ratio computed on January 1, 2016, was the highest of the seven years, while the 2012 PE ratio was the lowest.

**TRY EXERCISE 2-26** (pg. 56)

---

**BUSINESS APPLICATION**   **Constructing Bar Charts**

Excel Tutorial

**Bach, Lombard, & Wilson**   One of the most useful features of bar charts is that they can display multiple issues. Consider Bach, Lombard, & Wilson, a New England law firm. Recently, the firm handled a case in which a woman was suing her employer, a major electronics firm, claiming the company gave higher starting salaries to men than to women. Consequently, she stated, even though the company tended to give equal-percentage raises to women and men, the gap between the two groups widened.

Attorneys at Bach, Lombard, & Wilson had their staff assemble massive amounts of data. Table 2.9 provides an example of the type of data they collected. A bar chart is a more effective way to convey this information, as Figure 2.9 shows. From this graph we can quickly see that in all years except 2013, the starting salaries for males did exceed those for females. The bar chart also illustrates that the general trend in starting salaries for both groups has been increasing, though with a slight downturn in 2015. Do you think the information in Figure 2.9 alone is sufficient to rule in favor of the claimant in this lawsuit? Bar charts like the one in Figure 2.9 that display two or more variables are referred to as *cluster bar charts*.

**TABLE  2.9**   Starting Salary Data

| Year | Males: Average Starting Salaries | Females: Average Starting Salaries |
|------|----------------------------------|------------------------------------|
| 2010 | $45,000 | $41,000 |
| 2011 | $48,000 | $46,000 |
| 2012 | $53,000 | $51,000 |
| 2013 | $58,000 | $59,000 |
| 2014 | $62,000 | $59,000 |
| 2015 | $59,000 | $54,000 |
| 2016 | $64,000 | $62,000 |

**FIGURE 2.9 Bar Chart of Starting Salaries**

**Starting Salaries — Males and Females**

**Pie Chart**
A graph in the shape of a circle. The circle is divided into "slices" corresponding to the categories or classes to be displayed. The size of each slice is proportional to the magnitude of the displayed variable associated with each category or class.

## Pie Charts

Another graphical tool that can be used to transform data into information is the **pie chart**.

---

**HOW TO DO IT** (Example 2-8)

**Constructing Pie Charts**

**1.** Define the categories for the variable of interest.

**2.** For each category, determine the appropriate measure or value. The value assigned to each category is the proportion the category is of the total for all categories.

**3.** Construct the pie chart by displaying one slice for each category that is proportional in size to the proportion the category value is of the total for all categories.

---

**EXAMPLE 2-8    Pie Charts**

**Golf Equipment**    A golf club manufacturer recently conducted a survey of 300 golfers. The survey asked questions about the impact of new technology on the game. One question asked the golfers to indicate which area of golf equipment is most responsible for improving an amateur golfer's game. The following data were obtained:

| Equipment | Frequency |
|---|---|
| Golf ball | 81 |
| Club head material | 66 |
| Shaft material | 63 |
| Club head size | 63 |
| Shaft length | 3 |
| Don't know | 24 |

To display these data in pie chart form, use the following steps:

STEP 1   **Define the categories.**
The categories are the six equipment-response categories.

STEP 2   **Determine the appropriate measure.**
The appropriate measure is the proportion of the golfers surveyed. The proportion for each category is determined by dividing the number of golfers in a category by the total sample size. For example, for the category "golf ball," the proportion is $81/300 = 0.27 = 27\%$.

STEP 3   **Construct the pie chart.**
The pie chart is constructed by dividing a circle into six slices (one for each category) such that each slice is proportional to the percentage of golfers in the category.

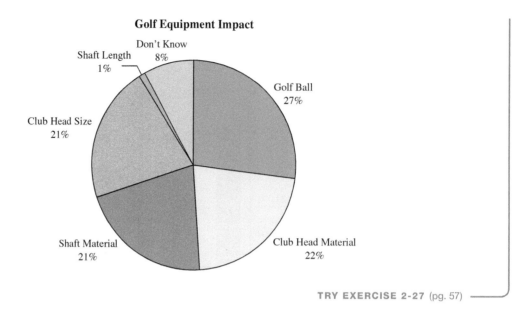

**Golf Equipment Impact**

**TRY EXERCISE 2-27** (pg. 57)

**FIGURE 2.10 Pie Chart: Per-Student Funding for Universities**

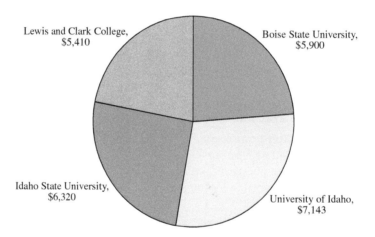

Pie charts are sometimes mistakenly used when a bar chart would be more appropriate. For example, a few years ago the student leaders at Boise State University wanted to draw attention to the funding inequities among the four public universities in Idaho. To do so, they rented a large billboard adjacent to a major thoroughfare through downtown Boise. The billboard contained a large pie chart like the one shown in Figure 2.10, where each slice indicated the funding per student at a given university. For a pie chart to be appropriate, the slices of the pie should represent parts of a total. But in the case of the billboard, that was not the case. The amounts merely represented the dollars of state money spent per student at each university. The sum of the four dollar amounts on the pie chart was a meaningless number. In this case, a bar chart like that shown in Figure 2.11 would have been more appropriate.

**OUTCOME 5**    ## Stem and Leaf Diagrams

Another graphical technique useful for doing an exploratory analysis of quantitative data is called the *stem and leaf diagram*. The stem and leaf diagram is similar to the histogram introduced in Section 2.1 in that it displays the distribution for the quantitative variable. However, unlike the histogram, in which the individual values of the data are lost if the variable of interest is broken into classes, the stem and leaf diagram shows the individual data values.

**FIGURE 2.11 Bar Chart: Per-Student Funding for Universities**

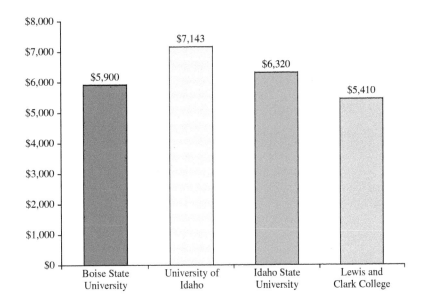

**HOW TO DO IT** (Example 2-9)

**Constructing Stem and Leaf Diagrams**

**1.** Sort the data from low to high.

**2.** Analyze the data for the variable of interest to determine how you wish to split the values into a stem and a leaf.

**3.** List all possible stems in a single column between the lowest and highest values in the data.

**4.** For each stem, list all leaves associated with the stem.

---

**EXAMPLE 2-9** | **Stem and Leaf Diagrams**

**Walk-In Health Clinic** The administrator for the Walk-In Health Clinic is interested in the number of patients who enter the clinic daily. One method for analyzing the data for a sample of 200 days is the stem and leaf diagram. The following data represent the number of patients on each of the 200 days:

| | | | | | | | | | |
|---|---|---|---|---|---|---|---|---|---|
| 113 | 112 | 63 | 127 | 110 | 129 | 142 | 115 | 192 | 94 |
| 165 | 121 | 105 | 140 | 85 | 93 | 105 | 140 | 93 | 126 |
| 183 | 118 | 67 | 104 | 162 | 110 | 76 | 109 | 91 | 132 |
| 88 | 96 | 132 | 80 | 144 | 112 | 57 | 139 | 123 | 124 |
| 172 | 149 | 198 | 114 | 88 | 111 | 133 | 117 | 138 | 134 |
| 53 | 147 | 108 | 109 | 153 | 89 | 159 | 99 | 130 | 93 |
| 161 | 118 | 115 | 117 | 128 | 98 | 125 | 184 | 134 | 132 |
| 117 | 127 | 166 | 72 | 122 | 109 | 124 | 92 | 82 | 69 |
| 110 | 128 | 151 | 67 | 142 | 177 | 135 | 121 | 143 | 89 |
| 160 | 115 | 138 | 79 | 104 | 76 | 89 | 110 | 44 | 140 |
| 117 | 103 | 59 | 109 | 145 | 117 | 162 | 108 | 141 | 139 |
| 148 | 175 | 107 | 117 | 87 | 87 | 150 | 152 | 80 | 168 |
| 88 | 127 | 131 | 85 | 143 | 101 | 137 | 111 | 128 | 147 |
| 110 | 81 | 111 | 149 | 154 | 90 | 150 | 117 | 101 | 116 |
| 153 | 176 | 112 | 147 | 87 | 177 | 190 | 66 | 62 | 154 |
| 143 | 122 | 176 | 153 | 97 | 106 | 86 | 62 | 146 | 98 |
| 134 | 135 | 127 | 118 | 109 | 143 | 146 | 152 | 140 | 95 |
| 102 | 137 | 158 | 69 | 122 | 135 | 136 | 129 | 91 | 136 |
| 135 | 86 | 131 | 154 | 132 | 59 | 136 | 85 | 142 | 137 |
| 155 | 190 | 120 | 154 | 102 | 109 | 97 | 157 | 144 | 149 |

The stem and leaf diagram is constructed using the following steps:

**STEP 1** **Sort the data from low to high.**

The lowest value is 44 patients and the highest value is 198 patients.

**STEP 2** **Split the values into a stem and leaf.**

Leaf = units place;   Stem = all digits left of the units place

For example, for the value 113, the stem is 11 and the leaf is 3. We are keeping one digit for the leaf.

STEP 3    List all possible stems from lowest to highest.

STEP 4    Itemize the leaves from lowest to highest and place next to the appropriate stems.

| | |
|---|---|
| 4 | 4 |
| 5 | 3 7 9 9 |
| 6 | 2 2 3 6 7 7 9 9 |
| 7 | 2 6 6 9 |
| 8 | 0 0 1 2 5 5 5 6 6 7 7 7 8 8 8 9 9 9 |
| 9 | 0 1 1 2 3 3 3 4 5 6 7 7 8 8 9 |
| 10 | 1 1 2 2 3 4 4 5 5 6 7 8 8 9 9 9 9 9 9 |
| 11 | 0 0 0 0 0 1 1 1 2 2 2 3 4 5 5 5 6 7 7 7 7 7 7 7 8 8 8 |
| 12 | 0 1 1 2 2 2 3 4 4 5 6 7 7 7 7 8 8 8 9 9 |
| 13 | 0 1 1 2 2 2 2 3 4 4 4 5 5 5 5 6 6 6 7 7 7 8 8 9 9 |
| 14 | 0 0 0 0 1 2 2 2 3 3 3 3 4 4 5 6 6 7 7 7 8 9 9 9 |
| 15 | 0 0 1 2 2 3 3 3 4 4 4 4 5 7 8 9 |
| 16 | 0 1 2 2 5 6 8 |
| 17 | 2 5 6 6 7 7 |
| 18 | 3 4 |
| 19 | 0 0 2 8 |

The stem and leaf diagram shows that most days have between 80 and 160 patients, with the most frequent value in the 110- to 120-patient range.

TRY EXERCISE 2-24 (pg. 56)

## 2.2 EXERCISES

### Skill Development

**2-23.** The following data reflect the percentages of employees with different levels of education:

| Education Level | Percentage |
|---|---|
| Less than high school graduate | 18 |
| High school graduate | 34 |
| Some college | 14 |
| College graduate | 30 |
| Graduate degree | 4 |
| | Total = 100 |

a. Develop a pie chart to illustrate these data.

b. Develop a horizontal bar chart to illustrate these data.

**2-24.** Given the following data, construct a stem and leaf diagram:

| | | | |
|---|---|---|---|
| 0.7 | 1.7 | 2.8 | 3.8 |
| 0.8 | 1.8 | 3.3 | 4.3 |
| 1.0 | 2.0 | 4.4 | 5.4 |
| 1.1 | 2.1 | 5.3 | 6.3 |
| 1.4 | 2.4 | 5.4 | 6.4 |
| 2.0 | 3.0 | | |

**2-25.** A university has the following number of students at each grade level:

| | |
|---|---|
| Freshman | 3,450 |
| Sophomore | 3,190 |
| Junior | 2,780 |
| Senior | 1,980 |
| Graduate | 750 |

a. Construct a bar chart that effectively displays these data.

b. Construct a pie chart to display these data.

c. Referring to the graphs constructed in parts a and b, indicate which you would favor as the more effective way of presenting these data. Discuss.

**2-26.** Given the following sales data for product type and sales region, construct at least two different bar charts that display the data effectively:

| | Region | | | |
|---|---|---|---|---|
| Product Type | East | West | North | South |
| XJ-6 Model | 200 | 300 | 50 | 170 |
| X-15-Y Model | 100 | 200 | 20 | 100 |
| Craftsman | 80 | 400 | 60 | 200 |
| Generic | 100 | 150 | 40 | 50 |

**2-27.** An annual report for Murphy Oil Corporation reports the following refinery yields, in barrels per day, by product category for the United States and the United Kingdom.

### United States

| Product Category | Refinery Yields— barrels per day |
|---|---|
| Gasoline | 61,128 |
| Kerosene | 11,068 |
| Diesel and Home Heating Oils | 41,305 |
| Residuals | 18,082 |
| Asphalt, LPG, and other | 14,802 |
| Fuel and Loss | 834 |

### United Kingdom

| Product Category | Refinery Yields— barrels per day |
|---|---|
| Gasoline | 20,889 |
| Kerosene | 11,374 |
| Diesel and Home Heating Oils | 25,995 |
| Residuals | 8,296 |
| Asphalt, LPG, and other | 14,799 |
| Fuel and Loss | 2,810 |

a. Construct a pie chart to display United States refinery yields by product per day. Display the refinery yields data for each product category as a percentage of total refinery yields for all product categories.

b. Construct a pie chart to display United Kingdom refinery yields by product per day. Display the refinery yields data for each product category as a percentage of total refinery yields for all product categories.

c. Construct a bar chart that effectively compares United States and United Kingdom refinery yields by product category.

**2-28.** Boston Properties is a real estate investment trust (REIT) that owns first-class office properties in selected markets. According to its website (www .bostonproperties.com), the 2014 occupancy rates for Boston Properties holdings in four major cities were:

| Region | 2014 Occupancy Rate |
|---|---|
| New York | 91.5% |
| Washington, D.C. | 92.9% |
| Boston | 91.1% |
| San Francisco | 88.6% |

a. Explain why a pie chart would not be appropriate to display these occupancy rate data.

b. Construct a bar chart to display the net operating income distribution by city.

c. Briefly comment on why a bar chart would be preferred over a pie chart for these data.

**2-29.** Hileman Services Company recently released the following data concerning its operating profits (in $billions) for the five years:

| Year | 2012 | 2013 | 2014 | 2015 | 2016 |
|---|---|---|---|---|---|
| Profit | 0.5 | 0.1 | 0.7 | 0.5 | 0.2 |

a. Construct a bar chart to graphically display these data.

b. Construct a pie chart to graphically display these data.

c. Referring to parts a and b, which graph would you recommend? Provide reasons for your choice.

**2-30.** National Football League (NFL) football continues to be arguably the most popular professional sport in the United States. The following data reflect the 2015 attendance at the eight regular-season home games for five NFL teams:

| Team | Home Games | Attendance |
|---|---|---|
| St. Louis | 8 | 419,220 |
| Minnesota | 8 | 419,440 |
| Oakland | 8 | 436,910 |
| Detroit | 8 | 490,782 |
| Cincinnati | 8 | 491,118 |

Produce a bar chart for these data.

## Business Applications

**2-31.** At the March meeting of the board of directors for the Graystone Services Company, one of the regional managers put the following data into a PowerPoint presentation to illustrate the ratio of the number of units manufactured to the number of employees at each of Graystone's five manufacturing plants:

| Plant Location | Units Manufactured/Employees |
|---|---|
| Bismarck, ND | 14.5 |
| Boulder, CO | 9.8 |
| Omaha, NE | 13.0 |
| Harrisburg, PA | 17.6 |
| Portland, ME | 5.9 |

a. Discuss whether a pie chart or a bar chart would be more appropriate to present these data graphically.

b. Construct the chart you recommended in part a.

**2-32.** The first few years after the turn of the century saw a rapid increase in housing values, followed by a rapid decline due in part to the sub-prime crisis. The following table indicates the increase in the number of homes valued at more than one million dollars before 2005:

| Year | Number of $1 Million Homes |
|------|----------------------------|
| 2000 | 394,878 |
| 2001 | 495,600 |
| 2002 | 595,441 |
| 2003 | 714,467 |
| 2004 | 1,034,386 |

Develop a horizontal bar chart to represent these data in graphical form.

**2-33.** The following data show the July 2011–June 2012 U.S. sales of antidepressant drugs by major brand:

| Antidepressant Drug | Sales (in Billions) |
|---------------------|---------------------|
| Cymbalta | 4.094 |
| Lexapro | 2.107 |
| Escitalopram | 0.65 |
| Prestig | 0.555 |

*Source:* "Top antidepressant drugs in the United States based on revenue in 2011–2012 (in million U.S. dollars)," www.statista.com

Construct an appropriate graph to display these data.

**2-34.** The 2014 Annual Report for Murphy Oil Corporation (see www.murphyoilcorp.com) shows the following numbers of barrels of crude oil produced per day in the United States annually from 2008 through 2014:

| 2008 | 2009 | 2010 | 2011 | 2012 | 2013 | 2014 |
|------|------|------|------|------|------|------|
| 10,668 | 17,053 | 19,579 | 13,936 | 12,804 | 11,943 | 14,366 |

Develop a bar chart that illustrates these data. Add appropriate labels.

**2-35.** The 2014 Annual Report of the Procter & Gamble Company (see www.pginvestor.com) reported the following percentages for its five global business segments:

| Global Segment | Percent of Total Sales |
|----------------|------------------------|
| Beauty | 24% |
| Grooming | 10% |
| Health Care | 9% |
| Fabric Care and Home Care | 32% |
| Baby Care and Family Care | 25% |

a. Construct a bar chart that displays this information by global business segment.
b. Construct a pie chart that displays each global business segment's percentage of total global segment net sales.

**2-36.** A fast-food restaurant monitors its drive-thru service times electronically to ensure that its speed of service is meeting the company's goals. A sample of 28 drive-thru times was recently taken and is shown here:

| Speed of Service (Time in Seconds) | | | |
|------|------|------|------|
| 83 | 138 | 145 | 147 |
| 130 | 79 | 156 | 156 |
| 90 | 85 | 68 | 93 |
| 178 | 76 | 73 | 119 |
| 92 | 146 | 88 | 103 |
| 116 | 134 | 162 | 71 |
| 181 | 110 | 105 | 74 |

a. Construct a stem and leaf diagram of the speed of service times.
b. What range of time might the restaurant say is the most frequent speed of service?

**2-37.** A random sample of 30 customer records for a physician's office showed the following times (in days) required to collect insurance payments:

| Number of Days to Collect Payment | | | | |
|----|----|----|----|----|
| 34 | 55 | 36 | 39 | 36 |
| 32 | 35 | 30 | 47 | 31 |
| 60 | 66 | 48 | 43 | 33 |
| 24 | 37 | 38 | 65 | 35 |
| 22 | 45 | 33 | 29 | 41 |
| 38 | 35 | 28 | 56 | 56 |

a. Construct a stem and leaf diagram of these data.
b. Within what range of days are most payments collected?

**2-38.** The following data represent the seat capacity for major airlines on an annual basis:

| Airline Seat Capacity (in Billions) | | | | |
|------------------|--------|--------|------------|-----|
| Airline Capacity | United | Delta | Northwest | U.S. Airways | ATA |
| | 145 | 130 | 92 | 54 | 21 |

a. Construct a bar graph representing the seat capacity of the major airlines for the five airlines indicated.
b. Produce a pie chart exhibiting the percentage of the total seat capacity for the five major airlines.

**2-39.** A survey conducted at a state university that asked students what brand of digital music device they owned yielded the following data:

| Corporation | Apple | Samsung | LG | Nokia | HTC |
|-------------|-------|---------|-----|-------|-----|
| Percentage | 74% | 6.4% | 3.9% | 3.6% | 2.6% |

a. Generate a bar chart to display these data.
b. Generate a pie chart to display these data.
c. Which of the two displays more effectively presents the data? Explain your answer.

## Computer Software Exercises

**2-40.** Suppose that cars with high consumer satisfaction ratings have a tendency to stay on dealers' lots a shorter period of time. As an example, suppose the Honda Ridgeline had stayed on dealers' lots an average of 24 days. The file titled **Honda** contains 50 lengths of stay on dealers' lots for Ridgeline trucks.
Construct a stem and leaf display for these data.

**2-41.** The manager for a credit union has selected a random sample of 300 of the credit union's credit card customers. The data are in a file called **Capital**. The manager is interested in graphically displaying the percentage of card holders of each gender.
a. Determine the appropriate type of graph to use in this application.
b. Construct the graph and interpret it.

**2-42.** A health insurance company selected a random sample of hospitals from each of four categories of hospitals: university related, religious related, community owned, and privately owned. At issue is the hospital charges associated with outpatient gall bladder surgery. The data are in the file called **Hospitals**.
a. Compute the average charge for each hospital category.
b. Construct a bar chart showing the averages by hospital category.
c. Discuss why a pie chart would not in this case be an appropriate graphical tool.

**2-43.** Amazon.com has become one of the success stories of the Internet age. Its growth can be seen by examining its increasing sales volume (in $billions) and the net income/loss during Amazon's operations. A file titled **Amazon** contains these data for its first 17 years (*source:* 2011 Amazon Annual Report).
a. Construct one bar graph illustrating the relationship between sales and income for each separate year of Amazon's existence.
b. Describe the type of relationship that exists between the years in business and Amazon's sales volume.
c. Amazon's sales rose sharply. However, its net income yielded losses, which increased during the first few years. In which year did this situation reverse itself and show improvement in the net income balance sheet?

**2-44.** In your capacity as assistant to the administrator at Freedom Hospital, you have been asked to develop a graphical presentation that focuses on the insurance carried by the geriatric patients at the hospital. The data file **Patients** contains data for a sample of geriatric patients. In developing your presentation, please do the following:
a. Construct a pie chart that shows the percentage of patients with each health insurance payer.
b. Develop a bar chart that shows total charges for patients by insurance payer.
c. Develop a stem and leaf diagram for the length-of-stay variable.
d. Develop a bar chart that shows the number of males and females by insurance carrier.

---

| 2.3 | **Line Charts, Scatter Diagrams, and Pareto Charts** |

**OUTCOME 6**

## Line Charts

**Line Chart**
A two-dimensional chart showing time on the horizontal axis and the variable of interest on the vertical axis.

Most of the examples that have been presented thus far have involved *cross-sectional data*, or data gathered from many observations, all taken at the same time. However, if you have *time-series* data that are measured over time (e.g., monthly, quarterly, annually), an effective tool for presenting such data is a **line chart**. A line chart is also commonly called a trend chart.

Excel Tutorial

**BUSINESS APPLICATION**    Constructing Line Charts

**McGregor Vineyards**    McGregor Vineyards owns and operates a winery in the Sonoma Valley in northern California. At a recent company meeting, the financial manager expressed concern about the company's profit trend over the past 20 weeks. He presented weekly profit and sales data to McGregor management personnel. The data are in the file **McGregor**.

Initially, the financial manager developed two separate line charts for these data: one for sales, the other for profits. These are displayed in Figure 2.12. These line charts provide an indication that, although sales have been increasing, the profit trend is downward. But to fit both Excel graphs on one page, he had to compress the size of the graphs. This "flattened" the lines somewhat, masking the magnitude of the problem.

FIGURE 2.12 **Excel 2016 Output Showing McGregor Line Charts for Sales and Profits**

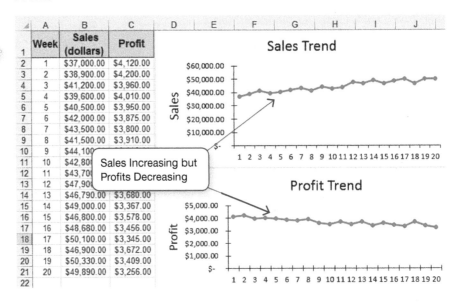

FIGURE 2.13 **Excel 2016 Line Chart of McGregor Profit and Sales Using a Single Vertical Axis**

What the financial manager needed is one graph with both profits and sales. Figure 2.13 shows his first attempt. This is better, but there still is a problem: the sales and profit variables are of different magnitudes. This results in the profit line being flattened out to almost a straight line. The profit trend is hidden.

To overcome this problem, the financial manager needed to construct his graph using two scales, one for each variable. Figure 2.14 shows the improved graph. We can now clearly see that although sales are moving steadily higher, profits are headed downhill. For some reason, costs are rising faster than revenues, and this graph should motivate McGregor Vineyards to look into the problem.

**FIGURE 2.14** Excel 2016 Line Chart for Sales and Profits Using Two Vertical Axes and Different Scales

---

**EXAMPLE 2-10** Line Charts

**Grogan Builders**   Grogan Builders produces mobile homes. The owners are planning to expand the manufacturing facilities. To do so requires additional financing. In preparation for the meeting with the bankers, the owners have assembled data on total annual sales for the past ten years. These data are shown as follows:

| 2008 | 2009 | 2010 | 2011 | 2012 | 2013 | 2014 | 2015 | 2016 | 2017 |
|------|------|------|------|------|------|------|------|------|------|
| 1,426 | 1,678 | 2,591 | 2,105 | 2,744 | 3,068 | 2,755 | 3,689 | 4,003 | 3,997 |

The owners wish to present these data in a line chart to effectively show the company's sales growth over the ten-year period. To construct the line chart, the following steps are used:

STEP **1**  **Identify the time-series variable.**
The time-series variable is units sold measured over ten years, with a maximum value of 4,003.

STEP **2**  **Construct the horizontal and vertical axes.**
The horizontal axis will have the ten time periods equally spaced. The vertical axis will start at zero and go to a value exceeding 4,003. We will use 4,500. The vertical axis will also be divided into 500-unit increments.

STEP **3**  **Plot the data values on the graph and connect the points with straight lines.**

TRY EXERCISE 2-46 (pg. 65)

**OUTCOME** **7**

**Scatter Diagram, or Scatter Plot**
A two-dimensional graph of plotted points in which the vertical axis represents values of one quantitative variable and the horizontal axis represents values of the other quantitative variable. Each plotted point has coordinates whose values are obtained from the respective variables.

**Dependent Variable**
A variable whose values are thought to be a function of, or dependent on, the values of another variable called the *independent variable*. On a scatter plot, the dependent variable is placed on the *y* axis and is often called the response variable.

**Independent Variable**
A variable whose values are thought to impact the values of the *dependent variable*. The independent variable, or explanatory variable, is often within the direct control of the decision maker. On a scatter plot, the independent variable, or explanatory variable, is graphed on the *x* axis.

**Excel Tutorial**

## Scatter Diagrams

In Section 2.1, we introduced a set of statistical procedures known as joint frequency distributions that allow the decision maker to examine two variables at the same time. Another procedure used to study two quantitative variables simultaneously is the **scatter diagram**, or the **scatter plot**.

There are many situations in which we are interested in understanding the *bivariate* relationship between two *quantitative* variables. For example, a company would like to know the relationship between sales and advertising. A bank might be interested in the relationship between savings account balances and credit card balances for its customers. A real estate agent might wish to know the relationship between the selling price of houses and the number of days that the houses have been on the market. The list of possibilities is almost limitless.

Regardless of the variables involved, we are looking for several key relationships when we develop a scatter diagram. Figure 2.15 shows scatter diagrams representing some key bivariate relationships that might exist between two quantitative variables.

Chapters 14 and 15 make extensive use of scatter diagrams. They introduce a statistical technique called *regression analysis* that focuses on the relationship between two variables. These variables are known as **dependent** and **independent variables**.

**FIGURE 2.15 Scatter Diagrams Showing Relationships Between *x* and *y***

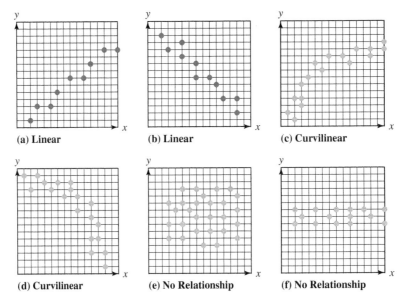

(a) Linear     (b) Linear     (c) Curvilinear

(d) Curvilinear     (e) No Relationship     (f) No Relationship

---

**BUSINESS APPLICATION**    **Creating Scatter Diagrams**

**Personal Computers**    Can you think of any product that has increased in quality and capability as rapidly as personal computers? Not that many years ago, an 8-MB RAM system with a 486 processor and a 640-KB hard drive sold in the mid-$2,500 range. Now the same money might buy a last-generation super computer!

The data file called **Personal Computers** contains historical data from the early 2000s on several personal computer characteristics, including processor speed, hard drive capacity, RAM, whether a monitor is included, and personal computer price. Of particular interest is the relationship between the computer price and processing speed. Our objective is to develop a scatter diagram to graphically depict what, if any, relationship exists between these two variables. The dependent variable is price and the independent variable is processor speed. Figure 2.16 shows the Excel scatter diagram output. The relationship between processor speed and price is somewhat curvilinear and positive.

**FIGURE 2.16 Excel 2016 Scatter Diagrams for Personal Computer Data**

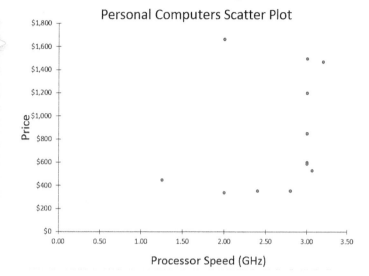

Personal Computers Scatter Plot

**HOW TO DO IT** (Example 2-11)

**Constructing Scatter Diagrams**

1. Identify the two quantitative variables and collect paired responses for the two variables.

2. Determine which variable will be placed on the vertical axis and which variable will be placed on the horizontal axis. Often the vertical axis can be considered the dependent variable ($y$) and the horizontal axis can be considered the independent variable ($x$).

3. Define the range of values for each variable and define the appropriate scale for the $x$ and $y$ axes.

4. Plot the joint values for the two variables by placing a point in the $x,y$ space. Do not connect the points.

**EXAMPLE 2-11**   **Scatter Diagrams**

**Main Drive Taxi Service**   The owner of Maine Drive Taxi Service has selected a sample of customer trips and has recorded the data on three variables: Distance in Miles, Trip Duration in Minutes, and Tip Amount to Driver. These data are as follows:

| Trip | Distance (Miles) | Trip Duration (Minutes) | Tip to Driver (% of Fare) |
|------|------------------|-------------------------|---------------------------|
| 1 | 4.2 | 7.9 | 14.2% |
| 2 | 8.7 | 16.3 | 10.0% |
| 3 | 0.9 | 3.0 | 28.0% |
| 4 | 3.3 | 7.9 | 20.0% |
| 5 | 13.5 | 24.6 | 15.5% |
| 6 | 5.3 | 13.0 | 20.2% |
| 7 | 4.0 | 9.3 | 22.0% |
| 8 | 11.7 | 19.3 | 14.6% |
| 9 | 2.1 | 5.2 | 23.0% |
| 10 | 1.7 | 4.7 | 20.0% |

To better understand these data, the manager is interested in knowing the relationship between trip distance and trip duration and also between trip distance and the percent tip. Scatter diagrams to visually illustrate these relationships can be constructed using the following steps:

**STEP 1**   **Identify the two variables of interest.**
In the first case, one variable is Distance (Miles) and the other is Trip Duration (Minutes). In the second case, one variable is Distance (Miles) and the other variable is Tip to Diver (% of Fare).

**STEP 2**   **Identify the dependent and independent variables.**
In each case, the independent ($x$) variable is Distance (Miles). Thus,

Case 1: $y$ = Trip Duration (Minutes)    $x$ = Distance (Miles)

Case 2: $y$ = Tip to Driver (% of Fare)    $x$ = Distance (Miles)

**STEP 3**   **Establish scales for the vertical and horizontal axes.**
The maximum value for each variable is:

Distance = 13.5 miles   Trip Duration = 24.6 minutes   Tip to Driver = 28.0%

The lower limit on the scale for each variable will be zero, and the upper limit must be as high or slightly higher than the above upper limits for each variable.

STEP 4 **Plot the joint values for the two variables by placing a point in the $x, y$ space.**

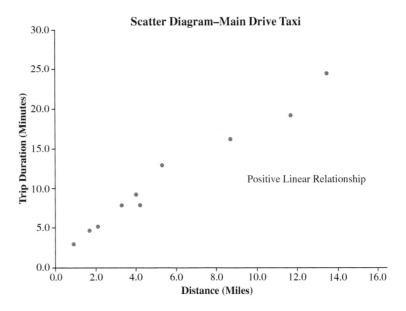

STEP 5 **Interpret the relationship.**

The first scatter diagram shows a strong positive linear relationship between the distance and the time the taxi trip takes. In the second case, the relationship between distance and the percentage amount tipped tends to be negative, which means that the longer the trip, the lower the percentage tip. The linear relationship in the second case does not appear to be as strong as it is in the first case.

**TRY EXERCISE 2-45** (pg. 65)

## Pareto Charts

Pareto charts are a commonly used graphical tool. They are used extensively when dealing with quality improvement applications.

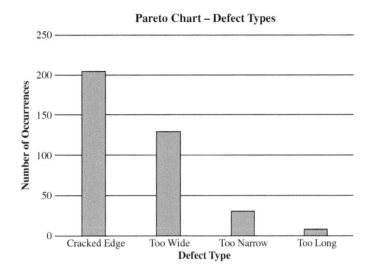

**Pareto Chart – Defect Types**

The Pareto Principle (often called the 80-20 rule), named after Alfredo Pareto, in quality applications indicates that the majority of quality problems are due to only a few causes. For example, a company that makes window trim may be having issues with defective trim pieces being produced. When they study a large number of these defects and assign a defect type to each one, a Pareto chart like that shown above could be developed. You should recognize this as a special type of bar chart where the bars have been arranged in order from high to low. The biggest defect problem based on the Pareto chart is cracked edge. They would want to fix this issue first.

## 2.3 EXERCISES

### Skill Development

**2-45.** The following data represent 11 observations of two quantitative variables:
$x$ = contact hours with client,
$y$ = profit generated from client.

| x | y | x | y | x | y | x | y |
|---|---|---|---|---|---|---|---|
| 45 | 2,345 | 54 | 3,811 | 34 | −700 | 24 | 1,975 |
| 56 | 4,200 | 24 | 2,406 | 45 | 3,457 | 32 | 206 |
| 26 | 278 | 23 | 3,250 | 47 | 2,478 | | |

  a. Construct a scatter diagram of the data. Indicate whether the diagram suggests a linear or nonlinear relationship between the dependent and independent variables.
  b. Determine how much influence extreme data points will have on your perception of the relationship between the independent and dependent variables by deleting the three data points with the smallest $x$ value. What appears to be the relationship between the dependent and independent variables?

**2-46.** You have the following sales data for the past 12 months. Develop a line graph for these data.

| Month | Jan | Feb | Mar | Apr | May | Jun |
|---|---|---|---|---|---|---|
| Sales | 200 | 230 | 210 | 300 | 320 | 290 |
| Month | Jul | Aug | Sep | Oct | Nov | Dec |
| Sales | 300 | 360 | 400 | 410 | 390 | 450 |

**2-47.** The following data have been selected for two variables, $y$ and $x$. Construct a scatter diagram for these two variables and indicate what type of relationship, if any, appears to be present.

| y | x |
|---|---|
| 100 | 23.5 |
| 250 | 17.8 |
| 70 | 28.6 |
| 130 | 19.3 |
| 190 | 15.9 |
| 250 | 19.1 |
| 40 | 35.3 |

**2-48.** The year-end net income (in millions) for a company for the years 2004–2015 are shown below:

| Year | 2004 | 2005 | 2006 | 2007 | 2008 | 2009 |
|------|------|------|------|------|------|------|
| Net Income | 25.9 | 31.5 | 31.7 | 31.7 | 34.5 | 36.8 |
| Year | 2010 | 2011 | 2012 | 2013 | 2014 | 2015 |
| Net Income | 64 | 101.3 | 77 | 87.1 | 118.6 | 172.5 |

Develop a chart that effectively displays the income data over time.

**2-49.** VanAuker Properties' controller collected the following data on annual sales and the years of experience of members of his sales staff:

| Sales $k: | 200 | 191 | 135 | 236 | 305 | 183 | 50 | 192 | 184 | 73 |
|-----------|-----|-----|-----|-----|-----|-----|-----|-----|-----|-----|
| Years: | 10 | 4 | 5 | 9 | 12 | 6 | 2 | 7 | 6 | 2 |

a. Construct a scatter diagram representing these data.
b. Determine the kind of relationship that exists (if any) between years of experience and sales.
c. Approximate the increase in sales that accrues with each additional year of experience for a member of the sales force.

## Business Applications

**2-50.** Amazon.com celebrated its 20th anniversary in July 2014. Its growth can be seen by examining its increasing sales volume (in $billions):

| | |
|------|-------|
| 2008 | 19.17 |
| 2009 | 24.51 |
| 2010 | 34.2 |
| 2011 | 48.08 |
| 2012 | 61.09 |
| 2013 | 74.45 |
| 2014 | 88.99 |

*Source:* 2014 Amazon Annual Report

a. Construct a line plot for Amazon's sales.
b. Describe the type of relationship that exists between the years in business and Amazon's sales volume.

**2-51.** The following data show worldwide computer/video game sales:

| Year | Sales (billions) |
|------|------------------|
| 2000 | $48.2 |
| 2001 | $45.4 |
| 2002 | $48.3 |
| 2003 | $41.9 |
| 2004 | $54.6 |
| 2005 | $44.3 |
| 2006 | $48.4 |
| 2007 | $62.9 |
| 2008 | $77.2 |

| Year | Sales (billions) |
|------|------------------|
| 2009 | $64.5 |
| 2010 | $65.9 |
| 2011 | $66.2 |
| 2012 | $63.0 |
| 2013 | $76.0 |

*Source:* http://vgsales.wikia.com/wiki/Video_game_industry

Construct a line chart showing these computer/video game sales data. Write a short statement describing the graph.

**2-52.** The recent performance of U.S. equity markets has increased the popularity of dividend-paying stocks for some investors. Shown below are the diluted net earnings per common share and the dividends per common share for the Procter & Gamble Company (P&G) for the years 1996–2015 (*source:* P&G Annual Reports, 1996–2015):

| Year | Diluted Net Earnings per Common Share | Dividends per Common Share |
|------|---------------------------------------|----------------------------|
| 1996 | $1.00 | $0.40 |
| 1997 | $1.14 | $0.45 |
| 1998 | $1.28 | $0.51 |
| 1999 | $1.29 | $0.57 |
| 2000 | $1.23 | $0.64 |
| 2001 | $1.03 | $0.70 |
| 2002 | $1.54 | $0.76 |
| 2003 | $1.85 | $0.82 |
| 2004 | $2.32 | $0.93 |
| 2005 | $2.66 | $1.03 |
| 2006 | $2.64 | $1.15 |
| 2007 | $3.04 | $1.28 |
| 2008 | $3.64 | $1.45 |
| 2009 | $4.26 | $1.64 |
| 2010 | $4.11 | $1.80 |
| 2011 | $3.93 | $1.97 |
| 2012 | $3.66 | $2.14 |
| 2013 | $3.86 | $2.29 |
| 2014 | $4.01 | $2.45 |
| 2015 | $2.44 | $2.59 |

a. Construct a line chart of diluted net earnings per common share for the years shown.
b. Construct a line chart of dividends per common share for the years shown.
c. Construct the appropriate chart for determining whether there is a relationship between diluted net earnings per common share and dividends per common share for the years shown. Briefly comment on the nature of any relationship you believe your chart reveals.

## Computer Software Exercises

**2-53.** On July 31, 2015, usatoday.com reported on big oil profits. The file **BIGOIL** contains the quarterly combined net incomes of the oil companies that are part of the S&P 500 Index for 2006 through the first half of 2015. The article discussed that after years of record profits, the oil companies were reporting their lowest profits in years due to the dramatic fall in crude oil prices during the latter part of 2014 and 2015. Develop the appropriate chart to display the oil company net incomes. Be sure to label the chart appropriately.

**2-54.** Having a Major League Baseball team in a city is generally considered to provide an economic boost to the community. The data file **MLB Attendance-2015** contains data for both home and road game attendance for all 30 MLB teams for 2015. Of interest is the relationship between average home attendance and average road attendance. Using the 2015 attendance data, construct the appropriate graph to help determine the relationship between these two variables and discuss the implications of the graph (*source:* http://espn.go.com/mlb/attendance).

**2-55.** Wilson Green started a lawn service business in 1997 with 15 customers. By 2015 his customer list had grown to 6,530 customers across three cities. The file **WilsonGreen** contains data showing the number of customers for each year between 1997 and 2015. Develop a line chart that shows the customer growth trend for Wilson's company.

**2-56.** As an analyst for one of the major airlines, you have been asked to analyze the relationship between passenger capacity and fuel consumption per hour. Data for 17 commonly flown planes are presented in the file called **Airplanes**. Develop the appropriate graph to illustrate the relationship between fuel consumption per hour and passenger capacity. Discuss.

**2-57.** Japolli Bakery tracks sales of its different bread products on a daily basis. The data are located in a data file called **Japolli Bakery**. Develop a line chart that displays these data. Discuss what, if any, conclusions you might be able to reach from the line chart.

**2-58.** The data in the file called **Gasoline Prices** reflect the average price of regular unleaded gasoline in the state of California for 36 consecutive years. The first price column is the actual average price of gasoline during each of those years. Construct an appropriate chart showing the actual average price of gasoline in California over the 15 years.

**2-59.** Federal flood insurance underwritten by the federal government was initiated in 1968. This federal flood insurance coverage has, according to *USA Today* ("How you pay for people to build in flood zones," Sep. 21, 2005), more than tripled between 1990 and 2004. A file titled **Flood** contains the amount of federal flood insurance coverage for each of the years from 1990 to 2004.
   a. Produce a line plot for these data.
   b. Describe the type of relationship between the year and the amount of federal flood insurance.
   c. Determine the average increase per year in federal flood insurance.

**2-60.** The G.H. Hillman Company sells designer clothes under various brand names. The data file called **Hillman** contains the unit sales (in thousands) for their swimsuit line in both the east and west regions for the years 2005 through 2015.
   a. Plot the sales data for the two regions on the same line graph.
   b. Discuss the relationship over time between sales in the west and east regions.

**2-61.** The sub-prime mortgage crisis that hit the world economy also affected the real estate market. Both new and existing home sales were affected. A file titled **EHSales** contains the number of existing homes sold (in millions) from September 2007 to September 2008.
   a. Construct a line plot for these data.
   b. The data file also contains data concerning the median selling price ($thousands). Construct a graph containing the line plot for both the number of sales (*millions*) and the median price of these sales for the indicated time period.
   c. Describe the relationship between the two line plots constructed in part b.

# 2 Overview

## Summary

**2.1** | **Frequency Distributions and Histograms** (pg. 29–49)

**OUTCOME 1** Construct frequency distributions both manually and with your computer.
**OUTCOME 2** Construct and interpret a frequency histogram.
**OUTCOME 3** Develop and interpret joint frequency distributions.

- A **frequency distribution** is used to determine the number of occurrences in your data that fall at each possible data value or within defined ranges of possible values. It represents a summary of the data, and from a frequency distribution you can create a graph called a histogram (see Figure 2.17).
- A **histogram** is a picture of the data's distribution. It shows the center, spread, and shape of the data. Options for histograms include converting the frequencies in a frequency distribution to **relative frequencies** and constructing a **relative frequency distribution** and a relative frequency histogram. Another option is to convert the frequency distribution to a **cumulative frequency distribution** and then a graph called an ogive.
- Finally, if you are analyzing two variables simultaneously, you may want to construct a **joint frequency distribution**.

**2.2** | **Bar Charts, Pie Charts, and Stem and Leaf Diagrams** (pg. 50–59)

**OUTCOME 4** Construct and interpret various types of bar charts.
**OUTCOME 5** Build a stem and leaf diagram.

- **Discrete data** or **nominal-** or **ordinal-level** data are described using bar charts or pie charts (see Figure 2.17).
- A **bar chart** can be arranged with the bars **vertical** or **horizontal**. A single bar chart can be used to describe two or more variables.
- A **pie chart** shows how the parts making up a total are distributed. The "slices" of the pie are often depicted as percentages of the total.
- A **stem and leaf diagram** provides a quick view of how the data are distributed.

**2.3** | **Line Charts, Scatter Diagrams, and Pareto Charts** (pg. 59–67)

**OUTCOME 6** Create a line chart and interpret the trend in the data.
**OUTCOME 7** Construct a scatter plot and interpret it.

- A **line chart** displays **time-series** data. The vertical axis displays the values of the time-series variable, and the horizontal axis contains the time increments. The points are plotted and are usually connected by straight lines (see Figure 2.17).
- A **scatter diagram** shows the relationship between two **quantitative** variables. The **dependent variable** is placed on the vertical axis, and the **independent variable** goes on the horizontal axis. The joint values are plotted as points in the two-dimensional space. The points are not connected with lines.

**FIGURE 2.17 Summary: Descriptive Statistical Techniques**

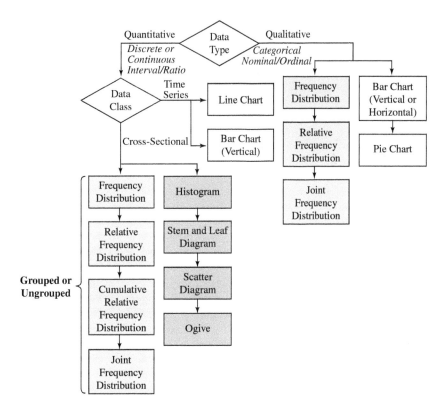

## Equations

**(2.1) Relative Frequency** pg. 32

$$\text{Relative frequency} = \frac{f_i}{n}$$

**(2.2) Class Width** pg. 36

$$W = \frac{\text{Largest value} - \text{Smallest value}}{\text{Number of classes}}$$

## Key Terms

**All-inclusive classes** pg. 35
**Bar chart** pg. 50
**Class boundaries** pg. 36
**Class width** pg. 36
**Continuous data** pg. 33
**Cumulative frequency distribution** pg. 37
**Cumulative relative frequency distribution** pg. 37

**Dependent variable** pg. 62
**Discrete data** pg. 29
**Equal-width classes** pg. 35
**Frequency distribution** pg. 29
**Frequency histogram** pg. 38

**Independent variable** pg. 62
**Line chart** pg. 59
**Mutually exclusive classes** pg. 35
**Ogive** pg. 41
**Pie chart** pg. 53
**Relative frequency** pg. 31
**Scatter diagram or scatter plot** pg. 62

## Chapter Exercises

### Conceptual Questions

**2-62.** Discuss the advantages of constructing a relative frequency distribution as opposed to a frequency distribution.

**2-63.** What are the characteristics of a data set that would lead you to construct a bar chart?

**2-64.** What are the characteristics of a data set that would lead you to construct a pie chart?

**2-65.** Discuss the differences in data that would lead you to construct a line chart as opposed to a scatter plot.

### Business Applications

**2-66.** As it has become more popular to stream movies over the Internet at home from such sources as Hulu and Netflix, ticket sales at movie theaters have been affected. The following data reflect annual ticket sales in billions at U.S. movie theaters between 1995 and 2015. Develop an appropriate chart to describe the trend in ticket sales over this time period.

| Year | 1995 | 1996 | 1997 | 1998 | 1999 | 2000 | 2001 | 2002 | 2003 | 2004 |
|---|---|---|---|---|---|---|---|---|---|---|
| Tickets (billions) | 1.22 | 1.27 | 1.42 | 1.45 | 1.44 | 1.39 | 1.44 | 1.58 | 1.55 | 1.47 |

| Year | 2005 | 2006 | 2007 | 2008 | 2009 | 2010 | 2011 | 2012 | 2013 | 2014 | 2015 |
|---|---|---|---|---|---|---|---|---|---|---|---|
| Tickets (billions) | 1.39 | 1.41 | 1.4 | 1.39 | 1.42 | 1.33 | 1.28 | 1.39 | 1.34 | 1.27 | 1.28 |

*Source:* www.the-numbers.com/market/

**2-67.** The following data represent the commuting distances for employees of the Pay-and-Carry Department store.
   a. The personnel manager for Pay-and-Carry would like you to develop a frequency distribution and histogram for these data.
   b. Develop a stem and leaf diagram for these data.
   c. Break the data into three groups (under 3 miles, 3 and under 6 miles, and 6 and over). Construct a pie chart to illustrate the proportion of employees in each category.

**Commuting Distance (miles)**

| | | | | | | | |
|---|---|---|---|---|---|---|---|
| 3.5 | 2.0 | 4.0 | 2.5 | 0.3 | 1.0 | 12.0 | 17.5 |
| 3.0 | 3.5 | 6.5 | 9.0 | 3.0 | 4.0 | 9.0 | 16.0 |
| 3.5 | 0.5 | 2.5 | 1.0 | 0.7 | 1.5 | 1.4 | 12.0 |
| 9.2 | 8.3 | 1.0 | 3.0 | 7.5 | 3.2 | 2.0 | 1.0 |
| 3.5 | 3.6 | 1.9 | 2.0 | 3.0 | 1.5 | 0.4 | 6.4 |
| 11.0 | 2.5 | 2.4 | 2.7 | 4.0 | 2.0 | 2.0 | 3.0 |

   d. Referring to part c, construct a bar chart to depict the proportion of employees in each category.
**2-68.** The data below show the per-semester tuition costs at two colleges in the Midwest over the past five years:

| | 2013 | 2014 | 2015 | 2016 | 2017 |
|---|---|---|---|---|---|
| Private College Tuition | $9,202 | $12,146 | $13,844 | $16,454 | $19,710 |
| Public College Tuition | $2,074 | $2,395 | $3,188 | $3,632 | $4,694 |

   a. Construct one bar graph illustrating the relationship between private and public university tuitions for the displayed years.
   b. Describe the tuition trend for both private and public college tuition.
**2-69.** A current emphasis in U.S. education is on the STEM disciplines (science, technology, engineering, and mathematics). The following data represent the percentages by gender of bachelor and doctorate degrees in physics:

| | Bachelor | | | Doctorate | | |
|---|---|---|---|---|---|---|
| | 2000 | 2007 | 2013 | 2000 | 2007 | 2013 |
| Men | 78 | 79 | 80 | 87 | 82 | 80 |
| Women | 22 | 21 | 20 | 13 | 18 | 20 |

*Source:* IPEDS Completion Survey & NSF-NIH Survey of Graduate Students & Postdoctorates in Science and Engineering, www.aps.org

   a. Use one graph that contains two bar charts, each of which represents the kind of degree received, to display the relationship between the percentages of men and women receiving each type of degree.
   b. Describe any trends that might be evident.
**2-70.** The Minnesota State Fishing Bureau has contracted with a university biologist to study the length of walleyes (fish) caught in Minnesota lakes. The biologist collected data on a sample of 1,000 fish caught and developed the following relative frequency distribution:

| Class Length (inches) | Relative Frequency $f_i$ |
|---|---|
| 8 to less than 10 | 0.22 |
| 10 to less than 12 | 0.15 |
| 12 to less than 14 | 0.25 |
| 14 to less than 16 | 0.24 |
| 16 to less than 18 | 0.06 |
| 18 to less than 20 | 0.05 |
| 20 to less than 22 | 0.03 |

   a. Construct a frequency distribution from this relative frequency distribution and then produce a histogram based on the frequency distribution.
   b. Construct a pie chart from the relative frequency distribution. Discuss which of the two graphs, the pie chart or the histogram, you think is more effective in presenting the fish length data.
**2-71.** A computer software company has been looking at the amount of time customers spend on hold after their call is answered by the central switchboard. The company would like to have at most 2% of the callers wait two minutes or more. The company's calling service has provided the following data showing how long each of last month's callers spent on hold:

| Class | Number |
|---|---|
| Less than 15 seconds | 456 |
| 15 to less than 30 seconds | 718 |
| 30 to less than 45 seconds | 891 |
| 45 to less than 60 seconds | 823 |
| 60 to less than 75 seconds | 610 |
| 75 to less than 90 seconds | 449 |
| 90 to less than 105 seconds | 385 |
| 105 to less than 120 seconds | 221 |
| 120 to less than 150 seconds | 158 |
| 150 to less than 180 seconds | 124 |
| 180 to less than 240 seconds | 87 |
| More than 240 seconds | 153 |

a. Develop a relative frequency distribution and ogive for these data.

b. The company estimates it loses an average of $30 in business from callers who must wait two minutes or more before receiving assistance. The company thinks that last month's distribution of waiting times is typical. Estimate how much money the company is losing in business per month because people have to wait too long before receiving assistance.

2-72. The regional sales manager for American Toys, Inc., recently collected data on weekly sales (in dollars) for the 15 stores in his region. He also collected data on the number of salesclerk work hours during the week for each of the stores. The data are as follows:

| Store | Sales | Hours |
|-------|-------|-------|
| 1 | 23,300 | 120 |
| 2 | 25,600 | 135 |
| 3 | 19,200 | 96 |
| 4 | 10,211 | 102 |
| 5 | 19,330 | 240 |
| 6 | 35,789 | 190 |
| 7 | 12,540 | 108 |
| 8 | 43,150 | 234 |
| 9 | 27,886 | 140 |
| 10 | 54,156 | 300 |
| 11 | 34,080 | 254 |
| 12 | 25,900 | 180 |
| 13 | 36,400 | 270 |
| 14 | 25,760 | 175 |
| 15 | 31,500 | 256 |

a. Develop a scatter diagram of these data. Determine which variable should be the dependent variable and which should be the independent variable.

b. Based on the scatter diagram, what, if any, conclusions might the sales manager reach with respect to the relationship between sales and number of clerk hours worked? Do any stores stand out as being different? Discuss.

## Computer Software Exercises

2-73. The file titled **Diesel$** contains the average on-highway diesel prices for the last Friday of the month in the years 2013–2015. Develop a line chart that shows the trend in diesel prices, and write a short statement that describes the price trend.

2-74. The data in the file titled **Digital** provide the brand of digital devices owned by a sample of consumers. Produce a pie chart that represents the market shares obtained from the referenced sample. Indicate the market shares and the identity of those manufacturers in the pie chart.

2-75. The file **Home-Prices** contains information about single-family housing prices in 100 metropolitan areas in the United States.

a. Construct a frequency distribution of median single-family home prices. Use the $2^k \geq n$ guideline to determine the appropriate number of classes.

b. Construct a cumulative relative frequency distribution for median single-family home prices.

2-76. Elliel's Department Store tracks its inventory on a monthly basis. Monthly data for 2008 through 2012 are in the file called **Elliels**.

a. Construct a line chart showing the monthly inventory over the five years. Discuss what this graph implies about inventory.

b. Sum the monthly inventory figures for each year. Then present the sums in bar chart form. Discuss whether you think this is an appropriate graph to describe the inventory situation at Elliels.

2-77. The file titled **Diesel$** contains the average nationwide diesel price on the last Friday of the month for the years 2013–2015.

a. Construct a histogram with 11 classes.

b. What are the possible reasons why one of the classes has a frequency count of zero?

## Case 2.1    Server Downtime

After getting outstanding grades in high school and scoring very high on his ACT and SAT tests, Clayton Haney had his choice of colleges but wanted to follow his parents' legacy and enrolled at Northwestern University. Clayton soon learned that there is a big difference between getting high grades in high school and being a good student. Although he was recognized as being quite bright and very quick to pick up on things, he had never learned how to study. As a result, after slightly more than two years at Northwestern, Clayton was asked to try his luck at another university. To the chagrin of his parents, Clayton decided that college was not for him.

After short stints working for a computer manufacturer and as a manager at a convenience store, Clayton landed an entry-level job working for EDS. EDS contracts to support information technology implementation and application for companies in the United States and throughout the world. Clayton received training in virtually all aspects of personal computers and local area networks and was assigned to work for a client in the Chicago area.

Clayton's first assignment was to research the downtime on one of the client's primary network servers. He was asked to study the downtime data for the month of April and to make a short presentation to the company's management. The downtime data are in a file called **Server Downtime**. Although Clayton is very good at solving computer problems, he has had no training or experience in analyzing data, so he is going to need some help.

### Required Tasks:

1. Construct a frequency distribution showing the number of times during the month that the server was down for each downtime cause category.

2. Develop a bar chart that displays the data from the frequency distribution in part a.
3. Develop a histogram that displays the downtime data.
4. Develop a pie chart that breaks down the percentage of total downtime that is attributed to each downtime cause during the month.

5. Prepare a short written report that discusses the downtime data. Make sure you include the graphs and charts in the report.

## Case 2.2  Hudson Valley Apples, Inc.

As a rule, Stacey Fredrick preferred to work in the field rather than do "office" work in her capacity as a midlevel manager with Hudson Valley Apples, Inc., a large grower and processor of apples in the state of New York. However, after just leaving a staff meeting at which she was asked to prepare a report of apple consumption in the United States, Stacey was actually looking forward to spending some time at her computer "crunching some numbers." Arden Golchein, senior marketing manager, indicated that he would e-mail her a data file that contained apple consumption data from 1977 through 2015 and told her that he wanted a report using graphs, charts, and tables to describe apple consumption.

When she got to her desk, the e-mail was waiting, and she saved the file under the name **Hudson Valley Apples**. Stacey had done quite a bit of descriptive analysis in her previous job with the New York State Department of Agriculture, so she had

several ideas for types of graphs and tables that she might construct. She began by creating a list of the tasks that she thought would be needed.

### Required Tasks:

1. Construct a line chart showing the total annual availability of apples.
2. Construct one line chart that shows two things: the annual availability of fresh apples and the annual availability of processed apples.
3. Construct a line chart that shows the annual availability for each type of processed apples.
4. Construct a histogram for the total annual availability of apples.
5. Write a short report that discusses the historical pattern of apple availability. The report will include all pertinent charts and graphs.

## Case 2.3  Pine River Lumber Company—Part 1

Martin Bledsoe wears several hats at the Pine River Lumber Company, including process improvement team leader, shipping manager, and assistant human resources manager. Pine River Lumber makes cedar fencing materials at its Naples, Idaho, facility, employing about 160 people.

More than 75% of the cost of the finished cedar fence boards is in the cedar logs that the company buys on the open market. Therefore, it is very important that the company get as much finished product as possible from each log. One of the most important steps in the manufacturing process is referred to as the head rig. The head rig is a large saw that breaks down the logs into slabs and cants. Figure C-2.3-A shows the concept. From small logs with diameters of 12 inches or less, one cant and four or fewer usable slabs are obtained. From larger logs, multiple cants and four slabs are obtained. Finished fence boards can be produced from both the slabs and the cants.

At some companies, the head rig cutting operation is automated and the cuts are made based on a scanner system and computer algorithms. However, at Pine River Lumber, the head rig is operated manually by operators who must look at a log as it arrives and determine how best to break the log down to get the most finished product. In addition, the operators are responsible for making sure that the cants are "centered" so that maximum product can be gained from them.

Recently, Martin Bledsoe headed up a study in which he videotaped 365 logs being broken down by the head rig. All three

operators, April, Sid, and Jim, were involved. Each log was marked as to its true diameter. Then Martin observed the way the log was broken down and the degree to which the cants were properly centered. He then determined the projected value of the finished product from each log given the way it was actually cut. In addition, he determined what the value would have been had the log been cut in the optimal way. Data for this study are in a file called **Pine River**.

You have been asked to assist Martin by analyzing these data using graphs, charts, and tables as appropriate. He wishes to focus on the lost profit to the company and whether there are differences among the operators. Also, do the operators tend to do a better job on small logs than on large logs? In general, he is hoping to learn as much as possible from this study and needs your help with the analysis.

**FIGURE C-2.3-A  Log Breakdown at the Head Rig**

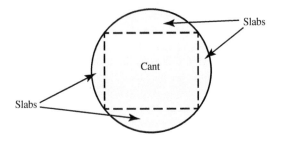

# 3 Describing Data Using Numerical Measures

## WHY YOU NEED TO KNOW

Suppose you are the advertising manager for a major airline and you want to develop an ad campaign touting how much cheaper your fares are than the competition's. You must be careful that your claims are valid. First, the Federal Trade Commission (FTC) is charged with regulating advertising and requires that advertising be truthful. Second, customers who can show that they were misled by an incorrect claim about prices could sue your company. You need to use statistical procedures to determine the validity of any claim you might want to make about your prices. Graphs and charts provide effective tools for transforming data into information; however, they are only a starting point. Graphs and charts do not reveal all the

**Quick Prep**

Review the definitions for nominal, ordinal, interval, and ratio data in Section 1.4.

Examine the statistical software, such as Excel, used during this course to identify the tools for computing descriptive measures. For instance, in Excel, look at the function wizard and the descriptive statistics tools on the Data tab under Data Analysis.

Review the material on frequency histograms in Section 2.1, paying special attention to how histograms help determine where the data are centered and how the data are spread around the center.

information contained in a set of data. To make your descriptive toolkit complete, you need to become familiar with key descriptive measures that are widely used to fully describe data.

You will need to combine the graphical tools discussed in Chapter 2 with the numerical measures introduced in this chapter.

---

**3.1** | ## Measures of Center and Location

You learned in Chapter 2 that frequency histograms are an effective way of converting quantitative data into useful information. The histogram provides a visual indication of where data are centered and how much spread there is in the data around the center. However, to fully describe a quantitative variable, we also need to compute measures of its center and spread. These measures can then be coupled with the histogram to give a clear picture of the variable's distribution. This section focuses on measures of the center of data. Section 3.2 introduces measures of the spread of data.

### Parameters and Statistics

Depending on whether we are working with a population or a sample, a numerical measure is known as either a **parameter** or a **statistic**.

### Population Mean

There are three important measures of the center of a set of data. The first of these is the **mean**, or average, of the data. To find the mean, we sum the values and divide the sum by the number of data values, as shown in Equation 3.1.

**Parameter**
A measure computed from the entire population. As long as the population does not change, the value of the parameter will not change.

**Statistic**
A measure computed from a sample that has been selected from a population. The value of the statistic will depend on which sample is selected.

**Mean**
A numerical measure of the center of a set of quantitative measures computed by dividing the sum of the values by the number of values in the data.

**Population Mean**

$$\mu = \frac{\sum_{i=1}^{N} x_i}{N} \tag{3.1}$$

where:

$\mu$ = Population mean (mu)
$N$ = Population size
$x_i$ = $i$th individual value of variable $x$

**Population Mean**
The average for all values in the population computed by dividing the sum of all values by the population size.

The **population mean** is represented by the Greek symbol $\mu$, pronounced "mu." The formal notation in the numerator for the sum of the $x$ values reads

$$\sum_{i=1}^{N} x_i \rightarrow \text{Sum all } x_i \text{ values where } i \text{ goes from 1 to } N$$

In other words, we are summing all $N$ values in the population.

Because you almost always sum all the data values, to simplify notation in this text, we generally will drop the subscripts after the first time we introduce a formula. Thus, the formula for the population mean will be written as

$$\mu = \frac{\sum x}{N}$$

**OUTCOME 1** | **BUSINESS APPLICATION** Population Mean

**The San Carlo Hotel** The manager of a small hotel in the wine country near Napa, California, was asked by the hotel's owner to analyze the Sunday night registration information for the past eight weeks. Data on three variables were collected:

**TABLE 3.1** San Carlo Hotel Data

| Week | Rooms Rented | Revenue | Complaints |
|------|-------------|---------|------------|
| 1 | 22 | $1,870 | 0 |
| 2 | 13 | $1,590 | 2 |
| 3 | 10 | $1,760 | 1 |
| 4 | 16 | $2,345 | 0 |
| 5 | 23 | $4,563 | 2 |
| 6 | 13 | $1,630 | 1 |
| 7 | 11 | $2,156 | 0 |
| 8 | 13 | $1,756 | 0 |

$x_1$ = Total number of rooms rented

$x_2$ = Total dollar revenue from the room rentals

$x_3$ = Number of customer complaints that came from guests each Sunday

These data are shown in Table 3.1. They are a population because they include all data that interest the owner.

Figure 3.1 shows the frequency histogram for the number of rooms rented. If the manager wants to describe the data further, she can locate the center of the data by finding the balance point for the histogram. Think of the horizontal axis as a plank and the histogram bars as weights proportional to their area. The center of the data would be the point at which the plank would balance. As shown in Figure 3.1, the balance point seems to be about 15 rooms.

Eyeing the histogram might yield a reasonable approximation of the center. However, computing a numerical measure of the center directly from the data is preferable. The most frequently used measure of the center is the mean. The population mean for number of rooms rented is computed using Equation 3.1 as follows:

> The Excel 2016 function for the mean is
>
> =**Average(22,13,10,16,23, 13,11,13)**

$$\mu = \frac{\sum x}{N} = \frac{22 + 13 + 10 + 16 + 23 + 13 + 11 + 13}{8}$$

$$= \frac{121}{8}$$

$$\mu = 15.125$$

Thus, the average number of rooms rented on Sunday for the past eight weeks is 15.125. This is the true balance point for the data. Take a look at Table 3.2, where we calculate what is called a *deviation* $(x_i - \mu)$ by subtracting the mean from each value, $x_i$.

**FIGURE 3.1** **Balance Point, Rooms Rented at San Carlo Hotel**

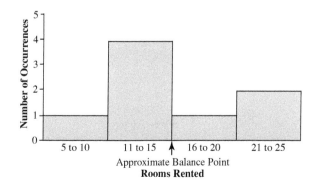

**TABLE 3.2** Deviations Around the Mean Using Hotel Data

| $x$ | $(x - \mu)$ = Deviation |
|---|---|
| 22 | $22 - 15.125 = 6.875$ |
| 13 | $13 - 15.125 = -2.125$ |
| 10 | $10 - 15.125 = -5.125$ |
| 16 | $16 - 15.125 = 0.875$ |
| 23 | $23 - 15.125 = 7.875$ |
| 13 | $13 - 15.125 = -2.125$ |
| 11 | $11 - 15.125 = -4.125$ |
| 13 | $13 - 15.125 = \underline{-2.125}$ |
| | $\sum(x - \mu) = \qquad 0.000 \leftarrow$ Sum of deviations equals zero. |

Note that the sum of the deviations of the data from the mean is zero. This is not a coincidence. *For any set of data, the sum of the deviations around the mean will be zero.*

---

**HOW TO DO IT** (Example 3-1)

**Computing the Population Mean (when the available data constitute the population of interest)**

**1.** Collect the data for the variable of interest for all items in the population. The data must be quantitative.

**2.** Sum all values in the population ($\Sigma x$).

**3.** Divide the sum ($\Sigma x$). by the number of values ($N$) in the population to get the population mean. The formula for the population mean is

$$\mu = \frac{\Sigma x}{N}$$

---

**EXAMPLE 3-1**   **Computing the Population Mean**

**United Airlines**   As the airline industry becomes increasingly competitive, in an effort to increase profits, many airlines are reducing flights. Therefore the supply of idled airplanes has increased. United Airlines, headquartered in Chicago, has decided to expand its fleet. Suppose United selects additional planes from a list of 17 possible planes, including such models as the Boeing 747-400, the Air Bus 300-B4, and the DC 9-10. At a recent meeting, the chief operating officer asked a member of his staff to determine the mean fuel consumption rate per hour of operation for the population of 17 planes.

STEP 1   **Collect data for the quantitative variable of interest.**
The staff member was able to determine, for each of the 17 planes, the hourly fuel consumption in gallons for a flight between Chicago and New York City. These data are recorded as follows:

| Airplane | Fuel Consumption (gal/hr) |
|---|---|
| B747-400 | 3,529 |
| L-1011-100/200 | 2,215 |
| DC-10-10 | 2,174 |
| A300-B4 | 1,482 |
| A310-300 | 1,574 |
| B767-300 | 1,503 |
| B767-200 | 1,377 |
| B757-200 | 985 |
| B727-200 | 1,249 |
| MD-80 | 882 |
| B737-300 | 732 |
| DC-9-50 | 848 |
| B727-100 | 806 |
| B737-100/200 | 1,104 |
| F-100 | 631 |
| DC-9-30-11 | 804 |
| DC-9-10 | 764 |

> The Excel 2016 function for the mean is
> =Average(3529,2215,2174, ...,764)

STEP 2   **Add the data values.**

$$\sum x = 3{,}529 + 2{,}215 + 2{,}174 + \cdots + 764 = 22{,}659$$

STEP 3   **Divide the sum by the number of values in the population using Equation 3.1.**

$$\mu = \frac{\sum x}{N} = \frac{22{,}659}{17} = 1{,}332.9$$

The mean number of gallons of fuel consumed per hour on these 17 planes is 1,332.9.

---

BUSINESS APPLICATION     Population Mean

Excel Tutorial

**The San Carlo Hotel** (*continued*)   In addition to collecting data on the number of rooms rented on Sunday nights, the San Carlo Hotel manager collected data on the room-rental revenue generated and the number of complaints on Sunday nights. Excel can quite easily be used to calculate numerical measures such as the mean. Because these data are the population of all nights of interest to the hotel manager, she can compute the population mean, $\mu$, revenue per night. The population mean is $\mu = \$2{,}208.75$, as shown in the Excel output in Figure 3.2. Likewise, the mean number of complaints is $\mu = 0.75$ per night. (Note that other measures are shown in the figure. We will discuss several of these later in the chapter.)

Now, for these eight Sunday nights, the manager can report to the hotel's owner that the mean number of rooms rented is 15.13 (rounded up from 15.125). This level of business generated a mean nightly revenue of \$2,208.75. The number of complaints averaged 0.75 (less than 1) per night. These values are the true means for the population and are, therefore, called parameters.

**FIGURE 3.2  Excel 2016 Output Showing Mean Revenue for the San Carlo Hotel**

> **Excel 2016 Instructions**
> 1. Open file: **San Carlo Hotel.**
> 2. Select the **Data** tab.
> 3. Click on **Data Analysis > Descriptive Statistics.**
> 4. Define data range for the desired variables.
> 5. Check **Summary Statistics.**
> 6. Name new Output Sheet.
> 7. On **Home** tab, adjust decimal places as desired.

|   | A | B | C | D | E | F |
|---|---|---|---|---|---|---|
| 1 | Rooms Rented | | Revenue | | Complaints | |
| 2 | | | | | | |
| 3 | Mean | 15.13 | Mean | 2208.75 | Mean | 0.75 |
| 4 | Standard Error | 1.73 | Standard Error | 348.54 | Standard Error | 0.31 |
| 5 | Median | 13 | Median | 1815 | Median | 0.5 |
| 6 | Mode | 13 | Mode | #N/A | Mode | 0 |
| 7 | Standard Deviation | 4.88 | Standard Deviation | 985.81 | Standard Deviation | 0.89 |
| 8 | Sample Variance | 23.84 | Sample Variance | 971821.93 | Sample Variance | 0.79 |
| 9 | Kurtosis | -0.60 | Kurtosis | 6.38 | Kurtosis | -1.48 |
| 10 | Skewness | 0.95 | Skewness | 2.46 | Skewness | 0.62 |
| 11 | Range | 13 | Range | 2,973 | Range | |
| 12 | Minimum | 10 | Minimum | 1,590 | Minimum | |
| 13 | Maximum | 23 | Maximum | 4,563 | Maximum | |
| 14 | Sum | 121 | Sum | 17,670 | Sum | |
| 15 | Count | 8 | Count | 8 | Count | |

> Mean rooms rented = 15.13
> Mean revenue = \$2,208.75
> Mean complaints = 0.75

---

OUTCOME **1**

**Sample Mean**
The average for all values in the sample computed by dividing the sum of all sample values by the sample size.

## Sample Mean

The data for the San Carlo Hotel constituted the population of interest. Thus, $\mu = 15.13$ rooms rented is the parameter measure. However, if the data constitute a sample rather than a population, the mean for the sample (**sample mean**) is computed using Equation 3.2.

**Sample Mean**

$$\bar{x} = \frac{\sum_{i=1}^{n} x_i}{n} \qquad (3.2)$$

where:

$\bar{x} =$ Sample mean (pronounced "$x$-bar")

$n =$ Sample size

Notice, Equation 3.2 is the same as Equation 3.1 *except* that we sum the sample values, not the population values, and divide by the sample size, not the population size.

The notation for the sample mean is $\bar{x}$. Sample descriptors (statistics) are usually assigned a Roman character. (Recall that population values usually are assigned a Greek character.)

---

**EXAMPLE 3-2**   **Computing a Sample Mean**

**Professor Salaries**   A newspaper reporter in Wisconsin collected a sample of seven university professors and determined their annual salaries. As part of her story, she wished to be able to report the mean salary. The following steps are used to calculate the sample mean salaries for professors in Wisconsin:

**STEP 1**   **Collect the sample data.**

$$\{x_i\} = \{\text{Professor salaries}\} = \{\$144,000; \$98,000; \$204,000;$$
$$\$177,000; \$155,000; \$316,000; \$100,000\}$$

**STEP 2**   **Add the values in the sample.**

$$\Sigma x = \$144,000 + \$98,000 + \$204,000 + \$177,000 + \$155,000$$
$$+ \$316,000 + \$100,000 = \$1,194,000$$

> The Excel 2016 function for the mean is
>
> =Average(144000,98000, ...,100000)

**STEP 3**   **Divide the sum by the sample size (Equation 3.2).**

$$\bar{x} = \frac{\Sigma x}{n} = \frac{\$1,194,000}{7} = \$170,571.43$$

Therefore, the mean salary for the sample of seven professors is $170,571.43.

---

## The Impact of Extreme Values on the Mean

The mean (population or sample) is the balance point for data, so using the mean as a measure of the center generally makes sense. However, the mean does have a potential disadvantage: *the mean can be affected by extreme values*. There are many instances in business when this may occur. For example, in a population or sample of income data, there likely will be extremes on the high end that will pull the mean upward from the center. Example 3-3 illustrates how an extreme value can affect the mean. In these situations, a second measure called the *median* may be more appropriate.

---

**EXAMPLE 3-3**   **Impact of Extreme Values**

**Professor Salaries** (*continued*)   Suppose the sample of professor salaries (see Example 3-2) had been slightly different. If the salary recorded as $316,000 had actually been $1,000,000 (must also be the college's football coach!), how would the mean be affected? We can see the impact as follows:

STEP **1**  **Collect the sample data.**

$$\{x_i\} = \{\text{Professor salaries}\} = \{\$144,000; \$98,000; \$204,000;$$
$$\$177,000; \$155,000; \$1,000,000; \$100,000\}$$

extreme value

STEP **2**  **Add the values.**

$$\Sigma x = \$144,000 + 98,000 + 204,000 + 177,000 + 155,000 + 1,000,000 + 100,000$$
$$= \$1,878,000$$

STEP **3**  **Divide the sum by the number of values in the sample.**

$$\bar{x} = \frac{\Sigma x}{n} = \frac{\$1,878,000}{7} = \$268,285.71$$

Recall, in Example 3-2, the sample mean was $170,571.43.

With only one value in the sample changed, the mean is now substantially higher than before. Because the mean is affected by extreme values, it may be a misleading measure of the data's center. In this case, the mean is larger than all but one of the professors' salaries.

**TRY EXERCISE 3-15** (pg. 93)

---

**OUTCOME 1**

## Median

**Median**

The median is a center value that divides a data array into two halves. We use $\tilde{u}$ to denote the population median and $M_d$ to denote the sample median.

**Data Array**

Data that have been arranged in numerical order.

Another measure of the center is called the **median**. The median is found by first arranging data in numerical order from smallest to largest. Data that are sorted in order are referred to as a **data array**.

Equation 3.3 is used to find the index point corresponding to the median value for a set of data placed in numerical order from low to high.

**Median Index**

$$i = \frac{1}{2}n \qquad\qquad (3.3)$$

where:

$i$ = The index of the point in the data array corresponding to the median value

$n$ = Sample size

If $i$ is not an integer, round its value up to the next higher integer. This next higher integer then is the position of the median in the data array.

If $i$ is an integer, the median is the average of the values in position $i$ and position $i + 1$.

For instance, suppose a personnel manager has analyzed the ages a sample of ten employees sorted from low to high as follows:

23   25   25   34   35   45   46   47   52   54

Using Equation 3.3 to find the median index, we get

$$i = \frac{1}{2}n = \frac{1}{2}(10) = 5$$

Since the index is an integer, the median value will be the average of the 5th and 6th values in the sorted data set. Thus, the median is

$$M_d = \frac{35 + 45}{2} = 40$$

Consider another case in which customers at a restaurant are asked to rate the service they received on a scale of 1 to 100. A sample of 15 customers were asked to provide the ratings. The data, sorted from low to high, are presented as follows:

60   68   75   77   80   80   80   85   88   90   95   95   95   95   99

Using Equation 3.3, we get the median index:

$$i = \frac{1}{2}n = \frac{1}{2}(15) = 7.5$$

Since the index is not an integer, we round 7.5 up to 8. Thus, the median ($M_d$) is the 8th data value from either end. In this case,

$$M_d = 85$$

---

**EXAMPLE 3-4    Computing the Median**

**Professor Salaries** (*continued*)    Consider again the example involving the newspaper reporter in Wisconsin and the sample salary data in Example 3-2. The median for these data is computed using the following steps:

STEP 1   **Collect the sample data.**

$$\{x_i\} = \{\text{Professor salaries}\} = \{\$144,000; \$98,000; \$204,000; \$177,000;$$
$$\$155,000; \$316,000; \$100,000\}$$

STEP 2   **Sort the data from smallest to largest, forming a data array.**

$$\{x_i\} = \{\$98,000; \$100,000; \$144,000; \$155,000; \$177,000; \$204,000; \$316,000\}$$

STEP 3   **Calculate the median index.**

Using Equation 3.3, we get $i = \frac{1}{2}(7) = 3.5$. Rounding up, the median is the fourth value from either end of the data array.

> The Excel 2016 function for the median is
>
> =Median(144000,98000,
>     ...,100000)
>
> *Note:* Excel sorts the data automatically when calculating the median.

STEP 4   **Find the median.**

$$\{x_i\} = \{\$98,000; \$100,000; \$144,000; \$155,000; \$177,000; \$204,000; \$316,000\}$$

$$\text{fourth value} = M_d$$

The median salary is $155,000. The notation for the sample median is $M_d$.

Note that if the number of data values in a sample or population is even, the median is the average of the two middle values.

TRY EXERCISE 3-2 (pg. 92)

---

**Symmetric Data**
Data sets whose values are evenly spread around the center. For symmetric data, the mean and median are equal.

**Skewed Data**
Data sets that are not symmetric. For skewed data, the mean is larger or smaller than the median.

**Right-Skewed Data**
A data distribution is right skewed if the mean for the data is larger than the median.

**Left-Skewed Data**
A data distribution is left skewed if the mean for the data is smaller than the median.

## Skewed and Symmetric Distributions

Data in a population or sample can be either **symmetric** or **skewed**, depending on how the data are distributed around the center.

In the original professor salary example (Example 3-2), the mean for the sample of seven managers was $170,571.43. In Example 3-4, the median salary was $155,000. Thus, for these data the mean and the median are not equal. This sample data set is **right skewed**, because $\bar{x} = \$170,571.43 > M_d = \$155,000$.

Figure 3.3 illustrates examples of right-skewed, **left-skewed**, and symmetric distributions. The greater the difference between the mean and the median, the more skewed the distribution. *Example 3-5 shows that an advantage of the median over the mean is that the*

**FIGURE 3.3 Skewed and Symmetric Distributions**

(a) **Right-Skewed**    (b) **Left-Skewed**    (c) **Symmetric**

*median is not affected by extreme values.* Thus, the median is particularly useful as a measure of the center when the data are highly skewed.[1]

---

**EXAMPLE 3-5    Impact of Extreme Values on the Median**

*Professor Salaries* (*continued*)    In Example 3-3, when we substituted a $1,000,000 salary for the professor who had a salary of $316,000, the sample mean salary increased from $170,571.43 to $268,285.71. What will happen to the median? The median is determined using the following steps:

**STEP 1    Collect the sample data.**

The sample salary data (including the extremely high salary) are

$$\{x_i\} = \{\text{Professor salaries}\} = \{\$144{,}000; \$98{,}000; \$204{,}000; \$177{,}000;$$
$$\$155{,}000; \$1{,}000{,}000; \$100{,}000\}$$

**STEP 2    Sort the data from smallest to largest, forming a data array.**

$$\{x_i\} = \{\$98{,}000; \$100{,}000; \$144{,}000; \$155{,}000; \$177{,}000; \$204{,}000; \$1{,}000{,}000\}$$

**STEP 3    Calculate the median index.**

Using Equation 3.3, we get $i = \dfrac{1}{2}(7) = 3.5$. Rounding up, the median is the fourth value from either end of the data array.

> The Excel 2016 function for the median is
>
> **=Median(144000,98000, ...,1000000)**
>
> *Note:* Excel sorts the data automatically when calculating the median.

**STEP 4    Find the median.**

$$\{x_i\} = \{\$98{,}000; \$100{,}000; \$144{,}000; \$155{,}000; \$177{,}000; \$204{,}000; \$1{,}000{,}000\}$$

$$\text{fourth value} = M_d$$

The median professor salary is $155,000, the same value as in Example 3-4, when the high salary was not included in the data. Thus, the median is not affected by extreme values in the data.

**TRY EXERCISE 3-2** (pg. 92)

---

**OUTCOME 1**

## Mode

The mean is the most commonly used measure of central location, followed closely by the median. However, the **mode** is a third value that is occasionally used as a measure of central location. The mode is the value in the data that occurs most frequently.

**Mode**
The mode is the value in a data set that occurs most frequently.

A data set may have more than one mode if two or more values tie for the most frequently occurring value. Example 3-6 illustrates this concept and shows how the mode is determined.

---

[1]Excel can be used to calculate a skewness statistic. The sign on the skewness statistic implies the direction of skewness. The higher the absolute value, the more the data are skewed. The 2016 Excel function for skewness is: = Skew(data range).

| EXAMPLE 3-6 | Determining the Mode |
| --- | --- |

**Smoky Mountain Pizza**   The owners of Smoky Mountain Pizza are planning to expand their restaurant to include an open-air patio. Before finalizing the design, the managers want to know what the most frequently occurring group size is so they can organize the seating arrangements to best meet demand. They wish to know the mode, which can be calculated using the following steps:

STEP 1   **Collect the sample data.**

A sample of 20 groups was selected at random. These data are

$$\{x_i\} = \{\text{People}\} = \{2, 4, 1, 2, 3, 2, 4, 2, 3, 6, 8, 4, 2, 1, 7, 4, 2, 4, 4, 3\}$$

STEP 2   **Organize the data into a frequency distribution.**

| $x_i$ | Frequency |
| --- | --- |
| 1 | 2 |
| 2 | 6 |
| 3 | 3 |
| 4 | 6 |
| 5 | 0 |
| 6 | 1 |
| 7 | 1 |
| 8 | 1 |
| | Total = 20 |

> The Excel 2016 function for the mode is
>
> =Mode.Sngl(2,4,1,2,3, 2,...,3)
>
> *Note:* The Mode function in Excel provides only one value of the mode even though there may be multiple modes in the data. If you are concerned about the possibility of multiple modes, you can use the function
>
> =Mode.Mult (2,4,1,2,3, 2,...,3)

STEP 3   **Determine the value(s) that occurs (occur) most frequently.**

In this case, there are two modes because the values 2 and 4 each occurred six times. Thus the modes are 2 and 4.

**TRY EXERCISE 3-2** (pg. 92)

A common mistake is to think that the mode is the frequency of the most frequently occurring value. In Example 3-6, you might be tempted to say that the mode = 6 because that was the highest frequency. Instead, there were two modes, 2 and 4, each of which occurred six times.

If no value occurs more frequently than any other, the data set is said to not have a mode. The mode might be particularly useful in describing the central location value for clothes sizes. For example, shoes come in full and half sizes. Consider the following sample data that have been sorted from low to high:

$$\{x\} = \{7.5, 8.0, 8.5, 9.0, 9.0, 10.0, 10.0, 10.0, 10.5, 10.5, 11.0, 11.5\}$$

The mean for these sample data is

$$\bar{x} = \frac{\sum x}{n} = \frac{7.5 + 8.0 + \cdots + 11.5}{12} = \frac{115.50}{12} = 9.63$$

Although 9.63 is the numerical average, the mode is 10 because more people wore that size shoe than any other. In making purchasing decisions, a shoe store manager would order more shoes at the modal size than at any other size. The mean isn't of any particular use in her purchasing decision.

Although the mode is considered to be a measure of the center, for some data the mode may not be a value near the center at all. For example, consider the following data that reflect the number of college credits that students are taking in a semester: {3, 3, 3, 3, 3, 8, 9, 12, 16, 16, 16}. If we use Excel to calculate the mean, median, and mode for these data, we get

$$=\text{Average}\ (3, 3, 3, 3, 3, 8, 9, 12, 16, 16, 16)\ =\ 8.4\ \text{credits}$$
$$=\text{Median}\ (3, 3, 3, 3, 3, 8, 9, 12, 16, 16, 16)\ =\ 8\ \text{credits}$$
$$=\text{Mode.Sngl}\ (3, 3, 3, 3, 3, 8, 9, 12, 16, 16, 16)\ =\ 3\ \text{credits}$$

In this case the mode happens to be the value 3, which also happens to be the lowest number of credits taken by any student and is not near the center of these data. Because of this feature, you need to be careful when using the mode as a measure of the center.

## Applying the Measures of Central Tendency

The cost of tuition is an important factor that most students and their families consider when deciding where to attend college. The data file **College Tuition** contains data on the average tuition at all colleges, by state, in the United States for the 2014–2015 academic year (*source:* http://trends.collegeboard.org/college-pricing/figures-tables). The cost of tuition is one of the variables in the data file. Suppose a guidance counselor who will be advising students about college choices wishes to conduct a descriptive analysis for this quantitative variable.

Figure 3.4 shows a frequency histogram generated using Excel. This histogram is a good place to begin the descriptive analysis, since it gives the analyst a good indication of the center value, the spread around the center, and the general shape of the distribution of the average tuition, by state, at U.S. colleges. Based on this histogram in Figure 3.4, what would you conclude about the distribution of college tuition? Is it skewed right or left?

We can extend the analysis by computing appropriate descriptive measures of central location for the tuition variable. Figure 3.5 shows the Excel output with descriptive measures for average tuition by state. First, focus on the primary measures of central location: mean and median. These are

$$\text{Mean} = \$9,011.96 \qquad \text{Median} = \$8,805.35$$

These statistics provide measures of the center of the tuition variable. The mean tuition value was $9,011.96, whereas the median was $8,805.35. Because the mean exceeds the median, we conclude that the data are right skewed—the same conclusion you should have reached by looking at the histogram in Figure 3.4.

**Issues with Excel**   In many instances, data files will have "missing values." That is, the values for one or more variables may not be available for some of the observations. The data may have been lost, or they were not measured when the data were collected. Many times when you receive data like this, the missing values will be coded in a special way. For example, the code "N/A" might be used or a "−99" might be entered to signify that the datum for that observation is missing.

Statistical software packages typically have flexible procedures for dealing with missing data. *However, Excel does not contain a missing-value option.* If you attempt to use certain data analysis options in Excel, such as Descriptive Statistics, in the presence of nonnumeric

**Excel 2016 Instructions**

1. Open file: **College Tuition.xlsx**.
2. Select the out-of-state tuition data.
3. On the **Insert** tab in the **Charts** group, click the **Insert Statistic Chart** arrow and click **Histogram**.
4. On the **Design** tab in the **Location** group, move the chart to a new sheet.
5. On the **Format Axis** sheet, change the **Number of bins** to 8.
6. On the **Design** tab, in the **Chart Layouts** group, set titles as desired and choose **Quick Layout 2**.

**FIGURE 3.4 Excel 2016 Frequency Histogram of College Tuition Costs**

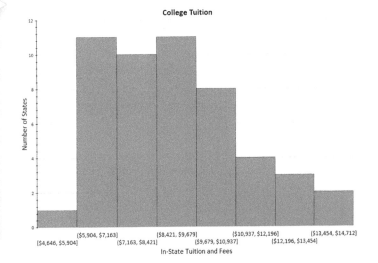

**FIGURE 3.5** **Excel 2016 Descriptive Statistics for College Tuition Data**

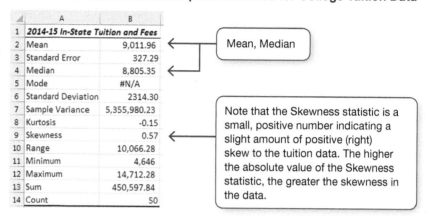

**Excel 2016 Instructions**

1. Open file: **College Tuition. xlsx**.
2. Select the **Data** tab.
3. Click on **Data Analysis > Descriptive Statistics**.
4. Define data range for the desired variables.
5. Check **Summary Statistics**.
6. Name new Output Sheet.
7. On **Home** tab, adjust decimal places and commas as desired.

| | A | B |
|---|---|---|
| 1 | *2014-15 In-State Tuition and Fees* | |
| 2 | Mean | 9,011.96 |
| 3 | Standard Error | 327.29 |
| 4 | Median | 8,805.35 |
| 5 | Mode | #N/A |
| 6 | Standard Deviation | 2314.30 |
| 7 | Sample Variance | 5,355,980.23 |
| 8 | Kurtosis | -0.15 |
| 9 | Skewness | 0.57 |
| 10 | Range | 10,066.28 |
| 11 | Minimum | 4,646 |
| 12 | Maximum | 14,712.28 |
| 13 | Sum | 450,597.84 |
| 14 | Count | 50 |

Mean, Median

Note that the Skewness statistic is a small, positive number indicating a slight amount of positive (right) skew to the tuition data. The higher the absolute value of the Skewness statistic, the greater the skewness in the data.

("N/A") data, you will get an error message. When that happens you must clear the missing values, generally by deleting all rows with missing values. In some instances, you can save the good data in the row by using **Edit-Clear-All** for the cell in question. However, a bigger problem exists when the missing value has been coded as an arbitrary numeric value (−99). In this case, unless you go into the data and clear these values, Excel will use the −99 values in the computations as if they are real values. The result will be incorrect calculations.

OUTCOME **1**

## Other Measures of Location

**Weighted Mean**    The arithmetic mean is the most frequently used measure of central location. Equations 3.1 and 3.2 are used when you have either a population or a sample. For instance, the sample mean is computed using

$$\bar{x} = \frac{\sum x}{n} = \frac{x_1 + x_2 + x_3 + \cdots + x_n}{n}$$

In this case, each $x$ value is given an equal weight in the computation of the mean. However, in some applications there is reason to weight the data values differently. In those cases, we

**Weighted Mean**
The mean value of data values that have been weighted according to their relative importance.

need to compute a **weighted mean**.

Equations 3.4 and 3.5 are used to find the weighted mean (or weighted average) for a population and for a sample, respectively.

**Weighted Mean for a Population**

$$\mu_w = \frac{\sum w_i x_i}{\sum w_i} \tag{3.4}$$

**Weighted Mean for a Sample**

$$\bar{x}_w = \frac{\sum w_i x_i}{\sum w_i} \tag{3.5}$$

where:

$w_i$ = The weight of the $i$th data value

$x_i$ = The $i$th data value

**EXAMPLE 3-7**  **Calculating a Weighted Population Mean**

**Myers & Associates**    The law firm of Myers & Associates was involved in litigating a discrimination suit concerning ski instructors at a ski resort in Colorado. One ski instructor from Germany had sued the operator of the ski resort, claiming he had not received equitable pay compared with the other ski instructors from Norway. In preparing a defense, the Myers attorneys planned to compute the mean annual income for all seven Norwegian ski instructors at the resort. However, because these instructors worked different numbers of days during the ski season, a weighted mean needed to be computed. This was done using the following steps:

STEP 1    **Collect the desired data and determine the weight to be assigned to each data value.**
In this case, the variable of interest was the income of the ski instructors. The population consisted of seven Norwegian instructors. The weights were the numbers of days that the instructors worked. The following data and weights were determined:

| $x_i$ = Income: | $7,600 | $3,900 | $5,300 | $4,000 | $7,200 | $2,300 | $5,100 |
|---|---|---|---|---|---|---|---|
| $w_i$ = Days: | 50 | 30 | 40 | 25 | 60 | 15 | 50 |

STEP 2    **Multiply each weight by the data value and sum these.**
$$\Sigma w_i x_i = (50)(\$7,600) + (30)(\$3,900) + \cdots + (50)(\$5,100) = \$1,530,500$$

STEP 3    **Sum the weights for all values (the weights are the days).**
$$\Sigma w_i = 50 + 30 + 40 + 25 + 60 + 15 + 50 = 270$$

STEP 4    **Compute the weighted mean.**
Divide the weighted sum by the sum of the weights. Because we are working with the population, the result will be the population weighted mean.
$$\mu_w = \frac{\Sigma w_i x_i}{\Sigma w_i} = \frac{\$1,530,500}{270} = \$5,668.52$$

Thus, taking into account the number of days worked, the Norwegian ski instructors had a mean income of $5,668.52.

**TRY EXERCISE 3-8** (pg. 92)

One weighted-mean example that you are probably very familiar with is your college grade point average (GPA). At most schools, A = 4 points, B = 3 points, and so forth. Each course has a certain number of credits (usually 1 to 5). The credits are the weights. Your GPA is computed by summing the product of points earned in a course and the credits for the course, and then dividing this sum by the total number of credits earned.

**Percentiles**    In some applications, we might wish to describe the location of the data in terms other than the center of the data. For example, prior to enrolling at your university, you took the SAT or ACT test and received a **percentile** score in math and verbal skills.
If you received word that your standardized exam score was at the 90th percentile, it means that you scored as high as or higher than 90% of the other students who took the exam. The score at the 50th percentile would indicate that you were at the median, where at least 50% scored at or below and at least 50% scored at or above your score.[2]

**Percentiles**
The $p$th percentile in a data array is a value that divides the data set into two parts. The lower segment contains at least $p$% and the upper segment contains at least $(100-p)$% of the data. The 50th percentile is the median.

---
[2]More rigorously, the percentile is that value (or set of values) such that at least $p$% of the data is as small as or smaller than that value and at least $(100 - p)$% of the data is at least as large as that value. For introductory courses, a convention has been adopted to average the largest and smallest values that qualify as a certain percentile. This is why the median was defined as it was earlier for data sets that have an even number of data values.

To illustrate how to manually approximate a percentile value, consider a situation in which 309 customers enter a Verizon store during the course of a day. The time (rounded to the nearest minute) that each customer spends is recorded. If we wish to approximate the 10th percentile, we would begin by first sorting the data in order from low to high, then assign each data value a location index from 1 to 309, and next determine the location index that corresponds to the 10th percentile using Equation 3.6.

**Percentile Location Index**

$$i = \frac{p}{100}(n)$$    (3.6)

where:

$$p = \text{Desired percent}$$
$$n = \text{Number of values in the data set}$$

If $i$ is not an integer, we round up to the next higher integer. The next integer greater than $i$ corresponds to the position of the $p$th percentile in the data set.

If $i$ is an integer, the $p$th percentile is the average of the values in position $i$ and position $i + 1$.

Thus, the index value associated with the 10th percentile is

$$i = \frac{p}{100}(n) = \frac{10}{100}(309) = 30.90$$

Because $i = 30.90$ is not an integer, we round to the next higher integer, which is 31. The 10th percentile corresponds to the value in the 31st position from the low end of the sorted data.

---

**HOW TO DO IT** (Example 3-8)

**Calculating Percentiles**

**1.** Sort the data in order from the lowest to highest value.

**2.** Determine the percentile location index, $i$, using Equation 3.6:

$$i = \frac{p}{100}(n)$$

where

$$p = \text{Desired percent}$$
$$n = \text{Number of values in the data set}$$

**3.** If $i$ is not an integer, then round to next higher integer. The $p$th percentile is located at the rounded index position. If $i$ is an integer, the $p$th percentile is the average of the values at location index positions $i$ and $i + 1$.

---

**EXAMPLE 3-8**   **Calculating Percentiles**

**Henson Trucking**   The Henson Trucking Company is a small company in the business of moving people from one home to another within the Dallas, Texas, area. Historically, the owners have charged the customers on an hourly basis, regardless of the distance of the move within the Dallas city limits. However, they are now considering adding a surcharge for moves over a certain distance. They have decided to base this charge on the 80th percentile. They have a sample of travel-distance data for 30 moves. These data are as follows:

| 13.5 | 8.6 | 16.2 | 21.4 | 21.0 | 23.7 | 4.1 | 13.8 | 20.5 | 9.6 |
|------|-----|------|------|------|------|-----|------|------|-----|
| 11.5 | 6.5 | 5.8 | 10.1 | 11.1 | 4.4 | 12.2 | 13.0 | 15.7 | 13.2 |
| 13.4 | 13.1 | 21.7 | 14.6 | 14.1 | 12.4 | 24.9 | 19.3 | 26.9 | 11.7 |

The 80th percentile can be computed using these steps.

**STEP 1   Sort the data from lowest to highest.**

| 4.1 | 4.4 | 5.8 | 6.5 | 8.6 | 9.6 | 10.1 | 11.1 | 11.5 | 11.7 |
|-----|-----|-----|-----|-----|-----|------|------|------|------|
| 12.2 | 12.4 | 13.0 | 13.1 | 13.2 | 13.4 | 13.5 | 13.8 | 14.1 | 14.6 |
| 15.7 | 16.2 | 19.3 | 20.5 | 21.0 | 21.4 | 21.7 | 23.7 | 24.9 | 26.9 |

STEP 2    **Determine percentile location index, *i*, using Equation 3.6.**
The 80th percentile location index is

$$i = \frac{p}{100}(n) = \frac{80}{100}(30) = 24$$

STEP 3    **Locate the appropriate percentile.**
Because $i = 24$ is an integer value, the 80th percentile is found by averaging the values in the 24th and 25th positions. These are 20.5 and 21.0. Thus, the 80th percentile is $(20.5 + 21.0)/2 = 20.75$; therefore, any distance exceeding 20.75 miles will be subject to a surcharge.

**TRY EXERCISE 3-7** (pg. 92)

The Excel 2016 function for a percentile is

= **Percentile.Inc (13.5, 8.6, ..., 11.7, .80)**

*Note:* Excel sorts the data automatically when calculating percentiles.

**Quartiles**    Another location measure that can be used to describe data is **quartiles**.

The first quartile corresponds to the 25th percentile. That is, it is the value at or below which there is at least 25% (one quarter) of the data and at or above which there is at least 75% of the data. The third quartile is also the 75th percentile. It is the value at or below which there is at least 75% of the data and at or above which there is at least 25% of the data. The second quartile is the 50th percentile and is also the median.

A quartile value can be approximated manually using the same method as for percentiles using Equation 3.6. For the 309 Verizon customer-service times mentioned earlier, the location of the first-quartile (25th percentile) index is found, after sorting the data, as

**Quartiles**
Quartiles in a data array are those values that divide the data set into four equal-sized groups. The median corresponds to the second quartile.

$$i = \frac{p}{100}(n) = \frac{25}{100}(309) = 77.25$$

Because 77.25 is not an integer value, we round up to 78. The first quartile is the 78th value from the low end of the sorted data.

The Excel 2016 function for a quartile is

= **Quartile.Inc (13.5, 8.6, ..., 11.7, 1)**

*Note:* The value for the quartile is entered as 1, 2, or 3 for first, second, or third quartile.

**Issues with Excel**    The quartile and percentile values from Excel will be slightly different from those we find manually using Equation 3.6. For example, referring to Example 3-8, when we use Excel to compute the 80th percentile for the moving distances, the value returned is 20.60 miles. This is slightly different from the 20.75 we found in Example 3-8.

OUTCOME **2**

## Box and Whisker Plots

**Box and Whisker Plots**
A graph that is composed of two parts: a box and the whiskers. The box has a width that ranges from the first quartile ($Q_1$) to the third quartile ($Q_3$). A vertical line through the box is placed at the median. Limits are located at a value that is 1.5 times the difference between $Q_1$ and $Q_3$ below $Q_1$ and above $Q_3$. The whiskers extend to the left to the lowest value within the limits and to the right to the highest value within the limits.

A descriptive tool that many decision makers like to use is called **box and whisker plot** (or a box plot). The box and whisker plot incorporates the five-number summary (minimum, first quartile, median, third quartile, and maximum) to graphically display quantitative data. It is also used to identify *outliers* that are unusually small or large data values that lie mostly by themselves.

**EXAMPLE 3-9**    **Constructing a Box And Whisker Plot**

**Rental Car Company**    A demand analyst for a rental car company has recently performed a study at one of the company's stores in which he determined the number of miles driven by rental car customers. He now wishes to construct a box and whisker plot as part of a presentation to describe customer driving patterns. The sorted sample data showing the miles driven are as follows. (The data are also listed in the file **Rental Car Miles**.)

| | | | | | | | | |
|---|---|---|---|---|---|---|---|---|
| 231 | 236 | 241 | 242 | 242 | 243 | 243 | 243 | 248 |
| 248 | 249 | 250 | 251 | 251 | 252 | 252 | 254 | 255 |
| 255 | 256 | 256 | 257 | 259 | 260 | 260 | 260 | 260 |
| 262 | 262 | 264 | 265 | 265 | 265 | 266 | 268 | 268 |
| 270 | 276 | 277 | 277 | 280 | 286 | 300 | 324 | 345 |

The box and whisker plot is constructed using the following steps:

**STEP 1**   **Sort the data values from low to high.**

**STEP 2**   **Calculate the 25th percentile ($Q_1$), the 50th percentile (median), and the 75th percentile ($Q_3$).**
The location index for $Q_1$ is

$$i = \frac{p}{100}(n) = \frac{25}{100}(45) = 11.25$$

Thus, $Q_1$ will be the 12th value, which is 250 miles. The median location is

$$i = \frac{p}{100}(n) = \frac{50}{100}(45) = 22.5$$

In the sorted data, the median is the 23rd value, which is 259 miles. The third-quartile location is

$$i = \frac{p}{100}(n) = \frac{75}{100}(45) = 33.75$$

Thus, $Q_3$ is the 34th data value. This is 266 miles.

**STEP 3**   **Draw the box so the ends correspond to $Q_1$ and $Q_3$.**

**STEP 4**   **Draw a vertical line through the box at the median.**

**STEP 5**   **Compute the upper and lower limits.**
The lower limit is computed as $Q_1 - 1.5(Q_3 - Q_1)$. This is

$$\text{Lower limit} = 250 - 1.5(266 - 250) = 226$$

The upper limit is $Q_3 + 1.5(Q_3 - Q_1)$. This is

$$\text{Upper limit} = 266 + 1.5(266 - 250) = 290$$

Any value outside these limits is identified as an outlier.

**STEP 6**   **Draw the whiskers.**
The whiskers are drawn to the smallest and largest values within the limits.

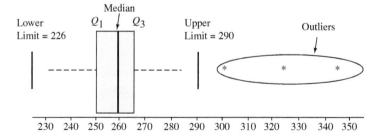

STEP 7   **Plot the outliers.**

The outliers are plotted as values outside the limits.

TRY EXERCISE 3-5 (pg. 92)

## Developing a Box and Whisker Plot in Excel 2016

A box and whisker plot can be developed quickly using Excel 2016. To illustrate, open the data file **MilestoFillup.xlsx**. The sampled data are in a single list. There is no need to sort the data. Figure 3.6 shows the options and the formatted and labeled box and whisker plot. Excel also makes it possible to modify the box and whisker display. The decision maker can choose to display the mean (**Show mean markers**) or not. The decision maker also has the ability to change how the quartiles are calculated (**Quartile Calculation**) by selecting either the **Inclusive median** or **Excusive median** option. (Note that by using the **Quick Layout** feature on the **Design** tab, the decision maker can alter the layout of the box and whisker plot, displaying the five number summary, if desired.)

Excel 2016 also makes it convenient to compare two or more variables using box and whisker plots. Open the data file **CampusATM.xlsx**, which contains data on daily ATM usage at two campus locations. The first column lists the location of the ATM, either North or South. The second column contains the daily usage. To develop a box and whisker plot using Excel that compares the two variables, select both columns and insert the box and whisker plot option. Excel automatically groups the data by location and produces the box and whisker plot of the daily ATM usage. Figure 3.7 shows the plot. A modified box and whisker plot of the same data is shown in Figure 3.8, where the mean marker and mean line options are used to show the mean daily usage of the respective locations. From the box and whisker figures it is evident that daily ATM usage is higher and more variable for the South location (note the higher median value, wider range, and wider distance between $Q_3$ and $Q_1$ for the South location). Daily usage at each location appears to be relatively symmetric, and no outliers are present.

**FIGURE 3.6** **Box and Whisker Plot**

### Excel 2016 Instructions

1. Open file: **MilestoFillup.xlsx**.
2. Select the data in Column A, excluding the label.
3. On the **Insert** tab in the **Charts** group, click the **Insert Statistic Chart** arrow and click **Box and Whisker**.
4. On the **Design** tab in the **Chart Layouts** group, add appropriate titles, delete the 1 label, and move the chart to a new sheet.
5. To view available options, select the plot, right mouse click, and click **Format Data Series** to open the **Format Data Series** sheet.
6. To increase the size of the box and whisker plot, select the vertical axis to display the **Format Axis** sheet and change the **Minimum Bound** to **200** and the **Maximum Bound** to **360**.

Miles Between Fill-ups Box and Whisker Plot

## Data-Level Issues

You need to be very aware of the level of data you are working with before computing the numerical measures introduced in this chapter. A common mistake is to compute means on nominal-level data. For example, a kitchen appliance maker surveyed a sample of customers to determine whether they preferred black, white, or stainless steel appliances. The data were coded as follows:

$$1 = black$$
$$2 = white$$
$$3 = stainless$$

FIGURE 3.7 **Box and Whisker Plot by Campus ATM Location**

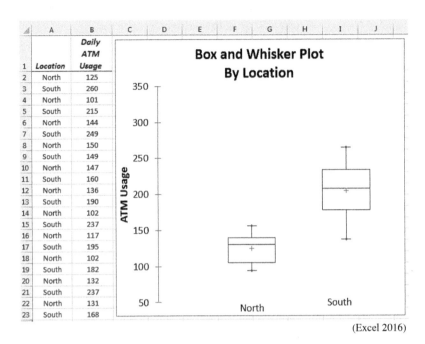

(Excel 2016)

A sample of the responses are

$$\text{Color code} = \{1, 1, 3, 2, 1, 2, 2, 2, 3, 1, 1, 1, 3, 2, 2, 1, 2\}$$

With these codes, the sample mean is

$$\bar{x} = \frac{\sum x}{n}$$

$$= \frac{30}{17} = 1.765$$

As you can see, reporting that customers prefer a color somewhere between black and white but closer to white would be meaningless. The mean should not be used with nominal data. This type of mistake tends to happen when people use computer software to perform their calculations. Asking Excel or other statistical software to compute the mean, median, and so on for all the variables in the data set is very easy. Then a table is created and, before long, the meaningless measures creep into your report. Don't let that happen.

There is also some disagreement about whether means should be computed on ordinal data. For example, in market research, a 5- or 7-point scale is often used to measure

FIGURE 3.8 **Box and Whisker Plot by Campus ATM Location**

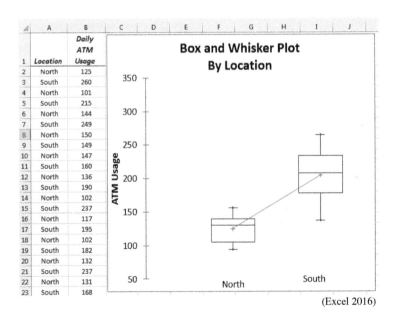

(Excel 2016)

customers' attitudes about products or TV commercials. The analyst might set up the following scale:

$$1 = \text{Strongly agree}$$
$$2 = \text{Agree}$$
$$3 = \text{Neutral}$$
$$4 = \text{Disagree}$$
$$5 = \text{Strongly disagree}$$

Customer responses to a particular question are obtained on this scale from 1 to 5. For a sample of $n = 10$ people, we might get the following responses to a question:

$$\text{Response} = \{2, 2, 1, 3, 3, 1, 5, 2, 1, 3\}$$

The mean rating is 2.3. We could then compute the mean for a second issue and compare the means. However, what exactly do we have? First, when we compute a mean for a scaled variable, we are making two basic assumptions:

1. We are assuming the distance between a rating of 1 and 2 is the same as the distance between 2 and 3. We are also saying these distances are exactly the same for the second issue's variable to which you wish to compare it. Although from a numerical standpoint this is true, in terms of what the scale is measuring, is the difference between strongly agree and agree the same as the difference between agree and neutral? If not, is the mean really a meaningful measure?
2. We are also assuming people who respond to the survey have the same definition of what "strongly agree" means or what "disagree" means. When you mark a 4 (disagree) on your survey, are you applying the same criteria as someone else who also marks a 4 on the same issue? If not, then the mean might be misleading.

Although these difficulties exist with ordinal data, we see many examples in which means are computed and used for decision purposes. In fact, we once had a dean who focused on one particular question on the course evaluation survey that was administered in every class each semester. This question was "Considering all factors of importance to you, how would you rate this instructor?"

$$1 = \text{Excellent} \quad 2 = \text{Good} \quad 3 = \text{Average} \quad 4 = \text{Poor} \quad 5 = \text{Very poor}$$

The dean then had his staff compute means for each class and for each professor. He then listed classes and faculty in order based on the mean values, and he based a major part of the performance evaluation on where a faculty member stood with respect to mean score on this one question. By the way, he carried the calculations for the mean out to three decimal places!

In general, the median is the preferred measure of central location for ordinal data instead of the mean.

Figure 3.9 summarizes the three measures of the center that have been discussed in this section.

**FIGURE 3.9 Descriptive Measures of the Center**

| Descriptive Measure | Computation Method | Data Level | Advantages/ Disadvantages |
|---|---|---|---|
| Mean | Sum of values divided by the number of values | Ratio Interval | • Numerical center of the data <br> • Sum of deviations from the mean is zero <br> • Sensitive to extreme values |
| Median | Middle value for data that have been sorted | Ratio Interval Ordinal | • Not sensitive to extreme values <br> • Computed only from the center values <br> • Does not use information from all the data |
| Mode | Value(s) that occur most frequently in the data | Ratio Interval Ordinal Nominal | • May not reflect the center <br> • May not exist <br> • Might have multiple modes |

## 3.1 EXERCISES

## Skill Development

**3-1.** A random sample of 15 articles in *Fortune* revealed the following word counts per article:

| | | | | |
|---|---|---|---|---|
| 5,176 | 6,005 | 5,052 | 5,310 | 4,188 |
| 4,132 | 5,736 | 5,381 | 4,983 | 4,423 |
| 5,002 | 4,573 | 4,209 | 5,611 | 4,568 |

Compute the mean, median, first quartile, and third quartile for these sample data.

**3-2.** The following data reflect the number of defects produced on an assembly line at the Dearfield Electronics Company for a period of 8 days:

| | | | | | | | |
|---|---|---|---|---|---|---|---|
| 3 | 0 | 2 | 0 | 1 | 3 | 5 | 2 |
| 5 | 1 | 3 | 0 | 0 | 1 | 3 | 3 |
| 4 | 3 | 1 | 8 | 4 | 2 | 4 | 0 |

a. Compute the mean number of defects for this population of days.
b. Compute the median number of defects produced for this population of days.
c. Determine if there is a mode number of defects and, if so, indicate the mode value.

**3-3.** A European cereal maker recently sampled 20 of its medium-size oat cereal packages to determine the weights of the cereal in each package. These sample data, measured in ounces, are as follows:

| | | | | | | | | | |
|---|---|---|---|---|---|---|---|---|---|
| 14.7 | 16.3 | 14.3 | 14.2 | 18.7 | 13.2 | 13.1 | 14.4 | 16.2 | 12.8 |
| 13.6 | 17.1 | 14.4 | 11.5 | 15.5 | 15.9 | 13.8 | 14.2 | 15.1 | 13.5 |

Calculate the first and third quartiles for these sample data.

**3-4.** The time (in seconds) that it took for each of 16 vehicles to exit a parking lot in downtown Cincinnati is:

| | | | |
|---|---|---|---|
| 106 | 153 | 169 | 116 |
| 135 | 78 | 51 | 129 |
| 100 | 141 | 72 | 101 |
| 130 | 125 | 128 | 139 |

Compute the mean, median, first quartile, and third quartile for the sample data.

**3-5.** A random sample of the miles driven by 20 rental car customers is shown as follows:

| | | | |
|---|---|---|---|
| 90 | 85 | 100 | 150 |
| 125 | 75 | 50 | 100 |
| 75 | 60 | 35 | 90 |
| 100 | 125 | 75 | 85 |
| 50 | 100 | 50 | 80 |

Develop a box and whisker plot for the sample data.

**3-6.** Examine the following data:

| | | | | | | | | | | |
|---|---|---|---|---|---|---|---|---|---|---|
| 23 | 65 | 45 | 19 | 35 | 28 | 39 | 100 | 50 | 26 | 25 | 27 |
| 24 | 17 | 12 | 106 | 23 | 19 | 39 | 70 | 20 | 18 | 44 | 31 |

a. Compute the quartiles.
b. Calculate the 90th percentile.
c. Develop a box and whisker plot.
d. Calculate the 20th and the 30th percentiles.

**3-7.** Consider the following data that represent the commute distances for students who attend Emory University:

| | | | | | | | | | |
|---|---|---|---|---|---|---|---|---|---|
| 3.1 | 4.7 | 8.4 | 11.6 | 12.1 | 13.0 | 13.4 | 16.1 | 17.3 | 20.8 |
| 22.8 | 24.3 | 26.2 | 26.6 | 26.7 | 31.2 | 32.2 | 35.8 | 35.8 | 39.8 |

a. Determine the 80th percentile.
b. Determine numbers that are the 25th and 75th percentiles.
c. Determine a number that qualifies as a median for these data.

**3-8.** A professor wishes to develop a numerical method for giving grades. He intends to base the grade on homework, two midterms, a project, and a final examination. He wishes the final exam to have the largest influence on the grade. He wants the project to have 10%, each midterm to have 20%, and the homework to have 10% of the influence on the semester grade.

a. Determine the weights the professor should use to produce a weighted average for grading purposes.
b. For a student with the following grades during the quarter, calculate a weighted average for the course:

| Instrument | Final | Project | Midterm 1 | Midterm 2 | Homework |
|---|---|---|---|---|---|
| Percentage Grade | 64 | 98 | 67 | 63 | 89 |

c. Calculate an (unweighted) average of these five scores and discuss why the weighted average would be preferable here.

## Business Applications

**3-9.** The manager for Jiffy Lube auto service center has collected data on the number of customers who agreed to purchase an air filter when they were also having their oil changed. The sample data are shown as follows:

| | | | | | | | | | | |
|---|---|---|---|---|---|---|---|---|---|---|
| 21 | 19 | 21 | 19 | 19 | 20 | 18 | 12 | 20 | 19 | 17 | 14 |
| 21 | 22 | 25 | 21 | 22 | 23 | 10 | 19 | 25 | 14 | 17 | 18 |

a. Compute the mean, median, and mode for these data.

b. Indicate whether the data are skewed or symmetric.

c. Construct a box and whisker plot for these data. Referring to your answer in part b, does the box plot support your conclusion about skewness? Discuss.

**3-10.** The following data reflect the tips received by a random sample of 20 restaurant servers in Denver, Colorado, during a 4-hour breakfast shift:

| | | | | |
|---|---|---|---|---|
| $33 | $32 | $42 | $31 | $31 |
| $37 | $37 | $34 | $47 | $31 |
| $42 | $29 | $36 | $32 | $25 |
| $29 | $47 | $26 | $32 | $40 |

a. Compute the mean tips received by the servers in the sample.

b. Compute the median tips received.

c. Determine the mode tips received.

d. Write a short paragraph discussing the statistics computed in parts a–c.

**3-11.** The La Costa Corporation operates five manufacturing plants in South America. The company tracks the medical costs that are directly attributed to accidents that occur at the plants each year. The following data show the medical costs (in U.S. dollars) at each plant for the most recent year along with the number of employees who work at each plant:

| Medical Cost | $7,400 | $14,400 | $12,300 | $6,200 | $3,100 |
|---|---|---|---|---|---|
| Employees | 123 | 402 | 256 | 109 | 67 |

a. Compute the weighted average for the annual medical costs at the five plants using the numbers of employees as the weights.

b. Explain why La Costa would want to use a weighted average in this situation rather than a simple average of the five medical cost values.

**3-12.** The Green Grass Lawn Company provides yard care services for residential customers. The company owner tracked the time his field employees spent at a sample of customer locations. He was hoping to use these data to help him with his scheduling and to establish billing rates. The following sample data, in minutes, were recorded:

| | | | | | | | | | |
|---|---|---|---|---|---|---|---|---|---|
| 31 | 27 | 29 | 22 | 24 | 30 | 28 | 21 | 29 | 26 |
| 22 | 17 | 17 | 20 | 38 | 10 | 38 | 25 | 27 | 23 |
| 23 | 13 | 17 | 34 | 25 | 29 | 22 | 22 | 14 | 11 |
| 29 | 26 | 29 | 29 | 37 | 32 | 27 | 26 | 18 | 22 |

Describe the central tendency of these data by computing the mean, median, and mode. Based on these measures, can you conclude that the distribution of time spent at customer locations is skewed or symmetric?

**3-13.** Eastern States Bank and Trust monitors its drive-thru service times electronically to ensure that its speed of service is meeting the company's goals. A sample of 28 drive-thru times was recently taken and is shown here:

| Speed of Service (time in seconds) | | | |
|---|---|---|---|
| 83 | 138 | 145 | 147 |
| 130 | 79 | 156 | 156 |
| 90 | 85 | 68 | 93 |
| 178 | 76 | 73 | 119 |
| 92 | 146 | 88 | 103 |
| 116 | 134 | 162 | 71 |
| 181 | 110 | 105 | 74 |

a. Compute the mean, median, and mode for these sample data.

b. Indicate whether the data are symmetric or skewed.

c. Construct a box and whisker plot for the sample data. Does the box and whisker plot support your conclusions in part b concerning the symmetry or skewness of these data?

**3-14.** Bahnson, English & Associates, a commercial real estate company, owns six office buildings that it leases to businesses. The lease price per square foot differs by building due to location and building amenities. Currently, all six buildings are fully leased at the prices shown here:

| | Price per Square Foot | Number of Square Feet |
|---|---|---|
| Building 1 | $ 75 | 125,000 |
| Building 2 | $ 85 | 37,500 |
| Building 3 | $ 90 | 77,500 |
| Building 4 | $ 45 | 35,000 |
| Building 5 | $ 55 | 60,000 |
| Building 6 | $110 | 130,000 |

a. Compute the weighted average (mean) price per square foot for these buildings.

b. Why is the weighted average price per square foot preferred to a simple average price per square foot in this case?

**3-15.** To obtain the average retail cost of a 55″ HD flat-screen TV in 2015, a survey yielded the following data (in U.S. dollars):

| | | | | |
|---|---|---|---|---|
| 606.70 | 558.12 | 625.82 | 533.70 | 464.37 |
| 511.15 | 400.56 | 538.20 | 531.64 | 632.14 |
| 474.86 | 567.46 | 588.39 | 528.78 | 610.32 |
| 564.71 | 912.68 | 475.87 | 545.25 | 589.15 |

a. Calculate the mean cost for these data.

b. Examine the data presented. Choose an appropriate measure of the center of the data, justify the choice, and calculate the measure.

**3-16.** The following table exhibits base salary data obtained from a survey of over 170 benchmark positions, including finance positions. It reports the salaries of a sample of 25 human resource directors for mid-sized firms. Assume the data are in thousands of dollars.

| | | | | | | | | |
|---|---|---|---|---|---|---|---|---|
| 173.1 | 171.2 | 141.9 | 112.6 | 211.1 | 156.5 | 145.4 | 134.0 | 192.0 |
| 185.8 | 168.3 | 131.0 | 214.4 | 155.2 | 164.9 | 123.9 | 161.9 | 162.7 |
| 178.8 | 161.3 | 182.0 | 165.8 | 213.1 | 177.4 | 159.3 | | |

a. Calculate the mean salary of the HR directors.
b. Based on measures of the center of the data, determine if the salary data are skewed.
c. Construct a box and whisker plot and summarize the characteristics of the salaries that it reveals.

**3-17.** According to the website www.fdic.gov, the Federal Deposit Insurance Corporation (FDIC) insures deposits in banks and thrift institutions for up to $250,000. In 2015, the FDIC insured 5,410 commercial banks with $14,727 billion in assets and 860 savings institutions with $1,074 billion in assets.

a. Calculate the average assets per entity for FDIC–insured institutions of both types.
b. Describe the relationship between the two averages calculated in part a. Can you provide a reason for the difference?
c. Would the two averages be considered parameters or statistics? Explain.

## Computer Software Exercises

**3-18.** Each year, *Bloomberg Businessweek* publishes information and rankings of master of business administration (MBA) programs. The data file **MBA Analysis** contains similar data on several variables for eight reputable MBA programs. The variables include pre– and post–MBA salary, percentage salary increase, undergraduate GPA, average Graduate Management Admission Test (GMAT) score, annual tuition, and expected annual student cost. Compute the mean and median for each of the variables in the database and write a short report that summarizes the data. Include any appropriate charts or graphs to assist in your report.

**3-19.** Dynamic random-access memory (DRAM) chips are routed through fabrication machines in an order that is referred to as a recipe. The data file **DRAM Chips** contains a sample of processing times, measured in fractions of hours, at a particular machine center for one chip recipe.

a. Compute the mean processing time.
b. Compute the median processing time.
c. Determine what the mode processing time is.
d. Calculate the 80th percentile for processing time.

**3-20.** Japolli Bakery tracks sales of its different bread products on a daily basis. The data for 22 consecutive days at one of its retail outlets in Nashville are in a file called **Japolli Bakery**. Calculate the mean, mode, and median sales for each of the bread categories and write

a short report that describes these data. Use any charts or graphs that may be helpful in more fully describing the data. Include a box and whisker plot comparing all the categories of breads.

**3-21.** In an effort to track housing prices, the National Association of Realtors developed the Pending Home Sales Index (PHSI), a leading indicator for the housing market. An index of 100 is equal to the average level of contract activity during 2001, the first year to be analyzed. The file titled **Pending** contains the PHSI from December 2014 to December 2015 (*source:* https://ycharts.com/indicators/pending_home_sales_index).

a. Determine the mean and median for the PHSI from December 2014 through December 2015. Specify the shape of the PHSI's distribution.
b. Calculate the first and third quartiles for the PHSI for the months December 2014 through December 2015.

**3-22.** A proposal in a South Carolina county calls for all homeowners over the age of 65 with incomes at or below the 20th percentile to get a reduction in property taxes. A random sample of 50 people over 65 was selected, and the household incomes (as reported on the most current federal tax return) were recorded. These data are in the file called **Property Tax Incomes**. Use these data to establish the income cutoff point to qualify for the property tax cut.

**3-23.** Suppose a random sample of 137 households in Detroit was taken as part of a study on annual household spending for food at home. The sample data are contained in the file **Detroit Eats**.

a. For the sample data, compute the mean and the median and construct a box and whisker plot.
b. Are the data skewed or symmetric?
c. Approximately what percent of the data values are between $2,900 and $3,250?

**3-24.** In order to practice law in Nebraska, individuals are required to pass the Nebraska bar exam. The exam is challenging, and most attorneys who attempt to take the exam spend a substantial amount of time studying. The file **Nebraska** contains data on the number of hours spent studying for a sample of 50 individuals who took and passed the bar exam.

a. Construct a stem and leaf display. Indicate the shape of data displayed by the stem and leaf display.
b. Use measures that indicate the shape of the distribution. Do these measures give results that agree with the shape shown in part a?
c. Considering your answers to parts a and b, indicate which measure you would recommend using to indicate the center of the data.

| 3.2 | **Measures of Variation** |
|---|---|

BUSINESS APPLICATION    ### Measuring Variation Using the Range

**Travel-Time Recreational Vehicles**   With the recent decrease in gasoline prices, the recreational vehicle market is improving. Consider the situation involving two manufacturing facilities for Travel-Time Recreational Vehicles. The division vice president asked the two plant managers to record the number of RVs produced weekly over a five-week period. The resulting sample data are shown in Table 3.3 to the left.

Instead of reporting these raw data, the managers reported only the mean and median for their data. The following are the computed statistics for the two plants:

**TABLE 3.3** Manufacturing Output for Travel-Time Recreational Vehicles

| Plant A | Plant B |
|---|---|
| 15 units | 23 units |
| 25 units | 26 units |
| 35 units | 25 units |
| 20 units | 24 units |
| 30 units | 27 units |

| Plant A | Plant B |
|---|---|
| $\bar{x} = 25$ units | $\bar{x} = 25$ units |
| $M_d = 25$ units | $M_d = 25$ units |

The division vice president looked at these statistics and concluded the following:

1. Average production is 25 units per week at both plants.
2. The median production is 25 units per week at both plants.
3. Because the mean and median are equal, the distribution of production output at the two plants is symmetric.
4. Based on these statistics, there is no reason to believe that the two plants are different in terms of their production output.

However, if he had taken a closer look at the raw data, he would have noticed a big difference between the two plants. The difference is the production **variation** from week to week. Plant B is very stable, producing almost the same number of units every week. Plant A varies considerably, with some high-output weeks and some low-output weeks. Thus, looking at only measures of the data's central location can be misleading. To fully describe a set of data, we need a measure of variation or spread.

**Variation**
A set of data exhibits variation if all the data are not the same value.

There is variation in everything that is made by humans or that occurs in nature. The variation may be small, but it is there. Given a fine enough measuring instrument, we can detect the variation. Variation either is a natural part of a process (or inherent to a product) or can be attributed to a special cause that is not considered random.

Business decision makers use several different measures of variation. In this section, we introduce four of these measures: range, interquartile range, variance, and standard deviation.

OUTCOME **3**

### Range

**Range**
The range is a measure of variation that is computed by finding the difference between the maximum and minimum values in a data set.

The simplest measure of variation is the **range**. It is both easy to compute and easy to understand.

The range is computed using Equation 3.7.

**Range**

$$R = \text{Maximum value} - \text{Minimum value} \tag{3.7}$$

BUSINESS APPLICATION    ### Calculating The Range

**Travel-Time Recreational Vehicles** (*continued*)   Table 3.3 showed the production-volume data for the two Travel-Time Recreational Vehicles plants. The range in production for each plant is determined using Equation 3.7 as follows:

|  | Plant A | Plant B |
|---|---|---|
|  | $R = $ Maximum $-$ Minimum | $R = $ Maximum $-$ Minimum |
|  | $R = 35 - 15$ | $R = 27 - 23$ |
|  | $R = 20$ | $R = 4$ |

> Excel 2016 does not have a Range function, but the range can be calculated using the Max and Min functions in an Excel equation as follows:
>
> = Max (data values)
> − Min (data values)

We see Plant A has a range that is five times as great as Plant B.

Although the range is quick and easy to compute, it does have some limitations. First, because we use only the high and low values to compute the range, it is very sensitive to extreme values in the data. Second, regardless of how many values are in the sample or population, the range is computed from only two of these values. For these reasons, it is considered a weak measure of variation.

OUTCOME **3**

## Interquartile Range

**Interquartile Range**

The interquartile range is a measure of variation that is determined by computing the difference between the third and first quartiles.

A measure of variation that tends to overcome the range's susceptibility to extreme values is called the **interquartile range**.

Equation 3.8 is used to compute the interquartile range.

**Interquartile Range**

$$\text{Interquartile range} = \text{Third quartile} - \text{First quartile} \tag{3.8}$$

---

**EXAMPLE 3-10** Computing the Interquartile Range

**Verizon Wireless** A systems capacity manager for Verizon Wireless is interested in better understanding the use of Verizon customer text messaging use. To do this, she has collected a random sample of 100 customers under the age of 25 and recorded the number of text messages they sent in a one-week period. She wishes to analyze the variation in these data by computing the range and the interquartile range. She can use the following steps to do so:

STEP 1 **Sort the data into a data array from lowest to highest.**
The 100 sorted values are as follows:

| | | | | | | | | | |
|---|---|---|---|---|---|---|---|---|---|
| 33 | 164 | 173 | 184 | 190 | 197 | 207 | 216 | 224 | 237 |
| 53 | 164 | 175 | 186 | 191 | 197 | 207 | 217 | 225 | 240 |
| 150 | 164 | 175 | 186 | 191 | 198 | 208 | 217 | 225 | 240 |
| 152 | 166 | 175 | 186 | 192 | 200 | 208 | 217 | 229 | 240 |
| 157 | 166 | 178 | 187 | 193 | 200 | 208 | 219 | 231 | 250 |
| 160 | 168 | 178 | 188 | 193 | 201 | 210 | 222 | 231 | 251 |
| 161 | 169 | 179 | 188 | 194 | 202 | 211 | 223 | 234 | 259 |
| 162 | 171 | 180 | 188 | 194 | 204 | 212 | 223 | 234 | 270 |
| 162 | 171 | 182 | 190 | 196 | 205 | 213 | 223 | 235 | 379 |
| 163 | 172 | 183 | 190 | 196 | 205 | 216 | 224 | 236 | 479 |

STEP 2 **Compute the range using Equation 3.7.**

$$R = \text{Maximum value} - \text{Minimum value}$$
$$R = 479 - 33 = 446$$

Note that the range is sensitive to extreme values. The low value of 33 and the high value of 479 cause the value of the range to be very large.

STEP 3 **Compute the first and third quartiles.**
Equation 3.6 can be used to find the locations of the third quartile (75th percentile) and the first quartile (25th percentile).

For $Q_3$, the location $i = \frac{75}{100}(100) = 75$. Thus, $Q_3$ is halfway between the 75th and 76th data values, which is found as follows:

$$Q_3 = (219 + 222)/2 = 220.50$$

For $Q_1$, the location is $i = \frac{25}{100}(100) = 25$. Then $Q_1$ is halfway between the 25th and 26th data values:

$$Q_1 = (178 + 178)/2 = 178$$

STEP 4  **Compute the interquartile range.**
The interquartile range overcomes the range's problem of sensitivity to extreme values. It is computed using Equation 3.8:

$$\text{Interquartile range} = Q_3 - Q_1$$
$$= 220.50 - 178 = 42.50$$

Note that the interquartile range would be unchanged even if the values on the high or low end of the distribution were even more extreme than those shown in these sample data.

> Excel 2016 can be used to calculate the interquartile range using the Quartile.Inc function in an equation as follows:
>
> **= Quartile.Inc (data values,3)**
> **− Quartile.Inc (data values,1)**

OUTCOME **3**

## Population Variance and Standard Deviation

Although the range is easy to compute and understand and the interquartile range is designed to overcome the range's sensitivity to extreme values, neither measure uses all the available data in its computation. Thus, both measures ignore potentially valuable information in data.

Two measures of variation that incorporate all the values in a data set are the **variance** and the **standard deviation**.

These two measures are closely related. The standard deviation is the positive square root of the variance. The standard deviation is in the original units (dollars, pounds, etc.), whereas the units of measure in the variance are squared. Because dealing with original units is easier than dealing with the square of the units, we usually use the standard deviation to measure variation in a population or sample.

**Variance**
The population variance is the average of the squared distances of the data values from the mean.

**Standard Deviation**
The standard deviation is the positive square root of the variance.

BUSINESS APPLICATION    Calculating the Variance and Standard Deviation

**Travel-Time Recreational Vehicles** (*continued*)   Recall the Travel-Time Recreational Vehicles application, in which we compared the weekly production output for two of the company's plants. Table 3.3 showed the data, which are considered a population for our purposes here.

Previously, we examined the variability in the output from these two plants by computing the ranges. Although those results gave us some sense of how much more variable Plant A is than Plant B, we also pointed out some of the deficiencies of the range. The variance and standard deviation offer alternatives to the range for measuring variation in data.

Equation 3.9 is the formula for the population variance. Like the population mean, the population variance and standard deviation are assigned Greek symbols.

**Population Variance**

$$\sigma^2 = \frac{\sum_{i=1}^{N}(x_i - \mu)^2}{N} \tag{3.9}$$

where:

$$\mu = \text{Population mean}$$
$$N = \text{Population size}$$
$$\sigma^2 = \text{Population variance (sigma squared)}$$

We begin by computing the variance for the output data from Plant A. The first step in manually calculating the variance is to find the mean using Equation 3.1:

> The Excel 2016 function for the population mean is
> =AVERAGE(15,25,35,20,30)

$$\mu = \frac{\sum x}{N} = \frac{15 + 25 + 35 + 20 + 30}{5} = \frac{125}{5} = 25$$

Next, we subtract the mean from each value, as shown in Table 3.4. Notice that the sum of the deviations from the mean is 0. Recall from Section 3.1 that this will be true for any set of data. The positive differences are canceled out by the negative differences. To overcome this fact when computing the variance, we square each of the differences and then sum the squared differences. These calculations are also shown in Table 3.4.

**TABLE 3.4 Computing the Population Variance: Squaring the Deviations**

| $x_i$ | $(x_i - \mu)$ | $(x_i - \mu)^2$ |
|---|---|---|
| 15 | $15 - 25 = -10$ | 100 |
| 25 | $25 - 25 = \phantom{-}0$ | 0 |
| 35 | $35 - 25 = \phantom{-}10$ | 100 |
| 20 | $20 - 25 = -5$ | 25 |
| 30 | $30 - 25 = \phantom{-}5$ | 25 |
| | $\sum(x_i - \mu) = \phantom{-}0$ | $\sum(x_i - \mu)^2 = 250$ |

The final step in computing the population variance is to divide the sum of the squared differences by the population size, $N = 5$:

> The Excel 2016 function for the population variance is
> =Var.P(15,25,35,20,30)

$$\sigma^2 = \frac{\sum(x - \mu)^2}{N} = \frac{250}{5} = 50$$

The population variance is 50 *RVs squared*.

Manual calculations for the population variance may be easier if you use an alternative formula for $\sigma^2$ that is the algebraic equivalent. This is shown as Equation 3.10.

**Population Variance Shortcut**

$$\sigma^2 = \frac{\sum x^2 - \frac{(\sum x)^2}{N}}{N} \tag{3.10}$$

Example 3-11 will illustrate using Equation 3.10 to find a population variance.

Because we squared the deviations to keep the positive values and negative values from canceling, the units of measure were also squared, but the term *RVs squared* doesn't have a meaning. To get back to the original units of measure, we take the square root of the variance. The result is the standard deviation. Equation 3.11 shows the formula for the population standard deviation.

**Population Standard Deviation**

$$\sigma = \sqrt{\sigma^2} = \sqrt{\frac{\sum_{i=1}^{N}(x_i - \mu)^2}{N}} \tag{3.11}$$

Therefore, the population standard deviation of Plant A's production output is

$$\sigma = \sqrt{50}$$
$$\sigma = 7.07 \text{ mobile homes}$$

The population standard deviation is a parameter and will not change unless the population values change.

We could repeat this process using the data for Plant B, which also had a mean output of 25 mobile homes. You should verify that the population variance is

$$\sigma^2 = \frac{\sum(x - \mu)^2}{N} = \frac{10}{5} = 2 \text{ mobile homes squared}$$

The standard deviation is found by taking the square root of the variance:

$$\sigma = \sqrt{2}$$
$$\sigma = 1.414 \text{ mobile homes}$$

Thus, Plant A has an output standard deviation that is five times larger than Plant B's. The fact that Plant A's range was also five times larger than the range for Plant B is merely a coincidence.

---

**EXAMPLE 3-11**  **Computing a Population Variance and Standard Deviation**

**Boydson Shipping Company**  Boydson Shipping Company owns and operates a fleet of ships that carry commodities between the countries of the world. In the past six months, the company has had seven contracts that called for shipments between Vancouver, Canada, and London, England. For many reasons, the travel time varies between these two locations. The scheduling manager is interested in knowing the variance and standard deviation in shipping times for these seven shipments. To find these values, he can follow these steps:

**HOW TO DO IT (Example 3-11)**

**Computing the Population Variance and Standard Deviation**

**1.** Collect quantitative data for the variable of interest for the entire population.

**2.** Use either Equation 3.9 or Equation 3.10 to compute the variance.

**3.** If Equation 3.10 is used, find the sum of the $x$ values ($\sum x$) and then square this sum ($\sum x)^2$.

**4.** Square each $x$ value and sum these squared values ($\sum x^2$).

**5.** Compute the variance using

$$\sigma^2 = \frac{\sum x^2 - \frac{(\sum x)^2}{N}}{N}$$

**6.** Compute the standard deviation by taking the positive square root of the variance:

$$\sigma = \sqrt{\sigma^2}$$

The Excel 2016 function for the population variance is

$$=\textbf{Var.P(5,7,5,9,7,4,6)}$$

STEP 1  **Collect the data for the population.**
The shipping times are shown as follows:

$$x = \text{Shipping weeks}$$
$$= \{5, 7, 5, 9, 7, 4, 6\}$$

STEP 2  **Select Equation 3.10 to find the population variance.**

$$\sigma^2 = \frac{\sum x^2 - \frac{(\sum x)^2}{N}}{N}$$

STEP 3  **Add the $x$ values and square the sum.**

$$\sum x = 5 + 7 + 5 + 9 + 7 + 4 + 6 = 43$$

$$(\sum x)^2 = (43)^2 = 1,849$$

STEP 4  **Square each of the $x$ values and sum these squares.**

$$\sum x^2 = 5^2 + 7^2 + 5^2 + 9^2 + 7^2 + 4^2 + 6^2 = 281$$

STEP 5  **Compute the population variance.**

$$\sigma^2 = \frac{\sum x^2 - \frac{(\sum x)^2}{N}}{N} = \frac{281 - \frac{1,849}{7}}{7} = 2.4082$$

The Excel 2016 function for the population standard deviation is

= Stdev.P (5,7,5,9,7,4,6)

The variance is in units squared, so in this example the population variance is 2.4082 weeks squared.

STEP 6 **Calculate the standard deviation as the positive square root of the variance.**

$$\sigma = \sqrt{\sigma^2} = \sqrt{2.4082} = 1.5518 \text{ weeks}$$

Thus, the standard deviation for the number of shipping weeks between Vancouver and London for the seven shipments is 1.5518 weeks.

**TRY EXERCISE 3-27** (pg. 103)

## Sample Variance and Standard Deviation

Equations 3.9, 3.10, and 3.11 are the equations for the population variance and standard deviation. Any time you are working with a population, these are the equations you will use. However, in many instances, you will be describing sample data that have been selected from the population. In addition to using different notations for the sample variance and sample standard deviation, the equations are also slightly different. Equations 3.12 and 3.13 can be used to find the sample variance. Note that Equation 3.13 is considered the shortcut formula for manual computations.

**Sample Variance**

$$s^2 = \frac{\sum_{i=1}^{n}(x_i - \bar{x})^2}{n - 1}$$

(3.12)

**Sample Variance Shortcut**

$$s^2 = \frac{\sum x^2 - \frac{\left(\sum x\right)^2}{n}}{n - 1}$$

(3.13)

where:

$$n = \text{Sample size}$$
$$\bar{x} = \text{Sample mean}$$
$$s^2 = \text{Sample variance}$$

The sample standard deviation is found by taking the square root of the sample variance, as shown in Equation 3.14.

**Sample Standard Deviation**

$$s = \sqrt{s^2} = \sqrt{\frac{\sum_{i=1}^{n}(x_i - \bar{x})^2}{n - 1}}$$

(3.14)

Take note in Equations 3.12, 3.13, and 3.14 that the denominator is $n - 1$ (sample size minus 1). This may seem strange, given that the denominator for the population variance and the standard deviation is simply $N$, the population size. The mathematical justification for the $n - 1$ divisor is outside the scope of this text. However, the general reason is that we want the average sample variance to equal the population variance. If we were to select all possible samples of size $n$ from a given population and for each sample compute the sample variance using Equation 3.12 or Equation 3.13, the average of all the sample variances would equal $\sigma^2$ (the population variance), provided we used $n - 1$ as the divisor. Using $n$ instead of $n - 1$ in the denominator would produce an average sample variance that is smaller than $\sigma^2$, the population variance. Because we want an estimator on average to equal the population variance, we use $n - 1$ in the denominator of $s^2$.

### EXAMPLE 3-12   Computing a Sample Variance and Standard Deviation

Fabco Industries    Based on the complaint by a major customer, the internal auditor for Fabco Industries, a manufacturer of equipment used in the food processing business, examined the transaction records related to ten of the company's customers. For each client, the auditor counted the number of incorrectly recorded entries (i.e., "defects"). The ten accounts can be considered to be samples of all possible Fabco customers that could be analyzed. To fully analyze the data, the auditor can calculate the sample variance and sample standard deviation using the following steps:

STEP 1  **Select the sample and record the data for the variable of interest.**

| Client | Defects = x | Client | Defects = x |
|--------|-------------|--------|-------------|
| 1 | 4 | 6 | 0 |
| 2 | 7 | 7 | 3 |
| 3 | 1 | 8 | 2 |
| 4 | 0 | 9 | 6 |
| 5 | 5 | 10 | 2 |

STEP 2  **Select either Equation 3.12 or Equation 3.13 to compute the sample variance.**
If we use Equation 3.12,

$$s^2 = \frac{\sum(x - \bar{x})^2}{n - 1}$$

STEP 3  **Compute $\bar{x}$.**
The sample mean number of defectives is

$$\bar{x} = \frac{\sum x}{n} = \frac{30}{10} = 3.0$$

STEP 4  **Determine the sum of the squared deviations of each $x$ value from $\bar{x}$.**

| Client | Defectives = x | $(x - \bar{x})$ | $(x - \bar{x})^2$ |
|--------|----------------|-----------------|-------------------|
| 1 | 4 | 1 | 1 |
| 2 | 7 | 4 | 16 |
| 3 | 1 | −2 | 4 |
| 4 | 0 | −3 | 9 |
| 5 | 5 | 2 | 4 |
| 6 | 0 | −3 | 9 |
| 7 | 3 | 0 | 0 |
| 8 | 2 | −1 | 1 |
| 9 | 6 | 3 | 9 |
| 10 | 2 | −1 | 1 |
| | $\sum = 30$ | $\sum = 0$ | $\sum = 54$ |

STEP 5 **Compute the sample variance using Equation 3.12.**

$$s^2 = \frac{\sum (x - \bar{x})^2}{n - 1} = \frac{54}{9} = 6$$

The sample variance is measured in squared units. Thus, the variance in this example is 6 defectives squared.

STEP 6 **Compute the sample standard deviation by taking the positive square root of the variance (see Equation 3.14).**

$$s = \sqrt{\frac{\sum (x - \bar{x})^2}{n - 1}} = \sqrt{\frac{54}{9}} = \sqrt{6}$$

$$s = 2.4495 \text{ defects}$$

This sample standard deviation measures the variation in the sample data for the number of incorrectly recorded entries.

**TRY EXERCISE 3-25** (pg. 103)

---

**BUSINESS APPLICATION**    Calculating Measures of Variation Using Excel

**Colleges**   In Section 3.1, the guidance counselor was interested in describing the data representing the average college tuition, by state, for all colleges in the United States. The data are in the file called **College Tuition**. Previously, we determined the following descriptive measures of the center for the variable:

$$\text{Mean} = \$9,011.96$$
$$\text{Median} = \$8,805.35$$

Next, the analyst will turn her attention to measures of variability. The range (maximum – minimum) is one measure of variability. Excel can be used to compute the range and the standard deviation of tuition, which is a more powerful measure of variation than the range. Figure 3.10 shows the Excel descriptive statistics results. We find the following measures of variation:

**Excel Tutorial**

$$\text{Range} = \$10,066.28$$
$$\text{Standard deviation} = \$2,314.30$$

These values are measures of the spread in the data. You should know that outlier values in a data set will increase both the range and standard deviation. One guideline for identifying outliers is the $\pm 3$ standard deviation rule. That is, if a value falls outside $\pm 3$ standard deviations from the mean, it is considered an outlier. Also, as shown in Section 3.1, outliers can be identified using box and whisker plots.

**FIGURE 3.10**   **Excel 2016 Descriptive Statistics for Colleges Data**

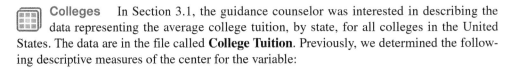

| 2014-15 In-State Tuition and Fees | |
| --- | --- |
| Mean | 9,011.96 |
| Standard Error | 327.29 |
| Median | 8,805.35 |
| Mode | #N/A |
| Standard Deviation | 2,314.30 |
| Sample Variance | 5,355,980.23 |
| Kurtosis | (0.15) |
| Skewness | 0.57 |
| Range | 10,066.28 |
| Minimum | 4646 |
| Maximum | 14,712.28 |
| Sum | 450,597.84 |
| Count | 50 |

## 3.2 EXERCISES

### Skill Development

**3-25.** Google is noted for its generous employee benefits. The following data reflect the number of vacation days that a sample of employees at Google have left to take before the end of the year:

| 3 | 0 | 2 | 0 | 1 | 3 | 5 | 2 |
|---|---|---|---|---|---|---|---|
| 5 | 1 | 3 | 0 | 0 | 1 | 3 | 3 |
| 4 | 3 | 1 | 8 | 4 | 2 | 4 | 0 |

a. Compute the range for these sample data.
b. Compute the variance for these sample data.
c. Compute the standard deviation for these sample data.

**3-26.** The following data reflect the number of times a population of business executives flew on business during the previous month:

| 4 | 6 | 9 | 4 | 5 | 7 |
|---|---|---|---|---|---|

a. Compute the range for these data.
b. Compute the variance and standard deviation.
c. Assuming that these data represent a sample rather than a population, compute the variance and standard deviation. Discuss the difference between the values computed here and in part b.

**3-27.** The following data are the population of ages of students who have recently purchased a sports video game:

| 16 | 15 | 17 | 15 | 15 | 15 |
|----|----|----|----|----|----|
| 14 | 9  | 16 | 15 | 13 | 10 |
| 8  | 18 | 20 | 17 | 17 | 17 |
| 18 | 23 | 7  | 15 | 20 | 10 |
| 14 | 14 | 12 | 12 | 24 | 21 |

a. Compute the population variance.
b. Compute the population standard deviation.

**3-28.** A county library in Minnesota reported the following numbers of books checked out in 15 randomly selected months:

| 5,176 | 6,005 | 5,052 | 5,310 | 4,188 |
|-------|-------|-------|-------|-------|
| 4,132 | 5,736 | 5,381 | 4,983 | 4,423 |
| 5,002 | 4,573 | 4,209 | 5,611 | 4,568 |

Determine the range, variance, and standard deviation for the sample data.

**3-29.** The following data show the number of hours spent watching television for 12 randomly selected freshmen attending a liberal arts college in the Midwest:

**Hours of Television Viewed Weekly**

| 7.5  | 11.5 | 14.4 | 7.8  |
|------|------|------|------|
| 13.0 | 10.3 | 5.4  | 12.0 |
| 12.2 | 8.9  | 8.5  | 6.6  |

Calculate the range, variance, standard deviation, and interquartile range for the sample data.

**3-30.** Consider the following two separate samples:

| 27 | 27 | 25 | 12 | 15 | 10 | 20 | 37 | 31 | 35 |
|----|----|----|----|----|----|----|----|----|----|

and

| 1 | 3 | 2 | 16 | 18 | 16 | 16 | 4 | 16 | 118 |
|---|---|---|----|----|----|----|---|----|-----|

a. Calculate the range, variance, standard deviation, and interquartile range for each data set.
b. Which data set is more spread out based on these statistics?
c. Now remove the largest number from each data set and repeat the calculations called for in part a.
d. Compare the results of parts a and c. Which statistic seems to be more affected by outliers?

**3-31.** The following set of data shows the number of alcoholic drinks that students at a Kansas university reported they had consumed in the past month:

| 24 | 16 | 23 | 26 | 30 | 21 | 15 | 9 |
|----|----|----|----|----|----|----|---|
| 18 | 27 | 14 | 6  | 14 | 10 | 12 |   |

a. Assume the data set is a sample. Calculate the range, variance, standard deviation, and interquartile range for the data set.
b. Assume the data set is a population. Calculate the range, variance, standard deviation, and interquartile range for the data set.
c. Indicate the relationship between the statistics and the respective parameters calculated in parts a and b.

### Business Applications

**3-32.** Easy Connect, Inc., provides access to computers for business uses. The manager monitors computer use to make sure that the number of computers is sufficient to meet the needs of the customers. The manager collected data on a sample of customers and tracked the times the customers started working at a computer until they were finished. The elapsed times, in minutes, are shown as follows:

| 40 | 42 | 18 | 32 | 43 | 35 | 11 | 39 | 36 | 37 |
|----|----|----|----|----|----|----|----|----|----|
| 8  | 34 | 34 | 50 | 20 | 39 | 31 | 75 | 33 | 17 |

Compute appropriate measures of the center and variation to describe the time customers spend on the computer.

**3-33.** A random sample of 20 pledges to a public radio fund-raiser revealed the following dollar pledges:

| 90  | 85  | 100 | 150 |
|-----|-----|-----|-----|
| 125 | 75  | 50  | 100 |
| 75  | 60  | 35  | 90  |
| 100 | 125 | 75  | 85  |
| 50  | 100 | 50  | 80  |

a. Compute the range, variance, standard deviation, and interquartile range for these sample data.

b. Briefly explain the difference between the range and the interquartile range as a measure of dispersion.

**3-34.** Gold's Gym selected a random sample of ten customers and monitored the number of times each customer used the workout facility in a one-month period. The following data were collected:

10   19   17   19   12   20   20   15   16   13

Gold's managers are considering a promotion in which they reward frequent users with a small gift. They have decided that they will only give gifts to those customers whose number of visits in a one-month period is 1 standard deviation above the mean. Find the minimum number of visits required to receive a gift.

**3-35.** A college registrar has been asked to prepare a report about the graduate students. Among other things, she wants to analyze the ages of the students. She has taken a sample of ten graduate students and has found the following ages:

32   22   24   27   27   33   28   23   24   21

a. Compute the range, interquartile range, and standard deviation for these data.

b. It is thought that the mean age of graduate students in U.S. colleges and universities is 37.8 years. Based on your calculations in part a, what might you conclude about the age of students in this college's programs?

**3-36.** The branch manager for the Commercial Savings & Trust Bank has been asked to prepare a presentation for next week's board meeting. At the presentation, she will discuss the status of her branch's loans issued for recreation vehicles (RVs). In particular, she will analyze the loan balances for a sample of ten RV loans. The following data were collected:

| | | | | |
|---|---|---|---|---|
| $11,509 | $8,088 | $13,415 | $17,028 | $16,754 |
| $18,626 | $4,917 | $11,740 | $16,393 | $ 8,757 |

a. Compute the mean loan balance.

b. Compute the loan balance standard deviation.

c. Write a one-paragraph statement that uses the statistics computed in parts a and b to describe the RV loan data at the branch.

**3-37.** A parking garage in Memphis monitors the time it takes customers to exit the parking structure from the time they get in their car until they are on the streets. A sample of 28 exits was recently taken and is shown here:

**Garage Exit (time in seconds)**

| | | | |
|---|---|---|---|
| 83 | 138 | 145 | 147 |
| 130 | 79 | 156 | 156 |
| 90 | 85 | 68 | 93 |
| 178 | 76 | 73 | 119 |
| 92 | 146 | 88 | 103 |
| 116 | 134 | 162 | 71 |
| 181 | 110 | 105 | 74 |

a. Calculate the range, interquartile range, variance, and standard deviation for these sample data.

b. If the minimum time and the maximum time in the sample data are both increased by 10 seconds, would this affect the value for the interquartile range that you calculated in part a? Why or why not?

c. Suppose the clock that electronically recorded the times was not working properly when the sample was taken and each of the sampled times needs to be increased by 10 seconds. How would adding 10 seconds to each of the sampled speed of service times change the sample variance of the data?

**3-38.** Students in a communications class studied network television programming and found the average nonprogramming minutes in an hour of prime-time broadcasting for network television was 15:48 minutes. For cable television, the average was 14:55 minutes.

a. Calculate the difference in the average clutter between network and cable television.

b. Suppose the standard deviation in the amount of clutter for both network and cable television was either 5 minutes or 15 seconds. Which standard deviation would lead you to conclude that there was a major difference in the two clutter averages? Comment.

**3-39.** The Bureau of Labor Statistics website (www.bls.gov) provides data on month-to-month changes in the U.S. export index for agricultural commodities during 2015. These data are reproduced here:

| Month | Change in Index |
|---|---|
| Jan | −1.1 |
| Feb | 0.1 |
| Mar | −1.4 |
| Apr | −2.5 |
| May | 1.0 |
| Jun | −1.6 |
| Jul | −1.3 |
| Aug | −0.9 |
| Sep | −1.6 |
| Oct | −2.2 |
| Nov | −1.4 |

a. Calculate the mean, standard deviation, and interquartile range for these data.

b. Consider the mean calculated in part a. What does this value indicate about the price index?

c. What does the standard deviation indicate about the price index?

**3-40.** The College Board website (www.collegeboard.org) reported that the average cost of textbooks and supplies for students attending four-year public colleges and universities in the 2015–16 academic year is $1,298. Suppose a sample of 15 students was selected and the following data reflect the amount they paid for textbooks and supplies during that year:

| | | | | |
|---|---|---|---|---|
| $1,517 | $1,371 | $1,216 | $1,200 | $1,075 |
| $1,310 | $1,366 | $1,569 | $1,503 | $1,422 |
| $1,388 | $1,236 | $1,372 | $1,531 | $1,185 |

a. Calculate the mean and standard deviation.

b. Determine the number of standard deviations the most extreme cost is away from the mean. If you were to advise a prospective student concerning the money the student should save to afford the cost of books and supplies for at least 90% of the colleges, determine the amount you would suggest. (*Hint:* Don't forget the yearly inflation of the cost of books and supplies.)

## Computer Software Exercises

**3-41.** The manager of a cell phone store in the Valley Mall recently collected data on a sample of 50 customers who purchased a cell phone and a monthly call plan. The data she recorded are in the data file called **Phone Survey**.

a. The manager is interested in describing the difference between male and female customers with respect to the price of the phone purchased. She wants to compute the mean and standard deviation of the phone purchase price for each group of customers.

b. The manager is also interested in an analysis of the phone purchase price based on whether the use will be for home or business. Again, she wants to compute the mean and standard deviation of the phone purchase price for each group of customers.

**3-42.** The file **MBA Analysis** contains data on several variables for eight MBA graduates. The variables include pre– and post–MBA salary, percentage salary increase, undergraduate GPA, average GMAT score, annual tuition, and expected annual student cost. Compute the mean, median, range, variance, and standard deviation for each of the variables in the database and write a short report that summarizes the data using these measures. Include any appropriate charts or graphs to assist in your report.

**3-43.** The First City Real Estate Company managing partner asked the office administrator to provide a descriptive analysis of the asking prices of the homes the company currently has listed. This list includes 319 homes; the price data, along with other home characteristics, are included in the data file called **First City Real Estate**. These data constitute a population.

a. Compute the mean listing price.

b. Compute the median listing price.

c. Compute the range in listing prices.

d. Compute the standard deviation in listing prices.

e. Write a short report using the statistics computed in parts a–d to describe the prices of the homes currently listed by First City Real Estate.

**3-44.** Suppose there is an investigation to determine whether the increased availability of generic drugs, Internet drug purchases, and cost controls have reduced out-of-pocket drug expenses. As a part of the investigation, a random sample of 196 privately insured adults with incomes above 200% of the poverty level was taken, and their 2015 out-of-pocket medical expenses for prescription drugs were collected. The data are in the file **Drug Expenses**.

a. Calculate the mean and median for the sample data.

b. Calculate the range, variance, standard deviation, and interquartile range for the sample data.

c. Construct a box and whisker plot for the sample data.

d. Write a short report that describes out-of-pocket drug expenses for privately insured adults whose incomes are greater than 200% of the poverty level.

**3-45.** The placement center at a southern university tracks starting salaries. The data file called **BI SALARIES** lists the starting salaries for the 25 graduates in Information Technology who completed one or more internships prior to graduation.

a. Calculate the mean, median, and standard deviation of the starting salaries for IT graduates.

b. Calculate the 80th percentile for these data.

c. Calculate the third-quartile salary for these data.

**3-46.** Efficient Hiring, Inc., assists companies in hiring people to fill jobs. The file called **Hired** contains data on a sample of the placements made by the company. The two variables in the file are the number of days required to fill the position and the costs incurred (including advertising, interviewing, etc.) to complete the hiring process.

a. Calculate the variance and standard deviation for the sample cost data.

b. Construct a box and whisker plot for the cost data. Does this plot indicate that extreme values (outliers) may be inflating the measures of spread calculated in part a?

c. Suggest and calculate a measure of spread that is not affected by outliers.

In the previous sections, we introduced several important descriptive measures that are useful for transforming data into meaningful information. Two of the most important of these measures are the mean and the standard deviation. In this section, we discuss several statistical tools that combine these two.

OUTCOME **4** ## Coefficient of Variation

The standard deviation measures the variation in a set of data. For decision makers, the standard deviation indicates how spread out a distribution is. For distributions that have the same mean, the distribution with the largest standard deviation has the greatest relative spread. When two or more distributions have different means, the relative spread cannot be determined by merely comparing standard deviations.

The **coefficient of variation** (*CV*) is used to measure the relative variation for distributions with different means.

The coefficient of variation for a population is computed using Equation 3.15, whereas Equation 3.16 is used for sample data.

**Coefficient of Variation**
The ratio of the standard deviation to the mean expressed as a percentage. The coefficient of variation is used to measure variation relative to the mean.

**Population Coefficient of Variation**

$$CV = \frac{\sigma}{\mu}(100)\% \qquad (3.15)$$

**Sample Coefficient of Variation**

$$CV = \frac{s}{\bar{x}}(100)\% \qquad (3.16)$$

When the coefficients of variation for two or more distributions are compared, the distribution with the largest *CV* is said to have the greatest relative spread.

In finance, the *CV* is one measure of the relative risk of a stock portfolio. Assume Portfolio A has a collection of stocks that average a 12% return with a standard deviation of 3% and Portfolio B has an average return of 6% with a standard deviation of 2%. We can compute the *CV* values for each as follows:

$$CV(A) = \frac{3}{12}(100)\% = 25\%$$

and

$$CV(B) = \frac{2}{6}(100)\% = 33\%$$

Even though Portfolio B has a lower standard deviation, it might be considered more risky than Portfolio A because B's *CV* is 33% and A's *CV* is 25%.

---

**EXAMPLE 3-13** ## Computing the Coefficient of Variation

**Agra-Tech Industries** Agra-Tech Industries has recently introduced feed supplements for both cattle and hogs that will increase the rate at which the animals gain weight. Three years of feedlot tests indicate that cattle fed the supplement will weigh an average of 125 pounds more than those not fed the supplement. However, not every steer on the supplement has the same weight gain; results vary. The standard deviation in weight-gain advantage for the steers in the three-year study has been 10 pounds.

Similar tests with hogs indicate those fed the supplement average 40 additional pounds compared with hogs not given the supplement. The standard deviation for the hogs was also 10 pounds. Even though the standard deviation is the same for both cattle and hogs, the mean weight gains differ. Therefore, the coefficient of variation is needed to compare relative variability. The coefficient of variation for each is computed using the following steps:

STEP 1 **Collect the sample (or population) data for the variable of interest.**
In this case, we have two samples: weight gain for cattle and weight gain for hogs.

STEP 2 **Compute the mean and the standard deviation.**
For the two samples in this example, we get

$$\text{Cattle: } \bar{x} = 125 \text{ lb and } s = 10 \text{ lb}$$
$$\text{Hogs: } \bar{x} = 40 \text{ lb and } s = 10 \text{ lb}$$

STEP 3 **Compute the coefficient of variation using Equation 3.15 (for populations) or Equation 3.16 (for samples).**
Because the data in this example are from samples, the $CV$ is computed using

$$CV = \frac{s}{\bar{x}}(100)\%$$

For each data set, we get

$$CV(\text{cattle}) = \frac{10}{125}(100)\% = 8\%$$

$$CV(\text{hogs}) = \frac{10}{40}(100)\% = 25\%$$

These results indicate that hogs exhibit much greater relative variability in weight gain compared with cattle.

**TRY EXERCISE 3-50** (pg. 111)

OUTCOME **5**

**Empirical Rule**
If the data distribution is bell-shaped, then the interval

$\mu \pm 1\sigma$ contains approximately 68% of the values

$\mu \pm 2\sigma$ contains approximately 95% of the values

$\mu \pm 3\sigma$ contains virtually all of the data values

**The Empirical Rule**    A tool that is helpful in describing data in certain circumstances is called the **Empirical Rule**. For the Empirical Rule to be used, the frequency distribution must be bell-shaped, such as the one shown in Figure 3.11.

**FIGURE 3.11  Illustrating the Empirical Rule for the Bell-Shaped Distribution**

**Empirical Rule**

**Eddey's Burgers** The standard deviation can be thought of as a measure of distance from the mean. Consider the Eddey's Burgers restaurant chain, which records the number of each hamburger option it sells each day at each location. The numbers of chili burgers sold each day for the past 365 days are in the file called **Eddey's Burgers**. Figure 3.12 shows the frequency histogram for those data. The distribution is nearly *symmetric* and is approximately *bell-shaped*. The mean number of chili burgers sold was 15.1, with a standard deviation of 3.1.

Excel Tutorial

The Empirical Rule is a very useful statistical concept for helping us understand the data in a bell-shaped distribution. In the Eddey's Burgers example, with $\bar{x} = 15.1$ and $s = 3.1$, if we move 1 standard deviation in each direction from the mean, approximately 68% of the data should lie within the following range:

$$15.1 \pm 1(3.1)$$
$$12.0 \text{------------------------} 18.2$$

The actual number of days Eddey's Burgers sold between 12 and 18 chili burgers is 263. Thus, out of 365 days, on 72% of the days, Eddey's Burgers sold between 12 and 18 chili burgers. (The reason that we didn't get exactly 68% is that the distribution in Figure 3.12 is not perfectly bell-shaped.)

If we look at the interval 2 standard deviations from either side of the mean, we would expect approximately 95% of the data. The interval is

$$15.1 \pm 2(3.1)$$
$$15.1 \pm 6.2$$
$$8.9 \text{------------------------} 21.30$$

Counting the values between these limits, we find 354 of the 365 values, or 97%. Again this is close to what the Empirical Rule predicted. Finally, according to the Empirical Rule, we would expect almost all of the data to fall within 3 standard deviations. The interval is

$$15.1 \pm 3(3.1)$$
$$15.1 \pm 9.3$$
$$5.80 \text{------------------------} 24.40$$

Looking at the data in Figure 3.12, we find that, in fact, all the data do fall within this interval. Therefore, if we know only the mean and the standard deviation for a set of data, the Empirical Rule gives us a tool for describing how the data are distributed if the distribution is bell-shaped.

**FIGURE 3.12 Excel 2016 Histogram for Eddey's Burgers Data**

Excel 2016 Instructions

1. Open file: **Eddey's Burgers. xlsx**.
2. See instructions from Chapter 2, page 40, for developing a histogram in Excel.

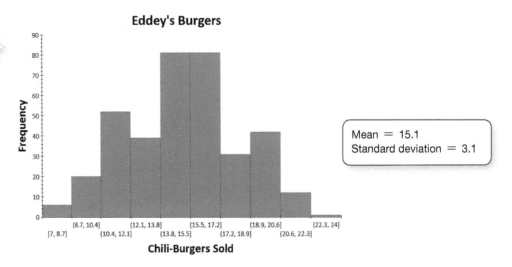

Mean = 15.1
Standard deviation = 3.1

OUTCOME **5**

# Tchebysheff's Theorem

The Empirical Rule applies when a distribution is bell-shaped. But what about the many situations in which a distribution is skewed and not bell-shaped? In these cases, we can use **Tchebysheff's theorem,** which is shown in the left margin.

Tchebysheff's theorem is conservative. It tells us nothing about the data within 1 standard deviation of the mean. Tchebysheff indicates that *at least* 75% of the data will fall within 2 standard deviations—it could be more. If we apply Tchebysheff's theorem to bell-shaped distributions, the percentage estimates are very low. The thing to remember is that Tchebysheff's theorem applies to *any distribution*. This gives it great flexibility.

OUTCOME **4**

# Standardized Data Values

When you are dealing with quantitative data, you will sometimes want to convert the measures to a form called **standardized data values**. This is especially useful when we wish to compare data from two or more distributions when the data scales for the two distributions are substantially different.

**Tchebysheff's Theorem**
Regardless of how data are distributed, *at least* $(1 - 1/k^2)$ of the values will fall within $k$ standard deviations of the mean. For example:

At least $\left(1 - \dfrac{1}{1^2}\right) = 0 = 0\%$ of the values will fall within $k = 1$ standard deviation of the mean.

At least $\left(1 - \dfrac{1}{2^2}\right) = \dfrac{3}{4} = 75\%$ of the values will lie within $k = 2$ standard deviations of the mean.

At least $\left(1 - \dfrac{1}{3^2}\right) = \dfrac{8}{9} = 89\%$ of the values will lie within $k = 3$ standard deviations of the mean.

**Standardized Data Values**
The number of standard deviations a value is from the mean. Standardized data values are sometimes referred to as z-scores.

| BUSINESS APPLICATION | Standardizing Data |

**Human Resources**   A local company uses placement exams as part of its hiring process. The company currently will accept scores from either of two tests: AIMS Hiring and BHS-Screen. The problem is that the AIMS Hiring test has an average score of 2,000 and a standard deviation of 200, whereas the BHS-Screen test has an average score of 80 with a standard deviation of 12. (These means and standard deviations were developed from a large number of people who have taken the two tests.) How can the company compare applicants when the average scores and measures of spread are so different for the two tests? One approach is to *standardize* the test scores.

The company is considering two applicants, John and Mary. John took the AIMS Hiring test and scored 2,344, whereas Mary took the BHS-Screen test and scored 95. Their scores can be standardized using Equation 3.17.

**Standardized Population Data**

$$z = \frac{x - \mu}{\sigma} \tag{3.17}$$

where:

$x$ = Original data value

$\mu$ = Population mean

$\sigma$ = Population standard deviation

$z$ = Standard score (number of standard deviations $x$ is from $\mu$)

If you are working with sample data rather than a population, Equation 3.18 can be used to standardize the values.

**Standardized Sample Data**

$$z = \frac{x - \bar{x}}{s} \tag{3.18}$$

where:

$x$ = Original data value

$\bar{x}$ = Sample mean

$s$ = Sample standard deviation

$z$ = Standard score

We can standardize the test scores for John and Mary using

$$z = \frac{x - \mu}{\sigma}$$

For the AIMS Hiring test, the mean, $\mu$, is 2,000 and the standard deviation, $\sigma$, equals 200. John's score of 2,344 converts to

$$z = \frac{2,344 - 2,000}{200}$$
$$z = 1.72$$

The BHS-Screen's $\mu = 80$ and $\sigma = 12$. Mary's score of 95 converts to

$$z = \frac{95 - 80}{12}$$
$$z = 1.25$$

> The Excel 2016 function to find the z-value is
>
> =Standardize(x,mean, standard deviation)
> =Standardize(95,80,12)

Compared to the average score on the AIMS Hiring test, John's score is 1.72 standard deviations higher. Mary's score is only 1.25 standard deviations higher than the average score on the BHS-Screen test. Therefore, even though the two tests used different scales, standardizing the data allows us to conclude that John scored relatively better on his test than Mary did on her test.

---

> **HOW TO DO IT** (Example 3-14)
>
> **Converting Data to Standardized Values**
>
> 1. Collect the population or sample values for the quantitative variable of interest.
>
> 2. Compute the population mean and standard deviation or the sample mean and standard deviation.
>
> 3. Convert the values to standardized z-values using Equation 3.17 or Equation 3.18. For populations,
>
> $$z = \frac{x - \mu}{\sigma}$$
>
> For samples,
>
> $$z = \frac{x - \bar{x}}{s}$$

**EXAMPLE 3-14** **Converting Data to Standardized Values**

**SAT and ACT Exams** Many colleges and universities require applicants to submit either SAT or ACT scores or both. One eastern university requires both exam scores. However, in assessing whether to admit a student, the university uses whichever exam score favors the student among all the applicants. Suppose the school receives 4,000 applications for admission. To determine which exam will be used for each student, the school will standardize the exam scores from both tests. To do this, it can use the following steps:

STEP 1 **Collect data.**
The university will collect the data for the 4,000 SAT scores and the 4,000 ACT scores for those students who applied for admission.

STEP 2 **Compute the mean and standard deviation.**
Assuming that these data reflect the population of interest for the university, the population mean is computed using

$$\text{SAT: } \mu = \frac{\sum x}{N} = 1,255 \qquad \text{ACT: } \mu = \frac{\sum x}{N} = 28.3$$

The standard deviation is computed using

$$\text{SAT: } \sigma = \sqrt{\frac{\sum (x - \mu)^2}{N}} = 72 \qquad \text{ACT: } \sigma = \sqrt{\frac{\sum (x - \mu)^2}{N}} = 2.4$$

STEP 3 **Standardize the data.**
Convert the $x$ values to $z$-values using

$$z = \frac{x - \mu}{\sigma}$$

Suppose a particular applicant has an SAT score of 1,228 and an ACT score of 27. These test scores can be converted to standardized scores:

$$\text{SAT: } z = \frac{x - \mu}{\sigma} = \frac{1,228 - 1,255}{72} = -0.375$$

$$\text{ACT: } z = \frac{x - \mu}{\sigma} = \frac{27 - 28.3}{2.4} = -0.542$$

> The Excel 2016 function to find the z-value is
>
> =Standardize(x,mean, standard deviation)

The negative $z$-values indicate that this student is below the mean on both the SAT and ACT exams. Because the university wishes to use the score that most favors the student, it will use the SAT score. The student is only 0.375 standard deviation below the SAT mean, compared with 0.542 standard deviation below the ACT mean.

TRY EXERCISE 3-52 (pg. 111)

## 3.3 EXERCISES

### Skill Development

**3-47.** A population of unknown shape has a mean of 3,000 and a standard deviation of 200.
  a. Find the minimum proportion of observations in the population that are in the range 2,600 to 3,400.
  b. Determine the maximum proportion of the observations that are above 3,600.
  c. What statement could you make concerning the proportion of observations that are smaller than 2,400?

**3-48.** The mean time that a certain model of light bulb will last is 400 hours, with a standard deviation equal to 50 hours.
  a. Calculate the standardized value for a light bulb that lasts 500 hours.
  b. Assuming that the distribution of hours that light bulbs last is bell-shaped, what percentage of bulbs could be expected to last longer than 500 hours?

**3-49.** Consider the following set of sample data:

78 121 143 88 110 107 62 122 130 95 78 139 89 125

  a. Compute the mean and standard deviation for these sample data.
  b. Calculate the coefficient of variation for these sample data and interpret its meaning.
  c. Using Tchebysheff's theorem, determine the range of values that should include at least 89% of the data. Count the number of data values that fall into this range and comment on whether your interval range was conservative.

**3-50.** You are given the following parameters for two populations:

| Population 1 | Population 2 |
|---|---|
| $\mu = 700$ | $\mu = 29,000$ |
| $\sigma = 50$ | $\sigma = 5,000$ |

  a. Compute the coefficient of variation for each population.
  b. Based on the answers to part a, which population has data values that are more variable relative to the size of the population mean?

**3-51.** Two distributions of data are being analyzed. Distribution A has a mean of 500 and a standard deviation equal to 100. Distribution B has a mean of 10 and a standard deviation equal to 4.0. Based on this information, use the coefficient of variation to determine which distribution has greater relative variation.

**3-52.** Given two distributions with the following characteristics:

| Distribution A | Distribution B |
|---|---|
| $\mu = 45,600$ | $\mu = 33.40$ |
| $\sigma = 6,333$ | $\sigma = 4.05$ |

If a value from Distribution A is 50,000 and a value from Distribution B is 40.0, convert each value to a standardized $z$-value and indicate which one is relatively closer to its respective mean.

**3-53.** If a sample mean is 1,000 and the sample standard deviation is 250, determine the standardized value for
  a. $x = 800$
  b. $x = 1,200$
  c. $x = 1,000$

**3-54.** The following data represent random samples taken from two different populations, A and B:

| A | 31 | 10 | 69 | 25 | 62 | 61 | 46 | 74 | 57 |
|---|---|---|---|---|---|---|---|---|---|
| B | 1,030 | 1,111 | 1,155 | 978 | 943 | 983 | 932 | 1,067 | 1,013 |

  a. Compute the mean and standard deviation for the sample data randomly selected from Population A.
  b. Compute the mean and standard deviation for the sample data randomly selected from Population B.
  c. Which sample has the greater spread when measured by the standard deviation?
  d. Compute the coefficient of variation for the sample data selected from Population A and from Population B. Which sample exhibits the greater relative variation?

**3-55.** Consider the following sample:

| 22 | 46 | 25 | 37 | 35 | 84 | 33 | 54 | 80 | 37 |
|---|---|---|---|---|---|---|---|---|---|
| 76 | 34 | 48 | 86 | 41 | 13 | 49 | 45 | 62 | 47 |
| 72 | 70 | 91 | 51 | 91 | 43 | 56 | 25 | 12 | 65 |

a. Calculate the mean and standard deviation for this data.
b. Determine the percentage of data values that fall in each of the following intervals:
$\bar{x} \pm s, \bar{x} \pm 2s, \bar{x} \pm 3s$.
c. Compare these with the percentages that should be expected from a bell-shaped distribution. Does it seem plausible that these data came from a bell-shaped population? Explain.

**3-56.** Consider the following population:

| 71 | 89 | 65 | 97 | 46 | 52 | 99 | 41 | 62 | 88 |
|----|----|----|----|----|----|----|----|----|----|
| 73 | 50 | 91 | 71 | 52 | 86 | 92 | 60 | 70 | 91 |
| 73 | 98 | 56 | 80 | 70 | 63 | 55 | 61 | 40 | 95 |

a. Determine the mean and variance.
b. Determine the percentage of data values that fall in each of the following intervals:
$\bar{x} \pm 2s, \bar{x} \pm 3s, \bar{x} \pm 4s$.
c. Compare these with the percentages specified by Tchebysheff's theorem.

## Business Applications

**3-57.** Pfizer, Inc., a major U.S. pharmaceutical company, is developing a new drug aimed at reducing the pain associated with migraine headaches. Two drugs are currently under development. One consideration in the evaluation of the medication is how long the pain-killing effects of the drugs last. A random sample of 12 tests for each drug revealed the following times (in minutes) until the effects of the drug were neutralized:

| Drug A | 258 214 243 227 235 222 240 245 245 234 243 211 |
|--------|--------------------------------------------------|
| Drug B | 219 283 291 277 258 273 289 260 286 265 284 266 |

a. Calculate the mean and standard deviation for each of the two drugs.
b. Based on the sample means calculated in part a, which drug appears to be effective longer?
c. Based on the sample standard deviations calculated in part a, which drug appears to have the greater variability in effect time?
d. Calculate the sample coefficient of variation for the two drugs. Based on the coefficient of variation, which drug has the greater variability in its time until the effect is neutralized?

**3-58.** The Jones and McFarlin Company provides call center support for a number of electronics manufacturers. Customers call in to receive assistance from a live technician for the products they have purchased from a retailer. Jones and McFarlin are experimenting with a new automated system that attempts to provide answers to customer questions. One issue of importance is the time customers have to spend on the line to get their question answered. The company has collected a sample of calls using each system and recorded the following times in seconds:

| Time to Complete the Call (in seconds) | |
|---|---|
| Automated Response | 131 80 140 118 79 94 103 145 113 100 122 |
| Live Response | 170 177 150 208 151 127 147 140 109 184 119 149 129 152 |

a. Compute the mean and standard deviation for the time to complete calls using the automated response.
b. Compute the mean and standard deviation for the time to complete calls to live service representatives.
c. Compute the coefficient of variation for the time to complete calls to the automated system and the live representatives. Which group has the greater relative variability in the time to complete calls?
d. Construct box and whisker plots for the time required to complete the two types of calls and briefly discuss.

**3-59.** Lockheed Martin is a supplier for the aerospace industry. Suppose Lockheed is considering switching to Cirus Systems, Inc., a new supplier for one of the component parts it needs for an assembly. At issue is the variability of the components supplied by Cirus Systems, Inc., compared to that of the existing supplier. The existing supplier makes the desired part with a mean diameter of 3.75 inches and a standard deviation of 0.078 inch. Unfortunately, Lockheed Martin does not have any of the exact same parts from the new supplier. Instead, the new supplier has sent a sample of 20 parts of a different size that it claims are representative of the type of work it can do. These sample data are shown here and in the data file called **Cirus**.

| Diameters (in inches) | | | | | |
|--------|--------|--------|--------|--------|--------|
| 18.018 | 17.856 | 18.095 | 17.992 | 18.086 | 17.812 |
| 17.988 | 17.996 | 18.129 | 18.003 | 18.214 | 18.313 |
| 17.983 | 18.153 | 17.996 | 17.908 | | |
| 17.948 | 18.219 | 18.079 | 17.799 | | |

Prepare a short letter to Lockheed Martin indicating which supplier you would recommend based on relative variability.

**3-60.** Suppose mortgage interest tax deductions average $8,268 for people with incomes between $50,000 and $200,000 and $365 for those with incomes of $40,000 to $50,000. Suppose also the standard deviations of the housing benefits in these two categories were equal to $2,750 and $120, respectively.
a. Examine the two standard deviations. What do these indicate about the range of benefits enjoyed by the two groups?
b. Repeat part a using the coefficient of variation as the measure of relative variation.

**3-61.** Thirty people were given an employment screening test, which is supposed to produce scores that are distributed according to a bell-shaped distribution. The following data reflect the scores of those 30 people:

| 76 | 75 | 74 | 56 | 61 | 76 |
|----|----|----|----|----|----|
| 62 | 96 | 68 | 62 | 78 | 76 |
| 84 | 67 | 60 | 96 | 77 | 59 |
| 67 | 81 | 66 | 71 | 69 | 65 |
| 58 | 77 | 82 | 75 | 76 | 67 |

The employment agency has in the past issued a rejection letter with no interview to the lower 16% taking the test. They also send the upper 2.5% directly to the company without an interview. Everyone else is interviewed. Based on the data and the assumption of a bell-shaped distribution, what scores should be used for the two cutoffs?

**3-62.** Each fall, the College Board releases survey results on how much colleges and universities are charging undergraduate students in the new academic year. The survey indicated that the average published tuition and fees for 2015–2016 were $9,410 at public four-year colleges and universities and $32,405 at private, nonprofit four-year colleges and universities (*source:* http://trends.collegeboard.org/college-pricing/). Assume that the standard deviation was approximately $4,500 at public four-year colleges and universities and approximately $12,000 for private colleges and universities.

  a. Do the private, nonprofit four-year colleges and universities have the larger relative variability? Provide statistical evidence to support your answer.

  b. If the data on published tuition and fees were bell-shaped, determine the largest and smallest amount paid at the four-year private, nonprofit colleges and universities.

## Computer Software Exercises

**3-63.** A West Coast newspaper has asked 50 certified public accountant (CPA) firms to complete the same tax return for a hypothetical head of household. The CPA firms have their tax experts complete the return with the objective of determining the total federal income tax liability. The data in the file **Taxes** show the taxes owed as figured by each of the 50 CPA firms. Theoretically, they should all come up with the same taxes owed.

  Based on these data, write a short article for the paper that describes the results of this experiment. Include in your article such descriptive statistics as the mean, median, and standard deviation. You might consider using percentiles, the coefficient of variation, and Tchebysheff's theorem to help describe the data.

**3-64.** Nike RZN Black is one of the golf balls Nike, Inc., produces. It must meet the specifications of the United States Golf Association (USGA). The USGA mandates that the diameter of the ball shall not be less than 1.682 inches (42.67 mm). To verify that this specification is met, sample golf balls are taken from the production line and measured. These data are found in the file titled **Diameter**.

  a. Calculate the mean and standard deviation of this sample.

  b. Examine the specification for the diameter of the golf ball again. Does it seem that the data could possibly be bell-shaped? Explain.

  c. Determine the proportion of diameters in the following intervals: $\bar{x} \pm 2s, \bar{x} \pm 3s, \bar{x} \pm 4s$. Compare these with the percentages specified by Tchebysheff's theorem.

**3-65.** A company in Silicon Valley periodically offers its employees a full health screen in which data are collected on several characteristics including percent body fat. Data for a sample of employees are in the file **Bodyfat**. Use percent body fat as the variable of interest in the following calculations:

  a. Calculate the mean and median for percent body fat.

  b. Calculate the standard deviation for percent body fat.

  c. Calculate the coefficient of variation for percent body fat.

  d. If an employee has a percent body fat equal to 29, would it be appropriate to say that this employee is in the 90th percentile or above? Explain.

**3-66.** Airfare prices were collected for a round trip from Los Angeles (LAX) to Salt Lake City (SLC). Airfare prices were also collected for a round trip from Los Angeles (LAX) to Barcelona, Spain (BCN). Airfares were obtained for the designated and nearby airports during high travel months. The passenger was to fly coach class round-trip, staying seven days. The data are contained in a file titled **Airfare**.

  a. Calculate the mean and standard deviation for each of the flights.

  b. Calculate an appropriate measure of the relative variability of these two flights.

  c. A British friend of yours is currently in Barcelona and wishes to fly to Los Angeles. If the flight fares are the same but priced in English pounds, determine his mean, standard deviation, and measure of relative dispersion for that data. (*Note:* $1 = 0.566 GBP.)

**3-67.** One factor that will be important for world trade is the growth rate of the population of the world's countries. The data file **Country Growth** contains the most recent United Nations data on the population and the growth rate for the last decade for 232 countries throughout the world as of 2015 (*source:* www.indexmundi.com/). Based on these data, which countries had growth rates more than 2 standard deviations higher than the mean growth rate? Which countries had growth rates more than 2 standard deviations below the mean growth rate?

# 3 Overview

## Summary

### Measures of Center and Location (pg. 74–94)

**OUTCOME 1** Compute the mean, median, mode, and weighted mean for a set of data and use these measures to describe data.

**OUTCOME 2** Construct a box and whisker graph and interpret it.

- The three numerical measures of the center of a data set are the mean, median, and mode.
- The **mean** is the **arithmetic average** and is the most frequently used measure. The mean is sensitive to extreme values in the data.
- Use the **median** if the data are **skewed** or **ordinal**-level data. The median is unaffected by extremes and is the middle value in the data array.
- The **mode** is the value in the data that occurs most frequently; it is less often used as a measure of the center.
- When the mean, median, or mode is computed from a **population**, the measure is a **parameter**. If the measure is computed from **sample** data, the measure is a **statistic**.
- Other measures of location that are commonly used are **percentiles** and **quartiles**.
- A **box and whisker plot** uses a box to display the range of the middle 50% of the data. The limits of whiskers are calculated based on the numerical distance between the first and third quartiles (see Figure 3.13).

### Measures of Variation (pg. 95–105)

**OUTCOME 3** Compute the range, interquartile range, variance, and standard deviation and use these measures to describe data.

- One of the major issues that business decision makers face is the **variation** that exists in their operations, processes, and people. Because virtually all data exhibit variation, it is important to measure it.
- The **range** is the difference between the highest value and the lowest value in the data.
- The **interquartile range** measures the numerical distance between the third and first quartiles; this alternative to the range ignores the extremes in the data.
- The two most frequently used measures of variation are the **variance** and the **standard deviation**. The equations for these two measures differ slightly depending on whether you are working with a population or a sample. The **standard deviation** is a measure of the average deviation of the individual data items around the mean; it is measured in the same units as the variable of interest (see Figure 3.13).

### Using the Mean and Standard Deviation Together (pg. 106–113)

**OUTCOME 4** Compute a $z$-score and the coefficient of variation and apply them in decision-making situations.

**OUTCOME 5** Use the Empirical Rule and Tchebysheff's theorem.

- The real power of statistical measures of the center and variation is seen when they are used together to fully describe the data.
- A measure that is used a great deal in business, especially in financial analysis, is the **coefficient of variation**. For two or more data sets, the larger the coefficient of variation, the greater the relative variation of the data.
- The **Empirical Rule** allows decision makers to better understand the data from a **bell-shaped** distribution.
- **Tchebysheff's theorem** helps describe data that are not bell-shaped.
- **z-values** for each individual data point measure the number of standard deviations a data value is from the mean (see Figure 3.13).

FIGURE 3.13 **Summary of Numerical Statistical Measures**

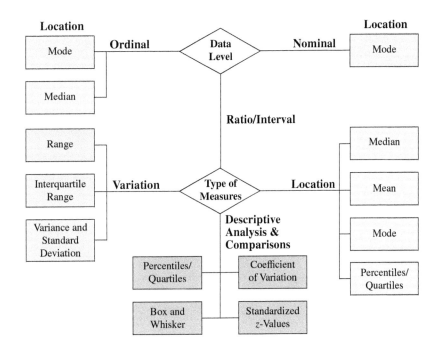

## Equations

**(3.1) Population Mean** pg. 74

$$\mu = \frac{\sum_{i=1}^{N} x_i}{N}$$

**(3.2) Sample Mean** pg. 78

$$\bar{x} = \frac{\sum_{i=1}^{n} x_i}{n}$$

**(3.3) Median Index** pg. 79

$$i = \frac{1}{2}n$$

**(3.4) Weighted Mean for a Population** pg. 84

$$\mu_w = \frac{\sum w_i x_i}{\sum w_i}$$

**(3.5) Weighted Mean for a Sample** pg. 84

$$\bar{x}_w = \frac{\sum w_i x_i}{\sum w_i}$$

**(3.6) Percentile Location Index** pg. 86

$$i = \frac{p}{100}(n)$$

**(3.7) Range** pg. 95

$$R = \text{Maximum value} - \text{Minimum value}$$

**(3.8) Interquartile Range** pg. 96

$$\text{Interquartile range} = \text{Third quartile} - \text{First quartile}$$

**(3.9) Population Variance** pg. 97

$$\sigma^2 = \frac{\sum_{i=1}^{N} (x_i - \mu)^2}{N}$$

**(3.10) Population Variance Shortcut** pg. 98

$$\sigma^2 = \frac{\sum x^2 - \frac{(\sum x)^2}{N}}{N}$$

**(3.11) Population Standard Deviation** pg. 98

$$\sigma = \sqrt{\sigma^2} = \sqrt{\frac{\sum_{i=1}^{N} (x_i - \mu)^2}{N}}$$

**(3.12) Sample Variance** pg. 100

$$s^2 = \frac{\sum_{i=1}^{n} (x_i - \bar{x})^2}{n - 1}$$

**(3.13) Sample Variance Shortcut** pg. 100

$$s^2 = \frac{\sum x^2 - \frac{(\sum x)^2}{n}}{n - 1}$$

**(3.14) Sample Standard Deviation** pg. 100

$$s = \sqrt{s^2} = \sqrt{\frac{\sum_{i=1}^{n} (x_i - \bar{x})^2}{n - 1}}$$

**(3.15) Population Coefficient of Variation** pg. 106

$$CV = \frac{\sigma}{\mu}(100)\%$$

**(3.16) Sample Coefficient of Variation** pg. 106

$$CV = \frac{s}{\bar{x}}(100)\%$$

**(3.17) Standardized Population Data** pg. 109

$$z = \frac{x - \mu}{\sigma}$$

**(3.18) Standardized Sample Data** pg. 109

$$z = \frac{x - \bar{x}}{s}$$

## Key Terms

Box and whisker plot pg. 87
Coefficient of variation pg. 106
Data array pg. 79
Empirical Rule pg. 107
Interquartile range pg. 96
Left-skewed data pg. 80
Mean pg. 74
Median pg. 79
Mode pg. 81

Parameter pg. 74
Percentiles pg. 85
Population mean pg. 74
Quartiles pg. 87
Range pg. 95
Right-skewed data pg. 80
Sample mean pg. 77
Skewed data pg. 80

Standard deviation pg. 97
Standardized data values pg. 109
Statistic pg. 74
Symmetric data pg. 80
Tchebysheff's theorem pg. 109
Variance pg. 97
Variation pg. 95
Weighted mean pg. 84

## Chapter Exercises

### Conceptual Questions

**3-68.** Consider the following questions concerning the sample variance:
   a. Is it possible for a variance to be negative? Explain.
   b. What is the smallest value a variance can be? Under what conditions does the variance equal this smallest value?
   c. Under what conditions is the sample variance smaller than the corresponding sample standard deviation?

**3-69.** For a continuous variable that has a bell-shaped distribution, determine the percentiles associated with the endpoints of the intervals specified in the Empirical Rule.

**3-70.** Consider that the Empirical Rule stipulates that virtually all of the data values are within the interval $\mu \pm 3\sigma$. Use this stipulation to determine an approximation for the standard deviation involving the range.

**3-71.** At almost every university in the United States, the university computes student grade point averages (GPAs). The following scale is typically used:

| | | |
|---|---|---|
| A = 4 points | B = 3 points | C = 2 points |
| D = 1 point | F = 0 points | |

Discuss what, if any, problems might exist when GPAs for two students are compared. What about comparing GPAs for students from two different universities?

**3-72.** Since the standard deviation of a set of data requires more effort to compute than the range does, what advantages does the standard deviation have when discussing the spread in a set of data?

**3-73.** The mode seems like a very simple measure of the location of a distribution. When would the mode be preferred over the median or the mean?

### Business Applications

**3-74.** Recently, a store manager tracked the time customers spent in the store from the time they took a number until they left. A sample of 16 customers was selected and the following data (measured in minutes) were recorded:

| | | | | | | | |
|---|---|---|---|---|---|---|---|
| 15 | 14 | 16 | 14 | 14 | 14 | 13 | 8 |
| 12 | 9 | 7 | 17 | 10 | 15 | 16 | 16 |

   a. Compute the mean, median, mode, range, interquartile range, and standard deviation.
   b. Develop a box and whisker plot for these data.

**3-75.** Suppose the mean age of video game players is 28, the standard deviation is 9 years, and the distribution is bell-shaped. To assist a video game company's marketing department in obtaining demographics to increase sales, determine the proportion of players who are
   a. between 19 and 28
   b. between 28 and 37
   c. older than 37

**3-76.** A traveler compared the costs for trips between Los Angeles and Denver, 1,016 miles one way. Los Angeles to Denver round-trip costs $159 by car and $218 by plane. Cost flexibility is greater with the flying trips because of greater airfare choices. The driving trip costs, except for the on-road lunches, are pretty much set in place. Assume the standard deviation for the cost of flying trips is approximately $54.

a. If a flight to Denver from Los Angeles was chosen at random, determine the proportion of the time that the cost would be less than $164. Assume the flight costs are bell-shaped.

b. Determine a flight cost that would qualify as the 25th percentile.

c. If nothing can be assumed about the distribution of the flight costs, determine the largest percentile that could be attributed to an airfare of $128.

**3-77.** A travel agency priced tickets between Spokane, Washington, and St. Louis, Missouri. The passenger was to fly coach class round-trip, staying seven days. Calls were made directly to airlines and to travel agents with the following results. Note that the data reflect round-trip airfare.

| | | | | | |
|---|---|---|---|---|---|
| $229.00 | $345.00 | $599.00 | $229.00 | $429.00 | $605.00 |
| $339.00 | $339.00 | $229.00 | $279.00 | $344.00 | $407.00 |

a. Compute the mean quoted airfare.

b. Compute the variance and standard deviation in airfares quoted. Treat the data as a sample.

**3-78.** The manager of a Miami restaurant recently selected a random sample of 18 customers and kept track of how long the customers were required to wait from the time they arrived at the restaurant until they were actually served dinner. This study resulted from several complaints the manager had received from customers saying that their wait time was unduly long and it appeared that the objective was to keep people waiting in the lounge for as long as possible to increase the lounge business. The following data were recorded, with time measured in minutes:

| | | | | | | | | |
|---|---|---|---|---|---|---|---|---|
| 34 | 24 | 43 | 56 | 74 | 20 | 19 | 33 | 55 |
| 43 | 54 | 34 | 27 | 34 | 36 | 24 | 54 | 39 |

a. Compute the mean waiting time for this sample of customers.

b. Compute the median waiting time for this sample of customers.

c. Compute the variance and standard deviation of waiting times for this sample of customers.

d. Develop a frequency distribution using six classes, each with a class width of 10. Make the lower limit of the first class 15.

e. Develop a frequency histogram for the frequency distribution.

f. Construct a box and whisker plot of these data.

g. The manager is considering giving a complimentary drink to customers whose waiting time is longer than the third quartile. Determine the minimum number of minutes a customer would have to wait in order to receive a complimentary drink.

**3-79.** Agri-Chemical Company has decided to implement a new incentive system for the managers of its three plants. The plan calls for a bonus to be paid next month

to the manager whose plant has the greatest relative improvement over the average monthly production volume. The following data reflect the historical production volumes at the three plants:

| Plant 1 | Plant 2 | Plant 3 |
|---|---|---|
| $\mu = 700$ | $\mu = 2,300$ | $\mu = 1,200$ |
| $\sigma = 200$ | $\sigma = 350$ | $\sigma = 30$ |

At the close of next month, the monthly output for the three plants was

Plant 1 = 810    Plant 2 = 2,600    Plant 3 = 1,320

Suppose the division manager has awarded the bonus to the manager of Plant 2 since her plant increased its production by 300 units over the mean, more than that for any of the other managers. Do you agree with the award of the bonus for this month? Explain, using the appropriate statistical measures to support your position.

**3-80.** According to the annual report issued by Wilson & Associates, an investment firm in Bowling Green, the stocks in its Growth Fund have generated an average return of 8% with a standard deviation of 2%. The stocks in the Specialized Fund have generated an average return of 18% with a standard deviation of 6%.

a. Based on the data provided, which of these funds has exhibited greater relative variability? Use the proper statistical measure to make your determination.

b. Suppose an investor who is very risk averse is interested in one of these two funds. Based strictly on relative variability, which fund would you recommend? Discuss.

c. Suppose the distributions for the two stock funds are bell-shaped with the means and standard deviations previously indicated. Which fund appears to be the better investment, assuming future returns will mimic past returns? Explain.

**3-81.** A large farm cooperative in North Dakota owns and leases prime farmland in the upper Midwest. Most of its 34,000 acres are planted in grain. The cooperative performs a substantial amount of testing to determine what seed types produce the greatest yields. Recently, the cooperative tested three types of corn seed on test plots. The following values were observed after the first test year:

| | Seed Type A | Seed Type B | Seed Type C |
|---|---|---|---|
| Mean Bushels/Acre | 88 | 56 | 100 |
| Standard Deviation | 25 | 15 | 16 |

a. Based on the results of this testing, which seed seems to produce the greatest average yield per acre? Comment on the type of testing controls that should have been used to make this study valid.

b. Suppose the company is interested in consistency. Which seed type shows the least relative variability?

c. Assuming the Empirical Rule applies, describe the production distribution for each of the three seed types.

d. Suppose you were a farmer and had to obtain at least 135 bushels per acre to escape bankruptcy. Which seed type would you plant? Explain your choice.

e. Rework your answer to part d assuming the farmer needed 115 bushels per acre instead.

**3-82.** The Hillcrest Golf and Country Club manager selected a random sample of the members and recorded the number of rounds of golf they played last season. The reason for his interest in these data is that the club is thinking of applying a discount to members who golf more than a specified number of rounds per year. The sample of eight people produced the following numbers of rounds played:

| 13 | 32 | 12 | 9 | 16 | 17 | 16 | 12 |
|----|----|----|---|----|----|----|----|

a. Compute the mean for these sample data.

b. Compute the median for these sample data.

c. Compute the mode for these sample data.

d. Calculate the variance and standard deviation for these sample data.

e. Note that one person in the sample played 32 rounds. What effect, if any, does this large value have on each of the three measures of location? Discuss.

f. For these sample data, which measure of location provides the best measure of the center of the data? Discuss.

g. Given this sample data, suppose the manager wishes to give discounts to golfers in the top quartile. What should the minimum number of rounds played be to receive a discount?

**3-83.** The Zagat Survey®, a leading provider of leisure-based survey results, released its New York City report that showed an average price per meal of $48.56 (*source:* www.zagat.com).

a. If the standard deviation of the price of meals in New York City was $10, determine the largest proportion of meal prices that could be higher than $60.

b. If the checks were paid in Chinese currency ($1 USD = 6.46 Chinese yuan), determine the mean and standard deviation of meal prices in New York City. How would this change of currency affect your answer to part a?

## Computer Software Exercises

**3-84.** The AJ Fitness Club is a full-service fitness center that offers its members the latest in weight machines, cardio equipment, and exercise classes. The company has surveyed its members to learn how satisfied they are with the services provided by the club. The data for this survey are in the file called **AJFitness**.

a. Calculate the mean and median age for those surveyed. Does the age distribution appear to be skewed or symmetric? Explain your answer.

b. Calculate the standard deviation for age.

c. Calculate the coefficient of variation for age.

d. Calculate the mean and standard deviation for the variable Typical Visits per Week broken down by gender.

**3-85.** The data in the file named **Fast100** were collected by D. L. Green & Associates. To aid the investment firm in locating appropriate investments for its clients, Sandra Williams, an assistant client manager, put together a database on 100 fast-growing companies. The database consists of data on eight variables for each of the 100 companies. Note that in some cases data are not available. A code of –99 has been used to signify missing data. These data will have to be omitted from any calculations.

a. Select the variable Sales. Develop a frequency distribution and histogram for Sales.

b. Compute the mean, median, and standard deviation for the Sales variable.

c. Determine the interquartile range for the Sales variable.

d. Construct a box and whisker plot for the Sales variable. Identify any outliers. Discard the outliers and recalculate the measures in part b.

e. Each year, a goal is set for sales. Next year's goal will be to have average sales that are at this year's 65th percentile. Identify next year's sales goal.

**3-86.** The Environmental Protection Agency (EPA) tests all new cars and provides a mileage rating for both city and highway driving conditions. For 2015, 32 electric cars were tested, and the data are contained in the data file **Electric Cars** (*source:* www.fueleconomy.gov). In this exercise, focus on the city and highway mileage data.

a. Calculate the sample mean miles per gallon (mpg) for both city and highway driving for the cars. Also calculate the sample standard deviation for the two mileage variables. Do the data tend to support the premise that cars get better mileage on the highway than around town? Discuss.

b. Referring to part a, what can the EPA conclude about the relative variability between car models for highway versus city driving? (*Hint:* Compute the appropriate measure to compare relative variability.)

**3-87.** A file titled **Halloween** contains sample data on Halloween spending for costumes.

a. Calculate the mean and standard deviation of these data.

b. Determine the following intervals for this data set: $\bar{x} \pm 1s, \bar{x} \pm 2s, \bar{x} \pm 3s$.

c. Suppose your responsibility as an assistant manager of a costume store was to determine the price of costumes to be sold. The manager has informed you to set the price of one costume so that it was beyond the budget of only 2.5% of the customers. Assume that the data set has a bell-shaped distribution.

**3-88.** The file titled **Payscale** contains data on salaries for a sample of MBA graduates in California and Florida.

a. Calculate the standard deviations of the salaries for both states' MBA graduates. Which state seems to have a wider spectrum of salaries for MBA graduates?

b. Calculate the average and median salary for each state's MBA graduates.

c. Examining the averages calculated in part b, determine which state's MBA graduates have the larger relative dispersion.

**3-89.** *Yahoo! Finance* makes available historical stock prices. It lists the opening, high, and low stock prices for each stock available on NYSE and NASDAQ. A file titled **GEstock** gives these data for General Electric (GE) for the period October 1, 2015, through December 24, 2015.

a. Calculate the difference between the opening and closing stock prices for GE over this time period. Then calculate the mean, median, and standard deviation of these differences.

b. Indicate what the mean in part a indicates about the relative opening and closing stock prices for GE.

**3-90.** Chad's Bakery makes a variety of bread types that it sells to supermarket chains in the area. One of Chad's problems is that the number of loaves of each type of bread sold each day by the chain stores varies considerably, making it difficult to know how many loaves to bake. A sample of daily demand data is contained in the file **Bakery**.

a. Which bread type has the highest average daily demand?

b. Develop a frequency distribution for each bread type.

c. Which bread type has the highest standard deviation in demand?

d. Which bread type has the greatest relative variability? Which type has the lowest relative variability?

e. Assuming that these sample data are representative of demand during the year, determine how many loaves of each type of bread should be made such that demand would be met on at least 75% of the days during the year.

f. Create a new variable called Total Loaves Sold. On which day of the week is the average for total loaves sold the highest?

**3-91.** For the past few years, one of the country's biggest tax-preparing companies has sponsored an event in which 50 certified public accountants from all sizes of CPA firms are asked to determine the tax owed for a fictitious citizen. The IRS is also asked to determine the "correct" tax owed. Last year, the "correct" figure stated by the IRS was $11,560. The file **Taxes** contains the data for the 50 accountants.

a. Compute a new variable that is the difference between the IRS number and the number determined by each accountant.

b. For this new variable computed in part a, develop a frequency distribution.

c. For the new variable computed in part a, determine the mean, median, and standard deviation.

d. Determine the percentile that would correspond to the "correct" tax figure if the IRS figure were one of the CPA firms' estimated tax figures. Describe what this implies about the agreement between the IRS and consultants' calculated tax.

**3-92.** The Cozine Corporation operates a garbage hauling business. Up to this point, the company has been charged a flat fee for each of the garbage trucks that enters the county landfill. The flat fee is based on the assumed truck weight of 45,000 pounds. In two weeks, the company is required to appear before the county commissioners to discuss a rate adjustment. In preparation for this meeting, Cozine has hired an independent company to weigh a sample of Cozine's garbage trucks just prior to their entering the landfill. The data file **Cozine** contains the data the company has collected.

a. Based on the sample data, what percentile does the 45,000-pound weight fall closest to?

b. Compute appropriate measures of central location for the data.

c. Construct a frequency histogram based on the sample data. Use the $2^k \geq n$ guideline (see Chapter 2) to determine the number of classes. Also, construct a box and whisker plot for these data. Discuss the relative advantages of histograms and box and whisker plots for presenting these data.

d. Use the information determined in parts a–c to develop a presentation to the county commissioners. Make sure the presentation attempts to answer the question of whether Cozine deserves a rate reduction.

**3-93.** A coffee shop would like to compare the dollar purchases of customers who use the drive-thru and those who come inside. A random sample of inside and drive-thru purchases is taken, and the results are contained in the file **Coffee Shop Sales**.

a. Develop a box and whisker plot to compare the dollar purchases of inside and drive-thru customers.

b. Are the purchases symmetric or skewed for each group?

c. Are the median purchase amounts for the two groups similar or not?

d. Does one customer group have more variation in its purchases?

e. Do any of the sampled values appear to be unusual?

f. Report the five-number summary for each of the two customer groups.

**3-94.** A manufacturing company operates three shifts daily. Each shift produces two models: A and B. A random sample of daily production has been taken, and the results are in the file **Shift Production**.

a. Develop a box and whisker plot that compares daily production by shift and model.

b. Is there a difference in the median output by shift? Is there a difference in the median output by model?

c. Does the variation in daily production appear to be symmetric or skewed across the shifts and model?

d. Are there any outliers in the data?

## Case 3.1    SDW—Human Resources

SDW is a large international construction company. The company was a major player in the reconstruction efforts in Iraq. However, the company is also involved in many small projects both in the United States and around the world. One of these is a sewer line installation project in a Midwest city. The contract is what is called a "cost plus" contract, meaning that the city will pay for all direct costs, including materials and labor, of the project plus an additional fee to SDW. Roberta Bernhart is the human resources (HR) manager for the project and is responsible for overseeing all aspects of employee compensation and HR issues.

SDW is required to produce a variety of reports to the city council on an annual basis. Recently, the council asked SDW to prepare a report showing the current hourly rates for the nonsalaried work crew on the project. Specifically, the council is interested in any proposed pay increases to the work crew that will ultimately be passed along to the city. In response to the city's request, Roberta put together a data file for all 19 nonsalaried work crew members, called **SDW**, which shows their current hourly pay rate and the proposed increase to take place the first of next month.

The city council expects the report to contain both graphic and numerical descriptive analyses. Roberta has outlined the following tasks and has asked you to help her.

### Required Tasks:

1. Develop and interpret histograms showing the distributions of current hourly rates and proposed new hourly rates for the crew members.
2. Compute and interpret key measures of central tendency and variation for the current and new hourly rates. Determine the coefficient of variation for each.
3. Compute a new variable called Pay Increase that reflects the difference between the proposed new pay rate and the current rate. Develop a histogram for this variable, and then compute key measures of the center and variation for the new variable.
4. Compute a new variable that is the percentage increase in hourly pay rate. Prepare a graphical and numerical description of this new variable.
5. Prepare a report to the city council that contains the results from tasks 1–4.

## Case 3.2    National Call Center

Candice Worthy and Philip Hanson are day-shift supervisors at National Call Center's Austin, Texas, facility. National provides contract call center services for a number of companies, including banks and major retail companies. Candice and Philip have both been with the company for slightly more than five years, having joined National right after graduating with bachelor's degrees from the University of Texas. As they walked down the hall together after the weekly staff meeting, the two friends were discussing the assignment they were just handed by Mark Gonzales, the division manager. The assignment came out of a discussion at the meeting in which one of National's clients wanted a report describing the calls being handled for them by National. Mark had asked Candice and Philip to describe the data in a file called **National Call Center** and produce a report that would both graphically and numerically analyze the data. The data are for a sample of 57 calls and for the following variables:

Account Number

Caller Gender

Account Holder Gender

Past Due Amount

Current Account Balance

Nature of Call (Billing Question or Other)

By the time they reached their office, Candice and Philip had outlined some of the key tasks that they needed to do.

### Required Tasks:

1. Develop bar charts showing the mean and median current account balance by gender of the caller.
2. Develop bar charts showing the mean and median current account balance by gender of the account holder.
3. Construct a scatter diagram showing current balance on the horizontal axis and past due amount on the vertical axis.
4. Compute the key descriptive statistics for the center and for the variation in current account balance broken down by gender of the caller, gender of the account holder, and nature of the call.
5. Repeat task 4 but compute the statistics for the past due balances.
6. Compute the coefficient of variation for current account balances for male and female account holders.
7. Develop frequency and relative frequency distributions for the gender of callers, gender of account holders, and nature of the calls.
8. Develop joint frequency and joint relative frequency distributions for the account holder gender by whether or not the account has a past due balance.
9. Write a report to National's client that contains the results for tasks 1–8 along with a discussion of these statistics and graphs.

## Case 3.3    Pine River Lumber Company—Part 2

Case 2.3 in Chapter 2 introduced you to the Pine River Lumber Company and to Martin Bledsoe, Pine River Lumber Company's process improvement team leader. Pine River Lumber makes cedar fencing materials at its Naples, Idaho, facility, employing about 160 people.

In Case 2.3 you were asked to help Martin develop a graphical descriptive analysis for data collected from the head rig. The head rig is a large saw that breaks down the logs into slabs and cants. Refer to Case 2.3 for more details involving a study that Martin recently conducted in which he videotaped 365 logs being broken down by the head rig. All three operators, April, Sid, and Jim, were involved. Each log was marked as to its true diameter. Then Martin observed the way the log was broken down and the degree to which

the cants were properly centered. He then determined the projected value of the finished product from each log given the way it was actually cut. In addition, he also determined what the value would have been had the log been cut in the optimal way. Data for this study are in a file called **Pine River Lumber**.

In addition to the graphical analysis that you helped Martin perform in Case 2.3, you have been asked to assist Martin by analyzing these data using appropriate measures of the center and variation. He wishes to focus on the lost profit to the company and whether there are differences among the operators. Also, do the operators tend to perform better on small logs than on large logs? In general, he is hoping to learn as much as possible from this study and needs your help with the analysis.

## Case 3.4    AJ's Fitness Center

When A. J. Reeser signed papers to take ownership of the fitness center previously known as the Park Center Club, he realized that he had just taken the biggest financial step in his life. Every asset he could pull together had been pledged against the mortgage. If the new AJ's Fitness Center didn't succeed, he would be in really bad shape financially. But A. J. didn't plan on failing. After all, he had never failed at anything.

As a high school football All-American, A. J. had been heavily recruited by major colleges around the country. Although he loved football, he and his family had always put academics ahead of sports. Thus, he surprised almost everyone other than those who knew him best when he chose to attend an Ivy League university not particularly noted for its football success. Although he excelled at football and was a member of two winning teams, he also succeeded in the classroom and graduated in four years. He spent six years working for McKinsey & Company, a major consulting firm, at which he gained significant experience in a broad range of business situations.

He was hired away from McKinsey & Company by the Dryden Group, a management services company that specializes in running health and fitness operations and recreational resorts throughout the world. After eight years of leading the Fitness Center section at Dryden, A. J. found that earning a high salary and the perks associated with corporate life were not satisfying him. Besides, the travel was getting old now that he had married and had two young children. When the opportunity to purchase the Park Center Club came, he decided that the time was right to control his own destiny.

A key aspect of the deal was that AJ's Fitness Club would keep its existing clientele, consisting of 1,833 memberships. One of the things A. J. was very concerned about was whether these members would stay with the club after the sale or move on to other fitness clubs in the area. He knew that keeping existing customers is a lot less expensive than attracting new customers.

Within days of assuming ownership, A. J. developed a survey that was mailed to all 1,833 members. The letter that accompanied the survey discussed A. J.'s philosophy and asked several key questions regarding the current level of satisfaction. Survey respondents were eligible to win a free lifetime membership in a drawing—an inducement that was no doubt responsible for the 1,214 usable responses.

To get help with the analysis of the survey data, A. J. approached the college of business at a local university with the

idea of having a senior student serve as an intern at AJ's Fitness Center. In addition to an hourly wage, the intern would get free use of the fitness facilities for the rest of the academic year.

The intern's first task was to key the data from the survey into a file that could be analyzed using a spreadsheet or a statistical software package. The survey contained eight questions that were keyed into eight columns, as follows:

Column 1:    Satisfaction with the club's weight- and exercise-equipment facilities

Column 2:    Satisfaction with the club's staff

Column 3:    Satisfaction with the club's exercise programs (aerobics, etc.)

Column 4:    Satisfaction with the club's overall service

Note that, columns 1 through 4 were coded on an ordinal scale as follows:

| 1 | 2 | 3 | 4 | 5 |
|---|---|---|---|---|
| Very unsatisfied | Unsatisfied | Neutral | Satisfied | Very satisfied |

Column 5:    Number of years that the respondent had been a member at this club

Column 6:    Gender (1 = Male, 2 = Female)

Column 7:    Typical number of visits to the club per week

Column 8:    Age

The data, saved in the file **AJFitness**, were clearly too much for anyone to comprehend in raw form. At yesterday's meeting, A. J. asked the intern to "make some sense of the data." When the intern asked for some direction, A. J.'s response was, "That's what I'm paying you the big bucks for. I just want you to develop a descriptive analysis of these data. Use whatever charts, graphs, and tables that will help us understand our customers. Also, use any pertinent numerical measures that will help in the analysis. For right now, give me a report that discusses the data. Why don't we set a time to get together next week to review your report?"

# 1-3 Special Review Section

This is the first of two special review sections in this text. The material in these sections, which is presented using block diagrams and flowcharts, is intended to help you tie together the material from several key chapters. These sections are not a substitute for reading and studying the chapters covered by the review. However, you can use this review material to add to your understanding of the individual topics in the chapters.

## Chapters 1–3

Chapters 1 through 3 introduce data, data collection, and statistical tools for describing data. The steps needed to gather "good" statistical data, transform them into usable information, and present the information in a manner that allows good decisions are outlined in the following figures.

**Transforming Data into Information**

# A Typical Application Sequence

**Determine a Need for Data**

- Research the issue
- Analyze business alternatives
- Respond to request for information

} Define your data requirements

**Define the Population**

- All items of interest—Who? What?
- Determine how to gain access to the population

**Determine What Data You Will Need**

- Identify the key variables (e.g., age, income, diameter, processing time, satisfaction rating)
- What categorical breakdowns will be needed?
  (e.g., analyze by gender, race, region, and class standing)

**Decide How the Data Will Be Collected**

- Experiment
- Observation
- Automation
- Telephone Survey
- Written Survey
- Personal Interview

**Decide on a Census or a Sample**

- Census: All items in the population
- Sample: A subset of the population

**Decide on Statistical or Nonstatistical Sampling**

- Statistical Sampling:
  - Simple Random Sample
  - Stratified Random Sample
  - Systematic Random Sample
  - Cluster Random Sample
- Nonstatistical Sampling:
  - Convenience Sample
  - Judgment Sample

**Determine Data Types and Measurement Level**

The method of descriptive statistical analysis that can be performed depends on the type of data and the level of data measurement for the variables in the data set. Typical studies will involve multiple types of variables and data levels.

- **Types of Data**

  | Quantitative | Qualitative |

- **Data Timing**

  | Cross-Sectional | Time-Series |

- **Data Level**

  *Lowest Level*  | Nominal |  Categories—no ordering implied

  *Mid-Level*  | Ordinal |  Categories—defined ordering

  *Highest Level*  | Interval/Ratio |  Measurements

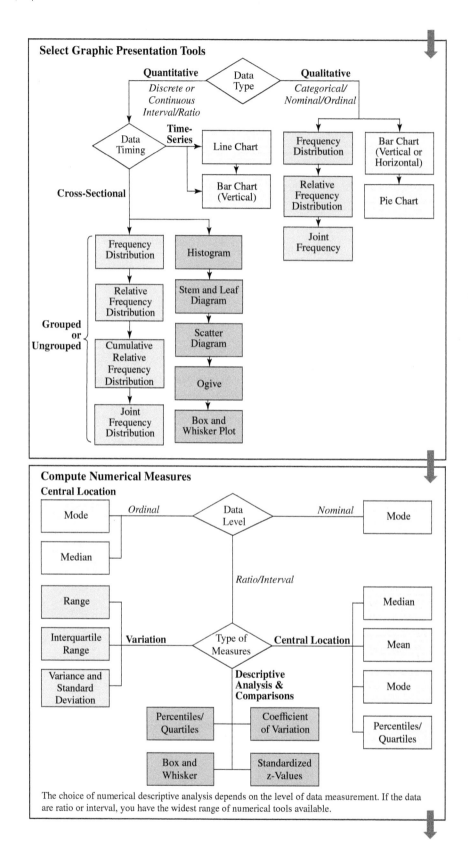

**Write the Statistical Report**

There is no one set format for writing a statistical report. However, there are a few suggestions you may find useful.

- *Lay the foundation:*
  Provide background and motivation for the analysis.

- *Describe the data collection methodology:*
  Explain how the data were gathered and the sampling techniques were used.

- *Use a logical sequence:*
  Follow a systematic plan for presenting your findings and analysis.

- *Label figures and tables by number:*
  Employ a consistent numbering and labeling format.

## EXERCISES

### Integrative Application Exercises

Chapters 1 to 3 have introduced you to the basics of descriptive statistics. Many of the business application problems, advanced business application problems, and cases in these chapters give you practice in performing descriptive statistical analysis. However, too often you are told which procedure you should use, or you can surmise which to use by the location of the exercise. It is important that you learn to identify the appropriate procedure on your own in order to solve problems for test purposes. But more important, this ability is essential throughout your career when you are required to select procedures for the tasks you will undertake. The following exercises will provide you with identification practice.

**SR.1.** Go to your university library and obtain the *Statistical Abstract of the United States*.
  a. Construct a frequency distribution for unemployment rate by state for the most current year available.
  b. Justify your choice of class limits and number of classes.
  c. Locate the unemployment rate for the state in which you are attending college. (1) What proportion of the unemployment rates are below that of your state? (2) Describe the distribution's shape with respect to symmetry. (3) If you were planning to build a new manufacturing plant, what state would you choose in which to build? Justify your answer. (4) Are there any unusual features of this distribution? Describe them.

**SR.2.** The State Industrial Development Council is currently working on a financial services brochure to send to out-of-state companies. It is hoped that the brochure will be helpful in attracting companies to relocate to your state. You are given the following frequency distribution on banks in your state:

| Deposit Size (in millions) | Number of Banks | Total Deposits (in millions) |
|---|---|---|
| Less than 5 | 2 | 7.2 |
| 5 to less than 10 | 7 | 52.1 |
| 10 to less than 25 | 6 | 111.5 |
| 25 to less than 50 | 3 | 95.4 |
| 50 to less than 100 | 2 | 166.6 |
| 100 to less than 500 | 2 | 529.8 |
| Over 500 | 2 | 1,663.0 |

  a. Does this frequency distribution violate any of the rules of construction for frequency distributions? If so, reconstruct the frequency distribution to remedy this violation.
  b. The council wishes to target companies that would require financial support from banks that have at least $25 million in deposits. Reconstruct the frequency distribution to attract such companies to relocate to your state. Do this by considering different classes that would accomplish such a goal.
  c. Reconstruct the frequency distribution to attract companies that require financial support from banks that have between $5 million and $25 million in deposits.
  d. Present an eye-catching, two-paragraph summary of what the data would mean to a company that is considering moving to the state. Your boss has said you need to include relative frequencies in this presentation.

**SR.3.** As an intern for Intel Corporation, suppose you have been asked to help the vice president prepare a newsletter to the shareholders. You have been given access to the data in a file called **Intel** that contains Intel Corporation financial data for the years 1987 through 1996. Go to the Internet or to Intel's annual report and update the file to include the same variables for the years 1997 to the present. Then use graphs to effectively present the data in a format that would be usable for the vice president's newsletter. Write a short article that discusses the information shown in your graphs.

**SR.4.** The Woodmill Company makes windows and door trim products. The first step in the process is to rip dimension ($2 \times 8, 2 \times 10$, etc.) lumber into narrower pieces. Currently, the company uses a manual process in which an experienced operator quickly looks at a board and determines what rip widths to use. The decision is based on the knots and defects in the wood.

A company in Oregon has developed an optical scanner that can be used to determine the rip widths. The scanner is programmed to recognize defects and to determine rip widths that will optimize the value of the board. A test run of 100 boards was put through the scanner and the rip widths were identified. However, the boards were not actually ripped. A lumber grader determined the resulting values for each of the 100 boards assuming that the rips determined by the scanner had been made. Next, the same 100 boards were manually ripped using the normal process. The grader then determined the value for each board after the manual rip process was completed. The resulting data, in the file **Woodmill**, consist of manual rip values and scanner rip values for each of the 100 boards.

a. Develop a frequency distribution for the board values for the scanner and the manual process.

b. Compute appropriate descriptive statistics for both manual and scanner values. Use these data along with the frequency distribution developed in part a

to prepare a written report that describes the results of the test. Be sure to include in your report a conclusion regarding whether the scanner outperforms the manual process.

c. Which process, scanner or manual, generated more values that were more than 2 standard deviations from the mean?

d. Which of the two processes has less relative variability?

**SR.5.** The commercial banking industry is undergoing rapid changes due to advances in technology and competitive pressures in the financial services sector. The data file **Banks** contains selected information tabulated by *Fortune* concerning the revenues, profitability, and number of employees for the 51 largest U.S. commercial banks. Use the information in this file to complete the following:

a. Compute the mean, median, and standard deviation for the three variables revenues, profits, and number of employees.

b. Convert the data for each variable to a *z*-value. Consider Mellon Bank Corporation headquartered in Pittsburgh. How does it compare to the average bank in the study on the three variables? Discuss.

c. As you can see by examining the data and by looking at the statistics computed in part a, not all banks had the same revenue, same profit, or same number of employees. Which variable had the greatest relative variation among the banks in the study?

d. Calculate a new variable: profits per employee. Develop a frequency distribution and a histogram for this new variable. Also compute the mean, median, and standard deviation for the new variable. Write a short report that describes the profits per employee for the banks.

e. Referring to part d, how many banks had a profit-per-employee ratio that was more than 2 standard deviations from the mean?

## Review Case 1    State Department of Insurance

This case study describes the efforts undertaken by the director of the Insurance Division to assess the magnitude of the uninsured motorist problem in a western state. The objective of the case study is to introduce you to a data collection application and show how one organization developed a database. The database **Liabins** contains a subset of the data actually collected by the state department.

The impetus for the case came from the legislative transportation committee, which heard much testimony during the recent legislative session about the problems that occur when an uninsured motorist is involved in a traffic accident in which damages to individuals and property occur. The state's law enforcement officers also testified that a large number of vehicles are not covered by liability insurance.

Because of both political pressure and a sense of duty to do what is right, the legislative committee spent many hours wrestling with what to do about drivers who do not carry the mandatory liability insurance. Because the actual magnitude of the problem was unknown,

the committee finally arrived at a compromise plan, which required the state Insurance Division to perform random audits of vehicles to determine whether the vehicle was covered by liability insurance. The audits are to be performed on approximately 1% of the state's 1 million registered vehicles each month. If a vehicle is found not to have liability insurance, the vehicle license and the owner's driver's license will be revoked for three months and a $250 fine will be imposed.

However, before actually implementing the audit process, which is projected to cost $1.5 million per year, Herb Kriner, director of the Insurance Division, was told to conduct a preliminary study of the uninsured motorists problem in the state and to report back to the legislative committee in six months.

### The Study

A random sample of 12 counties in the state was selected in a manner that gave the counties with higher numbers of registered vehicles proportionally higher chances of being selected. Two locations were selected in each county, and the state police set up roadblocks on a randomly selected day. Vehicles with in-state license plates were stopped at random until approximately 100 vehicles had been stopped at each location. The target total was about 2,400 vehicles statewide.

The issue of primary interest was whether the vehicle was insured. This was determined by observing whether the vehicle was carrying the required certificate of insurance. If so, the officer took down the insurance company name and address and the policy number. If the certificate was not in the car, but the owner stated that insurance was carried, the owner was given a postcard to return within five days supplying the required information. A vehicle was determined to be uninsured if no postcard was returned or if, subsequently, the insurance company reported that the policy was not valid on the day of the survey.

In addition to the issue of insurance coverage, Herb Kriner wanted to collect other information about the vehicle and the owner. This was done using a personal interview during which the police officer asked a series of questions and observed certain things such as seat belt usage and driver's and vehicle license expiration status. Also, the owners' driving records were obtained through the Transportation Department's computer division and added to the information gathered by the state police.

## The Data

The data are contained in the file **Liabins**. The sheet titled "Description" contains an explanation of the data set and the variables.

## Issues to Address

Herb has two weeks before he will give his presentation to the legislative subcommittee that has been dealing with the liability insurance issue. As Herb's chief analyst, your job is to perform a comprehensive analysis of the data and to prepare the report that Herb will deliver to the legislature. Remember, this report will go a long way in determining whether the state should spend the $1.5 million to implement a full liability insurance audit system.

## Term Project Assignments

For the project selected, you are to devise a sampling plan, collect appropriate data, and carry out a full descriptive analysis aimed at shedding light on the key issues for the project. The finished project will include a written report of a length and format specified by your professor.

## Project A

*Issue:* Your College of Business and Economics seeks input from business majors regarding class scheduling. Some potential issues are

- Day or evening
- Morning or afternoon
- One-day, two-day, or three-day schedules
- Weekend
- Location (on or off campus)

## Project B

*Issue:* Intercollegiate athletics is a part of most major universities. Revenue from attendance at major sporting events is one key to financing the athletic program. Investigate the drivers of attendance at your university's men's basketball and football games. Some potential issues:

- Game times
- Game days (basketball)
- Ticket prices
- Athletic booster club memberships
- Competition for entertainment dollars

## Project C

*Issue:* The department of your major is interested in surveying department alumni. Some potential issues are

- Satisfaction with degree
- Employment status
- Job satisfaction
- Suggestions for improving course content

## Capstone Project

### Project Objective

The objective of this business statistics capstone project is to provide you with an opportunity to integrate the statistical tools and concepts that you have learned thus far in your business statistics course. As with all real-world applications, completing this project will not require you to utilize every statistical technique covered in the first three chapters. Rather, an objective of the assignment is for you to determine which of the statistical tools and techniques are appropriate for the situation you have selected.

### Project Description

Assume that you are working as an intern for a financial management company. Your employer has a large number of clients who trust the company managers to invest their funds. In your position, you are responsible for producing reports for clients when they request information. Your company has a large data file with financial information for all 500 companies listed on the Standard & Poor's 500 Index for 2013. These data were compiled from various publicly available sources and are in a file called **S&P 500**.

Your role will be to perform certain statistical analyses that can be used to help convert these data into useful information that might aid the firm's clients in making investment decisions.

There are no firm guidelines on what the report should entail, but you are encouraged to blend appropriate descriptive measures with charts, graphs, and tables in order to transform the data into potentially useful information.

### Project Deliverables

To successfully complete this capstone project, you are required to deliver a management report that tells an interesting and informative story using correctly applied descriptive statistics tools.

The final report should be presented in a professional format using the style or format suggested by your instructor.

# 4 Introduction to Probability

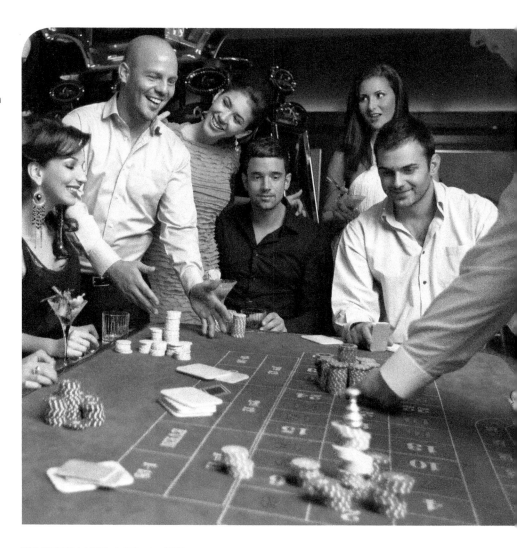

## WHY YOU NEED TO KNOW

The Powerball lottery raised the cost of buying a ticket from $1 to $2. With the higher ticket price, lottery officials expect the jackpot prize value to increase more rapidly and thereby entice even greater ticket sales. Most people recognize when they buy a lottery ticket that there is a very small *probability* of winning and that whether they win or lose is based on chance alone. In case you are not familiar with the Powerball lottery system, a drum contains 69 balls, numbered 1 to 69. The player must choose (or have a computer choose) five numbers between 1 and 69. The player also chooses a sixth number called the power ball

### Quick Prep

Review the discussion of statistical sampling in Section 1.3.

Examine recent business periodicals and newspapers, looking for examples in which probability concepts are discussed.

Think about how you determine what decision to make in situations in which you are uncertain about your choices.

from the numbers 1 to 26. On the night of the drawing, five balls are *randomly* selected and then placed in numerical order. Last, a sixth ball (the power ball) is randomly selected. To win the jackpot, the player must match all five numbers plus the power ball. The odds of winning are shown on the Powerball website to be 1 in 292,201,338, or about 0.00000000342. Later in the chapter, you will learn the method for computing probabilities like this. One analogy might put this in perspective. Suppose we take a college football field and cover it with 292,201,338 tiny red ants. One of these ants has a yellow dot on it. If you were blindfolded, your chances of picking the one ant with the yellow dot from the millions of ants on the football field would be the same as winning the Powerball jackpot! We suggest you come up with a different retirement strategy.

In business decision making, in many instances, chance is involved in determining the outcome of a decision. For instance, when a TV manufacturer establishes a warranty on its TV sets, there is a probability that any given TV will last less than the warranty life and customers will have to be compensated. Accountants perform audits on the financial statements of a client and sign off on the statements as accurate while realizing there is a chance that problems exist that were not uncovered by the audit. A food processor manufacturer recognizes that there is a chance that one or more of its products will be substandard and dissatisfy the customer. Airlines overbook flights to make sure that the plane is as full as possible because they know there is a certain probability that customers will not show for their flight. Professional poker players base their decisions to fold or play a hand based on their assessment of the chances that their hand beats those of their opponents.

If we always knew what the result of our decisions would be, our life as decision makers would be a lot less stressful. However, in most instances, uncertainty exists. To deal with this uncertainty, we need to know how to incorporate probability concepts into the decision process. Chapter 4 takes the first step in teaching you how to do this by introducing the basic concepts and rules of probability. You need to have a solid understanding of these basics before moving on to the more practical probability applications that you will encounter in business.

## 4.1    The Basics of Probability

The mathematical study of probability originated more than 300 years ago. The Chevalier de Méré, a French nobleman (who today would probably own a gaming house in Monte Carlo), began asking questions about games of chance. He was mostly interested in the probability of observing various outcomes when dice were repeatedly rolled. The French mathematician Blaise Pascal (you may remember studying Pascal's triangle in a mathematics class), with the help of his friend Pierre de Fermat, was able to answer de Méré's questions. Of course, Pascal began asking more and more complicated questions of himself and his colleagues, and the formal study of probability began.

### Important Probability Terms

Several explanations of what probability is have come out of this mathematical study. However, the definition of **probability** is quite basic.

**Probability**

The chance that a particular event will occur. The probability value will be in the range 0 to 1. A value of 0 means the event will not occur. A probability of 1 means the event will occur. Any number between 0 and 1 reflects the uncertainty of the event occurring. The definition given is for a countable number of events.

For instance, if we look out the window and see rain, we can say the probability of rain today is 1, since we know for sure that it will rain. If an airplane has a top speed of 450 mph, and the distance between City A and City B is 900 miles, we can say the probability the plane will make the trip in 1.5 hours is 0—it can't happen. These examples involve situations in which we are *certain* of the outcome, and our 1 and 0 probabilities reflect this.

However, uncertainty exists in most business situations. For instance, if a real estate investor has the option to purchase a small shopping mall, determining a rate of return on this investment involves uncertainty. The investor does not know with certainty whether she will make a profit, break even, or lose money. After looking closely at the situation, she might say the chance of making a profit is 0.30. This value between 0 and 1 reflects her uncertainty about whether she will make a profit from purchasing the shopping mall.

**Experiment**
A process that produces a single outcome whose result cannot be predicted with certainty.

**Events and Sample Space** In probability language, the process that produces the outcomes is an **experiment**. For instance, a very simple experiment might involve flipping a coin one time. When this experiment is performed, two possible experimental outcomes can occur: head or tail. If the coin-tossing experiment is expanded to involve two flips of the coin, the experimental outcomes are

Head on first flip and head on second flip, denoted by $(H,H)$

Head on first flip and tail on second flip, denoted by $(H,T)$

Tail on first flip and head on second flip, denoted by $(T,H)$

Tail on first flip and tail on second flip, denoted by $(T,T)$

In business situations, the experiment can be things like an investment decision, a personnel decision, or a choice of warehouse location.

The collection of possible experimental outcomes is called the **sample space**.

**Sample Space**
The collection of all outcomes that can result from a selection, decision, or experiment.

---

**EXAMPLE 4-1** | **Defining the Sample Space**

**West Side Drive-In** The sales manager for the West Side Drive-In is interested in analyzing the sales of its three best-selling hamburgers. As part of this analysis, he might be interested in determining the sample space (possible outcomes) for two randomly selected customers. To do this, he can use the following steps:

**STEP 1** **Define the experiment.**
The experiment is the sale. The item of interest is the product sold.

**STEP 2** **Define the outcomes for one trial of the experiment.**
The manager can define the outcomes to be

$$e_1 = \text{Hamburger}$$
$$e_2 = \text{Cheeseburger}$$
$$e_3 = \text{Bacon Burger}$$

**STEP 3** **Define the sample space.**
The sample space ($SS$) for an experiment involving a single sale is

$$SS = \{e_1, e_2, e_3\}$$

If the experiment is expanded to include two sales, the sample space is

$$SS = \{e_1, e_2, e_3, e_4, e_5, e_6, e_7, e_8, e_9\}$$

where the outcomes include what happens on both sales and are defined as

| Outcome | Sale 1 | Sale 2 |
|---------|--------|--------|
| $e_1$ | Hamburger | Hamburger |
| $e_2$ | Hamburger | Cheeseburger |
| $e_3$ | Hamburger | Bacon Burger |
| $e_4$ | Cheeseburger | Hamburger |
| $e_5$ | Cheeseburger | Cheeseburger |
| $e_6$ | Cheeseburger | Bacon Burger |
| $e_7$ | Bacon Burger | Hamburger |
| $e_8$ | Bacon Burger | Cheeseburger |
| $e_9$ | Bacon Burger | Bacon Burger |

**TRY EXERCISE 4-3** (pg. 138)

---

**Using Tree Diagrams** A tree diagram is often a useful way to define the sample space for an experiment to ensure that no outcomes are omitted or repeated. Example 4-2 illustrates how a tree diagram is used.

**EXAMPLE 4-2**    **Using a Tree Diagram to Define the Sample Space**

**Clearwater Research**    Clearwater Research is involved in a project in which television viewers were asked whether they objected to hard-liquor advertisements being shown on television. The analyst is interested in listing the sample space, using a tree diagram as an aid, when three viewers are interviewed. The following steps can be used:

STEP 1    **Define the experiment.**
Three people are interviewed and asked, "Would you object to hard-liquor advertisements on television?" Thus, the experiment consists of three trials.

STEP 2    **Define the outcomes for a single trial of the experiment.**
The possible outcomes when one person is interviewed are
no
yes

STEP 3    **Define the sample space for three trials using a tree diagram.**
Begin by determining the outcomes for a single trial. Illustrate these with tree branches beginning on the left side of the page:

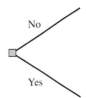

For each of these branches, add branches depicting the outcomes for a second trial. Continue until the tree has the number of sets of branches corresponding to the number of trials.

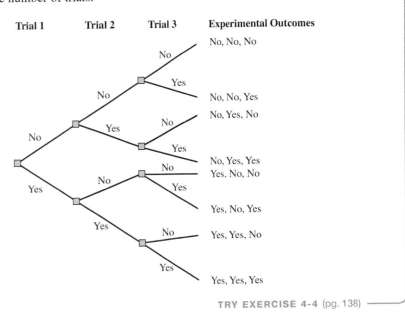

TRY EXERCISE 4-4 (pg. 138)

**Event**
A collection of experimental outcomes.

A collection of possible outcomes is called an **event**. An example will help clarify these terms.

**EXAMPLE 4-3**    **Defining an Event of Interest**

**KPMG Accounting**    The KPMG Accounting firm is interested in the sample space for an audit experiment in which the outcome of interest is the audit's completion status. The sample space is the list of all possible outcomes from the experiment. The accounting firm is also interested in specifying the outcomes that make up an event of interest. This can be done using the following steps:

STEP 1 **Define the experiment.**
The experiment consists of two randomly chosen audits.

STEP 2 **List the outcomes associated with one trial of the experiment.**
For a single audit, the following completion-status possibilities exist:

Audit done early
Audit done on time
Audit done late

STEP 3 **Define the sample space.**
For two audits (two trials), we define the sample space as follows:

| Experimental Outcome | | Audit 1 | Audit 2 |
|---|---|---|---|
| $e_1$ | = | Early | Early |
| $e_2$ | = | Early | On time |
| $e_3$ | = | Early | Late |
| $e_4$ | = | On time | Early |
| $e_5$ | = | On time | On time |
| $e_6$ | = | On time | Late |
| $e_7$ | = | Late | Early |
| $e_8$ | = | Late | On time |
| $e_9$ | = | Late | Late |

STEP 4 **Define the event of interest.**
The event of interest, at least one audit is completed late, is composed of all the outcomes in which one or more audits are late. This event ($E$) is

$$E = \{ e_3, e_6, e_7, e_8, e_9 \}$$

There are five ways in which one or more audits are completed late.

**TRY EXERCISE 4-3** (pg. 138)

**Mutually Exclusive Events** Keeping in mind the definitions for *experiment, sample space*, and *events*, we introduce two additional concepts. The first is **mutually exclusive events**.

**Mutually Exclusive Events**
Two events are mutually exclusive if the occurrence of one event precludes the occurrence of the other event.

**BUSINESS APPLICATION** Mutually Exclusive Events

**KPMG Accounting** Consider again the KPMG Accounting firm example. The possible outcomes for two audits are

| Experimental Outcomes | | Audit 1 | Audit 2 |
|---|---|---|---|
| $e_1$ | = | Early | Early |
| $e_2$ | = | Early | On time |
| $e_3$ | = | Early | Late |
| $e_4$ | = | On time | Early |
| $e_5$ | = | On time | On time |
| $e_6$ | = | On time | Late |
| $e_7$ | = | Late | Early |
| $e_8$ | = | Late | On time |
| $e_9$ | = | Late | Late |

Suppose we define one event as consisting of the outcomes in which at least one of the two audits is late:

$$E_1 = \{ e_3, e_6, e_7, e_8, e_9 \}$$

Further, suppose we define a second event as follows:

$$E_2 = \text{Neither audit is late} = \{ e_1, e_2, e_4, e_5 \}$$

Events $E_1$ and $E_2$ are mutually exclusive: if $E_1$ occurs, $E_2$ cannot occur; if $E_2$ occurs, $E_1$ cannot occur. That is, if at least one audit is late, then it is not possible for neither audit to be late. We can verify this fact by observing that no outcomes in $E_1$ appear in $E_2$. This observation provides another way of defining mutually exclusive events: two events are mutually exclusive if they have no common outcomes.

**EXAMPLE 4-4**    **Mutually Exclusive Events**

**Contract-Works, Inc.**    Contract-Works, Inc., does assembly work for high-tech companies. Each item produced on the assembly line can be thought of as an experimental trial. The managers at this facility can analyze their process to determine whether the events of interest are mutually exclusive using the following steps:

STEP 1    **Define the experiment.**
The experiment is producing a part on an assembly line.

STEP 2    **Define the outcomes for a single trial of the experiment.**
On each trial, the outcome is either a *good* or a *defective* item.

STEP 3    **Define the sample space.**
If two products are produced (two trials), the following sample space is defined:

| Experimental Outcomes | |
| --- | --- |
| **Product 1** | **Product 2** |
| $e_1$ = Good | Good |
| $e_2$ = Good | Defective |
| $e_3$ = Defective | Good |
| $e_4$ = Defective | Defective |

STEP 4    **Determine whether the events are mutually exclusive.**
Let event $E_1$ be defined as both products produced are good, and let event $E_2$ be defined as at least one product is defective:

$$E_1 = \text{Both good} = \{e_1\}$$
$$E_2 = \text{At least one defective} = \{e_2, e_3, e_4\}$$

Then events $E_1$ and $E_2$ are determined to be mutually exclusive because the two events have no outcomes in common. Having two good items and at the same time having at least one defective item is not possible.

**Independent and Dependent Events**    A second probability concept is that of **independent** versus **dependent events**.

**Independent Events**
Two events are independent if the occurrence of one event in no way influences the probability of the occurrence of the other event.

**Dependent Events**
Two events are dependent if the occurrence of one event impacts the probability of the other event occurring.

**BUSINESS APPLICATION**    **Independent and Dependent Events**

**Petroglyph**    Petroglyph is a natural gas exploration company with operations in the western United States and Canada. During the exploration phase, seismic surveys are conducted that provide information about the Earth's underground formations. Based on history, the company knows that if the seismic readings are favorable, gas will more likely be discovered than if the seismic readings are not favorable. However, the readings are not perfect indicators. Suppose the company currently is exploring in eastern Colorado. The possible outcomes for the seismic survey are defined as

$$e_1 = \text{Favorable}$$
$$e_2 = \text{Unfavorable}$$

If the company decides to drill, the outcomes are defined as

$$e_3 = \text{Strike gas}$$
$$e_4 = \text{Dry hole}$$

If we let the event $E_1$ be that the seismic survey is favorable and event $E_2$ be that the hole is dry, we can say that the two events are dependent because the probability of favorable report depends on whether there is gas underground. If the result of the survey was not related to the presence of gas, the events would be independent.

OUTCOME **1**

## Methods of Assigning Probability

Part of the confusion surrounding probability may be due to the fact that probability can be assigned to outcomes in more than one way. There are three common ways to assess the probabilities of outcomes: *classical probability assessment, relative frequency assessment*, and *subjective probability assessment*. The following notation is used when we refer to the probability of an event:

$$P(E_i) = \text{Probability of event } E_i \text{ occurring}$$

**Classical Probability Assessment**

The method of determining probability based on the ratio of the number of ways an outcome or event of interest can occur to the number of ways any outcome or event can occur when the individual outcomes are equally likely.

**Classical Probability Assessment**    The first method of probability assessment involves **classical probability**.

Consider again the experiment of flipping a coin one time. There are two possible outcomes: head and tail. If we assume the coin is fair, then each outcome is equally likely. Thus, with the classical assessment method, the probability of a head is the ratio of the number of ways a head can occur (1 way) to the total number of ways any outcome can occur (2 ways). Thus we get

$$P(\text{Head}) = \frac{1\,\text{way}}{2\,\text{ways}} = \frac{1}{2} = 0.50$$

The chance of a head occurring is 1 out of 2, or 0.50.

In those situations in which all possible outcomes are *equally likely*, the classical probability measurement is defined in Equation 4.1.

**Classical Probability Assessment**

$$P(E_i) = \frac{\text{Number of ways } E_i \text{ can occur}}{\text{Total number of possible outcomes}} \qquad \textbf{(4.1)}$$

---

**EXAMPLE 4-5**    **Classical Probability Assessment**

**King's Fine Clothes**    The managers at King's Fine Clothes plan to hold a special promotion over Labor Day weekend. Each customer who makes a purchase exceeding $100 will qualify to select an envelope from a large drum. Inside the envelope are coupons for percentage discounts off the purchase total. At the beginning of the weekend, there were 500 coupons. Four hundred of these were for a 10% discount, 50 were for 20%, 45 were for 30%, and 5 were for 50%. The probability of getting a particular discount amount can be determined using classical assessment with the following steps:

STEP **1** **Define the experiment.**
An envelope is selected from a large drum.

STEP **2** **Determine whether the possible outcomes are equally likely.**
In this case, the envelopes with the different discount amounts are unmarked from the outside and are thoroughly mixed in the drum. Thus, any one envelope

has the same probability of being selected as any other envelope. The outcomes are equally likely.

STEP 3  **Determine the total number of outcomes.**
There are 500 envelopes in the drum.

STEP 4  **Define the event of interest.**
We might be interested in assessing the probability that the first customer will get a 20% discount.

STEP 5  **Determine the number of outcomes associated with the event of interest.**
There are 50 coupons with a discount of 20% marked on them.

STEP 6  **Compute the classical probability using Equation 4.1.**

$$P(E_i) = \frac{\text{Number of ways } E_i \text{ can occur}}{\text{Total number of possible outcomes}}$$

$$P(20\% \text{ discount}) = \frac{\text{Number of ways } 20\% \text{ can occur}}{\text{Total number of possible outcomes}} = \frac{50}{500} = 0.10$$

*Note:* After the first customer removes an envelope from the drum, the probability that the next customer will get a particular discount will be different because the values in the denominator and possibly the numerator will change.

**TRY EXERCISE 4-10** (pg. 139)

As you can see, the classical approach to probability measurement is fairly straightforward. Many games of chance are based on classical probability assessment. However, classical probability assessment does not apply in many business situations. Rarely are the individual outcomes equally likely. For instance, you might be thinking of starting a business. The sample space is

$$SS = \{\text{Succeed, Fail}\}$$

Would it be reasonable to use classical assessment to determine the probability that your business will succeed? If so, we would make the following assessment:

$$P(\text{Succeed}) = \frac{1}{2}$$

If this were true, then the chance of any business succeeding would be 0.50. Of course, this is not true. Many factors go into determining the success or failure of a business. The possible outcomes (Succeed, Fail) are not equally likely. Instead, we need another method of probability assessment in these situations.

**Relative Frequency Assessment**

The method that defines probability as the number of times an event occurs divided by the total number of times an experiment is performed in a large number of trials.

## Relative Frequency Assessment

The **relative frequency assessment** approach is based on actual observations.

Equation 4.2 shows how the relative frequency assessment method is used to assess probabilities.

**Relative Frequency Assessment**

$$P(E_i) = \frac{\text{Number of times } E_i \text{ occurs}}{N} \tag{4.2}$$

where:

$$E_i = \text{The event of interest}$$

$$N = \text{Number of trials}$$

Relative Frequency Assessment

**Hathaway Heating & Air Conditioning** The sales manager at Hathaway Heating & Air Conditioning has recently developed the customer profile shown in Table 4.1. The profile is based on Hathaway's 500 customers. As a promotion for the company, the sales manager plans to randomly select a customer once a month and perform a free service on the customer's system. What is the probability that the first customer selected is a residential customer? What is the probability that the first customer has a Hathaway heating system?

To assess the probability that the customer selected is residential using the relative frequency of occurrence method, we determine from Table 4.1 the number of residential customers and divide by the total number of customers, both residential and commercial. We then apply Equation 4.2:

$$P(E_2) = P(\text{Residential}) = \frac{400}{500} = 0.80$$

Thus, there is an 80% chance the customer selected will be a residential customer.

The probability that the customer selected has a Hathaway heating system is determined by the ratio of the number of customers with Hathaway heating systems to the number of total customers:

$$P(E_3) = P(\text{Heating}) = \frac{200}{500} = 0.40$$

There is a 40% chance the randomly selected customer will have a Hathaway heating system.

The sales manager hopes the customer selected is a residential customer with a Hathaway heating system. Because there are 145 customers in this category, we can use the relative frequency method to assess the probability of this event occurring:

$$P(E_2 \text{ and } E_3) = P(\text{Residential with heating}) = \frac{145}{500} = 0.29$$

There is a 29% chance the customer selected will be a residential customer with a Hathaway heating system.

**TABLE 4.1** Hathaway Heating & Air Conditioning Co.

| | | Customer Category | | |
| | | $E_1$<br>Commercial | $E_2$<br>Residential | Total |
|---|---|---|---|---|
| $E_3$ | Heating Systems | 55 | 145 | 200 |
| $E_4$ | Air-Conditioning Systems | 45 | 255 | 300 |
| | Total | 100 | 400 | 500 |

EXAMPLE 4-6 Relative Frequency Probability Assessment

**Starbucks Coffee** The international coffee chain. Starbucks, sells caffeinated and decaffeinated drinks. One of the difficulties in this business is determining how much of a given product to prepare for the day. Suppose a store manager is interested in determining the probability that a customer will select a decaf versus a caffeinated drink. She has maintained records of customer purchases for the past three weeks. The probability can be assessed using relative frequency with the following steps:

STEP 1 **Define the experiment.**

A randomly chosen customer will select a drink.

STEP 2   **Define the event of interest.**
The manager is interested in the event $E_1$ customer selects a caffeinated drink.

STEP 3   **Determine the total number of occurrences.**
In this case, the manager has observed 2,250 drink sales in the past week. Thus, $N = 2,250$.

STEP 4   **For the event of interest, determine the number of occurrences.**
In the past week, 1,570 sales were for caffeinated drinks.

STEP 5   **Use Equation 4.2 to determine the probability assessment.**

$$P(E_1) = \frac{\text{Number of times } E_1 \text{ occurs}}{N} = \frac{1,570}{2,250} = 0.6978$$

Thus, based on past history, the chance that a customer will purchase a caffeinated drink is just under 0.70.

**TRY EXERCISE 4-9** (pg. 139)

**Potential Issues with the Relative Frequency Assessment Method**   You should be aware of a couple of concerns before applying the relative frequency assessment method. First, for this method to be useful, all of the observed frequencies must be comparable. For instance, consider again the case in which you are interested in starting a small business. Two outcomes can occur: business succeeds or business fails. If we are interested in the probability that the business will succeed, we might be tempted to study a sample of, say, 200 small businesses that have been started in the past and determine the number of those that have succeeded—say, 50. Using Equation 4.2 for the relative frequency method, we get

$$P(\text{Succeed}) = \frac{50}{200} = 0.25$$

However, before we can conclude that the chance your small business will succeed is 0.25, you must be sure that the conditions of each of the 200 businesses match your conditions (that is, location, type of business, management expertise and experience, financial standing, and so on). If not, then you should not use the relative frequency method.

Another issue involves the size of the denominator in Equation 4.2. If the number of possible occurrences is quite small, the probability assessment may be unreliable. For instance, suppose a basketball player took five free throws during the first ten games of the season and missed them all. The relative frequency method would determine the probability that he will make the next free throw to be

$$P(\text{Make}) = \frac{0 \text{ made}}{5 \text{ shots}} = \frac{0}{5} = 0.0$$

But do you think there is a zero chance that he will make his next free throw? No, even the notoriously poor free-throw shooter, Shaquille O'Neal, former National Basketball Association (NBA) star player, made some of his free throws. The problem is that the base of five free throws is too small to provide a reliable probability assessment.

**Subjective Probability Assessment**   Unfortunately, even though managers may have some past observations to guide their decision making, new factors will always be affecting each decision, making those observations only an approximate guide to the future. In other cases, managers may have little or no past data and, therefore, may not be able to use a relative frequency as even a starting point in assessing the desired probability. When past data are not available, decision makers must make a **subjective probability assessment**. A subjective probability is a measure of a personal conviction that an outcome will occur. Therefore, in this instance, probability represents a person's belief that an event will occur.

**Subjective Probability Assessment**
The method that defines the probability of an event as reflecting a decision maker's state of mind regarding the chances that the particular event will occur.

**Subjective Probability Assessment**

**URS Corporation** Because of a nationwide problem of deteriorating roads and bridges, many states have started increasing funding for highway-related projects. URS Corporation is preparing a bid for a major infrastructure construction project. The company's engineers are very good at defining all the elements of the projects (labor, materials, and so on) and know the costs of these with a great deal of certainty. In finalizing the bid amount, the managers add a profit markup to the projected costs. The problem is how much markup to add. If they add too much, they won't be the low bidder and may lose the contract. If they don't mark the bid up enough, they may get the project and make less profit than they might have made had they used a higher markup. The managers are considering four possible markup values, stated as percentages of base costs:

10%     12%     15%     20%

To make their decision, the managers need to assess the probability of winning the contract at each of these markup levels. Because they have never done a project exactly like this one, they can't rely on relative frequency assessment. Instead, they must subjectively assess the probability based on whatever information they currently have available, such as who the other bidders are, the rapport URS has with the potential client, and so forth.

After considering these values, the managers make the following assessments:

$$P(\text{Win at } 10\%) = 0.30$$
$$P(\text{Win at } 12\%) = 0.25$$
$$P(\text{Win at } 15\%) = 0.15$$
$$P(\text{Win at } 20\%) = 0.05$$

These assessments indicate the managers' state of mind regarding the chances of winning the contract. If new information (for example, a competitor drops out of the bidding) becomes available before the bid is submitted, these assessments could change. Note also that if different URS managers were involved, the subjective probability assessments would likely be different.

Each of the three methods by which probabilities are assessed has specific advantages and specific applications. Regardless of how decision makers arrive at a probability assessment, the rules by which people use these probabilities in decision making are the same. These rules will be introduced in Section 4.2.

## 4.1 EXERCISES

### Skill Development

**4-1.** A special roulette wheel, which has an equal number of red and black spots, has come up red four times in a row. Assuming that the roulette wheel is fair, what concept allows a player to know that the probability the next spin of the wheel will come up black is 0.5?

**4-2.** In a survey, respondents were asked to indicate their favorite brand of cereal (Post or Kellogg's). They were allowed only one choice. What is the probability concept that implies it is not possible for a single respondent to state both Post and Kellogg's are the favorite cereal?

**4-3.** If two customers are asked to list their choice of ice cream flavor from among vanilla, chocolate, and strawberry, list the sample space showing the possible outcomes.

**4-4.** Use a tree diagram to list the sample space for the number of movies rented by three customers at a Red Box video kiosk where customers are allowed to rent one, two, or three movies (assuming that each customer rents at least one movie).

**4-5.** In each of the following, indicate what method of probability assessment would most likely be used to assess the probability.
   a. What is the probability that a major earthquake will occur in California in the next three years?
   b. What is the probability that a customer will return a purchase for a refund?
   c. An inventory of appliances contains four white washers and one black washer. If a customer selects one at random, what is the probability that the black washer will be selected?

**4-6.** Long-time friends Pat and Tom agree on many things, but not the outcome of the American League pennant race and the World Series. Pat is originally from Boston, and Tom is from New York. They have a steak dinner bet on next year's race, with Pat betting on the Red Sox and Tom on the Yankees. Both are convinced they will win.

a. What probability assessment technique is being used by the two friends?

b. Why would the relative frequency technique not be appropriate in this situation?

**4-7.** Students who live on campus and purchase a meal plan are randomly assigned to one of three dining halls: the Commons, Northeast, and Frazier. What is the probability that the next student to purchase a meal plan will be assigned to the Commons?

**4-8.** The results of a census of 2,500 employees of a mid-sized company with 401(k) retirement accounts are as follows:

| Account Balance (to nearest $) | Male | Female |
|---|---|---|
| <$25,000 | 635 | 495 |
| $25,000–$49,999 | 185 | 210 |
| $50,000–$99,999 | 515 | 260 |
| ≥ $100,000 | 155 | 45 |

Suppose researchers are going to sample employees from the company for further study.

a. Based on the relative frequency assessment method, what is the probability that a randomly selected employee will be a female?

b. Based on the relative frequency assessment method, what is the probability that a randomly selected employee will have a 401(k) account balance between $25,000 and $49,999?

c. Compute the probability that a randomly selected employee will be a female with an account balance between $50,000 and $99,999.

**4-9.** Cross County Bicycles makes two mountain bike models, the XB-50 and the YZ-99, in three distinct colors. The following table shows the production volumes for last week:

| Model | Color | | |
|---|---|---|---|
| | Blue | Brown | White |
| XB-50 | 302 | 105 | 200 |
| YZ-99 | 40 | 205 | 130 |

a. Based on the relative frequency assessment method, what is the probability that a mountain bike is brown?

b. What is the probability that the mountain bike is a YZ-99?

c. What is the joint probability that a randomly selected mountain bike is a YZ-99 and brown?

d. Suppose a mountain bike is chosen at random. Consider the following two events: the event that model YZ-99 is chosen and the event that a white product is chosen. Are these two events mutually exclusive? Explain.

**4-10.** Cyber-Plastics, Inc., is in search of a CEO and a CFO. The company has a short list of candidates for each position. The four CEO candidates graduated from Chicago (C) and three Ivy League universities: Harvard (H), Princeton (P), and Yale (Y). The four CFO candidates graduated from MIT (M), Northwestern (N), and two Ivy League universities: Dartmouth (D) and Brown (B). One candidate from each of the respective lists will be chosen randomly to fill the positions. The event of interest is that both positions are filled with candidates from the Ivy League.

a. Determine whether the outcomes are equally likely.

b. Determine the number of equally likely outcomes.

c. Define the event of interest.

d. Determine the number of outcomes associated with the event of interest.

e. Compute the classical probability of the event of interest using Equation 4.1.

**4-11.** Three consumers go to a Best Buy to shop for high-definition televisions (HDTVs). Let $B$ indicate that one of the consumers buys an HDTV. Let $D$ be that the consumer doesn't buy an HDTV. Assume these events are equally likely. Consider the following: (1) only two consumers buy an HDTV, (2) at most two consumers buy HDTVs, and (3) at least two consumers buy HDTVs.

a. Determine whether outcomes 1, 2, and 3 are equally likely.

b. Determine the total number of equally likely outcomes for the three shoppers.

c. Define the events of interest in each of 1, 2, and 3. To define the events of interest, list the possible outcomes in each of the following events:

- Only two consumers buy an HDTV ($E_1$)
- At most two consumers buy HDTVs ($E_2$)
- At least two consumers buy HDTVs ($E_3$)

d. Determine the number of outcomes associated with each of the events of interest. Use the classical probability assessment approach to assign probabilities to each of the possible outcomes and calculate the probabilities of the events.

e. Compute the classical probabilities of each of the events in part d by using Equation 4.1.

## Business Applications

**4-12.** Cyber Communications, Inc., has a new cell phone product under development in the research and development (R&D) lab. It will increase the megapixel capability of cell phone cameras to the 12+ range. The head of R&D made a presentation to the company

CEO stating that the probability the company will earn a profit in excess of $20 million next year is 80%. What was the method used to make this probability assessment?

**4-13.** Five doctors work at the Evergreen Medical Clinic. The plan is to staff Saturdays with three doctors. The office manager has decided to make up Saturday schedules in such a way that no set of three doctors will be in the office together more than once. How many weeks can be covered by this schedule? (*Hint:* Use a tree diagram to list the sample space.)

**4-14.** Prince Windows, Inc., makes high-quality windows for the residential home market. Recently, three marketing managers were asked to assess the probability that sales for next year will be more than 15% higher than in the current year. One manager stated that the probability of this happening was 0.40. The second manager assessed the probability to be 0.60, and the third manager stated the probability was 0.90.
   a. What method of probability assessment is each of the three managers using?
   b. Which manager is expressing the least uncertainty in the probability assessment?
   c. Why is it that the three managers did not provide the same probability assessment?

**4-15.** The marketing manager for the *Charlotte Times* newspaper has commissioned a study of the advertisements in the classified section. The results for the Wednesday edition showed that 204 are help-wanted ads, 520 are real estate ads, and 306 are other ads.
   a. If the newspaper plans to select an ad at random each week to be published free, what is the probability that the ad for a specific week will be a help-wanted ad?
   b. What method of probability assessment is used to determine the probability in part a?
   c. Are the events that a help-wanted ad is chosen and that an ad for other types of products or services is chosen for this promotion on a specific week mutually exclusive? Explain.

**4-16.** One automobile dealership sells Buick, Cadillac, and Chevrolet automobiles. It also sells used cars that it gets as trade-ins on new car purchases. If two cars are sold on Tuesday by the dealership, what is the sample space for the type of cars that might be sold?

**4-17.** Welco Lumber manufactures cedar fencing products in Marysville, Washington. The company's quality manager inspected 5,900 boards and found that 4,100 could be rated as a #1 grade.
   a. If the manager wanted to assess the probability that a board being produced will be a #1 grade, what method of assessment would he likely use?
   b. Referring to your answer in part a, what would you assess the probability of a #1 grade board to be?

**4-18.** A survey of 490 workers included the question "In your opinion, is it OK for a company to monitor its employees' Internet use?" The possible responses were:

(1) Only after informing the employees, (2) Does not need to inform the employees, (3) Only when company believes an employee is misusing the Internet, (4) Company does not have the right, and (5) Only if an employee has previously misused the Internet. The following table contains the results for the respondents:

| Response | 1 | 2 | 3 | 4 | 5 |
|---|---|---|---|---|---|
| Number of Respondents | 150 | 206 | 109 | 11 | 14 |

   a. Calculate the probability that a randomly chosen respondent would indicate that there should be some restriction concerning the company's right to monitor Internet use.
   b. Indicate the method of probability assessment used to determine the probability in part a.
   c. Are the events that a randomly selected respondent chose response 1 and that another randomly selected respondent chose response 2 independent, mutually exclusive, or dependent events? Explain.

**4-19.** Famous Dave's is a successful barbeque chain and sells its beef, pork, and chicken items to three kinds of customers: dine-in, delivery, and carry-out. If last year's sales showed 12,753 orders were dine-in, 5,893 were delivery orders, and 3,122 orders were carry-out. Suppose an audit of last year's sales is being conducted.
   a. If a customer order is selected at random, what is the probability it will be a carry-out order?
   b. What method of probability assessment was used to determine the probability in part a?
   c. If two customer orders are selected at random, list the sample space indicating the type of order for both customers.

**4-20.** Based on a *USA Today* online article by Kevin McCoy ("IRS audit rate for individuals drops," www.usatoday.com, Feb. 24, 2015), an individual taxpayer's chance of being audited in 2014 was 0.0086, the lowest in ten years.
   a. Discuss the type of probability assessment used to arrive at the 0.0086 level.
   b. If a sample of 1,000 taxpayers is selected, what is the expected number that will be audited?

## Computer Software Exercises

**4-21.** The data in the file called **Babies** indicate whether the past 50 babies delivered at a local hospital were delivered using the caesarean method.
   a. Based on these data, what is the probability that a baby born in this hospital will be born using the caesarean method?
   b. What concerns might you have about using these data to assess the probability of a caesarean birth? Discuss.

**4-22.** Several years ago, a large state university conducted a survey of undergraduate students regarding their use of computers. The results of the survey are contained in the data file **ComputerUse**.

a. Based on the data from the survey, what was the probability that undergraduate students at this university will have a major that requires them to use a computer on a daily basis?

b. Based on the data from this survey, if a student was a business major, what was the probability of the student believing that the computer lab facilities are very adequate?

c. Select a sample of 20 students at your university and ask them what their major is and whether the major requires them to access the Internet for computational purposes, with a personal computer, notebook, tablet, or other device. Use these data to calculate the probability of each major requiring a computing device, and then compare your results with those from the older survey.

**4-23.** A scooter manufacturing company is notified whenever a scooter breaks down, and the problem is classified as being either mechanical or electrical. The company then matches the scooter to the plant where it was assembled. The file **Scooters** contains a random sample of 200 breakdowns. Use the data in the file and the relative frequency assessment method to find the following probabilities:

a. What is the probability a scooter was assembled at the Tyler plant?

b. What is the probability that a scooter breakdown was due to a mechanical problem?

c. What is the probability that a scooter was assembled at the Lincoln plant and had an electrical problem?

**4-24.** A survey on cell phone use asked, in part, what was the most important reason people give for not using a wireless phone exclusively. The responses were: (1) Like the safety of traditional phone, (2) Like having a land line for a backup phone, (3) Pricing not

attractive enough, (4) Weak or unreliable cell signal at home, (5) Need phone line for DSL Internet access, and (6) Other. The file titled **Wireless** contains the responses for the 1,088 respondents.

a. Calculate the probability that a randomly chosen respondent would not use a wireless phone exclusively because of some type of difficulty in placing and receiving calls with a wireless phone.

b. Calculate the probability that a randomly chosen person would not use a wireless phone exclusively because of some type of difficulty in placing and receiving calls with a wireless phone and is over the age of 55.

c. Determine the probability that a randomly chosen person would not use a wireless phone exclusively because of a perceived need for Internet access and the safety of a traditional phone.

d. Of those respondents under 36, determine the probability that an individual in this age group would not use a wireless phone exclusively because of some type of difficulty in placing and receiving calls with a wireless phone.

**4-25.** A selection of soft-drink users is asked to taste two disguised soft drinks and indicate which they prefer. The file titled **Challenge** contains the results of a study that was conducted on a college campus.

a. Determine the probability that a randomly chosen student prefers Pepsi.

b. Determine the probability that one of the students prefers Pepsi and is less than 20 years old.

c. Of those students who are less than 20 years old, calculate the probability that a randomly chosen student prefers (1) Pepsi and (2) Coke.

d. Of those students who are at least 20 years old, calculate the probability that a randomly chosen student prefers (1) Pepsi and (2) Coke.

## Measuring Probabilities

The probability attached to an event represents the likelihood the event will occur on a specified trial of an experiment. This probability also measures the perceived uncertainty about whether the event will occur.

**Possible Values and the Summation of Possible Values**    If we are certain about the outcome of an event, we will assign the event a probability of 0 or 1, where $P(E_i) = 0$ indicates the event $E_i$ will not occur and $P(E_i) = 1$ means $E_i$ will definitely occur. If we are uncertain about the result of an experiment, we measure this uncertainty by assigning a probability between 0 and 1.[1] Probability Rule 1 shows that the probability of an event occurring is always between 0 and 1.

---

[1]These statements are true only if the number of outcomes of an experiment is countable. They do not apply when the number of outcomes is infinitely uncountable. This will be discussed when continuous probability distributions are discussed in Chapter 6.

**Probability Rule 1**

For any event $E_i$,

$$0 \le P(E_i) \le 1 \quad \text{for all } i \tag{4.3}$$

All possible outcomes associated with an experiment form the sample space. Therefore, the sum of the probabilities of all possible outcomes is 1, as shown by Probability Rule 2.

**Probability Rule 2**

$$\sum_{i=1}^{k} P(e_i) = 1 \tag{4.4}$$

where:

$k$ = Number of outcomes in the sample

$e_i$ = $i$th outcome

**OUTCOME 2**

**Addition Rule for Individual Outcomes**   If a single event is made up of two or more individual outcomes, then the probability of the event is found by summing the probabilities of the individual outcomes. This is illustrated by Probability Rule 3.

**Probability Rule 3: Addition Rule for Individual Outcomes**

The probability of an event $E_i$ is equal to the *sum* of the probabilities of the individual outcomes that form $E_i$. For example, if

$$E_i = \{e_1, e_2, e_3\}$$

then

$$P(E_i) = P(e_1) + P(e_2) + P(e_3) \tag{4.5}$$

**BUSINESS APPLICATION**   **Addition Rule**

**Google**   Google has become synonymous with Web searches and is the leader in the search engine marketplace. Suppose officials at the northern California headquarters have recently performed a survey of computer users to determine how many Internet searches individuals do daily using Google. Table 4.2 shows the results of the survey of Internet users.

The sample space for the experiment for each respondent is

$$SS = \{e_1, e_2, e_3, e_4\}$$

where the possible outcomes are

$e_1$ = At least 10 searches
$e_2$ = 3 to 9 searches
$e_3$ = 1 to 2 searches
$e_4$ = 0 searches

**TABLE 4.2** Google's Survey Results

| Searches per Day | Frequency | Relative Frequency |
|---|---|---|
| $e_1$ = At least 10 | 400 | 0.08 |
| $e_2$ = 3 to 9 | 1,900 | 0.38 |
| $e_3$ = 1 to 2 | 1,500 | 0.30 |
| $e_4$ = 0 | 1,200 | 0.24 |
| Total | 5,000 | 1.00 |

Using the relative frequency assessment approach, we assign the following probabilities:

$$
\begin{aligned}
P(e_1) &= 400/5{,}000 = 0.08 \\
P(e_2) &= 1{,}900/5{,}000 = 0.38 \\
P(e_3) &= 1{,}500/5{,}000 = 0.30 \\
P(e_4) &= \underline{1{,}200/5{,}000 = 0.24} \\
&\hphantom{= 1{,}200/5{,}000} \Sigma = 1.00
\end{aligned}
$$

Assume we are interested in the event *respondent performs 1 to 9 searches per day*:

$$E = \text{Internet user performs 1 to 9 searches per day}$$

The outcomes that make up $E$ are

$$E = \{e_2, e_3\}$$

We can find the probability, $P(E)$, by using Probability Rule 3 (Equation 4.5), as follows:

$$
\begin{aligned}
P(E) &= P(e_2) + P(e_3) \\
&= 0.38 + 0.30 \\
&= 0.68
\end{aligned}
$$

---

**EXAMPLE 4-7** **The Addition Rule for Individual Outcomes**

**KQRT 1340 Radio** The KQRT 1340 radio station is a combination news/talk and "oldies" station. During a 24-hour day, a listener can tune in and hear any of the following four programs being broadcast:

"Oldies" music
News stories
Talk programming
Commercials

Recently, the station has been having trouble with its transmitter. Each day, the station's signal goes dead for a few seconds; it seems that these outages are equally likely to occur at any time during the 24-hour broadcast day. There seems to be no pattern regarding what is playing at the time the transmitter problem occurs. The station manager is concerned about the probability that these problems will occur during either a news story or a talk program.

STEP **1** **Define the experiment.**
The station conducts its broadcast starting at 12:00 midnight, extending until a transmitter outage is observed.

STEP **2** **Define the possible outcomes.**
The possible outcomes are each type of programming that is playing when the transmitter outage occurs. There are four possible outcomes:

$$
\begin{aligned}
e_1 &= \text{Oldies} \\
e_2 &= \text{News} \\
e_3 &= \text{Talk programs} \\
e_4 &= \text{Commercials}
\end{aligned}
$$

STEP **3** **Determine the probability of each possible outcome.**
The station manager has determined that out of the 1,440 minutes per day, 540 minutes are oldies, 240 minutes are news, 540 minutes are talk programs, and 120 minutes are commercials. Therefore, the probability of each type of programming being on at the moment the outage occurs is assessed as follows:

| Outcome | $P(e_i)$ |
|---|---|
| $e_1$ = Oldies | $P(e_1) = \dfrac{540}{1{,}440} = 0.375$ |
| $e_2$ = News | $P(e_2) = \dfrac{240}{1{,}440} = 0.167$ |
| $e_3$ = Talk programs | $P(e_3) = \dfrac{540}{1{,}440} = 0.375$ |
| $e_4$ = Commercials | $P(e_4) = \dfrac{120}{1{,}440} = 0.083$ |
| | $\Sigma = 1.000$ |

Note that based on Equation 4.4 (Probability Rule 2), the sum of the probabilities of the individual possible outcomes is 1.0.

**STEP 4** **Define the event of interest.**
The event of interest is a transmitter problem occurring during a news or talk program. This is

$$E = \{e_2, e_3\}$$

**STEP 5** **Use Probability Rule 3 (Equation 4.5) to compute the desired probability.**

$$P(E) = P(e_2) + P(e_3)$$
$$P(E) = 0.167 + 0.375$$
$$P(E) = 0.542$$

Thus, the probability is slightly higher than 0.5 that when a transmitter problem occurs, it will happen during either a news or talk program.

**TRY EXERCISE 4-26** (pg. 160)

**Complement**
The complement of an event $E$ is the collection of all possible outcomes not contained in event $E$.

**Complement Rule**   Closely connected with Probability Rules 1 and 2 is the **complement** of an event. The complement of event $E$ is represented by $\bar{E}$. The Complement Rule is a corollary to Probability Rules 1 and 2.

**Complement Rule**

$$P(\bar{E}) = 1 - P(E) \tag{4.6}$$

That is, the probability of the complement of event $E$ is 1 minus the probability of event $E$.

**EXAMPLE 4-8**   **The Complement Rule**

**Capital Consulting**   The managing partner for Capital Consulting is working on a proposal for a consulting project with a client in Sydney, Australia. The manager lists four possible net profits from the consulting engagement and his subjectively assessed probabilities related to each profit level.

| Outcome | P(Outcome) |
|---|---|
| $ 0 | 0.70 |
| $ 2,000 | 0.20 |
| $15,000 | 0.07 |
| $50,000 | 0.03 |
| | $\Sigma = 1.00$ |

Note that each probability is between 0 and 1 and that the sum of the probabilities is 1, as required by Probability Rules 1 and 2.

The manager plans to submit the proposal if the consulting engagement will have a positive profit, so he is interested in knowing the probability of an outcome greater than $0. This probability can be found using the Complement Rule with the following steps:

STEP 1  **Determine the probabilities for the outcomes.**

$$P(\$0) = 0.70$$
$$P(\$2,000) = 0.20$$
$$P(\$15,000) = 0.07$$
$$P(\$50,000) = 0.03$$

STEP 2  **Find the desired probability.**

Let $E$ be the consulting outcome event = $0. The probability of the $0 outcome is

$$P(E) = 0.70$$

The complement, $\overline{E}$, is all investment outcomes greater than $0. From the Complement Rule, the probability of profit greater than $0 is

$$P(\text{Profit} > \$0) = 1 - P(\text{Profit} = \$0)$$
$$P(\text{Profit} > \$0) = 1 - 0.70$$
$$P(\text{Profit} > \$0) = 0.30$$

Based on the manager's subjective probability assessment, there is a 30% chance the consulting project will have a positive profit.

TRY EXERCISE 4-32 (pg. 160)

OUTCOME **2**    **Addition Rule for Any Two Events**

**BUSINESS APPLICATION**    **Addition Rule**

Google (*continued*)    Suppose the staff who conducted the survey for Google discussed earlier also asked questions about the computer users' ages. The Google managers consider age important in designing their search engine methodologies. Table 4.3 shows the breakdown of the sample by age group and by the number of times a user performs a search each day.

Table 4.3 shows that seven events are defined. For instance, $E_1$ is the event that a computer user performs ten or more searches per day. This event is composed of three individual outcomes associated with the three age categories. These are

$$E_1 = \{e_1, e_2, e_3\}$$

In another case, event $E_5$ corresponds to a survey respondent being younger than 30 years of age. It is composed of four individual outcomes associated with the four levels of search activity. These are

$$E_5 = \{e_1, e_4, e_7, e_{10}\}$$

**TABLE 4.3** Google Search Study

| Searches per Day | | Age Group | | | Total |
|---|---|---|---|---|---|
| | | $E_5$ Under 30 | $E_6$ 30 to 50 | $E_7$ Over 50 | |
| $E_1$ | $\geq$ 10 searches | $e_1$ 200 | $e_2$ 100 | $e_3$ 100 | 400 |
| $E_2$ | 3 to 9 searches | $e_4$ 600 | $e_5$ 900 | $e_6$ 400 | 1,900 |
| $E_3$ | 1 to 2 searches | $e_7$ 400 | $e_8$ 600 | $e_9$ 500 | 1,500 |
| $E_4$ | 0 searches | $e_{10}$ 700 | $e_{11}$ 500 | $e_{12}$ 0 | 1,200 |
| Total | | 1,900 | 2,100 | 1,000 | 5,000 |

Table 4.3 illustrates two important concepts in data analysis: *joint frequencies* and *marginal frequencies*. Joint frequencies, which were discussed in Chapter 2, are the values inside the table. They provide information on age group and search activity jointly. Marginal frequencies are the row and column totals. These values give information on only the age group or only Google search activity.

For example, 2,100 people in the survey are in the 30- to 50-year age group. This column total is a marginal frequency for the age group 30 to 50 years, which is represented by $E_6$. Now notice that 600 respondents are younger than 30 years old and perform three to nine searches a day. The 600 is a joint frequency whose outcome is represented by $e_4$.

Table 4.4 shows the relative frequencies for the data in Table 4.3. These values are the probability assessments for the events and outcomes.

**TABLE 4.4** Google—Joint Probability Table

| Searches per Day | Age Group | | | |
|---|---|---|---|---|
| | $E_5$ Under 30 | $E_6$ 30 to 50 | $E_7$ Over 50 | Total |
| $E_1$  $\geq$ 10 searches | $e_1$ 200/5,000 = 0.04 | $e_2$ 100/5,000 = 0.02 | $e_3$ 100/5,000 = 0.02 | **400/5,000 = 0.08** |
| $E_2$  3 to 9 searches | $e_4$ 600/5,000 = 0.12 | $e_5$ 900/5,000 = 0.18 | $e_6$ 400/5,000 = 0.08 | **1,900/5,000 = 0.38** |
| $E_3$  1 to 2 searches | $e_7$ 400/5,000 = 0.08 | $e_8$ 600/5,000 = 0.12 | $e_9$ 500/5,000 = 0.10 | **1,500/5,000 = 0.30** |
| $E_4$  0 searches | $e_{10}$ 700/5,000 = 0.14 | $e_{11}$ 500/5,000 = 0.10 | $e_{12}$ 0/5,000 = 0.00 | **1,200/5,000 = 0.24** |
| **Total** | **1,900/5,000 = 0.38** | **2,100/5,000 = 0.42** | **1,000/5,000 = 0.20** | **5,000/5,000 = 1.00** |

Suppose we wish to find the probability of $E_4$ (0 searches) **or** $E_6$ (being in the 30-to-50 age group). That is,

$$P(E_4 \text{ or } E_6) = ?$$

**Probability Rule 4: Addition Rule for Any Two Events $E_1$ and $E_2$**

$$P(E_1 \text{ or } E_2) = P(E_1) + P(E_2) - P(E_1 \text{ and } E_2) \tag{4.7}$$

To find this probability, we must use Probability Rule 4.

The key word in knowing when to use Rule 4 is *or*. The word *or* indicates addition. Figure 4.1 is a Venn diagram that illustrates the application of the Addition Rule for Any Two Events. Notice that the probabilities of the outcomes in the overlap between the two events, $E_1$ and $E_2$, are double-counted when the probabilities of the outcomes in $E_1$ are added to those in $E_2$. Thus, the probabilities of the outcomes in the overlap, which is $E_1$ *and* $E_2$, need to be subtracted to avoid the double-counting.

**FIGURE 4.1 Venn Diagram—Addition Rule for Any Two Events**

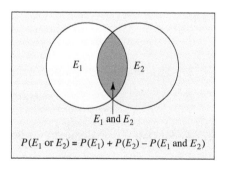

**TABLE 4.5** Google Searches—Addition Rule Example

| | Age Group | | | |
|---|---|---|---|---|
| Searches per Day | $E_5$ Under 30 | $E_6$ 30 to 50 | $E_7$ Over 50 | Total |
| $E_1$  ≥ 10 searches | $e_1$ 200/5,000 = 0.04 | $e_2$ 100/5,000 = 0.02 | $e_3$ 100/5,000 = 0.02 | 400/5,000 = 0.08 |
| $E_2$  3 to 9 searches | $e_4$ 600/5,000 = 0.12 | $e_5$ 900/5,000 = 0.18 | $e_6$ 400/5,000 = 0.08 | 1,900/5,000 = 0.38 |
| $E_3$  1 to 2 searches | $e_7$ 400/5,000 = 0.08 | $e_8$ 600/5,000 = 0.12 | $e_9$ 500/5,000 = 0.10 | 1,500/5,000 = 0.30 |
| $E_4$  0 searches | $e_{10}$ 700/5,000 = 0.14 | $e_{11}$ 500/5,000 = 0.10 | $e_{12}$ 0/5,000 = 0.00 | 1,200/5,000 = 0.24 |
| Total | 1,900/5,000 = 0.38 | 2,100/5,000 = 0.42 | 1,000/5,000 = 0.20 | 5,000/5,000 = 1.00 |

**TABLE 4.6** Google—Addition Rule Example

| | Age Group | | | |
|---|---|---|---|---|
| Searches per Day | $E_5$ Under 30 | $E_6$ 30 to 50 | $E_7$ Over 50 | Total |
| $E_1$  ≥ 10 searches | $e_1$ 200/5,000 = 0.04 | $e_2$ 100/5,000 = 0.02 | $e_3$ 100/5,000 = 0.02 | 400/5,000 = 0.08 |
| $E_2$  3 to 9 searches | $e_4$ 600/5,000 = 0.12 | $e_5$ 900/5,000 = 0.18 | $e_6$ 400/5,000 = 0.08 | 1,900/5,000 = 0.38 |
| $E_3$  1 to 2 searches | $e_7$ 400/5,000 = 0.08 | $e_8$ 600/5,000 = 0.12 | $e_9$ 500/5,000 = 0.10 | 1,500/5,000 = 0.30 |
| $E_4$  0 searches | $e_{10}$ 700/5,000 = 0.14 | $e_{11}$ 500/5,000 = 0.10 | $e_{12}$ 0/5,000 = 0.00 | 1,200/5,000 = 0.24 |
| Total | 1,900/5,000 = 0.38 | 2,100/5,000 = 0.42 | 1,000/5,000 = 0.20 | 5,000/5,000 = 1.00 |

Referring to the Google situation, the probability of $E_4$ (0 searches) *or* $E_6$ (being in the 30-to-50 age group) is

$$P(E_4 \text{ or } E_6) = ?$$

Table 4.5 shows the relative frequencies with the events of interest shaded. The overlap corresponds to the *joint occurrence* (intersection) of conducting 0 searches *and* being in the 30-to-50 age group. The probability of the outcomes in the overlap is represented by $P(E_4 \text{ and } E_6)$ and must be subtracted. This is done to avoid double-counting the probabilities of the outcomes that are in both $E_4$ and $E_6$ when calculating the $P(E_4 \text{ or } E_6)$. Thus,

$$\begin{aligned} P(E_4 \text{ or } E_6) &= P(E_4) + P(E_6) - P(E_4 \text{ and } E_6) \\ &= 0.24 + 0.42 - 0.10 \\ &= 0.56 \end{aligned}$$

Therefore, the probability that a respondent will either be in the 30-to-50 age group or perform 0 searches on a given day is 0.56.

What is the probability a respondent will perform 1 to 2 searches *or* be in the over-50 age group? Again, we can use Probability Rule 4:

$$P(E_3 \text{ or } E_7) = P(E_3) + P(E_7) - P(E_3 \text{ and } E_7)$$

Table 4.6 shows the relative frequencies for these events. We have

$$P(E_3 \text{ or } E_7) = 0.30 + 0.20 - 0.10 = 0.40$$

Thus, there is a 0.40 chance that a respondent will perform 1 to 2 searches **or** be in the over-50 age group.

---

**EXAMPLE 4-9**    **Addition Rule for Any Two Events**

**Greenfield Forest Products**    Greenfield Forest Products manufactures lumber for large material supply centers like Home Depot and Lowe's in the United States and Canada. A representative from Home Depot is due to arrive at the Greenfield plant for a meeting to discuss lumber quality. When the Home Depot representative arrives, he will ask Greenfield's managers to randomly select one board from the finished goods inventory for a quality check. Boards of three dimensions and three lengths are in the inventory. The following table shows the number of boards of each size and length:

| Length | Dimension $E_2 = 2'' \times 4''$ | $E_5 = 2'' \times 6''$ | $E_6 = 2'' \times 8''$ | Total |
|---|---|---|---|---|
| $E_1 = 8$ feet | 1,400 | 1,500 | 1,100 | 4,000 |
| $E_2 = 10$ feet | 2,000 | 3,500 | 2,500 | 8,000 |
| $E_3 = 12$ feet | 1,600 | 2,000 | 2,400 | 6,000 |
| Total | 5,000 | 7,000 | 6,000 | 18,000 |

The Greenfield manager will be selecting one board at random from the inventory to show the Home Depot representative. Suppose he is interested in the probability that the board selected will be 8 feet long or a 2″ × 6″. To find this probability, he can use the following steps:

STEP 1 **Define the experiment.**
One board is selected from the inventory and its dimensions are obtained.

STEP 2 **Define the events of interest.**
The manager is interested in boards that are 8 feet long.

$$E_1 = 8 \text{ feet}$$

He is also interested in the 2″ × 6″ dimension, so

$$E_5 = 2'' \times 6'' \text{ boards}$$

STEP 3 **Determine the probability for each event.**
There are 18,000 boards in inventory and 4,000 of these are 8 feet long, so

$$P(E_1) = \frac{4,000}{18,000} = 0.2222$$

Of the 18,000 boards, 7,000 are 2″ × 6″, so the probability is

$$P(E_5) = \frac{7,000}{18,000} = 0.3889$$

STEP 4 **Determine whether the two events overlap, and if so, compute the joint probability.**
Of the 18,000 total boards, 1,500 are 8 feet long and 2″ × 6″. Thus the joint probability is

$$P(E_1 \text{ and } E_5) = \frac{1,500}{18,000} = 0.0833$$

STEP 5 **Compute the desired probability using Probability Rule 4.**
$$P(E_1 \text{ or } E_5) = P(E_1) + P(E_5) - P(E_1 \text{ and } E_5)$$
$$P(E_1 \text{ or } E_5) = 0.2222 + 0.3889 - 0.0833$$
$$= 0.5278$$

The chance of selecting an 8-foot board or a 2″ × 6″ board is just under 0.53.

**TRY EXERCISE 4-28** (pg. 160)

**Addition Rule for Mutually Exclusive Events**   We indicated previously that when two events are mutually exclusive, both events cannot occur at the same time. Thus, for mutually exclusive events,

$$P(E_1 \text{ and } E_2) = 0$$

Therefore, when you are dealing with mutually exclusive events, the Addition Rule assumes a different form, shown as Probability Rule 5.

FIGURE 4.2 **Venn Diagram —Addition Rule for Two Mutually Exclusive Events**

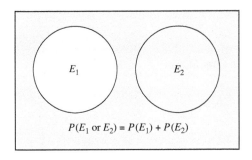

$$P(E_1 \text{ or } E_2) = P(E_1) + P(E_2)$$

**Probability Rule 5: Addition Rule for Mutually Exclusive Events**

For two mutually exclusive events $E_1$ and $E_2$,

$$P(E_1 \text{ or } E_2) = P(E_1) + P(E_2) \qquad (4.8)$$

Figure 4.2 is a Venn diagram illustrating the application of the Addition Rule for Mutually Exclusive Events.

## Conditional Probability

In dealing with probabilities, you will often need to determine the chances of two or more events occurring either at the same time or in succession. For example, a quality control manager for a manufacturing company may be interested in the probability of selecting two successive defective products from an assembly line. Or, the decision maker might know that an event has occurred and may then want to know the probability of a second event occurring. For instance, suppose that an oil company geologist who believes oil will be found at a certain drilling site makes a favorable report. Because oil is not always found at locations with a favorable report, the oil company's exploration vice president might well be interested in the probability of finding oil, given the favorable report.

Situations such as these relate to a probability concept known as **conditional probability**.

Probability Rule 6 offers a general rule for conditional probability. The notation $P(E_1 \mid E_2)$ reads "probability of event $E_1$ *given* event $E_2$ has occurred." Thus, the probability of one event is conditional upon another event having occurred.

**Conditional Probability**
The probability that an event will occur given that some other event has already happened.

**Probability Rule 6: Conditional Probability for Any Two Events**

For any two events $E_1$ and $E_2$,

$$P(E_1 \mid E_2) = \frac{P(E_1 \text{ and } E_2)}{P(E_2)} \qquad (4.9)$$

where:

$$P(E_2) > 0$$

Probability Rule 6 uses a *joint probability*, $P(E_1 \text{ and } E_2)$, and a *marginal probability*, $P(E_2)$, to calculate the conditional probability $P(E_1 \mid E_2)$.

**BUSINESS APPLICATION**    **Conditional Probability**

**R.C. Minch Networking**    R.C. Minch Networking is an Internet service provider to rural areas in the western United States. The company has studied its customers' Internet habits. Among the information collected are the data shown in Table 4.7.

The company is focusing on high-volume users, and one of the factors that will influence its marketing strategy is whether time spent using the Internet is related to a customer's

**TABLE 4.7** Joint Frequency Distribution for R.C. Minch Networking

| Hours per Month | Gender $E_4$ Female | $E_5$ Male | Total |
|---|---|---|---|
| $E_1 < 20$ | $e_1$ 450 | $e_2$ 500 | 950 |
| $E_2$ 20 to 40 | $e_3$ 300 | $e_4$ 800 | 1,100 |
| $E_3 > 40$ | $e_5$ 100 | $e_6$ 350 | 450 |
| Total | 850 | 1,650 | 2,500 |

gender. For example, suppose the company knows a user is female and wants to know the chances this woman will spend between 20 and 40 hours a month on the Internet. Let

$$E_2 = \{e_3, e_4\} = \text{Event: Person uses services 20 to 40 hours per month}$$

$$E_4 = \{e_1, e_3, e_5\} = \text{Event: User is female}$$

A marketing analyst needs to know the probability of $E_2$ given $E_4$.

One way to find the desired probability is as follows:

1. We know $E_4$ has occurred (customer is female). There are 850 females in the survey.
2. Of the 850 females, 300 use Internet services 20 to 40 hours per month.
3. Then,

$$P(E_2 \mid E_4) = \frac{300}{850}$$
$$= 0.35$$

However, we can also apply Probability Rule 6, as follows:

$$P(E_2 \mid E_4) = \frac{P(E_2 \text{ and } E_4)}{P(E_4)}$$

Table 4.8 shows the relative frequencies of interest. From Table 4.8, we get the joint probability

$$P(E_2 \text{ and } E_4) = 0.12$$

and

$$P(E_4) = 0.34$$

Then, applying Probability Rule 6, we have

$$P(E_2 \mid E_4) = \frac{0.12}{0.34} = 0.35$$

**TABLE 4.8** Joint Relative Frequency Distribution for R.C. Minch Networking

| Hours per Month | Gender $E_4$ Female | $E_5$ Male | Total |
|---|---|---|---|
| $E_1 < 20$ | $e_1$ 450/2,500 = 0.18 | $e_2$ 500/2,500 = 0.20 | 950/2,500 = 0.38 |
| $E_2$ 20 to 40 | $e_3$ 300/2,500 = 0.12 | $e_4$ 800/2,500 = 0.32 | 1,100/2,500 = 0.44 |
| $E_3 > 40$ | $e_5$ 100/2,500 = 0.04 | $e_6$ 350/2,500 = 0.14 | 450/2,500 = 0.18 |
| Total | 850/2,500 = 0.34 | 1,650/2,500 = 0.66 | 2,500/2,500 = 1.00 |

**EXAMPLE 4-10** **Computing Conditional Probabilities**

**Retirement Planning** Most financial publications suggest that the older the investor is, the more conservative his or her investment strategy should be. For example, younger investors might hold more in stocks, while older investors might hold more in bonds. Suppose the following table was created from a survey of investors. The table shows the number of people in the study by age group and percent of retirement funds in the stock market.

**Percent of Retirement Investments in the Stock Market**

| | Age of Investor | $E_5$ <5% | $E_6$ 5% < 10% | $E_7$ 10% < 30% | $E_8$ 30% < 50% | $E_9$ 50% or more | Total |
|---|---|---|---|---|---|---|---|
| $E_1$ | < 30 | 70 | 240 | 270 | 80 | 55 | 715 |
| $E_2$ | 30 < 50 | 90 | 300 | 630 | 1,120 | 1,420 | 3,560 |
| $E_3$ | 50 < 65 | 110 | 305 | 780 | 530 | 480 | 2,205 |
| $E_4$ | 65+ | 200 | 170 | 370 | 260 | 65 | 1,065 |
| | **Total** | 470 | 1,015 | 2,050 | 1,990 | 2,020 | 7,545 |

We are interested in knowing the probability that someone 65 or older will have 50% or more of retirement funds invested in the stock market. Assuming the data collected in this study reflect the population of investors, we can find this conditional probability using the following steps:

STEP 1 **Define the experiment.**
A randomly selected person age 65 or older has his or her portfolio analyzed for percent of retirement funds in the stock market.

STEP 2 **Define the events of interest.**
In this case, we are interested in two events:

$$E_4 = \text{At least 65 years old}$$
$$E_9 = 50\% \text{ or more in stocks}$$

STEP 3 **Define the probability statement of interest.**
We are interested in

$$P(E_9 | E_4) = \text{Probability of 50\% or more stocks } given \text{ at least 65 years}$$

STEP 4 **Convert the data to probabilities using the relative frequency assessment method.**
We begin with the event that is given to have occurred $(E_4)$. A total of 1,065 people in the study were at least 65 years of age. Of the 1,065 people, 65 had 50% or more of their retirement funds in the stock market.

$$P(E_9 | E_4) = \frac{65}{1,065} = 0.061$$

Thus, the conditional probability that someone at least 65 will have 50% or more of retirement assets in the stock market is 0.061. This value can also be found using Probability Rule 6 as shown in Step 5.

STEP 5 **Use Probability Rule 6 to find the conditional probability.**
You can also apply Probability Rule 6 to find the probability of a person having 50% or more of retirement assets in the stock market *given* that he or she is at least 65 years old.

$$P(E_9 | E_4) = \frac{P(E_9 \text{ and } E_4)}{P(E_4)}$$

The frequency assessment method is used to find the probability that the individual is 65 years or older:

$$P(E_4) = \frac{1,065}{7,545} = 0.1412$$

The joint probability of having 50% or more in stocks and being 65 years or older is

$$P(E_9 \text{ and } E_4) = \frac{65}{7,545} = 0.0086$$

Then, using Probability Rule 6, we get

$$P(E_9 | E_4) = \frac{P(E_9 \text{ and } E_4)}{P(E_4)} = \frac{0.0086}{0.1412} = 0.061$$

**TRY EXERCISE 4-34** (pg. 161)

**Tree Diagrams**   Another way of organizing the events of an experiment that aids in the calculation of probabilities is the *tree diagram.*

---

| BUSINESS APPLICATION | Using Tree Diagrams |

**R.C. Minch Networking** (*continued*)   Figure 4.3 illustrates the tree diagram for R.C. Minch Networking, the Internet service provider discussed earlier. Note that the branches at each node in the tree diagram represent mutually exclusive events. Moving from left to right, the first two branches indicate the two customer types (male and female—mutually exclusive events). Three branches grow from each of these original branches, representing the three possible categories for Internet use. The probabilities for the events male and female are shown on the first two branches. The probabilities shown on the right of the tree are the joint probabilities for each combination of gender and hours of use. These figures are found using Table 4.8, which was shown earlier. The probabilities on the branches following the male and female branches showing hours of use are conditional probabilities. For example, we can find the probability that a male customer ($E_5$) will spend more than 40 hours on the Internet ($E_3$) by

$$P(E_3 \mid E_5) = \frac{P(E_3 \text{ and } E_5)}{P(E_5)} = \frac{0.14}{0.66} = 0.2121$$

**Conditional Probability for Independent Events**   We previously discussed that two events are independent if the occurrence of one event has no bearing on the probability that the second event occurs. Therefore, when two events are independent, the rule for conditional probability takes a different form, as indicated in Probability Rule 7.

---

**Probability Rule 7: Conditional Probability for Independent Events**

For independent events $E_1$ and $E_2$,

$$P(E_1 \mid E_2) = P(E_1) \qquad P(E_2) > 0 \tag{4.10}$$

and

$$P(E_2 \mid E_1) = P(E_2) \qquad P(E_1) > 0$$

---

As Probablity Rule 7 shows, the conditional probability of one event occurring, given that a second independent event has already occurred, is simply the probability of the event occurring.

**FIGURE 4.3 Tree Diagram for R.C. Minch Networking**

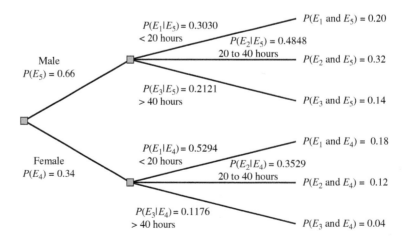

EXAMPLE 4-11    **Checking for Independence**

**Greenfield Forest Products** (*continued*)   In Example 4-9, the manager at the Greenfield Forest Products Company reported the following data on the boards in inventory:

| | Dimension | | | |
|---|---|---|---|---|
| Length | $E_4$ 2″ × 4″ | $E_5$ 2″ × 6″ | $E_6$ 2″ × 8″ | Total |
| $E_1 = 8$ feet | 1,400 | 1,500 | 1,100 | 4,000 |
| $E_2 = 10$ feet | 2,000 | 3,500 | 2,500 | 8,000 |
| $E_3 = 12$ feet | 1,600 | 2,000 | 2,400 | 6,000 |
| Total | 5,000 | 7,000 | 6,000 | 18,000 |

He will be selecting one board at random from the inventory to show a visiting customer. Of interest is whether the length of the board is independent of the dimension. This can be determined using the following steps:

STEP 1   **Define the experiment.**
A board is randomly selected and its dimensions determined.

STEP 2   **Define one event for length and one event for dimension.**
Let $E_2$ = Event that the board is 10 feet long and $E_5$ = Event that the board is 2″ × 6″.

STEP 3   **Determine the probability for each event.**

$$P(E_2) = \frac{8,000}{18,000} = 0.4444 \ \text{ and } \ P(E_5) = \frac{7,000}{18,000} = 0.3889$$

STEP 4   **Assess the joint probability of the two events occurring.**

$$P(E_2 \text{ and } E_5) = \frac{3,500}{18,000} = 0.1944$$

STEP 5   **Compute the conditional probability of one event given the other using Probability Rule 6.**

$$P(E_2 \mid E_5) = \frac{P(E_2 \text{ and } E_5)}{P(E_5)} = \frac{0.1944}{0.3889} = 0.50$$

STEP 6   **Check for independence using Probability Rule 7.**
Because $P(E_2 \mid E_5) = 0.50 > P(E_2) = 0.4444$, the two events, board length and board dimension, are not independent.

**TRY EXERCISE 4-42** (pg. 162)

OUTCOME 3   ## Multiplication Rule

We needed to find the joint probability of two events in the discussion on addition of two events and in the discussion on conditional probability. We were able to find $P(E_1 \text{ and } E_2)$ simply by examining the joint relative frequency tables. However, we often need to find $P(E_1 \text{ and } E_2)$ when we do not know the joint relative frequencies. When this is the case, we can use the Multiplication Rule for Any Two Events.

### Multiplication Rule for Any Two Events

**Probability Rule 8: Multiplication Rule for Any Two Events**

For two events $E_1$ and $E_2$,

$$P(E_1 \text{ and } E_2) = P(E_1)P(E_2 \mid E_1) \tag{4.11}$$

**Multiplication Rule**

**Hong Kong Fireworks** The Hong Kong Fireworks Company is a manufacturer of fireworks used by cities, fairs, and other commercial establishments for large-scale fireworks displays. The company uses two suppliers of material in making a particular product. The materials from the two suppliers are intermingled in the manufacturing process. When a case of fireworks is being made, the material is pulled randomly from inventory without regard to which company made it. Recently, a customer ordered two products. At the time of assembly, the material inventory contained 30 units of MATX and 50 units of Quinex. What is the probability that both fireworks products ordered by this customer will have MATX material?

To answer this question, we must recognize that two events are required to form the desired outcome:

$$E_1 = \text{Event: MATX material in first product}$$
$$E_2 = \text{Event: MATX material in second product}$$

The probability that both fireworks products contain MATX material is written as $P(E_1 \text{ and } E_2)$. The key word here is *and*, as contrasted with the Addition Rule, in which the key word is *or*. The *and* signifies that we are interested in the joint probability of two events, as noted by $P(E_1 \text{ and } E_2)$. To find this probability, we employ Probability Rule 8:

$$P(E_1 \text{ and } E_2) = P(E_1)P(E_2 \mid E_1)$$

We start by assuming that each unit of material in the inventory has the same chance of being selected for assembly. For the first fireworks product,

$$P(E_1) = \frac{\text{Number of MATX units}}{\text{Number of firework materials units in inventory}}$$

$$= \frac{30}{80} = 0.375$$

Then, because we are not replacing the first firework material, we find $P(E_2 \mid E_1)$ by

$$P(E_2 \mid E_1) = \frac{\text{Number of remaining MATX units}}{\text{Number of remaining firework materials units}}$$

$$= \frac{29}{79} = 0.3671$$

Now, by Probability Rule 8,

$$P(E_1 \text{ and } E_2) = P(E_1)P(E_2 \mid E_1) = (0.375)(0.3671)$$
$$= 0.1377$$

Therefore, there is a 13.77% chance the two fireworks products will contain the MATX material.

### Using a Tree Diagram

**Multiplication Rule**

**Hong Kong Fireworks** (*continued*) A tree diagram can be used to display the situation facing Hong Kong Fireworks. The company uses material from two suppliers, which is intermingled in the inventory. Recently a customer ordered two products and found that both contained the MATX material. Assuming that the inventory contains 30 MATX and 50 Quinex units, to determine the probability of both products containing the MATX material you can use a tree diagram. The two branches on the left side of the tree in Figure 4.4 show the possible material options for the first product. The two branches coming from each of the first branches show the possible material options for the second product. The probabilities at the

**FIGURE 4.4** **Tree Diagram for the Fireworks Product Example**

far right are the joint probabilities for the material options for the two products. As we determined previously, the probability that both products will contain a MATX unit is 0.1377, as shown on the top right on the tree diagram.

We can use the Multiplication Rule and the Addition Rule in one application when we determine the probability that two products will have different materials. Looking at Figure 4.4, we see there are two ways this can happen:

$$P[(\text{MATX } and \text{ Quinex}) \ or \ (\text{Quinex } and \text{ MATX})] \ = \ ?$$

If the first product is a MATX and the second one is a Quinex, then the first cannot be a Quinex and the second a MATX. These two events are mutually exclusive and, therefore, Probability Rule 5 can be used to calculate the required probability. The joint probabilities (generated from the Multiplication Rule) are shown on the right side of the tree. To find the desired probability, using Probability Rule 5, we can add the two joint probabilities:

$$
\begin{aligned}
P[(\text{MATX } and \text{ Quinex}) \ or \ (\text{Quinex } and \text{ MATX})] \ &= \\
0.2373 \ \ \ \ \ \ + \ \ \ \ \ \ \ \ 0.2373 \ \ \ &= \ 0.4746
\end{aligned}
$$

The chance that a customer buying two products will get two different materials is 47.46%.

---

**Multiplication Rule for Independent Events**    When we determined the probability that two products would have MATX material, we used the general multiplication rule (Rule 8). The general multiplication rule requires that conditional probability be used because the probability associated with the second product depends on the material selected for the first product. The chance of obtaining a MATX was lowered from 30/80 to 29/79, given that the first material was a MATX.

However, if the two events of interest are *independent*, the imposed condition does not alter the probability, and the Multiplication Rule takes the form shown in Probability Rule 9.

**Probability Rule 9: Multiplication Rule for Independent Events**

For independent events $E_1$ and $E_2$,

$$P(E_1 \text{ and } E_2) \ = \ P(E_1)P(E_2) \tag{4.12}$$

The joint probability of two independent events is simply the product of the probabilities of the two events. Rule 9 is the primary way that you can determine whether any two events are independent. If the product of the probabilities of the two events equals the joint probability, then the events are independent.

**EXAMPLE 4-12** Using the Multiplication Rule and the Addition Rule

**Christiansen Accounting**   Christiansen Accounting prepares tax returns for individuals and companies. Over the years, the firm has tracked its clients and has discovered that 12% of the individual returns have been selected for audit by the Internal Revenue Service. On one particular day, the firm signed two new individual tax clients. The firm is interested in the probability that at least one of these clients will be audited. This probability can be found using the following steps:

STEP 1   **Define the experiment.**
The IRS randomly selects a tax return to audit.

STEP 2   **Define the possible outcomes.**
For a single client, the following outcomes are defined:

$$A = \text{Audit}$$
$$N = \text{No audit}$$

For each of the clients, we define the outcomes as

Client 1: $A_1$; $N_1$
Client 2: $A_2$; $N_2$

STEP 3   **Define the overall event of interest.**
The event that Christiansen Accounting is interested in is

$$E = \text{At least one client is audited}$$

STEP 4   **List the outcomes for the events of interest.**
The possible outcomes for which at least one client will be audited are as follows:

| $E_1$: | $A_1$ | $A_2$ | both are audited |
|---|---|---|---|
| $E_2$: | $A_1$ | $N_2$ | only one client is audited |
| $E_3$: | $N_1$ | $A_2$ | |

STEP 5   **Compute the probabilities for the events of interest.**
Assuming the chances of the clients being audited are independent of each other, probabilities for the events are determined using Probability Rule 9 for independent events:

$$P(E_1) = P(A_1 \text{ and } A_2) = 0.12 \times 0.12 = 0.0144$$
$$P(E_2) = P(A_1 \text{ and } N_2) = 0.12 \times 0.88 = 0.1056$$
$$P(E_3) = P(N_1 \text{ and } A_2) = 0.88 \times 0.12 = 0.1056$$

STEP 6   **Determine the probability for the overall event of interest.**
Because events $E_1$, $E_2$, and $E_3$ are mutually exclusive, compute the probability of at least one client being audited using Rule 5, the Addition Rule for Mutually Exclusive Events:

$$P(E_1 \text{ or } E_2 \text{ or } E_3) = P(E_1) + P(E_2) + P(E_3)$$
$$= 0.0144 + 0.1056 + 0.1056$$
$$= 0.2256$$

The chance of one or both of the clients being audited is 0.2256.

**TRY EXERCISE 4-30** (pg. 160)

OUTCOME **4**   ## Bayes' Theorem

As decision makers, you will often encounter situations that require you to assess probabilities for events of interest. Your assessment may be based on relative frequency or subjectivity. However, you may then come across new information that causes you to revise the probability assessment. For example, a human resources manager who has interviewed a person for a sales job might assess a low probability that the person will succeed in sales. However, after seeing the person's very high score on the company's sales aptitude test, the manager might revise her assessment upward. A medical doctor might assign an 80% chance that a patient has a particular disease. However, after seeing positive results from a lab test, he might increase his assessment to 95%.

In these situations, you will need a way to formally incorporate the new information. One very useful tool for doing this is called *Bayes' Theorem*, named for the Reverend Thomas Bayes, who developed the special application of conditional probability in the 1700s. If we let event $B$ be an event that is given to have occurred, then the conditional probability of event $E_i$ occurring can be computed as shown earlier using Equation 4.9:

$$P(E_i|B) = \frac{P(E_i \text{ and } B)}{P(B)}$$

The numerator can be reformulated using the Multiplication Rule (Equation 4.11) as

$$P(E_i \text{ and } B) = P(E_i)P(B|E_i)$$

The conditional probability is then

$$P(E_i|B) = \frac{P(E_i)P(B|E_i)}{P(B)}$$

The denominator, $P(B)$, can be found by adding the probability of the $k$ possible ways that event $B$ can occur. This is

$$P(B) = P(E_1)P(B|E_1) + P(E_2)P(B|E_2) + \cdots + P(E_k)P(B|E_k)$$

Then, Bayes' Theorem is formulated as Equation 4.13.

**Bayes' Theorem**

$$P(E_i|B) = \frac{P(E_i)P(B|E_i)}{P(E_1)P(B|E_1) + P(E_2)P(B|E_2) + \cdots + P(E_k)P(B|E_k)} \tag{4.13}$$

where:

$$E_i = i\text{th event of interest of the } k \text{ possible events}$$
$$B = \text{Event that has occurred that might impact } P(E_i)$$

Events $E_1$ to $E_k$ are mutually exclusive and collectively exhaustive.

**BUSINESS APPLICATION**    Bayes' Theorem

**Tile Production**   The Glerum Tile and Flooring Company has two production facilities, one in Ohio and one in Virginia. The company makes the same type of tile at both facilities. The Ohio plant makes 60% of the company's total tile output and the Virginia plant 40%. All tiles from the two facilities are sent to regional warehouses, where they are intermingled. After extensive study, the quality assurance manager has determined that 5% of the tiles produced in Ohio and 10% of the tiles produced in Virginia are unusable due to quality problems. When the company sells a defective tile, it incurs not only the cost of replacing the item but also the loss of goodwill. The vice president for production would like to allocate these costs fairly between the two plants. To do so, he knows he must first determine the probability that a defective tile was produced by a particular production line. Specifically, he needs to answer these questions:

1. What is the probability that the tile was produced at the Ohio plant, given that the tile is defective?
2. What is the probability that the tile was produced at the Virginia plant, given that the tile is defective?

In notation form, with $D$ representing the event that an item is defective, what the manager wants to know is

$$P(\text{Ohio}\,|\,D) = ?$$
$$P(\text{Virginia}\,|\,D) = ?$$

We can use Bayes' Theorem (Equation 4.13) to determine these probabilities, as follows:

$$P(\text{Ohio}\,|\,D) = \frac{P(\text{Ohio})P(D\,|\,\text{Ohio})}{P(D)}$$

We know that event $D$(Defective tile) can happen if it is made in either Ohio or Virginia. Thus,

$$P(D) = P(\text{Ohio and Defective}) + P(\text{Virginia and Defective})$$
$$P(D) = P(\text{Ohio})P(D\,|\,\text{Ohio}) + P(\text{Virginia})P(D\,|\,\text{Virginia})$$

We already know that 60% of the tiles come from Ohio and 40% from Virginia. So, $P(\text{Ohio}) = 0.60$ and $P(\text{Virginia}) = 0.40$. These are called the *prior* probabilities. Without Bayes' Theorem, we would likely allocate the total cost of defects in a 60/40 split between Ohio and Virginia, based on total production. However, the new information about the quality from each line is

$$P(D\,|\,\text{Ohio}) = 0.05 \quad \text{and} \quad P(D\,|\,\text{Virginia}) = 0.10$$

which can be used to properly allocate the cost of defects. This is done using Bayes' Theorem:

$$P(\text{Ohio}\,|\,D) = \frac{P(\text{Ohio})P(D\,|\,\text{Ohio})}{P(\text{Ohio})P(D\,|\,\text{Ohio}) + P(\text{Virginia})P(D\,|\,\text{Virginia})}$$

Then

$$P(\text{Ohio}\,|\,D) = \frac{(0.60)(0.05)}{(0.60)(0.05) + (0.40)(0.10)} = 0.4286$$

and

$$P(\text{Virginia}\,|\,D) = \frac{P(\text{Virginia})P(D\,|\,\text{Virginia})}{P(\text{Virginia})P(D\,|\,\text{Virginia}) + P(\text{Ohio})P(D\,|\,\text{Ohio})}$$

$$P(\text{Virginia}\,|\,D) = \frac{(0.40)(0.10)}{(0.40)(0.10) + (0.60)(0.05)} = 0.5714$$

These probabilities are *revised* probabilities. The prior probabilities have been revised given the new quality information. We now see that 42.86% of the cost of defects should be allocated to the Ohio plant, and 57.14% should be allocated to the Virginia plant.

Note, the denominator $P(D)$ is the overall probability of a defective tile. This probability is

$$P(D) = P(\text{Virginia})P(D\,|\,\text{Virginia}) + P(\text{Ohio})P(D\,|\,\text{Ohio})$$
$$= (0.40)(0.10) + (0.60)(0.05)$$
$$= 0.04 + 0.03$$
$$= 0.07$$

Thus, 7% of all the tiles made by Glerum are defective.

You might prefer to use a tabular approach like that shown in Table 4.9 when you apply Bayes' Theorem.

**TABLE 4.9** Bayes' Theorem Calculations for Glerum Tile and Flooring

| Events | Prior Probabilities | Conditional Probabilities | Joint Probabilities | Revised Probabilities |
|--------|---------------------|---------------------------|---------------------|------------------------|
| Ohio | 0.60 | 0.05 | $(0.60)(0.05) = 0.03$ | $0.03/0.07 = 0.4286$ |
| Virginia | 0.40 | 0.10 | $(0.40)(0.10) = \underline{0.04}$ | $0.04/0.07 = \underline{0.5714}$ |
| | | | $0.07$ | $1.0000$ |

---

**EXAMPLE 4-13**  Bayes' Theorem

**Mammograms and Breast Cancer**   In 2012, a total of 224,147 women and 2,125 men in the United States were diagnosed with breast cancer (*source:* Centers for Disease Control and Prevention, www.cdc.gov/cancer/breast/statistics/). As with other forms of cancer, early detection is important for breast cancer survival. The most common method for detecting the presence of breast cancer is a mammogram. Suppose a 40-year-old woman wishes to know the probability that she has breast cancer. Further, suppose she decides to have a mammogram. In 75% of cases, if a 40-year-old woman actually has cancer, the mammogram will give a "positive" result (this is referred to as the *sensitivity* of the test). However, the mammogram will give a "positive" result 10% of the time when there actually is no cancer. This result is called a "false positive." Finally, suppose that 1.4% of all women 40 years old have breast cancer. Thus, prior to any test, this woman has a 0.014 chance of having cancer.

Now suppose the woman has the mammogram and her test comes back positive. What are her chances of having cancer? This probability can be determined using Bayes' Theorem as follows:

STEP 1  **Define the events.**
There are two events:

$$C = \text{Cancer} \qquad NC = \text{No cancer}$$

STEP 2  **Define the prior probabilities.**
The probabilities of the events prior to having any result from the mammogram are

$$P(C) = 0.014 \qquad P(NC) = 0.986$$

STEP 3  **Define an event that if it occurs could alter the prior probabilities.**
In this case, the altering event is a positive test result from the mammogram. We label this event $MP$.

STEP 4  **Determine the conditional probabilities.**
The conditional probabilities are associated with the positive test results:

$$P(MP|C) = 0.75 \qquad P(MP|NC) = 0.10$$

STEP 5  **Use the tabular approach for Bayes' Theorem.**

| Event | Prior Probabilities | Conditional Probabilities | Joint Probabilities | Revised Probabilities |
|-------|---------------------|---------------------------|---------------------|------------------------|
| $C$ Cancer | $P(C) = 0.014$ | $P(MP|C) = 0.75$ | $P(C)P(MP|C) =$ $0.014 \times 0.75 = 0.0105$ | $0.0105/0.1091 = 0.0962$ |
| $NC$ No cancer | $P(NC) = 0.986$ | $P(MP|NC) = 0.10$ | $P(NC)\,P(MP|NC) =$ $0.986 \times 0.10 = 0.0986$ | $0.0986/0.1091 = 0.9038$ |
| | $\overline{\phantom{xxx}}$ $1.000$ | | $P(MP) = 0.1091$ | |

Thus, even though the mammogram gave a positive test result, this woman's probability of actually having breast cancer is 0.0962.

**TRY EXERCISE 4-33** (pg. 160)

## 4.2 EXERCISES

### Skill Development

**4-26.** Based on weather data collected in Racine, Wisconsin, on Christmas Day, the weather had the following distribution:

| Event | Relative Frequency |
|---|---|
| Clear & dry | 0.20 |
| Cloudy & dry | 0.30 |
| Rain | 0.40 |
| Snow | 0.10 |

a. Based on these data, what is the probability that next Christmas will be dry?

b. Based on the data, what is the probability that next Christmas will be rainy or cloudy and dry?

c. Supposing next Christmas is dry, determine the probability that it will also be cloudy.

**4-27.** A Jack in the Box franchise manager has determined that the chance a customer will order a soft drink is 0.90. The probability that a customer will order a hamburger is 0.60. The probability that a customer will order french fries is 0.50.

a. If a customer places an order, what is the probability that the order will include a soft drink and no fries if these two events are independent?

b. The restaurant has also determined that if a customer orders a hamburger, the probability the customer will also order fries is 0.80. Determine the probability that the order will include a hamburger and fries.

**4-28.** Ponderosa Paint and Glass carries three brands of paint. A customer wants to buy another gallon of paint to match paint she purchased at the store previously. She can't recall the brand name and does not wish to return home to find the old can of paint. So she selects two of the three brands of paint at random and buys them.

a. What is the probability that she matched the paint brand?

b. Her husband also goes to the paint store and fails to remember what brand to buy. So he also purchases two of the three brands of paint at random. Determine the probability that both the woman and her husband fail to get the correct brand of paint. (*Hint:* Are the husband's selections independent of his wife's selections?)

**4-29.** A college basketball team has 10 players; 5 are seniors, 2 are juniors, and 3 are sophomores. Two players are randomly selected to serve as captains for the next game. What is the probability that both players selected are seniors?

**4-30.** Martin Technology has sales offices in four cities: Dallas, Seattle, Boston, and Los Angeles. An analysis of the company's accounts receivables reveals the number of overdue invoices by days, as shown here.

| Days Overdue | Dallas | Seattle | Boston | Los Angeles |
|---|---|---|---|---|
| Under 30 days | 137 | 122 | 198 | 287 |
| 30–60 days | 85 | 46 | 76 | 109 |
| 61–90 days | 33 | 27 | 55 | 48 |
| Over 90 days | 18 | 32 | 45 | 66 |

Assume the invoices are stored and managed from a central database.

a. What is the probability that a randomly selected invoice from the database is from the Boston sales office?

b. What is the probability that a randomly selected invoice from the database is between 30 and 90 days overdue?

c. What is the probability that a randomly selected invoice from the database is over 90 days old and from the Seattle office?

d. If a randomly selected invoice is from the Los Angeles office, what is the probability that it is 60 or fewer days overdue?

**4-31.** Three events occur with probabilities $P(E_1) = 0.35, P(E_2) = 0.15, P(E_3) = 0.40$. If the event $B$ occurs, the probability becomes $P(E_1|B) = 0.25, P(B) = 0.30$.

a. Calculate $P(E_1 \text{ and } B)$.

b. Compute $P(E_1 \text{ or } B)$.

c. Assume that $E_1, E_2,$ and $E_3$ are independent events. Calculate $P(E_1 \text{ and } E_2 \text{ and } E_3)$.

**4-32.** The URS Corporation construction company has submitted two bids, one to build a large hotel in London and the other to build a commercial office building in New York City. The company believes it has a 40% chance of winning the hotel bid and a 25% chance of winning the office building bid. The company also believes that winning the hotel bid is independent of winning the office building bid.

a. What is the probability the company will win both contracts?

b. What is the probability the company will win at least one contract?

c. What is the probability the company will lose both contracts?

**4-33.** Suppose a quality manager for Dell Computers has collected the following data on the quality status of disk drives by supplier. She inspected a total of 700 disk drives.

| Supplier | Drive Status | |
|---|---|---|
| | Working | Defective |
| Company A | 120 | 10 |
| Company B | 180 | 15 |
| Company C | 50 | 5 |
| Company D | 300 | 20 |

a. Based on these inspection data, what is the probability of randomly selecting a disk drive from Company B?

b. What is the probability of a defective disk drive being received by the computer company?

c. What is the probability of a defect given that Company B supplied the disk drive?

**4-34.** Three events occur with probabilities $P(E_1) = 0.35$, $P(E_2) = 0.25$, $P(E_3) = 0.40$. Other probabilities are $P(B|E_1) = 0.25$, $P(B|E_2) = 0.15$, $P(B|E_3) = 0.60$.

a. Compute $P(E_1|B)$.

b. Compute $P(E_2|B)$.

c. Compute $P(E_3|B)$.

**4-35.** A hair salon caters to both men and women. In addition to hairstyles, the salon provides back and shoulder massages. In the past, 42% of men have requested a massage and 61% of women have requested a massage. The salon's customers are 48.2% men and 51.8% women.

a. If a customer arrives for a hair appointment, what is the probability that the customer is a woman who will also request a massage?

b. Calculate the probability that a customer will ask for a massage.

c. Given that the customer does ask for a massage, what is the probability that the customer is a man?

## Business Applications

**4-36.** A local FedEx Office has three black-and-white copy machines and two color copiers. Based on historical data, the chance that each black-and-white copier will be down for repairs is 0.10. The color copiers are more of a problem and are down 20% of the time each.

a. Based on this information, what is the probability that if a customer needs a color copy, both color machines will be down for repairs?

b. If a customer wants both a color copy and a black-and-white copy, what is the probability that the necessary machines will be available? (Assume that the color copier can also be used to make a black-and-white copy if needed.)

c. If the manager wants to have at least a 99% chance of being able to furnish a black-and-white copy on demand, is the present configuration sufficient? (Assume that the color copier can also be used to make a black-and-white copy if needed.) Back up your answer with appropriate probability computations.

d. What is the probability that all five copiers will be up and running at the same time? Suppose the manager added a fourth black-and-white copier. How would the probability of all copiers being ready at any one time be affected?

**4-37.** Suppose the managers at FedEx Office wish to meet the increasing demand for color photocopies and to have more reliable service. (Refer to Exercise 4-36.) As a goal, they would like to have at least a 99.9% chance

of being able to furnish a black-and-white copy or a color copy on demand. They also wish to purchase only four copiers. They have asked for your advice regarding the mix of black-and-white and color copiers. Supply them with your advice. Provide calculations and reasons to support your advice.

**4-38.** The Snappy Service gas station manager is thinking about a promotion that she hopes will bring in more business to the full-service island. She is considering the option that when a customer requests a fill-up, if the pump stops with the dollar amount at $19.99, the customer will get the gasoline free. Previous studies show that 70% of the customers require more than $20.00 when they fill up, so would not be eligible for the free gas. What is the probability that a customer will get free gas at this station if the promotion is implemented?

**4-39.** Suppose the manager in Exercise 4-38 is concerned about alienating customers who buy more than $20.00 of gas, since they would not be eligible to win the free gas under the original concept. To overcome this, she is thinking about changing the contest. The customer will get free gas for any of the following dollar amounts:

$21.11, $22.22, $23.33, $24.44, $25.55, $26.66, $27.77, $28.88, $29.99

Past data show that only 5% of all customers require more than $30.00. If one of these big-volume customers arrives, he will get to blindly draw a ball from a box containing 100 balls (99 red, 1 white). If the white ball is picked, the customer gets gas free. Considering this new promotion, what is the probability that a customer will get free gas?

**4-40.** A government-sponsored subsidized housing construction contract is to be awarded on the basis of a blind drawing from those who have bid. In addition to the Hubble Company, five other companies have submitted bids.

a. What is the probability of the Hubble Construction Company winning the bid?

b. Suppose two contracts are to be awarded by a blind draw. What is the probability of Hubble winning both contracts? Assume sampling with replacement.

c. Referring to part b, what is the probability of Hubble not winning either contract?

d. Referring to part b, what is the probability of Hubble winning exactly one contract?

**4-41.** Drake Marketing and Promotions has randomly surveyed 200 men who watch professional sports. The men were separated according to their educational level (college degree or not) and whether they preferred the NBA or the National Football League (NFL). The results of the survey are shown:

| Sports Preference | College Degree | No College Degree |
|---|---|---|
| NBA | 40 | 55 |
| NFL | 10 | 95 |

a. What is the probability that a randomly selected survey participant prefers the NFL?

b. What is the probability that a randomly selected survey participant has a college degree and prefers the NBA?

c. Suppose a survey participant is randomly selected and you are told that he has a college degree. What is the probability that this man prefers the NFL?

d. Is a survey participant's preference for the NBA independent of having a college degree?

4-42. Online marketplaces such as Craigslist have become a popular way for individuals to buy and sell miscellaneous items. The table below shows the numbers of days products stayed active (not sold) on one of these sites and also the price range of the items.

| | Days on the Website | | |
|---|---|---|---|
| Price Range | 1–7 | 8–30 | Over 30 |
| Under $200 | 125 | 15 | 30 |
| $200–$500 | 200 | 150 | 100 |
| $501–$1,000 | 400 | 525 | 175 |
| Over $1,000 | 125 | 140 | 35 |

a. Using the relative frequency approach to probability assessment, what is the probability that a product will be on the website more than 7 days?

b. Is the event 1–7 *days on the website* independent of the price $200–$500?

c. Suppose an item has just sold and was on the website less than 8 days, what is the most likely price range for that item?

4-43. Vegetables from the summer harvest are processed at a food-processing plant. The manager has found a case of cans that have not been properly sealed. There are three lines that processed cans of this type, and the manager wants to know which line is most likely to be responsible for this mistake. Provide the manager this information.

| Line | Contribution to Total | Proportion Defective |
|---|---|---|
| 1 | 0.40 | 0.05 |
| 2 | 0.35 | 0.10 |
| 3 | 0.25 | 0.07 |

4-44. A corporation has 11 manufacturing plants. Of these, seven are domestic and four are outside the United States. Each year a performance evaluation is conducted for four randomly selected plants. What is the probability that a performance evaluation will include at least one plant outside the United States? (*Hint:* Begin by finding the probability that only domestic plants are selected.)

4-45. Parts and materials for the skis made by the Downhill Adventures Company are supplied by two suppliers. Supplier A's materials make up 30% of what is used, with Supplier B providing the rest. Past records

indicate that 15% of Supplier A's materials are defective and 10% of B's are defective. Since it is impossible to tell which supplier the materials came from once they are in inventory, the manager wants to know which supplier more likely supplied the defective materials the supervisor has brought to his attention. Provide the manager this information.

4-46. A major electronics manufacturer has determined that when one of its televisions is sold, there is 0.08 chance that it will need service before the warranty period expires. It has also assessed a 0.05 chance that a digital video recorder (DVR) will need service prior to the expiration of the warranty.

a. Suppose a customer purchases one of the company's televisions and one of the DVRs. What is the probability that at least one of the products will require service prior to the warranty expiring?

b. Suppose a retailer sells four televisions on a particular Saturday. What is the probability that none of the four will need service prior to the warranty expiring?

c. Suppose a retailer sells four televisions on a particular Saturday. What is the probability that at least one will need repair?

4-47. A Bipartisan Policy Center article indicated that 57.5% of the voting-age voters cast ballots in the 2012 presidential election. It also indicated that 70% of registered voters voted in the election. The percent of those who voted for President Obama was 51.01%. (*Source:* "2012 Election turnout dips below 2008 and 2004 levels: Number of eligible voters increases by eight million, five million fewer votes cast," http://bipartisanpolicy.org, Nov. 2012.)

a. Determine the proportion of voting-age voters who voted for President Obama.

b. Determine the proportion of voting-age voters who were registered to vote.

4-48. A distributor of outdoor yard lights has four suppliers. This past season, she purchased 40% of the lights from Franklin Lighting, 30% from Wilson & Sons, 20% from Evergreen Supply, and the rest from A. L. Scott. In prior years, 3% of Franklin's lights were defective, 6% of the Wilson lights were defective, 2% of Evergreen's were defective, and 8% of the Scott lights were defective. When the lights arrive at the distributor, she puts them in inventory without identifying the supplier. Suppose that a defective light string has been pulled from inventory; what is the probability that it was supplied by Franklin Lighting?

4-49. Trucks that travel on U.S. highways have to stop at various locations to be weighed and inspected for safe brakes and light systems. Of these trucks, 77% are on interstate commerce business while 23% are intrastate. Of the intrastate trucks, 3.1% are flagged for safety defects compared to 0.7% of those that are on interstate business.

a. Calculate the probability that a randomly chosen truck is an interstate truck and is not flagged for a safety violation.

b. What is the probability that a randomly chosen truck does not have a safety defect?

c. If two trucks are chosen randomly, determine the probability that only one of the two does not have a safety defect.

**4-50.** Suppose an auditor has 18 tax returns, 12 of which are for physicians. If three of the 18 tax returns are randomly selected, then what is the probability that at least one of the three selected will be a physician's tax return?

**4-51.** A box of 50 remote control devices contains three that have a defective power button. If devices are randomly sampled from the box and inspected one at a time, determine

a. The probability that the first control device is defective.

b. The probability that the first control device is good and the second control device is defective.

c. The probability that the first three sampled devices are all good.

## Computer Software Exercises

**4-52.** ECCO is a company that makes backup alarms for commercial vehicles and heavy-duty equipment like forklifts. The data file **ECCO** contains warranty claim data as well as the associated manufacturing plant, complaint code, and shift for each alarm that is manufactured.

a. Based on the data provided, what is the probability that a warranty claim comes from a product made in the Salt Lake City plant?

b. What is the probability that a warranty claim was made on the swing shift at the Toronto plant?

c. What is the probability that a warranty claim is for a cracked lens or for a wiring complaint?

d. If two warranty claims are examined, what is the probability that both are from a product made in Boise?

**4-53.** A hotel chain conducted a survey of its guests. Sixty-two surveys were completed. Based on the data from the survey, found in the file **HotelSurvey**, answer the following questions using the relative frequency assessment method.

a. Of two customers selected, what is the probability that both will be on a business trip?

b. What is the probability that a customer will be on a business trip or will experience a hotel problem during a stay at the hotel?

c. What is the probability that a customer on business has an in-state area code phone number?

d. Based on the data in the survey, can the hotel manager conclude that a customer's rating regarding staff attentiveness is independent of whether he or she is traveling on business, pleasure, or both? Use the rules of probability to make this determination.

**4-54.** Continuing with the survey done by the managers of the hotel chain, based on the data from the survey, found in the file **HotelSurvey**, determine the following probabilities using the relative frequency assessment method.

a. Of two customers selected, what is the probability that neither will be on a business trip?

b. What is the probability that a customer will be on a business trip or will not experience a hotel problem during a stay at the hotel?

c. What is the probability that a customer on a pleasure trip has an in-state area code phone number?

**4-55.** A survey asked what the most important reason was that people give for not using a wireless phone exclusively. The responses were: (1) Like the safety of traditional phone, (2) Like having a land line for a backup phone, (3) Pricing not attractive enough, (4) Weak or unreliable cell signal at home, (5) Need phone line for DSL Internet access, and (6) Other. The file titled **Wireless** contains the responses for 1,088 respondents.

a. Of those respondents 36 or older, determine the probability that an individual in this age group would not use a wireless phone exclusively because of concern about a weak or unreliable cell signal.

b. If three respondents were selected at random from those respondents younger than 36, calculate the probability that at least one of the respondents stated the most important reason for not using a wireless exclusively was that he or she liked having a land line for a backup phone.

**4-56.** The loan manager for High Desert Bank selected a sample of the loans her bank has made in the past year. The file **High Desert Banking** contains information on the loan amount, the type of loan, and the region of the state where the loan was made.

a. What is the probability of randomly selecting a loan from this file that has a loan amount of at least $73,500 and was made in Region 2?

b. Find the probability that a commercial loan is less than $90,000.

c. If two loans are selected, what is the probability that both are commercial loans?

**4-57.** A report indicated the average number of days it takes for an American company to fill a job vacancy is 48 days. Sample data similar to those used in the study are in a file titled **Hired**. Categories for the days and hire cost are provided under the headings "Time" and "Cost," respectively.

a. Calculate the probability that a company vacancy took at most 100 days or cost at most $4,000 to fill.

b. Of the vacancies that took at most 100 days to fill, calculate the probability that the cost was at most $4,000.

c. If three of the vacancies were chosen at random, calculate the probability that two of the vacancies cost at most $4,000 to fill.

4-58. A scooter manufacturing company is notified whenever a scooter breaks down, and the problem is classified as being either mechanical or electrical. The company then matches the scooter to the plant where it was assembled. The file **Scooters** contains a random sample of 200 breakdowns. Use the data in the file to answer these questions.

a. If a scooter was assembled in the Tyler plant, what is the probability its breakdown was due to an electrical problem?

b. Is the probability of a scooter having a mechanical problem independent of the scooter being assembled at the Lincoln plant?

c. If mechanical problems are assigned a cost of $75 and electrical problems are assigned a cost of $100, how much cost would be budgeted for the Lincoln and Tyler plants next year if a total of 500 scooters were expected to be returned for repair?

# 4 Overview

## Summary

**OUTCOME 1** Identify situations for which each of the three approaches to assessing probabilities applies.

- **Probability** is the way we measure uncertainty about events. To properly use probability, you need to know the probability rules and the terms associated with probability.
- Important probability concepts and terminology include **sample space, dependent** and **independent events**, and **mutually exclusive events**.
- Probabilities are assessed in three main ways: **classical assessment, relative frequency assessment**, and **subjective assessment**.

**OUTCOME 2** Be able to apply the Addition Rule.

**OUTCOME 3** Know how to use the Multiplication Rule.

**OUTCOME 4** Know how to use Bayes' Theorem for applications involving conditional probabilities.

- Section 4.2 introduces nine probability rules, including three **addition rules** and two **multiplication rules**.
- Rules for **conditional probability** and the **complement rule** are also very useful.
- **Bayes' Theorem** is used to calculate conditional probabilities in situations where the probability of the given event is not provided and must be calculated. Bayes' Theorem is also used to revise prior probabilities after new information affecting the prior becomes available.

## Equations

**(4.1) Classical Probability Assessment** pg. 134

$$P(E_i) = \frac{\text{Number of ways } E_i \text{ can occur}}{\text{Total number of possible outcomes}}$$

**(4.2) Relative Frequency Assessment** pg. 135

$$P(E_i) = \frac{\text{Number of times } E_i \text{ occurs}}{N}$$

**(4.3) Probability Rule 1** pg. 142

$$0 \le P(E_i) \le 1 \text{ for all } i$$

**(4.4) Probability Rule 2** pg. 142

$$\sum_{i=1}^{k} P(e_i) = 1$$

**(4.5) Probability Rule 3** pg. 142

Addition Rule for Individual Outcomes:

The probability of an event $E_i$ is equal to the *sum* of the probabilities of the individual outcomes that form $E_i$. For example, if

$$E_i = \{e_1, e_2, e_3\}$$

then

$$P(E_i) = P(e_1) + P(e_2) + P(e_3)$$

**(4.6) Complement Rule** pg. 144

$$P(\overline{E}) = 1 - P(E)$$

**(4.7) Probability Rule 4** pg. 146
Addition Rule for Any Two Events $E_1$ and $E_2$:

$$P(E_1 \text{ or } E_2) = P(E_1) + P(E_2) - P(E_1 \text{ and } E_2)$$

**(4.8) Probability Rule 5** pg. 149
Addition Rule for Mutually Exclusive Events $E_1$ and $E_2$:

$$P(E_1 \text{ or } E_2) = P(E_1) + P(E_2)$$

**(4.9) Probability Rule 6** pg. 149
Conditional Probability for Any Two Events $E_1$ and $E_2$:

$$P(E_1 \mid E_2) = \frac{P(E_1 \text{ and } E_2)}{P(E_2)}$$

**(4.10) Probability Rule 7** pg. 152
Conditional Probability for Independent Events $E_1$ and $E_2$:

$$P(E_1 \mid E_2) = P(E_1) \quad P(E_2) > 0$$

and

$$P(E_2 | E_1) = P(E_2) \quad P(E_1) > 0$$

**(4.11) Probability Rule 8** pg. 153

Multiplication Rule for Any Two Events $E_1$ and $E_2$:

$$P(E_1 \text{ and } E_2) = P(E_1)P(E_2 | E_1)$$

**(4.12) Probability Rule 9** pg. 155

Multiplication Rule for Independent Events $E_1$ and $E_2$:

$$P(E_1 \text{ and } E_2) = P(E_1)P(E_2)$$

**(4.13) Bayes' Theorem** pg. 157

$$P(E_i | B) = \frac{P(E_i)P(B | E_i)}{P(E_1)P(B | E_1) + P(E_2)P(B | E_2) + \cdots + P(E_k)P(B | E_k)}$$

## Key Terms

**Classical probability assessment** pg. 134
**Complement** pg. 144
**Conditional probability** pg. 149
**Dependent events** pg. 133

**Event** pg. 131
**Experiment** pg. 130
**Independent events** pg. 133
**Mutually exclusive events** pg. 132

**Probability** pg. 129
**Relative frequency assessment** pg. 135
**Sample space** pg. 130
**Subjective probability assessment** pg. 137

## Chapter Exercises

### Conceptual Questions

**4-59.** Discuss what is meant by *classical probability assessment* and indicate why classical assessment is not often used in business applications.

**4-60.** Discuss what is meant by the *relative frequency assessment approach* to probability assessment. Provide a business-related example, other than the one given in the text, in which this method of probability assessment might be used.

**4-61.** Discuss what is meant by *subjective probability*. Provide a business-related example in which subjective probability assessment would likely be used. Also, provide an example of when you have personally used subjective probability assessment.

**4-62.** Examine the relationship between independent, dependent, and mutually exclusive events. Consider two events $A$ and $B$ that are mutually exclusive such that $P(A) \neq 0$.
  a. Calculate $P(A | B)$.
  b. What does your answer to part a say about whether two mutually exclusive events are dependent or independent?
  c. Consider two events $C$ and $D$ such that $P(C) = 0.4$ and $P(C | D) = 0.15$. (1) Are events $C$ and $D$ mutually exclusive? (2) Are events $C$ and $D$ independent or dependent? Are dependent events necessarily mutually exclusive events?

**4-63.** Consider the following table:

|  | **A** | **Ā** | **Totals** |
|---|---|---|---|
| **B** | 800 | 200 | 1,000 |
| **B̄** | 600 | 400 | 1,000 |
| **Totals** | 1,400 | 600 | 2,000 |

Explore the complements of conditional events:
  a. Calculate the following probabilities: $P(A | B), P(\bar{A} | \bar{B}), P(\bar{A} | B), P(A | \bar{B})$.
  b. Now determine which pair of events are complements of each other. (*Hint:* Use the probabilities calculated in part a and the Complement Rule.)

**4-64.** Examine the following table:

|  | **A** | **Ā** | **Totals** |
|---|---|---|---|
| **B** | 200 | 800 | 1,000 |
| **B̄** | 300 | 700 | 1,000 |
| **Totals** | 500 | 1,500 | 2,000 |

  a. Calculate the following probabilities: $P(A), P(\bar{A}), P(A | B), P(\bar{A} | B), P(A | \bar{B})$, and $P(\bar{A} | \bar{B})$.
  b. Show that (1) $A$ and $B$, (2) $A$ and $\bar{B}$ (3) $\bar{A}$ and $B$, and (4) $\bar{A}$ and $\bar{B}$ are dependent events.

### Business Applications

**4-65.** An accounting professor at a state university in Vermont recently gave a three-question multiple-choice quiz. Each question had four optional answers.
  a. What is the probability of getting a perfect score if you were forced to guess at each question?
  b. Suppose it takes at least two correct answers out of three to pass the test. What is the probability of passing if you are forced to guess at each question? What does this indicate about studying for such an exam?
  c. Suppose through some late-night studying you are able to correctly eliminate two answers on each question. Now answer parts a and b.

**4-66.** When customers call into a bank's customer service phone line, they are asked to complete a survey on service quality at the end of the call. Six out of ten customers agree to do the survey. Of those who agree, 58% actually respond to all the questions.

a. Calculate the probability that a randomly chosen customer who calls the service phone line will actually complete the survey.

b. Calculate the probability that a randomly chosen customer will not complete the survey.

c. If three customers are selected at random, what is the probability that only two of them will complete the survey?

**4-67.** An April 14, 2015, article by Kara Brandeisky on *Money* magazine's website states that overall the chance of being audited by the IRS in 2014 was 0.86%. If a person's adjusted gross income was between $50,000 and $74,999, the chance of being audited drops to 0.53%, and for those with incomes higher than $10 million, the chance goes up to 16.22%. (*Source:* "These are the people who are most likely to get audited," http://time.com/money, Apr. 14, 2015.)

a. Considering these probabilities, discuss which probability assessment method you believe was most likely used to arrive at the probabilities.

b. Based on chance alone and not taking income into account, what is the overall probability that you and your best friend will both be audited? (Assume the 2014 chances still apply.)

c. Suppose Harry made $13 million in 2014 and his daughter, Sarah, made $55,000. What is the probability that neither of them was audited by the IRS based on chance alone?

**4-68.** Simmons Furniture Company is considering changing its starting hour from 8:00 A.M. to 7:30 A.M. A census of the company's 1,200 office and production workers shows that 370 of its 750 production workers favor the change and a total of 715 workers favor the change. To further assess worker opinion, the region manager decides to talk with random workers.

a. What is the probability a randomly selected worker will be in favor of the change?

b. What is the probability a randomly selected worker will be against the change *and* be an office worker?

c. Are the events *job type* and *opinion* independent? Explain.

**4-69.** A national convenience store chain determines that 70% of gas purchases are made with a credit or debit card.

a. Indicate the type of probability assessment method that the gas station manager would use to assess this probability.

b. In one local store, 10 randomly chosen customers were observed. All 10 of these customers used a credit or a debit card. If the 70% statistic applies to this area, determine the probability that 10 out of 10 customers would use a credit or debit card.

c. If 90% of gas purchases paid for at the pump were made with a credit or debit card, determine the probability that 10 out of 10 customers would use a credit or debit card.

d. Based on your answers to parts b and c, does it appear that a larger percentage of local individuals use credit or debit cards than is true for the nation as a whole? Explain.

**4-70.** Ponderosa Paint and Glass makes paint at three plants. It then ships the unmarked paint cans to a central warehouse. Plant A supplies 50% of the paint, and past records indicate that the paint is incorrectly mixed 10% of the time. Plant B contributes 30%, with paint mixed incorrectly 5% of the time. Plant C supplies 20%, with paint mixed incorrectly 20% of the time. If Ponderosa guarantees its product and spent $10,000 replacing improperly mixed paint last year, how should the cost be distributed among the three plants?

**4-71.** Recently, several long-time customers at the Sweet Haven Chocolate Company have complained about the quality of the chocolates. It seems there are several partially covered chocolates being found in boxes. The defective chocolates should have been caught when the boxes were packed. The manager is wondering which of the three packers is not doing the job properly. Clerk 1 packs 40% of the boxes and usually has a 2% defective rate. Clerk 2 packs 30%, with a 2.5% defective rate. Clerk 3 boxes 30% of the chocolates, and her defective rate is 1.5%. Which clerk is most likely responsible for the boxes that raised the complaints?

**4-72.** An investment firm is thinking of opening a skiing facility in Colorado. It is trying to decide whether to open an area catering to family skiers or to some other group. To help make its decision, it gathers the following information. Let

$A_1$ = Family will ski
$A_2$ = Family will not ski
$B_1$ = Family has children but none in the 8–16 age group
$B_2$ = Family has children in the 8–16 age group
$B_3$ = Family has no children

Then, for this location,

$$P(A_1) = 0.40$$
$$P(B_2) = 0.35$$
$$P(B_1) = 0.25$$
$$P(A_1 | B_2) = 0.70$$
$$P(A_1 | B_1) = 0.30$$

a. Use the probabilities given to construct a joint probability distribution table.

b. What is the probability a family will ski *and* have children who are not in the 8–16 age group? How do you write this probability?

c. What is the probability a family with children in the 8–16 age group will not ski?

d. Are the categories *skiing* and *family composition* independent?

**4-73.** Fifty chief executive officers of small to medium-sized companies were classified according to their gender and functional background as shown in the table below:

| Functional Background | Male | Female | Total |
|---|---|---|---|
| Marketing | 4 | 10 | 14 |
| Finance | 11 | 5 | 16 |
| Operations | 17 | 3 | 20 |
| Total | 32 | 18 | 50 |

a. If a chief executive is randomly selected from this group, what is the probability that the executive is a female?

b. What is the probability that a randomly selected executive from this group is a male whose functional background is marketing?

c. Assume that an executive is selected and you are told that the executive's functional background was in operations. What is the probability that this executive is a female?

d. Assume that an executive is selected and you are told that the executive is a female. What is the probability the executive's functional area is marketing?

e. Are gender and functional background independent for this set of executives?

4-74. A manufacturing firm has two suppliers for an electrical component used in its process: one in Mexico and one in China. The supplier in Mexico ships 82% of all the electrical components used by the firm and has a defect rate of 4%. The Chinese supplier ships 18% of the electrical components used by the firm and has a defect rate of 6%.

a. Calculate the probability that an electrical component is defective.

b. Suppose an electrical component is defective. What is the probability that component was shipped from Mexico? (*Hint:* Use Bayes' Theorem.)

4-75. Five hundred smartphone owners were asked if they use their phones to download sports scores. The responses are summarized in the following table:

| | Yes | No | Total |
|---|---|---|---|
| **Male** | 195 | 105 | 300 |
| **Female** | 70 | 130 | 200 |
| **Total** | 265 | 235 | 500 |

a. What is the probability that a randomly sampled respondent from this group has used his or her smartphone to download sports scores?

b. What is the probability that a randomly sampled respondent from this group is female?

c. Given that a randomly sampled respondent has not used the smartphone to download sports scores, what is the probability that he or she is male?

d. Are the responses Yes and No independent of the respondent's gender?

4-76. Thirty-five percent of all customers of a large, national financial institution manage their accounts online. What is the probability that in a random sample of three customers from the financial institution's database,

a. All three manage their accounts online?

b. Exactly two manage their accounts online?

c. None manage online?

d. At least one manages online?

e. The third sampled customer is the first one to manage online?

4-77. Based on a survey conducted by students in the Recreational Management degree program, 65% of students at the university own a mountain bike, 25% own a road bike, and 6% own both a mountain bike and a road bike.

a. What is the probability that a randomly selected student at this university has a mountain bike but not a road bike?

b. What is the probability that a randomly selected student at this university has a road bike but not a mountain bike?

c. What is the probability that a randomly selected student at this university has neither a road bike nor a mountain bike?

d. Given that a randomly selected student at this university has a road bike, what is the probability he or she also has a mountain bike?

4-78. After spending some time studying company e-mail accounts, the manager of computer security for a large insurance company has determined that 25% of all e-mail messages sent to employee computers are spam. If 80% of all spam e-mails sent to personal accounts contain the word *guarantee* in the Subject line, and only 4% of nonspam e-mails contain the word *guarantee*, what is the probability that an employee's next e-mail that contains the word *guarantee* in the Subject line is spam?

4-79. A large coffeehouse has found that 50% of all customers order food, 80% order coffee, and 40% order both. What is the probability that a customer at the coffeehouse

a. Orders food but no coffee?

b. Orders coffee but no food?

c. Orders neither coffee nor food?

4-80. In a large shipment of electronic parts, 0.015 are bad. The parts are tested using a machine that correctly identifies bad parts as defective with a probability of 0.98, and correctly identifies good parts as nondefective with a probability of 0.95.

a. If a part is randomly sampled from the shipment and tested, what is the probability that the testing machine identifies the part as defective?

b. Given that the test indicates a nondefective part, what is the probability the part is truly good?

4-81. In a large metropolitan area, 65% of commuters ride the train only, 18% ride the bus only, and 6% ride both the train and the bus to get to the central business district.

a. What is the probability that a randomly selected commuter does not ride the train to the central business district?

b. What is the probability that a randomly selected commuter rides neither the train nor the bus to the central business district?

c. If a randomly selected commuter rides the bus, what is the probability that he or she also rides the train?

## Computer Software Exercises

**4-82.** A survey of 300 CEOs was conducted in which the CEOs were to list their corporation's geographical location: Northeast (NE), Southeast (SE), Midwest (MW), Southwest (SW), and West (W). They were also requested to indicate their company's industrial type: Communications (C), Electronics (E), Finance (F), and Manufacturing (M). The file titled **CEOInfo** contains sample data similar to those used in this study.

a. Determine the probability that a randomly chosen CEO would have a corporation in the West.

b. Compute the probability that a randomly chosen CEO would have a corporation in the West and head an electronics corporation.

c. Calculate the probability that a randomly chosen CEO would have a corporation in the East or head a communications corporation.

d. Of the corporations located in the East, calculate the probability that a randomly selected CEO would head a communications corporation.

**4-83.** The ECCO company makes backup alarms for machinery like forklifts and commercial trucks. When a customer returns one of the alarms under warranty, the quality manager logs data on the product. From the data available in the file named **Ecco,** use relative frequency to answer the following questions.

a. What is the probability the product was made at the Salt Lake City plant?

b. What is the probability the reason for the return was a wiring problem?

c. What is the joint probability the returned item was from the Salt Lake City plant and had a wiring-related problem?

d. What is the probability a returned item was made on the day shift at the Salt Lake City plant and had a cracked lens problem?

e. If an item was returned, what is the most likely profile for the item, including plant location, shift, and cause of problem?

**4-84.** Continuing with the ECCO company from Exercise 4-83, when a customer returns one of the alarms under warranty, the quality manager logs data on the product. From the data available in the **Ecco** file, use relative frequency to answer these questions.

a. If a part was made in the Salt Lake City plant, what is the probability the cause of the returned part was wiring?

b. If the company incurs a $30 cost for each returned alarm, what percentage of the cost should be assigned to each plant if it is known that 70% of all production is done in Boise, 20% in Salt Lake City, and the rest in Toronto?

**4-85.** The Employee Benefit Research Institute (EBRI) issued a report in 2015 that stated 55% of workers are confident they have enough money for a comfortable retirement. (*Source:* R. Helman, C. Copeland, and J. VanDerhei, "The 2015 Retirement Confidence Survey: Having a retirement savings plan a key factor in Americans' retirement confidence," Apr. 2015, no. 413.) The file titled **Retirement** contains sample data similar to those used in this study.

a. Construct a frequency distribution of the total savings and investments using the intervals (1) Less than $25,000, (2) $25,000–$49,999, (3) $50,000–$99,999, (4) $100,000–$249,999, and (5) $250,000 or more.

b. Determine the probability that an individual who has saved for retirement has saved less than $50,000. Use relative frequencies.

c. Determine the probability that a randomly chosen individual has saved less than $50,000 toward retirement.

d. Calculate the probability that at least two of four individuals have saved less than $50,000 toward retirement.

**4-86.** The Center for Generational Kinetics had categorized the American workforce as follows: (1) Traditionalists (born before 1945); (2) Baby Boomers (born 1945 to 1964); (3) Generation X (born 1965 to 1976); (4) Millennials–Generation Y (born 1977 to 1995); (5) Generation Z (born 1996 or after) (*source:* "Five generations of employees in today's workforce, http://genhq.com, Apr. 27, 2015). The file called **Communication** contains data on the preferred method of communication for the first four categories in the workforce.

a. Construct a frequency distribution for each of the generations. Use the communication categories (1) Gp Meeting, (2) Face to Face, (3) E-mail, and (4) Other.

b. Calculate the probability that a randomly chosen member of the workforce prefers communicating face to face.

c. Given that an individual in the workforce prefers to communicate face to face, determine the generation of which the individual is most likely a member.

## Case 4.1    Great Air Commuter Service

The Great Air Commuter Service Company started in 1984 to provide efficient and inexpensive commuter travel between Boston and New York City. People in the airline industry know Peter Wilson, the principal owner and operating manager of the company, as "a real promoter." Before founding Great Air, Peter operated a small regional airline in the Rocky Mountains with varying success. When Cascade Airlines offered to buy his company, Peter decided to sell and return to the East.

Peter arrived at his office near Fenway Park in Boston a little later than usual this morning. He had stopped to have a business breakfast with Aaron Little, his long-time friend and sometime partner in various business deals. Peter needed some advice and through the years has learned to rely on Aaron as a ready source, no matter what the subject.

Peter explained to Aaron that his commuter service needed a promotional gimmick to improve its visibility in the business communities in Boston and New York. Peter was thinking of running a contest on each flight and awarding the winner a prize. The idea would be that travelers who commute between Boston and New York might just as well have fun on the way and have a chance to win a nice prize.

As Aaron listened to Peter outlining his contest plans, his mind raced through contest ideas. Aaron thought that a large variety of contests would be needed, because many of the passengers would likely be repeat customers and might tire of the same old thing. In addition, some of the contests should be chance-type contests, whereas others should be skill based.

"Well, what do you think?" asked Peter. Aaron finished his scrambled eggs before responding. When he did, it was completely in character. "I think it will fly," Aaron said, and proceeded to offer a variety of suggestions.

Peter felt good about the enthusiastic response Aaron had given to the idea and thought that the ideas discussed at breakfast presented a good basis for the promotional effort. Now back at the office, Peter does have some concerns with one part of the plan. Aaron thought that in addition to the regular in-flight contests for prizes (such as free flights, dictation equipment, and business periodical subscriptions), each month on a randomly selected day a major prize should be offered on all Great Air flights. This would encourage regular business fliers to fly Great Air all the time. Aaron proposed that the prize could be a trip to the Virgin Islands or somewhere similar, or the cash equivalent.

Great Air has three flights daily to New York and three flights returning to Boston, for a total of six flights. Peter is concerned that the cost of funding six prizes of this size each month plus six daily smaller prizes might be excessive. He also believes that it might be better to increase the size of the large prize to something such as a new car but use a contest that will not guarantee a winner.

But what kind of a contest can be used? Just as he is about to dial Aaron's number, Margaret Runyon, Great Air's marketing manager, enters Peter's office. He has been waiting for her to return from a meeting so he can run the contest idea past her and get her input.

Margaret's response is not as upbeat as Aaron's, but she does think the idea is worth exploring. She offers an idea for the large-prize contest that she thinks might be workable. She outlines the contest as follows.

On the first of each month, she and Peter will randomly select a day for that month on which the major contest will be run. That date will not be disclosed to the public. Then, on each flight that day, the flight attendant will have passengers write down their birthdays (month and day). If any two people on the plane have the same birthday, they will place their names in a hat and one name will be selected to receive the grand prize.

Margaret explains that because the capacity of each flight is 40 passengers plus the crew, there is a very low chance of a birthday match and, therefore, the chance of giving away a grand prize on any one flight is small. Peter likes the idea, but when he asks Margaret what the probability is that a match will occur, her response does not sound quite right. She believes the probability for a match will be 40/365 for a full plane and less than that when there are fewer than 40 passengers aboard.

After Margaret leaves, Peter decides that it would be useful to know the probability of one or more birthday matches on flights with 20, 30, and 40 passengers. He realizes that he will need some help from someone with knowledge of statistics.

### Required Tasks:

1. Assume that there are 365 days in a year (in other words, there is no leap year). Also assume there is an equal probability of a passenger's birthday falling on any one of the 365 days. Calculate the probability that there will be at least one birthday match for a flight containing exactly 20 passengers. (*Hint:* This calculation is made easier if you will first calculate the probability that there are no birthday matches for a flight containing 20 passengers.)
2. Repeat task 1 for a flight containing 30 passengers and a flight containing 40 passengers. Again, it will be easier to compute the probabilities of one or more matches if you first compute the probability of no birthday matches.
3. Assuming that each of the six daily flights carries 20 passengers, calculate the probability that the airline will have to award two or more major prizes that month. (*Hint:* It will be easier to calculate the probability of interest by first calculating the probability that the airline will award one or fewer prizes in a month.)

---

### Case 4.2    Pittsburg Lighting

Melinda Wilson and Tony Green left their boss's office, heading for the conference room to discuss the task they were given. The task is to select one of two suppliers of outdoor light bulbs. Ordinarily this would be pretty easy—just pick the one that offers the best price. But in this case, their boss is planning to provide a warranty on the bulbs that will require the company to replace all defects at no cost to the customer. Thus, product quality is an important consideration.

As Melinda and Tony settled into the plush leather conference room seats, they quickly realized that this task was more

challenging than they first thought. To get started, Melinda went to the white board and began listing the information they already knew. The total purchase will be 100,000 bulbs. Regardless of which supplier is selected, the bulbs will be sold to Pittsburg customers at the same price. The first supplier, Altman Electronics, was one that Pittsburg had used for years, so they had solid quality information. Both Melinda and Tony agreed that this supplier could be counted on for a 5% defect rate. Altman's bid was $3.00 per light. The second supplier, Crestwell Lights and Fixtures, was new to Pittsburg but offered an attractive bid of $2.70 per unit.

However, neither Melinda nor Tony had any information about Crestwell's likely defect rate. They realized that they needed to get some more information before they could proceed.

After lunch, Tony made a few calls to other lighting companies in different parts of the country to see what their experience was with Crestwell. When Tony met Melinda back in the conference room, he went to the board and wrote his findings:

| Crestwell Defect Rate | Probability |
|---|---|
| 0.02 | 0.05 |
| 0.03 | 0.10 |
| 0.04 | 0.20 |
| 0.05 | 0.25 |
| 0.06 | 0.20 |
| 0.07 | 0.15 |
| 0.08 | 0.05 |

Tony explained to Melinda that Crestwell's defect performance differed depending on which person he talked to, but these numbers summarized his findings. Thus, Crestwell's defect rate is uncertain, and the probability of any specific defect level is really a subjective assessment made by Tony. Because customers would be refunded for defective bulbs, both Melinda and Tony were concerned that they could not determine a particular defect rate. Melinda suggested that they place a small order for 100 bulbs and see what the defect rate for those turned out to be. She figured that they could use this added information to learn more about Crestwell's quality.

A few days later, Pittsburg received the 100 bulbs from Crestwell, and after the warehouse crew tested them, six were found to be defective. Tony did some quick calculations and found the following:

| Possible Defect Rate | Probability of Six Defects *Given* the Possible Defect Rate |
|---|---|
| 0.02 | 0.01 |
| 0.03 | 0.05 |
| 0.04 | 0.11 |
| 0.05 | 0.15 |
| 0.06 | 0.17 |
| 0.07 | 0.15 |
| 0.08 | 0.12 |

Melinda and Tony returned to the conference room to consider all this information and make a decision regarding which supplier to select. They agreed that they would recommend the supplier that would provide the lowest overall cost, including purchasing and warranty replacement costs.

## Required Tasks:

1. Calculate the total purchase cost for each supplier.
2. Calculate the warranty replacement cost for Altman Electronics, and calculate the total cost if Altman is selected.
3. Calculate the total cost (purchase cost plus defective replacement) for each possible defect level if Crestwell is selected.
4. Based on the findings from the order of 100 bulbs from Crestwell, use Bayes' Theorem to revise Tony's subjective probability assessments for Crestwell's possible defect rates.
5. Using the revised probabilities, calculate the weighted total costs for Crestwell, and compare this to the total costs calculated for Altman. Which supplier should Melinda and Tony recommend?

# 5 Discrete Probability Distributions

## WHY YOU NEED TO KNOW

How many toll stations should be constructed when a new toll bridge is built? If there are four toll stations, will drivers have to wait too long, or will there be too many toll stations and excess employees? To help answer these questions, decision makers use a probability distribution known as the *Poisson distribution*.

Pak-Sense, a manufacturer of temperature sensor equipment for the food industry, receives component parts for its sensors weekly from suppliers. When a batch of parts arrives, the quality-assurance section randomly samples a fixed number of parts and tests them to see

**Quick Prep**

Review the concepts of simple random sampling discussed in Chapter 1.

Review the discussion of weighted averages in Chapter 3.

Review the basic rules of probability in Chapter 4, including the Addition and Multiplication Rules.

if any are defective. Suppose in one such test, a sample of 50 parts is selected from a supplier whose contract calls for at most 2% defective parts. How many defective parts in the sample of 50 should Pak-Sense expect if the contract is being satisfied? What should be concluded if the sample contains three defects? Answers to these questions require calculations based on a probability distribution known as the *binomial distribution*.

A personnel manager has a chance to promote three people from ten equally qualified candidates. Suppose none of six women are selected by the manager. Is this evidence of gender bias, or would we expect to see this type of result? A distribution known as the *hypergeometric distribution* would be very helpful in addressing this issue.

The *binomial*, *Poisson*, and *hypergeometric* distributions are three *discrete* probability distributions used in business decision making. This chapter introduces discrete probability distributions and shows how they are used in business settings. Through the use of these well-established discrete probability distributions, you will be better prepared for making decisions in an uncertain environment.

## Introduction to Discrete Probability Distributions

### Random Variables

**Random Variable**
A variable that takes on different numerical values based on chance.

As discussed in Chapter 4, when a random experiment is performed, some outcome must occur. When the experiment has a quantitative characteristic, we can associate a number with each outcome. For example, an inspector who examines three HD televisions can judge each television as "acceptable" or "unacceptable." The outcome of the experiment defines the specific number of acceptable televisions. The possible outcomes are

$$x = \{0, 1, 2, 3\}$$

The value $x$ is called a **random variable**, since the numerical values it takes on are random and vary from trial to trial. Although the inspector knows these are the possible values for the variable before she samples, she does not know which value will occur in any given trial. Further, the value of the random variable may be different each time three HD televisions are inspected.

**Discrete Random Variable**
A random variable that can assume only a finite number of values or an infinite sequence of values such as 0, 1, 2, . . . .

Two classes of random variables exist: **discrete random variables** and **continuous random variables**. For instance, if a bank auditor randomly examines 15 accounts to verify the accuracy of the balances, the number of inaccurate account balances can be represented by a discrete random variable with the following values:

$$x = \{0, 1, \ldots, 15\}$$

**Continuous Random Variables**
Random variables that can assume an uncountably infinite number of values.

In another situation, ten employees were recently hired by a major electronics company. The number of females in that group can be described as a discrete random variable with possible values equal to

$$x = \{0, 1, 2, 3, \ldots, 10\}$$

Notice that the value for a discrete random variable is often determined by counting. In the bank auditing example, the value of variable $x$ is determined by counting the number of accounts with errors. In the hiring example, the value of variable $x$ is determined by counting the number of females hired.

In other situations, the random variable is said to be continuous. For example, the exact time it takes a city bus to complete its route may be any value between two points, say 30 minutes to 35 minutes. If $x$ is the time required, then $x$ is continuous because, if measured precisely enough, the possible values, $x$, can be any value in the interval 30 to 35 minutes. Other examples of continuous variables include measures of distance and measures of weight when measured precisely. A continuous random variable is generally defined by measuring, which is contrasted with a discrete random variable, whose value is typically determined by counting. Chapter 6 focuses on some important probability distributions for continuous random variables.

**Displaying Discrete Probability Distributions Graphically** The probability distribution for a discrete random variable is composed of the values the variable can assume and the probabilities for each of the possible values. For example, if three parts are tested to determine whether they are defective, the probability distribution for the number of defectives might be

| $x$ = Number of Defectives | $P(x)$ |
|:---:|:---:|
| 0 | 0.10 |
| 1 | 0.30 |
| 2 | 0.40 |
| 3 | 0.20 |
| | $\sum = \overline{1.00}$ |

Graphically, the discrete probability distribution associated with these defectives can be represented by the areas of rectangles in which the base of each rectangle is one unit wide and the height corresponds to the probability. The areas of the rectangles sum to 1. Figure 5.1 illustrates two examples of discrete probability distributions. Figure 5.1(a) shows a discrete random variable with only three possible outcomes. Figure 5.1(b) shows the probability distribution for a discrete variable that has 21 possible outcomes. Note that as the number of possible outcomes increases, the distribution becomes smoother and the individual probability of any particular value tends to be reduced. In all cases, the sum of the probabilities is 1.

Discrete probability distributions have many applications in business decision-making situations. In the remainder of this section, we discuss several issues that are of particular importance to discrete probability distributions.

**FIGURE 5.1 Discrete Probability Distributions**

**(a) Discrete Probability Distribution (3 possible outcomes)**

**(b) Discrete Probability Distribution (21 possible outcomes)**

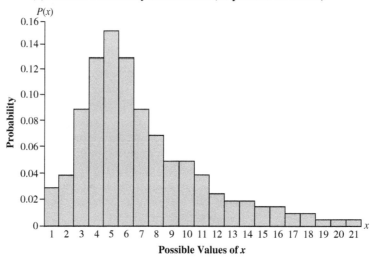

## Mean and Standard Deviation of Discrete Distributions

A probability distribution, like a frequency distribution, can be only partially described by a graph. To aid in a decision situation, you may need to calculate the distribution's *mean* and *standard deviation*. These values measure the central location and spread, respectively, of the probability distribution.

**OUTCOME 1**

**Expected Value**
The mean of a probability distribution. The average value when the experiment that generates values for the random variable is repeated over the long run.

**Calculating the Mean**   The mean of a discrete probability distribution is also called the **expected value** of the random variable from an experiment. The expected value is actually a *weighted average* of the random variable values in which the weights are the probabilities assigned to the values. The expected value is calculated using Equation 5.1.

**Expected Value of a Discrete Probability Distribution**

$$E(x) = \sum xP(x) \tag{5.1}$$

where:

$$E(x) = \text{Expected value of } x$$
$$x = \text{Values of the random variable}$$
$$P(x) = \text{Probability of the random variable taking on the value } x$$

**Calculating the Standard Deviation**   The standard deviation measures the spread, or dispersion, in a set of data. The standard deviation also measures the spread in the values of a random variable. To calculate the standard deviation for a discrete probability distribution, use Equation 5.2.

**Standard Deviation of a Discrete Probability Distribution**

$$\sigma_x = \sqrt{\sum [x - E(x)]^2 P(x)} \tag{5.2}$$

where:

$$x = \text{Values of the random variable}$$
$$E(x) = \text{Expected value of } x$$
$$P(x) = \text{Probability of the random variable taking on the value } x$$

An example will help you understand how to apply Equations 5.1 and 5.2.

**EXAMPLE 5-1**

### Computing the Mean and Standard Deviation of a Discrete Random Variable

**Swenson Security Sales**   Swenson Security Sales is a company that sells and installs home security systems throughout the eastern United States. Each week, the company's quality managers examine one randomly selected security installation in states where the company operates to see whether the installers made errors. The discrete random variable, $x$, is the number of errors discovered on each installation examined, ranging from 0 to 3. The following frequency distribution was developed after an examination of 400 security installations:

| $x$ | Frequency |
|---|---|
| 0 | 150 |
| 1 | 110 |
| 2 | 50 |
| 3 | 90 |
| $\sum$ = | 400 |

Assuming that these data reflect typical performance by the installers, the company leadership wishes to develop a discrete probability distribution for the number of install errors and compute the mean and standard deviation for the distribution. This can be done using the following steps:

**STEP 1** **Convert the frequency distribution into a probability distribution using the relative frequency assessment method.**

| x | Frequency |
|---|---|
| 0 | 150/400 = 0.375 |
| 1 | 110/400 = 0.275 |
| 2 | 50/400 = 0.125 |
| 3 | 90/400 = 0.225 |
| | $\Sigma$ = 1.000 |

**STEP 2** **Compute the expected value using Equation 5.1.**

$$E(x) = \sum xP(x)$$

We can calculate the expected value using the following table:

| x | P(x) | xP(x) |
|---|---|---|
| 0 | 0.375 | 0.000 |
| 1 | 0.275 | 0.275 |
| 2 | 0.125 | 0.250 |
| 3 | 0.225 | 0.675 |
| | Sum = | 1.200 |

$$E(x) = 1.20$$

The expected value is 1.20 errors per install. Thus, if we assume the distribution of the number of errors is representative of that of each individual installation, the long-run average number of errors per installation will be 1.20.

**STEP 3** **Compute the standard deviation using Equation 5.2.**

$$\sigma_x = \sqrt{\sum [x - E(x)]^2 P(x)}$$

| x | P(x) | [x − E(x)] | [x − E(x)]² | [x − E(x)]²P(x) |
|---|---|---|---|---|
| 0 | 0.375 | 0 − 1.2 = −1.20 | 1.44 | 0.540 |
| 1 | 0.275 | 1 − 1.2 = −0.20 | 0.04 | 0.011 |
| 2 | 0.125 | 2 − 1.2 = 0.80 | 0.64 | 0.080 |
| 3 | 0.225 | 3 − 1.2 = 1.80 | 3.24 | 0.729 |
| | | | | $\Sigma$ = 1.360 |

$$\sigma_x = \sqrt{1.36} = 1.17$$

The standard deviation of the discrete probability distribution is 1.17 errors per security system installed.

**TRY EXERCISE 5-4** (pg. 177)

---

**BUSINESS APPLICATION** **Expected Values**

**Morrison Batteries** Morrison Batteries imports batteries from Taiwan used in electric toothbrushes. For one particular product line, Morrison currently has two suppliers. Both suppliers have poor records when it comes to quality. Morrison is planning to purchase 100,000

**TABLE 5.1** Probability Distribution—Defect Rate for Supplier B

| Defect Rate x | Probability P(x) |
|---|---|
| 0.01 | 0.3 |
| 0.05 | 0.4 |
| 0.10 | 0.2 |
| 0.15 | 0.1 |

batteries and wants to use the least-cost supplier for the entire purchase. Supplier A is less expensive by $1.20 per fixture and has an ongoing record of supplying 10% defects. Supplier B is more expensive but may be a higher-quality supplier. Morrison records indicate that the rate of defects from Supplier B varies. Table 5.1 shows the probability distribution for the defect percentages for Supplier B. Each defect is thought to cost the company $9.50.

For Supplier A, at a defect rate of 0.10, out of 100,000 units the number of defects is expected to be 10,000. The cost of these is $9.50 × 10,000 = $95,000. For Supplier B, the expected defect rate is found using Equation 5.1 as follows:

$$E(\text{Defect rate}) = \sum x\,P(x)$$
$$E(\text{Defect rate}) = (0.01)(0.3) + (0.05)(0.4) + (0.10)(0.2) + (0.15)(0.1)$$
$$E(\text{Defect rate}) = 0.058$$

Thus, Supplier B is expected to supply 5.8% defects, or 5,800 out of the 100,000 units ordered, for an expected cost of $9.50 × 5,800 = $55,100. Based on defect cost alone, Supplier B is less expensive ($55,100 versus $95,000). However, recall that Supplier B's product sells for $1.20 per unit more. Thus, on a 100,000-unit order, Supplier B costs an extra $1.20 × 100,000 = $120,000 more than Supplier A. The relative costs are

Supplier A = $95,000     Supplier B = $55,100 + $120,000 = $175,100

Therefore, based on expected costs, Supplier A should be selected to supply the 100,000 light fixtures.

# 5.1 EXERCISES

## Skill Development

**5-1.** An economics quiz contains six multiple-choice questions. Let x represent the number of questions a student answers correctly.
  a. Is x a continuous or discrete random variable?
  b. What are the possible values of x?

**5-2.** Two numbers are randomly drawn without replacement from a list of five. If the five numbers are 2, 2, 4, 6, 8, what is the probability distribution of the sum of the two numbers selected? Show the probability distribution graphically.

**5-3.** If the Prudential Insurance Company surveys its customers to determine the number of children under age 22 living in each household,
  a. What is the random variable for this survey?
  b. Is the random variable discrete or continuous?

**5-4.** Given the following discrete probability distribution:

| x | P(x) |
|---|---|
| 50 | 0.375 |
| 65 | 0.15 |
| 70 | 0.225 |
| 75 | 0.05 |
| 90 | 0.20 |

  a. Calculate the expected value of x.
  b. Calculate the variance of x.
  c. Calculate the standard deviation of x.

**5-5.** Because of weather issues, the number of days next week that the captain of a charter fishing boat can leave port is uncertain. Let x = number of days the boat is able to leave port per week. The following probability distribution for the variable, x, was determined based on historical data when the weather was poor:

| x | P(x) |
|---|---|
| 0 | 0.05 |
| 1 | 0.10 |
| 2 | 0.10 |
| 3 | 0.20 |
| 4 | 0.20 |
| 5 | 0.15 |
| 6 | 0.15 |
| 7 | 0.05 |

Based on the probability distribution, what is the expected number of days per week the captain can leave port?

**5-6.** Consider the following discrete probability distribution:

| x | P(x) |
|---|---|
| 3 | 0.13 |
| 6 | 0.12 |
| 9 | 0.15 |
| 12 | 0.60 |

a. Calculate the variance and standard deviation of the random variable.

b. Let $y = x + 7$. Calculate the variance and standard deviation of the random variable $y$.

c. Let $z = 7x$. Calculate the variance and standard deviation of the random variable $z$.

d. From your calculations in part a and part b, indicate the effect that adding a constant to a random variable has on its variance and standard deviation.

e. From your calculations in part a and part c, indicate the effect that multiplying a random variable with a constant has on the variance and the standard deviation of the random variable.

5-7. Given the following discrete probability distribution:

| x | P(x) |
|---|------|
| 100 | 0.25 |
| 125 | 0.30 |
| 150 | 0.45 |

a. Calculate the expected value of $x$.
b. Calculate the variance of $x$.
c. Calculate the standard deviation of $x$.

5-8. Two six-sided dice are rolled and the random variable $x$ is the sum of the values produced by each die.

a. List the possible outcomes for the variable $x$, determine the probability for each outcome, and calculate the expected value of $x$.

b. Calculate the variance of $x$.

c. Calculate the standard deviation of $x$.

5-9. Consider the following discrete probability distribution:

| x | P(x) |
|---|------|
| 5 | 0.10 |
| 10 | 0.15 |
| 15 | 0.25 |
| 20 | 0.50 |

a. Calculate the expected value of the random variable.

b. Let $y = x + 5$. Calculate the expected value of the new random variable $y$.

c. Let $z = 5x$. Calculate the expected value of the new random variable $z$.

d. From your calculations in part a and part b, indicate the effect that adding a constant to a random variable has on the expected value of the random variable.

e. From your calculations in part a and part c, indicate the effect that multiplying a random variable by a constant has on the expected value of the random variable.

5-10. Examine the following probability distribution:

| x | 5 | 10 | 15 | 20 | 25 | 30 | 35 | 40 | 45 | 50 |
|---|---|----|----|----|----|----|----|----|----|----|
| P(x) | 0.01 | 0.05 | 0.14 | 0.20 | 0.30 | 0.15 | 0.05 | 0.04 | 0.01 | 0.05 |

a. Calculate the expected value and standard deviation for this random variable.

b. Denote the expected value as $\mu$. Calculate $\mu - \sigma$ and $\mu + \sigma$.

c. Determine the proportion of the distribution that is contained within the interval $\mu \pm \sigma$.

d. Repeat part c for $(1)\mu \pm 2\sigma$ and $(2)\mu \pm 3\sigma$.

## Business Applications

5-11. Assume the distribution of the number of people per household in the United States is shown in the following table:

| x | P(x) |
|---|------|
| 2 | 0.27 |
| 3 | 0.25 |
| 4 | 0.28 |
| 5 | 0.13 |
| 6 | 0.04 |
| 7 | 0.03 |

a. Calculate the expected number of people per household in the United States.

b. Compute the standard deviation of the number of people per household.

5-12. Using historical records, a manufacturing firm has developed the following probability distribution for the number of days required to get components from its suppliers. The distribution is shown here, where the random variable $x$ is the number of days.

| x | P(x) |
|---|------|
| 2 | 0.15 |
| 3 | 0.45 |
| 4 | 0.30 |
| 5 | 0.075 |
| 6 | 0.025 |

a. What is the average lead time for the component?

b. What is the coefficient of variation for delivery lead time?

c. How might the manufacturing firm use this information?

5-13. Marque Electronics is a family-owned electronics repair business. The owner has read an advertisement from a local competitor that guarantees all television repairs within four days. Based on his company's experience, he wants to know if he can offer a similar guarantee. His past service records are used to determine the following probability distribution:

| Number of Days | Probability |
|----------------|-------------|
| 1 | 0.15 |
| 2 | 0.25 |

| Number of Days | Probability |
|:---:|:---:|
| 3 | 0.30 |
| 4 | 0.18 |
| 5 | 0.12 |

a. Calculate the mean number of days his customers wait for a repair.
b. Calculate the variance and standard deviation.
c. Based on the calculations in parts a and b, what conclusion should the manager reach regarding his company's repair times?

**5-14.** Cramer's Bar and Grille in Dallas can seat 130 people at a time. The manager has been gathering data on the number of minutes a party of four spends in the restaurant from the moment they are seated to when they pay the check. What is the mean number of minutes for a dinner party of four? What are the variance and standard deviation?

| Number of Minutes | Probability |
|:---:|:---:|
| 60 | 0.05 |
| 70 | 0.15 |
| 80 | 0.20 |
| 90 | 0.45 |
| 100 | 0.10 |
| 110 | 0.05 |

**5-15.** Rossmore Brothers, Inc., sells plumbing supplies for commercial and residential applications. The company currently has only one supplier for a particular type of faucet. Based on historical data, the company has assessed the following probability distribution for the proportion of defective faucets it receives from this supplier:

| Proportion Defective | Probability |
|:---:|:---:|
| 0.01 | 0.4 |
| 0.02 | 0.3 |
| 0.05 | 0.2 |
| 0.10 | 0.1 |

This supplier charges Rossmore Brothers, Inc., $29.00 per unit for this faucet. Although the supplier will replace any defects free of charge, Rossmore managers figure the cost of dealing with the defects is about $5.00 each.

a. Assuming that Rossmore Brothers is planning to purchase 2,000 of these faucets from the supplier, what is the total expected cost to Rossmore Brothers for the deal?
b. Suppose that Rossmore Brothers has an opportunity to buy the same faucets from another supplier at a cost of $28.50 per unit. However, based on its investigations, Rossmore Brothers has assessed the following probability distribution for the proportion of defective faucets that will be delivered by the new supplier:

| Proportion Defective | Probability |
|:---:|:---:|
| 0.01 | 0.1 |
| 0.02 | 0.1 |
| 0.05 | 0.7 |
| 0.10 | 0.1 |

Assuming that the defect cost is still $5.00 each and based on the total expected cost for an order of 2,000 faucets, should Rossmore buy from the new supplier or stick with its original supplier?

**5-16.** A consumer electronics store stocks four alarm clock radios. If it has fewer than four clock radios available at the end of a week, the store restocks the item to bring the in-stock level up to four. If weekly demand is greater than the four units in stock, the store loses sales. The radio sells for $25 and costs the store $15. The manager estimates that the probability distribution of weekly demand for the radio is as follows:

| Weekly Demand | Probability |
|:---:|:---:|
| 0 | 0.05 |
| 1 | 0.05 |
| 2 | 0.10 |
| 3 | 0.20 |
| 4 | 0.40 |
| 5 | 0.10 |
| 6 | 0.05 |
| 7 | 0.05 |

a. What is the expected weekly demand for the alarm clock radio?
b. What is the probability that weekly demand will be greater than the number of available radios?
c. What is the expected weekly profit from the sale of the alarm clock radio? (Remember: There are only four clock radios available in any week to meet demand.)
d. On average, how much profit is lost each week because the radio is not available when demanded?

**5-17.** Fiero Products, LTD, of Bologna, Italy, makes a variety of footwear, including indoor slippers, children's shoes, and flip-flops. To keep up with increasing demand, it is considering three expansion plans: (1) a small factory with yearly fixed costs of $150,000 that will increase the production of flip-flops to 400,000; (2) a mid-sized factory with yearly fixed costs of $250,000 that will increase the production of flip-flops to 600,000; and (3) a large factory with yearly fixed costs of $350,000 that will increase the production of flip-flops to 900,000. The profit per flip-flop is projected to be $0.75. The probability distribution of demand for flip-flops is considered to be

| Demand | 300,000 | 700,000 | 900,000 |
|:---|:---:|:---:|:---:|
| Probability | 0.2 | 0.5 | 0.3 |

a. Compute the expected profit for each of the expansion plans. Assume that the company will produce to its capacity regardless of demand.

b. Calculate the standard deviation for each of the expansion plans.

c. Which expansion plan would you recommend? Provide the statistical reasoning behind your selection.

**5-18.** A large corporation in search of a CEO and a CFO has narrowed the fields for each position to a short list. The CEO candidates graduated from Chicago (C) and three Ivy League universities: Harvard (H), Princeton (P), and Yale (Y). The four CFO candidates graduated from MIT (M), Northwestern (N), and two Ivy League universities: Dartmouth (D) and Brown (B). The personnel director wishes to determine the distribution of the number of Ivy League graduates who could fill these positions.

a. Assume the selections were made randomly. Construct the probability distribution of the number of Ivy League graduates who could fill these positions.

b. Would it be surprising if both positions were filled with Ivy League graduates?

c. Calculate the expected value and standard deviation of the number of Ivy League graduates who could fill these positions.

## Computer Software Exercises

**5-19.** A national retailer has entered into an agreement with a publisher to begin selling a food and beverage magazine on a trial basis. The retailer pays $1.95 and sells the magazine for $3.49. During the trial period, the retailer placed 10 copies of the magazine in each of 150 stores throughout the country. The file titled **Sold** contains the number of magazines sold in each of the stores.

a. Produce a frequency distribution for these data. Convert the frequency distribution into a probability distribution using the relative frequency assessment method.

b. Calculate the expected profit from the sale of these 10 magazines.

c. The retailer is negotiating returning all unsold magazines for a salvage price. Determine the salvage price that will be needed to yield a positive profit.

**5-20.** Pfizer Inc. is the manufacturer of Revolution (Selamectin), a topical parasiticide used for the treatment, control, and prevention of flea infestation, heartworm, and ear mites for dogs and cats. One of its selling points is that it provides protection for an entire month. The file titled **Fleafree** contains data similar to those obtained in Pfizer's research. It presents the number of days Revolution could remain effective when applied to mature cats.

a. Produce a frequency distribution for these data. Convert the frequency distribution into a probability distribution using the relative frequency assessment method.

b. Calculate the expected value and standard deviation for the number of days Revolution could remain effective.

c. If the marketing department wished to advertise the number of days that 90% of the cats remain protected while using Revolution, what would this number of days be?

**5-21.** Fiber Fabrics, Inc., makes boat tops for a number of boat manufacturers. Its fabric has a limited two-year warranty. Periodic testing is done to determine if the warranty policy should be changed. One such study examined those covers that became unserviceable while still under warranty. Data produced by the study are contained in the file titled **Covers**. The data represent assessment of the number of months a cover was used until it became unserviceable.

a. Produce a frequency distribution for these data. Convert the frequency distribution into a probability distribution using the relative frequency assessment method.

b. Calculate the expected value and standard deviation for the time until the covers became unserviceable.

c. The quality-control department thinks that among those covers that do become unserviceable while still under warranty, the majority last longer than 19 months. Produce the relevant statistic to verify this assumption.

| 5.2 | **The Binomial Probability Distribution** |

Several theoretical discrete distributions have extensive applications in business decision making. A probability distribution is called *theoretical* when the mathematical properties of its random variable are used to produce its probabilities. Such distributions are different from the distributions that are obtained subjectively or from observation. Sections 5.2 and 5.3 focus on theoretical discrete probability distributions. Chapter 6 will introduce important theoretical continuous probability distributions.

**OUTCOME 2**

**Binomial Distribution Characteristics**
A distribution that gives the probability of $x$ successes in $n$ trials in a process that meets the following conditions:

1. A trial has only two possible outcomes: a success or a failure.
2. There is a fixed number, $n$, of identical trials.
3. The trials of the experiment are independent of each other. This means that if one outcome is a success, this does not influence the chance of another outcome being a success.
4. The process must be consistent in generating successes and failures. That is, the probability, $p$, associated with a success remains constant from trial to trial.
5. If $p$ represents the probability of a success, then $(1 - p) = q$ is the probability of a failure.

# The Binomial Distribution

The first theoretical probability distribution we will consider is the **binomial distribution** that describes processes whose trials have only two possible outcomes. The physical events described by this type of process are widespread. For instance, a quality-control system in a manufacturing plant labels each tested item as either defective or acceptable. A firm bidding for a contract either will or will not get the contract. A marketing research firm may receive responses to a questionnaire in the form of "Yes, I will buy" or "No, I will not buy." The personnel manager in an organization is faced with two possible outcomes each time he offers a job—the applicant either accepts the offer or rejects it.

# Characteristics of the Binomial Distribution

The binomial distribution requires that the experiment's trials be independent. This can be assured if the sampling is performed with replacement from a finite population. This means that an item is sampled from a population and returned to the population, after its characteristic(s) have been recorded, before the next item is sampled. However, sampling with replacement is the exception rather than the rule in business applications. Most often, the sampling is performed without replacement. Strictly speaking, when sampling is performed without replacement, the conditions for the binomial distribution cannot be satisfied. However, the conditions are approximately satisfied if the sample selected is quite small relative to the size of the population from which the sample is selected. A commonly used rule of thumb is that the binomial distribution can be applied if the sample size is at most 5% of the population size.

---

**BUSINESS APPLICATION**   ## Using the Binomial Distribution

**Lake City Automotive**   Lake City Automotive performs 300 automobile transmission repairs every week. Transmission repair work is not an exact science, and a percentage of the work done must be reworked. Lake City shop managers have determined that even when the mechanics perform the work in a proper manner, 10% of the time the car will have to be worked on again to fix a problem. The binomial distribution applies to this situation because the following conditions exist:

1. There are only two possible outcomes when a car is repaired: either it needs rework or it doesn't. We are interested in the cars that need rework, so we will consider a reworked car to be a "success." A success occurs when we observe the outcome of interest.
2. All repair jobs are considered to be equivalent in the work required.
3. The outcome of a repair (rework or no rework) is independent of whether the preceding repair required rework or not.
4. The probability of rework being needed, $p = 0.10$, remains constant from car to car.
5. The probability of a car not needing rework, $q = 1 - p = 0.90$, remains constant from car to car.

To determine whether the Lake City mechanics are continuing to function at the standard level of performance, the shop supervisor randomly selects four cars from the week's list of repaired vehicles and tracks those to see whether they need rework or not (note that the need for rework is known within a few hours of the work being completed). Because the sample size is small relative to the size of the population ($4/300 = 0.0133$ or $1.33\%$), the conditions of independence and constant probability will be approximately satisfied because the sample is less than 5% of the population.

The number of reworked cars is the random variable of interest. The number of reworked units is limited to discrete values, $x = 0, 1, 2, 3$, or $4$. One way to determine the probability that the random variable will have any of the discrete values is to list the sample space, as shown in Table 5.2. We can find the probability of zero cars needing rework, for instance, by employing the Multiplication Rule for Independent Events:

$$P(x = 0 \text{ reworks}) = P(G \text{ and } G \text{ and } G \text{ and } G)$$

**TABLE 5.2 Sample Space for Lake City Automotive**

| Results | No. of Reworked Cars | No. of Ways |
|---------|---------------------|-------------|
| G,G,G,G | 0 | 1 |
| G,G,G,D | 1 | |
| G,G,D,G | 1 | 4 |
| G,D,G,G | 1 | |
| D,G,G,G | 1 | |
| G,G,D,D | 2 | |
| G,D,G,D | 2 | |
| D,G,G,D | 2 | 6 |
| G,D,D,G | 2 | |
| D,G,D,G | 2 | |
| D,D,G,G | 2 | |
| D,D,D,G | 3 | |
| D,D,G,D | 3 | 4 |
| D,G,D,D | 3 | |
| G,D,D,D | 3 | |
| D,D,D,D | 4 | 1 |

where:

$$G = \text{Car does not require rework}$$

Here,

$$P(G) = 0.90$$

and we have assumed the repair jobs are independent. Using the Multiplication Rule for Independent Events introduced in Chapter 4 (Rule 9),

$$P(G \text{ and } G \text{ and } G \text{ and } G) = P(G)P(G)P(G)P(G) = (0.90)(0.90)(0.90)(0.90)$$
$$= 0.90^4$$
$$= 0.6561$$

We can also find the probability of exactly one reworked car in a sample of four. This is accomplished using both the Multiplication Rule for Independent Events and the Addition Rule for Mutually Exclusive Events, which was also introduced in Chapter 4 (Rule 5):

$$D = \text{Car needs rework}$$
$$P(x = 1 \text{ rework}) = P(G \text{ and } G \text{ and } G \text{ and } D) + P(G \text{ and } G \text{ and } D \text{ and } G)$$
$$+ P(G \text{ and } D \text{ and } G \text{ and } G) + P(D \text{ and } G \text{ and } G \text{ and } G)$$

where:

$$P(G \text{ and } G \text{ and } G \text{ and } D) = P(G)P(G)P(G)P(D) = (0.90)(0.90)(0.90)(0.10)$$
$$= (0.90^3)(0.10)$$

Likewise:

$$P(G \text{ and } G \text{ and } D \text{ and } G) = (0.90^3)(0.10)$$
$$P(G \text{ and } D \text{ and } G \text{ and } G) = (0.90^3)(0.10)$$
$$P(D \text{ and } G \text{ and } G \text{ and } G) = (0.90^3)(0.10)$$

Then:

$$P(1 \text{ rework}) = (0.90^3)(0.10) + (0.90^3)(0.10) + (0.90^3)(0.10) + (0.90^3)(0.10)$$
$$= (4)(0.90^3)(0.10)$$
$$= 0.2916$$

Note that each of the four possible ways of finding one reworked car has the same probability $[(0.90^3)(0.10)]$. We determine the probability of one of the ways to obtain one reworked car and multiply this value by the number of ways (four) of obtaining one reworked car. This produces the overall probability of one reworked car.

**Combinations**   In this relatively simple application, we can fairly easily list the sample space and from that count the number of ways that each possible number of reworked cars can occur. However, for examples with larger sample sizes, this approach is inefficient. A more effective method for counting the number of ways binomial events can occur is called the *counting rule for combinations*. This rule is used to find the number of outcomes from an experiment in which $x$ objects are to be selected from a group of $n$ objects. Equation 5.3 is used to find the combinations.

**Counting Rule for Combinations**

$$C_x^n = \frac{n!}{x!(n-x)!}$$  **(5.3)**

where:

$C_x^n$ = Number of combinations of $x$ objects selected from $n$ objects
$n! = n(n-1)(n-2) \ldots (2)(1)$
$0! = 1$ by definition

The Excel 2016 function for combinations is

= **COMBIN(number, number chosen)**
= **COMBIN(4,2)**

Using Equation 5.3, we find the number of ways that $x = 2$ reworked cars can occur in a sample of $n = 4$ as

$$C_x^n = \frac{n!}{x!(n-x)!} = \frac{4!}{2!(4-2)!} = \frac{(4)(3)(2)(1)}{(2)(1)(2)(1)} = \frac{24}{4} = 6 \text{ ways}$$

Refer to Table 5.2 to see that this is the same value for two reworked cars in a sample of four that was obtained by listing the sample space.

Now we can find the probabilities of two reworked cars:

$$P(2 \text{ reworks}) = (6)(0.90^2)(0.10^2)$$
$$= 0.0486$$

Use this method to verify the following:

$$P(3 \text{ reworks}) = (4)(0.90)(0.10^3)$$
$$= 0.0036$$
$$P(4 \text{ reworks}) = (1)(0.10^4)$$
$$= 0.0001$$

**TABLE 5.3** Binomial Distribution for Lake City Automotive: $n = 4, p = 0.10$

| $x$ = # of Reworks | $P(x)$ |
|---|---|
| 0 | 0.6561 |
| 1 | 0.2916 |
| 2 | 0.0486 |
| 3 | 0.0036 |
| 4 | 0.0001 |
| $\Sigma$ = | 1.0000 |

The key to developing the probability distribution for a binomial process is first to determine the probability of any one way the event of interest can occur and then to multiply this probability by the number of ways that event can occur. Table 5.3 shows the binomial probability distribution for the number of reworked cars in a sample size of four when the probability of any individual car requiring rework is 0.10. The probability distribution is graphed in Figure 5.2. Most samples would contain zero or one reworked car when the mechanics at Lake City are performing the work to standard.

**FIGURE 5.2** Binomial Distribution for Lake City Automotive

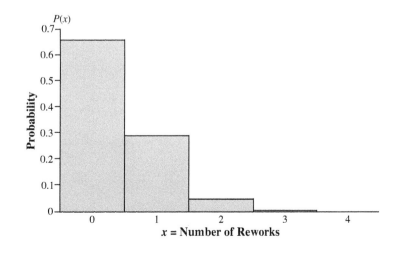

$x$ = Number of Reworks

**Binomial Formula**    The steps that we have taken to develop this binomial probability distribution can be summarized through a formula called the *binomial formula*, shown as Equation 5.4. Note that this formula is composed of two parts: the combinations of $x$ items selected from $n$ items and the probability that $x$ items can occur.

**Binomial Formula**

$$P(x) = \frac{n!}{x!(n-x)!} p^x q^{n-x} \tag{5.4}$$

where:

$$n = \text{Random sample size}$$
$$x = \text{Number of successes (when a success is defined as what we are looking for)}$$
$$n - x = \text{Number of failures}$$
$$p = \text{Probability of a success}$$
$$q = 1 - p = \text{Probability of a failure}$$
$$n! = n(n-1)(n-2)(n-3)\ldots(2)(1)$$
$$0! = 1 \text{ by definition}$$

Applying Equation 5.4 to the Lake City Automotive example for $n = 4, p = 0.10$, and $x = 2$ reworked cars, we get

$$P(x) = \frac{n!}{x!(n-x)!} p^x q^{n-x}$$

$$P(2) = \frac{4!}{2!2!}(0.10^2)(0.90^2) = 6(0.10^2)(0.90^2) = 0.0486$$

> The Excel 2016 function for the binomial distribution formula is
>
> =BINOM.DIST(x,n,p,False)
> =BINOM.DIST(2,4,.10,False)

This is the same value we calculated earlier when we listed out the sample space above.

---

**EXAMPLE 5-2**    **Using the Binomial Formula**

**Creative Style and Cut**    Creative Style and Cut, an upscale beauty salon in San Francisco, offers a full refund to anyone who is not satisfied with the way his or her hair looks after it has been cut and styled. The owners believe the hairstyle satisfaction from customer to customer is independent and that the probability a customer will ask for a refund is 0.20. Suppose a random sample of six customers is observed. In four instances, the customer has asked for a refund. The owners might be interested in the probability of four refund requests from six customers. If the binomial distribution applies, the probability can be found using the following steps:

STEP 1    **Define the characteristics of the binomial distribution.**
In this case, the characteristics are

$$n = 6, \qquad p = 0.20, \qquad q = 1 - p = 0.80$$

STEP 2    **Determine the probability of $x$ successes in $n$ trials using the binomial formula, Equation 5.4.**
In this case, $n = 6, p = 0.20, q = 0.80$, and we are interested in the probability of $x = 4$ successes.

$$P(x) = \frac{n!}{x!(n-x)!} p^x q^{n-x}$$

> The Excel 2016 function for the binomial distribution formula is
>
> =BINOM.DIST(x,n,p,False)
> =BINOM.DIST(4,6,.20,False)

$$P(4) = \frac{6!}{4!(6-4)!}(0.20^4)(0.80^{6-4})$$
$$P(4) = 15(0.20^4)(0.80^2)$$
$$P(4) = 0.0154$$

There is only a 0.0154 chance that exactly four customers will want a refund in a sample of six if the chance that any one of the customers will want a refund is 0.20.

TRY EXERCISE 5-24 (pg. 189)

**Using the Binomial Distribution Table**   Using Equation 5.4 to develop the binomial distribution is not difficult, but it can be time consuming. To make binomial probabilities easier to find, you can use the binomial table in Appendix B. This table is constructed to give *cumulative* probabilities for different sample sizes and probabilities of success. Each column is headed by a probability, *p*, which is the probability associated with a success. The column headings correspond to probabilities of success ranging from 0.01 to 1.00. Down the left side of the table are integer values that correspond to the number of successes, *x*, for the specified sample size, *n*. The values in the body of the table are the cumulative probabilities of *x* or fewer successes in a random sample of size *n*.

---

**BUSINESS APPLICATION**    Binomial Distribution Table

**Big O Tire Company**   Big O Tire Company operates tire and repair stores throughout the western United States. Upper management is considering offering a money-back warranty if one of their tires fails in the first 20,000 miles of use. The managers are willing to make this warranty if 2% or fewer of the tires they sell fail within the 20,000-mile limit.

The company plans to test 10 randomly selected tires over 20,000 miles of use in test conditions. The number of possible failures will be $x = 0, 1, 2, \ldots, 10$. We can use the binomial table in Appendix B to develop the probability distribution. This table is called a *cumulative* probability table. Go to the table for $n = 10$ and the column for $p = 0.02$. The values of *x* are listed down the left side of the table. For example, the cumulative probability of $x \leq 2$ occurrences is 0.9991. This means that it is extremely likely that 2 or fewer tires in a sample of 10 would fail in the first 20,000 miles of use if the overall fraction of tires that will fail is 0.02. The probability of 3 or more failures in the sample of $n = 10$ is

$$P(x \geq 3) = 1 - P(x \leq 2) = 1 - 0.9991 = 0.0009$$

There are about 9 chances in 10,000 that 3 or more tires in a sample of 10 will fail if the probability of a failure for any one tire is $p = 0.02$. If the test did show that 3 tires failed, the true rate of tire failures *likely* exceeds 2%, and the company should have serious doubts about making the warranty.

---

**EXAMPLE 5-3**    Using the Binomial Table

**Television Ratings**   On Monday, February 1, 2015, the day after the 2015 Super Bowl between the Seahawks and the Patriots, SB Nation reported that the game was the most-watched program in history, with over 45% of all people with TVs in the United States tuned to the game on Sunday (*source:* Adam Stites, "Super Bowl 2015 ratings: Patriots, Seahawks duel sets TV ratings record," www.sbnation.com, Feb. 2, 2015). Assuming that the 45% rating is correct, what is the probability that in a random sample of 20 people, 2 or fewer would have been tuned to the Super Bowl? This question can be answered, assuming that the binomial distribution applies, using the following steps:

STEP 1   **Define the characteristics of the binomial distribution.**
In this case, the characteristics are

$$n = 20, \qquad p = 0.45 \qquad q = 1 - p = 0.55$$

The Excel 2016 function for the binomial distribution formula is

**= BINOM.DIST(x,n,p,True)**
**= BINOM.DIST(2,20,.45,True)**

*Note:* The "True" argument returns the cumulative probability of x or fewer successes.

STEP **2** **Define the event of interest.**
We are interested in knowing

$$P(x \le 2) = P(0) + P(1) + P(2)$$

STEP **3** **Go to the binomial table in Appendix B to find the desired probability.**
In this case, we locate the section of the table corresponding to sample size equal to $n = 20$ and go to the column headed $p = 0.45$ and the row labeled $x = 2$. The cumulative probability $P(x \le 2)$ listed in the table is 0.0009.

Thus, there is only a 0.0009 chance that 2 or fewer people in a random sample of 20 were tuned to the Super Bowl. Thus, it is unlikely that 2 or fewer in a sample of 20 would have been tuned to the Super Bowl in 2015.

**TRY EXERCISE 5-28** (pg. 190)

---

**EXAMPLE 5-4** | **Using the Binomial Distribution**

**Anderson Research** Anderson Research is a full-service marketing research consulting firm. Recently it was retained to do a project for a major U.S. airline. The airline was considering changing from an assigned-seating reservation system to one in which fliers would be able to take any seat they wished on a first-come, first-served basis. The airline believes that 80% of its fliers would like this change if it was accompanied with a reduction in ticket prices. Anderson Research will survey a large number of customers on this issue, but prior to conducting the full research, it has selected a random sample of 20 customers and determined that 12 like the proposed change. What is the probability of finding 12 or fewer who like the change if the probability is 0.80 that a customer will like the change?

If we assume the binomial distribution applies, we can use the following steps to answer this question:

STEP **1** **Define the characteristics of the binomial distribution.**
In this case, the characteristics are

$$n = 20, \quad p = 0.80, \quad q = 1 - p = 0.20$$

STEP **2** **Define the event of interest.**
We are interested in knowing

$$P(x \le 12)$$

The Excel 2016 function for the binomial distribution formula is

**= BINOM.DIST(x,n,p,True)**
**= BINOM.DIST(12,20,.80,True)**

STEP **3** **Go to the binomial table in Appendix B to find the desired probability.**
Locate the table for the sample size, $n$. Locate the column for $p = 0.80$. Go to the row corresponding to $x = 12$ and the column for $p = 0.80$ in the section of the table for $n = 20$ to find

$$P(x \le 12) = 0.0321$$

Thus, it is quite unlikely that if 80% of customers like the new seating plan, 12 or fewer in a sample of 20 would like it. The airline may want to rethink its plan.

**TRY EXERCISE 5-29** (pg. 190)

**Mean and Standard Deviation of the Binomial Distribution** In Section 5.1, we stated that the mean of a discrete probability distribution is also referred to as the *expected value*. The expected value of a discrete random variable, $x$, is found using Equation 5.1:

$$\mu_x = E(x) = \sum x P(x)$$

**Mean of a Binomial Distribution** This equation for the expected value can be used with any discrete probability distribution, including the binomial. However, if we are working with a binomial distribution, the expected value can be found more easily by using Equation 5.5.

### Expected Value of a Binomial Distribution

$$\mu_x = E(x) = np \tag{5.5}$$

where:

$$n = \text{Sample size}$$
$$p = \text{Probability of a success}$$

---

**BUSINESS APPLICATION** **Binomial Distribution**

Excel Tutorial

**Catalog Sales**   Catalog sales have been a part of the U.S. economy for many years, and companies such as Lands' End, L.L. Bean, and Eddie Bauer have enjoyed increased business. One feature that has made mail-order buying so popular is the ease with which customers can return merchandise. Nevertheless, one mail-order catalog has the goal of no more than 11% of all purchased items returned.

The binomial distribution can describe the number of items returned. For instance, in a given hour, the company shipped 300 items. If the probability of an item being returned is $p = 0.11$, the expected number of items (mean) to be returned is

$$\mu_x = E(x) = np$$
$$\mu_x = E(x) = (300)(0.11) = 33$$

Thus, the average number of returned items for each 300 items shipped is 33.

Suppose the company sales manager wants to know if the return rate is stable at 11%. To test this, she monitors a random sample of 300 items and finds that 44 have been returned. This return rate exceeds the mean of 33 units, which concerns her. However, before reaching a conclusion, she will be interested in the probability of observing 44 or more returns in a sample of 300:

$$P(x \geq 44) = 1 - P(x \leq 43)$$

The binomial table in Appendix B does not contain sample sizes as large as 300. Instead, we can use Excel's BINOM.DIST function to find the probability. The Excel output in Figure 5.3 shows the cumulative probability of 43 or fewer is equal to

$$P(x \leq 43) = 0.97$$

Then the probability of 44 or more returns is

$$P(x \geq 44) = 1 - 0.97 = 0.03$$

There is only a 3% chance of 44 or more items being returned if the 11% return rate is still in effect. This small probability suggests that the return rate may have increased above 11%, because we would not expect to see 44 returned items.

**FIGURE 5.3** **Excel 2016 Binomial Distribution Output for Catalog Sales**

---

**EXAMPLE 5-5**   **Finding the Mean of the Binomial Distribution**

**Anderson Research** (*continued*)   In Example 5-4, Anderson Research was hired to do a study for a major airline that is planning to change from a designated-seat assignment plan to an open-seating system. The company believes that 80% of its customers approve of the idea. Anderson Research interviewed a sample of $n = 20$ and found 12 who like the proposed change. If the airline is correct in its assessment of the probability, what is the expected number of people in a sample of $n = 20$ who will like the change? We can find this using the following steps:

STEP 1   **Define the characteristics of the binomial distribution.**
In this case, the characteristics are

$$n = 20, \quad p = 0.80, \quad q = 1 - p = 0.20$$

STEP 2   **Use Equation 5.5 to find the expected value.**

$$\mu_x = E(x) = np$$
$$E(x) = 20(0.80) = 16$$

The average number who would say they like the proposed change is 16 in a sample of 20.

**TRY EXERCISE 5-33a** (pg. 190)

**Standard Deviation of a Binomial Distribution**   The standard deviation for any discrete probability distribution can be calculated using Equation 5.2. We show this again as

$$\sigma_x = \sqrt{\sum [x - E(x)]^2 P(x)}$$

If a discrete probability distribution meets the binomial distribution conditions, the standard deviation is more easily computed by Equation 5.6.

**Standard Deviation of the Binomial Distribution**

$$\sigma = \sqrt{npq} \tag{5.6}$$

where:

$n$ = Sample size
$p$ = Probability of a success
$q = 1 - p$ = Probability of a failure

---

**EXAMPLE 5-6**   **Finding the Standard Deviation of a Binomial Distribution**

**Anderson Research** (*continued*)   Refer to Examples 5-4 and 5-5, in which Anderson Research surveyed a sample of $n = 20$ airline customers about changing the way seats are assigned on flights. The airline believes that 80% of its customers approve of the proposed change. Example 5-5 showed that if the airline is correct in its assessment, the expected number in a sample of 20 who would like the change is 16. However, there are other possible outcomes if 20 customers are surveyed. What is the standard deviation of the random variable, $x$, in this case? We can find the standard deviation for the binomial distribution using the following steps:

STEP 1   **Define the characteristics of the binomial distribution.**
In this case, the characteristics are

$$n = 20, \quad p = 0.80, \quad q = 1 - p = 0.20$$

STEP 2    **Use Equation 5.6 to calculate the standard deviation.**

$$\sigma = \sqrt{npq} = \sqrt{20(0.80)(0.20)} = 1.7889$$

TRY EXERCISE 5-33b (pg. 190)

**Additional Information about the Binomial Distribution**   At this point, several comments about the binomial distribution are worth making. If $p$, the probability of a success, is 0.50, the binomial distribution is *symmetric* and bell-shaped, regardless of the sample size. This is illustrated in Figure 5.4, which shows frequency histograms for samples of $n = 5, n = 10$, and $n = 50$. Notice that all three distributions are centered at the expected value, $E(x) = np$.

When the value of $p$ differs from 0.50 in either direction, the binomial distribution is skewed. The skewness will be most pronounced when $n$ is small and $p$ approaches 0 or 1. However, the binomial distribution becomes more symmetric and bell-shaped as $n$ increases. The frequency histograms shown in Figure 5.5 bear this out.

**FIGURE 5.4  The Binomial Distribution with Varying Sample Sizes ($p = 0.50$)**

(a) **Number of Successes ($x$)**

(b) **Number of Successes ($x$)**

(c) **Number of Successes ($x$)**

**FIGURE 5.5  The Binomial Distribution with Varying Sample Sizes ($p = 0.05$)**

(a) **Number of Successes ($x$)**

(b) **Number of Successes ($x$)**

(c) **Number of Successes ($x$)**

## 5.2 EXERCISES

### Skill Development

**5-22.** Suppose that 20% of credit cards have an outstanding balance at the credit card limit. A bank manager randomly selects 15 customers and finds 4 that have balances at the limit. Assume that the properties of the binomial distribution apply.
  a. What is the probability of finding 4 customers in a sample of 15 who have "maxed out" their credit cards?
  b. What is the probability that 4 or fewer customers in the sample will have balances at the limit of the credit card?

**5-23.** For a binomial distribution with a sample size equal to 10 and a probability of a success equal to 0.30, what is the probability that the sample will contain exactly three successes? Use the binomial formula to determine the probability.

**5-24.** Use the binomial formula to calculate the following probabilities for an experiment in which $n = 5$ and $p = 0.4$:
  a. the probability that $x$ is at most 1
  b. the probability that $x$ is at least 4
  c. the probability that $x$ is less than 1

**5-25.** If a binomial distribution applies with a sample size of $n = 20$, find
   a. the probability of 5 successes if the probability of a success is 0.40
   b. the probability of at least 7 successes if the probability of a success is 0.25
   c. the expected value, $n = 20, p = 0.20$
   d. the standard deviation, $n = 20, p = 0.20$

**5-26.** Suppose that 40% of all home buyers will do some remodeling to their home within the first five years of home ownership. Assuming this is true, use the binomial distribution to determine the probability that in a random sample of 20 homeowners, 2 or fewer will remodel their homes. Use the binomial table.

**5-27.** Find the probability of exactly 5 successes in a sample of $n = 10$ when the probability of a success is 0.70.

**5-28.** Assuming the binomial distribution applies with a sample size of $n = 15$, find
   a. the probability of 5 or more successes if the probability of a success is 0.30
   b. the probability of fewer than 4 successes if the probability of a success is 0.75
   c. the expected value of the random variable if the probability of success is 0.40
   d. the standard deviation of the random variable if the probability of success is 0.40

**5-29.** A random variable follows a binomial distribution with a probability of success equal to 0.65. For a sample size of $n = 7$, find
   a. the probability of exactly 3 successes
   b. the probability of 4 or more successes
   c. the probability of exactly 7 successes
   d. the expected value of the random variable

**5-30.** A random variable follows a binomial distribution with a probability of success equal to 0.45. For $n = 11$, find
   a. the probability of exactly 1 success
   b. the probability of 4 or fewer successes
   c. the probability of at least 8 successes

**5-31.** Use the binomial distribution table to determine the following probabilities:
   a. $n = 6, p = 0.08$; find $P(x = 2)$
   b. $n = 9, p = 0.80$; determine $P(x < 4)$
   c. $n = 11, p = 0.65$; calculate $P(2 < x \le 5)$
   d. $n = 14, p = 0.95$; find $P(x \ge 13)$
   e. $n = 20, p = 0.50$; compute $P(x > 3)$

**5-32.** Use the binomial distribution in which $n = 6$ and $p = 0.3$ to calculate the following probabilities:
   a. $x$ is at most 1
   b. $x$ is at least 2
   c. $x$ is more than 5
   d. $x$ is less than 6

**5-33.** Given a binomial distribution with $n = 8$ and $p = 0.40$, obtain the following:
   a. the mean
   b. the standard deviation
   c. the probability that the number of successes is larger than the mean

   d. the probability that the number of successes is within $\pm 2$ standard deviations of the mean

## Business Applications

**5-34.** Magic Valley Memorial Hospital administrators have recently received an internal audit report that indicates that 15% of all patient bills contain an error of one form or another. After spending considerable effort to improve the hospital's billing process, the administrators are convinced that things have improved. They believe that the new error rate is somewhere closer to 0.05.
   a. Suppose that recently the hospital randomly sampled 10 patient bills and conducted a thorough study to determine whether an error exists. It found 3 bills with errors. Assuming that managers are correct that they have improved the error rate to 0.05, what is the probability that they would find 3 or more errors in a sample of 10 bills?
   b. Referring to part a, what conclusion would you reach based on the probability of finding 3 or more errors in the sample of 10 bills?

**5-35.** According to NBCNews.com, 38% of the voters in California voted for Mitt Romney, the Republican candidate for President in 2012 (*source:* "Presidential election results," http://nbcnews.com, updated Feb. 9, 2016). A small startup company in San Jose, California, has ten employees.
   a. How many of the employees would you expect to have voted for Romney?
   b. All of the employees indicated that they voted for Obama. Determine the probability of this assuming they are representative of all California voters.
   c. Eight of the employees voted for Obama. Determine the probability that at least 8 of the employees would vote for Obama if they are representative of all California voters.
   d. Based on your calculations in parts b and c, do the employees reflect the California trend? Support your answer with statistical calculations and reasoning.

**5-36.** Hewlett-Packard receives large shipments of microprocessors from Intel Corp. It must try to ensure the proportion of microprocessors that are defective is small. Suppose HP decides to test five microprocessors out of a shipment of thousands of these microprocessors. Suppose that if at least one of the microprocessors is defective, the shipment is returned.
   a. If Intel Corp.'s shipment contains 10% defective microprocessors, calculate the probability the entire shipment will be returned.
   b. If Intel and HP agree that Intel will not provide more than 5% defective chips, calculate the probability that the entire shipment will be returned even though only 5% are defective.
   c. Calculate the probability that the entire shipment will be kept by HP even though the shipment has 10% defective microprocessors.

**5-37.** Consider the case of Dragon Wood Furniture in Bismarck, North Dakota, which believes that when the drilling process is operating at an acceptable rate, the upper limit on the percent of incorrectly drilled dowel holes is 4%. To monitor its drilling process, Dragon Wood Furniture randomly samples 20 products each hour and determines if the dowel hole in each product is correctly drilled or not. If, in the sample of 20 holes, Dragon Wood discovers 1 or more incorrectly drilled holes, the production process is stopped and the drilling process is recalibrated.

a. If the process is really operating correctly ($p = 0.04$), what is the probability that the sampling effort will produce $x = 0$ defective holes and thus the process will properly be left to continue running?

b. Suppose the true defect rate has risen to 0.10. What is the probability the sample will produce results that properly tell the managers to halt production to recalibrate the drilling machine?

c. Prepare a short letter to the manufacturing manager at Dragon Wood Furniture discussing the effectiveness of the sampling process that her company is using. Base your response on the results to parts a and b.

**5-38.** Assume that for an ad campaign to be successful, at least 80% of those seeing a television commercial must be able to recall the name of the company featured in the commercial one hour after viewing the commercial. Before distributing an ad campaign nationally, an advertising company plans to show the commercial to a random sample of 20 people. It will also show the same people two additional commercials for different products or businesses.

a. Assuming that the advertisement will be successful (80% will be able to recall the name of the company in the ad), what is the expected number of people in the sample who will recall the company featured in the commercial one hour after viewing the three commercials?

b. Suppose that in the sample of 20 people, 11 were able to recall the name of the company in the commercial one hour after viewing. Based on the premise that the advertising campaign will be successful, what is the probability of 11 or fewer people being able to recall the company name?

c. Based on your responses to parts a and b, what conclusion might the advertising executives make about this particular advertising campaign?

**5-39.** A national retailer has tracked customers who visit its website and determined that 37% will make a purchase from the website within six months of their first website visit. Eight people who visit the website are selected at random as a sample.

a. Determine the expected number in the sample who will make a purchase within six months.

b. Calculate the standard deviation in the number who will make a purchase.

c. Calculate the probability of 2 or fewer making a purchase within six months.

**5-40.** Recent ratings for the popular *NCIS* television show indicate that 7.5% of all television households tune in to the show. The ratings also show that 9.7% of all television households tune in to *The Big Bang Theory*. (*Source:* "Top Ten," www.nielsen.com, week of Jan. 25, 2016.)

a. Suppose a consultant submits a report saying that it found no one in a sample of 25 homes who claimed to have watched the program *NCIS*. What is the probability of this happening, assuming that the Nielsen ratings for the show are accurate?

b. Assume the producers for *The Big Bang Theory* planned to survey 1,000 people on the day following the broadcast of the program. The purpose of the survey was to determine what the reaction would be if one of the leading characters was retired from the show. Based on the ratings, what would be the expected number of people who would end up being included in the analysis, assuming that all 1,000 people could be reached?

**5-41.** A small hotel in a popular resort area has 20 rooms. The hotel manager estimates that 15% of all confirmed reservations are "no-shows." Consequently, the hotel accepts confirmed reservations for as many as 25 rooms. If more confirmed reservations arrive than there are rooms, the overbooked guests are sent to another hotel and given a complimentary dinner. If the hotel currently has 25 confirmed reservations, find

a. the probability that no customers will be sent to another hotel

b. the probability that exactly 2 guests will be sent to another hotel

c. the probability that 3 or more guests will be sent to another hotel

**5-42.** A manufacturing firm produces a product that has a ceramic coating. The coating is baked on to the product, and the baking process is known to produce 15% defective items (for example, cracked or chipped finishes). Every hour, 20 products from the thousands that are baked hourly are sampled from the ceramic-coating process and inspected.

a. What is the probability that 5 defective items will be found in the next sample of 20?

b. On average, how many defective items would be expected to occur in each sample of 20?

c. How likely is it that 15 or more nondefective (good) items would occur in a sample due to chance alone?

**5-43.** The Employee Benefit Research Institute reports that 58% of workers say they are confident they have saved enough for retirement (*source:* "The 2015 Retirement Confidence Survey: Having a retirement savings plan a key factor in Americans' retirement confidence, EBRI Issue Brief #413, April 2015).

a. If a random sample of 30 workers is taken, what is the probability that fewer than 17 workers are confident?

b. If a random sample of 50 workers is taken, what is the probability that more than 40 workers are confident about their retirement savings?

**5-44.** Radio frequency identification (RFID) is an electronic scanning technology that can be used to identify items in a number of ways. One advantage of RFID is that it can eliminate the need to manually count inventory, which can help improve inventory management. The technology is not infallible, however, and sometimes errors occur when items are scanned. If the probability that a scanning error occurs is 0.0065, use Excel to find

  a. the probability that exactly 20 items will be scanned incorrectly from the next 5,000 items scanned

  b. the probability that more than 20 items will be scanned incorrectly from the next 5,000 items scanned

  c. the probability that the number of items scanned incorrectly is between 10 and 25 from the next 5,000 items scanned

  d. the expected number of items scanned incorrectly from the next 5,000 items scanned

**5-45.** Managers at an electronics retailer have tracked the frequency with which product rebates are redeemed and found that for their company, 40% of all rebates are actually redeemed. Further, they found the following redemption rates: 50% for a $30 rebate on a $100 product, 10% for a $10 rebate on a $100 product, and 35% for a $50 rebate on a $200 product.

  a. Calculate the weighted average proportion of redemption rates for this retailer using the size of the rebate to establish the weights.

  b. A random sample of 20 individuals who purchased an item accompanied by a rebate were asked if they submitted their rebate. Suppose four of the questioned individuals said they did redeem their rebate. If the overall 40% estimate of the redemption rate is correct, determine the expected number of rebates that would be redeemed.

**5-46.** A large grocery store chain believes that 80% of its customers use the store's "Savings Card" when they make purchases. The manager in charge of the "Savings Card" program has randomly sampled 20 customers and observed whether or not they used the card.

  a. Assuming that the 80% use rate is correct, what is the probability that 17 or more people in the sample used the card?

  b. Repeat the calculation in part a assuming that the use rate is 70%; that it is 90%.

  c. Based on your calculations, if 17 people in the sample used the card, which of the three use rates do you think is most accurate? Explain.

## Computer Software Exercises

**5-47.** According to the most recent study conducted by the U.S. Bureau of Labor Statistics, 28.1% of female spouses outearn their male counterparts (*source:* "Highlights of women's earnings in 2014," BLS Reports, Report 1058,

November 2015). The file titled **Gendergap** contains the incomes of 150 married couples in Utah.

  a. Determine the number of families in which the female outearns her husband.

  b. Calculate the expected number of female spouses who outearn their male counterparts in the sample of 150 married couples based on the Bureau of Labor Statistics study.

  c. If the percent of married women in Utah who outearn their male spouses is the same as that indicated by the Bureau of Labor Statistics, determine the probability that at least the number found in part a would occur.

  d. Based on your calculation in part c, does the Bureau of Labor Statistics' percentage seem plausible if Utah is not different from the rest of the United States?

**5-48.** An online shoe retailer has the goal of keeping the percent of returns due to incorrect size to no more than 5%. The file titled **Shoesize** contains a sample of 125 shoe sizes that were sent to customers and the sizes that were actually ordered.

  a. Determine the number of pairs of wrong-size shoes that were delivered to customers.

  b. Calculate the probability of obtaining at least that many pairs of wrong-sized shoes delivered to customers if the proportion of incorrect sizes is actually 0.05.

  c. On the basis of your calculation, determine whether the retailer has kept the percent of returns due to incorrect size to no more than 5%. Support your answer with statistical reasoning.

  d. If the company sells 5 million pairs of shoes in one year and it costs an average of $4.75 a pair to return them, calculate the expected cost associated with wrong-sized shoes being returned using the probability calculated from the sample data.

**5-49.** International Data Corp. (IDC) has shown that the average return on business analytics projects was almost four-and-a-half times the initial investment (*source:* "IDC study finds analytics projects yield 431% average ROI," www.idg.com/, October 2002). Analytics consists of tools and applications that present better metrics to the user and to the probable future outcome of an event. IDC looked at how long it takes a typical company to recoup its investment in analytics. It determined that 63% of the U.S. corporations that adopted analytics took two years or less to recoup their investment. The file titled **Analytics** contains a sample of the time it might have taken 35 corporations to recoup their investment in analytics.

  a. Determine the number of corporations that recovered their investment in analytics in 24 months or less.

  b. Calculate the probability of obtaining at most the number of corporations that you determined in part a if the percent of those recovering their investment is as indicated by IDC.

## Other Probability Distributions

The binomial distribution is very useful in many business situations, as indicated by the examples and applications presented in the previous section. In this section, we introduce two other useful discrete probability distributions: the Poisson distribution and the hypergeometric distribution.

**OUTCOME 3**

## The Poisson Distribution

To use the binomial distribution, we must be able to count the number of successes and the number of failures. Although in many situations you may be able to count the number of successes, you might not be able to count the number of failures. For example, suppose a company builds freeways in Vermont. The company could count the number of potholes that develop per mile, but how could it count the number of nonpotholes? Or what about a hospital supplying emergency medical services in Los Angeles? It could easily count the number of emergency calls it receives in one hour, but how could it determine how many calls it did not receive? Obviously, in these cases, the number of possible outcomes (successes + failures) is difficult, if not impossible, to determine. If the total number of possible outcomes cannot be determined, the binomial distribution cannot be applied. In these cases you may be able to use the Poisson distribution.

**Poisson Distribution**

The Poisson distribution describes a process that extends over space, time, or any well-defined interval or unit of inspection in which the outcomes of interest occur at random and we can count the number of outcomes that occur in any given interval. The Poisson distribution, rather than the binomial distribution, is used when the total number of possible outcomes cannot be determined.

**Characteristics of the Poisson Distribution**   The **Poisson distribution**[1] describes a process that extends over time, space, or any well-defined unit of inspection. The outcomes of interest, such as emergency calls or potholes, occur at random, and we count the number of outcomes that occur in a given segment of time or space. We might count the number of emergency calls in a one-hour period or the number of potholes in a one-mile stretch of freeway. As we did with the binomial distribution, we will call these outcomes *successes* even though (like potholes) they might be undesirable.

The possible numbers of successes are the integers $0, 1, 2, \ldots$, and we would like to know the probability of each of these values. For example, what is the chance of getting exactly four emergency calls in a particular hour? What is the chance that a chosen one-mile stretch of freeway will contain zero potholes?

We can use the Poisson probability distribution to answer these questions if we make the following assumptions:

1. We know $\lambda$, the average number of successes in one segment. For example, we might know that there is an average of 8 emergency calls per hour ($\lambda = 8$) or an average of 15 potholes per mile of freeway ($\lambda = 15$).
2. The probability of $x$ successes in a segment is the same for all segments of the same size. For example, the probability distribution of emergency calls is the same for any one-hour period of time at the hospital.
3. What happens in one segment has no influence on any nonoverlapping segment. For example, the number of calls arriving between 9:30 P.M. and 10:30 P.M. has no influence on the number of calls between 11:00 P.M. and 12:00 midnight.

Once $\lambda$ has been determined, we can calculate the average occurrence rate for any number of segments ($t$). This is $\lambda t$. Note that $\lambda$ and $t$ must be in compatible units. If we have $\lambda = 8$ calls per hour, the segments must be in hours or fractional parts of an hour. That is, if we have $\lambda = 8$ per hour and we wish to work with half-hour time periods, the segment would be

$$t = \frac{1}{2}\text{hour} = 0.50 \text{ hour}$$

not $t = 30$ minutes. In this case, $\lambda t$ will be $(8)(0.5) = 4$ per half hour.

Although the Poisson distribution is often used to describe situations such as the number of customers who arrive at a hospital emergency room per hour or the number of calls the

---

[1]The Poisson distribution can be derived as the limiting distribution of the binomial distribution as the number of trials $n$ tends to infinity and the probability of success decreases to zero. It serves as a good approximation to the binomial when $n$ is large.

Hewlett-Packard LaserJet printer service center receives in a 30-minute period, the segments need not be time intervals. Poisson distributions are also used to describe such random variables as the number of knots in a sheet of plywood or the number of contaminants in a gallon of lake water. The segments would be the sheet of plywood and the gallon of water.

Another important point is that $\lambda t$, the average number in $t$ segments, is not necessarily the number we will see if we observe the process for $t$ segments. We might expect an average of 20 people to arrive at a checkout stand in any given hour, but we do not expect to find exactly that number arriving every hour. The actual arrivals will form a distribution with an expected value, or mean, equal to $\lambda t$. So, for the Poisson distribution,

$$E[x] = \mu_x = \lambda t$$

Once $\lambda$ and $t$ have been specified, the probability for any discrete value in the Poisson distribution can be found using Equation 5.7.

**Poisson Probability Distribution**

$$P(x) = \frac{(\lambda t)^x e^{-\lambda t}}{x!} \tag{5.7}$$

where:

$t$ = Number of segments of interest
$x$ = Number of successes in $t$ segments
$\lambda$ = Expected number of successes in one segment
$e$ = Base of the natural logarithm system (2.71828 . . . )

---

**BUSINESS APPLICATION** **Poisson Distribution**

**Whole Foods Grocery** A study conducted at Whole Foods Grocery shows that the average number of arrivals to the checkout section of the store per hour is 16. Further, the distribution for the number of arrivals is considered to be Poisson distributed. Figure 5.6 shows the shape of the Poisson distribution for $\lambda = 16$. The probability of each possible number of customers arriving can be computed using Equation 5.7. For example, we can find the probability of $x = 12$ customers in one hour ($t = 1$) as follows:

The Excel 2016 function for the Poisson distribution formula is

= POISSON.DIST(x,Mean,False)
= POISSON.DIST(12,16,False)

$$P(x = 12) = \frac{(\lambda t)^x e^{-\lambda t}}{x!} = \frac{16^{12} e^{-16}}{12!} = 0.0661$$

**FIGURE 5.6 Poisson Distribution for Whole Foods Checkout Arrivals with $\lambda = 16$**

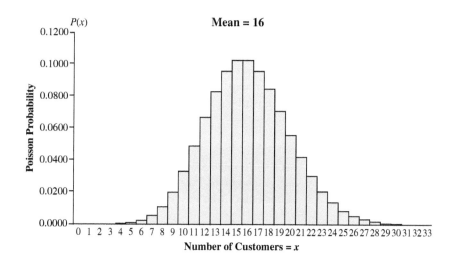

**Poisson Probability Distribution Table**    As was the case with the binomial distribution, a table of probabilities exists for the Poisson distribution. (The Poisson table appears in Appendix C.) The Poisson table shows the cumulative probabilities for $x$ or fewer occurrences for different $\lambda t$ values. We can use the following business application to illustrate how to use the Poisson table.

---

**BUSINESS APPLICATION**    Using the Poisson Distribution Table

**Whole Foods Grocery** (*continued*)    At Whole Foods Grocery, customers are thought to arrive at the checkout section according to a Poisson distribution with $\lambda = 16$ customers per hour. (See Figure 5.6.) Based on previous studies, the store manager believes that the service time for each customer is quite constant at six minutes. Suppose, during each six-minute time period, the store has three checkers available. This means that three customers can be served during each six-minute segment. The manager is interested in the probability that one or more customers will have to wait for service during a six-minute period.

To determine this probability, you will need to convert the mean arrivals from $\lambda = 16$ customers per hour to a new average for a six-minute segment. Six minutes corresponds to 0.10 hour, so the segment size is $t = 0.10$. Then the mean number of arrivals in six minutes is $\lambda t = 16(0.10) = 1.6$ customers.

Now, because there are three checkers, any time four or more customers arrive in a six-minute period, at least one customer will have to wait for service. Thus,

$$P(4 \text{ or more customers}) = P(4) + P(5) + P(6) + \cdots$$

or you can use the *Complement Rule*, discussed in Chapter 4, as follows:

$$P(4 \text{ or more customers}) = 1 - P(x \leq 3)$$

> The Excel 2016 function for the cumulative Poisson distribution is
>
> = **POISSON.DIST**(x,Mean,True)
> = **POISSON.DIST**(3,1.6,True)
>
> *Note:* The "True" argument signifies that the POISSON.DIST function will return the cumulative probability of $x$ or fewer successes.

The Poisson table in Appendix C can be used to find the necessary probabilities. To use the table, first go across the top of the table until you find the desired value of $\lambda t$. In this case, look for $\lambda t = 1.6$. Next, go down the left-hand side to find the value of $x$ corresponding to the number of occurrences of interest. For example, consider $x = 3$ customer arrivals. Because Appendix C is a cumulative Poisson table, the probabilities are for $x$ or fewer occurrences. The probability of $x \leq 3$ is given in the table as 0.9212. Thus,

$$P(x \leq 3) = 0.9212$$

Then the probability of four or more customers arriving is

$$P(4 \text{ or more customers}) = 1 - P(x \leq 3)$$
$$P(4 \text{ or more customers}) = 1 - 0.9212 = 0.0788$$

Given the store's capacity to serve three customers in a six-minute period, the probability of one or more customers having to wait is 0.0778.

---

**EXAMPLE 5-7**    Using the Poisson Distribution

**Fashion Leather Products**    Fashion Leather Products, headquartered in Argentina, makes leather clothing for export to many other countries around the world. Before shipping, quality managers perform tests on the leather products. The industry standards call for the average number of defects per square meter of leather to not exceed five. During a recent test, the inspector selected 3 square meters and found 18 defects. To determine the probability of this event occurring if the leather meets the industry standards, assuming that the Poisson distribution applies, the company can perform the following steps:

STEP 1  **Define the segment unit.**
Because the mean was stated as five defects per square meter, the segment unit in this case is one meter.

**HOW TO DO IT** (Example 5-7)

**Using the Poisson Distribution**

1. Define the segment units. The segment units are usually blocks of time, areas of space, or volume.

2. Determine the mean of the random variable. The mean is the parameter that defines the Poisson distribution and is referred to as $\lambda$. It is the average number of successes in a segment of unit size.

3. Determine $t$, the number of the segments to be considered, and then calculate $\lambda t$.

4. Define the event of interest and use the Poisson formula, the Poisson table, or Excel to find the probability.

The Excel 2016 function for the cumulative Poisson distribution is

=**POISSON.DIST(x,Mean,True)**
=**POISSON.DIST(17,15,True)**

**STEP 2**   **Determine the mean of the random variable.**

In this case, if the company meets the industry standards, the mean will be

$$\lambda = 5$$

**STEP 3**   **Determine the segment size $t$.**

The company quality inspectors analyzed 3 square meters, which is equal to 3 units. So $t = 3.0$. Then

$$\lambda t = (5)(3.0) = 15.0$$

When looking at 3 square meters, the company would expect to find at most 15.0 defects if the industry standards are being met.

**STEP 4**   **Define the event of interest and use the Poisson formula, the Poisson table, or Excel to find the probability.**

In this case, 18 defects were observed. Because 18 exceeds the expected number ($\lambda t = 15.0$) the company would want to find:

$$P(x \geq 18) = P(x = 18) + P(x = 19) + \cdots$$

The Poisson table in Appendix C is used to determine these probabilities. Locate the desired probability under the column headed $\lambda t = 15.0$. Then find the values of $x$ down the left-hand column:

$$P(x \geq 18) = 1 - P(x \leq 17)$$
$$= 1 - 0.7489$$
$$= 0.2511$$

There is about a 0.25 chance of finding 18 or more defects in 3 square meters of leather products made by Fashion Leather if they are meeting the quality standard.

**TRY EXERCISE 5-50** (pg. 202)

**The Mean and Standard Deviation of the Poisson Distribution**   The mean of the Poisson distribution is $\lambda t$. This is the value we use to specify which Poisson distribution we are using. We must know the mean before we can find probabilities for a Poisson distribution.

Figure 5.6 illustrated that the outcome of a Poisson distributed variable is subject to variation. Like any other discrete probability distribution, the standard deviation for the Poisson can be computed using Equation 5.2:

$$\sigma_x = \sqrt{\sum [x - E(x)]^2 P(x)}$$

However, for a Poisson distribution, the standard deviation also can be found using Equation 5.8.

**Standard Deviation of the Poisson Distribution**

$$\sigma = \sqrt{\lambda t} \tag{5.8}$$

The standard deviation of the Poisson distribution is simply the square root of the mean. Therefore, if you are working with a Poisson process, reducing the mean will reduce the variability also.

**Excel Tutorial**

**BUSINESS APPLICATION**    **The Poisson Probability Distribution**

**Fashion Tile**   To illustrate the importance of the relationship between the mean and standard deviation of the Poisson distribution, consider Fashion Tile in New York City. The company makes ceramic tile for kitchens and bathrooms. The quality standards call for the number of imperfections in a tile to average 3 or fewer. The distribution of imperfections is thought to

**FIGURE 5.7  Excel 2016 Output for Fashion Tile Example**

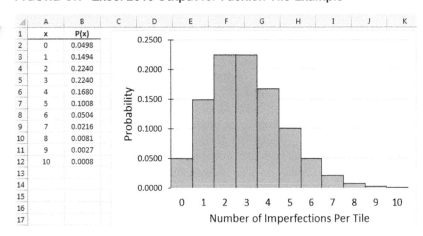

be Poisson. Many software packages, including Excel, will generate Poisson probabilities in much the same way as for the binomial distribution, which was discussed in Section 5.2. If we assume that the company is meeting the standard, Figure 5.7 shows the Poisson probability distribution generated using Excel when $\lambda t = 3.0$. Even though the average number of defects is 3, the manager is concerned about the high probabilities associated with the number of tile imperfections equal to 4, 5, 6, or more on a tile. The variability is too great. From Equation 5.8, the standard deviation for this distribution is

$$\sigma = \sqrt{3.0} = 1.732$$

This large standard deviation means that although some tiles will have few, if any, imperfections, others will have several, causing problems for installers and unhappy customers.

A quality-improvement effort directed at reducing the average number of imperfections to 2.0 would also reduce the standard deviation to

$$\sigma = \sqrt{2.0} = 1.414$$

Further reductions in the average would also reduce variation in the number of imperfections between tiles. This would mean more consistency for installers and higher customer satisfaction.

**OUTCOME 3**

## The Hypergeometric Distribution

Although the binomial and Poisson distributions are very useful in many business decision-making situations, both require that the trials be independent. For instance, in binomial applications, the probability of a success in one trial must be the same as the probability of a success in any other trial. Although there are certainly times when this assumption can be satisfied, or at least approximated, in instances in which the population is fairly small and we are sampling without replacement, the condition of independence will not hold. In these cases, a discrete probability distribution referred to as the *hypergeometric distribution* can be useful.

**BUSINESS APPLICATION**    The Hypergeometric Probability Distribution

**Dolby Industries**  Dolby Industries contracts with a manufacturer to make women's handbags. Because of the intense competition in the marketplace for handbags, Dolby has made every attempt to provide high-quality products. However, a recent production run of 20 handbags of a particular model contained 2 units that tested out as defective. The problem was traced to a shipment of defective latches that Dolby's Chinese partner received shortly before the production run started.

The production manager ordered that the entire batch of 20 handbags be isolated from other production output until further testing could be completed. Unfortunately, a new shipping clerk packaged 10 of these isolated handbags and shipped them to a California retailer to fill an order that was already overdue. By the time the production manager noticed what had happened, the handbags were already in transit.

The immediate concern was whether one or more of the defectives had been included in the shipment. The new shipping clerk thought there was a good chance that no defectives were included. Short of reinspecting the remaining handbags, how might Dolby Industries determine the probability that no defectives were actually shipped?

At first glance, it might seem that the question could be answered by employing the binomial distribution with $n = 10, p = 2/20 = 0.10$, and $x = 0$. Using the binomial distribution table in Appendix B, we get

$$P(x = 0) = 0.3487$$

There is a 0.3487 chance that no defectives were shipped, assuming the selection process satisfied the requirements of a binomial distribution. However, for the binomial distribution to be applicable, the trials must be independent, and the probability of a success, $p$, must remain constant from trial to trial. In order for this to occur when the sampling is from a "small," *finite* population as is the case here, the sampling must be performed with *replacement*. This means that after each item is selected, it is returned to the population and, therefore, may be selected again later in the sampling.

In the Dolby example, the sampling was performed without replacement because each handbag could be shipped only one time. Also, the population of handbags is finite with size $N = 20$, which is a "small" population. Thus, $p$, the probability of a defective handbag, does not remain equal to 0.10 on each trial. The value of $p$ on any particular trial depends on what has already been selected on previous trials.

The event of interest is

$$G\ G\ G\ G\ G\ G\ G\ G\ G\ G$$

The probability that the first item selected for shipment was good is 18/20, because there were 18 good handbags in the batch of 20. Now, assuming the first unit selected was good, the probability the second unit was good is 17/19, because we then had only 19 handbags to select from and 17 of those would be good. The probability that all 10 items selected were good is

$$\frac{18}{20} \times \frac{17}{19} \times \frac{16}{18} \times \frac{15}{17} \times \frac{14}{16} \times \frac{13}{15} \times \frac{12}{14} \times \frac{11}{13} \times \frac{10}{12} \times \frac{9}{11} = 0.2368$$

This value is not the same as the 0.3847 probability we got when we used the binomial distribution. This demonstrates that when sampling is performed without replacement from small, finite populations, the binomial distribution can produce inaccurate probabilities. To protect against large inaccuracies, the binomial distribution should be used only when the sample is small relative to the size of the population. Under that circumstance, the value of $p$ will not change very much as the sample is selected, and the binomial distribution will be a reasonable approximation to the actual probability distribution.

In cases in which the sample is large relative to the size of the population, a discrete probability distribution, called the **hypergeometric distribution**, is the correct distribution for computing probabilities for the random variable of interest.

We use combinations (see Section 5.2) to form the equation for computing probabilities for the hypergeometric distribution. When each trial has two possible outcomes (success and failure), hypergeometric probabilities are computed using Equation 5.9.

Notice that the numerator of Equation 5.9 is the product of the number of ways you can select $x$ successes in a random sample out of the $X$ successes in the population and the

**Hypergeometric Distribution**
The hypergeometric distribution is formed as the ratio of the number of ways an event of interest can occur to the total number of ways any event can occur.

### Hypergeometric Distribution (Two Possible Outcomes per Trial)

$$P(x) = \frac{C_{n-x}^{N-X} \cdot C_x^X}{C_n^N} \qquad (5.9)$$

where:

$N$ = Population size
$X$ = Number of successes in the population
$n$ = Sample size
$x$ = Number of successes in the sample
$n - x$ = Number of failures in the sample

number of ways you can select $n - x$ failures in a sample from the $N - X$ failures in the population. The denominator in the equation is the number of ways the sample can be selected from the population.

In the earlier Dolby Industries example, the probability of zero defectives being shipped ($x = 0$) is

> The Excel 2016 function for the hypergeometric distribution is
>
> = **HYPGEOM.DIST(x,n, X,N,False)**
> = **HYPGEOM.DIST(0,10,2,20,False)**

$$P(x = 0) = \frac{C_{10-0}^{20-2} \cdot C_0^2}{C_{10}^{20}}$$

$$P(x = 0) = \frac{C_{10}^{18} \cdot C_0^2}{C_{10}^{20}}$$

Carrying out the arithmetic, we get

$$P(x = 0) = \frac{(43{,}758)(1)}{184{,}756} = 0.2368$$

As we found before, the probability that zero defectives were included in the shipment is 0.2368, or approximately 24%.

The probabilities of $x = 1$ and $x = 2$ defectives can also be found by using Equation 5.9, as follows:

$$P(x = 1) = \frac{C_{10-1}^{20-2} \cdot C_1^2}{C_{10}^{20}} = 0.5264$$

> The Excel 2016 function for the hypergeometric distribution is
>
> = **HYPGEOM.DIST(1,10,2,20,False)**
>
> and
>
> = **HYPGEOM.DIST(2,10,2,20,False)**

and

$$P(x = 2) = \frac{C_{10-2}^{20-2} \cdot C_2^2}{C_{10}^{20}} = 0.2368$$

Thus, the hypergeometric probability distribution for the number of defective handbags in a random selection of 10 is

| x | P(x) |
|---|------|
| 0 | 0.2368 |
| 1 | 0.5264 |
| 2 | 0.2368 |
| | $\sum P(x) = 1.0000$ |

Recall that when we introduced the hypergeometric distribution, we said that it is used in situations when we are sampling without replacement from a finite population. However, when the population size is large relative to the sample size, decision makers typically use the binomial distribution as an approximation of the hypergeometric. This eases the computational burden and provides useful approximations in those cases. Although there is no exact rule for when the binomial approximation can be used, we suggest that the sample should be less than 5% of the population size. Otherwise, use the hypergeometric distribution when sampling is done without replacement from the finite population.

EXAMPLE 5-8

### The Hypergeometric Distribution (One of Two Possible Outcomes per Trial)

**Gender Equity** One of the positive changes in U.S. business practice in the past few decades has been the inclusion of more women in the management ranks of companies. Tom Peters, management consultant and author, stated in his 2006 book *In Search of Excellence* that one of the reasons Middle Eastern countries have suffered economically compared with countries such as the United States is that they have not included women in their economic system. However, there are still concerns in U.S. business. Consider a situation in which a Maryland company needed to downsize one department that had 30 people—12 women and 18 men. Ten people were laid off, and upper management said the layoffs were done randomly. By chance alone, 40% (12/30) of the layoffs would be women. However, of the 10 laid off, 8 were women. This is 80%, not the 40% due to chance. A labor attorney is interested in the probability of eight or more women being laid off due to chance alone. This can be determined using the following steps:

STEP 1 **Determine the population size and the combined sample size.**
The population size and sample size are

$$N = 30 \quad \text{and} \quad n = 10$$

STEP 2 **Define the event of interest.**
The attorney is interested in the event:

$$P(x \geq 8) = ?$$

What are the chances that eight or more women would be selected?

STEP 3 **Determine the number of successes in the population and the number of successes in the sample.**
In this situation, a success is the event that a woman is selected. There are $X = 12$ women in the population and $x \geq 8$ in the sample. We will break this down as $x = 8, x = 9, x = 10$.

STEP 4 **Compute the desired probabilities using Equation 5.9.**

$$P(x) = \frac{C_{n-x}^{N-X} \cdot C_x^X}{C_n^N}$$

We want:

$$P(x \geq 8) = P(x = 8) + P(x = 9) + P(x = 10)$$

$$P(x = 8) = \frac{C_{10-8}^{30-12} \cdot C_8^{12}}{C_{10}^{30}} = \frac{C_2^{18} \cdot C_8^{12}}{C_{10}^{30}} = 0.0025$$

$$P(x = 9) = \frac{C_1^{18} \cdot C_9^{12}}{C_{10}^{30}} = 0.0001$$

$$P(x = 10) = \frac{C_0^{18} \cdot C_{10}^{12}}{C_{10}^{30}} \approx 0.0000$$

Therefore, $P(x \geq 8) = 0.0025 + 0.0001 + 0.0000 = 0.0026$

The chances that 8 or more women would have been selected among the 10 people chosen for layoff strictly due to chance is 0.0026. The attorney will likely wish to challenge the layoffs based on this extremely low probability.

**TRY EXERCISE 5-53** (pg. 202)

> The Excel 2016 function for the cumulative hypergeometric distribution is
>
> =HYPGEOM.DIST(x,n,X, N,True)
>
> For $P(x \geq 8)$, use $1 - P(x \leq 7)$:
>
> =1-HYPGEOM.DIST(7,10, 12,30,True)
> =0.002655

### The Hypergeometric Distribution with More Than Two Possible Outcomes per Trial
Equation 5.9 assumes that on any given sample selection or trial, only one of two possible outcomes will occur. However, the hypergeometric distribution can easily be extended to consider any number of possible categories of outcomes on a given trial by employing Equation 5.10.

**Hypergeometric Distribution (*k* Possible Outcomes per Trial)**

$$P(x_1, x_2, \ldots, x_k) = \frac{C_{x_1}^{X_1} \cdot C_{x_2}^{X_2} \cdot C_{x_3}^{X_3} \cdot \ldots \cdot C_{x_k}^{X_k}}{C_n^N} \tag{5.10}$$

where:

$$\sum_{i=1}^{k} X_i = N$$

$$\sum_{i=1}^{k} x_i = n$$

$N$ = Population size
$n$ = Total sample size
$X_i$ = Number of items in the population with outcome $i$
$x_i$ = Number of items in the sample with outcome $i$

---

**EXAMPLE 5-9** **The Hypergeometric Distribution for Multiple Outcomes**

**Breakfast Cereal Preferences** Consider a marketing study that involves placing breakfast cereal made by four different companies in a basket at the exit to a food store. A sign on the basket invites customers to take one box of cereal free of charge. At the beginning of the study, the basket contains the following:

5 brand A
4 brand B
6 brand C
4 brand D

The researchers were interested in the brand selection patterns for customers who could select without regard to price. Suppose six customers were observed and three selected brand B, two selected brand D, and one selected brand C. No one selected brand A. The probability of this selection mix, assuming the customers were selecting entirely at random without replacement from a finite population, can be found using the following steps:

**STEP 1** **Determine the population size and the combined sample size.**
The population size and sample size are

$$N = 19 \quad \text{and} \quad n = 6$$

**STEP 2** **Define the event of interest.**
The event of interest is

$$P(x_1 = 0; x_2 = 3; x_3 = 1; x_4 = 2) = ?$$

**STEP 3** **Determine the number in each category in the population and the number in each category in the sample.**

| | | |
|---|---|---|
| $X_1 = 5$ | (brand A) | $x_1 = 0$ |
| $X_2 = 4$ | (brand B) | $x_2 = 3$ |
| $X_3 = 6$ | (brand C) | $x_3 = 1$ |
| $X_4 = 4$ | (brand D) | $x_4 = 2$ |
| $N = 19$ | | $n = 6$ |

**STEP 4** **Compute the desired probability using Equation 5.10.**

$$P(x_1, x_2, x_3, \ldots, x_k) = \frac{C_{x_1}^{X_1} \cdot C_{x_2}^{X_2} \cdot C_{x_3}^{X_3} \cdot \ldots \cdot C_{x_k}^{X_k}}{C_n^N}$$

$$P(0,3,1,2) = \frac{C_0^5 \cdot C_3^4 \cdot C_1^6 \cdot C_2^4}{C_6^{19}}$$

$$= \frac{(1)(4)(6)(6)}{27,132} = \frac{144}{27,132}$$

$$= 0.0053$$

There are slightly more than 5 chances in 1,000 of this exact selection occurring by random chance.

**TRY EXERCISE 5-52** (pg. 202)

## 5.3 EXERCISES

### Skill Development

**5-50.** The mean number of errors per page missed by a proofreader at a major publishing company is thought to be 1.5, with the number of errors distributed according to a Poisson distribution. If three pages are examined, what is the probability that more than three errors will be observed?

**5-51.** Arrivals to a bank automated teller machine (ATM) are distributed according to a Poisson distribution with a mean equal to three per 15 minutes.
   a. Determine the probability that in a given 15-minute segment, no customers will arrive at the ATM.
   b. What is the probability that fewer than four customers will arrive in a 30-minute segment?

**5-52.** Consider a situation in which a used-car lot contains five Fords, four General Motors (GM) cars, and five Toyotas. If five cars are selected at random to be placed on a special sale, what is the probability that three are Fords and two are GMs?

**5-53.** A population of 10 items contains 3 that are red and 7 that are green. What is the probability that in a random sample of 3 items selected without replacement, 2 red and 1 green items are selected?

**5-54.** If a random variable follows a Poisson distribution with $\lambda = 20$ and $t = \frac{1}{2}$, find the
   a. expected value, variance, and standard deviation of this Poisson distribution
   b. probability of exactly 8 successes

**5-55.** A corporation has 11 manufacturing plants. Of these, seven are domestic and four are located outside the United States. Each year, a performance evaluation is conducted for 4 randomly selected plants.
   a. What is the probability that a performance evaluation will include exactly 1 plant outside the United States?
   b. What is the probability that a performance evaluation will contain 3 plants from the United States?
   c. What is the probability that a performance evaluation will include 2 or more plants from outside the United States?

**5-56.** Determine the following values associated with a Poisson distribution with $\lambda t$ equal to 3:
   a. $P(x \le 3)$
   b. $P(x > 3)$
   c. $P(2 < x \le 5)$
   d. Find the smallest $x'$ so that $P(x \le x') > 0.50$.

**5-57.** A random variable, $x$, has a hypergeometric distribution with $N = 10$, $X = 7$, and $n = 4$. Calculate the following quantities:
   a. $P(x = 3)$
   b. $P(x = 5)$
   c. $P(x \ge 4)$
   d. Find the largest $x'$ so that $P(x > x') > 0.25$.

### Business Applications

**5-58.** A new phone-answering system installed by the Ohio Power Company is capable of handling five calls every 10 minutes. Prior to installing the new system, company analysts determined that the incoming calls to the system are Poisson distributed with a mean equal to two every 10 minutes. If the analysts are correct about this incoming call distribution, what is the probability that in a 10-minute period more calls will arrive than the system can handle? Based on this probability, comment on the adequacy of the new answering system.

**5-59.** Plywood contains minor imperfections that can be repaired with small "plugs." One customer will accept plywood with a maximum of 3.5 plugs per sheet on average. Suppose a shipment was sent to this customer, and when the customer inspected two sheets at random, 10 plugged defects were counted. What is the probability of observing 10 or more plugged defects if in fact the 3.5 average per sheet is being satisfied? Comment on what this probability implies about whether you think the company is meeting the 3.5 per sheet defect rate.

**5-60.** When things are operating properly, E-Bank United, an Internet bank, can process a maximum of 25 electronic transfers every minute during the busiest periods of the day. If it receives more transfer requests than this, then the bank's computer system will become so overburdened that it will slow to the point that no electronic transfers can be handled. If during the busiest periods of the day requests for electronic transfers arrive at the rate of 170 per 10-minute period on average, what is the probability that the system will be overwhelmed by requests? Assume that the process can be described using a Poisson distribution.

**5-61.** A stock portfolio contains 20 stocks. Of these stocks, 10 are considered "large-cap" stocks, 5 are "mid cap," and 5 are "small cap." The portfolio manager has been asked by his client to develop a report that highlights 7 randomly selected stocks. When she presents her report to the client, all 7 of the stocks are large-cap stocks. The client is very suspicious that the manager has not randomly selected the stocks. She believes that the chances of all 7 of the stocks being large cap must be very low. Compute the probability of all 7 being large cap and comment on the concerns of the client.

**5-62.** College-Pro Painting does home interior and exterior painting. The company uses inexperienced painters that do not always do a high-quality job. It believes that its painting process can be described by a Poisson distribution with an average of 4.8 defects per 400 square feet of painting.
   a. What is the probability that a 400-square-foot painted section will have fewer than 6 blemishes?
   b. What is the probability that six randomly sampled sections of size 400 square feet will each have 7 or fewer blemishes?

**5-63.** A shipment of 20 hard drives was shipped to The Tech-Crew computer repair company. Four of them were defective. One of the technicians selected 5 of the hard drives to put in his parts inventory and then went on three service calls.
   a. Determine the probability that only 1 of the 5 hard drives is defective.
   b. Determine the probability that 3 of the 5 hard drives are not defective.
   c. Determine the probability that the technician will have enough hard drives to replace 3 defective hard drives at the repair sites.

**5-64.** John Thurgood founded a company that translates Chinese books into English. His company is currently testing a computer-based translation service. Since Chinese symbols are difficult to translate, John assumes the computer program will make some errors, but then so do human translators. The computer error rate is supposed to be an average of 3 per 400 words of translation. Suppose John randomly selects a 1,200-word passage. Assuming that the Poisson distribution applies and that the computer error rate is actually 3 errors per 400 words,
   a. determine the probability that no errors will be found
   b. calculate the probability that more than 14 errors will be found

   c. find the probability that fewer than 9 errors will be found
   d. If 15 errors are found in the 1,200-word passage, what would you conclude about the computer company's claim? Why?

**5-65.** Beacon Hill Trees & Shrubs currently has an inventory of 10 fruit trees, 8 pine trees, and 14 maple trees. It plans to give 4 trees away at next Saturday's lawn and garden show in the city park. The 4 winners can select which type of tree they want. Assume they select randomly.
   a. What is the probability that all 4 winners will select the same type of tree?
   b. What is the probability that 3 winners will select pine trees and the other tree will be a maple?
   c. What is the probability that no fruit trees and 2 of each of the others will be selected?

**5-66.** Fasteners used in a manufacturing process are shipped by the supplier to the manufacturer in boxes that contain 20 fasteners. Because the fasteners are critical to the production process, their failure will cause the product to fail. The manufacturing firm and the supplier have agreed that a random sample of 4 fasteners will be selected from every box and tested to see if the fasteners meet the manufacturer's specifications. The nature of the testing process is such that tested fasteners become unusable and must be discarded. The supplier and the manufacturer have agreed that if 2 or more fasteners fail the test, the entire box will be rejected as being defective. Assume that a new box has just been received for inspection. If the box has 5 defective fasteners, what is the probability that a random sample of 4 will have 2 or more defective fasteners? What is the probability the box will be accepted?

**5-67.** The owner of Lucky Dogs sells spicy hot dogs from a pushcart. Lucky Dogs is open every day between 11:00 A.M. and 1:00 P.M. Assume the demand for spicy hot dogs follows a Poisson distribution with a mean of 50 per hour.
   a. What is the probability the owner will run out of spicy hot dogs over the two-hour period if he stocks his cart with 115 spicy hot dogs every day?
   b. How many spicy hot dogs should the owner stock if he wants to limit the probability of being out of stock to less than 2.5%?

   (*Hint:* To solve this problem use Excel's POISSON. DIST function.)

**5-68.** The U.S. Department of Labor reported that about one-third of eligible workers haven't enrolled in their employers' 401(k) plans (*source:* "Automatic enrollment 401(k) plans for small businesses," www.dol.gov/ebsa/publications/). Costco has been contemplating new incentives to encourage more participation from its employees. Of the 12 employees in one of Costco's automotive departments, 5 have enrolled in Costco's 401(k) plan. The store manager has randomly selected 7 of the automotive department employees to receive investment training.

a. Calculate the probability that all of the employees currently enrolled in the 401(k) program are selected for the investment training.

b. Calculate the probability that none of the employees currently enrolled in the 401(k) program is selected for the investment training.

c. Compute the probability that more than half of the employees currently enrolled in the 401(k) program are selected for the investment training.

**5-69.** The 2014 State of Women-Owned Businesses report commissioned by American Express indicated 31% of private firms had female owners, 53% had male owners, and 16% had male and female co-owners (*source:* www.womenable.com). In one community, there are 100 privately owned firms. Ten privately owned firms are selected to receive assistance in marketing their products. Assume the percentages indicated by American Express apply to this community.

a. Calculate the probability that one-half of the firms selected will be solely owned by a woman, 3 owned by men, and the rest co-owned by women and men.

b. Calculate the probability that all of the firms selected will be solely owned by women.

c. Calculate the probability that 6 will be owned by a woman and the rest co-owned.

## Computer Software Exercises

**5-70.** An employment agency survey contacted 130 small firms. One of the many inquiries was to determine the number of employees the firms had. The file titled **Employees** contains the responses by the firms. The number of employees was grouped into the following categories: (1) fewer than 20; (2) 20–99; (3) 100–499; and (4) 500 or more.

a. Determine the number of firms in each of these categories.

b. If the agency contacts 25 of these firms to gather more information, determine the probability that it will choose the following number of firms in each category: (1) 22, (2) 2, (3) 1, and (4) 0.

c. Calculate the probability that it will choose all of the firms from those businesses that have fewer than 20 workers.

**5-71.** Cliff Summey is the quality-assurance engineer for Sticks and Stones Billiard Supply, a manufacturer of billiard supplies. One of the items that Sticks and Stones produces is sets of pocket billiard balls. Cliff has been monitoring the finish of the pocket billiard balls. He is concerned that sets of billiard balls have been shipped with an increasing number of scratches. The company's goal is to have no more than an average of one scratch per set of pocket billiard balls. A set contains 16 balls. Over the last week, Cliff selected a sample of 48 billiard balls and inspected them to determine the number of scratches. The data collected by Cliff are displayed in the file called **Poolball**.

a. Determine the number of scratches in the sample.

b. Calculate the average number of scratches for 48 pocket billiard balls if Sticks and Stones has met its goal.

c. Determine the probability that there would be at least as many scratches observed per set of pocket billiard balls if Sticks and Stones has met its goal.

d. Based on the sample evidence, does it appear that Sticks and Stones has met its goal? Provide statistical reasons for your conclusion.

# 5 Overview

## Summary

**OUTCOME 1** Be able to calculate and interpret the expected value of a discrete random variable.

- A **discrete random** variable can assume only a finite number of values or an infinite sequence of values, such as 0, 1, 2, . . . .
- The **expected value** is the mean of a discrete random variable. It represents the long-run average value for the random variable.
- The graph of a discrete random variable looks like a histogram, with the values of the random variable on the horizontal axis and the bars above the values having heights corresponding to the probability of the outcome occurring. The sum of the individual probabilities is 1.

**OUTCOME 2** Be able to apply the binomial distribution to business decision-making situations.

- The **binomial distribution** applies when an experimental trial has only two possible outcomes called *success* and *failure*, the probability of success remains constant from trial to trial, the trials are **independent**, and there is a fixed number of identical trials.
- The probabilities for a binomial distribution can be calculated using Equation 5.4, from the **binomial table** in Appendix B, or found using Excel.
- We can find the expected value of the binomial distribution by multiplying $n$, the number of trials, by $p$, the probability of a success on any one trial.
- The shape of a binomial distribution depends on the sample size (number of trials) and $p$, the probability of a success. When $p$ is close to 0.50, the binomial distribution is fairly symmetric and bell–shaped. Even when $p$ is near 0 or 1, if $n$, the sample size, is large, the binomial distribution is still fairly symmetric and bell–shaped.

**OUTCOME 3** Be able to compute probabilities for the Poisson and hypergeometric distributions and apply these distributions to decision-making situations.

- Although the **binomial distribution** may be the discrete distribution that business decision makers use most often, the Poisson distribution and the hypergeometric distribution are also frequently employed.
- The **Poisson distribution** is used when the value of the random variable is found by counting the number of occurrences within a defined segment of time or space. If you know the mean number of occurrences per segment, you can use the Poisson formula, the Poisson tables in Appendix C, or Excel to find the probability of any specific number of occurrences within the segment. The Poisson distribution is often used to describe the number of customers who arrive at a service facility in a specific amount of time.
- Decision makers use the **hypergeometric** distribution in situations where the sample size is large relative to the size of the population and the sampling is done without replacement.

## Equations

**(5.1) Expected Value of a Discrete Probability Distribution** pg. 175

$$E(x) = \sum xP(x)$$

**(5.2) Standard Deviation of a Discrete Probability Distribution** pg. 175

$$\sigma_x = \sqrt{\sum [x - E(x)]^2 P(x)}$$

**(5.3) Counting Rule for Combinations** pg. 183

$$C_x^n = \frac{n!}{x!(n-x)!}$$

**(5.4) Binomial Formula** pg. 184

$$P(x) = \frac{n!}{x!(n-x)!}p^x q^{n-x}$$

**(5.5) Expected Value of a Binomial Distribution** pg. 187

$$\mu_x = E(x) = np$$

**(5.6) Standard Deviation of the Binomial Distribution** pg. 188

$$\sigma = \sqrt{npq}$$

**(5.7) Poisson Probability Distribution** pg. 194

$$P(x) = \frac{(\lambda t)^x e^{-\lambda t}}{x!}$$

**(5.8) Standard Deviation of the Poisson Distribution** pg. 196

$$\sigma = \sqrt{\lambda t}$$

**(5.9) Hypergeometric Distribution (Two Possible Outcomes per Trial)** pg. 199

$$P(x) = \frac{C_{n-x}^{N-X} \cdot C_x^X}{C_n^N}$$

**(5.10) Hypergeometric Distribution (k Possible Outcomes per Trial)** pg. 201

$$P(x_1, x_2, \ldots, x_k) = \frac{C_{x_1}^{X_1} \cdot C_{x_2}^{X_2} \cdot C_{x_3}^{X_3} \cdot \ldots \cdot C_{x_k}^{X_k}}{C_n^N}$$

## Key Terms

Binomial distribution characteristics pg. 181
Continuous random variable pg. 173
Counting rule for combinations pg. 183

Discrete random variable pg. 173
Expected value pg. 175
Hypergeometric distribution pg. 198

Random variable pg. 173
Poisson distribution pg. 193

## Chapter Exercises

### Conceptual Questions

**5-72.** Three discrete distributions were discussed in this chapter. Each was defined by a random variable that measured the number of successes. To apply these distributions, you must know which one to use. Describe the distinguishing characteristics for each distribution.

**5-73.** How is the shape of the binomial distribution changed for a given value of $p$ as the sample size is increased? Discuss.

**5-74.** Discuss the basic differences and similarities between the binomial distribution and the Poisson distribution.

**5-75.** Beginning statistics students are often puzzled by two characteristics of distributions in this chapter: (1) the trials are independent, and (2) the probability of a success remains constant from trial to trial. Students often think these two characteristics are the same. The questions in this exercise point out the difference. Consider a hypergeometric distribution where $N = 3, X = 2$, and $n = 2$.
   a. Mathematically demonstrate that the trials for this experiment are dependent by calculating the probability of obtaining a success on the second trial if the first trial resulted in a success. Repeat this calculation if the first trial was a failure. Use these two probabilities to prove that the trials are dependent.
   b. Now calculate the probability that a success is obtained on each of the three respective trials and, therefore, demonstrate that the trials are dependent

but that the probability of a success is constant from trial to trial.

**5-76.** Consider an experiment in which a sample of size $n = 5$ is taken from a binomial distribution.
   a. Calculate the probability of each value of the random variable for the probability of a success equal to (1) 0.1, (2) 0.25, (3) 0.50, (4) 0.75, and (5) 0.9.
   b. Which probabilities produced a right-skewed distribution? Why?
   c. Which probability of a success yielded a symmetric distribution? Why?
   d. Which probabilities produced a left-skewed distribution? Discuss why.

### Business Applications

**5-77.** The McMillan Newspaper Company sometimes makes printing errors in its advertising and is forced to provide corrected advertising in the next issue of the paper. The managing editor has done a study of this problem and found the following data:

| No. of Errors $x$ | Relative Frequency |
|---|---|
| 0 | 0.56 |
| 1 | 0.21 |
| 2 | 0.13 |
| 3 | 0.07 |
| 4 | 0.03 |

a. Using the relative frequencies as probabilities, what is the expected number of errors? Interpret what this value means to the managing editor.

b. Compute the variance and standard deviation for the number of errors and explain what these values measure.

5-78. The Ziteck Corporation buys parts from international suppliers. One part is currently being purchased from a Malaysian supplier under a contract that calls for at most 5% of the 10,000 parts to be defective. When a shipment arrives, Ziteck randomly samples 10 parts. If it finds 2 or fewer defectives in the sample, it keeps the shipment; otherwise, it returns the entire shipment to the supplier.

a. Assuming that the conditions for the binomial distribution are satisfied, what is the probability that the sample will lead Ziteck to keep the shipment if the defect rate is actually 0.05?

b. Suppose the supplier is actually sending Ziteck 10% defects. What is the probability that the sample will lead Ziteck to accept the shipment anyway?

c. Comment on this sampling plan (sample size and accept/reject point). Do you think it favors either Ziteck or the supplier? Discuss.

5-79. California-based Wagner Foods, Inc., has a process that inserts fruit juice into 24-ounce containers. When the process is in control, half the cans actually contain more than 24 ounces and half contain less. Suppose a quality inspector has just randomly sampled nine cans and found that all nine had more than 24 ounces. Calculate the probability that this result would occur if the filling process was actually still in control. Based on this probability, what conclusion might be reached? Discuss.

5-80. Your company president has told you that the company experiences product returns at the rate of two per month with the number of product returns distributed as a Poisson random variable. Determine the probability that next month there will be

a. no returns

b. one return

c. two returns

d. more than two returns

e. In the last three months your company has had only one month in which the number of returns was at most two. Calculate the probability of this event occurring. What will you tell the president of your company concerning the return rate? Make sure you support your statement with something other than opinion.

5-81. The Defense Department has recently advertised for bids for producing a new night-vision binocular. Vista Optical has decided to submit a bid for the contract. The first step was to supply a sample of binoculars for the army to test at its Kentucky development grounds. Vista makes a superior night-vision binocular. However, the 4 sent to the army for testing were taken from a development-lab project of 20 units that contained 4 defectives. The army has indicated it will reject any

manufacturer that submits 1 or more defective binoculars. What is the probability that this mistake has cost Vista any chance for the contract?

5-82. *USA Today* reported that only 0.8% of all big-business partnerships are audited by the IRS (*source:* Gregory Korte, "IRS audits less than 1% of big-business partnerships," *USA Today*, Apr. 17, 2014).

a. One member of a partnership's board of directors is on the board of directors of four other large partnerships. Calculate the expected number of these five partnerships that should get audited, assuming selection is random.

b. Three of the five partnerships were actually audited. Determine the probability that at least three of the five partnerships would be audited if 0.8% of large partnerships are audited. (Assume random selection.)

c. The board member is concerned that the partnerships have been singled out to be audited by the Internal Revenue Service (IRS). Respond to these thoughts using probability and statistical logic.

5-83. Stafford Production, Inc., is concerned with the quality of the parts it purchases that will be used in the end items it assembles. Part number 34-78D is used in the company's new laser printer. The parts are sensitive to dust and can easily be damaged in shipment even if they are acceptable when they leave the vendor's plant. In a shipment of four parts, the purchasing agent has assessed the following probability distribution for the number of defective products:

| x | P(x) |
|---|------|
| 0 | 0.20 |
| 1 | 0.20 |
| 2 | 0.20 |
| 3 | 0.20 |
| 4 | 0.20 |

a. What is the expected number of defectives in a shipment of four parts? Discuss what this value really means to Stafford Production, Inc.

b. Compute and interpret the standard deviation of the number of defective parts in a shipment of four.

c. Examine the probabilities as assessed and indicate why this probability distribution might be called a uniform distribution. Provide some reasons why the probabilities might all be equal, as they are in this case.

5-84. Bach Photographs takes school pictures and charges only $0.99 for a sitting, which consists of six poses. The company then makes up three packages that are offered to the parents, who have a choice of buying 0, 1, 2, or all 3 of the packages. Based on his experience in the business, Bill Bach has assessed the following probabilities of the number of packages that a parent purchases:

| No. of Packages x | P(x) |
|---|---|
| 0 | 0.30 |
| 1 | 0.40 |
| 2 | 0.20 |
| 3 | 0.10 |

a. What is the expected number of packages purchased by each parent?
b. What is the standard deviation for the random variable, $x$?
c. Suppose all of the picture packages are to be priced at the same level. How much should they cost if Bach Photographs wants to break even? Assume that the production costs are $3.00 per package. Remember that the sitting charge is $0.99.

**5-85.** The managing partner for Westwood One Investment Managers, Inc., gave a public seminar in which she discussed investment risk analysis. She reminded people that the coefficient of variation often can be used as a measure of risk of an investment. (See Chapter 3 for a review of the coefficient of variation.) To demonstrate her point, she used two hypothetical stocks as examples. She let $x$ equal the change in assets for a $1,000.00 investment in Stock 1 and $y$ reflect the change in assets for a $1,000.00 investment in Stock 2. She showed the seminar participants the following probability distributions:

| x | P(x) | y | P(y) |
|---|---|---|---|
| −$1,000.00 | 0.10 | −$1,000.00 | 0.20 |
| 0.00 | 0.10 | 0.00 | 0.40 |
| 500.00 | 0.30 | 500.00 | 0.30 |
| 1,000.00 | 0.30 | 1,000.00 | 0.05 |
| 2,000.00 | 0.20 | 2,000.00 | 0.05 |

a. Compute the expected values for random variables $x$ and $y$.
b. Compute the standard deviations for random variables $x$ and $y$.
c. Recalling that the coefficient of variation is determined by the ratio of the standard deviation to the mean, compute the coefficient of variation for each random variable.
d. Referring to part c, suppose the seminar director said that the first stock was riskier, since its standard deviation was greater than the standard deviation of the second stock. How would you respond? (*Hint:* What do the coefficients of variation imply?)

**5-86.** A company located in the southern part of the United States has slightly more than 2,200 employees. Of these, 35% have a college degree. Twenty employees were selected at random to potentially receive special training in quality-improvement techniques. Half of these 20 have college degrees. Four of these 20 were then randomly selected to actually attend the training program.

a. What is the probability that all four of those selected have a college degree?
b. Assuming that the four are selected at random, what is the expected number among those four who have college degrees?
c. What is the probability that fewer than 2 of those in the final group have a college degree?

**5-87.** On a busy weekend, a hotel has 20 rooms that are not occupied. Suppose that smoking is allowed in 8 of the rooms. A small tour group arrives that has four smokers and six nonsmokers. The desk clerk randomly selects 10 rooms and gives the keys to the tour guide to distribute to the travelers.

a. Compute the probability that the tour guide will have the correct mix of rooms so that all members of the tour group will receive a room that accommodates their smoking preferences.
b. Determine the probability that the tour guide will assign at least one nonsmoker to a smoking room.

**5-88.** If 70% of all Millennials are currently saving for retirement, calculate the probabilities of the following events from a random sample of 15 Millennials:

a. exactly 5 are saving for retirement
b. more than 7 are saving for retirement
c. between 3 and 8 are saving for retirement
d. fewer than 6 are saving for retirement
e. exactly 3 are not saving for retirement

**5-89.** If 80% of all business airline travelers prefer to carry on baggage rather than check it, calculate the probabilities of the following events from a random sample of 25 business travelers:

a. exactly 15 prefer to carry on baggage
b. fewer than 5 prefer to check baggage
c. more than 10 prefer to carry on baggage

**5-90.** If 80% of all college students in the United States text during classes, calculate the probabilities of the following events from a random sample of 15 college students:

a. exactly 9 text during class
b. more than 6 text during class
c. between 3 and 12 text during class

**5-91.** A machine produces parts continuously throughout the day with an overall defect rate of 0.80%. Periodically a random sample of 50 parts is sampled, and if 3 or more are found to be defective the machine is stopped and adjusted.

a. What is the probability that in the next sample of 50 parts the machine is stopped for adjustment?
b. What is the probability that in the next sample of 50 parts the machine is allowed to keep processing parts even though its defect rate has increased to 1.1%?
c. How many parts would need to be sampled, on average, in order to find one defective part if the overall defect rate is 0.80%?

**5-92.** A large mail sorting center finds that, on average, 0.5% of all letters have an incorrect address.

a. What is the probability that in a random sample of 500 letters at the center exactly 5 letters have an incorrect address?

b. What is the probability that the random variable $x$, where $x = $ Number of letters with an incorrect address, will exceed its mean by more than 2 standard deviations in a random sample of 500 letters?

**5-93.** The manager of an online shopping website has determined that an average of 11 customers per minute make a purchase on Saturdays. The manager believes that each customer's purchase is independent and has decided to model the number of purchases per minute using a Poisson probability distribution.

a. What is the probability that during a one-minute interval on Saturday exactly 8 purchases will be made?

b. What is the probability that more than 15 purchases will be made during a one-minute interval on Saturday?

## Computer Software Exercises

**5-94.** A 23-mile stretch of a two-lane highway east of Paso Robles, California, was once considered a "death trap" by residents of San Luis Obispo County. Formerly known as "Blood Alley," Highway 46 gained notoriety for the number of fatalities (29) and crashes over a 240-week period. More than two-thirds involved head-on collisions. The file titled **Crashes** contains the simulated number of fatal crashes during this time period.

a. Determine the average number of crashes in the 240 weeks.

b. Calculate the probability that at least 19 crashes would occur over the 240-week period if the average number of crashes per week was as calculated in part a.

c. Calculate the probability that at least 19 crashes would occur over a five-year period if the average number of crashes per week was as calculated in part a.

d. A coalition of state, local, and private organizations devised a coordinated and innovative approach to dramatically reduce deaths and injuries on this road. During the 16 months before and after completion of the project, fatal crashes were reduced to zero. Calculate the probability that there would be no fatal crashes if the mean number of fatal crashes was not changed by the coalition. Does it appear that the average number of fatal accidents has indeed decreased?

**5-95.** Darlington Industries produces a wide array of home safety and security products, including smoke alarms. As part of its quality-control program, it constantly tests to assure that the alarms work. A change in the manufacturing process requires the company to determine the proportion of alarms that fail the quality-control tests. Each day, 20 smoke alarms are taken from the production line and tested, and the number of defectives is recorded. A file titled **Smokeless** contains the results from the last 90 days of testing.

a. Compute the proportion of defective smoke alarms.

b. Calculate the expected number and the standard deviation of defectives for each day's testing. Assume the proportion of defectives is what was computed in part a. (*Hint:* Recall the formulas for the mean and the standard deviation for a binomial distribution.)

c. To make sure that the proportion of defectives does not change, the quality-control manager wants to establish control limits that are 3 standard deviations above the mean and 3 standard deviations below the mean. Calculate these limits.

d. Determine the probability that a randomly chosen set of 20 smoke alarms would have a number of defectives that was beyond the control limits established in part c.

**5-96.** Covercraft manufactures covers to protect automobile and boat interiors and finishes. Its Block-It 200 Series fabric has a limited two-year warranty. Periodic testing is done to determine if the warranty policy should be changed. One such study examined those covers that became unserviceable while still under warranty. Data that could be produced by such a study are contained in the file titled **Covers**. The data represent the number of months a cover was used until it became unserviceable. Covercraft might want to examine more carefully the covers that became unserviceable while still under warranty. Specifically, it wants to examine those that became unserviceable before they had been in use one year.

a. Determine the number of covers that became unserviceable before they had been in use less than a year and a half.

b. If Covercraft quality-control staff selects 20 of the covers at random, determine the probability that none of them will have failed before they had been in service a year and a half.

c. If Covercraft quality-control staff needs to examine at least 5 of the failed covers, determine the probability that they will obtain this many.

## Case 5.1    SaveMor Pharmacies

A common practice now is for large retail pharmacies to buy the customer base from smaller, independent pharmacies. The way this works is that the buyer requests to see the customer list along with the buying history. The buyer then makes an offer based on its projection of how many of the seller's customers will move their business to the buyer's pharmacy and on how many dollars of new business will come to the buyer as a result of the purchase. Once the deal is made, the buyer and seller usually send out a joint letter to the seller's customers explaining the transaction and informing them that their prescription files have been transferred to the purchasing company.

The problem is that there is no guarantee regarding what proportion of the existing customers will make the switch to the buying company. That is the issue facing Heidi Fendenand, acquisitions manager for SaveMor Pharmacies. SaveMor has the opportunity to purchase the 6,780-person customer base from Hubbard Pharmacy in San Jose, California. Based on previous acquisitions, Heidi believes that if 70% or more of the customers will make the switch, then the deal is favorable to SaveMor. However, if 60% or less make the move to SaveMor, then the deal will be a bad one and she would recommend against it.

Quincy Kregthorpe, a research analyst who works for Heidi, has suggested that SaveMor take a new approach to this acquisition decision. He has suggested that SaveMor contact a random sample of 20 Hubbard customers telling them of the proposed sale and asking them if they will be willing to switch their business to SaveMor. Quincy has suggested that if 15 or more of the 20 customers indicate that they would make the switch, then SaveMor should go ahead with the purchase. Otherwise, it should decline the deal or negotiate a lower purchase price.

Heidi liked this idea and contacted Cal Hubbard, Hubbard's owner, to discuss the idea of surveying 20 randomly selected customers. Cal was agreeable as long as only these 20 customers would be told about the potential sale.

Before taking the next step, Heidi met with Quincy to discuss the plan one more time. She was concerned that the proposed sampling plan might have too high a probability of rejecting the purchase deal even if it was a positive one from SaveMor's viewpoint. On the other hand, she was concerned that the plan might also have a high probability of accepting the purchase deal when in fact it would be unfavorable to SaveMor. After discussing these concerns for over an hour, Quincy finally offered to perform an evaluation of the sampling plan.

## Required Tasks:

1. Compute the probability that the sampling plan will provide a result that suggests that SaveMor should reject the deal even if the true proportion of all customers who would switch is actually 0.70.
2. Compute the probability that the sampling plan will provide a result that suggests that SaveMor should accept the deal even if the true proportion of all customers who would switch is actually only 0.60.
3. Write a short report to Heidi outlining the sampling plan, the assumptions on which the evaluation of the sampling plan has been based, and the conclusions regarding the potential effectiveness of the sampling plan. The report should make a recommendation about whether Heidi should go through with using the sampling plan.

| Case 5.2 | Arrowmark Vending |
|---|---|

Arrowmark Vending has the contract to supply pizza at all home football games for a university in the Big 12 athletic conference. It is a constant challenge at each game to determine how many pizzas to have available at the games. Tom Kealey, operations manager for Arrowmark, has determined that his fixed cost of providing pizzas, whether he sells 1 pizza or 4,000 pizzas, is $1,000. This cost includes hiring employees to work at the concession booths, hiring extra employees to cook the pizzas the day of the game, delivering them to the game, and advertising during the game. He believes that this cost should be equally allocated between two types of pizzas.

Tom has determined that he will supply only two types of pizzas: plain cheese and pepperoni-and-cheese combo. His cost to make a plain cheese pizza is $4.50 each, and his cost to make a pepperoni-and-cheese combo is $5.00 each. Both pizzas will sell for $9.00 at the game. Unsold pizzas have no value and are donated to a local shelter for the homeless.

Experience has shown the following demand distributions for the two types of pizza at home games:

| Plain Cheese Demand | Probability | Pepperoni-and-Cheese Demand | Probability |
|---|---|---|---|
| 200 | 0.10 | 300 | 0.10 |
| 300 | 0.15 | 400 | 0.20 |
| 400 | 0.15 | 500 | 0.25 |
| 500 | 0.20 | 600 | 0.25 |
| 600 | 0.20 | 700 | 0.15 |
| 700 | 0.10 | 800 | 0.05 |
| 800 | 0.05 | | |
| 900 | 0.05 | | |

## Required Tasks:

1. For each type of pizza, determine the profit (or loss) associated with producing at each possible demand level. For instance, determine the profit if 200 plain cheese pizzas are produced and 200 are demanded. What is the profit if 200 plain cheese pizzas are produced but 300 were demanded, and so on?

2. Compute the expected profit associated with each possible production level (assuming Tom will produce at only one of the possible demand levels) for each type of pizza.
3. Prepare a short report that provides Tom with the information regarding how many of each type of pizza he should produce if he wants to achieve the highest expected profit from pizza sales at the game.

## Case 5.3 Boise Cascade Corporation

At the Boise Cascade Corporation, lumber mill logs arrive by truck and are scaled (measured to determine the number of board feet) before they are dumped into a log pond. Figure C-5.3 illustrates the basic flow. The mill manager must determine how many scale stations to have open during various times of the day. If he has too many stations open, the scalers will have excessive idle time and the cost of scaling will be unnecessarily high. On the other hand, if too few scale stations are open, some log trucks will have to wait.

**FIGURE C-5.3 Truck Flow for Boise Cascade Mill Example**

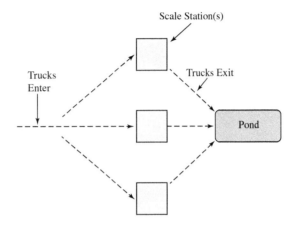

The manager has studied the truck arrival patterns and has determined that during the first open hour (7:00 A.M.–8:00 A.M.), the trucks randomly arrive at 12 per hour on average. Each scale station can scale 6 trucks per hour (10 minutes each). If the manager knew how many trucks would arrive during the hour, he would know how many scale stations to have open.

| 0 to 6 trucks: | open 1 scale station |
| 7 to 12 trucks: | open 2 scale stations, etc. |

However, the number of trucks is a random variable and is uncertain. Your task is to provide guidance for the decision.

# 6 Introduction to Continuous Probability Distributions

### 6.1 The Normal Probability Distribution (pg. 213–226)

**OUTCOME 1** Convert a normal distribution to a standard normal distribution.

**OUTCOME 2** Determine probabilities using the standard normal distribution.

**OUTCOME 3** Calculate values of the random variable associated with specified probabilities from a normal distribution.

### 6.2 Other Continuous Probability Distributions (pg. 226–232)

**OUTCOME 4** Calculate probabilities associated with a uniformly distributed random variable.

**OUTCOME 5** Determine probabilities using an exponential probability distribution.

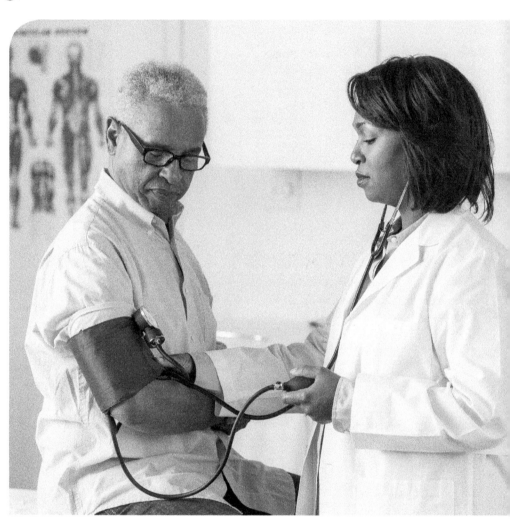

## WHY YOU NEED TO KNOW

You will encounter many business situations in which the random variable of interest is discrete and in which probability distributions such as the binomial, the Poisson, or the hypergeometric will be useful in making decisions. Chapter 5 introduced these distributions and provided many examples and applications. But there will also be many situations in which the random variable of interest is *continuous* rather than discrete. For instance, a pharmaceutical company may be interested in the probability that a new drug will reduce blood pressure by more than 20 points for patients. Blood pressure is the continuous random variable of interest. Kellogg Company could be interested in the probability that cereal boxes labeled as containing 16 ounces will

### Quick Prep

Review the methods for determining the probability for a discrete random variable in Chapter 5.

Review the discussion of the mean and standard deviation in Sections 3.1 and 3.2.

Review the concept of *z*-scores outlined in Section 3.3.

actually contain at least that much cereal. Here, the variable of interest is weight, which can be measured on a continuous scale.

The managers at Harley Davidson might be interested in a measure called throughput time, which is the time it takes from when a motorcycle is started on the manufacturing line until it is completed. Lots of factors can affect the throughput time, including breakdowns, need for rework, the type of accessories added to the motorcycle, and worker productivity. The managers might be interested in determining the probability that the throughput time will be between 3.5 and 5.0 hours. In this case, time is the random variable of interest and is continuous.

In each of these examples, the value of the variable of interest is determined by measuring (measuring the blood pressure reading, measuring the weight of cereal in a box, measuring the time required to make a motorcycle). In every instance, the number of possible values for the variable is limited only by the capacity of the measuring device. Chapter 6 introduces the *normal distribution*, which is important because a great many applications involve random variables that possess the characteristics of the normal distribution. In addition, many of the topics in the remaining chapters are based on the normal distribution. In addition to the normal distribution, you will be introduced to the uniform distribution and the exponential distribution. Both are important continuous probability distributions and have many applications in business decision making. You need to have a firm understanding and working knowledge of all three continuous probability distributions introduced in this chapter.

| 6.1 | **The Normal Distribution** |

A Pepsi can is supposed to contain 12 ounces, but it might actually contain any amount between 11.90 and 12.10 ounces, such as 11.9853 ounces. When the variable of interest, such as the volume of soda in a can, is approximately continuous, the probability distribution associated with the random variable is called a *continuous probability distribution*.

One important difference between discrete and continuous probability distributions involves the calculation of probabilities associated with specific values of the random variable. For instance, in a market research example in which 100 people are surveyed and asked whether they have a positive view of a product, we could use the binomial distribution introduced in Chapter 5 to find the probability of any specific number of positive reviews, such as $P(x = 75)$ or $P(x = 76)$. Although these individual probabilities may be small values, they can be computed because the random variable is discrete. However, if the random variable is continuous, as in the Pepsi example, there is an uncountable infinite number of possible outcomes for the random variable. Theoretically, the probability of any one of these individual outcomes is 0. That is, $P(x = 11.92) = 0$ or $P(x = 12.05) = 0$. Thus, when you are working with continuous distributions, you need to find the probability for a range of possible values such as $P(x \leq 11.92)$ or $P(11.92 \leq x \leq 12.0)$. You can also conclude that

$$P(x \leq 11.92) = P(x < 11.92)$$

because we assume that $P(x = 11.92) = 0$.

There are many different continuous probability distributions, but the most important of these is the **normal distribution**.

### The Normal Distribution[1]

Figure 6.1 illustrates a typical normal distribution and highlights the normal distribution's characteristics. All normal distributions have the same general bell shape as the one shown in Figure 6.1. However, they can differ in their mean value and their variation, depending on the situation being considered. The process being represented determines the scale of the horizontal axis. It may be pounds, inches, dollars, or any other attribute with a continuous measurement. Figure 6.2 shows three normal distributions with different centers and different spreads. Note that the total area (probability) under each normal curve equals 1.

The normal distribution is described by the rather complicated-looking probability density function, Equation 6.1.

**Normal Distribution**

The normal distribution is a bell-shaped distribution with the following properties:

1. It is *unimodal*; that is, the normal distribution peaks at a single value.

2. It is *symmetric*; this means that the two areas under the curve between the mean and any two points equidistant on either side of the mean are identical. One side of the distribution is the mirror image of the other side.

3. The mean, median, and mode are equal.

4. The normal distribution approaches the horizontal axis on either side of the mean toward plus and minus infinity ($\infty$). In more formal terms, the normal distribution is *asymptotic* to the *x* axis.

5. The amount of variation in the random variable determines the height and spread of the normal distribution.

---

[1]It is common to refer to the very large family of normal distributions as "*the* normal distribution."

**FIGURE 6.1 Characteristics of the Normal Distribution**

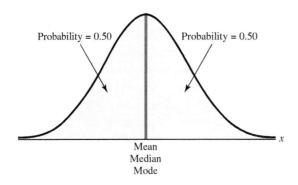

**FIGURE 6.2 Different Normal Distribution Centers and Spreads**

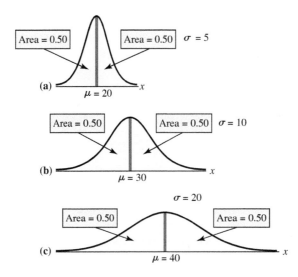

**Normal Probability Density Function**

$$f(x) = \frac{1}{\sigma\sqrt{2\pi}}e^{-(x-\mu)^2/2\sigma^2}$$

(6.1)

where:

$x =$ Any value of the continuous random variable

$\sigma =$ Population standard deviation

$\pi = 3.14159$

$e =$ Base of the natural log $= 2.71828$

$\mu =$ Population mean

**Standard Normal Distribution**

A normal distribution that has a mean = 0.0 and a standard deviation = 1.0. The horizontal axis is scaled in z-values that measure the number of standard deviations a point is from the mean. Values above the mean have positive z-values. Values below the mean have negative z-values.

To graph the normal distribution, we need to know the mean, $\mu$, and the standard deviation, $\sigma$. Placing $\mu, \sigma$, and a value of the variable, $x$, into the probability density function, we can calculate a height, $f(x)$, of the density function. If we could try enough $x$ values, we could construct curves like those shown in Figures 6.1 and 6.2.

The area under the normal curve corresponds to probability. Because $x$ is a continuous random variable, the probability, $P(x)$, is equal to 0 for any particular $x$. However, we can find the probability for a range of values between $x_1$ and $x_2$ by finding the area under the curve between these two values. A special normal distribution called the *standard normal distribution* is used to find areas (probabilities) for all normal distributions.

**OUTCOME 1**

## The Standard Normal Distribution

The trick to finding probabilities for a normal distribution is to convert the normal distribution to a **standard normal distribution**.

To convert a normal distribution to a standard normal distribution, the values ($x$) of the random variable are standardized as outlined in Chapter 3. The conversion formula is shown as Equation 6.2.

### Standardized Normal z-Value

$$z = \frac{x - \mu}{\sigma} \tag{6.2}$$

where:

> $z$ = Scaled value (the number of standard deviations a point $x$ is from the mean)
> $x$ = Any point on the horizontal axis
> $\mu$ = Mean of the specific normal distribution
> $\sigma$ = Standard deviation of the specific normal distribution

Equation 6.2 *scales* any normal distribution axis from its true units (time, weight, dollars, volume, and so forth) to the standard measure referred to as a *z-value*. Thus, any value of the normally distributed continuous random variable can be represented by a unique *z*-value. Positive *z*-values represent corresponding values of the random variable, $x$, that are higher than the population mean. Values of $x$ that are less than the population mean have *z*-values that are negative.

---

**BUSINESS APPLICATION**    ## Standard Normal Distribution

**EMT Response Times**    The time it takes an emergency response team to arrive at the scene where assistance is needed is very important to the medical outcome. In a particular city, data show that the mean response time is 4 minutes with a standard deviation of 2 minutes. The response times are also shown to be normally distributed as shown in Figure 6.3.

Recently, the EMT director pulled data for three emergency calls and found the following response times:

> Call 1: 4 minutes
> Call 2: 5.5 minutes
> Call 3: 2.3 minutes

We use Equation 6.2 to convert these values from a normally distributed population with $\mu = 4$ and $\sigma = 2$ to corresponding *z*-values in a standard normal distribution. For Call 1, we get

The Excel 2016 function for *z*-values is

=STANDARDIZE(x,mean, standard_dev)
=STANDARDIZE(4,4,2)

$$z = \frac{x - \mu}{\sigma} = \frac{4 - 4}{2} = 0$$

Note that, Call 1 took 4 minutes, which happens to be equal to the population mean for all calls. The standardized *z*-value corresponding to the population mean is 0. This indicates that the population mean is 0 standard deviations from itself.

For Call 2, we get

The Excel 2016 function for *z*-values is

=STANDARDIZE(x,mean, standard_dev)
=STANDARDIZE(5.5,4,2)

$$z = \frac{x - \mu}{\sigma} = \frac{5.5 - 4}{2} = 0.75$$

**FIGURE 6.3 Distribution of EMT Call Times**

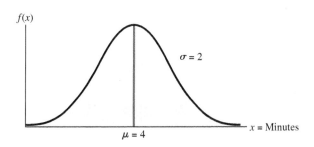

**FIGURE 6.4 Standard Normal Distribution**

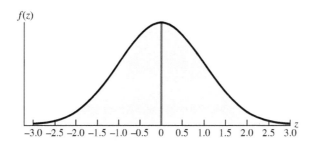

Thus, a call that takes 5.5 minutes is 0.75 standard deviation higher than the mean for all calls. The standardized $z$-value for Call 3 is

$$z = \frac{x - \mu}{\sigma} = \frac{2.3 - 4}{2} = -0.85$$

This means a call that takes only 2.3 minutes is 0.85 standard deviation below the population mean. *Note:* A negative $z$-value always indicates the $x$-value is less than the mean, $\mu$.

The $z$-value represents the number of standard deviations a point is above or below the population mean. Equation 6.2 can be used to convert any specified value, $x$, from the population distribution to a corresponding $z$-value. If the population distribution is normally distributed as shown in Figure 6.3, then the distribution of $z$-values is also normally distributed and is called the *standard normal distribution*. Figure 6.4 shows a standard normal distribution where the horizontal axis represents $z$-*values*. Note, virtually the entire area under a normal distribution falls within 3 standard deviations either side of the mean.

You can convert the normal distribution to a standard normal distribution and use the standard normal table to find the desired probability. Example 6-1 shows the steps required to do this.

OUTCOME **2**

## Using the Standard Normal Table

The *standard normal table* in Appendix D provides probabilities (or areas under the normal curve) associated with many different $z$-values. The standard normal table is constructed so that the probabilities provided represent the chance of a value being between a positive $z$-value and its population mean, 0.

The standard normal table is also reproduced in Table 6.1. This table provides probabilities for $z$-values between $z = 0.00$ and $z = 3.09$. Note that because the normal distribution is symmetric, the probability of a value being between a positive $z$-value and its population mean, 0, is the same as the probability of a value being between a negative $z$-value and its population mean, 0. So we can use one standard normal table for both positive and negative $z$-values.

**HOW TO DO IT** (Example 6-1)

**Using the Normal Distribution**

**1.** Determine the mean, $\mu$, and the standard deviation, $\sigma$.

**2.** Define the event of interest, such as $P(x \geq x_1)$.

**3.** Convert the normal distribution to the standard normal distribution using Equation 6.2:

$$z = \frac{x - \mu}{\sigma}$$

**4.** Use the standard normal distribution table to find the probability associated with the calculated $z$-value. The table gives the probability between the $z$-value and the mean.

**5.** Determine the desired probability using the knowledge that the probability of a value being on either side of the mean is 0.50 and the total probability under the normal distribution is 1.0.

| **EXAMPLE 6-1** | **Using the Standard Normal Table** |

**Airline Passenger Loading Times**    After completing a study, Chicago O'Hare Airport managers have concluded that the time needed to get passengers loaded onto an airplane is normally distributed with a mean equal to 15 minutes and a standard deviation equal to 3.5 minutes. Recently one airplane required 22 minutes to get passengers on board and ready for takeoff. To find the probability that a flight will take 22 or more minutes to get passengers loaded, you can use the following steps:

STEP 1    **Determine the mean and standard deviation for the random variable.**
The parameters of the probability distribution are

$$\mu = 15 \quad \text{and} \quad \sigma = 3.5$$

STEP 2    **Define the event of interest.**
The flight load time is 22 minutes. We wish to find

$$P(x \geq 22) = ?$$

> The Excel 2016 function for normal distribution probabilities is
>
> = **NORM.DIST(x, mean, standard_dev,cumulative)**
> = **NORM.DIST(22,15,3.5, True) = .9772**
>
> *Note:* The NORM.DIST function gives the cumulative probability for all values equal to or less than *x* when cumulative is set to TRUE.

STEP 3 **Convert the random variable to a standardized value using Equation 6.2.**

$$z = \frac{x - \mu}{\sigma} = \frac{22 - 15}{3.5} = 2.00$$

STEP 4 **Find the probability associated with the z-value in the standard normal distribution table (Table 6.1 or Appendix D).**

To find the probability associated with $z = 2.00$ [ i.e., $P(0 \le z \le 2.00)$ ], do the following:

1. Go down the left-hand column of the table to $z = 2.0$.
2. Go across the top row of the table to the column 0.00 for the second decimal place in $z = 2.00$.
3. Find the value where the row and column intersect.

The value, 0.4772, is the probability that a value in a normal distribution will lie between the mean and 2.00 standard deviations above the mean.

> The Excel 2016 function for normal distribution probabilities $P(x \ge 22)$ is found using
>
> = **1 − NORM.DIST(22,15,3.5, True)**

STEP 5 **Determine the probability for the event of interest.**

$$P(x \ge 22) = ?$$

We know that the area on each side of the mean under the normal distribution is equal to 0.50. In Step 4, we computed the probability associated with $z = 2.00$ to be 0.4772, which is the probability of a value falling between the mean and 2.00 standard deviations above the mean. Then, the probability we are looking for is

$$P(x \ge 22) = P(z \ge 2.00) = 0.5000 - 0.4772 = 0.0228$$

**TRY EXERCISE 6-2** (pg. 223)

---

**BUSINESS APPLICATION**    ## The Normal Distribution

**Fruit Production**   Consider a situation involving a fruit grower in Florida. The grapefruit from this grower's orchard have a mean weight of 16 ounces. The grapefruit weight is normally distributed with $\mu = 16$ and $\sigma = 4$. A local television station that runs a consumer advocacy program reported that a grapefruit from this grower was selected and weighed only 14 ounces. The reporter said she thought it should have been heavier if the mean weight is supposed to be 16 ounces. The grower, when interviewed, said that he thought the probability was quite high that a grapefruit would weigh 14 or more ounces. To check his statement out, we want to find

$$P(x \ge 14) = ?$$

This probability corresponds to the area under a normal distribution to the right of $x = 14$ ounces. This will be the sum of the area between $x = 14$ and $\mu = 16$ and the area to the right of $\mu = 16$. Refer to Figure 6.5.

To find this probability, we first convert $x = 14$ ounces to its corresponding $z$-value. This is equivalent to determining the number of standard deviations $x = 14$ is from the population mean of $\mu = 16$. Equation 6.2 is used to do this as follows:

> The Excel 2016 function for z-values is
>
> = **STANDARDIZE(x,mean, standard_dev)**
> = **STANDARDIZE(14,16,4)**

$$z = \frac{x - \mu}{\sigma} = \frac{14 - 16}{4} = -0.50$$

Because the normal distribution is symmetric, even though the $z$-value is –0.50, we find the desired probability by going to the standard normal distribution table for a positive $z = 0.50$. The probability in the table for $z = 0.50$ corresponds to the probability of a $z$-value occurring between $z = 0.50$ and $z = 0.00$. This is the same as the probability of a $z$-value falling between $z = -0.50$ and $z = 0.00$. Thus, from the standard normal table (Table 6.1 or Appendix D), we get

$$P(-0.50 \le z \le 0.00) = 0.1915$$

**TABLE 6.1** Standard Normal Distribution Table

0.1985
Example:
$z = 0.52$ (or $-0.52$)
$P(0 \leq z \leq 0.52) = 0.1985$, or 19.85%

| z | .00 | .01 | .02 | .03 | .04 | .05 | .06 | .07 | .08 | .09 |
|---|-----|-----|-----|-----|-----|-----|-----|-----|-----|-----|
| 0.0 | .0000 | .0040 | .0080 | .0120 | .0160 | .0199 | .0239 | .0279 | .0319 | .0359 |
| 0.1 | .0398 | .0438 | .0478 | .0517 | .0557 | .0596 | .0636 | .0675 | .0714 | .0753 |
| 0.2 | .0793 | .0832 | .0871 | .0910 | .0948 | .0987 | .1026 | .1064 | .1103 | .1141 |
| 0.3 | .1179 | .1217 | .1255 | .1293 | .1331 | .1368 | .1406 | .1443 | .1480 | .1517 |
| 0.4 | .1554 | .1591 | .1628 | .1664 | .1700 | .1736 | .1772 | .1808 | .1844 | .1879 |
| 0.5 | .1915 | .1950 | .1985 | .2019 | .2054 | .2088 | .2123 | .2157 | .2190 | .2224 |
| 0.6 | .2257 | .2291 | .2324 | .2357 | .2389 | .2422 | .2454 | .2486 | .2517 | .2549 |
| 0.7 | .2580 | .2611 | .2642 | .2673 | .2704 | .2734 | .2764 | .2794 | .2823 | .2852 |
| 0.8 | .2881 | .2910 | .2939 | .2967 | .2995 | .3023 | .3051 | .3078 | .3106 | .3133 |
| 0.9 | .3159 | .3186 | .3212 | .3238 | .3264 | .3289 | .3315 | .3340 | .3365 | .3389 |
| 1.0 | .3413 | .3438 | .3461 | .3485 | .3508 | .3531 | .3554 | .3577 | .3599 | .3621 |
| 1.1 | .3643 | .3665 | .3686 | .3708 | .3729 | .3749 | .3770 | .3790 | .3810 | .3830 |
| 1.2 | .3849 | .3869 | .3888 | .3907 | .3925 | .3944 | .3962 | .3980 | .3997 | .4015 |
| 1.3 | .4032 | .4049 | .4066 | .4082 | .4099 | .4115 | .4131 | .4147 | .4162 | .4177 |
| 1.4 | .4192 | .4207 | .4222 | .4236 | .4251 | .4265 | .4279 | .4292 | .4306 | .4319 |
| 1.5 | .4332 | .4345 | .4357 | .4370 | .4382 | .4394 | .4406 | .4418 | .4429 | .4441 |
| 1.6 | .4452 | .4463 | .4474 | .4484 | .4495 | .4505 | .4515 | .4525 | .4535 | .4545 |
| 1.7 | .4554 | .4564 | .4573 | .4582 | .4591 | .4599 | .4608 | .4616 | .4625 | .4633 |
| 1.8 | .4641 | .4649 | .4656 | .4664 | .4671 | .4678 | .4686 | .4693 | .4699 | .4706 |
| 1.9 | .4713 | .4719 | .4726 | .4732 | .4738 | .4744 | .4750 | .4756 | .4761 | .4767 |
| 2.0 | .4772 | .4778 | .4783 | .4788 | .4793 | .4798 | .4803 | .4808 | .4812 | .4817 |
| 2.1 | .4821 | .4826 | .4830 | .4834 | .4838 | .4842 | .4846 | .4850 | .4854 | .4857 |
| 2.2 | .4861 | .4864 | .4868 | .4871 | .4875 | .4878 | .4881 | .4884 | .4887 | .4890 |
| 2.3 | .4893 | .4896 | .4898 | .4901 | .4904 | .4906 | .4909 | .4911 | .4913 | .4916 |
| 2.4 | .4918 | .4920 | .4922 | .4925 | .4927 | .4929 | .4931 | .4932 | .4934 | .4936 |
| 2.5 | .4938 | .4940 | .4941 | .4943 | .4945 | .4946 | .4948 | .4949 | .4951 | .4952 |
| 2.6 | .4953 | .4955 | .4956 | .4957 | .4959 | .4960 | .4961 | .4962 | .4963 | .4964 |
| 2.7 | .4965 | .4966 | .4967 | .4968 | .4969 | .4970 | .4971 | .4972 | .4973 | .4974 |
| 2.8 | .4974 | .4975 | .4976 | .4977 | .4977 | .4978 | .4979 | .4979 | .4980 | .4981 |
| 2.9 | .4981 | .4982 | .4982 | .4983 | .4984 | .4984 | .4985 | .4985 | .4986 | .4986 |
| 3.0 | .4987 | .4987 | .4987 | .4988 | .4988 | .4989 | .4989 | .4989 | .4990 | .4990 |

To illustrate: 19.85% of the area under a normal curve lies between the mean, $\mu$, and a point 0.52 standard deviation units away.

> The Excel 2016 function for normal distribution probabilities $P(x \geq 14)$ is found using
>
> $= 1 - $ **NORM.DIST(14,16,4,True)**

This is the area between $x = 14$ and $\mu = 16$ in Figure 6.5. We now add 0.1915 to 0.5000 $[P(x > 16 = 0.5000)]$. Therefore, the probability that a grapefruit will weigh 14 or more ounces is

$$P(x \geq 14) = 0.1915 + 0.5000 = 0.6915$$

This is illustrated in Figure 6.5. Thus, there is nearly a 70% chance that a grapefruit will weigh at least 14 ounces.

**FIGURE 6.5 Probabilities from the Normal Curve for Fruit Production**

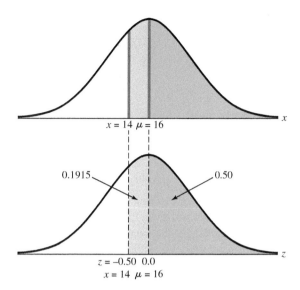

---

**EXAMPLE 6-2**    Using the Normal Distribution

**Lockheed Martin**   Lockheed Martin, the defense contractor, designs and builds communication satellite systems to be used by the U.S. military. Because of the very high cost (more than $1 billion each), the company performs numerous tests on every component. These tests tend to extend the component assembly time. For example, suppose the time required to construct and test (called build time) a particular component part is thought to be normally distributed, with a mean equal to 30 hours and a standard deviation equal to 4.7 hours. To keep the assembly flow moving on schedule, this component needs to have a build time between 26 and 35 hours. To determine the probability of this happening, use the following steps:

STEP 1   **Determine the mean, $\mu$, and the standard deviation, $\sigma$.**
The mean build time for this step in the process is thought to be 30 hours, and the standard deviation is thought to be 4.7 hours.

STEP 2   **Define the event of interest.**
We are interested in determining the following:

$$P(26 \leq x \leq 35) = ?$$

STEP 3   **Convert values of the specified normal distribution to corresponding values of the standard normal distribution using Equation 6.2.**

$$z = \frac{x - \mu}{\sigma}$$

We need to find the $z$-value corresponding to $x = 26$ and to $x = 35$.

$$z = \frac{x - \mu}{\sigma} = \frac{26 - 30}{4.7} = -0.85 \quad \text{and} \quad z = \frac{35 - 30}{4.7} = 1.06$$

STEP 4   **Use the standard normal table to find the probabilities associated with each $z$-value.**

For $z = -0.85$, the probability is 0.3023.
For $z = 1.06$, the probability is 0.3554.

> The Excel 2016 function for normal distribution probabilities $P(26 \leq x \leq 35)$ is found using
>
> **=NORM.DIST(35,30,4.7,True)**
> **− NORM.DIST(26,30,4.7,True)**

STEP 5   **Determine the desired probability for the event of interest.**

$$P(26 \leq x \leq 35) = 0.3023 + 0.3554 = 0.6577$$

Thus, there is a 0.6577 chance that the build time will be such that assembly will stay on schedule.

**TRY EXERCISE 6-13** (pg. 224)

Using the Normal Distribution

**General Electric Company** By the end of 2015, ten states, including California, had passed legislation requiring automakers to sell a certain percentage of zero-emissions cars within their borders. One current alternative is battery-powered cars. The major problem with battery-operated cars is the limited time they can be driven before the batteries must be recharged. Suppose that General Electric (GE) has developed a Longlife battery pack it claims will power a car at a sustained speed of 45 miles per hour for an average of 8 hours. But of course there will be variations: Some battery packs will last longer and some less than 8 hours. Current data indicate that the standard deviation of battery operation time before a charge is needed is 0.4 hour. Data show a normal distribution of uptime on these battery packs. Automakers are concerned that batteries may run short. For example, drivers might find an "8-hour" battery that lasts 7.5 hours or less unacceptable. What are the chances of this happening with the Longlife battery pack?

To calculate the probability the batteries will last 7.5 hours or less, we find the appropriate area under the normal curve shown in Figure 6.6. There is just over 1 chance in 10 that a battery will last 7.5 hours or less when the vehicle is driven at 45 miles per hour.

**FIGURE 6.6 Longlife Battery Uptimes**

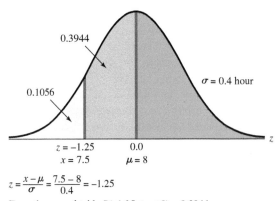

$$z = \frac{x-\mu}{\sigma} = \frac{7.5-8}{0.4} = -1.25$$

From the normal table $P(-1.25 \le z \le 0) = 0.3944$
Then we find $P(x \le 7.5 \text{ hours}) = 0.5000 - 0.3944 = 0.1056$

Suppose this level of reliability is unacceptable to the automakers. Instead of a 10% chance of an "8-hour" battery lasting 7.5 hours or less, the automakers will accept no more than a 2% chance. GE managers ask what the mean uptime would have to be to meet the 2% requirement.

Assuming that uptime is normally distributed, we can answer this question by using the standard normal distribution. However, instead of using the standard normal table to find a probability, we use it in reverse to find the $z$-value that corresponds to a known probability. Figure 6.7 shows the uptime distribution for the battery packs. Note that the 2% probability is shown in the left tail of the distribution. This is the allowable chance of a battery lasting 7.5 hours or less. We must solve for $\mu$, the mean uptime that will meet this requirement.

**FIGURE 6.7 Longlife Battery, Solving for the Mean Uptime**

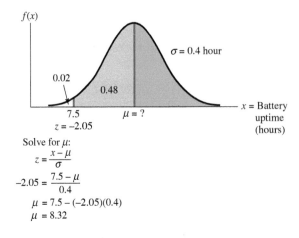

Solve for $\mu$:
$$z = \frac{x-\mu}{\sigma}$$
$$-2.05 = \frac{7.5-\mu}{0.4}$$
$$\mu = 7.5 - (-2.05)(0.4)$$
$$\mu = 8.32$$

1. Go to the body of the standard normal table, where the probabilities are located, and find the probability as close to 0.48 as possible. This is 0.4798.
2. Determine the $z$-value associated with 0.4798. This is $z = 2.05$. Because we are below the mean, the $z$ is negative. Thus, $z = -2.05$.
3. The formula for $z$ is

$$z = \frac{x - \mu}{\sigma}$$

4. Substituting the known values, we get

$$-2.05 = \frac{7.5 - \mu}{0.4}$$

5. Solve for $\mu$:

$$\mu = 7.5 - (-2.05)(0.4) = 8.32 \text{ hours}$$

General Electric will need to increase the mean life of the battery pack to 8.32 hours to meet the automakers' requirement that no more than 2% of the batteries fail in 7.5 hours or less.

---

OUTCOME **3**

Excel Tutorial

**BUSINESS APPLICATION**    Using the Normal Distribution

**State Bank and Trust**    The director of operations for the State Bank and Trust recently performed a study of the time bank customers spent from when they walk into the bank until they complete their banking. The file **State Bank** contains the data for a sample of 1,045 customers randomly observed over a four-week period. The customers in the survey were limited to those who were there for basic bank business, such as making a deposit or a withdrawal or cashing a check. The histogram in Figure 6.8 shows that the service times have an approximate normal distribution.[2]

The mean service time for the 1,045 customers was 22.14 minutes, with a standard deviation equal to 6.09 minutes. On the basis of these data, the manager assumes that the service times are normally distributed with $\mu = 22.14$ and $\sigma = 6.09$. Given these assumptions, the manager is considering providing a gift certificate to a local restaurant to any customer who is required to spend more than 30 minutes to complete basic bank business. Before doing this, she is interested in the probability of having to pay off on this offer.

**Excel 2016 Instructions**

1. Open file: **State Bank.xlsx**.
2. Create a histogram using the Excel instructions from Chapter 2.
3. Delete the gridlines and label axes and title appropriately.

**FIGURE 6.8** **Excel 2016 Output for State Bank and Trust Service Times**

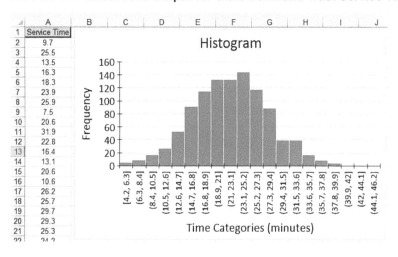

---

[2]A statistical technique known as the chi-square goodness-of-fit test, introduced in Chapter 13, can be used to determine statistically whether the data have a normal distribution.

**FIGURE 6.9** **Normal Distribution for the State Bank and Trust Example**

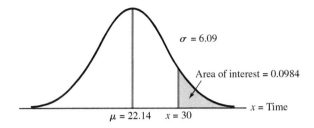

**FIGURE 6.10** **Excel 2016 Output for State Bank and Trust**

**Excel 2016 Instructions**

1. Open a blank worksheet.
2. Select **Formulas**.
3. Click on $f_x$ *(function wizard)*.
4. Select the **Statistical** category.
5. Select the **NORM.DIST** function.
6. Fill in the requested information in the template.
7. **True** indicates cumulative probabilities.
8. Click **OK**.

Figure 6.9 shows the theoretical distribution, with the area of interest identified. The manager is interested in finding

$$P(x > 30 \text{ minutes})$$

This can be done manually or with Excel. Figure 6.10 shows the Excel output. The cumulative probability is

$$P(x \leq 30) = 0.9016$$

Then to find the probability of interest, we subtract this value from 1.0, giving

$$P(x > 30 \text{ minutes}) = 1.0 - 0.9016 = 0.0984$$

Thus, there are just under 10 chances in 100 that the bank would have to give out a gift certificate. Suppose the manager believes this policy is too liberal. She wants to set the time limit so that the chance of giving out the gift is at most only 5%. We can use the standard normal table or Excel to find the new limit. To use the table, we first consider that the manager wants a 5% area in the upper tail of the normal distribution. This will leave

$$0.50 - 0.05 = 0.45$$

between the new time limit and the mean. Now look at the body of the standard normal table, where the probabilities are, and locate the value as close to 0.45 as possible (0.4495 or 0.4505). Next, we find the *z*-value that corresponds to this probability. Because 0.45 lies midway between 0.4495 and 0.4505, we interpolate halfway between $z = 1.64$ and $z = 1.65$ to get

$$z = 1.645$$

Now, we know

$$z = \frac{x - \mu}{\sigma}$$

The Excel 2016 function to find values of *x* corresponding to a normal distribution probability is found using

**=NORM.INV(probability,mean, standard_dev)**
**=NORM.INV(.95,22.14,6.09)**

We then substitute the known values and solve for *x*:

$$1.645 = \frac{x - 22.14}{6.09}$$
$$x = 22.14 + 1.645(6.09)$$
$$x = 32.16 \text{ minutes}$$

Therefore, any customer required to spend more than 32.16 minutes will receive the gift. This should result in no more than 5% of the customers getting the restaurant certificate. Obviously, the bank will work to reduce the average service time or standard deviation so even fewer customers will have to be in the bank for more than 32.16 minutes.

**Approximate Areas under the Normal Curve**   In Chapter 3 we introduced the Empirical Rule for probabilities with bell-shaped distributions. For the normal distribution we can make this rule more precise. Knowing the areas under the normal curve between $\pm 1\sigma$, $\pm 2\sigma$, and $\pm 3\sigma$ provides useful benchmarks for estimating probabilities and checking reasonableness of results. Figure 6.11 shows these benchmark areas for any normal distribution.

**FIGURE 6.11 Approximate Areas under the Normal Curve**

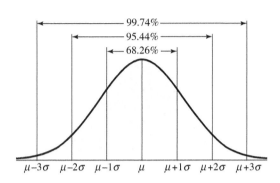

## 6.1 EXERCISES

### Skill Development

**6-1.** For a normally distributed population with $\mu = 200$ and $\sigma = 20$, determine the standardized $z$-value for each of the following:
   a. $x = 225$
   b. $x = 190$
   c. $x = 240$

**6-2.** For a standardized normal distribution, calculate the following probabilities:
   a. $P(z < 1.5)$
   b. $P(z \geq 0.85)$
   c. $P(-1.28 < z < 1.75)$

**6-3.** For a standardized normal distribution, calculate the following probabilities:
   a. $P(0.00 < z \leq 2.33)$
   b. $P(-1.00 < z \leq 1.00)$
   c. $P(1.78 < z < 2.34)$

**6-4.** For a standardized normal distribution, determine a value, say $z_0$, so that
   a. $P(0 < z < z_0) = 0.4772$
   b. $P(-z_0 \leq z < 0) = 0.45$
   c. $P(-z_0 \leq z \leq z_0) = 0.95$
   d. $P(z > z_0) = 0.025$
   e. $P(z \leq z_0) = 0.01$

**6-5.** Consider a random variable, $z$, that has a standardized normal distribution. Determine the following probabilities:

   a. $P(0 < z < 1.96)$
   b. $P(z > 1.645)$
   c. $P(1.28 < z \leq 2.33)$
   d. $P(-2 \leq z \leq 3)$
   e. $P(z > -1)$

**6-6.** A random variable, $x$, has a normal distribution with $\mu = 13.6$ and $\sigma = 2.90$. Determine a value, $x_0$, so that
   a. $P(x > x_0) = 0.05$.
   b. $P(x \leq x_0) = 0.975$.
   c. $P(\mu - x_0 \leq x \leq \mu + x_0) = 0.95$.

**6-7.** For the following normal distributions with parameters as specified, calculate the required probabilities:
   a. $\mu = 5, \sigma = 2$; calculate $P(0 < x < 8)$.
   b. $\mu = 5, \sigma = 4$; calculate $P(0 < x < 8)$.
   c. $\mu = 3, \sigma = 2$; calculate $P(0 < x < 8)$.
   d. $\mu = 4, \sigma = 3$; calculate $P(x > 1)$.
   e. $\mu = 0, \sigma = 3$; calculate $P(x > 1)$.

**6-8.** A population is normally distributed with $\mu = 100$ and $\sigma = 20$.
   a. Find the probability that a value randomly selected from this population will be greater than 130.
   b. Find the probability that a value randomly selected from this population will be less than 90.
   c. Find the probability that a value randomly selected from this population will be between 90 and 130.

**6-9.** A random variable is known to be normally distributed with the following parameters:

$$\mu = 5.5 \quad \text{and} \quad \sigma = 0.50$$

a. Determine the value of $x$ such that the probability of a value from this distribution exceeding $x$ is at most 0.10.

b. Referring to your answer in part a, what must the population mean be changed to if the probability of exceeding the value of $x$ found in part a is reduced from 0.10 to 0.05?

**6-10.** A randomly selected value from a normal distribution is found to be 2.1 standard deviations above its mean.

a. What is the probability that a randomly selected value from the distribution will be greater than 2.1 standard deviations above the mean?

b. What is the probability that a randomly selected value from the distribution will be less than 2.1 standard deviations from the mean?

**6-11.** Assume that a random variable is normally distributed with a mean of 1,500 and a variance of 324.

a. What is the probability that a randomly selected value will be greater than 1,550?

b. What is the probability that a randomly selected value will be less than 1,485?

c. What is the probability that a randomly selected value will be either less than 1,475 or greater than 1,535?

**6-12.** A random variable is normally distributed with a mean of 25 and a standard deviation of 5. If an observation is randomly selected from the distribution,

a. What value will be exceeded 10% of the time?

b. What value will be exceeded 85% of the time?

c. Determine two values of which the smaller has 25% of the values below it and the larger has 25% of the values above it.

d. What value will 15% of the observations be below?

**6-13.** A random variable is normally distributed with a mean of 60 and a standard deviation of 9.

a. What is the probability that a randomly selected value from the distribution will be less than 46.5?

b. What is the probability that a randomly selected value from the distribution will be greater than 78?

c. What is the probability that a randomly selected value will be between 51 and 73.5?

## Business Applications

**6-14.** A global financial institution transfers a large data file every evening from offices around the world to its London headquarters. Once the file is received, it must be cleaned and partitioned before being stored in the company's data warehouse. Each file is the same size and the times required to transfer, clean, and partition a file are normally distributed, with a mean of 1.5 hours and a standard deviation of 15 minutes.

a. If one file is selected at random, what is the probability that it will take longer than 1 hour and 55 minutes to transfer, clean, and partition the file?

b. If a manager must be present until 85% of the files are transferred, cleaned, and partitioned, how long will the manager need to be there?

c. What percentage of the data files will take between 63 minutes and 110 minutes to be transferred, cleaned, and partitioned?

**6-15.** Doggie Nuggets Inc. (DNI) sells large bags of dog food to warehouse clubs. DNI uses an automatic filling process to fill the bags. Weights of the filled bags are approximately normally distributed with a mean of 50 kilograms and a standard deviation of 1.25 kilograms.

a. What is the probability that a filled bag will weigh less than 49.5 kilograms?

b. What is the probability that a randomly sampled filled bag will weigh between 48.5 and 51 kilograms?

c. What is the minimum weight a bag of dog food could be and remain in the top 15% of all bags filled?

d. DNI is unable to adjust the mean of the filling process. However, it is able to adjust the standard deviation of the filling process. What would the standard deviation need to be so that no more than 2% of all filled bags weigh more than 52 kilograms?

**6-16.** An expensive watch is powered by a 3-volt lithium battery expected to last three years. Suppose the life of the battery has a standard deviation of 0.3 year and is normally distributed.

a. Determine the probability that the watch's battery will last longer than 3.5 years.

b. Calculate the probability that the watch's battery will last more than 2.75 years.

c. Compute the length-of-life value for which 10% of the watch's batteries last longer.

**6-17.** The average number of acres burned by forest and range fires in a large county in the West is 5,300 acres per year, with a standard deviation of 750 acres. The distribution of the number of acres burned is normal.

a. Compute the probability that more than 6,000 acres will be burned in any year.

b. Determine the probability that fewer than 5,000 acres will be burned in any year.

c. What is the probability that between 3,500 and 5,200 acres will be burned?

d. In those years when more than 6,500 acres are burned, help is needed from eastern-region fire teams. Determine the probability help will be needed in any year.

**6-18.** An Internet retailer stocks a popular electronic toy at a central warehouse that supplies the eastern United States. Every week, the retailer makes a decision about how many units of the toy to stock. Suppose that weekly demand for the toy is approximately normally distributed with a mean of 2,500 units and a standard deviation of 300 units.

a. If the retailer wants to limit the probability of being out of stock of the electronic toy to no more than 2.5% in a week, how many units should the central warehouse stock?

b. If the retailer has 2,750 units on hand at the start of the week, what is the probability that weekly demand will be greater than inventory?

c. If the standard deviation of weekly demand for the toy increases from 300 units to 500 units, how many more toys would have to be stocked to ensure that the probability of weekly demand exceeding inventory is no more than 2.5%?

**6-19.** F&G Industries manufactures a motor that is marketed with a warranty that guarantees it will be replaced free of charge if it fails within the first 13,000 hours of operation. On average, F&G motors operate for 15,000 hours with a standard deviation of 1,250 hours before failing. The number of operating hours before failure is approximately normally distributed.

a. What is the probability that a motor will have to be replaced free of charge?

b. What percentage of F&G motors can be expected to operate for more than 17,500 hours?

c. If F&G wants to design a motor so that no more than 1% are replaced free of charge, what would the average hours of operation before failure have to be if the standard deviation remains at 1,250 hours?

**6-20.** A private equity firm is evaluating two alternative investments. Although the returns are random, each investment's return can be described using a normal distribution. The first investment has a mean return of $2,000,000 with a standard deviation of $125,000. The second investment has a mean return of $2,275,000 with a standard deviation of $500,000.

a. How likely is it that the first investment will return $1,900,000 or less?

b. How likely is it that the second investment will return $1,900,000 or less?

c. If the firm would like to limit the probability of a return being less than $1,750,000, which investment should it make?

**6-21.** The lead analyst at L.J. Raney & Associates has recently analyzed the company's 403(b) clients and determined that the annual contribution is approximately normally distributed with a mean equal to $6,400. Further, he has determined that the probability a customer will contribute more than $13,000 is 0.025. Based on this information, what is the standard deviation of contributions to the 403(b) program?

**6-22.** Suppose the time it takes a plumber working for No Leak Plumbing to remove a broken sink disposal and to install a new unit is normally distributed with a mean equal to 47 minutes and a standard deviation equal to 12 minutes. The company bills at $75.00 for the first 30 minutes and $2.00 per minute for anything beyond 30 minutes.

Suppose the going rate for this procedure by other plumbing shops in the area is $85.00, not including the cost of the new equipment. If No Leak bids the disposal job at $85, on what percentage of such jobs will the actual time required exceed the time for which it will be getting paid?

**6-23.** A maternity wear designer sells dresses and pants priced around $150 each for an average total sale of $1,200. The total sale has a normal distribution with a standard deviation of $350.

a. Calculate the probability that a randomly selected customer will have a total sale of more than $1,500.

b. Compute the probability that the total sale will be within $\pm 2$ standard deviations of the mean total sales.

c. Determine the median total sale.

**6-24.** A Coca-Cola bottling plant's product line includes 12-ounce cans of Coke products. The cans are filled by an automated filling process that can be adjusted to any mean fill volume and that will fill cans according to a normal distribution. However, not all cans will contain the same volume due to variation in the filling process. Historical records show that regardless of what the mean is set at, the standard deviation in fill will be 0.035 ounce. Operations managers at the plant know that if they put too much Coke in a can, the company loses money. If too little is put in the can, customers are short changed, and the State Department of Weights and Measures may fine the company.

a. Suppose the industry standards for fill volume call for each 12-ounce can to contain between 11.98 and 12.02 ounces. Assuming that the manager sets the mean fill at 12 ounces, what is the probability that a can will contain a volume of Coke product that falls in the desired range?

b. Assume that the manager is focused on an upcoming audit by the Department of Weights and Measures. She knows the process is to select one Coke can at random and that if it contains less than 11.97 ounces, the company will be reprimanded and potentially fined. Assuming that the manager wants at most a 5% chance of this happening, at what level should she set the mean fill level? Comment on the ramifications of this step, assuming that the company fills tens of thousands of cans each week.

**6-25.** MP-3 players are wonderful devices for people who want to have access to their favorite music and videos in a portable format. They can store massive numbers of songs and videos. Although owners have the potential to store lots of data, suppose the actual disk storage being used is normally distributed with a mean equal to 1.95 GB and a standard deviation of 0.48 GB. Consider a new company that is thinking of entering the market with a low-cost device that has only 1.0 GB of storage. The marketing slogan will be "Why Pay for Storage Capacity That You Don't Need?"

Based on the usage data provided above, what percentage of owners, given their current usage, would have enough capacity with the new 1-GB player?

**6-26.** An article by Erin El Issa titled "2015 American household credit card debt study" (www.nerdwallet.com) indicated that for households that had credit card debt, the average debt in 2015 was $15,355. Assume that credit card debt per household has a normal distribution with a standard deviation of $3,000.

  a. Determine the percentage of households that have a credit card debt of more than $15,000.

  b. One household has a credit card debt that is at the 95th percentile. Determine its credit card debt.

**6-27.** At one Georgia-Pacific plywood plant, the operations manager has been struggling to make sure that the plywood thickness meets quality standards. Specifically, all sheets of their 3/4-inch plywood must fall within the range 0.747 to 0.753 inch in thickness. Suppose the current process produces plywood that has thicknesses that are normally distributed with a mean of 0.751 inch and a standard deviation equal to 0.004 inch.

  a. Use Excel to determine the proportion of plywood sheets that will meet quality specifications (0.747 to 0.753), given how the current process is performing.

  b. Referring to part a, suppose the manager is unhappy with the proportion of product meeting specifications. Assuming that he can get the mean adjusted to 0.75 inch, what must the standard deviation be if he is going to have 98% of his product meet specifications?

**6-28.** A senior loan officer for Whitney National Bank has recently studied the bank's real estate loan portfolio and found that the loan balances are approximately normally distributed with a mean of $155,600 and a standard deviation equal to $33,050. As part of an internal audit, bank auditors recently randomly selected 100 real estate loans from the portfolio of all loans and found that 80 of these loans had balances below $170,000. The senior loan officer is interested in knowing the expected proportion of loans in the portfolio that would have balances below $170,000.

You are asked to conduct an appropriate analysis and write a short report to the senior loan officers with your conclusion about the sample.

## Computer Software Exercises

**6-29.** A November 13, 2015, report titled "The cost of hiring a new employee" by Annie Mueller (www.investopedia.com) indicated that the average cost for an American company to fill a job vacancy was $3,500. Sample data showing costs to hire employees are in a file titled **Hired**.

  a. Produce a relative frequency histogram for these data. Does it seem plausible the data were sampled from a normally distributed population?

  b. Calculate the mean and standard deviation of the cost of filling a job vacancy.

  c. Determine the probability that the cost of filling a job vacancy would be between $2,000 and $3,000.

**6-30.** The file titled **Drug$** contains daily cost data for a drug used to treat lung disease.

  a. Produce a relative frequency histogram for these data. Does it seem plausible the data were sampled from a population that was normally distributed?

  b. Compute the mean and standard deviation for the sample data in the file.

  c. Assuming the sample came from a normally distributed population and the sample standard deviation is a good approximation for the population standard deviation, determine the probability that a randomly chosen transaction would yield a price of $2.12 or less even though the population mean was $2.51.

**6-31.** A file titled **Tuition** contains the percent change in in-state tuition for 50 universities (*source:* The College Board, "Annual survey of colleges," October 2015).

  a. Produce a relative frequency histogram for these data. Does it seem plausible that the data are from a population that has a normal distribution?

  b. Find the percent change in tuition that corresponds to the 80th percentile.

---

| 6.2 | **Other Continuous Probability Distributions** |
| --- | --- |

This section introduces two additional continuous probability distributions that are used in business decision making: the uniform distribution and the exponential distribution.

**OUTCOME 4**

## The Uniform Distribution

**Uniform Distribution**
The probability distribution where the probability over any interval is the same as the probability over any other interval of the same width.

The **uniform distribution** is sometimes referred to as the *distribution of little information* because the probability over any interval of the continuous random variable is the same as the probability over any other interval of the same width.

**FIGURE 6.12 Uniform Distributions**

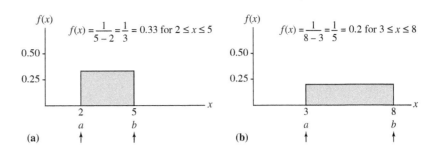

Equation 6.3 defines the *continuous uniform density function.*

**Continuous Uniform Density Function**

$$f(x) = \begin{cases} \dfrac{1}{b-a} & \text{if } a \le x \le b \\ 0 & \text{otherwise} \end{cases} \qquad (6.3)$$

where:

$f(x)$ = Value of the density function at any $x$-value
$a$ = The smallest value assumed by the uniform random variable of interest
$b$ = The largest value assumed by the uniform random variable of interest

Figure 6.12 shows two examples of uniform probability distributions with different $a$ to $b$ intervals. Note the height of the probability density function is the same for all values of $x$ between $a$ and $b$ for a given distribution. The graph of the uniform distribution is a rectangle.

---

**EXAMPLE 6-3** | **Using the Uniform Distribution**

**Georgia-Pacific**   The Georgia-Pacific Company owns and operates several tree farms in different parts of the United States and South America. The lead botanist for the company has stated that pine trees on one parcel of land will increase in diameter between 1 and 4 inches per year according to a uniform distribution. Suppose the company is interested in the probability that a given tree will have an increased diameter of more than 2 inches. The probability can be determined using the following steps:

STEP 1   **Define the density function.**
The height of the probability rectangle, $f(x)$, for the tree growth interval of 1 to 4 inches is determined using Equation 6.3, as follows:

$$f(x) = \frac{1}{b-a}$$

$$f(x) = \frac{1}{4-1} = \frac{1}{3} = 0.33$$

STEP 2   **Define the event of interest.**
The botanist is specifically interested in a tree that has increased by more than 2 inches in diameter. This event of interest is $x > 2.0$.

STEP 3   **Calculate the required probability.**
We determine the probability as follows:

$$\begin{aligned} P(x > 2.0) &= 1 - P(x \le 2.0) \\ &= 1 - f(x)(2.0 - 1.0) \\ &= 1 - 0.33(1.0) \\ &= 1 - 0.33 \\ &= 0.67 \end{aligned}$$

Thus, there is a 0.67 probability that a tree will increase by more than 2 inches in diameter.

TRY EXERCISE 6-32 (pg. 230)

Like the normal distribution, the uniform distribution can be further described by specifying the mean and the standard deviation. We can compute these values using Equations 6.4 and 6.5.

**Mean and Standard Deviation of a Uniform Distribution**

Mean (expected value):

$$E(x) = \mu = \frac{a + b}{2} \tag{6.4}$$

Standard deviation:

$$\sigma = \sqrt{\frac{(b - a)^2}{12}} \tag{6.5}$$

where:

$a$ = The smallest value assumed by the uniform random variable of interest
$b$ = The largest value assumed by the uniform random variable of interest

---

**EXAMPLE 6-4** **The Mean and Standard Deviation of a Uniform Distribution**

 **Surgery Recovery** The chief administrator of a San Francisco–area surgical center has analyzed data from a large number of shoulder surgeries conducted by her center and others in a medical association in California. The analysis shows that the recovery time for shoulder surgery ranges between 15 and 45 weeks. Without any further information, the administrator will apply a uniform distribution to surgery times. Based on this, she can determine the mean and standard deviation for the recovery duration using the following steps:

STEP 1 **Define the density function.**
Equation 6.3 is used to define the distribution:

$$f(x) = \frac{1}{b - a} = \frac{1}{45 - 15} = \frac{1}{30} = 0.0333$$

STEP 2 **Compute the mean of the probability distribution using Equation 6.4.**

$$\mu = \frac{a + b}{2} = \frac{15 + 45}{2} = 30$$

Thus, the mean recovery time is 30 weeks.

STEP 3 **Compute the standard deviation using Equation 6.5.**

$$\sigma = \sqrt{\frac{(b - a)^2}{12}} = \sqrt{\frac{(45 - 15)^2}{12}} = \sqrt{75} = 8.66$$

The standard deviation is 8.66 weeks.

**TRY EXERCISE 6-34** (pg. 231)

---

**Exponential Distribution**
The exponential distribution is a continuous distribution that is appropriate to use if the following conditions are true.

- $x$ is the time (or distance) between events, with $x > 0$.

- The occurrence of one event does not affect the probability that a second event will occur. That is, events occur independently.

- The rate at which events occur is constant. The rate cannot be higher in some intervals and lower in other intervals.

- Two events cannot occur at exactly the same instant.

OUTCOME 5 **The Exponential Distribution**

Another continuous probability distribution frequently used in business situations is the **exponential distribution**. The exponential distribution is often used to measure the time that elapses between two occurrences of an event, such as the time between "hits" on an Internet home page. Analysts might also use the exponential distribution to describe the time between arrivals of customers at a bank drive-in teller window or the time between failures of an electronic component. Equation 6.6 shows the probability density function for the exponential distribution.

### Exponential Density Function

A continuous random variable that is exponentially distributed has the probability density function given by

$$f(x) = \lambda e^{-\lambda x}, \qquad x \geq 0 \tag{6.6}$$

where:

$$e = 2.71828 \ldots$$
$$1/\lambda = \text{The mean time between events } (\lambda > 0)$$

Note that the parameter that defines the exponential distribution is $\lambda$ (lambda). You should recall from Chapter 5 that $\lambda$ is the mean value for the Poisson distribution. If the number of occurrences per time period is known to be Poisson distributed with a mean of $\lambda$, then the time between occurrences will be exponentially distributed with a mean time of $1/\lambda$.

If we select a value for $\lambda$, we can graph the exponential distribution by substituting $\lambda$ and different values for $x$ into Equation 6.6. For instance, Figure 6.13 shows exponential density functions for $\lambda = 0.5$, $\lambda = 1.0$, $\lambda = 2.0$, and $\lambda = 3.0$. Note in Figure 6.13 that for any exponential density function, $f(x)$, $f(0) = \lambda$, as $x$ increases, $f(x)$ approaches zero. It can also be shown that *the standard deviation of any exponential distribution is equal to the mean, $1/\lambda$.*

As with any continuous probability distribution, the probability that a value will fall within an interval is the area under the graph between the two points defining the interval. Equation 6.7 is used to find the probability that a value will be equal to or less than a particular value for an exponential distribution.

### Exponential Probability

$$P(0 \leq x \leq a) = 1 - e^{-\lambda a} \tag{6.7}$$

where:

$$a = \text{The value of interest}$$
$$1/\lambda = \text{Mean}$$
$$e = \text{Base of natural log; 2.71828}$$

Appendix E contains a table of $e^{-\lambda a}$ values for different values of $\lambda a$. You can use this table and Equation 6.7 to find the probabilities when the $\lambda a$ of interest is contained in the table. You can also use Excel to find exponential probabilities, as the following application illustrates.

**FIGURE 6.13 Exponential Distributions**

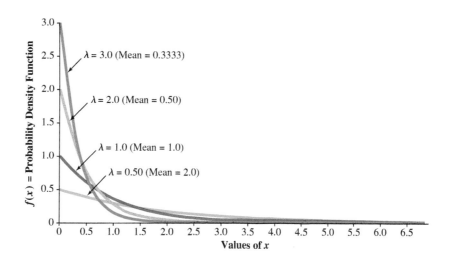

**Using Excel to Calculate Exponential Probabilities**

**Excel Tutorial**

**Haines Services**   The Haines Services Company provides online chat support for small business customers who have purchased accounting software. It has been determined that the number of customers who attempt to connect to the chat service per hour is Poisson distributed with $\lambda = 30$ per hour. The time between connect requests is exponentially distributed with a mean time between requests of 2.0 minutes, computed as follows:

$$\lambda = 30 \text{ attempts per 60 minutes} = 0.50 \text{ attempt per minute}$$

The mean time between attempted connects, then, is

$$1/\lambda = \frac{1}{0.50} = 2.0 \text{ minutes}$$

At the current staffing level, if customer requests are too close together—45 seconds (0.75 minute) or less—the chat connection will fail and the customer will get a message saying to try again later. The managers at Haines are analyzing whether they should add additional service representatives to eliminate this problem. They need to know the probability that a customer will fail to connect. Thus, they want to solve

$$P(x \le 0.75 \text{ minute}) = ?$$

To find this probability using a calculator, we need to first determine $\lambda a$. In this example, $\lambda = 0.50$ and $a = 0.75$. Then,

$$\lambda a = (0.50)(0.75) = 0.3750$$

We find that the desired probability is

$$1 - e^{-\lambda a} = 1 - e^{-0.3750}$$
$$= 0.3127$$

> The Excel 2016 function for finding probabilities for an exponential distribution is
>
> =**EXPON.DIST(x,lambda,True)**
> =**EXPON.DIST(0.75,0.5,True)**

The managers can also use the EXPON.DIST function in Excel to compute the precise value for the desired probability. Using Excel, Figure 6.14 shows that the chance of failing to connect is 0.3127. This means that nearly one-third of the customers will get the try-again message at the current staffing level.

**FIGURE 6.14 Excel 2016 Exponential Probability Output for Haines Services**

> **Excel 2016 Instructions**
> 1. Open a blank worksheet.
> 2. On the **Formula** tab, click on $f_x$ *(function wizard)*.
> 3. Select **Statistical** category.
> 4. Select **EXPON.DIST** function.
> 5. Supply $x$ and $\lambda$.
> 6. Set **Cumulative = TRUE** for cumulative probability.

# 6.2 EXERCISES

## Skill Development

**6-32.** A continuous random variable is uniformly distributed between 100 and 150.
  a. What is the probability a randomly selected value will be greater than 135?

  b. What is the probability a randomly selected value will be less than 115?
  c. What is the probability a randomly selected value will be between 115 and 135?

**6-33.** Determine the following:
  a. the probability that a uniform random variable whose range is between 10 and 30 assumes a value in the interval (10 to 20) or (15 to 25)
  b. the quartiles for a uniform random variable whose range is from 4 to 20
  c. the mean time between events for an exponential random variable that has a median equal to 10
  d. the 90th percentile for an exponential random variable that has the mean time between events equal to 0.4

**6-34.** Suppose a random variable, $x$, has a uniform distribution with $a = 5$ and $b = 9$.
  a. Calculate $P(5.5 \leq x \leq 8)$.
  b. Determine $P(x > 7)$.
  c. Compute the mean, $\mu$, and standard deviation, $\sigma$, of this random variable.
  d. Determine the probability that $x$ is in the interval $(\mu \pm 2\sigma)$.

**6-35.** Let $x$ be an exponential random variable with $\lambda = 0.5$. Calculate the following probabilities:
  a. $P(x < 5)$
  b. $P(x > 6)$
  c. $P(5 \leq x \leq 6)$
  d. $P(x \geq 2)$
  e. the probability that $x$ is at most 6

**6-36.** The useful life of an electrical component is exponentially distributed with a mean of 2,500 hours.
  a. What is the probability the circuit will last more than 3,000 hours?
  b. What is the probability the circuit will last between 2,500 and 2,750 hours?
  c. What is the probability the circuit will fail within the first 2,000 hours?

**6-37.** The time between telephone calls to a cable television service call center follows an exponential distribution with a mean of 1.5 minutes.
  a. What is the probability that the time between the next two calls will be 45 seconds or less?
  b. What is the probability that the time between the next two calls will be greater than 112.5 seconds?

## Business Applications

**6-38.** Suppose buses stop in front of your hotel every 20 minutes throughout the day. If you show up at the bus stop at a random moment during the day, determine the probability that
  a. you will have to wait more than 10 minutes
  b. you will have to wait only 6 minutes or less
  c. you will have to wait between 8 and 15 minutes

**6-39.** Assume that the time it takes to build a laser printer is uniformly distributed between 8 and 15 hours.
  a. What are the chances that it will take more than 10 hours to build a printer?
  b. How likely is it that a printer will require less than 9 hours?

  c. Suppose a single customer orders two printers. Determine the probability that the first and second printer each will require less than 9 hours to complete.

**6-40.** The time required to prepare a dry cappuccino using whole milk at the Daily Grind Coffee House is uniformly distributed between 25 and 35 seconds. Assuming a customer has just ordered a whole-milk dry cappuccino,
  a. What is the probability that the preparation time will be more than 29 seconds?
  b. What is the probability that the preparation time will be between 28 and 33 seconds?
  c. What percentage of whole-milk dry cappuccinos will be prepared within 31 seconds?
  d. What is the standard deviation of preparation times for a dry cappuccino using whole milk at the Daily Grind Coffee House?

**6-41.** The time to failure for a power supply unit used in a particular brand of personal computer (PC) is thought to be exponentially distributed with a mean of 4,000 hours as per the contract between the vendor and the PC maker. The PC manufacturer has just had a warranty return from a customer who had the power supply fail after 2,100 hours of use.
  a. What is the probability that the power supply would fail at 2,100 hours or less? Based on this probability, do you feel the PC maker has a right to require that the power supply maker refund the money on this unit?
  b. Assuming that the PC maker has sold 100,000 computers with this power supply, approximately how many should be returned due to failure at 2,100 hours or less?

**6-42.** A delicatessen is open 24 hours a day every day of the week. If, on the average, 20 orders are received by fax every two hours throughout the day, find the
  a. probability that a faxed order will arrive within the next 9 minutes
  b. probability that the time between two faxed orders will be between 3 and 6 minutes
  c. probability that 12 or more minutes will elapse between faxed orders

**6-43.** According to the website www.zanebenefits.com, the average cost of a stay in a U.S. hospital in 2013 for a Medicare patient was $11,600. Assume that this cost is exponentially distributed.
  a. Determine the probability that a randomly selected patient's stay in an American hospital will cost more than $14,000.
  b. Calculate the probability that a randomly selected patient's stay in an American hospital will cost less than $9,000.

**6-44.** During the busiest time of the day at the Daily Grind Coffee House, an average of 15 customers arrive per 20-minute period.
  a. What is the probability that a customer will arrive within the next 3 minutes?

b. What is the probability that the time between the arrivals of customers is 12 minutes or more?

c. What is the probability that the next customer will arrive within 4 and 6 minutes?

**6-45.** A California water company has determined that the average customer billing is $1,250 per year and the amounts billed have an exponential distribution.

a. Calculate the probability that a randomly chosen customer would spend more than $5,000.

b. Compute the probability that a randomly chosen customer would spend more than the average amount spent by all customers of this company.

**6-46.** Charter Southeast Airlines states that the flight between Fort Lauderdale, Florida, and Los Angeles takes 5 hours and 37 minutes. Assume that the actual flight times are uniformly distributed between 5 hours and 20 minutes and 5 hours and 50 minutes.

a. Determine the probability that the flight will be more than 10 minutes late.

b. Calculate the probability that the flight will be more than 5 minutes early.

c. Compute the average flight time between these two cities.

d. Determine the variance in the flight times between these two cities.

**6-47.** Currently, a manufacturing company experiences an average of 0.75 equipment-related shutdowns per shift and the downtimes are exponentially distributed. In testing, a new maintenance program reduced the frequency of equipment-related shutdowns to an average of 0.20 per shift, exponentially distributed. The new maintenance program is expensive, but the company will install it if it can help achieve the target of four consecutive shifts without an equipment-related shutdown.

a. Under the current system, what is the probability that the plant would run four or more consecutive shifts without an equipment-related shutdown?

b. Using the computer decision model, what is the probability that the plant could run four or more consecutive shifts without an equipment-related shutdown? Has the decision model helped the company achieve its goal?

c. What would be the maximum average number of shutdowns allowed per day such that the probability

of experiencing four or more consecutive shifts without an equipment-related shutdown is greater than or equal to 0.70?

## Computer Software Exercises

**6-48.** The Schoonover Manufacturing Company monitors the downtimes on its key fabricating machines. Data for the downtimes on one particular machine are in the file called **Schoonover**. Assuming that the downtimes are exponentially distributed, use the data in the file to find the probability that a specific downtime will exceed 1.0 minute.

**6-49.** An article titled "You're paying record ATM fees" by Kathryn Vasel states that the average ATM fee banks charge nonbank customers is $4.52 per transaction (*source:* http://money.cnn.com, Oct. 5, 2015). The file titled **ATM Fees** contains the ATM fees paid by a sample of individuals.

a. Calculate the mean and standard deviation of the ATM fees.

b. Assume that the ATM fees are exponentially distributed with the same mean as that of the sample. Determine the probability that a randomly chosen bank's ATM fee would be greater than $3.00.

**6-50.** The San Luis Obispo, California, Transit Program provides daily fixed-route transit service to the general public within the city limits and to Cal Poly State University's staff and students. The most heavily traveled route schedules a city bus to arrive at Cal Poly at 8:54 A.M. The file titled **Late** lists plausible differences between the actual and scheduled times of arrival rounded to the nearest minute for this route.

a. Produce a relative frequency histogram for these data. Does it seem plausible the data came from a population that has a uniform distribution?

b. Provide the density for this uniform distribution.

c. Classes start 10 minutes after the hour and classes are a 5-minute walk from the drop-off point. Determine the probability that a randomly chosen bus on this route would cause the students on board to be late for class. Assume the differences form a continuous uniform distribution with a range the same as the sample.

d. Determine the median difference between the actual and scheduled arrival times.

# 6 Overview

## Summary

**6.1** | **The Normal Probability Distribution** (pg. 213–226)

**OUTCOME 1** Convert a normal distribution to a standard normal distribution.

**OUTCOME 2** Determine probabilities using the standard normal distribution.

**OUTCOME 3** Calculate values of the random variable associated with specified probabilities from a normal distribution.

- The **normal distribution** is a symmetric, bell-shaped probability distribution. Half the probability lies to the right and half lies to the left of the mean.
- To find probabilities associated with a normal distribution, we convert to a **standard normal distribution** by first converting values of the random variables to **standardized z-values**. The probabilities associated with a range of values for the random variable are found using the normal distribution table in Appendix D or using Excel.

**6.2** | **Other Continuous Probability Distributions** (pg. 226–232)

**OUTCOME 4** Calculate probabilities associated with a uniformly distributed random variable.

**OUTCOME 5** Determine probabilities using an exponential probability distribution.

- Although the normal distribution is by far the most frequently used continuous probability distribution, two other continuous distributions are the uniform distribution and the exponential distribution.
- With the **uniform distribution**, the probability over any interval is the same as the probability over any other interval of the same width. We can compute the probabilities for the uniform distribution using Equation 6.3.
- The **exponential distribution** is based on a single parameter, $\lambda$, and is often used to describe random service times or the time between customer arrivals in waiting-line applications. We can compute the probability over a range of values for an exponential distribution using either Equation 6.7 or the exponential table in Appendix E. Excel also has a function for calculating exponential probabilities.
- Figure 6.15 summarizes the discrete probability distributions introduced in Chapter 5 and the continuous probability distributions introduced in this chapter.

**FIGURE 6.15 Probability Distribution Summary**

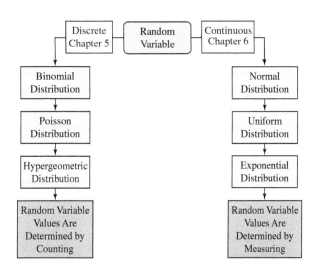

## Equations

**(6.1) Normal Probability Density Function** pg. 214

$$f(x) = \frac{1}{\sigma\sqrt{2\pi}}e^{-(x-\mu)^2/2\sigma^2}$$

**(6.2) Standardized Normal z-Value** pg. 215

$$z = \frac{x - \mu}{\sigma}$$

**(6.3) Continuous Uniform Density Function** pg. 227

$$f(x) = \begin{cases} \dfrac{1}{b - a} & \text{if } a \le x \le b \\ 0 & \text{otherwise} \end{cases}$$

**(6.4) Mean of a Uniform Distribution** pg. 228

$$E(x) = \mu = \frac{a + b}{2}$$

**(6.5) Standard Deviation of a Uniform Distribution** pg. 228

$$\sigma = \sqrt{\frac{(b - a)^2}{12}}$$

**(6.6) Exponential Density Function** pg. 229

$$f(x) = \lambda e^{-\lambda x}, \quad x \ge 0$$

**(6.7) Exponential Probability** pg. 229

$$P(0 \le x \le a) = 1 - e^{-\lambda a}$$

## Key Terms

**Exponential distribution** pg. 228
**Normal distribution** pg. 213
**Standard normal distribution** pg. 214

**Standardized z-value** pg. 215
**Uniform distribution** pg. 226

## Chapter Exercises

### Conceptual Questions

**6-51.** Discuss the difference between discrete and continuous probability distributions. Discuss two situations in which a variable of interest may be considered either continuous or discrete.

**6-52.** Recall the Empirical Rule from Chapter 3. It states that if the data distribution is bell-shaped, then the interval $\mu \pm \sigma$ contains approximately 68% of the values, $\mu \pm 2\sigma$ contains approximately 95%, and $\mu \pm 3\sigma$ contains virtually all of the data values. The bell-shaped distribution referenced is the normal distribution.

    a. Verify that a standard normal distribution contains approximately 68% of the values in the interval $\mu \pm \sigma$.

    b. Verify that a standard normal distribution contains approximately 95% of the values in the interval $\mu \pm 2\sigma$.

    c. Verify that a standard normal distribution contains virtually all of the data in the interval $\mu \pm 3\sigma$.

**6-53.** The probability that a value from a normally distributed random variable will exceed the mean is 0.50. The same is true for the uniform distribution. Why is this not necessarily true for the exponential distribution? Discuss and show examples to illustrate your point.

**6-54.** Suppose you tell one of your fellow students that when working with a continuous distribution, it does not make sense to try to compute the probability of any specific value, since it will be zero. She says that when the experiment is performed, some value must occur; the probability can't be zero. Your task is to respond to her statement and, in doing so, explain why it is appropriate to find the probability for specific ranges of values for a continuous distribution.

**6-55.** The exponential distribution has a characteristic that is called the "memoryless" property. This means $P(X > x) = P(X > x + x_0 \mid X > x_0)$. To illustrate this, consider the calls coming into 911. Suppose that the distribution of the time between occurrences has an exponential distribution with a mean of one half hour ($=0.5$).

    a. Calculate the probability that no calls come in during the first hour.

    b. Now suppose that you are monitoring the call frequency, and you note that a call does not come in during the first two hours. Determine the probability that no calls will come in during the next hour.

**6-56.** Revisit Exercise 6-55, but examine whether it would matter when you started monitoring the 911 calls if the time between occurrences had a uniform distribution with a mean of 2 and a range of 4.

    a. Calculate the probability that no call comes in during the first hour.

    b. Now suppose that you are monitoring the call frequency, and you note that no call comes in during the first two hours. Determine the probability that no calls will arrive during the next hour.

**6-57.** Suppose that, on average, 20 customers arrive every hour at a twenty-four-hour coffee shop. Assume that the time between customer arrivals is exponentially distributed. Determine

a. the probability that a customer arrives within the next 2 minutes

b. the probability that the time between two arriving customers will be between 1 and 4 minutes

c. the probability that 5 or more minutes will pass between customer arrivals

**6-58.** Assume that the time required to receive a confirmation that an electronic transfer has occurred is uniformly distributed between 30 and 90 seconds.

a. What is the probability that a randomly selected transfer will take between 30 and 45 seconds?

b. What is the probability that a randomly selected transfer will take between 50 and 90 seconds?

c. What proportion of transfers will take between 40 and 75 seconds?

## Business Applications

**6-59.** The manager for Select-a-Seat, a company that sells tickets to athletic games, concerts, and other events, has determined that the number of people arriving at the Broadway location on a typical day is Poisson distributed with a mean of 12 per hour. It takes approximately four minutes to process a ticket request. Thus, if customers arrive in intervals that are less than four minutes, they will have to wait. Assuming that a customer has just arrived and the ticket agent is starting to serve that customer, what is the probability that the next customer who arrives will have to wait in line?

**6-60.** The Four Brothers Lumber Company is considering buying a machine that planes lumber to the correct thickness. The machine is advertised to produce "6-inch lumber" having a thickness that is normally distributed, with a mean of 6 inches and a standard deviation of 0.1 inch.

a. If building standards in the industry require a 99% chance of a board being between 5.85 and 6.15 inches, should Four Brothers purchase this machine? Why or why not?

b. To what level would the company that manufactures the machine have to reduce the standard deviation for the machine to conform to industry standards?

**6-61.** Two automatic dispensing machines are being considered for use in a fast-food chain. The first dispenses an amount of liquid that has a normal distribution with a mean of 11.9 ounces and a standard deviation of 0.07 ounce. The second dispenses an amount of liquid that has a normal distribution with a mean of 12.0 ounces and a standard deviation of 0.05 ounce. Acceptable amounts of dispensed liquid are between 11.9 and 12.0 ounces. Calculate the relevant probabilities and determine which machine should be selected.

**6-62.** A small private ambulance service in Kentucky has determined that the time between emergency calls is exponentially distributed with a mean of 41 minutes. When a unit goes on call, it is out of service for 60 minutes. If a unit is busy when an emergency call is received, the call is immediately routed to another

service. The company is considering buying a second ambulance. However, before doing so, the owners are interested in determining the probability that a call will come in before the ambulance is back in service. Without knowing the costs involved in this situation, does this probability tend to support the need for a second ambulance? Discuss.

**6-63.** Assume that after the first 12 hours, the average remaining useful life of a particular battery (before recharging is required) is 9 hours and that the remaining time is exponentially distributed. What is the probability that a randomly sampled battery of this type will last between 15 and 17 hours?

**6-64.** Suppose eye shadow and eyeliner have a shelf life of up to three years, and the shelf lives of these two products are exponentially distributed with an average of one year.

a. Calculate the probability that the shelf life of eye shadow will be longer than three years.

b. Determine the probability that at least one of these products will have a shelf life of more than three years.

c. Determine the probability that a purchased eyeliner that is useful after one year will be useful after three years.

**6-65.** The Shadow Mountain Golf Course is preparing for a major LPGA golf tournament. Since parking near the course is extremely limited (room for only 500 cars), the course officials have contracted with the local community to provide parking and a bus shuttle service. Sunday, the final day of the tournament, will have the largest crowd, and the officials estimate there will be between 8,000 and 12,000 cars needing parking spaces but think no value is more likely than another. The tournament committee is discussing how many parking spots to contract from the city. If they want to limit the chance of not having enough provided parking to 10%, how many spaces do they need from the city on Sunday?

**6-66.** A product made by a window manufacturer is constructed of two pieces of glass 0.125 inch thick, with a thin layer of vinyl sandwiched between them. The average thickness of the laminated safety glass is 0.25 inch. The thickness of the glass does not vary from the mean by more than 0.10 inch. Assume the thickness of the glass has a uniform distribution.

a. Provide the density for this uniform distribution.

b. If the glass has a thickness that is more than 0.05 inch below the mean, it must be discarded for safety considerations. Determine the probability that a randomly selected glass is discarded due to safety considerations.

c. If the glass thickness is more than 0.075 inch above the mean, it will create installation problems and must be discarded. Calculate the probability that a randomly selected glass will be rejected due to installation concerns.

**6-67.** A traffic control camera at a busy intersection records, on average, 5 traffic violations per hour. Assume that the random variable number of recorded traffic violations follows a Poisson distribution.

a. What is the probability that the next recorded violation will occur within 5 minutes?

b. How likely is it that no traffic violations will be recorded within the next 7 minutes?

**6-68.** A manager of a plywood plant organized a study of the diameters of trees that are shipped to the mill. After collecting a large amount of data on diameters, he concluded that the diameter is approximately normally distributed with a mean of 14.25 inches and a standard deviation of 2.92 inches. Because of the way plywood is made, there is a certain amount of waste on each log because the peeling process leaves a core that is approximately 3 inches in diameter. For this reason, the manager feels that any log less than 10 inches in diameter is not profitable for making plywood. Based on the data collected, what is the probability that a log will be unprofitable?

**6-69.** The personnel manager for a large company is interested in the distribution of sick-leave hours for employees of her company. A recent study revealed the distribution to be approximately normal, with a mean of 58 hours per year and a standard deviation of 14 hours.

An office manager in one division has reason to believe that during the past year, two of his employees have taken excessive sick leave relative to everyone else. The first employee used 74 hours of sick leave, and the second used 90 hours. What would you conclude about the office manager's claim and why?

**6-70.** If the number of hours between servicing required for a particular snowmobile engine is exponentially distributed with an average of 118 hours, determine the probability that a randomly selected engine

a. will run at least 145 hours before servicing is needed

b. will run at most 161 hours before servicing is needed

**6-71.** Assume that the amount of time eighth-graders take to complete an assessment examination is 78 minutes with a standard deviation of 12 minutes.

a. What proportion of eighth-graders complete the assessment examination in 72 minutes or less?

b. What proportion of eighth-graders complete the assessment examination in 82 minutes or more?

c. For what number of minutes would 90% of all eighth-graders complete the assessment examination?

**6-72.** If the time required to assemble a product is normally distributed with a mean of 180 minutes and a standard deviation of 15 minutes, find

a. the probability that the product's assembly time would be less than 168 minutes

b. the probability that the product's assembly time would be greater than 190 minutes

c. the probability that the product's assembly time would be between 174 and 186 minutes

d. the number of minutes within which 95% of all such products would be assembled

**6-73.** If the average amount of time customers spent online at an Internet retailer's site is assumed to be normally distributed with mean of 36 minutes and a standard deviation of 6 minutes, determine

a. the probability a customer spends less than 30 minutes on the site

b. the probability a customer spends more than 44 minutes on the site

c. the probability a customer spends between 33 and 42 minutes on the site

d. the amount of time in minutes spent on the site by no more than 5% of all customers

**6-74.** A manufacturing process applies a coating to a product in order to protect it from scratches. If the thickness of the coating follows a uniform distribution between 15 and 25 microns, determine

a. the mean thickness of the protective coating

b. the standard deviation of the thickness of the protective coating

c. the probability that the protective coating is between 18 and 22 microns thick

d. the probability that the protective coating is less than 23 microns thick

## Computer Software Exercises

**6-75.** The Cozine Corporation runs the landfill operation outside Little Rock, Arkansas. Each day, each of the company's trucks makes several trips from the city to the landfill. On each entry, the truck is weighed. The data file **Cozine** contains a sample of 200 truck weights. Determine the mean and standard deviation for the garbage truck weights. Assume that these sample values are representative of the population of all Cozine garbage trucks, and that the weights are normally distributed.

a. Determine the probability that a truck will arrive at the landfill weighing in excess of 46,000 pounds.

b. Compare the probability in part a to the proportion of trucks in the sample that weighed more than 46,000 pounds. What does this imply to you?

c. Suppose the managers are concerned that trucks are returning to the landfill before they are fully loaded. If they have set a minimum weight of 38,000 pounds before the truck returns to the landfill, what is the probability that a truck will fail to meet the minimum standard?

**6-76.** The Hydronics Company's research and development department came up with two weight-loss products. To determine whether these products are effective, the company has conducted a test. A total of 300 people who were 30 pounds or more overweight were recruited to participate in the study. Of these, 100 people were given a placebo supplement, 100 people were given Product 1, and 100 people were given Product 2. As might be expected, some people dropped out of the study before the four-week study period was completed.

The weight loss (or gain) for each individual is listed in the data file called **Hydronics**.

a. Develop a frequency histogram for the weight loss (or gain) for those people on Product 1. Does it appear from this graph that weight loss is approximately normally distributed?

b. Referring to part a and assuming that a normal distribution does apply, compute the mean and standard deviation weight loss for the Product 1 users.

c. Referring to parts a and b, assume that the weight-change distribution for Product 1 users is normally distributed and that the sample mean and standard deviation are used to directly represent the population mean and standard deviation. What is the probability that a Product 1 user will lose over 12 pounds in a four-week period?

d. Referring to your answer in part c, would it be appropriate for the company to claim that Product 1 users can expect to lose as much as 12 pounds in four weeks? Discuss.

**6-77.** Midwest Fan Manufacturing Inc. is a distributor of quality ventilation equipment. Midwest Fan's products include the AXC range hood exhaust fans. The file titled **Fan Life** contains the length of life of 125 randomly chosen AXC fans that were used in an accelerated life-testing experiment.

a. Produce a relative frequency histogram for the data. Does it seem plausible the data came from a population that has an exponential distribution?

b. Calculate the mean and standard deviation of the fans' length of life.

c. Calculate the median length of life of the fans.

d. Determine the probability that a randomly chosen fan will have a life longer than 25,000 hours.

**6-78.** Team Marketing Report (TMR) produces the Fan Cost Index™ (FCI) survey, which tracks the cost of attendance for a family of four at National Football League (NFL) games. The FCI includes four average-price tickets, four small soft drinks, two small beers, four hot dogs, two game programs, parking, and two adult-size caps. The league's average FCI in 2015 was $480.89 (*source:* www.teammarketing.com). The file **NFL Price** shows the actual costs for 175 randomly chosen families of four.

a. Produce a relative frequency histogram for these data. Does it seem plausible the data were sampled from a population that was normally distributed?

b. Calculate the mean and standard deviation of the league's FCI.

c. Calculate the 90th percentile of the league's FCI.

d. The San Francisco 49ers had an FCI of $640.00. Determine the percentile of the FCI of a randomly chosen family whose FCI is the same as that of the 49ers' average FCI.

**6-79.** The Future-Vision Digital and Satellite TV Company recently surveyed its customers. A total of 548 responses were received. Among other things, the respondents were asked to indicate their household income. The data from the survey are found in a file named **Future-Vision**.

a. Develop a frequency histogram for the income variable. Does it appear from the graph that income is approximately normally distributed? Discuss.

b. Compute the mean and standard deviation for the income variable.

c. Referring to parts a and b and assuming that income is normally distributed and the sample mean and standard deviation are good substitutes for the population values, what is the probability that a Future-Vision customer will have an income exceeding $40,000?

d. Suppose that Future-Vision managers are thinking about offering a monthly discount to customers who have a household income below a certain level. If the management wants to grant discounts to no more than 7% of the customers, what income level should be used for the cutoff?

**6-80.** Championship Billiards, owned by D & R Industries, in Lincolnwood, Illinois, provides some of the finest billiard fabrics, cushion rubber, and component parts in the industry. It sells billiard cloth in bolts and half-bolts. A half-bolt of billiard cloth has an average length of 35 yards with widths of either 62 or 66 inches. The file titled **Half Bolts** contains the lengths of 120 randomly selected half-bolts.

a. Produce a relative frequency histogram for these data. Does it seem plausible the data came from a population that has a uniform distribution?

b. Provide the density, $f(x)$, for this uniform distribution.

c. A billiard retailer, Sticks & Stones Billiard Supply, is going to recover the pool tables in the local college pool hall, which has eight tables. It takes approximately 3.8 yards per table. If Championship ships a randomly chosen half-bolt, determine the probability that it will contain enough cloth to recover the eight tables.

# Case 6.1    State Entitlement Programs

Franklin Joiner, director of health, education, and welfare, had just left a meeting with the state's newly elected governor and several of the other recently appointed department heads. One of the governor's campaign promises was to try to halt the rising cost of a certain state entitlement program. In several speeches, the governor indicated the

state of Idaho should allocate funds only to those individuals ranked in the bottom 10% of the state's income distribution. Now the governor wants to know how much one could earn before being disqualified from the program, and he also wants to know the range of incomes for the middle 95% of the state's income distribution.

Frank had mentioned in the meeting that he thought incomes in the state could be approximated by a normal distribution and that mean per capita income was about $33,000 with a standard deviation of nearly $9,000. The governor was expecting a memo in his office by 3:00 P.M. that afternoon with answers to his questions.

## Required Tasks:

1. Assuming that incomes can be approximated using a normal distribution with the specified mean and standard deviation, calculate the income that cut off the bottom 10% of incomes.

2. Assuming that incomes can be approximated using a normal distribution with the specified mean and standard deviation, calculate the middle 95% of incomes. *Hint:* This requires calculating two values.

3. Write a short memo describing your results and how they were obtained. Your memo should clearly state the income that would disqualify people from the program, as well as the range of incomes in the middle 95% of the state's income distribution.

## Case 6.2    Credit Data, Inc.

Credit Data, Inc., has been monitoring the amount of time its bill collectors spend on calls that produce contacts with consumers. Management is interested in the distribution of time a collector spends on each call in which he or she initiates contact, informs a consumer about an outstanding debt, discusses a payment plan, and receives payments by phone. Credit Data is mostly interested in how quickly a collector can initiate and end a conversation to move on to the next call. For employees of Credit Data, time is money in the sense that one account may require one call and 2 minutes to collect, whereas another account may take five calls and 10 minutes per call to collect. The company has discovered that the time collectors spend talking to consumers about accounts is approximated by a normal distribution with a mean of 8 minutes and a standard deviation of 2.5 minutes. The managers believe that the mean is too high and should be reduced by more efficient phone call methods. Specifically, they wish to have no more than 10% of all calls require more than 10.5 minutes.

## Required Tasks:

1. Assuming that training can affect the average time but not the standard deviation, the managers are interested in knowing to what level the mean call time needs to be reduced in order to meet the 10% requirement.

2. Assuming that the standard deviation can be affected by training but the mean time will remain at 8 minutes, to what level must the standard deviation be reduced in order to meet the 10% requirement?

3. If nothing is done, what percent of all calls can be expected to require more than 10.5 minutes?

## Case 6.3    National Oil Company—Part 1

Chad Williams, field geologist for the National Oil Company, settled into his first-class seat on the Sun-Air flight between Los Angeles and Oakland, California. Earlier that afternoon, he had attended a meeting with the design engineering group at the Los Angeles New Product Division. He was now on his way to the home office in Oakland. He was looking forward to the one-hour flight because it would give him a chance to reflect on a problem that surfaced during the meeting. It would also give him a chance to think about the exciting opportunities that lay ahead in Australia.

Chad works with a small group of highly trained people at National Oil who literally walk the Earth looking for new sources of oil. They make use of the latest in electronic equipment to take a wide range of measurements from many thousands of feet below the Earth's surface. It is one of these electronic machines that is the source of Chad's current problem. Engineers in Los Angeles have designed a sophisticated enhancement that will greatly improve the equipment's ability to detect oil. The enhancement requires 800 capacitors, which must operate within ±0.50 micron from the specified standard of 12 microns.

The problem is that the supplier can provide capacitors that operate according to a normal distribution, with a mean of 12 microns and a standard deviation of 1 micron. Thus, Chad knows that not all capacitors will meet the specifications required by the new piece of exploration equipment. This will mean that to have at least 800 usable capacitors, National Oil will have to order more than 800 from the supplier. However, these items are very expensive, so he wants to order as few as possible to meet their needs. At the meeting, the group agreed that they wanted a 98% chance that any order of capacitors would contain the sufficient number of usable items. If the project is to remain on schedule, Chad must place the order by tomorrow. He wants the new equipment ready to go by the time he leaves for an exploration trip in Australia. As he reclined in his seat, sipping a cool lemonade, he wondered whether a basic statistical technique could be used to help determine how many capacitors to order.

# 7 Introduction to Sampling Distributions

## WHY YOU NEED TO KNOW

The director of tourism in a Caribbean country conducted a study that showed that the mean daily expenditure for adult visitors to the country is $318.69. The mean value is based on a statistical sample of 780 adult visitors. The $318.69 is a *statistic*, not a *parameter*, because it is based on a sample rather than an entire population. If you were this official, you might have some questions:

### Quick Prep

**Review** the discussion of random sampling in Chapter 1.

**Review** the steps for computing means and standard deviations in Chapter 3.

**Make sure** you are familiar with the normal distribution and how to compute standardized *z*-values as introduced in Chapter 6.

**Review** the concepts associated with finding probabilities with a standard normal distribution as discussed in Chapter 6.

- Is the actual population mean equal to $318.69?
- If the population mean is not $318.69, how close is $318.69 to the true population mean?
- Is a sample of 780 taken from a population of almost 2 million annual visitors to the country sufficient to provide a "good" estimate of the population mean?

A furniture manufacturer that makes made-to-assemble furniture kits selects a random sample of kits boxed and ready for shipment to customers. These kits are unboxed and inspected to see whether what is in the box matches exactly what is supposed to be in the box. This past week, 150 kits were sampled and 15 had one or more discrepancies. This is a 10% defect rate. Should the quality engineer conclude that exactly 10% of the 6,900 furniture kits made since the first of the year reached the customer with one or more order discrepancies? Is the actual percentage higher or lower than 10% and, if so, by how much? Should the quality engineer request that more furniture kits be sampled?

The questions facing the tourism director and the furniture quality engineer are common to people in business everywhere. You will almost assuredly find yourself in a similar situation many times in the future. To help answer these questions, you need to have an understanding of *sampling distributions*. Whenever decisions are based on samples rather than an entire population, questions about the sample results exist. Anytime we sample from a population, many, many samples could be selected. Each sample will contain different items. Because of this, the sample means for each possible sample can be different, or the sample percentages can be different. The sampling distribution describes the distribution of possible sample outcomes. Knowing what this distribution looks like will help you understand the specific result you obtained from the one sample you selected.

This chapter introduces you to the important concepts of sampling error and sampling distributions and discusses how you can use this knowledge to help answer the questions facing the tourism director and the quality engineer. The information presented here provides an essential building block to understanding statistical estimation and hypothesis testing, which will be covered in upcoming chapters.

## 7.1 | Sampling Error: What It Is and Why It Happens

As discussed in previous chapters, you will encounter many situations in business in which you take a sample from a population and then must analyze the sample data. Chapter 1 introduced several different statistical sampling techniques, including simple random sampling. The objective of random sampling is to gather data that accurately represent a population.

OUTCOME **1**

## Calculating Sampling Error

Regardless of how careful we are in using random sampling methods, the sample may not be a perfect representation of the population. For example, we might compute a *statistic* such as $\bar{x}$ for sample data. Unless the sample is a perfect replication of the population, the statistic will likely not equal the *parameter*, $\mu$. In this case, the difference between the sample mean and the population mean is called **sampling error**. The sampling error is computed using Equation 7.1.

**Sampling Error**
The difference between a measure computed from a sample (a statistic) and corresponding measure computed from the population (a parameter).

**Sampling Error of the Sample Mean**

$$\text{Sampling error} = \bar{x} - \mu \qquad (7.1)$$

where:

$\bar{x}$ = Sample mean
$\mu$ = Population mean

**TABLE 7.1** Square Feet for Office Complex Projects

| Complex | Square Feet |
|---------|-------------|
| 1 | 114,560 |
| 2 | 202,300 |
| 3 | 78,600 |
| 4 | 156,700 |
| 5 | 134,600 |
| 6 | 88,200 |
| 7 | 177,300 |
| 8 | 155,300 |
| 9 | 214,200 |
| 10 | 303,800 |
| 11 | 125,200 |
| 12 | 156,900 |

The Excel 2016 function for the mean is

=AVERAGE(data values)
=AVERAGE(114560,202300, ...,125200,156900)

**Parameter**
A measure computed from the entire population. As long as the population does not change, the value of the parameter will not change.

**Simple Random Sample**
A sample selected in such a manner that each possible sample of a given size has an equal chance of being selected.

**BUSINESS APPLICATION** **Sampling Error**

**Hummel Development Corporation** The Hummel Development Corporation has built 12 office complexes. Table 7.1 shows a list of the 12 projects and the total square footage of each project.

Because these 12 projects are all the office complexes the company has worked on, the square-feet area for all 12 projects, shown in Table 7.1, is a population. Equation 7.2 is used to compute the mean square feet in the population of projects.

**Population Mean**

$$\mu = \frac{\sum x}{N}$$

(7.2)

where:

$\mu$ = Population mean
$x$ = Values in the population
$N$ = Population size

The mean square feet for the 12 office complexes is

$$\mu = \frac{114{,}560 + 202{,}300 + \cdots + 125{,}200 + 156{,}900}{12}$$

$\mu = 158{,}972$ square feet

The average area of the complexes built by the firm is 158,972 square feet. This value is a **parameter**. No matter how many times we compute the value, assuming no arithmetic mistakes, we will get the same value for the population mean.

Hummel is a finalist to be the developer of a new office building in Madison, Wisconsin. The client who will hire the firm plans to select a **simple random sample** of $n = 5$ projects from those the finalists have completed. The client plans to travel to these office buildings to see the quality of the construction and to interview owners and occupants. (You may want to refer to Chapter 1 to review the material on simple random samples.)

Referring to the office complex data in Table 7.1, suppose the client randomly selects the following five Hummel projects from the population:

| Complex | Square Feet |
|---------|-------------|
| 5 | 134,600 |
| 4 | 156,700 |
| 1 | 114,560 |
| 8 | 155,300 |
| 9 | 214,200 |

The key factor in the selection process is the finalists' past performance on large projects, so the client might be interested in the mean size of the office buildings that the Hummel Development Company has built. Equation 7.3 is used to compute the sample mean.

**Sample Mean**

$$\bar{x} = \frac{\sum x}{n}$$

(7.3)

where:

$\bar{x}$ = Sample mean
$x$ = Sample values selected from the population
$n$ = Sample size

> The Excel 2016 function for the mean is
>
> =AVERAGE(data values)
> =AVERAGE(134600,156700, 114560,155300,214200)

The sample mean is

$$\bar{x} = \frac{134{,}600 + 156{,}700 + 114{,}560 + 155{,}300 + 214{,}200}{5} = \frac{775{,}360}{5} = 155{,}072$$

The average number of square feet in the random sample of five Hummel office buildings selected by the client is 155,072. This value is a *statistic* based on the sample.

Recall the mean for the population:

$$\mu = 158{,}972 \text{ square feet}$$

The sample mean is

$$\bar{x} = 155{,}072 \text{ square feet}$$

As you can see, the sample mean does not equal the population mean. This difference is called the *sampling error*. Using Equation 7.1, we compute the sampling error as follows:

$$\text{Sampling error} = \bar{x} - \mu$$
$$= 155{,}072 - 158{,}972 = -3{,}900 \text{ square feet}$$

The sample mean for the random sample of $n = 5$ office buildings is 3,900 square feet less than the population mean. Regardless of how carefully you construct your sampling plan, you can expect to see sampling error. A random sample will almost never be a perfect representation of its population. The sample value and the population value will most likely be different.

Suppose the client who selected the random sample throws these five projects back into the stack and selects a second random sample of five as follows:

| Complex | Square Feet |
| --- | --- |
| 9 | 214,200 |
| 6 | 88,200 |
| 5 | 134,600 |
| 12 | 156,900 |
| 10 | 303,800 |

The mean for this sample is

$$\bar{x} = \frac{214{,}200 + 88{,}200 + 134{,}600 + 156{,}900 + 303{,}800}{5} = \frac{897{,}700}{5} = 179{,}540 \text{ square feet}$$

This time, the sample mean is larger than the population mean. The sampling error is

$$\bar{x} - \mu = 179{,}540 - 158{,}972$$
$$= 20{,}568 \text{ square feet}$$

This example illustrates some useful fundamental concepts:

- The size of the sampling error depends on which sample is selected.
- The sampling error may be positive or negative.
- There is potentially a different $\bar{x}$ for each possible sample.

If the client wanted to use these sample means to estimate the population mean, in one case they would be 3,900 square feet too small, and in the other they would be 20,568 square feet too large.

**EXAMPLE 7-1**  **Computing the Sampling Error**

Smart Televisions    The website for a major seller of consumer electronics has 10 different brands of Smart televisions available. The stated prices for the 55-inch size for the 10 brands are listed as follows:

| $479 | $569 | $599 | $649 | $649 | $699 | $699 | $749 | $799 | $799 |
| --- | --- | --- | --- | --- | --- | --- | --- | --- | --- |

Suppose a competitor who is monitoring this company has randomly sampled $n = 4$ Smart TV brands and recorded the prices from the population of $N = 10$. The selected Smart TV prices are

| | | | |
|---|---|---|---|
| $569 | $649 | $799 | $799 |

The sampling error can be computed using the following steps:

STEP 1 **Determine the population mean using Equation 7.2.**

$$\mu = \frac{\sum x}{N} = \frac{479 + 569 + 599 + \cdots + 799 + 799}{10} = \frac{6,690}{10} = \$669$$

STEP 2 **Compute the sample mean using Equation 7.3.**

$$\bar{x} = \frac{\sum x}{n} = \frac{569 + 649 + 799 + 799}{4} = \frac{2,816}{4} = \$704$$

STEP 3 **Compute the sampling error using Equation 7.1.**

$$\bar{x} - \mu = 704 - 669 = \$35$$

This sample of four has a sampling error of $35. The sample of TV prices has a slightly higher mean price than the mean for the population.

TRY EXERCISE 7-1 (pg. 245)

### The Role of Sample Size in Sampling Error

BUSINESS APPLICATION **Sampling Error**

**Hummel Development Corporation** (*continued*)   Previously, we selected random samples of 5 office complexes from the 12 projects Hummel Development Corporation has built. We then computed the resulting sampling error. There are actually 792 possible samples of size 5 that may be taken from 12 projects. This value is found using the counting rule for combinations, which was discussed in Chapter 5.[1]

In actual situations, the decision maker selects only one sample and then uses the sample measure to estimate the population measure. A "small" sampling error may be acceptable. However, if the sampling error is too "large," conclusions about the population could be misleading.

We can look at the extremes on either end to evaluate the potential for extreme sampling error. The population of square feet for the 12 projects is

| Complex | Square Feet | Complex | Square Feet |
|---|---|---|---|
| 1 | 114,560 | 7 | 177,300 |
| 2 | 202,300 | 8 | 155,300 |
| 3 | 78,600 | 9 | 214,200 |
| 4 | 156,700 | 10 | 303,800 |
| 5 | 134,600 | 11 | 125,200 |
| 6 | 88,200 | 12 | 156,900 |

Suppose, by chance, the developers ended up with the five smallest office complexes in their sample. These are

| Complex | Square Feet |
|---|---|
| 3 | 78,600 |
| 6 | 88,200 |
| 1 | 114,560 |
| 11 | 125,200 |
| 5 | 134,600 |

---

[1]The number of combinations of $x$ items from a sample of size $n$ is $\dfrac{n!}{x!(n-x)!}$.

The mean of this sample is

$$\bar{x} = 108{,}232 \text{ square feet}$$

Of all the possible random samples of five, this one provides the smallest sample mean. The sampling error is

$$\bar{x} - \mu = 108{,}232 - 158{,}972 = -50{,}740 \text{ square feet}$$

On the other extreme, suppose the sample contains the five largest office complexes, as follows:

| Complex | Square Feet |
|---|---|
| 10 | 303,800 |
| 9 | 214,200 |
| 2 | 202,300 |
| 7 | 177,300 |
| 12 | 156,900 |

The mean for this sample is $\bar{x} = 210{,}900$. This is the largest possible sample mean from all the possible samples of five complexes. The sampling error in this case is

$$\bar{x} - \mu = 210{,}900 - 158{,}972 = 51{,}928 \text{ square feet}$$

The potential for extreme sampling error ranges from

$$-50{,}740 \text{ to } +51{,}928 \text{ square feet}$$

The remaining possible random samples of five will provide sampling errors between these limits.

What happens if the size of the sample selected is larger or smaller? Suppose the client scales back his sample size to $n = 3$ office complexes. Table 7.2 shows the extremes.

By reducing the sample size from five to three, the range of potential sampling error has increased from

$$(-50{,}740 \text{ to } +51{,}928 \text{ square feet})$$

to

$$(-65{,}185.33 \text{ to } +81{,}128 \text{ square feet})$$

This illustrates that the potential for extreme sampling error is greater when smaller sample sizes are used. Likewise, larger sample sizes reduce the range of potential sampling error.

Although larger sample sizes reduce the potential for extreme sampling error, there is no guarantee that a larger sample size will always give a smaller sampling error. For example, Table 7.3 shows two further applications of the office complex data. As shown, this random sample of three has a sampling error of $-2{,}672$ square feet, whereas the larger random sample of size five has a sampling error of 16,540 square feet. In this case, the smaller sample

**TABLE 7.2** Hummel Office Building Example for $n = 3$ (Extreme Samples)

| Smallest Office Buildings | | Largest Office Buildings | |
|---|---|---|---|
| Complex | Square Feet | Complex | Square Feet |
| 3 | 78,600 | 10 | 303,800 |
| 6 | 88,200 | 9 | 214,200 |
| 1 | 114,560 | 2 | 202,300 |
| $\bar{x} = 93{,}786.67$ sq. feet | | $\bar{x} = 240{,}100$ sq. feet | |
| Sampling Error: | | Sampling Error: | |
| $93{,}786.67 - 158{,}972 = -65{,}185.33$ square feet | | $240{,}100 - 158{,}972 = 81{,}128$ square feet | |

**TABLE 7.3** Hummel Office Building Example with Different Sample Sizes

| n = 5 | | n = 3 | |
|---|---|---|---|
| Complex | Square Feet | Complex | Square Feet |
| 4 | 156,700 | 12 | 156,900 |
| 1 | 114,560 | 8 | 155,300 |
| 7 | 177,300 | 4 | 156,700 |
| 11 | 125,200 | | |
| 10 | 303,800 | | |

$\bar{x} = 175,512$ sq. feet

Sampling Error:

$175,512 - 158,972 = 16,540$ square feet

$\bar{x} = 156,300$ sq. feet

Sampling Error:

$156,300 - 158,972 = -2,672$ square feet

is "better" than the larger sample. However, in Section 7.2, you will learn that the average sampling error produced by large samples is less than the average sampling error from small samples.

## 7.1 EXERCISES

### Skill Development

**7-1.** A population has a mean of 125. If a random sample of 8 items from the population results in the following sampled values, what is the sampling error for the sample?

| 103 | 123 | 99 | 107 | 121 | 100 | 100 | 99 |
|---|---|---|---|---|---|---|---|

**7-2.** The following data are the 16 values in a population:

| 10 | 5 | 19 | 20 | 10 | 8 | 10 | 2 |
|---|---|---|---|---|---|---|---|
| 14 | 18 | 7 | 8 | 14 | 2 | 3 | 10 |

a. Compute the population mean.
b. Suppose a simple random sample of 5 values from the population is selected with the following results:

| 5 | 10 | 20 | 2 | 3 |
|---|---|---|---|---|

Compute the mean of this sample.
c. Based on the results for parts a and b, compute the sampling error for the sample mean.

**7-3.** The following population is provided:

| 17 | 15 | 8 | 12 | 9 | 7 | 9 | 11 |
|---|---|---|---|---|---|---|---|
| 12 | 14 | 16 | 9 | 5 | 10 | 14 | 13 |
| 12 | 12 | 11 | 9 | 14 | 8 | 14 | 12 |

Further, a simple random sample from this population gives the following values:

| 12 | 9 | 5 | 10 | 14 | 11 |
|---|---|---|---|---|---|

Compute the sampling error for the sample mean in this situation.

**7-4.** Consider the following population:

| 18 | 26 | 32 | 17 | 34 | 17 | 17 | 22 |
|---|---|---|---|---|---|---|---|
| 29 | 24 | 24 | 35 | 13 | 29 | 38 | |

The following sample was drawn from this population:

| 35 | 18 | 24 | 17 | 24 | 32 | 17 | 29 |
|---|---|---|---|---|---|---|---|

a. Determine the sampling error for the sample mean.
b. Determine the largest possible sampling error for this sample of $n = 8$.

**7-5.** Assume that the following represent a population of $N = 24$ values:

| 10 | 14 | 32 | 9 | 34 | 19 | 31 | 24 |
|---|---|---|---|---|---|---|---|
| 33 | 11 | 14 | 30 | 6 | 27 | 33 | 32 |
| 28 | 30 | 10 | 31 | 19 | 13 | 6 | 35 |

a. If a random sample of $n = 10$ items includes the following values, compute the sampling error for the sample mean:

| 32 | 19 | 6 | 11 | 10 |
|---|---|---|---|---|
| 19 | 28 | 9 | 13 | 33 |

b. For a sample of size $n = 6$, compute the range for the possible sampling error. (*Hint:* Find the sampling error for the 6 smallest sample values and the 6 largest sample values.)

c. For a sample of size $n = 12$, compute the range for the possible sampling error. How does sample size affect the potential for extreme sampling error?

**7-6.** Assume that the following represent a population of $N = 16$ values:

| 25 | 12 | 21 | 13 | 19 | 17 | 15 | 18 |
|----|----|----|----|----|----|----|----|
| 23 | 16 | 18 | 15 | 22 | 14 | 23 | 17 |

a. Compute the population mean.
b. If a random sample of $n = 9$ includes the following values:

| 12 | 18 | 13 | 17 | 23 | 14 | 16 | 25 | 15 |
|----|----|----|----|----|----|----|----|----|

compute the sample mean and calculate the sampling error for this sample.

c. Determine the range of extreme sampling error for a sample of size $n = 4$. (*Hint:* Calculate the lowest possible sample mean and highest possible sample mean.)

**7-7.** Consider the following population:

| 3 | 6 | 9 |
|---|---|---|

a. Calculate the population mean.
b. Select, with replacement, and list each possible sample of size 2. Also, calculate the sample mean for each sample.
c. Calculate the sampling error associated with each sample mean.
d. Assuming that each sample is equally likely, produce the distribution of the sampling errors.

## Business Applications

**7-8.** Hillman Management Services manages apartment complexes in Tulsa, Oklahoma. They currently have 30 units available for rent. The monthly rental prices (in dollars) for this population of 30 units are

| 455 | 690 | 450 | 495 | 550 | 780 | 800 | 395 | 500 | 405 |
|-----|-----|-----|-----|-----|-----|-----|-----|-----|-----|
| 675 | 550 | 490 | 495 | 700 | 995 | 650 | 550 | 400 | 750 |
| 600 | 780 | 650 | 905 | 415 | 600 | 600 | 780 | 575 | 750 |

a. What is the range of possible sampling error if a random sample of size $n = 6$ is selected from the population?
b. What is the range of possible sampling error if a random sample of size $n = 10$ is selected? Compare your answers to parts a and b and explain why the difference exists.

**7-9.** Findings from the National Health and Nutrition Examination Survey (NHANES) 1999–2002, the National Health Interview Survey (NHIS) 2006, and the Tobacco Use Supplement to the Current Population Survey (TUS-CPS) 2003 in the Pfizer Facts report titled "Smoking in the United States workforce"

(www.pfizer.com) indicate that smokers miss an average of 6 days of work per year due to smoking-related health issues. At a large company, the following sample data were collected:

| 13 | 4  | 4  | 5  | 12 | 8  | 9  | 11 | 1  | 5  |
|----|----|----|----|----|----|----|----|----|----|
| 6  | 9  | 14 | 6  | 3  | 5  | 10 | 7  | 0  | 14 |
| 6  | 15 | 0  | 2  | 5  | 3  | 10 | 8  | 6  | 7  |
| 0  | 0  | 15 | 14 | 6  | 2  | 2  | 1  | 4  | 15 |
| 10 | 12 | 3  | 0  | 14 | 10 | 0  | 1  | 9  | 14 |

Determine the sampling error of this sample, assuming that the 6.0-day average is correct.

**7-10.** An Internet service provider states that the average number of hours its customers are online each day is 3.75. Suppose a random sample of 14 of the company's customers is selected and the average number of hours that they are online each day is measured. The sample results are

| 3.11 | 1.97 | 3.52 | 4.56 | 7.19 | 3.89 | 7.71 |
|------|------|------|------|------|------|------|
| 2.12 | 4.68 | 6.78 | 5.02 | 4.28 | 3.23 | 1.29 |

Based on the sample of 14 customers, how much sampling error exists? Would you expect the sampling error to increase or decrease if the sample size was increased to 40?

**7-11.** The Anasazi Real Estate Company has 20 listings for homes in Santa Fe, New Mexico. The number of days each house has been on the market without selling is as follows:

| 26 | 45 | 16 | 77 | 33 | 50 | 19 | 23 | 55 | 107 |
|----|----|----|----|----|----|----|----|----|-----|
| 88 | 15 | 7  | 19 | 30 | 60 | 80 | 66 | 31 | 17  |

a. If we consider these 20 values to be the population of interest, what is the mean of the population?
b. The company is making a sales brochure and wishes to feature 5 homes selected at random from the list. The number of days the 5 sampled homes have been on the market is

| 77 | 60 | 15 | 31 | 23 |
|----|----|----|----|----|

If these 5 houses were used to estimate the mean for all 20, what would the sampling error be?

c. What is the range of possible sampling error if 5 homes are selected at random from the population?

**7-12.** The administrator at Saint Frances Hospital is concerned about the amount of overtime the nursing staff is incurring and wonders whether so much overtime is really necessary. The hospital employs 60 nurses. Following is the number of hours of overtime reported by each nurse last week. These data are the population of interest.

| Nurse | Overtime | Nurse | Overtime | Nurse | Overtime |
|-------|----------|-------|----------|-------|----------|
| 1 | 2 | 21 | 4 | 41 | 3 |
| 2 | 1 | 22 | 2 | 42 | 3 |
| 3 | 7 | 23 | 3 | 43 | 2 |
| 4 | 0 | 24 | 5 | 44 | 1 |
| 5 | 4 | 25 | 5 | 45 | 3 |
| 6 | 2 | 26 | 6 | 46 | 3 |
| 7 | 6 | 27 | 2 | 47 | 3 |
| 8 | 4 | 28 | 2 | 48 | 3 |
| 9 | 2 | 29 | 7 | 49 | 4 |
| 10 | 5 | 30 | 4 | 50 | 6 |
| 11 | 5 | 31 | 4 | 51 | 0 |
| 12 | 4 | 32 | 3 | 52 | 3 |
| 13 | 5 | 33 | 3 | 53 | 4 |
| 14 | 0 | 34 | 4 | 54 | 6 |
| 15 | 6 | 35 | 5 | 55 | 0 |
| 16 | 0 | 36 | 5 | 56 | 3 |
| 17 | 2 | 37 | 0 | 57 | 3 |
| 18 | 4 | 38 | 0 | 58 | 7 |
| 19 | 2 | 39 | 4 | 59 | 5 |
| 20 | 5 | 40 | 3 | 60 | 7 |

Devise a method to select a random sample of 6 nurses. Determine the mean hours of overtime for these 6 nurses and calculate the sampling error associated with this particular sample mean.

**7-13.** Princess Cruises offered a 9-day voyage from Beijing to Bangkok during the time period from May to August. The announced price, excluding airfare, for a room with an ocean view or a balcony was listed as $3,475. Cruise fares usually are quite variable due to discounting by the cruise line and travel agents. A sample of 20 passengers who purchased this cruise paid the following amounts (in dollars):

3,559 3,005 3,389 3,505 3,605 3,545 3,529 3,709 3,229 3,419
3,439 3,375 3,349 3,559 3,419 3,569 3,559 3,575 3,449 3,119

a. Calculate the sample mean cruise fare.
b. Determine the sampling error for this sample.
c. Would the results obtained in part b indicate that the average cruise fare during this period for this cruise is different from the listed price? Explain your answer from a statistical point of view.

**7-14.** An investment advisor has worked with 24 clients for the past five years. Following are the percentage rates of average five-year returns that these 24 clients experienced over this time on their investments:

11.2 11.2 15.9  2.7  4.6  7.6  15.6  1.3   3.3  4.8  12.8  14.9
10.1 10.9  4.9 −2.1 12.5  3.7   7.6  4.9  10.2  0.4   9.6 −0.5

This investment advisor plans to introduce a new investment program to a sample of his customers this year. Because this is experimental, he plans to randomly select 5 of the customers to be part of the

program. However, he would like those selected to have a mean return rate close to the population mean for the 24 clients. Suppose the following 5 values represent the average five-year annual return for the clients that were selected in the random sample:

11.2    −2.1   12.5    1.3    3.3

Calculate the sampling error associated with the mean of this random sample. What would you tell this advisor regarding the sample he has selected?

**7-15.** A computer lab at a small college has 25 computers. Twice during the day, a full scan for viruses is performed on each computer. Because of differences in the configuration of the computers, the times required to complete the scan are different for each machine. Records for the scans are kept and indicate that the time (in seconds) required to perform the scan for each machine is as shown here.

**Time in Seconds to Complete Scan**

| | | | | |
|---|---|---|---|---|
| 1,500 | 1,347 | 1,552 | 1,453 | 1,371 |
| 1,362 | 1,447 | 1,362 | 1,216 | 1,378 |
| 1,647 | 1,093 | 1,350 | 1,834 | 1,480 |
| 1,522 | 1,410 | 1,446 | 1,291 | 1,601 |
| 1,365 | 1,575 | 1,134 | 1,532 | 1,534 |

a. What is the mean time required to scan all 25 computers?
b. Suppose a random sample of 5 computers is taken and the scan times for each are as follows: 1,534, 1,447, 1,371, 1,410, and 1,834. If these 5 randomly sampled computers are used to estimate the mean scan time for all 25 computers, what would the sampling error be?
c. What is the range of possible sampling error if a random sample size of 7 computers is taken to estimate the mean scan time for all 25 machines?

## Computer Software Exercises

**7-16.** The file **2015 Cardinals** shows the player positions and annual salaries for the Arizona Cardinals NFL players (*source*: www.spotrac.com/nfl).
a. Calculate the average total salary for the Arizona Cardinals.
b. Calculate the smallest sample mean for total salary and the largest sample mean for total salary using a sample size of 10. Calculate the sampling error for each sample mean.
c. Repeat the calculations in part b for a sample size of 5.
d. What effect does a change in the sample size appear to have on the dispersion of the sampling errors?

**7-17.** The file titled **Clothing** contains the monthly retail sales ($millions) of U.S. women's clothing stores for 70 months. A sample taken from this population to estimate the average sales in this time period follows:

| | | | | | | |
|---|---|---|---|---|---|---|
| 2,942 | 2,574 | 2,760 | 2,939 | 2,642 | 2,905 | 2,568 |
| 2,677 | 2,572 | 3,119 | 2,697 | 2,884 | 2,632 | 2,742 |
| 2,671 | 2,884 | 2,946 | 2,825 | 2,987 | 2,729 | 2,676 |
| 2,846 | 3,112 | 2,924 | 2,676 | | | |

a. Calculate the population mean.

b. Calculate the sample mean.

c. How much sampling error is present in this sample?

d. Determine the range of possible sampling error if 25 sales figures are sampled at random from this population.

**7-18.** The file **Dow Jones** contains date, open, high, low, close, and volume for the Dow Jones Industrial Average (DJIA) for almost eight years of trading days.

a. Assuming that the data in the file constitute the population of interest, what is the population mean closing value for the DJIA?

b. Using Excel, select a random sample of 10 days' closing values (make certain not to include duplicate days) and calculate the sample mean and the sampling error for the sample.

c. Repeat part b with a sample size of 50 days' closing values.

d. Repeat part b with a sample size of 100 days' closing values.

e. Write a short statement describing your results. Were they as expected? Explain.

**7-19.** Wilson Lumber Company is a privately held company that makes cedar siding, cedar lumber, and cedar fencing products. The major cost of production is the cedar logs that are the raw material necessary to make the finished cedar products. The dollar value to be achieved from a log depends initially on the diameter of the log. The file called **Wilson** contains a random sample of logs of various diameters and the potential value of the finished products that could be developed from the log if it is made into fence boards.

a. Calculate the sample mean potential value for each diameter of logs in the sample.

b. Discuss whether there is a way to determine how much sampling error exists for a given diameter log based on the sample. Can you determine whether the sampling error will be positive or negative? Discuss.

**7-20.** Maher, Barney, and White LLC is a legal firm with 40 employees. All of the firm's employees are eligible to participate in the company's 401(k) plan, and the firm is proud of its 100% participation rate.

The file **MBW 401** contains last year's year-end 401(k) account balance for each of the firm's 40 employees.

a. Compute the population mean and population standard deviation for the year-end 401(k) account balances at Maher, Barney, and White.

b. Suppose that an audit of the firm's 401(k) plan is being conducted and 12 randomly selected employee account balances are to be examined. If the following employees (indicated by employee number) are randomly selected to be included in the study, what is the estimate for the year-end mean 401(k) account balance? How much sampling error is present in this estimate?

| Employee # | | | | | | | | | | | |
|---|---|---|---|---|---|---|---|---|---|---|---|
| 26 | 8 | 31 | 3 | 38 | 30 | 17 | 9 | 21 | 39 | 18 | 11 |

c. Calculate the range of possible sampling error if a random sample of 15 employees is used to estimate the year-end mean 401(k) account balance.

**7-21.** The Badke Foundation was set up by the Fred Badke family following his death in 2001. To help in the solicitation of contributions, the foundation was considering the idea of hiring a consulting company that specialized in this activity. Leaders of the consulting company maintained in their presentation that the mean contribution from families who actually contribute after receiving a specially prepared letter would be $20.00.

Before actually hiring the company, the Badke Foundation sent out the letter and request materials to many people in the San Diego area. They received contributions from 166 families. The contribution amounts are in the data file called **Badke**.

a. Assuming that these data reflect a random sample of the population of contributions that would be received, compute the sampling error based on the claim made by the consulting firm.

b. Comment on any issues you have with the assumption that the data represent a random sample. Does the calculation of the sampling error matter if the sample is not a random sample? Discuss.

## 7.2 Sampling Distribution of the Mean

Section 7.1 introduced the concept of sampling error. A random sample selected from a population will not perfectly match the population. Thus, the sample statistic likely will not equal the population parameter. If this difference arises because the random sample is not a perfect representation of the population, it is called sampling error.

In business applications, decision makers select a single random sample from a population. They compute a sample measure and use it to make decisions about the entire population. For example, Nielsen Media Research takes a single random sample of television viewers to determine the percentage of the population who are watching a particular program during a particular week. Of course, the sample selected is only one of many possible samples that could have been selected from the same population. The sampling error will differ depending

**Sampling Distribution**
The distribution of all possible values of a statistic for a given sample size that has been randomly selected from a population.

on which sample is selected. If, in theory, you were to select all possible random samples of a given size and compute the sample means for each one, these means would vary above and below the true population mean. If we graphed these values as a histogram, the graph would be the **sampling distribution**.

In this section, we introduce the basic concepts of sampling distributions. We will use an Excel tool to select repeated samples from the same population for demonstration purposes only.

**OUTCOME 2**

## Simulating the Sampling Distribution for $\bar{x}$

| BUSINESS APPLICATION | Sampling Distributions |
| --- | --- |

**Aims Investment Company**  Aims Investment Company handles employee retirement funds, primarily for small companies. The file named **AIMS** contains data on the number of mutual funds in each client's portfolio. The file contains data for all 200 Aims customers, so it is considered a population. Figure 7.1 shows a histogram for the population.

The mean number of mutual funds in a portfolio is 2.505 funds. The standard deviation is 1.503 funds. The graph in Figure 7.1 indicates that the population is spread between zero and six funds, with more customers owning two funds than any other number.

Suppose the controller at Aims plans to select a random sample of 10 accounts. In Excel, we can use the Sampling tool to generate the random sample. Figure 7.2 shows the number of mutual funds owned for a random sample of 10 clients. The sample mean of 2.5 is also shown. To illustrate the concept of a sampling distribution, we repeat this process 500 times, generating 500 different random samples of 10. For each sample, we compute the sample mean.

**FIGURE 7.1 Distribution of Mutual Funds for the Aims Investment Company**

**Population of Funds Owned**

(histogram: Number of Customers vs. Number of Mutual Funds)

**FIGURE 7.2 Excel 2016 Output for the Aims Investment Company First Sample Size $n = 10$**

**Excel 2016 Instructions**

1. Open file: **Aims.xlsx**.
2. Select **Data > Data Analysis**.
3. Select **Sampling**.
4. Define the population data range {B2:B201}.
5. Select **Random, Number of Samples: 10**.
6. Select **Output Range**: D2.
7. Compute sample mean using Excel = **Average** function, using the range (D2:D11).

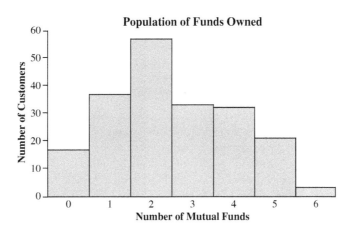

| | A | B | C | D |
| --- | --- | --- | --- | --- |
| 1 | Cutomer Number | Number of Mutual Fund Accounts | | |
| 2 | 19100 | 4 | | 2 |
| 3 | 5034 | 4 | | 4 |
| 4 | 29824 | 1 | | 2 |
| 5 | 44955 | 0 | | 2 |
| 6 | 44230 | 5 | | 3 |
| 7 | 47923 | 5 | | 1 |
| 8 | 725 | 2 | | 2 |
| 9 | 20371 | 3 | | 4 |
| 10 | 43162 | 4 | | 1 |
| 11 | 6929 | 1 | | 4 |
| 12 | 12252 | 4 | Mean= | 2.5 |

**FIGURE 7.3 Aims Investment Company, Histogram of 500 Sample Means from Sample Size $n = 10$**

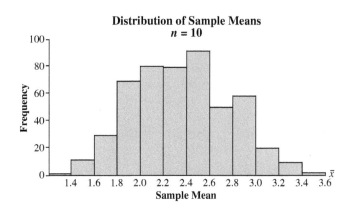

Figure 7.3 shows the frequency histogram for these sample means. Note that the horizontal axis represents the $\bar{x}$-values. The graph in Figure 7.3 is not a complete sampling distribution because it is based on only 500 samples out of the many possible samples that could be selected. However, this simulation gives us an idea of what the sampling distribution looks like.

Look again at the population distribution in Figure 7.1 and compare it with the shape of the frequency histogram in Figure 7.3. Although the population distribution is somewhat skewed, the distribution of sample means is taking the shape of a normal distribution.

Note also the population mean number of mutual funds owned by the 200 Aims Investment customers in the population is 2.505. If we average the 500 sample means in Figure 7.3, we get 2.41. This value is the *mean of the 500 sample means*. It is reasonably close to the population mean.

Had we selected all possible random samples of size 10 from the population and computed all possible sample means, the average of all the possible sample means would be equal to the population mean. This concept is expressed as Theorem 1.

### Theorem 1

For any population, the average value of all possible sample means computed from all possible random samples of a given size from the population equals the population mean. This is expressed as

$$\mu_{\bar{x}} = \mu$$

**Unbiased Estimator**
A characteristic of certain statistics in which the average of all possible values of the sample statistic equals a parameter, no matter the value of the parameter.

When the average of all possible values of the sample statistic equals the corresponding parameter, no matter the value of the parameter, we call that statistic an **unbiased estimator** of the parameter.

Also, the population standard deviation is 1.503 mutual funds. This measures the variation in the number of mutual funds between individual customers. When we compute the standard deviation of the 500 sample means, we get 0.421, which is considerably smaller than the population standard deviation. If all possible random samples of size $n$ are selected from the population, the distribution of possible sample means will have a standard deviation that is equal to the population standard deviation divided by the square root of the sample size, as Theorem 2 states.

### Theorem 2

For any population, the standard deviation of the possible sample means computed from all possible random samples of size $n$ is equal to the population standard deviation divided by the square root of the sample size. This is shown as

$$\sigma_{\bar{x}} = \frac{\sigma}{\sqrt{n}}$$

Recall the population standard deviation is $\sigma = 1.503$. Then, based on Theorem 2, had we selected all possible random samples of size $n = 10$ rather than only 500 samples, the standard deviation for the possible sample means would be

$$\sigma_{\bar{x}} = \frac{\sigma}{\sqrt{n}} = \frac{1.503}{\sqrt{10}} = 0.475$$

Our simulated value of 0.421 is fairly close to 0.475.

**FIGURE 7.4 Aims Investment Company, Histogram of Sample Means from Sample Size $n = 20$**

The standard deviation of the sampling distribution will be less than the population standard deviation. To further illustrate, suppose we increased the sample size from $n = 10$ to $n = 20$ and selected 500 new samples of size 20. Figure 7.4 shows the distribution of the 500 different sample means.

The distribution in Figure 7.4 is even closer to a normal distribution than what we observed in Figure 7.3. As sample size increases, the distribution of sample means more closely resembles a normal distribution. The average sample mean for these 500 samples is 2.53, and the standard deviation of the different sample means is 0.376. Based on Theorems 1 and 2, for a sample size of 20, we would expect the following:

$$\mu_{\bar{x}} = \mu = 2.505 \quad \text{and} \quad \sigma_{\bar{x}} = \frac{\sigma}{\sqrt{n}} = \frac{1.503}{\sqrt{20}} = 0.336$$

Thus, our simulated values are quite close to the theoretical values we would expect had we selected all possible random samples of size 20.

**Sampling from Normal Populations**  The previous discussion began with the population of mutual funds shown in Figure 7.1. The population was not normally distributed, but as we increased the sample size, the sampling distribution of possible sample means began to approach a normal distribution. We will return to this situation shortly, but what happens if the population itself is normally distributed? To help answer this question, we can apply Theorem 3.

### Theorem 3

If a population is normally distributed, with mean $\mu$ and standard deviation $\sigma$, the sampling distribution of the sample mean $\bar{x}$ is also normally distributed with a mean equal to the population mean ($\mu_{\bar{x}} = \mu$) and a standard deviation equal to the population standard deviation divided by the square root of the sample size ($\sigma_{\bar{x}} = \sigma/\sqrt{n}$).

In Theorem 3, the quantity ($\sigma_{\bar{x}} = \sigma/\sqrt{n}$) is the *standard deviation of the sampling distribution*. Another term that is given to this is the *standard error of $\bar{x}$*, because it is the measure of the standard deviation of the potential sampling error.

We can again use simulation to demonstrate Theorem 3. We begin by using Excel to generate a normally distributed population. Figure 7.5 shows a simulated population that is approximately normally distributed with a mean equal to 1,000 and a standard deviation equal to 200. The data range is from 250 to 1,800.

Next, we have simulated the selection of 2,000 random samples of size $n = 10$ from the normally distributed population and computed the sample mean for each sample. These sample means can then be graphed as a frequency histogram, as shown in Figure 7.6. This histogram represents the sampling distribution. Note that it, too, is approximately normally distributed.

**FIGURE 7.5 Simulated Normal Population Distribution**

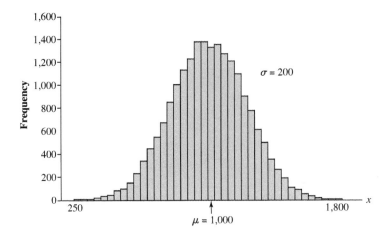

**FIGURE 7.6 Approximated Sampling Distribution ($n = 10$)**

We next compute the average of the 2,000 sample means and use it to approximate $\mu_{\bar{x}}$, as follows:

$$\mu_{\bar{x}} \approx \frac{\sum \bar{x}}{2,000} = \frac{2,000,178}{2,000} \approx 1,000$$

The mean of these sample means is approximately 1,000. This is the same value as the population mean.

We also approximate the standard deviation of the sample means as follows:

$$\sigma_{\bar{x}} \approx \sqrt{\frac{\sum (\bar{x} - \mu_{\bar{x}})^2}{2,000}} = 62.10$$

We see the standard deviation of the sample means is 62.10. This is much smaller than the population standard deviation, which is 200. The smallest sample mean was just less than 775, and the largest sample mean was just higher than 1,212. Recall, however, that the population ranged from 250 to 1,800. The variation in the sample means always will be less than the variation for the population as a whole. Using Theorem 3, we would expect the sample means to have a standard deviation equal to

$$\sigma_{\bar{x}} = \frac{\sigma}{\sqrt{n}} = \frac{200}{\sqrt{10}} = 63.25$$

Our simulated standard deviation of 62.10 is quite close to the theoretical value of 63.25.

Suppose we again use the simulated population shown in Figure 7.5, with $\mu = 1,000$ and $\sigma = 200$. We are interested in seeing what the sampling distribution will look like for different sample sizes. For a sample size of 5, Theorem 3 indicates that the sampling distribution will be normally distributed and have a mean equal to 1,000 and a standard deviation equal to

$$\sigma_{\bar{x}} = \frac{200}{\sqrt{5}} = 89.44$$

If we were to take a random sample of 10 (as simulated earlier), Theorem 3 indicates the sampling distribution would be normal, with a mean equal to 1,000 and a standard deviation equal to

$$\sigma_{\bar{x}} = \frac{200}{\sqrt{10}} = 63.25$$

For a sample size of 20, the sampling distribution will be centered at $\mu_{\bar{x}} = 1,000$, with a standard deviation equal to

$$\sigma_{\bar{x}} = \frac{200}{\sqrt{20}} = 44.72$$

Notice that as we increase the sample size, the standard deviation of the sampling distribution is reduced. This means the potential for extreme sampling error is reduced when we use larger sample sizes. Figure 7.7 shows sampling distributions for sample sizes of 5, 10, and 20. When the population is normally distributed, the sampling distribution of $\bar{x}$ is always normal and centered at the population mean. Only the spread in the distribution changes as the sample size changes.

This illustrates a very important statistical concept referred to as *consistency*. Earlier we defined a statistic as *unbiased* if the average value of the statistic equals the parameter to be estimated. Theorem 1 asserted that the sample mean is an *unbiased* estimator of the population mean no matter the value of the parameter. However, just because a statistic is unbiased does not tell us whether the statistic will be close in value to the parameter. But if, as the sample size is increased, we can expect the value of the statistic to become closer to the parameter, then we say that the statistic is a **consistent estimator** of the parameter. Figure 7.7 illustrates that the sample mean is a consistent estimator of the population mean.

The sampling distribution is composed of all possible sample means of the same size. Half the sample means lie above the center of the sampling distribution and half lie below. The relative distance that a given sample mean is from the center can be determined by *standardizing* the sampling distribution. As discussed in Chapter 6, a standardized value is determined by converting the value from its original units into a $z$-value. A $z$-value measures the number of standard deviations a value is from the mean. We can use this same concept when working with a sampling distribution. Equation 7.4 shows how the $z$-values are computed.

**Consistent Estimator**

An unbiased estimator is said to be a consistent estimator if the difference between the estimator and the parameter tends to become smaller as the sample size becomes larger.

**FIGURE 7.7 Theorem 3 Examples**

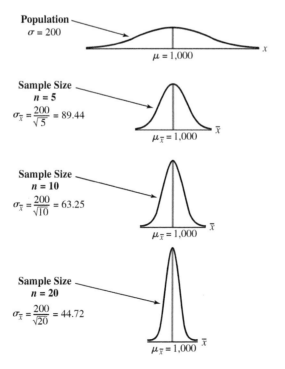

**z-Value for Sampling Distribution of $\bar{x}$**

$$z = \frac{\bar{x} - \mu}{\dfrac{\sigma}{\sqrt{n}}}$$    (7.4)

where:

$\bar{x}$ = Sample mean
$\mu$ = Population mean
$\sigma$ = Population standard deviation
$n$ = Sample size

Note that if the sample being selected is large relative to the size of the population (greater than 5% of the population size), we need to modify how we compute the standard deviation of the sampling distribution and z-value using what is known as the *finite population correction factor*, as shown in Equation 7.5.

**z-Value Adjusted for the Finite Population Correction Factor**

$$z = \frac{\bar{x} - \mu}{\dfrac{\sigma}{\sqrt{n}}\sqrt{\dfrac{N - n}{N - 1}}}$$    (7.5)

where:

$N$ = Population size
$n$ = Sample size

$\sqrt{\dfrac{N - n}{N - 1}}$ = Finite population correction factor

The finite population correction factor is used to calculate the standard deviation of the sampling distribution when the sample size is greater than 5% of the population size.

---

**EXAMPLE 7-2**    **Finding the Probability That $\bar{x}$ Is in a Given Range**

**Scribner Products**    Scribner Products manufactures flooring materials for the residential and commercial construction industries. One item they make is a mosaic tile for bathrooms and showers. When the production process is operating according to specifications, the diagonal dimension of a tile used for decorative purposes is normally distributed with a mean equal to 1.5 inches and a standard deviation of 0.05 inch. Before shipping a large batch of these tiles, Scribner quality analysts have selected a random sample of eight tiles with the following diameters:

| 1.57 | 1.59 | 1.48 | 1.60 | 1.59 | 1.62 | 1.55 | 1.52 |

The analysts want to use these measurements to determine if the process is no longer operating within the specifications. The following steps can be used:

**STEP 1**    **Determine the mean for this sample.**

$$\bar{x} = \frac{\sum x}{n} = \frac{12.52}{8} = 1.565 \text{ inches}$$

**STEP 2**    **Define the sampling distribution for $\bar{x}$ using Theorem 3.**
Theorem 3 indicates that if the population is normally distributed, the sampling distribution for $\bar{x}$ is also normally distributed, with

$$\mu_{\bar{x}} = \mu \quad \text{and} \quad \sigma_{\bar{x}} = \frac{\sigma}{\sqrt{n}}$$

Thus, in this case, the mean of the sampling distribution should be 1.50 inches, and the standard deviation should be $0.05/\sqrt{8} = 0.0177$ inch.

STEP 3 **Define the probability statement of interest.**
Because the sample mean is $\bar{x} = 1.565$, which is greater than the mean of the sampling distribution, we want to find

$$P(\bar{x} \geq 1.565 \text{ inches}) = ?$$

STEP 4 **Convert the sample mean to a standardized z-value, using Equation 7.4.**

$$z = \frac{\bar{x} - \mu}{\dfrac{\sigma}{\sqrt{n}}} = \frac{1.565 - 1.50}{\dfrac{0.05}{\sqrt{8}}} = \frac{0.065}{0.0177} = 3.67$$

STEP 5 **Use the standard normal distribution table to determine the desired probability.**

$$P(z \geq 3.67) = ?$$

The standard normal distribution table in Appendix D does not show $z$-values as high as 3.67. This implies that $P(z \geq 3.67) \approx 0.00$.

**TRY EXERCISE 7-26** (pg. 259)

So, if the production process is working properly, there is virtually no chance that a random sample of eight tiles will have a mean diameter of 1.565 inches or greater. Because the analysts at Scribner Products did find this sample result, there is a very good chance that something is wrong with the process.

OUTCOME **3**

## The Central Limit Theorem

Theorem 3 applies when the population is normally distributed. Although this is the case in many business situations, there are also many situations in which the population is not normal. For example, incomes in a region tend to be right skewed. Some distributions, such as people's weight, are bimodal (a peak weight group for males and another peak weight group for females).

What does the sampling distribution of $\bar{x}$ look like when a population is not normally distributed? The answer is . . . it depends. It depends on what the shape of the population is and what size sample is selected. To illustrate, suppose we have a U-shaped population, such as the one in Figure 7.8, with mean $= 14.00$ and standard deviation $= 3.00$. Now, we select 3,000 simple random samples of size 3 and compute the mean for each sample. These $\bar{x}$-values are graphed in the histogram shown in Figure 7.9.

The average of these 3,000 sample means is

$$\frac{\sum \bar{x}}{3,000} \approx \mu_{\bar{x}} = 14.02$$

Notice this value is approximately equal to the population mean of 14.00, as Theorem 1 would suggest.[2]

**FIGURE 7.8 Simulated Nonnormal Population**

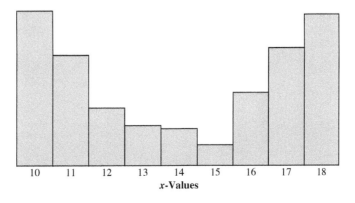

*x*-Values

---

[2]Note that if we had selected all possible samples of 3, the average of the sample means would have been equal to the population mean.

**FIGURE 7.9 Frequency Histogram of $\bar{x}(n = 3)$**

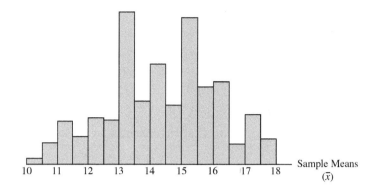

Next we compute the standard deviation as

$$\sigma_{\bar{x}} \approx \sqrt{\frac{\Sigma(\bar{x} - \mu_{\bar{x}})^2}{3{,}000}} = 1.82$$

The standard deviation of the sampling distribution is less than the standard deviation for the population, which was 3.00. This is always the case.

The frequency histogram of $\bar{x}$-values for the 3,000 samples of 3 looks different from the population distribution, which is U-shaped. Suppose we increase the sample size to 10 and take 3,000 samples from the same U-shaped population. The resulting frequency histogram of $\bar{x}$-values is shown in Figure 7.10. Now the frequency distribution looks much like a normal distribution. The average of the sample means is still equal to 14.02, which is virtually equal to the population mean. The standard deviation for this sampling distribution is now reduced to 0.97.

This example is not a special case. Instead, it illustrates a very important statistical concept called the Central Limit Theorem.

### Theorem 4: The Central Limit Theorem

For simple random samples of $n$ observations taken from a population with mean $\mu$ and standard deviation $\sigma$, regardless of the population's distribution, provided the sample size is sufficiently large, the distribution of the sample means, $\bar{x}$, is approximately normal with a mean equal to the population mean ($\mu_{\bar{x}} = \mu$) and a standard deviation equal to the population standard deviation divided by the square root of the sample size ($\sigma_{\bar{x}} = \sigma/\sqrt{n}$). The larger the sample size, the better the approximation to the normal distribution.

The Central Limit Theorem is very important because with it, we know the shape of the sampling distribution even though we may not know the shape of the population distribution. The one catch is that the sample size must be "sufficiently large." What is a sufficiently large sample size?

The answer depends on the shape of the population. If the population is quite symmetric, then sample sizes as small as 2 or 3 can provide a normally distributed sampling distribution.

**FIGURE 7.10 Frequency Histogram of $\bar{x}(n = 10)$**

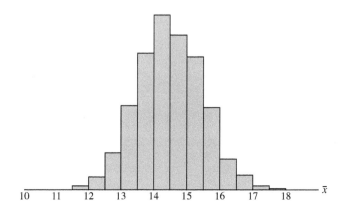

If the population is highly skewed or otherwise irregularly shaped, the required sample size will be larger. Recall the example of the U-shaped population. The frequency distribution obtained from samples of 3 was shaped differently than the population, but not like a normal distribution. However, for samples of 10, the frequency distribution of the sample means was a very close approximation to a normal distribution. Figures 7.11, 7.12, and 7.13 show some examples of the Central Limit Theorem concept. Simulation studies indicate that even for very strange-looking populations, samples of 25 to 30 produce sampling distributions that are approximately normal. Thus, *a conservative definition of a sufficiently large sample size is* $n \geq 30$. The Central Limit Theorem is illustrated in the following example.

**FIGURE 7.11 Central Limit Theorem with Uniform Population Distribution**

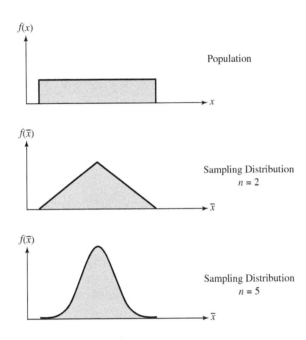

**FIGURE 7.12 Central Limit Theorem with Triangular Population**

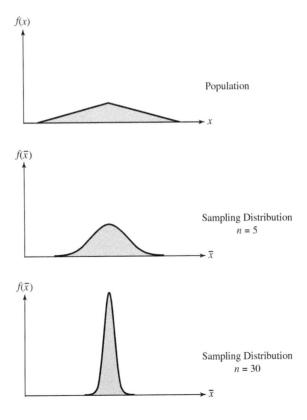

**FIGURE 7.13 Central Limit Theorem with Skewed Population**

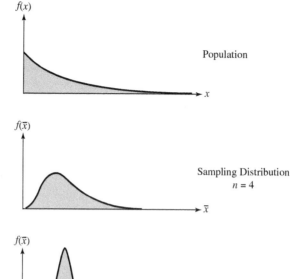

**EXAMPLE 7-3**    Finding the Probability That $\bar{x}$ Is in a Given Range

**Westside Drive-In**    Past sales records indicate that the dollar value of lunch orders at the Westside Drive-In are right skewed, with a population mean of $12.50 per customer and a standard deviation of $5.50. The manager has selected a random sample of 100 lunch receipts. She is interested in determining the probability that the mean lunch order for this sample from this population will fall between $12.25 and $13.00. To find this probability, she can use the following steps:

**STEP 1  Determine the sample mean.**
In this case, two sample means are being considered:

$$\bar{x} = \$12.25 \quad \text{and} \quad \bar{x} = \$13.00$$

**STEP 2  Define the sampling distribution.**
We can use the Central Limit Theorem because the sample size is large enough ($n = 100$) to determine that the sampling distribution will be approximately normal (even though the population is right skewed), with

$$\mu_{\bar{x}} = \$12.50 \quad \text{and} \quad \sigma_{\bar{x}} = \frac{\$5.50}{\sqrt{100}} = \$0.55$$

**STEP 3  Define the probability statement of interest.**
The manager is interested in

$$P(\$12.25 \leq \bar{x} \leq \$13.00) = ?$$

**STEP 4  Use the standard normal distribution to find the probability of interest.**
Assuming the population of lunch receipts is quite large, we use Equation 7.4 to convert the sample means to corresponding z-values.

$$z = \frac{\bar{x} - \mu}{\frac{\sigma}{\sqrt{n}}} = \frac{12.25 - 12.50}{\frac{5.50}{\sqrt{100}}} = -0.46 \quad \text{and} \quad z = \frac{\bar{x} - \mu}{\frac{\sigma}{\sqrt{n}}} = \frac{13.00 - 12.50}{\frac{5.50}{\sqrt{100}}} = 0.91$$

From the standard normal table in Appendix D, the probability associated with $z = -0.46$ is 0.1772, and the probability for $z = 0.91$ is 0.3186. Therefore,

$$P(\$12.25 \leq \bar{x} \leq \$13.00) = P(-0.46 \leq z \leq 0.91) = 0.1772 + 0.3186 = 0.4958$$

There is nearly a 0.50 chance that the sample mean will fall in the range $12.25 to $13.00.

**TRY EXERCISE 7-30** (pg. 259)

# 7.2 EXERCISES

## Skill Development

**7-22.** A population with a mean of 1,250 and a standard deviation of 400 is known to be highly skewed to the right. If a random sample of 64 items is selected from the population, what is the probability that the sample mean will be less than 1,325?

**7-23.** Suppose that a population is known to be normally distributed with $\mu = 2{,}000$ and $\sigma = 230$. If a random sample of size $n = 8$ is selected, calculate the probability that the sample mean will exceed 2,100.

**7-24.** A normally distributed population has a mean of 500 and a standard deviation of 60.
   a. Determine the probability that a random sample of size 16 selected from this population will have a sample mean less than 475.
   b. Determine the probability that a random sample of size 25 selected from the population will have a sample mean greater than or equal to 515.

**7-25.** If a population is known to be normally distributed with $\mu = 250$ and $\sigma = 40$, what will be the characteristics of the sampling distribution for $\bar{x}$ based on a random sample of size 25 selected from the population?

**7-26.** Suppose nine items are randomly sampled from a normally distributed population with a mean of 100 and a standard deviation of 20. The nine randomly sampled values are

| | | | | |
|---|---|---|---|---|
| 125 | 95 | 66 | 116 | 99 |
| 91 | 102 | 51 | 110 | |

Calculate the probability of getting a sample mean that is smaller than the sample mean for these nine sampled values.

**7-27.** A random sample of 100 items is selected from a population of size 350. What is the probability that the sample mean will exceed 200 if the population mean is 195 and the population standard deviation equals 20? (*Hint:* Use the finite population correction factor since the sample size is more than 5% of the population size.)

**7-28.** Given a distribution that has a mean of 40 and a standard deviation of 13, calculate the probability that a sample of 49 has a sample mean that is

   a. greater than 37
   b. at most 43
   c. between 37 and 43
   d. between 43 and 45
   e. no more than 35

**7-29.** Consider a normal distribution with mean = 12 and standard deviation = 90. Calculate $P(\bar{x} > 36)$ for each of the following sample sizes:
   a. $n = 1$
   b. $n = 9$
   c. $n = 16$
   d. $n = 25$

## Business Applications

**7-30.** Andersen Windows makes windows for use in homes and commercial buildings. The standards for glass thickness call for the glass to average 0.375 inch with a standard deviation equal to 0.050 inch. Suppose a random sample of $n = 50$ windows yields a sample mean of 0.392 inch.
   a. What is the probability of $\bar{x} \geq 0.392$ if the windows meet the standards?
   b. Based on your answer to part a, what would you conclude about the population of windows? Is it meeting the standards?

**7-31.** Tax Professionals, Inc., is a tax preparation service with offices located throughout the western United States. Suppose the average number of returns processed by employees of Tax Professionals during tax season is 12 per day with a standard deviation of 3 per day. A random sample of 36 employees taken during tax season revealed the following number of returns processed daily:

| | | | | | | | | |
|---|---|---|---|---|---|---|---|---|
| 11 | 17 | 13 | 9 | 13 | 13 | 13 | 12 | 15 |
| 15 | 13 | 15 | 10 | 15 | 13 | 13 | 15 | 13 |
| 9 | 9 | 9 | 15 | 13 | 14 | 9 | 14 | 11 |
| 11 | 17 | 16 | 9 | 8 | 10 | 15 | 12 | 11 |

   a. What is the probability of having a sample mean equal to or smaller than the sample mean for this sample if the population mean is 12 processed returns daily with a standard deviation of 3 returns per day?

b. What is the probability of having a sample mean larger than the one obtained from this sample if the population mean is 12 processed returns daily with a standard deviation of 3 returns per day?

c. Explain how it is possible to answer parts a and b when the population distribution of daily tax returns at Tax Professionals is not known.

**7-32.** SeaFair Fashions relies on its sales force of 220 to do an initial screening of all new fashions. The company is bringing out a new line of swimwear and has invited 40 salespeople to its Orlando home office. An issue of constant concern to the SeaFair sales office is the volume of orders generated by each salesperson. Last year, the overall company average was $417,330 with a standard deviation of $45,285. (*Hint:* The finite population correction factor, Equation 7.5, is required.)

a. Determine the probability the sample of 40 will have a sales average less than $400,000.

b. What shape do you think the distribution of all possible sample means of 40 will have? Discuss.

c. Determine the value of the standard deviation of the distribution of the sample mean of all possible samples of size 40.

d. How would the answers to parts a, b, and c change if the home office brought 60 salespeople to Orlando? Provide the respective answers for this sample size.

e. Each year SeaFair invites the sales personnel with sales above the 85th percentile to enjoy a complimentary vacation in Hawaii. Determine the smallest average salary for the sales personnel who were in Hawaii last year. (Assume the sales were normally distributed last year.)

**7-33.** Suppose the life of a particular brand of calculator battery is approximately normally distributed with a mean of 75 hours and a standard deviation of 10 hours.

a. What is the probability that a single battery randomly selected from the population will have a life between 70 and 80 hours?

b. What is the probability that 16 randomly sampled batteries from the population will have a sample mean life of between 70 and 80 hours?

c. If the manufacturer of the battery is able to reduce the standard deviation of battery life from 10 to 9 hours, what would be the probability that 16 batteries randomly sampled from the population will have a sample mean life of between 70 and 80 hours?

**7-34.** Sands, Inc., makes particleboard for the building industry. Particleboard is built by mixing wood chips and resins together and pressing the sheets under extreme heat and pressure to form a 4-feet $\times$ 8-feet sheet that is used as a substitute for plywood. The strength of the particleboard is tied to the board's weight. Boards that are too light are brittle and do not meet the quality standard for strength. Boards that are too heavy are strong but are difficult for customers to

handle. The company knows that there will be variation in the boards' weight. Product specifications call for the weight per sheet to average 10 pounds with a standard deviation of 1.75 pounds. During each shift, Sands employees select and weigh a random sample of 25 boards. The boards are thought to have a normally distributed weight.

If the average of the sample slips below 9.60 pounds, an adjustment is made to the process to add more moisture and resins to increase the weight (and, Sands hopes, the strength).

a. Assuming that the process is operating correctly according to specifications, what is the probability that a sample will indicate that an adjustment is needed?

b. Assume the population mean weight per sheet slips to 9 pounds. Determine the probability that the sample will indicate an adjustment is not needed.

c. Assuming that 10 pounds is the mean weight, what should the cutoff be if the company wants no more than a 5% chance that a sample of 25 boards will have an average weight less than 9.6 lbs?

**7-35.** The branch manager for United Savings and Loan in Seaside, Virginia, has worked with her employees in an effort to reduce the waiting time for customers at the bank. She and the team concluded that average waiting time is now down to 3.5 minutes with a standard deviation equal to 1.0 minute. However, before making a statement at a managers' meeting, this branch manager wanted to double-check that the process was working as thought. To make this check, she randomly sampled 25 customers and recorded the time they had to wait. She discovered that mean wait time for this sample of customers was 4.2 minutes. Based on the team's claims about waiting time, what is the probability that a sample mean for $n = 25$ people would be as large as or larger than 4.2 minutes? What should the manager conclude based on these data?

**7-36.** Mileage ratings for cars and trucks generally come with a qualifier stating actual mileage will depend on driving conditions and habits. Ford states the Ecoboost F-150 3.5-L V6 will average 20 miles per gallon with combined town and country driving (*source:* www.ford.com/trucks). Assume the mean stated by Ford is the actual average, and the distribution has a standard deviation of 3 mpg.

a. Given the above mean and standard deviation, what is the probability that 100 drivers will average more than 19.2 miles per gallon?

b. Suppose 1,000 drivers were randomly selected. What is the probability the average obtained by these drivers will exceed 19.2 mpg?

**7-37.** Many airlines limit the weight of checked bags to 50 pounds on domestic U.S. flights. They will carry heavier bags, but at an additional fee. Suppose one major airline has stated in an internal memo to employees that the mean weight for bags checked last year on the airline was 34.3 pounds with a standard deviation of 5.7 pounds. Further, it stated that the distribution of weights was approximately normally distributed. This memo was leaked to a consumers' group in Atlanta. This group selected and weighed a random sample of 14 bags to be checked on a flight departing from Atlanta. The following data (pounds) were recorded:

| 29 | 27 | 40 | 34 | 30 | 30 | 35 |
|----|----|----|----|----|----|----|
| 44 | 33 | 28 | 36 | 33 | 30 | 40 |

What is the probability that a sample mean as small as or smaller than the one for this sample would occur if the airline's claims about the population of baggage weight is accurate? Comment on the results.

**7-38.** Carter & Whitmore is a New York–based marketing research company. One of the items that the company tracks is the expenditure on over-the-counter (OTC) cough medicines. They indicated that consumers spent $620 million on OTC cough medicines in the United States. The company also indicated that nearly 30 million visits for coughs were made to doctors' offices in the United States.

a. Determine the average cost of OTC cough medicines per doctor's office visit based on 30 million purchases.

b. Assuming that the average cost indicated in part a is the true average cost of OTC cough medicines per doctor's visit and the standard deviation is $10, determine the probability that the average cost for a random selection of 30 individuals will result in an average expenditure of more than $25 in OTC cough medicines.

c. Determine the 90th percentile for the average cost of OTC cough medicines for a sample of 36 individuals, all of whom have visited a doctor's office for cough symptoms.

## Computer Software Exercises

**7-39.** One of the top-selling video games is *Call of Duty: Black Ops*. While prices vary widely depending on store or website, the suggested retail price for this video game is $59.95. The file titled **Black Ops** contains a random sample of the retail prices paid for *Call of Duty: Black Ops*.

a. Calculate the sample mean and standard deviation of retail prices paid for *Call of Duty: Black Ops*.

b. To determine if the average retail price has fallen, assume the population mean is $59.95 and calculate the probability that a sample of size 200 would result in a sample mean no larger than the one calculated in part a. Assume that the sample standard deviation is representative of the population standard deviation.

c. In part b, you used $59.95 as the population mean. Calculate the probability required in part b assuming that the population mean is $59.50.

d. On the basis of your calculations in parts b and c, does it seem likely that the average retail price for *Call of Duty: Black Ops* has decreased? Explain.

**7-40.** Acee Bottling and Distributing bottles and markets Pepsi products in southwestern Vermont. The average fill volume for Pepsi cans is supposed to be 12 ounces. The filling machine has a known standard deviation of 0.05 ounce. Each week, the company selects a simple random sample of 60 cans and carefully measures the volume in each can. The results of the latest sample are shown in the file called **Acee Bottling**. Based on the data in the sample, what would you conclude about whether the filling process is working as expected? Base your answer on the probability of observing the sample mean you compute for these sample data.

**7-41.** Bruce Leichtman is president of Leichtman Research Group, Inc. (LRG), which specializes in research and consulting on broadband, media, and entertainment industries. In a recent survey, the company determined the cost of HDMI Digital Antenna systems ranged from about $25 to $200 (*source:* "Everything you need to know about replacing cable with an HDTV antenna," www.fool.com, June 27, 2015). The file titled **HDCosts** contains a sample of the total costs including installation for HDMI Digital Antenna systems. Assume that the population average cost is $150 and the standard deviation is $50.

a. Create a box and whisker plot and use it and the sample average to determine if the population from which this sample was obtained could be normally distributed.

b. Determine the probability that the mean of a random sample of size 150 costs would be more than $5 different from the mean of the sample described above.

c. Given your response to part a, do you believe the results obtained in part b are valid? Explain.

**7-42.** The file **CEO Compensation** contains data on the top paid executives in 2014 (*source:* "Fortunate 50 data," *Pittsburgh Post-Gazette*, www.post-gazette.com, June 14, 2015).

a. Treating the data in the file as the population of interest, compute the population mean and standard deviation for CEO compensation.

b. Use Excel to select a simple random sample of 20 executive compensation amounts. Compute the sample mean for this sample. Find the probability of getting a sample mean as extreme as or more extreme than the one you got. (*Hint:* Use the finite population correction factor because the sample is large relative to the size of the population.)

**7-43.** Referring to Exercise 7-42, the data file called **CEO Compensation** contains data for highly paid CEOs. Treat the values as the population. If a random sample of size 5 was selected from this population, what is the largest possible sampling error that could occur when estimating the population mean?

**7-44.** The file **Salaries** contains the annual salary for all faculty at a small state college in the Midwest. Assume that these faculty salaries represent the population of interest.

   a. Compute the population mean and population standard deviation.

   b. Develop a frequency distribution of these data using 10 classes. Do the population data appear to be normally distributed?

   c. What is the probability that a random sample of 16 faculty selected from the population would have a sample mean annual salary greater than or equal to $56,650?

   d. Suppose the following 25 faculty were randomly sampled from the population and used to estimate the population mean annual salary:

| Faculty ID Number | | | | |
|---|---|---|---|---|
| 137 | 040 | 054 | 005 | 064 |
| 134 | 013 | 199 | 168 | 027 |
| 095 | 065 | 193 | 059 | 192 |
| 084 | 176 | 029 | 143 | 182 |
| 009 | 033 | 152 | 068 | 044 |

What would the sampling error be?

   e. Referring to part d, what is the probability of obtaining a sample mean smaller than the one obtained from this sample?

---

| 7.3 | **Sampling Distribution of a Proportion** |
|---|---|

## Working with Proportions

In many instances, the objective of sampling is to estimate a population proportion. For instance, an accountant may be interested in determining the proportion of accounts payable balances that are correct. A production supervisor may wish to determine the percentage of product that is defect-free. A marketing research department might want to know the proportion of potential customers who will purchase a particular product. In all these instances, the decision makers could select a sample, compute the sample proportion, and make their decision based on the sample results.

Sample proportions are subject to sampling error, just as are sample means. The concept of sampling distributions provides us a way to assess the potential magnitude of the sampling error for proportions in given situations.

---

**BUSINESS APPLICATION**    Sampling Distributions for Proportions

**Population Proportion**

The fraction of values in a population that have a specific attribute.

**Waste Management**    Waste Management is a national garbage and hazardous waste disposal company. Customer service managers at its Florida division surveyed every customer who used Waste Management's services in a suburb of Tampa during the month of September last year. The key question in the survey was "Are you satisfied with the service received?"

The population size was 80 customers. The number of customers who answered "Yes" to the question was 72. The value of interest in this example is the **population proportion**. Equation 7.6 is used to compute a population proportion.

**Population Proportion**

$$p = \frac{X}{N} \tag{7.6}$$

where:

    $p$ = Population proportion
    $X$ = Number of items in the population having the attribute of interest
    $N$ = Population size

The proportion of customers in the population who are satisfied with the service by Waste Management is

$$p = \frac{72}{80} = 0.90$$

Therefore, 90% of the population responded "Yes" to the survey question. This is the *parameter*. It is a measurement taken from the population. It is the "true value."

Now, suppose that Waste Management wishes to do a follow-up survey for a simple random sample of $n = 20$ from the same 80 customers. What fraction of this sample will be people who had previously responded "Yes" to the satisfaction question?

The answer depends on which sample is selected. There are many ($3.5353 \times 10^{18}$ to be precise) possible random samples of 20 that could be selected from 80 people. However, the company will select only one of these possible samples. At one extreme, suppose the 20 people selected for the sample included all 8 who answered "No" to the satisfaction question and 12 others who answered "Yes." The **sample proportion** is computed using Equation 7.7.

**Sample Proportion**

The fraction of items in a sample that have the attribute of interest.

**Sample Proportion**

$$\bar{p} = \frac{x}{n} \tag{7.7}$$

where:

$\bar{p}$ = Sample proportion
$x$ = Number of items in the sample with the attribute of interest
$n$ = Sample size

For the Waste Management example, the sample proportion of "Yes" responses is

$$\bar{p} = \frac{12}{20} = 0.60$$

The sample proportion of "Yes" responses is 0.60, whereas the population proportion is 0.90. The difference between the sample value and the population value is sampling error. Equation 7.8 is used to compute the sampling error involving a single proportion.

**Single-Proportion Sampling Error**

$$\text{Sampling error} = \bar{p} - p \tag{7.8}$$

where:

$p$ = Population proportion
$\bar{p}$ = Sample proportion

Then for this extreme situation we get

$$\text{Sampling error} = 0.60 - 0.90 = -0.30$$

If a sample on the other extreme had been selected and all 20 people came from the original list of 72 who had responded "Yes," the sample proportion would be

$$\bar{p} = \frac{20}{20} = 1.00$$

For this sample, the sampling error is

$$\text{Sampling error} = 1.00 - 0.90 = 0.10$$

Thus, the range of sampling error in this example is from −0.30 to 0.10. As with any sampling situation, you can expect some sampling error. The sample proportion will probably not equal the population proportion because the sample selected will not be a perfect replica of the population.

**EXAMPLE 7-4** **Sampling Error for a Proportion**

**AT&T Acquisition of DirecTV** In the summer of 2015, the Federal Communications Commission (FCC) approved AT&T's plan to acquire DirecTV for $48.5 billion. Projections showed that the new firm would be the largest pay TV company in the United States, with approximately 26 million subscribers. Based on a statement by DirecTV's executives, its shareholders gave overwhelming support to the deal, with 99% of those voting favoring the deal (*sources:* Jacob Kastrenakes, "FCC approves AT&T–DirecTV merger," www.theverge.com, July 24, 2015; and Aaron Mamiit, "DirecTV shareholders unanimously approves AT&T takeover," www.techtimes.com, Sep. 27, 2015). Suppose a shareholder advocacy group decides to randomly sample 40 shareholder proxies to see whether the share owners had voted in favor of the sale. Among these 40 proxies, 36 had "Approve" votes. The advocacy group can now use the following steps to assess the sampling error associated with the sample of proxies they selected:

**STEP 1** **Determine the population proportion.**

In this case, the proportion of votes cast in favor of the merger is

$$p = 0.99$$

This is the number of approval votes divided by the total number of shares.

**STEP 2** **Compute the sample proportion using Equation 7.7.**

The sample proportion is

$$\bar{p} = \frac{x}{n} = \frac{36}{40} = 0.90$$

**STEP 3** **Compute the sampling error using Equation 7.8.**

$$\text{Sampling error} = \bar{p} - p = 0.90 - 0.99 = -0.09$$

The proportion of "Approval" votes from the shares in this sample is 0.09 less than the reported proportion.

**TRY EXERCISE 7-47** (pg. 267)

**OUTCOME 4**

## Sampling Distribution of $\bar{p}$

In many applications, you will be interested in determining the proportion, $p$, of all items in a population that possess a particular attribute. The best estimate of this population proportion will be $\bar{p}$, the sample proportion. However, any inference about how close your estimate is to the true population value will be based on the distribution of the sample proportion, $\bar{p}$, whose underlying distribution is the binomial. However, if the sample size is sufficiently large such that

$$np \geq 5 \quad \text{and} \quad n(1 - p) \geq 5$$

then the normal distribution can be used as a reasonable approximation to the discrete binomial distribution. If the sample size is large enough, the distribution of all possible sample proportions will be approximately normally distributed. In addition to being normally distributed, the sampling distribution will have a mean and standard error as indicated in Equations 7.9 and 7.10.[3]

---

[3]An application of the Central Limit Theorem provides the rationale for this statement. Recall that $\bar{p} = x/n$, where $x$ is the sum of random variables $(x_i)$ whose values are 0 and 1. Therefore, $\bar{p}$ is in reality just a sample mean. Each of these $x_i$ can be thought of as binomial random variables from a sample of size $n = 1$. Thus, they each have a mean of $\mu = np = p$ and a variance of $\sigma^2 = np(1 - p) = p(1 - p)$. As we have seen from the Central Limit Theorem, the sample mean has an expected value of $\mu$ and a variance of $\sigma^2/n$. Thus, the sample proportion has an expected value of $\mu = p$ and a variance of $\sigma^2 = \dfrac{p(1 - p)}{n}$.

**Mean and Standard Error of the Sampling Distribution of $\bar{p}$**

$$\text{Mean} = \mu_{\bar{p}} = p \qquad\qquad (7.9)$$

and

$$\text{Standard error} = \sigma_{\bar{p}} = \sqrt{\frac{p(1-p)}{n}} \qquad\qquad (7.10)$$

where:

$$p = \text{Population proportion}$$
$$n = \text{Sample size}$$
$$\bar{p} = \text{Sample proportion}$$

In Section 7.2, we introduced Theorem 4, the Central Limit Theorem, which indicates that regardless of the shape of the population distribution, the distribution of possible sample means is approximately normal as long as the sample size is sufficiently large. Theorem 5 is similar but pertains to the sampling distribution for the sample proportion.

**Theorem 5: Sampling Distribution of $\bar{p}$**

Regardless of the value of the population proportion, $p$ (with the obvious exceptions of $p = 0$ and $p = 1$), the sampling distribution for the sample proportion, $\bar{p}$, is approximately normally distributed with $\mu_{\bar{p}} = p$ and

$$\sigma_{\bar{p}} = \sqrt{\frac{p(1-p)}{n}},$$

provided $np \geq 5$ and $n(1-p) \geq 5$. The approximation to the normal distribution improves as the sample size increases and $p$ approaches 0.50.

**BUSINESS APPLICATION**     ## Sampling Distribution for Proportions

**On-Line Coupons, Inc.**   On-Line Coupons, Inc., sells discount coupons online. Buyers get discounts at local retail and service establishments. The On-Line Coupon marketing managers have analyzed coupon redemption patterns and have determined that 15% of coupons purchased are not used by the buyer. There appears to be no particular pattern to the failure to use coupons, and one coupon going unused seems independent of whether any other coupon is not used.

Suppose that the company recently received an e-mail from a retailer who says that 18% of the 500 coupons sold were not used by the expiration date. Assume the nonuse rate for the population of all coupons sold is

$$p = 0.15$$

How likely is it that a sample of

$$n = 500$$

coupons sold will have 18% or more unused? To answer this question, we first check to determine if the sample size is sufficiently large. Because both

$$n(p) = 500(0.15) = 75 \geq 5 \quad \text{and} \quad n(1-p) = 500(0.85) = 425 \geq 5$$

we can safely conclude that the sampling distribution of sample proportions will be approximately normal. Using Equations 7.9 and 7.10, we can compute the mean and standard error for the sampling distribution as follows:

$$\mu_{\bar{p}} = 0.15$$

and

$$\sigma_{\bar{p}} = \sqrt{\frac{(0.15)(0.85)}{500}} = 0.016$$

Equation 7.11 is used to convert the sample proportion to a standardized $z$-value.

**$z$-Value for Sampling Distribution of $\bar{p}$**

$$z = \frac{\bar{p} - p}{\sigma_{\bar{p}}} \qquad\qquad (7.11)$$

where:

$z$ = Number of standard errors $\bar{p}$ is from $p$

$\bar{p}$ = Sample proportion

$\sigma_{\bar{p}} = \sqrt{\dfrac{p(1 - p)}{n}}$ = Standard error of the sampling distribution[4]

$p$ = Population proportion

From Equation 7.11, we get

$$z = \frac{\bar{p} - p}{\sigma_{\bar{p}}} = \frac{0.18 - 0.15}{\sqrt{\dfrac{(0.15)(0.85)}{500}}} = 1.88$$

Therefore, the 0.18 nonredemption rate reported by the retailer is 1.88 standard errors above the population proportion of 0.15. Figure 7.14 illustrates that the chances of a defect rate of 0.18 or higher is

$$P(\bar{p} \geq 0.18) = 0.0301$$

Because this is a low probability, the On-Line Coupons managers would want to investigate whether there was something unusual about this coupon offering.

---

> **HOW TO DO IT** (Example 7-5)
>
> **Sampling Distribution of $\bar{p}$**
>
> **1.** Determine the population proportion, $p$, using
>
> $$p = \frac{X}{N}$$
>
> **2.** Calculate the sample proportion using
>
> $$\bar{p} = \frac{x}{n}$$
>
> **3.** Determine the mean and standard deviation of the sampling distribution using
>
> $$\mu_{\bar{p}} = p \quad \text{and} \quad \sigma_{\bar{p}} = \sqrt{\frac{p(1 - p)}{n}}$$
>
> **4.** Define the event of interest. For example:
>
> $$P(\bar{p} \geq 0.30) = ?$$
>
> **5.** If $np$ and $n(1 - p)$ are both $\geq 5$, then convert $\bar{p}$ to a standardized $z$-value using
>
> $$z = \frac{\bar{p} - p}{\sigma_{\bar{p}}}$$
>
> **6.** Use the standard normal distribution table in Appendix D to determine the required probability.

| **EXAMPLE 7-5** | **Finding the Probability That $\bar{p}$ Is in a Given Range** |

**Craigslist** An analysis is performed to determine the proportion of "apartment for rent" ads placed online that result in a rental within two weeks. A reporter for a national publication about these types of services wishes to make the claim that the proportion of apartments rented within two weeks using Craigslist is 0.80. Before doing this, she has selected a simple random sample of 100 "apartment for rent" ads and tracked down the rental result for each. Of these, 73 resulted in a rental within the two-week period. To determine the probability of 73 or fewer apartments being rented in two weeks or less if the 0.80 proportion is correct, we can use the following steps:

**STEP 1** **Determine the population proportion, $p$.**

The population proportion is believed to be $p = 0.80$ based on the analysis.

**FIGURE 7.14 Standard Normal Distribution for On-Line Coupons, Inc.**

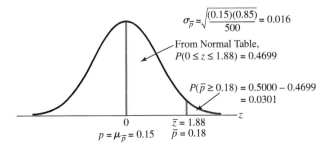

---

[4]If the sample size $n$ is greater than 5% of the population size, the standard error of the sampling distribution should be computed using the finite population correction factor as $\sigma_{\bar{p}} = \sqrt{\dfrac{p(1 - p)}{n}} \sqrt{\dfrac{N - n}{N - 1}}$.

STEP 2    **Calculate the sample proportion.**
In this case, a random sample of $n = 100$ ads was selected, with 73 having the attribute of interest (rented within two weeks). Thus,

$$\bar{p} = \frac{x}{n} = \frac{73}{100} = 0.73$$

STEP 3    **Determine the mean and standard deviation of the sampling distribution.**
The mean of the sampling distribution is equal to $p$, the population proportion. So

$$\mu_{\bar{p}} = p = 0.80$$

The standard deviation of the sampling distribution for $\bar{p}$ is computed using

$$\sigma_{\bar{p}} = \sqrt{\frac{p(1 - p)}{n}} = \sqrt{\frac{0.80(1 - 0.80)}{100}} = 0.04$$

STEP 4    **Define the event of interest.**
In this case, because 0.73 is less than 0.80, we are interested in

$$P(\bar{p} \le 0.73) = ?$$

STEP 5    **If $np$ and $n(1 - p)$ are both $\ge 5$, then convert $\bar{p}$ to a standardized $z$-value.**
Checking, we get

$$np = 100(0.80) = 80 \ge 5 \quad \text{and} \quad n(1 - p) = 100(0.20) = 20 \ge 5$$

Then we convert to a standardized $z$-value using

$$z = \frac{\bar{p} - p}{\sigma_{\bar{p}}} = \frac{0.73 - 0.80}{\sqrt{\dfrac{0.80(1 - 0.80)}{100}}} = -1.75$$

STEP 6    **Use the standard normal distribution table in Appendix D to determine the probability for the event of interest.**
We want

$$P(\bar{p} \le 0.73) \quad \text{or} \quad P(z \le -1.75)$$

From the normal distribution table for $z = -1.75$, we get 0.4599, which corresponds to the probability of a $z$-value between $-1.75$ and $0.0$. To get the probability of interest, we subtract 0.4599 from 0.5000, giving 0.0401.

There is only a 4% chance that a random sample of $n = 100$ apartments would produce a sample proportion of $\bar{p} \le 0.73$ that are rented within two weeks if the true population proportion is 0.80. The author might want to use caution before making this claim.

**TRY EXERCISE 7-45** (pg. 267)

## 7.3 EXERCISES

### Skill Development

**7-45.** A population has a proportion equal to 0.30. Calculate the following probabilities with $n = 100$:
   a. $P(\bar{p} \le 0.35)$
   b. $P(\bar{p} > 0.40)$
   c. $P(0.25 < \bar{p} \le 0.40)$
   d. $P(\bar{p} \ge 0.27)$.

**7-46.** If a random sample of 200 items is taken from a population in which the proportion of items having a desired attribute is $p = 0.30$, what is the probability that the proportion of successes in the sample will be less than or equal to 0.27?

**7-47.** The proportion of items in a population that possess a specific attribute is known to be 0.70.
   a. If a simple random sample of size $n = 100$ is selected and the proportion of items in the sample that contain the attribute of interest is 0.65, what is the sampling error?
   b. Referring to part a, what is the probability that a sample of size 100 would have a sample proportion of 0.65 or less if the population proportion is 0.70?

**7-48.** Given a population in which the proportion of items with a desired attribute is $p = 0.25$, if a sample of 400 is taken,
   a. What is the standard deviation of the sampling distribution of $\bar{p}$?
   b. What is the probability the proportion of successes in the sample will be greater than 0.22?

**7-49.** Given a population in which the probability of success is $p = 0.20$, if a sample of 500 items is taken, then
   a. Calculate the probability the proportion of successes in the sample will be between 0.18 and 0.23.
   b. Calculate the probability the proportion of successes in the sample will be between 0.18 and 0.23 if the sample size is 200.

**7-50.** Given a population in which the proportion of items with a desired attribute is $p = 0.50$, if a sample of 200 is taken,
   a. Find the probability the proportion of successes in the sample will be between 0.47 and 0.51.
   b. Referring to part a, what would the probability be if the sample size were 100?

**7-51.** Given a population in which the probability of a success is $p = 0.40$, if a sample of 1,000 is taken,
   a. Calculate the probability the proportion of successes in the sample will be less than 0.42.
   b. What is the probability the proportion of successes in the sample will be greater than 0.44?

**7-52.** A random sample of size 100 is to be taken from a population that has a proportion equal to 0.35. The sample proportion will be used to estimate the population proportion.
   a. Calculate the probability that the sample proportion will be within $\pm 0.05$ of the population proportion.
   b. Calculate the probability that the sample proportion will be within $\pm 1$ standard error of the population proportion.
   c. Calculate the probability that the sample proportion will be within $\pm 0.10$ of the population proportion.

**7-53.** A survey is conducted from a population of people of whom 40% have a college degree. The following sample data were recorded for a question asked of each person sampled, "Do you have a college degree?"

| YES | NO  | NO  | YES | YES |
| --- | --- | --- | --- | --- |
| YES | YES | YES | YES | YES |
| YES | NO  | NO  | NO  | YES |
| NO  | YES | YES | NO  | NO  |
| NO  | YES | YES | YES | NO  |
| YES | NO  | YES | NO  | NO  |
| YES | NO  | NO  | NO  | YES |
| YES | NO  | NO  | NO  | NO  |
| NO  | NO  | YES | NO  | NO  |
| NO  | YES | NO  | YES | YES |
| NO  | NO  | NO  | YES | NO  |
| NO  | NO  | NO  | YES | YES |

   a. Calculate the sample proportion of respondents who have a college degree.
   b. What is the probability of getting a sample proportion as extreme as or more extreme than the one observed in part a if the population has 40% with college degrees?

## Business Applications

**7-54.** United Manufacturing and Supply makes sprinkler valves for use in residential sprinkler systems. United entered into a contract to supply 40,000 sprinkler valves. The contract called for at least 97% of the valves to be free of defects. Before shipping the valves, United managers tested 200 randomly selected valves and found 190 defect-free valves in the sample. The managers wish to know the probability of finding 190 or fewer defect-free valves if in fact the population of 40,000 valves is 97% defect-free. Discuss how they could use this information to determine whether to ship the valves to the customer.

**7-55.** The J R Simplot Company is one of the world's largest privately held agricultural companies in North America. One of its major products is french fries that are sold primarily on the commercial market to customers such as McDonald's and Burger King. French fries have numerous quality attributes that are important to customers. One of these is called "dark ends," which are the dark-colored ends that can result when the fries are cooked. Suppose a major customer will accept no more than 0.06 of the fries having dark ends.

   The customer called the Simplot Company saying that a recent random sample of 300 fries was tested from a shipment and 27 fries had dark ends. Assuming that the population does meet the 0.06 standard, what is the probability of getting a sample of 300 with 27 or more dark ends? Comment on this result.

**7-56.** The National Association of Realtors released a survey indicating that a surprising 66% of first-time home buyers purchased their homes with low down payment loans (*source:* Scholastica Cororaton, "66 Percent of first-time home buyers make a low down payment," http://economistsoutlook.blogs.realtor.org, Apr. 8, 2015).
   a. Suppose 12 out of the 20 first-time home buyers sampled received low down payment loans. Calculate the probability that at least 12 in a sample of 20 first-time buyers would take out a such a loan if the population proportion is 0.66.
   b. Determine the probability requested in part a if the nationwide proportion is 0.53.
   c. Determine the probability that between 8 and 12 of a sample of 20 first-time home buyers would take low down payment loans if the 66% value applies.

**7-57.** Ten percent of engineers (electrical, mechanical, civil, and industrial) are women (*source:* Rebecca Adams, "40 Percent of female engineers are leaving the field.

This might be why," www.huffingtonpost.com, Aug. 12, 2014). Suppose a random sample of 50 engineers is selected.

a. How likely is it that the random sample of 50 engineers will contain 8 or more women in these positions?

b. How likely is it that the random sample will contain fewer than 5 women in these positions?

c. If the random sample included 200 engineers, how would this change your answer to part b?

7-58. The federal government's Home Affordable Refinance Program (HARP) is designed to help homeowners who owe more on their mortgages than their houses are worth. A lending agency advertises that 33% of those holding "underwater" mortgages would be helped by the HARP program.

a. In samples of mortgages of size 75, what proportion of the samples would you expect to produce sample proportions equal to or larger than 0.25?

b. If a sample of size 75 produced a sample proportion of 0.45, what would you infer that the population proportion of mortgages "underwater" really is?

c. In a sample of 90 mortgages, what would the probability be that the number of mortgages "underwater" would be between 25 and 35?

7-59. Airgistics provides air consolidation and freight-forwarding services for companies that ship their products internationally. Airgistics currently believes that it achieves a 96% on-time performance for its customers. A random sample of 200 customer shipments was selected for study, and 188 of them were found to have met the on-time delivery promise.

a. What is the probability that a random sample of 200 customer shipments would contain 188 or fewer on-time deliveries if the true population of on-time deliveries is 96%?

b. Would you be surprised if the random sample of 200 customer shipments had 197 on-time deliveries?

c. Suppose the random sample of 200 customer shipments revealed that 178 were on time. Would such a finding cause you to question Airgistics's claim that its on-time performance is 96%? Support your answer with a probability calculation.

## Computer Software Exercises

7-60. Several years ago, Procter & Gamble (P&G) merged with Gillette. One of the concerns after the merger was the increasing burden of retirement expenditures. An effort was made to encourage employees to participate in 401(k) accounts. Nationwide, 66% of eligible workers participated in these accounts. The file titled **Gillette** contains responses of 200 P&G workers when asked if they were currently participating in a 401(k) account.

a. Determine the sample proportion of P&G workers who participate in 401(k) accounts.

b. Determine the sampling error if in reality P&G workers have the same proportion of participants in 401(k) accounts as does the rest of the nation.

c. Determine the probability that a sample proportion at least as large as that obtained in the sample would be obtained if P&G workers have the same proportion of participants in 401(k) accounts as does the rest of the nation.

d. Does it appear that a larger proportion of P&G workers participate in 401(k) accounts than do the workers of the nation as a whole? Support your response.

7-61. The Bureau of Transportation Statistics releases information concerning the monthly percentage of U.S. airline flights that land no later than 15 minutes after scheduled arrival. Suppose that last year, the proportion of flights that met this criterion was 0.785. Each year, these data become available as soon as feasible. However, the airlines can provide preliminary results by obtaining a sample. The file titled **Ontime** contains the sample data indicating the number of minutes after scheduled arrival time that the aircraft arrived. Note that a negative entry indicates the minutes earlier than the scheduled arrival time that the aircraft arrived.

a. Calculate the proportion of sampled airline flights that landed within 15 minutes of scheduled arrival.

b. Calculate the probability that a sample proportion of on-time flights would be within $\pm 0.06$ of a population proportion equal to 0.785.

c. If the airlines' goal was to attain the same proportion of on-time arrivals as last year, do the preliminary results indicate that they have met this goal? Support your assertions.

7-62. In 2015, the average hourly compensation (salary plus benefits) for federal workers was $40.32 (*source:* Chris Edwards, "Reducing the costs of federal worker pay and benefits," www.downsizinggovernment.org, Oct. 1, 2015). The file titled **Paychecks** contains a random sample of the hourly amounts paid to state and local government workers.

a. If hourly pay is thought to be normally distributed, what proportion of state and local employees receive pay higher than the average for federal employees?

b. Calculate the sample proportion of the state and local government workers whose hourly compensation is less than $40.32.

c. Determine the probability that the sample proportion would be equal to or greater than the answer obtained in part b.

d. On the basis of your work in parts a, b, and c, would you conclude that the proportion of state and local workers whose hourly compensation is less than $40.32 is the same as that of the federal workers? Explain.

7-63. A popular magazine aimed at IT professionals states that 65% of its subscribers have an annual income greater than $100,000. The magazine's staff uses this high proportion of subscribers earning more than $100,000 as a selling point when trying to get companies to place advertisements with the magazine.

Before committing to advertise in this magazine, a computer hardware firm's market research staff has decided to randomly sample 196 subscribers to verify the claim that 65% of its subscribers earn more than $100,000 annually.

a. The file **High Tech** contains the responses of 196 randomly sampled subscribers of the magazine. The responses are coded "Yes" if the subscriber earns more than $100,000 annually and "No" if the subscriber does not earn more than $100,000 annually. Open the file and create a new variable that has a value of 1 for "Yes" and 0 for "No."

b. Calculate the sample proportion of subscribers who have annual incomes greater than $100,000.

c. How likely is it that a sample proportion greater than the one calculated in part b would be obtained if the population proportion of subscribers earning more than $100,000 annually is 65%?

d. Based on the probability you calculated in part c, should the hardware firm advertise in this magazine?

**7-64.** A study conducted by Reuters revealed that 14% of Fortune 1000 companies either terminated or froze their defined-benefit pension plans (*source:* Mark

Miller, "The vanishing defined-benefit pension and its discontents," www.reuters.com, May 6, 2014). As part of a study to evaluate how well its benefits package compares to other Fortune 1000 companies, a retail firm randomly samples 36 Fortune 1000 companies in odd-numbered years and asks them to complete a benefits questionnaire. One question asked is whether the company has changed its defined-benefits pension plan by either freezing it or terminating it during the survey year. The results of the survey are contained in the file **Pension Survey**.

a. Open the file and create a new variable that has a value equal to 1 if the firm has either terminated or frozen its defined-benefits pension plan and equal to 0 if the firm has not significantly altered its pension plan. Determine the sample proportion of companies that either terminated or froze their defined-benefits pension plan.

b. How likely is it that a sample proportion greater than or equal to the one found in the survey would occur if the true population of firms that have terminated or frozen their defined-benefits pension plan is as reported by Reuters?

# 7 Overview

## Summary

**7.1** Sampling Error: What It Is and Why It Happens (pg. 240–248)

**OUTCOME 1** Define and compute sampling error.

- A **simple random sample** is a sample selected in such a manner that each possible sample of a given size has an equal chance of being selected.
- Regardless of how carefully we select a random sample, the sample may not perfectly represent the population, and the sample statistic does not likely equal the population parameter.
- The **sampling error** is the difference between a measure computed from a sample (a **statistic**) and the corresponding measure computed from the population (a **parameter**).
- Some fundamental concepts associated with sampling errors are:
  - The size of the sampling error depends on which sample is selected.
  - The sampling error may be positive or negative.
  - There is potentially a different statistic for each sample selected.

**7.2** Sampling Distribution of the Mean (pg. 248–262)

**OUTCOME 2** Determine the mean and standard deviation for the sampling distribution of the sample mean, $\bar{x}$.

**OUTCOME 3** Apply concepts associated with the Central Limit Theorem.

- In business applications, decision makers select a single sample from a population, from which they compute a sample measure and use it to make decisions about the population.
- The **sampling distribution** is the distribution of all possible values of a statistic for a given sample size that has been randomly selected from a population.
- A statistic is an **unbiased estimator** of a parameter when the average of all possible values of the sample statistic equals the corresponding parameter. An unbiased estimator is **consistent** if the difference between the estimator and the parameter becomes smaller as the sample size increases.
- The **Central Limit Theorem** states that for a sufficiently large sample size, the distribution of the sample means is approximately normal, with a mean equal to the population mean and a standard deviation equal to the population standard deviation divided by the square root of the sample size. The larger the sample size, the better the approximation to the normal distribution.

**7.3** Sampling Distribution of a Proportion (pg. 262–270)

**OUTCOME 4** Determine the mean and standard deviation for the sampling distribution of the sample proportion, $\bar{p}$.

- The objective of sampling is often to estimate a **population proportion**. In situations where we are interested in determining the proportion of all items that possess a particular attribute, the **sample proportion** is the best estimate of the true population proportion.
- The distribution of the sample proportion is binomial; however, if the sample size is sufficiently large, then the Central Limit Theorem tells us that we can use the normal distribution as an approximation.

## Equations

**(7.1) Sampling Error of the Sample Mean** pg. 240

$$\text{Sampling error} = \bar{x} - \mu$$

**(7.2) Population Mean** pg. 241

$$\mu = \frac{\sum x}{N}$$

**(7.3) Sample Mean** pg. 241

$$\bar{x} = \frac{\sum x}{n}$$

**(7.4) z-Value for Sampling Distribution of $\bar{x}$** pg. 254

$$z = \frac{\bar{x} - \mu}{\dfrac{\sigma}{\sqrt{n}}}$$

**(7.5) z-Value Adjusted for the Finite Population Correction Factor** pg. 254

$$z = \frac{\bar{x} - \mu}{\dfrac{\sigma}{\sqrt{n}}\sqrt{\dfrac{N-n}{N-1}}}$$

**(7.6) Population Proportion** pg. 262

$$p = \frac{X}{N}$$

**(7.7) Sample Proportion** pg. 263

$$\bar{p} = \frac{x}{n}$$

**(7.8) Single-Proportion Sampling Error** pg. 263

$$\text{Sampling error} = \bar{p} - p$$

**(7.9) Mean of the Sampling Distribution of $\bar{p}$** pg. 265

$$\text{Mean} = \mu_{\bar{p}} = p$$

**(7.10) Standard Error of the Sampling Distribution of $\bar{p}$** pg. 265

$$\text{Standard error} = \sigma_{\bar{p}} = \sqrt{\frac{p(1-p)}{n}}$$

**(7.11) z-Value for the Sampling Distribution of $\bar{p}$** pg. 266

$$z = \frac{\bar{p} - p}{\sigma_{\bar{p}}}$$

## Key Terms

**Consistent estimator** pg. 253
**Parameter** pg. 241
**Population proportion** pg. 262

**Sample proportion** pg. 263
**Sampling distribution** pg. 249
**Sampling error** pg. 240

**Simple random sample** pg. 241
**Unbiased estimator** pg. 250

## Chapter Exercises

### Conceptual Questions

**7-65.** Under what conditions should the finite population correction factor be used in determining the standard error of a sampling distribution?

**7-66.** A sample of size 30 is obtained from a population that has a proportion of 0.34. Determine the range of sampling errors possible when the sample proportion is used to estimate the population proportion. (*Hint:* Review the Empirical Rule.)

**7-67.** Discuss why the sampling distribution is less variable than the population distribution. Give a short example to illustrate your answer.

**7-68.** Discuss the similarities and differences between a standard deviation and a standard error.

**7-69.** A researcher has collected all possible samples of size 150 from a population and listed the sample means for each of these samples.

    a. If the average of the sample means is 450.55, what would be the numerical value of the true population mean? Discuss.

    b. If the standard deviation of the sample means is 12.25, determine the standard deviation of the model from which the samples came. To perform this calculation, assume the population has a size of 1,250.

**7-70.** Consider the standard error of a sample proportion obtained from a sample of size 100.

    a. Determine the standard error obtained from a population with $p = 0.1$.

    b. Repeat part a for a population proportion equal to (1) 0.5 and (2) 0.9.

    c. Which population proportion results in the largest standard error?

    d. Given your responses to parts a, b, and c, which value of a population proportion would produce the largest sampling error?

**7-71.** If a population is known to be normally distributed, what size sample is required to ensure that the sampling distribution of $\bar{x}$ is normally distributed?

**7-72.** Suppose we are told the sampling distribution developed from a sample of size 400 has a mean of 56.78 and a standard error of 9.6. If the population is known to be normally distributed, what are the population mean and population standard deviation? Discuss how these values relate to the values for the sampling distribution.

## Business Applications

**7-73.** The Baily Hill Bicycle Shop sells mountain bikes and offers a maintenance program to its customers. The manager has found the average repair bill during the maintenance program's first year to be $15.30 with a standard deviation of $7.00.
  a. What is the probability a random sample of 40 customers will have a mean repair cost exceeding $16.00?
  b. What is the probability the mean repair cost for a sample of 100 customers will be between $15.10 and $15.80?
  c. The manager has decided to offer a spring special. He is aware of the mean and standard deviation for repair bills last year. Therefore, he has decided to randomly select and repair the first 50 bicycles for $14.00 each. He notes this is not even 1 standard deviation below the mean price to make such repairs. He asks your advice. Is this a risky thing to do? Based upon the probability of a repair bill being $14.00 or less, what would you recommend? Discuss.

**7-74.** When its ovens are working properly, the time required to bake fruit pies at Ellardo Bakeries is normally distributed with a mean of 45 minutes and a standard deviation of 5 minutes. Yesterday, a random sample of 16 pies had an average baking time of 50 minutes.
  a. If Ellardo's ovens are working correctly, how likely is it that a sample of 16 pies would have an average baking time of 50 minutes or more?
  b. Would you recommend that Ellardo inspect its ovens to see if they are working properly? Justify your answer.

**7-75.** A company in Hawaii manages properties for owners and arranges for the properties to be rented. A property owner who is considering using the company is told that 10% of the bookings result in a cancellation.
  a. Assuming that the company's number is correct, what is the probability that a random sample of 50 bookings ends up with 9 or more cancellations?
  b. If an audit of 200 of the company's bookings showed 25 cancellations, what might you conclude about the company's 10% claim? Base your answer on probability.

**7-76.** Suppose at your university, some administrators believe that the proportion of students preferring to take classes at night exceeds 0.30. The president is skeptical and so has an assistant take a simple random sample of 200 students. Of these, 66 indicate that they prefer night classes. What is the probability of finding a sample proportion equal to or greater than that found if the president's skepticism is justified? Assume $n \leq 5\%$ of $N$.

**7-77.** A year-old study found that the service time for all drive-thru customers at the Stardust Coffee Shop is uniformly distributed between 3 and 6 minutes. Assuming the service time distribution has not changed, a random sample of 49 customers is taken and the service time for each is recorded.
  a. Calculate the mean and standard deviation of service times for all drive-thru customers at the Stardust Coffee Shop. (*Hint:* Review the uniform distribution from Chapter 6.)
  b. What is the probability that a sample of 49 customers would have a sample mean of 4.25 minutes or more if the true population mean and standard deviation for service times are as calculated in part a?
  c. How can the probability in part b be determined when the population of service times is not normally distributed?

**7-78.** The time it takes a mechanic to tune an engine is known to be normally distributed with a mean of 45 minutes and a standard deviation of 14 minutes.
  a. Determine the mean and standard error of a sampling distribution for a sample size of 20 tune-ups. Draw a picture of the sampling distribution.
  b. Calculate the largest sampling error you would expect to make in estimating the population mean with the sample size of 20 tune-ups.

**7-79.** Suppose Frito-Lay managers set the fill volume on Cheetos to an average of 16 ounces. The filling machine is known to fill with a standard deviation of 0.25 ounce with a normal distribution around the mean fill level.
  a. What is the probability that a single bag of Cheetos will have a fill volume that exceeds 16.10 ounces?
  b. What is the probability that a random sample of 12 bags of Cheetos will have a mean fill volume that exceeds 16.10 ounces?
  c. Compare your answers to parts a and b and discuss why they are different.

**7-80.** Frank N. Magid Associates conducted a telephone survey of 1,109 consumers to obtain the number of cell phones in houses. The survey obtained the following results:

| Cell phones | 0 | 1 | 2 | 3 | $\geq 4$ |
|---|---|---|---|---|---|
| Percent | 19 | 26 | 33 | 13 | 9 |

  a. Determine the sample proportion of households that own two or more cell phones.
  b. If the proportion of households that own two or more cell phones is equal to 0.50, determine the probability that in a sample of 1,109 consumers at least 599 would own two or more cell phones.
  c. The margin of error of the survey was $\pm 3$ percentage points. This means that the sample proportion would be within 3 percentage points of the population proportion. If the population proportion of households that had two or more cell phones was 0.50, determine the probability that the margin of error would be $\pm 3$ percentage points.

**7-81.** The Bendbo Corporation has a total of 300 employees in its two manufacturing locations and the headquarters office. A study conducted five years ago showed the average commuting distance to work for Bendbo employees was 6.2 miles with a standard deviation of 3 miles. Recently, a follow-up study based on a random sample of 100 employees indicated an average travel distance of 5.9 miles.

   a. Assuming that the mean and standard deviation of the original study hold, what is the probability of obtaining a sample mean of 5.9 miles or less?

   b. Based on this probability, do you think the average travel distance may have decreased?

   c. A second random sample of 40 was selected. This sample produced a mean travel distance of 5.9 miles. If the mean for all employees is 6.2 miles and the standard deviation is 3 miles, what is the probability of observing a sample mean of 5.9 miles or less?

   d. Discuss why the probabilities differ even though the sample results were the same in each case.

**7-82.** One of the most critical factors in the success of PC makers is how fast they can turn their inventory of parts. Faster inventory turns mean lower average inventory cost. The vice president (VP) of manufacturing for a computer manufacturing company said there is no reason to continue offering hard disk drives that have less than a 250-GB storage capacity since only 10% of customers ask for the smaller hard disks. After much discussion and debate about the accuracy of the VP's figure, it was decided to sample 100 orders from the past week's sales. This sample revealed 14 requests for drives with less than 250-GB capacity.

   a. Determine the probability of finding 14 or more requests like this if the VP's assertion is correct. Do you believe that the proportion of customers requesting hard drives with storage capacity is as small as 0.10? Explain.

   b. Suppose a second sample of 100 customers was selected. This sample again yielded 14 requests for a hard drive with less than 250 GB of storage. Combining this sample information with that found in part a, what conclusion would you now reach regarding the VP's 10% claim? Base your answer on probability.

**7-83.** While the average starting salary for graduates with an MBA degree varies depending on the university attended and field of emphasis, one source shows a 2015 median annual starting base salary of $100,000 (*source:* Claire Zillman, "Congrats, MBA grads! You're getting a $45,000 raise," http://fortune.com, May 19, 2015). Assuming that starting salaries are normally distributed and the standard deviation is $20,000, what is the probability that a random sample of 100 MBA graduates will have a mean starting salary of $96,000 or less?

**7-84.** Suppose executives at Netflix believe that at least 70% of customers would return a movie 2 days after it arrives

at their home. A sample of 500 customers found 68% returned the movie prior to the third day. Given the executives' estimate, what would be the probability of a sample result with 68% or fewer returns prior to the third day?

**7-85.** A computer scanner can read a bar code on a package correctly 97% of the time. One company monitors the accuracy of the scanner by randomly sampling packages and verifying that each package has been correctly scanned. Random samples of size $n = 25, 50, 100$, and 200 have recently been taken with the following results:

| Sample Size, $n$ | Number Correctly Scanned |
|---|---|
| 25 | 24 |
| 50 | 49 |
| 100 | 95 |
| 200 | 193 |

   a. For each sample size, calculate the sample proportion.

   b. For each sample size, what are the mean and standard deviation of the sampling distribution?

   c. Calculate the single-proportion sampling error for each sample size.

   d. For a sample of size $n = 200$, what is the probability of finding 198 correctly scanned packages?

## Computer Software Exercises

**7-86.** The **Patients** file contains information for a random sample of geriatric patients. During a meeting, one hospital administrator indicated that 70% of the geriatric patients are males.

   a. Based on the data contained in the file, would you conclude the administrator's assertion concerning the proportion of male geriatric patients is correct? Justify your answer.

   b. The administrator also believes 80% of all geriatric patients are covered by Medicare (Code = CARE). Again, based on the data contained in the file, what conclusion should the hospital administrator reach concerning the proportion of geriatric patients covered by Medicare? Discuss.

**7-87.** A business professor stated that 74% of adult Internet users are confident making purchases online. As part of a class project, the students in the class were required to survey adults in the community. The file titled **Online** contains "Yes" and "No" responses to the survey conducted by the students.

   a. Calculate the sample proportion of adults who are confident.

   b. If the professor's value for the proportion of confident adults is correct, determine the sampling error for the sample in the file.

   c. Determine the probability that a sample proportion would be at most as far away as the **Online** sample if the 74% value was correct.

d. The students indicated that their sample had a margin of error of $\pm 2\%$. The sample size used by the students' poll was 2,022. Calculate the probability that the sample proportion from a sample size of 2,022 would be at most as far away from the population proportion as suggested by the margin of error.

7-88. The file **High Desert Banking** contains information regarding consumer, real estate, and small commercial loans made last year by the bank. Use your computer software to do the following:

   a. Construct a frequency histogram using eight classes for dollar value of loans made last year. Does the population distribution appear to be normally distributed?

   b. Compute the population mean for all loans made last year.

   c. Compute the population standard deviation for all loans made last year.

   d. Select a simple random sample of 36 loans. Compute the sample mean. By how much does the sample mean differ from the population mean? Use the Central Limit Theorem to determine the probability that you would have a sample mean this small or smaller and the probability that you would have a sample mean this large or larger.

7-89. Covercraft manufactures covers to protect automobile interiors and finishes. Its Block-It 380 Series fabric has a limited three-year warranty. Periodic testing is commonly done to determine if warranty policies should be changed. One possible study was to examine those covers that became unserviceable while still under warranty. Data that could be produced by such a study are contained in the file titled **Covers**. The data represent the number of months a cover was used until it became unserviceable. Covercraft might want to examine more carefully the covers that became unserviceable while still under warranty. Specifically, it wants to examine those that became unserviceable before they had been in use one year.

   a. Covercraft has begun to think that it should lower its warranty period to perhaps 20 months. It believes that in doing this, 20% of the covers that now fail before the warranty is up will have surpassed the 20-month warranty rate. Calculate the proportion of the sample that became unserviceable after 20 months of service.

   b. Determine the probability of obtaining a sample proportion at least as large as that calculated in part a if the true proportion was equal to 0.20.

   c. Based on your calculation in part b, should Covercraft lower its warranty period to 20 months? Support your answer.

7-90. The data file **Trucks** contains data on a sample of 200 trucks that were weighed on two scales. The WIM (weigh-in-motion) scale weighs the trucks as they drive down the highway. The POE scale weighs the trucks while they are stopped at the port-of-entry station. The maker of the WIM scale believes that its scale will weigh heavier than the POE scale 60% of the time when gross weight is considered.

   a. Create a new variable that has a value $= 1$ when the WIM gross weight $>$ POE gross weight, and 0 otherwise.

   b. Determine the sample proportion of times the WIM gross weight exceeds the POE gross weight.

   c. Based on this sample, what is the probability of finding a proportion less than that found in part b? For this calculation, assume the WIM maker's assertion is correct.

   d. Based on the probability found in part c, what should the WIM maker conclude? Is his 60% figure reasonable?

## Case 7.1    Carpita Bottling Company—Part 1

Don Carpita owns and operates Carpita Bottling Company in Lakeland, Wisconsin. The company bottles soda and beer and distributes the products in the counties surrounding Lakeland.

The company has four bottling machines, which can be adjusted to fill bottles at any mean fill level between 2 ounces and 72 ounces. The machines exhibit some variation in actual fill from the mean setting. For instance, if the mean setting is 16 ounces, the actual fill may be slightly more or less than that amount.

Three of the four filling machines are relatively new, and their fill variation is not as great as that of the older machine. Don has observed that the standard deviation in fill for the three new machines is about 1% of the mean fill level when the mean fill is set at 16 ounces or less, and it is 0.5% of the mean at settings exceeding 16 ounces. The older machine has a standard deviation of about 1.5% of the mean setting regardless of the mean fill setting. However, the older machine tends to underfill bottles more than overfill, so the older machine is set at a mean fill slightly in excess of the desired mean to compensate for the propensity to underfill. For example, when 16-ounce bottles are to be filled, the machine is set at a mean fill level of 16.05 ounces.

The company can simultaneously fill bottles with two brands of soda using two machines, and it can use the other two machines to bottle beer. Although each filling machine has its own warehouse and the products are loaded from the warehouse directly onto a truck, products from two or more filling machines may be loaded on the same truck. However, an individual store almost always receives bottles on a particular day from just one machine.

On Saturday morning, Don received a call at home from the J. R. Summers grocery store manager. She was very upset because the shipment of 16-ounce bottles of beer received yesterday contained several bottles that were not adequately filled. The manager wanted Don to replace the entire shipment at once.

Don gulped down his coffee and prepared to head to the store to check out the problem. He started thinking how he could determine which machine was responsible for the problem. If he could at least determine whether it was the old machine or one of the new ones, he could save his maintenance people a lot of time and effort checking all the machines.

His plan was to select a sample of 64 bottles of beer from the store and measure the contents. Don figured that he might be able to determine, on the basis of the average contents, whether it was more likely that the beer was bottled by a new machine or by the old one.

The results of the sampling showed an average of 15.993 ounces. Now Don needs some help in determining whether a sample mean of 15.993 ounces or less is more likely to come from the new machines or the older machine.

## Case 7.2    Truck Safety Inspection

The Idaho Department of Law Enforcement, in conjunction with the federal government, recently began a truck inspection program in Idaho. The current inspection effort is limited to an inspection of only those trucks that visually appear to have some defect when they stop at one of the weigh stations in the state. The proposed inspection program will not be limited to the trucks with visible defects, but will potentially subject all trucks to a comprehensive safety inspection.

Jane Lund of the Department of Law Enforcement is in charge of the new program. She has stated that the ultimate objective of the new truck inspection program is to reduce the number of trucks with safety defects operating in Idaho. Ideally, all trucks passing through or operating within Idaho would be inspected once a month, and substantial penalties would be applied to operators if safety defects were discovered. Ms. Lund is confident that such an inspection program would, without fail, reduce the number of

defective trucks operating on Idaho's highways. However, each safety inspection takes about an hour, and because of limited money to hire inspectors, she realizes that all trucks cannot be inspected. She also knows it is unrealistic to have trucks wait to be inspected until trucks ahead of them have been checked. Such delays would cause problems with the drivers.

In meetings with her staff, Jane has suggested that before the inspection program begins, the number of defective trucks currently operating in Idaho should be estimated. This estimate can be compared with later estimates to see if the inspection program has been effective. To arrive at this initial estimate, Jane thinks that some sort of sampling plan to select representative trucks from the population of all trucks in the state must be developed. She has suggested that this sampling be done at the eight weigh stations near Idaho's borders, but she is unsure how to establish a statistically sound sampling plan that is practical to implement.

# 8 Estimating Single Population Parameters

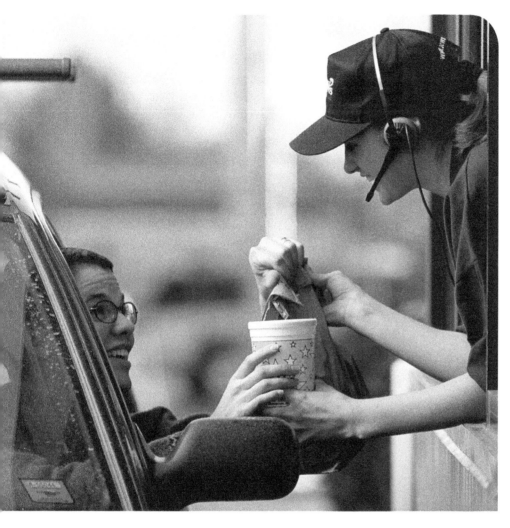

## WHY YOU NEED TO KNOW

Chapter 1 discussed various sampling techniques, including statistical and nonstatistical methods. Chapter 7 introduced the concepts of sampling error and sampling distributions. Chapter 8 builds on these concepts and introduces the steps needed to develop and interpret statistical estimations of various population values. You will undoubtedly need to estimate population parameters as a regular part of your managerial decision-making activities. In addition, you will receive estimates that other people have developed that you will need to evaluate before relying on them as inputs to your decision-making process. Was the sample

**Quick Prep**

Review material on calculating and interpreting sample means and standard deviations in Chapter 3.

Review the normal distribution in Section 6.1.

Make sure you understand the concepts associated with sampling distributions for $\bar{x}$ and $\bar{p}$ by reviewing Sections 7.1, 7.2, and 7.3.

size large enough to provide valid estimates of the population parameter? How confident can you be that the estimate matches the population parameter of interest? You can answer these and similar questions using the concepts and procedures presented in this chapter.

## Point and Confidence Interval Estimates for a Population Mean

**OUTCOME 1**

## Point Estimates and Confidence Intervals

Every election year, political parties and news agencies conduct polls. These polls attempt to determine the percentage of voters who favor a particular candidate or a particular issue. For example, suppose a poll indicates that 82% of the people older than 18 in your state favor banning texting while driving a motor vehicle. The pollsters have not contacted every person in the state; rather, they have sampled only a relatively few people to arrive at the 82% figure. In statistical terminology, the 82% is the **point estimate** of the true population percentage of people who favor a ban on texting and driving.

The Environmental Protection Agency (EPA) tests the mileage of automobiles sold in the United States. Since the EPA cannot test every car, the resulting mileage rating is actually a point estimate for the true average mileage of all cars of a given model.

Fast-food restaurants, such as McDonald's, Burger King, and Jack in the Box, operate drive-thru windows so that customers can order and receive their food without getting out of the car. To make sure that this process runs smoothly, these companies collect sample data on the time vehicles spend in the drive-thru line. After they collect the sample data, they compute the average waiting time. This average for the sample data is a point estimate for the true average waiting time for all cars that arrive at the drive-thru during a certain time period.

Which point estimator the decision maker uses depends on the population characteristic the decision maker wishes to estimate. However, regardless of the population value being estimated, we always expect **sampling error**.

Chapter 7 discussed sampling error. You cannot eliminate sampling error, but you can deal with it in your decision process. For example, if cost accountants use $\bar{x}$, the average cost to make a sample of items, to establish the true average per unit cost of production, the point estimate, $\bar{x}$, will most likely not equal the population mean, $\mu$. In fact, the probability of $\bar{x} = \mu$ is essentially zero. With $\bar{x}$ as their only information, the cost accountants have no way of determining exactly how far $\bar{x}$ is from $\mu$.

To overcome this problem with point estimates, the most common procedure is to calculate an interval estimate known as a **confidence interval**. An application will help to make this definition clear.

**Point Estimate**
A single statistic, determined from a sample, that is used to estimate the corresponding population parameter.

**Sampling Error**
The difference between a measure (a statistic) computed from a sample and the corresponding measure (a parameter) computed from the population.

**Confidence Interval**
An interval developed from sample values such that if all possible intervals of a given width are constructed, a percentage of these intervals, known as the confidence level, will include the true population parameter.

Excel Tutorial

**BUSINESS APPLICATION**  ## Calculating a Confidence Interval Estimate

**Health Star Energy Drink**  The Health Star Energy Drink Company makes and markets a popular energy-enhancing drink. The company has installed a new machine that automatically fills the energy drink bottles. The machine allows the operator to adjust the mean fill quantity. However, no matter what the mean setting, the actual volume of the energy drink liquid will vary from bottle to bottle. The machine has been carefully tested and is known to fill bottles with an amount of liquid that has a standard deviation of $\sigma = 0.2$ ounce.

The filling machine has been adjusted to fill cans at an average of 12 ounces. After running the machine for several hours, a manager selects a simple random sample of 100 bottles and measures the volume of liquid in each bottle in the company's quality lab. Figure 8.1 shows the frequency histogram of the sample data. (The data are in a file called **Health Star Energy**.) Notice that the distribution is centered at a point larger than 12 ounces. The manager

FIGURE 8.1 **Excel 2016 Histogram for Health Star Energy Drink**

Excel 2016 Instructions
1. Open File: **Health Star Energy.xlsx**.
2. See instructions in Chapter 2 for developing a histogram in Excel.

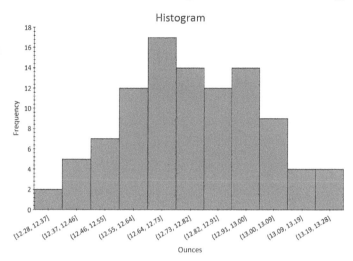

FIGURE 8.2 **Sampling Distribution of $\bar{x}$**

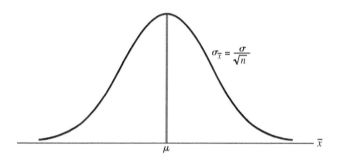

wishes to use the sample data to estimate the mean fill amount for all bottles filled by this machine.

The sample mean computed from 100 bottles is $\bar{x} = 12.79$ ounces. This is the *point estimate* of the population mean, $\mu$. Because of the potential for sampling error, the manager should not expect a particular $\bar{x}$ to equal $\mu$. However, as discussed in Chapter 7, the Central Limit Theorem indicates that the distribution of all possible sample means for samples of size $n = 100$ will be approximately normally distributed around the population mean with its spread measured by $\sigma/\sqrt{n}$, as illustrated in Figure 8.2.

Although the sample mean is 12.79 ounces, the manager knows the true population mean may be larger or smaller than this number. To account for the potential for sampling error, the manager can develop a *confidence interval estimate* for $\mu$. This estimate will take the following form:

Lower confidence limit ↑ Upper confidence limit

$\bar{x} = 12.79$

The key is to determine the upper and lower limits of the interval. The specific method for computing these values depends on whether the population standard deviation, $\sigma$, is known or unknown. We first take up the case in which $\sigma$ is known.

## Confidence Interval Estimate for the Population Mean, $\sigma$ Known

**Standard Error**
A value that measures the spread of the sample means around the population mean. The standard error is reduced when the sample size is increased.

There are two cases that must be considered. The first is the case in which the simple random sample is drawn from a normal distribution. Given a population mean of $\mu$ and a population standard deviation of $\sigma$, the sampling distribution of the sample mean is a normal distribution with a mean of $\mu_{\bar{x}} = \mu$ and a standard deviation (or **standard error**) of $\sigma_{\bar{x}} = \sigma/\sqrt{n}$. This is true for any sample size.

FIGURE 8.3 **Critical Value for a 95% Confidence Interval**

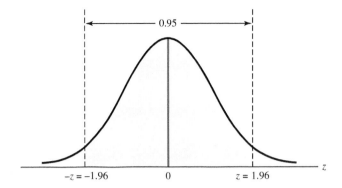

The second case is one in which the population may not be normally distributed. Chapter 7 addressed these specific circumstances. Recall that in such cases, the Central Limit Theorem can be invoked if the sample size is sufficiently large ($n \geq 30$). Then, the sampling distribution is an approximately normal distribution, with a mean of $\mu_{\bar{x}} = \mu$ and a standard deviation of $\sigma_{\bar{x}} = \sigma / \sqrt{n}$. The approximation becomes more precise as the sample size increases. The standard deviation, $\sigma / \sqrt{n}$, is known as the *standard error* of the sample mean.

In both these cases, the sampling distribution for $\bar{x}$ is assumed to be normally distributed. When we look at the sampling distribution in Figure 8.2, it is apparent that the probability that any $\bar{x}$ will exceed $\mu$ is the same as the probability that any $\bar{x}$ will be less than $\mu$. We also know from our discussion in Chapter 7 that we can calculate the percentage of sample means in the interval formed by a specified distance above and below $\mu$. This percentage corresponds to the probability that the sample mean will be in the specified interval. For example, the probability of obtaining a value for $\bar{x}$ that is within 1.96 standard errors on either side of $\mu$ is 0.95. To verify this, recall from Chapter 7 that the standardized $z$-value measures the number of standard errors $\bar{x}$ is from $\mu$. The probability from the standard normal distribution table that corresponds to $z = 1.96$ is 0.4750. Likewise, the probability corresponding to $z = -1.96$ is equal to 0.4750. Therefore,

$$P(-1.96 \leq z \leq 1.96) = 0.4750 + 0.4750 = 0.95$$

This is illustrated in Figure 8.3.

Because the standard error is $\sigma / \sqrt{n}$, 95% of all sample means will fall in the range

$$\mu - 1.96 \frac{\sigma}{\sqrt{n}} \qquad \mu + 1.96 \frac{\sigma}{\sqrt{n}}$$

This is illustrated in Figure 8.4.

In a like manner, we can determine that 80% of all sample means will fall in the range

$$\mu - 1.28 \frac{\sigma}{\sqrt{n}} \qquad \mu + 1.28 \frac{\sigma}{\sqrt{n}}$$

Also, we can determine that 90% of all sample means will fall in the range

$$\mu - 1.645 \frac{\sigma}{\sqrt{n}} \qquad \mu + 1.645 \frac{\sigma}{\sqrt{n}}$$

This concept can be generalized to any probability by substituting the appropriate $z$-value from the standard normal distribution.

Now, given that our objective is to estimate $\mu$ based on a random sample of size $n$, if we form an interval estimate using

$$\bar{x} \pm z \frac{\sigma}{\sqrt{n}}$$

then the proportion of all possible intervals that contain $\mu$ will equal the probability associated with the specified $z$-value. In estimation terminology, the $z$-value is referred to as the *critical value*.

**FIGURE** 8.4 **95% Confidence Intervals from Selected Random Samples**

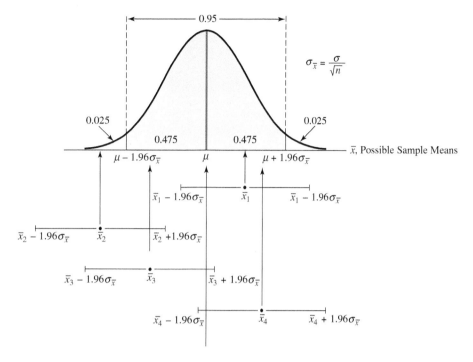

*Note*: Most intervals include $\mu$ but some do not. Those intervals that do not contain the population mean are developed from sample means that fall in either tail of the sampling distribution. If all possible intervals were constructed from a given sample size, 95% would include $\mu$.

OUTCOME 2    **Confidence Interval Calculation**   Confidence interval estimates can be constructed using the general format shown in Equation 8.1.

**Confidence Interval General Format**

$$\text{Point estimate } \pm \text{ (Critical value)(Standard error)} \qquad \textbf{(8.1)}$$

**Confidence Level**
The percentage of all possible confidence intervals that will contain the true population parameter.

The first step in developing a confidence interval estimate is to specify the **confidence level** that is needed to determine the critical value.

Once you decide on the confidence level, the next step is to determine the critical value. If the population standard deviation is known and either the population is normally distributed or the sample size is large enough to comply with the Central Limit Theorem requirements, then the critical value is a $z$-value from the standard normal table. Table 8.1 shows some of the most frequently used critical values.

The next step is to compute the standard error for the sampling distribution, shown in Chapter 7 and also earlier in this chapter to be $\sigma_{\bar{x}} = \sigma/\sqrt{n}$. Then, Equation 8.2 is used to compute the confidence interval estimate for $\mu$.

**TABLE** 8.1 **Critical Values for Commonly Used Confidence Levels**

| Confidence Level | Critical Value |
|---|---|
| 80% | $z = 1.28$ |
| 90% | $z = 1.645$ |
| 95% | $z = 1.96$ |
| 99% | $z = 2.575$ |

*Note:* Instead of using the standard normal table, you can find the critical $z$-value using Excel's NORM.S.INV function.

**Confidence Interval Estimate for $\mu$, $\sigma$ Known**

$$\bar{x} \pm z \frac{\sigma}{\sqrt{n}} \qquad \textbf{(8.2)}$$

where:

$z$ = Critical value from the standard normal table for a specified confidence level
$\sigma$ = Population standard deviation
$n$ = Sample size

**Confidence Interval Estimate for $\mu$**

Health Star Energy Drink (*continued*)   Recall that the sample of 100 energy drink bottles produced a sample mean of $\bar{x} = 12.79$ ounces and the Health Star manager knows that $\sigma = 0.2$ ounce. Thus, the 95% confidence interval estimate for the population mean is

$$\bar{x} \pm z\frac{\sigma}{\sqrt{n}}$$

$$12.79 \pm 1.96\frac{0.2}{\sqrt{100}}$$

$$12.79 \pm 0.04$$

$$12.75 \text{ ounces} \text{————} 12.83 \text{ ounces}$$

Based on this sample information, the Health Star manager believes that the true mean fill for all bottles is somewhere within the following interval:

12.75 ounces                                                    12.83 ounces

$\bar{x} = 12.79$ ounces

Because this interval does not contain the target mean of 12 ounces, the manager should conclude that the filling equipment is out of adjustment and is putting in too much energy drink, on average.

---

**HOW TO DO IT** (EXAMPLE 8-1)

**Confidence Interval Estimate for $\mu$ with $\sigma$ Known**
(when either the population distribution is normal or the sample size *n* is $\geq$ 30)

**1.** Define the population of interest and select a simple random sample of size *n*.

**2.** Specify the confidence level.

**3.** Compute the sample mean using

$$\bar{x} = \frac{\sum x}{n}$$

**4.** Determine the standard error of the sampling distribution using

$$\sigma_{\bar{x}} = \frac{\sigma}{\sqrt{n}}$$

**5.** Determine the critical value, *z*, from the standard normal table.

**6.** Compute the confidence interval estimate using

$$\bar{x} \pm z\frac{\sigma}{\sqrt{n}}$$

**EXAMPLE 8-1** **Confidence Interval Estimate for $\mu$, $\sigma$ Known**

**Textbook Purchases**   One contributor to the cost of a college education is the purchase of textbooks. Administrators at a well-known private university are interested in estimating the average amount students spend on textbooks during the first four years at the university. To estimate this value, they can use the following steps:

STEP 1   **Define the population of interest and select a simple random sample of size *n*.**
The population is the amount students spend on textbooks over four years. A simple random sample of 200 students will be selected, and the amount spent on textbooks will be recorded.

STEP 2   **Specify the confidence level.**
The administrators want to develop a 90% confidence interval estimate. Thus, 90% of all possible interval estimates will contain the population mean.

STEP 3   **Compute the sample mean.**
After administrators select the sample and record the dollars spent on textbooks by each of the 200 students sampled, the sample mean is computed using

$$\bar{x} = \frac{\sum x}{n}$$

Assume the sample mean is $5,230.

STEP 4   **Determine the standard error of the sampling distribution.**
Suppose past studies have indicated that the population standard deviation for the amounts students spend on textbooks at this university is

$$\sigma = \$500$$

Then the standard error of the sampling distribution is computed using

$$\sigma_{\bar{x}} = \frac{\sigma}{\sqrt{n}} = \frac{\$500}{\sqrt{200}} = \$35.36$$

STEP 5    **Determine the critical value, z, from the standard normal table.**
Because the sample size is large, the Central Limit Theorem applies. The sampling distribution will be normally distributed, and the critical value will be a z-value from the standard normal distribution. The administrators want a 90% confidence level, so the z-value is 1.645.

STEP 6    **Compute the confidence interval estimate.**
The 90% confidence interval estimate for the population mean is

$$\bar{x} \pm z\frac{\sigma}{\sqrt{n}}$$

$$\$5,230 \pm 1.645\frac{500}{\sqrt{200}}$$

$$\$5,230 \pm \$58.16$$

$$\$5,171.84 \text{——} \$5,288.16$$

Thus, based on the sample results, with 90% confidence, the university administrators believe that the true population mean amount students spend on textbooks is between $5,171.84 and $5,288.16.

TRY EXERCISE 8-2 (pg. 299)

### Interpreting Confidence Intervals

There is a subtle distinction to be made here. Beginning students often wonder if it is permissible to say, "There is a 0.90 probability that the population mean is between $5,171.84 and $5,288.16." This may seem to be the logical consequence of constructing a confidence interval. However, we must be very careful to attribute probability only to random events or variables. Because the population mean is a fixed value, there can be no probability statement about the population mean. The confidence interval we have computed either will contain the population mean or will not. If you were to produce all the possible 90% confidence intervals using the mean of each possible sample of a given size from the population, then 90% of these intervals would contain the population mean.

### Impact of the Confidence Level on the Interval Estimate

**BUSINESS APPLICATION**    **Margin of Error**

**Health Star Energy Drink** (*continued*)    In the Health Star Energy Drink example, the manager specified a 95% confidence level. The resulting confidence interval estimate for the population mean was

$$\bar{x} \pm z\frac{\sigma}{\sqrt{n}}$$

$$12.79 \pm 1.96\frac{0.2}{\sqrt{100}}$$

$$12.79 \pm 0.04$$

$$12.75 \text{ ounces} \text{———} 12.83 \text{ ounces}$$

**Margin of Error**

The amount that is added to and subtracted from the point estimate to determine the endpoints of the confidence interval. Also, a measure of how close we expect the point estimate to be to the population parameter with the specified level of confidence.

The quantity 0.04 on the right of the $\pm$ sign is called the **margin of error**. This is illustrated in Equation 8.3. The margin of error defines the relationship between the sample mean and the population mean. Thus, the sample mean of 12.79 ounces is thought to be within $\pm 0.04$ of the true population mean.

Now suppose the manager at Health Star is willing to settle for 80% confidence. This will affect the critical value. To determine the new value, we divide 0.80 by 2, giving 0.40. Then we go to the standard normal table and locate a probability value (area under the curve)

**Margin of Error for Estimating $\mu$, $\sigma$ Known**

$$e = z\frac{\sigma}{\sqrt{n}} \tag{8.3}$$

where:

$e$ = Margin of error
$z$ = Critical value
$\dfrac{\sigma}{\sqrt{n}}$ = Standard error of the sampling distibution

that is as close to 0.40 as possible. The corresponding $z$-value is 1.28. The 80% confidence interval estimate is

$$\bar{x} \pm z\frac{\sigma}{\sqrt{n}}$$

$$12.79 \pm (1.28)\frac{0.2}{\sqrt{100}}$$

$$12.79 \pm 0.0256$$

12.7644 ounces ——————— 12.8156 ounces

Based on this sample information and the 80% confidence interval, we believe that the true average fill level is between approximately 12.76 ounces and 12.82 ounces.

By lowering the confidence level, we are less likely to obtain an interval that contains the population mean. However, on the positive side, the margin of error has been reduced from $\pm 0.04$ ounce to $\pm 0.0256$ ounce. For equivalent samples from a population:

1. If the confidence level is decreased, the margin of error is reduced.
2. If the confidence level is increased, the margin of error is increased.

The Health Star Energy Drink manager will need to decide which is more important, a higher confidence level or a lower margin of error.

---

**EXAMPLE 8-2** **Impact of Changing the Confidence Level**

**National Recycling**   National Recycling operates a garbage hauling company in a southern Maine city. Each year, the company must apply for a new contract with the state. The contract is based in part on the pounds of recycled materials collected. Part of the analysis that goes into contract development is an estimate of the mean pounds of recycled material submitted by each customer in the city on a quarterly basis. The city has asked for both 99% and 90% confidence interval estimates for the mean. If, after the contract has been signed, the actual mean pounds deviates from the estimate over time, an adjustment will be made (up or down) in the amount National Recycling receives. The steps used to generate these estimates follow:

STEP 1   **Define the population of interest and select a simple random sample of size $n$.**
The population is the collection of all of National Recycling's customers, and a simple random sample of $n = 100$ customers is selected.

STEP 2   **Specify the confidence level.**
The city requires 99% and 90% confidence interval estimates.

STEP 3   **Compute the sample mean.**
After the sample has been selected and the pounds of recycled materials have been determined for each of the 100 customers sampled, the sample mean is computed using

$$\bar{x} = \frac{\sum x}{n}$$

Suppose the sample mean is 40.78 pounds.

STEP 4 **Determine the standard error of the sampling distribution.**

Suppose, from past years, the population standard deviation is known to be $\sigma = 12.6$ pounds. Then the standard error of the sampling distribution is computed using

$$\sigma_{\bar{x}} = \frac{\sigma}{\sqrt{n}} = \frac{12.6}{\sqrt{100}} = 1.26 \text{ pounds}$$

> The Excel 2016 function to determine the critical z-value is
>
> **= ABS(NORM.S.INV(.005))**
>
> where the argument (.005) is $(1 - 0.99)/2$ and
>
> **= ABS(NORM.S.INV(.05))**
>
> where the argument (.05) is $(1 - 0.90)/2$.

STEP 5 **Determine the critical value, z, from the standard normal table.**

First, the state wants a 99% confidence interval estimate, so the z-value is determined by finding a probability in Appendix D corresponding to $0.99/2 = 0.495$. The correct z-value is between $z = 2.57$ and $z = 2.58$. We split the difference to get the critical value: $z = 2.575$. For 90% confidence, the critical $z$ is determined to be 1.645.

STEP 6 **Compute the confidence interval estimate.**

The 99% confidence interval estimate for the population mean is

$$\bar{x} \pm z \frac{\sigma}{\sqrt{n}}$$

$$40.78 \pm 2.575 \frac{12.6}{\sqrt{100}}$$

$$40.78 \pm 3.24$$

37.54 pounds ————— 44.02 pounds

> The Excel 2016 function to compute the margin of error at 99% confidence when the population standard deviation is known is
>
> **= CONFIDENCE. NORM(.01,12.6,100)**
>
> To determine the margin of error at 90% confidence, use
>
> **= CONFIDENCE. NORM(.10,12.6,100).**

The margin of error at 99% confidence is $\pm 3.24$ pounds.

The 90% confidence interval estimate for the population mean is

$$\bar{x} \pm z \frac{\sigma}{\sqrt{n}}$$

$$40.78 \pm 1.645 \frac{12.6}{\sqrt{100}}$$

$$40.78 \pm 2.07$$

38.71 pounds ————— 42.85 pounds

The margin of error is only 2.07 pounds when the confidence level is reduced from 99% to 90%. The margin of error is smaller when the confidence level is smaller.

**TRY EXERCISE 8-5** (pg. 292)

Lowering the confidence level is one way to reduce the margin of error. However, by examining Equation 8.3, you will note there are two other values that affect the margin of error. One of these is the population standard deviation. The more the population's standard deviation, $\sigma$, can be reduced, the smaller the margin of error will be. Businesses do not want large standard deviations for measurements related to the quality of a product. In fact, corporations spend considerable effort to decrease the variation in their products either by changing their process or by controlling variables that cause the variation. Typically, decision makers pursue all avenues for reducing the standard deviation before they entertain thoughts of reducing the confidence level.

Unfortunately, there are many situations in which reducing the population standard deviation is not possible. In these cases, another step that can be taken to reduce the margin of error is to increase the sample size. As you learned in Chapter 7, an increase in sample size reduces the standard error of the sampling distribution. This can be the most direct way of reducing the margin of error as long as obtaining an increased sample size is not prohibitively costly or unattainable for other reasons.

## Impact of the Sample Size on the Interval Estimate

| BUSINESS APPLICATION | Understanding the Value of a Larger Sample Size |

**Health Star Energy Drink** (*continued*)    Suppose the Health Star Energy Drink production manager decided to increase the sample to 400 bottles. This is a fourfold increase over the original sample size. We learned in Chapter 7 that an increase in sample size reduces the standard error of the sampling distribution because the standard error is computed as $\sigma / \sqrt{n}$. Thus, without adversely affecting his confidence level, the manager can reduce the margin of error by increasing his sample size.

Assume that the sample mean for the larger sample size also happens to be $\bar{x} = 12.79$ ounces. The new 95% confidence interval estimate is

$$12.79 \pm 1.96 \frac{0.2}{\sqrt{400}}$$
$$12.79 \pm 0.02$$
$$12.77 \text{ ounces} \text{ ——— } 12.81 \text{ ounces}$$

Notice that by increasing the sample size to 400 bottles, the margin of error is reduced from the original 0.04 ounce to 0.02 ounce. The production manager now believes that his sample mean is within $\pm 0.02$ ounce of the true population mean.

The production manager was able to reduce the margin of error without reducing the confidence level. However, the downside is that sampling 400 bottles instead of 100 bottles will cost more money and take more time. That's the trade-off. Absent the possibility of reducing the population standard deviation, if he wants to reduce the margin of error, he must either reduce the confidence level or increase the sample size, or some combination of each. If he is unwilling to do so, he will have to accept the larger margin of error.

## Confidence Interval Estimates for the Population Mean, $\sigma$ Unknown

In the Health Star Energy Drink application, the manager was dealing with a filling machine that had a known standard deviation in fill volume. You may encounter situations in which the standard deviation is known. However, in most cases, if you do not know the population mean, you also will not know the population standard deviation. When this occurs, you need to make a minor, but important, modification to the confidence interval estimation process.

### Student's *t*-Distribution

When the population standard deviation is known, the sampling distribution of the mean has only one unknown parameter: its mean, $\mu$. This is estimated by $\bar{x}$. However, when the population standard deviation is unknown, there are two unknown parameters, $\mu$ and $\sigma$, which can be estimated by $\bar{x}$ and $s$, respectively. This estimation doesn't affect the general format for a confidence interval, as shown earlier in Equation 8.1:

$$\text{Point estimate} \pm (\text{Critical value})(\text{Standard error})$$

However, not knowing the population standard deviation does affect the critical value. Recall that when $\sigma$ is known and the population is normally distributed or the Central Limit Theorem applies, the critical value is a *z*-value taken from the standard normal table. But when $\sigma$ is not known, the critical value is a *t*-value taken from a family of distributions called the **Student's *t*-distributions**.

Because the specific *t*-distribution chosen is based on its *degrees of freedom*, it is important to understand what *degrees of freedom* means. Recall that the sample standard deviation is an estimate of the population's standard deviation and is defined as

$$s = \sqrt{\frac{\sum (x - \bar{x})^2}{n - 1}}$$

**Student's *t*-Distributions**
A family of distributions that is bell-shaped and symmetric like the standard normal distribution but with greater area in the tails. Each distribution in the *t*-family is defined by its degrees of freedom. As the degrees of freedom increase, the *t*-distribution approaches the normal distribution.

Therefore, if we wish to estimate the population standard deviation, we must first calculate the sample mean. The sample mean is itself an estimator of a parameter, namely, the population mean. We can determine the sample mean from a sample of $n$ randomly and independently chosen data values. Then, once we know the sample mean, there are only $n - 1$ independent pieces of data information left in the sample.

To illustrate, examine a sample of size $n = 3$ in which the sample mean is calculated to be 12. This implies that the sum of the three data values equals $36 (3 \times 12)$. If you know that the first two data values are 10 and 8, then the third data value is determined to be 18. Similarly, if you know that the first two data values are 18 and 7, the third data value must be 11. You are free to choose any two of the three data values. In general, if you must estimate $k$ parameters before you are able to estimate the population's standard deviation from a sample of $n$ data values, you have the freedom to choose any $n - k$ data values before the remaining $k$ values are determined. This value, $n - k$, is called the **degrees of freedom**.

**Degrees of Freedom**

The number of independent data values available to estimate the population's standard deviation. If $k$ parameters must be estimated before the population's standard deviation can be calculated from a sample of size $n$, the degrees of freedom are equal to $n - k$.

When the population is normally distributed, the $t$-value represents the number of standard errors $\bar{x}$ is from $\mu$, as shown in Equation 8.4. Appendix F contains a table of standardized $t$-values that correspond to specified tail areas and different degrees of freedom. The $t$-table is used to determine the critical value when we do not know the population standard deviation. The $t$-table is reproduced on the inside back endsheet of your text and also in Table 8.2. Note that in Equation 8.4, we use the sample standard deviation, $s$, to estimate the population standard deviation, $\sigma$. The fact that we are estimating $\sigma$ is the reason the $t$-distribution is more spread out (i.e., has a larger standard deviation) than the normal distribution (see Figure 8.5). By estimating $\sigma$, we are introducing more uncertainty into the estimation process; therefore, achieving the same level of confidence requires a $t$-value larger than the $z$-value for the same confidence level. As the sample size increases, our estimate of $\sigma$ becomes better and the $t$-distribution converges to the $z$-distribution.

**$t$-Value for $\bar{x}$**

$$t = \frac{\bar{x} - \mu}{\dfrac{s}{\sqrt{n}}}$$

(8.4)

where:

$\bar{x}$ = Sample mean
$\mu$ = Population mean
$s$ = Sample standard deviation
$n$ = Sample size

**Assumption**

The $t$-distribution is based on the assumption that the population is normally distributed. However, as long as the population is reasonably symmetric, we can use the $t$-distribution.

**FIGURE 8.5 $t$-Distribution and Normal Distribution**

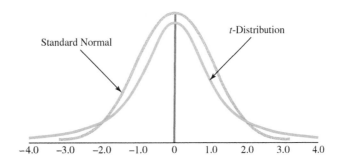

**TABLE 8.2** Values of *t* for Selected Probabilities

| | Probabilities (or Areas under *t*-Distribution Curve) | | | | | | | | |
|---|---|---|---|---|---|---|---|---|---|
| **Conf. Level** | 0.1 | 0.3 | 0.5 | 0.7 | 0.8 | 0.9 | 0.95 | 0.98 | 0.99 |
| **One Tail** | 0.45 | 0.35 | 0.25 | 0.15 | 0.1 | 0.05 | 0.025 | 0.01 | 0.005 |
| **Two Tails** | 0.9 | 0.7 | 0.5 | 0.3 | 0.2 | 0.1 | 0.05 | 0.02 | 0.01 |
| *d.f.* | | | | | *Values of t* | | | | |
| 1 | 0.1584 | 0.5095 | 1.0000 | 1.9626 | 3.0777 | 6.3137 | 12.7062 | 31.8210 | 63.6559 |
| 2 | 0.1421 | 0.4447 | 0.8165 | 1.3862 | 1.8856 | 2.9200 | 4.3027 | 6.9645 | 9.9250 |
| 3 | 0.1366 | 0.4242 | 0.7649 | 1.2498 | 1.6377 | 2.3534 | 3.1824 | 4.5407 | 5.8408 |
| 4 | 0.1338 | 0.4142 | 0.7407 | 1.1896 | 1.5332 | 2.1318 | 2.7765 | 3.7469 | 4.6041 |
| 5 | 0.1322 | 0.4082 | 0.7267 | 1.1558 | 1.4759 | 2.0150 | 2.5706 | 3.3649 | 4.0321 |
| 6 | 0.1311 | 0.4043 | 0.7176 | 1.1342 | 1.4398 | 1.9432 | 2.4469 | 3.1427 | 3.7074 |
| 7 | 0.1303 | 0.4015 | 0.7111 | 1.1192 | 1.4149 | 1.8946 | 2.3646 | 2.9979 | 3.4995 |
| 8 | 0.1297 | 0.3995 | 0.7064 | 1.1081 | 1.3968 | 1.8595 | 2.3060 | 2.8965 | 3.3554 |
| 9 | 0.1293 | 0.3979 | 0.7027 | 1.0997 | 1.3830 | 1.8331 | 2.2622 | 2.8214 | 3.2498 |
| 10 | 0.1289 | 0.3966 | 0.6998 | 1.0931 | 1.3722 | 1.8125 | 2.2281 | 2.7638 | 3.1693 |
| 11 | 0.1286 | 0.3956 | 0.6974 | 1.0877 | 1.3634 | 1.7959 | 2.2010 | 2.7181 | 3.1058 |
| 12 | 0.1283 | 0.3947 | 0.6955 | 1.0832 | 1.3562 | 1.7823 | 2.1788 | 2.6810 | 3.0545 |
| 13 | 0.1281 | 0.3940 | 0.6938 | 1.0795 | 1.3502 | 1.7709 | 2.1604 | 2.6503 | 3.0123 |
| 14 | 0.1280 | 0.3933 | 0.6924 | 1.0763 | 1.3450 | 1.7613 | 2.1448 | 2.6245 | 2.9768 |
| 15 | 0.1278 | 0.3928 | 0.6912 | 1.0735 | 1.3406 | 1.7531 | 2.1315 | 2.6025 | 2.9467 |
| 16 | 0.1277 | 0.3923 | 0.6901 | 1.0711 | 1.3368 | 1.7459 | 2.1199 | 2.5835 | 2.9208 |
| 17 | 0.1276 | 0.3919 | 0.6892 | 1.0690 | 1.3334 | 1.7396 | 2.1098 | 2.5669 | 2.8982 |
| 18 | 0.1274 | 0.3915 | 0.6884 | 1.0672 | 1.3304 | 1.7341 | 2.1009 | 2.5524 | 2.8784 |
| 19 | 0.1274 | 0.3912 | 0.6876 | 1.0655 | 1.3277 | 1.7291 | 2.0930 | 2.5395 | 2.8609 |
| 20 | 0.1273 | 0.3909 | 0.6870 | 1.0640 | 1.3253 | 1.7247 | 2.0860 | 2.5280 | 2.8453 |
| 21 | 0.1272 | 0.3906 | 0.6864 | 1.0627 | 1.3232 | 1.7207 | 2.0796 | 2.5176 | 2.8314 |
| 22 | 0.1271 | 0.3904 | 0.6858 | 1.0614 | 1.3212 | 1.7171 | 2.0739 | 2.5083 | 2.8188 |
| 23 | 0.1271 | 0.3902 | 0.6853 | 1.0603 | 1.3195 | 1.7139 | 2.0687 | 2.4999 | 2.8073 |
| 24 | 0.1270 | 0.3900 | 0.6848 | 1.0593 | 1.3178 | 1.7109 | 2.0639 | 2.4922 | 2.7970 |
| 25 | 0.1269 | 0.3898 | 0.6844 | 1.0584 | 1.3163 | 1.7081 | 2.0595 | 2.4851 | 2.7874 |
| 26 | 0.1269 | 0.3896 | 0.6840 | 1.0575 | 1.3150 | 1.7056 | 2.0555 | 2.4786 | 2.7787 |
| 27 | 0.1268 | 0.3894 | 0.6837 | 1.0567 | 1.3137 | 1.7033 | 2.0518 | 2.4727 | 2.7707 |
| 28 | 0.1268 | 0.3893 | 0.6834 | 1.0560 | 1.3125 | 1.7011 | 2.0484 | 2.4671 | 2.7633 |
| 29 | 0.1268 | 0.3892 | 0.6830 | 1.0553 | 1.3114 | 1.6991 | 2.0452 | 2.4620 | 2.7564 |
| 30 | 0.1267 | 0.3890 | 0.6828 | 1.0547 | 1.3104 | 1.6973 | 2.0423 | 2.4573 | 2.7500 |
| 40 | 0.1265 | 0.3881 | 0.6807 | 1.0500 | 1.3031 | 1.6839 | 2.0211 | 2.4233 | 2.7045 |
| 50 | 0.1263 | 0.3875 | 0.6794 | 1.0473 | 1.2987 | 1.6759 | 2.0086 | 2.4033 | 2.6778 |
| 60 | 0.1262 | 0.3872 | 0.6786 | 1.0455 | 1.2958 | 1.6706 | 2.0003 | 2.3901 | 2.6603 |
| 70 | 0.1261 | 0.3869 | 0.6780 | 1.0442 | 1.2938 | 1.6669 | 1.9944 | 2.3808 | 2.6479 |
| 80 | 0.1261 | 0.3867 | 0.6776 | 1.0432 | 1.2922 | 1.6641 | 1.9901 | 2.3739 | 2.6387 |
| 90 | 0.1260 | 0.3866 | 0.6772 | 1.0424 | 1.2910 | 1.6620 | 1.9867 | 2.3685 | 2.6316 |
| 100 | 0.1260 | 0.3864 | 0.6770 | 1.0418 | 1.2901 | 1.6602 | 1.9840 | 2.3642 | 2.6259 |
| 250 | 0.1258 | 0.3858 | 0.6755 | 1.0386 | 1.2849 | 1.6510 | 1.9695 | 2.3414 | 2.5956 |
| 500 | 0.1257 | 0.3855 | 0.6750 | 1.0375 | 1.2832 | 1.6479 | 1.9647 | 2.3338 | 2.5857 |
| ∞ | 0.1257 | 0.3853 | 0.6745 | 1.0364 | 1.2816 | 1.6449 | 1.9600 | 2.3264 | 2.5758 |

**BUSINESS APPLICATION**    Using the *t*-Distribution

Excel Tutorial

**Kapalua Software**    Kapalua Software develops educational and business software and operates a service center, where employees respond to customer calls about questions and problems with the company's software packages. Recently, a team of Kapalua employees was asked to study the average length of time service representatives spend with customers. The team decided to collect a simple random sample of 25 calls and then estimate the population mean call time based on the sample data. Not only did the team not know the average length of time, $\mu$, but it also didn't know the standard deviation of length of the service time, $\sigma$.

Table 8.3 shows the sample data for 25 calls. (These data are in a file called **Kapalua**.) The managers at Kapalua Software are willing to assume the population of call times is approximately normal.

Kapalua's sample mean and standard deviation are

$$\bar{x} = 7.088 \text{ minutes}$$
$$s = 4.64 \text{ minutes}$$

**TABLE 8.3** Sample Call Times for Kapalua Software

| 7.1 | 11.6 | 12.4 | 8.5 | 0.4 |
|------|------|------|------|------|
| 13.6 | 1.7 | 11.0 | 6.1 | 11.0 |
| 1.4 | 16.9 | 3.7 | 3.3 | 0.8 |
| 3.6 | 2.6 | 14.6 | 6.1 | 6.4 |
| 1.9 | 7.7 | 8.8 | 6.9 | 9.1 |

If the managers need a single-valued estimate of the population mean, they would use the point estimate, $\bar{x} = 7.088$ minutes. However, they should realize that this point estimate is subject to sampling error. To take the sampling error into account, the managers can construct a confidence interval estimate. Equation 8.5 shows the formula for the confidence interval estimate for the population mean when the population standard deviation is unknown.

**OUTCOME 2**

**Confidence Interval Estimate for $\mu$, $\sigma$ Unknown**

The Excel 2016 functions to determine the sample mean and standard deviation are

= AVERAGE(7.1,11.6,...,9.1)
= STDEV.S(7.1,11.6,...,9.1)

$$\bar{x} \pm t\frac{s}{\sqrt{n}} \qquad (8.5)$$

where:

$\bar{x}$ = Sample mean
$t$ = Critical value from the *t*-distribution with $n - 1$ degrees of freedom for the desired confidence level
$s$ = Sample standard deviation
$n$ = Sample size

The Excel 2016 function to determine the critical *t*-value is

**T.INV.2T (Probability,Deg_freedom)**

where Probability equals 1 − Confidence level and Deg_freedom is $n - 1$.

= **T.INV.2T(0.05,24)**

where the argument (0.05) is 1 − 0.95.

The first step is to specify the desired confidence level. For example, suppose the Kapalua team specifies a 95% confidence level. To get the critical *t*-value from the *t*-table in Appendix F or Table 8.2, go to the top of the table to the row labeled "Conf. Level." Locate the column headed "0.95." Next, go to the row corresponding to

$$n - 1 = 25 - 1 = 24 \text{ degrees of freedom}$$

The critical *t*-value for 95% confidence and 24 degrees of freedom is

$$t = 2.0639$$

The Kapalua team can now compute the 95% confidence interval estimate using Equation 8.5 as follows:

The Excel 2016 function to compute the margin of error when the population standard deviation is unknown is

**CONFIDENCE.T(Alpha, Standard_dev,Size)**

where Alpha = 1 − Confidence level, Standard_Dev is the sample standard deviation, and Size is the sample size.

**CONFIDENCE.T(.05,4.64,25)**

$$\bar{x} \pm t\frac{s}{\sqrt{n}}$$
$$7.088 \pm 2.0639\frac{4.64}{\sqrt{25}}$$
$$7.088 \pm 1.915$$
$$5.173 \text{ min} \longrightarrow 9.003 \text{ min}$$

Therefore, based on the random sample of 25 calls and the 95% confidence interval, the Kapalua Software team has estimated the true average time per call to be between 5.173 minutes and 9.003 minutes.

FIGURE 8.6 **Excel 2016 Output for the Kapalua Example**

| | A | B | C | D |
|---|---|---|---|---|
| 1 | Minutes | | Minutes | |
| 2 | 7.1 | | | |
| 3 | 13.6 | | Mean | 7.088 |
| 4 | 1.4 | | Standard Error | 0.928 |
| 5 | 3.6 | | Median | 6.9 |
| 6 | 1.9 | | Mode | 11 |
| 7 | 11.6 | | Standard Deviation | 4.640 |
| 8 | 1.7 | | Sample Variance | 21.534 |
| 9 | 16.9 | | Kurtosis | -0.757 |
| 10 | 2.6 | | Skewness | 0.347 |
| 11 | 7.7 | | Range | 16.5 |
| 12 | 12.4 | | Minimum | 0.4 |
| 13 | 11.0 | | Maximum | 16.9 |
| 14 | 3.7 | | Sum | 177.2 |
| 15 | 14.6 | | Count | 25 |
| 16 | 8.8 | | Confidence Level(95.0%) | 1.915 |

Mean = 7.088
Confidence Interval:
7.088 ± 1.915
5.173 ——— 9.003

Margin of Error

Excel has a procedure for computing the confidence interval estimate of the population mean. The Excel output is shown in Figure 8.6. Note that the margin of error is printed. You use it and the sample mean to compute the upper and lower limits.

---

**EXAMPLE 8-3** Confidence Interval Estimate for $\mu$, $\sigma$ Unknown

**Internal Audit** An internal auditor for a retail grocery store is interested in the degree to which the customer bill matches the true cost of the items purchased. In particular, the auditor wishes to develop a 90% confidence interval estimate for the population mean difference between customer bill and true cost. Errors can be positive, negative, or zero when no error occurred. To do so, the auditor can use the following steps:

STEP 1 **Define the population and select a simple random sample of size $n$ from the population.**

In this case, the population consists of the individual sales receipt errors made by the grocery clerks in a given week. A simple random sample of $n = 20$ transactions is selected, with the following data:

| | | | | | | | | | |
|---|---|---|---|---|---|---|---|---|---|
| $0.00 | $1.20 | $0.43 | $1.00 | $1.47 | $0.83 | $0.50 | $3.34 | $1.58 | $1.46 |
| −$0.36 | −$1.10 | $2.60 | $0.00 | $0.00 | −$1.70 | $0.83 | $1.99 | $0.00 | $1.34 |

STEP 2 **Specify the confidence level.**

A 90% confidence interval estimate is desired.

STEP 3 **Compute the sample mean and sample standard deviation.**

After the auditors have selected the sample and determined the tax amounts for each of the 20 customers sampled, the sample mean is computed using

$$\bar{x} = \frac{\sum x}{n} = \frac{\$15.41}{20} = \$0.77$$

The sample standard deviation is computed using

$$s = \sqrt{\frac{\sum (x - \bar{x})^2}{n - 1}} = \sqrt{\frac{(0.00 - 0.77)^2 + (1.20 - 0.77)^2 + \cdots + (1.34 - 0.77)^2}{20 - 1}} = \$1.19$$

STEP 4 **Determine the standard error of the sampling distribution.**

Because the population standard deviation is unknown, the standard error of the sampling distribution is estimated using

$$\sigma_{\bar{x}} = \frac{s}{\sqrt{n}} = \frac{\$1.19}{\sqrt{20}} = \$0.27$$

STEP 5 **Determine the critical value for the desired level of confidence.**

Because we do not know the population standard deviation and the sample size is reasonably small, the critical value will come from the *t*-distribution, providing we can assume that the population is normally distributed. A box and whisker plot can give some insight about how the population might look.

> The Excel 2016 function to determine the critical *t*-value is
>
> **= T.INV.2T(1-conf. level, deg. of freedom)**
> **= T.INV.2T(0.10,19)**
>
> where the argument (0.10) is 1 − 0.90.

| Minimum | −1.7 |
| First Quartile | 0 |
| Median | 0.83 |
| Third Quartile | 1.47 |
| Maximum | 3.34 |

This diagram does not indicate that there is any serious skewness or other abnormality in the data, so we will continue with the normal distribution assumption.

The critical value for 90% confidence and $20 - 1 = 19$ degrees of freedom is found in the *t*-distribution table as $t = 1.7291$.

STEP 6 **Compute the confidence interval estimate.**

The 90% confidence interval estimate for the population mean is

> The Excel 2016 function for computing the 90% confidence interval margin of error when the population standard deviation is unknown is
>
> **= CONFIDENCE.T(.10,1.19,20)**

$$\bar{x} \pm t\frac{s}{\sqrt{n}}$$

$$0.77 \pm 1.7291\frac{1.19}{\sqrt{20}}$$

$$0.77 \pm 0.46$$

$$\$0.31 \text{———} \$1.23$$

Thus, based on the sample data, with 90% confidence, the auditors can conclude that the population mean error per transaction is between $0.31 and $1.23.

**TRY EXERCISE 8-7** (pg. 292)

**Estimation with Larger Sample Sizes**  We saw earlier that a change in sample size can affect the margin of error in a statistical estimation situation when the population standard deviation is known. This is also true in applications in which the standard deviation is not known. In fact, the effect of a change is compounded because the change in sample size affects both the calculation of the standard error and the critical value from the *t*-distribution.

The *t*-distribution table in Appendix F shows degrees of freedom up to 30 and then incrementally to 500. Observe that for any confidence level, as the degrees of freedom increase, the *t*-value gets smaller as it approaches a limit equal to the *z*-value from the standard normal table in Appendix D for the same confidence level. If you need to estimate the population mean with a sample size that is not listed in the *t*-table, you can use the Excel T.INV.2T function to get the critical *t*-value for any specified degrees of freedom and then use Equation 8.5.

You should have noticed that the format for confidence interval estimates for $\mu$ is essentially the same, regardless of whether the population standard deviation is known. The basic format is

$$\text{Point estimate} \pm (\text{Critical value})(\text{Standard error})$$

Later in this chapter, we introduce estimation examples in which the population value of interest is *p*, the population proportion. The same confidence interval format is used. In addition, the trade-offs between margin of error, confidence level, and sample size that were discussed in this section apply to every other estimation situation.

## 8.1 EXERCISES

### Skill Development

**8-1.** Assuming the population of interest is approximately normally distributed, construct a 95% confidence interval estimate for the population mean given the following values:

$$\bar{x} = 18.4 \quad s = 4.2 \quad n = 13$$

**8-2.** Construct a 90% confidence interval estimate for the population mean given the following values:

$$\bar{x} = 70 \quad \sigma = 15 \quad n = 65$$

**8-3.** Construct a 95% confidence interval estimate for the population mean given the following values:

$$\bar{x} = 300 \quad \sigma = 55 \quad n = 250$$

**8-4.** Construct a 98% confidence interval estimate for the population mean given the following values:

$$\bar{x} = 120 \quad \sigma = 20 \quad n = 50$$

**8-5.** Determine the 90% confidence interval estimate for the population mean of a normal distribution given $n = 100$, $\sigma = 121$, and $\bar{x} = 1,200$.

**8-6.** Determine the margin of error for a confidence interval estimate for the population mean of a normal distribution given the following information:
a. confidence level $= 0.98$, $n = 13$, $s = 15.68$
b. confidence level $= 0.99$, $n = 25$, $\sigma = 3.47$
c. confidence level $= 0.98$, standard error $= 2.356$

**8-7.** The following sample data have been collected based on a simple random sample from a normally distributed population:

| 2 | 8 | 0 | 2 | 3 |
|---|---|---|---|---|
| 5 | 3 | 1 | 4 | 2 |

a. Compute a 90% confidence interval estimate for the population mean.
b. Show what the impact would be if the confidence level is increased to 95%. Discuss why this occurs.

**8-8.** A random sample of size 20 yields $\bar{x} = 3.13$ and $s^2 = 1.45$. Calculate a confidence interval for the population mean whose confidence level is as follows:
a. 0.99
b. 0.98
c. 0.95
d. 0.90
e. 0.80
f. What assumptions were necessary to establish the validity of the confidence intervals calculated in parts a through e?

**8-9.** A random sample of $n = 12$ values taken from a normally distributed population resulted in the following sample values:

| 107 | 109 | 99 | 91 | 103 | 105 |
|-----|-----|-----|-----|-----|-----|
| 105 | 94 | 107 | 94 | 97 | 113 |

Use the sample information to construct a 95% confidence interval estimate for the population mean.

**8-10.** A random sample of $n = 9$ values taken from a normally distributed population with a population variance of 25 resulted in the following sample values:

| 53 | 46 | 55 | 45 | 44 | 52 | 46 | 60 | 49 |
|----|----|----|----|----|----|----|----|----|

Use the sample values to construct a 90% confidence interval estimate for the population mean.

**8-11.** A random sample was selected from a population having a normal distribution. Calculate a 90% confidence interval estimate for $\mu$ for each of the following situations:
a. $\Sigma x = 134$, $n = 10$, $s = 3.1$
b. $\Sigma x = 3,744$, $n = 120$, $s = 8.2$
c. $\Sigma x = 40.5$, $n = 9$, $\sigma = 2.9$
d. $\Sigma x = 585.9$, $\Sigma x^2 = 15,472.37$, $n = 27$
(*Hint:* Refer to Equation 3.13.)

### Business Applications

**8-12.** Allante Pizza delivers pizzas throughout its local market area at no charge to the customer. However, customers often tip the driver. The owner is interested in estimating the mean tip income per delivery. To do this, she has selected a simple random sample of 12 deliveries and has recorded the tips that the drivers received. These data are

| $2.25 | $2.50 | $2.25 | $2.00 | $2.00 | $1.50 |
|-------|-------|-------|-------|-------|-------|
| $0.00 | $2.00 | $1.50 | $2.00 | $3.00 | $1.50 |

a. Based on these sample data, what is the best point estimate to use as an estimate of the true mean tip per delivery?
b. Suppose the owner is interested in developing a 90% confidence interval estimate. Given the fact that the population standard deviation is unknown, what distribution will be used to obtain the critical value?
c. Referring to part b, what assumption is required to use the specified distribution to obtain the critical value? Develop a box and whisker plot to illustrate whether this assumption seems to be reasonably satisfied.
d. Referring to parts b and c, construct and interpret the 90% confidence interval estimate for the population mean.

**8-13.** The BelSante Company operates retail pharmacies in 10 eastern states. The company's internal audit department selected a random sample of 300 prescriptions issued throughout the system. The objective of the sampling was to estimate the average dollar value of all prescriptions issued by the company. The following data were collected:

$$\bar{x} = \$14.23$$
$$s = 3.00$$

a. Determine the 90% confidence interval estimate for the true average sales value for prescriptions issued by the company. Interpret the interval estimate.

b. One of its retail outlets reported that it had monthly revenue of $7,392 from 528 prescriptions. Are such results to be expected? Do you believe that the retail outlet should be audited? Support your answer with calculations and logic.

**8-14.** A consumer group conducted a study of SUV owners to estimate the mean highway mileage for their vehicles. A simple random sample of 91 SUV owners was selected, and the owners were asked to report their highway mileage. The following results were summarized from the sample data:

$$\bar{x} = 21.3 \text{ mpg}$$
$$s = 6.3 \text{ mpg}$$

Based on these sample data, compute and interpret a 90% confidence interval estimate for the mean highway mileage for SUVs.

**8-15.** Car buyers often add accessories to their new cars. A sample of 179 Mini Cooper purchasers yielded a sample mean of $5,000 worth of accessories added to the purchase above the $20,200 base sticker price. Suppose the cost of accessories purchased for all Minis has a standard deviation of $1,500.

a. Calculate a 95% confidence interval for the average cost of accessories on Minis.

b. Determine the margin of error in estimating the average cost of accessories on Minis.

c. What sample size would be required to reduce the margin of error by 50%?

**8-16.** Bolton, Inc., an Internet service provider (ISP), has experienced rapid growth in the past five years. As a part of its marketing strategy, the company promises fast connections and dependable service. To achieve its objectives, the company constantly evaluates the capacity of its servers. One component of its evaluation is an analysis of the average amount of time a customer is connected and actively using the Internet daily. A random sample of 12 customer records shows the following daily usage times, in minutes:

| 268 | 336 | 296 | 311 | 306 | 335 |
| 301 | 278 | 290 | 393 | 373 | 329 |

a. Using the sample data, compute the best point estimate of the population mean for daily usage times for Bolton's customers.

b. The managers of Bolton's marketing department would like to develop a 99% confidence interval estimate for the population mean daily customer usage time. Because the population standard deviation of daily customer usage time is unknown and the sample size is small, what assumption must the marketing managers make concerning the population of daily customer usage times?

c. Construct and interpret a 99% confidence interval for the mean daily usage time for Bolton's customers.

d. Assume that before the sample was taken, Bolton's marketing staff believed that mean daily usage for its customers was 267 minutes. Does their assumption concerning mean daily usage seem reasonable based on the confidence interval developed in part c?

**8-17.** In a study conducted by a national credit card company, corporate clients were surveyed to determine the extent to which hotel room rates quoted by central reservation systems differ from the rates negotiated by the companies. The study found that the mean overcharge by hotels was $11.35 per night. Suppose a follow-up study was done in which a random sample of 30 corporate hotel bookings was analyzed. Only those cases in which an error occurred were included in the study. The following data show the amounts by which the quoted rate differs from the negotiated rate. Positive values indicate an overcharge and negative values indicate an undercharge.

| $15.45 | $24.81 | $6.02 | $14.00 | $25.60 | $8.29 |
| -$17.34 | -$5.72 | $11.61 | $3.48 | 18.91 | $7.14 |
| $6.64 | $12.48 | $6.31 | -$4.85 | $5.72 | $12.72 |
| $5.23 | $4.57 | $15.84 | $2.09 | -$4.56 | $3.00 |
| $23.60 | $30.86 | $9.25 | $0.93 | $20.73 | $12.45 |

a. Compute a 95% confidence interval estimate for the mean error in hotel charges. Interpret the confidence interval estimate.

b. Based on the interval computed in part a, do these sample data tend to support the results of the study? Explain.

**8-18.** A regional U.S. commercial bank issues both Visa and MasterCard credit cards. As a part of its annual review of the profitability of each type of credit card, the bank randomly samples 36 customers to measure the average quarterly charges per card. It has completed its analysis of the Visa card accounts and is now focused on its MasterCard customers. A random sample of 36 MasterCard accounts shows the following spending per account (rounded to the nearest dollar):

| $2,869 | $3,770 | $2,854 | $2,750 | $2,574 | $2,972 |
| $2,549 | $3,267 | $3,013 | $2,707 | $2,794 | $1,189 |
| $2,230 | $2,178 | $3,032 | $3,485 | $2,679 | $2,010 |
| $1,994 | $2,768 | $3,853 | $2,064 | $3,244 | $2,738 |
| $2,807 | $2,395 | $3,405 | $3,006 | $3,368 | $2,691 |
| $1,996 | $3,008 | $2,730 | $2,518 | $2,710 | $3,719 |

a. Based on these randomly sampled accounts, what is the best point estimate of the true mean quarterly spending for MasterCard account holders?

b. If the bank is interested in developing a 95% confidence interval estimate of mean quarterly spending, what distribution will be used to determine the critical value?

c. Determine the standard error of the sampling distribution.

d. Construct the 95% confidence interval estimate for the population mean of quarterly MasterCard spending for the bank's customers.

e. If the bank desires to have a higher level of confidence in its interval estimate, what will happen to the margin of error?

**8-19.** Nielsen's Advertising and Audiences report indicated that during the 2013–2014 season, more people in the United States watched TV on Sundays during prime time (average of 125 million) than any other night of the week. Monday ranked second with 120 million U.S. viewers. Fridays and Saturdays had the fewest viewers (107 million and 108 million, respectively). (*Source:* Jack Loechner, "Primetime TV data with Nielsen," www.mediapost.com, May 30, 2014.) Suppose that in 2015, the company sampled 10 Sundays and found the following number of viewers (numbers are in millions) on each Sunday:

| 119 | 132 | 114 | 133 | 141 | 127 | 138 | 140 | 130 | 121 |
|-----|-----|-----|-----|-----|-----|-----|-----|-----|-----|

a. Calculate the sample standard deviation and mean number of minutes spent viewing television.

b. Calculate a 95% confidence interval for the average number of viewers on Sundays in 2015.

**8-20.** The concession managers for the Arkansas Travelers (a minor league baseball team located in Little Rock) are interested in estimating the average amount spent on food by fans attending the team's Friday night home games. Suppose a random sample of 36 receipts for food orders was taken from last year's receipts for Friday night home games with the following food expenditures recorded:

| $30.50 | $10.63 | $3.77 | $21.90 | $21.95 | $9.65 |
|--------|--------|-------|--------|--------|-------|
| $14.31 | $11.39 | $25.36 | $15.79 | $30.88 | $12.20 |
| $8.48 | $20.70 | $28.54 | $9.13 | $15.54 | $14.95 |
| $11.96 | $11.91 | $8.28 | $12.87 | $24.26 | $21.04 |
| $20.08 | $10.08 | $25.37 | $12.02 | $11.61 | $11.22 |
| $25.36 | $28.07 | $17.71 | $23.00 | $31.79 | $17.70 |

a. Based on the sampled receipts, what is the best point estimate for the mean food expenditures for Friday night home games?

b. Use the sample information to construct a 95% confidence interval estimate for the true mean expenditures for Friday night home games.

c. Before the sample was taken, the food concessions manager stated that mean food expenditures were about $19.00 per order. Does his statement seem consistent with the results obtained in part b?

## Computer Software Exercises

**8-21.** Suppose a study of 196 randomly sampled privately insured adults with incomes more than 200% of the current poverty level will be used to measure out-of-pocket medical expenses for prescription drugs for this income class. The sample data are in the file **Drug Expenses**.

a. Based on the sample data, construct a 95% confidence interval estimate for the mean annual out-of-pocket expenditures on prescription drugs for this income class. Interpret this interval.

b. The study's authors hope to use the information to make recommendations concerning insurance reimbursement guidelines and patient copayment recommendations. If the margin of error calculated in part a is considered to be too large for this purpose, what options are available to the study's authors?

**8-22.** One of the reasons for multiple car accidents on highways is thought to be poor visibility. One of this text's authors participated in a study of one rural section of highway in Idaho that had been the scene of several multiple-car accidents (*source:* Michael Kyte, Patrick Shannon, and Fred Kitchener, "Idaho Storm Warning System operational test - final report," Report No. IVH9316 (601), National Institute for Advanced Transportation Technology, University of Idaho, December 2000). Two visibility sensors were located near the site for the purposes of recording the number of miles of visibility each time the visibility reading is performed. The two visibility sensors are made by different companies, Scorpion and Vanguard. Develop a 95% confidence interval estimates of the mean visibility at this location as recorded by each of the visibility sensors. The random sample data are in a file called **Visibility**. Comment on whether there appears to be a difference between the two sensors in terms of average visibility readings.

**8-23.** The file **Danish Coffee** contains a random sample of 144 Danish coffee drinkers and measures the annual coffee consumption in kilograms for each sampled coffee drinker. A marketing research firm wants to use this information to develop an advertising campaign to increase Danish coffee consumption.

a. Based on the sample's results, what is the best point estimate of average annual coffee consumption for Danish coffee drinkers?

b. Develop and interpret a 90% confidence interval estimate for the mean annual coffee consumption of Danish coffee drinkers.

**8-24.** The manager at a new tire and repair shop in  Hartford, Connecticut, wants to establish guidelines for the time it should take to put new brakes on vehicles. In particular, he is interested in estimating the mean installation time for brakes for passenger cars and SUVs made by three different manufacturers. To help with this process, he set up an experiment in his shop in which five brake jobs were performed for each manufacturer and each class of vehicle. He recorded the number of minutes that it took to complete the jobs. These data are in a file called **Brake-test**.

a. Use Excel to compute the point estimate for the population mean installation time for each category.

b. Use Excel to compute the necessary sample statistics needed to construct 95% confidence interval estimates for the population mean installation times. What assumption is required?

c. Based on the results from part b, what conclusions might you reach about the three companies in terms of the time it takes to install their brakes for passenger cars and for SUVs? Discuss.

**8-25.** The TSA PreCheck program is intended to be a way to shorten airline security lines for "trusted travelers." Suppose 13,000 people living in a major southern city paid an annual $80 fee to participate in the program. They spent an average of 200 seconds in security lines. For comparison purposes, suppose a sample was obtained of the time it took the other passengers to pass through security a TSA check station. The file titled **PASSTIME** contains these data. Assume the distribution of time required to pass through security for those flyers in the PreCheck program is normally distributed.

a. Calculate the sample mean and the sample standard deviation for this sample of passenger times.

b. Assume that the distribution of the time required to pass through security is normally distributed. Use the sample data to construct a 95% confidence interval for the average time required to pass through security.

c. What is the margin of error for the confidence interval constructed in part b?

**8-26.** The per capita consumption of chicken rose from 36.9 pounds in 1965 to 90.1 pounds in 2015 according to the National Chicken Council (www.nationalchickencouncil.org). Suppose the organization wishes to estimate the mean chicken consumption for the year 2016. A file titled **Chickwt** contains the data.

a. Calculate the mean and standard deviation of the amount of chicken consumed by the individuals in the sample.

b. Calculate a 99% confidence interval for the 2016 per capita consumption of chicken in the United States.

---

**OUTCOME 3**    **8.2**    **Determining the Required Sample Size for Estimating a Population Mean**

We have discussed the basic trade-offs that are present in all statistical estimations: the desire is to have a high confidence level, a low margin of error, and a small sample size. The problem is that these three objectives conflict. For a given sample size, a high confidence level will tend to generate a large margin of error. For a given confidence level, a small sample size will result in an increased margin of error. Reducing the margin of error requires either reducing the confidence level or increasing the sample size, or both.

A common question from business decision makers who are planning an estimation application is: How large a sample size do I really need? To answer this question, we usually begin by asking a couple of questions of our own:

1. How much money do you have budgeted to do the sampling?
2. How much will it cost to select each item in the sample?

The answers to these questions provide the upper limit on the sample size that can be selected. For instance, if the decision maker indicates that she has a $2,000 budget for selecting the sample and the cost will be about $10 per unit to collect the sample, the sample size's upper limit is $2,000 ÷ $10 = 200 units.

Keeping in mind the estimation trade-offs discussed earlier, the issue should be fully discussed with the decision maker. For instance, is a sample of 200 sufficient to give the desired margin of error at a specified confidence level? Is 200 more than is needed, or not enough, to achieve the desired margin of error?

Therefore, before we can give a definite answer about what sample size is needed, the decision maker must specify her confidence level and a desired margin of error. Then the required sample size can be computed.

# Determining the Required Sample Size for Estimating $\mu$, $\sigma$ Known

**BUSINESS APPLICATION**    Calculating the Required Sample Size

**Highway Planning**    The California Transportation Administration is interested in knowing the mean number of miles that male adults in San Diego commute to and from work on a weekly basis. Because of the size of the population of interest, the only way to get this number is to select a random sample of adult males and develop a statistical estimation of the mean. Officials at the Transportation Administration have specified that the estimate must be based on a 95% confidence level. Further, the margin of error for the population mean must not exceed $\pm 30$ miles. Given these requirements, what size sample is needed?

To answer this question, if the population standard deviation is known, we start with Equation 8.3, the equation for calculating the margin of error:

$$e = z\frac{\sigma}{\sqrt{n}}$$

We next substitute into this equation the values we know. For example, the margin of error was specified to be

$$e = 30 \text{ miles}$$

> The Excel 2016 function to determine the critical $z$-value is
>
> **= NORM.S.INV(.025)**
>
> where the argument (.025) is $(1 - 0.95)/2$.
> *Note:* This function will return a negative $z$-value of $-1.96$ because 0.025 is in the left-hand side of the distribution.

The confidence level was specified to be 95%. The $z$-value for 95% is 1.96. (Refer to the standard normal table in Appendix D or use Excel.) This gives us

$$30 = 1.96\frac{\sigma}{\sqrt{n}}$$

We need to know the population standard deviation. The Transportation Administration officials might know this value from other studies that it has conducted in the past. Assume for this example that $\sigma$, the population standard deviation, is 200 miles. We can now substitute

$$\sigma = 200$$

into the equation for $e$, as follows:

$$30 = 1.96\frac{200}{\sqrt{n}}$$

We now have a single equation with one unknown, $n$, the sample size. Doing the algebra to solve for $n$, we get

$$n = \left(\frac{(1.96)(200)}{30}\right)^2 = 170.73 \approx 171 \text{ adult males}$$

Thus, to meet the requirements of the California Transportation Administration, a sample of $n = 171$ adult males should be selected. Equation 8.6 is used to determine the required sample size for estimating a single population mean when $\sigma$ is known.

### Sample Size Requirement for Estimating $\mu$, $\sigma$ Known

$$n = \left(\frac{z\sigma}{e}\right)^2 = \frac{z^2\sigma^2}{e^2} \tag{8.6}$$

where:

$z = $ Critical value for the specified confidence level
$e = $ Desired margin of error
$\sigma = $ Population standard deviation

If the officials conducting this study feel that the cost of sampling 171 adult males will be too high, it might allow for a higher margin of error or a lower confidence level. For example, if the confidence level is lowered to 90%, the z-value is lowered to 1.645, as found in the standard normal table.

We can now use Equation 8.6 to determine the revised sample-size requirement:

$$n = \frac{(1.645)^2(200)^2}{30^2} = 120.27 = 121$$

The Transportation Administration will need to sample only 121 (120.27 rounded up) adult males for a confidence level of 90% rather than 95%.

> The Excel 2016 function to determine the critical z-value is
>
> **= NORM.S.INV(.05)**
>
> where the argument (.05) is (1 − 0.90)/2.
> *Note:* Again, the z-value will be negative.

---

| EXAMPLE 8-4 | Determining the Required Sample Size, $\sigma$ Known |

**United Meat Producers**   The general manager for United Meat Producers is interested in estimating the mean pounds of hamburger that are purchased per month by households in the Honolulu, Hawaii, area. He would like his estimate to be within plus or minus 0.50 pound per month, and he would like the estimate to be at the 99% confidence level. Assume that the standard deviation for the hamburger purchase amount is 4.0 pounds. To determine the required sample size, he can use the following steps:

STEP 1   **Specify the desired margin of error.**
   The manager wishes to have his estimate be within $\pm 0.50$ pound, so the margin of error is

$$e = 0.50 \text{ pound}$$

STEP 2   **Determine the population standard deviation.**
   Based on other studies, the manager is willing to conclude that the population standard deviation is known. Thus,

$$\sigma = 4.0$$

> The Excel 2016 function to determine the critical z-value is
>
> **= NORM.S.INV(.005)**
>
> where the argument (.005) is (1 − 0.99)/2.
> *Note:* Again, the z-value will be negative.

STEP 3   **Determine the critical value for the desired level of confidence.**
   The critical value will be a z-value from the standard normal table for 99% confidence. This is

$$z = 2.575$$

STEP 4   **Compute the required sample size using Equation 8.6.**
   The required sample size is

$$n = \frac{z^2\sigma^2}{e^2} = \frac{(2.575)^2(4.0)^2}{0.50^2} = 424.36 \approx 425 \text{ households}$$

*Note:* The sample size is always rounded up to the next integer value.

**TRY EXERCISE 8-27** (pg. 299)

## Determining the Required Sample Size for Estimating $\mu$, $\sigma$ Unknown

Equation 8.6 assumes you know the population standard deviation. Although this may be the case in some situations, most likely we won't know the population standard deviation. To get around this problem, we can use three approaches. One is to use a value for $\sigma$ that is considered to be at least as large as the true $\sigma$. This will provide a conservatively large sample size.

The second option is to select a **pilot sample**, a sample from the population that is used explicitly to estimate $\sigma$.

The third option is to use the range of the population to estimate the population's standard deviation. Recall the Empirical Rule in Chapter 3 and the examination in Chapter 6 of the normal distribution. Both sources suggest that $\mu \pm 3\sigma$ contains virtually all of the data values of a normal distribution. If this were the case, then $\mu - 3\sigma$ would be approximately

**Pilot Sample**
A sample taken from the population of interest of a size smaller than the anticipated sample size that will be used to provide an estimate for the population standard deviation.

the smallest number and $\mu + 3\sigma$ would be approximately the largest number. Remember that Range $= R =$ Maximum value $-$ Minimum value. So, $R \approx (\mu + 3\sigma) - (\mu - 3\sigma) = 6\sigma$. We can, therefore, obtain an estimate of the standard deviation as

$$\sigma \approx \frac{R}{6}$$

We seldom know the standard deviation of the population. However, very often we have a good idea about the largest and smallest values of the population. Therefore, this third method can be used in many instances in which you do not wish to, or cannot, obtain a pilot sample or you are unable to offer a conjecture concerning a conservatively large value of the standard deviation.

---

| EXAMPLE 8-5 | **Determining the Required Sample Size, $\sigma$ Unknown** |

**Oceanside Petroleum**  Consider a situation in which the regional manager for Oceanside Petroleum in Oregon wishes to know the mean gallons of gasoline customers purchase each time they fill up their cars. Not only does he not know $\mu$, but he also does not know the population standard deviation. He wants a 90% confidence level and is willing to have a margin of error of 0.50 gallon in estimating the true mean gallons purchased. The required sample size can be determined using the following steps:

STEP 1 **Specify the desired margin of error.**
The manager wants the estimate to be within $\pm 0.50$ gallon of the true mean. Thus,

$$e = 0.50$$

> The Excel 2016 function to determine the sample standard deviation is
> **=STDEV.S(18.9,22.4,...,28.8)**

STEP 2 **Determine an estimate for the population standard deviation.**
The manager will select a pilot sample of $n = 20$ fill-ups and record the number of gallons for each. These values are

| 18.9 | 22.4 | 24.6 | 25.7 | 26.3 | 28.4 | 21.7 | 31.0 | 19.0 | 31.7 |
| 17.4 | 25.5 | 20.1 | 34.3 | 25.9 | 20.3 | 21.6 | 25.8 | 31.6 | 28.8 |

The estimate for the population standard deviation is the sample standard deviation for the pilot sample. This is computed using

$$s = \sqrt{\frac{\sum (x - \bar{x})^2}{n - 1}} = \sqrt{\frac{(18.9 - 25.05)^2 + (22.4 - 25.05)^2 + \cdots + (28.8 - 25.05)^2}{20 - 1}} = 4.85$$

We will use

$$\sigma \approx 4.85$$

STEP 3 **Determine the critical value for the desired level of confidence.**
The critical value will be a $z$-value from the standard normal table. The 90% confidence level gives

$$z = 1.645$$

STEP 4 **Calculate the required sample size using Equation 8.6.**
Using the pilot sample's standard deviation, we can calculate the required sample size as

$$n = \frac{z^2 \sigma^2}{e^2} = \frac{(1.645^2)(4.85^2)}{0.50^2} = 254.61 = 255$$

The required sample size is 255 fill-ups, but we can use the pilot sample as part of this total. Thus, the net required sample size in this case is $255 - 20 = 235$.

**TRY EXERCISE 8-34** (pg. 299)

## 8.2 EXERCISES

### Skill Development

**8-27.** What sample size is needed to estimate a population mean within $\pm 50$ of the true mean value using a confidence level of 95%, if the true population variance is known to be 122,500?

**8-28.** An advertising company wishes to estimate the mean household income for all single working professionals who own a foreign automobile. If the advertising company wants a 90% confidence interval estimate with a margin of error of $\pm \$2,500$, what sample size is needed if the population standard deviation is known to be $27,500$?

**8-29.** A manager wishes to estimate a population mean using a 95% confidence interval estimate that has a margin of error of $\pm 44.0$. If the population standard deviation is thought to be 680, what is the required sample size?

**8-30.** A sample size must be determined for estimating a population mean given that the confidence level is 90% and the desired margin of error is 0.30. The largest value in the population is thought to be 15 and the smallest value is thought to be 5. Calculate the sample size required to estimate the population using a generously large sample size. (*Hint:* Use the range/6 option.)

**8-31.** Suppose a study estimated the population mean for a variable of interest using a 99% confidence interval. If the width of the estimated confidence interval (the difference between the upper limit and the lower limits) is 600 and the sample size used in estimating the mean is 1,000, what is the population standard deviation?

**8-32.** Determine the smallest sample size required to estimate the population mean under the following specifications:
a. $e = 2.4$, confidence level = 80%, data between 50 and 150
b. $e = 2.4$, confidence level = 90%, data between 50 and 150
c. $e = 1.2$, confidence level = 90%, data between 50 and 150
d. $e = 1.2$, confidence level = 90%, data between 25 and 175

**8-33.** Calculate the smallest sample size required to estimate the population mean under the following specifications:
a. confidence level 95%, $\sigma = 16$, and $e = 4$
b. confidence level 90%, $\sigma = 23$, and $e = 0.5$
c. confidence level 99%, $\sigma = 0.5$, and $e = 1$
d. confidence level 98%, $\sigma = 1.5$, and $e = 0.2$
e. confidence level 95%, $\sigma = 6$, and $e = 2$

**8-34.** A decision maker is interested in estimating the mean of a population based on a random sample. She wants the confidence level to be 90% and the margin of error to be $\pm 0.30$. She does not know what the population standard deviation is, so she has selected the following pilot sample:

| | | | | |
|---|---|---|---|---|
| 8.80 | 4.89 | 10.98 | 15.11 | 14.79 |
| 16.93 | 1.27 | 9.06 | 14.38 | 5.65 |
| 7.24 | 3.24 | 2.61 | 6.09 | 6.91 |

Based on this pilot sample, how many more items must be sampled so that the decision maker can make the desired confidence interval estimate?

### Business Applications

**8-35.** A production process that fills 12-ounce cereal boxes is known to have a population standard deviation of 0.009 ounce. If a consumer protection agency would like to estimate the mean fill, in ounces, for 12-ounce cereal boxes with a confidence level of 92% and a margin of error of 0.001, what size sample must be used?

**8-36.** A public policy research group is conducting a study of health care plans and would like to estimate the average dollars contributed annually to health savings accounts by participating employees. A pilot study conducted a few months earlier indicated that the standard deviation of annual contributions to such plans was $1,225. The research group wants the study's findings to be within $100 of the true mean with a confidence level of 90%. What sample size is required?

**8-37.** With the high cost of fuel and intense competition, the major airline companies have had a very difficult time financially in recent years. Many carriers are now charging for checked bags. One carrier is considering charging a two-tiered rate based on the weight of checked bags. Before deciding at what weight to increase the charge, the airline wishes to estimate the mean weight per bag checked by passengers. It wants the estimate to be within $\pm 0.25$ pound of the true population mean. A pilot sample of bags checked gives the following results:

| | | | | | | | | | |
|---|---|---|---|---|---|---|---|---|---|
| 35 | 33 | 37 | 33 | 36 | 40 | 34 | 40 | 39 | 40 |
| 39 | 41 | 35 | 42 | 43 | 46 | 34 | 41 | 38 | 44 |

a. What size sample should the airline use if it wants to have 95% confidence?
b. Suppose the airline managers do not want to take as large a sample as the one you determined in part a. What general options do they have to lower the required sample size?

**8-38.** The Northwest Pacific Phone Company wishes to estimate the average number of minutes its customers spend on long-distance calls per month. The company wants the estimate made with 99% confidence and a margin of error of no more than 5 minutes.
a. Assume that the standard deviation for long-distance calls is 21 minutes per month. What should the sample size be?

b. Determine the required sample size if the confidence level were changed from 99% to 90%.

c. What would the required sample size be if the confidence level was 95% and the margin of error was 8 minutes?

**8-39.** In the past, minimum payments on outstanding credit card balances were set at about 2% of the outstanding balance. Recently, banks have started to increase the minimum payments. For example, Bank of America Corporation in Charlotte, North Carolina, went from a minimum of 2.2% of the balance to a flat amount of $10 plus all fees and interest due.

Suppose that a large California-based bank is in the process of considering what changes to make. It wishes to survey its customers to estimate the mean percent payment on their outstanding balance that customers would like to see. The bank wants to construct a 99% confidence interval estimate with a margin of error of ± 0.2%. A pilot sample of $n = 50$ customers showed a sample standard deviation of 1.4%.

How many more customers does the bank need to survey in order to construct the interval estimate?

**8-40.** The Longmont Computer Leasing Company wishes to estimate the mean number of pages that will be printed in a month on its leased printers. The plan is to select a random sample of printers and record the number on each printer's counter at the beginning of May. Then, at the end of May, the number on the counter will be recorded again and the difference will be the number of copies on that printer for the month. The company wants the estimate to be within ± 100 pages of the true mean with a 95% confidence level.

a. The standard deviation in pages printed is thought to be about 1,400 pages. How many printers should be sampled?

b. Suppose that the conjecture concerning the size of the standard deviation is off (plus or minus) by as much as 10%. What percent change in the required sample size would this produce?

**8-41.** The average monthly cable television bill is estimated to be $99.10 (*source:* Yoni Heisler, "The average cable bill has increased by nearly 40% in 5 years," http://bgr.com, Sep. 9, 2015). Suppose the study used to arrive at this estimate had a margin of error of $1.25 and a confidence level of 95%. Suppose the standard deviation of the monthly cost of cable television bills is $10.00.

a. Determine the sample size of the study.

b. Calculate the sample size required to decrease the margin of error by a dollar.

c. A typical sample size used in national surveys is 1,500 to 2,000. Determine a range for the margin of error corresponding to this range of sample sizes.

**8-42.** More than ever before, film production companies are gravitating toward Eastern Europe. Studios there have a reputation of skilled technical work at a relatively low cost. Bulgaria is one country with costs below those of Hollywood. As an example, weekly soundstage rental prices are quoted as $3.77 per

square meter in Bulgaria and $16.50 for its U.S. counterparts. To verify these figures, a sample of 50 rentals was taken, producing the average quoted for Bulgaria.

a. Determine the standard deviation of the weekly rental prices if the margin of error associated with the estimate was $2 using a 95% confidence level.

b. How much would the sample size have to be increased to decrease the margin of error by $1?

c. Calculate the change in the standard deviation that would have to be realized to produce the same decrease in the margin of error as realized in part b.

**8-43.** A regional chain of fast-food restaurants would like to estimate the mean time required to serve its drive-thru customers. Because speed of service is a critical factor in the success of its restaurants, the chain wants to have as accurate an estimate of the mean service time as possible.

a. If the population standard deviation of service times is known to be 30 seconds, how large a sample must be used to estimate the mean service time for drive-thru customers if a margin of error of no more than ± 10 seconds of the true mean with a 99% level of confidence is required?

b. Suppose the manager believes that a margin of error of ± 10 seconds is too high and has decided it should be cut in half to a margin of error of ± 5 seconds. He is of the opinion that by cutting the margin of error in half, the required sample size will double over what was required for a margin of error of ± 10. Is the manager correct concerning the sample-size requirement for the reduced margin of error? (Provide supporting calculations.)

## Computer Software Exercises

**8-44.** Bruce Leichtman is president of Leichtman Research Group, Inc. (LRG), which specializes in research and consulting on broadband, media, and entertainment industries. In a survey, the company determined the cost of HDMI Digital Antenna systems ranged from about $25 to $200 (*source:* "Everything you need to know about replacing cable with an HDTV antenna," www.fool.com, June 27, 2015). The file titled **HDcosts** contains a sample of the total costs including installation for HDMI Digital Antenna systems required to watch television in HD.

a. Produce a 95% confidence interval for the population mean cost of the purchases required to watch television in HD.

b. Calculate the margin of error for this experiment.

c. If you were to view the sample used in part a as a pilot sample, how many additional data values would be required to produce a margin of error of 5? Assume the population standard deviation is 50.353.

**8-45.** Suppose a random sample of 137 households in Detroit was selected to determine the average annual household spending on food at home for Detroit residents.

The sample results are contained in the file **Detroit Eats**.

a. Using the sample standard deviation as an estimate for the population standard deviation, calculate the sample size required to estimate the true population mean to within $\pm \$25$ with 95% confidence. How many additional samples must be taken?

b. Using the sample standard deviation as an estimate for the population standard deviation, calculate the sample size required to estimate the true population mean to within $\pm \$25$ with 90% confidence. How many additional samples must be taken?

**8-46.** According to the Office of Personnel Management (www.OPM.gov), in 2015 the average hourly wage rate for federal employees was slightly higher than $43. The file titled **Paychecks** contains a random sample of the hourly amounts paid to state and local government workers.

a. Generate the margin of error for estimating the average hourly amounts paid to state and local government workers with a 98% confidence level.

b. Determine the sample size required to produce a margin of error equal to 1.40 with a confidence level of 98%. Assume the population standard deviation equals 6.22.

c. Does it appear that state and local government workers have a lower average hourly compensation than federal workers? Support your opinion.

**8-47.** Premium Information Solutions, Inc., specializes in providing call center services for companies that wish to outsource their call center activities. Premium Information Solutions is currently negotiating with a new client who wants to be billed for the number of minutes that Premium Information Solutions is on the phone with customers. Before a contract is written, Premium Information Solutions plans to receive a random sample of calls and keep track of the minutes spent on the phone with the customer. From this, it plans to estimate the mean call time. It wishes to develop a 95% confidence interval estimate for the population mean call time and wants this estimate to be within $\pm 0.15$ minute. The question is, how many calls should Premium Information Solutions use in its sample?

Since the population standard deviation is unknown, a pilot sample was taken by having three call centers operated by Premium Information Solutions each take 50 calls for a total pilot sample of 150 calls. The minutes for each of these calls are listed in the file called **Premium Information**.

a. How many additional calls will be needed to compute the desired confidence interval estimate for the population mean?

b. In the event that the managers at Premium Information Solutions want a smaller sample size, what options do they have? Discuss in general terms.

## 8.3 Estimating a Population Proportion

The previous sections have illustrated the methods for developing confidence interval estimates when the population value of interest is the mean. However, you will encounter many situations in which the value of interest is the proportion of items in the population that possess a particular attribute. For example, you may wish to estimate the proportion of customers who are satisfied with the service provided by your company. The notation for the *population proportion* is $p$. The point estimate for $p$ is the *sample proportion*, $\bar{p}$, which is computed using Equation 8.7.

**Sample Proportion**

$$\bar{p} = \frac{x}{n}$$

(8.7)

where:

$\bar{p}$ = Sample proportion
$x$ = Number of items in the sample with the attribute of interest
$n$ = Sample size

In Chapter 7, we introduced the sampling distribution for proportions. We indicated then that when the sample size is sufficiently large $[np \geq 5$ and $n(1 - p) \geq 5]$, the sampling distribution can be approximated by a normal distribution centered at $p$ with a standard error for $\bar{p}$ computed using Equation 8.8.

**Standard Error for $\bar{p}$**

$$\sigma_{\bar{p}} = \sqrt{\frac{p(1-p)}{n}} \tag{8.8}$$

where:

$p$ = Population proportion
$n$ = Sample size

Notice that in Equation 8.8, the population proportion, $p$, is required. But if we already knew the value for $p$, we would not need to determine its estimate. If $p$ is unknown, we can estimate the value for the standard error by substituting $\bar{p}$ for $p$ as shown in Equation 8.9, providing that $n\bar{p} \geq 5$ and $n(1 - \bar{p}) \geq 5$.

**Estimate for the Standard Error of $\bar{p}$**

$$\sigma_{\bar{p}} \approx \sqrt{\frac{\bar{p}(1-\bar{p})}{n}} \tag{8.9}$$

where:

$\bar{p}$ = Sample proportion
$n$ = Sample size

Figure 8.7 illustrates the sampling distribution for $\bar{p}$.

**OUTCOME 4**

## Confidence Interval Estimate for a Population Proportion

The confidence interval estimate for a population proportion is formed using the same general format that we used to estimate a population mean. This was shown originally as Equation 8.1:

Point estimate $\pm$ (Critical value)(Standard error)

Equation 8.10 shows the specific format for *confidence intervals involving population proportions*.

**Confidence Interval Estimate for $p$**

$$\bar{p} \pm z\sqrt{\frac{\bar{p}(1-\bar{p})}{n}} \tag{8.10}$$

where:

$\bar{p}$ = Sample proportion
$n$ = Sample size
$z$ = Critical value from the standard normal distribution for the
      desired confidence level

**FIGURE 8.7 Sample Distribution for $\bar{p}$**

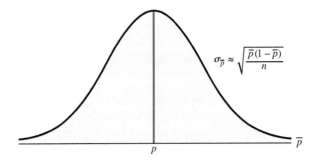

The critical value for a confidence interval estimate of a population proportion will always be a $z$-value from the standard normal distribution. Recall from Table 8.1 that the most commonly used critical values are

| Critical Value | Confidence Level |
|---|---|
| $z = 1.28$ | 80% |
| $z = 1.645$ | 90% |
| $z = 1.96$ | 95% |
| $z = 2.575$ | 99% |

For other confidence levels, you can find the critical $z$-value in the standard normal distribution table in Appendix D.

---

## HOW TO DO IT (Example 8-6)

**Developing a Confidence Interval Estimate for a Population Proportion**

**1.** Define the population and variable of interest for which to estimate the population proportion.

**2.** Determine the sample size and select a random sample. Note that the sample must be large enough so that $np \geq 5$ and $n(1 - p) \geq 5$.

**3.** Specify the level of confidence and obtain the critical value from the standard normal distribution table.

**4.** Calculate $\bar{p}$, the sample proportion.

**5.** Construct the interval estimate using Equation 8.10:

$$\bar{p} \pm z\sqrt{\frac{\bar{p}(1 - \bar{p})}{n}}$$

---

The Excel 2016 function to determine the critical $z$-value is

**= NORM.S.INV(.025)**

where the argument (.025) is $(1 - 0.95)/2$.
*Note:* Again, the $z$-value will be negative.

---

**EXAMPLE 8-6** | **Confidence Interval for a Population Proportion**

**Royal Haciendas Resort** Suppose the Royal Haciendas Resort in Playa del Carmen, Mexico, is thinking of starting a new promotion. When a customer checks out of the resort after spending five or more days, the customer would be given a voucher that is good for two free nights on the next stay of five or more nights at the resort.

The marketing manager might be interested in estimating the *proportion* of customers who will take advantage of the voucher. Assume that from a simple random sample of 100 customers, 62 returned the voucher. A confidence interval estimate for the true population proportion is found using the following steps:

STEP 1 **Define the population and the variable of interest.**
The population is all customers who received a voucher from Royal Haciendas, and the variable of interest is the number who use the voucher for two free nights.

STEP 2 **Determine the sample size.**
The manager uses a simple random sample of $n = 100$ customers who received vouchers. (Note that as long as $p \geq 0.05$ and $p \leq 0.95$, a sample size of 100 will meet the requirements that $np \geq 5$ and $n(1 - p) \geq 5$.)

STEP 3 **Specify the desired level of confidence and determine the critical value.**
Assuming that a 95% confidence level is desired, the critical value from the standard normal distribution table (Appendix D) is $z = 1.96$.

STEP 4 **Compute the sample proportion based on the sample data.**
Equation 8.7 is used to compute the sample proportion:

$$\bar{p} = \frac{x}{n} = \frac{62}{100} = 0.62$$

STEP 5 **Compute the confidence interval estimate using Equation 8.10.**
The 95% confidence interval estimate is

$$\bar{p} \pm z\sqrt{\frac{\bar{p}(1 - \bar{p})}{n}}$$

$$0.62 \pm 1.96\sqrt{\frac{0.62(1 - 0.62)}{100}}$$

$$0.62 \pm 0.095$$

$$0.525 \text{ ——— } 0.715$$

Using the sample of 100 customers and a 95% confidence interval, the manager estimates that the true percentage of customers who will take advantage of the two-free-night option will be between 52.5% and 71.5%.

**TRY EXERCISE 8-50** (pg. 306)

**OUTCOME 5**

# Determining the Required Sample Size for Estimating a Population Proportion

Changing the confidence level affects the interval width. Likewise, changing the sample size affects the interval width. An increase in sample size reduces the standard error and reduces the interval width. A decrease in the sample size has the opposite effect. For many applications, decision makers would like to determine a required sample size before doing the sampling. As was the case for estimating the population mean, the required sample size in a proportion application is based on the desired margin of error, the desired confidence level, and the variation in the population. The margin of error, $e$, is computed using Equation 8.11.

**Margin of Error for Estimating $p$**

$$e = z\sqrt{\frac{p(1 - p)}{n}} \qquad\qquad (8.11)$$

where:

$p$ = Value used to represent the population proportion
$z$ = Critical value from standard normal distribution for the desired confidence level
$n$ = Sample size

Equation 8.12 is used to determine the required sample size for a given confidence level and margin of error.

**Sample Size for Estimating $p$**

$$n = \frac{z^2 p(1 - p)}{e^2} \qquad\qquad (8.12)$$

where:

$p$ = Value used to represent the population proportion
$e$ = Desired margin of error
$z$ = Critical value from standard normal distribution for the desired confidence level

**BUSINESS APPLICATION**    Calculating the Required Sample Size

**Royal Haciendas Resort** (*continued*)    Referring to Example 8-6, recall that the marketing manager developed a confidence interval estimate for the proportion of customers who would redeem the voucher for two free nights at the resort. This interval was

$$0.62 \pm 1.96\sqrt{\frac{0.62(1 - 0.62)}{100}}$$

$$0.62 \pm 0.095$$

$$0.525 \text{———— } 0.715$$

The calculated margin of error in this situation is 0.095. Suppose the marketing manager wants the margin of error reduced to $e = \pm 0.04$ at a 95% confidence level. This will require an increase in the sample size. To apply Equation 8.12, the decision maker must specify the margin of error and the confidence level. However, the population proportion, $p$, is not something that can be controlled. In fact, if you already knew the value for $p$, you wouldn't need to estimate it and the sample-size issue wouldn't come up.

Two methods overcome this problem. First, you can select a *pilot sample* and compute the sample proportion, $\bar{p}$, and substitute $\bar{p}$ for $p$. Then, once the sample size is computed, you can use the pilot sample as part of the overall required sample.

Second, you can select a conservative value for $p$. The closer $p$ is to 0.50, the greater the variation because $p(1 - p)$ is greatest when $p = 0.50$. If the manager has reason to believe that the population proportion, $p$, will be about 0.60, he could use a value for $p$ a little closer to 0.50—say, 0.55. If he doesn't have a good idea of what $p$ is, he could conservatively use $p = 0.50$, which will give a sample size at least large enough to meet requirements.

Suppose the Royal Haciendas manager selects a pilot sample of $n = 100$ customers and provides them with the vouchers. Further, suppose $x = 62$ of these customers respond to the mailing. Then,

$$\bar{p} = \frac{62}{100} = 0.62$$

is substituted for $p$ in Equation 8.12. For a 95% confidence level, the $z$-value is

$$z = 1.96$$

and the margin of error is equal to

$$e = 0.04$$

We substitute these values into Equation 8.12 and solve for the required sample size:

$$n = \frac{1.96^2(0.62)(1 - 0.62)}{0.04^2} = 565.676 = 566$$

Because the pilot sample of 100 can be included, the Royal Haciendas Resort manager needs to give out an additional 466 vouchers to randomly selected customers. If this is more than the company can afford or wishes to include in the sample, the margin of error can be increased or the confidence level can be reduced.

---

**EXAMPLE 8-7**   **Sample Size Determination for Estimating $p$**

**Naumann Research**   The customer account manager for Naumann Research, a marketing research company located in Cincinnati, Ohio, is interested in estimating the proportion of a client's customers who like a new television commercial. She wishes to develop a 90% confidence interval estimate and would like to have the estimate be within $\pm 0.05$ of the true population proportion. To determine the required sample size, she can use the following steps:

STEP 1  **Define the population and variable of interest.**
The population is all potential customers in the market area. The variable of interest is the number of customers who like the new television commercial.

STEP 2  **Determine the level of confidence and find the critical $z$-value using the standard normal distribution table.**
The desired confidence level is 90%. The $z$-value for 90% confidence is 1.645.

STEP 3  **Determine the desired margin of error.**
The account manager wishes the margin of error to be 0.05.

STEP 4  **Arrive at a value to use for $p$.**
Two options can be used to obtain a value for $p$:

1. Use a pilot sample and compute $\bar{p}$, the sample proportion. Use $\bar{p}$ to approximate $p$.
2. Select a value for $p$ that is closer to 0.50 than you actually believe the value to be. If you have no idea what $p$ might be, use $p = 0.50$, which will give the largest possible sample size for the stated confidence level and margin of error.

In this case, suppose the account manager has no idea what $p$ is but wants to make sure that her sample is sufficiently large to meet her estimation requirements. Then she will use $p = 0.50$.

STEP 5  **Use Equation 8.12 to determine the sample size.**

$$n = \frac{z^2 p(1 - p)}{e^2} = \frac{1.645^2(0.50)(1 - 0.50)}{0.05^2} = 270.6025 = 271$$

The account manager should randomly survey 271 customers. (Always round up.)

TRY EXERCISE 8-49 (pg. 306)

---

## 8.3 EXERCISES

### Skill Development

**8-48.** Compute the 90% confidence interval estimate for the population proportion, $p$, based on a sample size of 100 when the sample proportion, $\bar{p}$, is equal to 0.40.

**8-49.** A pilot sample of 75 items was taken, and the number of items with the attribute of interest was found to be 15. How many more items must be sampled to construct a 99% confidence interval estimate for $p$ with a 0.025 margin of error?

**8-50.** A decision maker is interested in estimating a population proportion. A sample of size $n = 150$ yields 115 successes. Based on these sample data, construct a 90% confidence interval estimate for the true population proportion.

**8-51.** At issue is the proportion of people in a particular county who do not have health care insurance coverage. A simple random sample of 240 people was asked if they have insurance coverage, and 66 replied that they did not have coverage. Based on these sample data, determine the 95% confidence interval estimate for the population proportion.

**8-52.** A magazine company is planning to survey customers to determine the proportion who will renew their subscription for the coming year. The magazine wants to estimate the population proportion with 95% confidence and a margin of error equal to $\pm 0.02$. What sample size is required?

**8-53.** A random sample of size 150 taken from a population yields a proportion equal to 0.35.
 a. Determine if the sample size is large enough so that the sampling distribution can be approximated by a normal distribution.
 b. Construct a 90% confidence interval for the population proportion.
 c. Interpret the confidence interval calculated in part b.
 d. Produce the margin of error associated with this confidence interval.

**8-54.** A random sample of 200 items reveals that 144 of the items have the attribute of interest.

 a. What is the point estimate for the population proportion for all items having this attribute?
 b. Use the information from the random sample to develop a 95% confidence interval estimate for the population proportion, $p$, of all items having this attribute of interest.

**8-55.** A random sample of 40 television viewers was asked if they had watched the current week's *The Voice*. The following data represent their responses:

| no | no | no | yes | no | no | no | yes | no | yes |
|----|----|----|-----|----|----|----|-----|----|-----|
| no | no | no | yes | no | no | no | no | yes | no |
| yes | no | no | no | no | yes | no | no | no | no |
| no | no | no | no | no | no | no | no | no | no |

 a. Calculate the proportion of viewers in the sample who indicated they watched the current week's episode of *The Voice*.
 b. Compute a 95% confidence interval for the proportion of viewers in the sample who indicated they watched the current week's episode of *The Voice*.
 c. Calculate the smallest sample size that would produce a margin of error of 0.025 if the population proportion is well represented by the sample proportion in part a.

### Business Applications

**8-56.** Suppose that Drive for Life, a safety group sponsored by Volvo of North America, surveyed 1,100 drivers. Among those surveyed, 74% said that careless or aggressive driving was the biggest threat on the road. One-third of the respondents said that cell phone usage by other drivers was the driving behavior that annoyed them the most.

 Based on these data and assuming that the sample was a simple random sample, construct and interpret a 95% confidence interval estimate for the true proportion in the population of all drivers who are annoyed by cell phone users.

**8-57.** A survey of 499 women revealed that 38% wear flats to work.

a. Use this sample information to develop a 99% confidence interval for the population proportion of women who wear flats to work.

b. Suppose we also wish to estimate the proportion of women who wear athletic shoes to work with a margin of error of 0.01 with 95% confidence. Determine the sample size required.

**8-58.** The television landscape has changed as networks compete with streaming services such as Netflix and Hulu. Suppose that CBS plans to conduct interviews with television viewers in an attempt to estimate the proportion of viewers in the 18–49 age group who watch "most" of their television on network television as opposed to streaming services. CBS wishes to have 95% confidence and a margin of error in its estimate of $\pm 0.03$. A pilot sample of size 50 was selected, and the sample proportion was 0.61. To achieve these results with a simple random sample, how many additional viewers should be sampled?

**8-59.** Most major airlines allow passengers to carry one luggage item (of a certain maximum size) and one personal item onto the plane. However, suppose the more carry-on baggage passengers have, the longer it takes to unload and load passengers. One regional airline is considering changing its policy to allow only one carry-on item per passenger. Before doing so, it decided to collect some data. Specifically, a random sample of 1,000 passengers was selected. The passengers were observed, and the number of bags carried on the plane was noted. Out of the 1,000 passengers, 345 had two carry-on items.

a. Based on this sample, develop and interpret a 95% confidence interval estimate for the proportion of the traveling population that would have been affected if the one-bag limit had been in effect. Discuss your result.

b. The domestic version of Boeing's 747 has a capacity for 568 passengers. Determine an interval estimate of the number of passengers that you would expect to carry more than one piece of luggage on the plane. Assume the plane is at its passenger capacity.

c. Suppose the airline also noted whether each passenger was male or female. Out of the 1,000 passengers observed, 690 were males. Of this group, 280 had more than one bag. Using these data, obtain and interpret a 95% confidence interval estimate for the proportion of male passengers in the population who would have been affected by the one-bag limit. Discuss.

d. Suppose the airline decides to conduct a survey of its customers to determine their opinion of the proposed one-bag limit. Airline managers expect that only about 15% will approve of the proposal. Based on this assumption, what size sample should

the airline take if it wants to develop a 95% confidence interval estimate for the population proportion who will approve with a margin of error of $\pm 0.02$?

**8-60.** Suppose the Akron Chamber of Commerce has decided to conduct a survey to estimate the proportion of adults between the ages of 25 and 35 living in the metropolitan area who have a college degree in a high-technology field. The chamber hopes to use the survey's results to attract more high-technology firms to the region. The chamber wants the survey to estimate the population proportion within a margin of error of 0.03 percentage points with a level of confidence of 95%.

a. If the chamber has no information concerning the proportion of adults between the ages of 25 and 35 who have a college degree in a high-technology field before the survey is taken, how large a sample size must be used?

b. Suppose the chamber conducted a pilot study of 200 adults between the ages of 25 and 35 that indicated 28 with the desired attribute. How large a sample would be needed for the survey to estimate the population proportion within a margin of error of $\pm 0.03$ with a 95% level of confidence?

**8-61.** Many companies have started to feature female models who are not the traditional thin women who have graced magazine covers for the last 20 to 25 years. Suppose Calvin Klein conducted a survey of 1,000 women in which 91% of the respondents said they were satisfied with what they see in the mirror. The managers would like to use these data to develop a 90% confidence interval estimate for the true proportion of all women who are satisfied with their bodies. Develop and interpret the 90% confidence interval estimate.

**8-62.** A multinational corporation employing several thousand workers at its campus in a large city in the southwestern United States would like to estimate the proportion of its employees who commute to work by any means other than an automobile. The company hopes to use the information to develop a proposal to encourage more employees to forgo their automobiles as a part of their commute. A pilot study of 100 randomly sampled employees found that 14 commute to work by means other than an automobile.

a. How many more employees must the company randomly sample to be able to estimate the true population of employees who commute to work by means other than an automobile with a margin of error of $\pm 0.03$ and a level of confidence of 90%?

b. Suppose that after the full sample is taken, it was found that 50 employees commute to work by means other than an automobile. Construct a 90% confidence interval estimate for the true population of employees who commute to work using means other than an automobile. (*Hint:* Your sample size will be the total sample size required for part a.)

**8-63.** A survey of 725 teenagers between the ages of 13 and 18 conducted by JA Worldwide/Deloitte & Touche USA LLP found that 69% agree that people who practice good business ethics are more successful than those who do not (*source:* "2007 Executive Summary: JA Worldwide/Deloitte Teen Ethics Survey").

a. Calculate the 90% confidence interval estimate for the true population proportion, *p*, given the survey information.

b. What is the largest possible sample size needed to estimate the true population proportion, *p*, within a margin of error of $\pm 0.02$ with a confidence level of 95% if there was no prior knowledge concerning the proportion of respondents who would agree with the survey's question?

c. If the survey in part a had a margin of error of $\pm 0.04$ percentage points, determine the level of confidence that was used in estimating the population proportion if there was no prior knowledge concerning the percentage of teenagers who would respond as they did.

**8-64.** Suppose a survey of individuals between the ages of 26 and 82 indicated that 66% of seniors, 61% of Baby Boomers, and 58% of Generation X expect IRAs to be their primary source of income in retirement. The margin of error was given as $\pm 5$ percentage points.

a. Calculate a 95% confidence interval for the proportion of seniors who expect IRAs to be their primary source of income in retirement.

b. Although the sample size for the entire survey was listed, the sample size for each of the three generations was not given. Assuming the confidence level was 95%, determine the sample size for each of the three generations.

**8-65.** The website www.totalregistration.net tracks Advanced Placement (AP) exam registrations and performance by U.S. high school students. In 2015, the website reported that 573,000 AP exams were taken by over 290,000 students. Of the students who took the AP microeconomics exam, 36% scored a 4 or a 5. In an attempt to determine if the proportion of those passing the math and science AP exams is equal to the 36% success rate for microneconomics, a random sample of 300 students enrolled in AP math and science classes has been selected.

a. If 35 of the students in the sample passed at least one AP math or science exam, calculate the proportion of those students who passed at least one AP math or science exam. Does this statistic indicate that the proportion of students who passed at least one AP math or science exam is less than the 36% success rate for microeconomics? Support your assertion.

b. Calculate a 98% confidence interval for the proportion of those students who passed at least one AP math or science exam.

## Computer Software Exercises

**8-66.** Suppose that 34% of workers between the ages of 35 and 44 own a 401(k)-type retirement plan. Suppose the Atlanta Chamber of Commerce conducted a survey to determine the participation rate of 35- to 44-year-old working adults in the Atlanta metropolitan area who have 401(k)-type retirement plans. The Atlanta survey randomly sampled 144 working adults in Atlanta between the ages of 35 and 44. The results of the survey can be found in the file **Atlanta Retirement**.

a. Use the information in the file to compute a 95% confidence interval estimate for the true population proportion of working adults in Atlanta between the ages of 35 and 44 who have 401(k)-type retirement plans.

b. Based on the confidence interval calculated in part a, can the Atlanta Chamber of Commerce advertise that a greater percentage of working adults in Atlanta between the ages of 35 and 44 have 401(k) plans than in the nation as a whole for the same age group? Support your answer with the confidence level you calculated above.

**8-67.** According to the Retirement Confidence Survey, in 2015, 36% of individuals expected to begin retirement after the age of 65 (*source:* 2015 RCS Fact Sheet # 2, Expectations about retirement, www.ebri.org/). Suppose a recent random sample of 90 households in the greater Miami area was taken and respondents were asked whether they plan to retire after the age of 65. The results are in the file **RetireAfter65**.

a. Based on the random sample of Miami households, what is the best point estimate for the proportion of all Miami respondents who expect to retire after age 65?

b. Construct a 99% confidence interval estimate for the true population proportion of Miami respondents who expect to retire after age 65.

c. If the sponsors of the Miami study found that the margin of error was too high, what could they do to reduce it if they were not willing to change the level of confidence?

**8-68.** Neverslip, Inc., produces belts for industrial use. As part of its continuous process-improvement program, Neverslip has decided to monitor on-time shipments of its products. Suppose a random sample of 140 shipments was taken from shipping records for the last quarter and the shipment was recorded as being either "on time" or "late." The results of the sample are contained in the file **Neverslip**.

a. Using the randomly sampled data, calculate a 90% confidence interval estimate for the true population proportion, *p*, for on-time shipments for Neverslip.

b. What is the margin of error for the confidence interval calculated in part a?

c. One of Neverslip's commitments to its customers is that 95% of all shipments will arrive on time. Based on the confidence interval calculated in part a, is Neverslip meeting its on-time commitment?

**8-69.** Several states including Colorado, Idaho, Kansas, Mississippi, Oregon, Texas, Utah, and Wisconsin have passed legislation that allows holders of gun permits to carry weapons on college campuses in those states. The file **Guns** contains data from a survey conducted on a college campus in another state that is considering similar legislation. The data indicate whether the students interviewed approve of the legislation.

a. Calculate and interpret a 99% confidence interval estimate for the proportion of males and the proportion of females who approve of the legislation.

b. Based on the interval estimates in part a, can you infer which gender has a higher proportion of approval? Explain.

**8-70.** The Emerging Workforce Study conducted by Harris Interactive on behalf of Spherion, a leader in providing value-added staffing, recruiting, and workforce solutions, utilized a random sample of 225 senior human resources executives (*source:* www.spherion.com). The survey asked which methods the executives felt led them to find their best candidates. The file titled **Referrals** contains the responses indicating those who chose "referrals" as their best method.

a. Determine the margin of error that would accrue with a confidence level of 95%.

b. Calculate a 95% confidence interval for the proportion of executives who chose "referrals" as their best method.

c. Determine the sample size required to decrease the margin of error by 25%.

# 8 Overview

## Summary

**8.1** **Point and Confidence Interval Estimates for a Population Mean** (pg. 278–295)

**OUTCOME 1** Distinguish between a point estimate and a confidence interval estimate.

**OUTCOME 2** Construct and interpret a confidence interval estimate for a single population mean using both the standard normal and *t*-distributions. (See Figure 8.8.)

- A **point estimate** is a single statistic determined from a sample that is used to estimate the corresponding population parameter. Decision makers rely on point estimates when it is impossible to know the true population parameter.
- A **standard error** is a value that measures the spread of the sample means around the population mean. The standard error is reduced when the sample size is increased.
- A **confidence interval** is an interval developed from sample values such that if all possible intervals of a given width are constructed, a certain percentage of these intervals (known as the **confidence level**) will include the true population parameter.
- A confidence interval can be calculated using this general format: Point estimate $\pm$ (Critical value) (Standard error). The size of the interval and the confidence level chosen have an impact on the interval estimate.
- The **margin of error** is the amount that is subtracted from and added to the point estimate to determine the endpoints of the confidence interval. To lower the margin of error, we can lower the confidence level or increase the sample size.
- When estimating a population mean, we must distinguish between those cases where the standard deviation is known and those where it is not known. When it is known, the population mean is estimated using a critical value from the standard normal table for a specified confidence interval. When the population standard deviation is not known, the critical value is a *t-value* taken from a family of distributions called the **Student's *t*-distributions**.
- The specific *t*-distribution chosen depends on the number of independent data values available to estimate the population's standard deviation—a value known as the **degrees of freedom**.

**8.2** **Determining the Required Sample Size for Estimating a Population Mean** (pg. 295–301)

**OUTCOME 3** Determine the required sample size for estimating a single population mean.

- When conducting estimations, decision makers need to answer the question How large a sample size do I need?
- The answer depends on the resources available for sampling and the cost to select and measure each item sampled. The confidence level and desired margin of error must also be determined.
- A **pilot sample** is a sample taken from the population of interest smaller than the anticipated sample size and used to provide an estimate of the population standard deviation.

**FIGURE 8.8 Flow Diagram for Confidence Interval Estimation Alternatives**

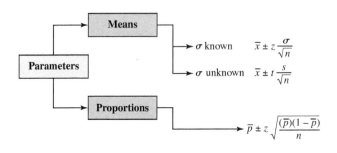

310

**8.3**    **Estimating a Population Proportion** (pg. 301–309)

**OUTCOME 4** Establish and interpret a confidence interval estimate for a single population proportion.

**OUTCOME 5** Determine the required sample size for estimating a single population proportion.

- The objective of sampling is frequently to estimate a population proportion. The confidence interval estimate for a **population proportion** is formed using the same general format as a population mean:

    Point estimate ± (Critical value) (Standard error)

- The critical value for a confidence interval estimate of a population proportion is always a *z*-value from the standard normal distribution.
- Changing the confidence level or sample size affects the interval width.
- As was the case for estimating the population mean, the required sample size for estimating a population proportion is based on the desired margin of error.

# Equations

**(8.1) Confidence Interval General Format** pg. 281

$$\text{Point estimate } \pm \text{ (Critical value)(Standard error)}$$

**(8.2) Confidence Interval Estimate for $\mu$, $\sigma$ Known** pg. 281

$$\bar{x} \pm z\frac{\sigma}{\sqrt{n}}$$

**(8.3) Margin of Error for Estimating $\mu$, $\sigma$ Known** pg. 284

$$e = z\frac{\sigma}{\sqrt{n}}$$

**(8.4) *t*-Value for $\bar{x}$** pg. 287

$$t = \frac{\bar{x} - \mu}{\dfrac{s}{\sqrt{n}}}$$

**(8.5) Confidence Interval Estimate for $\mu$, $\sigma$ Unknown** pg. 289

$$\bar{x} \pm t\frac{s}{\sqrt{n}}$$

**(8.6) Sample Size Requirement for Estimating $\mu$, $\sigma$ Known** pg. 296

$$n = \left(\frac{z\sigma}{e}\right)^2 = \frac{z^2\sigma^2}{e^2}$$

**(8.7) Sample Proportion** pg. 301

$$\bar{p} = \frac{x}{n}$$

**(8.8) Standard Error for $\bar{p}$** pg. 302

$$\sigma_{\bar{p}} = \sqrt{\frac{p(1-p)}{n}}$$

**(8.9) Estimate for the Standard Error of $\bar{p}$** pg. 302

$$\sigma_{\bar{p}} \approx \sqrt{\frac{\bar{p}(1-\bar{p})}{n}}$$

**(8.10) Confidence Interval Estimate for $p$** pg. 302

$$\bar{p} \pm z\sqrt{\frac{\bar{p}(1-\bar{p})}{n}}$$

**(8.11) Margin of Error for Estimating $p$** pg. 304

$$e = z\sqrt{\frac{p(1-p)}{n}}$$

**(8.12) Sample Size for Estimating $p$** pg. 304

$$n = \frac{z^2 p(1-p)}{e^2}$$

# Key Terms

# Chapter Exercises

## Conceptual Questions

**8-71.** Explain why the critical value for a given confidence level when the population variance is not known is always greater than the critical value for the same confidence level when the population variance is known.

**8-72.** When we need to estimate the population mean, and the population standard deviation is unknown, we are hit with a "double whammy" when it comes to the margin of error. Explain what the "double whammy" is and why it occurs. (*Hint:* Consider the sources of variation in the margin of error.)

**8-73.** An insurance company in Iowa recently conducted a survey of its automobile policy customers to estimate the mean miles these customers commute to work each day. The result based on a random sample of 300 policyholders indicated the population mean was between 3.5 and 6.7 miles. This interval estimate was constructed using 95% confidence.

   After receiving this result, one of the managers was overheard telling a colleague that 95% of all customers commute between 3.5 and 6.7 miles to work each day. How would you respond to this statement? Is it correct? Why or why not? Discuss.

**8-74.** Examine the equation for the margin of error when estimating a population mean

$$e = z\frac{\sigma}{\sqrt{n}}$$

Indicate the effect on the margin of error resulting from an increase in each of the following items:
a. confidence level
b. z-value
c. standard deviation
d. sample size
e. standard error

## Business Applications

**8-75.** A random sample of 64 bicycle-riding adults in Portland indicated that 24 always wore a helmet while riding. Use the sample information to develop a 95% confidence interval estimate for the true population proportion of bicycle-riding adults in Portland who wear a helmet while riding.

**8-76.** Suppose a random sample of 197 accounts from a corporate credit card database revealed a sample average balance of $2,325 with a standard deviation of $144. Use the sample information to develop a 95% confidence interval for the true population of all credit card balances for this corporate credit card.

**8-77.** A random sample of 441 shoppers revealed that 76% made at least one purchase at a discount store last month.
a. Based on this sample information, what is the 90% confidence interval for the population proportion of shoppers who made at least one discount store purchase last month?
b. The city of San Luis Obispo, California, has a population of 35,000 people. Referring to part a, determine a 90% confidence interval for the number of shoppers who made at least one discount store purchase last month.

**8-78.** PR Newswire reported that in 2015, the mean number of text messages sent by Americans was 32 (*source:* "No time to talk: Americans sending/receiving five times as many texts compared to phone calls each day, according to new report," www.prnewswire.com, Mar. 25, 2015). A major cell phone service provider conducted a study of a random sample of 400 customers in the age group between 18 and 35. The objective of the survey was to estimate the average number of text messages sent per day. The 400 people in the study sent 10,320 text messages on the day the study was conducted. The standard deviation in the number of texts sent per day by the people in this study was 9.5 messages.
a. What is the point estimate for the mean number of text messages sent per day by customers in this age group?
b. Calculate and interpret the 95% confidence interval estimate for the mean number of text messages sent per day by customers in this age group.
c. If the company would like to have a smaller margin of error, indicate what options are available.
d. Based on the result in part b, what conclusion might you reach about the PR Newswire statement about the mean being 32 messages per person?

**8-79.** Arco Manufacturing makes electronic video recorders. As part of its quality efforts, the company wishes to estimate the mean number of days a video recorder is used before it needs repair. A pilot sample of 40 video recorders indicates a sample standard deviation of 200 days. The company wishes its estimate to have a margin of error of no more than 50 days, and the confidence level must be 95%.
a. Given this information, how many additional video recorders should be sampled?
b. The pilot study was initiated because of the costs involved in sampling. Each sampled observation costs approximately $10 to obtain. Originally, it was thought that the population's standard deviation might be as large as 300. Determine the amount of money saved by obtaining the pilot sample. (*Hint:* Determine the total cost of obtaining the required samples for both methods.)

**8-80.** A random sample of 25 4-wheel-drive pickups of the same year and model revealed the following miles per gallon (mpg) values:

| | | | | |
|------|-------|------|------|------|
| 12.4 | 13.0  | 12.6 | 12.1 | 13.1 |
| 13.0 | 12.0  | 13.1 | 11.4 | 12.6 |
| 9.5  | 13.25 | 12.4 | 10.7 | 11.7 |
| 10.0 | 14.0  | 10.9 | 9.9  | 10.2 |
| 11.0 | 11.9  | 9.9  | 12.0 | 11.3 |

Assume that the population for mpg for this model year is normally distributed.
a. Use the sample results to develop a 95% confidence interval estimate for the population mean miles per gallon.
b. Determine the average number of gallons of gasoline the pickups described here would use to travel between Los Angeles and San Francisco—a distance of approximately 400 miles.
c. Another sample of the same size is to be obtained. If you know that the average miles per gallon in the second sample will be larger than the one obtained in part a, determine the probability that the sample

mean will be larger than the upper confidence limit of the confidence interval you calculated.

**8-81.** In an article titled "Airport screeners' strains, sprains highest among workers," Thomas Frank reported that the injury rate for airport screeners was 29%, far exceeding the 4.5% injury rate for the rest of the federal workforce. The 48,000 full- and part-time screeners were reported to have missed nearly a quarter-million days because of injuries in the recent fiscal year. (*Source:* Thomas Frank, *USA Today,* http://usatoday30.usatoday.com, Jan. 10, 2006.)

  a. Calculate the average number of days airport screeners missed.

  b. If one were to estimate the average number of days missed to within 1 hour in 2006 with a confidence level of 90%, determine the smallest sample size that would be required. Assume the standard deviation of the number of days missed is 1.5 days and that a work day consists of 8 hours.

  c. How close could the estimate get if a sample of size 100 was used?

## Computer Software Exercises

**8-82.** On its first day on the stock market, the Chinese Internet search engine, Baidu, increased its share price from $27.00 to $122.54, an increase of 454%. This was larger than any other Chinese initial public offering (IPO) and the second biggest for a foreign IPO. However, of the nine other biggest foreign IPOs with the largest first-day gains, all are trading below their IPO prices by an average of 88%. To determine the relationship between the IPOs with the largest first-day gains and the other IPOs, a sample might be taken to determine the average percentage decrease in the share prices of those IPOs not in the group of the nine IPOs with the largest first-day gains. A file titled **BigIPO$** contains such a sample. Note that an increase in share prices is represented as a negative decrease.

  a. Calculate a 95% confidence interval for the average percentage decrease after the first-day offering in the share of those IPOs not in the 9 IPOs with the largest first-day gains.

  b. Does it appear that there is a difference in the average percentage decrease in the share prices of the two groups? Support your assertions.

**8-83.** The Future-Vision Company is considering applying for a franchise to market satellite television dish systems in a Florida market area. As part of the company's research into this opportunity, staff in the new acquisitions department conducted a survey of 548 homes selected at random in the market area. They asked a number of questions on the survey. The data for some of the variables are in a file called **Future-Vision.** One key question asked whether the household was currently connected to cable TV.

  a. Using the sample information, what is the 95% confidence interval estimate for the true proportion of households in the market area that subscribe to cable television?

  b. Based on the sample data, develop a 95% confidence interval estimate for the mean income and interpret this estimate.

**8-84.** The quality manager for a major automobile manufacturer is interested in estimating the mean number of paint defects in cars produced by the company. She wishes to have her estimate be within $\pm 0.10$ of the true mean and wants 98% confidence in the estimate. The file called **CarPaint** contains data from a pilot sample that was conducted for the purpose of determining a value to use for the population standard deviation. How many additional cars need to be sampled to provide the estimate required by the quality manager?

**8-85.** In 2015, *Mortal Kombat X* was one of the most popular video games (*source:* Chris Morris, "'Mortal Kombat' secrets discovered after 24 years," http://fortune.com, Feb. 26, 2016). The file titled **KOMBATX** contains a sample of the retail prices paid for *Mortal Kombat X* in 2015.

  a. Calculate a 95% confidence interval for the population mean retail price paid for *Mortal Kombat X* in 2015.

  b. What sample size would be required to generate a margin of error of $0.25?

**8-86.** The Jordeen Bottling Company did an extensive sampling of its soft-drink inventory in which 5,000 cans were sampled. Employees weighed each can and used these weights to determine the fluid ounces in the cans. The data are in a file called **Jordeen.** Based on this sample data, should the company conclude that the mean volume is 12 ounces? Base your conclusion on a 95% confidence interval estimate and discuss.

**8-87.** National Products is a distributor of printer and copier paper for commercial use. The data file called **Sales** contains the annual, year-to-date sales values for each of the company's customers. Suppose the internal audit department has decided to audit a sample of these accounts. Specifically, they have decided to sample 36 accounts. However, before they actually conduct the in-depth audit (a process that involves tracking all transactions for each sampled account), they want to be sure that the sample they have selected is representative of the population.

  a. Compute the population mean.

  b. Use all the data in the population to develop a frequency distribution and histogram.

  c. Calculate the proportion of accounts for customers in each region of the country.

  d. Select a random sample of accounts. Develop a frequency distribution for these sample data. Compare this distribution to that of the population. (*Hint:* You might want to consider using relative frequencies for comparison purposes.)

  e. Construct a 95% confidence interval estimate for the population mean sales per customer. Discuss how you would use this interval estimate to help determine whether the sample is a good representation of the population. (*Hint:* You may want to use the finite population correction factor, since the sample is large relative to the size of the population.)

f. Use the information developed in parts a through e to draw a conclusion about whether the sample is a representative sample of the population. What other information would be desirable? Discuss.

**8-88.** A company's internal audit department randomly sampled 121 employee business travel reimbursement requests as part of an ongoing program to control and manage travel expenses. The dollar amounts for the sampled travel reimbursement requests are contained in the file titled **Travel**. As a part of its analysis, the internal audit department would like to estimate the population mean reimbursement request for the company.

a. Calculate the sample mean for the sampled travel reimbursement requests.

b. Calculate the sample standard deviation for the sampled travel reimbursement requests.

c. Compute the margin of error for a 95% confidence interval estimate for the true population mean for travel reimbursement requests for this company.

d. State the 95% confidence interval estimate for the true population mean for travel reimbursement requests for this company.

e. In its final report, the internal audit department stated that based on its analysis of the sampled travel reimbursement requests, "95% of all

submitted requests for travel reimbursement were for values between the limits found in part d." Is this a correct interpretation of the confidence interval that was calculated? If this statement is not correct, rewrite it in a form that is correct.

**8-89.** A marketing research firm conducted a survey to determine how adults in Philadelphia use their cell phones. The survey randomly sampled 250 adults. One question asked in the survey was "Over the past month have you used your cell phone to pay a credit card bill?" The sampled responses to this question are contained in the file **CellPhone**.

a. How many of the 250 randomly sampled adults in Philadelphia used their cell phone last month to pay a credit card bill?

b. What point estimate should the marketing research firm use to estimate the true population proportion of adults in Philadelphia who used their cell phones last month to pay a credit card bill?

c. Construct a 95% confidence interval estimate for the true proportion of adults in Philadelphia who used their cell phones last month to pay a credit card bill.

d. For a 90% confidence interval estimate of the true proportion of adults who used their cell phone last month to pay a credit card bill, what sample size is required if the marketing research firm wants a margin of error that is no higher than $\pm 0.02$?

## Case 8.1    Management Solutions, Inc.

The round trip to the "site" was just under 360 miles, which gave Fred Kitchener and Mike Kyte plenty of time to discuss the next steps in the project. The site is a rural stretch of highway in Idaho where two visibility sensors are located. The project is part of a contract Fred's company, Management Solutions, Inc., has with the state of Idaho and the Federal Highway Administration. Under the contract, among other things, Management Solutions is charged with evaluating the performance of a new technology for measuring visibility. The larger study involves determining whether visibility sensors can be effectively tied to electronic message signs that would warn motorists of upcoming visibility problems in rural areas.

Mike Kyte, a transportation engineer and professor at the University of Idaho, has been involved with the project as a consultant to Fred's company since the initial proposal. Mike is very knowledgeable about visibility sensors and traffic systems. Fred's expertise is in managing projects like this one, in which it is important to get people from multiple organizations to work together effectively.

As the pair headed back toward Boise from the site, Mike was more excited than Fred had seen him in a long time. Fred reasoned that the source of excitement was that they had finally been successful in getting solid data to compare the two visibility sensors in a period of low visibility. The previous day at the site had been very foggy. The Scorpion Sensor is a tested technology that Mike has worked with for some time in urban applications. However, it has never before been installed in such a remote location as this

stretch of Highway I-84, which connects Idaho and Utah. The other sensor produced by the Vanguard Company measures visibility in a totally new way using laser technology.

The data that had excited Mike so much were collected by the two sensors and fed back to a computer system at the port of entry near the test site. The measurements were collected every five minutes for the 24-hour day. As Fred took advantage of the 75-mph speed limit through southern Idaho, Mike kept glancing at the data on the printout he had made of the first few five-minute time periods. The Scorpion system had not only provided visibility readings, but it also had provided other weather-related data, such as temperature, wind speed, wind direction, and humidity.

Mike's eyes went directly to the two visibility columns. Ideally, the visibility readings for the two sensors would be the same at any five-minute period, but they weren't. After a few exclamations of surprise from Mike, Fred suggested that they come up with an outline for the report they would have to make from these data for the project team meeting next week. Both agreed that a full descriptive analysis of all the data, including graphs and numerical measures, was necessary. In addition, Fred wanted to use these early data to provide an estimate for the mean visibility provided by the two sensors. They agreed that estimates were needed for the day as a whole and also for only those periods when the Scorpion system showed visibility under 1.0 mile. They also felt that the analysis should look at the other weather factors, too, but they weren't sure just what was needed.

As the lights in the Boise Valley became visible, Mike agreed to work up a draft of the report, including a narrative based on the data in the file called **Visibility**. Fred said that he would set up the project team meeting agenda, and Mike could make the presentation. Both men agreed that the data were strictly a sample and that more low-visibility data would be collected when conditions occurred.

## Case 8.2    Federal Aviation Administration

In January 2003, the FAA ordered that passengers be either weighed or surveyed about their weight prior to boarding 10- to 19-seat passenger planes. The order was instituted in response to a crash that occurred on January 8, 2003, in Charlotte, North Carolina, in which all 21 passengers, including the pilot and copilot, of a 19-seat Beech 1900 turboprop died. One possible cause of the crash was that the plane may have been carrying too much weight. (*Source:* Matthew L. Wald, "Weight estimates on air passengers will be increased," www.nytimes.com, May 13, 2003.)

The airlines were asked to weigh adult passengers and carry-on bags randomly over a one-month period to estimate the mean weight per passenger (including luggage). A total of 426 people and their luggage were weighed, and the sample data are contained in a data file called **FAA**.

### Required Tasks:

1. Prepare a descriptive analysis of the data using charts, graphs, and numerical measures.
2. Construct and interpret a 95% confidence interval estimate for the mean weight for male passengers.
3. Construct and interpret a 95% confidence interval estimate for the mean weight for female passengers.
4. Construct and interpret a 95% confidence interval estimate for the mean weight for all passengers.
5. Indicate what sample size would be required if the margin of error in the estimate for the mean of all passengers is to be reduced by half.

## Case 8.3    Cell Phone Use

Helen Hutchins and Greg Haglund took the elevator together to the fourth-floor meeting room, where they were scheduled to meet the rest of the market research team at the Franklin Company. On the way up, Helen mentioned that she had terminated her contract for the land-line telephone in her apartment and was going to be using her cell phone exclusively to save money. "I rarely use my house phone anymore and about the only calls I get are from organizations wanting donations or doing surveys," she said. Greg said that he and his wife were thinking about doing the same thing.

As Helen and Greg walked toward the meeting room, Helen suddenly stopped. "If everyone did what I am doing, wouldn't that affect our marketing research telephone surveys?" she asked. "I mean, when we make calls the numbers are all to land-line phones. Won't we be missing out on some people we should be talking to when we do our surveys?" Helen continued. Greg indicated that it could be a problem if very many people were using cell phones exclusively like Helen. "Maybe we need to discuss this at the meeting today," Greg said.

When Helen and Greg brought up the subject to the market research team, several others indicated that they had been having similar concerns. It was decided that a special study was needed among the Franklin customer base to estimate the proportion of customers who were now using only a cell phone for telephone service. It was decided to randomly sample customers using personal interviews at their business outlets, but no one had any idea of how many customers they needed to interview. One team member mentioned that he had read an Associated Press article recently that said about 8% of all households have only a cell phone. Greg mentioned that any estimate they came up with should have a margin of error of $\pm 0.03$, and the others at the meeting agreed.

### Required Tasks:

1. Assuming that the group wishes to develop a 95% confidence interval estimate, determine the required sample size if the population proportion of cell phone–only users is 8%.
2. Supposing the group is unwilling to use the 8% baseline proportion and wants to have the sample size be conservatively large enough to provide a margin of error of no greater than $\pm 0.03$ with 95% confidence, determine the sample size that will be needed.

# 9 Introduction to Hypothesis Testing

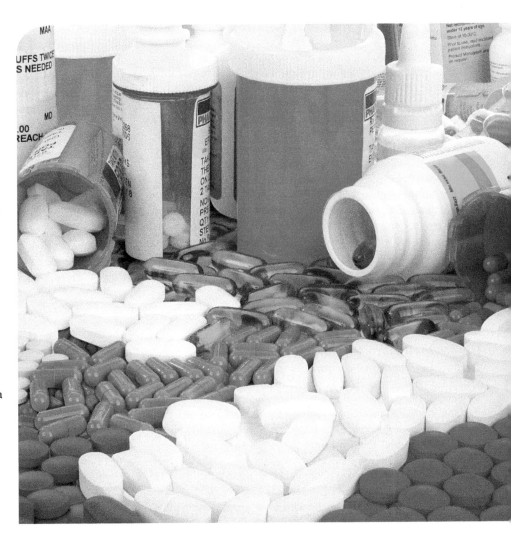

## WHY YOU NEED TO KNOW

Estimating a population parameter based on a sample statistic is one area of business statistics called *statistical inference*. The basic tools for estimation were introduced in Chapter 8. Another important application of statistical inference is *hypothesis testing*. In hypothesis testing, we make a hypothesis (or statement) concerning a population parameter. We then use sample data to either deny or confirm the validity of the proposed hypothesis.

Hypothesis testing is performed regularly in many industries. Companies in the pharmaceutical industry must perform many hypothesis tests on new drug products before they are deemed to be safe and effective by the U.S. Food and Drug Administration (FDA). In

---

### Quick Prep

Review the concepts associated with the Central Limit Theorem in Section 7.2.

Examine Section 7.3 on the sampling distribution for proportions.

Familiarize yourself with the Student's *t*-distributions in Section 8.1 and normal probability distributions in Section 6.1.

Review the standard normal distribution and the Student's *t*-distribution tables, making sure you know how to find critical values in both tables.

these instances, the drug is hypothesized to be both unsafe and ineffective. The FDA does not wish to certify that the drug is safe and effective unless sufficient evidence is obtained that this is the case. Then, if the sample results from the studies performed provide "significant" evidence that the drug is safe and effective, the FDA will allow the company to market the drug.

As another example, suppose an orange juice plant in Orlando, Florida, produces approximately 120,000 bottles of orange juice daily. Each bottle is supposed to contain 32 fluid ounces. However, like all processes, the automated filling machine is subject to variation, and each bottle will contain either slightly more or less than the 32-ounce target. The important thing is that the mean fill is 32 fluid ounces. The manager might state the *hypothesis* that the mean for all bottles filled is equal to 32 ounces. Every two hours, the plant quality manager selects a random sample of bottles and computes the sample mean. If the sample mean is "significantly" greater or less than the desired 32 ounces, then the sample data will have provided sufficient evidence that the average for the population of bottles is not 32 ounces and the manager's *hypothesis* would be rejected. The machine would then be stopped until repairs or adjustments had been made. However, if the sample mean is "close" to 32 ounces, the data will not refute the hypothesis and the machine would be allowed to continue filling bottles.

Hypothesis testing is the basis of the legal system in which judges and juries hear evidence in court cases. In a criminal case, the hypothesis in the American legal system is that the defendant is innocent. Based on the totality of the evidence presented in the trial, if the jury concludes that "beyond a reasonable doubt" the defendant committed the crime, the hypothesis of innocence will be rejected and the defendant will be found guilty. If the evidence is not strong enough, the defendant will be judged not guilty.

Hypothesis testing is a major part of business statistics. Chapter 9 introduces the fundamentals involved in conducting hypothesis tests. Many of the remaining chapters in this text will introduce additional hypothesis-testing techniques, so you need to gain a solid understanding of concepts presented in this chapter.

## 9.1 Hypothesis Tests for Means

By now you know that the information in a sample is subject to sampling error. The sample mean almost certainly does not equal the population mean. Therefore, in situations in which you need to test a claim about a population mean by using the sample mean, you can't simply compare the sample mean to the claim and reject the claim if $\bar{x}$ and the claimed value of $\mu$ are different. Instead, you need a testing procedure that incorporates the potential for sampling error.

*Statistical hypothesis testing* provides managers with a structured analytical method for making decisions of this type. It lets them make decisions in such a way that the probability of decision errors can be controlled, or at least measured. Even though statistical hypothesis testing does not eliminate the uncertainty in the managerial environment, the techniques involved often allow managers to identify and control the level of uncertainty.

The techniques presented in this chapter assume that the data are selected using an appropriate statistical sampling process and that the data are interval or ratio level. In short, we assume we are working with good data.

### Formulating the Hypotheses

**Null Hypothesis**
The statement about the population parameter that will be assumed to be true during the conduct of the hypothesis test. The null hypothesis will be rejected only if the sample data provide substantial contradictory evidence.

**Alternative Hypothesis**
The hypothesis that includes all population values not included in the null hypothesis. The alternative hypothesis will be selected only if there is strong enough sample evidence to support it. The alternative hypothesis is deemed to be true if the null hypothesis is rejected.

**Null and Alternative Hypotheses** In hypothesis testing, two hypotheses are formulated. One is the **null hypothesis**. The null hypothesis is represented by $H_0$ and contains an equality sign, such as $=$, $\leq$, or $\geq$. The second hypothesis is the **alternative hypothesis** (represented by $H_A$). Based on the sample data, either we reject $H_0$ or we do not reject $H_0$.

Correctly specifying the null and alternative hypotheses is important. If this is done incorrectly, the results obtained from the hypothesis test may be misleading. Unfortunately, how you should formulate the null and alternative hypotheses is not always obvious. As you gain experience with hypothesis-testing applications, the process becomes easier. To help you get started, we have developed some general guidelines you should find helpful.

**Testing the Status Quo** In many cases, you will be interested in whether a situation has changed. We refer to this as testing the status quo, and this is a common application of hypothesis testing. For example, Kellogg's makes many food products, including a variety of breakfast cereals. At the company's Battle Creek, Michigan, plant, Frosted Mini-Wheats are produced and packaged for distribution around the world. If the packaging process is working properly, the mean fill per box is 16 ounces. Suppose every hour, quality analysts at the plant select a random sample of filled boxes and measure their contents. They do not wish to unnecessarily stop the packaging process, since doing so can be quite costly. Thus, the packaging process will not be stopped unless there is sufficient evidence that the average fill is different from 16 ounces—that is, $H_A: \mu \neq 16$. The analysts use the sample data to test the following null and alternative hypotheses:

$$H_0: \mu = 16 \text{ ounces (status quo)}$$
$$H_A: \mu \neq 16 \text{ ounces}$$

The null hypothesis is reasonable because the line supervisor would assume the process is operating correctly before starting production. As long as the sample mean is "reasonably" close to 16 ounces, the analysts will assume the filling process is working properly. Only when the sample mean is seen to be too large or too small will the analysts reject the null hypothesis (the status quo) and take action to identify and fix the problem.

As another example, the Transportation Security Administration (TSA), which is responsible for screening passengers at U.S. airports, publishes on its website the average waiting times for customers to pass through security. For example, suppose on Mondays between 9:00 A.M. and 10:00 A.M., the average waiting time at Atlanta's Hartsfield International Airport is supposed to be 15 minutes or less at the main checkpoint. Periodically, TSA staff will select a random sample of passengers during this time slot and measure their actual waiting times to determine if the average time is longer than the guidelines require. The alternative hypothesis is, therefore, stated as $H_A: \mu > 15$ minutes. The sample data will be used to test the following null and alternative hypotheses:

$$H_0: \mu \leq 15 \text{ minutes (status quo)}$$
$$H_A: \mu > 15 \text{ minutes}$$

Only if the sample mean waiting time is "substantially" greater than 15 minutes will TSA employees reject the null hypothesis and conclude there is a problem with staffing levels. Otherwise, they will assume that the 15-minute standard (the status quo) is being met, and no action will be taken.

**Testing a Research Hypothesis** Many business and scientific decision-making situations involve research applications. For example, companies such as Intel, Procter & Gamble, Apple, Pfizer, and 3M continually introduce new and hopefully improved products. However, before introducing a new product, the companies want to determine whether the new product is superior to the original. In the case of drug companies like Pfizer, the government requires them to show their products are both safe and effective. Because statistical evidence is needed to indicate that the new product is safe and effective, the default position (or null hypothesis) is that it is unsafe and ineffective. The burden of proof is placed on the new product, and the alternative hypothesis is formulated as the **research hypothesis**.

**Research Hypothesis**
The hypothesis the decision maker attempts to demonstrate to be true. Because this is the hypothesis deemed to be the most important to the decision maker, it will be declared true only if the sample data strongly indicate that it is true.

For example, suppose Goodyear Tire and Rubber Company has a new tread design that its engineers claim will outlast its competitor's leading tire. New technology is able to produce tires whose longevity is better than the competitors' tires but are less expensive. Thus, if Goodyear can be sure that the new tread design will last longer than the competition's, it will realize a profit that will justify the introduction of the tire with the new tread design. The competitor's tire has been demonstrated to provide an average of 60,000 miles of use. Therefore, the research hypothesis for Goodyear is that its tire will last longer than its competitor's, meaning that the tire will last an average of *more than* 60,000 miles. The research hypothesis becomes the alternative hypothesis:

$$H_0: \mu \leq 60,000$$
$$H_A: \mu > 60,000 \text{ (research hypothesis)}$$

The burden of proof is on Goodyear. Only if the sample data show a sample mean that is "substantially" greater than 60,000 miles will the null hypothesis be rejected and Goodyear's position be affirmed.

In another example, suppose Nunhems Brothers Seed Company has developed a new variety of bean seed. Nunhems Brothers will introduce this seed variety on the market only if the seed provides yields superior to the current seed variety. Experience shows the current seed provides a mean yield of 60 bushels per acre. To test the new variety of beans, Nunhems Brothers researchers will set up the following null and alternative hypotheses:

$$H_0: \mu \leq 60 \text{ bushels}$$

$$H_A: \mu > 60 \text{ bushels (research hypothesis)}$$

The alternative hypothesis is the research hypothesis. If the null hypothesis is rejected, then Nunhems Brothers will have statistical evidence to show that the new variety of beans is superior to the existing product.

**Testing a Claim about the Population**  Analyzing claims using hypothesis tests can be complicated. Sometimes you will want to give the benefit of the doubt to the claim, but in other instances you will be skeptical about the claim and will want to place the burden of proof on the claim. For consistency purposes, the rule adopted in this text is that if the claim contains the equality, the claim becomes the null hypothesis. If the claim does not contain the equality, the claim is the alternative hypothesis.

A radio commercial stated the average waiting time at a medical clinic is less than 15 minutes. A claim like this can be tested using hypothesis testing. The null and alternative hypotheses should be formulated such that one contains the claim and the other reflects the opposite position. Since, in this example, the claim that the average wait time *is less* than 15 minutes does not contain the equality ($\mu < 15$), the claim should be the alternative hypothesis. The appropriate null and alternative hypotheses are, then, as follows:

$$H_0: \mu \geq 15$$

$$H_A: \mu < 15 \text{ (claim)}$$

In cases like this where the claim corresponds to the alternative hypothesis, the *burden of proof* is on the claim. If the sample mean is "substantially" less than 15 minutes, the null hypothesis would be rejected and the alternative hypothesis (and the claim) would be accepted. Otherwise, the null hypothesis would not be rejected and the claim could not be accepted.

---

**OUTCOME** **1**

**EXAMPLE 9-1**    Formulating the Null and Alternative Hypotheses

**Student Work Hours**    In today's economy, university students often work many hours to pay for the high costs of a college education. Suppose a university in the Midwest is considering changing its class schedule to accommodate students working long hours. The registrar has stated a change is needed because the mean number of hours worked by undergraduate students at the university is more than 20 per week. The first step in testing this claim is to establish the appropriate null and alternative hypotheses:

**HOW TO DO IT** (Example 9-1)

**Formulating the Null and Alternative Hypotheses**

1. Identify the population parameter of interest (e.g., $\mu$, $p$, or $\sigma$).

2. Identify the hypothesis of interest to the researcher or analyst. It could involve testing a status quo, a research hypothesis, or a claim.

3. The null hypothesis will contain the equal sign; the alternative hypothesis will not. The range of all possible values for the parameter must be included in the null and alternative hypotheses. Therefore, if $H_0: \mu \leq 15$, the alternative hypothesis must be $H_A: \mu > 15$.

STEP 1    **Determine the population parameter of interest.**
The population parameter of interest is the mean hours worked, $\mu$. The null and alternative hypotheses must be stated in terms of the population parameter.

STEP 2    **Identify the hypothesis of interest.**
The registrar has made a claim that the mean hours worked "is more than 20" per week. Because changing the class scheduling system would be expensive and time consuming, the claim should not be declared true unless the sample data strongly indicate that it is true. Thus, the burden of proof is placed on the registrar to justify her claim that the mean is greater than 20 hours.

STEP 3    **Formulate the null and alternative hypotheses.**
Keep in mind that the equality goes in the null hypothesis.

$$H_0: \mu \leq 20 \text{ hours}$$

$$H_A: \mu > 20 \text{ hours (claim)}$$

**TRY EXERCISE 9-13a** (pg. 336)

Example 9-2 illustrates another example of how the null and alternative hypotheses are formulated.

---

**EXAMPLE 9-2** **Formulating the Null and Alternative Hypotheses**

**Nabisco Foods**   One of the leading products made by Nabisco is the snack cracker called Wheat Thins. Nabisco uses an automatic filling machine to fill the Wheat Thins boxes with the desired weight. For instance, when the company is running the product for Costco on the fill line, the machine is set to fill the oversized boxes with 20 ounces. If the machine is working properly, the mean fill will be equal to 20 ounces. Each hour, a sample of boxes is collected and weighed, and the technicians determine whether the machine is still operating correctly or whether it needs adjustment. The following steps can be used to establish the null and alternative hypotheses to be tested:

STEP 1 **Determine the population parameter of interest.**
The population parameter of interest is the mean weight per box, $\mu$.

STEP 2 **Identify the hypothesis of interest.**
The status quo is that the machine is filling the boxes with the proper amount, which is $\mu = 20$ ounces. We will believe this to be true unless we find evidence to suggest otherwise. If such evidence exists, then the filling process needs to be adjusted.

STEP 3 **Formulate the null and alternative hypotheses.**
The null and alternative hypotheses are

$$H_0: \mu = 20 \text{ ounces (status quo)}$$

$$H_A: \mu \neq 20 \text{ ounces}$$

TRY EXERCISE 9-14a (pg. 336)

---

OUTCOME **2**

**Type I Error**
*Rejecting* the null hypothesis when it is, in fact, true.

**Type II Error**
*Failing to reject* the null hypothesis when it is, in fact, false.

**Types of Statistical Errors**   Because of the potential for extreme sampling error, two possible errors can occur when a hypothesis is tested: **Type I** and **Type II errors**. These errors show the relationship between what actually exists (a state of nature) and the decision made based on the sample information.

Figure 9.1 shows the possible actions and states of nature associated with any hypothesis-testing application. As you can see, there are three possible outcomes: no error (correct decision), Type I error, and Type II error. *Only one of these results can occur when you test a hypothesis.* From Figure 9.1, you can see that if the null hypothesis is true and an error is made, it must be a Type I error. On the other hand, if the null hypothesis is false and an error is made, it must be a Type II error.

Many statisticians argue that you should never use the phrase "accept the null hypothesis." Instead, you should use "*do not reject* the null hypothesis." Thus, the only two hypothesis-testing decisions would be *reject* $H_0$ or *do not reject* $H_0$. This is why in a jury verdict to acquit a defendant, the verdict is "not guilty" rather than innocent. Just because the evidence is insufficient to convict does not necessarily mean that the defendant is innocent. The same is true with hypothesis testing. Just because the sample data do not lead to rejecting the null hypothesis, we cannot be sure that the null hypothesis is true.

This thinking is appropriate when hypothesis testing is employed in situations in which some future action is not dependent on the results of the hypothesis test. However, in most business applications, the purpose of the hypothesis test is to direct the decision maker to take one action or another based on the test results. So, in this text, when we apply hypothesis

**FIGURE 9.1 The Relationship between Decisions and States of Nature**

| | | State of Nature | |
|---|---|---|---|
| | | Null Hypothesis True | Null Hypothesis False |
| Decision | Conclude Null True (Don't Reject $H_0$) | Correct Decision | Type II Error |
| | Conclude Null False (Reject $H_0$) | Type I Error | Correct Decision |

testing to decision-making situations, *not rejecting* the null hypothesis is essentially the same as *accepting it*. The same action will be taken whether we conclude that the null hypothesis is not rejected or that it is accepted.[1]

**BUSINESS APPLICATION**    Type I and Type II Statistical Errors

**Price & Associates Construction**   Price & Associates is a residential home developer in the Phoenix, Arizona, area. They build single-family homes in the $300,000 to $500,000 price range. Because of the volume of homes they build, they have refined their processes to be very efficient. For example, they have a company standard that a home should not require more than 25 days for framing and roofing. The managing partner at Price & Associates wishes to test whether the mean framing and roofing times have changed since the mean time of 25 days was established a few years ago. If he treats the average framing and roofing time of 25 days or less as the status quo, the null and alternative hypotheses to be tested are

$$H_0: \mu \leq 25 \text{ days (status quo)}$$

$$H_A: \mu > 25 \text{ days}$$

The managing partner will select a random sample of homes built in 2016. In this application, a Type I error would occur if the sample data lead the manager to conclude that the mean framing and roofing time exceeds 25 days ($H_0$ is rejected) when in fact $\mu \leq 25$ days. If a Type I error occurred, the manager would needlessly spend time and resources trying to speed up a process that already meets the original time frame.

Alternatively, a Type II error would occur if the sample evidence leads the manager to incorrectly conclude that $\mu \leq 25$ days ($H_0$ is not rejected) when the mean framing and roofing time exceeds 25 days. Now the manager would take no action to improve framing and roofing times at Price & Associates when changes are needed to improve the building time.

**OUTCOME** ▌3    # Significance Level and Critical Value

The objective of a hypothesis test is to use sample information to decide whether to reject the null hypothesis about a population parameter. How do decision makers determine whether the sample information supports or refutes the null hypothesis? The answer to this question is the key to understanding statistical hypothesis testing.

In hypothesis tests for a single population mean, the sample mean, $\bar{x}$, is used to test the hypotheses under consideration. Depending on how the null and alternative hypotheses are formulated, certain values of $\bar{x}$ will tend to support the null hypothesis, whereas other values will appear to support the alternative hypothesis. In the Price & Associates example, the null and alternative hypotheses were formulated as

$$H_0: \mu \leq 25 \text{ days}$$

$$H_A: \mu > 25 \text{ days}$$

Values of $\bar{x}$ less than or equal to 25 days would tend to support the null hypothesis. By contrast, values of $\bar{x}$ greater than 25 days would tend to refute the null hypothesis. The larger the value of $\bar{x}$, the greater the evidence that the null hypothesis should be rejected. However, because we expect some sampling error, do we want to reject $H_0$ for any value of $\bar{x}$ that is greater than 25 days? Probably not. But should we reject $H_0$ if $\bar{x} = 26$ days, or $\bar{x} = 30$ days, or $\bar{x} = 35$ days? At what point do we stop attributing the result to sampling error?

To perform the hypothesis test, we need to select a *cutoff* point that is the demarcation between rejecting and not rejecting the null hypothesis. Our *decision rule* for the Price & Associates application is then:

---

[1]Whichever language you use, you should make an effort to understand both arguments and make an informed choice. If your instructor requests that you reference the action in a particular way, it behooves you to follow the instructions. Having gone through this process ourselves, we prefer to state the choice as "don't reject the null hypothesis." We will use this terminology throughout this text.

**FIGURE 9.2 Sampling Distribution of $\bar{x}$ for Price & Associates**

If $\bar{x} >$ Cutoff, reject $H_0$.

If $\bar{x} \leq$ Cutoff, do not reject $H_0$.

If $\bar{x}$ is greater than the cutoff, we will reject $H_0$ and conclude that the average framing and roofing time *does* exceed 25 days. If $\bar{x}$ is less than or equal to the cutoff, we will not reject $H_0$; in this case our test does not give sufficient evidence that the framing and roofing time exceeds 25 days.

Recall from the Central Limit Theorem (see Chapter 7) that, for large samples, the distribution of the possible sample means is approximately normal, with a center at the population mean, $\mu$. The null hypothesis in our example is $\mu \leq 25$ days. Figure 9.2 shows the sampling distribution for $\bar{x}$ assuming that $\mu = 25$. The shaded region on the right is called the *rejection region*. The area of the rejection region gives the probability of getting an $\bar{x}$ larger than the cutoff when $\mu$ is really 25, so it is the probability of making a Type I statistical error. This probability is called the **significance level** of the test and is given the symbol $\alpha$ (alpha).

The decision maker carrying out the test specifies the significance level, $\alpha$. The value of $\alpha$ is chosen based on the costs involved in committing a Type I error. If making a Type I error is costly, we want the probability of a Type I error to be small. If a Type I error is less costly, then we can allow a higher probability of a Type I error.

However, in choosing $\alpha$, we must also take into account the probability of making a Type II error, which is given the symbol $\beta$ (beta). The two error probabilities, $\alpha$ and $\beta$, are inversely related. That is, if we reduce $\alpha$, then $\beta$ will increase.[2] Thus, in setting $\alpha$, we must consider both sides of the issue.[3]

Calculating the specific dollar costs associated with making Type I and Type II errors is often difficult and may require a subjective management decision. Therefore, any two managers might well arrive at different alpha levels. However, in the end, the choice for alpha must reflect the decision maker's best estimate of the costs of these two errors.

Having chosen a significance level, $\alpha$, the decision maker then must calculate the corresponding cutoff point, which is called a **critical value**.

**Significance Level**
The maximum allowable probability of committing a Type I statistical error. The probability is denoted by the symbol $\alpha$.

**Critical Value**
The value corresponding to a significance level that determines those test statistics that lead to rejecting the null hypothesis and those that lead to a decision not to reject the null hypothesis.

OUTCOME **4**

## Hypothesis Test for $\mu$, $\sigma$ Known

**Calculating Critical Values** To calculate critical values corresponding to a chosen $\alpha$, we need to know the sampling distribution of the sample mean $\bar{x}$. If our sampling conditions satisfy the Central Limit Theorem requirements or if the population is normally distributed and we know the population standard deviation $\sigma$, then the sampling distribution of $\bar{x}$ is a normal distribution centered at the population mean $\mu$ with a standard deviation equal to $\sigma/\sqrt{n}$. (For many population distributions, the Central Limit Theorem applies for sample sizes as small as 4 or 5. Sample sizes $n \geq 30$ assure us that the sampling distribution will be approximately normal regardless of population distribution.) With this information, we can calculate a critical $z$-value, called $z_\alpha$, or a critical $\bar{x}$-value, called $\bar{x}_\alpha$. We illustrate both calculations in the Price & Associates business application that follows.

---

[2] The sum of $\alpha$ and $\beta$ may coincidentally equal 1. However, in general, the sum of these two error probabilities does not equal 1 since they are not complements.

[3] We will discuss Type II errors more fully later in this chapter. Contrary to the Type I error situation in which we are able to choose the desired $\alpha$ level, $\beta$ is computed based on certain assumptions. Methods for computing $\beta$ are shown in Section 9.3.

**BUSINESS APPLICATION**    Conducting the Hypothesis Test

**Price & Associates Construction** (*continued*)    Recall that the managing partner at Price & Associates was interested in testing whether the mean number of days required to frame and roof a new house is less than or equal to 25. Suppose the managing partner decides he is willing to incur a 0.10 probability of committing a Type I error. Assume also that the population standard deviation, $\sigma$, for framing and roofing homes is three days and the sample size is 64 homes. Given that the sample size is large ($n \geq 30$) and the population standard deviation is known ($\sigma = 3$ days), we can state the critical value in two ways. First, we can establish the critical value as a $z$-value.

Figure 9.3 shows that if the rejection region on the upper end of the sampling distribution has an area of 0.10, the critical $z$-value, $z_\alpha$, from the standard normal table (or by using Excel's NORM.S.INV function) corresponding to the critical value is 1.28. Thus, $z_{0.10} = 1.28$. If the sample mean lies more than 1.28 standard deviations above $\mu = 25$ days, $H_0$ should be rejected; otherwise, we will not reject $H_0$.

Second, we can also express the critical value in the same units as the sample mean. In the Price & Associates example, we can calculate a critical $\bar{x}$-value, $\bar{x}_\alpha$, so that if $\bar{x}$ is greater than the critical value, we should reject $H_0$. If $\bar{x}$ is less than or equal to $\bar{x}_\alpha$, we should not reject $H_0$. Equation 9.1 shows how $\bar{x}_\alpha$ is computed. Figure 9.4 illustrates the use of Equation 9.1 for computing the critical value, $\bar{x}_\alpha$.

> The Excel 2016 function for determining the critical $z$-value for an upper-tailed test with alpha = 0.10 is
>
> = **NORM.S.INV(1-alpha)**
>
> = **NORM.S.INV(.90)**

**$\bar{x}_\alpha$ for Hypothesis Tests, $\sigma$ Known**

$$\bar{x}_\alpha = \mu + z_\alpha \frac{\sigma}{\sqrt{n}} \tag{9.1}$$

where:

$\mu$ = Hypothesized value for the population mean

$z_\alpha$ = Critical value from the standard normal distribution

$\sigma$ = Population standard deviation

$n$ = Sample size

Applying Equation 9.1, we determine the value for $\bar{x}_\alpha$ as follows:

$$\bar{x}_\alpha = \mu + z_\alpha \frac{\sigma}{\sqrt{n}}$$

$$\bar{x}_{0.10} = 25 + 1.28 \frac{3}{\sqrt{64}}$$

$$\bar{x}_{0.10} = 25.48 \text{ days}$$

> The Excel 2016 function for the critical value is
>
> = **NORM.INV(1-alpha,mean, standard error)**
>
> = **NORM.INV(0.9,25,3/sqrt(64))**

If $\bar{x} > 25.48$ days, $H_0$ should be rejected and changes should be made in the construction process; otherwise, $H_0$ should not be rejected and the process should not be changed. Any sample mean between 25.48 and 25 days would be attributed to sampling error, and the null hypothesis would not be rejected. A sample mean of 25.48 or fewer days will support the null hypothesis.

**FIGURE 9.3** **Determining the Critical Value as a z-Value**

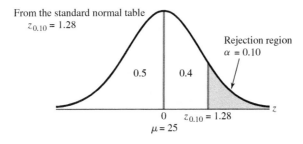

**FIGURE 9.4 Determining the Critical Value as an $\bar{x}$-Value for the Price & Associates Example**

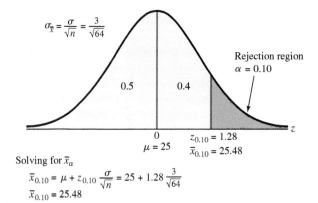

$$\sigma_{\bar{x}} = \frac{\sigma}{\sqrt{n}} = \frac{3}{\sqrt{64}}$$

Rejection region
$\alpha = 0.10$

0.5     0.4

0
$\mu = 25$

$z_{0.10} = 1.28$
$\bar{x}_{0.10} = 25.48$

Solving for $\bar{x}_{\alpha}$

$$\bar{x}_{0.10} = \mu + z_{0.10}\frac{\sigma}{\sqrt{n}} = 25 + 1.28\frac{3}{\sqrt{64}}$$

$$\bar{x}_{0.10} = 25.48$$

**Decision Rules and Test Statistics**   To conduct a hypothesis test, you can use three equivalent approaches. You can calculate a $z$-value and compare it to the critical value, $z_{\alpha}$. Alternatively, you can calculate the sample mean, $\bar{x}$, and compare it to the critical value, $\bar{x}_{\alpha}$. Finally, you can use a method called the $p$-value approach, to be discussed later in this section. It makes no difference which approach you use; each method yields the same conclusion.

Suppose $\bar{x} = 26$ days. How we test the null hypothesis depends on the procedure we used to establish the critical value. First, using the $z$-value method, we establish the following decision rule:

**Hypotheses**

$$H_0: \mu \leq 25 \text{ days}$$

$$H_A: \mu > 25 \text{ days}$$

$$\alpha = 0.10$$

**Decision Rule**

If $z > z_{0.10}$, reject $H_0$.

If $z \leq z_{0.10}$, do not reject $H_0$.

where:

$$z_{0.10} = 1.28$$

Recall that the number of homes sampled is 64 and the population standard deviation is assumed known at three days. The calculated $z$-value is called the **test statistic**.

The $z$-test statistic is computed using Equation 9.2.

**Test Statistic**

A function of the sampled observations that provides a basis for testing a statistical hypothesis.

***z*-Test Statistic for Hypothesis Tests for $\mu$, $\sigma$ Known**

$$z = \frac{\bar{x} - \mu}{\dfrac{\sigma}{\sqrt{n}}} \tag{9.2}$$

where:

$\bar{x} =$ Sample mean
$\mu =$ Hypothesized value for the population mean
$\sigma =$ Population standard deviation
$n =$ Sample size

Given that $\bar{x} = 26$ days, we can apply Equation 9.2 to get

$$z = \frac{\bar{x} - \mu}{\dfrac{\sigma}{\sqrt{n}}} = \frac{26 - 25}{\dfrac{3}{\sqrt{64}}} = 2.67$$

Thus, $\bar{x} = 26$ is 2.67 standard deviations above the hypothesized mean. Because,

$$z = 2.67 > z_{0.10} = 1.28, \text{ reject } H_0.$$

Because the sample mean is 2.67 standard deviations above the hypothesized mean, which exceeds our rejection cutoff of 1.28 standard deviations, we have sufficient evidence to reject the null hypothesis.

Now we use the second approach, which established (see Figure 9.4) a decision rule, as follows:

**Decision Rule**

$$\text{If } \bar{x} > \bar{x}_{0.10}, \text{ reject } H_0.$$

$$\text{Otherwise, do not reject } H_0.$$

Then,

$$\text{If } \bar{x} > 25.48 \text{ days, reject } H_0.$$

$$\text{Otherwise, do not reject } H_0.$$

Because

$$\bar{x} = 26 > \bar{x}_{0.10} = 25.48, \text{ reject } H_0.$$

Note that the two approaches yield the same conclusion, as they always will if you perform the calculations correctly. We have found that academic applications of hypothesis testing tend to use the $z$-value method, whereas business applications of hypothesis testing often use the $\bar{x}$ approach.

You will often come across a different language used to express the outcome of a hypothesis test. For instance, a statement for the hypothesis test just presented might be "The hypothesis test was significant at an $\alpha$ (or significance level) of 0.10." This simply means that the null hypothesis was rejected using a significance level of 0.10.

---

**HOW TO DO IT** (Example 9-3)

**One-Tailed Test for a Hypothesis about a Population Mean, $\sigma$ Known**

1. Specify the population parameter of interest.

2. Formulate the null hypothesis and the alternative hypothesis in terms of the population mean, $\mu$.

3. Specify the desired significance level ($\alpha$).

4. Construct the rejection region. (We strongly suggest you draw a picture showing where in the distribution the rejection region is located.)

5. Compute the test statistic:

$$\bar{x} = \frac{\sum x}{n} \quad \text{or} \quad z = \frac{\bar{x} - \mu}{\dfrac{\sigma}{\sqrt{n}}}$$

6. Reach a decision. Compare to the test statistic $\bar{x}_\alpha$ or $z_\alpha$.

7. Draw a conclusion regarding the null hypothesis.

---

**EXAMPLE 9-3**    One-Tailed Hypothesis Test for $\mu$, $\sigma$ Known

**Mountain States Surgery Center**    Mountain States Surgery Center in Denver, Colorado, performs many knee-replacement procedures each year. Research physicians at Mountain States have developed a surgery procedure they believe will reduce the average patient recovery time. The hospital board will not recommend the new procedure unless there is substantial evidence to suggest that it is better than the existing procedure. Records indicate that the current mean recovery time for the existing procedure is 142 days, with a standard deviation of 15 days. To test whether the new procedure actually results in a lower mean recovery time, the new procedure was performed on a random sample of 36 patients.

**STEP 1    Specify the population parameter of interest.**
We are interested in the mean recovery time, $\mu$.

**STEP 2    Formulate the null and alternative hypotheses.**

$$H_0: \mu \geq 142 \text{ (status quo)}$$

$$H_A: \mu < 142$$

**STEP 3    Specify the desired significance level ($\alpha$).**
The researchers wish to test the hypothesis using a 0.05 level of significance.

**STEP 4    Determine the critical value.**
This will be a one-tailed test, with the rejection region in the lower (left-hand) tail of the sampling distribution. The critical value is $-z_{0.05} = -1.645$. Therefore, the decision rule becomes:

$$\text{If } z < -1.645, \text{ reject } H_0; \text{ otherwise, do not reject } H_0.$$

**FIGURE 9.5 Mountain States Surgery Hypothesis Test**

$$\sigma_{\bar{x}} = \frac{\sigma}{\sqrt{n}} = \frac{15}{\sqrt{36}}$$

Rejection region
$\alpha = 0.05$

0.45

0.50

$-z_{0.05} = -1.645$

0

$\mu = 142$

$z$

> The value of $z$ can also be determined here from the Excel function
> **= NORM.S.INV(alpha)**
> For this example:
> **= NORM.S.INV(0.05)**

STEP 5    **Compute the test statistic.**
For this example, we will use $z$. Assume the sample mean, computed using $\bar{x} = \dfrac{\sum x}{n}$, is 140.2 days. Then,

$$z = \frac{\bar{x} - \mu}{\dfrac{\sigma}{\sqrt{n}}} = \frac{140.2 - 142}{\dfrac{15}{\sqrt{36}}} = -0.72$$

STEP 6    **Reach a decision.** (See Figure 9.5.)
The decision rule is:

If $z < -1.645$, reject $H_0$.

Otherwise, do not reject.

Because $-0.72 > -1.645$, do not reject $H_0$.

STEP 7    **Draw a conclusion.**
There is not sufficient evidence to conclude that the new knee replacement procedure results in a shorter average recovery period. Thus, Mountain States will not be able to recommend the new procedure on the grounds that it reduces recovery time.

**TRY EXERCISE 9-5** (pg. 335)

---

**EXAMPLE 9-4**    **Hypothesis Test for $\mu$, $\alpha$ Known**

**Quality Car Care, Inc.**    Quality Car Care, Inc., performs vehicle maintenance services for car owners in Vancouver, Canada. The company has advertised that the mean time for a complete routine maintenance (lube, oil change, tire rotation, etc.) is 40 minutes or less. The company has received complaints from several individuals saying the mean time required to complete the service exceeds the advertised mean of 40 minutes. Before responding, employees at Quality Car Care plan to test this claim using an alpha level equal to 0.05 and a random sample size of $n = 100$ past services. Suppose the population standard deviation is known to be $\sigma = 8$ minutes. The hypothesis test can be conducted using the following steps:

STEP 1    **Specify the population parameter of interest.**
The population parameter of interest is the mean test time, $\mu$.

STEP 2    **Formulate the null and alternative hypotheses.**
The claim made by the company is $\mu \leq 40$. Thus, the null and alternative hypotheses are

$H_0$: $\mu \leq 40$ minutes (claim)

$H_A$: $\mu > 40$ minutes

STEP 3  **Specify the significance level.**
The alpha level is specified to be 0.05.

STEP 4  **Determine the critical value.**
Alpha is the area under the standard normal distribution to the right of the critical value. The population standard deviation is known and the sample size is large, so the test statistic comes from the standard normal distribution. The critical $z$-value, $z_{0.05}$, is found by locating the $z$-value that corresponds to an area equal to $0.50 - 0.05 = 0.45$. The critical $z$-value from the standard normal table is 1.645.
We can calculate $\bar{x}_\alpha = \bar{x}_{0.05}$ using Equation 9.1 as follows:

$$\bar{x}_\alpha = \mu + z_\alpha \frac{\sigma}{\sqrt{n}}$$

$$\bar{x}_{0.05} = 40 + 1.645\left(\frac{8}{\sqrt{100}}\right)$$

$$\bar{x}_{0.05} = 41.32$$

> The Excel 2016 function for the critical value is
> = **NORM.INV(0.95,40,8/ sqrt(100))**

STEP 5  **Compute the sample mean.**
Suppose that the sample of 100 service records produced a sample mean of 43.5 minutes.

STEP 6  **Reach a decision.**
The decision rule is:

If $\bar{x} > 41.32$, reject $H_0$.

Otherwise, do not reject.

Because $\bar{x} = 43.5 > 41.32$, we reject $H_0$.

STEP 7  **Draw a conclusion.**
There is sufficient evidence to conclude that the mean time required to perform the maintenance service exceeds the advertised time of 40 minutes. Quality Car Care will likely want to modify its service process to shorten the average completion time or change its advertisement.

**TRY EXERCISE 9-7** (pg. 335)

**p-Value Approach**  In addition to the two hypothesis-testing approaches discussed previously, there is a third approach for conducting hypothesis tests. This approach uses a **p-value** instead of a critical value.

*p-*Value
The probability (assuming the null hypothesis is true) of obtaining a test statistic at least as extreme as the test statistic we calculated from the sample. The *p*-value is also known as the *observed significance level.*

If the calculated *p*-value is smaller than the probability in the rejection region ($\alpha$), then the null hypothesis is rejected. If the calculated *p*-value is greater than or equal to $\alpha$, then the hypothesis will not be rejected. The *p*-value approach is popular today because *p*-values are usually computed by statistical software packages, including Excel. The advantage to reporting test results using a *p*-value is that it provides more information than simply stating whether the null hypothesis is rejected. The decision maker is presented with a measure of the degree of significance of the result (i.e., the *p*-value). This offers the reader the opportunity to evaluate the *extent* to which the data disagree with the null hypothesis, not just whether they disagree.

**EXAMPLE 9-5**   **Hypothesis Test Using *p*-Values, $\sigma$ Known**

**Dodger Stadium Parking**   Suppose the parking manager for the Los Angeles Dodgers baseball team has studied the exit times for cars leaving the ballpark after a game and believes that recent changes to the traffic flow leaving the stadium have increased, rather than decreased, average exit times. Assume that prior to the changes, the mean exit time per vehicle was 36 minutes, with a population standard deviation equal to 11 minutes. To test the parking manager's belief that the mean time now exceeds 36 minutes, a simple random sample of $n = 200$ vehicles is selected, and a sample mean of 36.8 minutes is calculated. Using an alpha level of 0.05, the manager can use the following steps to conduct the hypothesis test:

STEP 1 **Specify the population parameter of interest.**
The Dodger Stadium parking manager is interested in the mean exit time per vehicle, $\mu$.

STEP 2 **Formulate the null and alternative hypotheses.**
Based on the manager's claim that the current mean exit time is longer than before the traffic flow changes, the null and alternative hypotheses are

$$H_0: \mu \leq 36 \text{ minutes}$$

$$H_A: \mu > 36 \text{ minutes (claim)}$$

STEP 3 **Specify the significance level.**
The alpha level specified for this test is $\alpha = 0.05$.

STEP 4 **Construct the decision rule.**
The decision rule is:

If $p$-value $< \alpha = 0.05$, reject $H_0$.

Otherwise, do not reject $H_0$.

STEP 5 **Compute the test statistic (find the $p$-value).**
Because the sample size is large and the population standard deviation is assumed known, the test statistic will be a $z$-value, which is computed as follows:

$$z = \frac{\bar{x} - \mu}{\frac{\sigma}{\sqrt{n}}} = \frac{36.8 - 36}{\frac{11}{\sqrt{200}}} = 1.0285 = 1.03$$

In this example, the $p$-value is the probability of a $z$-value from the standard normal distribution being at least as large as 1.03. This is stated as

$$p\text{-value} = P(z \geq 1.03)$$

From the standard normal distribution table in Appendix D,

$$P(z \geq 1.03) = 0.5000 - 0.3485 = 0.1515$$

STEP 6 **Reach a decision.**
Because $p$-value $= 0.1515 > \alpha = 0.05$, do not reject the null hypothesis.

STEP 7 **Draw a conclusion.**
The difference between the sample mean and the hypothesized population mean is not large enough to attribute the difference to anything but sampling error.

**TRY EXERCISE 9-6** (pg. 335)

> The Excel 2016 function for finding the $p$-value for an upper-tail test is:
>
> = 1 – NORM.S.DIST(z,True)
>
> = 1 – NORM.S.DIST(1.03,True)
>
> or alternatively:
>
> = 1 – NORM.DIST (x,mean, standard deviation,true)
>
> where $x$ is the sample mean
>
> = 1 – NORM.DIST(36.8,36,11/ SQRT(200),TRUE)

Why do we need three methods to test the same hypothesis when they all give the same result? The answer is that we don't. However, you need to be aware of all three methods because you will encounter each in business situations. The $p$-value approach is especially important because many statistical software packages provide a $p$-value that you can use to test a hypothesis quite easily, and a $p$-value provides a measure of the degree of significance associated with the hypothesis test. In this text, we will use both test-statistic approaches as well as the $p$-value approach to hypothesis testing.

## Types of Hypothesis Tests

Hypothesis tests are formulated as either **one-tailed tests** or **two-tailed tests** depending on how the null and alternative hypotheses are presented.

For instance, in the Price & Associates application, the null and alternative hypotheses are

| | |
|---|---|
| Null hypothesis | $H_0: \mu \leq 25$ days |
| Alternative hypothesis | $H_A: \mu > 25$ days |

**One-Tailed Test**

A hypothesis test in which the entire rejection region is located in one tail of the sampling distribution. In a one-tailed test, the entire alpha level is located in one tail of the distribution.

**Two-Tailed Test**

A hypothesis test in which the rejection region is split into the two tails of the sampling distribution. In a two-tailed test, the alpha level is split evenly between the two tails.

This hypothesis test is one tailed because the entire rejection region is located in one tail (the upper tail) and the null hypothesis will be rejected only if the sample mean falls in the extreme upper tail of the sampling distribution (see Figure 9.4). In this application, it will take a sample mean substantially larger than 25 days to reject the null hypothesis.

In Example 9-2 involving Nabisco, the null and alternative hypotheses for the mean fill of Wheat Thins boxes are

$$H_0: \mu = 20 \text{ ounces (status quo)}$$

$$H_A: \mu \neq 20 \text{ ounces}$$

In this two-tailed hypothesis test, the null hypothesis will be rejected if the sample mean is extremely large (upper tail) or extremely small (lower tail). The rejection region, or the alpha level, is split evenly between the two tails.

### *p*-Value for Two-Tailed Tests

In the *p*-value example about Dodger Stadium, the rejection region was located in one tail of the sampling distribution. In those cases, the null hypothesis has the $\geq$ or $\leq$ format. However, sometimes the null hypothesis is stated as a direct equality. The following application shows how to use the *p*-value approach for a two-tailed test.

---

**BUSINESS APPLICATION**    Using *p*-Values to Test a Null Hypothesis

**Golden Peanut Company**    Consider the Golden Peanut Company in Alpharetta, Georgia, which packages salted and unsalted unshelled peanuts in 16-ounce sacks. The company's filling process strives for an average fill amount equal to 16 ounces. Therefore, Golden will test the following null and alternative hypotheses:

$$H_0: \mu = 16 \text{ ounces (status quo)}$$

$$H_A: \mu \neq 16 \text{ ounces}$$

The null hypothesis will be rejected if the test statistic falls in either tail of the sampling distribution. The size of the rejection region is determined by $\alpha$. Each tail has an area equal to $\alpha/2$.

The *p*-value for the two-tailed test is computed in a manner similar to that for a one-tailed test. First, we determine the *z*-test statistic as follows:

$$z = \frac{\bar{x} - \mu}{\dfrac{\sigma}{\sqrt{n}}}$$

Suppose for this situation, Golden managers calculated $z = 3.32$. In a one-tailed test, the area calculated to form the *p*-value is determined by the direction in which the inequality is pointing in the alternative hypotheses. However, in a two-tailed test, the tail area in which the test statistic is located is initially calculated. In this case, we find $P(z > 3.32)$ using either the standard normal table in Appendix D or Excel's NORM.S.DIST function (= NORM.S.DIST(3.32,True) = 0.9995). In this case, because $z = 3.32$ exceeds the table values, we use Excel to obtain

$$P(z \leq 3.32) = 0.9995$$

Then

$$P(z > 3.32) = 1 - 0.9995 = 0.0005$$

However, because this is a two-tailed hypothesis test, the *p*-value is found by multiplying the 0.0005 value by 2 (to account for the chance that our sample result could have been on either side of the distribution). Thus

$$p\text{-value} = 2(0.0005) = 0.0010$$

**FIGURE 9.6** **Two-Tailed Test for the Golden Peanut Example**

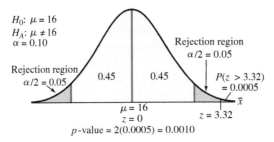

$H_0: \mu = 16$
$H_A: \mu \neq 16$
$\alpha = 0.10$

Rejection region
$\alpha/2 = 0.05$

Rejection region
$\alpha/2 = 0.05$

0.45      0.45

$P(z > 3.32)$
$= 0.0005$

$\bar{x}$

$\mu = 16$
$z = 0$

$z = 3.32$

$p\text{-value} = 2(0.0005) = 0.0010$

**Decision Rule:**
If $p$-value $< \alpha = 0.10$, reject $H_0$.
Otherwise, do not reject $H_0$.
Because $p$-value $= 0.0010 < \alpha = 0.10$, reject $H_0$.

Assuming an alpha $= 0.10$ level, then because

$$p\text{-value} = 0.0010 < \alpha = 0.10, \text{ we reject } H_0.$$

Figure 9.6 illustrates the two-tailed test for the Golden Peanut Company example.

---

**HOW TO DO IT** (Example 9-6)

**Two-Tailed Test for a Hypothesis about a Population Mean, $\sigma$ Known**

**1.** Specify the population parameter of interest.

**2.** Formulate the null and alternative hypotheses in terms of the population mean, $\mu$.

**3.** Specify the desired significance level, $\alpha$.

**4.** Construct the rejection region. Determine the critical values for each tail, $z_{\alpha/2}$ and $-z_{\alpha/2}$, from the standard normal table. If needed, calculate $\bar{x}_{(\alpha/2)L}$ and $\bar{x}_{(\alpha/2)U}$.

Define the two-tailed decision rule using one of the following:

- If $z > z_{\alpha/2}$ or if $z < -z_{\alpha/2}$, reject $H_0$; otherwise, do not reject $H_0$.

- If $\bar{x} < \bar{x}_{(\alpha/2)/L}$ or $\bar{x} > \bar{x}_{(\alpha/2)/U}$, reject $H_0$; otherwise, do not reject $H_0$.

- If $p$-value $< a$, reject $H_0$; otherwise, do not reject $H_0$.

**5.** Compute the test statistic, $z = (\bar{x} - \mu)/(\sigma/\sqrt{n})$, or $\bar{x}$, or find the $p$-value.

**6.** Reach a decision.

**7.** Draw a conclusion.

---

**EXAMPLE 9-6**    **Two-Tailed Hypothesis Test for $\mu$, $\sigma$ Known**

**Hargrove Wood Products**    Hargrove Wood Products has lumber, plywood, and paper plants in several areas of the United States. At its La Grande, Oregon, plywood plant, the company makes plywood used in residential and commercial building. One product made at the La Grande plant is 3/8-inch plywood, which must have a mean thickness of 0.375 inch. The standard deviation, $\sigma$, is known to be 0.05 inch. Before sending a shipment to customers, Hargrove managers test whether they are meeting the 0.375-inch requirements by selecting a random sample of $n = 100$ sheets of plywood and collecting thickness measurements.

**STEP 1**    **Specify the population parameter of interest.**
The mean thickness of plywood is of interest.

**STEP 2**    **Formulate the null and the alternative hypotheses.**
The null and alternative hypotheses are

$$H_0: \mu = 0.375 \text{ inch (status quo)}$$

$$H_A: \mu \neq 0.375 \text{ inch}$$

Note that the test is two tailed because the company is concerned that the plywood could be too thick or too thin.

**STEP 3**    **Specify the desired significance level ($\alpha$).**
The managers wish to test the hypothesis using $\alpha = 0.05$.

**STEP 4**    **Construct the rejection region.**
This is a two-tailed test. The critical $z$-values for the upper and lower tails are found in the standard normal table:

$$-z_{\alpha/2} = -z_{0.05/2} = -z_{0.025} = -1.96$$

and

$$z_{\alpha/2} = z_{0.05/2} = z_{0.025} = 1.96$$

The two-tailed decision rule is:

If $z > 1.96$ or if $z < -1.96$, reject $H_0$; otherwise, do not reject $H_0$.

**STEP 5**    **Compute the test statistic.**
Select the random sample and calculate the sample mean.
Suppose that the sample mean for the random sample of 100 measurements is

$$\bar{x} = \frac{\sum x}{n} = 0.378 \text{ inch}$$

The $z$-test statistic is

$$z = \frac{\bar{x} - \mu}{\dfrac{\sigma}{\sqrt{n}}} = \frac{0.378 - 0.375}{\dfrac{0.05}{\sqrt{100}}} = 0.60$$

STEP 6   **Reach a decision.**
Because $z = 0.60 < 1.96$, do not reject the null hypothesis.
We can also compute the $p$-value and compare it to alpha. This is done as follows:

$$P(z > 0.60) = \mathbf{1 - NORM.S.\ DIST(0.60, True)} = 1 - 0.7257 = 0.2743$$

Now, because this is a two-tailed test, we find the $p$-value by doubling this probability: $p$-value $- 2(0.2743) = 0.5486$.

Since the $p$-value $0.5486 > 0.05$, we do not reject the null hypothesis.

STEP 7   **Draw a conclusion.**
The Hargrove Wood Products Company does not have sufficient evidence to reject the null hypothesis. Thus, it will ship the plywood.

**TRY EXERCISE 9-5** (pg. 335)

## Hypothesis Test for $\mu$, $\sigma$ Unknown

In Chapter 8, we introduced situations in which the objective was to estimate a population mean when we did not know the population standard deviation. In those cases, the critical value is a $t$-value from the $t$-distribution rather than a $z$-value from the standard normal distribution. The same logic is used in hypothesis testing when $\sigma$ is unknown (which will usually be the case). Equation 9.3 is used to compute the test statistic for testing hypotheses about a population mean when the population standard deviation is unknown.

**$t$-Test Statistic for Hypothesis Tests for $\mu$, $\sigma$ Unknown**

$$t = \frac{\bar{x} - \mu}{\dfrac{s}{\sqrt{n}}} \tag{9.3}$$

where:

$\bar{x}$ = Sample mean

$\mu$ = Hypothesized value for the population mean

$s$ = Sample standard deviation, $s = \sqrt{\dfrac{\sum (x - \bar{x})^2}{n - 1}}$

$n$ = Sample size

To employ the $t$-distribution, we must make the following assumption:

**Assumption**    The population is normally distributed.

If the population from which we select the simple random sample is approximately normal, then the $t$-test statistic computed using Equation 9.3 will be distributed according to a $t$-distribution with $n - 1$ degrees of freedom.

**HOW TO DO IT** (Example 9-7)

**One- or Two-Tailed Tests for $\mu$, $\sigma$ Unknown**

**1.** Specify the population parameter of interest, $\mu$.

**2.** Formulate the null hypothesis and the alternative hypothesis.

**3.** Specify the desired significance level ($\alpha$).

**4.** Construct the rejection region. If it is a two-tailed test, determine the critical values for each tail, $t_{\alpha/2}$ and $-t_{\alpha/2}$, from the $t$-distribution table. If the test is a one-tailed test, find either $t_\alpha$ or $-t_\alpha$, depending on the tail of the rejection region. Degrees of freedom are $n - 1$. If desired, the critical $t$-values can be used to find the appropriate $\bar{x}_\alpha$ or the $\bar{x}_{(\alpha/2)L}$ and $\bar{x}_{(\alpha/2)U}$ values. Define the decision rule.

   **a.** If the test statistic is in the rejection region, reject $H_0$; otherwise, do not reject $H_0$.

   **b.** If the $p$-value is less than $\alpha$, reject $H_0$; otherwise, do not reject $H_0$.

**5.** Compute the test statistic or find the $p$-value.

Select the random sample and calculate the sample mean, $\bar{x} = \Sigma x/n$, and the sample standard deviation:

$$s = \sqrt{\frac{\Sigma(x - \bar{x})^2}{n - 1}}$$

Then calculate

$$t = \frac{\bar{x} - \mu}{\dfrac{s}{\sqrt{n}}}$$

or the $p$-value.

**6.** Reach a decision.

**7.** Draw a conclusion.

---

**EXAMPLE 9-7** **Hypothesis Test for $\mu$, $\sigma$ Unknown**

**Dairy Fresh Ice Cream** The Dairy Fresh Ice Cream plant in Greensboro, Alabama, uses a filling machine for its 64-ounce cartons. There is some variation in the actual amount of ice cream that goes into the carton. The machine can go out of adjustment and put a mean amount either less or more than 64 ounces in the cartons. To monitor the filling process, the production manager selects a simple random sample of 16 filled ice cream cartons each day. If he can assume that the population is approximately normally distributed, he can test whether the machine is still in adjustment using the following steps:

**STEP 1** **Specify the population parameter of interest.**
The manager is interested in the mean amount of ice cream.

**STEP 2** **Formulate the appropriate null and alternative hypotheses.**
The status quo is that the machine continues to fill ice cream cartons with a mean equal to 64 ounces. Thus, the null and alternative hypotheses are

$$H_0\text{: } \mu = 64 \text{ ounces (Machine is in adjustment.)}$$
$$H_A\text{: } \mu \neq 64 \text{ ounces (Machine is out of adjustment.)}$$

**STEP 3** **Specify the desired level of significance.**
The test will be conducted using an alpha level equal to 0.05.

**STEP 4** **Construct the decision rule.**
Now we determine the critical values from the $t$-distribution. Based on the null and alternative hypotheses, this test is two tailed. Thus, we split the alpha into two tails and determine the critical values from the $t$-distribution with $n - 1$ degrees of freedom. Using Appendix F, we find the critical $t$'s for a two-tailed test with $\alpha = 0.05$ and $16 - 1 = 15$ degrees of freedom are $t = \pm 2.1315$. The Excel function, $= \text{T.INV.2T(Probability,Deg\_Freedom)}$ can also be used to find the critical $t$-value for a two-tailed test:

$$= \text{T.INV.2T}(0.05,15) = 2.1315.$$

The decision rule for this two-tailed test is:

If $t < -2.1315$ or $t > 2.1315$, reject $H_0$.

Otherwise, do not reject $H_0$.

**STEP 5** **Compute the $t$-test statistic.**
The sample data are

| | | | | | | | |
|---|---|---|---|---|---|---|---|
| 62.7 | 64.7 | 64.0 | 64.5 | 64.6 | 65.0 | 64.4 | 64.2 |
| 64.6 | 65.5 | 63.6 | 64.7 | 64.0 | 64.2 | 63.0 | 63.6 |

The sample mean is

$$\bar{x} = \frac{\Sigma x}{n} = \frac{1,027.3}{16} = 64.2$$

The sample standard deviation is

$$s = \sqrt{\frac{\Sigma(x - \bar{x})^2}{n - 1}} = 0.72$$

The $t$-test statistic, using Equation 9.3, is

$$t = \frac{\bar{x} - \mu}{\dfrac{s}{\sqrt{n}}} = \frac{64.2 - 64}{\dfrac{0.72}{\sqrt{16}}} = 1.11$$

**STEP 6** **Reach a decision.**
Because $t = 1.11$ is not less than $-2.1315$ and not greater than $2.1315$, we do not reject the null hypothesis.

STEP 7 **Draw a conclusion.**
Based on these sample data, the company does not have sufficient evidence to conclude that the filling machine is out of adjustment.

TRY EXERCISE 9-12 (pg. 336)

---

**EXAMPLE 9-8** **Testing the Hypothesis for $\mu$ Unknown**

**American Southern Fried Chicken**   An American Southern Fried Chicken store in New Orleans has studied its service operations and determined that the time required for a customer to be served is normally distributed, with a mean equal to 540 seconds. However, the manager overseeing all of American Southern's facilities in the Southeast has charged his staff with improving service times. After undertaking some process changes and training, managers at the New Orleans store have selected a random sample of 16 customers and wish to determine whether the mean service time is now less than 540 seconds.

STEP 1 **Specify the population parameter of interest.**
The mean service time for all customers at the New Orleans store is the population parameter of interest.

STEP 2 **Formulate the null and alternative hypotheses.**
The null and alternative hypotheses are

$$H_0: \mu \geq 540 \text{ seconds (status quo)}$$

$$H_A: \mu < 540 \text{ seconds}$$

STEP 3 **Specify the significance level.**
The test will be conducted at the 0.01 level of significance. Thus, $\alpha = 0.01$.

STEP 4 **Construct the rejection region.**
Because this is a one-tailed test and the rejection region is in the lower tail, as indicated in $H_A$, the critical value from the $t$-distribution with $16 - 1 = 15$ degrees of freedom is $-t_\alpha = -t_{0.01} = -2.6025$.
The decision rule for this one-tailed test is:

If $t < -2.6025$, reject $H_0$.

Otherwise, do not reject $H_0$.

> The Excel 2016 function for determining a left-tailed critical $t$-value is
>
> = T.INV(alpha,deg. of freedom)
>
> = T.INV(0.01,15)
>
> For a right-tailed critical value, use
>
> = 1 − T.INV(1 − alpha,deg. of freedom)

STEP 5 **Compute the test statistic.**
The sample mean for the random sample of 16 customers is $\bar{x} = \Sigma x/n = 510$ seconds, and the sample standard deviation is $\sqrt{\dfrac{\Sigma(x - x)^2}{n - 1}} = 45$ seconds.
If we assume that the population distribution is approximately normal, the test statistic is

$$t = \frac{\bar{x} - \mu}{\dfrac{s}{\sqrt{n}}} = \frac{510 - 540}{\dfrac{45}{\sqrt{16}}} = -2.67$$

STEP 6 **Reach a decision.**
Because $t = -2.67 < -2.6025$, the null hypothesis is rejected.
Alternatively, the left-tailed $p$-value can be computed using an Excel function:

$$= \text{T.DIST}(-2.67, 15, \text{True}) = 0.0087$$

Since the $p$-value $0.0087 < 0.01$, we reject the null hypothesis.

STEP 7 **Draw a conclusion.**
There is sufficient evidence to conclude that the mean service time has been reduced below 540 seconds.

TRY EXERCISE 9-15 (pg. 336)

Excel Tutorial

**BUSINESS APPLICATION**   Hypothesis Tests Using Excel

**Franklin Tire and Rubber Company**   The Franklin Tire and Rubber Company conducted a test on a new tire design to determine whether the company could make the claim that the mean tire mileage exceeds 60,000 miles. A simple random sample of 100 tires was tested, and the number of miles each tire lasted until it no longer met the federal government minimum tread thickness was recorded. The data (shown in thousands of miles) are in the file called **Franklin**.

The null and alternative hypotheses to be tested are

$$H_0: \mu \leq 60$$
$$H_A: \mu > 60 \,(\text{research hypothesis})$$
$$\alpha = 0.05$$

Excel does not have a special procedure for testing hypotheses for single population means. However, the Excel add-ins software called XLSTAT has the necessary hypothesis-testing tools.[4] Figure 9.7 shows the Excel XLSTAT output.

We denote the critical value of an upper- or lower-tailed test with a significance level of $\alpha$ as $t_\alpha$ or $-t_\alpha$. The critical value for $\alpha = 0.05$ and 99 degrees of freedom is $t_{0.05} = 1.6604$. Using the critical value approach, the decision rule is:

If the $t$-test statistic $> 1.6604$, reject $H_0$; otherwise, do not reject $H_0$.

The sample mean, based on a sample of 100 tires, is $\bar{x} = 60.17$ (60,170 miles), and the sample standard deviation is $s = 4.701$ (4,701 miles). The $t$-test statistic shown in Figure 9.7 is computed as follows:

$$t = \frac{\bar{x} - \mu}{\frac{s}{\sqrt{n}}} = \frac{60.17 - 60}{\frac{4.701}{\sqrt{100}}} = 0.3616$$

Because

$$t = 0.3616 < t_{0.05} = 1.6604, \text{ do not reject the null hypothesis.}$$

**Excel 2016 (XLSTAT) Instructions**

1. Open **XLSTAT**. Open File: **Franklin.xlsx**.
2. Select **Parametric tests**. From the drop-down options, choose **One-sample *t*-test and *z*-test**.
3. In **Data input box**, select data from the file including the column label. Check the box for **Column labels** and check the box for **Student's *t*-test**. For the **Data format**, check **One sample**. To place the output on a new worksheet, select **Sheet**.
4. Select **Options**. Choose **Alternative hypothesis** drop-down box. Select **Mean 1 > Theoretical mean**. Theoretical mean (hypothesized mean) input is **60**. Significance level (%) is 5.
5. Select **Outputs**. Check box for **Descriptive statistics** if desired.
6. Click **OK**.

**FIGURE 9.7**   **Excel 2016 (XLSTAT) Output for Franklin Tire Hypothesis Test Results**

_____

[4] This test can be done in Excel without the benefit of the XLSTAT add-ins by using Excel equations. Please refer to the Excel tutorial for the specifics.

Thus, based on the sample data, the evidence is insufficient to conclude that the new tires have an average life exceeding 60,000 miles. Based on this test, the company would not be justified in making the claim.

Franklin managers could also use the $p$-value approach to test the null hypothesis because the output shown in Figure 9.7 provides the $p$-value. In this case, the $p$-value $= 0.3592$. The decision rule for a test is

If $p$-value $< \alpha$, reject $H_0$; otherwise, do not reject $H_0$.

Because

$$p\text{-value} = 0.3592 > \alpha = 0.05$$

we do not reject the null hypothesis. This is the same conclusion we reached using the $t$-test statistic approach.

This section has introduced the basic concepts of hypothesis testing. There are several ways to test a null hypothesis. Each method will yield the same result; however, computer software such as Excel shows the $p$-values automatically. Therefore, decision makers increasingly use the $p$-value approach.

# 9.1 EXERCISES

## Skill Development

**9-1.** Determine the appropriate critical value(s) for each of the following tests concerning the population mean:
a. upper-tailed test: $\alpha = 0.025; n = 25; \sigma = 3.0$
b. lower-tailed test: $\alpha = 0.05; n = 30; s = 9.0$
c. two-tailed test: $\alpha = 0.02; n = 51; s = 6.5$
d. two-tailed test: $\alpha = 0.10; n = 36; \sigma = 3.8$

**9-2.** Determine whether each of the following pairs of hypotheses is valid for a hypothesis test. Explain reasons for any pair that is indicated to be invalid.
a. $H_0: \mu = 15, H_A: \mu > 15$
b. $H_0: \mu = 20, H_A: \mu \neq 20$
c. $H_0: \mu < 30, H_A: \mu > 30$
d. $H_0: \mu \leq 40, H_A: \mu \geq 40$
e. $H_0: \bar{x} \leq 45, H_A: \bar{x} > 45$
f. $H_0: \mu \leq 50, H_A: \mu > 55$

**9-3.** Provide the relevant critical value(s) for each of the following circumstances:
a. $H_A: \mu > 13, n = 15, \sigma = 10.3, \alpha = 0.05$
b. $H_A: \mu \neq 21, n = 23, s = 35.40, \alpha = 0.02$
c. $H_A: \mu \neq 35, n = 41, \sigma = 35.407, \alpha = 0.01$
d. $H_A: \mu < 49$; data: 12.5, 15.8, 44.3, 22.6, 18.4; $\alpha = 0.10$
e. $H_A: \bar{x} > 15, n = 27, \sigma = 12.4$

**9-4.** For each of the following $z$-test statistics, compute the $p$-value assuming that the hypothesis test is a one-tailed test:
a. $z = 1.34$
b. $z = 2.09$
c. $z = -1.55$

**9-5.** For the following hypothesis test:

$$H_0: \mu = 200$$
$$H_A: \mu \neq 200$$
$$\alpha = 0.01$$

with $n = 64, \sigma = 9$, and $\bar{x} = 196.5$, state

a. the decision rule in terms of the critical value of the test statistic
b. the calculated value of the test statistic
c. the conclusion

**9-6.** For the following hypothesis test:

$$H_0: \mu \leq 45$$
$$H_A: \mu > 45$$
$$\alpha = 0.02$$

with $n = 80, \sigma = 9$, and $\bar{x} = 47.1$, state

a. the decision rule in terms of the critical value of the test statistic
b. the calculated value of the test statistic
c. the appropriate $p$-value
d. the conclusion

**9-7.** For the following hypothesis test:

$$H_0: \mu \geq 23$$
$$H_A: \mu < 23$$
$$\alpha = 0.025$$

with $n = 25, s = 8$, and $\bar{x} = 20$, state

a. the decision rule in terms of the critical value of the test statistic
b. the calculated value of the test statistic
c. the conclusion

**9-8.** For the following hypothesis test:

$$H_0: \mu = 60.5$$
$$H_A: \mu \neq 60.5$$
$$\alpha = 0.05$$

with $n = 15, s = 7.5$, and $\bar{x} = 62.2$, state

a. the decision rule in terms of the critical value of the test statistic
b. the calculated value of the test statistic
c. the conclusion

**9-9.** For the following hypothesis test:

$$H_0: \mu \leq 70$$
$$H_A: \mu > 70$$

with $n = 20$, $\bar{x} = 71.2$, $s = 6.9$, and $\alpha = 0.1$, state
a. the decision rule in terms of the critical value of the test statistic
b. the calculated value of the test statistic
c. the conclusion

**9-10.** A sample taken from a population yields a sample mean of 58.4. Calculate the $p$-value for each of the following circumstances:
a. $H_A: \mu > 58$, $n = 16$, $\sigma = 0.8$
b. $H_A: \mu \neq 45$, $n = 41$, $s = 35.407$
c. $H_A: \mu \neq 45$, $n = 41$, $\sigma = 35.407$
d. $H_A: \mu < 69$; data: 60.1, 54.3, 57.1, 53.1, 67.4

**9-11.** For each of the following scenarios, indicate which type of statistical error could have been committed or, alternatively, that no statistical error was made. When warranted, provide a definition for the indicated statistical error.
a. Unknown to the statistical analyst, the null hypothesis is actually true.
b. The statistical analyst fails to reject the null hypothesis.
c. The statistical analyst rejects the null hypothesis.
d. Unknown to the statistical analyst, the null hypothesis is actually true and the analyst fails to reject the null hypothesis.
e. Unknown to the statistical analyst, the null hypothesis is actually false.
f. Unknown to the statistical analyst, the null hypothesis is actually false and the analyst rejects the null hypothesis.

## Business Applications

**9-12.** According to the website www.truecar.com, the average price of a new sports car sold in the United States in 2015 was $31,262. Suppose that, during the first three months of 2016, a random sample of prices paid for sports cars is as follows:

| | | | | | |
|---|---|---|---|---|---|
| $29,121 | $31,472 | $28,054 | $31,005 | $36,295 | $32,771 |
| $26,205 | $33,299 | $25,602 | $33,726 | $39,731 | $27,816 |

Based on the sample information, can you conclude at the $\alpha = 0.05$ level of significance that the average sports car price in 2016 is lower than in 2015?

**9-13.** The director of a state agency believes that the average starting salary for clerical employees in the state is less than $30,000 per year. To test her hypothesis, she has collected a simple random sample of 100 starting clerical salaries from across the state and found that the sample mean is $29,750.
a. State the appropriate null and alternative hypotheses.

b. Assuming the population standard deviation is known to be $2,500 and the significance level for the test is to be 0.05, what is the critical value (stated in dollars)?
c. Referring to your answer in part b, what conclusion should she reach with respect to the null hypothesis?
d. Referring to your answer in part c, which of the two statistical errors might have been made in this case? Explain.

**9-14.** A mail-order business prides itself in its ability to fill customers' orders in six days or less on average. Periodically, the operations manager selects a random sample of customer orders and determines the number of days required to fill the orders. Based on this sample information, he decides if the desired standard is being met. He will assume that the average number of days to fill customers' orders is six or less unless the data strongly suggest otherwise.
a. Establish the appropriate null and alternative hypotheses.
b. On one occasion when a sample of 40 customers was selected, the average number of days was 6.65, with a sample standard deviation of 1.5 days. Can the operations manager conclude that his mail-order business is achieving its goal? Use a significance level of 0.025 to answer this question.
c. Calculate the $p$-value for this test. Conduct the test using this $p$-value.
d. The operations manager wishes to monitor the efficiency of his mail-order service often. Therefore, he does not wish to repeatedly calculate $t$-values to conduct the hypothesis tests. Obtain the critical value, $\bar{x}_\alpha$, so that the manager can simply compare the sample mean to this value to conduct the test.

**9-15.** An internal report issued by the marketing manager for a national oil-change franchise indicated that the mean number of miles between oil changes for franchise customers is at least 3,600. One Texas franchise owner conducted a study to determine whether the marketing manager's statement was accurate for his franchise's customers. He selected a simple random sample of 10 customers and determined the number of miles each had driven the car between oil changes. He obtained the following sample data:

| | | | | |
|---|---|---|---|---|
| 3,655 | 4,204 | 1,946 | 2,789 | 3,555 |
| 3,734 | 3,208 | 3,311 | 3,920 | 3,902 |

a. State the appropriate null and alternative hypotheses.
b. Use the test statistic approach with $\alpha = 0.05$ to test the null hypothesis.

**9-16.** The makers of Mini-Oats Cereal have an automated packaging machine that can be set at any targeted fill level between 12 and 32 ounces. At the end of every shift (eight hours), 16 boxes are selected at random. Based on these sample results, the production control manager determines whether the filling machine needs to be readjusted or whether it remains all right to operate. Use $\alpha = 0.05$.

a. Establish the appropriate null and alternative hypotheses to be tested for boxes that are supposed to contain an average of 24 ounces.

b. At the end of a particular shift during which the machine was filling 24-ounce boxes of Mini-Oats, the sample mean of 16 boxes was 24.42 ounces, with a standard deviation of 0.72 ounce. Assist the production control manager in determining if the machine is achieving its targeted average.

c. Why do you suppose the production control manager would prefer to make this hypothesis test a two-tailed test? Discuss.

d. Conduct the test using a *p*-value. (*Hint:* Use Excel's T.DIST.2T function.)

e. Considering the result of the test, which of the two types of errors in hypothesis testing could you have made?

**9-17.** The owner of an Axion health club franchise believes that the average age of her members is under 55 years. To test this, a random sample of 30 was selected; the following sample data are the ages of the sampled individuals:

| 40 | 55 | 78 | 27 | 55 | 33 |
| 51 | 76 | 54 | 67 | 40 | 31 |
| 60 | 61 | 50 | 42 | 78 | 80 |
| 25 | 38 | 74 | 46 | 48 | 57 |
| 30 | 65 | 80 | 26 | 46 | 49 |

a. State the appropriate null and alternative hypotheses.

b. Use the test statistic approach to test the null hypothesis with $\alpha = 0.01$.

## Computer Software Exercises

**9-18.** At a recent meeting, the manager of a national call center for a major Internet bank made the statement that the average past-due amount for customers who have been called previously about their bills is now no larger than $20.00. Other bank managers at the meeting suggested that this statement may be in error and that it might be worthwhile to conduct a test to see if there is statistical support for the call center manager's statement. The file called **Bank Call Center** contains data for a random sample of 67 customers from the call center population. Assuming that the population standard deviation for past due amounts is known to be $60.00, what should be concluded based on the sample data? Test using $\alpha = 0.10$.

**9-19.** The *Consumer Expenditures* report released by the U.S. Bureau of Labor Statistics found the average annual household spending on food at home in 2014 was $3,971. Suppose a random sample of 137 households in Detroit was taken to determine whether the average annual expenditure on food at home was less for consumer units in Detroit than in the nation as a whole. The sample results are in the file **Detroit Eats**. Based on the sample results, can it be concluded at the $\alpha = 0.02$ level of significance that average consumer-unit spending for food at home in Detroit is less than the national average?

**9-20.** The 2015 Milliman Medical Index reported on its website (www.milliman.com) that in 2015 the average annual out-of-pocket medical costs for a typical family were $4,065. A health care consumer group believes that the actual average is higher than this. Suppose the group has selected a random sample of 120 "typical" households and asked them to go back through their 2015 medical payments and record the total out-of-pocket expenses for the year. The data are in a file called **Medical Expenses**. Based on these data, what should the consumer group conclude, using an alpha level of 0.01?

**9-21.** A key factor in the world's economic condition is the population growth of countries. The file called **Country Growth** contains data for 232 countries. Consider these countries to be all the countries in the world. (*Source: The World Factbook*, 2015 population estimates, www.cia.gov.)

a. From this population, suppose a systematic random sample of every fifth country is selected starting with the fifth country on the list. From this sample, test the null hypothesis that the mean population growth between the years 1990 and 2000 is equal to 1.5%. Test using $\alpha = 0.05$.

b. Now compute the average population growth rate for all 232 countries. After examining the result of the hypothesis test in part a, what type of statistical error, if any, was committed? Explain your answer.

**9-22.** Hono Golf is a manufacturer of golf products in Taiwan and China. One of the golf accessories it produces at its plant in Tainan Hsing, Taiwan, is plastic golf tees. The injector molder produces golf tees that are designed to have an average height of 66 mm. To determine if this specification is met, random samples are taken from the production floor. One sample is contained in the file labeled **THeight**.

a. Determine if the process is not producing the tees to specification. Use a significance level of 0.01.

b. If the hypothesis test determines the specification is not being met, the production process will be shut down while causes and remedies are determined. At times, this occurs even though the process is functioning to specification. What type of statistical error is this?

**Hypothesis Tests for a Proportion**

So far, this chapter has focused on hypothesis tests about a single population mean. Although many decision problems involve a test of a population mean, there are also cases in which the parameter of interest is the population proportion. For example, a production manager might consider the proportion of defective items produced on an assembly line to determine whether the line should be restructured. Likewise, a life insurance salesperson's performance assessment might include the proportion of existing clients who renew their policies.

OUTCOME **5**

## Testing a Hypothesis about a Single Population Proportion

The basic concepts of hypothesis testing for proportions are the same as those for means:

1. The null and alternative hypotheses are stated in terms of a population parameter, now $p$ instead of $\mu$, and the sample statistic is $\bar{p}$ instead of $\bar{x}$.
2. The null hypothesis should be a statement concerning the parameter that includes the equality.
3. The significance level of the hypothesis determines the size of the rejection region.
4. The test can be one or two tailed, depending on how the alternative hypothesis is formulated.

---

**BUSINESS APPLICATION** | Testing a Hypothesis for a Population Proportion

**Sampson and Koenig Financial Center** The Sampson and Koenig Financial Center purchases installment loans that were originally made by independent appliance dealers and heating and air conditioning installers. Ideally, all loans purchased by Sampson and Koenig will be fully documented. However, the company's internal auditors periodically need to check to make sure the internal controls are being followed. The audit manager examined the documentation on the company's portfolio of 9,460 installment loans. The internal control procedures require that the file on each installment loan account contain certain specific documentation, such as a list of applicant assets, statement of monthly income, list of liabilities, and certificate of automobile insurance. If an account contains all the required documentation, then it complies with company procedures.

The audit manager has established a 1% noncompliance rate as the company's standard. If more than 1% of the 9,460 loans do not have appropriate documentation, then the internal controls are not effective and the company needs to improve the situation. The audit staff does not have enough time to examine all 9,460 files to determine the true population noncompliance rate. As a result, the audit staff selects a random sample of 600 files, examines them, and determines the number of files not in compliance with bank documentation requirements. The sample findings will tell the manager if the bank is exceeding the 1% noncompliance rate for the population of all 9,460 loan files. The manager will not act unless the noncompliance rate exceeds 1%. The default position is that the internal controls are effective. Thus, the null and alternative hypotheses are

$$H_0: p \leq 0.01 \,(\text{internal controls are effective})(\text{Status Quo})$$

$$H_A: p > 0.01 \,(\text{internal controls are not effective})$$

Suppose the sample of 600 accounts uncovered 9 files with inadequate loan documentation. The question is whether 9 out of 600 is sufficient to conclude that the company has a problem. To answer this question statistically, we need to recall a lesson from Chapter 7.

Requirement

> The sample size, $n$, is large such that $np \geq 5$ and $n(1 - p) \geq 5$.[5]

---

[5] A paper published in *Statistical Science* by L. Brown et al. titled "Interval Estimation for a Binomial Proportion" in 2001, pp. 101–133, suggests that the requirement should be $np \geq 15$ and $n(1 - p) > 15$. However, most sources still use the $\geq 5$ limit.

If this requirement is satisfied, the sampling distribution is approximately normal with mean $= p$ and standard deviation $= \sqrt{p(1 - p)/n}$.

The auditors have a general policy of performing these tests with a significance level of

$$\alpha = 0.02$$

They are willing to reject a true null hypothesis 2% of the time. In this case, if a Type I statistical error is committed, the internal controls will be considered ineffective when, in fact, they are working as intended.

Once the null and alternative hypotheses and the significance level have been specified, we can formulate the decision rule for this test. Figure 9.8 shows how the decision rule is developed. Notice the critical value, $\bar{p}_{0.02}$, is 2.05 standard deviations above $p = 0.01$. Thus, the decision rule is:

If $\bar{p} > \bar{p}_{0.02} = 0.0182$, reject $H_0$.

There were 9 deficient files in the sample of 600 files. This means that

$$\bar{p} = 9/600 = 0.015$$

Because

$$\bar{p} = 0.015 < 0.0182, \text{do not reject } H_0.$$

The null hypothesis, $H_0$, should not be rejected, based on these sample data. Therefore, the auditors will conclude the system of internal controls is working effectively.

Alternatively, we could have based the test on a test statistic $(z)$ with a standard normal distribution. This test statistic is calculated using Equation 9.4.

### z-Test Statistic for Proportions

$$z = \frac{\bar{p} - p}{\sqrt{\dfrac{p(1 - p)}{n}}} \tag{9.4}$$

where:

$$\bar{p} = \text{Sample proportion}$$
$$p = \text{Hypothesized population proportion}$$
$$n = \text{Sample size}$$

The $z$-value for this test statistic is

$$z = \frac{0.015 - 0.01}{0.004} = 1.25$$

**FIGURE 9.8 Decision Rule for Sampson and Koenig Example**

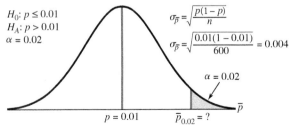

$H_0: p \leq 0.01$
$H_A: p > 0.01$
$\alpha = 0.02$

$\sigma_{\bar{p}} = \sqrt{\dfrac{p(1-p)}{n}}$

$\sigma_{\bar{p}} = \sqrt{\dfrac{0.01(1-0.01)}{600}} = 0.004$

$\alpha = 0.02$

$p = 0.01$    $\bar{p}_{0.02} = ?$    $\bar{p}$

Sample results:
$x = 9$ bad files
$\bar{p} = x/n = 9/600$
$= 0.015$

Solving for $\bar{p}_\alpha$:  $\bar{p}_\alpha = p + z_\alpha \sigma_{\bar{p}}$
From the normal table $z_{0.02} = 2.05$; then:
$\bar{p}_{0.02} = 0.01 + (2.05)(0.004)$
$\bar{p}_{0.02} = 0.0182$

**Decision Rule:**
If $\bar{p} > 0.0182$, reject $H_0$; otherwise, do not reject.
Because $\bar{p} = 0.015 < 0.0182$, do not reject $H_0$.

As was established in Figure 9.8 , the critical value is

$$z_{0.02} = 2.05$$

We reject the null hypothesis only if $z > z_{0.02}$. Because

$$z = 1.25 < 2.05$$

we don't reject the null hypothesis. This, of course, is the same conclusion we reached when we used $\bar{p}$ as the test statistic. Both test statistics must yield the same decision.

---

**HOW TO DO IT** (Example 9-9)

**Testing Hypotheses about a Single Population Proportion**

1. Specify the population parameter of interest.

2. Formulate the null and alternative hypotheses.

3. Specify the significance level for testing the null hypothesis.

4. Construct the rejection region. For a one-tailed test, determine the critical value, $z_\alpha$, from the standard normal distribution table or

$$\bar{p}_\alpha = p + z_\alpha \sqrt{\frac{p(1 - p)}{n}}$$

For a two-tailed test, determine the critical values:

$z_{(\alpha/2)L}$ and $z_{(\alpha/2)U}$ from the standard normal table or

$\bar{p}_{(\alpha/2)L}$ and $\bar{p}_{(\alpha/2)U}$

5. Compute the test statistic:

$$\bar{p} = \frac{x}{n} \text{ or } z = \frac{\bar{p} - p}{\sqrt{\frac{p(1 - p)}{n}}}$$

or determine the $p$-value.

6. Reach a decision by comparing $z$ to $z_\alpha$ or $\bar{p}$ to $\bar{p}_\alpha$ or by comparing the $p$-value to $\alpha$.

7. Draw a conclusion.

---

**EXAMPLE 9-9** **Testing Hypotheses for a Single Population Proportion**

**The Developmental Basketball League** The Developmental League, or D-League for short, is for players who are not on NBA rosters but who want to fine-tune their skills in hopes of getting called up to the NBA. Suppose one of the D-League's teams is considering increasing the season ticket prices for basketball games. The marketing manager is concerned that some people will cancel their ticket orders if this change occurs. If more than 10% of the season ticket orders would be canceled, the marketing manager does not want to implement the price increase. To test this, a random sample of ticket holders is surveyed and asked what they would do if the prices were increased.

STEP 1 **Specify the population parameter of interest.**
The parameter of interest is the population proportion of season ticket holders who would cancel their orders.

STEP 2 **Formulate the null and alternative hypotheses.**
The null and alternative hypotheses are

$$H_0: p \leq 0.10$$
$$H_A: p > 0.10$$

STEP 3 **Specify the significance level.**
The alpha level for this test is $\alpha = 0.05$.

STEP 4 **Determine the critical value.**
1. Using the critical $z$ value:
   The critical value from the standard normal table for this upper-tailed test is $z_\alpha = z_{0.05} = 1.645$.
   The decision rule is:

   If $z > 1.645$, reject $H_0$; otherwise, do not reject.

2. Using the critical $\bar{p}$ value:
   As you learned in Section 9.1, there are alternative approaches to testing a hypothesis. In addition to the $z$-test statistic approach, you could compute the critical value, $\bar{p}_\alpha$, and compare $\bar{p}$ to $\bar{p}_\alpha$. The critical value is computed as follows:

   $$\bar{p}_\alpha = p + z_\alpha \sqrt{\frac{p(1 - p)}{n}}$$

   $$\bar{p}_{0.05} = 0.10 + 1.645 \sqrt{\frac{0.10(1 - 0.10)}{100}} = 0.149$$

   The decision rule is:

   If $\bar{p} > \bar{p}_{0.05} = 0.149$, reject $H_0$. Otherwise, do not reject.

3. Using the $p$-value:
   The decision rule is:

   If $p$-value $< \alpha = 0.05$, reject $H_0$. Otherwise, do not reject.

STEP 5 **Compute the test statistic or the $p$-value.**
The random sample of $n = 100$ season ticket holders showed that 14 would cancel their ticket orders if the price change were implemented.

1. The sample proportion and $z$-test statistic are

$$\bar{p} = \frac{x}{n} = \frac{14}{100} = 0.14$$

$$z = \frac{\bar{p} - p}{\sqrt{\dfrac{p(1-p)}{n}}} = \frac{0.14 - 0.10}{\sqrt{\dfrac{0.10(1-0.10)}{100}}} = 1.33$$

> The Excel 2016 function for the finding a p-value for an upper tail test is
>
> = 1 – Norm.S.Dist(z,True)
>
> = 1 – Norm.S.Dist(1.33,True)

2. Using the $\bar{p}$-critical value:
   The critical $\bar{p}$ value was previously calculated to be $p = 0.149$.
3. Using the $p$-value:
   To find the $p$-value for a one-tailed test, we use the calculated $z$-value shown previously in Step 5 to be $z = 1.33$. Then,

$$p\text{-value} = P(z > 1.33)$$

From the standard normal table, the probability associated with $z = 1.33$, that is, $P(0 \leq z \leq 1.33)$, is 0.4082. Then

$$p\text{-value} = 0.5 - 0.4082 = 0.0918$$

**STEP 6  Reach a decision.**
1. Using the $z$-test statistic:
   The decision rule is:

$$\text{If } z > z_{0.05}, \text{ reject } H_0.$$

Because $z = 1.33 < 1.645$, do not reject $H_0$.
2. Using the $\bar{p}$-critical value:
   The decision rule is:

$$\text{If } \bar{p} > \bar{p}_{0.05}, \text{ reject } H_0.$$

Because $0.14 < 0.149$, do not reject $H_0$. This is the same decision we reached using the $z$-test statistic approach.
3. Using the $p$-value:
   The decision rule is:

$$\text{If } p\text{-value} < \alpha = 0.05, \text{ reject } H_0.$$

Because $p$-value $= 0.0918 > 0.05$, do not reject $H_0$.
All three hypothesis-testing approaches provide the same decision.

**STEP 7  Draw a conclusion.**
Based on the sample data, the marketing manager does not have sufficient evidence to conclude that more than 10% of the season ticket holders will cancel their ticket orders.

TRY EXERCISE 9-28 (pg. 342)

## 9.2 EXERCISES

### Skill Development

**9-23.** Determine the appropriate critical value(s) for each of the following tests concerning the population proportion:
   a. upper-tailed test: $\alpha = 0.025, n = 48$
   b. lower-tailed test: $\alpha = 0.05, n = 30$
   c. two-tailed test: $\alpha = 0.02, n = 129$
   d. two-tailed test: $\alpha = 0.10, n = 36$

**9-24.** Calculate the $z$-test statistic for a hypothesis test in which the null hypothesis states that the population proportion, $p$, equals 0.40 if the following sample information is present:

$$n = 150, x = 30$$

**9-25.** Given the following null and alternative hypotheses

$$H_0: p \geq 0.60$$
$$H_A: p < 0.60$$

test the hypothesis using $\alpha = 0.01$ assuming that a sample of $n = 200$ yielded $x = 105$ items with the desired attribute.

**9-26.** For the following hypothesis test:

$$H_0: p = 0.40$$
$$H_A: p \neq 0.40$$
$$\alpha = 0.01$$

with $n = 64$ and $\bar{p} = 0.42$, state

a. the decision rule in terms of the critical value of the test statistic
b. the calculated value of the test statistic
c. the conclusion

**9-27.** For the following hypothesis test:

$$H_0: p \geq 0.75$$
$$H_A: p < 0.75$$
$$\alpha = 0.025$$

with $n = 100$ and $\bar{p} = 0.66$, state

a. the decision rule in terms of the critical value of the test statistic
b. the calculated value of the test statistic
c. the conclusion

**9-28.** A test has the following hypotheses:

$$H_0: p \leq 0.45$$
$$H_A: p > 0.45$$

The sample size is 30, and the sample proportion is 0.55.

a. For an $\alpha = 0.025$, determine the critical value.
b. Calculate the numerical value of the test statistic.
c. State the test's conclusion.
d. Determine the $p$-value.

**9-29.** A sample of size 25 was obtained to test the hypotheses

$$H_0: p = 0.30$$
$$H_A: p \neq 0.30$$

Calculate the $p$-value for each of the following sample results:

a. $\bar{p} = 0.12$
b. $\bar{p} = 0.35$
c. $\bar{p} = 0.42$
d. $\bar{p} = 0.5$

## Business Applications

**9-30.** Suppose a recent random sample of employees nationwide who have a 401(k) retirement plan found that 18% of them had borrowed against it in the last year. A random sample of 100 employees from a local company who have a 401(k) retirement plan found that 14 had borrowed from their plan. Based on the sample results, is it possible to conclude, at the $\alpha = 0.025$ level of significance, that the local company had a lower proportion of borrowers from its 401(k) retirement plan than the 18% reported nationwide?

**9-31.** An issue that faces individuals investing for retirement is allocating assets among different investment choices. Suppose a study conducted 10 years ago showed that

65% of investors preferred stocks to real estate as an investment. In a recent random sample of 900 investors, 360 preferred real estate to stocks. Are these new data sufficient to allow you to conclude that the proportion of investors preferring stocks to real estate has declined from 10 years ago? Conduct your analysis at the $\alpha = 0.02$ level of significance.

**9-32.** A major issue facing many states is whether to legalize casino gambling. Suppose the governor of one state believes that more than 55% of the state's registered voters would favor some form of legal casino gambling. However, before backing a proposal to allow such gambling, the governor has instructed his aides to conduct a statistical test on the issue. To do this, the aides have hired a consulting firm to survey a simple random sample of 300 voters in the state. Of these 300 voters, 175 actually favored legalized gambling.

a. State the appropriate null and alternative hypotheses.
b. Assuming that a significance level of 0.05 is used, what conclusion should the governor reach based on these sample data? Discuss.

**9-33.** An article on the Federal Trade Commission's website stated that 13% of complaints received were about identity theft (*source:* "Identity theft tops FTC's consumer complaint categories again in 2014," www.ftc.gov, Feb. 27, 2015). Suppose a random sample of 90 complaints is obtained. Of these complaints, 14 were regarding identity theft. Based on these sample data, what conclusion should be reached about the statement made by the article? (Test using $\alpha = 0.10$.)

**9-34.** Suppose the general manager for a company that produces a complicated electronics component used in the navigation systems in automobiles has stated that 69% of the items made pass through the final inspection without needing any rework. The quality manager wished to verify this statement. To do so, she selected a random sample of $n = 180$ parts at the final inspection station and found that 105 passed and did not require rework.

a. State the appropriate null and alternative hypotheses for testing the general manager's claim.
b. Using alpha $= 0.05$, from the sample data, what conclusion should the quality manager reach about the general manager's claim? Discuss.

**9-35.** An ABC/Washington Post poll indicates that 45% of U.S. adults favor a ban on assault weapons (*source:* Gary Langer, "Most now oppose an assault weapons ban; doubts about stopping a lone wolf run high (poll), http://abcnews.go.com, Dec. 16, 2015). Suppose 300 adults in California are randomly surveyed and 162 say they favor a ban. Based on this sample, test using alpha $= 0.025$ whether the proportion of California residents who favor the ban is different from the proportion given in the poll for the country as a whole. Report the $p$-value for this hypothesis test.

**9-36.** Assume the sports page of your local newspaper reported that 65% of males over the age of 17 in the

United States would skip an important event such as a birthday party or an anniversary dinner to watch their favorite professional sports team play. A random sample of 676 adult males over the age of 17 in the Dallas-Fort Worth market reveals that 507 would be willing to skip an important event to watch their favorite team play. Given the results of the survey, can you conclude that the proportion of adult males who would skip an important event to watch their favorite team play is greater in the Dallas-Fort Worth area than in the nation as a whole? Conduct your test at the $\alpha = 0.01$ level of significance.

**9-37.** Many banks and other financial institutions have raised the minimum payment customers must pay on their outstanding credit card balances. Suppose a claim is made that more than 40% of all credit card holders pay the minimum payment. To test this claim, a random sample of payments made by credit card customers was collected. The sample contained data for 400 customers, of which 174 paid the minimum payment.
   a. State the appropriate null and alternative hypotheses.
   b. Based on the sample data, test the null hypothesis using an alpha level equal to 0.05. Discuss the results of the test.

**9-38.** A Harris poll indicated that 57% of Americans prefer to watch movies at home instead of going to a theater (*source:* Pamela Gocobachi, "Adults prefer staying home over going to the movies, survey says," www. ew.com, Jan. 10, 2014). Suppose students in a communications class at a major university polled 400 individuals in the region in which the university is located and found that 220 people preferred watching movies at home. Based on these sample data, can we conclude that the people in this region have a lower preference for watching movies at home than the Harris poll reported? Test using an alpha = 0.05 level. State the *p*-value for this test.

**9-39.** The practice of "phishing," or using the Internet to pilfer personal information, has become an increasing concern, not only for individual computer users but also for online retailers and financial institutions. Suppose a credit union in Oregon instituted an extensive online security and education program in an effort to combat phishing. The credit union's managers believe that if more than 10% of their customers have cut back on their Internet banking transactions because of fears about phishing, they will be required to take more stringent action to lower this percentage.

   The credit union's information technology department analyzed 200 randomly selected accounts and determined that 24 had cut back on their Internet banking transactions.
   a. State the appropriate null and alternative hypotheses for this situation.
   b. Using $\alpha = 0.05$ and the *p*-value approach, indicate whether the sample data support taking more stringent action.

**9-40.** A large number of complaints have been received in the past six months regarding airlines losing fliers' baggage. Suppose one airline spokesman claimed that less than 1% of all bags fail to arrive at the destination with the passenger. To test this claim, 800 bags were randomly selected at various airports in the United States when they were checked with this airline. Of these, 6 failed to reach the destination when the passenger (owner) arrived.
   a. Is this sufficient evidence to support the airline spokesman's claim? Test using a significance level of 0.05. Discuss.
   b. Estimate the proportion of bags that fail to arrive at the proper destination using a technique for which 95% confidence applies.

**9-41.** Suppose a national survey indicated that 56% of women have had experience with a global positioning system (GPS) device. The survey indicated that 66% of the men surveyed have used a GPS device.
   a. If the survey was based on a sample size of 200 men, do these data indicate that the proportion of men is the same as the proportion of women who have had experience with a GPS device? Use a significance level of 0.05.
   b. Obtain the *p*-value for the test indicated in part a.

## Computer Software Exercises

**9-42.** An article on the BBC website states that of more than 25 billion e-mail messages monitored by the company Symantec, 46.4% were classified as junk (*source:* "Spam email levels at 12-year low," www.bbc.com, July 17, 2015). Suppose a company installed a new system for identifying junk messages and then diverting them to the spam folder. A random sample of 1,000 incoming messages was selected, and of these, 488 were declared to be junk. If we assume that the 46.4% figure is the true percentage, then based on the sample data, does it look like this company receives a higher proportion of junk messages than the population in general? Test using an alpha = 0.01 level.

**9-43.** Suppose 40% of investors have funded a Roth IRA rollover from an employer-sponsored retirement account. The HR manager for a Florida company selected a random sample of 90 households in the greater Miami area, and respondents were asked whether they had ever funded a Roth IRA account with a rollover from an employer-sponsored retirement plan. The results are in the file **Miami Rollover**. Based on the sample data, can you conclude at the 0.10 level of significance that the proportion of households in the greater Miami area that have funded a Roth IRA with a rollover is different from the proportion for all households reported in the study?

**9-44.** According to the Forbes website (www.forbes.com), in 2012, more than 32% of all U.S. employees chose to participate in an employer-sponsored 401(k) retirement plan. Suppose researchers at a university in Georgia sampled employees in Atlanta with the aim of testing the Forbes claim. The results of the survey can be found in the file **Atlanta Retirement**. Based on the survey results, can the researchers conclude that the participation rate for employees in Atlanta is higher than 32%? Conduct your analysis at the 0.025 level of significance.

**9-45.** The Electronic Controls Company (ECCO) is one of the largest makers of backup alarms in the world. Backup alarms are the safety devices that emit a high-pitched beeping sound when a truck, forklift, or other equipment is operated in reverse. ECCO is well known in the industry for its high quality and excellent customer service, but some products are returned under warranty due to quality problems. ECCO's operations manager recently stated that less than half of the warranty returns are wiring-related problems. To verify if she is correct, a company intern was asked to select a random sample of warranty returns and determine the proportion that were returned due to wiring problems. The data the intern collected are shown in the data file called **ECCO**.

a. State the appropriate null and alternative hypotheses.

b. Conduct the hypothesis test using $\alpha = 0.02$ and provide an interpretation of the result of the hypothesis test in terms of the operation manager's claim.

**9-46.** A study of cell phone users found cell phones give people personal power, allow unprecedented mobility, and give people the ability to conduct their business on the go. Interesting enough, gender differences can be found in phone use. Women see their cell phone as a means of expression and social communication, whereas males tend to use it as an interactive toy. According to the Pew Research Institute (www.pewinternet.org), in 2015, 66% of women owned a cell phone.

a. State the appropriate null and alternative hypotheses for testing the Pew Research Institute claim.

b. Based on a random sample of women shown in the data file called **Phone Survey**, test the null hypothesis. (Use $\alpha = 0.05$.)

**9-47.** Suppose that the Seadoo of Canada company standard for customer satisfaction is 95% "highly satisfied." Company managers recently completed a customer survey of 700 customers from around the world. The responses to the question "What is your overall level of satisfaction with Seadoo?" are provided in the file called **Seadoo**.

a. State the appropriate null and alternative hypotheses to be tested.

b. Using an alpha level of 0.05, conduct the hypothesis test and discuss the conclusions.

---

**9.3** | **Type II Errors**

Sections 9.1 and 9.2 provided several examples that illustrated how hypotheses and decision rules for tests of the population mean or population proportion are formulated. In these examples, we determined the critical values by first specifying the significance level, alpha: the maximum allowable probability of committing a Type I error. As we indicated, if the cost of committing a Type I error is high, the decision maker will want to specify a small significance level.

This logic provides a basis for establishing the critical value for the hypothesis test. However, it ignores the possibility of committing a Type II error. Recall that a Type II error occurs if a false null hypothesis is "accepted." The probability of a Type II error is given by the symbol $\beta$, the Greek letter *beta*. We discussed in Section 9.1 that $\alpha$ and $\beta$ are inversely related. That is, if we make $\alpha$ smaller, $\beta$ will increase. However, the two are not proportional. A case in point: Cutting $\alpha$ in half will not necessarily double $\beta$.

**OUTCOME 6**    **Calculating Beta**

Once $\alpha$ has been specified for a hypothesis test involving a particular sample size, $\beta$ cannot also be specified. Rather, the $\beta$ value is fixed for any specified value in that alternative hypothesis, and all the decision maker can do is calculate it. Therefore, $\beta$ is not a single value; it depends on the selected value taken from the range of values in the alternative hypothesis. Because a Type II error occurs when a false null hypothesis is "accepted" (refer to Figure 9.1, "do not reject $H_0$" block), there is a $\beta$ value for each possible population value for which the alternative hypothesis is true. To calculate beta, we must first specify a "what-if" value for a true population parameter taken from the alternative hypothesis. Then, $\beta$ is computed conditional on that parameter being true. Keep in mind that $\beta$ is computed before the sample is taken, so its value is not dependent on the sample outcome.

For instance, if the null hypothesis is that the mean income for a population is equal to or greater than \$30,000, then $\beta$ could be calculated for any value of $\mu$ less than \$30,000. We would get a different $\beta$ for each value of $\mu$ in that range. An application will help clarify this concept.

**BUSINESS APPLICATION**   Calculating the Probability of a Type II Error

**Westberg Products**   Westberg Products designs, manufactures, and distributes products for customers such as Lowe's and Home Depot. Westberg has developed a new low-cost, energy-saving light bulb to last more than 700 hours on average. If a hypothesis test could confirm this, the company would use the "greater than 700 hours" statement in its advertising. The null and alternative hypotheses are

$$H_0: \mu \leq 700 \text{ hours}$$
$$H_A: \mu > 700 \text{ hours (research hypothesis)}$$

A type II error will occur if a false null hypothesis is "accepted." The null hypothesis is false for all possible values of $\mu > 700$ hours. Thus, for each of the infinite number of possibilities for $\mu > 700$, a value of $\beta$ can be determined. (*Note:* $\sigma$ is assumed to be a known value of 15 hours.)

Figure 9.9 shows how $\beta$ is calculated if the true value of $\mu$ is 701 hours. By specifying the significance level to be 0.05 and a sample size of 100 bulbs, the chance of committing a Type II error is approximately 0.8365. This means that if the true population mean is 701 hours, there is nearly an 84% chance that the sampling plan Westberg Products is using will not reject the null hypothesis that the mean is 700 hours or less.

Figure 9.10 shows that if the "what-if" mean value ($\mu = 704$) is farther from the hypothesized mean ($\mu = 700$), beta becomes smaller. The greater the difference between the mean specified in $H_0$ and the mean selected from $H_A$, the easier it is to tell the two apart, and the less likely we are to fail to reject the null hypothesis when it is actually false. Of course, the opposite is also true. As the mean selected from $H_A$ moves increasingly closer to the mean specified in $H_0$, the harder it is for the hypothesis test to distinguish between the two and the higher $B$ will be.

**FIGURE 9.9 Beta Calculation for "True" $\mu = 701$**

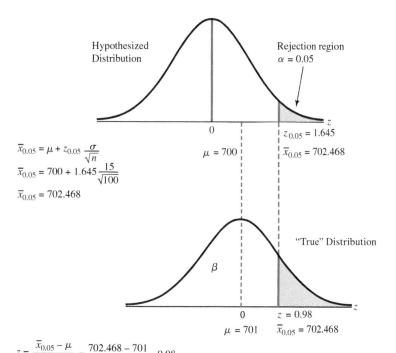

$$\bar{x}_{0.05} = \mu + z_{0.05}\frac{\sigma}{\sqrt{n}}$$
$$\bar{x}_{0.05} = 700 + 1.645\frac{15}{\sqrt{100}}$$
$$\bar{x}_{0.05} = 702.468$$

$$z = \frac{\bar{x}_{0.05} - \mu}{\frac{\sigma}{\sqrt{n}}} = \frac{702.468 - 701}{\frac{15}{\sqrt{100}}} = 0.98$$

From the standard normal table, $P(0 \leq z \leq 0.98) = 0.3365$
$\beta = 0.5000 + 0.3365 = 0.8365$

**FIGURE 9.10 Beta
Calculation for "True"
$\mu = 704$**

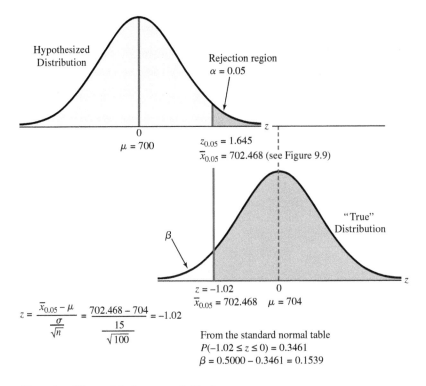

$$z = \frac{\overline{x}_{0.05} - \mu}{\frac{\sigma}{\sqrt{n}}} = \frac{702.468 - 704}{\frac{15}{\sqrt{100}}} = -1.02$$

From the standard normal table
$P(-1.02 \le z \le 0) = 0.3461$
$\beta = 0.5000 - 0.3461 = 0.1539$

## Controlling Alpha and Beta

Ideally, we want both alpha and beta to be as small as possible. Although we can set alpha at any desired level, for a specified sample size and standard deviation, the calculated value of beta depends on the population mean chosen from the alternative hypothesis and on the significance level. For a specified sample size, reducing alpha will increase beta. However, we can simultaneously control the size of both alpha and beta if we are willing to change the sample size.

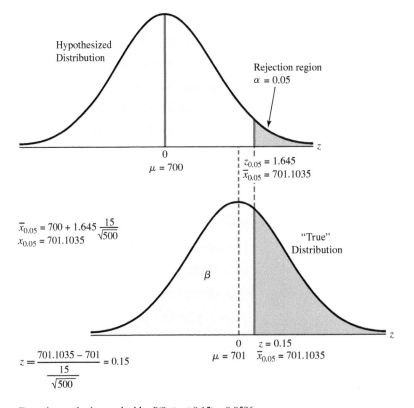

$$z = \frac{701.1035 - 701}{\frac{15}{\sqrt{500}}} = 0.15$$

From the standard normal table, $P(0 \le z \le 0.15) = 0.0596$
$\beta = 0.5000 + 0.0596 = 0.5596$

Westberg Products planned to take a sample of 100 light bulbs. In Figure 9.9, we showed that beta = 0.8365 when the "true" population mean was 701 hours. This is a very large probability and would be unacceptable to the company. However, if the company is willing to incur the cost associated with a sample size of 500 bulbs, the probability of a Type II error could be reduced to 0.5596, as shown in Figure 9.11. This is a big improvement and is due to the fact that the standard error $(\sigma/\sqrt{n})$ is reduced because of the increased sample size.

---

**HOW TO DO IT** (Example 9-10)

**Calculating Beta**

1. Specify the population parameter of interest.

2. Formulate the null and alternative hypotheses.

3. Specify the significance level. (*Hint:* Draw a picture of the hypothesized sampling distribution showing the rejection region(s) and the "acceptance" region found by specifying the significance level.)

4. Determine the critical value, $z_\alpha$, from the standard normal distribution.

5. Determine the critical value, $\bar{x}_\alpha = \mu + z_\alpha\sigma/\sqrt{n}$ for an upper-tailed test, or $\bar{x}_\alpha = \mu - z_\alpha\sigma/\sqrt{n}$ for a lower-tailed test.

6. Specify the stipulated value for $\mu$, the "true" population mean for which you wish to compute $\beta$. (*Hint:* Draw the "true" distribution immediately below the hypothesized distribution.)

7. Compute the z-value based on the stipulated population mean as

$$z = \frac{\bar{x}_\alpha - \mu}{\dfrac{\sigma}{\sqrt{n}}}$$

8. Use the standard normal table to find $\beta$, the probability associated with "accepting" (not rejecting) the null hypothesis when it is false.

---

**EXAMPLE 9-10** Calculating Beta

**Goldman Tax Software** Goldman Tax Software develops software for use by individuals and small businesses to complete federal and state income tax forms. The company has claimed its customers save an average of more than $200 each by using the Goldman software. A consumer group plans to randomly sample 64 customers to test this claim. The standard deviation of the amount saved is assumed to be $100. Before testing, the consumer group is interested in knowing the probability that it will make a Type II error by mistakenly concluding that the mean savings is less than or equal to $200 when, in fact, it does exceed $200, as the company claims. To find beta if the true population mean is $210, the company can use the following steps:

**STEP 1** **Specify the population parameter of interest.**
The consumer group is interested in the mean savings of Goldman Tax Software clients, $\mu$.

**STEP 2** **Formulate the null and alternative hypotheses.**
The null and alternative hypotheses are

$$H_0: \mu \leq \$200$$
$$H_A: \mu > \$200 \text{ (claim)}$$

**STEP 3** **Specify the significance level.**
The one-tailed hypothesis test will be conducted using $\alpha = 0.05$.

**STEP 4** **Determine the critical value, $z_\alpha$, from the standard normal distribution.**
The critical value from the standard normal is $z_\alpha = z_{0.05} = 1.645$.

**STEP 5** **Calculate the $\bar{x}_\alpha$ critical value.**

$$\bar{x}_{0.05} = \mu + z_{0.05}\frac{\sigma}{\sqrt{n}} = 200 + 1.645\frac{100}{\sqrt{64}} = 220.56$$

Thus, the null hypothesis will be rejected if $\bar{x} > 220.56$.

**STEP 6** **Specify the stipulated value for $\mu$.**
The null hypothesis is false for all values greater than $200. What is beta if the stipulated population mean is $210?

**STEP 7** **Compute the z-value based on the stipulated population mean.**

$$z = \frac{\bar{x}_{0.05} - \mu}{\dfrac{\sigma}{\sqrt{n}}} = \frac{220.56 - 210}{\dfrac{100}{\sqrt{64}}} = 0.84$$

**STEP 8** **Determine beta.**
From the standard normal table, the probability associated with $z = 0.84$ is 0.2995. Then $\beta = 0.5000 + 0.2995 = 0.7995$. There is a 0.7995 probability that the hypothesis test will lead the consumer group to mistakenly believe that the mean tax savings is less than or equal to $200 when, in fact, the mean savings is $210.

**TRY EXERCISE 9-49** (pg. 351)

**EXAMPLE 9-11** **Calculating Beta for a Two-Tailed Test**

**Davidson Tree and Landscape** Davidson Tree and Landscape provides trees to home and garden centers in the United States and Canada. One product is the Norway maple. The ideal tree diameter for shipment to the nurseries is 2.25 inches. If the diameter is too large or small, it seems to affect the trees' ability to survive transport and planting. Before each large shipment is sent out, quality managers for Davidson randomly sample 20 Norway maple trees and measure the diameter of each. The standard deviation is tightly controlled as well. Assume $\sigma = 0.005$ inch. Suppose the quality manager is interested in how likely it is that if he conducts a hypothesis test, he will conclude that the mean diameter is equal to 2.25 inches when, in fact, the mean equals 2.255 inches. Thus, he wants to know the probability of a Type II error. To find beta for this test procedure under these conditions, the engineers can use the following steps:

STEP 1 **Specify the population parameter of interest.**
The quality manager is interested in the mean diameter of Norway maple trees, $\mu$.

STEP 2 **Formulate the null and alternative hypotheses.**
The null and alternative hypotheses are

$$H_0: \mu = 2.25 \text{ (status quo)}$$
$$H_A: \mu \neq 2.25$$

STEP 3 **Specify the significance level.**
The two-tailed hypothesis test will be conducted using $\alpha = 0.05$.

STEP 4 **Determine the critical values, $z_{(\alpha/2)L}$ and $z_{(\alpha/2)U}$, from the standard normal distribution.**
The critical value from the standard normal is $z_{(\alpha/2)L}$ and $z_{(\alpha/2)U} = \pm z_{0.025} = \pm 1.96$.

STEP 5 **Calculate the $\bar{x}_{(\alpha/2)L}$ and $\bar{x}_{(\alpha/2)U}$ critical values.**

$$\bar{x}_{L,U} = \mu \pm z_{0.025}\frac{\sigma}{\sqrt{n}} = 2.25 \pm 1.96\frac{0.005}{\sqrt{20}} \rightarrow \bar{x}_L = 2.2478; \bar{x}_U = 2.2522$$

Thus, the null hypothesis will be rejected if $\bar{x} < 2.2478$ or $\bar{x} > 2.2522$.

STEP 6 **Specify the stipulated value of $\mu$.**
The stipulated value of $\mu$ is 2.255.

STEP 7 **Compute the $z$-values based on the stipulated population mean.**

$$z = \frac{\bar{x}_L - \mu}{\frac{\sigma}{\sqrt{n}}} = \frac{2.2478 - 2.255}{\frac{0.005}{\sqrt{20}}} = -6.44 \text{ and } z = \frac{\bar{x}_U - \mu}{\frac{\sigma}{\sqrt{n}}} = \frac{2.2522 - 2.255}{\frac{0.005}{\sqrt{20}}} = -2.50$$

STEP 8 **Determine beta and reach a conclusion.**
Beta is the probability from the standard normal distribution between $z = -6.44$ and $z = -2.50$. From the standard normal table, we get

$$(0.5000 - 0.5000) + (0.5000 - 0.4938) = 0.0062$$

Thus, beta = 0.0062. There is a very small chance (only 0.0062) that this hypothesis test will fail to detect that the mean diameter has shifted to 2.255 inches from the desired mean of 2.25 inches. This low beta will give the quality manager confidence that his test can detect problems when they occur.

**TRY EXERCISE 9-59** (pg. 352)

As shown in Section 9.2, many business applications involve hypotheses tests about population proportions rather than population means. Example 9-12 illustrates the steps needed to compute beta for a hypothesis test involving proportions.

---

**EXAMPLE 9-12**  **Calculating Beta for a Test of a Population Proportion**

**Small Business Startups**    A Wall Street Journal online article stated that more than 30% of all new businesses in 2014 were started by women (*source:* Ruth Simon, "Women started smaller percentage of U.S. businesses in 2014," www.wsj.com, May 13, 2015). Suppose the Economic Development Office in an eastern state said the percentage was closer to 25%. As a result, the director planned to conduct a test to determine if the percentage of small businesses started by women was less than 30%. Additionally, he was quite interested in determining the probability (beta) of "accepting" the claim ($\geq 30\%$) of the online article if in fact the true percentage was 25%. A simple random sample of 500 small businesses will be selected.

**STEP 1**  **Specify the population parameter of interest.**
  The director is interested in the proportion of female-started small businesses, $p$.

**STEP 2**  **Formulate the null and alternative hypotheses.**

$$H_0: p \geq 0.30 \text{ (claim)}$$
$$H_A: p < 0.30$$

**STEP 3**  **Specify the significance level.**
  The one-tailed hypothesis test will be conducted using $\alpha = 0.025$.

**STEP 4**  **Determine the critical value, $z_\alpha$, from the standard normal distribution.**

$$z_\alpha = z_{0.025} = -1.96$$

**STEP 5**  **Calculate the $\bar{p}_\alpha$ critical value.**

$$\bar{p}_{0.025} = p - z_{0.025}\sqrt{\frac{p(1-p)}{n}} = 0.30 - 1.96\sqrt{\frac{0.30(1-0.30)}{500}} = 0.2598$$

  Therefore, the null hypothesis will be rejected if $\bar{p} < 0.2598$.

**STEP 6**  **Specify the stipulated value for the "true" $p$.**
  The stipulated value is 0.25.

**STEP 7**  **Compute the $z$-value based on the stipulated population proportion.**

$$z = \frac{\bar{p}_\alpha - p}{\sqrt{\dfrac{p(1-p)}{n}}} = \frac{0.2598 - 0.25}{\sqrt{\dfrac{0.25(1-0.25)}{500}}} = 0.51$$

**STEP 8**  **Determine beta.**
  From the standard normal table, the probability associated with $z = 0.51$ is 0.1950. Then $\beta = 0.5000 - 0.1950 = 0.3050$. Thus, there is a 0.3050 chance that the hypothesis test will "accept" the null hypothesis that the percentage of women-started small businesses is $\geq 30\%$ if in fact the true percentage is only 25%. The Economic Development Office may wish to increase the sample size to reduce beta.

**TRY EXERCISE 9-57** (pg. 352)

---

As you now know, hypothesis tests are subject to error. The two potential statistical errors are:

  Type I (rejecting a true null hypothesis)
  Type II (failing to reject or "accepting" a false null hypothesis)

In most business applications, adverse consequences are associated with each type of error. In some cases, the errors can mean dollar costs to a company. For instance, suppose a health insurance company is planning to set its premium rates based on a hypothesized mean annual claims amount per participant as follows:

$$H_0: \mu \leq \$1,700$$
$$H_A: \mu > \$1,700$$

If the company tests the hypothesis and "accepts" the null, it will institute the planned premium rate structure. However, if a Type II error is committed, the actual average claim will exceed $1,700, and the company will incur unexpected payouts and suffer reduced profits. On the other hand, if the company rejects the null hypothesis, it will probably increase its premium rates. But if a Type I error is committed, there would be no justification for the rate increase, and the company may not be competitive in the marketplace and could lose customers.

In other cases, the costs associated with either a Type I or a Type II error may be even more serious. If a drug company's hypothesis tests for a new drug incorrectly conclude that the drug is safe when in fact it is not (Type I error), the company's customers may become ill or even die as a result. You might refer to reports dealing with pain medications such as Vicodin. On the other hand, a Type II error would mean that a potentially useful and safe drug would most likely not be made available to people who need it if the hypothesis tests incorrectly determined that the drug was not safe.

In the U.S. legal system where a defendant is hypothesized to be innocent, a Type I error by a jury would result in a conviction of an innocent person. DNA evidence has recently resulted in a number of convicted people being set free. A case in point is Hubert Geralds, who was convicted of killing Rhonda King in 1994 in the state of Illinois. On the other hand, Type II errors in our court system result in guilty people being set free to potentially commit other crimes.

The bottom line is that as a decision maker using hypothesis testing, you need to be aware of the potential costs associated with both Type I and Type II statistical errors and conduct your tests accordingly.

## Power of the Test

In the previous examples, we have been concerned about the chance of making a Type II error. We would like beta to be as small as possible. If the null hypothesis is false, we want to reject it. Another way to look at this is that we would like the hypothesis test to have a high probability of rejecting a false null hypothesis. This concept is expressed by what is called the **power** of the test. When the alternative hypothesis is true, the power of the test is computed using Equation 9.5.

**Power**

The probability that the hypothesis test will correctly reject the null hypothesis when the null hypothesis is false.

**Power**

$$\text{Power} = 1 - \beta \tag{9.5}$$

Refer again to the business application involving Westberg Products. Beta calculations were presented in Figures 9.9, 9.10, and 9.11. For example, in Figure 9.9, the company was interested in the probability of a Type II error if the "true" population mean was 701 hours instead of the hypothesized mean of 700 hours. This probability, called beta, was shown to be 0.8365. Then for this same test,

$$\text{Power} = 1 - \beta$$
$$\text{Power} = 1 - 0.8365 = 0.1635$$

Thus, in this situation, there is only a 0.1635 chance that the hypothesis test will correctly reject the null hypothesis that the mean is 700 or fewer hours when in fact it really is 701 hours.

In Figure 9.10, when a "true" mean of 704 hours was considered, the value of beta dropped to 0.1539. Likewise, the power is increased:

$$\text{Power} = 1 - 0.1539 = 0.8461$$

So the probability of correctly rejecting the null hypothesis increases to 0.8461 when the "true" mean is 704 hours.

We also saw in Figure 9.11 that an increase in the sample size resulted in a lower beta value. For a "true" mean of 701 but with a sample size increase from 100 to 500, the value of beta dropped from 0.8365 to 0.5596. That means that the power increases from 0.1635 to 0.4404 due to the increased size of the sample.

**FIGURE 9.12 Westberg Products—Power Curve**

$H_0: \mu \leq 700$
$H_A: \mu > 700$
$a = 0.05$
$\bar{x}_{0.05} = 702.468$

| "True" $\mu$ | $z = \dfrac{702.468 - \mu}{\dfrac{15}{\sqrt{100}}}$ | $\beta$ | Power $(1-\beta)$ |
|---|---|---|---|
| 700.5 | 1.31 | 0.9049 | 0.0951 |
| 701 | 0.98 | 0.8365 | 0.1635 |
| 702 | 0.31 | 0.6217 | 0.3783 |
| 703 | −0.36 | 0.3594 | 0.6406 |
| 704 | −1.02 | 0.1539 | 0.8461 |
| 705 | −1.69 | 0.0455 | 0.9545 |
| 706 | −2.36 | 0.0091 | 0.9909 |

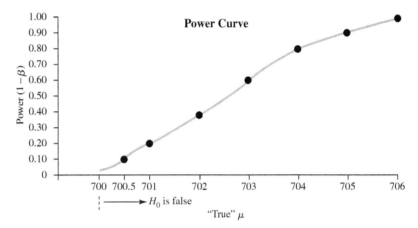

**Power Curve**

A graph showing the probability that the hypothesis test will correctly reject a false null hypothesis for a range of possible "true" values for the population parameter.

A graph called a **power curve** can be created to show the power of a hypothesis test for various levels of the "true" population parameter. Figure 9.12 shows the power curve for the Westberg Products application for a sample size of $n = 100$.

---

## 9.3 EXERCISES

### Skill Development

**9-48.** You are given the following null and alternative hypotheses:

$$H_0: \mu = 200$$
$$H_A: \mu \neq 200$$
$$\alpha = 0.10$$

Calculate the probability of committing a Type II error when the population mean is 197, the sample size is 36, and the population standard deviation is known to be 24.

**9-49.** You are given the following null and alternative hypotheses:

$$H_0: \mu = 1.20$$
$$H_A: \mu \neq 1.20$$
$$\alpha = 0.10$$

a. If the true population mean is 1.25, determine the value of beta. Assume the population standard deviation is known to be 0.50 and the sample size is 60.

b. Referring to part a, calculate the power of the test.

c. Referring to parts a and b, what could be done to increase the power and reduce beta when the true population mean is 1.25? Discuss.

d. Indicate clearly the decision rule that would be used to test the null hypothesis, and determine what decision should be made if the sample mean were 1.23.

**9-50.** You are given the following null and alternative hypotheses:

$$H_0: \mu \geq 4{,}350$$
$$H_A: \mu < 4{,}350$$
$$\alpha = 0.05$$

a. If the true population mean is 4,345, determine the value of beta. Assume the population standard deviation is known to be 200 and the sample size is 100.

b. Referring to part a, calculate the power of the test.

c. Referring to parts a and b, what could be done to increase the power and reduce beta when the true population mean is 4,345? Discuss.

d. Indicate clearly the decision rule that would be used to test the null hypothesis, and determine what decision should be made if the sample mean were 4,337.50.

**9-51.** You are given the following null and alternative hypotheses:

$$H_0: \mu \leq 500$$
$$H_A: \mu > 500$$
$$\alpha = 0.01$$

Calculate the probability of committing a Type II error when the population mean is 505, the sample size is 64, and the population standard deviation is known to be 36.

**9-52.** Consider the following hypotheses:

$$H_0: \mu \geq 103$$
$$H_A: \mu < 103$$

A sample of size 20 is to be taken from a population with a mean of 100 and a standard deviation of 4. Determine the probability of committing a Type II error for each of the following significance levels:

a. $\alpha = 0.01$

b. $\alpha = 0.025$

c. $\alpha = 0.05$

**9-53.** Solve for beta when the "true" population mean is 103 and the following information is given:

$$H_0: \mu \leq 100$$
$$H_A: \mu > 100$$
$$\alpha = 0.05$$
$$\sigma = 10$$
$$n = 49$$

**9-54.** For each of the following situations, indicate what the general impact on the Type II error probability will be:

a. The alpha level is increased.

b. The "true" population mean is moved farther from the hypothesized population mean.

c. The alpha level is decreased.

d. The sample size is increased.

**9-55.** Consider the following hypotheses:

$$H_0: \mu = 30$$
$$H_A: \mu \neq 30$$

A sample of size 50 is to be taken from a population with a standard deviation of 13. The hypothesis test is to be conducted using a significance level of 0.05. Determine the probability of committing a Type II error when

a. $\mu = 22$

b. $\mu = 25$

c. $\mu = 29$

**9-56.** Consider the following hypotheses:

$$H_0: \mu \leq 201$$
$$H_A: \mu > 201$$

A sample is to be taken from a population with a mean of 203 and a standard deviation of 3. The hypothesis test is to be conducted using a significance level of 0.05. Determine the probability of committing a Type II error when

a. $n = 10$

b. $n = 20$

c. $n = 50$

**9-57.** The following hypotheses are to be tested:

$$H_0: p \leq 0.65$$
$$H_A: p > 0.65$$

A random sample of 500 is taken. Using each set of information following, compute the power of the test.

a. $\alpha = 0.01$, true $p = 0.68$

b. $\alpha = 0.025$, true $p = 0.67$

c. $\alpha = 0.05$, true $p = 0.66$

**9-58.** The following hypotheses are to be tested:

$$H_0: p \geq 0.35$$
$$H_A: p < 0.35$$

A random sample of 400 is taken. Using each set of information following, compute the power of the test.

a. $\alpha = 0.01$, true $p = 0.32$

b. $\alpha = 0.025$, true $p = 0.33$

c. $\alpha = 0.05$, true $p = 0.34$

## Business Applications

**9-59.** According to data from the Environmental Protection Agency (www3.epa.gov), in 2015 the average daily water consumption for a household of four people in the United States was more than 400 gallons. Suppose a state agency plans to test this claim using an alpha level equal to 0.05 and a random sample of 100 households with four people.

a. State the appropriate null and alternative hypotheses.

b. Calculate the probability of committing a Type II error if the true population mean is 385 gallons. Assume that the population standard deviation is known to be 40 gallons.

**9-60.** Swift is the holding company for Swift Transportation Co., Inc., a truckload carrier headquartered in Phoenix, Arizona. Suppose, before Swift switched to its current computer-based billing system, the average payment time from customers was approximately 40 days. Before purchasing the present billing system, Swift performed a test by examining a random sample of 24 invoices to see if the system would reduce the average billing time. The sample indicates that the average payment time is 38.7 days.

a. The company that created the billing system indicates that the system would reduce the average billing time to less than 40 days. Conduct a hypothesis test to determine if the new computer-based billing system would reduce the average billing time to less than 40 days. Assume the standard deviation is known to be 6 days. Use a significance level of 0.025.

b. If the billing system actually reduced the average billing time to 36 days, determine the probability that a wrong decision was made in part a.

**9-61.** Waiters at Finegold's Restaurant and Lounge earn most of their income from tips. Each waiter is required to "tip-out" a portion of tips to the table bussers and hostesses. The manager has based the "tip-out" rate on the assumption that the mean tip is at least 15% of the customer bill. To make sure that this is the correct assumption, he has decided to conduct a test by randomly sampling 60 bills and recording the actual tips.

a. State the appropriate null and alternative hypotheses.

b. Calculate the probability of a Type II error if the true mean is 14%. Assume that the population standard deviation is known to be 2% and that a significance level equal to 0.01 will be used to conduct the hypothesis test.

**9-62.** An Ohio-based company provides a full range of insurance and financial services. Suppose the company reported the results of a survey of 1,097 identity theft victims. The survey shows victims spend an average of 81 hours trying to resolve their cases. If the true average time spent was 81 hours, determine the probability that a test of hypothesis designed to test that the average was less than 85 hours would select the research hypothesis. Use $\alpha = 0.05$ and a standard deviation of 50.

**9-63.** According to accountingweb.com, in 2015 the average starting salary for accounting graduates was at least $53,300. Suppose the American Society for Certified Public Accountants planned to test this claim by randomly sampling 200 accountants who recently graduated.

a. State the appropriate null and alternative hypotheses.

b. Compute the power of the hypothesis test to reject the null hypothesis if the true average starting salary is only $52,000. Assume that the population standard deviation is known to be $4,600 and the test is to be conducted using an alpha level equal to 0.01.

**9-64.** According to an Internet source, per-capita beer consumption in North Dakota is 43.6 gallons per year (*source:* Thomas C. Frohlich, "States drinking the most beer," http://247wallst.com, June 30, 2015). Suppose a survey is designed to determine if the per-capita consumption has changed in 2016. A

hypothesis test is to be conducted using a sample size of 1,500, a significance level of 0.01, and a standard deviation of 40. Determine the probability that the test will be able to correctly detect that the per-capita consumption has changed if it has declined by an average of 4 gallons.

**9-65.** According to an April 28, 2015, report on the website newsroom.aaa.com, the average cost of owning and operating a sedan automobile in the United States in 2015 was $8,968. Suppose the California AAA plans to conduct a survey to determine whether the average cost of owning a sedan in California is less than the $8,968 for the nation as a whole. Assuming that the population standard deviation for the cost of owning a sedan is $750, what is the probability that the hypothesis test will lead to accepting the null hypothesis if the true mean cost in California is actually $8,700.

a. State the null and alternative hypotheses.

b. Calculate the probability of a Type II error. Assume that the test is to be conducted using an alpha level of 0.01 and a sample size of 100 sedan owners.

**9-66.** The union negotiations between labor and management at the Stone Container paper mill in Minnesota hit a snag when management asked labor to take a cut in health insurance coverage. As part of its justification, management claimed that the average insurance claims filed by union employees did not exceed $250 per employee. The union's chief negotiator requested that a sample of 100 employees' records be selected and that this claim be tested statistically. The claim would be accepted if the sample data did not strongly suggest otherwise. The significance level for the test was set at 0.10.

a. State the null and alternative hypotheses.

b. Before the sample was selected, the negotiator was interested in knowing the power of the test if the mean insurance claim was $260. (Assume the standard deviation in claims is $70.00, as determined in a similar study at another plant location.) Calculate this probability for the negotiator.

c. Referring to part b, how would the power of the test change if $\alpha = 0.05$ is used?

## Computer Software Exercises

**9-67.** Suppose, consultants for a national hotel chain conducted a review of hotel bills and found that on average, errors in hotel bills resulted in overpayment of $11.35 per night. To determine if such mistakes are being made at a major hotel chain, the CEO requested a survey yielding the following data:

| | | | | | | | | | |
|---|---|---|---|---|---|---|---|---|---|
| 9.99 | 9.87 | 11.53 | 12.40 | 12.36 | 11.68 | 12.52 | 9.34 | 13.13 | 10.78 |
| 9.76 | 10.88 | 10.61 | 10.29 | 10.23 | 9.29 | 8.82 | 8.70 | 8.22 | 11.01 |
| 12.40 | 9.55 | 11.30 | 10.21 | 8.19 | 10.56 | 8.49 | 7.99 | 8.03 | 10.53 |

The file **OverPay** contains these data.

a. Conduct a hypothesis test with $\alpha = 0.05$ to determine if the average overpayment is smaller than that indicated by the consultants.

b. If the actual average overpayment at the hotel chain was $11 with an actual standard deviation of $1.50, determine the probability that the hypothesis test would correctly indicate that the actual average is less than $11.35.

**9-68.** An industry expert claimed that sporting goods stores average sales of $260 a square foot. A hypothesis test was requested to determine if the data supported the statement made by the expert using an $\alpha = 0.05$ and a sample size of 41. Calculate the probability that the data will indicate that sporting goods stores do average sales of $260 per square foot when the true average is $300 per square foot. Assume a population standard deviation of 100. The file called **Apparel** contains data for a random sample of several competitors' sales per square foot. Use $\alpha = 0.05$.

# 9 Overview

## Summary

### Hypothesis Tests for Means (pg. 317–337)

**OUTCOME 1** Formulate the null and alternative hypotheses for applications involving a single population mean or proportion.

**OUTCOME 2** Identify what Type I and Type II errors are in the context of a business application.

**OUTCOME 3** Correctly formulate a decision rule for testing a hypothesis.

**OUTCOME 4** Use the test statistic, critical value, and $p$-value approaches to test a hypothesis.

- In hypothesis testing, two hypotheses are formulated: the null hypothesis and the alternative hypothesis.
- The **null hypothesis** is a statement about the population parameter that will be rejected only if the sample data provide substantial contradictory evidence. It always contains an equal sign.
- The **alternative hypothesis** is a statement that contains all population values not included in the null hypothesis. If the null hypothesis is rejected, then the alternative hypothesis is deemed to be true.
- The null and alternative hypotheses must be specified correctly so that the results obtained from the test are not misleading.
- Because of sampling error, two possible errors can occur when a hypothesis is tested: Type I and Type II errors.
- A **Type I error** occurs when the null hypothesis is rejected and in fact, it is true.
- The maximum allowable probability of committing a Type I statistical error is called the **significance level**. The significance level is specified by the decision maker who is conducting the test.
- A **Type II error** occurs when the decision maker fails to reject the null hypothesis and it is, in fact, false. Controlling for a Type II error is more difficult than controlling for a Type I error.
- Once the null and alternative hypotheses have been stated and the significance level specified, the decision maker must determine the critical value. The **critical value** is the value corresponding to a significance level that determines those test statistics that lead to rejecting the null hypothesis and those that lead to not rejecting the null hypothesis. The decision maker calculates a test statistic from the sample data and compares it to the critical value. A decision is then made whether to reject or to not reject the null hypothesis.
- A $p$-value is often used to test hypotheses, especially when hypothesis testing is done by computer. The **$p$-value** is the probability (assuming that the null hypothesis is true) of obtaining a test statistic at least as extreme as the test statistic calculated from the sample. If the $p$-value is smaller than the significance level, then the null hypothesis is rejected.
- Hypothesis tests may be either one-tailed or two-tailed.
- A **one-tailed test** is a hypothesis test in which the entire rejection region is located in one tail of the sampling distribution.
- A **two-tailed test** is a hypothesis test in which the entire rejection region is divided evenly into the two tails of the sampling distribution.

### Hypothesis Tests for a Proportion (pg. 338–344)

**OUTCOME 5** Formulate null and alternative hypothesis for applications involving a single population proportion.

- Hypothesis tests for a single population proportion follow the same steps as hypothesis tests for a single population mean. Those steps are:
  - State the null and alternative hypotheses in terms of the population parameter, now $p$ instead of $\mu$.
  - The null hypothesis is a statement concerning the parameter that includes the equal sign.
  - The significance level specified by the decision maker determines the size of the rejection region.

**FIGURE 9.13 Deciding Which Hypothesis-Testing Procedure to Use**

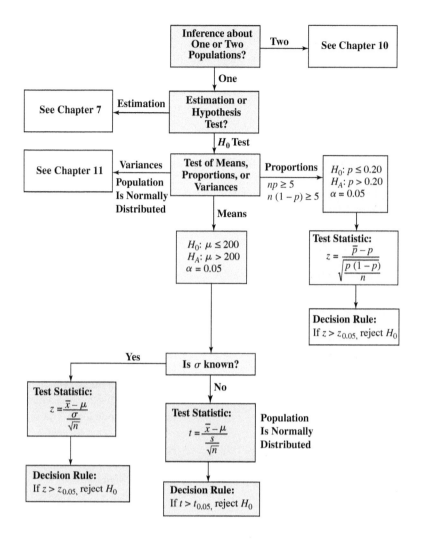

- The test can be a one- or two-tailed test, depending on how the alternative hypothesis is formulated.
- For help deciding which hypothesis-testing procedure to use, refer to Figure 9.13.

## 9.3   Type II Errors (pg. 344–354)

**OUTCOME 6** Compute the probability of a Type II error.

- The probability of committing a Type II error is denoted by $\beta$.
- Once the significance level for a hypothesis test has been specified, $\beta$ cannot also be specified. Rather, $\beta$ is a fixed value and all the decision maker can do is calculate it.
- $\beta$ is not a single value. Because a Type II error occurs when a false null hypothesis is "accepted," there is a $\beta$ value for each possible population value for which the null hypothesis is false.
- To calculate $\beta$, the decision maker must first specify a "what-if" value for the true population parameter. Then $\beta$ is computed before the sample is taken, so its value is not dependent on the sample outcome.
- The size of both $\alpha$ and $\beta$ can be simultaneously controlled if the decision maker is willing to increase the sample size.
- The probability that the hypothesis test will correctly reject the null hypothesis when the null hypothesis is false is referred to as the **power** of the test. The power of the test is computed as $1 - \beta$.
- A **power curve** is a graph showing the probability that the hypothesis test will correctly reject a false null hypothesis for a range of possible "true" values for the population parameter.

# Equations

**(9.1) $\bar{x}_\alpha$ for Hypothesis Tests, $\sigma$ Known** pg. 323

$$\bar{x}_\alpha = \mu + z_\alpha \frac{\sigma}{\sqrt{n}}$$

**(9.2) $z$-Test Statistic for Hypothesis Tests for $\mu$, $\sigma$ Known** pg. 324

$$z = \frac{\bar{x} - \mu}{\frac{\sigma}{\sqrt{n}}}$$

**(9.3) $t$-Test Statistic for Hypothesis Tests for $\mu$, $\sigma$ Unknown** pg. 331

$$t = \frac{\bar{x} - \mu}{\frac{s}{\sqrt{n}}}$$

**(9.4) $z$-Test Statistic for Proportions** pg. 339

$$z = \frac{\bar{p} - p}{\sqrt{\frac{p(1 - p)}{n}}}$$

**(9.5) Power** pg. 350

$$\text{Power} = 1 - \beta$$

# Key Terms

Alternative hypothesis pg. 317
Critical value pg. 322
Null hypothesis pg. 317
One-tailed test pg. 328
$p$-Value pg. 327

Power pg. 350
Power curve pg. 351
Research hypothesis pg. 318
Significance level pg. 322
Test statistic pg. 324

Two-tailed test pg. 328
Type I error pg. 320
Type II error pg. 320

# Chapter Exercises

## Conceptual Questions

**9-69.** What is meant by the term *critical value* in a hypothesis-testing situation? Illustrate what you mean with a business example.

**9-70.** Discuss the issues a decision maker should consider when determining the significance level to use in a hypothesis test.

**9-71.** Discuss the two types of statistical errors that can occur when a hypothesis is tested. Illustrate what you mean by using a business example for each.

**9-72.** Discuss why it is necessary to use an estimate of the standard error for a confidence interval and not for a hypothesis test concerning a population proportion.

**9-73.** Examine the test statistic used in testing a population proportion. Why is it impossible to test the hypothesis that the population proportion equals zero using such a test statistic? Try to determine a way that such a test could be conducted.

**9-74.** Recall that the power of the test is the probability the null hypothesis is rejected when $H_0$ is false. Explain whether power is definable if the given parameter is the value specified in the null hypothesis.

**9-75.** What is the maximum probability of committing a Type I error called? How is this probability determined? Discuss.

**9-76.** In a hypothesis test, indicate the type of statistical error that can be made if
   a. the null hypothesis is rejected
   b. the null hypothesis is not rejected

   c. the null hypothesis is true
   d. the null hypothesis is not true

**9-77.** While conducting a hypothesis test, indicate the effect on
   a. $\beta$ when $\alpha$ is decreased while the sample size remains constant
   b. $\beta$ when $\alpha$ is held constant and the sample size is increased
   c. the power when $\alpha$ is held constant and the sample size is increased
   d. the power when $\alpha$ is decreased and the sample size is held constant

**9-78.** The Mortimor Chemical Company develops and manufactures pharmaceutical drugs for distribution and sale in the United States. The pharmaceutical business can be very lucrative when useful and safe drugs are introduced into the market. Whenever the Mortimor research lab considers putting a drug into production, the company must actually establish the following sets of null and alternative hypotheses:

|  Set 1 | Set 2 |
| --- | --- |
| $H_0$: The drug is not safe. | $H_0$: The drug is not effective. |
| $H_A$: The drug is safe. | $H_A$: The drug is effective. |

Take each set of hypotheses separately.
   a. Discuss the considerations that should be made in establishing alpha and beta.
   b. For each set of hypotheses, describe what circumstances suggest that a Type I error would be of more concern.

c. For each set of hypotheses, describe what circumstances suggest that a Type II error would be of more concern.

9-79. For each of the following scenarios, indicate which test statistic would be used or which test could not be conducted using the materials from this chapter:

a. testing a mean when $\sigma$ is known and the population sampled from has a normal distribution

b. testing a mean when $\sigma$ is unknown and the population sampled from has a normal distribution

c. testing a proportion in which $np = 12$ and $n(1 - p) = 4$

d. testing a mean when $s$ is obtained from a small sample and the population sampled from has a skewed distribution

## Business Applications

9-80. Fairfield Automotive is the local dealership selling Honda automobiles. It recently stated in an advertisement that Honda owners average more than 85,000 miles before trading in or selling their Hondas. To test this, an independent agency selected a simple random sample of 80 Honda owners who have either traded or sold their Hondas and determined the number of miles on the car when the owner parted with the car. It plans to test Fairfield's claim at the $\alpha = 0.05$ level.

a. State the appropriate null and alternative hypotheses.

b. If the sample mean is 86,200 miles and the sample standard deviation is 12,000 miles, what conclusion should be reached about the claim?

9-81. Sanchez Electronics sells electronic components for car stereos. It claims that the average life of a component exceeds 4,000 hours. To test this claim, it has selected a random sample of $n = 12$ of the components and traced the life between installation and failure. The following data were obtained:

| 1,973 | 4,838 | 3,805 | 4,494 | 4,738 | 5,249 |
| 4,459 | 4,098 | 4,722 | 5,894 | 3,322 | 4,800 |

a. State the appropriate null and alternative hypotheses.

b. Assuming that the test is to be conducted using a 0.05 level of significance, what conclusion should be reached based on these sample data? Be sure to examine the required normality assumption.

9-82. The Utah State Tax Commission attempts to set up payroll tax–withholding tables such that by the end of the year, an employee's income tax withholding is about $100 below his actual income tax owed to the state. The commission director claims that when all the Utah tax returns are in, the average additional payment will be less than $100. A random sample of 50 accounts revealed an average additional payment of $114 with a sample standard deviation of $50.

a. Testing at a significance level of 0.10, do the sample data refute the director's claim?

b. Determine the largest sample mean (with the same sample size and standard deviation) that would fail to refute the director's claim.

9-83. Technological changes in golf clubs have affected the way golfers of all abilities are able to play the game. The impact seems to be especially great for professionals. Suppose there is a controversy over whether professional and amateur players should have different rules for what constitutes legal golf clubs and equipment in general. One golf organization has indicated that more than 50% of all golfers believe the rules should be different for professionals and amateurs. To test this, a survey of 300 randomly selected golfers found that 200 said there should be different equipment rules.

a. To test the organization's "more than 50%" claim, what should the null and alternative hypotheses be?

b. Based on the sample data and alpha = 0.05, use the $p$-value approach to test whether the null hypothesis should be rejected.

9-84. An advertising agency on the West Coast formulated a media advertising campaign for an online retailer and claims that its ads received "high approval" by at least 21% of adults with incomes over $50,000 per year. Suppose the online retailer wishes to test this claim. It surveys 883 adults with incomes exceeding $50,000 and finds that 168 give the ads "high approval."

a. State the appropriate null and alternative hypotheses to be tested.

b. Based on an alpha level of 0.05, do the sample data provide evidence to reject the ad agency's claim?

9-85. The college of business at a state university has a computer literacy requirement for all graduates. Students must show proficiency with a computer spreadsheet software package and with a word-processing software package. To assess whether students are computer literate, a test is given at the end of each semester. The test is designed so that at least 70% of all students who have taken a special microcomputer course will pass. The college does not wish to declare that fewer than 70% of the students pass the test unless there is strong sample evidence to indicate this. Suppose that, in a random sample of 100 students who have recently finished the microcomputer course, 63 pass the proficiency test.

a. Using a significance level of 0.05, what conclusions should the administrators make regarding the difficulty of the test?

b. Describe a Type II error in the context of this problem.

9-86. The makers of Free Run Dog Food have an automated filling machine that can be set at any targeted fill level between 10 and 40 pounds. At the end of every shift (eight hours), 16 bags are selected at random and the mean and standard deviation of the sample are computed. Based on these sample results, the production control manager determines whether the filling machine needs to be readjusted or whether it remains all right to operate. Previous data suggest the fill level has a normal distribution with a standard deviation of 0.65 pound. Use $\alpha = 0.05$. At the end of a run of 20-pound bags, a sample of 16 bags was taken and tested using a two-tailed test to determine if the mean fill level was equal to 20 pounds.

a. Calculate the probability that the test procedure will detect that the average fill level is not equal to 20 pounds when in fact it equals 20.5 pounds.

b. On the basis of your calculation in part a, would you suggest a change in the test procedure? Explain what change you would make and the reasons you would make this change.

**9-87.** The Nielsen Company reported that 30% of Millennials (ages 21–34) order groceries online (*source:* "The future of grocery: E-commerce, digital technology and changing shopper preferences around the world," www.nielsen.com, Apr. 29, 2015). Suppose a U.S. grocery company wishes to test whether the Nielsen figure applies in its market area. To do this, it surveys 400 Millennials and finds that 132 currently order groceries online.

a. State the appropriate null and alternative hypotheses.

b. Based on an alpha $= 0.05$ level, do the data from the sample indicate that the Millennials in this store's market area order groceries online to a greater or lesser extent than the 30% that the Nielson study indicated?

**9-88.** The Sledge Tire and Rubber Company plans to warranty its new mountain bike tire for 12 months. However, before it does this, the company wants to be sure that the mean lifetime of the tires is at least 18 months under normal operations. The company plans to test this statistically using a random sample of tires. The test will be conducted using an alpha level of 0.03.

a. If the population mean is actually 16.5 months, determine the probability the hypothesis test will lead to incorrectly failing to reject the null hypothesis. Assume that the population standard deviation is known to be 2.4 months and the sample size is 60.

b. If the population mean is actually 17.3 months, calculate the chance of committing a Type II error. This is a specific example of a generalization relating the probability of committing a Type II error and the parameter being tested. State this generalization.

**9-89.** The personnel manager for a large airline has claimed that, on average, workers are asked to work no more than 3 hours overtime per week. Assume that the standard deviation in overtime hours per worker is 1.2 hours.

Suppose the union negotiators wish to test this claim by sampling payroll records for 250 employees. They believe that the personnel manager's claim is untrue but want to base their conclusion on the sample results.

a. State the research, null, and alternative hypotheses and discuss the meaning of Type I and Type II errors in the context of this case.

b. Establish the appropriate decision rule if the union wishes to have no more than a 0.01 chance of a Type I error.

c. The payroll records produced a sample mean of 3.15 hours. Do the union negotiators have a basis for a grievance against the airline? Support your answer with a relevant statistical procedure.

**9-90.** The Lazer Company has a contract to produce a part for Boeing Corporation that must have an average diameter of 6 inches and a standard deviation of 0.10 inch. The Lazer Company has developed the process that will meet the specifications with respect to the standard deviation, but it is still trying to meet the mean specifications. A test run (considered a random sample) of parts was produced, and the company wishes to determine whether this latest process that produced the sample will produce parts meeting the requirement of an average diameter equal to 6 inches.

a. Specify the appropriate research, null, and alternative hypotheses.

b. Develop the decision rule assuming that the sample size is 200 parts and the significance level is 0.01.

c. What should the Lazer Company conclude if the sample mean diameter for the 200 parts is 6.03 inches? Discuss.

**9-91.** The senior vice president for human resources at a high-tech company often initiates surveys concerning the company's personnel. A typical survey asked, "Do you feel it's OK for your company to monitor your Internet use?" Of the 405 respondents, 223 chose "Only after informing me." The company would consider monitoring if more than 50% of its workers wouldn't mind if informed beforehand that the company was going to monitor their Internet usage. Test statistically to determine if the proportion of workers who wouldn't object to the company monitoring their Internet use after they were informed is more than 50%. Use a significance level of 0.05.

**9-92.** Many companies have changed their employee retirement plans to ones based on 401(k) savings. An article in a business magazine states that the average 401(k) balance is $91,800 (*source:* Donna Rosato, "401(k) savings hit a record high. How do you stack up?" http://time.com/money, Apr. 29, 2015). A regional brokerage firm wishes to test whether its clients' accounts average more than this amount. A firm manager selects a random sample of 55 clients and finds a sample average of $94,785.

a. If the standard deviation for the amounts in the clients' accounts is $17,340, determine if the manager is correct in her assumption that the firm's clients have more than the reported average amount. Use a significance level of 0.025 and discuss any assumptions you made to answer this question.

b. Based on your answer to part a, what type of statistical error could have been made?

## Computer Software Exercises

**9-93.** The Air Frame Company sells cellular phones and airtime in several northwestern states. The marketing manager stated that the average age of its customers is under 40. This came up in conjunction with a proposed advertising plan that is to be directed toward a young audience. Before actually completing the advertising plan, Air Frame decided to randomly sample customers. Among the questions asked in the survey of 50 customers from the northwest was the customer's age. The data are available in a data file called **Cell Phone Survey**.

  a. Based on the statement made by the marketing manager, formulate the appropriate null and alternative hypotheses.

  b. The marketing manager must support his statement concerning average customer age in an upcoming board meeting. Using a significance level of 0.10, provide this support for the marketing manager.

  c. Consider the result of the hypothesis test you conducted in part b. Which of the two types of hypothesis test errors could you have committed? How could you discover if you had, indeed, made this error?

  d. Calculate the critical value, $\bar{x}_\alpha$.

  e. Determine the $p$-value and conduct the test using the $p$-value approach.

**9-94.** The AJ Fitness Center has surveyed 1,214 of its customers. Of particular interest is whether more than 60% of the customers who express overall service satisfaction with the club (represented by codes 4 or 5) are female. If this is not the case, the promotions director feels she must initiate new exercise programs that are designed specifically for women. Should the promotions director initiate the new exercise programs? Support your answer with the relevant hypothesis test utilizing a $p$-value to perform the test. The data are found in a data file called **AJ Fitness** ($\alpha = 0.05$).

**9-95.** To determine whether the Wilson Company is complying with federal requirements for cleaning its wastewater, sample measurements are taken every so often. One requirement is that the average pH level not exceed 7.4. Suppose a sample of 95 pH measurements has been taken. The data for these measurements are shown in a file called **Wilson Water**.

  a. Considering the requirement for pH level, state the appropriate null and alternative hypotheses. Discuss why it is appropriate to form the hypotheses with the federal standard as the alternative hypothesis.

  b. Based on the sample data of pH levels, what should the company conclude about its current status on meeting the federal requirement? Test the hypothesis at the 0.05 level. Discuss your results in a memo to the company's environmental relations manager.

**9-96.** The Haines Lumber Company makes plywood for the furniture industry. One of its products is 3/4-inch oak veneer panels. It is very important that the panels conform to specifications. One specification calls for the panels to be made to an average thickness of 0.75 inch. Suppose

each hour, 5 panels are selected at random and measured. After 20 hours, a total of 100 panels have been measured. The thickness measures are in a file called **Haines**.

  a. Formulate the appropriate null and alternative hypotheses relative to the thickness specification.

  b. Based on the sample data, what should the company conclude about the status of its product meeting the thickness specification? Test at a significance level of 0.01. Discuss your results in a report to the production manager.

**9-97.** The Inland Empire Food Store Company has stated in its advertising that the average shopper will save more than $5.00 per week by shopping at Inland stores. A consumer group has decided to test this assertion by sampling 50 shoppers who currently shop at other stores. Suppose the company selects the customers and then notes each item purchased at their regular stores. These same items are then priced at the Inland store, and the total bills are compared. The data in the file **Inland Foods** reflect the savings at Inland for the 50 shoppers. Note that those cases where the bill was higher at Inland are marked with a minus sign.

  a. Set up the appropriate null and alternative hypotheses to test Inland's claim.

  b. Using a significance level of 0.05, develop the decision rule and test the hypothesis. Can Inland support its advertising claim?

  c. Which type of hypothesis error would the consumer group be most interested in controlling? Which type of hypothesis test error would the company be most interested in controlling? Explain your reasoning.

**9-98.** Suppose a bank that issues credit cards claims that its customers' average annual spending per active account is $6,920. An analysis of a random sample of credit card accounts provided the data on annual spending. This sample is in the file labeled **ASpending**.

  a. Conduct a hypothesis test to determine if the average annual spending per active account actually exceeds the $6,920 value. Use a $p$-value approach and a significance level of 0.025.

  b. If the annual spending per active account is actually normally distributed with a mean of $5,560 and a standard deviation of $1,140, determine the probability that a randomly chosen account would have an annual spending larger than $6,920.

**9-99.** At the annual meeting of the Golf Equipment Manufacturer's Association, a speaker made the claim that at least 30% of all golf clubs being used are knock-off clubs that look very much like the more expensive originals. This claim prompted the association to conduct a study to see if the problem was as big as the speaker said. Suppose a random sample of 400 golfers was selected. Out of the 400 golfers, data were collected from 294. Based on the golfer's response to the question about the brand of clubs used, a determination was made whether the club was "original" or a "copy." The data are in a file called **Golf Survey**.

a. Based on the sample data, what conclusion should be reached if the hypothesis is tested at a significance level of 0.05? Show the decision rule.
b. Determine whether a Type I or Type II error for this hypothesis test would be more serious. Given your determination, would you advocate raising or lowering the significance level for this test? Explain your reasoning.
c. Confirm that the sample proportion's distribution can be approximated by a normal distribution.
d. Based on the sample data, what conclusion should be drawn about the use of knock-off clubs? Is the speaker's statement justified?

**9-100.** TOMRA Systems ASA is a Norwegian company that manufactures reverse vending machines (RVMs). In most cases, RVMs are used in markets that have deposits on beverage containers, offering an efficient and convenient method of identifying the deposit amount of each container returned and providing a refund to the customer. For a single-container machine to pay for itself in one year, it would need to generate an average monthly income of more than $750. A random sample of single-machine monthly incomes was obtained to determine if that goal could be reached.

This sample is contained in the file labeled **RVM-Income**. Conduct a hypothesis test to determine if the goal can be reached. Use a significance level of 0.05 and the *p*-value approach.

---

| **Case 9.1** | **Carpita Bottling Company—Part 2** |
| --- | --- |

Don Carpita and his younger brother, Edward, purchased Carpita Bottling Company from their father in 1983. The company bottles and distributes soda and beer in Wisconsin. Since purchasing the company, Don has been instrumental in modernizing operations.

One of the latest acquisitions is a filling machine that can be adjusted to fill at any average level desired. Because the bottles and cans filled by the company are exclusively the 12-ounce size, when they received the machine, Don set the fill level to 12 ounces and left it that way. According to the manufacturer's specifications, the machine will fill bottles or cans around the average, with a standard deviation of 0.15 ounce.

Don just returned from a convention at which he attended a panel discussion related to problems with filling machines. One bottling company representative discussed a problem her company had. It failed to learn that its machine's average fill went out of adjustment until several months later, when its cost accounting department reported some problems with beer production in bulk not matching output in bottles and cans. It turns out that the machine's average fill had increased from 12 ounces to 12.07 ounces. With large volumes of production, this deviation meant a substantial loss in profits.

Another company reported the same type of problem, but in the opposite direction. Its machine began filling bottles with slightly less than 12 ounces on the average. Although the consumers could not detect the shortage in a given bottle, the state and federal agencies responsible for checking the accuracy of packaged products discovered the problem in their testing and substantially fined the company for the underfill.

These problems were a surprise to Don. He had not considered the possibility that the machine might go out of adjustment and pose these types of problems. In fact, he became very concerned because the problems of losing profits and potentially being fined by the government were ones that he wished to avoid, if possible. After the convention, Don and Ed decided to hire a consulting firm with expertise in these matters to assist them in setting up a procedure for monitoring the performance of the filling machine.

The consultant suggested that they set up a sampling plan in which once a month, they would sample some number of bottles and measure their volumes precisely. If the average of the sample deviated too much from 12 ounces, they would shut the machine down and make the necessary adjustments. Otherwise, they would let the filling process continue. The consultant identified two types of problems that could occur from this sort of sampling plan:

1. They might incorrectly decide to adjust the machine when it was not really necessary to do so.
2. They might incorrectly decide to allow the filling process to continue when, in fact, the true average had deviated from 12 ounces.

After carefully considering what the consultant told them, Don indicated that he wanted no more than a 0.02 chance of the first problem occurring because of the costs involved. He also decided that if the true average fill had slipped to 11.99 ounces, he wanted no more than a 0.05 chance of not detecting this with his sampling plan. He wanted to avoid problems with state and federal agencies. Finally, if the true average fill had actually risen to 12.007 ounces, he wanted to be able to detect this 98% of the time with his sampling plan. Thus, he wanted to avoid the lost profits that would result from such a problem.

In addition, Don needs to determine how large a sample size is necessary to meet his requirements.

---

| **Case 9.2** | **Wings of Fire** |
| --- | --- |

Following his graduation from college, Tony Smith wanted to continue to live and work in Oxford. However, the community was small and there were not a lot of readily available opportunities for a new college graduate. Fortunately, Tony had some experience working in the food service industry gained in the summers and throughout high school at his uncle's restaurant in Buffalo. When Tony decided to leverage his experience into a small delivery and take-out restaurant located close to the university, he thought he had hit on a great idea. Tony would offer limited fare consisting of the buffalo wings his uncle had perfected at his restaurant. Tony called his restaurant Wings of Fire. Although success came slowly, the uniqueness of Tony's offering coupled with the growth of the university community made Wings of Fire a success.

Tony's business was pretty simple. Tony purchased wings locally. The wings were then seasoned and prepared in Tony's restaurant. Once an order was received, Tony cooked the wings, which were then delivered or picked up by the customer. Tony's establishment was small, and there was no place for customers to dine in the restaurant. However, his wings proved so popular that over time, Tony hired several employees, including three delivery drivers. Business was steady and predictable during the week, with the biggest days being home-football Saturdays.

A little over a year ago, Oxford really began to grow and expand. Tony noticed that his business was beginning to suffer when other fast-food delivery restaurants opened around campus. Some of these restaurants were offering guarantees such as "30 minutes or it's free." Tony's Wings of Fire now had to compete with fish tacos, specialty pizzas, and gourmet burgers. Most of these new restaurants, however, were dine-in establishments that provided carry-out and delivery as a customer convenience. However, Tony was certain that he would need to offer a delivery guarantee to remain competitive with the newer establishments.

Tony was certain that a delivery guarantee of "30 minutes or it's free" could easily be accomplished every day except on football Saturdays. Tony thought that if he could offer a 30-minute guarantee on his busiest day, he would be able to hold onto and perhaps even recover market share from the competition. However, before he was willing to commit to such a guarantee, Tony wanted to ensure that it was possible to meet the 30-minute promise.

Tony knew it would be no problem for customers to pick up orders within 30 minutes of phoning them in. However, he was less confident about delivering orders to customers in 30 minutes or less. Not only would the wings need to be cooked and packaged, but the delivery time might be affected by the availability of drivers. Tony decided that he needed to analyze the opportunity further.

As a part of his analysis, Tony decided to take a random sample of deliveries over five different football weekends. He did not consider cooking time and packaging time in his analysis because wings were not cooked for individual orders. Rather, large numbers of wings were cooked at a single time and then packaged in boxes of 12. Tony therefore decided to focus his analysis on the time required to deliver cooked and packaged wings. He collected information on the amount of time an order had to wait for a driver (the pick-up time) as well as the amount of time required to transport the wings to the customer (the drive time). The sampled information is in the file **Wings of Fire**. Tony is not willing to offer the guarantee on football Saturdays unless he can be reasonably sure that the total time to deliver a customer's order is less than 30 minutes, on average. Tony would also like to have an estimate of the actual time required to deliver a customer's order on football Saturdays. Finally, Tony would like to know how likely it is that the total time to make a delivery would be longer than 30 minutes. Based on the sampled data, should Tony offer the guarantee? What percent of the Saturday deliveries would result in a customer receiving a free order? What recommendations might help Tony improve his Saturday delivery times?

## Required Tasks:

1. Use the sample information to compute a measure of performance that Tony can use to analyze his delivery performance.
2. State a hypothesis test that would help Tony decide whether or not to offer the delivery guarantee.
3. Calculate sample statistics and formally test the hypothesis stated in (2).
4. Estimate the probability of an order taking longer than 30 minutes.
5. Summarize your findings and make a recommendation in a short report.

# 10 Estimation and Hypothesis Testing for Two Population Parameters

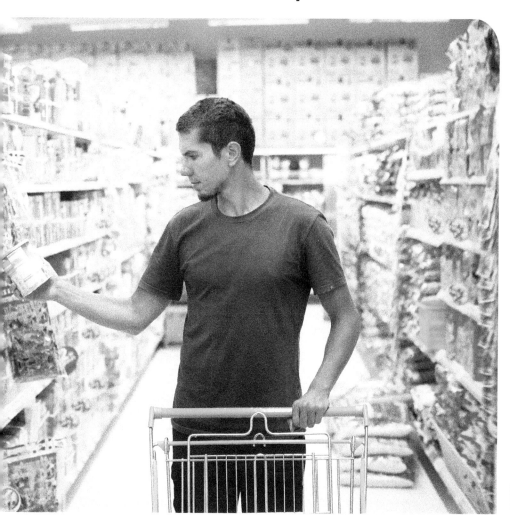

## WHY YOU NEED TO KNOW

Chapter 9 introduced the concepts of hypothesis testing and illustrated its application through examples involving a single population parameter. However, in many business decision-making situations, managers must decide between two or more alternatives. For example, every day consumers make choices between products from two or more competing brands. Fleet managers in large companies must decide which model and make of car to purchase next year. Airlines must decide whether to purchase replacement planes from Boeing or Airbus. When deciding on a new advertising campaign, a company may need to evaluate proposals from competing advertising agencies. Hiring decisions may require a personnel

### Quick Prep

Review material on calculating and interpreting sample means and standard deviations in Chapter 3.

Review the normal distribution in Section 6.1.

Make sure you understand the concepts associated with sampling distributions for $\bar{x}$ and $\bar{p}$ by reviewing Sections 7.1, 7.2, and 7.3.

Review the steps for developing confidence interval estimates for a single population

mean and a single population proportion in Sections 8.1 and 8.3.

Review the methods for testing hypotheses about single population means and single population proportions in Chapter 9.

director to select one employee from a list of applicants. Production managers are often confronted with decisions concerning whether to change a production process or leave it alone.

In this chapter, we introduce techniques that you can use to make statistical comparisons between two populations. In a later chapter, we will discuss methods to extend this comparison to more than two populations. Whether we are discussing cases involving two populations or more than two populations, the techniques we present are all extensions of the statistical tools involving a single population parameter introduced in Chapters 8 and 9.

| 10.1 | Estimation for Two Population Means Using Independent Samples |
|---|---|

In this section, we examine situations in which we are interested in the difference between two population means, looking first at the case in which the samples from the two populations are **independent**.

**Independent Samples**

Samples selected from two or more populations in such a way that the occurrence of values in one sample has no influence on the occurrence of values in the other sample(s).

We will introduce techniques for estimating the difference between the means of two populations in the following situations:

1. The population standard deviations are known and the samples are independent.
2. The population standard deviations are unknown and the samples are independent.

OUTCOME **1**

### Estimating the Difference between Two Population Means When $\sigma_1$ and $\sigma_2$ Are Known, Using Independent Samples

Recall that in Chapter 8 we used the standard normal distribution $z$-values to establish the critical value and develop the interval estimate when the population standard deviation is assumed known and the population distribution is assumed to be normally distributed.[1] The general format for a confidence interval estimate is shown in Equation 10.1.

**Confidence Interval, General Format**

$$\text{Point estimate} \pm (\text{Critical value})(\text{Standard error})$$ **(10.1)**

In business situations, you will often need to estimate the difference between two population means. For instance, you may wish to estimate the difference in mean starting salaries between male and female accounting graduates, the difference in mean production output in union and nonunion factories, or the difference in mean service times at two different fast-food businesses. In these situations, the best point estimate for $\mu_1 - \mu_2$ is

$$\text{Point estimate} = \bar{x}_1 - \bar{x}_2$$

In situations in which you know the population standard deviations, $\sigma_1$ and $\sigma_2$, and when the samples selected from the two populations are independent, an extension of the Central Limit Theorem tells us that the sampling distribution for all possible differences between $\bar{x}_1$ and $\bar{x}_2$ is approximately normally distributed with a standard error computed as shown in Equation 10.2.

**Standard Error of $\bar{x}_1 - \bar{x}_2$ When $\sigma_1$ and $\sigma_2$ Are Known**

$$\sigma_{\bar{x}_1 - \bar{x}_2} = \sqrt{\frac{\sigma_1^2}{n_1} + \frac{\sigma_2^2}{n_2}}$$ **(10.2)**

where:

$$\sigma_1^2 = \text{Variance of Population 1}$$
$$\sigma_2^2 = \text{Variance of Population 2}$$
$$n_1 \text{ and } n_2 = \text{Sample sizes of Populations 1 and 2}$$

---

[1]If the samples from the two populations are large ($n \geq 30$), the normal distribution assumption is not required.

Further, the critical value for determining the confidence interval is a $z$-value from the standard normal distribution. In these circumstances, the confidence interval estimate for $\mu_1 - \mu_2$ is found by using Equation 10.3.

**Confidence Interval Estimate for $\mu_1 - \mu_2$ When $\sigma_1$ and $\sigma_2$ Are Known, Independent Samples**

$$(\bar{x}_1 - \bar{x}_2) \pm z\sqrt{\frac{\sigma_1^2}{n_1} + \frac{\sigma_2^2}{n_2}}$$    **(10.3)**

> The Excel 2016 function for the critical $z$-value is
>
> = ABS(NORM.S.INV((1 −
>    Confidence Level)/2))
> *Note:* The ABS function provides a positive $z$-value.

The $z$-values for several of the most commonly used confidence levels are

| Confidence Level | Critical $z$-value |
|---|---|
| 80% | $z = 1.28$ |
| 90% | $z = 1.645$ |
| 95% | $z = 1.96$ |
| 99% | $z = 2.575$ |

**EXAMPLE 10-1**

## Confidence Interval Estimate for $\mu_1 - \mu_2$ When $\sigma_1$ and $\sigma_2$ Are Known, Using Independent Samples

**Healthy Body**    Healthy Body is a small chain of fitness centers located primarily in the South but with some clubs scattered in other parts of the United States and Canada. The club in Winston-Salem, North Carolina, worked with a business class from a local university on a project in which a team of students observed Healthy Body customers with respect to their club usage. As part of the study, the students measured the time that customers spent in the club during a visit. The objective is to estimate the difference in mean times spent per visit for male and female customers. Previous studies indicate that the standard deviation is 11 minutes for males and 16 minutes for females. To develop a 95% confidence interval estimate for the difference in mean times, the following steps are taken:

STEP 1   **Define the population parameter of interest and select independent samples from the two populations.**

In this case, the company is interested in estimating the difference in mean times spent in the club for males and females. The measure of interest is $\mu_1 - \mu_2$.

The student team has selected simple random samples of 100 males and 100 females at different times in the Winston-Salem club.

STEP 2   **Specify the desired confidence level.**
The plan is to develop a 95% confidence interval estimate.

STEP 3   **Compute the point estimate.**
The resulting sample means are

Males: $\bar{x}_1 = 34.5$ minutes        Females: $\bar{x}_2 = 42.4$ minutes

The point estimate is

$$\bar{x}_1 - \bar{x}_2 = 34.5 - 42.4 = -7.9 \text{ minutes}$$

Women in the sample spent an average of 7.9 minutes longer in the club.

STEP 4   **Determine the standard error of the sampling distribution.**
The standard error is calculated as

$$\sigma_{\bar{x}_1 - \bar{x}_2} = \sqrt{\frac{\sigma_1^2}{n_1} + \frac{\sigma_2^2}{n_2}} = \sqrt{\frac{11^2}{100} + \frac{16^2}{100}} = 1.9416$$

The Excel 2016 function for the critical $z$-value is

$=$ABS(NORM.S.INV((1 −

  Confidence Level)/2))

$=$ABS(NORM.S.INV((1 − .95)/

  2))

STEP 5 **Determine the critical value, $z$, from the standard normal table.**

The interval estimate will be developed using a 95% confidence interval. Because the population standard deviations are known, the critical value is a $z$-value from the standard normal table. The critical value is

$$z = 1.96$$

STEP 6 **Develop the confidence interval estimate using Equation 10.3.**

$$(\bar{x}_1 - \bar{x}_2) \pm z\sqrt{\frac{\sigma_1^2}{n_1} + \frac{\sigma_2^2}{n_2}}$$

$$-7.9 \pm 1.96\sqrt{\frac{11^2}{100} + \frac{16^2}{100}}$$

$$-7.9 \pm 3.8056$$

The 95% confidence interval estimate for the difference in mean times spent in the Winston-Salem Healthy Body center for men and women is

$$-11.7056 \le (\mu_1 - \mu_2) \le -4.0944$$

Thus, based on the sample data and the specified confidence level, women spend on average between 4.09 and 11.71 minutes longer at this Healthy Body Fitness Center.

TRY EXERCISE 10-4 (pg. 372)

OUTCOME **1** ## Estimating the Difference between Two Population Means When $\sigma_1$ and $\sigma_2$ Are Unknown, Using Independent Samples

In Chapter 8, you learned that, when estimating a single population mean when the population standard deviation is unknown, the critical value is a $t$-value from the $t$-distribution. This is also the case when you are interested in estimating the difference between two population means, if the following assumptions hold:

**Assumptions**

- The populations are normally distributed.
- The populations have equal variances.
- The samples are independent.

The following application illustrates how a confidence interval estimate is developed using the $t$-distribution.

**BUSINESS APPLICATION**

### Estimating the Difference between Two Population Means

**Retirement Investing** Many people who have entered the workforce in the past 20 years are actively investing in their own retirement accounts. One investment alternative is a tax-sheltered annuity (TSA) marketed by life insurance companies. Certain people, depending on occupation, qualify to invest part of their paychecks in a TSA and pay no federal income tax on this money until it is withdrawn. The insurance companies invest this money in either stock or bond portfolios. A second alternative open to many people is a plan known as a 401(k), in which employees contribute a portion of their pretax paychecks to purchase stocks, bonds, or mutual funds. In some cases, employers match all or part of the employee contributions. In many 401(k) systems, the employees can control how their funds are invested.

A recent study was conducted in North Dakota to estimate the difference in mean annual contributions for individuals covered by the two plans [TSA or 401(k)]. A simple random sample of 15 people from the population of adults who are eligible for a TSA investment was selected. A second sample of 15 people was selected from the population of adults in North Dakota who have 401(k) plans. The variable of interest is the dollar amount of money invested in the retirement plan during the previous year. Specifically, we are interested in estimating $\mu_1 - \mu_2$ using a 95% confidence interval estimate, where:

$\mu_1$ = Mean dollars invested by the TSA-eligible population during the past year

$\mu_2$ = Mean dollars invested by the 401(k)-eligible population during the past year

| TSA-Eligible | 401(k)-Eligible |
|---|---|
| $n_1 = 15$ | $n_2 = 15$ |
| $\bar{x}_1 = \$2,119.70$ | $\bar{x}_2 = \$1,777.70$ |
| $s_1 = \$709.70$ | $s_2 = \$593.90$ |

The values above were computed from the two samples. Before applying the $t$-distribution, we need to determine whether the assumptions are likely to be satisfied. First, the samples are considered independent because the amount invested by one group should have no influence on the likelihood that any specific amount will be invested for the second sample.

Next, Figure 10.1 shows the sample data and the box and whisker plots for the two samples. These plots exhibit characteristics that are reasonably consistent with normal distributions and approximately equal variances. You should know that the $t$-distribution can still be used even when there are small violations of the assumptions. This is particularly the case when the sample sizes are approximately equal.[2]

Equation 10.4 can be used to develop the confidence interval estimate for the difference between two population means when you have small independent samples.

**Confidence Interval Estimate for $\mu_1 - \mu_2$ When $\sigma_1$ and $\sigma_2$ Are Unknown, Independent Samples**

$$(\bar{x}_1 - \bar{x}_2) \pm ts_p\sqrt{\frac{1}{n_1} + \frac{1}{n_2}} \tag{10.4}$$

where:

$$s_p = \sqrt{\frac{(n_1 - 1)s_1^2 + (n_2 - 1)s_2^2}{n_1 + n_2 - 2}} = \text{Pooled standard deviation}$$

$t$ = Critical $t$-value from the $t$-distribution, with degree of freedom equal to $n_1 + n_2 - 2$

**FIGURE 10.1 Sample Information for the Investment Study: Dollars Invested in Each Retirement Plan**

**Box and Whisker Plot**

| TSA | 401(k) |
|---|---|
| 3,122 | 1,781 |
| 3,253 | 2,594 |
| 2,021 | 1,615 |
| 2,479 | 334 |
| 2,318 | 2,322 |
| 1,407 | 2,234 |
| 2,641 | 2,022 |
| 1,648 | 1,603 |
| 2,439 | 1,395 |
| 1,059 | 1,604 |
| 2,799 | 2,676 |
| 1,714 | 1,773 |
| 951 | 1,156 |
| 2,372 | 2,092 |
| 1,572 | 1,465 |

Box and Whisker Plot
Five-Number Summary

| | TSA | 401(k) |
|---|---|---|
| Minimum | 951 | 334 |
| First Quartile | 1,572 | 1,465 |
| Median | 2,318 | 1,773 |
| Third Quartile | 2,641 | 2,234 |
| Maximum | 3,253 | 2,676 |

[2]Chapter 11 introduces a statistical procedure for testing whether two populations have equal variances. Chapter 13 provides a statistical procedure for testing whether a population is normally distributed.

To use Equation 10.4, we must compute the pooled standard deviation, $s_p$. If the equal-variance assumption holds, then both $s_1^2$ and $s_2^2$ are estimators of the same population variance, $\sigma^2$. To use only one of these, say $s_1^2$, to estimate $\sigma^2$ would be disregarding the information obtained from the other sample. To use the average of $s_1^2$ and $s_2^2$, if the sample sizes were different, would ignore the fact that more information about $\sigma^2$ is obtained from the sample having the larger sample size. We therefore use a weighted average of $s_1^2$ and $s_2^2$, denoted as $s_p^2$, to estimate $\sigma^2$, where the weights are the degrees of freedom associated with each sample. The square root of $s_p^2$ is known as the *pooled standard deviation* and is computed using

$$s_p = \sqrt{\frac{(n_1 - 1)s_1^2 + (n_2 - 1)s_2^2}{n_1 + n_2 - 2}}$$

For the retirement investing example, the pooled standard deviation is

$$s_p = \sqrt{\frac{(n_1 - 1)s_1^2 + (n_2 - 1)s_2^2}{n_1 + n_2 - 2}} = \sqrt{\frac{(15 - 1)(709.7)^2 + (15 - 1)(593.9)^2}{15 + 15 - 2}} = 654.37$$

For situations involving two independent samples, the degrees of freedom for the $t$-distribution is equal to $n_1 + n_2 - 2$. Using the $t$-distribution table, we find that the critical $t$-value for

$$n_1 + n_2 - 2 = 15 + 15 - 2 = 28$$

degrees of freedom and 95% confidence is

$$t = 2.0484$$

> The Excel 2016 function for determining the critical $t$-value is
>
> =T.INV.2T(1−Confidence Level, degrees of freedom)
> =T.INV.2T(1−0.95,28)

Now we can develop the interval estimate using Equation 10.4:

$$(\bar{x}_1 - \bar{x}_2) \pm t s_p \sqrt{\frac{1}{n_1} + \frac{1}{n_2}}$$

$$(2{,}119.70 - 1{,}777.70) \pm 2.0484(654.37)\sqrt{\frac{1}{15} + \frac{1}{15}}$$

$$342 \pm 489.45$$

Thus, the 95% confidence interval estimate for the difference in mean dollars for people who invest in a TSA versus those who invest in a 401(k) is

$$-\$147.45 \leq (\mu_1 - \mu_2) \leq \$831.45$$

This confidence interval estimate crosses zero and therefore indicates there may be no difference between the mean contributions to TSA accounts and to 401(k) accounts by adults in North Dakota. The implication of this result is that the average amount invested by those individuals who invest in pretax TSA programs is no more or no less than that invested by those participating in after-tax 401(k) programs. Thus, based on the sample data there is no reason to believe that one investment is more popular than another.

---

**EXAMPLE 10-2**

## Confidence Interval Estimate for $\mu_1 - \mu_2$ when $\sigma_1$ and $\sigma_2$ Are Unknown, Using Independent Samples

**Andreason Marketing, Inc.**  Andreason Marketing, Inc., has been hired by a major U.S. newspaper to estimate the difference in mean times that newspaper subscribers spend reading the Saturday newspaper when they read a hard copy version or read the paper online. Simple random samples of six print copy readers and eight online readers participated in the study. The estimate can be developed using the following steps:

**STEP 1  Define the population parameter of interest and select independent samples from the two populations.**

The objective here is to estimate the difference between the two groups with respect to the mean time spent reading the Saturday newspaper. The parameter of interest is $\mu_1 - \mu_2$.

The marketing company has selected simple random samples of six print copy readers and eight online readers. Because the reading time for one person does not influence the reading time for any other person, the samples are independent.

**STEP 2  Specify the confidence level.**

The marketing firm wishes to have a 95% confidence interval estimate.

**STEP 3  Compute the point estimate.**

The resulting sample means and sample standard deviations for the two groups are

$$\text{Hard copy:} \quad \bar{x}_1 = 13.6 \text{ minutes} \qquad \text{Online:} \quad \bar{x}_2 = 11.2 \text{ minutes}$$
$$s_1 = 3.1 \text{ minutes} \qquad\qquad\qquad s_2 = 5.0 \text{ minutes}$$
$$n_1 = 6 \qquad\qquad\qquad\qquad n_2 = 8$$

The point estimate is

$$\bar{x}_1 - \bar{x}_2 = 13.6 - 11.2 = 2.4 \text{ minutes}$$

**STEP 4  Determine the standard error of the sampling distribution.**

The pooled standard deviation is computed using

$$s_p = \sqrt{\frac{(n_1 - 1)s_1^2 + (n_2 - 1)s_2^2}{n_1 + n_2 - 2}} = \sqrt{\frac{(6 - 1)3.1^2 + (8 - 1)5^2}{6 + 8 - 2}} = 4.31$$

The standard error is then calculated as

$$s_p\sqrt{\frac{1}{n_1} + \frac{1}{n_2}} = 4.31\sqrt{\frac{1}{6} + \frac{1}{8}} = 2.3277$$

> The Excel 2016 function for determining the critical $t$-value is
>
> $= \text{T.INV.2T}(1 - 0.95, 12)$

**STEP 5  Determine the critical value, $t$, from the $t$-distribution table.**

Because the population standard deviations are unknown, the critical value will be a $t$-value from the $t$-distribution as long as the population variances are equal and the populations are assumed to be normally distributed.

The critical $t$ for 95% confidence and $6 + 8 - 2 = 12$ degrees of freedom is

$$t = 2.1788$$

**STEP 6  Develop a confidence interval using Equation 10.4.**

$$(\bar{x}_1 - \bar{x}_2) \pm ts_p\sqrt{\frac{1}{n_1} + \frac{1}{n_2}}$$

where:

$$s_p = \sqrt{\frac{(n_1 - 1)s_1^2 + (n_2 - 1)s_2^2}{n_1 + n_2 - 2}} = \sqrt{\frac{(6 - 1)3.1^2 + (8 - 1)5^2}{6 + 8 - 2}} = 4.31$$

Then the confidence interval estimate is

$$2.4 \pm 2.1788(4.31)\sqrt{\frac{1}{6} + \frac{1}{8}}$$
$$2.4 \pm 5.0715$$
$$-2.6715 \le (\mu_1 - \mu_2) \le 7.4715$$

Because the interval crosses zero, we cannot conclude that a difference exists between the two types of readers with respect to the mean reading time for the Saturday edition. Thus, with respect to this factor, it does not seem to matter whether the person is a hard copy reader or an online reader.

**TRY EXERCISE 10-1** (pg. 371)

**What If the Population Variances Are Not Equal?**    If you have reason to believe that the population variances are substantially different, Equation 10.4 is not appropriate for computing the confidence interval. Instead of computing the pooled standard deviation as part of the confidence interval formula, we use Equations 10.5 and 10.6.

---

**Confidence Interval for $\mu_1 - \mu_2$ When $\sigma_1$ and $\sigma_2$ Are Unknown and Not Equal, Independent Samples**

$$(\bar{x}_1 - \bar{x}_2) \pm t\sqrt{\frac{s_1^2}{n_1} + \frac{s_2^2}{n_2}} \tag{10.5}$$

where:

$t$ is from the $t$-distribution with degrees of freedom computed using:

**Degrees of Freedom for Estimating Difference between Population Means When $\sigma_1$ and $\sigma_2$ Are Not Equal**

$$df = \frac{(s_1^2/n_1 + s_2^2/n_2)^2}{\left(\dfrac{(s_1^2/n_1)^2}{n_1 - 1} + \dfrac{(s_2^2/n_2)^2}{n_2 - 1}\right)} \tag{10.6}$$

---

**EXAMPLE 10-3**

## Estimating $\mu_1 - \mu_2$ When the Population Variances Are Not Equal

**Citibank**    The marketing managers at Citibank are planning to roll out a new marketing campaign addressed at increasing bank card use. As one part of the campaign, the company will be offering a low-interest-rate incentive to induce people to spend more money using its charge cards. However, the company is concerned whether this plan will have a different impact on married card holders than on unmarried card holders. So, prior to starting the marketing campaign nationwide, the company tests it on a random sample of 30 unmarried and 25 married customers. The managers wish to estimate the difference in mean credit card spending for unmarried versus married for a two-week period immediately after being exposed to the marketing campaign. Based on past data, the managers have reason to believe the spending distributions for unmarried and married will be approximately normally distributed, but they are unwilling to conclude that the population variances for spending are equal for the two populations.

A 95% confidence interval estimate for the difference in population means can be developed using the following steps:

STEP 1  **Define the population parameter of interest.**
The parameter of interest is the difference between the mean dollars spent on credit cards by unmarried versus married customers in the two-week period after being exposed to Citi's new marketing program.

STEP 2  **Specify the confidence level.**
The research manager wishes to have a 95% confidence interval estimate.

STEP 3  **Compute the point estimate.**
Independent samples of 30 unmarried and 25 married customers were taken, and the credit card spending for each sampled customer during the two-week period was recorded. The following sample results were observed:

|  | Unmarried | Married |
|---|---|---|
| Mean | $455.10 | $268.90 |
| St. Dev. | $102.40 | $ 77.25 |

The point estimate is the difference between the two sample means:

$$\text{Point estimate} = \bar{x}_1 - \bar{x}_2 = 455.10 - 268.90 = 186.20$$

STEP 4  **Determine the standard error of the sampling distribution.**
The standard error is calculated as

$$\sqrt{\frac{s_1^2}{n_1} + \frac{s_2^2}{n_2}} = \sqrt{\frac{102.40^2}{30} + \frac{77.25^2}{25}} = 24.25$$

> The Excel 2016 function for determining the exact critical $t$-value is
> $= \text{T.INV.2T}(1 - 0.95, 52)$

STEP 5  **Determine the critical value, $t$, from the $t$-distribution table.**
Because we are unable to assume the population variances are equal, we must first use Equation 10.6 to calculate the degrees of freedom for the $t$-distribution. This is done as follows:

$$df = \frac{(s_1^2/n_1 + s_2^2/n_2)^2}{\left(\dfrac{(s_1^2/n_1)^2}{n_1 - 1} + \dfrac{(s_2^2/n_2)^2}{n_2 - 1}\right)}$$

$$= \frac{(102.40^2/30 + 77.25^2/25)^2}{\left(\dfrac{(102.40^2/30)^2}{29} + \dfrac{(77.25^2/25)^2}{24}\right)} = \frac{346{,}011.98}{6{,}586.81} = 52.53$$

Thus, the degrees of freedom (rounded down) is 52. At the 95% confidence level, using the $t$-distribution table, the approximate $t$-value is 2.0086. Note that since there is no entry for 52 degrees of freedom in the table, we have selected the $t$-value associated with 95% confidence and 50 degrees of freedom, which provides a slightly larger $t$-value than would have been the case for 52 degrees of freedom. Thus, the interval estimate will be generously wide.

STEP 6  **Develop the confidence interval estimate using Equation 10.5.**
The confidence interval estimate is computed using

$$(\bar{x}_1 - \bar{x}_2) \pm t\sqrt{\frac{s_1^2}{n_1} + \frac{s_2^2}{n_2}}$$

Then the interval estimate is

$$(\$455.10 - \$268.90) \pm 2.0086\sqrt{\frac{102.40^2}{30} + \frac{77.25^2}{25}}$$

$$\$186.20 \pm \$48.72$$

$$\$137.48 \le (\mu_1 - \mu_2) \le \$234.92$$

$$\$137.48 \text{————} \$234.92$$

The test provides evidence to conclude that unmarried customers, after being introduced to the marketing program, spend more than married customers, on average, by anywhere from $137.48 to $234.92 in the two weeks following the marketing campaign. But before concluding that the campaign is more effective for unmarried than married customers, the managers would want to compare these results with data from customer accounts prior to the marketing campaign.

TRY EXERCISE 10-6 (pg. 372)

## 10.1 EXERCISES

### Skill Development

**10-1.** The following information is based on independent random samples taken from two normally distributed populations having equal variances:

| | |
|---|---|
| $n_1 = 15$ | $n_2 = 13$ |
| $\bar{x}_1 = 50$ | $\bar{x}_2 = 53$ |
| $s_1 = 5$ | $s_2 = 6$ |

Based on the sample information, determine the 90% confidence interval estimate for the difference between the two population means.

**10-2.** The following information is based on independent random samples taken from two normally distributed populations having equal variances:

$$n_1 = 24 \qquad n_2 = 28$$
$$\bar{x}_1 = 130 \qquad \bar{x}_2 = 125$$
$$s_1 = 19 \qquad s_2 = 17.5$$

Based on the sample information, determine the 95% confidence interval estimate for the difference between the two population means.

**10-3.** Construct a 90% confidence interval estimate for the difference between two population means given the following sample data selected from two normally distributed populations with equal variances:

| Sample 1 | | | Sample 2 | | |
|---|---|---|---|---|---|
| 29 | 25 | 31 | 42 | 39 | 38 |
| 35 | 35 | 37 | 42 | 40 | 43 |
| 21 | 29 | 34 | 46 | 39 | 35 |

**10-4.** Construct a 95% confidence interval estimate for the difference between two population means based on the following information:

| Population 1 | Population 2 |
|---|---|
| $\bar{x}_1 = 355$ | $\bar{x}_2 = 320$ |
| $\sigma_1 = 34$ | $\sigma_2 = 40$ |
| $n_1 = 50$ | $n_2 = 80$ |

**10-5.** Construct a 95% percent confidence interval for the difference between two population means using the following sample data that have been selected from normally distributed populations with different population variances:

| Sample 1 | | | | Sample 2 | | | |
|---|---|---|---|---|---|---|---|
| 473 | 386 | 406 | 379 | 349 | 359 | 346 | 395 |
| 346 | 438 | 391 | 328 | 398 | 401 | 411 | 384 |
| 388 | 388 | 456 | 429 | 363 | 437 | 388 | 273 |

**10-6.** Two random samples were selected independently from populations having normal distributions. The following statistics were extracted from the samples:

$$\bar{x}_1 = 42.3 \qquad \bar{x}_2 = 32.4$$

a. If $\sigma_1 = 3$ and $\sigma_2 = 2$ and the sample sizes are $n_1 = 50$ and $n_2 = 50$, construct a 95% confidence interval for the difference between the two population means.

b. If $\sigma_1 = \sigma_2$, $s_1 = 3$, and $s_2 = 2$, and the sample sizes are $n_1 = 10$ and $n_2 = 10$, construct a 95%

confidence interval for the difference between the two population means.

c. If $\sigma_1 \neq \sigma_2$, $s_1 = 3$, and $s_2 = 2$, and the sample sizes are $n_1 = 10$ and $n_2 = 10$, construct a 95% confidence interval for the difference between the two population means.

## Business Applications

**10-7.** Barton Industries operates two manufacturing facilities that specialize in doing custom manufacturing work for the semiconductor industry. The facility in Denton, Texas, is highly automated, whereas the facility in Lincoln, Nebraska, has more manual functions. For the past few months, both facilities have been working on a large order for a specialized product. The vice president of operations is interested in estimating the difference in mean times it takes to complete a part on the two lines. To do this, he has requested that a random sample of 15 parts at each facility be tracked from start to finish and the times required be recorded. The following sample data were recorded:

| Denton, Texas | Lincoln, Nebraska |
|---|---|
| $\bar{x}_1 = 56.7$ hours | $\bar{x}_2 = 70.4$ hours |
| $s_1 = 7.1$ hours | $s_1 = 8.3$ hours |

Assuming that the populations are normally distributed with equal population variances, construct and interpret a 95% confidence interval estimate.

**10-8.** A credit card company operates two customer service centers: one in Boise and one in Richmond. Callers to the service centers dial a single number, and a computer program routes callers to the center having the fewest calls waiting. As part of a customer service review program, the credit card center would like to determine whether the average length of a call (not including hold time) is different for the two centers. The managers of the customer service centers are willing to assume that the populations of interest are normally distributed with equal variances. Suppose a random sample of phone calls to the two centers is selected and the following results are reported:

| | Boise | Richmond |
|---|---|---|
| Sample Size | 120 | 135 |
| Sample Mean (seconds) | 195 | 216 |
| Sample St. Dev. (seconds) | 35.10 | 37.80 |

a. Using the sample results, develop a 90% confidence interval estimate for the difference between the two population means.

b. Based on the confidence interval constructed in part a, what can be said about the difference between the average call times at the two centers?

**10-9.** A pet food producer manufactures and then fills 25-pound bags of dog food on two different production lines located in separate cities. In an effort to determine whether differences exist between the average fill rates for the two lines, a random sample of 19 bags from line 1 and a random sample of 23 bags from line 2 were recently selected. Each bag's weight was measured and the following summary measures from the samples were reported:

|  | Production Line 1 | Production Line 2 |
| --- | --- | --- |
| Sample Size, $n$ | 19 | 23 |
| Sample Mean, $\bar{x}$ | 24.96 | 25.01 |
| Sample Standard Deviation, $s$ | 0.07 | 0.08 |

Management believes that the fill rates of the two lines are normally distributed with equal variances.

a. Calculate the point estimate for the difference between the population means of the two lines.

b. Develop a 95% confidence interval estimate of the true mean difference between the two lines.

c. Based on the 95% confidence interval estimate calculated in part b, what can the managers of the production lines conclude about the differences between the average fill rates for the two lines?

**10-10.** Two companies that manufacture batteries for electronics products have submitted their products to an independent testing agency. The agency tested 200 of each company's batteries and recorded the lengths of time the batteries lasted before failure. The following results were determined:

| Company A | Company B |
| --- | --- |
| $\bar{x} = 41.5$ hours | $\bar{x} = 39.0$ hours |
| $s = 3.6$ | $s = 5.0$ |

Based on these data, determine the 95% confidence interval to estimate the difference in average life of the batteries for the two companies. Do these data indicate that one company's batteries will outlast the other company's batteries on average? Explain.

**10-11.** One of the key factors in concrete work is the time it takes for the concrete to "set up." Suppose a concrete supplier is considering a new additive that can be put in the concrete mix to help reduce the setup time. Before going ahead with the additive, the company plans to test it against the current additive. To do this, 14 batches of concrete are mixed using each of the additives. The following results are observed:

| Old Additive | New Additive |
| --- | --- |
| $\bar{x} = 17.2$ hours | $\bar{x} = 15.9$ hours |
| $s = 2.5$ hours | $s = 1.8$ hours |

a. Use these sample data to construct a 90% confidence interval estimate for the difference in mean setup times for the two concrete additives. On the basis of the confidence interval produced, do you agree that the

new additive helps reduce the setup time for cement? (Assume the populations are normally distributed with equal variances.) Explain your answer.

b. Assuming that the new additive is slightly more expensive than the old additive, do the data support switching to the new additive if the managers of the company are primarily interested in reducing the average setup time?

**10-12.** According to the Bureau of Labor Statistics (www. bls.gov), in 2014 men spent an average of 42 hours per week on leisure activities, while women averaged 36.4 hours. Suppose the sample standard deviations were 11.0 and 10.2 hours, respectively. These results were obtained from samples of women and men of size 5,848 and 5,752, respectively. In this study, *leisure* refers to time spent socializing, in passive leisure, in active leisure, volunteering, in pet care, gardening, and recreational child care. Assume that the amounts of leisure time spent by men and women have normal distributions with equal population variances.

a. Determine the pooled estimate of the common populations' standard deviation.

b. Produce the margin of error to estimate the difference of the two population means with a confidence level of 95%.

c. Calculate a 95% confidence interval for the difference in the average leisure time between women and men.

d. Does your result in part c indicate that the average amount of men's leisure time was larger than that of women in 2014? Support your assertions.

**10-13.** Delfonte's Restaurant is a popular Italian eatery in the Bay Area in northern California. Recently, there has been a dispute among the wait staff regarding the amount of tips waiters receive on the lunch shift and on the dinner shift. Before making any adjustments in schedules or other policies, the restaurant manager wants to estimate the difference in the average tips earned by waiters on the lunch schedule and on the dinner schedule. To do this, she has randomly selected 40 tip reports for lunch servers over the past year and 50 tip reports from dinner servers. The following results were observed:

|  | Lunch | Dinner |
| --- | --- | --- |
| Mean Tips | $62.40 | $102.50 |
| St. Dev. Tips | $11.30 | $12.33 |

Assuming equal population variances, develop and interpret a 95% confidence interval estimate for the difference between the population mean tips earned by lunch servers and by dinner servers.

## Computer Software Exercises

**10-14.** Recently, Logston Enterprises was notified by a law firm representing several female employees that a lawsuit was going to be filed claiming that the company gave males preferential treatment when it came to pay raises. The Logston human resources manager has requested that an estimate be made of the difference between mean percentage raises granted to males versus females. Sample data are contained in the file **Logston Enterprises**. She wants you to develop and interpret a 95% confidence interval estimate. She further states that the distribution of percentage raises can be assumed approximately normal, and she expects the population variances to be about equal.

**10-15.** The owner of the A.J. Fitness Center is interested in estimating the difference in mean years that female members have been with the club compared with male members. He wishes to develop a 95% confidence interval estimate. Sample data are in the file called **AJ Fitness**. Assuming that the sample data are approximately normal and that the two populations have equal variances, develop and interpret the confidence interval estimate. Discuss the result.

**10-16.** Berhhart Billiards, Inc., conducted an experiment to measure the speed a cue ball attains when struck by various weighted pool cues. The conjecture is that a light cue generates faster speeds while breaking the balls at the beginning of a game of pool. Anecdotal experience has indicated that a billiard cue weighing less than 19 ounces generates faster speeds. Berhhart used a robotic arm to investigate this claim. The research generated the data given in the file titled **Breakcue**.

a. Calculate the sample standard deviation and mean speed produced by cues in the two weight categories: (1) under 19 ounces and (2) at or above 19 ounces.

b. Calculate a 95% confidence interval for the difference in the average speed of a cue ball generated by each of the weight categories.

c. Is the anecdotal evidence correct? Support your assertion.

d. What assumptions are required so that your results in part b would be valid?

**10-17.** The Federal Reserve reported in its comprehensive Survey of Consumer Finances, released every three years, that the average income of families in the United States increased between in 2010 and 2013. A sample of incomes was taken in 2010 and again in 2013. After being adjusted for inflation, the data that arise from these samples are given in a file titled **Federal Reserve**.

a. Determine the percentage increase indicated by the two samples.

b. Using these samples, produce a 90% confidence interval for the difference in the average family incomes between 2010 and 2013.

c. Is it plausible that there has been no increase in the average income of U.S. families? Support your assertion.

d. How large an error could you have made by using the difference in the sample means to estimate the difference in the population means?

---

OUTCOME **2** | **10.2**

## Hypothesis Tests for Two Population Means Using Independent Samples

You are going to encounter situations that will require you to test whether two populations have equal means or whether one population mean is larger (or smaller) than another. These hypothesis-testing applications are just an extension of the hypothesis-testing process introduced in Chapter 9 for a single population mean. They also build directly on the estimation process introduced in Section 10.1.

In this section, we will introduce hypothesis-testing techniques for the difference between the means of two populations in the following situations:

1. The population standard deviations are known and the samples are independent.
2. The population standard deviations are unknown and the samples are independent.

The remainder of this section presents examples of hypothesis tests for these different situations.

### Testing for $\mu_1 - \mu_2$ When $\sigma_1$ and $\sigma_2$ Are Known, Using Independent Samples

Samples are considered to be *independent* when the samples from the two populations are taken in such a way that the occurrence of values in one sample has no influence on the occurrence of values in the second sample. In special cases in which we know the population standard deviations and the samples are independent, the test statistic is a $z$-value computed using Equation 10.7.

**z-Test Statistic for $\mu_1 - \mu_2$ When $\sigma_1$ and $\sigma_2$ Are Known, Independent Samples**

$$z = \frac{(\bar{x}_1 - \bar{x}_2) - (\mu_1 - \mu_2)}{\sqrt{\dfrac{\sigma_1^2}{n_1} + \dfrac{\sigma_2^2}{n_2}}}$$    **(10.7)**

where:

$(\mu_1 - \mu_2) = $ Hypothesized difference in population means

If the calculated $z$-value using Equation 10.7 exceeds the critical $z$-value from the standard normal distribution, the null hypothesis is rejected. Example 10-4 illustrates the use of this test statistic.

---

**EXAMPLE 10-4**

### Hypothesis Test for $\mu_1 - \mu_2$ When $\sigma_1$ and $\sigma_2$ Are Known, Independent Samples

**HOW TO DO IT** (Example 10-4)

**The Hypothesis-Testing Process for Two Population Means**

1. Specify the population parameter of interest.

2. Formulate the appropriate null and alternative hypotheses. The null hypothesis should contain the equality. Possible formats for hypotheses testing concerning two populations means are

$H_0: \mu_1 - \mu_2 = c$
$H_A: \mu_1 - \mu_2 \neq c$ two-tailed test

$H_0: \mu_1 - \mu_2 \leq c$
$H_A: \mu_1 - \mu_2 > c$ one-tailed test

$H_0: \mu_1 - \mu_2 \geq c$
$H_A: \mu_1 - \mu_2 < c$ one-tailed test

where $c = $ any specified number.

3. Specify the significance level ($\alpha$) for testing the hypothesis. Alpha is the maximum allowable probability of committing a Type I statistical error.

4. Determine the rejection region and develop the decision rule.

5. Compute the test statistic or the $p$-value. Of course, you must first select simple random samples from each population and compute the sample means.

6. Reach a decision. Apply the decision rule to determine whether to reject the null hypothesis.

7. Draw a conclusion.

**Brooklyn Brick, Inc.**    Brooklyn Brick, Inc., is a Pennsylvania-based company that makes bricks and concrete blocks for the building industry. One product is a brick facing material that looks like a real brick but is much thinner. The ideal thickness is 0.50 inch. The bricks that the company makes must be very uniform in their dimension so that brickmasons can build straight walls. The company has two plants that produce brick facing products, and the technology used at the two plants is slightly different. At Plant 1, the standard deviation in the thickness of brick facing products is known to be 0.025 inch, and the standard deviation at Plant 2 is 0.034 inch. These are known values. However, the company is interested in determining whether there is a difference in the average thickness of brick facing products made at the two plants. Specifically, the company wishes to know whether Plant 2 provides brick facing products that have a greater mean thickness than the products produced at Plant 1. If the test determines that Plant 2 does provide thicker materials than Plant 1, the managers will have the maintenance department attempt to adjust the process to reduce the mean thickness. To test this, you can use the following steps:

**STEP 1**    **Specify the population parameter of interest.**
This is $\mu_1 - \mu_2$, the difference in the two population means.

**STEP 2**    **Formulate the appropriate null and alternative hypotheses.**
We are interested in determining whether the mean thickness for Plant 2 exceeds that for Plant 1. The following null and alternative hypotheses are specified:

$$H_0: \mu_1 - \mu_2 \geq 0.0 \qquad\qquad H_0: \mu_1 \geq \mu_2$$
$$\text{or}$$
$$H_A: \mu_1 - \mu_2 < 0.0 \qquad\qquad H_A: \mu_1 < \mu_2$$

**STEP 3**    **Specify the significance level for the test.**
The test will be conducted using $\alpha = 0.05$.

**STEP 4**    **Determine the rejection region and state the decision rule.**
Because the population standard deviations are assumed to be known, the critical value is a $z$-value from the standard normal distribution. This test is a one-tailed lower-tail test, with $\alpha = 0.05$. From the standard normal distribution, the critical $z$-value is

$$-z_{0.05} = -1.645$$

The decision rule compares the test statistic found in Step 5 to the critical $z$-value:

If $z < -1.645$, reject the null hypothesis.

Otherwise, do not reject the null hypothesis.

Alternatively, you can state the decision rule in terms of a $p$-value, as follows:

If $p$-value $< \alpha = 0.05$, reject the null hypothesis.

Otherwise, do not reject the null hypothesis.

STEP **5** **Compute the test statistic.**

Select simple random samples of brick facing pieces from the two populations and compute the sample means. A simple random sample of 100 brick facing pieces is selected from Plant 1's production, and another simple random sample of 100 brick facing pieces is selected from Plant 2's production. The samples are independent because the thicknesses of the brick pieces made by one plant can in no way influence the thicknesses of the bricks made by the other plant. The means computed from the samples are

$$\bar{x}_1 = 0.501 \text{ inch} \quad \text{and} \quad \bar{x}_2 = 0.509 \text{ inch}$$

The test statistic is obtained using Equation 10.7:

$$z = \frac{(\bar{x}_1 - \bar{x}_2) - (\mu_1 - \mu_2)}{\sqrt{\dfrac{\sigma_1^2}{n_1} + \dfrac{\sigma_2^2}{n_2}}}$$

$$z = \frac{(0.501 - 0.509) - 0}{\sqrt{\dfrac{0.025^2}{100} + \dfrac{0.034^2}{100}}} = -1.90$$

STEP **6** **Reach a decision.**

The critical $-z_{0.05} = -1.645$, and the test statistic value was computed to be $z = -1.90$. Applying the decision rule, we have:

Because $z = -1.90 < -1.645$, reject the null hypothesis.

Figure 10.2 illustrates this hypothesis test.

STEP **7** **Draw a conclusion.**

There is statistical evidence to conclude that the brick facings made by Plant 2 have a larger mean thickness than those made by Plant 1. Thus, the managers of Brooklyn Brick, Inc., need to take action to reduce the mean thicknesses from Plant 2.

**FIGURE 10.2 Example 10-4 Hypothesis Test**

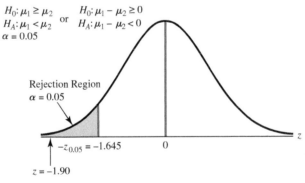

$H_0: \mu_1 \geq \mu_2$ $\quad$ $H_0: \mu_1 - \mu_2 \geq 0$
$H_A: \mu_1 < \mu_2$ $\quad$ or $\quad$ $H_A: \mu_1 - \mu_2 < 0$
$\alpha = 0.05$

Rejection Region
$\alpha = 0.05$

$-z_{0.05} = -1.645$ $\qquad$ 0

$z = -1.90$

**Test Statistic:**

$$z = \frac{(\bar{x}_1 - \bar{x}_2) - (\mu_1 - \mu_2)}{\sqrt{\dfrac{\sigma_1^2}{n_1} + \dfrac{\sigma_2^2}{n_2}}} = \frac{(0.501 - 0.509) - 0}{\sqrt{\dfrac{0.025^2}{100} + \dfrac{0.034^2}{100}}} = -1.90$$

**Decision Rule:**

Since $z = -1.90 < z = -1.645$, reject $H_0$.
Conclude that the brick facings made by Plant 2 have a larger mean thickness than those made by Plant 1.

**TRY EXERCISE 10-21** (pg. 384)

**Using _p_-Values**   The $z$-test statistic computed in Example 10-4 indicates that the difference in sample means is 1.90 standard errors below the hypothesized difference of zero. Because this falls below the $z$ critical level of $-1.645$, we rejected the null hypothesis. You could have also tested this hypothesis using the $p$-value approach introduced in Chapter 9. The $p$-value for this one-tailed test is the probability of a $z$-value in a standard normal distribution being less than $-1.90$. From the standard normal table, the probability associated with $z = -1.90$ is 0.4713. Then the $p$-value is

$$p\text{-value} = 0.5000 - 0.4713 = 0.0287$$

The decision rule to use with $p$-values is:

> If $p$-value $< \alpha$, reject the null hypothesis.
>
> Otherwise, do not reject the null hypothesis.

Because

$$p\text{-value} = 0.0287 < \alpha = 0.05$$

reject the null hypothesis and conclude that the mean brick facing thickness for Plant 2 is greater than the mean thickness for products produced by Plant 1.

> The Excel 2016 function for finding the $p$-value for a lower-tailed test is
>
> $=$ **NORM.S.DIST(z,TRUE)**
>
> $=$ **NORM.S.DIST($-$1.90,TRUE)**

## Testing for $\mu_1 - \mu_2$ When $\sigma_1$ and $\sigma_2$ Are Unknown, Using Independent Samples

In Section 10.1 we showed that to develop a confidence interval estimate for the difference between two population means when the standard deviations are unknown, we can use the $t$-distribution to obtain the critical value. As you might suspect, we can use this same approach for hypothesis-testing situations. Equation 10.8 shows the $t$-test statistic that will be used when $\sigma_1$ and $\sigma_2$ are unknown.

**$t$-Test Statistic for $\mu_1 - \mu_2$ When $\sigma_1$ and $\sigma_2$ Are Unknown and Assumed Equal, Independent Samples**

$$t = \frac{(\bar{x}_1 - \bar{x}_2) - (\mu_1 - \mu_2)}{s_p \sqrt{\dfrac{1}{n_1} + \dfrac{1}{n_2}}}, \quad df = n_1 + n_2 - 2 \qquad \textbf{(10.8)}$$

where:

$$\bar{x}_1 \text{ and } \bar{x}_2 = \text{ Sample means from Populations 1 and 2}$$
$$\mu_1 - \mu_2 = \text{ Hypothesized difference between population means}$$
$$n_1 \text{ and } n_2 = \text{ Sample sizes from the two populations}$$
$$s_p = \text{ Pooled standard deviation (see Equation 10.4)}$$

The test statistic in Equation 10.8 is based on three assumptions:

**Assumptions**
- Each population has a normal distribution.[3]
- The two population variances, $\sigma_1^2$ and $\sigma_2^2$, are equal.
- The samples are independent.

Notice that in Equation 10.8, we are using the pooled estimate for the common population standard deviation that we developed in Section 10.1.

---

[3]In Chapter 13, we will introduce a technique called the goodness-of-fit test, which we can use to test whether the sample data come from a population that is normally distributed.

## Hypothesis Test for the Difference between Two Population Means

**Retirement Investing** (*continued*)   Recall the earlier example discussing a study in North Dakota involving retirement investing. We were interested in determining whether there is a difference in mean annual contributions for individuals covered by TSAs and those with 401(k) retirement programs. A simple random sample of 15 people from the population of adults who are eligible for a TSA investment was selected. A second sample of 15 people was selected from the population of adults in North Dakota who have 401(k) plans. The variables of interest are the dollars invested in the two retirement plans during the previous year.

Specifically, we are interested in testing the following null and alternative hypotheses:

$$H_0: \mu_1 - \mu_2 = 0.0 \qquad\qquad H_0: \mu_1 = \mu_2$$
$$\text{or}$$
$$H_A: \mu_1 - \mu_2 \neq 0.0 \qquad\qquad H_A: \mu_1 \neq \mu_2$$

$\mu_1$ = Mean dollars invested by the TSA-eligible population during the past year
$\mu_2$ = Mean dollars invested by the 401(k)-eligible population during the past year

We select a significance level of $\alpha = 0.05$. The sample results are

| TSA-Eligible | 401(k)-Eligible |
|---|---|
| $n_1 = 15$ | $n_2 = 15$ |
| $\bar{x}_1 = \$2{,}119.70$ | $\bar{x}_2 = \$1{,}777.70$ |
| $s_1 = \$709.70$ | $s_2 = \$593.90$ |

Because the investments by individuals with TSA accounts are in no way influenced by investments by individuals with 401(k) accounts, the samples are considered independent. The box and whisker plots shown earlier in Figure 10.1 are consistent with what might be expected if the populations have equal variances and are approximately normally distributed.

We are now in a position to complete the hypothesis test to determine whether the mean dollar amount invested by TSA employees is different from the mean amount invested by 401(k) employees. We first determine the critical values from the *t*-distribution table in Appendix F with degrees of freedom equal to

$$n_1 + n_2 - 2 = 15 + 15 - 2 = 28$$

and $\alpha = 0.05$ for the two-tailed test. The appropriate critical *t*-values are

$$t_{0.025} = \pm 2.0484$$

The Excel 2016 function to find the critical *t*-value for a two-tailed test is

= T.INV.2T(alpha,Degrees of Freedom)
= T.INV.2T(0.05,28)

To continue the hypothesis test, we compute the pooled standard deviation:

$$s_p = \sqrt{\frac{(n_1 - 1)s_1^2 + (n_2 - 1)s_2^2}{n_1 + n_2 - 2}} = \sqrt{\frac{(15 - 1)(709.7)^2 + (15 - 1)(593.9)^2}{15 + 15 - 2}} = 654.37$$

Note that the pooled standard deviation is partway between the two sample standard deviations. Now, keeping in mind that the hypothesized difference between $\mu_1$ and $\mu_2$ is zero, we compute the *t*-test statistic using Equation 10.8, as follows:

$$t = \frac{(\bar{x}_1 - \bar{x}_2) - (\mu_1 - \mu_2)}{s_p \sqrt{\dfrac{1}{n_1} + \dfrac{1}{n_2}}} = \frac{(2{,}119.70 - 1{,}777.70) - 0.0}{654.37 \sqrt{\dfrac{1}{15} + \dfrac{1}{15}}} = 1.4313$$

This indicates that the difference in sample means is 1.4313 standard errors above the hypothesized difference of zero. Because

$$t = 1.4313 < t_{0.025} = 2.0484$$

the null hypothesis should not be rejected.

**FIGURE 10.3 Hypothesis Test for the Equality of the Two Population Means for the North Dakota Investment Study**

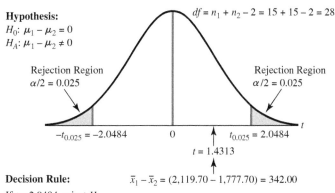

**Hypothesis:**
$H_0: \mu_1 - \mu_2 = 0$
$H_A: \mu_1 - \mu_2 \neq 0$

$df = n_1 + n_2 - 2 = 15 + 15 - 2 = 28$

Rejection Region
$\alpha/2 = 0.025$

Rejection Region
$\alpha/2 = 0.025$

$-t_{0.025} = -2.0484$    0    $t_{0.025} = 2.0484$

$t = 1.4313$

$\bar{x}_1 - \bar{x}_2 = (2{,}119.70 - 1{,}777.70) = 342.00$

**Decision Rule:**
If $t > 2.0484$, reject $H_0$.
If $t < -2.0484$, reject $H_0$.
Otherwise, do not reject $H_0$.

**Test Statistic:**
$$t = \frac{(\bar{x}_1 - \bar{x}_2) - (\mu_1 - \mu_2)}{s_p\sqrt{\dfrac{1}{n_1} + \dfrac{1}{n_2}}} = \frac{(2{,}119.70 - 1{,}777.70) - 0.0}{654.37\sqrt{\dfrac{1}{15} + \dfrac{1}{15}}} = 1.4313$$

where:
$$s_p = \sqrt{\frac{(n_1 - 1)s_1^2 + (n_2 - 1)s_2^2}{n_1 + n_2 - 2}} = \sqrt{\frac{(15-1)(709.7)^2 + (15-1)(593.9)^2}{15 + 15 - 2}} = 654.37$$

Figure 10.3 summarizes this hypothesis test. The difference in sample means is attributed to sampling error. Based on the sample data, there is no statistical justification to believe that the mean annual investment by individuals eligible for the TSA option is different from those individuals eligible for the 401(k) plan.

---

**BUSINESS APPLICATION**

## Using Excel to Test for the Difference between Two Population Means

Excel Tutorial

**SUV Vehicle Mileage**   Excel has a procedure for performing the necessary calculations to test hypotheses involving two population means. Consider a national car rental company that is interested in testing whether there is a difference in mean mileage for sport utility vehicles (SUVs) driven in town versus those driven on the highway. Based on its experience with regular automobiles, the company believes the mean highway mileage will exceed the mean city mileage.

To test this belief, the company has randomly selected 25 SUV rentals driven only on the highway and another random sample of 25 SUV rentals driven only in the city. The vehicles were filled with 14 gallons of gasoline. The company then asked each customer to drive the car until it ran out of gasoline. At that point, the elapsed miles were noted and the miles per gallon (mpg) were recorded. For their trouble, the customers received free use of the SUV and a coupon valid for one week's free rental. The results of the experiment are contained in the file **Mileage**.

We can use Excel to perform the calculations required to determine whether the manager's belief about SUV highway mileage is justified. We first formulate the null and alternative hypotheses to be tested:

$$H_0: \mu_1 - \mu_2 \leq 0.0 \qquad\qquad H_0: \mu_1 \leq \mu_2$$
$$\text{or}$$
$$H_A: \mu_1 - \mu_2 > 0.0 \qquad\qquad H_A: \mu_1 > \mu_2$$

Population 1 represents highway mileage, and Population 2 represents city mileage. The test is conducted using a significance level of $0.05 = \alpha$.

Figure 10.4 shows the descriptive statistics for the two independent samples.

**FIGURE 10.4** **Excel 2016 Output—SUV Mileage Descriptive Statistics**

| | A | B | C | D |
|---|---|---|---|---|
| 1 | *Highway Mileage* | | *City Mileage* | |
| 2 | | | | |
| 3 | Mean | 19.6468 | Mean | 16.146 |
| 4 | Standard Error | 0.8594 | Standard Error | 1.0882 |
| 5 | Median | 19.5400 | Median | 16.6200 |
| 6 | Mode | #N/A | Mode | #N/A |
| 7 | Standard Deviation | 4.2969 | Standard Deviation | 5.4412 |
| 8 | Sample Variance | 18.4637 | Sample Variance | 29.6064 |
| 9 | Kurtosis | -0.4179 | Kurtosis | -0.7224 |
| 10 | Skewness | 0.1929 | Skewness | -0.3829 |
| 11 | Range | 15.66 | Range | 18.43 |
| 12 | Minimum | 12.27 | Minimum | 5.89 |
| 13 | Maximum | 27.93 | Maximum | 24.32 |
| 14 | Sum | 491.17 | Sum | 403.65 |
| 15 | Count | 25 | Count | 25 |

**FIGURE 10.5** **Excel 2016 Output Box and Whisker Plot—SUV Mileage Test**

SUV Mileage Box and Whisker Plot

☐ Highway Mileage  ■ City Mileage

Figure 10.5 displays the Excel box and whisker plots for the two samples. Based on these plots, the normal distribution and equal variance assumptions appear reasonable. We will proceed with the test of means assuming normal distributions and equal variances.

Figure 10.6 shows the Excel output for the hypothesis test. The mean highway mileage is 19.6468 mpg, whereas the mean for city driving is 16.146. At issue is whether this difference in sample means (19.6468 − 16.146 = 3.5008 mpg) is sufficient to conclude that the mean highway mileage exceeds the mean city mileage. The one-tailed critical *t*-value for $\alpha = 0.05$ is shown in Figure 10.6 to be

$$t_{0.05} = 1.6772$$

Figure 10.6 shows that the "*t* Stat" value from Excel, which is the calculated test statistic (or *t*-value, based on Equation 10.8), is

$$t = 2.52$$

The difference in sample means (3.5008 mpg) is 2.52 standard errors larger than the hypothesized difference of zero. Because the test statistic

$$t = 2.52 > t_{0.05} = 1.6772$$

we reject the null hypothesis. Thus, the sample data do provide sufficient evidence to conclude that mean SUV highway mileage exceeds mean SUV city mileage, and this study confirms the expectations of the rental company managers. This will factor into the company's fuel pricing.

The output shown in Figure 10.6 also provides the *p*-value for the one-tailed test, which can also be used to test the null hypothesis. Recall that if the calculated *p*-value is less than alpha, the null hypothesis should be rejected. The decision rule is:

**Excel 2016 Instructions**

1. Open file: **Mileage.xlsx**.
2. Select **Data > Data Analysis**.
3. Select **t-test: Two Sample Assuming Equal Variances**.
4. Define data ranges for the two variables of interest.
5. Set **Hypothesized Difference** to 0.0.
6. Set **Alpha** at 0.05.
7. Specify Output Location.
8. Click **OK**.
9. Click the **Home** tab and adjust decimal points in output.

**FIGURE 10.6 Excel 2016 Output for the SUV Mileage *t*-Test for Two Population Means**

| | A | B | C |
|---|---|---|---|
| 1 | t-Test: Two-Sample Assuming Equal Variances | | |
| 2 | | | |
| 3 | | Highway Mileage | City Mileage |
| 4 | Mean | 19.6468 | 16.146 |
| 5 | Variance | 18.4637 | 29.6064 |
| 6 | Observations | 25 | 25 |
| 7 | Pooled Variance | 24.0351 | |
| 8 | Hypothesized Mean Difference | 0 | |
| 9 | df | 48 | |
| 10 | t Stat | 2.5246 | |
| 11 | P(T<=t) one-tail | 0.0075 | |
| 12 | t Critical one-tail | 1.6772 | |
| 13 | P(T<=t) two-tail | 0.0149 | |
| 14 | t Critical two-tail | 2.0106 | |

If $p$-value $< 0.05$, reject $H_0$.

Otherwise, do not reject $H_0$.

The $p$-value for the one-tailed test is 0.0075. Because $0.0075 < 0.05$, the null hypothesis is rejected. This is the same conclusion we reached using the test statistic approach.

---

**EXAMPLE 10-5**

## Hypothesis Test for $\mu_1 - \mu_2$ When $\sigma_1$ and $\sigma_2$ Are Unknown, Using Independent Samples

**Color Printer Ink Cartridges**    Companies such as Canon, Hewlett-Packard, and Epson are the printer market leaders and make most of their printer-related profits by selling replacement ink cartridges. Suppose an independent test agency wishes to conduct a test to determine whether name-brand ink cartridges generate more color pages on average than competing generic ink cartridges. The test can be conducted using the following steps:

STEP 1    **Specify the population parameter of interest.**
We are interested in determining whether the mean number of pages printed by name-brand cartridges (Population 1) exceeds the mean pages printed by generic cartridges (Population 2).

STEP 2    **Formulate the appropriate null and alternative hypotheses.**
The null and alternative hypotheses are:

$$H_0: \mu_1 - \mu_2 \leq 0.0 \qquad\qquad H_0: \mu_1 \leq \mu_2$$
$$\text{or}$$
$$H_A: \mu_1 - \mu_2 > 0.0 \qquad\qquad H_A: \mu_1 > \mu_2$$

STEP 3    **Specify the significance level for the test.**
The test will be conducted using $\alpha = 0.05$.
When the populations have standard deviations that are unknown, the critical value is a $t$-value from the $t$-distribution if the populations are assumed to be normally distributed and the population variances are assumed to be equal.
A simple random sample of 10 users was selected, and the users were given a name-brand cartridge. A second sample of 8 users were given generic cartridges. Both groups used their printers until the ink ran out and then recorded number of pages printed. The samples are independent because the pages printed by users in one group did not in any way influence the pages printed by users in the second group. The means computed from the samples are

$$\bar{x}_1 = 322.5 \text{ pages} \quad \text{and} \quad \bar{x}_2 = 298.3 \text{ pages}$$

Because we do not know the population standard deviations, we compute the sample standard deviations from the sample data:

$$s_1 = 48.3 \text{ pages} \quad \text{and} \quad s_2 = 53.3 \text{ pages}$$

Suppose previous studies have shown that the number of pages printed by both types of cartridges tends to be approximately normally distributed with equal variances.

STEP 4 **Construct the rejection region.**

Based on a one-tailed test with $\alpha = 0.05$, the critical value is a $t$-value from the $t$-distribution with $10 + 8 - 2 = 16$ degrees of freedom. From the $t$-table, the critical $t$-value is

$$t_{0.05} = 1.7459 = \text{Critical value}$$

We compare the calculated test statistic from Step 5 to the critical $t$-value to form the decision rule. The decision rule is:

If $t > 1.7459$, reject the null hypothesis.
Otherwise, do not reject the null hypothesis.

> The Excel 2016 function for finding the critical $t$-value for a one-tailed, upper-tailed test is
>
> =T.INV(1−alpha,deg. of freedom)
> =T.INV(1−0.05,16)

STEP 5 **Determine the test statistic using Equation 10.8.**

$$t = \frac{(\bar{x}_1 - \bar{x}_2) - (\mu_1 - \mu_2)}{s_p \sqrt{\dfrac{1}{n_1} + \dfrac{1}{n_2}}}$$

The pooled standard deviation is

$$s_p = \sqrt{\frac{(n_1 - 1)s_1^2 + (n_2 - 1)s_2^2}{n_1 + n_2 - 2}} = \sqrt{\frac{(10 - 1)48.3^2 + (8 - 1)53.3^2}{10 + 8 - 2}} = 50.55$$

Then the $t$-test statistic is

$$t = \frac{(322.5 - 298.3) - 0.0}{50.55\sqrt{\dfrac{1}{10} + \dfrac{1}{8}}} = 1.0093$$

STEP 6 **Reach a decision.**

Because

$$t = 1.0093 < t_{0.05} = 1.7459$$

do not reject the null hypothesis.
Figure 10.7 illustrates the hypothesis test.

STEP 7 **Draw a conclusion.**

Based on these sample data, there is insufficient evidence to conclude that the mean number of pages printed by name-brand ink cartridges exceeds the mean for generic cartridges.

**TRY EXERCISE 10-20** (pg. 384)

**What If the Population Variances Are Not Equal?** In the previous examples, we assumed that the population variances were equal, and we carried out the hypothesis test for two population means using Equation 10.8. Even in cases in which the population variances are not equal, the $t$-test as specified in Equation 10.8 is generally considered to be appropriate as long as the sample sizes are equal.[4] However, if the sample sizes are not equal and if the sample data lead us to suspect that the variances are not equal, the $t$-test statistic must be approximated using Equation 10.9.[5] In cases in which the variances are not equal, the degrees of freedom are computed using Equation 10.10.

---

[4]Studies show that when the sample sizes are equal or almost equal, the $t$-distribution is appropriate even when one population variance is twice the size of the other.
[5]Chapter 11 introduces a statistical procedure for testing whether two populations have equal variances.

**FIGURE 10.7 Example 10-5 Hypothesis Test**

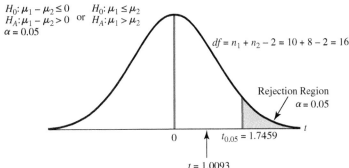

$H_0: \mu_1 - \mu_2 \leq 0$    $H_0: \mu_1 \leq \mu_2$
$H_A: \mu_1 - \mu_2 > 0$   or   $H_A: \mu_1 > \mu_2$
$\alpha = 0.05$

$df = n_1 + n_2 - 2 = 10 + 8 - 2 = 16$

Rejection Region
$\alpha = 0.05$

$0$    $t_{0.05} = 1.7459$

$t = 1.0093$

where:

$$s_p = \sqrt{\frac{(n_1 - 1)s_1^2 + (n_2 - 1)s_2^2}{n_1 + n_2 - 2}} = \sqrt{\frac{(10 - 1)\,48.3^2 + (8 - 1)\,53.3^2}{10 + 8 - 2}} = 50.55$$

**Test Statistic:**

$$t = \frac{(\bar{x}_1 - \bar{x}_2) - (\mu_1 - \mu_2)}{s_p\sqrt{\dfrac{1}{n_1} + \dfrac{1}{n_2}}} = \frac{(322.5 - 298.3) - 0}{50.55\sqrt{\dfrac{1}{10} + \dfrac{1}{8}}} = 1.0093$$

**Decision Rule:**

Because $t = 1.0093 < t_{0.05} = 1.7459$, do not reject $H_0$.

**$t$-Test Statistic for $\mu_1 - \mu_2$ When Population Variances Are Unknown and Not Assumed Equal**

$$t = \frac{(\bar{x}_1 - \bar{x}_2) - (\mu_1 - \mu_2)}{\sqrt{\dfrac{s_1^2}{n_1} + \dfrac{s_2^2}{n_2}}} \qquad (10.9)$$

**Degrees of Freedom for $t$-Test Statistic When Population Variances Are Not Equal**

$$df = \frac{(s_1^2/n_1 + s_2^2/n_2)^2}{\left(\dfrac{(s_1^2/n_1)^2}{n_1 - 1} + \dfrac{(s_2^2/n_2)^2}{n_2 - 1}\right)} \qquad (10.10)$$

## 10.2 EXERCISES

### Skill Development

**10-18.** A decision maker wishes to test the following null and alternative hypotheses using an alpha level equal to 0.05:

$$H_0: \mu_1 - \mu_2 = 0$$
$$H_A: \mu_1 - \mu_2 \neq 0$$

The population standard deviations are assumed to be known. After the sample data are collected, the test statistic is computed to be

$$z = 1.78$$

a. Using the test statistic approach, what conclusion should be reached about the null hypothesis?

b. Using the $p$-value approach, what decision should be reached about the null hypothesis?

c. Will the two approaches (test statistic and $p$-value) ever provide different conclusions based on the same sample data? Explain.

**10-19.** The following null and alternative hypotheses have been stated:

$$H_0: \mu_1 - \mu_2 = 0$$
$$H_A: \mu_1 - \mu_2 \neq 0$$

To test the null hypothesis, random samples have been selected from the two normally distributed populations with equal variances. The following sample data were observed:

| Sample from Population 1 | | | Sample from Population 2 | | |
|---|---|---|---|---|---|
| 33 | 29 | 35 | 46 | 43 | 42 |
| 39 | 39 | 41 | 46 | 44 | 47 |
| 25 | 33 | 38 | 50 | 43 | 39 |

Test the null hypothesis using an alpha level equal to 0.05.

**10-20.** You are given the following null and alternative hypotheses:

$$H_0: \mu_1 \geq \mu_2$$
$$H_A: \mu_1 < \mu_2$$

together with the following sample information:

| Sample 1 | Sample 2 |
|---|---|
| $n_1 = 14$ | $n_2 = 18$ |
| $\bar{x}_1 = 565$ | $\bar{x}_2 = 578$ |
| $s_1 = 28.9$ | $s_2 = 26.3$ |

a. Assuming that the populations are normally distributed with equal variances, test at the 0.10 level of significance whether you would reject the null hypothesis based on the sample information. Use the test statistic approach.

b. Assuming that the populations are normally distributed with equal variances, test at the 0.05 level of significance whether you would reject the null hypothesis based on the sample information. Use the test statistic approach.

**10-21.** Given the following null and alternative hypotheses, conduct a hypothesis test using an alpha equal to 0.05. (*Note:* The population standard deviations are assumed to be known.)

$$H_0: \mu_1 \leq \mu_2$$
$$H_A: \mu_1 > \mu_2$$

The sample means for the two populations are as follows:

$$\bar{x}_1 = 144 \quad \bar{x}_2 = 129$$
$$\sigma_1 = 11 \quad \sigma_2 = 16$$
$$n_1 = 40 \quad n_2 = 50$$

**10-22.** The following statistics were obtained from independent samples from populations that have normal distributions:

| | 1 | 2 |
|---|---|---|
| $n_i$ | 41 | 51 |
| $\bar{x}_i$ | 25.4 | 33.2 |
| $s_i$ | 5.6 | 7.4 |

a. Use these statistics to conduct a test of hypothesis if the alternative hypothesis is $\mu_1 - \mu_2 < -4$. Use a significance level of 0.01.

b. Determine the *p*-value for the test described in part a.

c. Describe the type of statistical error that could have been made as a result of your hypothesis test.

**10-23.** You are given the following null and alternative hypotheses:

$$H_0: \mu_1 - \mu_2 = 0$$
$$H_A: \mu_1 - \mu_2 \neq 0$$

and the following sample information:

| Sample 1 | Sample 2 |
|---|---|
| $n_1 = 125$ | $n_2 = 120$ |
| $s_1 = 31$ | $s_2 = 38$ |
| $\bar{x}_1 = 130$ | $\bar{x}_2 = 105$ |

a. Develop the appropriate decision rule, assuming a significance level of 0.05 is to be used.

b. Test the null hypothesis and indicate whether the sample information leads you to reject or fail to reject the null hypothesis. Use the test statistic approach.

**10-24.** Consider the following two independently chosen samples whose population variances are not equal to each other:

| Sample 1 | 12.1 | 13.4 | 11.7 | 10.7 | 14.0 |
|---|---|---|---|---|---|
| Sample 2 | 10.5 | 9.5 | 8.2 | 7.8 | 11.1 |

a. Using a significance level of 0.025, test the null hypothesis that $\mu_1 - \mu_2 \leq 0$.

b. Calculate the *p*-value.

## Business Applications

**10-25.** High Mountain produces a variety of climbing and mountaineering equipment. One of its products is a traditional three-strand climbing rope. An important characteristic of any climbing rope is its tensile strength. High Mountain produces the three-strand rope on two separate production lines: one in Bozeman and the other in Challis. The Bozeman line has recently installed new production equipment. High Mountain regularly tests the tensile strength of its ropes by randomly selecting ropes from production and subjecting them to various tests. The most recent random sample of ropes, taken after the new equipment was installed at the Bozeman plant, revealed the following:

| Bozeman | Challis |
|---|---|
| $\bar{x}_1 = 7{,}200$ lb | $\bar{x}_2 = 7{,}087$ lb |
| $s_1 = 425$ lb | $s_2 = 415$ lb |
| $n_1 = 25$ | $n_2 = 20$ |

High Mountain's production managers are willing to assume that the population of tensile strengths for each plant is approximately normally distributed with equal variances. Based on the sample results, can High

Mountain's managers conclude that there is a difference between the mean tensile strengths of ropes produced in Bozeman and Challis? Conduct the appropriate hypothesis test at the 0.05 level of significance.

**10-26.** The management of the Seaside Golf Club regularly monitors the golfers on its course for speed of play. Suppose a random sample of golfers was taken in 2015 and another random sample of golfers was selected in 2016. The results of the two samples (in minutes) are as follows:

| 2015 | 2016 |
|------|------|
| $\bar{x}_1 = 225$ | $\bar{x}_2 = 219$ |
| $s_1 = 20.25$ | $s_2 = 21.70$ |
| $n_1 = 36$ | $n_2 = 31$ |

Based on the sample results, can the management of the Seaside Golf Club conclude that average speed of play was different in 2016 than in 2015? Conduct the appropriate hypothesis test at the 0.10 level of significance. Assume that the management of the club is willing to accept the assumption that the populations of playing times for each year are approximately normally distributed with equal variances.

**10-27.** A grocery store manager believes that the mean amount spent by customers on dairy products per visit is higher in stores in which the dairy section is in the central part of the store compared with stores that have the dairy section at the rear of the store. To consider relocating the dairy products, the manager feels that the increase in the mean amount spent by customers must be at least 25 cents. To determine whether relocation is justified, her staff selected a random sample of 25 customers at stores in which the dairy section is central in the store. She selected a second sample of 25 customers in stores with the dairy section at the rear of the store. The following sample results were observed:

| Central Dairy | Rear Dairy |
|---------------|------------|
| $\bar{x}_1 = \$3.74$ | $\bar{x}_2 = \$3.26$ |
| $s_1 = \$0.87$ | $s_2 = \$0.79$ |

a. Conduct a hypothesis test with a significance level of 0.05 to determine if the manager should relocate the dairy products in those stores that have their dairy products in the rear of the store. Assume equal population variances.

b. If a statistical error associated with hypothesis testing was made in this hypothesis test, what error could it have been? Explain.

**10-28.** A major paint manufacturer came out with a new paint product designed to be used in areas that are particularly prone to periodic moisture and hot sun. The R&D department set up a test in which two random samples of paint were selected. The first sample consisted of 25 one-gallon containers of the company's best-selling

paint, and the second sample consisted of 15 one-gallon containers of the new paint under consideration. The following statistics were computed from each sample and refer to the number of square feet that each gallon will cover:

| Best-Selling Paint | New Paint Product |
|--------------------|-------------------|
| $\bar{x}_1 = 423$ sq. feet | $\bar{x}_2 = 406$ sq. feet |
| $s_1 = 22.4$ sq. feet | $s_2 = 16.8$ sq. feet |
| $n_1 = 25$ | $n_2 = 15$ |

The R&D managers are concerned that the average area covered per gallon will be less for the new paint than for the existing product. Based on the sample data, what should they conclude if they base their conclusion on a significance level equal to 0.01?

**10-29.** Suppose a major retailer replaced some of the traditional checkout stands with automated self-checkout facilities. After making the change in some test stores, the company performed a study to determine whether the average purchase through a self-checkout facility was less than the average purchase at the traditional checkout stand.

To conduct the test, a random sample of 125 customer transactions at the self-checkout was obtained, and a second random sample of 125 transactions from customers using the traditional checkout process was obtained. The following statistics were computed from each sample:

| Self-Checkout | Traditional Checkout |
|---------------|----------------------|
| $\bar{x}_1 = \$45.68$ | $\bar{x}_2 = \$78.49$ |
| $s_1 = \$58.20$ | $s_2 = \$62.45$ |
| $n_1 = 125$ | $n_2 = 125$ |

Based on these sample data, assuming equal population variances, what should be concluded with respect to the average transaction amount for the two checkout processes? Test using an $\alpha = 0.05$ level.

**10-30.** Being "connected" is very important to people of all ages these days. One way people stay connected is by sending text messages. Suppose a study was conducted at a major state university to determine whether the mean number of texts sent per day differed between male and female undergraduate students. Forty female students and 35 male students participated in the study. These students were randomly selected, and the number of text messages they sent on a particular day was recorded. The students did not know which day was going to be used in the study. The following statistics were computed:

| | Males | Females |
|---------|-------|---------|
| Mean | 32.4 | 54.3 |
| St. Dev | 13.9 | 14.8 |

Using an alpha level equal to 0.05, conduct the hypothesis test to determine whether there is a difference in the average numbers of text messages sent by males and females.

**10-31.** Suppose a study has been conducted to determine whether there is a difference between student debt at private, for-profit universities and public universities. It was found from random samples that the private, for-profit colleges' average debt was $38,300 and the public colleges average was $11,800. The respective standard deviations were $2,050 and $2,084. The sample sizes were 75 and 205, respectively.

Conduct a hypothesis test to determine if the average student debt is at least $25,000 more for graduates from private colleges than from public colleges. Use a 0.01 significance level and a *p*-value approach for this hypothesis test. Assume equal population variances.

## Computer Software Exercises

**10-32.** Suppose a job-placement firm would like to know if the 2015 average starting salary for chemical engineering majors is higher than the 2015 average starting salary for electrical engineering majors. To conduct its test, the job-placement firm has selected a random sample of 124 electrical engineering majors and 110 chemical engineering majors who graduated and received jobs in 2015. Each graduate was asked to report his or her starting salary. The results of the survey are contained in the file **Starting Salaries**.

   a. Conduct a hypothesis test to determine whether the mean starting salary for 2015 graduates in chemical engineering is higher than the mean starting salary for 2015 graduates in electrical engineering. Conduct the test at the 0.05 level of significance. Be sure to state a conclusion. (Assume that the firm believes the two populations from which the samples were taken are approximately normally distributed with equal variances.)

   b. Suppose the job-placement firm is unwilling to assume that the two populations are normally distributed with equal variances. Conduct the appropriate hypothesis test to determine whether a difference exists between the mean starting salaries for the two groups. Use a level of significance of

0.05. What conclusion should the job-placement firm reach based on the hypothesis test?

**10-33.** A fitness journal is interested in knowing whether there is a difference in percentage body fat for people who are 30 and under versus those who are over 30 years old. It is thought that the older group will have a higher average body fat percentage. A file titled **BodyFat** contains data for a sample of people tested at fitness centers in the St. Louis area. Using an alpha level equal to 0.05, conduct the appropriate hypothesis test. Assume equal population variances.

**10-34.** The Fairfield Inn and the Residence Inn are two of the hotel brands owned by Marriott. At a recent meeting, a manager posed the question whether the average length of stay was different at these two properties in the United States. A summer intern was assigned the task of testing to see if there is a difference. She started by selecting a simple random sample of 100 hotel reservations from Fairfield Inn. Next, she selected a simple random sample of 100 hotel reservations from Residence Inn. In both cases, she recorded the number of nights' stay for each reservation. The resulting data are in the file called **Marriott**.

   a. State the appropriate null and alternative hypotheses.

   b. Based on these sample data and a 0.05 level of significance, what conclusion should be made about the average length of stay at these two hotel chains?

**10-35.** Suppose that one national airline had an objective of getting an additional $5 to $10 per trip in revenue from its customers through the implementation of various fees. Surveys could be used to determine the success of the company's actions. The file titled **AirRevenue** contains results of samples gathered before and after the company implemented its changes.

   a. Produce a 95% confidence interval for the difference in the average fares paid by passengers before and after the change in policy. Based on the confidence interval, is it possible that revenue per passenger increased by at least $10? Explain your response.

   b. Conduct a test of hypothesis to answer the question posed in part a. Use a significance level of 0.025.

   c. Did you reach the same conclusion in both parts a and b? Is this a coincidence or will it always be so? Explain your response.

---

**OUTCOME 3**    **10.3**    **Interval Estimation and Hypothesis Tests for Paired Samples**

**Paired Samples**
Samples that are selected in such a way that values in one sample are matched with the values in the second sample for the purpose of controlling for extraneous factors. Another term for paired samples is *dependent samples*.

Sections 10.1 and 10.2 introduced the methods by which decision makers can estimate and test the hypotheses for the difference between the means for two populations when the two samples are independent. In each example, the samples were independent because the sample values from one population did not have the potential to influence the values that would be selected from the second population. However, there are instances in business in which you will want to use **paired samples** to control for sources of variation that might otherwise distort the conclusions of a study.

## Why Use Paired Samples?

Using paired samples should be considered in many business situations. For instance, a paint manufacturer might be interested in comparing the number of square feet that a new paint mix will cover per gallon with the area for an existing paint mixture. One approach would be to have one random sample of painters apply a gallon of the new paint mixture. A second sample of painters would be given the existing mixture. In both cases, the number of square feet covered by the gallon of paint would be recorded. In this case, the samples would be independent because the area covered by painters using the new mixture would not be in any way affected by the area covered by painters using the existing mixture.

This would be a fine way to do the study unless the painters themselves could influence the area that the paint will cover. For instance, suppose some painters, because of their technique or experience, are able to cover more area with a gallon of paint than other painters regardless of the type of paint used. Then, if by chance most of these "good" painters happened to get assigned to the new mix, the results might show that the new mix covers more area, not because it is a better paint, but because the painters who used it during the test were better.

To combat this potential problem, the company might want to use paired samples. To do this, one group of painters would be selected and each painter would use one gallon of each paint mix. We would measure the area covered by each painter for both paint mixes. Doing this controls for the effect of the painters' ability or technique. The following application involving gasoline supplemented with ethanol is one in which paired samples are most likely warranted.

---

**BUSINESS APPLICATION**    Estimation Using Paired Samples

**Testing Ethanol Mixed Gasoline**    A major oil company wanted to estimate the difference in average mileage for cars using regular gasoline compared with cars using a gasoline-and-ethanol mixture. The company used a paired-sample approach to control for any variation in mileage arising because of different cars and drivers. A random sample of 10 motorists (and their cars) was selected. Each car was filled with regular gasoline. The car was driven 200 miles on a specified route. The car then was filled again with gasoline and the miles per gallon were computed. After the 10 cars completed this process, the same steps were performed using the gasoline mixed with ethanol. Because the same cars and drivers tested both types of fuel, the miles-per-gallon measurements for the ethanol mixture and regular gasoline are related. The two samples are not independent but are instead considered paired samples. Thus, we will compute the *paired difference* between the values from each sample, using Equation 10.11.

**FIGURE 10.8** Excel 2016 Worksheet for Ethanol Mixed Gasoline Study

| | A | B | C |
|---|---|---|---|
| 1 | Ethanol | Regular | d |
| 2 | 19.8 | 20.7 | -0.9 |
| 3 | 28.8 | 25.8 | 3 |
| 4 | 20.4 | 27.8 | -7.4 |
| 5 | 18.7 | 14.9 | 3.8 |
| 6 | 23.4 | 21.6 | 1.8 |
| 7 | 27.1 | 21.1 | 6 |
| 8 | 28.4 | 28 | 0.4 |
| 9 | 21.4 | 13 | 8.4 |
| 10 | 26.4 | 24.4 | 2 |
| 11 | 19.9 | 14.3 | 5.6 |

**Paired Difference**

$$d = x_1 - x_2 \tag{10.11}$$

where:

$$d = \text{Paired difference}$$
$$x_1 \text{ and } x_2 = \text{Values from Samples 1 and 2, respectively}$$

Figure 10.8 shows the Excel worksheet for this mileage study with the paired differences computed. The data are in the file called **Ethanol-Gas**.

The first step to develop the interval estimate is to compute the *mean paired difference*, $\bar{d}$, using Equation 10.12. This value is the best point estimate for the population mean paired difference, $\mu_d$.

**Excel 2016 Instructions**
1. Open file: **Ethanol-Gas.xlsx**.

**Point Estimate for the Population Mean Paired Difference, $\mu_d$**

$$\bar{d} = \frac{\sum_{i=1}^{n} d_i}{n} \tag{10.12}$$

where:

$\bar{d}_i = i$th paired difference value

$n = $ Number of paired differences

Using Equation 10.12, we determine $\bar{d}$ as follows:

$$\bar{d} = \frac{\sum d}{n} = \frac{22.7}{10} = 2.27$$

The next step is to compute the *sample standard deviation for the paired differences* using Equation 10.13.

**Sample Standard Deviation for Paired Differences**

$$s_d = \sqrt{\frac{\sum_{i=1}^{n} (d_i - \bar{d})^2}{n - 1}} \tag{10.13}$$

where:

$d_i = i$th paired difference

$\bar{d} = $ Mean paired difference

The sample standard deviation for the paired differences is

$$s_d = \sqrt{\frac{\sum (d - \bar{d})^2}{n - 1}} = \sqrt{\frac{172.8}{10 - 1}} = 4.38$$

Assuming that the population of paired differences is normally distributed, we compute the confidence interval estimate for the population mean paired difference $\mu_d$ using Equation 10.14.

**Confidence Interval Estimate for Population Mean Paired Difference, $\mu_d$**

$$\bar{d} \pm t\frac{s_d}{\sqrt{n}} \tag{10.14}$$

where:

$t = $ Critical $t$-value from $t$-distribution with $n - 1$ degrees of freedom

$\bar{d} = $ Sample mean paired difference

$s_d = $ Sample standard deviation of paired differences

$n = $ Number of paired differences (sample size)

> The Excel 2016 function for determining the critical $t$-value is
>
> =T.INV.2T(1 − Conf. Level, degrees of freedom)
> =T.INV.2T(0.05,9)

For a 95% confidence interval with $10 - 1 = 9$ degrees freedom, we use a critical $t$-value from the $t$-distribution of

$$t = 2.2622$$

The interval estimate obtained from Equation 10.14 is

$$\bar{d} \pm t\frac{s_d}{\sqrt{n}}$$

$$2.27 \pm 2.2622 \frac{4.38}{\sqrt{10}}$$

$$2.27 \pm 3.13$$

$$-0.86 \text{ mpg} \text{———} 5.40 \text{ mpg}$$

Because the interval estimate contains zero, there may be no difference in the average mileage with regular gasoline versus the ethanol mixture.

**EXAMPLE 10-6**

## Confidence Interval Estimate for the Difference between Population Means, Paired Samples

**PGA of America Testing Center**   Technology has done more to change golf than possibly any other sport in recent years. Titanium woods, hybrid irons, and new golf ball designs have had an impact on professional and amateur golfers alike. Suppose a golf ball maker has developed a new ball technology, and PGA of America is interested in estimating the mean difference in driving distance for this new ball versus the existing best-seller. To conduct the test, the PGA of America staff selected six professional golfers and had each golfer hit each ball one time. The following are the steps used to develop a confidence interval estimate for the difference in population means for paired samples:

STEP 1   **Define the population parameter of interest.**
Because the same golfers hit each golf ball, the company is controlling for the variation in the golfers' ability to hit a golf ball. The samples are paired, and the population value of interest is $\mu_d$, the mean paired difference in distance. We assume that the population of paired differences is normally distributed.

STEP 2   **Specify the desired confidence level and determine the appropriate critical value.**
The research director wishes to have a 95% confidence interval estimate.

STEP 3   **Collect the sample data and compute the point estimate, $\bar{d}$.**
The sample data, paired differences, are shown as follows:

| Golfer | Existing Ball | New Ball | d |
|--------|---------------|----------|-----|
| 1 | 280 | 276 | 4 |
| 2 | 299 | 301 | −2 |
| 3 | 278 | 285 | −7 |
| 4 | 301 | 299 | 2 |
| 5 | 268 | 273 | −5 |
| 6 | 295 | 300 | −5 |

The point estimate is computed using Equation 10.12:

$$\bar{d} = \frac{\sum d}{n} = \frac{-13}{6} = -2.17 \text{ yards}$$

The Excel 2016 functions for the mean and sample standard deviation are
= **AVERAGE(4,−2,−7,2,−5,−5)**
= **STDEV.S(4,−2,−7,2,−5,−5)**

STEP 4   **Calculate the standard deviation, $s_d$.**
The standard deviation for the paired differences is computed using Equation 10.13:

$$s_d = \sqrt{\frac{\sum (d - \bar{d})^2}{n - 1}} = 4.36 \text{ yards}$$

The Excel 2016 function for determining the critical *t*-value is

= T.INV.2T(1 − Conf. Level, degrees of freedom)

= T.INV.2T(0.05,5)

STEP 5 **Determine the critical value, *t*, from the *t*-distribution table.**

The critical *t*-value for 95% confidence and $6 - 1 = 5$ degrees of freedom is

$$t = 2.5706$$

STEP 6 **Compute the confidence interval estimate using Equation 10.14.**

$$\bar{d} \pm t \frac{s_d}{\sqrt{n}}$$

$$-2.17 \pm 2.5706 \frac{4.36}{\sqrt{6}}$$

$$-2.17 \pm 4.58$$

$$-6.75 \text{ yards} \underline{\hspace{2cm}} 2.41 \text{ yards}$$

Based on these sample data and the confidence interval estimate, which contains zero, PGA of America must conclude that the new ball's average distance may not be any longer than that of the existing best-seller. This may affect whether the company that developed the new ball will continue to make it.

TRY EXERCISE 10-38 (pg. 392)

The key in deciding when to use paired samples is whether a factor exists that might adversely influence the results of the estimation. In the ethanol mixed gasoline test example, we controlled for potential outside influence by using the same cars and drivers to test both gasolines. In Example 10-6, we controlled for golfer ability by having the same golfers hit both golf balls. If you determine there is no need to control for an outside source of variation, then you should use independent samples, as discussed earlier in this chapter.

OUTCOME **2**     ## Hypothesis Testing for Paired Samples

As we just illustrated, paired samples can be used in some cases to control for an outside source of variation. For instance, in Example 10-5 involving the ink cartridges, the original test of whether name-brand cartridges yield a higher mean number of printed pages than generic cartridges involved different users for the two types of cartridges, so the samples were independent. However, different users may use more or less ink as a rule; therefore, we could control for that source of variation by having the same people use both types of cartridges in a paired test format.

If a paired-sample experiment is used, the test statistic is computed using Equation 10.15.

*t*-Test Statistic for Paired-Sample Test

$$t = \frac{\bar{d} - \mu_d}{\frac{s_d}{\sqrt{n}}}, \qquad df = n - 1 \qquad \textbf{(10.15)}$$

where:

$\bar{d}$ = Mean paired difference = $\dfrac{\sum d}{n}$

$\mu_d$ = Hypothesized population mean paired difference

$s_d$ = Sample standard deviation for paired differences = $\sqrt{\dfrac{\sum (d - \bar{d})^2}{n - 1}}$

$n$ = Number of paired values in the sample

| EXAMPLE 10-7 | Hypothesis Test for $\mu_d$, Paired Samples |

**Color Printer Ink Cartridges** (*continued*)    Referring to Example 10-5, suppose the experiment regarding ink cartridges is conducted differently. Instead of having different samples of people use name-brand and generic cartridges, the test is done using paired samples. This means that the same people will use both types of cartridges, and the pages printed in each case will be recorded. The test under this paired-sample scenario can be conducted using the following steps. Six randomly selected people have agreed to participate.

STEP 1 **Specify the population value of interest.**
In this case, we will form paired differences by subtracting the generic pages from the name-brand pages. We are interested in determining whether name-brand cartridges produce more printed pages, on average, than generic cartridges, so we expect the paired difference to be positive. We assume that the paired differences are normally distributed.

STEP 2 **Formulate the null and alternative hypotheses.**
The null and alternative hypotheses are

$$H_0: \mu_d \leq 0.0$$
$$H_A: \mu_d > 0.0$$

STEP 3 **Specify the significance level for the test.**
The test will be conducted using $\alpha = 0.01$.

STEP 4 **Determine the rejection region and the critical value.**
The critical value is a $t$-value from the $t$-distribution, with $\alpha = 0.01$ and $6 - 1 = 5$ degrees of freedom. The critical value is

$$t_{0.01} = 3.3649$$

The decision rule is:

> If $t > 3.3649$, reject the null hypothesis.
> Otherwise, do not reject the null hypothesis.

> The Excel 2016 function for the critical $t$-value for an upper-tailed test is
>
> = T.INV(1 – alpha,Degrees of Freedom)
> = T.INV(0.99,5)

STEP 5 **Compute the test statistic.**
Select the random sample and compute the mean and standard deviation for the paired differences.

In this case, a random sample of six people tests each type of cartridge. The following data and paired differences were observed:

| Printer User | Name-Brand | Generic | d |
|---|---|---|---|
| 1 | 306 | 300 | 6 |
| 2 | 256 | 260 | −4 |
| 3 | 402 | 357 | 45 |
| 4 | 299 | 286 | 13 |
| 5 | 306 | 290 | 16 |
| 6 | 257 | 260 | −3 |

The mean paired difference is

$$\bar{d} = \frac{\sum d}{n} = \frac{73}{6} = 12.17$$

The standard deviation for the paired differences is

$$s_d = \sqrt{\frac{\sum (d - \bar{d})^2}{n - 1}} = 18.02$$

The test statistic is calculated using Equation 10.15:

$$t = \frac{\bar{d} - \mu_d}{\frac{s_d}{\sqrt{n}}} = \frac{12.17 - 0.0}{\frac{18.02}{\sqrt{6}}} = 1.6543$$

STEP 6 **Reach a decision.**
Because $t = 1.6543 < t_{0.01} = 3.3649$, do not reject the null hypothesis.

STEP 7 **Draw a conclusion.**
Based on these sample data, there is insufficient evidence to conclude that name-brand ink cartridges print more pages on average than generic brands.

**TRY EXERCISE 10-39** (pg. 392)

# 10.3 EXERCISES

## Skill Development

**10-36.** The following dependent samples were randomly selected. Use the sample data to construct a 95% confidence interval estimate for the population mean paired difference.

| Sample 1 | Sample 2 | Sample 1 | Sample 2 |
|----------|----------|----------|----------|
| 22 | 31 | 23 | 31 |
| 25 | 24 | 25 | 27 |
| 27 | 25 | 28 | 31 |
| 26 | 32 | 27 | 31 |
| 22 | 25 | 23 | 26 |
| 21 | 27 | | |

**10-37.** The following paired sample data have been obtained from normally distributed populations. Construct a 90% confidence interval estimate for the mean paired difference between the two population means.

| Sample # | Population 1 | Population 2 |
|----------|--------------|--------------|
| 1 | 3,693 | 4,635 |
| 2 | 3,679 | 4,262 |
| 3 | 3,921 | 4,293 |
| 4 | 4,106 | 4,197 |
| 5 | 3,808 | 4,536 |
| 6 | 4,394 | 4,494 |
| 7 | 3,878 | 4,094 |

**10-38.** You are given the following results of a paired-difference test:

$$\bar{d} = -4.6$$
$$s_d = 0.25$$
$$n = 16$$

a. Construct and interpret a 99% confidence interval estimate for the paired difference in mean values.
b. Construct and interpret a 90% confidence interval estimate for the paired difference in mean values.

**10-39.** The following sample data have been collected from a paired sample from two populations. The claim is that the first population mean will be at least as large as the mean of the second population. This claim will be assumed to be true unless the data strongly suggest otherwise.

| Sample 1 | Sample 2 | Sample 1 | Sample 2 |
|----------|----------|----------|----------|
| 4.4 | 3.7 | 2.6 | 4.2 |
| 2.7 | 3.5 | 2.4 | 5.2 |
| 1.0 | 4.0 | 2.0 | 4.4 |
| 3.5 | 4.9 | 2.8 | 4.3 |
| 2.8 | 3.1 | | |

a. State the appropriate null and alternative hypotheses.
b. Based on the sample data, what should you conclude about the null hypothesis? Test using $\alpha = 0.10$.
c. Calculate a 90% confidence interval for the difference in the population means. Are the results from the confidence interval consistent with the outcome of your hypothesis test? Explain why or why not.

**10-40.** A paired sample study has been conducted to determine whether two populations have equal means. Twenty paired samples were obtained with the following sample results:

$$\bar{d} = 12.45 \qquad s_d = 11.0$$

Based on these sample data and a significance level of 0.05, what conclusion should be made about the population means?

**10-41.** The following samples are observations taken from the same elements at two different times:

| Unit | Sample 1 | Sample 2 |
|------|----------|----------|
| 1 | 15.1 | 4.8 |
| 2 | 12.3 | 5.7 |
| 3 | 14.9 | 6.2 |
| 4 | 17.5 | 9.4 |
| 5 | 18.1 | 2.3 |
| 6 | 18.4 | 4.7 |

a. Assume that the populations are normally distributed and construct a 90% confidence interval for the difference in the means of the distribution at the times in which the samples were taken.

b. Perform a test of hypothesis to determine if the difference in the means of the distribution at the first time period is 10 units larger than at the second time period. Use a level of significance equal to 0.10.

**10-42.** Consider the following samples obtained from two normally distributed populations whose variances are equal:

Sample 1:  11.2   11.2   7.4   8.7   8.5   13.5   4.5   11.9

Sample 2:  11.7   9.5   15.6   16.5   11.3   17.6   17.0   8.5

a. Suppose that the samples were independent. Perform a test of hypothesis to determine if there is a difference in the two population means. Use a significance level of 0.05.

b. Now suppose that the samples were paired samples. Perform a test of hypothesis to determine if there is a difference in the two population means.

c. How do you account for the difference in the outcomes of part a and part b? Support your assertions with a statistical rationale.

## Business Applications

**10-43.** One of the advances that helped to diminish carpal tunnel syndrome is ergonomic keyboards. The ergonomic keyboards may also increase typing speed. Ten administrative assistants were chosen to type on both standard and ergonomic keyboards. The resulting word-per-minute typing speeds follow:

Ergonomic:   69   80   60   71   73   64   63   70   63   74

Standard:    70   68   54   56   58   64   62   51   64   53

a. Were the two samples obtained independently? Support your assertion.

b. Conduct a hypothesis test to determine if the ergonomic keyboards increase the average words per minute attained while typing. Use a $p$-value approach with a significance level of 0.01. Assume equal population variances.

**10-44.** Production engineers at Sinotron believe that a modified layout on its assembly lines might increase average worker productivity (measured in the number of units produced per hour). However, before the engineers are ready to install the revised layout officially across the entire firm's production lines, they would like to study the modified line's effects on output. The following data represent the average hourly production output of 12 randomly sampled employees before and after the line was modified:

| Employee | 1 | 2 | 3 | 4 | 5 | 6 | 7 | 8 | 9 | 10 | 11 | 12 |
|----------|----|----|----|----|----|----|----|----|----|----|----|----|
| Before | 49 | 45 | 43 | 44 | 48 | 42 | 46 | 46 | 49 | 42 | 46 | 44 |
| After | 49 | 46 | 48 | 50 | 46 | 50 | 45 | 46 | 47 | 51 | 51 | 49 |

At the 0.05 level of significance, can the production engineers conclude that the modified (after) layout has increased average worker productivity?

**10-45.** The United Way raises money for community charity activities. In one community, the fundraising committee was concerned about whether there is a difference in the proportion of employees who give to United Way depending on whether the employer is a private business or a government agency. A random sample of people who had been contacted about contributing last year was selected. Of those contacted, 70 worked for a private business and 50 worked for a government agency. For the 70 private-sector employees, the mean contribution was $230.25 with a standard deviation equal to $55.52. For the 50 government employees in the sample, the mean and standard deviation were $309.45 and $61.75, respectively. Assume equal population variances.

a. Based on these sample data and $\alpha = 0.05$, what should the committee conclude? Be sure to show the decision rule.

b. Construct a 95% confidence interval for the difference between the mean contributions of private business and government agency employees who contribute to United Way. Do the hypothesis test and the confidence interval produce compatible results? Explain and give reasons for your answer.

**10-46.** Added to your car's fuel tank in tiny amounts, an additive called acetone is thought to increase the miles per gallon for vehicles that have gasoline or diesel engines. To test whether this product actually does increase fuel efficiency in passenger cars, a consumer group has randomly selected 10 people to participate in the study. The following procedure is used:

1. The car's odometer reading is recorded at the time of fill-up.

2. When the tank is nearly empty, the driver is to bring the car to the same gas station and pump and have it refilled with gasoline. The odometer is read again and the miles per gallon are recorded. This time, a prescribed quantity of acetone is added to the fuel.

3. When the tank is nearly empty, the driver is to bring the car back to the same station and pump to have it filled. The miles per gallon will be recorded.

The following miles per gallon (mpg) were recorded:

| Driver | MPG: No Additive | MPG: Acetone Added |
|--------|------------------|--------------------|
| 1 | 18.4 | 19.0 |
| 2 | 23.5 | 22.8 |
| 3 | 31.4 | 30.9 |
| 4 | 26.5 | 26.9 |
| 5 | 27.2 | 28.4 |
| 6 | 16.3 | 18.2 |
| 7 | 19.4 | 19.2 |
| 8 | 20.1 | 21.4 |
| 9 | 14.2 | 16.1 |
| 10 | 22.1 | 21.5 |

a. Discuss the appropriateness of the way this study was designed and conducted. Why didn't the consumer group select two samples with different drivers in each and have one group use the acetone and the other group not use it? Discuss.

b. Using a significance level of 0.05, what conclusion should the consumer group reach based on these sample data? Discuss.

**10-47.** An article in *The American Statistician* (M. L. R. Ernst, et al., "Scatterplots for unordered pairs," 50 (1996), pp. 260–265) reports on the difference in the measurements by two evaluators of the cardiac output of 23 patients using Doppler echocardiography. Both observers took measurements from the same patients. The measured outcomes were as follows:

| Patient | 1 | 2 | 3 | 4 | 5 | 6 | 7 | 8 | 9 | 10 | 11 | 12 |
|---------|---|---|---|---|---|---|---|---|---|----|----|----|
| Evaluator 1 | 4.8 | 5.6 | 6.0 | 6.4 | 6.5 | 6.6 | 6.8 | 7.0 | 7.0 | 7.2 | 7.4 | 7.6 |
| Evaluator 2 | 5.8 | 6.1 | 7.7 | 7.8 | 7.6 | 8.1 | 8.0 | 8.21 | 6.6 | 8.1 | 9.5 | 9.6 |

| Patient | 13 | 14 | 15 | 16 | 17 | 18 | 19 | 20 | 21 | 22 | 23 |
|---------|----|----|----|----|----|----|----|----|----|----|----|
| Evaluator 1 | 7.7 | 7.7 | 8.2 | 8.2 | 8.3 | 8.5 | 9.3 | 10.2 | 10.4 | 10.6 | 11.4 |
| Evaluator 2 | 8.5 | 9.5 | 9.1 | 10.0 | 9.1 | 10.8 | 11.5 | 11.5 | 11.2 | 11.5 | 12.0 |

Conduct a hypothesis test to determine if the average cardiac outputs measured by the two evaluators differ. Use a significance level of 0.02. Assume the population variances to be equal.

## Computer Software Exercises

**10-48.** The file titled **House** contains data for a sample showing the average and median housing prices for selected areas in the United States in November 2015

and November 2016. Assume the data can be viewed as samples of the relevant populations.

a. Discuss whether the two samples are independent or dependent.

b. Based on your answer to part a, calculate a 90% confidence interval for the difference between the means of the average and median selling prices for houses during November 2015.

c. Noting your answer to part b, would it be plausible to assert that the mean of the average selling prices for houses during November 2015 is more than the average of the median selling prices during this period? Support your assertions.

d. Using a *p*-value approach and a significance level of 0.05, conduct a hypothesis test to determine if the mean of the average selling prices for houses during November 2015 is more than $30,000 larger than the mean of the median selling prices during this period.

**10-49.** A treadmill manufacturer has developed a new machine with softer tread and better fans than its current model. The manufacturer believes these new features will enable runners to run for longer times than they can on its current machines. To determine whether the desired result is achieved, the manufacturer randomly sampled 35 runners. Each runner was measured for one week on the current machine and for one week on the new machine. The weekly total number of minutes for each runner on the two types of machines was collected. The results are contained in the file **Treadmill**. At the 0.02 level of significance, can the treadmill manufacturer conclude that the new machine has the desired result?

**10-50.** A study addressed the per-person spending at selected airports for merchandise, excluding food, gifts, and news items. A file titled **Revenues** contains sample data selected from airport retailers in 2012 and again in 2015.

a. Conduct a hypothesis test to determine if the average amount of retail spending by air travelers increased at least as much as approximately $0.10 a year from 2012 to 2015. Use a significance level of 0.025.

b. Using the appropriate analysis (that of part a or other appropriate methodology), substantiate the statement that average retail purchases in airports increased over the time period between 2012 and 2015. Support your assertions.

c. Parts a and b give what seems to be a mixed message. Is there a way to determine what values are plausible for the difference between the average revenues in 2012 and 2015? If so, conduct the appropriate procedure.

## 10.4 Estimation and Hypothesis Tests for Two Population Proportions

The previous sections illustrated the methods for estimating and testing hypotheses involving two population means. These methods can be applied in many business situations. However, there are other instances involving two populations in which the measures of interest are not the population means. For example, Chapter 9 introduced the methodology for testing hypotheses involving a single population proportion. This section extends that methodology to tests involving hypotheses about the difference between two population proportions. First, we will look at a confidence interval estimation involving two population proportions.

**OUTCOME 4**

## Estimating the Difference between Two Population Proportions

**BUSINESS APPLICATION**

### Estimating the Difference between Two Population Proportions

**Bicycle Design**    An outdoor magazine conducted an interesting study involving a prototype bicycle made by a Swiss manufacturer. The prototype had no identification on it to indicate the name of the manufacturer. Of interest was the difference in the proportions of men and women who rated the bicycle as high quality.

Obviously, there was no way to gauge the attitudes of the entire population of men and women who could eventually judge the quality of the bicycle. Instead, the reporter for the magazine asked a random sample of 425 men and 370 women to rate the bicycle's quality. In the results that follow, the variable $x$ indicates the number in the sample who said the bicycle was high quality.

| Men | Women |
|---|---|
| $n_1 = 425$ | $n_2 = 370$ |
| $x_1 = 240$ | $x_2 = 196$ |

Based on these sample data, the sample proportions are

$$\bar{p}_1 = \frac{240}{425} = 0.565 \quad \text{and} \quad \bar{p}_2 = \frac{196}{370} = 0.530$$

The point estimate for the difference in population proportions is

$$\bar{p}_1 - \bar{p}_2 = 0.565 - 0.530 = 0.035$$

So, the single best estimate for the difference in the proportions of men and women who rated the bicycle prototype as high quality is 0.035. However, all point estimates are subject to sampling error. We can develop a confidence interval estimate for the difference in population proportions using Equation 10.16, providing the sample sizes are sufficiently large. A rule of thumb for "sufficiently large" is that $n\bar{p}$ and $n(1 - \bar{p})$ are greater than or equal to 5 for each sample.

**Confidence Interval Estimate for $p_1 - p_2$**

$$(\bar{p}_1 - \bar{p}_2) \pm z \sqrt{\frac{\bar{p}_1(1 - \bar{p}_1)}{n_1} + \frac{\bar{p}_2(1 - \bar{p}_2)}{n_2}} \tag{10.16}$$

where:

$\bar{p}_1$ = Sample proportion from Population 1

$\bar{p}_2$ = Sample proportion from Population 2

$z$ = Critical value from the standard normal table

The analysts can substitute the sample results into Equation 10.16 to establish a 95% confidence interval estimate, as follows:

$$(0.565 - 0.530) \pm 1.96 \sqrt{\frac{0.565(1 - 0.565)}{425} + \frac{0.530(1 - 0.530)}{370}}$$

$$0.035 \pm 0.069$$

$$-0.034 \le (p_1 - p_2) \le 0.104$$

Thus, based on the sample data and using a 95% confidence interval, the analysts estimate that the true difference in the proportions of men and women who rate the prototype as high quality is between −0.034 and 0.104. At one extreme, 3.4% more women rate the bicycle as high in quality. At the other extreme, 10.4% more men rate the bicycle as high quality. Because zero is included in the interval, there may be no difference between the proportions of men and women who rate the prototype as high quality based on these data. Consequently, the reporter is not able to conclude that one group or the other is more likely to rate the prototype bicycle high in quality.

## Hypothesis Tests for the Difference between Two Population Proportions

### Testing for the Difference between Two Population Proportions

**Excel Tutorial**

**Pomona Fabrications**   Pomona Fabrications, Inc., produces handheld hair dryers that several major retailers sell as in-house brands. A critical component of a handheld hair dryer is the motor-heater unit, which accounts for most of the dryer's cost and for most of the product's reliability problems. Product reliability is important to Pomona because the company offers a one-year warranty. Of course, Pomona is also interested in reducing production costs.

Pomona's R&D department has created a new motor-heater unit with fewer parts than the current unit, which would lead to a 15% cost savings per hair dryer. However, the company's vice president of product development is unwilling to authorize the new component unless it is more reliable than the current motor-heater.

The R&D department has decided to test samples of both units to see which motor-heater is more reliable. Of each type, 250 will be tested under conditions that simulate one year's use, and the proportion of each type that fails within that time will be recorded. This leads to the formulation of the following null and alternative hypotheses:

$$H_0\!: p_1 - p_2 \ge 0.0 \qquad\qquad H_0\!: p_1 \ge p_2$$
$$\text{or}$$
$$H_A\!: p_1 - p_2 < 0.0 \qquad\qquad H_A\!: p_1 < p_2$$

where:

$p_1$ = Population proportion of new dryer type that fails in simulated one-year period

$p_2$ = Population proportion of existing dryer type that fails in simulated one-year period

The null hypothesis states that the new motor-heater is no better than the old, or current, motor-heater. The alternative states that the new unit has a smaller proportion of failures within one year than the current unit. In other words, the alternative states that the new unit is more reliable. The company wants clear evidence before changing units. If the null hypothesis is rejected, the company will conclude that the new motor-heater unit is more reliable than the old unit and should be used in producing the hair dryers. To test the null hypothesis, we can use the test statistic approach.

The test statistic is based on the sampling distribution of $\bar{p}_1 - \bar{p}_2$. In Chapter 7, we showed that when $np \ge 5$ and $n(1 - p) \ge 5$, the sampling distribution of the sample proportion is approximately normally distributed, with a mean equal to $p$ and a variance equal to $p(1 - p)/n$.

Likewise, in the two-sample case, the sampling distribution of $\bar{p}_1 - \bar{p}_2$ is also approximately normal if

**Assumptions**

$$n_1 p_1 \geq 5,\; n_1(1 - p_1) \geq 5, \quad \text{and} \quad n_2 p_2 \geq 5, n_2(1 - p_2) \geq 5$$

Because $p_1$ and $p_2$ are unknown, we substitute the sample proportions, $\bar{p}_1$ and $\bar{p}_2$, to determine whether the sample size requirements are satisfied.

The mean of the sampling distribution of $\bar{p}_1 - \bar{p}_2$ is the difference of the population proportions, $p_1 - p_2$. The variance is, however, the sum of the variances, $p_1(1 - p_1)/n_1 + p_2(1 - p_2)/n_2$. Because the test is conducted using the assumption that the null hypothesis is true, we assume that $p_1 = p_2 = p$ and estimate their common value, $p$, using a pooled estimate, as shown in Equation 10.17. The $z$-test statistic for the difference between two proportions is given as Equation 10.18.

**Pooled Estimator for Overall Proportion**

$$\bar{p} = \frac{n_1\bar{p}_1 + n_2\bar{p}_2}{n_1 + n_2} = \frac{x_1 + x_2}{n_1 + n_2} \tag{10.17}$$

where:

$x_1$ and $x_2$ = Number from Samples 1 and 2 with the characteristic of interest

**$z$-Test Statistic for Difference between Population Proportions**

$$z = \frac{(\bar{p}_1 - \bar{p}_2) - (p_1 - p_2)}{\sqrt{\bar{p}(1 - \bar{p})\left(\dfrac{1}{n_1} + \dfrac{1}{n_2}\right)}} \tag{10.18}$$

where:

$(p_1 - p_2)$ = Hypothesized difference in proportions from Populations 1 and 2, respectively

$\bar{p}_1$ and $\bar{p}_2$ = Sample proportions for samples selected from Populations 1 and 2, respectively

$\bar{p}$ = Pooled estimator for the overall proportion for both populations combined

The reason for taking a weighted average in Equation 10.17 is to give more weight to the larger sample. Note that the numerator is the total number of items with the characteristic of interest in the two samples, and the denominator is the total sample size. Again, the pooled estimator, $\bar{p}$, is used when the null hypothesis states that there is no difference between the population proportions.

Assume that Pomona is willing to use a significance level of 0.05 and that 55 of the new motor-heaters and 75 of the originals failed the one-year test. Figure 10.9 illustrates the development of the decision rule and the hypothesis test. As you can see, Pomona should reject the null hypothesis based on the sample data. Thus, the firm should conclude that the new motor-heater is more reliable than the old one. Because the new one is also less costly, the company should now use the new unit in the production of hair dryers.

The $p$-value approach to hypothesis testing could also have been used to test Pomona's hypothesis. In this case, the calculated value of the test statistic, $z = -2.04$, results in a $p$-value of $0.5 - 0.4793 = 0.0207$ from the standard normal table. Because this $p$-value is smaller than the significance level of 0.05, we would reject the null hypothesis. Remember, whenever your $p$-value is smaller than the alpha value, your sample contains evidence to reject the null hypothesis.

**FIGURE 10.9 Hypothesis Test of Two Population Proportions for Pomona Fabrications**

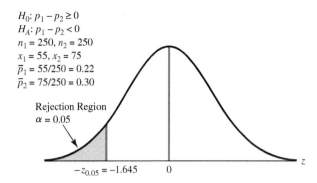

$H_0: p_1 - p_2 \geq 0$
$H_A: p_1 - p_2 < 0$
$n_1 = 250, n_2 = 250$
$x_1 = 55, x_2 = 75$
$\bar{p}_1 = 55/250 = 0.22$
$\bar{p}_2 = 75/250 = 0.30$

Rejection Region
$\alpha = 0.05$

$-z_{0.05} = -1.645 \qquad 0$

**Decision Rule:**
If $z < -1.645$, reject $H_0$.
If $z \geq -1.645$, do not reject $H_0$.

**Test Statistic:**

$$z = \frac{(\bar{p}_1 - \bar{p}_2) - (p_1 - p_2)}{\sqrt{\bar{p}(1-\bar{p})\left(\frac{1}{n_1} + \frac{1}{n_2}\right)}} = \frac{(0.22 - 0.30) - 0}{\sqrt{0.26(1-0.26)\left(\frac{1}{250} + \frac{1}{250}\right)}} = -2.04$$

where:

$$\bar{p} = \frac{250(0.22) + 250(0.30)}{250 + 250} = \frac{55 + 75}{500} = 0.26$$

Since $z = -2.04 < -1.645$, reject $H_0$.

---

**EXAMPLE 10-8**

## Hypothesis Test for the Difference between Two Population Proportions

**Transportation Security Administration** The Transportation Security Administration (TSA) is responsible for transportation security at all U.S. airports. The TSA is evaluating two suppliers of a body-scanning system it is considering purchasing. Both companies have designed scanners with the ability to detect weapons that are made of nonmetallic materials. The TSA is interested in determining whether there is a difference in the proportions of nonmetallic weapons detected by scanners from the two suppliers. To conduct this test, use the following steps:

STEP 1 **Specify the population parameter of interest.**
In this case, the population parameter of interest is the population proportion of detected weapons. At issue is whether there is a difference between the two suppliers in terms of the proportion of weapons detected.

STEP 2 **Formulate the appropriate null and alternative hypotheses.**
The null and alternative hypotheses are

$$H_0: p_1 - p_2 = 0.0$$
$$H_A: p_1 - p_2 \neq 0.0$$

The Excel 2016 functions for the upper and lower-tailed critical z-value are

= NORM.S.INV(1 − alpha/2)
= NORM.S.INV(.99) and
= NORM.S.INV(alpha/2)
= NORM.S.INV(.01)

STEP 3 **Specify the significance level.**
The test will be conducted using an $\alpha = 0.02$.

STEP 4 **Determine the rejection region and the critical value.**
For a two-tailed test, the critical values for each side of the distribution are

$$-z_{0.01} = -2.33 \text{ and } z_{0.01} = 2.33$$

The decision rule based on the $z$-test statistic is:

If $z < -2.33$ or $z > 2.33$, reject the null hypothesis.
Otherwise, do not reject the null hypothesis.

STEP 5  **Compute the z-test statistic using Equation 10.18 and apply it to the decision rule.**

Two hundred nonmetallic weapons of different types have been scanned by systems from each supplier. For Supplier 1, 186 forgeries are detected, and for Supplier 2, 168 are detected. The sample proportions are

$$\bar{p}_1 = \frac{x_1}{n_1} = \frac{186}{200} = 0.93 \qquad \bar{p}_2 = \frac{x_2}{n_2} = \frac{168}{200} = 0.84$$

The test statistic is then calculated using Equation 10.18:

$$z = \frac{(\bar{p}_1 - \bar{p}_2) - (p_1 - p_2)}{\sqrt{\bar{p}(1 - \bar{p})\left(\dfrac{1}{n_1} + \dfrac{1}{n_2}\right)}}$$

where:

$$\bar{p} = \frac{n_1\bar{p}_1 + n_2\bar{p}_2}{n_1 + n_2} = \frac{200(0.93) + 200(0.84)}{200 + 200} = 0.885 \quad \text{(see Equation 10.17)}$$

Then

$$z = \frac{(0.93 - 0.84) - 0.0}{\sqrt{0.885(1 - 0.885)\left(\dfrac{1}{200} + \dfrac{1}{200}\right)}} = 2.8211$$

STEP 6  **Reach a decision.**

Because $z = 2.8211 > z_{0.01} = 2.33$, reject the null hypothesis.

STEP 7  **Draw a conclusion.**

The difference between the two sample proportions provides sufficient evidence to allow us to conclude that a difference exists between the two suppliers. The TSA can infer that Supplier 1 provides the better scanner for this purpose.

**TRY EXERCISE 10-54 (PG. 399)**

---

## 10.4 EXERCISES

### Skill Development

**10-51.** In each of the following cases, determine if the sample sizes are large enough so that the sampling distribution of the differences in the sample proportions can be approximated with a normal distribution:
a. $n_1 = 15$, $n_2 = 20$, $x_1 = 6$, and $x_2 = 16$
b. $n_1 = 10$, $n_2 = 30$, $\bar{p}_1 = 0.6$, and $x_2 = 19$
c. $n_1 = 25$, $n_2 = 16$, $x_1 = 6$, and $\bar{p}_2 = 0.40$
d. $n_1 = 100$, $n_2 = 75$, $\bar{p}_1 = 0.05$, and $\bar{p}_2 = 0.05$

**10-52.** Given the following sample information randomly selected from two populations

| Sample 1 | Sample 2 |
|----------|----------|
| $n_1 = 200$ | $n_2 = 150$ |
| $x_1 = 40$ | $x_2 = 27$ |

a. Determine if the sample sizes are large enough so that the sampling distribution for the difference between the sample proportions is approximately normally distributed.

b. Calculate a 95% confidence interval for the difference between the two population proportions.

**10-53.** Given the following null and alternative hypotheses and level of significance

$$H_0: p_1 = p_2$$
$$H_A: p_1 \neq p_2$$
$$\alpha = 0.10$$

together with the sample information

| Sample 1 | Sample 2 |
|----------|----------|
| $n_1 = 120$ | $n_2 = 150$ |
| $x_1 = 42$ | $x_2 = 57$ |

conduct the appropriate hypothesis test using the p-value approach. What conclusion should be reached concerning the null hypothesis?

**10-54.** Given the following null and alternative hypotheses

$$H_0: p_1 - p_2 \leq 0.0$$
$$H_A: p_1 - p_2 > 0.0$$

and the following sample information

| Sample 1 | Sample 2 |
|----------|----------|
| $n_1 = 60$ | $n_2 = 80$ |
| $x_1 = 30$ | $x_2 = 24$ |

a. Based on $\alpha = 0.02$ and the sample information, what should be concluded with respect to the null and alternative hypotheses? Be sure to clearly show the decision rule.

b. Calculate the $p$-value for this hypothesis test. Based on the $p$-value, would the null hypothesis be rejected? Support your answer with calculations and/or reasons.

**10-55.** Independent random samples of size 50 and 75 are selected. The sampling results in 35 and 35 successes, respectively. Test the following hypotheses:

a. $H_0: p_1 - p_2 = 0$ vs. $H_A: p_1 - p_2 \neq 0$. Use $\alpha = 0.05$.

b. $H_0: p_1 - p_2 \geq 0$ vs. $H_A: p_1 - p_2 < 0$. Use $\alpha = 0.05$.

c. $H_0: p_1 - p_2 \leq 0$ vs. $H_A: p_1 - p_2 > 0$. Use $\alpha = 0.025$.

d. $H_0: p_1 - p_2 = 0.05$ vs. $H_A: p_1 - p_2 \neq 0.05$. Use $\alpha = 0.02$.

## Business Applications

**10-56.** A survey conducted at a midwestern university and involving 1,500 students found that 68% had taken out a student loan to help pay for the cost of their college education. A similar study at a southeastern university in which 2,000 students were surveyed found that 47% had taken out a student loan.

a. Develop and interpret a 95% confidence interval estimate for the difference in the proportions of students who have loans at these two universities.

b. Using an alpha level of 0.05, test the hypothesis that there is no difference in the proportions of students at these two universities who have loans. How does this test compare with the result you found in part a?

**10-57.** Suppose, as part of a national study of economic competitiveness, a marketing research firm randomly sampled 200 adults between the ages of 27 and 35 living in metropolitan Seattle and 180 adults between the ages of 27 and 35 living in metropolitan Minneapolis. Each adult selected in the sample was asked, among other things, whether he or she had a college degree. From the Seattle sample, 66 adults answered yes, and from the Minneapolis sample, 63 adults answered yes. Based on the sample data, can we conclude that there is a difference between the population proportions of adults between the ages of 27 and 35 in the two cities with college degrees? Use a

level of significance of 0.01 to conduct the appropriate hypothesis test.

**10-58.** Suppose a random sample of 100 U.S. companies taken in 2015 showed that 21 offered high-deductible health insurance plans to their workers. A separate random sample of 120 firms taken in 2016 showed that 30 offered high-deductible health insurance plans to their workers. Based on the sample results, can you conclude that there is a higher proportion of U.S. companies offering high-deductible health insurance plans to their workers in 2016 than in 2015? Conduct your hypothesis test at a level of significance $\alpha = 0.05$.

**10-59.** Suppose a project was undertaken by the Department of Labor in a southern state to determine whether the most recent economic recession affected men and women differently. Assume that random samples of 485 adult males and 242 adult females were selected. The subjects were asked whether they had been involuntarily unemployed for four or more consecutive weeks at any time during the past 8 years. Fifty of the men responded "Yes" to the question, and 36 women responded "Yes." Based on these sample data, what should the analysis by the Department of Labor conclude about whether there is a statistical difference in the proportions of men and women who were unemployed? Test at an alpha $= 0.05$ level.

## Computer Software Exercises

**10-60.** As part of a nationwide study on home Internet use, suppose that researchers randomly sampled 150 urban households and 150 rural households. Among the questions asked of the sampled households was whether they used the Internet to download and play computer games. The survey results for this question are contained in the file **Internet Games**. Based on the sample results, can the researchers conclude that there is a difference between the proportions of urban households and rural households that use the Internet to download and play computer games? Conduct your test using a level of significance $\alpha = 0.01$.

**10-61.** One of the questions on a survey of business managers in different parts of the world was whether executives ranked innovation as the top priority for the coming year. Suppose the responses from 400 executives in the United States and 300 in Asia are given in the file titled **Priority**.

a. Determine if the sample sizes are large enough to provide assurance that the sampling distribution of the difference in the sample proportions of executives who feel innovation is their top priority is normally distributed.

b. Determine if the same proportion of U.S. and Asian executives feel that innovation is their top priority for the coming year. Use a significance value of 0.05 and the *p*-value approach.

**10-62.** A marketing research firm is interested in determining whether there is a difference between the proportion of households in Chicago and the proportion of households in Milwaukee who purchase groceries online. The research firm decided to randomly sample households earning more than $50,000 a year in the two cities and ask them if they purchased any groceries online last year. The random sample involved 150 Chicago households and 135 Milwaukee households.

The results of the sample can be found in the file **Online Groceries**.

a. Construct a 95% confidence interval estimate for the difference between the two population proportions.

b. At the 0.10 level of significance, can the marketing research firm conclude that a greater proportion of households in Chicago earning more than $50,000 annually buys more groceries online than do similar households in Milwaukee? Support your answer with the appropriate hypothesis test.

# 10 Overview

## Summary

### 10.1 Estimation for Two Population Means Using Independent Samples (pg. 364–374)

**OUTCOME 1** Be able to apply the techniques for using independent samples to develop interval estimates for the difference between two population means.

- **Independent samples** are samples selected from two or more populations in such a way that the occurrence of values in one sample has no influence on the occurrence of values in the other sample(s).
- There are different techniques for estimating the difference between two population means of independent samples for each of the following situations:
  - The population standard deviations are known and the samples are independent.
  - The population standard deviations are unknown and the samples are independent.

### 10.2 Hypothesis Tests for Two Population Means Using Independent Samples (pg. 374–386)

**OUTCOME 2** Be able to apply the techniques for using independent samples to test hypotheses for the difference between two population means.

- As in Section 10.1, the hypothesis-testing techniques for the difference between the means of two normally distributed populations vary based on whether the population standard deviations are known or unknown.
- Just as with hypothesis tests involving a single population parameter, two possible errors can occur:
  - Type I error: Rejecting $H_0$ when it is true (alpha error).
  - Type II error: Not rejecting $H_0$ when it is false (beta error).

### 10.3 Interval Estimation and Hypothesis Tests for Paired Samples (pg. 386–394)

**OUTCOME 3** Develop confidence interval estimates and conduct hypothesis tests for the difference between two population means for paired samples.

- **Paired samples** are samples that are selected in such a way that the values in one sample are matched with the values in the second sample for the purpose of controlling for extraneous factors. Paired samples are sometimes used to control for sources of variation that might otherwise distort the estimations or hypothesis tests.
- Paired samples are used when it is necessary to control for an outside source of variation.

### 10.4 Estimation and Hypothesis Tests for Two Population Proportions (pg. 395–401)

**OUTCOME 4** Carry out hypothesis tests and establish interval estimates for the difference between two population proportions.

- Many business applications involve confidence intervals and hypothesis tests for two population proportions.
- The general format for confidence interval estimates for the difference between two population proportions is, as always:

  Point estimate $\pm$ (Critical value)(Standard error)

- Confidence intervals that involve the difference between two population proportions always have a $z$-value as the critical value.
- Hypothesis tests for the difference between two population proportions require the calculation of a pooled estimator for the overall proportion.
- For help in selecting the appropriate procedures for establishing a confidence interval and conducting the hypothesis test when the decision-making situation involves two population parameters, see Figure 10.10.

402

FIGURE 10.10 **Estimation and Hypothesis Testing Flow Diagram**

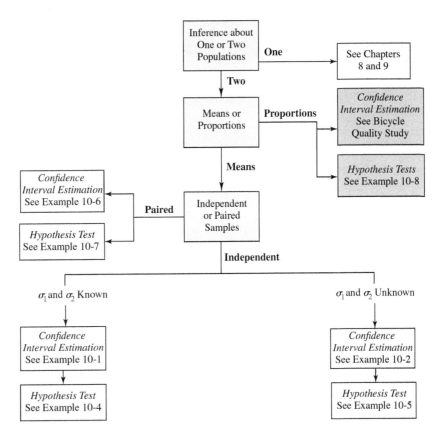

# Equations

**(10.1) Confidence Interval, General Format** pg. 364

Point estimate $\pm$ (Critical value)(Standard error)

**(10.2) Standard Error of $\bar{x}_1 - \bar{x}_2$ When $\sigma_1$ and $\sigma_2$ Are Known** pg. 364

$$\sigma_{\bar{x}_1 - \bar{x}_2} = \sqrt{\frac{\sigma_1^2}{n_1} + \frac{\sigma_2^2}{n_2}}$$

**(10.3) Confidence Interval Estimate for $\mu_1 - \mu_2$ When $\sigma_1$ and $\sigma_2$ Are Known, Independent Samples** pg. 365

$$(\bar{x}_1 - \bar{x}_2) \pm z\sqrt{\frac{\sigma_1^2}{n_1} + \frac{\sigma_2^2}{n_2}}$$

**(10.4) Confidence Interval Estimate for $\mu_1 - \mu_2$ When $\sigma_1$ and $\sigma_2$ Are Unknown, Independent Samples** pg. 367

$$(\bar{x}_1 - \bar{x}_2) \pm ts_p\sqrt{\frac{1}{n_1} + \frac{1}{n_2}}$$

**(10.5) Confidence Interval Estimate for $\mu_1 - \mu_2$ When $\sigma_1$ and $\sigma_2$ Are Unknown and Not Equal, Independent Samples** pg. 370

$$(\bar{x}_1 - \bar{x}_2) \pm t\sqrt{\frac{s_1^2}{n_1} + \frac{s_2^2}{n_2}}$$

**(10.6) Degrees of Freedom for Estimating Difference between Population Means When $\sigma_1$ and $\sigma_2$ Are Not Equal** pg. 370

$$df = \frac{(s_1^2/n_1 + s_2^2/n_2)^2}{\left(\dfrac{(s_1^2/n_1)^2}{n_1 - 1} + \dfrac{(s_2^2/n_2)^2}{n_2 - 1}\right)}$$

**(10.7) z-Test Statistic for $\mu_1 - \mu_2$ When $\sigma_1$ and $\sigma_2$ Are Known, Independent Samples** pg. 375

$$z = \frac{(\bar{x}_1 - \bar{x}_2) - (\mu_1 - \mu_2)}{\sqrt{\dfrac{\sigma_1^2}{n_1} + \dfrac{\sigma_2^2}{n_2}}}$$

**(10.8) t-Test Statistic for $\mu_1 - \mu_2$ When $\sigma_1$ and $\sigma_2$ Are Unknown and Assumed Equal, Independent Samples** pg. 377

$$t = \frac{(\bar{x}_1 - \bar{x}_2) - (\mu_1 - \mu_2)}{s_p\sqrt{\dfrac{1}{n_1} + \dfrac{1}{n_2}}}, \qquad df = n_1 + n_2 - 2$$

**(10.9) t-Test Statistic for $\mu_1 - \mu_2$ When Population Variances Are Unknown and Not Assumed Equal** pg. 383

$$t = \frac{(\bar{x}_1 - \bar{x}_2) - (\mu_1 - \mu_2)}{\sqrt{\dfrac{s_1^2}{n_1} + \dfrac{s_2^2}{n_2}}}$$

**(10.10) Degrees of Freedom for t-Test Statistic When Population Variances Are Not Equal** pg. 383

$$df = \frac{(s_1^2/n_1 + s_2^2/n_2)^2}{\left(\dfrac{(s_1^2/n_1)^2}{n_1 - 1} + \dfrac{(s_2^2/n_2)^2}{n_2 - 1}\right)}$$

**(10.11) Paired Difference** pg. 387

$$d = x_1 - x_2$$

**(10.12) Point Estimate for the Population Mean Paired Difference, $\mu_d$** pg. 388

$$\bar{d} = \frac{\sum\limits_{i=1}^{n} d_i}{n}$$

**(10.13) Sample Standard Deviation for Paired Differences** pg. 388

$$s_d = \sqrt{\frac{\sum\limits_{i=1}^{n}(d_i - \bar{d})^2}{n - 1}}$$

**(10.14) Confidence Interval Estimate for Population Mean Paired Difference, $\mu_d$** pg. 388

$$\bar{d} \pm t\frac{s_d}{\sqrt{n}}$$

**(10.15) $t$-Test Statistic for Paired-Sample Test** pg. 390

$$t = \frac{\bar{d} - \mu_d}{\dfrac{s_d}{\sqrt{n}}}, \quad df = (n - 1)$$

**(10.16) Confidence Interval Estimate for $p_1 - p_2$** pg. 395

$$(\bar{p}_1 - \bar{p}_2) \pm z\sqrt{\frac{\bar{p}_1(1 - \bar{p}_1)}{n_1} + \frac{\bar{p}_2(1 - \bar{p}_2)}{n_2}}$$

**(10.17) Pooled Estimator for Overall Proportion** pg. 397

$$\bar{p} = \frac{n_1\bar{p}_1 + n_2\bar{p}_2}{n_1 + n_2} = \frac{x_1 + x_2}{n_1 + n_2}$$

**(10.18) $z$-Test Statistic for Difference between Population Proportions** pg. 397

$$z = \frac{(\bar{p}_1 - \bar{p}_2) - (p_1 - p_2)}{\sqrt{\bar{p}(1 - \bar{p})\left(\dfrac{1}{n_1} + \dfrac{1}{n_2}\right)}}$$

## Key Terms

Independent samples pg. 364

Paired samples pg. 386

## Chapter Exercises

### Conceptual Questions

**10-63.** Why, when dealing with two independent samples where you cannot assume the population variances are equal, should the degrees of freedom be adjusted?

**10-64.** Explain, in nontechnical terms, why pairing observations, if possible, is often a more effective tool than taking independent samples.

**10-65.** Consider the following samples obtained from two normally distributed populations whose variances are equal:

| Sample 1: | 11.2 | 11.2 | 7.4 | 8.7 | 8.5 | 13.5 | 4.5 | 11.9 |
|---|---|---|---|---|---|---|---|---|
| Sample 2: | 11.7 | 9.5 | 15.6 | 16.5 | 11.3 | 17.6 | 17.0 | 8.5 |

a. Suppose that the samples were independent. Perform a test of hypothesis to determine if there is a difference in the two population means. Use a significance level of 0.05.

b. Now suppose that the samples were paired samples. Perform a test of hypothesis to determine if there is a difference in the two population means.

c. How do you account for the difference in the outcomes of part a and part b? Support your assertions with a statistical rationale.

**10-66.** Examine the following information:

|  | Sample 1 | Sample 2 |
|---|---|---|
| $n_i$ | 15 | 10 |
| $\bar{x}_i$ | 12 | 7 |
| $s_i$ | 3 | 4 |

a. Construct a 95% confidence interval for $\mu_1 - \mu_2$.

b. Calculate the test statistic that would be used to determine if $\mu_1 - \mu_2 = $ the upper confidence limit. Given this, what would the result of the test be if you wish to test $\mu_1 - \mu_2 = $ Any value larger than the upper confidence limit?

c. Calculate the test statistic that would be used to determine if $\mu_1 - \mu_2 = $ the lower confidence limit. Given this, what would the result of the test be if you wish to test $\mu_1 - \mu_2 = $ Any value smaller than the lower confidence limit?

d. Given your answers to parts b and c, how could you use the confidence interval to conduct a two-tailed hypothesis test?

## Business Applications

**10-67.** A student group at a university in the Midwest is conducting a study to determine whether there is a difference between textbook prices for used textbooks sold by the on-campus bookstore and by an Internet retailer. A random sample of 10 textbooks was taken, and the campus price and the Internet retailer's price for each textbook were collected. At the 0.01 level of significance, can the student group conclude that there is a difference in average textbook prices for used textbooks sold on campus and over the Internet?

| Textbook | Physics | Chemistry | Spanish | Accounting | Calculus |
|---|---|---|---|---|---|
| Campus | $108 | $114 | $114 | $110 | $118 |
| Internet | $124 | $120 | $107 | $112 | $122 |

| Textbook | Economics | Art | Biology | History | English |
|---|---|---|---|---|---|
| Campus | $108 | $119 | $115 | $119 | $114 |
| Internet | $123 | $125 | $108 | $117 | $119 |

**10-68.** Bach Photographs is a photography business with studios in two locations. The owner is interested in monitoring business activity closely at the two locations. Among the factors in which he is interested is whether the mean customer orders per day for the two locations are the same. A random sample of 11 days' orders for the two locations showed the following data:

| Location A | | | Location B | | |
|---|---|---|---|---|---|
| $444 | $478 | $501 | $233 | $127 | $230 |
| $200 | 400 | 350 | 299 | 250 | 300 |
| $167 | 250 | 300 | 800 | 340 | 400 |
| $300 | 600 | | 780 | 370 | |

The owner wishes to know the difference between the average amounts in customer orders per day for the two locations. He has no idea what this difference might be. What procedure would you use under these circumstances? Explain your reasoning.

**10-69.** Allstate Insurance is one of the major automobile insurance companies in the country. The western region claims manager instructed an intern in her department to develop a confidence interval estimate of the difference between the mean years that male customers have been insured by Allstate versus female customers. The intern randomly selected 13 male and 13 female customers from the account records and recorded the number of years that each customer had been insured by Allstate. These data (rounded to the nearest year) are as follows:

| Males | | | | Females | | | |
|---|---|---|---|---|---|---|---|
| 14 | 9 | 9 | 16 | 3 | 10 | 4 | 7 |
| 14 | 8 | 12 | 11 | 10 | 5 | 4 | 1 |
| 6 | 9 | 10 | 7 | 4 | 4 | 6 | 9 |
| 3 | | | | 2 | | | |

Based on these data, construct and interpret a 90% confidence interval estimate for the difference between the mean years for male and female customers. Assume equal population variances.

**10-70.** The Eaton Company sells breakable china through a mail-order system that has been very profitable. One of its major problems is freight damage. It insures the items at shipping, but the inconvenience to the customer when a piece gets broken can cause the customer not to place another order in the future. Thus, packaging is important to the Eaton Company.

In the past, the company has purchased two different packaging materials from two suppliers. The assumption was that there would be no difference in the proportion of damaged shipments resulting from the use of either packaging material. The sales manager decided a study of this issue should be done. Therefore, a random sample of 300 orders using shipping material 1 and a random sample of 250 orders using material 2 were pulled from the files. The number of damaged parcels, $x$, was recorded for each material as follows:

| Material 1 | Material 2 |
|---|---|
| $n_1 = 300$ | $n_2 = 250$ |
| $x_1 = 19$ | $x_2 = 12$ |

a. Is the normal distribution a good approximation for the distribution of the difference between the sample proportions? Provide support for your answer.
b. Based on the sample information and $\alpha = 0.03$, what should the Eaton Company conclude?

**10-71.** Home purchases by single women in the United States were expected to increase in 2016 due to the increase in the number of women who earn over $100,000 (*source: Laura Lorenzetti, "There may be more single women buying houses in 2016," http://fortune.com, Dec. 28, 2015*). Suppose a survey of 200 male and 302 female home buyers showed that their mean incomes were $58,939 and $47,315, respectively, while the standard deviations were $10,000 and $9,000. Assume equal population variances.

a. Conduct a hypothesis test to determine if the mean income for men buying homes is more than $10,000 larger than that of women. Use a significance level of 0.05.
b. Describe any additional assumptions required to validate the results of the hypothesis test in part a.

**10-72.** TurboTax and H&R Block are two of the best-selling tax preparation software packages. Suppose a consumer group conducted a test to estimate the difference in the mean times it would take an individual to complete his or her federal income tax using these two products. To run the test, the group selected a random sample of 16 people. All 16 people used both software packages to complete their taxes. Half of the people were assigned TurboTax first, followed by H&R Block. The other half used the products in the opposite order. The following values represent the time in minutes that it took to complete the returns:

| Individual | TurboTax | H&R Block |
|:---:|:---:|:---:|
| 1 | 70 | 88 |
| 2 | 56 | 71 |
| 3 | 79 | 89 |
| 4 | 94 | 66 |
| 5 | 93 | 78 |
| 6 | 101 | 64 |
| 7 | 42 | 74 |
| 8 | 71 | 99 |
| 9 | 91 | 79 |
| 10 | 59 | 68 |
| 11 | 65 | 93 |
| 12 | 50 | 93 |
| 13 | 47 | 86 |
| 14 | 60 | 86 |
| 15 | 63 | 81 |
| 16 | 43 | 83 |

Based on these sample data, construct and interpret a 95% confidence interval estimate for the difference between the two population means.

**10-73.** Suppose a random sample taken in 2016 of 900 adults between the ages of 25 and 34 years of age indicated that 396 owned a home. In 1984, a random sample of 900 adults in the same age group showed that 432 owned a home. At the 0.02 level of significance, is there a difference between the proportions of homeowners in 2016 and 1984 for the 25 to 34 age group?

**10-74.** A random sample of 64 bicycle-riding adults in Portland indicated that 24 always wore a helmet while riding. A random sample of 80 bicycle-riding adults in Amsterdam indicated that 18 always wore a helmet while riding.
   a. Use the sample information to develop a 95% confidence interval estimate for the difference between the proportions of bicycle-riding adults who wear a helmet in the two cities.
   b. Based on the sample results, can you conclude that the population proportion of bicycle-riding adults in Portland who always wear a helmet is greater than the population proportion of bicycle-riding adults in

Amsterdam who always wear a helmet? Use a 0.025 level of significance to conduct the test.

**10-75.** Surprisingly, a 2004 report indicated that injuries such as strains and sprains were higher among airport screeners than any other federal work group (*source: Associated Press, "Airport screeners have highest injury rate," www.nbcnews.com, May 3, 2004*). Suppose a 2016 study found the injury rate for airport screeners was 29%, far exceeding the 4.5% injury rate for the rest of the federal workforce. Assume the sample sizes required to obtain these percentages were 75 and 125, respectively.
   a. Determine if the sample sizes were large enough to allow the distribution of the difference of the sample proportions to be approximated with a normal distribution.
   b. Conduct a hypothesis test to determine if the injury rate of airport screeners is more than 10% larger than that of the rest of the federal workforce. Use a significance level of 0.05.

## Computer Software Exercises

**10-76.** Reviewers from the Oregon Evidence-Based Practice Center at the Oregon Health and Science University investigated the effectiveness of prescription drugs in assisting people to fall asleep and stay asleep (*source: S. Carson, M.S. McDonagh, et al., "Drug class review: Newer drugs for insomnia," Oregon Health and Science University, 2008, http://www.ncbi.nlm.nih.gov*). Suppose a study of two drugs, Sonata and Ambien, was conducted and the resulting data are contained in a file titled **Shuteye**. The samples reflect an experiment in which individuals were randomly given the two brands of pills on separate evenings. Their time spent sleeping was recorded for each of the brands of sleeping pills.
   a. Does the experiment seem to have dependent or independent samples? Explain your reasoning.
   b. Do the data indicate that there is a statistical difference in the mean time spent sleeping for people using the two drugs? Test using alpha = 0.05. Conduct a statistical procedure to determine this.
   c. Develop and interpret a 95% confidence interval for the difference in mean sleeping time for people using the two drugs.

**10-77.** Suppose that as part of a study on student loan debt, a national agency that underwrites student loans is examining the differences in student loan debt for undergraduate students. One question the agency would like to address specifically is whether the mean undergraduate debt of nonminority students is less than the mean undergraduate debt of minority students. To conduct the study, suppose that a random sample of 92 nonminority students and a random sample of 110 minority students who completed an undergraduate degree were taken. The undergraduate debt incurred for financing college for each sampled student was collected. The sample results can be found in the file **Student Debt**.

a. Assume that the agency believes the two populations from which the samples were taken are approximately normally distributed with equal variances. Conduct a hypothesis test at the 0.01 level of significance to determine whether the mean undergraduate debt for nonminority students is less than the mean undergraduate debt for minority students.

b. For what values of alpha would your decision in part a change?

c. Suppose the agency is unwilling to assume the two populations from which the samples are taken are approximately normally distributed with equal variances. Conduct the appropriate test to determine whether the mean undergraduate debt for nonminority students is less than the mean undergraduate debt for minority students. Use the *p*-value approach to conduct the test. State a conclusion.

**10-78.** The file titled **Tuition and Fees** shows the in-state and out-of-state tuition and fees for a sample of public universities in the United States. Based on these data, develop a 90% confidence interval estimate for the difference in average tuition between in state and out of state.

**10-79.**  Vintner Mortgage Company's board of directors asked the president whether the company had a difference in the proportions of loan defaults between residential and commercial customers. To prepare an answer to this question, company officials selected a random sample of 200 residential loans and 105 commercial loans. The loans were analyzed to determine their status. A loan that is still being paid was labeled "Active" and a default loan was labeled "Default." The resulting data are in a file called **Vintner**.

a. Based on the sample data and a significance level equal to 0.05, does there appear to be a difference in the proportion of loan defaults between residential and commercial customers?

b. Prepare a short response to the Vintner board of directors. Include in your report a graph of the data that supports your statistical analysis.

c. Consider the outcome of the hypothesis test in part a. In the last five audits, 10 residential and 10 commercial customers were selected. In three of the audits, there were more residential than commercial loan defaults. Determine the probability of such an occurrence.

---

## Case 10.1  Larabee Engineering—Part 1

Cregg Hart is manufacturing manager for Larabee Engineering, a locomotive engine and rail car manufacturer. The company has been very successful in recent years, and in July 2016 signed two major contracts totaling nearly $200 million. A key to the company's success has been its focus on quality. Customers from around the world have been very pleased with the attention to detail put forth by Larabee Engineering.

One of the things Cregg has been adamant about is that Larabee Engineering's suppliers also provide high quality. As a result, when the company finds good suppliers, it stays with them and tries to establish a long-term relationship. However, Cregg must also factor in the costs of parts and materials and has instructed his purchasing staff to be on the lookout for "better deals."

Recently, Sheryl Carleson, purchasing manager, identified a new rivet supplier in Europe that claims its rivets are as good or better quality than the current supplier's but at a much lower cost. One key quality factor is the rivet diameter. When Sheryl approached Cregg about the possibility of going with the new supplier for rivets, he suggested they conduct a test to determine if there is any difference in the average diameter of the rivets from the two companies. Sheryl requested that the new company send

100 rivets, and she pulled a random sample of 100 rivets from her inventory of rivets from the original supplier. She then asked an intern to measure the diameters to three decimal places using a micrometer. The resulting data from both suppliers are given in the file called **Larabee**.

### Required Tasks:

1. Develop histograms showing the distribution of rivet diameters from the two suppliers.
2. Compute the sample means and standard deviations for the two sets of data.
3. Comment on whether it appears the assumptions required for testing if the two populations have equal mean diameters are satisfied.
4. Select a level of significance for testing whether the two suppliers have rivets with equal mean diameters. Discuss the factors you used in arriving at the level of significance you have selected.
5. Perform the appropriate null and alternative hypothesis test. Discuss the results.
6. Prepare a short report outlining your recommendation.

---

## Case 10.2  Hamilton Marketing Services

Alex Hamilton founded Hamilton Marketing Services after leaving a major marketing consulting firm in Chicago. Hamilton Marketing Services focuses on small- to medium-sized retail firms and has been quite successful in providing a wide range of marketing and advertising services.

A few weeks ago, a relatively new customer that Alex himself has been working with for the past several months called with an idea. This customer, a pet-grooming company, is interested in changing the way it prices its full-service dog-grooming service. The customer is considering two options: (1) a flat $40.00 per-visit price and (2) a $30.00 per-visit price if the dog owner signs up for a series of four groomings. However, the pet-grooming service is unsure how these options would be received by its customers. The owner was hoping there was some type of study Alex could have his company do that would provide information on what the difference in response rates would be for the two pricing options. He was interested in determining if one option would bring in more revenue than the other.

At the time, Alex suggested that a flier with an attached coupon be sent to a random sample of potential customers. One sample of customers would receive the coupon listing the $40.00 price. A second sample of customers would receive the coupon listing the $30.00 price and the requirement for signing up for a series of four visits. Each coupon would have an expiration date of one month from the date of issue. Then the pet-grooming store owner could track the responses to these coupon offers and bring the data back to Alex for analysis.

Yesterday, the pet store owner e-mailed an Excel file called **Grooming Price Test** to Alex. Alex has now asked you to assist with the analysis. He has mentioned using a 95% confidence

interval and wants a short report describing the data and summarizing which pricing strategy is preferred both from a proportion-response standpoint and from a revenue-producing standpoint.

### Required Tasks:

1. Compute a sample proportion for the responses for the two coupon options under consideration.
2. Develop a 95% confidence interval for the difference between the proportions of responses between the two options.
3. Use the confidence interval developed in (2) to draw a conclusion regarding whether or not there is any statistical evidence that there is a difference in response rates between the two coupon options.
4. Determine whether or not there is a difference between the two coupon options in terms of revenue generated.
5. Identify any other issues or factors that should be considered in deciding which coupon option to use.
6. Develop a short report summarizing your analysis and conclusions.

---

### Case 10.3    Green Valley Assembly Company

The Green Valley Assembly Company assembles consumer electronics products for manufacturers that need temporary extra production capacity. As such, it has periodic product changes. Because the products Green Valley assembles are marketed under the label of well-known manufacturers, high quality is a must.

Tom Bradley, of the Green Valley personnel department, has been very impressed by recent research concerning job-enrichment programs. In particular, he has been impressed with the increases in quality that seem to be associated with these programs. However, some studies have shown no significant increase in quality, and they imply that the money spent on such programs has not been worthwhile.

Tom has talked to Sandra Hansen, the production manager, about instituting a job-enrichment program in the assembly operation at Green Valley. Sandra was somewhat pessimistic about the potential, but she agreed to introduce the program. The plan was to implement the program in one wing of the plant and continue with the current method in the other wing. The procedure was to be in effect for six months. After that period, a test would be made to determine the effectiveness of the job-enrichment program.

After the six-month trial period, a random sample of employees from each wing produced the following output measures:

| Old | Job-Enriched |
|---|---|
| $n_1 = 50$ | $n_2 = 50$ |
| $\bar{x}_1 = 11/\text{hr}$ | $\bar{x}_2 = 9.7/\text{hr}$ |
| $s_1 = 1.2/\text{hr}$ | $s_2 = 0.9/\text{hr}$ |

Both Sandra and Tom wonder whether the job-enrichment program has affected production output. They would like to use these sample results to determine whether the average output has changed and to determine whether the employees' consistency has been affected by the new program.

A second sample from each wing was selected. The measure was the quality of the products assembled. In the "old" wing, 79 products were tested and 12% were found to be defectively assembled. In the "job-enriched" wing, 123 products were examined and 9% were judged defectively assembled.

With all these data, Sandra and Tom are beginning to get a little confused. However, they realize that there must be some way to use the information to make a judgment about the effectiveness of the job-enrichment program.

---

### Case 10.4    U-Need-It Rental Agency

Richard Fundt has operated the U-Need-It rental agency in a northern Wisconsin city for the past five years. One of the biggest rental items has always been chainsaws; lately, the demand for these saws has increased dramatically. Richard buys chainsaws at a special industrial rate and then rents them for $10 per day. The chainsaws are used an average of 50 to 60 days per year. Although Richard makes money on any chainsaw, he obviously makes more on those saws that last the longest.

Richard worked for a time as a repairperson and can make most repairs on the equipment he rents, including chainsaws. However, he would also like to limit the time he spends making repairs. U-Need-It is currently stocking two types of saws: North Woods and Accu-Cut. Richard has an impression that one of the models, Accu-Cut, does not seem to break down as much as the other. Richard currently has 8 North Woods saws and 11 Accu-Cut saws. He decides to keep track of the number of hours each is used between major repairs. He finds the following values, in hours:

| Accu-Cut | | North Woods | |
|---|---|---|---|
| 48 | 46 | 48 | 78 |
| 39 | 88 | 44 | 94 |
| 84 | 29 | 72 | 59 |
| 76 | 52 | 19 | 52 |
| 41 | 57 | | |
| 24 | | | |

The North Woods sales representative has stated that the company may be raising the price of its saws in the near future. This will make them slightly more expensive than the Accu-Cut models. However, the prices have tended to move with each other in the past.

## Required Tasks:

1. Compute and interpret a 95% confidence interval estimate for the difference in mean hours between repairs for the two chainsaw brands.
2. Based on your findings in part 1, is there justification for the North Woods proposed price increase based on chainsaw performance?

# 11 Hypothesis Tests and Estimation for Population Variances

## WHY YOU NEED TO KNOW

In addition to hypothesis-testing applications involving one and two population means and proportions like those presented in Chapters 9 and 10, there are also business situations in which decision makers must reach a conclusion about the value of a single population variance or about the relationship between two population variances. For example, knowing that a crate of strawberries weighs 4 pounds on average may not be enough. The produce manager for a supermarket chain may also be concerned about the variability in weight of the strawberries. If there is too much variability, then some crates may be underfilled, which will cause problems with consumers who believe they have been cheated. If the crate is overfilled,

**Quick Prep**

Review material on calculating and interpreting sample means and variances in Chapter 3.

Examine Section 9.1 on formulating null and alternative hypotheses.

Make sure you understand the concepts of Type I and Type II error discussed in Chapter 9.

the store is giving away strawberries and incurring unnecessary costs. The manager must monitor both the average weight and the variation in weight of the strawberry crates.

A manager may also be required to decide if there is a difference in the variability of sales between two different sales territories or if the output of one production process is more or less variable than another. In this chapter, we discuss methods that can be used to make inferences concerning one and two population variances. The techniques presented in this chapter will also introduce new distributions that will be used in later chapters. When reading this chapter, keep in mind that the techniques discussed here are extensions of the estimation and hypothesis-testing concepts introduced in Chapters 8, 9, and 10.

## 11.1 Hypothesis Tests and Estimation for a Single Population Variance

You can expect to encounter many cases in which you will be as interested in the spread or variation of a population as in its central location. For instance, military planes designed to penetrate enemy defenses have a ground-following radar system. The radar tells the pilot exactly how far the plane is above the ground. A radar unit that is correct *on the average* is useless if the readings are distributed widely around the average value. Many airport shuttle systems have stopping sensors to deposit passengers at the correct spot in a terminal. A sensor that, *on the average*, lets passengers off at the correct point could leave many irritated people long distances up and down the track if it has too much variability. Therefore, many product specifications involve both an average value and some limit on the dispersion that the individual values can have. For example, the specification for a steel push pin may be an average length of 1.78 inches plus or minus 0.01 inch. A company using these pins would be interested in both the average length and how much these pins vary in length.

Business applications in which the population variance is important use one of two statistical procedures: hypothesis tests or confidence interval estimates. In hypothesis-testing applications, a null hypothesis is formulated in terms of the population variance, $\sigma^2$. For example, a bank manager might hypothesize that the population variance in customer service times, $\sigma^2$, is no greater than 36 minutes squared (remember, variance is in squared units). Then, based on sample data from the population of bank customers, the manager will either reject or not reject the null hypothesis.

In other cases, the application might require an analyst to estimate the population variance. For instance, a regional transportation manager is planning to conduct a survey of residents in a suburb to determine the mean number of times per week they use their cars for purposes other than commuting to work. Before conducting the survey, she needs to determine the required sample size. One key factor in determining the sample size (see Chapter 9) is the value of the population variance. Thus, before she can determine the required sample size, she might want to estimate the population variance by taking a *pilot sample* and constructing a confidence interval estimate for $\sigma^2$.

This section introduces the methods for testing hypotheses and for constructing confidence interval estimates for a single population variance.

### Chi-Square Test for One Population Variance

Usually when we think of measuring variation, we use the standard deviation as the measure because it is measured in the same units as the mean. Ideally, in the ground-following radar example mentioned earlier, we would want to test to see whether the standard deviation exceeds a certain level, as determined by the product specifications. Unfortunately, there is no statistical test that directly tests the standard deviation. However, there is a procedure called the chi-square test that can be used to test the population variance. We can convert any hypothesis test for a population standard deviation into one involving the population variance, as shown in the following example.

**BUSINESS APPLICATION**   ## Hypothesis for a Population Variance

**Valley Appliance Repair**   Valley Appliance Repair is a company that repairs equipment and appliances for businesses and private residential customers. Looking at past records, the operations manager has determined the mean job time for a properly trained crew is 2 hours, with a standard deviation not to exceed 0.5 hour. Recent data indicate that the 2-hour average is being achieved. However, the manager is worried that variability in repair times may have increased. The company's job schedule is built around the assumptions of $\mu = 2$ hours and $\sigma = 0.5$ hour. If the repair-time standard deviation exceeds 0.5 hour, the job schedule gets disrupted.

The operations manager has decided to select a random sample of recent jobs and use the sample data to determine whether the repair-time standard deviation exceeds 0.5 hour.

Ideally, the manager would like to test the following null and alternative hypotheses:

$$H_0: \sigma \leq 0.5 \text{ (service standard)}$$
$$H_A: \sigma > 0.5$$

Because there is no statistical technique for directly testing hypotheses about a population standard deviation, we will use a test for a population variance. We first convert the standard deviation to a variance by squaring the standard deviation. Then we restate the null and alternative hypotheses as follows:

$$H_0: \sigma^2 \leq 0.25$$
$$H_A: \sigma^2 > 0.25$$

As with all hypothesis tests, the decision to reject or not reject the null hypothesis will be based on the statistic computed from the sample. In testing hypotheses about a single population variance, the appropriate sample statistic is $s^2$, the *sample variance*.

To test a null hypothesis about a population variance, we compare $s^2$ with the hypothesized population variance, $\sigma^2$. To do this, we need to standardize the distribution of the sample variance in much the same way as we did to use the $z$-distribution and the $t$-distribution when testing hypotheses about the population mean.

**Assumption** { When the random sample is from a normally distributed population, the distribution for the standardized sample variance is a *chi-square distribution*.

The chi-square distribution is a continuous distribution of a standardized random variable, computed using Equation 11.1.

**Chi-Square Test for a Single Population Variance**

$$\chi^2 = \frac{(n-1)s^2}{\sigma^2} \tag{11.1}$$

where:

$$\chi^2 = \text{Standardized chi-square variable}$$
$$n = \text{Sample size}$$
$$s^2 = \text{Sample variance}$$
$$\sigma^2 = \text{Hypothesized variance}$$

The distribution of $\chi^2$ is a chi-square distribution with $n - 1$ degrees of freedom.

**FIGURE 11.1 Chi-Square Distributions**

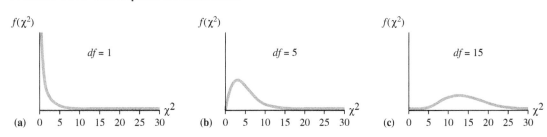

The central location and shape of the chi-square distribution depend only on the degrees of freedom, $n - 1$. Figure 11.1 illustrates chi-square distributions for various degrees of freedom. Note that as the degrees of freedom increase, the chi-square distribution becomes more symmetric.

---

**BUSINESS APPLICATION**    **Testing a Single Population Variance**

**Valley Appliance Repair** (*continued*)   Returning to the Valley Appliance Repair application, suppose the operations manager took a random sample of 20 service calls and found a variance of 0.33 hour squared. Figure 11.2 illustrates the hypothesis test at a significance level of 0.10.

Appendix G contains a table of upper-tailed chi-square critical values for various probabilities and degrees of freedom. The use of the chi-square table is similar to the use of the $t$-distribution table. For example, to find the critical value, $\chi^2_{0.10}$, for the Valley Appliance Repair application, we determine the degrees of freedom, $n - 1 = 20 - 1 = 19$, and the desired significance level, 0.10. Because this is an upper-tailed, one-tailed test, we go to the chi-square table under the column headed 0.10 and find the $\chi^2$ value in this column that intersects the row corresponding to the appropriate degrees of freedom. We find the critical value of $\chi^2_{0.10} = 27.2036$.

As you can see in Figure 11.2, the chi-square test statistic, calculated using Equation 11.1, is

$$\chi^2 = \frac{(n - 1)s^2}{\sigma^2} = \frac{(19)(0.33)}{0.25} = 25.08$$

> The Excel 2016 function for finding an upper-tailed chi-square critical value is
> = **CHISQ.INV.RT(alpha,degrees of freedom)**
> = **CHISQ.INV.RT(0.10,19)**

**FIGURE 11.2 Chi-Square Test for One Population Variance for the Valley Appliance Repair Example**

**Hypothesis:**

$H_0: \sigma^2 \leq 0.25$

$H_A: \sigma^2 > 0.25$

$\alpha = 0.10$

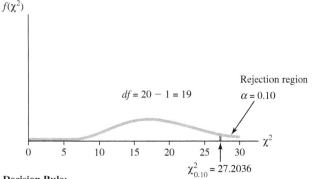

$df = 20 - 1 = 19$

Rejection region

$\alpha = 0.10$

$\chi^2_{0.10} = 27.2036$

**Decision Rule:**

If $\chi^2 > \chi^2_{0.10} = 27.2036$, reject $H_0$.

Otherwise, do not reject $H_0$.

The calculated chi-square test statistic is

$$\chi^2 = \frac{(n - 1)s^2}{\sigma^2} = \frac{19(0.33)}{0.25} = 25.08$$

Because $25.08 < 27.2036$, do not reject $H_0$.

This falls to the left of the rejection region, meaning the manager should not reject the null hypothesis based on these sample data. Thus, based on these results, there is insufficient evidence to conclude that the repair teams are completing their work with a standard deviation of more than 0.5 hour.

---

**HOW TO DO IT** (Example 11-1)

**Hypotheses Tests for a Single Population Variance**

**1.** Specify the population parameter of interest.

**2.** Formulate the null and alternative hypotheses in terms of $\sigma^2$, the population variance.

**3.** Specify the level of significance for the hypothesis test.

**4.** Construct the rejection region and define the decision rule. Obtain the critical value, $\chi_\alpha^2$, from the chi-square distribution table.

**5.** Compute the test statistic.

Select a random sample and compute the sample variance,

$$s^2 = \frac{\Sigma(x - \bar{x})^2}{n - 1}$$

Based on the sample variance, determine $\chi^2 = \frac{(n - 1)s^2}{\sigma^2}$

**6.** Reach a decision.

**7.** Draw a conclusion.

---

The Excel 2016 function for finding an upper-tailed chi-square critical value is

$= \textbf{CHISQ.INV.RT(0.05,19)}$

---

### EXAMPLE 11-1 One-Tailed Hypothesis Tests for a Population Variance

**Lockheed Martin Corporation** The Lockheed Martin Corporation is a major defense contractor as well as the maker of commercial products such as space satellite systems. The quality specialist at the Sunnyvale, California, Space Systems facility has been informed that one specification listed in the contract between Lockheed Martin and the Department of Defense concerns the variability in the diameter of the part that will be installed on a satellite. Hundreds of these parts are used on each satellite. Before installing these parts, Lockheed Martin quality specialists will take a random sample of 20 parts from the batch and test to see whether the standard deviation exceeds the 0.05-inch specification. This can be done using the following steps:

**STEP 1** **Specify the population parameter of interest.**
The standard deviation for the diameter of a part is the parameter of interest.

**STEP 2** **Formulate the null and alternative hypotheses.**
The null and alternative hypotheses must be stated in terms of the population variance, so we convert the specification, $\sigma = 0.05$, to the variance, $\sigma^2 = 0.0025$. The null and alternative hypotheses are

$$H_0: \sigma^2 \leq 0.0025$$
$$H_A: \sigma^2 > 0.0025$$

**STEP 3** **Specify the significance level.**
The hypothesis test will be conducted using $\alpha = 0.05$.

**STEP 4** **Construct the rejection region and define the decision rule.**
Note that this hypothesis test is a one-tailed, upper-tailed test. Thus we obtain the critical value from the chi-square table, where the area in the upper tail corresponds to $\alpha = 0.05$. The critical value from the chi-square distribution with $20 - 1 = 19$ degrees of freedom and 0.05 level of significance is

$$\chi_\alpha^2 = \chi_{0.05}^2 = 30.1435$$

The decision rule is stated as:

If $\chi^2 > \chi_{0.05}^2 = 30.1435$, reject $H_0$; otherwise, do not reject.

**STEP 5** **Compute the test statistic.**
The random sample of $n = 20$ parts gives a sample variance for part diameter of $s^2 = \frac{\Sigma(x - \bar{x})^2}{n - 1} = 0.0108$.

The test statistic is

$$\chi^2 = \frac{(n - 1)s^2}{\sigma^2} = \frac{(20 - 1)0.0108}{0.0025} = 82.08$$

**STEP 6** **Reach a decision.**
Because $\chi^2 = 82.08 > 30.1435$, reject the null hypothesis.

**STEP 7** **Draw a conclusion.**
Conclude that the variance of the population does exceed the 0.0025 limit. The company appears to have a problem with the variation of this part. The quality specialist will likely contact the supplier to discuss the issue.

**TRY EXERCISE 11-3** (pg. 417)

### EXAMPLE 11-2    Two-Tailed Hypothesis Tests for a Population Variance

**Genesis Technology**    The research and development manager for Genesis Technology, a clean-tech startup company headquartered in Pittsburgh, has spent the past several months overseeing a project in which the company has been experimenting with different designs of storage devices for solar energy. One important attribute of a storage device for electricity is the variability in storage capacity. Consistent capacity is desirable so that consumers can more accurately predict the amount of time they can expect the "battery" system to last under normal conditions. Genesis Technology engineers have determined that one particular storage design will yield an average of 88 minutes per cell with a standard deviation of 6 minutes. During the past few weeks, the engineers have made some modifications to the design and are interested in determining whether this change has affected the standard deviation either up or down. The test was conducted on a random sample of 12 individual storage cells containing the modified design. The following data show the minutes of use that were recorded:

| 89 | 85 | 91 | 95 |
|----|----|----|----|
| 95 | 97 | 81 | 89 |
| 94 | 86 | 87 | 83 |

This data can be analyzed using the following steps:

STEP 1 **Specify the population parameter of interest.**
The engineers are interested in the standard deviation of the time (in minutes) that the storage cells last under normal use.

STEP 2 **Formulate the null and alternative hypotheses.**
The null and alternative hypotheses are stated in terms of the population variance, since there is no test that deals specifically with the population standard deviation. Thus, we must convert the population standard deviation, $\sigma = 6$, to a variance, $\sigma^2 = 36$. Because the engineers are interested in whether there has been a change (up or down), the test will be a two-tailed test with the null and alternative hypotheses formulated as follows:

$$H_0: \sigma^2 = 36$$
$$H_A: \sigma^2 \neq 36$$

The Excel 2016 functions for finding upper- and lower-tailed chi-square critical values are
Upper tail:
= **CHISQ.INV.RT(0.05,11)**
Lower tail:
= **CHISQ.INV.RT(0.95,11)**

STEP 3 **Specify the significance level.**
The hypothesis test will be conducted using an $\alpha = 0.10$.

STEP 4 **Construct the rejection region and define the decision rule.**
Because this is a two-tailed test, two critical values from the chi-square distribution in Appendix G are required, one for the upper (right) tail and one for the lower (left) tail. The alpha will be split evenly between the two tails with $\alpha/2 = 0.05$ in each tail. The degrees of freedom for the chi-square distribution are $n - 1 = 12 - 1 = 11$. The upper-tailed critical value is found by locating the column headed 0.05 and going to the row for 11 degrees of freedom. This gives $\chi^2_{0.05} = 19.6752$. The lower-tailed critical value is found by going to the column headed 0.95 and to the row for 11 degrees of freedom. (Refer to Figure 11.3.) This gives $\chi^2_{0.95} = 4.5748$. Thus, the decision rule is:

If $\chi^2 > \chi^2_{0.05} = 19.6752$, or if
$\chi^2 < \chi^2_{0.95} = 4.5748$, reject the null hypothesis.

Otherwise, do not reject the null hypothesis.

**FIGURE 11.3 Chi-Square Rejection Regions for Two-Tailed Test of One Population Variance**

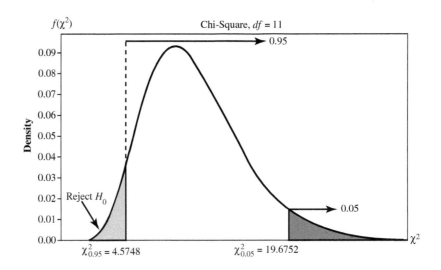

STEP 5 **Compute the test statistic.**
The random sample of $n = 12$ cells gives a sample standard variance computed as

$$s^2 = \frac{\Sigma(x - \bar{x})^2}{n - 1} = 26.6$$

Then the test statistic is

$$\chi^2 = \frac{(n - 1)s^2}{\sigma^2} = \frac{(12 - 1)26.6}{36} = 8.13$$

STEP 6 **Reach a decision.**
Because $\chi^2 = 8.13 > \chi^2_{0.95} = 4.5748$ and $\chi^2 = 8.13 < \chi^2_{0.05} = 19.6752$, do not reject the null hypothesis based on these sample data.

STEP 7 **Draw a conclusion.**
Even though the $s^2$ of 26.6 is less than the hypothesized value of 36, the engineers at Genesis Technology can state there is insufficient evidence to conclude that the modified design has had any effect on the variability of storage life.

**TRY EXERCISE 11-2** (pg. 417)

OUTCOME **2** ## Interval Estimation for a Population Variance

Chapter 8 introduced confidence interval estimation for a single population mean and a single population proportion. We now extend those concepts to situations in which we are interested in estimating a population variance. Although the basic concepts are the same when we interpret a confidence interval estimate for a variance, the methodology for computing the interval estimate is slightly different. Equation 11.2 is used to construct the interval estimate.

**Confidence Interval Estimate for a Population Variance**

$$\frac{(n - 1)s^2}{\chi^2_U} \leq \sigma^2 \leq \frac{(n - 1)s^2}{\chi^2_L}$$ 
(11.2)

where:

$$s^2 = \text{Sample variance}$$
$$n = \text{Sample size}$$
$$\chi^2_L = \text{Lower critical value}$$
$$\chi^2_U = \text{Lower critical value}$$

FIGURE 11.4 **Critical Values for Estimating s²**

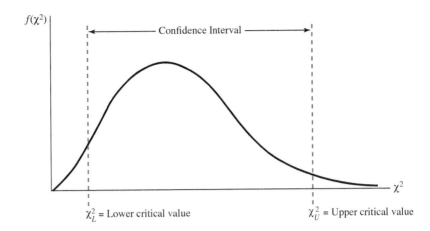

$f(\chi^2)$

Confidence Interval

$\chi^2$

$\chi_L^2$ = Lower critical value          $\chi_U^2$ = Upper critical value

The logic of Equation 11.2 is best demonstrated using Figure 11.4. In a manner similar to the discussion associated with Figure 8.4 in Chapter 8 for estimating a population mean, when we estimate a population variance, for a 95% confidence interval estimate, 95% of the area in the distribution will be between the lower and upper critical values. But since the chi-square distribution is not symmetric and contains only positive numbers, we must find two values from the table in Appendix G.

In Equation 11.2, the denominators come from the chi-square distribution with $n - 1$ degrees of freedom. For example, in an application in which the sample size is $n = 10$ and the desired confidence level is 95%, there is 0.025 in both the lower and upper tails of the distribution. Then from the chi-square table in Appendix G, we get the critical value

$$\chi_U^2 = \chi_{0.025}^2 = 19.0228$$

Likewise, we get

$$\chi_L^2 = \chi_{0.975}^2 = 2.7004$$

Now suppose that the sample variance computed from the sample of $n = 10$ values is $s^2 = 44$. Then, using Equation 11.2, we construct the 95% confidence interval as follows:

$$\frac{(10 - 1)44}{19.0228} \leq \sigma^2 \leq \frac{(10 - 1)44}{2.7004}$$

$$20.82 \leq \sigma^2 \leq 146.64$$

Thus, at the 95% confidence level, we conclude that the population variance is in the range 20.82 to 146.64. By taking the square root, we convert the interval estimate to one for the population standard deviation:

$$4.56 \leq \sigma \leq 12.11$$

> The Excel 2016 functions for finding the upper- and lower-tailed chi-square critical values are
> Upper tail:
> = **CHISQ.INV.RT(0.025,9)**
> Lower tail:
> = **CHISQ.INV.RT(0.975,9)**

## 11.1 EXERCISES

### Skill Development

**11-1.** A random sample of 20 values was selected from a population, and the sample standard deviation was computed to be 360. Based on this sample result, compute a 95% confidence interval estimate for the true population standard deviation.

**11-2.** You are given the following null and alternative hypotheses:

$$H_0: \sigma^2 = 100$$
$$H_A: \sigma^2 \neq 100$$

a. Test when $n = 27$, $s = 9$, and $\alpha = 0.10$. Be certain to state the decision rule.

b. Test when $n = 17$, $s = 6$, and $\alpha = 0.05$. Be certain to state the decision rule.

**11-3.** A manager is interested in determining if the population standard deviation has dropped below 130. Based on a sample of $n = 20$ items selected randomly from the population, conduct the appropriate hypothesis test at a 0.05 significance level. The sample standard deviation is 105.

**11-4.** The following sample data have been collected for testing whether a population standard deviation is equal to 40. Conduct the appropriate hypothesis test using $\alpha = 0.05$.

| | | | | |
|---|---|---|---|---|
| 318 | 255 | 323 | 325 | 334 |
| 354 | 266 | 308 | 321 | 297 |
| 316 | 272 | 346 | 266 | 309 |

**11-5.** You are given the following null and alternative hypotheses:

$$H_0: \sigma^2 = 50$$
$$H_A: \sigma^2 \neq 50$$

a. Test when $n = 12$, $s = 9$, and $\alpha = 0.10$. Be certain to state the decision rule.

b. Test when $n = 19$, $s = 6$, and $\alpha = 0.05$. Be certain to state the decision rule.

**11-6.** Suppose a random sample of 22 items produces a sample standard deviation of 16.

a. Use the sample results to develop a 90% confidence interval estimate for the population variance.

b. Use the sample results to develop a 95% confidence interval estimate for the population variance.

**11-7.** Historical data indicate that the standard deviation of a process is 6.3. A recent sample of size

a. 28 produced a variance of 66.2. Test to determine if the variance has increased using a significance level of 0.05.

b. 8 produced a variance of 9.02. Test to determine if the variance has decreased using a significance level of 0.025. Use the test statistic approach.

c. 18 produced a variance of 62.9. Test to determine if the variance has changed using a significance level of 0.10.

**11-8.** Examine the sample obtained from a normally distributed population:

| | | | | | | |
|---|---|---|---|---|---|---|
| 5.2 | 10.4 | 5.1 | 2.1 | 4.8 | 15.5 | 10.2 |
| 8.7 | 2.8 | 4.9 | 4.7 | 13.4 | 15.6 | 14.5 |

a. Calculate the variance.

b. Calculate the probability that a randomly chosen sample would produce a sample variance at least as large as that produced in part a if the population variance was equal to 20.

c. What is the statistical term used to describe the probability calculated in part b?

d. Conduct a hypothesis test to determine if the population variance is larger than 15.3. Use a significance level equal to 0.05.

## Business Applications

**11-9.** In an effort to increase public acceptance of a light rail system, the manager for City Transit Services in Seattle is interested in estimating the standard deviation for the time it takes a bus to travel between the University of Washington and the downtown bus terminal. To develop an estimate for the standard deviation, he has collected a random sample of the times required for 15 trips. The sample standard deviation is 6.2 minutes. Based on these data, what is the 90% confidence interval estimate for the true population standard deviation?

**11-10.** The consulting firm of Winston & Associates has been retained by an electronic component assembly company in Phoenix to design and program a computer simulation model that will include probability distributions for things like machine downtime, defect rates, daily demand, and so forth. For one machine center, Winston consultants have assumed that when the center goes down for repairs, the time that it will be down will be normally distributed with an average of 30 minutes and a standard deviation equal to 10 minutes. Before finalizing the model, the consultants will collect a random sample of downtimes and test whether their downtime assumptions are valid. The following sample data reflect 10 randomly selected downtimes at this machine center from records over the past four weeks:

| | | | | |
|---|---|---|---|---|
| 25 | 11 | 34 | 49 | 48 |
| 56 | 2 | 26 | 46 | 14 |

a. Using $\alpha = 0.05$, conduct the appropriate test for the mean downtime at this machine center.

b. Using $\alpha = 0.05$, conduct the appropriate test for the standard deviation in downtime.

c. Based on the hypothesis tests in parts a and b, what conclusions should the consulting company reach? Discuss.

**11-11.** The Flicks is a small, independent theater that shows foreign and limited-release films. One attraction of the Flicks is that it sells beer and wine as well as premium coffee and pastries at its snack bar. Friday and Saturday nights are the busiest nights at the Flicks, and the owners of the theater are interested in estimating the variance in sales at the snack bar on Friday nights. Suppose a random sample of 14 Friday evenings is selected and the snack bar sales for each evening are recorded. The results of the sample in dollars are as follows:

| | | | | | | |
|---|---|---|---|---|---|---|
| 279.66 | 329.91 | 314.99 | 358.08 | 341.14 | 303.28 | 325.88 |
| 369.29 | 336.90 | 316.54 | 356.57 | 313.49 | 351.04 | 295.36 |

Use the random sample of sales to develop a 95% confidence interval estimate of the variance of Friday night snack bar sales at the Flicks.

**11-12.** A key measure for airlines is the number of minutes a plane deviates from the scheduled arrival time. Experience indicates that even under the best of circumstances, there will be inherent variability. Suppose one major airline has set standards that require the planes to have a standard deviation for late (or early) arrival times of 2 minutes or less. To determine whether this requirement is being met, each month the airline selects a random sample of 12 airplane arrivals and determines the number of minutes early or late the flight is. For last month, the times, rounded to the nearest minute, were

| 3 | −7 | 4 | 2 | −2 | 5 | 11 | −3 | 4 | 6 | −4 | 1 |
|---|----|---|---|----|---|----|----|---|---|----|---|

State the appropriate null and alternative hypotheses regarding the standard deviation. Use the sample data to conduct the hypothesis test with $\alpha = 0.05$.

What should the airline conclude regarding its arrival standards?

**11-13.** A software design firm has developed a prototype educational computer game for children. One of the important factors in the success of a game like this is the time it takes the child to play the game. Two factors are important: the mean time it takes to play and the variability in time required from child to child. Experience indicates that the mean time should be 10 minutes or less and the standard deviation should not exceed 4 minutes. The company has decided to test this prototype with 10 children selected at random from the local school district. The following values represent the time (rounded to the nearest minute) each child spent until completing the game:

| 9 | 14 | 11 | 8 | 13 | 15 | 11 | 10 | 7 | 12 |
|---|----|----|---|----|----|----|----|---|----|

a. State the appropriate null and alternative hypotheses for testing the requirement regarding the mean value.

b. Referring to part a, test the hypotheses using a significance level equal to 0.10. What assumption will be required?

c. State the appropriate null and alternative hypotheses regarding the standard deviation. Use the sample data to conduct the hypothesis test with a significance level = 0.10.

**11-14.** The consistency of the diameters of wheel bearings is vital to the operation of the wheel. The specifications require that the variance of these diameters be no more than 0.0015 centimeter squared. The diameter is continually monitored by the quality-control team. Twenty subsamples of size 10 are obtained every day. One of these subsamples produced bearings that had a variance of 0.00317 centimeter squared.

Conduct a hypothesis test to determine if the quality control team should advise management to stop production and search for causes of the inconsistency of the bearing diameters. Use a significance level of 0.05.

**11-15.** A sample of 25 construction workers in San Antonio, Texas, yielded an average wage of $17.55 and a standard deviation of $1.46. Suppose the population standard deviation for hourly construction wages is $3.50. Test to determine whether the standard deviation of the construction workers' wages in San Antonio is less than $3.50. Use a significance level of 0.05 to make your determination.

## Computer Software Exercises

**11-16.** Many homes are heated with oil furnaces. The file titled **Homeheat** contains a sample of heating costs for selected households. Assume that historical data indicate that household heating costs have a standard deviation of about $100.

Conduct a test of hypothesis to see if the variability in heating costs in the sample was larger than that indicated by historical data. Use both the test statistic and a significance level of 0.025.

**11-17.** The Pet Health and Food Corporation is a producer of pet food. One of its products is Wonderfood, which is a dry cat food that comes in various sizes. The company guarantees that 32% of this cat food is crude protein. In the 6.6-pound (3-kilogram) size, this would mean that 2.11 pounds are crude protein. Of course, these figures are averages. The amount of crude protein varies with each sack of cat food. The file titled **Catfood** contains the amounts of crude protein found in sacks randomly sampled from the production line. The company would like the standard deviation for protein to be no greater than 0.0167.

Conduct hypothesis test to determine if the standard deviation of the weight of crude protein in the cat food is too large to meet the company's wishes. Use a significance level of 0.01.

**11-18.** The Fillmore Institute has established a service designed to help charities increase the amount of money they collect from their direct-mail solicitations. Its consulting is aimed at increasing the mean dollar amount returned from each giver and also at reducing the variation in amount contributed from giver to giver. The Badke Foundation collects money for heart disease research. Over the last eight years, records show that the average contribution per returned envelope is $14.25 with a standard deviation of $6.44. The Badke Foundation directors decided to try the Fillmore services on a test basis. They used the recommended letters and other request materials and sent out 1,000 requests. From these, 166 were returned. The data showing the dollars returned per giver are in the file called **Badke**.

Based on the sample data, what conclusions should the Badke Foundation reach regarding the Fillmore consulting services? Use appropriate hypothesis tests with a significance level = 0.05 to reach your conclusions.

OUTCOME **3**    ## F-Test for Two Population Variances

Section 11.1 introduced a method for testing hypotheses involving a single population standard deviation. Recall that to conduct the test, we had to first convert the standard deviation to a variance. Then we used the chi-square distribution to determine whether the sample variance led us to reject the null hypothesis. However, decision makers are often faced with decision problems involving two population standard deviations. Although there is no hypothesis test that directly tests two standard deviations, there is a procedure we can use to test two population variances. We typically formulate null and alternative hypotheses using one of the following forms:

| Two-Tailed Test | Upper One-Tailed Test | Lower One-Tailed Test |
|---|---|---|
| $H_0: \sigma_1^2 = \sigma_2^2$ | $H_0: \sigma_1^2 \leq \sigma_2^2$ | $H_0: \sigma_1^2 \geq \sigma_2^2$ |
| $H_A: \sigma_1^2 \neq \sigma_2^2$ | $H_A: \sigma_1^2 > \sigma_2^2$ | $H_0: \sigma_1^2 < \sigma_2^2$ |

To test a hypothesis involving two population variances, we first compute the sample variances. We then compute the test statistic shown as Equation 11.3.

### F-Test Statistic for Testing whether Two Populations Have Equal Variances

$$F = \frac{s_i^2}{s_j^2} \quad (df: D_1 = n_i - 1 \quad \text{and} \quad D_2 = n_j - 1) \qquad (11.3)$$

where:

$n_i$ = Sample size from the $i$th population
$n_j$ = Sample size from the $j$th population
$s_i^2$ = Sample variance from the $i$th population
$s_j^2$ = Sample variance from the $j$th population

Analyzing this test statistic requires that we introduce the $F$-distribution. Although it is beyond the scope of this book, statistical theory shows the $F$-distribution is formed by the ratio of two independent chi-square variables. Like the chi-square and $t$-distributions, the appropriate $F$-distribution is determined by its degrees of freedom. However, the $F$-distribution has two sets of degrees of freedom, $D_1$ and $D_2$, which depend on the sample sizes for the variances in the numerator and denominator, respectively, in Equation 11.3.

To apply the $F$-distribution to test two population variances, we must be able to assume the following are true:

Assumptions

- The populations are normally distributed.
- The samples are randomly and independently selected.

**Independent Samples**

Samples selected from two or more populations in such a way that the occurrence of values in one sample has no influence on the probability of the occurrence of values in the other sample(s).

**Independent Samples** will occur when the sample data are obtained in such a way that the values in one sample do not influence the probability that the values in the second sample will be selected.

The test statistic shown in Equation 11.3 is formed as the ratio of two sample variances. There are two key points to remember when formulating this ratio.

1. To use the *F*-distribution table in this text, for a two-tailed test, always place the larger sample variance in the numerator. This will make the calculated *F*-value greater than 1.0 and push the *F*-value toward the upper tail of the *F*-distribution.
2. For the one-tailed test, examine the alternative hypothesis. For the population that is *predicted* (based on the alternative hypothesis) to have the larger variance, place that sample variance in the numerator.

The following applications and examples will illustrate the specific methods used for testing for a difference between two population variances.

---

**BUSINESS APPLICATION**    ## Testing Two Population Variances

*E. Coli* Bacteria Testing   Recent years have seen several national scares involving food contaminated with *E. coli* bacteria. The recommended preventive measure when dealing with hamburger patties is to cook the meat at a required temperature. However, different meat patties, despite being cooked for the same amount of time, will have different final internal temperatures because of variations in the patties and variations in burner temperatures. To address this problem, a regional hamburger chain plans to replace its current burners with one of two new digitally controlled models.

The chain's purchasing agents have arranged to randomly sample 11 meat patties cooked by Model 1 and 13 meat patties cooked by Model 2 to learn if there is a difference in temperature variation between the two models. If a difference exists, the chain's managers have decided to select the model that provides the smaller variation in final internal meat temperature. Ideally, they would like a test that compares standard deviations, but no such test exists. Instead, they must convert the standard deviations to variances. The hypotheses are

$$H_0: \sigma_1^2 = \sigma_2^2$$
$$H_A: \sigma_1^2 \neq \sigma_2^2$$

The null and alternative hypotheses are formulated for a two-tailed test. Intuitively, you might reason that if the two population variances are actually equal, the sample variances should be approximately equal also. That would mean that the ratio of the two sample variances should be approximately 1. The managers will reject the null hypothesis if one sample variance is significantly larger than the other and if the ratio of sample variances is significantly greater than 1. The managers will use a significance level of $\alpha = 0.10$.

The next step is to collect the sample data. Figure 11.5 shows the sample data and the box and whisker plot. The assumption of independence is met because the two burners were

**FIGURE 11.5 *E. coli* Bacteria Testing Sample Data**

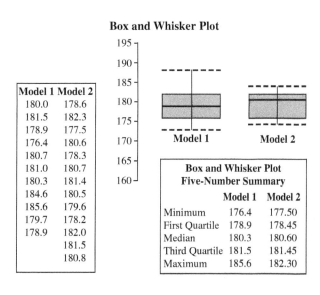

| Model 1 | Model 2 |
|---------|---------|
| 180.0 | 178.6 |
| 181.5 | 182.3 |
| 178.9 | 177.5 |
| 176.4 | 180.6 |
| 180.7 | 178.3 |
| 181.0 | 180.7 |
| 180.3 | 181.4 |
| 184.6 | 180.5 |
| 185.6 | 179.6 |
| 179.7 | 178.2 |
| 178.9 | 182.0 |
|  | 181.5 |
|  | 180.8 |

**Box and Whisker Plot Five-Number Summary**

|  | Model 1 | Model 2 |
|---|---------|---------|
| Minimum | 176.4 | 177.50 |
| First Quartile | 178.9 | 178.45 |
| Median | 180.3 | 180.60 |
| Third Quartile | 181.5 | 181.45 |
| Maximum | 185.6 | 182.30 |

used to cook different meat patties and the temperature measures are not related. The box and whisker plots provide no evidence to suggest that the distributions are highly skewed, so the assumption that the populations are normally distributed is assumed to hold.

The sample variances are computed using Equation 11.4.

**Sample Variance**

$$s^2 = \frac{\Sigma(x - \bar{x})^2}{n - 1}$$ (11.4)

where:

$$\bar{x} = \frac{\Sigma x}{n} = \text{Sample mean}$$

$$n = \text{Sample size}$$

Based on the sample data shown in Figure 11.5, the sample variances are

$$s_1^2 = 6.7 \quad \text{and} \quad s_2^2 = 2.5$$

The null hypothesis is that the two population variances are equal, making this a two-tailed test. Thus, we form the test statistic using Equation 11.3 by placing the larger sample variance in the numerator. The calculated $F$-value is

$$F = \frac{s_1^2}{s_2^2} = \frac{6.7}{2.5} = 2.68$$

If the calculated $F$-value exceeds the critical value, then the managers will reject the null hypothesis. The critical $F$-value is determined by locating the appropriate $F$-distribution table for the desired alpha level and the correct degrees of freedom. This requires the following thought process:

1. If the test is two tailed, use the table corresponding to $\alpha/2$. For example, if $\alpha = 0.10$ for a two-tailed test, the appropriate $F$ table is the one with the upper tail equal to 0.05.
2. If the test is one tailed, use the $F$ table corresponding to the significance level. If $\alpha = 0.05$ for a one-tailed test, use the table with the upper-tailed area equal to 0.05.

In this example, the test is two tailed and $\alpha$ is 0.10. Thus, we go to the $F$-distribution table in Appendix H for the upper-tailed area equal to 0.05.

The next step is to determine the appropriate degrees of freedom. In Chapter 8, we stated that the degrees of freedom of any test statistic are equal to the number of independent data values available to estimate the population variance. We lose 1 degree of freedom for each parameter we are required to estimate. For both the numerator and denominator in Equation 11.3, we must estimate the population mean, $\bar{x}$, before we calculate $s^2$. In each case, we lose 1 degree of freedom. Therefore, we have two distinct degrees of freedom, $D_1$ and $D_2$, where $D_1$ is equal to $n_1 - 1$ for the variance in the numerator of the $F$-test statistic and $D_2$ is equal to $n_2 - 1$ for the variance in the denominator. Recall that for a two-tailed test, the larger sample variance is placed in the numerator. In this example, Model 1 has the larger sample variance, so we place Model 1 in the numerator with a sample size of 11, so $D_1 = 11 - 1 = 10$ and $D_2 = 13 - 1 = 12$.

Then we locate the page of the $F$ table corresponding to the desired upper-tailed area. In this text, we have three options (0.05, 0.025, and 0.01). The $F$ table is arranged in columns and rows. The columns correspond to the $D_1$ degrees of freedom, and the rows correspond to the $D_2$ degrees of freedom. For this example, the critical $F$-value at the intersection of $D_1 = 10$ and $D_2 = 12$ degrees of freedom is 2.753.

Figure 11.6 summarizes the hypothesis test. Note that the decision rule is:

> The Excel 2016 function for the critical $F$-value for a two-tailed test is
>
> = F.INV.RT(Alpha/2,D1,D2)
> = F.INV.RT(0.05,10,12)

If calculated $F > 2.753$, reject $H_0$.
Otherwise, do not reject $H_0$.

**FIGURE 11.6** *F*-Test for the *E. coli* Example

**Hypothesis:**
$H_0: \sigma_1^2 = \sigma_2^2$
$H_A: \sigma_1^2 \neq \sigma_2^2$
$\alpha = 0.10$

*df*: $D_1 = 10, D_2 = 12$

Rejection region
$\alpha/2 = 0.05$

$F_{0.05} = 2.753$

**Decision Rule:**

If $F > 2.753$, reject $H_0$.

Otherwise, do not reject $H_0$.

The *F*-test is

$$F = \frac{s_1^2}{s_2^2} = \frac{6.7}{2.5} = 2.68$$

Because F = 2.68 < $F_{0.05}$ = 2.753, do not reject $H_0$.

*Note*: The right-hand tail of the *F*-distribution always contains an area of $\alpha/2$ if the hypothesis is two-tailed.

Because $F = 2.68 < 2.753$, the conclusion is that the null hypothesis is not rejected based on these sample data; that is, there is not sufficient evidence to support a conclusion that there is a difference in the population variances of internal meat temperatures.

---

**EXAMPLE 11-3**   **Two-Tailed Test for Two Population Variances**

**Mass Transit**   To accommodate population growth, city leaders in a southwestern city formed the City Mass Transit Agency, which has been responsible for expanding public transportation options, including bus service and light rail. However, mass transit will work only if the service is reliable. Suppose city leaders are concerned about the waiting times for passengers who use the downtown transit center between 4:00 P.M. and 6:00 P.M. Monday through Friday. Of particular interest is whether there is a difference in the standard deviations in waiting times at concourses A and B. The following steps can be used to test whether there is a difference in population standard deviations:

STEP 1   **Specify the population parameter of interest.**
The population parameter of interest is the standard deviation in waiting times at the two concourses.

STEP 2   **Formulate the appropriate null and alternative hypotheses.**
Because we are interested in determining if a difference exists in standard deviation and because neither concourse is predicted to have a higher variance, the test is two tailed, and the hypotheses are established as

$$H_0: \sigma_A^2 = \sigma_B^2$$
$$H_A: \sigma_A^2 \neq \sigma_B^2$$

*Note:* The hypotheses are stated in terms of the population variances.

STEP 3   **Specify the level of significance.**
The test will be conducted using an $\alpha = 0.02$.

STEP 4 **Construct the rejection region.**

To determine the critical value from the *F*-distribution, we can use either Excel or the *F* table in Appendix H. The degrees of freedom are $D_1 =$ Numerator sample size $- 1$ and $D_2 =$ Denominator sample size $- 1$. As shown in the statistics section of the stem and leaf display in Step 5, concourse B has the larger sample standard deviation. Thus we get

$$D_1 = n_B - 1 = 31 - 1 = 30 \quad \text{and} \quad D_2 = n_A - 1 = 25 - 1 = 24$$

Then for $\alpha/2 = 0.01$, we get a critical $F_{0.01} = 2.577$. The null hypothesis is rejected if $F > F_{0.01} = 2.577$. Otherwise, do not reject the null hypothesis.

STEP 5 **Compute the test statistic.**

The test statistic is formed by the ratio of the two sample variances. Because this is a two-tailed test, the larger sample variance is placed in the numerator.

We select random samples from each population of interest, determine whether the assumptions have been satisfied, and compute the test statistic.

Random samples of 25 passengers from concourse A and 31 passengers from concourse B were selected, and the waiting time for each passenger was recorded. There is no connection between the two samples, so the assumption of independence is satisfied. The stem and leaf diagrams do not dispute the assumption of normality.

**Stem and Leaf Display for Concourse A**
Stem unit: 1

| Statistics | |
|---|---|
| Sample Size | 25 |
| Mean | 14.58 |
| Median | 14.16 |
| Std. Deviation | 3.77 |
| Minimum | 8.89 |
| Maximum | 22.16 |

```
 8| 9
 9| 0
10| 2 3 7 9
11| 9
12| 2 4
13| 2
14| 0 0 2 2
15| 5 6 9
16| 2
17| 0 4
18| 2 4
19|
20| 8
21| 5
22| 2
```

**Stem and Leaf Display for Concourse B**
Stem unit: 1

| Statistics | |
|---|---|
| Sample Size | 31 |
| Mean | 16.25 |
| Median | 15.77 |
| Std. Deviation | 4.79 |
| Minimum | 4.70 |
| Maximum | 24.38 |

```
 4| 7
 5|
 6| 3
 7|
 8|
 9|
10| 8
11| 1 4 7
12| 2 4
13|
14| 2 2 2 8
15| 1 3 7 8
16|
17| 0 5
18| 3 6
19| 0 1 3 3
20| 4
21| 8 8
22| 1 4 9
23|
24| 4
```

The test statistic is

$$F = \frac{4.79^2}{3.77^2} = 1.614$$

STEP 6 **Reach a decision.**

We compare the test statistic to the critical value and reach a conclusion with respect to the null hypothesis.

Because $F = 1.614 < F_{0.01} = 2.577$, do not reject the null hypothesis.

STEP 7 **Draw a conclusion.**

There is no reason to conclude that there is a difference in the variability of waiting time at concourses A and B.

**TRY EXERCISE 11-24** (pg. 427)

Excel Tutorial

**BUSINESS APPLICATION**    One-Tailed Test for Two Population Variances

**Bank ATMs**    One-tailed tests on two population variances are performed much like two-tailed tests. Consider the systems development group for a midwestern bank that has developed a new software algorithm for its automatic teller machines (ATMs). Although reducing the average transaction time is an objective, the systems programmers also want to reduce the variability in transaction speed. They believe the standard deviation for transaction time will be less with the new software (Population 2) than it was with the old algorithm (Population 1). For their analysis, the programmers have performed 7 test runs using the original software and 11 test runs using the new system. Although the managers want to determine the standard deviation of transaction time, they must perform the test as a test of variances because no method exists for testing standard deviations directly. Thus, the null and alternative hypotheses are

$$H_0: \sigma_1^2 \leq \sigma_2^2 \quad \text{or} \quad \sigma_1^2 - \sigma_2^2 \leq 0$$
$$H_A: \sigma_1^2 > \sigma_2^2 \quad \text{or} \quad \sigma_1^2 - \sigma_2^2 > 0$$

The hypothesis is to be tested using a significance level of $\alpha = 0.01$.

In order to use the $F$-test to test whether these sample variances come from populations with equal variances, we need to make sure that the sample variances are independent and the populations are approximately normally distributed. Because the test runs using the two algorithms were unique, the variances are independent.

Figure 11.7 illustrates the one-tailed hypothesis test for this situation using a significance level of 0.01. Recall that in a two-tailed test, the $F$-ratio has the larger sample variance in the numerator and the smaller variance in the denominator. In a one-tailed test, we look to the alternative hypothesis to determine which sample variance should go in the numerator. In this example, Population 1 (the original software) is thought to have the larger variance. Then the sample variance from Population 1 forms the numerator, regardless of the size of the sample variances. Excel correctly computes the calculated $F$-ratio.

If you are performing the test manually, the $F$-ratio needs to be formed correctly for two reasons. First, the correct $F$-ratio will be computed. Second, the correct degrees of freedom will be used to determine the critical value to test the null hypothesis. In this one-tailed example, the numerator represents Population 1 and the denominator represents Population 2. This means that the degrees of freedom are

$$D_1 = n_1 - 1 = 7 - 1 = 6 \quad \text{and} \quad D_2 = n_2 - 1 = 11 - 1 = 10$$

Using the $F$-distribution table in Appendix H, you can find

$$F_{0.01} = 5.386$$

for this one-tailed test with $\alpha = 0.01$.

The Excel 2016 function for the critical $F$-value for a one-tailed test is

= F.INV.RT(Alpha,D1,D2)
= F.INV.RT(0.01,6,10)

**FIGURE 11.7 Excel 2016 Output—$F$-Test Example of ATM Transaction Time**

**Excel 2016 Instructions**

1. Open file: **ATM.xlsx**.
2. Select **Data > Data Analysis**.
3. Select *F*-Test Two Sample for Variances.
4. Define the data range for the two variables.
5. Specify **Alpha** equal to 0.01.
6. Specify output location.
7. Click **OK**.
8. Click the **Home** tab and adjust decimal places in output.

| | A | B | C | D |
|---|---|---|---|---|
| 1 | F-Test Two-Sample for Variances | | | |
| 3 | | Original Software | New System | |
| 4 | Mean | 46.5571 | 29.8 | |
| 5 | Variance | 612.6762 | 51.494 | |
| 6 | Observations | 7 | 11 | |
| 7 | df | 6 | 10 | |
| 8 | F | 11.898 | | |
| 9 | P(F<=f) one-tail | 0.0005 | | |
| 10 | F Critical one-tail | 5.3858 | | |

| 11 | Because the calculated F = 11.898 > F₀.₀₁ = 5.3858, |
| 12 | reject the null hypothesis and conclude that the |
| 13 | population variance for the Original Software |
| 14 | exceeds the population variance for the New |
| 15 | System. |
| 16 | |

The sample data for the test runs are in a file called **ATM**. The sample variances are

$$s_1^2 = 612.68$$
$$s_2^2 = 51.49$$

Thus, the calculated $F$-ratio is

$$F = \frac{612.68}{51.49} = 11.898$$

As shown in Figure 11.7, the calculated $F = 11.898 > F_{0.01} = 5.386$, so the null hypothesis, $H_0$, is rejected. Based on the sample data, the systems programmers have evidence to support their claim that the new ATM algorithm will result in reduced transaction-time variability.

There are many business decision-making applications in which you need to test whether two populations have unequal variances.

---

### EXAMPLE 11-4 One-Tailed Test for Two Population Variances

**Goodyear Tire Company**   Suppose the Goodyear Tire Company has entered into a contract to supply tires for a leading Japanese automobile manufacturer. Goodyear executives were originally planning to make all the tires at their Ohio plant, but they also have an option to build some tires at their Michigan plant. A critical quality characteristic for the tires is tread thickness, and the automaker wants to know if the standard deviation in tread thickness of tires produced at the Ohio plant (Population 1) exceeds the standard deviation for tires produced at the Michigan plant (Population 2). If so, the automaker will specify that Goodyear use the Michigan plant for all tires because a high standard deviation is not desirable. The following steps can be used to conduct a test for the two suppliers:

STEP 1   **Specify the population parameter of interest.**
Goodyear is concerned with the standard deviation in tread thickness. Therefore, the population parameter of interest is the standard deviation, $\sigma$.

STEP 2   **Formulate the appropriate null and alternative hypotheses.**
Because the Japanese automaker's customers are concerned with whether the Ohio plant's tread standard deviation exceeds that for the Michigan plant, the test will be one tailed. The null and alternative hypotheses are as follows:

$$H_0: \sigma_1^2 \leq \sigma_2^2$$
$$H_A: \sigma_1^2 > \sigma_2^2$$

*Note:* The hypotheses must be stated in terms of the population variances.

STEP 3   **Specify the significance level.**
The test will be conducted using an alpha level equal to 0.05.

> The Excel 2016 function for the critical $F$-value for a one-tailed test is
>
> = F.INV.RT(Alpha,D1,D2)
> = F.INV.RT(0.05,10,10)

STEP 4   **Construct the rejection region.**
Based on sample sizes of 11 tires from each Goodyear plant, the critical value for a one-tailed test with $\alpha = 0.05$ and $D_1 = 10$ and $D_2 = 10$ degrees of freedom is 2.978. The null hypothesis is rejected if $F > F_{0.05} = 2.978$. Otherwise, do not reject the null hypothesis.

STEP 5   **Compute the test statistic.**
Assume that a simple random sample of 11 tires was selected from each Goodyear plant with the sample variances of

$$s_1^2 = 0.799 \quad \text{and} \quad s_2^2 = 0.547$$

The assumptions of independence and normal populations are believed to be satisfied in this case.

The test statistic is an $F$-ratio formed by placing the variance that is predicted to be larger (as shown in the alternative hypothesis) in the numerator. The Ohio plant is predicted to have the larger variance in the alternative hypothesis. Thus the test statistic is

$$F = \frac{0.799}{0.547} = 1.4607$$

STEP 6  **Reach a decision.**
Because $F = 1.4607 < F_{0.05} = 2.978$, do not reject the null hypothesis.

STEP 7  **Draw a conclusion.**
Based on the sample data, there is insufficient evidence to conclude that the variance of tread thickness from the Ohio plant (Population 1) is greater than that for the Michigan plant (Population 2). Therefore, Goodyear managers are free to produce tires at either manufacturing plant.

TRY EXERCISE 11-19 (pg. 427)

## 11.2 EXERCISES

### Skill Development

**11-19.** You are given the following null and alternative hypotheses

$$H_0: \sigma_1^2 \leq \sigma_2^2$$
$$H_A: \sigma_1^2 > \sigma_2^2$$

and sample information

| Sample 1 | Sample 2 |
| --- | --- |
| $n_1 = 13$ | $n_2 = 21$ |
| $s_1^2 = 1,450$ | $s_2^2 = 1,320$ |

a. If $\alpha = 0.05$, state the decision rule for the hypothesis.
b. Test the hypothesis and indicate whether the null hypothesis should be rejected.

**11-20.** You are given the following null and alternative hypotheses

$$H_0: \sigma_1^2 \leq \sigma_2^2$$
$$H_A: \sigma_1^2 > \sigma_2^2$$

and sample information

| Sample 1 | Sample 2 |
| --- | --- |
| $n_1 = 21$ | $n_2 = 12$ |
| $s_1^2 = 345.7$ | $s_2^2 = 745.2$ |

a. If $\alpha = 0.01$, state the decision rule for the hypothesis. (Be careful to pay attention to the alternative hypothesis to construct this decision rule.)
b. Test the hypothesis and indicate whether the null hypothesis should be rejected.

**11-21.** Find the appropriate critical $F$-value, from the $F$-distribution table, for each of the following:
a. $D_1 = 16, D_2 = 14, \alpha = 0.01$
b. $D_1 = 5, D_2 = 12, \alpha = 0.05$
c. $D_1 = 16, D_2 = 20, \alpha = 0.01$

**11-22.** You are given the following null and alternative hypotheses

$$H_0: \sigma_1^2 = \sigma_2^2$$
$$H_A: \sigma_1^2 \neq \sigma_2^2$$

and sample information

| Sample 1 | Sample 2 |
| --- | --- |
| $n_1 = 11$ | $n_2 = 21$ |
| $s_1 = 15$ | $s_2 = 33$ |

a. If $\alpha = 0.02$, state the decision rule for the hypothesis.
b. Test the hypothesis and indicate whether the null hypothesis should be rejected.

**11-23.** Consider the following two independently chosen samples:

| Sample 1 | Sample 2 |
| --- | --- |
| 12.1 | 10.5 |
| 13.4 | 9.5 |
| 11.7 | 8.2 |
| 10.7 | 7.8 |
| 14.0 | 11.1 |

Use a significance level of 0.05 for testing the null hypothesis that $\sigma_1^2 \leq \sigma_2^2$.

**11-24.** You are given two random samples with the following information:

| Item | Sample 1 | Sample 2 |
| --- | --- | --- |
| 1 | 19.6 | 21.3 |
| 2 | 22.1 | 17.4 |
| 3 | 19.5 | 19.0 |
| 4 | 20.0 | 21.2 |
| 5 | 21.5 | 20.1 |
| 6 | 20.2 | 23.5 |
| 7 | 17.9 | 18.9 |
| 8 | 23.0 | 22.4 |
| 9 | 12.5 | 14.3 |
| 10 | 19.0 | 17.8 |

Based on these samples, test at $\alpha = 0.10$ whether the true difference in population variances is equal to zero.

## Business Applications

**11-25.** In an effort to better understand current donation patterns, the manager of a public broadcast radio station is studying past data. A staff member from a neighboring state has speculated that male viewers' donations have greater variability in amount than do those of females. To test this, the manager has selected random samples of 25 men and 25 women from people who donated during last year's telethon. The following statistics were computed from the sample data:

| Males | Females |
|---|---|
| $\bar{x} = \$12.40$ | $\bar{x} = \$8.92$ |
| $s = \$2.50$ | $s = \$1.34$ |

Based on a significance level of 0.05, does it appear that male viewers' donations have greater variability in amount than do those of female viewers?

**11-26.** As purchasing agent for the Horner-Williams Company, you have primary responsibility for securing high-quality raw materials at the best possible price. One particular material that the Horner-Williams Company uses a great deal of is aluminum. After careful study, you have been able to reduce the prospective vendors to two. It is unclear whether these two vendors produce aluminum that is equally durable.

To compare durability, the recommended procedure is to put pressure on the aluminum until it cracks. The vendor whose aluminum requires the greatest average pressure will be judged to be the one that provides the most durable product.

For this test, 14 pieces from Vendor 1 and 14 pieces from Vendor 2 are selected at random. The following results in pounds per square inch (psi) are noted:

| Vendor 1 | Vendor 2 |
|---|---|
| $n_1 = 14$ | $n_2 = 14$ |
| $\bar{x}_1 = 2{,}345$ psi | $\bar{x}_2 = 2{,}411$ psi |
| $s_1 = 300$ | $s_2 = 250$ |

Before you test the hypothesis about a difference in population means, suppose you are concerned about whether the assumption of equal population variances is satisfied.

a. Based on the sample data, what would you conclude from a test at the significance level of 0.10?

b. Would your conclusion differ if you tested at the significance level of 0.02? Discuss.

**11-27.** The production control manager at Ashmore Manufacturing is interested in determining whether there is a difference in the standard deviation of product diameters for part #XC-343 for units made at the Trenton, New Jersey, plant versus those made at the Atlanta plant. The Trenton plant is highly automated and thought to provide better quality control. Thus, the parts produced in the Atlanta plant should be more variable than those made at the Trenton plant.

A random sample of 15 parts was selected from those produced last week at Trenton. The standard deviation for these parts was 0.14 inch. A sample of 13 parts was selected from those made in Atlanta. The sample standard deviation for these parts was 0.202 inch.

Based on these sample data, is there sufficient evidence to conclude that the Trenton plant produces parts that are less variable than those of the Atlanta plant? Test using $\alpha = 0.05$.

**11-28.** A national fitness chain is considering being a representative for a weight loss program. The company has narrowed the choice of programs down to two options. Before making the selection, the company has asked each weight loss program to submit a random sample of 40 individuals who were on the program for six months. Of particular interest to the fitness chain is whether there is a difference in the standard deviations of weight loss between the two programs. They will choose the program that has the least amount of variability. The following data were obtained from the two programs:

| | Program A | Program B |
|---|---|---|
| Mean Weight Loss | 17 lb | 20.2 lb |
| St. Deviation | 4.7 lb | 7.3 lb |
| Sample Size | 40 | 40 |

a. State the appropriate null and alternative hypotheses.

b. Using an alpha level equal to 0.10, test whether the two weight loss programs have equal population standard deviations.

**11-29.** The human resources director of a Fortune 500 company recently asked her staff to conduct a study of the airline miles traveled by executives of the company during the most recent calendar year. The analyst assigned to the project randomly sampled 20 male executives and 20 female executives. Of particular interest to the HR director is whether males have greater variability in miles flown than females. The following statistics were computed from the samples:

| | Males | Females |
|---|---|---|
| Mean Miles | 73,201 | 67,890 |
| St. Deviation Miles | 2,636 | 1,513 |
| Sample Size | 20 | 20 |

Based on these data, what should the HR director conclude if an alpha level equal to 0.05 is used?

## Computer Software Exercises

**11-30.** The Celltone company is in the business of providing cellular phone coverage. Recently, it conducted a study of its customers who have purchased either the

Basic Plan or the Business Plan service. At issue is the number of minutes of use by the customers during the midnight-to-7:00 A.M. time period Monday through Friday over a four-week period. Celltone managers believe that the standard deviation in minutes used by Basic Plan customers will be more than that for the Business Plan customers. Data for this study are in a file called **Celltone**. Assume that the managers wish to test this using a 0.05 level of significance. Determine if the standard deviation in minutes used by Basic Plan customers is more than that for the Business Plan customers using an alpha level equal to 0.05.

**11-31.** The First Night Stage Company operates a small, nonprofit theater group in Milwaukee. This company's representatives proposed two different solicitation brochures. They are interested in determining whether there is a difference in the standard deviation of dollars returned between the two brochures. To test this, a random sample of 20 people was selected to receive Brochure A and another random sample of 20 people was selected to receive Brochure B. The data are contained in the file called **First-Night**. Based on these sample data, what should the First Night Company conclude about the two brochures with

respect to their variability? Test using a significance level of 0.02.

**11-32.** The vehicle fleet manager for a large agricultural company realizes that passenger cars get higher average mileage than pickup trucks. However, he is interested in determining whether there is a difference in the variability in miles per gallon for the two types of vehicles. To test this, he has selected a random sample of trucks and a random sample of passenger cars and divided the total miles driven in the past year for each vehicle by the total gallons of gasoline used by each vehicle. The overall mpg values for the two samples are in the file called **MPG**. Based on the sample data, what conclusions should the manager reach about whether there is a difference in the variability in miles per gallon for the two types of vehicles? Test using an alpha level equal to 0.05.

**11-33.** A state income tax audit was recently conducted at a restaurant in St. Louis. The audit team selected a random sample of 100 lunch and 100 dinner sales receipts and recorded the state sales tax collected on each receipt. These values are recorded in the file called **Tax Receipts**. Of interest is whether the standard deviation in tax receipts differs for lunches and dinners. Test at the 0.02 alpha level.

# 11 Overview

## Summary

**11.1** | **Hypothesis Tests and Estimation for a Single Population Variance** (pg. 411–419)

**OUTCOME 1** Formulate and carry out hypothesis tests for a single population variance.

**OUTCOME 2** Develop and interpret confidence interval estimates for a population variance.

- A **chi-square test for one population variance** is used to test whether the value of the sample variance supports or refutes a null hypothesis for a population variance.
- If there is no hypothesis of interest concerning the population variance, the **confidence interval estimate for $\sigma^2$** is used to obtain a plausible range of values for the population variance.
- Often a standard deviation is the parameter of interest. In such cases, the standard deviation is converted into one involving the variance.

**11.2** | **Hypothesis Tests for Two Population Variances** (pg. 420–429)

**OUTCOME 3** Formulate and carry out hypothesis tests for the difference between two population variances.

- The **F-test for two population variances** is the hypothesis test for determining whether two populations have equal variances.

## Equations

**(11.1) Chi-Square Test for a Single Population Variance** pg. 412

$$\chi^2 = \frac{(n-1)s^2}{\sigma^2}$$

**(11.2) Confidence Interval Estimate for a Population Variance** pg. 416

$$\frac{(n-1)s^2}{\chi_U^2} \le \sigma^2 \le \frac{(n-1)s^2}{\chi_L^2}$$

**(11.3) F-Test Statistic for Testing whether Two Populations Have Equal Variances** pg. 420

$$F = \frac{s_i^2}{s_j^2} \quad (df: D_1 = n_i - 1 \quad \text{and} \quad D_2 = n_j - 1)$$

**(11.4) Sample Variance** pg. 422

$$s^2 = \frac{\Sigma(x - \bar{x})^2}{n - 1}$$

## Key Term

Independent samples pg. 420

## Chapter Exercises

### Conceptual Questions

**11-34.** Identify three situations in which the measured output of a process has a small variation and three situations in which the variation is larger (for instance, the time for a phone call to ring the other phone versus the time to speak to a person when making a service call).

**11-35.** In a journal related to your major, locate an article in which a hypothesis about one or two population variances is tested. Discuss why population variation was an important issue in the article, how the hypothesis was formulated, and the results of the test.

**11-36.** Much of the emphasis of modern quality-control efforts involves identifying sources of variation in the processes associated with service or manufacturing operations. Discuss why variation in a process may affect quality.

**11-37.** Identify three situations in which organizations would be interested in limiting the standard deviation of a process.

**11-38.** Consider testing the hypothesis $H_A: \sigma_1^2 > \sigma_2^2$ using samples of respective sizes 6 and 11. Note that we could represent this hypothesis as $H_A: \sigma_1^2/\sigma_2^2 > 1$. The samples yield $s_1^2 = 42.2$ and $s_2^2 = 1.1$. It hardly seems worth the time to conduct this test, since the test statistic will obviously be very large, leading us to reject the null hypothesis. However, we might wish to test a hypothesis such as $H_A: \sigma_1^2/\sigma_2^2 > 10$. The test is conducted exactly the same way as the $F$-test of this chapter except that you use the test statistic $F = (s_1^2/s_2^2)(1/k)$, where $k = 10$ for this hypothesis. Conduct the indicated test using a significance level of 0.05.

## Business Applications

**11-39.** In response to customer surveys, a candy company has recently increased the size of the chopped nuts used in its caramel nut bar. The marketing manager is concerned that this change will affect the variability in the cooking temperature and compromise the taste and consistency of the candy. The head cook took a random sample of 27 batches and found the standard deviation of the temperature to be 1.15°F. Realizing this is only a point estimate, the marketing manager has requested a 98% confidence interval estimate for the population variance in the cooking temperature.

**11-40.** Maher Saddles, Inc., produces bicycle seats. Maher's operations manager has recently made a change in the production process for the company's high-end seat. After waiting for the process to become stable, he took a random sample of 25 assembly times and found a standard deviation of 47 seconds. He recognizes this is only a point estimate of the variation in completion times and so wants to report both a 90% and a 95% confidence interval estimate for the population variance.

**11-41.** A medical research group is investigating what differences might exist between two pain-killing drugs, Azerlieve and Zynumbic. The researchers have already established there is no difference between the two drugs in terms of the average amount of time required before they take effect. However, they are also interested in knowing if there is any difference between the variability of time until pain relief occurs. A random sample of 24 patients using Azerlieve and 32 patients using Zynumbic yielded the following results:

| Azerlieve | Zynumbic |
|---|---|
| $n_A = 24$ | $n_Z = 32$ |
| $s_A = 37.5$ seconds | $s_Z = 41.3$ seconds |

Based on these sample data, can the researchers conclude a difference exists between the two drugs?

**11-42.** The Littleton Corporation manufactures high-speed electronic cables and focuses on products for the specialty electronics and data networking markets, including connectivity. One of its products is a fiber-optic cable. It is designed to have an overall diameter of 0.440 inch. A standard deviation greater than 0.05 inch would be unacceptable. A sample of size 20 was taken from the production line and yielded a standard deviation of 0.070.

   a. Determine if the standard deviation does not meet specifications using a significance level of 0.01.

   b. Describe a Type II error in the context of this exercise.

**11-43.** Soft drink bottling companies such as Coca-Cola use quality-control techniques to assure that the average amount of cola in the cans is 12 ounces. Suppose that a small standard deviation of 0.05 ounce is acceptable. Any significant difference from this standard deviation would reflect badly on the quality-control measures concerning the average amount of cola in the 12-ounce cans. A sample of size 20 indicated that the standard deviation was 0.070.

   a. Determine if the standard deviation of the amount of cola in the cans differs from the standard deviation specified by the quality control division. Use a significance level of 0.10.

   b. The quality-control sampling occurs several times a day. In one day, seven samples were taken, and three indicated that the standard deviation was not 0.05. If the seven samples were taken at a time in which the standard deviation met specifications, determine the probability of having at least three out of seven samples indicate the specification was not being met.

**11-44.** The 2013 graduating class from St. Francis University in Pennsylvania had the highest average student debt of all U.S. college graduates: $50,275. Second on the list was The Citadel in South Carolina, at $48,862. (*Source: Susannah Snider, "10 Colleges that leave graduates with the most student loan debt," www.usnews.com, Feb. 17, 2015.*) Suppose random samples of 75 St. Francis students and 205 Citadel students showed student loan standard deviations of $2,050 and $2,084, respectively. Using an alpha level of 0.10, test to determine whether the standard deviations for student loans at these two universities are equal.

## Computer Software Exercises

**11-45.** Suppose the California State Highway Patrol conducted a study on a stretch of interstate highway south of San Francisco to determine what differences, if any, existed in driving speeds of cars licensed in California and cars licensed out of state. One of the issues to be examined was whether there was a difference in the variability of driving speeds for California and out-of-state cars. The data file **Speed-Test** contains speeds of 140 randomly selected California cars and 75 randomly selected out-of-state cars. Based on these sample results, can you conclude at the 0.05 level of significance that there is a difference between the variations in driving speeds?

**11-46.** The operations manager for Cozine Corporation is concerned with variation in the number of pounds of garbage collected per truck. The manager believes the current truck routing system provides for consistent garbage pickup per truck and is unwilling to reroute the trucks unless the variability, measured by the standard deviation in pounds per truck, is greater than 3,900 pounds. The data file **Cozine** contains 200 truck weights. Assuming the data represent a random sample of 200 trucks selected from Cozine's daily operations, is there evidence the manager needs to change the routes to better balance the loads? Conduct your analysis using a 0.10 level of significance. Be sure to state your conclusion in terms of the operations manager's decision.

**11-47.** Recently, the X-John Company developed a new cell phone battery it believes will be less expensive to produce. There is concern about the consistency in the lasting power of the battery. If there is too much variability in battery life, cellular phone users will be unwilling to buy X-John batteries even if they are less expensive. Engineers have specified the standard deviation of battery life must be less than 5 hours. Treat the measurements in the file **X-John** as a random sample of 100 of the new batteries. Based on this sample, is there evidence that the standard deviation of battery life is less than 5 hours? Conduct the appropriate hypothesis test using a level of significance of 0.01. Report the $p$-value for this test and be sure to state a conclusion in business terms.

**11-48.** Freedom Hospital is concerned that there is a greater variability in the total charges for men than in the total charges for women. To investigate this issue, the hospital collected a random sample of data for 137 patients. The data are contained in the file **Patients**. Using the data for total charges, conduct the appropriate test to respond to the hospital's concern. Use a significance level of 0.05. State your conclusion in terms that address the issue raised by the hospital.

**11-49.** The Transportation Security Administration (TSA) has instituted the PreCheck program for some airline passengers. Individuals who pay a fee, and who are approved, can receive a TSA PreCheck message on their boarding pass that allows expedited screening at many U.S. airports. This program is intended to reduce the

screening times for these passengers (*source:* www.tsa.gov/tsa-precheck). Suppose for comparison purposes, a sample of the time it took the other passengers to pass through security at Orlando was obtained. The file titled **Passtime** contains these data.

a. Although the average time to pass through security is of importance, the standard deviation is also important. Conduct a hypothesis test to determine if the standard deviation is greater than $1\frac{1}{2}$ minutes (i.e., 90 seconds). Use a significance level of 0.01.

b. Considering the results of your hypothesis tests in part a, determine and define the type of statistical error that could have been made.

**11-50.** Suppose staff members at the U.S. Department of Education conducted a study to determine whether there is a difference in the variation in teacher salaries between Oregon and Nebraska. The sample data are in the file called **Teacher Salaries**. Perform the appropriate statistical test using a significance level of 0.05.

**11-51.** Phone Solutions provides assistance to users of a personal finance software package. Users of the software call with their questions, and trained consultants provide answers and information. One concern that Phone Solutions must deal with is the staffing of its call centers. As part of the staffing issue, it seeks to reduce the average variability in the time each consultant spends with each caller. A study of this issue is currently under way at the company's three call centers. Each call center manager has randomly sampled 50 days of calls, and the collected times, in minutes, are in the file **Phone Solutions**.

a. Call Center 1 has set the goal that the variation in phone calls, measured by the standard deviation of length of calls in minutes, should be less than 3.5 minutes. Using the data in the file, can the operations manager of Call Center 1 conclude that her consultants are meeting the goal? Use a 0.10 level of significance.

b. Can the manager conclude that there is greater variability in the average length of phone calls for Call Center 3 than for Call Center 2? Again, use a 0.10 level of significance to conduct the appropriate test.

---

## Case 11.1    Larabee Engineering—Part 2

Cregg Hart is manufacturing manager for Larabee Engineering, a locomotive engine and rail car manufacturer (see Case 10.1). The company has been very successful in recent years, and in July 2016 signed two major contracts totaling nearly $200 million. A key to the company's success has been its focus on quality. Customers from around the world have been very pleased with the attention to detail put forth by Larabee Engineering.

In Case 10.1, Sheryl Carleson came to Cregg with a new supplier of rivets that would provide a substantial price advantage over

the current supplier. Cregg asked Sheryl to conduct a study in which samples of rivets were selected from both suppliers. (Data are in the file called **Larabee Engineering**.) In Case 10.1, the focus was on the mean diameter and a test to determine whether the population means were the same for the two suppliers. However, Cregg reminds Sheryl that not only is the mean diameter important, so too is the variation in diameter. Too much variation in rivet diameter adversely affects quality. Cregg showed Sheryl the following table to emphasize what he meant:

| Diameter (Inches) | | |
| --- | --- | --- |
| Company A | | Company B |
| 0.375 | | 0.375 |
| 0.376 | | 0.400 |
| 0.374 | | 0.350 |
| 0.375 | | 0.325 |
| 0.375 | | 0.425 |
| 0.376 | | 0.340 |
| 0.374 | | 0.410 |
| 0.375 | Mean | 0.375 |
| 0.00082 | St. Dev. | 0.03808 |

As Sheryl examined this example that Cregg had prepared, she was quickly convinced that looking at the mean diameters would not be enough to fully compare the rivet suppliers. She told Cregg that she would also ask her intern to perform the following tasks.

## Required Tasks:

1. Review the results from Case 10.1.
2. Conduct the appropriate hypothesis test to determine whether the two suppliers have equal standard deviations. (Test using a significance level equal to 0.05.)
3. Prepare a short report that ties together the results from Cases 10.1 and 11.1 to present to Cregg along with a conclusion as to whether the new supplier seems viable based on rivet diameters.

# 12 Analysis of Variance

## WHY YOU NEED TO KNOW

After completing Chapters 9 through 11, you understand that regardless of the population parameter in question, hypothesis-testing steps are basically the same:

1. Specify the population parameter of interest.
2. Formulate the null and alternative hypotheses.
3. Specify the level of significance.
4. Determine a decision rule defining the rejection and "acceptance" regions.

### Quick Prep

**Review** the computational methods for the sample mean and the sample variance in Chapter 3.

**Review** the basics of hypothesis testing discussed in Section 9.1.

**Re-examine** the material on hypothesis testing for the difference between two population variances in Section 11.2.

5. Select a random sample of data from the population(s). Compute the appropriate sample statistic(s). Finally, calculate the test statistic.

6. Reach a decision. Reject the null hypothesis, $H_0$, if the sample statistic falls in the rejection region; otherwise, do not reject the null hypothesis. If the test is conducted using the $p$-value approach, reject $H_0$ when the $p$-value is smaller than the significance level; otherwise, do not reject $H_0$.

7. Draw a conclusion. State the result of your hypothesis test in the context of the exercise or analysis of interest.

Chapter 9 focused on hypothesis tests for a single population. Chapters 10 and 11 expanded the hypothesis-testing process to include applications that involve differences between two populations. However, you will encounter many instances involving more than two populations. For example, the vice president of operations at Farber Rubber, Inc., oversees production at Farber's six U.S. manufacturing plants. Because each plant uses slightly different manufacturing processes, the vice president needs to know if there are any differences in the average strength of the products made at the different plants.

Similarly, *Golf Digest*, a publisher of articles about golf, might wish to determine which of five major brands of golf balls has the greatest mean distance off the tee. The Environmental Protection Agency (EPA) might conduct a test to determine if there is a difference in the average miles-per-gallon performance of cars manufactured by the Big Three U.S. automobile producers. In each of these cases, analysts may be required to test a hypothesis involving more than two population means.

This chapter introduces a tool called analysis of variance (ANOVA), which can be used to test whether there are differences among three or more population means. There are several ANOVA procedures, depending on the type of test being conducted. Our aim in this chapter is to introduce you to ANOVA and to illustrate how to use Excel to perform the calculations involved in hypothesis tests with three or more population means.

OUTCOME **1**    **12.1**

## One-Way Analysis of Variance

In Chapter 10, we introduced the *t*-test for testing whether two populations have equal means when the samples from the two populations are independent. However, you will often encounter situations in which you are interested in determining whether three or more populations have equal means. To handle these situations, you will need a new tool called *analysis of variance (ANOVA)*. There are many different analysis of variance designs to fit different situations; the simplest is known as **one-way analysis of variance**.

**One-Way Analysis of Variance**
An analysis of variance design in which independent samples are obtained from two or more levels of a single factor for the purpose of testing whether the levels have equal means.

## Introduction to One-Way ANOVA

**BUSINESS APPLICATION**    Applying One-Way Analysis of Variance

**Chicago Connection Sandwich Company**  The Chicago Connection Sandwich Company is a privately held company that operates in four locations in Columbus, Ohio. The VP of sales for the company is interested in knowing whether the dollar values for orders made by individual customers differ, on average, between the four locations.

To answer this question, staff at the VP's office have selected a random sample of eight customers at each of the four locations and recorded the order amounts. These are shown in Table 12.1.

In this example, we are interested in whether the different locations generate different mean order sizes. In other words, we are trying to determine if location is one of the possible causes of the variation in the dollar values of the orders placed by customers (the response variable). In this case, location is called a **factor**.

**Factor**
A quantity under examination in an experiment as a possible cause of variation in the response variable.

TABLE 12.1 Chicago Connection Sandwich Company Order Data

| Customer | Store Locations | | | |
|---|---|---|---|---|
|  | 1 | 2 | 3 | 4 |
| 1 | $4.10 | $6.90 | $4.60 | $12.50 |
| 2 | 5.90 | 9.10 | 11.40 | 7.50 |
| 3 | 10.45 | 13.00 | 6.15 | 6.25 |
| 4 | 11.55 | 7.90 | 7.85 | 8.75 |
| 5 | 5.25 | 9.10 | 4.30 | 11.15 |
| 6 | 7.75 | 13.40 | 8.70 | 10.25 |
| 7 | 4.78 | 7.60 | 10.20 | 6.40 |
| 8 | 6.22 | 5.00 | 10.80 | 9.20 |

|  | | | | | Grand Mean |
|---|---|---|---|---|---|
| Mean | $\bar{x}_1 = \$7.00$ | $\bar{x}_2 = \$9.00$ | $\bar{x}_3 = \$8.00$ | $\bar{x}_4 = \$9.00$ | $\bar{\bar{x}} = \$8.25$ |
| Variance | $s_1^2 = 7.341$ | $s_2^2 = 8.423$ | $s_3^2 = 7.632$ | $s_4^2 = 5.016$ | |

*Note:* Data are the dollar values of the orders by customers at the four locations.

**Levels**
The categories, measurements, or strata of a factor of interest in a particular experiment.

The *single factor* of interest is location. This factor has four **levels**. These four levels are the four locations: 1, 2, 3, and 4. Each level is a population of interest, and the values seen in Table 12.1 are sample values taken from those populations.

The null and alternative hypotheses to be tested are

$$H_0: \mu_1 = \mu_2 = \mu_3 = \mu_4 \text{ (mean order sizes are equal)}$$
$$H_A: \text{At least two of the population means are different}$$

The appropriate statistical tool for conducting the hypothesis test related to this experimental design is analysis of variance. Because this ANOVA addresses an experiment that has only one factor, it is a one-way ANOVA, or a single factor ANOVA. Because the sample size for each location (level) is the same, the experiment has a **balanced design**.

**Balanced Design**
An experiment has a balanced design if the factor levels have equal sample sizes.

ANOVA tests the null hypothesis that three or more populations have the same mean. The test is based on four assumptions:

**Assumptions**

1. All populations are normally distributed.
2. The population variances are equal.
3. The observations are independent—that is, the occurrence of any one individual value does not affect the probability that any other observation will occur.
4. The data are interval or ratio level.

If the null hypothesis is true, the populations have identical distributions. If so, the sample means for random samples from each population should be close in value. The basic logic of ANOVA is the same as the two-sample *t*-test introduced in Chapter 10. The null hypothesis should be rejected only if the sample means are substantially different.

## Partitioning the Sum of Squares

To understand the logic of ANOVA, you should note several things about the data in Table 12.1. First, the dollar values of the orders are different throughout the data table. Some values are higher; others are lower. Thus, variation exists across all customer orders. This variation is called the **total variation** in the data.

**Total Variation**
The aggregate dispersion of the individual data values across the various factor levels.

Next, within any particular location (i.e., factor level), not all customers had the same dollar order. For instance, within Level 1, order amounts ranged from $4.10 to $11.55.

**Within-Sample Variation**
The dispersion among the data values within a particular factor level.

**Between-Sample Variation**
The dispersion among the factor sample means.

Similar differences occur within the other levels. The variation within the factor levels is called the **within-sample variation**.

Finally, the sample means for the four restaurant locations are not all equal. Thus, variation exists between the four averages. This variation between the factor level means is referred to as the **between-sample variation**.

Recall that the sample variance is computed as

$$s^2 = \frac{\sum (x - \bar{x})^2}{n - 1}$$

The sample variance is the sum of the squared deviations from the sample mean divided by its degrees of freedom. When all the data from all the samples are included, $s^2$ is the estimator of the *total variation*. The numerator of this estimator is called the *total sum of squares* (*SST*) and can be partitioned into the sum of squares associated with the estimators of the between-sample variation and the within-sample variation, as shown in Equation 12.1.

**Partitioned Sum of Squares**

$$SST = SSB + SSW \qquad \qquad \textbf{(12.1)}$$

where:

$$SST = \text{Total sum of squares}$$
$$SSB = \text{Sum of squares between}$$
$$SSW = \text{Sum of squares within}$$

After we partition the sum of squares, we divide *SSB* and *SSW* by their respective degrees of freedom to produce two estimates for the overall population variance. If the between-sample variance estimate is large relative to the within-sample estimate, the ANOVA procedure will lead us to reject the null hypothesis and conclude that the population means are different. The question is, how can we determine at what point any difference is statistically significant?

## The ANOVA Assumptions

OUTCOME **2**    **BUSINESS APPLICATION**    Understanding the ANOVA Assumptions

**Chicago Connection Sandwich Company** (*continued*)   Recall that the VP of sales for the Chicago Connection Sandwich Company is interested in testing whether the four locations generate orders of equal dollar value. The null and alternative hypotheses are

$$H_0: \mu_1 = \mu_2 = \mu_3 = \mu_4$$
$$H_A: \text{At least two population means are different}$$

Before we jump into the ANOVA calculations, recall the four basic assumptions of ANOVA:

1. All populations are normally distributed.
2. The population variances are equal.
3. The sampled observations are independent.
4. The data's measurement level is interval or ratio.

Figure 12.1 illustrates the first two assumptions. The populations are normally distributed and the spread (variance) is the same for each population. However, this figure shows that the populations have different means—and therefore the null hypothesis is false. Figure 12.2 illustrates the same assumptions but in a case in which the population means are equal; therefore, the null hypothesis is true.

FIGURE 12.1 **Normal Populations with Equal Variances and Unequal Means**

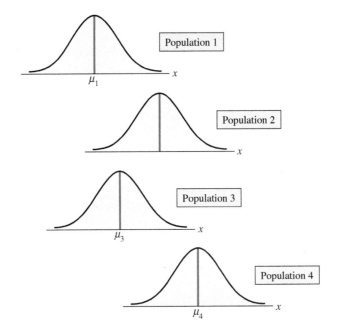

FIGURE 12.2 **Normal Populations with Equal Variances and Equal Means**

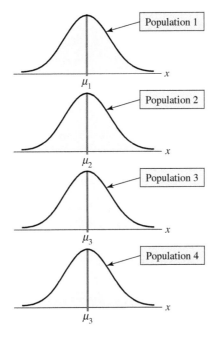

You can do a rough check to determine whether the normality assumption is satisfied by developing graphs of the sample data from each population. Histograms are probably the best graphical tool for checking the normality assumption, but they require a fairly large sample size. The stem and leaf diagram and box and whisker plot are alternatives when sample sizes are smaller. If the graphical tools show plots consistent with a normal distribution, then that evidence suggests the normality assumption is satisfied.[1] Figure 12.3 illustrates the box and whisker plot for the Chicago Connection data. Note that when the sample sizes are very small, as they are here, the graphical techniques may not be very effective.

Examining the sample data to see whether the basic assumptions are satisfied is always a good idea, but you should be aware that the analysis of variance procedures discussed in

---

[1]Chapter 13 introduces a goodness-of-fit approach to testing whether sample data come from a normally distributed population.

**FIGURE 12.3 Box and Whisker Plot for Chicago Connection Sandwich Company**

Box and Whisker Plot

| Box and Whisker Plot Five-Number Summary | | | | |
|---|---|---|---|---|
| | **1** | **2** | **3** | **4** |
| Minimum | 4.1 | 5.0 | 4.3 | 6.25 |
| First Quartile | 4.78 | 6.9 | 4.6 | 6.4 |
| Median | 6.06 | 8.5 | 8.275 | 8.975 |
| Third Quartile | 10.45 | 13.0 | 10.8 | 11.15 |
| Maximum | 11.55 | 13.4 | 11.4 | 12.5 |

this chapter are *robust*, in the sense that the analysis of variance test is relatively unperturbed when the equal-variance assumption is not met. This is especially so when all samples are the same size, as in the Chicago Connection Company example. Hence, for one-way analysis of variance, or any other ANOVA design, try to have equal sample sizes when possible. Recall that we earlier referred to an analysis of variance design that has equal sample sizes as a *balanced design*. If for some reason you are unable to use a balanced design, the rule of thumb is that the ratio of the largest sample size to the smallest sample size should not exceed 1.5.

When the samples are the same size (or meet the 1.5 ratio rule), the analysis of variance procedures are also robust with respect to the assumption that the populations are normally distributed. So, in brief, the one-way ANOVA for independent samples can be applied to virtually any set of interval- or ratio-level data.

Finally, if the data are not interval or ratio level, or if they do not satisfy the normal distribution assumption, Chapter 17 introduces an ANOVA procedure called the Kruskal-Wallis one-way ANOVA, which does not require that these assumptions be met.

## Applying One-Way ANOVA

Although the previous discussion covers the essence of ANOVA, to determine whether the null hypothesis should be rejected requires that we actually determine values of the estimators for the total variation, between-sample variation, and within-sample variation. Most ANOVA tests are done using a computer, but we will illustrate the manual computational approach one time to show you how it is done. Because software such as Excel can be used to perform all calculations, we will use the computer in future examples. The software packages will do all the computations while we focus on interpreting the results.

**BUSINESS APPLICATION**    Developing the ANOVA Table

**Chicago Connection Sandwich Company** (*continued*)    Now we are ready to perform the one-way ANOVA computations for the Chicago Connection Sandwich example. Recall from Equation 12.1 that we can partition the total sum of squares into two components:

$$SST = SSB + SSW$$

The *total sum of squares* is computed as shown in Equation 12.2.

### Total Sum of Squares

$$SST = \sum_{i=1}^{k} \sum_{j=1}^{n_i} (x_{ij} - \bar{\bar{x}})^2 \qquad (12.2)$$

where:

$SST$ = Total sum of squares
$k$ = Number of populations (treatments)
$n_i$ = Sample size from population $i$
$x_{ij}$ = $j$th measurement from population $i$
$\bar{\bar{x}}$ = Grand mean (mean of all the data values)

Equation 12.2 is not as complicated as it appears. Manually applying Equation 12.2 to the data shown in Table 12.1 on page 436 (*Grand mean* = $\bar{\bar{x}}$ = 8.25), we can compute the *SST* as follows:

$$SST = (4.10 - 8.25)^2 + (5.90 - 8.25)^2 + (10.45 - 8.25)^2 + \cdots + (9.20 - 8.25)^2$$
$$SST = 220.88$$

Thus, the sum of the squared deviations of all values from the grand mean is 220.88. Equation 12.2 can also be restated as

$$SST = \sum_{i=1}^{k} \sum_{j=1}^{n_i} (x_{ij} - \bar{\bar{x}})^2 = (n_T - 1)s^2$$

where $s^2$ is the sample variance for all data combined, and $n_T$ is the sum of the combined sample sizes.

We now need to determine how much of this total sum of squares is due to between-sample sum of squares and how much is due to within-sample sum of squares. The between-sample portion is called the *sum of squares between* and is found using Equation 12.3.

### Sum of Squares Between

$$SSB = \sum_{i=1}^{k} n_i (\bar{x}_i - \bar{\bar{x}})^2 \qquad (12.3)$$

where:

$SSB$ = Sum of squares between samples
$k$ = Number of populations
$n_i$ = Sample size from population $i$
$\bar{x}_i$ = Sample mean from population $i$
$\bar{\bar{x}}$ = Grand mean

We can use Equation 12.3 to manually compute SSB for the Chicago Connection data, as follows:

$$SSB = 8(7 - 8.25)^2 + 8(9 - 8.25)^2 + 8(8 - 8.25)^2 + 8(9 - 8.25)^2$$
$$SSB = 22$$

Once both the *SST* and *SSB* have been computed, we can easily compute the *sum of squares within* (also called the sum of squares error, *SSE*) using Equation 12.4. The sum of squares within can also be computed directly using Equation 12.5.

**Sum of Squares Within**

$$SSW = SST - SSB \tag{12.4}$$

or

**Sum of Squares Within**

$$SSW = \sum_{i=1}^{k} \sum_{j=1}^{n_i} (x_{ij} - \bar{x}_i)^2 \tag{12.5}$$

where:

$SSW$ = Sum of squares within samples

$k$ = Number of populations

$n_i$ = Sample size from population $i$

$\bar{x}_i$ = Sample mean from population $i$

$x_{ij}$ = $j$th measurement from population $i$

For the Chicago Connection Sandwich example, the $SSW$ is

$$SSW = 220.88 - 22.00$$
$$= 198.88$$

These computations are the essential first steps in performing the ANOVA test to determine whether the population means are equal. Table 12.2 illustrates the ANOVA table format used to conduct the test. The format shown in Table 12.2 is the standard ANOVA table layout. For the Chicago Connection example, we substitute the numerical values for $SSB$, $SSW$, and $SST$ and complete the ANOVA table, as shown in Table 12.3. The mean square column contains the $MSB$ (mean square between samples) and the $MSW$ (mean square within samples).[2] We compute these values by dividing the sums of squares by their respective degrees of freedom, as shown in Table 12.3.

**TABLE 12.2    One-Way ANOVA Table: The Basic Format**

| Source of Variation | SS | df | MS | F-ratio |
|---|---|---|---|---|
| Between samples | $SSB$ | $k - 1$ | $MSB$ | $\dfrac{MSB}{MSW}$ |
| Within samples | $SSW$ | $n_T - k$ | $MSW$ | |
| Total | $SST$ | $n_T - 1$ | | |

where:

$k$ = Number of populations

$n_T$ = Sum of the sample sizes from all populations

$df$ = Degrees of freedom

$MSB$ = Mean square between $= \dfrac{SSB}{k - 1}$

$MSW$ = Mean square within $= \dfrac{SSW}{n_T - k}$

---

[2]$MSW$ is also known as the *mean square error (MSE)*.

**TABLE 12.3** One-Way ANOVA Table for the Chicago Connection Sandwich Company

| Source of Variation | SS | df | MS | F-ratio |
|---|---|---|---|---|
| Between samples | 22.00 | 3 | 7.33 | $\frac{7.33}{7.10} = 1.03$ |
| Within samples | 198.88 | 28 | 7.10 | |
| Total | 220.88 | 31 | | |

where:

$$MSB = \text{Mean square between} = \frac{SSB}{k-1} = \frac{22}{3} = 7.33$$

$$MSW = \text{Mean square within} = \frac{SSW}{n_T - k} = \frac{198.88}{28} = 7.10$$

We restate the null and alternative hypotheses for the Chicago Connection example:

$$H_0: \mu_1 = \mu_2 = \mu_3 = \mu_4$$
$$H_A: \text{At least two population means are different}$$

Now glance back at Figures 12.1 and 12.2. If the null hypothesis is true (that is, all the means are equal—Figure 12.2), the $MSW$ and $MSB$ will be equal, except for the presence of sampling error. However, the more the sample means differ (Figure 12.1), the larger the $MSB$ becomes. As the $MSB$ increases, it will tend to get larger than the $MSW$. When this difference gets too large, we can conclude that the population means must not be equal, and we reject the null hypothesis. But how do we determine what "too large" is? How do we know when the difference is due to more than just sampling error?

To answer these questions, recall from Chapter 11 that the $F$-distribution is used to test whether two populations have the same variance. In the ANOVA test, if the null hypothesis is true, the ratio of $MSB$ over $MSW$ forms an $F$-distribution with $D_1 = k - 1$ and $D_2 = n_T - k$ degrees of freedom. If the calculated $F$-ratio in Table 12.3 gets too large, we reject the null hypothesis.

Figure 12.4 illustrates the hypothesis test for a significance level of 0.05. Because the calculated $F$-ratio $= 1.03$ is less than the critical $F_{0.05} = 2.95$ (found using Excel's F.INV.RT function) with 3 and 28 degrees of freedom, we cannot reject the null hypothesis. The $F$-ratio indicates that the between-levels estimate and the within-levels estimate are not different enough to conclude that the population means are different. Thus, there is insufficient statistical evidence to conclude that any one of the four locations will generate higher average-dollar lunch orders than any of the other locations. Therefore, the VP of sales has no reason to think that in Columbus, Ohio, the location of his restaurants is influencing the size of customer orders.

> The Excel 2016 function for the ANOVA critical $F$-value is
>
> = F.INV.RT(alpha, k-1,$n_T$-k)
> = F.INV.RT(0.05,3,28)

**FIGURE 12.4** Chicago Connection Sandwich Company Hypothesis Test

$H_0: \mu_1 = \mu_2 = \mu_3 = \mu_4$
$H_A$: At least two population means are different
$\alpha = 0.05$

Degrees of Freedom:
$D_1 = k - 1 = 4 - 1 = 3$
$D_2 = n_T - k = 32 - 4 = 28$

Rejection Region

$F = 1.03$          $F_{0.05} = 2.95$

Decision Rule:

If: $F > F_{0.05}$, reject $H_0$; otherwise, do not reject $H_0$.

Then: $F = \dfrac{MSB}{MSW} = \dfrac{7.33}{7.10} = 1.03$

Because: $F = 1.03 < F_{0.05} = 2.95$, we do not reject $H_0$.

**OUTCOME** 2    **EXAMPLE 12-1** One-Way Analysis of Variance

**Roderick, Wilterding & Associates** Roderick, Wilterding & Associates (RWA) operates automobile dealerships in three regions: the West, Southwest, and Northwest. Recently, RWA's general manager questioned whether the company's mean profit margin per vehicle sold differed by region. To determine this, the following steps can be performed:

**STEP 1** **Specify the parameter(s) of interest.**
The parameter of interest is the mean profit margin per vehicle in each region.

**STEP 2** **Formulate the null and alternative hypotheses.**
The appropriate null and alternative hypotheses are

$$H_0: \mu_W = \mu_{SW} = \mu_{NW}$$
$H_A$: At least two populations have different means

**STEP 3** **Specify the significance level ($\alpha$) for testing the hypothesis.**
The test will be conducted using an $\alpha = 0.05$.

**STEP 4** **Select independent simple random samples from each population, and compute the sample means and the grand mean.**
There are three regions. Simple random samples of vehicles sold in these regions have been selected: 10 in the West, 8 in the Southwest, and 12 in the Northwest. Note that even though the sample sizes are not equal, the largest sample is not more than 1.5 times as large as the smallest sample size. The following sample data were collected (in dollars):

| West | Southwest | Northwest | West | Southwest | Northwest |
|------|-----------|-----------|------|-----------|-----------|
| 3,700 | 3,300 | 2,900 | 5,300 | 2,700 | 3,300 |
| 2,900 | 2,100 | 4,300 | 2,200 | 4,500 | 3,700 |
| 4,100 | 2,600 | 5,200 | 3,700 | 2,400 | 2,400 |
| 4,900 | 2,100 | 3,300 | 4,800 | | 4,400 |
| 4,900 | 3,600 | 3,600 | 3,000 | | 3,300 |
| | | | | | 4,400 |
| | | | | | 3,200 |

The sample means are

$$\bar{x}_W = \frac{\sum x}{n} = \frac{\$39,500}{10} = \$3,950$$

$$\bar{x}_{SW} = \frac{\$23,300}{8} = \$2,912.50$$

$$\bar{x}_{NW} = \frac{\$44,000}{12} = \$3,666.67$$

and the grand mean, the mean of the data from all samples, is

$$\bar{\bar{x}} = \frac{\sum \sum x}{n_T} = \frac{\$3,700 + \$2,900 + \cdots + \$3,200}{30}$$

$$= \frac{\$106,800}{30}$$

$$= \$3,560$$

The Excel 2016 function for the ANOVA critical *F*-value is

= F.INV.RT(alpha,k-1,n$_T$-k)

= F.INV.RT(0.05,2,27)

STEP 5 **Determine the decision rule.**

The critical *F*-value from the *F*-distribution table in Appendix H for $D_1 = 2$ and $D_2 = 27$ degrees of freedom is between 3.316 and 3.403. The exact value $F_{0.05} = 3.354$ can be found using Excel's F.INV.RT function.

The decision rule is:

> If $F > 3.354$, reject the null hypothesis.
>
> Otherwise, do not reject the null hypothesis.

STEP 6 **Create the ANOVA table.**

Compute the total sum of squares, sum of squares between, and sum of squares within, and complete the ANOVA table.

**Total Sum of Squares**

$$SST = \sum_{i=1}^{k} \sum_{j=1}^{n_i} (x_{ij} - \bar{\bar{x}})^2$$
$$= (3,700 - 3,560)^2 + (2,900 - 3,560)^2 + \cdots + (3,200 - 3,560)^2$$
$$= 26,092,000$$

**Sum of Squares Between**

$$SSB = \sum_{i=1}^{k} n_i (\bar{x}_i - \bar{\bar{x}})^2$$
$$= 10(3,950 - 3,560)^2 + 8(2,912.50 - 3,560)^2 + 12(3,666.67 - 3,560)^2$$
$$= 5,011,583$$

**Sum of Squares Within**

$$SSW = SST - SSB = 26,092,000 - 5,011,583 = 21,080,417$$

The ANOVA table is

| Source of Variation | SS | df | MS | F-ratio |
|---|---|---|---|---|
| Between samples | 5,011,583 | 2 | 2,505,792 | $\frac{2,505,792}{780,756} = 3.209$ |
| Within samples | 21,080,417 | 27 | 780,756 | |
| Total | 26,092,000 | 29 | | |

STEP 7 **Reach a decision.**

Because the *F*-test statistic $= 3.209 < F_{0.05} = 3.354$, we do not reject the null hypothesis based on these sample data.

STEP 8 **Draw a conclusion.**

We are not able to detect a difference in the mean profit margin per vehicle sold by region.

TRY EXERCISE 12-2 (pg. 450)

---

**BUSINESS APPLICATION**  **Using Excel to Perform One-Way ANOVA**

Excel Tutorial

**Best Health Corporation**  The Best Health Corporation makes and distributes health products. The company's research department is experimenting with two new herb-based weight loss–enhancing products. To gauge their effectiveness, suppose that researchers at the company conducted a test using 300 human subjects over a six-week period. All the people in the study were between 30 and 40 pounds overweight.

One-third of the subjects were randomly selected to receive a placebo—in this case, a pill containing only vitamin C. One-third of the subjects were randomly selected and given

Product 1. The remaining 100 people received Product 2. The subjects did not know which pill they had been assigned. Each person was asked to take the pill regularly for six weeks and otherwise observe his or her normal routine. At the end of six weeks, the subjects' weight loss was recorded. The company was hoping to find statistical evidence that at least one of the products is an effective weight-loss aid.

The file **Best Health** contains the study data. Positive values indicate that the subject lost weight, whereas negative values indicate that the subject gained weight during the six-week study period. As often happens in studies involving human subjects, people drop out. Thus, at the end of six weeks, only 89 placebo subjects, 91 Product 1 subjects, and 83 Product 2 subjects with valid data remained. Consequently, this experiment resulted in an unbalanced design. Although the sample sizes are not equal, they are close to being the same size and do not violate the 1.5-ratio rule of thumb mentioned earlier.

The null and alternative hypotheses to be tested using a significance level of 0.05 are

$$H_0: \mu_1 = \mu_2 = \mu_3$$
$$H_A: \text{At least two population means are different}$$

The factor is the weight-loss aid, which has three levels: placebo, Product 1, and Product 2. We will use a significance level of $\alpha = 0.05$.

Figure 12.5 shows the Excel analysis of variance results. The top section of the Excel ANOVA output provides descriptive information for the three levels. The ANOVA table is shown in the other section of the output. These tables look like the ones we generated manually in the Chicago Connection Sandwich Company application and in the Roderick, Wilterding & Associates Example 12-1. Excel also computes the p-value and displays the critical F-value from the F-distribution table. Thus, you can test the null hypothesis by comparing the calculated F to the critical F or by comparing the p-value to the significance level.

The decision rule is

$$\text{If } F > F_{0.05} = 3.03, \text{ reject } H_0.$$
$$\text{Otherwise, do not reject } H_0.$$

or

$$\text{If } p\text{-value} < \alpha = 0.05, \text{ reject } H_0.$$
$$\text{Otherwise, do not reject } H_0.$$

Because

$$F = 20.48 > F_{0.05} = 3.03 \text{ (or } p\text{-value} = 0.0000 < \alpha = 0.05)$$

we reject the null hypothesis and conclude there is a difference in the mean weight loss for people on the three treatments. At least two of the populations have different means. The top portion of Figure 12.5 shows the descriptive measures for the sample data. For example, the subjects who took the placebo actually gained an average of 1.75 pounds. Subjects on Product 1 lost an average of 2.45 pounds, and subjects on Product 2 lost an average of 2.58 pounds.

**FIGURE 12.5 Excel 2016 Output: Best Health Weight Loss ANOVA Results**

**Excel 2016 Instructions**

1. Open File: **Best Health.xlsx**.
2. Select **Data > Data Analysis**.
3. Select **ANOVA: Single Factor**.
4. Define data range (columns B, C, and D).
5. Specify **Alpha** equal 0.05.
6. Indicate output choice.
7. Click **OK**.

**Anova: Single Factor**

**SUMMARY**

| Groups | Count | Sum | Average | Variance |
|---|---|---|---|---|
| Placebo | 89 | -156 | -1.7506 | 31.4782 |
| Product # 1 | 91 | 223.2 | 2.4527 | 22.8247 |
| Product # 2 | 83 | 214.5 | 2.5843 | 24.1845 |

**ANOVA**

| Source of Variation | SS | df | MS | F | P-value | F crit |
|---|---|---|---|---|---|---|
| Between Groups | 1,072.37 | 2 | 536.186 | 20.479 | 0.0000 | 3.0305 |
| Within Groups | 6,807.44 | 260 | 26.182 | | | |
| Total | 7,879.81 | 262 | | | | |

**OUTCOME 3**   # The Tukey-Kramer Procedure for Multiple Comparisons

What does the Best Health conclusion imply about which treatment results in greater weight loss? One method for testing which populations have different means after the one-way ANOVA has led us to reject the null hypothesis is called the *Tukey-Kramer procedure for multiple comparisons*.[3]

The Tukey-Kramer procedure allows us to simultaneously examine all pairs of populations *after* the ANOVA test has been completed without increasing the true alpha level. Because these comparisons are made after the ANOVA *F*-test, the procedure is called a post-test (or post-hoc) procedure.

The first step in using the Tukey-Kramer procedure is to compute the absolute differences between each pair of sample means. Using the results shown in Figure 12.5, we get the following absolute differences:

$$|\bar{x}_1 - \bar{x}_2| = |-1.75 - 2.45| = 4.20$$
$$|\bar{x}_1 - \bar{x}_3| = |-1.75 - 2.58| = 4.33$$
$$|\bar{x}_2 - \bar{x}_3| = |2.45 - 2.58| = 0.13$$

The Tukey-Kramer procedure requires us to compare these absolute differences to the *critical range* that is computed using Equation 12.6.

**Tukey-Kramer Critical Range**

$$\text{Critical range} = q_{1-\alpha}\sqrt{\frac{MSW}{2}\left(\frac{1}{n_i} + \frac{1}{n_j}\right)} \qquad \textbf{(12.6)}$$

where:

$q_{1-\alpha} =$ Value from studentized range table (Appendix I), with $D_1 = k$ and $D_2 = n_T - k$ degrees of freedom for the desired level of $1 - \alpha$ [$k =$ Number of populations or factor levels, and $n_T =$ Total number of data values from all populations (levels) combined]

$MSW =$ Mean square within

$n_i$ and $n_j =$ Sample sizes from populations (levels) $i$ and $j$, respectively

We compute a critical range for each pairwise comparison, but if the sample sizes are equal, only one critical-range calculation is necessary because the quantity under the radical in Equation 12.7 will be the same for all comparisons. If the calculated pairwise comparison value is greater than the critical range, we conclude that the difference is significant.

To determine the $q$-value from the studentized range table in Appendix I for a significance level equal to

$$\alpha = 0.05$$

and

$$D_1 = k = 3 \text{ and } D_2 = n_T - k = 260 \text{ degrees of freedom.}$$

For $D_2 = n_T - k = 260$ degrees of freedom, we use the row labeled $\infty$. The studentized range value for $1 - 0.05 = 0.95$ is approximately

$$q_{0.95} = 3.31$$

Then, for the placebo versus Product 1 comparison,

$$n_1 = 89 \quad \text{and} \quad n_2 = 91$$

---

[3]There are other methods for making these comparisons. Statisticians disagree over which method to use. Later, we introduce alternative methods.

**TABLE 12.4** Best Health Pairwise Comparisons—Tukey-Kramer Test

| | $\lvert \bar{x}_i - \bar{x}_j \rvert$ | Critical Range | Significant? |
|---|---|---|---|
| Placebo vs. Product 1 | 4.20 | 1.785 | Yes |
| Placebo vs. Product 2 | 4.33 | 1.827 | Yes |
| Product 1 vs. Product 2 | 0.13 | 1.818 | No |

we use Equation 12.6 to compute one critical range, as follows:

$$\text{Critical range} = q_{1-\alpha}\sqrt{\frac{MSW}{2}\left(\frac{1}{n_i} + \frac{1}{n_j}\right)}$$

$$\text{Critical range} = 3.31\sqrt{\frac{26.18}{2}\left(\frac{1}{89} + \frac{1}{91}\right)} = 1.785$$

Because

$$\lvert \bar{x}_1 - \bar{x}_2 \rvert = 4.20 > 1.785$$

we conclude that

$$\mu_1 \neq \mu_2$$

The mean weight loss for the placebo group is not equal to the mean weight loss for the Product 1 group. Table 12.4 summarizes the results for the three pairwise comparisons. From the table, we see that Product 1 and Product 2 both offer significantly higher average weight loss than the placebo. However, the sample data do not indicate a difference in the average weight loss between Product 1 and Product 2. Thus, the company can only conclude that both Product 1 and Product 2 are superior to a placebo.

**OUTCOME** 3

Excel Tutorial

**EXAMPLE 12-2** The Tukey-Kramer Procedure for Multiple Comparison

**Digitron, Inc.** Digitron, Inc., makes disc brakes for automobiles. Digitron's research and development (R&D) department tested four brake systems to determine if there is a difference in the average stopping distances among them. Forty identical mid-sized cars were driven on a test track. Ten cars were fitted with Brake A, 10 with Brake B, and so forth. An electronic, remote switch was used to apply the brakes at exactly the same point on the road. The number of feet required to bring the car to a full stop was recorded. The data are in the file titled **Digitron**. Because we care to determine whether the four brake systems have the same or different mean stopping distances, the test is a one-way (single-factor) test with four levels and can be completed using the following steps:

**STEP 1** **Specify the parameter(s) of interest.**
The parameter of interest is the mean stopping distance for each brake type. The company is interested in knowing whether a difference exists in the mean stopping distances for the four brake types.

**STEP 2** **Formulate the appropriate null and alternative hypotheses.**
The appropriate null and alternative hypotheses are

$$H_0: \mu_1 = \mu_2 = \mu_3 = \mu_4$$
$$H_A: \text{At least two population means are different}$$

**STEP 3** **Specify the significance level for the test.**
The test will be conducted using $\alpha = 0.05$.

**STEP 4** **Select independent simple random samples from each population.**

**STEP 5** **Check to see that the normality and equal-variance assumptions have been satisfied.**
Because of the small sample size, the box and whisker plot is used.

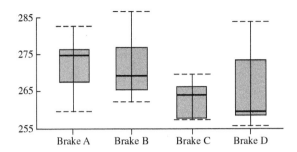

The box plots indicate some skewness in the samples and question the assumption of equality of variances.

Recall our earlier discussion that when the sample sizes are equal, as they are in this example, the ANOVA test is robust in regard to both the equal-variance and normality assumptions.

STEP 6  **Determine the decision rule.**

Because $k - 1 = 3$ and $n_T - k = 36$, from Excel we find $F_{0.05} = 2.866$. The decision rule is:

> If calculated $F > F_{0.05} = 2.866$, reject $H_0$, or
> if $p$-value $< \alpha = 0.05$, reject $H_0$. Otherwise, do not reject $H_0$.

> The Excel 2016 function for the ANOVA critical F value is
> = **F.INV.RT(alpha,k,n$_T$-k)**
> = **F.INV.RT(0.05,3,36)**

STEP 7  **Use Excel to construct the ANOVA table.**

Figure 12.6 shows the Excel output for the ANOVA.

STEP 8  **Reach a decision.**

From Figure 12.6, we see that

$$F = 3.89 > F_{0.05} = 2.866, \text{ and } p\text{-value} = 0.017 < 0.05$$

We reject the null hypothesis.

STEP 9  **Draw a conclusion.**

We conclude that not all population means are equal. But which systems are different? Is one system superior to all the others?

STEP 10  **Use the Tukey-Kramer test to determine which populations have different means.**

Because we have rejected the null hypothesis of equal means, we need to perform a post–ANOVA multiple comparisons test. Using Equation 12.6 to

**FIGURE 12.6** **Excel 2016 One-Way ANOVA Output for the Digitron Example**

> **Excel 2016 Instructions**
> 1. Open file: **Digitron.xlsx**.
> 2. Select **Data > Data Analysis**.
> 3. Select **ANOVA: Single Factor**.
> 4. Define data range (columns B, C, D, and E).
> 5. Specify **Alpha** equal 0.05.
> 6. Indicate output location.
> 7. Click **OK**.

| | A | B | C | D | E | F | G |
|---|---|---|---|---|---|---|---|
| 1 | Anova: Single Factor | | | | | | |
| 2 | | | | | | | |
| 3 | SUMMARY | | | | | | |
| 4 | *Groups* | *Count* | *Sum* | *Average* | *Variance* | | |
| 5 | Brake A | 10 | 2,723.59 | 272.36 | 49.90 | | |
| 6 | Brake B | 10 | 2,713.30 | 271.33 | 61.86 | | |
| 7 | Brake C | 10 | 2,623.14 | 262.31 | 21.74 | | |
| 8 | Brake D | 10 | 2,652.36 | 265.24 | 106.44 | | |
| 9 | | | | | | | |
| 10 | ANOVA | | | | | | |
| 11 | *Source of Variation* | *SS* | *df* | *MS* | *F* | *P-value* | *F crit* |
| 12 | Between Groups | 699.16 | 3.000 | 233.055 | 3.885 | 0.017 | 2.866 |
| 13 | Within Groups | 2,159.37 | 36.000 | 59.982 | | | |
| 14 | | | | | | | |
| 15 | Total | 2,858.53 | 39.000 | | | | |
| 16 | | | | | | | |
| 17 | Because calculated F = 3.885 > F critical = 2.866, reject the null hypothesis | | | | | | |
| 18 | and conclude that the population means are not equal. | | | | | | |

construct the critical range to compare to the absolute differences in all possible pairs of sample means, we find the critical range is[4]

$$\text{Critical range} = q_{1-\alpha}\sqrt{\frac{MSW}{2}\left(\frac{1}{n_i} + \frac{1}{n_j}\right)} = 3.85\sqrt{\frac{59.98}{2}\left(\frac{1}{10} + \frac{1}{10}\right)}$$

$$\text{Critical range} = 9.43$$

Only one critical range is necessary because the sample sizes are equal. If any pair of sample means has an absolute difference $|\bar{x}_i - \bar{x}_j|$ greater than the critical range, we can infer that a difference exists in those population means. The possible pairwise comparisons (part of a family of comparisons called *contrasts*) are

| Contrast | Significant Difference? |
|---|---|
| $|\bar{x}_1 - \bar{x}_2| = |272.3590 - 271.3299| = 1.0291 < 9.43$ | No |
| $|\bar{x}_1 - \bar{x}_3| = |272.3590 - 262.3140| = 10.0450 > 9.43$ | Yes |
| $|\bar{x}_1 - \bar{x}_4| = |272.3590 - 265.2357| = 7.1233 < 9.43$ | No |
| $|\bar{x}_2 - \bar{x}_3| = |271.3299 - 262.3140| = 9.0159 < 9.43$ | No |
| $|\bar{x}_2 - \bar{x}_4| = |271.3299 - 265.2357| = 6.0942 < 9.43$ | No |
| $|\bar{x}_3 - \bar{x}_4| = |262.3140 - 265.2357| = 2.9217 < 9.43$ | No |

Therefore, based on the Tukey-Kramer procedure, we can infer that Population 1 (Brake A) and Population 3 (Brake C) have different mean stopping distances. Because short stopping distances are preferred, Brake C would be preferred over Brake A, but no other differences are supported by these sample data. For the other contrasts, the difference between the two sample means is insufficient to conclude that a difference in population means exists.

**TRY EXERCISE 12-6** (pg. 451)

## Fixed Effects Versus Random Effects in Analysis of Variance

In the Digitron brake example, the company was testing four brake systems, and these were the only brake systems under consideration. The ANOVA was intended to determine whether there was a difference in these four brake systems only. In the Best Health weight-loss example, the company was interested in determining whether there was a difference in mean weight loss for two supplements and the placebo. In the Chicago Connection Sandwich example involving locations, the company analyzed orders at four selected locations, and the ANOVA test was used to determine whether there was a difference in means for these four locations only. Thus, in each of these examples, the inferences extend only to the factor levels being analyzed, and the levels are assumed to be the only levels of interest. This type of test is called a *fixed effects analysis of variance test*.

Suppose in the Chicago Connection example that instead of reducing the list of possible restaurant locations to the four in Columbus, Ohio, the company had selected a random sample of four of its restaurants from all its Midwest locations. In that case, the factor levels included in the test would be a random sample of the possible levels. Then, if the ANOVA leads to rejecting the null hypothesis, the conclusion applies to all possible Chicago Connection Sandwich locations rather than just the four in Columbus. The assumption is that the possible levels have a normal distribution and the tested levels are a random sample from this distribution. When the factor levels are selected through random sampling, the analysis of variance test is called a *random effects test*.

---

[4]The $q$-value from the studentized range table with $\alpha = 0.05$ and degrees of freedom equal to $k = 4$ and $n_T - k = 36$ must be approximated using degrees of freedom 4 and 30 because the table does not show degrees of freedom of 4 and 36. This value is 3.85. Rounding down to 30 will give a larger $q$ value and a conservatively large critical range.

## 12.1 EXERCISES

### Skill Development

**12-1.** A startup cell phone applications company is interested in determining whether household incomes are different for subscribers to three different service providers. A random sample of 25 subscribers to each of the three service providers was taken, and the annual household income for each subscriber was recorded. The partially completed ANOVA table for the analysis is shown here:

**ANOVA**

| Source of Variation | SS | df | MS | F |
|---|---|---|---|---|
| Between Groups | 2,949,085,157 | | | |
| Within Groups | | | | |
| Total | 9,271,678,090 | | | |

a. Complete the ANOVA table by filling in the missing sums of squares, the degrees of freedom for each source, the mean square, and the calculated $F$-test statistic.
b. Based on the sample results, can the startup firm conclude that there is a difference in household incomes for subscribers to the three service providers? You may assume normal distributions and equal variances. Conduct your test at the $\alpha = 0.10$ level of significance. Be sure to state a critical $F$-statistic, a decision rule, and a conclusion.

**12-2.** An analyst is interested in testing whether four populations have equal means. The following sample data have been collected from populations that are assumed to be normally distributed with equal variances:

| Sample 1 | Sample 2 | Sample 3 | Sample 4 |
|---|---|---|---|
| 9 | 12 | 8 | 17 |
| 6 | 16 | 8 | 15 |
| 11 | 16 | 12 | 17 |
| 14 | 12 | 7 | 16 |
| 14 | 9 | 10 | 13 |

Conduct the appropriate hypothesis test using a significance level equal to 0.05.

**12-3.** A manager is interested in testing whether three populations have equal population means. Simple random samples of size 10 were selected from each population. The following ANOVA table and related statistics were computed:

**ANOVA: Single Factor**

**Summary**

| Groups | Count | Sum | Average | Variance |
|---|---|---|---|---|
| Sample 1 | 10 | 507.18 | 50.72 | 35.06 |
| Sample 2 | 10 | 405.79 | 40.58 | 30.08 |
| Sample 3 | 10 | 487.64 | 48.76 | 23.13 |

**ANOVA**

| Source | SS | df | MS | F | p-value | F-ratio |
|---|---|---|---|---|---|---|
| Between Groups | 578.78 | 2 | 289.39 | 9.84 | 0.0006 | 3.354 |
| Within Groups | 794.36 | 27 | 29.42 | | | |
| Total | 1,373.14 | 29 | | | | |

a. State the appropriate null and alternative hypotheses.
b. Based on your answer to part a, what conclusions can you reach about the null and alternative hypotheses? Use a 0.05 level of significance.
c. If warranted, use the Tukey-Kramer procedure for multiple comparisons to determine which populations have different means. (Assume $\alpha = 0.05$.)

**12-4.** Respond to each of the following questions using this partially completed one-way ANOVA table:

| Source of Variation | SS | df | MS | F-ratio |
|---|---|---|---|---|
| Between Samples | 1,745 | | | |
| Within Samples | | 240 | | |
| Total | 6,504 | 246 | | |

a. How many different populations are being considered in this analysis?
b. Fill in the ANOVA table with the missing values.
c. State the appropriate null and alternative hypotheses.
d. Based on the analysis of variance $F$-test, what conclusion should you reach regarding the null hypothesis? Test using a significance level of 0.01.

**12-5.** Respond to each of the following questions using this partially completed one-way ANOVA table:

| Source of Variation | SS | df | MS | F-ratio |
|---|---|---|---|---|
| Between Samples | | 3 | | |
| Within Samples | 405 | | | |
| Total | 888 | 31 | | |

a. How many different populations are being considered in this analysis?

b. Fill in the ANOVA table with the missing values.
c. State the appropriate null and alternative hypotheses.
d. Based on the analysis of variance $F$-test, what conclusion should you reach regarding the null hypothesis? Test using $\alpha = 0.05$.

**12-6.** You are given the following sample data:

| Item | Group 1 | Group 2 | Group 3 | Group 4 |
|------|---------|---------|---------|---------|
| 1 | 20.9 | 28.2 | 17.8 | 21.2 |
| 2 | 27.2 | 26.2 | 15.9 | 23.9 |
| 3 | 26.6 | 21.6 | 18.4 | 19.5 |
| 4 | 22.1 | 29.7 | 20.2 | 17.4 |
| 5 | 25.3 | 30.3 | 14.1 | |
| 6 | 30.1 | 25.9 | | |
| 7 | 23.8 | | | |

a. Based on the computations for the within- and between-sample variation, develop the ANOVA table and test the appropriate null hypothesis using $\alpha = 0.05$. Use the $p$-value approach.
b. If warranted, use the Tukey-Kramer procedure to determine which populations have different means. Use $\alpha = 0.05$.

**12-7.** Examine the three samples obtained independently from three populations:

| Item | Group 1 | Group 2 | Group 3 |
|------|---------|---------|---------|
| 1 | 14 | 17 | 17 |
| 2 | 13 | 16 | 14 |
| 3 | 12 | 16 | 15 |
| 4 | 15 | 18 | 16 |
| 5 | 16 | | 14 |
| 6 | | | 16 |

a. Conduct a one-way analysis of variance on the data. Use alpha $= 0.05$.
b. If warranted, use the Tukey-Kramer procedure to determine which populations have different means. Use an experiment-wide error rate of 0.05.

## Business Applications

**12-8.** A major U.S. bank wishes to determine if the average credit card balance depends on the type of credit card used. Under consideration are Visa, MasterCard, Discover, and American Express. The sample sizes to be used for each level are 25, 25, 26, and 23, respectively.
a. Describe the parameter of interest for this analysis.
b. Determine the factor associated with this experiment.
c. Describe the levels of the factor associated with this analysis.
d. State the number of degrees of freedom available for determining the between-sample variation.
e. State the number of degrees of freedom available for determining the within-sample variation.

f. State the number of degrees of freedom available for determining the total variation.

**12-9.** EverRun Incorporated produces treadmills for use in exercise clubs and recreation centers. The treadmill motors are purchased from an outside vendor. Currently, EverRun is considering which motor to include in its new ER1500 series. Three potential suppliers have been identified: Venetti, Madison, and Edison. EverRun has decided to make its decision based on how long a motor operates at a high level of speed and incline before it fails. A random sample of 10 motors of each type is selected, and each motor is tested to determine how many minutes (rounded to the nearest minute) it operates before it needs to be repaired. The sample information for each motor is as follows:

| Venetti | Madison | Edison |
|---------|---------|--------|
| 14,722 | 13,649 | 13,296 |
| 14,699 | 13,592 | 13,262 |
| 12,627 | 11,788 | 11,552 |
| 13,010 | 12,623 | 11,036 |
| 13,570 | 14,552 | 12,978 |
| 14,217 | 13,441 | 12,170 |
| 13,687 | 13,404 | 12,674 |
| 13,465 | 13,427 | 11,851 |
| 14,786 | 12,049 | 12,342 |
| 12,494 | 11,672 | 11,557 |

a. At the $\alpha = 0.01$ level of significance, is there a difference in the average times before failure for the three different suppliers' motors?
b. Is it possible for EverRun to decide on a single motor supplier based on the analysis of the sample results? Support your answer by conducting the appropriate post-test analysis.

**12-10.** Lanhu Cement Corporation is a leading North American cement producer, with more than 6.5 million metric tons of annual capacity. One of its products is Portland cement. Portland cement's properties and performance standards are defined by its type designation. Each type is designated by a Roman numeral. Ninety-two percent of the Portland cement produced in North America is Type I, II, or I/II. One characteristic of the type of cement is its compressive strength. Sample data for the compressive strength (in psi) are shown as follows:

| Type | Compressive Strength | | | |
|------|------|------|------|------|
| I | 4,972 | 4,983 | 4,889 | 5,063 |
| II | 3,216 | 3,399 | 3,267 | 3,357 |
| I/II | 4,073 | 3,949 | 3,936 | 3,925 |

a. Develop the appropriate ANOVA table to determine if there is a difference in the average compressive strengths among the three types of Portland cement. Use a significance level of 0.01.

b. If warranted, use the Tukey-Kramer procedure to determine which populations have different mean compressive strengths. Use an experiment-wide error rate of 0.01.

**12-11.** The Weidmann Group Companies are in the insulation systems technology business. One facet of its expertise is the development of dielectric fluids in electrical equipment. Mineral oil–based dielectric fluids have been used more extensively than other dielectric fluids. Their only shortcomings are their relatively low flash and fire points. One study examined the fire point of mineral oil, high-molecular-weight hydrocarbon (HMWH), and silicone. The fire points for each of these fluids were as follows:

| Fluid | Fire Points (°C) | | | | |
|---|---|---|---|---|---|
| Mineral Oil | 162 | 151 | 168 | 165 | 169 |
| HMWH | 312 | 310 | 300 | 311 | 308 |
| Silicone | 343 | 337 | 345 | 345 | 337 |

a. Develop the appropriate ANOVA table to determine if there is a difference in the average fire points among the types of dielectric fluids. Use a significance level of 0.05.
b. If warranted, use the Tukey-Kramer procedure to determine which populations have different mean fire points. Use an experiment-wide error rate of 0.05.

**12-12.** The manager at the Lawrence National Bank is interested in determining whether there is a difference in the mean time that customers spend completing their transactions depending on which of four tellers they use. To conduct the test, the manager has selected simple random samples of 15 customers for each of the tellers and has timed them (in seconds) from the moment they start their transaction to the time the transaction is completed and they leave the teller station. The manager then asked one of her assistants to perform the appropriate statistical test. The assistant returned with the following partially completed ANOVA table:

| Summary | | | | |
|---|---|---|---|---|
| Groups | Count | Sum | Average | Variance |
| Teller 1 | 15 | 3,043.9 | | 827.4 |
| Teller 2 | 15 | 3,615.5 | | 472.2 |
| Teller 3 | 15 | 3,427.7 | | 445.6 |
| Teller 4 | 15 | 4,072.4 | | 619.4 |

| ANOVA | | | | | | |
|---|---|---|---|---|---|---|
| Source of Variation | SS | df | MS | F-ratio | p-value | Critical F-value |
| Between Groups | 36,530.6 | | | | 0.0000 | 2.7694 |
| Within Groups | | | | | | |
| Total | 69,633.7 | 59 | | | | |

a. State the appropriate null and alternative hypotheses.
b. Fill in the missing parts of the ANOVA table and perform the statistical hypothesis test using $\alpha = 0.05$.
c. Based on the result of the test in part b, if warranted, use the Tukey-Kramer method with $\alpha = 0.05$ to determine which teller requires the most time on average to complete a customer's transaction.

**12-13.** Suppose as part of your job you are responsible for installing emergency lighting in a series of state office buildings. Bids have been received from four manufacturers of battery-operated emergency lights. The costs are about equal, so the decision will be based on the length of time the lights last before failing. A sample of four lights from each manufacturer has been tested, with the following values (time in hours) recorded for each manufacturer:

| Type A | Type B | Type C | Type D |
|---|---|---|---|
| 1,024 | 1,270 | 1,121 | 923 |
| 1,121 | 1,325 | 1,201 | 983 |
| 1,250 | 1,426 | 1,190 | 1,087 |
| 1,022 | 1,322 | 1,122 | 1,121 |

a. Using a significance level equal to 0.01, what conclusion should you reach about the four manufacturers' battery-operated emergency lights? Explain.
b. If the test conducted in part a reveals that the null hypothesis should be rejected, what manufacturer should be used to supply the lights? Can you eliminate one or more manufacturers based on these data? Use the appropriate test and $\alpha = 0.01$ for multiple comparisons. Discuss.

## Computer Software Exercises

**12-14.** Automatic shutoff valves can prevent extensive water damage from plumbing failures. The valves contain sensors that cut off water flow in the event of a leak, thereby preventing flooding. One important characteristic is the time (in milliseconds) required for the sensor to detect the water leak. Sample data obtained for four different shutoff valves are contained in the file titled **Waterflow**.
a. Produce the appropriate ANOVA table and conduct a hypothesis test to determine if the mean detection times differ among the four shutoff valve models. Use a significance level of 0.05.
b. Use the Tukey-Kramer multiple comparison technique to discover any differences in the average detection times. Use a significance level of 0.05.
c. Which of the four shutoff valves would you recommend? State your criterion for your selection.

**12-15.** A regional package delivery company is considering changing from full-size vans to minivans. The company sampled minivans from each of three manufacturers. The number sampled represents the number the manufacturer was able to provide for the test. Each minivan was driven for 5,000 miles, and the operating cost per mile was computed. The operating costs, in cents per mile, for the 12 are provided in the data file called **Delivery**:

| Mini 1 | Mini 2 | Mini 3 |
|--------|--------|--------|
| 13.3 | 12.4 | 13.9 |
| 14.3 | 13.4 | 15.5 |
| 13.6 | 13.1 | 15.2 |
| 12.8 | | 14.5 |
| 14.0 | | |

a. Perform an analysis of variance on these data. Assume a significance level of 0.05. Do the experimental data provide evidence that the average operating costs per mile for the three types of minivans are different? Use a $p$-value approach.

b. Referring to part a, based on the sample data and the appropriate test for multiple comparisons, what conclusions should you reach concerning which type of car the delivery company should adopt? Discuss and prepare a report to the company CEO. Use $\alpha = 0.05$.

c. Provide an estimate of the maximum and minimum difference in average savings per year if the CEO chooses the "best" versus the "worst" minivan using operating costs as a criterion. Assume that minivans are driven 30,000 miles a year. Use a 90% confidence interval.

**12-16.**  The owner of Ranch Style Burgers wants to determine if its speed of service is different across its four outlets. Orders at the Ranch Style restaurants are tracked electronically, and the chain is able to determine the speed with which every order is filled. The owner decided to randomly sample 20 orders from each of the four restaurants it operates. The speed of service for each randomly sampled order was noted and is contained in the file **Ranch Style**.

a. At the $\alpha = 0.05$ level of service, can the Ranch Style owner conclude that the speed of service is different across the four restaurants in the chain?

b. If the chain concludes that there is a difference in speed of service, is there a particular restaurant the chain should focus its attention on? Use the appropriate test for multiple comparisons to support your decision. Use $\alpha = 0.05$.

**12-17.** To determine whether which company makes auto batteries affects the average length of life of the battery, the sample data in the file titled **Start** were obtained. The data represent the lengths of life (months) for batteries of the same specifications for each of three manufacturers.

a. Determine if the average length of battery life is different among the batteries produced by the three manufacturers. Use a significance level of 0.05.

b. Which manufacturer produces the battery with the longest average length of life? If warranted, conduct the Tukey-Kramer procedure to determine this. Use a significance level of 0.05. (*Note:* You will need to manipulate the data columns to obtain the appropriate factor levels.)

---

**OUTCOME 4**    **12.2**    **Randomized Complete Block Analysis of Variance**

Section 12.1 introduced one-way ANOVA for testing hypotheses that involve three or more population means. This ANOVA method is appropriate as long as we are interested in analyzing one factor at a time and we select independent random samples from the populations. For instance, Example 12-2 involving brake systems at Digitron, Inc. (Figure 12.6) illustrated a situation in which we were interested in only one factor: type of brake system. The measurement of interest was the stopping distance with each brake system. To test the hypothesis that the four brake systems were equal with respect to average stopping distance, four groups of the same make and model cars were assigned to each brake system independently. Thus, the one-way ANOVA design was appropriate.

There are, however, situations in which we know that a second factor is impinging on the response variable of interest. Chapter 10 introduced the concept of paired samples and indicated that there are instances when you will want to test for differences in two population means by controlling for sources of variation that might adversely affect the analysis. For instance, in the Digitron example, we might be concerned that, even though we used the same make and model of car in the study, the cars themselves may interject a source of variability that could affect the result. To control for this, we could use the concept of *paired samples* by using the same 10 cars for each of the four brake systems. When an additional factor with two or more levels is involved, we can use a design technique called *blocking* to eliminate the additional factor's effect on the statistical analysis of the main factor of interest.

# Randomized Complete Block ANOVA

Excel Tutorial

**BUSINESS APPLICATION** | **A Randomized Block Design**

**Citizen's State Bank** At Citizen's State Bank, homeowners can borrow money against the equity they have in their homes. To determine equity, the bank determines the home's value and subtracts the mortgage balance. The maximum loan is 90% of the equity.

The bank outsources the home appraisals to three companies: Allen & Associates, Heist Appraisal, and Appraisal International. The bank managers know that appraisals are not exact. Some appraisal companies may overvalue homes on average, whereas others might undervalue homes.

Bank managers wish to test the hypothesis that there is no difference in the average house appraisals among the three different companies. The managers could select a random sample of homes for Allen & Associates to appraise, a second sample of homes for Heist Appraisal to work on, and a third sample of homes for Appraisal International. They would then use one-way ANOVA to compare the sample means. Obviously a problem could occur if, by chance, one company received larger, higher-quality homes located in better neighborhoods than the other companies. This company's appraisals would naturally be higher on average, not because it tended to appraise higher, but because the homes were simply more expensive.

Citizen's State Bank managers need to control for the variation in size, quality, and location of homes to fairly test that the three companies' appraisals are equal on average. To do this, they select a random sample of properties and have each company appraise the same properties. In this case, the properties are called *blocks*, and the test design is called a *randomized complete block design*.

The data in Table 12.5 were obtained when each appraisal company was asked to appraise the same five properties. The bank managers wish to test the following hypothesis:

$$H_0: \mu_1 = \mu_2 = \mu_3$$

$H_A$: At least two populations have different means

The randomized block design requires the following assumptions:

**Assumptions**

> 1. The populations are normally distributed.
> 2. The populations have equal variances.
> 3. The observations within samples are independent.
> 4. The data measurement must be interval or ratio level.

Because the managers have chosen to have the same properties appraised by each company (block on property), a method known as *randomized complete block ANOVA* must be employed to test the hypothesis. This method is similar to the one-way ANOVA in Section 12.1. However, there is one more source of variation to be accounted for: the block variation. As was the case in Section 12.1, we must find estimators for each source of variation.

**TABLE 12.5** Citizen's State Bank Property Appraisals (in thousands of dollars)

| Property (Block) | Appraisal Company | | | Block Mean |
| --- | --- | --- | --- | --- |
| | Allen & Associates | Heist Appraisal | Appraisal International | |
| 1 | 78 | 82 | 79 | 79.67 |
| 2 | 102 | 102 | 99 | 101.00 |
| 3 | 68 | 74 | 70 | 70.67 |
| 4 | 83 | 88 | 86 | 85.67 |
| 5 | 95 | 99 | 92 | 95.33 |
| Factor-Level Mean | $\bar{x}_1 = 85.2$ | $\bar{x}_2 = 89$ | $\bar{x}_3 = 85.2$ | $\bar{\bar{x}} = 86.47 =$ Grand mean |

Identifying the appropriate sums of squares and then dividing each by its degrees of freedom does this. As was the case in the one-way ANOVA, we obtain the sums of squares by partitioning the total sum of squares (*SST*). However, in this case the *SST* is divided into three components instead of two, as shown in Equation 12.7.

**Sum of Squares Partitioning for Randomized Complete Block Design**

$$SST = SSB + SSBL + SSW \tag{12.7}$$

where:

$$SST = \text{Total sum of squares}$$
$$SSB = \text{Sum of squares between factor levels}$$
$$SSBL = \text{Sum of squares between blocks}$$
$$SSW = \text{Sum of squares within levels}$$

Both *SST* and *SSB* are computed just as we did with one-way ANOVA, using Equations 12.2 and 12.3. The *sum of squares for blocking (SSBL)* is computed using Equation 12.8.

**Sum of Squares for Blocking**

$$SSBL = \sum_{j=1}^{b} k(\bar{x}_j - \bar{\bar{x}})^2 \tag{12.8}$$

where:

$$k = \text{Number of levels for the factor}$$
$$b = \text{Number of blocks}$$
$$\bar{x}_j = \text{The mean of the } j\text{th block}$$
$$\bar{\bar{x}} = \text{Grand mean}$$

Finally, we compute the *sum of squares within (SSW)* using Equation 12.9. This sum of squares is what remains (the residual) after the variation for all factors and blocks has been removed. This residual sum of squares may be due to the inherent variability of the data, measurement error, or other unidentified sources of variation. Therefore, the sum of squares within is also known as the sum of squares error, *SSE*.

**Sum of Squares Within**

$$SSW = SST - (SSB + SSBL) \tag{12.9}$$

Table 12.6 shows the completely randomized block ANOVA table format and equations for degrees of freedom, mean squares, and *F*-ratios. As you can see, we now have two *F*-ratios. The reason for this is that we test not only to determine whether the population (factor) means are equal but also to obtain an indication of whether the blocking was necessary by examining the ratio of the mean square for blocks to the mean square within.

Although you could manually compute the necessary values for the randomized block design, Excel has a procedure that will do all the computations and build the ANOVA table. The Citizen's State Bank appraisal data are included in the file **Citizens**. (Note that the first column contains labels for each block.)

Figure 12.7 shows the ANOVA output. Using Excel to perform the computations frees the decision maker to focus on interpreting the results. Note that Excel refers to the randomized block ANOVA as two-factor ANOVA without replication.

**TABLE 12.6** Basic Format for the Randomized Block ANOVA Table

| Source of Variation | SS | df | MS | F-ratio |
|---|---|---|---|---|
| Between blocks | SSBL | $b - 1$ | MSBL | $\dfrac{MSBL}{MSW}$ |
| Between samples | SSB | $k - 1$ | MSB | $\dfrac{MSB}{MSW}$ |
| Within samples | SSW | $(k - 1)(b - 1)$ | MSW | |
| Total | SST | $n_T - 1$ | | |

where:

$$k = \text{Number of levels}$$
$$b = \text{Number of blocks}$$
$$df = \text{Degrees of freedom}$$
$$n_T = \text{Combined sample size}$$
$$MSBL = \text{Mean square blocking} = \frac{SSBL}{b - 1}$$
$$MSB = \text{Mean square between} = \frac{SSB}{k - 1}$$
$$MSW = \text{Mean square within} = \frac{SSW}{(k - 1)(b - 1)}$$

*Note:* Some randomized block ANOVA tables put *SSB* first, followed by *SSBL*.

The main issue is to determine whether the three appraisal companies differ in average appraisal values. The primary test is

$$H_0: \mu_1 = \mu_2 = \mu_3$$
$$H_A: \text{At least two populations have different means}$$
$$\alpha = 0.05$$

Using the output presented in Figure 12.7, we can test this hypothesis two ways. First, we can use the *F*-distribution approach. Figure 12.8 shows the results of this test. Based on the sample data, because $F = 8.54 > 4.46$ we reject the null hypothesis and conclude that the three appraisal companies do not provide equal average values for properties.

The second approach to testing the null hypothesis is the *p*-value approach. The decision rule in an ANOVA application for *p*-values is

If *p*-value $< \alpha$, reject $H_0$; otherwise, do not reject $H_0$.

In this case, $\alpha = 0.05$ and the *p*-value in Figure 12.7 is 0.01. Because *p*-value $= 0.01 < \alpha = 0.05$, we reject the null hypothesis.

**FIGURE 12.7** Excel 2016 Output: Citizen's State Bank Analysis of Variance

**Excel 2016 Instructions**
1. Open file: **Citizens.xlsx**.
2. Select **Data > Data Analysis**.
3. Select **ANOVA: Two Factor Without Replication**.
4. Define data range (A1:D6).
5. Specify **Alpha** equal 0.05 and check **Labels**.
6. Indicate output location.
7. Click **OK**.

**Anova: Two-Factor Without Replication**

| | A | B | C | D | E | F | G |
|---|---|---|---|---|---|---|---|
| 3 | SUMMARY | Count | Sum | Average | Variance | | |
| 4 | 1 | 3 | 239 | 79.67 | 4.33 | | |
| | 2 | 3 | 303 | 101.00 | 3.00 | | |
| | 3 | 3 | 212 | 70.67 | 9.33 | | |
| | 4 | 3 | 257 | 85.67 | 6.33 | | |
| 8 | 5 | 3 | 286 | 95.33 | 12.33 | | |
| 9 | | | | | | | |
| 10 | Allen & Associates | 5 | 426 | 85.2 | 182.7 | | |
| 11 | Heist Appraisal | 5 | 445 | 89 | 136 | | |
| 12 | Appraisal Int'l | 5 | 426 | 85.2 | 126.7 | | |
| 13 | | | | | | | |
| 14 | ANOVA | | | | | | |
| 15 | Source of Variation | SS | df | MS | F | P-value | F crit |
| 16 | Rows | 1,759.07 | 4 | 439.77 | 156.13 | 0.0000 | 3.84 |
| 17 | Columns | 48.13 | 2 | 24.07 | 8.54 | 0.01 | 4.46 |
| 18 | Error | 22.53 | 8 | 2.82 | | | |
| 19 | | | | | | | |
| 20 | Total | 1829.733 | 14 | | | | |

Blocks

Blocking Test

Main Factor Test

Within Variation

**FIGURE 12.8** **Appraisal Company Hypothesis Test For Citizen's State Bank**

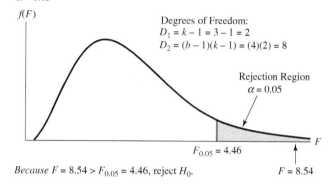

$H_0: \mu_1 = \mu_2 = \mu_3$
$H_A$: At least two population means are different
$\alpha = 0.05$

$f(F)$

Degrees of Freedom:
$D_1 = k - 1 = 3 - 1 = 2$
$D_2 = (b - 1)(k - 1) = (4)(2) = 8$

Rejection Region
$\alpha = 0.05$

$F_{0.05} = 4.46$

$F$

*Because $F = 8.54 > F_{0.05} = 4.46$, reject $H_0$.*

$F = 8.54$

Both the $F$-distribution approach and the $p$-value approach give the same result, as they must.

**Was Blocking Necessary?**   Before we take up the issue of determining which company provides the highest mean property values, we need to discuss one other issue. Recall that the bank managers chose to control for variation between properties by having each appraisal company evaluate the same five properties. This restriction is called blocking, and the properties are the blocks. The ANOVA output in Figure 12.7 contains information that allows us to test whether blocking was necessary.

If blocking was necessary, it would mean that appraisal values are in fact influenced by the particular property being appraised. The blocks then form a second factor of interest, and we formulate a secondary hypothesis test for this factor, as follows:

$$H_0: \mu_{b1} = \mu_{b2} = \mu_{b3} = \mu_{b4} = \mu_{b5}$$
$$H_A: \text{Not all block means are equal}$$

Note that we are using $\mu_{bj}$ to represent the mean of the $j$th block.

We can test the blocking hypothesis by examining the $F$-ratio formed by $MSBL/MSW$ in the ANOVA table. If the $F$-ratio is large, it implies that the blocks had a large effect on the response variable and that they were probably helpful in improving the precision of the $F$-test for the primary factor's means. The output in Figure 12.7 provides the $F$-value and $p$-value for this test to determine if the blocking was a necessity. Because $F = 156.13 > F_{0.05} = 3.84$, we conclude that the blocking design was necessary.

If a hypothesis test indicates that blocking is not necessary, then the chance of a Type II error for the primary hypothesis has been unnecessarily increased by the use of blocking. The reason is that by blocking, we not only partition the sum of squares but also partition the degrees of freedom. Therefore, the denominator of $MSW$ is decreased, and $MSW$ will most likely increase. If blocking isn't needed, the $MSW$ will tend to be relatively larger than if we had run a one-way design with independent samples. This can lead to "accepting" the null hypothesis for the primary test when it actually should have been rejected.

Therefore, if blocking is indicated to be unnecessary, follow these rules:

1. If the primary $H_0$ is rejected, proceed with your analysis and decision making. There is no concern.
2. If the primary $H_0$ is not rejected, redo the study without using blocking. Run a one-way ANOVA with independent samples.

**OUTCOME** **4**    **EXAMPLE 12-3**   **Performing a Randomized Block Analysis of Variance**

**Online Course Exams**   A business statistics professor at a Philadelphia, Pennsylvania, university has developed an online course for business statistics that is available through the school's Division of Extended Studies. To reduce the possibility of students sharing exam questions with one another, the professor has developed three different midterm exams that are to be graded on a 1,000-point scale. However, before she uses the exams in a live class, she wants to determine if the tests will yield the same mean scores. To test this, she selects a random sample of 14 students who have been through the course material. Each student will take each test. The order in which the tests are taken is randomized and the scores are recorded. A randomized block analysis of variance test can be performed using the following steps:

STEP 1    **Specify the parameter of interest and formulate the appropriate null and alternative hypotheses.**

The parameter of interest is the mean test score for the three different exams, and the question is whether there is a difference among the mean scores for the three. The appropriate null and alternative hypotheses are

$$H_0: \mu_1 = \mu_2 = \mu_3$$
$$H_A: \text{At least two populations have different means}$$

In this case, the professor wants to control for variation in student ability by having the same students take all three tests. The test scores will be independent because the scores achieved by one student do not influence the scores achieved by other students. Here, the students are the blocks.

STEP 2    **Specify the level of significance for conducting the tests.**
The tests will be conducted using $\alpha = 0.05$.

STEP 3    **Select simple random samples from each population and compute treatment means, block means, and the grand mean.**
The following sample data were observed:

| Student | Exam 1 | Exam 2 | Exam 3 | Block Means |
|---------|--------|--------|--------|-------------|
| 1 | 830 | 647 | 630 | 702.33 |
| 2 | 743 | 840 | 786 | 789.67 |
| 3 | 652 | 747 | 730 | 709.67 |
| 4 | 885 | 639 | 617 | 713.67 |
| 5 | 814 | 943 | 632 | 796.33 |
| 6 | 733 | 916 | 410 | 686.33 |
| 7 | 770 | 923 | 727 | 806.67 |
| 8 | 829 | 903 | 726 | 819.33 |
| 9 | 847 | 760 | 648 | 751.67 |
| 10 | 878 | 856 | 668 | 800.67 |
| 11 | 728 | 878 | 670 | 758.67 |
| 12 | 693 | 990 | 825 | 836.00 |
| 13 | 807 | 871 | 564 | 747.33 |
| 14 | 901 | 980 | 719 | 866.67 |
| Treatment means | 793.57 | 849.50 | 668.00 | 770.36 = Grand mean |

STEP 4    **Use Excel's Two Factor Without Replication procedure in the Data Analysis tools to compute the sums of squares and complete the ANOVA table.**
Four sums of squares are required:

**Total Sum of Squares** (Equation 12.2)

$$SST = \sum_{i=1}^{k} \sum_{j=1}^{n_i} (x_{ij} - \bar{\bar{x}})^2 = 614{,}641.6$$

**Sum of Squares Between** (Equation 12.3)

$$SSB = \sum_{i=1}^{k} n_i (\bar{x}_i - \bar{\bar{x}})^2 = 241{,}912.7$$

**Sum of Squares for Blocking** (Equation 12.8)

$$SSBL = \sum_{j=1}^{b} k(\bar{x}_j - \bar{\bar{x}})^2 = 116{,}605.0$$

**Sum of Squares Within** (Equation 12.9)

$$SSW = SST - (SSB + SSBL) = 256{,}123.9$$

Here is the ANOVA table (see Table 12.6 format):

| Source | SS | df | MS | F-ratio |
|---|---|---|---|---|
| Between blocks | 116,605.0 | 13 | 8,969.6 | 0.9105 |
| Between samples | 241,912.7 | 2 | 120,956.4 | 12.2787 |
| Within samples | 256,123.9 | 26 | 9,850.9 | |
| Total | 614,641.6 | 41 | | |

STEP 5 **Test to determine whether blocking is effective.**

Fourteen students were used to evaluate the three tests. These students constitute the blocks, so if blocking is effective, the mean test scores across the three tests will not be the same for all 14 students. The null and alternative hypotheses are

$$H_0: \mu_{b1} = \mu_{b2} = \mu_{b3} = \cdots = \mu_{b14}$$
$$H_A: \text{Not all means are equal (blocking is effective)}$$

As shown in Step 3, the $F$-test statistic to test this null hypothesis is formed by

$$F = \frac{MSBL}{MSW} = \frac{8{,}969.6}{9{,}850.9} = 0.9105$$

> The Excel 2016 function for the critical $F$-value is
>
> = F.INV.RT(alpha,D1,D2)
> = F.INV.RT(0.05,13,26)

The critical $F$-value from the $F$-distribution, with $\alpha = 0.05$ and $D_1 = 13$ and $D_2 = 26$ degrees of freedom, can be found using the F.INV.RT function in Excel as $F_{0.05} = 2.119$. Then, because

$$F = 0.9105 < F_{\alpha=0.05} = 2.119, \text{do not reject the null hypothesis.}$$

This means that based on these sample data we cannot conclude that blocking was effective.

STEP 6 **Conduct the main hypothesis test to determine whether the populations have equal means.**

We are considering three different statistics exams. At issue is whether the mean scores are equal for the three exams. The appropriate null and alternative hypotheses are

$$H_0: \mu_1 = \mu_2 = \mu_3$$
$$H_A: \text{At least two populations have different means}$$

As shown in the ANOVA table in Step 3, the $F$-test statistic for this null hypothesis is formed by

$$F = \frac{MSB}{MSW} = \frac{120{,}956.4}{9{,}850.9} = 12.2787$$

> The Excel 2016 function for the critical $F$-value is
>
> = F.INV.RT(alpha,D1,D2)
> = F.INV.RT(0.05,2,26)

The critical $F$-value from the $F$-distribution, with $\alpha = 0.05$ and $D_1 = 2$ and $D_2 = 26$ degrees of freedom, can be found using the F.INV.RT function in Excel as $F = 3.369$. Then, because

$$F = 12.2787 > F_{\alpha=0.05} = 3.369, \text{reject the null hypothesis.}$$

Even though in Step 5 we concluded that blocking was not effective, the sample data still lead us to reject the primary null hypothesis and conclude that the three tests do not all have the same mean score. The professor will now be interested in looking into the issue in more detail to determine which tests yield higher or lower average scores.

TRY EXERCISE 12-21 (pg. 462)

**OUTCOME 3**

# Fisher's Least Significant Difference Test

We can use an analysis of variance test to determine whether the populations of interest have different means. However, even if we reject the null hypothesis of equal population means, the ANOVA does not specify which population means are different. In Section 12.1, we showed how to use the Tukey-Kramer multiple comparisons procedure to determine where the population differences occur for a one-way ANOVA design. Likewise, *Fisher's least significant difference test* is one test for multiple comparisons that we can use for a randomized block ANOVA design.

If we reject the primary null hypothesis, then we can compare the absolute differences in sample means from any two populations to the *least significant difference (LSD)*, as computed using Equation 12.10.

**Fisher's Least Significant Difference**

$$LSD = t_{\alpha/2} \sqrt{MSW} \sqrt{\frac{2}{b}}$$

**(12.10)**

where:

$t_{\alpha/2}$ = One-tailed value from Student's $t$-distribution for $\alpha/2$ and $(k - 1)(b - 1)$ degrees of freedom

$MSW$ = Mean square within from ANOVA table

$b$ = Number of blocks

$k$ = Number of levels of the main factor

---

**EXAMPLE 12-4** **Applying Fisher's Least Significant Difference Test**

**Online Course Exams** (*continued*)  Recall that in Example 12-3, the business statistics professor used a randomized block ANOVA design to conclude that the midterm exams do not all have the same mean score. To determine which populations (tests) have different means, you can use the following steps:

**STEP 1** **Compute the *LSD* statistic using Equation 12.10.**

$$LSD = t_{\alpha/2} \sqrt{MSW} \sqrt{\frac{2}{b}}$$

For a significance level equal to 0.05, the critical $t$-value for $(3 - 1)(14 - 1) = 26$ degrees of freedom is

$$t_{0.05/2} = 2.0555$$

The mean square within from the ANOVA table (see Example 12-3, Step 4) is

$$MSW = 9,850.9$$

The *LSD* is

$$LSD = t_{\alpha/2} \sqrt{MSW} \sqrt{\frac{2}{b}} = 2.0555 \sqrt{9,850.9} \sqrt{\frac{2}{14}} = 77.11$$

> The Excel 2016 function for the critical *t*-value is
>
> = T.INV.2T(alpha,df)
> = T.INV.2T(0.025,26)

**STEP 2** **Compute the sample means from each population.**

$$\bar{x}_1 = \frac{\sum x}{n} = 793.57 \qquad \bar{x}_2 = \frac{\sum x}{n} = 849.50 \qquad \bar{x}_3 = \frac{\sum x}{n} = 668$$

**STEP 3** **Form all possible contrasts by finding the absolute differences between all pairs of sample means. Compare these to the *LSD* value.**

| Absolute Difference | Comparison | Significant Difference? |
|---|---|---|
| $\lvert \bar{x}_1 - \bar{x}_2 \rvert = \lvert 793.57 - 849.50 \rvert = 55.93$ | $55.93 < 77.11$ | No |
| $\lvert \bar{x}_1 - \bar{x}_3 \rvert = \lvert 793.57 - 668 \rvert = 125.57$ | $125.57 > 77.11$ | Yes |
| $\lvert \bar{x}_2 - \bar{x}_3 \rvert = \lvert 849.50 - 668 \rvert = 181.50$ | $181.50 > 77.11$ | Yes |

We infer, based on the sample data, that the mean score for Test 1 exceeds the mean for Test 3, and the mean for Test 2 exceeds the mean for Test 3. Now the professor may wish to evaluate Test 3 to see why the scores are lower than for the other two tests. No difference is detected between Tests 1 and 2.

TRY EXERCISE 12-22 (pg. 462)

## 12.2 EXERCISES

### Skill Development

**12-18.** A student club wishes to determine whether there are differences in new textbook prices at on-campus bookstores, off-campus bookstores, and Internet bookstores. To control for differences in textbook prices that might exist across disciplines, suppose the students randomly selected 12 textbooks and recorded the price of each of the 12 books at each of the three retailers. You are given that normality and equal-variance assumptions have been met. The partially completed ANOVA table based on the study's findings is shown here:

#### ANOVA

| Source of Variation | SS | df | MS | F |
|---|---|---|---|---|
| Textbooks | 16,624 | | | |
| Retailer | 2.4 | | | |
| Error | | | | |
| Total | 17,477.6 | | | |

a. Complete the ANOVA table by filling in the missing sums of squares, the degrees of freedom for each source, the mean square, and the calculated $F$-test statistic for each possible hypothesis test.
b. Based on the study's findings, was it correct to block for differences in textbooks? Conduct the appropriate test at the $\alpha = 0.10$ level of significance.
c. Based on the study's findings, can you conclude that there is a difference in the average price of textbooks across the three retail outlets? Conduct the appropriate hypothesis test at the $\alpha = 0.10$ level of significance.

**12-19.** The following data were collected for a randomized block analysis of variance design with four populations and eight blocks:

| | Group 1 | Group 2 | Group 3 | Group 4 |
|---|---|---|---|---|
| **Block 1** | 56 | 44 | 57 | 84 |
| **Block 2** | 34 | 30 | 38 | 50 |
| **Block 3** | 50 | 41 | 48 | 52 |
| **Block 4** | 19 | 17 | 21 | 30 |
| **Block 5** | 33 | 30 | 35 | 38 |
| **Block 6** | 74 | 72 | 78 | 79 |
| **Block 7** | 33 | 24 | 27 | 33 |
| **Block 8** | 56 | 44 | 56 | 71 |

a. State the appropriate null and alternative hypotheses for the treatments and determine whether blocking is necessary.
b. Construct the appropriate ANOVA table.
c. Using a significance level equal to 0.05, can you conclude that blocking was necessary in this case? Use a test-statistic approach.
d. Based on the data and a significance level equal to 0.05, is there a difference in population means for the four groups? Use a $p$-value approach.
e. If you found that a difference exists in part d, use the *LSD* approach to determine which populations have different means.

**12-20.** The following ANOVA table and accompanying information are the result of a randomized block ANOVA test.

| Summary | Count | Sum | Average | Variance |
|---|---|---|---|---|
| 1 | 4 | 443 | 110.8 | 468.9 |
| 2 | 4 | 275 | 68.8 | 72.9 |
| 3 | 4 | 1,030 | 257.5 | 1,891.7 |
| 4 | 4 | 300 | 75.0 | 433.3 |
| 5 | 4 | 603 | 150.8 | 468.9 |
| 6 | 4 | 435 | 108.8 | 72.9 |
| 7 | 4 | 1,190 | 297.5 | 1,891.7 |
| 8 | 4 | 460 | 115.0 | 433.3 |
| Sample 1 | 8 | 1,120 | 140.0 | 7,142.9 |
| Sample 2 | 8 | 1,236 | 154.5 | 8,866.6 |
| Sample 3 | 8 | 1,400 | 175.0 | 9,000.0 |
| Sample 4 | 8 | 980 | 122.5 | 4,307.1 |

ANOVA

| Source of Variation | SS | df | MS | F | p-value | F-crit |
|---|---|---|---|---|---|---|
| Rows | 199,899 | 7 | 28,557.0 | 112.8 | 0.0000 | 2.488 |
| Columns | 11,884 | 3 | 3,961.3 | 15.7 | 0.0000 | 3.073 |
| Error | 5,317 | 21 | 253.2 | | | |
| Total | 217,100 | 31 | | | | |

    a. How many blocks were used in this study?
    b. How many populations are involved in this test?
    c. Test to determine whether blocking is effective using an alpha level equal to 0.05.
    d. Test the main hypothesis of interest using $\alpha = 0.05$.
    e. If warranted, conduct an *LSD* test with $\alpha = 0.05$ to determine which population means are different.

**12-21.** The following sample data were recently collected in the course of conducting a randomized block analysis of variance. Based on these sample data, what conclusions should you reach about blocking effectiveness and about the means of the three populations involved? Test using a significance level equal to 0.05.

| Block | Sample 1 | Sample 2 | Sample 3 |
|---|---|---|---|
| 1 | 30 | 40 | 40 |
| 2 | 50 | 70 | 50 |
| 3 | 60 | 40 | 70 |
| 4 | 40 | 40 | 30 |
| 5 | 80 | 70 | 90 |
| 6 | 20 | 10 | 10 |

**12-22.** A randomized complete block design is carried out, resulting in the following statistics:

| Source | $\bar{x}_1$ | $\bar{x}_2$ | $\bar{x}_3$ | $\bar{x}_4$ |
|---|---|---|---|---|
| Primary Factor | 237.15 | 315.15 | 414.01 | 612.52 |
| Block | 363.57 | 382.22 | 438.33 | |

$SST = 364,428$

    a. Determine if blocking was effective for this design.
    b. Using a significance level of 0.05, produce the relevant ANOVA and determine if the average responses of the factor levels are equal to each other.
    c. If you discovered that there were differences among the average responses of the factor levels, use the *LSD* approach to determine which populations have different means.

## Business Applications

**12-23.** The Goodson Company manufactures four different products that it ships to customers throughout Canada.

Breakage cost is very expensive, and Goodson would like to select a mode of delivery that reduces the amount of product breakage. The managers have decided to examine the dollar amounts of breakage incurred by the three alternative modes of transportation under consideration. The executives conducting the study wish to control for differences due to type of product. The company randomly assigns each product to each carrier and monitors the dollar amounts of breakage that occur over the course of 100 shipments. The dollar amounts of breakage per shipment (to the nearest dollar) are as follows:

| | Rail | Plane | Truck |
|---|---|---|---|
| **Product 1** | $7,960 | $8,053 | $8,818 |
| **Product 2** | $8,399 | $7,764 | $9,432 |
| **Product 3** | $9,429 | $9,196 | $9,260 |
| **Product 4** | $6,022 | $5,821 | $5,676 |

    a. Was the Goodson Company correct in its decision to block for type of product? Conduct the appropriate hypothesis test using a level of significance of 0.01.
    b. Is there a difference due to carrier type? Conduct the appropriate hypothesis test using a level of significance of 0.01.

**12-24.** Suppose the California Lettuce Research Board conducted a study that followed a randomized complete block design where variety of lettuce (Salinas and Sniper) was the main factor and salinity levels (SAR) were the blocks. The measurements (the number of lettuce heads from each plot) of the kind observed were

| SAR | Salinas | Sniper |
|---|---|---|
| 3 | 104 | 109 |
| 5 | 160 | 163 |
| 7 | 142 | 146 |
| 10 | 133 | 156 |

    a. Determine if blocking was effective for this design.
    b. Using a significance level of 0.05, produce the relevant ANOVA and determine if the average number of lettuce heads among the SARs are equal to each other.
    c. If you discovered that there were differences among the average number of lettuce heads among the SARs, use the *LSD* approach to determine which populations have different means.

**12-25.** D&G Industries operates three shifts every day of the week. Each shift includes full-time hourly workers, nonsupervisory salaried employees, and supervisors/ managers. D&G Industries would like to know if there is a difference among the shifts in terms of the number of hours of work missed due to employee illness. To control for differences that might exist across employee groups, D&G Industries randomly selects one employee from each employee group and shift and records the number of hours missed for one year. The results of the study are shown here:

|  | Shift 1 | Shift 2 | Shift 3 |
|---|---|---|---|
| Hourly | 48 | 54 | 60 |
| Nonsupervisory | 31 | 36 | 55 |
| Supervisors/Managers | 25 | 33 | 40 |

a. Develop the appropriate test to determine whether blocking is effective or not. Conduct the test at the $\alpha = 0.05$ level of significance.
b. Develop the appropriate test to determine whether there are differences in the average number of hours missed due to illness across the three shifts. Conduct the test at the $\alpha = 0.05$ level of significance.
c. If it is determined that the average hours of work missed due to illness are not the same for the three shifts, use the *LSD* approach to determine which shifts have different means.

12-26. Assume that a large accounting firm provides auditing training for its employees in three different auditing methods. Auditors were grouped into four blocks according to the education they had received: (1) high school, (2) bachelor's, (3) master's, and (4) doctorate. Three auditors at each education level were used—one assigned to each method. They were given a post-training examination consisting of complicated auditing scenarios. The scores for the 12 auditors were as follows:

|  | Method 1 | Method 2 | Method 3 |
|---|---|---|---|
| Doctorate | 83 | 81 | 82 |
| Master's | 77 | 75 | 79 |
| Bachelor's | 74 | 73 | 75 |
| High School | 72 | 70 | 69 |

a. Indicate why blocking was employed in this design.
b. Determine if blocking was effective for this design by producing the relevant ANOVA.
c. Using a significance level of 0.05, determine if the average post-training examination scores among the auditing methods are equal to each other.
d. If you discovered that there were differences in the average post-training examination scores among the auditing methods, use the *LSD* approach to determine which populations have different means.

## Computer Software Exercises

12-27. Applebee's International, Inc., is the largest chain of casual dining restaurants in the country, with more than

2,000 restaurants across the United States (*source: Applebees.com*). Suppose the company is interested in determining if mean weekly revenues differ among three restaurants in a particular city. The file titled **Applebees** contains revenue data for a sample of weeks for each of the three locations.

a. Test to determine if blocking the week on which the testing was done was necessary. Use a significance level of 0.05.
b. Based on the data gathered by Applebee's, can it be concluded that there is a difference in the average revenues among the three restaurants?
c. If you did conclude that there was a difference in the average revenues, use Fisher's *LSD* approach to determine which restaurant has the lowest mean sales.

12-28. In a local community, there are three grocery chain stores. A local news station recently sent a reporter to the three stores to check prices on several items. She found that each store had the lowest price for certain specific items. The station set up a test in which 20 shoppers were given a list of grocery items and were sent to each of the three chain stores. The sales receipts from each of the three stores are recorded in the data file **Groceries**.

a. Why should this price test be conducted using the design that the television station used? What was it attempting to achieve by having the same shopping lists used at each of the three grocery stores?
b. Based on a significance level of 0.05 and these sample data, test to determine whether blocking was necessary in this example. State the null and alternative hypotheses. Use a test-statistic approach.
c. Based on these sample data, can you conclude the three grocery stores have different sample means? Test using a significance level of 0.05. State the appropriate null and alternative hypotheses. Use a *p*-value approach.
d. Based on the sample data, which store has the highest average prices? Use Fisher's *LSD* test if appropriate.

12-29. Although natural fibers like manila, sisal, and cotton were once the predominant rope materials, industrial synthetic fibers dominate the marketplace today, with most ropes made of nylon, polyester, or polypropylene. One of the principal traits of rope material is its breaking strength. Assume that a research project generated the data given in the file titled **Knots**. The data listed were gathered on 10 different days from $\frac{1}{2}''$-diameter ropes.

a. Test to determine if inserting the day on which the testing was done was necessary. Use a significance level of 0.05.
b. Based on the data, can it be concluded that there is a difference in the average breaking strengths of nylon, polyester, and polypropylene?
c. If you concluded that there was a difference in the average breaking strength of the rope material, use

Fisher's *LSD* approach to determine which material has the highest breaking strength.

**12-30.** Students in a business statistics class were asked to do a project in which a "market basket" of food products was selected at random from those items offered in three grocery stores in Boise, Idaho: Walmart, WinCo, and Albertsons. At issue was whether the mean prices at the three stores are equal or whether there is a difference in prices. The sample data are in the data file called **Food Price Comparisons**. Using an alpha level equal to 0.05, test to determine whether the three stores have equal population mean prices. If you conclude that there are differences in the mean prices, perform the appropriate post-test to determine which stores have different means.

---

OUTCOME **5**     **12.3**     ### Two-Factor Analysis of Variance with Replication

Section 12.2 introduced an ANOVA procedure called the randomized complete block ANOVA. We use this method when we are interested in testing whether the means for the populations (levels) for a factor of interest are equal and we want to control for potential variation due to a second factor. The second factor is called the blocking factor. Consider again the Citizen's State Bank property appraisal application, in which the bank managers wanted to determine whether the mean property valuations were the same for three different appraisal companies. The company used the same five properties to test each appraisal company in an attempt to reduce any variability that might exist due to the properties involved in the test. The properties were the blocks in that example, but we were not really interested in knowing whether the mean appraisals were the same for all properties. The single factor of interest was the appraisal companies.

However, you will encounter many situations in which there are actually two or more factors of interest in the same study. In this section, we limit our discussion to situations involving only two factors. The technique that we use when we wish to analyze two factors is called *two-factor ANOVA with replications*.

## Two-Factor ANOVA with Replications

**BUSINESS APPLICATION**     ### Using Software for Two-Factor ANOVA

**Excel Tutorial**

**Fly High Airlines**     Like other major U.S. airlines, Fly High Airlines is concerned because many of its frequent flier program members have accumulated large quantities of free miles.[5] The airline worries that at some point in the future, there will be a big influx of customers wanting to use their miles and the airline will have difficulty satisfying all the requests at once. Suppose Fly High conducted an experiment in which each of three methods for redeeming frequent flier miles was offered to a sample of 16 customers. Each customer had accumulated more than 100,000 frequent flier miles. The customers were equally divided into four age groups. The variable of interest was the number of miles redeemed by the customers during the six-week trial. Table 12.7 shows the number of miles redeemed for each person. These data are also contained in the **Fly High** file.

Method 1 offered cash inducements to use miles, Method 2 offered discount vacation options, and Method 3 offered access to a discount-shopping program through the Internet. The airline wants to know if the mean numbers of miles redeemed under the three redemption methods are equal and whether the mean miles redeemed are the same across the four age groups.

A *two-factor ANOVA* design is the appropriate method to use in this case because the airline has two factors of interest. Factor A is the redemption offer type with three levels. Factor B is the age group of each customer with four levels. As shown in Table 12.7, there are $3 \times 4 = 12$ cells in the study and four customers in each cell. The measurements are called *replications* because we get four measurements (miles redeemed) at each combination of redemption offer level (Factor A) and age level (Factor B).

Two-factor ANOVA follows the same logic as all other ANOVA designs. Each factor of interest introduces variability into the experiment. As was the case in Sections 12.1 and 12.2,

---

[5]Name changed at request of the airline.

**TABLE 12.7** Fly High Airlines Frequent Flier Miles Data

|  | Cash Option | Vacation | Shopping |
|---|---|---|---|
| **Under 25 years** | 30,000 | 40,000 | 25,000 |
|  | 0 | 25,000 | 25,000 |
|  | 25,000 | 0 | 75,000 |
|  | 0 | 0 | 5,000 |
| **25 to 40 years** | 60,000 | 40,000 | 30,000 |
|  | 0 | 25,000 | 25,000 |
|  | 0 | 5,000 | 50,000 |
|  | 25,000 | 25,000 | 0 |
| **41 to 60 years** | 40,000 | 25,000 | 25,000 |
|  | 25,000 | 50,000 | 50,000 |
|  | 25,000 | 0 | 0 |
|  | 0 | 25,000 | 0 |
| **Over 60 years** | 0 | 45,000 | 30,000 |
|  | 5,000 | 25,000 | 25,000 |
|  | 25,000 | 0 | 25,000 |
|  | 50,000 | 50,000 | 50,000 |

we must find estimators for each source of variation. Identifying the appropriate sums of squares and then dividing each by its degrees of freedom does this. As in the one-way ANOVA, the total sum of squares (*SST*) in two-factor ANOVA can be partitioned. The *SST* is partitioned into four parts as follows:

1. One part is due to differences in the levels of Factor A ($SS_A$).
2. Another part is due to the levels of Factor B ($SS_B$).
3. Another part is due to the *interaction* between Factor A and Factor B ($SS_{AB}$). (We will discuss the concept of interaction between factors later.)
4. The final component making up the total sum of squares is the sum of squares due to the inherent random variation in the data (*SSE*).

Figure 12.9 illustrates this partitioning concept. The variations due to each of these components will be estimated using the respective *mean squares* obtained by dividing the sums of squares by their degrees of freedom. If the variation accounted for by Factor A and Factor B is large relative to the error variation, we will tend to conclude that the factor levels have different means.

Table 12.8 illustrates the format of the two-factor ANOVA. We can test three different hypotheses from the information in this ANOVA table. First, for Factor A (redemption options), we have

$$H_0: \mu_{A1} = \mu_{A2} = \mu_{A3}$$
$$H_A: \text{Not all Factor A means are equal}$$

**FIGURE 12.9** **Two-Factor ANOVA—Partitioning of Total Sums of Squares**

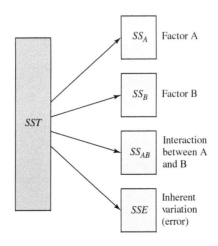

For Factor B (age levels):

$$H_0: \mu_{B1} = \mu_{B2} = \mu_{B3} = \mu_{B4}$$
$$H_A: \text{Not all Factor B means are equal}$$

**TABLE 12.8 Basic Format of the Two-Factor ANOVA Table**

| Source of Variation | SS | df | MS | F-ratio |
|---|---|---|---|---|
| Factor A | $SS_A$ | $a - 1$ | $MS_A$ | $\dfrac{MS_A}{MSE}$ |
| Factor B | $SS_B$ | $b - 1$ | $MS_B$ | $\dfrac{MS_B}{MSE}$ |
| AB interaction | $SS_{AB}$ | $(a - 1)(b - 1)$ | $MS_{AB}$ | $\dfrac{MS_{AB}}{MSE}$ |
| Error | SSE | $n_T - ab$ | $MSE$ | |
| Total | SST | $n_T - 1$ | | |

where:

$$a = \text{Number of levels of Factor A}$$
$$b = \text{Number of levels of Factor B}$$
$$n_T = \text{Total number of observations in all cells}$$
$$MS_A = \text{Mean square Factor A} = \frac{SS_A}{a - 1}$$
$$MS_B = \text{Mean square Factor B} = \frac{SS_B}{b - 1}$$
$$MS_{AB} = \text{Mean square interaction} = \frac{SS_{AB}}{(a - 1)(b - 1)}$$
$$MSE = \text{Mean square error} = \frac{SSE}{n_T - ab}$$

We test to determine whether interaction exists between the two factors:

$H_0$: Factors A and B do not interact to affect the mean response

$H_A$: Factors A and B do interact

Here are the assumptions required to use two-factor ANOVA:

**Assumptions**

1. The population values for each combination of pairwise factor levels are normally distributed.
2. The variances for each population are equal.
3. The samples are independent.
4. The data measurement is interval or ratio level.

Although we could manually compute all the necessary values to complete Table 12.8 using the equations shown in Table 12.9, this would be a time-consuming task for even a small example because the equations for the various sum-of-squares values are quite complicated. Instead, you will want to use software such as Excel to perform the two-factor ANOVA.

**Interaction Explained** Before we share the ANOVA results for the Fly High Airlines example, a few comments regarding the concept of *factor interaction* are needed. Consider our example involving the two factors: miles-redemption-offer type and age category of customer. The response variable is the number of miles redeemed in the six weeks after the offer. Suppose one redemption-offer type is really better and results in a higher average miles being redeemed. If there is no interaction between age and offer type, then customers of all ages will have uniformly higher average miles redeemed for this offer type compared with the other

**TABLE 12.9** Two-Factor ANOVA Equations

Total Sum of Squares

$$SST = \sum_{i=1}^{a} \sum_{j=1}^{b} \sum_{k=1}^{n} (x_{ijk} - \bar{\bar{x}})^2 \tag{12.11}$$

Sum of Squares Factor A

$$SS_A = bn \sum_{i=1}^{a} (\bar{x}_{i..} - \bar{\bar{x}})^2 \tag{12.12}$$

Sum of Squares Factor B

$$SS_B = an \sum_{j=1}^{b} (\bar{x}_{.j.} - \bar{\bar{x}})^2 \tag{12.13}$$

Sum of Squares Interaction between Factors A and B

$$SS_{AB} = n \sum_{i=1}^{a} \sum_{j=1}^{b} (\bar{x}_{ij.} - \bar{x}_{i..} - \bar{x}_{.j.} + \bar{\bar{x}})^2 \tag{12.14}$$

Sum of Squares Error

$$SSE = \sum_{i=1}^{a} \sum_{j=1}^{b} \sum_{k=1}^{n} (x_{ijk} - \bar{x}_{ij.})^2 \tag{12.15}$$

where:

$$\bar{\bar{x}} = \frac{\sum_{i=1}^{a} \sum_{j=1}^{b} \sum_{k=1}^{n} x_{ijk}}{abn} = \text{Grand mean}$$

$$\bar{x}_{i..} = \frac{\sum_{j=1}^{b} \sum_{k=1}^{n} x_{ijk}}{bn} = \text{Mean of each level of Factor A}$$

$$\bar{x}_{.j.} = \frac{\sum_{i=1}^{a} \sum_{k=1}^{n} x_{ijk}}{an} = \text{Mean of each level of Factor B}$$

$$\bar{x}_{ij.} = \sum_{k=1}^{n} \frac{x_{ijk}}{n} = \text{Mean of each cell}$$

$a$ = Number of levels of Factor A

$b$ = Number of levels of Factor B

$n$ = Number of replications in each cell

offer types. If another offer type yields lower average miles, and if there is no interaction, all age groups receiving this offer type will redeem uniformly lower miles on average than the other offer types. Figure 12.10 illustrates a situation with no interaction between the two factors.

**FIGURE 12.10** **Differences between Factor-Level Mean Values: No Interaction**

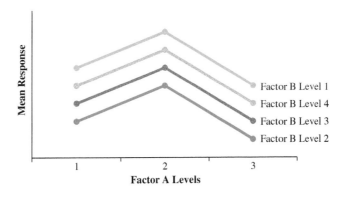

**FIGURE 12.11 Differences between Factor-Level Mean Values: Interaction Is Present**

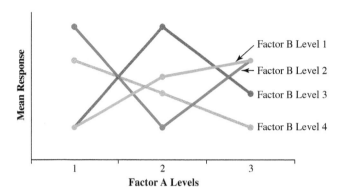

However, if interaction exists between the factors, we would see a graph similar to the one shown in Figure 12.11. Interaction would be indicated if one age group redeemed higher average miles than the other age groups with one program but lower average miles than the other age groups on the other mileage-redemption programs. In general, interaction occurs if the differences in the averages of the response variable for the various levels of one factor— say, Factor A—are not the same for each level of the other factor—say, Factor B. The general idea is that interaction between two factors means that the effect due to one of them is not uniform across all levels of the other factor.

Another example in which potential interaction might exist occurs in plywood manufacturing, where thin layers of wood called veneer are glued together to form plywood. One of the important qualities of plywood is its strength. However, plywood is made from different species of wood (pine, fir, hemlock, etc.), and different types of glue are available. If some species of wood work better (stronger plywood) with certain glues, whereas other species work better with different glues, we say that the wood species and the glue type interact.

If interaction is suspected, it should be accounted for by subtracting the interaction term ($SS_{AB}$) from the total sum-of-squares term in the ANOVA. From a strictly arithmetic point of view, the effect of computing $SS_{AB}$ and subtracting it from $SST$ is that $SSE$ is reduced. Also, if the corresponding variation due to interaction is significant, the variation within the factor levels (error) is significantly reduced. This can make it easier to detect a difference in the population means if such a difference actually exists. If so, $MSE$ will most likely be reduced. This will produce a larger $F$-test statistic, which will more likely lead to correctly rejecting the null hypothesis. Thus, considering potential interaction improves your chances of finding a difference in the Factor A and Factor B mean values, if such a difference exists. This will depend, of course, on the relative size of $SS_{AB}$ and the respective changes in the degrees of freedom. We will comment later on the appropriateness of testing the factor hypotheses if interaction is present. *Note that to measure the interaction effect, the sample size for each combination of Factor A and Factor B must be 2 or greater (replicated design).*

**FIGURE 12.12 Excel 2016 Data Format for Two-Factor ANOVA for Fly High Airlines**

Excel 2016 Instructions
1. Open file: **Fly High.xlsx**.

| | A | B<br>Cash<br>Option | C<br>Vacation | D<br>Shopping | |
|---|---|---|---|---|---|
| 1 | | | | | ← Factor A Names |
| 2 | under 25 yrs | 30,000 | 40,000 | 25,000 | |
| 3 | | 0 | 25,000 | 25,000 | |
| 4 | | 25,000 | 0 | 75,000 | |
| 5 | | 0 | 0 | 5,000 | |
| 6 | 25-40 yrs | 60,000 | 40,000 | 30,000 | |
| 7 | | 0 | 25,000 | 25,000 | |
| 8 | | 0 | 5,000 | 50,000 | |
| 9 | | 25,000 | 25,000 | 0 | |
| 10 | 41-60 yrs | 40,000 | 25,000 | 25,000 | |
| 11 | | 25,000 | 50,000 | 50,000 | |
| 12 | | 25,000 | 0 | 0 | |
| 13 | | 0 | 25,000 | 0 | |
| 14 | over 60 yrs | 0 | 45,000 | 30,000 | |
| 15 | | 5,000 | 25,000 | 25,000 | |
| 16 | | 25,000 | 0 | 25,000 | |
| 17 | | 50,000 | 50,000 | 50,000 | |

Factor B Names

Excel has a tool for performing two-factor ANOVA with replications. It can be used to compute the different sums of squares and complete the ANOVA table. However, Excel requires that the data be organized in a special way, as shown in Figure 12.12. (Note that Excel's first row must contain the names for each level of Factor A. Also, Excel's column A contains the Factor B level names. These must be in the row corresponding to the first sample item value for each Factor B level.)

The Excel two-factor ANOVA output for this example is actually too big to fit on one screen. The top portion of the printout shows summary information for each cell, including means and variances (see Figure 12.13). At the bottom of the output (scroll down) is the ANOVA table shown in Figure 12.14. Excel changes a few labels. For example, Factor A (the miles redemption options) is now referred to as *Columns*. Factor B (age groups) is referred to as *Sample*. In Figure 12.14, we see all the information necessary to test whether the three redemption offers (Factor A) result in different mean miles redeemed.

$$H_0: \mu_{A1} = \mu_{A2} = \mu_{A3}$$
$$H_A: \text{Not all Factor A means are equal}$$
$$\alpha = 0.05$$

Both the $p$-value and $F$-distribution approaches can be used. Because

$$p\text{-value (columns)} = 0.56 > \alpha = 0.05$$

we do not reject the null hypothesis $H_0$. (Also, $F = 0.59 < F_{0.05} = 3.259$; the null hypothesis is not rejected.) This means the test data do not indicate that a difference exists between the average numbers of miles redeemed for the three types of offers. None seems superior to the others.

We can also test to determine if age level makes a difference in frequent flier miles redeemed:

$$H_0: \mu_{B1} = \mu_{B2} = \mu_{B3} = \mu_{B4}$$
$$H_A: \text{Not all Factor B means are equal}$$
$$\alpha = 0.05$$

In Figure 12.14, we see that the

$$p\text{-value} = 0.88 > \alpha = 0.05$$
$$(\text{Also, } F = 0.22 < F_{0.05} = 2.866)$$

**FIGURE 12.13 Excel 2016 Output (Part 1) for Two-Factor ANOVA for Fly High Airlines**

Excel 2016 Instructions

1. Open file: **Fly High.xlsx**.
2. Select **Data > Data Analysis**.
3. Select **ANOVA: Two Factor With Replication**.
4. Define data range (include Factor A and B labels).
5. Specify the number of **rows per sample:** 4.
6. Specify **Alpha** equal 0.05.
7. Indicate output range.

| | A | B | C | D | E |
|---|---|---|---|---|---|
| 1 | Anova: Two-Factor With Replication | | | | |
| 2 | | | | | |
| 3 | SUMMARY | Cash Option | Vacation | Shopping | Total |
| 4 | <25 years | | | | |
| 5 | Count | 4 | 4 | 4 | 12 |
| 6 | Sum | 55,000 | 65,000 | 130,000 | 250,000 |
| 7 | Average | 13,750 | 16,250 | 32,500 | 20,833.33 |
| 8 | Variance | 256,250,000 | 389,583,333 | 891,666,667 | 494,696,970 |
| 9 | | | | | |
| 10 | 25-40 years | | | | |
| 11 | Count | 4 | 4 | 4 | 12 |
| 12 | Sum | 85,000 | 95,000 | 105,000 | 285,000 |
| 13 | Average | 21,250 | 23,750 | 26,250 | 23,750 |
| 14 | Variance | 806,250,000 | 206,250,000 | 422,916,667 | 396,022,727 |
| 15 | | | | | |
| 16 | 41-60 years | | | | |
| 17 | Count | 4 | 4 | 4 | 12 |
| 18 | Sum | 90,000 | 100,000 | 75,000 | 265,000 |
| 19 | Average | 22,500 | 25,000 | 18,750 | 22,083.33 |
| 20 | Variance | 275,000,000 | 416,666,667 | 572,916,667 | 352,083,333 |
| 21 | | | | | |
| 22 | over 60 years | | | | |
| 23 | Count | 4 | 4 | 4 | 12 |
| 24 | Sum | 80,000 | 120,000 | 130,000 | 330,000 |
| 25 | Average | 20,000 | 30,000 | 32,500 | 27,500 |
| 26 | Variance | 516,666,667 | 516,666,667 | 141,666,667 | 352,272,727 |
| 27 | | | | | |
| 28 | Total | | | | |
| 29 | Count | 16 | 16 | 16 | |
| 30 | Sum | 310,000 | 380,000 | 440,000 | |
| 31 | Average | 19,375 | 23,750 | 27,500 | |
| 32 | Variance | 382,916,667 | 331,666,667 | 440,000,000 | |

FIGURE 12.14 **Excel 2016 Output (Part 2) for Two-Factor ANOVA for Fly High Airlines**

| 35 | ANOVA | | | | | | |
|---|---|---|---|---|---|---|---|
| 36 | *Source of Variation* | *SS* | *df* | *MS* | *F* | *P-value* | *F crit* |
| 37 | Sample | 302,083,333 | 3 | 100,694,444 | 0.22 | 0.88 | 2.866 |
| 38 | Columns | 529,166,667 | 2 | 264,583,333 | 0.59 | 0.56 | 3.259 |
| 39 | Interaction | 779,166,667 | 6 | 129,861,111 | 0.29 | 0.94 | 2.364 |
| 40 | Within | 16,237,500,000 | 36 | 451,041,667 | | | |
| 41 | Total | 17,847,916,667 | 47 | | | | |

Excel terminology:
Sample = Factor B (age)
Columns = Factor A (program)
Within = Error

F-statistics and p-values for testing the three different hypotheses of interest in the two-factor test

Thus, we do not reject the null hypothesis. The test data do not indicate that customer age significantly influences the average number of frequent flier miles that will be redeemed.

Finally, we can also test for interaction. The null hypothesis is that no interaction exists. The alternative is that interaction does exist between the two factors. The ANOVA table in Figure 12.14 shows a $p$-value of 0.94, which is greater than $\alpha = 0.05$. Based on these data, interaction between the two factors does not appear to exist. This would indicate that the differences in the average mileage redeemed between the various age categories are the same for each redemption-offer type.

## A Caution about Interaction

In this example, the sample data indicate that no interaction between Factors A and B is present. Based on the sample data, we were unable to conclude that the three redemption offers resulted in different average frequent flier miles redeemed. Finally, we were unable to conclude that a difference in average miles redeemed occurred over the four different age groups.

The appropriate approach is to first test for interaction. If we do not reject the interaction null hypothesis, we proceed to test the Factor A and Factor B hypotheses. However, if we conclude that interaction is present between the two factors, hypothesis tests for Factors A and B generally *should not* be performed. The reason is that findings of significance for either factor might be due only to interactive effects when the two factors are combined and not to the fact that the levels of the factor differ significantly. It is also possible that interactive effects might mask differences between means of one of the factors for at least some of the levels of the other factor. If significant interaction is present, the experimenter may conduct a one-way ANOVA to test the levels of one of the factors—for example, Factor A—using only one level of the other factor—Factor B.

Thus, when conducting hypothesis tests for a two-factor ANOVA:

1. Test for interaction.
2. If interaction is present, conduct a one-way ANOVA to test the levels of one of the factors using only one level of the other factor.
3. If no interaction is found, test Factor A and Factor B.

## 12.3 EXERCISES

### Skill Development

**12-31.** Consider the following data from a two-factor experiment:

| | Factor A | | |
|---|---|---|---|
| Factor B | Level 1 | Level 2 | Level 3 |
| Level 1 | 43 | 25 | 37 |
| | 49 | 26 | 45 |
| Level 2 | 50 | 27 | 46 |
| | 53 | 31 | 48 |

a. Determine if there is interaction between Factor A and Factor B. Use the $p$-value approach and a significance level of 0.05.
b. Does the average response vary among the levels of Factor A? Use the test-statistic approach and a significance level of 0.05.
c. Determine if there are differences in the average response between the levels of Factor B. Use the $p$-value approach and a significance level of 0.05.

**12-32.** Examine the following two-factor analysis of variance table:

| Source | SS | df | MS | F-ratio |
|---|---|---|---|---|
| Factor A | 162.79 | 4 | | |
| Factor B | | | 28.12 | |
| AB Interaction | 262.31 | 12 | | |
| Error | _____ | — | | |
| Total | 1,298.74 | 84 | | |

   a. Complete the analysis of variance table.
   b. Determine if interaction exists between Factor A and Factor B. Use $\alpha = 0.05$.
   c. Determine if the levels of Factor A have equal means. Use a significance level of 0.05.
   d. Does the ANOVA table indicate that the levels of Factor B have equal means? Use a significance level of 0.05.

**12-33.** Consider the following data for a two-factor experiment:

| | Factor A | | |
|---|---|---|---|
| Factor B | Level 1 | Level 2 | Level 3 |
| Level 1 | 33 | 30 | 21 |
| | 31 | 42 | 30 |
| | 35 | 36 | 30 |
| Level 2 | 23 | 30 | 21 |
| | 32 | 27 | 33 |
| | 27 | 25 | 18 |

   a. Based on the sample data, do Factors A and B have significant interaction? State the appropriate null and alternative hypotheses and test using a significance level of 0.05.
   b. Based on these sample data, can you conclude that the levels of Factor A have equal means? Test using a significance level of 0.05.
   c. Do the data indicate that the levels of Factor B have different means? Test using a significance level equal to 0.05.

**12-34.** Consider the following partially completed two-factor analysis of variance table, which is an outgrowth of a study in which Factor A has four levels and Factor B has three levels. The number of replications was 11 in each cell.

| Source of Variation | SS | df | MS | F-ratio |
|---|---|---|---|---|
| Factor A | 345.1 | 4 | | |
| Factor B | | | 28.12 | |
| AB Interaction | 1,123.2 | 12 | | |
| Error | 256.7 | | | |
| Total | 1,987.3 | 84 | | |

   a. Complete the analysis of variance table.
   b. Based on the sample data, can you conclude that the two factors have significant interaction? Test using a significance level equal to 0.05.
   c. Based on the sample data, should you conclude that the means for Factor A differ across the four levels or the means for Factor B differ across the three levels? Discuss.
   d. Considering the outcome of part b, determine what can be said concerning the differences of the levels of Factors A and B. Use a significance level of 0.10 for any hypothesis tests required. Provide a rationale for your response to this question.

**12-35.** A two-factor experiment yielded the following data:

| | Factor A | | |
|---|---|---|---|
| Factor B | Level 1 | Level 2 | Level 3 |
| Level 1 | 375 | 402 | 395 |
| | 390 | 396 | 390 |
| Level 2 | 335 | 336 | 320 |
| | 342 | 338 | 331 |
| Level 3 | 302 | 485 | 351 |
| | 324 | 455 | 346 |

   a. Determine if there is interaction between Factor A and Factor B. Use the $p$-value approach and a significance level of 0.05.
   b. Given your findings in part a, determine any significant differences among the response means of the levels of Factor A for level 1 of Factor B.
   c. Repeat part b at levels 2 and 3 of Factor B, respectively.

## Business Applications

**12-36.** Suppose that a national health club chain investigated how many pounds heavier club members were than their perceived ideal weight. It investigated whether these perceptions differed among different regions of the country and gender of the respondents. The following data (pounds) reflect the survey results:

| | Region | | | |
|---|---|---|---|---|
| Gender | West | Midwest | South | Northeast |
| Men | 14 | 18 | 15 | 16 |
| | 13 | 16 | 15 | 14 |
| Women | 16 | 20 | 17 | 17 |
| | 13 | 18 | 17 | 13 |

   a. Determine if there is interaction between region and gender. Use the $p$-value approach and a significance level of 0.05.

b. Given your findings in part a, determine any significant differences among the discrepancies between the average existing and desired weights in the regions.

c. Repeat part b for the gender factor.

**12-37.** The Lanier Company produces a single product on three production lines that use different equipment. The production manager is considering changing the layouts of the lines and would like to know what effects different layouts would have on production output. Suppose that the manager measured the average output for each line over four randomly selected weeks using each of the three layouts under consideration. The output (in hundreds of units produced) was measured for each line for each of the four weeks for each layout being evaluated. The results of the study are as follows:

|  | Line 1 | Line 2 | Line 3 |
|---|---|---|---|
| **Layout 1** | 12 | 12 | 11 |
|  | 10 | 14 | 10 |
|  | 12 | 10 | 14 |
|  | 12 | 11 | 12 |
| **Layout 2** | 17 | 16 | 18 |
|  | 18 | 15 | 18 |
|  | 15 | 16 | 17 |
|  | 17 | 17 | 18 |
| **Layout 3** | 12 | 10 | 11 |
|  | 12 | 11 | 11 |
|  | 11 | 11 | 10 |
|  | 11 | 11 | 12 |

a. Based on the sample data, can the production manager conclude that there is an interaction effect between the type of layout and the production line? Conduct the appropriate test at the 0.05 level of significance.

b. At the 0.05 level of significance, can the production manager conclude that there is a difference in mean output across the three production lines?

c. At the 0.05 level of significance, can the production manager conclude that there is a difference in mean output due to the type of layout used?

**12-38.** A popular consumer food item was displayed in different locations in the same aisle of a grocery store to determine what, if any, effect different placements might have on its sales. The product was placed at one of three heights on the aisle—low, medium, and high—and at one of three locations in the store—at the front of the store, at the middle of the store, or at the rear of the store. The numbers of units sold of the product at the various height and distance combinations were recorded each week for five weeks. The following results were obtained:

|  | Front | Middle | Rear |
|---|---|---|---|
| **Low** | 125 | 195 | 126 |
|  | 143 | 150 | 136 |
|  | 150 | 160 | 129 |
|  | 138 | 195 | 136 |
|  | 149 | 162 | 147 |
| **Medium** | 141 | 186 | 128 |
|  | 137 | 161 | 133 |
|  | 145 | 157 | 148 |
|  | 150 | 165 | 145 |
|  | 130 | 194 | 141 |
| **High** | 129 | 157 | 149 |
|  | 141 | 152 | 137 |
|  | 148 | 186 | 138 |
|  | 130 | 164 | 126 |
|  | 137 | 176 | 138 |

a. At the 0.10 level of significance, is there an interaction effect?

b. At the 0.10 level of significance, does the height of the product's placement have an effect on the product's mean sales?

c. At the 0.10 level of significance, does the location in the store have an effect on the product's mean sales?

## Computer Software Exercises

**12-39.** Mt. Jumbo Plywood Company makes plywood for use in furniture production. The first major step in the plywood process is peeling the logs into thin layers of veneer. Ideally, when a log is reduced to a 4-inch core diameter, the lathe releases the core and a new log is loaded onto the lathe. However, a problem called "spinouts" occurs if the lathe kicks out a core that has more than 4 inches left. This wastes wood and costs the company money.

Two factors are believed to affect the core diameter: the vat temperature and the time the logs spend in the vat prior to peeling. The lathe supervisor has recently conducted a test during which logs were peeled at each combination of temperature and time. The sample data for this experiment are in the data file called **Mt Jumbo**. The data are the core diameters in inches.

a. Based on the sample data, is there an interaction between water temperature and vat hours? Test using a significance level of 0.01. Discuss what interaction would mean in this situation. Use a $p$-value approach.

b. Based on the sample data, is there a difference in the mean core diameters at the three water temperatures? Test using a significance level of 0.01.

c. Do the sample data indicate a difference in the mean core diameters across the three vat times analyzed in this study? Use a significance level of 0.10 and a $p$-value approach.

**12-40.** Assume that a psychologist conducted a study to determine whether there are differences between the ability of history majors and mathematics majors to solve various types of puzzles. Twenty-five mathematics majors and 25 history majors were randomly selected from the students at a liberal arts college in Maine. Five students from each major were randomly assigned one of the following puzzles: a crossword puzzle, a cryptogram, a logic problem, a maze, and a cross sums. The time in minutes (rounded to the nearest minute) was recorded for each student in the study. If a student could not complete a puzzle in the maximum time allowed, or completed a puzzle incorrectly, then a penalty of 10 minutes was added to his or her time. The results are shown in the file **Puzzle**.

a. Plot the mean time to complete a puzzle for each puzzle type by major. What conclusion would you reach about the interaction between major and puzzle type?

b. At the 0.05 level of significance, is there an interaction effect?

c. If interaction is present, conduct a one-way ANOVA to test whether the mean time to complete a puzzle for history majors depends on the type of puzzle. Does the mean time to complete a puzzle for mathematics majors depend on the type of puzzle? Conduct the one-way ANOVA tests at a level of significance of 0.05.

**12-41.** The Iams Company sells dog and cat foods (dry and canned) in countries throughout the world. Suppose that one of the company's dry dog food products comes in five formulas. One of the ingredients is of particular importance: crude fat. To discover if there is a difference in the average percentage of crude fat among the five formulas and among the production sites, the sample data found in the file titled **Iams** were obtained.

a. Determine if there is interaction between the five formulas and the plant sites where they are produced. Use the $p$-value approach and a significance level of 0.025.

b. Given your findings in part a, determine if there is a difference in the average percentage of crude fat in the five formulas. Use a test-statistic approach with a significance level of 0.025.

c. One important finding will be whether the average percent of crude fat for the "Reduced Fat" formula is equal to the advertised 9%. Conduct a relevant hypothesis test to determine this using a significance level of 0.05.

**12-42.** As part of a class project, students chose beers and obtained the sodium content from the labels on the bottle or can. The file titled **Sodium** contains the amounts of sodium (mg) found in 12 fluid ounces of beer produced by the four major producers: Anheuser-Busch Inc., Miller Brewing Co., Coors Brewing Co., and Pabst Brewing Co. The types of beer (ales, lager, and specialty beers) were also scrutinized in the analysis.

a. Determine if there is interaction between the producer and the type of beer. Use a significance level of 0.05.

b. Given your findings in part a, determine if there is a difference in the average amounts of sodium in 12 ounces of beer among the producers of the beer. Use a significance level of 0.05.

c. Repeat part b for the types of beer. Use a significance level of 0.05.

# 12 Overview

## Summary

**OUTCOME 1** Be able to apply the basic logic of analysis of variance.

**OUTCOME 2** Perform a hypothesis test for a single-factor design using analysis of variance manually and with the aid of Excel.

**OUTCOME 3** Conduct and interpret post–analysis of variance pairwise comparisons procedures.

- Sometimes independent samples are obtained from two or more levels of a single factor to determine if the levels have equal means.
- Use analysis of variance to conduct a hypothesis test for this kind of experimental design.
- This procedure is called a **one-way analysis of variance** because it addresses an experiment that involves only one factor.
- The concept acknowledges that the data produced by the completely randomized design are not all the same value, which indicates that there is variation in the data. This variation is referred to as the **total variation**.
- Each level's data exhibits dispersion as well, which is called the **within-sample variation**. The dispersion between the factor levels is called the **between-sample variation**.
- The ratio between estimators of these two variances forms the test statistic used to detect differences in the means of the levels. If the null hypothesis of equal means is rejected, we can use the Tukey-Kramer procedure to determine which pairs of populations have different means.

**OUTCOME 3** Conduct and interpret post–analysis of variance pairwise comparisons procedures.

**OUTCOME 4** Recognize when randomized block analysis of variance is useful and be able to perform analysis of variance on a randomized block design.

- In some situations, an additional known factor with at least two levels may impinge on the response variable of interest.
- A technique called blocking is used in such cases to eliminate the effects of the levels of the additional known factor on the analysis of variance.
- A multiple comparisons procedure known as **Fisher's least significant difference** can be used to determine any difference among the population means of a randomized block ANOVA design.

**OUTCOME 5** Perform analysis of variance on a two-factor design of experiments with replications using Excel and interpret the output.

- Two-factor ANOVA follows the same logic used in the one-way and complete block ANOVA designs.
- In the latter two procedures, there is only one factor of interest. In two-factor ANOVA, there are two factors of interest. Each factor of interest introduces variability into the experiment.
- Sometimes the presence of a level of one factor affects the relationship between the response variable and the levels of the other factor.
- This effect is called interaction and, if present, is another source of variation. As in Sections 12.1 and 12.2, we must find estimators for each source of variation.
- We find these estimators by identifying the appropriate sums of squares and then dividing each by its degrees of freedom. If the variation accounted for by Factor A, Factor B, and the interaction is large relative to the error variation, we tend to conclude that the factor levels have different means.
- When we wish to analyze two factors as described above, we use the **two-factor ANOVA with replications** technique.

## Equations

**(12.1) Partitioned Sum of Squares** pg. 437

$$SST = SSB + SSW$$

**(12.2) Total Sum of Squares** pg. 440

$$SST = \sum_{i=1}^{k} \sum_{j=1}^{n_i} (x_{ij} - \bar{\bar{x}})^2$$

**(12.3) Sum of Squares Between** pg. 440

$$SSB = \sum_{i=1}^{k} n_i (\bar{x}_i - \bar{\bar{x}})^2$$

**(12.4) Sum of Squares Within** pg. 441

$$SSW = SST - SSB$$

**(12.5) Sum of Squares Within** pg. 441

$$SSW = \sum_{i=1}^{k} \sum_{j=1}^{n_i} (x_{ij} - \bar{x}_i)^2$$

**(12.6) Tukey-Kramer Critical Range** pg. 446

$$\text{Critical range} = q_{1-\alpha} \sqrt{\frac{MSW}{2} \left( \frac{1}{n_i} + \frac{1}{n_j} \right)}$$

**(12.7) Sum of Squares Partitioning for Randomized Complete Block Design** pg. 455

$$SST = SSB + SSBL + SSW$$

**(12.8) Sum of Squares for Blocking** pg. 455

$$SSBL = \sum_{j=1}^{b} k (\bar{x}_j - \bar{\bar{x}})^2$$

**(12.9) Sum of Squares Within** pg. 455

$$SSW = SST - (SSB + SSBL)$$

**(12.10) Fisher's Least Significant Difference** pg. 460

$$LSD = t_{\alpha/2} \sqrt{MSW} \sqrt{\frac{2}{b}}$$

**(12.11) Total Sum of Squares** pg. 467

$$SST = \sum_{i=1}^{a} \sum_{j=1}^{b} \sum_{k=1}^{n} (x_{ijk} - \bar{\bar{x}})^2$$

**(12.12) Sum of Squares Factor A** pg. 467

$$SS_A = bn \sum_{i=1}^{a} (\bar{x}_{i..} - \bar{\bar{x}})^2$$

**(12.13) Sum of Squares Factor B** pg. 467

$$SS_B = an \sum_{j=1}^{b} (\bar{x}_{.j.} - \bar{\bar{x}})^2$$

**(12.14) Sum of Squares Interaction between Factors A and B** pg. 467

$$SS_{AB} = n \sum_{i=1}^{a} \sum_{j=1}^{b} (\bar{x}_{ij.} - \bar{x}_{i..} - \bar{x}_{.j.} + \bar{\bar{x}})^2$$

**(12.15) Sum of Squares Error** pg. 467

$$SSE = \sum_{i=1}^{a} \sum_{j=1}^{b} \sum_{k=1}^{n} (x_{ijk} - \bar{x}_{ij.})^2$$

## Key Terms

Balanced design pg. 436
Between-sample variation pg. 437
Factor pg. 435

Levels pg. 436
One-way analysis of variance pg. 435

Total variation pg. 436
Within-sample variation pg. 437

## Chapter Exercises

### Conceptual Questions

**12-43.** A one-way analysis of variance has just been performed. The conclusion reached is that the null hypothesis stating the population means are equal has not been rejected. What would you expect the Tukey-Kramer procedure for multiple comparisons to show if it were performed for all pairwise comparisons? Discuss.

**12-44.** In journals related to your major, locate two articles in which tests of three or more population means were important. Discuss the issue being addressed, how the data were collected, the results of the statistical test, and any conclusions drawn based on the analysis.

**12-45.** Discuss why in some circumstances it is appropriate to use the randomized complete block design. Give an example other than those discussed in the text in which this design could be used.

**12-46.** A two-way analysis of variance experiment is to be conducted to examine CEO salaries ($K) as a function of the number of years the CEO has been with the company and the size of the company's sales. The years

spent with the company are categorized into
$0-3$, $4-6$, $7-9$, and $>9$ years. The size of the company
is categorized using sales ($million) per year into three
categories: $0-50$, $51-100$, and $>100$.

a. Describe the factors associated with this experiment.
b. List the levels of each of the factors identified in
   part a.
c. List the treatment combinations of the experiment.
d. Indicate the components of the ANOVA table that
   will be used to explain the variation in the CEOs'
   salaries.
e. Determine the degrees of freedom for each of the
   components in the ANOVA if two replications are
   used.

**12-47.** In any of the multiple comparison techniques (Tukey-
Kramer, *LSD*), the estimate of the within-sample
variance uses data from the entire experiment. However,
if one were to do a two-sample *t*-test to determine if
there were a difference between any two means, the
estimate of the population variances would include only
data from the two specific samples under consideration.
Explain this seeming discrepancy.

## Business Applications

**12-48.** Your company provides computers, printers, and
Internet service to a number of engineers and
programmers so that they can telecommute from home.
Because of the cost of name-brand printer replacement
cartridges, several companies have entered the
secondary market. You are currently considering offers
from four companies. The prices are equivalent, so you
will make your decision based on number of pages
printed. You have given samples of four cartridges to 16
programmers and engineers and have received the
following numbers of pages:

| Supplier A | Supplier B | Supplier C | Supplier D |
|---|---|---|---|
| 424 | 650 | 521 | 323 |
| 521 | 725 | 601 | 383 |
| 650 | 826 | 590 | 487 |
| 422 | 722 | 522 | 521 |

a. Using a significance level equal to 0.01, what
   conclusion should you reach about the four
   manufacturers' printer cartridges? Explain.
b. If the test conducted in part a reveals that you
   should reject the null hypothesis, which supplier
   should you choose? Is there one or more you can
   eliminate based on these data? Use the appropriate
   test for multiple comparisons. Discuss.

**12-49.** The Promise Seed Co. is owned by Tim Carlton Jr. One
of Promise's most demanded seeds is corn. Promise
continues to increase production to meet the growing
demand. To this end, it conducts an experiment such as
the one presented here to determine the combination of
fertilizer and seed type that produces the largest number
of kernels per ear.

| | Fert. 1 | Fert. 2 | Fert. 3 | Fert. 4 |
|---|---|---|---|---|
| Seed A | 807 | 995 | 894 | 903 |
| | 800 | 909 | 907 | 904 |
| Seed B | 1,010 | 1,098 | 1,000 | 1,008 |
| | 912 | 987 | 801 | 912 |
| Seed C | 1,294 | 1,286 | 1,298 | 1,199 |
| | 1,097 | 1,099 | 1,099 | 1,201 |

a. Determine if there is interaction between the type of
   seed and the type of fertilizer. Use a significance
   level of 0.05.
b. Given your findings in part a, determine if there is a
   difference in the average number of kernels per ear
   among the seeds.
c. Repeat part b for the types of fertilizer. Use a
   significance level of 0.05.

**12-50.** Many people rely on local accountants to handle their
tax work. A local television station, which prides itself
on doing investigative reporting, decided to do a study to
determine the extent to which tax errors are made by the
accounting firms.

The station selected eight people to have their taxes
figured at each of three accounting offices in its market
area. The following data show the tax bills (in dollars) as
figured by each of the three accounting offices:

| Return | Office 1 | Office 2 | Office 3 |
|---|---|---|---|
| 1 | 4,376.20 | 5,100.10 | 4,988.03 |
| 2 | 5,678.45 | 6,234.23 | 5,489.23 |
| 3 | 2,341.78 | 2,242.60 | 2,121.90 |
| 4 | 9,875.33 | 10,300.30 | 9,845.60 |
| 5 | 7,650.20 | 8,002.90 | 7,590.88 |
| 6 | 1,324.80 | 1,450.90 | 1,356.89 |
| 7 | 2,345.90 | 2,356.90 | 2,345.90 |
| 8 | 15,468.75 | 16,080.70 | 15,376.70 |

a. Discuss why this test was conducted as a
   randomized block design. Why did the station think
   it was important to have all three offices do the
   returns for each of the eight people?
b. Test to determine whether blocking was necessary
   in this situation. Use a significance level of 0.01.
   State the null and alternative hypotheses.
c. Based on the sample data, can the station report
   statistical evidence that there is a difference in
   the mean taxes due on tax returns? Test using a
   significance level of 0.01. State the appropriate null
   and alternative hypotheses.
d. Referring to part c, if you did conclude that a
   difference exists, use the appropriate test to
   determine which office has the highest mean tax due.

**12-51.** A senior analyst working for a stock brokerage firm has
reviewed purchases his customers have made over the
last six months. He has categorized the mutual funds
purchased into eight categories: (1) Aggressive Growth

(AG), (2) Growth (G), (3) Growth-Income (G-I), (4) Income Funds (IF), (5) International (I), (6) Asset Allocation (AA), (7) Precious Metal (PM), and (8) Bond (B). The percentage gains accrued by three randomly selected customers in each group are as follows:

| Mutual Fund | AG | G | G-I | IF | I | AA | PM | B |
|---|---|---|---|---|---|---|---|---|
| | 6 | 7 | 5 | 1 | 14 | −3 | 5 | −1 |
| | 7 | −2 | 6 | 0 | 13 | 7 | 7 | 3 |
| | 12 | 0 | 2 | 6 | 10 | 7 | 5 | 2 |

a. Develop the appropriate ANOVA table to determine if there is a difference in the average percentage gains accrued by his customers among the mutual fund types. Use a significance level of 0.05.

b. Use the Tukey-Kramer procedure to determine which mutual fund type has the highest average percentage gain. Use an experiment-wide error rate of 0.05.

12-52. A manufacturer of tortilla chips has recently developed a new product, a blue corn tortilla chip. The manufacturer has arranged with a regional supermarket chain to display the chips at the end of the aisle at four different locations in stores that have had similar weekly sales in snack foods. The dollar volumes of sales for the last six weeks in the four stores are as follows:

| | Store | | | |
|---|---|---|---|---|
| Week | 1 | 2 | 3 | 4 |
| 1 | $1,430 | $ 980 | $1,780 | $2,300 |
| 2 | $2,200 | $1,400 | $2,890 | $2,680 |
| 3 | $1,140 | $1,200 | $1,500 | $2,000 |
| 4 | $ 880 | $1,300 | $1,470 | $1,900 |
| 5 | $1,670 | $1,300 | $2,400 | $2,540 |
| 6 | $ 990 | $ 550 | $1,600 | $1,900 |

a. If the assumptions of a one-way ANOVA design are satisfied in this case, what should the manufacturer conclude about the average sales at the four stores? Use a significance level of 0.05.

b. Discuss whether you think the assumptions of a one-way ANOVA are satisfied in this case and indicate why or why not. If they are not, what design is appropriate? Discuss.

c. Perform a randomized block analysis of variance test using a significance level of 0.05 to determine whether the mean sales for the four stores are different.

d. Comment on any differences between the means in parts b and c.

e. Suppose blocking was necessary and the researcher chooses not to use blocks. Discuss what impact this could have on the results of the analysis of variance.

f. Use Fisher's least significant difference procedure to determine which, if any, stores have different true average weekly sales.

## Computer Software Exercises

12-53. Gordon Manufacturing produces golf balls. Recently, Gordon developed a golf ball made from a space-age material. This new golf ball promises greater distance off the tee. To test Gordon Manufacturing's claim, a test was set up to measure the average distance of four different golf balls (the New Gordon, Competitor 1, Competitor 2, Competitor 3) hit by a driving machine using three different types of drivers (Driver 1, Driver 2, Driver 3). The results (rounded to the nearest yard) are listed in the data file called **Gordon**. Conduct a test to determine if there are significant differences due to type of golf ball.

a. Does there appear to be interaction between type of golf ball and type of driver?

b. Conduct a test to determine if there is a significant effect due to the type of driver used.

c. How could Gordon Manufacturing use the results of the tests?

12-54. Although retail store scanners can save customers a great deal of time, scanners sometimes misread an item's price code. Before investing in one of three new systems, a store would like to determine if there is a difference in scanner accuracy. To investigate possible differences in scanner accuracy, 30 shopping carts were randomly selected from customers at the store. The 30 carts differed from each other in both the number and types of items each contained. The items in each cart were then scanned by the three new scanning systems under consideration as well as by the current scanner used in the store at a specially designed test facility for the purposes of the analysis. Each item was also checked manually, and a count was kept of the number of scanning errors made by each scanner for each basket. Each of the scannings was repeated 30 times, and the average number of scanning errors was determined. The sample data are in the file called **Scanners**.

a. What type of experimental design did the store use to test for differences among scanning systems? Why was this type of design selected?

b. State the primary hypotheses of interest for this test.

c. At the 0.01 level of significance, is there a difference in the average number of errors among the four different scanners?

d. (1) Is there a difference in the average number of errors by cart? (2) Was the store correct in blocking by cart?

e. If you determined that there is a difference in the average number of errors among the four different scanners, identify where those differences exist.

f. Do you think that the store should upgrade from its existing scanner to Scanner A, Scanner B, or Scanner C? What other factors may it want to consider before making a decision?

12-55. PhoneEx provides call center services for many different companies. A large increase in its business has made it necessary to establish a new call center. Four cities are being considered—Little Rock, Wichita,

Tulsa, and Memphis. The new center will employ approximately 1,500 workers, and PhoneEx will transfer 75 people from its Omaha center to the new location. One concern in the choice of where to locate the new center is the cost of housing for the employees who will be moving there. To help determine whether significant housing cost differences exist across the competing sites, PhoneEx has asked a real estate broker in each city to randomly select a list of 33 homes between 5 and 15 years old and ranging in size between 1,975 and 2,235 square feet. The prices (in dollars) that were recorded for each city are contained in the file called **PhoneEx**.

a. At the 0.05 level of significance, is there evidence to conclude that the average prices of houses between 5 and 15 years old and ranging in size between 1,975 and 2,235 square feet are not the same in the four cities? Use the *p*-value approach.

b. At the 0.05 level of significance, is there a difference in average housing price between Wichita and Little Rock? Between Little Rock and Tulsa? Between Tulsa and Memphis?

c. Determine the sample size required to estimate the average housing price in Wichita to within $500 with a 95% confidence level. Assume the required parameters' estimates are sufficient for this calculation.

**12-56.** An investigation into the effects of various levels of nitrogen conducted by a the agricultural department at a state university addressed the pounds per acre of

nitrogen required to produce certain yield levels of corn on fields that had previously been planted with other crops. Suppose the data in the file titled **Nitrogen** indicate the amounts of nitrogen required to produce given quantities of corn planted.

a. Determine if there is interaction between the yield levels of corn and the crop that had been previously planted in the field. Use a significance level of 0.05.

b. Given your findings in part a, determine any significant differences among the average pounds per acre of nitrogen required to produce yield levels of corn on fields that had been planted with corn as the previous crop.

c. Repeat part b for soybeans and grass sod, respectively.

**12-57.** Customer orders placed online at an e-retailer are filled and shipped from four different warehouses in Cleveland, Atlanta, Denver, and Sacramento. A random sample of 20 customer orders was chosen from all shipments filled in a week at the four warehouses and weighed. The weights of each sampled shipment (in ounces) are contained in the file **Shipments**.

a. Based on the sampled shipments, can you conclude that the average weights of shipped customer orders are different at the four warehouses? Use an alpha of 0.05 to conduct the appropriate test analysis.

b. If a difference is found in the average weights of customer orders, conduct a post-test analysis to determine where the difference exists. Use an alpha of 0.05 to perform the comparison.

## Case 12.1    Agency for New Americans

Denise Walker was physically exhausted after her first outing as a volunteer for the Agency for New Americans in Raleigh, North Carolina. Denise had a fairly good career going with various federal agencies after graduating with a degree in accounting. She decided to stay at home after she and her husband started a family. Now that their youngest was in high school, Denise decided she needed something more to do than manage the household. She decided on volunteer work and joined the Agency for New Americans.

The purpose of the Agency for New Americans is to help new arrivals become comfortable with the basic activities necessary to function in American society. One of the major activities, of course, is shopping for food and other necessities. Denise had just returned from her first outing to a supermarket with a recently arrived Somali Bantu family. It was their first time also, and they were astonished by both the variety and selection. Since the family was on a very limited budget, Denise spent much time talking about comparison shopping, and for someone working with a new currency, this was hard. She didn't even want to tell them the store they were in was only one of four possible chains within a mile of their apartment. Denise realized the store she started with would be the one they would automatically return to when on their own.

Next week, Denise and the family are scheduled to go to a discount store. Denise typically goes to a national chain close to

her house but hasn't felt the need to be primarily a value shopper for some time. Since she feels the Somali family will automatically return to the store she picks, and she has her choice of two national chains and one regional chain, she decides to not automatically take them to "her" store. Because each store advertises low prices and meeting all competitors' prices, she also doesn't want to base her decision on what she hears on commercials. Instead, she picks a random selection of items and finds the prices in each store. The items and prices are shown in the file **New Americans**. In looking at the data, Denise sees there are differences in some prices but wonders if there is any way to determine which store to take the family to.

### Required Tasks:

1. Identify the major issue in the case.
2. Identify the appropriate statistical test that could be conducted to address the case's major issue.
3. Explain why you selected the test you choose in (2).
4. State the appropriate null and alternative hypotheses for the statistical test you identified.
5. Perform the statistical test(s). Be sure to state your conclusion(s).
6. If possible, identify the stores that Denise should recommend to the family.
7. Summarize your analysis and findings in a short report.

## Case 12.2    McLaughlin Salmon Works

John McLaughlin's father correctly predicted that a combination of declining wild populations of salmon and an increase in demand for fish in general would create a growing market for salmon grown in "fish farms." Over recent years, an increasing percentage of salmon, trout, and catfish, for example, come from commercial operations. At first, operating a fish farm consisted of finding an appropriate location, installing the pens, putting in smelt, and feeding the fish until they grew to the appropriate size. However, as the number of competitors increased, successful operation required taking a more scientific approach to raising fish.

Over the past year, John has been looking at the relationship between food intake and weight gain. Since food is a major cost of the operation, the higher the weight gain for a given amount of food, the more cost-effective the food. John's most recent effort involved trying to determine the relationship between four component mixes and three size progressions for the food pellets. Since smaller fish require smaller food pellets but larger pellets contain more food, one question John was addressing was at what rate to move from smaller to larger pellets. Also, since fish are harder to individually identify than livestock, the study involved constructing small individual pens and giving fish in each pen a different

combination of pellet mix and size progression. This involved a reasonable cost but a major commitment of time, and John's father wasn't sure the cost and time were justified. John has just gathered his first set of data and has started to analyze it. The data are shown in the file called **McLaughlin Salmon Works**. John not only is interested in whether one component mix, or one pellet size progression, seemed to lead to maximum weight gain but would really like to find one combination of mix and size progression that proved to be superior.

### Required Tasks:
1. Identify the major issues in the case.
2. Identify an appropriate statistical analysis to perform.
3. Explain why you selected the test you choose in (2).
4. State the appropriate null and alternative hypotheses for the statistical test you identified.
5. Perform the statistical test(s). Be sure to state your conclusion(s).
6. Is there one combination of mix and size progression that is superior to the others?
7. Summarize your analysis and findings in a short report.

## Case 12.3    NW Pulp and Paper

Cassie Coughlin had less than a week to finish her presentation to the CEO of NW Pulp and Paper. Cassie had inherited a project started by her predecessor as head of the new-product development section of the company, and by the nature of the business, dealing with wood products, projects tended to have long lifetimes. Her predecessor had successfully predicted the consequences of a series of events that, in fact, had occurred:

1. The western United States, where NW Pulp and Paper had its operations, was running out of water, caused by a combination of population growth and increased irrigation. The situation had currently been made worse by several years of drought. This meant many farming operations were becoming unprofitable.
2. The amount of timber harvesting from national forests continued to be limited.
3. At least some of the land that had been irrigated would become less productive due to alkaline deposits caused by taking water from rivers.

Based on these three factors, Cassie's predecessor had convinced the company to purchase a 2,000-acre farm that had four types of soil commonly found in the West and also had senior

water rights. Water rights in the West are given by the state, and senior rights are those that will continue to be able to get irrigation water after those with junior rights are cut off. His idea had been to plant three types of genetically modified poplar trees (these are generally fast-growing trees) on the four types of soil and assess growth rates. His contention was it might be economically feasible for the company to purchase more farms that were becoming less productive and to become self-sufficient in its supply of raw material for making paper.

The project had been started 15 years ago, and since her predecessor had since retired, Cassie was now in charge of the project. The primary focus of the 15-year review was tree growth. Growth in this case did not refer to height but wood volume. Volume is assessed by measuring the girth of the tree three feet above the ground. She had just received data from the foresters who had been managing the experiment. They had taken a random sample of measurements from each of the tree types. The data are shown in the file **NW Pulp and Paper**. Cassie knew the CEO would at least be interested in whether one type of tree was generally superior and whether there was some unique combination of soil type and tree type that stood out.

## Case 12.4    Quinn Restoration

Last week John Quinn sat back in a chair with his feet on his deck and nodded at his wife, Kate. They had just finished a conversation that would likely influence the direction of their lives for the next several years or longer.

John retired a little less than a year ago after 25 years in the Lake Oswego police department. He had steadily moved up the

ranks and retired as a captain. Although his career had, in his mind, gone excellently, he had been working much more than he had been home. Initially upon retiring, he had reveled in the ability to spend time doing things he was never able to do while working: complete repairs around the house, travel with his wife, spend time with the children still at home, and visit those who had moved out.

He was even able to knock five strokes off his golf handicap. However, he had become increasingly restless, and both he and Kate agreed he needed something to do, but that something did not involve a full-time job.

John had, over the years, bought, restored, and sold a series of older Corvettes. Although this had been entirely a hobby, it also had been a profitable one. The discussion John and Kate just concluded involved expanding this hobby, not into a full-time job, but into a part-time business. John would handle the actual restoration, which he enjoyed, and Kate would cover the paperwork, ordering parts, keeping track of expenses, and billing clients, which John did not like. The last part of their conversation involved ordering parts.

In the past, John had ordered parts for old Corvettes from one of three possible sources: Weckler's, American Auto Parts, or Corvette Central. Kate, however, didn't want to call all three any time John needed a part but instead wanted to set up an account with one of the three and be able to order parts over the Internet. The question was which company, if any, would be the appropriate choice. John agreed to develop a list of common parts. Kate would then call each of the companies asking for their prices, and, based on this information, determine with which company to establish the account. Kate spent time over the last week on the phone developing the data located in the data file called **Quinn Restoration**. The question John now faced is whether the prices he found could lead him to conclude that one company will be less expensive, on average, than the other two.

## Business Statistics Capstone Project

## Theme: Analysis of Variance

### Project Objective

The objective of this business statistics capstone project is to provide you with an opportunity to integrate the statistical tools and concepts you have learned in your business statistics course. As in all real-world applications, it is not expected through the completion of this project that you will have utilized every statistical technique you have been taught in this course. Rather, an objective of the assignment will be for you to determine which of the statistical tools and techniques are appropriate to employ for the situation you have selected.

### Project Description

You are to identify a business or organizational issue that is appropriately addressed using analysis of variance or experimental design. You will need to specify one or more sets of null and alternative hypotheses to be tested in order to reach conclusions pertaining to the business or organizational issue you have selected. You are responsible for designing and carrying out an "experiment" or otherwise collecting appropriate data required to test the hypotheses using one or more of the analysis of variance designs introduced in your text and statistics course.

There is no minimum sample size. The sample size should depend on the design you choose and the cost and difficulty in obtaining the data. You are responsible for making sure that the data are accurate. All methods (or sources) for data collection should be fully documented.

### Project Deliverables

To successfully complete this capstone project, you are required to deliver, at a minimum, the following items in the context of a management report:

- A complete description of the central issue of your project and of the background of the company or organization you have selected as the basis for the project.
- A clear and concise explanation of the data-collection method used. Included should be a discussion of your rationale for selecting the analysis of variance technique(s) used in your analysis.
- A complete descriptive analysis of all variables in the data set, including both numerical and graphical analysis. You should demonstrate the extent to which the basic assumptions of the analysis of variance designs have been satisfied.
- Provide a clear and concise review of the hypotheses tests that formed the objective of your project. Show any post–ANOVA multiple comparison tests where appropriate.
- Offer a summary and conclusion section that relates back to the central issue(s) of your project and discusses the results of the hypothesis tests.
- All pertinent appendix materials.

The final report should be presented in a professional format using the style or format suggested by your instructor.

# 8–12 Special Review Section

This review section, which is presented using block diagrams and flowcharts, is intended to help you tie together the material from several key chapters. This section is not a substitute for reading and studying the chapters covered by the review. However, you can use this review material to add to your understanding of the individual topics in the chapters.

## Chapters 8–12

Statistical inference is the process of reaching conclusions about a population based on a random sample selected from the population. Chapters 8–12 introduced the fundamental concepts of statistical inference involving two major categories of inference: *estimation* and *hypothesis testing*. These chapters have covered a fairly wide range of different situations that, for beginning students, can sometimes seem overwhelming. The following diagrams will, we hope, help you better identify which specific estimation or hypothesis-testing technique to use in a given situation. These diagrams form something resembling a decision support system that you should be able to use as a guide through the estimation and hypothesis-testing processes.

To use the diagrams, start with the one labeled A. From the exercise, determine whether it asks you to estimate a parameter or test a hypothesis. Next, determine how many populations are involved and then proceed to the appropriate diagram. Continuing in that manner, you will be led to the correct equations to use for any situation covered in Chapters 8–12.

**Estimate $\mu$, $\sigma$ Unknown**

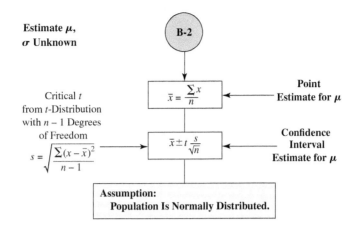

Critical $t$ from $t$-Distribution with $n - 1$ Degrees of Freedom

$$s = \sqrt{\frac{\sum(x - \bar{x})^2}{n - 1}}$$

**B-2**

$$\bar{x} = \frac{\sum x}{n}$$

$$\bar{x} \pm t \frac{s}{\sqrt{n}}$$

**Point Estimate for $\mu$**

**Confidence Interval Estimate for $\mu$**

**Assumption: Population Is Normally Distributed.**

**Estimate $p$, Population Proportion**

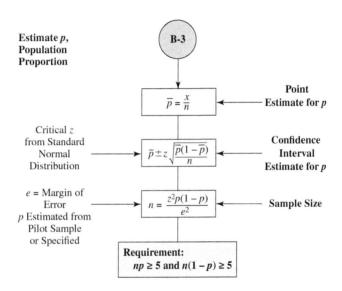

**B-3**

Critical $z$ from Standard Normal Distribution

$e$ = Margin of Error
$p$ Estimated from Pilot Sample or Specified

$$\bar{p} = \frac{x}{n}$$

$$\bar{p} \pm z \sqrt{\frac{\bar{p}(1 - \bar{p})}{n}}$$

$$n = \frac{z^2 p(1 - p)}{e^2}$$

**Point Estimate for $p$**

**Confidence Interval Estimate for $p$**

**Sample Size**

**Requirement: $np \geq 5$ and $n(1 - p) \geq 5$**

**Estimating 2 Population Parameters**

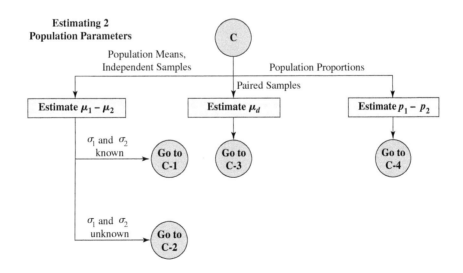

**C**

Population Means, Independent Samples

Population Proportions

Paired Samples

**Estimate $\mu_1 - \mu_2$**

**Estimate $\mu_d$**

**Estimate $p_1 - p_2$**

$\sigma_1$ and $\sigma_2$ known → **Go to C-1**

**Go to C-3**

**Go to C-4**

$\sigma_1$ and $\sigma_2$ unknown → **Go to C-2**

Estimate $\mu_1 - \mu_2$, $\sigma_1$ and $\sigma_2$ Known

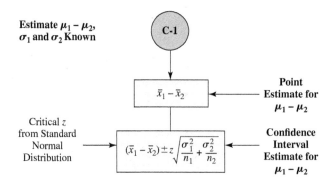

Estimate $\mu_1 - \mu_2$, $\sigma_1$ and $\sigma_2$ Unknown

Estimate $\mu_d$, Paired Samples

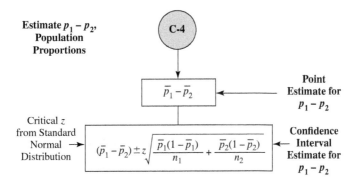

**Estimate $p_1 - p_2$, Population Proportions**

C-4

$\bar{p}_1 - \bar{p}_2$ ← **Point Estimate for $p_1 - p_2$**

**Critical $z$ from Standard Normal Distribution** → $(\bar{p}_1 - \bar{p}_2) \pm z \sqrt{\dfrac{\bar{p}_1(1 - \bar{p}_1)}{n_1} + \dfrac{\bar{p}_2(1 - \bar{p}_2)}{n_2}}$ ← **Confidence Interval Estimate for $p_1 - p_2$**

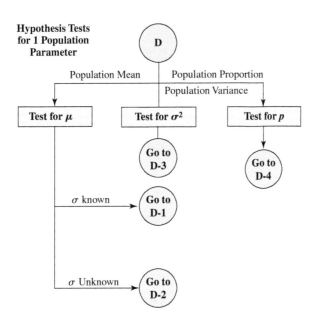

**Hypothesis Tests for 1 Population Parameter**

D

Population Mean | Population Proportion

Population Variance

**Test for $\mu$**    **Test for $\sigma^2$**    **Test for $p$**

Go to D-3    Go to D-4

$\sigma$ known → Go to D-1

$\sigma$ Unknown → Go to D-2

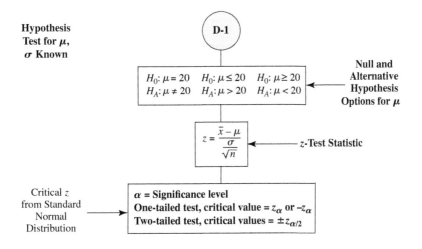

**Hypothesis Test for $\mu$, $\sigma$ Known**

D-1

$H_0: \mu = 20$   $H_0: \mu \leq 20$   $H_0: \mu \geq 20$
$H_A: \mu \neq 20$   $H_A: \mu > 20$   $H_A: \mu < 20$ ← **Null and Alternative Hypothesis Options for $\mu$**

$z = \dfrac{\bar{x} - \mu}{\dfrac{\sigma}{\sqrt{n}}}$ ← $z$-**Test Statistic**

**Critical $z$ from Standard Normal Distribution** → $\alpha$ = **Significance level**
**One-tailed test, critical value = $z_\alpha$ or $-z_\alpha$**
**Two-tailed test, critical values = $\pm z_{\alpha/2}$**

**Hypothesis
Test for $\mu$,
$\sigma$ Unknown**

**Hypothesis
Test for $\sigma^2$**

**Hypothesis Test for $p$**

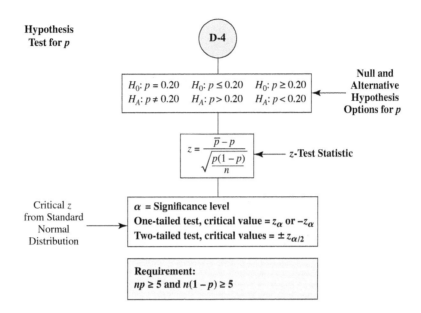

D-4

$H_0$: $p = 0.20$    $H_0$: $p \leq 0.20$    $H_0$: $p \geq 0.20$
$H_A$: $p \neq 0.20$    $H_A$: $p > 0.20$    $H_A$: $p < 0.20$

**Null and Alternative Hypothesis Options for $p$**

$$z = \frac{\bar{p} - p}{\sqrt{\dfrac{p(1-p)}{n}}}$$

$z$-**Test Statistic**

Critical $z$ from Standard Normal Distribution

$\alpha$ = **Significance level**
**One-tailed test, critical value** = $z_\alpha$ or $-z_\alpha$
**Two-tailed test, critical values** = $\pm z_{\alpha/2}$

**Requirement:**
$np \geq 5$ and $n(1-p) \geq 5$

**Hypothesis Tests for 2 Population Parameters**

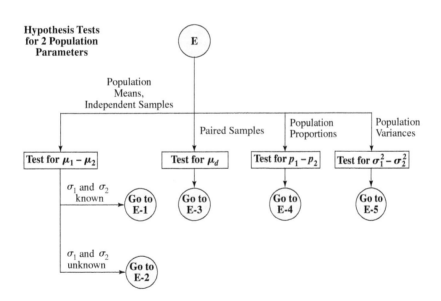

E

Population Means, Independent Samples

Paired Samples

Population Proportions

Population Variances

Test for $\mu_1 - \mu_2$

Test for $\mu_d$

Test for $p_1 - p_2$

Test for $\sigma_1^2 - \sigma_2^2$

$\sigma_1$ and $\sigma_2$ known

Go to E-1

Go to E-3

Go to E-4

Go to E-5

$\sigma_1$ and $\sigma_2$ unknown

Go to E-2

**Hypothesis Test for $\mu_1 - \mu_2$, $\sigma_1$ and $\sigma_2$ Known**

E-1

$$H_0: \mu_1 - \mu_2 = 0 \quad H_0: \mu_1 - \mu_2 \geq 0 \quad H_0: \mu_1 - \mu_2 \leq 0$$
$$H_A: \mu_1 - \mu_2 \neq 0 \quad H_A: \mu_1 - \mu_2 < 0 \quad H_A: \mu_1 - \mu_2 > 0$$

Null and Alternative Hypothesis Options for Testing $\mu_1 - \mu_2$

$$z = \frac{(\bar{x}_1 - \bar{x}_2) - (\mu_1 - \mu_2)}{\sqrt{\dfrac{\sigma_1^2}{n_1} + \dfrac{\sigma_2^2}{n_2}}}$$

$z$-Test Statistic for $\mu_1 - \mu_2$

Critical $z$ from Standard Normal Distribution

$\alpha$ = Significance level
One-tailed test, critical value = $z_\alpha$ or $-z_\alpha$
Two-tailed test, critical values = $\pm z_{\alpha/2}$

**Hypothesis Test for $\mu_1 - \mu_2$, $\sigma_1$ and $\sigma_2$ Unknown**

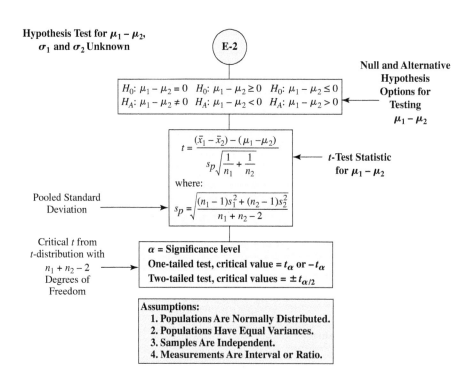

E-2

$$H_0: \mu_1 - \mu_2 = 0 \quad H_0: \mu_1 - \mu_2 \geq 0 \quad H_0: \mu_1 - \mu_2 \leq 0$$
$$H_A: \mu_1 - \mu_2 \neq 0 \quad H_A: \mu_1 - \mu_2 < 0 \quad H_A: \mu_1 - \mu_2 > 0$$

Null and Alternative Hypothesis Options for Testing $\mu_1 - \mu_2$

$$t = \frac{(\bar{x}_1 - \bar{x}_2) - (\mu_1 - \mu_2)}{s_p\sqrt{\dfrac{1}{n_1} + \dfrac{1}{n_2}}}$$

where:

$$s_p = \sqrt{\frac{(n_1 - 1)s_1^2 + (n_2 - 1)s_2^2}{n_1 + n_2 - 2}}$$

$t$-Test Statistic for $\mu_1 - \mu_2$

Pooled Standard Deviation

Critical $t$ from $t$-distribution with $n_1 + n_2 - 2$ Degrees of Freedom

$\alpha$ = Significance level
One-tailed test, critical value = $t_\alpha$ or $-t_\alpha$
Two-tailed test, critical values = $\pm t_{\alpha/2}$

**Assumptions:**
1. Populations Are Normally Distributed.
2. Populations Have Equal Variances.
3. Samples Are Independent.
4. Measurements Are Interval or Ratio.

**Hypothesis Test for $\mu_d$, Paired Samples**

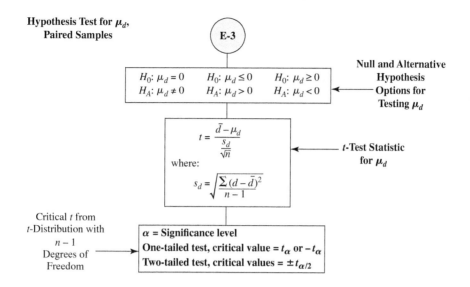

**E-3**

$H_0: \mu_d = 0$    $H_0: \mu_d \leq 0$    $H_0: \mu_d \geq 0$
$H_A: \mu_d \neq 0$    $H_A: \mu_d > 0$    $H_A: \mu_d < 0$

**Null and Alternative Hypothesis Options for Testing $\mu_d$**

$$t = \frac{\bar{d} - \mu_d}{\frac{s_d}{\sqrt{n}}}$$

where:

$$s_d = \sqrt{\frac{\sum(d - \bar{d})^2}{n - 1}}$$

**$t$-Test Statistic for $\mu_d$**

Critical $t$ from $t$-Distribution with $n - 1$ Degrees of Freedom

$\alpha$ = **Significance level**
**One-tailed test, critical value = $t_\alpha$ or $-t_\alpha$**
**Two-tailed test, critical values = $\pm t_{\alpha/2}$**

**Hypothesis Test for Difference Between Proportions, $p_1 - p_2$**

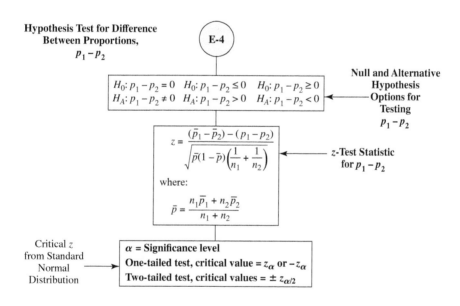

**E-4**

$H_0: p_1 - p_2 = 0$    $H_0: p_1 - p_2 \leq 0$    $H_0: p_1 - p_2 \geq 0$
$H_A: p_1 - p_2 \neq 0$    $H_A: p_1 - p_2 > 0$    $H_A: p_1 - p_2 < 0$

**Null and Alternative Hypothesis Options for Testing $p_1 - p_2$**

$$z = \frac{(\bar{p}_1 - \bar{p}_2) - (p_1 - p_2)}{\sqrt{\bar{p}(1 - \bar{p})\left(\frac{1}{n_1} + \frac{1}{n_2}\right)}}$$

where:

$$\bar{p} = \frac{n_1 \bar{p}_1 + n_2 \bar{p}_2}{n_1 + n_2}$$

**$z$-Test Statistic for $p_1 - p_2$**

Critical $z$ from Standard Normal Distribution

$\alpha$ = **Significance level**
**One-tailed test, critical value = $z_\alpha$ or $-z_\alpha$**
**Two-tailed test, critical values = $\pm z_{\alpha/2}$**

**Hypothesis Test for Difference Between Population Variances, $\sigma_1^2 - \sigma_2^2$**

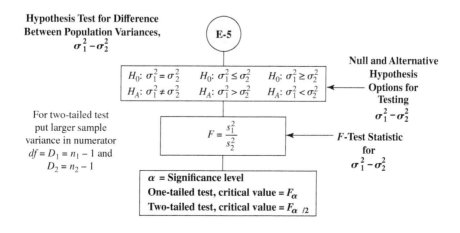

**E-5**

$H_0: \sigma_1^2 = \sigma_2^2$    $H_0: \sigma_1^2 \leq \sigma_2^2$    $H_0: \sigma_1^2 \geq \sigma_2^2$
$H_A: \sigma_1^2 \neq \sigma_2^2$    $H_A: \sigma_1^2 > \sigma_2^2$    $H_A: \sigma_1^2 < \sigma_2^2$

**Null and Alternative Hypothesis Options for Testing $\sigma_1^2 - \sigma_2^2$**

For two-tailed test put larger sample variance in numerator $df = D_1 = n_1 - 1$ and $D_2 = n_2 - 1$

$$F = \frac{s_1^2}{s_2^2}$$

**$F$-Test Statistic for $\sigma_1^2 - \sigma_2^2$**

$\alpha$ = **Significance level**
**One-tailed test, critical value = $F_\alpha$**
**Two-tailed test, critical value = $F_{\alpha/2}$**

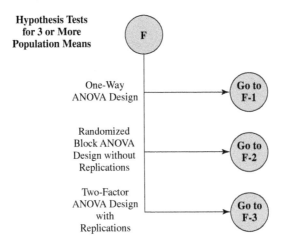

**Hypothesis Tests for 3 or More Population Means**

One-Way ANOVA Design → Go to F-1

Randomized Block ANOVA Design without Replications → Go to F-2

Two-Factor ANOVA Design with Replications → Go to F-3

**One-Way ANOVA Design for 3 or More Population Means**

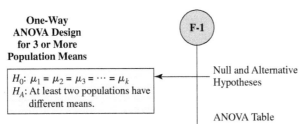

$H_0$: $\mu_1 = \mu_2 = \mu_3 = \cdots = \mu_k$
$H_A$: At least two populations have different means.

Null and Alternative Hypotheses

ANOVA Table

| Source of Variation | SS | df | MS | F-Ratio |
|---|---|---|---|---|
| Between samples | $SSB$ | $k - 1$ | $MSB$ | $\dfrac{MSB}{MSW}$ |
| Within samples | $\underline{SSW}$ | $n_T - k$ | $MSW$ | |
| Total | $SST$ | $n_T - 1$ | | |

Critical $F$ from $F$-Distribution with $D_1 = k - 1$ and $D_2 = n_T - k$ Degrees of Freedom

$\alpha$ = **Significance level**
**Critical value = $F_\alpha$**

If null hypothesis is rejected, compare all possible pairs $|\bar{x}_i - \bar{x}_j|$ to Tukey-Kramer critical range.

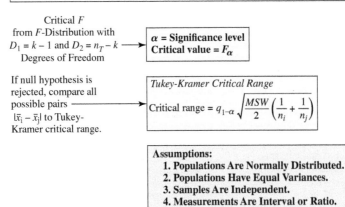

*Tukey-Kramer Critical Range*

Critical range = $q_{1-\alpha} \sqrt{\dfrac{MSW}{2}\left(\dfrac{1}{n_i} + \dfrac{1}{n_j}\right)}$

**Assumptions:**
1. **Populations Are Normally Distributed.**
2. **Populations Have Equal Variances.**
3. **Samples Are Independent.**
4. **Measurements Are Interval or Ratio.**

**Randomized Block ANOVA Design for 3 or More Population Means**

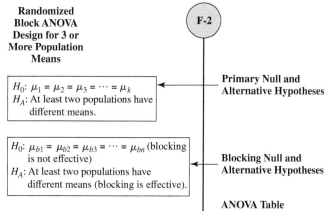

F-2

$H_0: \mu_1 = \mu_2 = \mu_3 = \cdots = \mu_k$
$H_A$: At least two populations have different means.

← **Primary Null and Alternative Hypotheses**

$H_0: \mu_{b1} = \mu_{b2} = \mu_{b3} = \cdots = \mu_{bn}$ (blocking is not effective)
$H_A$: At least two populations have different means (blocking is effective).

← **Blocking Null and Alternative Hypotheses**

**ANOVA Table**

| Source of Variation | SS | df | MS | F-Ratio |
|---|---|---|---|---|
| Between blocks | SSBL | $b-1$ | MSBL | $\dfrac{MSBL}{MSW}$ |
| Between samples | SSB | $k-1$ | MSB | $\dfrac{MSB}{MSW}$ |
| Within samples | SSW | $(k-1)(b-1)$ | MSW | |
| Total | SST | $n_T - 1$ | | |

Critical $F$ from $F$-Distribution →

$\alpha$ = **Significance level**
**Blocking critical value** = $F_\alpha$
$df = D_1 = b-1$ and $D_2 = (k-1)(b-1)$
**Primary critical value** = $F_\alpha$, $df = D_1 = k-1$
and $D_2 = (k-1)(b-1)$

**Assumptions:**
1. **Populations Are Normally Distributed.**
2. **Populations Have Equal Variances.**
3. **Observations within Samples Are Independent.**
4. **Measurements Are Interval or Ratio.**

If the null hypothesis is rejected, compare all $|\bar{x}_i - \bar{x}_j|$ to
Fisher's $LSD = t_{\alpha/2}\sqrt{MSW}\sqrt{\dfrac{2}{b}}$

**Two-Factor
ANOVA Design
with Replications**

F-3

**Factor A Null and
Alternative Hypotheses**

$H_0$: $\mu_{A1} = \mu_{A2} = \mu_{A3} = \cdots = \mu_{Ak}$
$H_A$: Not all Factor A means
are equal.

**Factor B Null and
Alternative Hypotheses**

$H_0$: $\mu_{B1} = \mu_{B2} = \mu_{B3} = \cdots = \mu_{Bn}$
$H_A$: Not all Factor B means
are equal.

$H_0$: Factors A and B do not interact
to affect the mean response.
$H_A$: Factors A and B do interact.

**Null and Alternative
Hypotheses for Testing
Whether the Two
Factors Interact**

**ANOVA Table**

| Source of Variation | SS | df | MS | F-Ratio |
|---|---|---|---|---|
| Factor A | $SS_A$ | $a - 1$ | $MS_A$ | $\dfrac{MS_A}{MSE}$ |
| Factor B | $SS_B$ | $b - 1$ | $MS_B$ | $\dfrac{MS_B}{MSE}$ |
| AB Interaction | $SS_{AB}$ | $(a-1)(b-1)$ | $MS_{AB}$ | $\dfrac{MS_{AB}}{MSE}$ |
| Error | $SSE$ | $n_T - ab$ | $MSE$ | |
| Total | $SST$ | $n_T - 1$ | | |

$\alpha$ = Significance level
**Factor A critical value** = $F_\alpha$, $df = D_1 = a - 1$ and $D_2 = n_T - ab$
**Factor B critical value** = $F_\alpha$, $df = D_1 = b - 1$ and $D_2 = n_T - ab$
**Interaction critical value** = $F_\alpha$, $df = D_1 = (a-1)(b-1)$ and $D_2 = n_T - ab$

**Assumptions:**
  1. **The Population Values for Each Combination of Pairwise Factor
     Levels Are Normally Distributed.**
  2. **The Variances for Each Population Are Equal.**
  3. **The Samples Are Independent.**
  4. **Measurements Are Interval or Ratio.**

## Using the Flow Diagrams

**Example Problem:** A travel agent in Florida is interested in determining whether there is a difference in the mean out-of-pocket costs incurred by customers on two major cruise lines. To test this, she has selected a simple random sample of 20 customers who have taken cruise line I and has asked these people to track their costs over and above the fixed price of the cruise. She did the same for a second simple random sample of 15 people who took cruise line II.

You can use the flow diagrams to direct you to the appropriate statistical tool.

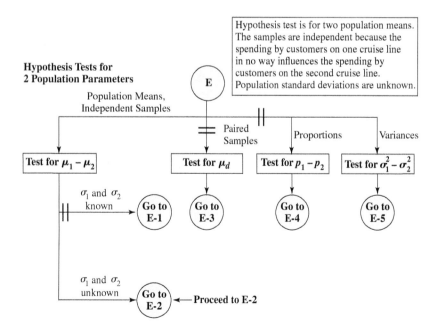

At E-2, we determine the null hypothesis to be

$$H_0: \mu_1 - \mu_2 = 0$$
$$H_A: \mu_1 - \mu_2 \neq 0$$

Next, we establish the test statistic as

$$t = \frac{(\bar{x}_1 - \bar{x}_2) - (\mu_1 - \mu_2)}{s_p \sqrt{\dfrac{1}{n_1} + \dfrac{1}{n_2}}}$$

where:

$$s_p = \sqrt{\frac{(n_1 - 1)s_1^2 + (n_2 - 1)s_2^2}{n_1 + n_2 - 2}}$$

Finally, the critical value is a $t$-value from the $t$-distribution with $20 + 15 - 2 = 33$ degrees of freedom. Note that if the degrees of freedom are not shown in the $t$ table, use Excel's T.INV.2T function to determine the $t$-value.

Thus, by using the flow diagrams and answering a series of basic questions, you should be successful in identifying the statistical tools required to address any problem or application covered in Chapters 8–12. You are encouraged to apply this process to the application problems and projects listed here.

# EXERCISES

## Integrative Application Problems

**SR.1.** Brandon Outdoor Advertising supplies neon signs to retail stores. A major complaint from its clients is that letters in the signs can burn out and leave the signs looking silly, depending on which letters stop working. The primary cause of neon letters not working is the failure of the starter unit attached to each letter. Starter units fail primarily based on turn-on/turn-off cycles. The present unit bought by Brandon averages 1,000 cycles before failure. A new manufacturer has approached Brandon claiming to have a model that is superior to the current unit. Brandon is skeptical but agrees to sample 50 starter units. It says it will buy from the new supplier if the sample results indicate the new unit is better. The sample of 50 gives the following values:

> Sample mean = 1,010 cycles
> Sample standard deviation = 48 cycles

Would you recommend changing suppliers?

**SR.2.** PestFree Chemicals has developed a new fungus preventative that may have a significant market among potato growers. Unfortunately, the actual extent of the fungus problem in any year depends on rainfall, temperature, and many other factors. To test the new chemical, PestFree has used it on 500 acres of potatoes and has used the leading competitor on an additional 500 acres. At the end of the season, 120 acres treated by the new chemical show significant levels of fungus infestation, whereas 160 of the acres treated by the leading chemical show significant infestation. Do these data provide statistical proof that the new product is superior to the leading competitor?

**SR.3.** Tucker Electronics has decided to try to do something about turnover among assembly-line workers at its plants. It implemented two trial personnel policies, one based on an improved hiring policy and the other based on increasing worker responsibility. These policies were put into effect at two different plants, with the following results:

|  | Plant 1 | Plant 2 |
|---|---|---|
|  | Improved Hiring | Increased Responsibility |
| Workers in trial group | 800 | 900 |
| Turnover proportion | 0.05 | 0.09 |

Do these data provide evidence that there is a difference between the turnover rates for the two trial policies?

**SR.4.** A Big 10 university has been approached by Wilson Sporting Goods. Wilson has developed a football designed specifically for practice sessions. Wilson would like to claim the ball will last for 500 practice hours before it needs to be replaced. Wilson has supplied six balls for use during spring and fall practice. The following data have been gathered on the hours of use before the ball must be replaced:

| Hours | |
|---|---|
| 551 | 511 |
| 479 | 435 |
| 440 | 466 |

Do you see anything wrong with Wilson claiming the ball will last 500 hours?

**SR.5.** The management of a chain of movie theaters believes the average weekend attendance at its downtown theater is greater than at its suburban theater. The following sample results were found from their accounting data:

|  | Downtown | Suburban |
|---|---|---|
| Number of weekends | 11 | 10 |
| Average attendance | 855 | 750 |
| Sample variance | 1,684 | 1,439 |

Do these data provide sufficient evidence to indicate there is a difference in average attendance? The company is also interested in whether there is a significant difference in the variability of attendance.

**SR.6.** A large mail-order company has placed an order for 5,000 thermal-powered fans to sit on wood-burning stoves from a supplier in Canada, with the stipulation that no more than 2% of the units will be defective. To check the shipment, the company tests a random sample of 400 fans and finds 11 defective. Should this sample evidence lead the company to conclude the supplier has violated the terms of the contract?

**SR.7.** A manufacturer of automobile shock absorbers is interested in comparing the durability of its shocks with that of its two biggest competitors. To make the comparison, a set of one each of the manufacturer's and of the competitor's shocks were randomly selected and installed on the rear wheels of each of six randomly selected cars of the same type. After the cars had been driven 20,000 miles, the strength of each test shock was measured, coded, and recorded.

| Car number | Manufacturer's | Competitor 1 | Competitor 2 |
|---|---|---|---|
| 1 | 8.8 | 9.3 | 8.6 |
| 2 | 10.5 | 9.0 | 13.7 |
| 3 | 12.5 | 8.4 | 11.2 |
| 4 | 9.7 | 13.0 | 9.7 |
| 5 | 9.6 | 12.0 | 12.2 |
| 6 | 13.2 | 10.1 | 8.9 |

Do these data present sufficient evidence to conclude there is a difference in the mean strengths of the three types of shocks after 20,000 miles?

**SR.8.** AstraZeneca is the maker of the stomach medicine Prilosec. Suppose the company's management is concerned about the cost of its medicines. The company's internal audit department selected a random sample of 300 purchases of Prilosec at drugstores and pharmacies in the United States. They wished to determine how much customers are spending on this medicine. In the sample, the mean price per 20-milligram tablet of Prilosec was $0.30. The sample

had a standard deviation of $0.08. Determine an estimate for the average range of costs for a tablet of Prilosec.

**SR.9.** A manufacturer of PC monitors is interested in the effects that the type of glass and the type of phosphor used in the manufacturing process have on the brightness of the monitors. The director of research and development has received anecdotal evidence that the type of glass does not affect the brightness of the monitor as long as phosphor type 2 is used. However, the evidence seems to indicate that the type of glass does make a difference if two other phosphor types are used. Here are data to validate this anecdotal evidence.

|  |  | Phosphor Type | |
|---|---|---|---|
| Glass Type | 1 | 2 | 3 |
| 1 | 279 | 307 | 287 |
|  | 254 | 313 | 290 |
|  | 297 | 294 | 285 |
| 2 | 243 | 253 | 252 |
|  | 245 | 232 | 236 |
|  | 267 | 223 | 278 |

Conduct a procedure to verify or repudiate the anecdotal evidence.

**SR.10.** The Vilmore Corporation is considering two word-processing programs for its PCs. One factor that will influence its decision is the ease of use in preparing a business report. Consequently, Jody Vilmore selected a random sample of nine typists from the clerical pool and asked them to type a typical report using both word processors. The typists then were timed (in seconds) to determine how quickly they could type one of the frequently used forms. The results were as follows:

| Typist | Processor 1 | Processor 2 |
|---|---|---|
| 1 | 82 | 75 |
| 2 | 76 | 80 |
| 3 | 90 | 70 |
| 4 | 55 | 58 |
| 5 | 49 | 53 |
| 6 | 82 | 75 |
| 7 | 90 | 80 |
| 8 | 45 | 45 |
| 9 | 70 | 80 |

Jody wishes to have an estimate of the smallest and biggest differences that might exist in the average times required for typing the business form using the two programs. Provide this information.

**SR.11.** The research department of an appliance manufacturing firm has developed a solid-state switch for its blender that the department claims will reduce the percentage of appliances being returned under the one-year full warranty by a range of 3% to 6%. To determine if the claim can be supported, the testing department selects a group of the blenders manufactured with the new switch and the old switch and subjects them to a normal year's worth of wear. Out of 250 blenders tested with the new switch, 9 would have been returned. Sixteen would have been returned out of the 250 blenders with the old switch. Use a statistical procedure to verify or refute the department's claim.

**SR-12.** The Ecco Company makes electronics products for distribution throughout the world. As a member of the quality department, you are interested in the warranty claims made by customers who have experienced problems with Ecco products. The file called **Ecco** contains data for a random sample of warranty claims. Large warranty claims not only cost the company money but also provide adverse publicity. The quality manager has asked you to provide her with a range of values that represent the percentage of warranty claims filed for more than $300. Provide this information for your manager.

# 13 Goodness-of-Fit Tests and Contingency Analysis

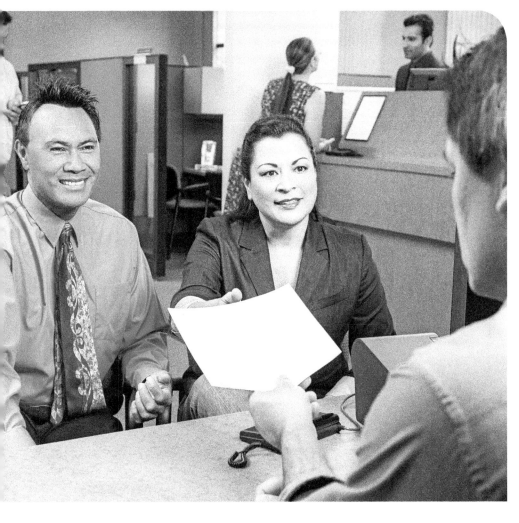

## WHY YOU NEED TO KNOW

So far, we have discussed numerous descriptive tools and techniques, as well as estimation and hypothesis tests for one and two populations, hypothesis tests using the *t*-distribution, and analysis of variance. However, as we have often mentioned, these statistical tools are limited to use only in those conditions for which they were originally developed. For example, to use the tests that employ the *t*-distribution, we assume that the sampled populations are normally distributed. In other situations, we assume that the binomial or Poisson distribution applies. How do we know which distribution applies to our situation? Fortunately, a statistical technique called *goodness of fit* exists that can help us answer this question. Using goodness-of-fit tests, we can decide whether a set of data comes from a specific hypothesized distribution.

You will also encounter many business situations in which the level of data measurement for the variable of interest is either nominal or ordinal, not interval or ratio. For example, a bank may use a code to indicate whether a customer is a good or poor credit risk. The bank may also have data for these customers that indicate, by a code, whether each person is buying or renting a home. The loan officer may be interested in determining whether credit-risk status is independent of homeownership. Because both credit risk and homeownership are qualitative, or categorical, variables, their measurement level is nominal and we cannot use the statistical techniques introduced in Chapters 8 through 12 to analyze this problem. We therefore need a new statistical tool to assist the manager in reaching an inference about the customer population. This statistical tool is contingency analysis. Contingency analysis is a widely used tool for analyzing the relationship between qualitative variables, one that decision makers in all business areas find helpful for data analysis.

**OUTCOME** **1** **13.1**

## Introduction to Goodness-of-Fit Tests

Many of the statistical procedures introduced in earlier chapters require that the sample data come from populations that are normally distributed. For example, when we use the *t*-distribution in confidence interval estimation or hypothesis testing about one or two population means, we assume that the population(s) of interest is (are) normally distributed. The *F*-test introduced in Chapters 11 and 12 is based on the assumption that the populations are normally distributed. But how can you determine whether these assumptions are satisfied? In other instances, you may wish to employ a particular probability distribution to help solve a problem related to an actual business process. To solve the problem, you may find it necessary to know whether the actual data from the process fit the probability distribution being considered. In such instances, you can use a statistical technique known as a *goodness-of-fit test*.

The term *goodness of fit* aptly describes the technique. Suppose Nordstrom's, a major retail department store, believes the proportions of customers who use each of the four entrances to its Seattle, Washington, store are the same. This would mean that customer arrivals are *uniformly* distributed across the four entrances. Suppose a sample of 1,000 customers is observed entering the store, and the entrance (East, West, North, South) selected by each customer is recorded. Table 13.1 shows the results of the sample. If the assumption about the entrances being used uniformly holds and if no sampling error was involved, we would expect one-fourth of the customers, or 250, to enter through each door. When we allow for the potential of sampling error, we would still expect close to 250 customers to enter through each entrance. The question is, how "good is the fit" between the sample data in Table 13.1 and the expected number of 250 people at each entrance? At what point do we no longer believe that the differences between what is actually observed at each entrance and what we expected can be attributed to sampling error? If these differences get too big, we will reject the uniformity assumption and conclude that customers prefer some entrances to others.

**TABLE 13.1 Customer Door Entrance Data**

| Entrance | Number of Customers |
|----------|---------------------|
| East | 260 |
| West | 290 |
| North | 230 |
| South | 220 |

### Chi-Square Goodness-of-Fit Test

The *chi-square goodness-of-fit test* is one of the statistical tests that we can use to determine whether the sample data come from any hypothesized distribution. Consider the following application.

**BUSINESS APPLICATION** **Conducting a Goodness-of-Fit Test**

**Town & Country Taxi** Town & Country Taxi operates taxi service in the city seven days a week. The operations manager is interested in matching the number of cabs in service with customer demand throughout the week.

Currently, the company runs the same number of taxis Monday through Friday, with reduced staffing on Saturday and Sunday.

This is because the operations manager believes that demand for taxicabs is fairly level throughout the week and about 25% less on weekends. The manager has decided to study demand to see whether the assumed demand pattern still applies.

The operations manager requested a random sample of 20 days for each day of the week that showed the number of customers using Town & Country cabs on each of the sample days. A portion of those data follows:

| Day | Customer Count | Day | Customer Count |
|---|---|---|---|
| Monday, May 6 | 325 | Monday, July 15 | 323 |
| Monday, October 7 | 379 | Wednesday, April 3 | 467 |
| Tuesday, July 2 | 456 | | |

**TABLE 13.2** Customer Count Data for the Town & Country Taxi Example

| Day | Total Customer Count |
|---|---|
| Sunday | 4,502 |
| Monday | 6,623 |
| Tuesday | 8,308 |
| Wednesday | 10,420 |
| Thursday | 11,032 |
| Friday | 10,754 |
| Saturday | 4,361 |
| Total | 56,000 |

For the 140 days observed, the total count was 56,000 customers. The total customer counts for each day of the week are shown in Table 13.2 and are graphed in Figure 13.1.

Recall that the operations manager at Town & Country Taxi based his cab supply on the premise that from Monday to Friday the customer demand is essentially the same, and on Saturdays and Sundays, it is down 25%. If this is so, how many of the 56,000 taxi customers would we expect on Monday? How many on Tuesday, and so forth? To figure out this demand, we determine weighting factors by allocating four units each to days Monday through Friday and three units each (representing the 25% reduction) to Saturday and Sunday. The total number of units is then $(5 \times 4) + (2 \times 3) = 26$. The proportion of total customers expected on each weekday is $4/26$, and the proportion expected on each weekend day is $3/26$. The expected number of customers on a weekday is $(4/26) \times 56,000 = 8,615.38$, and the expected number on each weekend day is $(3/26) \times 56,000 = 6,461.54$.

Figure 13.2 shows a graph with the actual sample data and the expected values. With the exception of what might be attributed to sampling error, if the taxi demand distribution assumed by the operations manager is correct, the actual frequencies for each day of the week should fit quite closely with the expected frequencies. As you can see in Figure 13.2, the actual data and the expected data do not match perfectly. However, is the difference enough to warrant changing how the manager schedules the taxis? The situation facing Town & Country Taxi is one for which a number of statistical tests have been developed. One of the most frequently used is the *chi-square goodness-of-fit test*. What we need to examine is how well the sample data fit the hypothesized distribution. The following null and alternative hypotheses can represent this:

$H_0$: The customer demand distribution is evenly spread through the weekdays and is 25% lower on the weekend.

$H_A$: The customer demand follows some other distribution.

**FIGURE 13.1 Graph of Actual Frequencies for Town & Country Taxi**

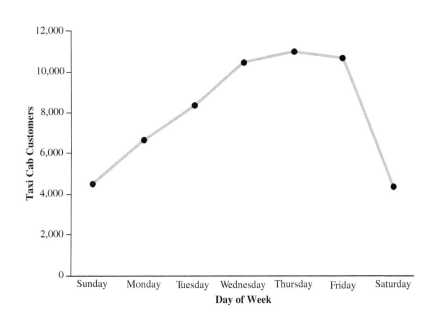

**FIGURE 13.2 Actual and Expected Frequencies for Town & Country Taxi**

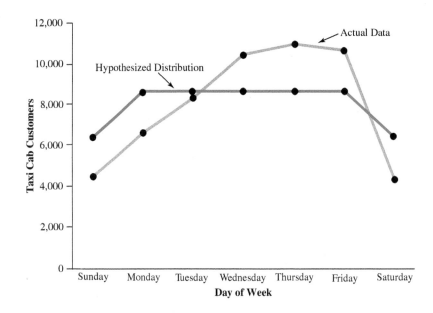

Equation 13.1 is for the chi-square goodness-of-fit test statistic. The logic behind this test is to determine how far the actual *observed frequency* is from the *expected frequency*. Because we are interested in whether a difference exists, positive or negative, we remove the effect of negative values by squaring the differences. In addition, how important this difference is really depends on the magnitude of the expected frequency (e.g., a difference of 5 is more important if the expected frequency is 10 than if the expected frequency is 1,000), so we divide the squared difference by the expected frequency. Finally, we sum these difference ratios for all days. This sum is a statistic that has an approximate chi-square distribution.

**Chi-Square Goodness-of-Fit Test Statistic**

$$\chi^2 = \sum_{i=1}^{k} \frac{(o_i - e_i)^2}{e_i}$$

(13.1)

where:

$o_i$ = Observed frequency for category $i$

$e_i$ = Expected frequency for category $i$

$k$ = Number of categories

The $\chi^2$ statistic is distributed approximately as a chi-square only if the sample size is large.

**Special Note**

A sample size of at least 30 is sufficient in most cases provided that none of the expected frequencies are less than 5.

We will discuss this issue of expected cell frequencies later. If the calculated chi-square statistic gets large, this is evidence to suggest that the fit of the actual data to the hypothesized distribution is not good and that the null hypothesis should be rejected.

Figure 13.3 shows the hypothesis-test process and results for this chi-square goodness-of-fit test. Note that the degrees of freedom for the chi-square test are equal to $k - 1$, where $k$ is the number of categories or observed cell frequencies. In this example, we have 7 categories corresponding to the days of the week, so the degrees of freedom are $7 - 1 = 6$. The critical value of 12.5916 is found in Appendix G for an upper-tailed test with 6 degrees of freedom and a significance level of 0.05.

The Excel 2016 function to determine the chi-square critical value is

= CHISQ.INV.RT(alpha,df)

= CHISQ.INV.RT(.05,6)

**FIGURE 13.3 Chi-Square Goodness-of-Fit Test for Town & Country Taxi**

**Hypotheses:**

$H_0$: Customer demand is evenly spread through the weekdays and is 25% lower on weekends.
$H_A$: Customer demand follows some other distribution.
$\alpha = 0.05$

|  | Total Customer Count | |
|---|---|---|
|  | Observed | Expected |
| Day | $o_i$ | $e_i$ |
| Sunday | 4,502 | 6,461.54 |
| Monday | 6,623 | 8,615.38 |
| Tuesday | 8,308 | 8,615.38 |
| Wednesday | 10,420 | 8,615.38 |
| Thursday | 11,032 | 8,615.38 |
| Friday | 10,754 | 8,615.38 |
| Saturday | 4,361 | 6,461.54 |
| Total | 56,000 | 56,000 |

**Test Statistic:**

$$\chi^2 = \sum_{i=1}^{k} \frac{(o_i - e_i)^2}{e_i} = \frac{(4{,}502 - 6{,}461.54)^2}{6{,}461.54} + \frac{(6{,}623 - 8{,}615.38)^2}{8{,}615.38} + \cdots + \frac{(4{,}361 - 6{,}461.54)^2}{6{,}461.54}$$

$$\chi^2 = 594.2 + 460.8 + \cdots + 682.9$$

$$\chi^2 = 3{,}335.6$$

$df = k - 1 = 7 - 1 = 6$

Rejection Region

$\alpha = 0.05$

$\chi^2_{0.05} = 12.5916$

**Decision Rule:**

If $\chi^2 > 12.5916$, reject $H_0$.
Otherwise, do not reject $H_0$.

Because $3{,}335.6 > 12.5916$, reject $H_0$.

Based on the sample data, we can conclude that the customer distribution is not the same as previously indicated.

The Excel 2016 function for a chi-square $p$-value is =CHISQ.TEST(Actual_range,Expected_range)

Small $p$-values provide evidence that the null hypothesis should be rejected. The actual_range refers to the observed values, and the expected_range to the expected values.

As Figure 13.3 indicates, $\chi^2 = 3{,}335.6 > 12.5916$, so the null hypothesis is rejected and the manager should conclude that the demand pattern does not match the previously assumed distribution. The data in Figure 13.3 indicate that demand is heavier than expected Wednesday through Friday and lighter than expected on the other days. The operations manager may now wish to increase the number of taxicabs in service on Wednesday, Thursday, and Friday to more closely approximate current demand patterns.

**EXAMPLE 13-1    Chi-Square Goodness-of-Fit Test**

**Central University Food Service**    One of the challenges in managing a food service on a college campus is predicting how many students will show up for meals. Central University operates three different food service facilities on campus. All three are the "all-you-can-eat" style where students pay a fixed fee to enter the dining hall and then can eat as much as they want from the various food stations in the hall. One of these is the River View Café on campus, which is open five days a week—

Monday through Friday—for breakfast, lunch, and dinner. The manager has been ordering food products for the three meals under the assumption that there is no difference in demand for various food products by day of the week. For example, at dinner, she provides pie as one of the dessert choices. She orders the same number of pies from a local bakery each day. Recently, the manager has received complaints that the café frequently runs out of pie. As a first step in analyzing this issue, the manager decides to conduct a test to determine if student demand for pie is *uniformly distributed* across the five days of the week or whether some other demand distribution applies. She starts by ordering enough pie every day so that there is no possible chance of running out and counts the number of pieces that are selected each day. A chi-square goodness-of-fit test can then be conducted using the following steps:

STEP 1 **Formulate the appropriate null and alternative hypotheses.**
Because the number of pieces of pie demanded is supposed to be the same across the five days of the week, the following null and alternative hypotheses are formed:

$H_0$: Distribution of pie demand is uniform across the five days.

$H_A$: Distribution of pie demand is not uniform.

STEP 2 **Specify the significance level.**
The test will be conducted using $\alpha = 0.05$.

> The Excel 2016 function to determine the chi-square critical value is
>
> $=$ **CHISQ.INV.RT(alpha,df)**
> $=$ **CHISQ.INV.RT(.05,4)**

STEP 3 **Determine the critical value.**
The critical value for the goodness-of-fit test is a chi-square value from the chi-square distribution, with $k - 1 = 5 - 1 = 4$ degrees of freedom, and $\alpha = 0.05$ is 9.4877.

STEP 4 **Collect the sample data and compute the chi-square test statistic.**
The manager collects the data over the course of a representative four-week period. The following data represent the number of pieces of pie used on each day of the week:

| Day | Mon | Tue | Wed | Thu | Fri |
|---|---|---|---|---|---|
| **Pieces of Pie** | 358 | 402 | 577 | 403 | 380 |

Students took a total of 2,120 pieces of pie over the test period. Under the hypothesis of a uniform distribution, 20% of this total (424) should be selected on each day of the week. Thus, the expected number of pieces needed each day, if the null hypothesis of a uniform distribution is true, is 424. This number is the expected cell frequency.

We use Equation 13.1 to form the test statistic based on these sample data:

$$\chi^2 = \sum_{i=1}^{k} \frac{(o_i - e_i)^2}{e_i} = \frac{(358 - 424)^2}{424} + \frac{(402 - 424)^2}{424}$$
$$+ \frac{(577 - 424)^2}{424} + \frac{(403 - 424)^2}{424} + \frac{(380 - 424)^2}{424}$$
$$= 72.231$$

STEP 5 **Reach a decision.**
The decision rule is:

If $\chi^2 > 9.4877$, reject $H_0$.

Otherwise, do not reject $H_0$.

Because $\chi^2 = 72.231 > 9.4877$, reject the null hypothesis.

The test can also be conducted using the Excel function CHISQ.TEST: $=$CHISQ.TEST(actual_range, expected_range). The function returns the *p*-value for the hypothesis test, which is 0.000000 to six decimal places in this example. Small *p*-values provide evidence to reject the null hypothesis and conclude that the demand for pie is not equal across the five days of the week.

STEP 6 **Draw a conclusion.**

The manager of the River View Café should conclude that the demand for pie is not equal across the five days of the week. As a result, she will need to alter the ordering process to account for a nonuniform demand distribution for this food item.

TRY EXERCISE 13-2 (pg. 508)

---

**BUSINESS APPLICATION**

## Using Software to Conduct a Goodness-of-Fit Test

**Excel Tutorial**

**Woodtrim Products, Inc.** Woodtrim Products, Inc., makes wood moldings, door-frames, and window frames. It purchases lumber from mills throughout New England and eastern Canada. The first step in the production process is to rip the lumber into narrower strips. Different widths are used for different products. For example, wider pieces with no imperfections are used to make door and window frames. Once an operator decides on the appropriate width, that information is locked into a computer and a ripsaw automatically cuts the board to the desired size. The manufacturer of the saw claims that the ripsaw cuts an average deviation of zero from target and that the differences from target are normally distributed, with a standard deviation of 0.01 inch.

Woodtrim has recently become concerned that the ripsaw may not be cutting to the manufacturer's specifications because operators at other machines downstream in the production process are finding excessive numbers of ripped pieces that are too wide or too narrow.

A quality improvement team (QIT) has started to investigate the problem. Team members selected a random sample of 300 boards just as they came off the ripsaw. To provide a measure of control, the only pieces sampled in the initial study had stated widths of $2^{7}/_{8}$ (2.875) inches. Each piece's width was measured halfway from its end. A portion of the differences between the target 2.875 inches and the actual measured width are shown in Figure 13.4. The full data set is in the file **Woodtrim**. The team can use these data and the chi-square goodness-of-fit testing procedure to test the following null and alternative hypotheses:

$H_0$: The differences are normally distributed, with $\mu = 0$ and $\sigma = 0.01$.

$H_A$: The differences are not normally distributed, with $\mu = 0$ and $\sigma = 0.01$.

This example differs slightly from the previous examples because the hypothesized distribution is continuous rather than discrete. Thus, we must organize the data into a grouped-data frequency distribution (see Chapter 2), as shown in the Excel output in Figure 13.5. Our choice of classes requires careful consideration. The chi-square goodness-of-fit test compares the actual cell frequencies with the expected cell frequencies. The test statistic from Equation 13.1,

$$\chi^2 = \sum_{i=1}^{k} \frac{(o_i - e_i)^2}{e_i}$$

is approximately chi-square distributed if the expected cell frequencies are large. Because the expected cell frequencies are used in computing the test statistic, the general recommendation is that the goodness-of-fit test be performed only when all expected cell frequencies are

**FIGURE 13.4 Excel 2016 Woodtrim Products Test Data**

Excel 2016 Instructions
1. Open file: **Woodtrim.xlsx**.

| | A | B | C | D |
|---|---|---|---|---|
| 1 | Sample | Actual Width | Target Width | Difference |
| 2 | 1 | 2.870 | 2.875 | -0.005 |
| 3 | 2 | 2.863 | 2.875 | -0.012 |
| 4 | 3 | 2.885 | 2.875 | 0.010 |
| 5 | 4 | 2.872 | 2.875 | -0.003 |
| 6 | 5 | 2.891 | 2.875 | 0.016 |
| 7 | 6 | 2.893 | 2.875 | 0.018 |
| 8 | 7 | 2.868 | 2.875 | -0.007 |
| 9 | 8 | 2.861 | 2.875 | -0.014 |
| 10 | 9 | 2.889 | 2.875 | 0.014 |

**FIGURE 13.5** **Excel 2016 Results—Goodness-of-Fit Test for the Woodtrim Example**

| Classes | Observed Frequencies | Normal Distribution Probability | Expected Frequency | $(o-e)^2/e$ |
|---|---|---|---|---|
| less than -0.02 | 0 | 0.02275 | 6.83 | 6.825 |
| -0.02 and under -0.01 | 42 | 0.13591 | 40.77 | 0.037 |
| -0.01 and under 0.00 | 133 | 0.34134 | 102.40 | 9.142 |
| 0.00 and under 0.01 | 75 | 0.34134 | 102.40 | 7.333 |
| 0.01 and under 0.02 | 47 | 0.13591 | 40.77 | 0.951 |
| 0.02 and over | 3 | 0.02275 | 6.83 | 2.144 |
| | | | $X^2$ calc = | 26.432 |
| | | | p-value = | 0.0001 |

at least 5. If any of the cells have expected frequencies less than 5, the cells should be combined in a meaningful way such that the expected frequencies are at least 5. We have chosen to use $k = 6$ classes. The number of classes is your choice. You can perform the chi-square goodness-of-fit test using Excel. (The tutorials that accompany this text take you through the specific steps required to complete this example.) Figure 13.5 shows the normal distribution probabilities, expected cell frequencies, and chi-square calculation. The calculated chi-square statistic is $\chi^2 = 26.432$. The p-value associated with $\chi^2 = 26.432$ and $6 - 1 = 5$ degrees of freedom is 0.0001. Therefore, because p-value = 0.0001 is less than any reasonable level of alpha, we reject the null hypothesis and conclude that the ripsaw is not currently meeting the manufacturer's specification. The saw errors are not normally distributed with mean equal to 0 and standard deviation equal to 0.01.

**Special Note**

Note that in this case, because the null hypothesis specified both the mean and the standard deviation, we computed the normal distribution probabilities using these values. However, if the mean and/or the standard deviation had not been specified, we would use the sample mean and standard deviation in the probability computation. We would lose 1 additional degree of freedom for each parameter that was estimated from the sample data. This is true any time sample statistics are specified in place of population parameters in the hypothesis.

---

**EXAMPLE 13-2** **Goodness-of-Fit Test**

**Call Center Support, Discount Electronics**   Discount Electronics sells a wide variety of electronics products over the Internet. One of the features that differentiates Discount Electronics from other online electronics retailers is the 24-7 call center support it provides to customers. A key component is that the answers provided by the call center representatives are accurate. Because of the high volume of calls received each week, the call center quality-assurance manager can't analyze every call for accuracy of assistance provided. Instead, every day, 10 calls are randomly recorded for quality purposes. The manager has a policy that if two or more of the 10 calls provide incorrect answers, all call center representatives are required to come in after hours for a special training session. If the review finds one or fewer incorrect answers, no training session is required that week. The manager believes that if the call center representatives are doing their job, at most 10% of the customers who call in will get a wrong answer. At issue is whether the manager can evaluate this sampling plan using a binomial distribution with $n = 10$ and $p = 0.10$. To test this, a goodness-of-fit test can be performed using the following steps:

**STEP 1** **Formulate the appropriate null and alternative hypotheses.**
In this case, the null and alternative hypotheses are

$H_0$: Distribution of incorrect answers is binomial, with $n = 10$ and $p = 0.10$.
$H_A$: Distribution is not binomial, with $n = 10$ and $p = 0.10$.

> The Excel 2016 function to determine the chi-square critical value is
>
> **=CHISQ.INV.RT(alpha,df)**
> **=CHISQ.INV.RT(.025,3)**

**STEP 2  Specify the level of significance.**
The test will be conducted using $\alpha = 0.025$.

**STEP 3  Determine the critical value.**
The critical value depends on the degrees of freedom and the level of significance. The degrees of freedom is equal to $k - 1$, where $k$ is the number of categories for which observed and expected frequencies are recorded. In this case, the managers have set up the following four groups:

Incorrect answers: 0, 1, 2, 3 and over

Therefore, $k = 4$, and the degrees of freedom is $4 - 1 = 3$. The chi-square critical value for $\alpha = 0.025$ found in Appendix G is 9.3484.

**STEP 4  Collect the sample data and compute the chi-square test statistic using Equation 13.1.**
The company selected a simple random sample of 100 days' test results from past recorded calls and counted the number of incorrect answers in each sample of 10 calls. The following table shows the computations for the chi-square statistic:

| Incorrect Answers | $o$ Observed Defects | Binomial Probability $n = 10, p = 0.10$ | $e$ Expected Frequency | $\dfrac{(o_i - e_i)^2}{e_i}$ |
|---|---|---|---|---|
| 0 | 30 | 0.3487 | 34.87 | 0.6802 |
| 1 | 40 | 0.3874 | 38.74 | 0.0410 |
| 2 | 20 | 0.1937 | 19.37 | 0.0205 |
| 3 and over | 10 | 0.0702 | 7.02 | 1.2650 |
| Total | 100 | | | 2.0067 |

The calculated chi-square test statistic is $\chi^2 = 2.0067$.

**STEP 5  Reach a decision.**
Because $\chi^2 = 2.0067$ is less than the critical value of 9.3484, we do not reject the null hypothesis.

The $p$-value for the example can be found using the Excel function =CHISQ.TEST(actual_range,expected_range) = 0.5713. Because the $p$-value is greater than alpha, we do not reject the null hypothesis.

**STEP 6  Draw a conclusion.**
The binomial distribution may be the appropriate distribution to describe the company's sampling plan.

**TRY EXERCISE 13-1** (pg. 508)

---

**EXAMPLE 13-3  Goodness-of-Fit Test**

**University Internet Service**    Students in a computer information systems class at a major university have established an Internet service provider (ISP) company for the university's students, faculty, and staff. Customers of this ISP connect via a wireless signal available throughout the campus and surrounding business area. Capacity is always an issue for an ISP, and the students had to estimate the capacity demands for their service. Suppose before they opened for business, the students conducted a survey of likely customers. Based on this survey, they estimated that demand during the late afternoon and evening hours is Poisson distributed (refer to Chapter 5) with a mean equal to 10 users per hour. Based on this assumption, the students developed the ISP with the capacity to handle 20 users simultaneously. However, they have lately been receiving complaints from customers who say they have been denied access to the system because 20 users are already online. The students are now interested in determining whether the demand is still Poisson distributed with a mean equal to 10 per hour. To test this, they have collected data on the number of user requests for ISP access for 225 randomly selected time periods during the heavy-use hours. The following steps can be used to conduct the statistical test:

STEP 1   **State the appropriate null and alternative hypotheses.**
The null and alternative hypotheses are

$H_0$: The demand is Poisson distributed with mean equal to 10 users per time period.

$H_A$: The demand is not distributed as a Poisson distribution with mean equal to 10 per period.

STEP 2   **Specify the level of significance.**
The hypothesis test will be conducted using $\alpha = 0.05$.

STEP 3   **Determine the critical value.**
The critical value depends on the level of significance and the number of degrees of freedom. The degrees of freedom is equal to $k - 1$, where $k$ is the number of categories. In this case, after collapsing the categories to get the expected frequencies to be at least 5, we have 13 categories (see Step 4). Thus, the degrees of freedom for the chi-square critical value is $13 - 1 = 12$. For 12 degrees of freedom and a level of significance equal to 0.05, from Appendix G we find a critical value of $\chi^2 = 21.0261$. Thus the decision rule is:

If $\chi^2 > 21.0261$, reject the null hypothesis.
Otherwise, do not reject.

> The Excel 2016 function to determine the chi-square critical value is
>
> = CHISQ.INV.RT(alpha,df)
> = CHISQ.INV.RT(.05,12)

STEP 4   **Collect the sample data and compute the chi-square test statistic using Equation 13.1.**
A random sample of 225 time periods was selected, and the number of users requesting access to the ISP in each time period was recorded. The observed frequencies based on the sample data are as follows:

| Number of Requests | Observed Frequency | Number of Requests | Observed Frequency |
|---|---|---|---|
| 0 | 0 | 10 | 18 |
| 1 | 2 | 11 | 14 |
| 2 | 1 | 12 | 17 |
| 3 | 3 | 13 | 18 |
| 4 | 4 | 14 | 25 |
| 5 | 3 | 15 | 28 |
| 6 | 8 | 16 | 23 |
| 7 | 6 | 17 | 17 |
| 8 | 11 | 18 | 9 |
| 9 | 7 | 19 and over | 11 |
|  |  | Total | 225 |

To compute the chi-square test statistic, we must determine the expected frequencies. We start by determining the probability for each number of user requests based on the hypothesized distribution. (Poisson with $\lambda t = 10$.)

We then calculate the expected frequencies by multiplying each probability by the total observed frequency of 225. These results are as follows:

| Number of Requests | Observed Frequency | Poisson Probability $\lambda t = 10$ | Expected Frequency |
|---|---|---|---|
| 0 | 0 | 0.0000 | 0.00 |
| 1 | 2 | 0.0005 | 0.11 |
| 2 | 1 | 0.0023 | 0.52 |
| 3 | 3 | 0.0076 | 1.71 |
| 4 | 4 | 0.0189 | 4.25 |
| 5 | 3 | 0.0378 | 8.51 |
| 6 | 8 | 0.0631 | 14.20 |
| 7 | 6 | 0.0901 | 20.27 |
| 8 | 11 | 0.1126 | 25.34 |
| 9 | 7 | 0.1251 | 28.15 |
| 10 | 18 | 0.1251 | 28.15 |
| 11 | 14 | 0.1137 | 25.58 |
| 12 | 17 | 0.0948 | 21.33 |
| 13 | 18 | 0.0729 | 16.40 |
| 14 | 25 | 0.0521 | 11.72 |
| 15 | 28 | 0.0347 | 7.81 |
| 16 | 23 | 0.0217 | 4.88 |
| 17 | 17 | 0.0128 | 2.88 |
| 18 | 9 | 0.0071 | 1.60 |
| 19 and over | 11 | 0.0072 | 1.62 |
| Total | 225 | 1.0000 | 225.00 |

Now we need to check if any of the expected cell frequencies are less than 5. In this case, there are several instances where this is so. To deal with this, we collapse the number of categories so that all expected frequencies are at least 5. Doing this gives the following:

| Number of Requests | Observed Frequency | Poisson Probability $\lambda t = 10$ | Expected Frequency |
|---|---|---|---|
| 4 or fewer | 10 | 0.0293 | 6.59 |
| 5 | 3 | 0.0378 | 8.51 |
| 6 | 8 | 0.0631 | 14.20 |
| 7 | 6 | 0.0901 | 20.27 |
| 8 | 11 | 0.1126 | 25.34 |
| 9 | 7 | 0.1251 | 28.15 |
| 10 | 18 | 0.1251 | 28.15 |
| 11 | 14 | 0.1137 | 25.58 |
| 12 | 17 | 0.0948 | 21.33 |
| 13 | 18 | 0.0729 | 16.40 |
| 14 | 25 | 0.0521 | 11.72 |
| 15 | 28 | 0.0347 | 7.81 |
| 16 or more | 60 | 0.0488 | 10.98 |
| Total | 225 | 1.0000 | 225.00 |

Now we can compute the chi-square test statistic using Equation 13.1 as follows:

$$\chi^2 = \sum_{i=1}^{k} \frac{(o_i - e_i)^2}{e_i}$$

$$= \frac{(10 - 6.59)^2}{6.59} + \frac{(3 - 8.51)^2}{8.51} + \cdots + \frac{(60 - 10.98)^2}{10.98}$$

$$= 338.5$$

STEP 5 **Reach a decision.**
Because $\chi^2 = 338.5 > 21.0261$, we reject the null hypothesis.

STEP 6 **Draw a conclusion.**
The demand is not Poisson distributed with a mean of 10. The students should conclude that either the mean demand per period has increased from 10 or the distribution is not Poisson or both. They may need to add more capacity to the ISP business.

TRY EXERCISE 13-3 (pg. 508)

## 13.1 EXERCISES

### Skill Development

**13-1.** A large retailer receives shipments of batteries for consumer electronic products in packages of 50 batteries. The packages are held at a distribution center and are shipped to retail stores as requested. Because some packages may contain defective batteries, the retailer randomly samples 400 packages from its distribution center and tests to determine whether the batteries are defective or not. The most recent sample of 400 packages revealed the following observed frequencies for defective batteries per package:

| # of Defective Batteries per Package | Frequency of Occurrence |
|---|---|
| 0 | 165 |
| 1 | 133 |
| 2 | 65 |
| 3 | 28 |
| 4 or more | 9 |

The retailer's managers would like to know if they can evaluate this sampling plan using a binomial distribution with $n = 50$ and $p = 0.02$. Test at the $\alpha = 0.01$ level of significance.

**13-2.** The following frequency distribution shows the number of times an outcome was observed from the toss of a die. Based on the frequencies that were observed from 2,400 tosses of the die, can it be concluded at the 0.05 level of significance that the die is fair?

| Outcome | Frequency |
|---|---|
| 1 | 352 |
| 2 | 418 |
| 3 | 434 |
| 4 | 480 |
| 5 | 341 |
| 6 | 375 |

**13-3.** Based on the sample data in the following frequency distribution, conduct a test to determine whether the population from which the sample data were selected is Poisson distributed with mean equal to 6. Test using $\alpha = 0.05$.

| x | Frequency | x | Frequency |
|---|---|---|---|
| 2 or less | 7 | 9 | 53 |
| 3 | 29 | 10 | 35 |
| 4 | 26 | 11 | 28 |
| 5 | 52 | 12 | 18 |
| 6 | 77 | 13 | 13 |
| 7 | 77 | 14 or more | 13 |
| 8 | 72 | Total | 500 |

**13-4.** A chi-square goodness-of-fit test will be conducted to test whether a population is normally distributed. No statement has been made regarding the value of the population mean and standard deviation. A frequency distribution has been formed based on a random sample of 1,000 values. The frequency distribution has $k = 8$ classes. Assuming that the test is to be conducted at the $\alpha = 0.10$ level, determine the correct decision rule to be used.

**13-5.** An experiment is run that is claimed to have a binomial distribution with $p = 0.15$ and $n = 18$, and the number of successes is recorded. The experiment is conducted 200 times with the following results:

| Number of Successes | 0 | 1 | 2 | 3 | 4 | 5 |
|---|---|---|---|---|---|---|
| Observed Frequency | 80 | 75 | 39 | 6 | 0 | 0 |

Using a significance level of 0.01, is there sufficient evidence to conclude that the distribution is binomial with $p = 0.15$ and $n = 18$?

**13-6.** Data collected from a hospital emergency room reflect the number of patients per day that visited the emergency room due to cardiac-related symptoms. Suppose it is believed that the number of cardiac patients entering the emergency room per day over a two-month period has a Poisson distribution with a mean of 8 patients per day.

| 6 | 7 | 9 | 7 | 5 | 6 | 7 | 7 | 5 | 10 |
| 9 | 9 | 7 | 2 | 8 | 5 | 7 | 10 | 6 | 7 |
| 12 | 12 | 10 | 8 | 8 | 14 | 7 | 9 | 10 | 7 |
| 4 | 9 | 6 | 4 | 11 | 9 | 10 | 7 | 5 | 10 |
| 8 | 8 | 10 | 7 | 9 | 2 | 10 | 12 | 10 | 9 |
| 8 | 11 | 7 | 9 | 11 | 7 | 16 | 7 | 9 | 10 |

Use a chi-square goodness-of-fit test to determine if the data come from a Poisson distribution with mean of 8. Test using a significance level of 0.01.

## Business Applications

**13-7.** The loan manager at a New York City branch bank keeps track of the number of loan applicants who visit his branch's loan department per week. Having enough loan officers available is one of the ways of providing excellent service. Over the last year, the loan manager accumulated the following data:

| Number of Customers | 0 | 1 | 2 | 3 | 4 | 5 | ≥6 |
|---|---|---|---|---|---|---|---|
| Frequency | 1 | 2 | 9 | 11 | 14 | 6 | 9 |

From previous years, the manager believes that the number of customer arrivals has a Poisson distribution with an average of 3.5 loan applicants per week. Determine if the loan officer's belief is correct using a significance level of 0.025.

**13-8.** Suppose that managers of a major book publisher believe that the occurrence of typographical errors in the books the company publishes is Poisson distributed with a mean of 0.2 per page. Because of some customer quality complaints, the managers have arranged to conduct a test to determine if the error distribution still holds. A total of 400 pages were randomly selected, and the number of errors per page was counted. These data are summarized in the following frequency distribution:

| Errors | Frequency |
|---|---|
| 0 | 335 |
| 1 | 56 |
| 2 | 7 |
| 3 | 2 |
| Total | 400 |

Conduct the appropriate hypothesis test using a significance level equal to 0.01. Discuss the results.

**13-9.** The Baltimore Steel and Pipe Company recently developed a new pipe product for a customer. According to specifications, the pipe is supposed to have an average outside diameter of 2.00 inches with a standard deviation equal to 0.10 inch, and the outside diameters are to be normally distributed. Before going into full-scale production, the company selected a random sample of 30 sections of pipe from the initial test run. The following data were recorded:

| Pipe Section | Diameter (inches) | Pipe Section | Diameter (inches) |
|---|---|---|---|
| 1 | 2.04 | 16 | 1.96 |
| 2 | 2.13 | 17 | 1.89 |
| 3 | 2.07 | 18 | 1.99 |
| 4 | 1.99 | 19 | 2.13 |
| 5 | 1.90 | 20 | 1.90 |
| 6 | 2.06 | 21 | 1.91 |
| 7 | 2.19 | 22 | 1.95 |
| 8 | 2.01 | 23 | 2.18 |
| 9 | 2.05 | 24 | 1.94 |
| 10 | 1.98 | 25 | 1.93 |
| 11 | 1.95 | 26 | 2.08 |
| 12 | 1.90 | 27 | 1.82 |
| 13 | 2.10 | 28 | 1.94 |
| 14 | 2.02 | 29 | 1.96 |
| 15 | 2.11 | 30 | 1.81 |

a. Using a significance level of 0.01, perform the appropriate test.
b. Based on these data, should the company conclude that it is meeting the product specifications? Explain your reasoning.

**13-10.** The Cincinnati Dowel & Wood Products Co. manufactures wood dowels and wood turnings. Four-inch-diameter boat plugs are one of its products. The quality-control procedures aimed at maintaining the 4-inch diameter are valid only if the diameters have a normal distribution. The quality-control manager obtained the following summary diameters taken from randomly selected boat plugs on the production line:

| Interval | Frequency | Interval | Frequency |
|---|---|---|---|
| <3.872 | 4 | 4.001–4.025 | 8 |
| 3.872–3.916 | 6 | 4.026–4.052 | 2 |
| 3.917–3.948 | 11 | 4.053–4.084 | 4 |
| 3.949–3.975 | 9 | 4.085–4.128 | 0 |
| 3.976–4.000 | 5 | > 4.128 | 1 |

The boat plug diameters are specified to have a normal distribution with a mean of 4 inches and a standard deviation of 0.10 inch. Determine if the distribution of the 4-inch boat plugs is currently adhering to specification. Use a chi-square goodness-of-fit test and a significance level of 0.05.

## Computer Software Exercises

**13-11.** The owners of Big Boy Burgers are considering remodeling their facility to include a drive-thru window. A key factor in determining how to design the drive-thru is the distribution of customer service times. The Big Boy owners have received service time data from a similar operation in another city. The data in the file called **Big Boy** reflect the service time per car.

Based on these sample data, is there sufficient evidence to conclude that the service times are not normally distributed? Test using the chi-square distribution and $\alpha = 0.05$.

**13-12.** It has been reported that revenues at Disneyland and Disney World have been increasing faster than attendance, which means it costs more to spend time in the Magic Kingdom (*source:* Jason Garcia, "Disney parks visitor spending is growing faster than attendance," *Orlando Sentinel*, https://skift.com, Mar. 16, 2014). Suppose executives at the Walt Disney Company are interested in estimating the mean spending per capita for people who visit Disney World in Orlando, Florida. Since they do not know the population standard deviation, they plan to use the *t*-distribution (see Chapter 9) to conduct the test. However, they realize that the *t*-distribution requires that the population be normally distributed. Assume that 600 customers were randomly surveyed, and the amount they spent during their stay at Disney World was recorded. These data are in the file called **Disney**. Before using these sample data to estimate the population mean, the Disney executives wish to test to determine whether the population is normally distributed.

a. State the appropriate null and alternative hypotheses.

b. Organize the data into six classes and form the grouped data frequency distribution (refer to Chapter 2).

c. Using the sample mean and sample standard deviation, calculate the expected frequencies, assuming that the null hypothesis is true.

d. Conduct the test statistic and compare it to the appropriate critical value for a significance level equal to 0.05. What conclusion should the executives reach? Discuss.

**13-13.** Again, working with the data in Exercise 13-11, the number of cars that arrive in each 10-minute period is a factor that will determine whether Big Boy has the capacity to handle the drive-thru business. In addition to studying the service times, the owners counted the number of cars that arrived at a deli in the nearby town in a sample of 10-minute time periods. These data are in a file called **Drive-Thru Traffic**.

Based on these data, is there evidence to conclude that the arrivals are not Poisson distributed? State the appropriate null and alternative hypotheses, and test using a significance level of 0.025.

**13-14.** Ellison Mechanical manufactures automatic shutoff valves that can prevent extensive water damage from plumbing failures. The valves contain sensors that cut off water flow in the event of a leak, thereby preventing flooding. One important characteristic is the time (in milliseconds) required for the sensor to detect the water flow. The data obtained for four different shutoff valves are contained in the file titled **Waterflow**. Based on these data, test to determine if the sensor response times are normally distributed. Use a chi-square goodness-of-fit test and a significance level of 0.05. Use five groups of equal width to conduct the test.

**13-15.** Drivers who use the bridges in the San Francisco area have the option of paying the tolls using a system called FasTrak (*source:* "All electronic tolling on the Golden Gate Bridge," www.bayareafastrak.org). Suppose the regional transportation director is interested in increasing the percentage of drivers who use this method as opposed to stopping at the tollbooth to pay cash. Before beginning a comprehensive advertising campaign to promote FasTrak, the manager wishes to estimate the percentage of drivers who are currently using the system. To do this, 15 vehicles each day are tracked through the toll lanes of the bridges. The numbers of drivers who use FasTrak to pay their toll for a period of three months appear in the file titled **Fastrak**.

Determine if the distribution of the number of FasTrak users could be described as binomial with a population proportion equal to 0.50. Use a chi-square goodness-of-fit test and a significance level of 0.05.

---

**OUTCOME 2**     **13.2**     **Introduction to Contingency Analysis**

In Chapters 9 and 10, you were introduced to hypothesis tests involving one and two population proportions. Although these techniques are useful in many cases, you will also encounter many situations that involve multiple population proportions. For example, a mutual fund company offers six different mutual funds. The president of the company may wish to determine if the proportion of customers who select each mutual fund is related to the four sales regions in which the customers reside. A hospital administrator who collects service-satisfaction data from patients might be interested in determining whether there is a significant difference in patient rating by hospital department. A personnel manager for a large corporation might be interested in determining whether there is a relationship between level of employee job satisfaction and job classification. In each of these cases, the proportions relate to characteristic categories of the variable of interest. The six mutual funds, four sales regions, hospital departments, and job classifications are all specific categories.

These situations involving categorical data call for a new statistical tool known as *contingency analysis* to help analysts make decisions when multiple proportions are involved. Contingency analysis can be used when a level of data measurement is either nominal or ordinal and the values are determined by counting the number of occurrences in each category.

## 2 × 2 Contingency Tables

BUSINESS APPLICATION    Applying Contingency Analysis

**Dalgarno Photo, Inc.**    Dalgarno Photo, Inc., gets much of its business taking photographs for college yearbooks. Dalgarno hired a first-year masters of business administration (MBA) student to develop the survey it mailed to 850 yearbook editors at the colleges and universities in its market area. The representatives were unaware that Dalgarno Photo had developed the survey.

The survey asked about the photography and publishing activities associated with yearbook development. For instance, what photographer and publisher services did the schools use, and what factors were most important in selecting services? The survey instrument contained 30 questions, which were coded into 137 separate variables.

Among his many interests in this study, Dalgarno's marketing manager questioned whether college funding source and gender of the yearbook editor were related in some manner. To analyze this issue, we examine these two variables more closely. Source of university funding is a categorical variable, coded as follows:

$$1 = \text{Private funding}$$
$$2 = \text{State funding}$$

Of the 221 respondents who provided data for this variable, 155 came from privately funded colleges or universities and 66 were from state-funded institutions.

The second variable, gender of the yearbook editor, is also a categorical variable, with two response categories, coded as follows:

$$1 = \text{Male}$$
$$2 = \text{Female}$$

Of the 221 responses to the survey, 164 were from females and 57 were from males.

In cases in which both the variables of interest are categorical and the decision maker is interested in determining whether a relationship exists between the two, analysts use a statistical technique known as contingency analysis. We first set up a two-dimensional table called a **contingency table**. The contingency table for these two variables is shown in Table 13.3.

Table 13.3 shows that 14 of the respondents were males from schools that are privately funded. The numbers at the extreme right and along the bottom are called the *marginal frequencies*. For example, 57 respondents were males, and 155 respondents were from privately funded institutions.

The issue of whether there is a relationship between responses to these two variables is formally addressed through a hypothesis test, in which the null and alternative hypotheses are stated as follows:

$H_0$: Gender of yearbook editor is independent of the college's funding source.

$H_A$: Gender of yearbook editor *is not* independent of the college's funding source.

If the null hypothesis is true, the population proportion of yearbook editors from private institutions who are males should be equal to the proportion of male editors from state-funded institutions. These two proportions should also equal the population proportion of male editors without regard to a school's funding source. To illustrate, we can use the sample data to determine the sample proportion of male editors as follows:

$$p_M = \frac{\text{Number of male editors}}{\text{Number of respondents}} = \frac{57}{221} = 0.2579$$

Then, if the null hypothesis is true, we would expect 25.79% of the 155 privately funded schools, or 39.98 schools, to have a male yearbook editor. We would also expect 25.79% of

**Contingency Table**

A table used to classify sample observations according to two or more identifiable characteristics. It is also called a *cross-tabulation table*.

**TABLE 13.3** Contingency Table (Observed Data) for Dalgarno Photo

| Source of Funding | | | |
|---|---|---|---|
| Gender | Private | State | |
| Male | 14 | 43 | 57 |
| Female | 141 | 23 | 164 |
| | 155 | 66 | 221 |

**TABLE 13.4** Contingency Table for Dalgarno Photo

| | Source of Funding | | |
|---|---|---|---|
| Gender | Private | State | Total |
| Male | $o_{11} = 14$ | $o_{12} = 43$ | 57 |
| | $e_{11} = 39.98$ | $e_{12} = 17.02$ | |
| Female | $o_{21} = 141$ | $o_{22} = 23$ | 164 |
| | $e_{21} = 115.02$ | $e_{22} = 48.98$ | |
| Total | 155 | 66 | 221 |

the 66 state-funded schools, or 17.02 schools, to have male yearbook editors. (Note that the expected frequencies need not be integer values. Note also that the sum of expected frequencies in any column or row add up to the marginal frequency of the row or column.) We can use this reasoning to determine the expected number of respondents in each cell of the contingency table, as shown in Table 13.4.

We can simplify the calculations needed to produce the expected values for each cell. Note that the first cell's expected value, 39.98, was obtained by the following calculation:

$$e_{11} = 0.2579(155) = 39.98$$

However, because the probability, 0.2579, is calculated by dividing the row total, 57, by the grand total, 221, the expected cell value calculation can also be represented as

$$e_{11} = \frac{(\text{Row total})(\text{Column total})}{\text{Grand total}} = \frac{(57)(155)}{221} = 39.98$$

We can calculate the expected value for the next cell in the same row. The expected number of male yearbook editors in state-funded schools is

$$e_{12} = \frac{(\text{Row total})(\text{Column total})}{\text{Grand total}} = \frac{(57)(66)}{221} = 17.02$$

Keep in mind that the row and column totals (the marginal frequencies) must be the same for the expected values as for the observed values. Therefore, when there is only one cell left in a row or a column for which we must calculate an expected value, we can obtain it by subtraction. So, as an example, we could have calculated the expected value $e_{12}$ as

$$e_{12} = 57 - 39.98 = 17.02$$

Allowing for sampling error, we would expect the actual frequencies in each cell to approximately match the corresponding expected cell frequencies when the null hypothesis is true. The greater the difference between the actual and the expected frequencies, the more likely that the null hypothesis of independence is false and should be rejected. The statistical test to determine whether the sample data support or refute the null hypothesis is given by Equation 13.2. Do not be confused by the double summation in Equation 13.2; it merely indicates that all rows and columns must be used in calculating $\chi^2$. As was the case in the goodness-of-fit tests, the degrees of freedom are the number of independent data values obtained from the experiment. In any given row, once you know $c - 1$ of the data values,

**Chi-Square Contingency Test Statistic**

$$\chi^2 = \sum_{i=1}^{r} \sum_{j=1}^{c} \frac{(o_{ij} - e_{ij})^2}{e_{ij}} \quad \text{with } df = (r - 1)(c - 1) \tag{13.2}$$

where:

$o_{ij}$ = Observed frequency in cell $(i, j)$

$e_{ij}$ = Expected frequency in cell $(i, j)$

$r$ = Number of rows

$c$ = Number of columns

**FIGURE 13.6 Chi-Square Contingency Analysis Test for Dalgarno Photo**

**Hypotheses:**

$H_0$: Gender of yearbook editor is independent of college's funding source.
$H_A$: Gender of yearbook editor is not independent of college's funding source.
$\alpha = 0.05$

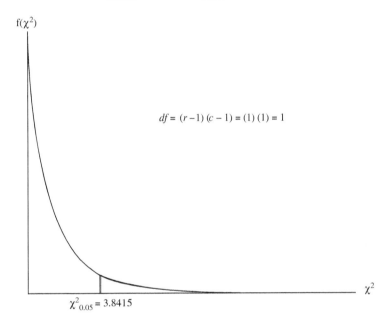

|  | Private | State |
|---|---|---|
| Male | $o_{11} = 14$ <br> $e_{11} = 39.98$ | $o_{12} = 43$ <br> $e_{12} = 17.02$ |
| Female | $o_{21} = 141$ <br> $e_{21} = 115.02$ | $o_{22} = 23$ <br> $e_{22} = 48.98$ |

**Test Statistic:**

$$\chi^2 = \sum_{i=1}^{r} \sum_{j=1}^{c} \frac{(o_{ij} - e_{ij})^2}{e_{ij}} = \frac{(14 - 39.98)^2}{39.98} + \frac{(43 - 17.02)^2}{17.02}$$
$$+ \frac{(141 - 115.02)^2}{115.02} + \frac{(23 - 48.98)^2}{48.98} = 76.19$$

$f(\chi^2)$

$df = (r - 1)(c - 1) = (1)(1) = 1$

$\chi^2_{0.05} = 3.8415$

$\chi^2$

**Decision Rule:**

If $\chi^2 > 3.8415$, reject $H_0$.
Otherwise, do not reject $H_0$.
Because $76.19 > 3.8415$, reject $H_0$. Thus, the gender of the yearbook editor and the school's source of funding are not independent.

the remaining data value is determined. For instance, once you know that 14 of the 57 male editors were from privately funded institutions, you know that 43 were from state-funded institutions. Similarly, once $r - 1$ data values in a column are known, the remaining data value is determined. Therefore, the degrees of freedom are obtained by the expression $(r - 1)(c - 1)$.

Figure 13.6 presents the hypotheses and test results for this example. As was the case in the goodness-of-fit tests, the test statistic has a distribution that can be approximated by the chi-square distribution if the expected values are larger than 5. Note that the calculated chi-square statistic is compared to the tabled value of chi-square for an $\alpha = 0.05$ and degrees of freedom $= (2 - 1)(2 - 1) = 1$. Because $\chi^2 = 76.19 > 3.8415$, the null hypothesis of independence should be rejected. Dalgarno Photo representatives should conclude that the gender of the yearbook editor and each school's source of funding are not independent. By examining the data in Figure 13.6, they can see that private schools are more likely to have female editors, whereas state schools are more likely to have male yearbook editors.

The Excel 2016 function for determining the chi-square critical value is

= CHISQ.INV.RT(alpha,df)
= CHISQ.INV.RT(.05,1)

**EXAMPLE 13-4** 2 × 2 Contingency Analysis

**Jury Selection**    In major cases, trial attorneys often spend significant time and resources trying to understand how potential jurors may respond to particular testimony. Suppose that in a civil case involving a high-profile client in Pittsburgh, Pennsylvania, the lawyers were interested in understanding whether potential jurors might recall a DUI arrest of their client a year ago. They are specifically concerned about whether there is a relationship between gender and the ability to recall the DUI incident. They conducted a test in which they randomly called 100 people (potential jurors) and asked them to indicate whether they recalled a DUI conviction by the client. The lawyers are interested in determining whether there is a relationship between gender and a person's ability to recall the DUI arrest. To test this, the following steps can be used:

STEP 1  **Specify the null and alternative hypotheses.**
The law firm is interested in testing whether a relationship exists between gender and recall ability. Here are the appropriate null and alternative hypotheses.

$H_0$: Ability to recall the client's arrest is independent of gender.
$H_A$: Recall ability and gender are not independent.

STEP 2  **Determine the significance level.**
The test will be conducted using a 0.01 level of significance.

> The Excel 2016 function for determining the chi-square critical value is
> = CHISQ.INV.RT(alpha,df)
> = CHISQ.INV.RT(.01,1)

STEP 3  **Determine the critical value.**
The critical value for this test is the chi-square value, with
$(r - 1)(c - 1) = (2 - 1)(2 - 1) = 1$ degree of freedom and $\alpha = 0.01$.
From Appendix G, the critical value is 6.6349.

STEP 4  **Collect the sample data and compute the chi-square test statistic using Equation 13.2.**
The following contingency table shows the results of the sampling:

|  | Female | Male | Total |
|---|---|---|---|
| Recall | 33 | 25 | 58 |
| Do Not Recall | 22 | 20 | 42 |
| Total | 55 | 45 | 100 |

Note that 58% of all the data values result in a recall. If the ability to recall the client's DUI arrest is independent of gender, you would expect the same percentage (58%) would occur for each gender. Thus, 58% of the males $[\,0.58(45) = 26.10\,]$ would be expected to recall. In general, the expected cell frequencies are determined by multiplying the row total by the column total and dividing by the overall sample size. For example, for the cell corresponding to female and recall, we get

$$\text{Expected} = \frac{58 \times 55}{100} = 31.90$$

The expected cell values for all cells are

|  | Female | Male | Total |
|---|---|---|---|
| Recall | $o = 33$<br>$e = 31.90$ | $o = 25$<br>$e = 26.10$ | 58 |
| Do Not Recall | $o = 22$<br>$e = 23.10$ | $o = 20$<br>$e = 18.90$ | 42 |
| Total | 55 | 45 | 100 |

After checking to make sure all the expected cell frequencies $\geq 5$, we compute the test statistic using Equation 13.2:

$$\chi^2 = \sum_{i=1}^{r} \sum_{j=1}^{c} \frac{(o_{ij} - e_{ij})^2}{e_{ij}}$$

$$= \frac{(33 - 31.90)^2}{31.90} + \frac{(25 - 26.10)^2}{26.10} + \frac{(22 - 23.10)^2}{23.10} + \frac{(20 - 18.90)^2}{18.90} = 0.20$$

STEP 5 **Reach a decision.**

Because $\chi^2 = 0.20 < 6.6349$, do not reject the null hypothesis.

The $p$-value for the test can be found using the function =CHISQ.TEST(actual_range, expected_range).

For this example, the function returns a $p$-value of 0.6542, which does not support rejecting the null hypothesis.

STEP 6 **Draw a conclusion.**

Based on the sample data, there is no reason to believe that being able to recall that the client was previously arrested for DUI is related to gender. Thus, based on this issue, the lawyers have no reason to prefer male or female jurors.

**TRY EXERCISE 13-17** (pg. 517)

## $r \times c$ **Contingency Tables**

**BUSINESS APPLICATION**    **Larger Contingency Tables**

Excel Tutorial

**Benton Stone & Tile**    Benton Stone & Tile makes a wide variety of products for the building industry. It pays market wages, provides competitive benefits, and offers attractive options for employees in an effort to create a satisfied workforce and reduce turnover. However, suppose that several supervisors have complained that employee absenteeism is becoming a problem. In response to these complaints, the human resources manager studied a random sample of 500 employees. One aim of this study was to determine whether there is a relationship between absenteeism and marital status. Absenteeism during the past year was broken down into three levels:

1. 0 absences
2. 1 to 5 absences
3. Over 5 absences

Marital status was divided into four categories:

1. Single    2. Married
3. Divorced    4. Widowed

Table 13.5 shows the contingency table for the sample of 500 employees. The table is also shown in the file **Benton**. The null and alternative hypotheses to be tested are

$H_0$: Absentee behavior is independent of marital status.

$H_A$: Absentee behavior is not independent of marital status.

As with $2 \times 2$ contingency analysis, the test for independence can be made using the chi-square test, where we compare the expected cell frequencies to the actual cell frequencies and use the test statistic shown as Equation 13.2. The logic of the test says that if the actual and expected frequencies closely match, then the null hypothesis of independence is not rejected. However, if the actual and expected cell frequencies are substantially different overall, the null hypothesis of independence is rejected. The calculated chi-square statistic is compared to

**TABLE 13.5 Contingency Table for Benton Stone & Tile**

| | Absences | | | |
| --- | --- | --- | --- | --- |
| Marital Status | 0 | 1–5 | Over 5 | Row Totals |
| Single | 84 | 82 | 34 | 200 |
| Married | 50 | 64 | 36 | 150 |
| Divorced | 50 | 34 | 16 | 100 |
| Widowed | 16 | 20 | 14 | 50 |
| Column Totals | 200 | 200 | 100 | 500 |

an Appendix G critical value for the desired significance and degrees of freedom equal to $(r - 1)(c - 1)$.

The expected cell frequencies are determined assuming that the row and column variables are independent. This means, for example, that the probability of a married person being absent more than 5 days during the year is the same as the probability of any employee being absent more than 5 days. An easy way to compute the expected cell frequencies, $e_{ij}$, is given by Equation 13.3.

**Expected Cell Frequencies**

$$e_{ij} = \frac{(i\text{th row total})(j\text{th column total})}{\text{Total sample size}} \tag{13.3}$$

For example, the expected cell frequency for row 1, column 1, is

$$e_{11} = \frac{(200)(200)}{500} = 80$$

and the expected cell frequency for row 2, column 3, is

$$e_{23} = \frac{(150)(100)}{500} = 30$$

Figure 13.7 shows the completed contingency table with the actual and expected cell frequencies that were developed using Excel. The calculated chi-square test value is computed as follows:

$$
\begin{aligned}
\chi^2 &= \sum_{i=1}^{r} \sum_{j=1}^{c} \frac{(o_{ij} - e_{ij})^2}{e_{ij}} \\
&= \frac{(84 - 80)^2}{80} + \frac{(82 - 80)^2}{80} + \cdots + \frac{(20 - 20)^2}{20} + \frac{(14 - 10)^2}{10} \\
&= 10.88
\end{aligned}
$$

**Excel 2016 Instructions**

1. Open file: **Benton.xlsx**.
2. Compute expected cell frequencies using Excel formula.
3. Compute chi-square statistic using Excel formula.

**FIGURE 13.7 Excel 2016 Output—Benton Stone & Tile Contingency Analysis Test**

| | A | B | C | D E F |
| --- | --- | --- | --- | --- |
| 1 | | | Absentee Rate | |
| 2 | Marital Status | Zero | 1 - 5 | Over 5 | Row Totals |
| 4 | Single | 84 | 82 | 34 | 200 |
| 5 | | 80 | 80 | 40 | |
| 7 | Married | 50 | 64 | 36 | 150 |
| 8 | | 60 | 60 | 30 | |
| 10 | Divorced | 50 | 34 | 16 | 100 |
| 11 | | 40 | 40 | 20 | |
| 13 | Widowed | 16 | 20 | 14 | 50 |
| 14 | | 20 | 20 | 10 | |
| 16 | Column Total | 200 | 200 | 100 | 500 |
| 18 | Chi-Square = | 10.8833 | | |

Expected Frequency found using =(D$16*$F13)/$F$16

The Excel 2016 function for determining the chi-square critical value is

$= \text{CHISQ.INV.RT(alpha,df)}$
$= \text{CHISQ.INV.RT(.05,6)}$

The degrees of freedom are $(r - 1)(c - 1) = (4 - 1)(3 - 1) = 6$. You can use the chi-square table in Appendix G to get the chi-square critical value for $\alpha = 0.05$ and 6 degrees of freedom, which equals 12.5916. Because the calculated chi-square value (10.883) shown in Figure 13.7 is less than 12.5916, we cannot reject the null hypothesis. Based on these sample data, there is *insufficient evidence* to conclude that absenteeism and marital status are not independent.

## Chi-Square Test Limitations

The chi-square distribution is only an approximation of the true distribution for contingency analysis. We use the chi-square approximation because in most instances it is impractical to compute the true distribution. However, the approximation (and, therefore, the conclusion reached) is quite good when all expected cell frequencies are at least 5.0. When expected cell frequencies drop below 5.0, the calculated chi-square value tends to be inflated and may inflate the true probability of a Type I error beyond the stated significance level. As a rule, if the null hypothesis is not rejected, you do not need to worry when the expected cell frequencies drop below 5.0.

There are two alternatives that you can use to overcome the small expected-cell-frequency problem. The first is to increase the sample size. This may increase the marginal frequencies in each row and column enough to increase the expected cell frequencies. The second option is to combine the categories of the row and/or column variables. If you do decide to group categories together, there should be some logic behind the resulting categories. You don't want to lose the meaning of the results through poor groupings. You will need to examine each situation individually to determine whether the option of grouping classes to increase expected cell frequencies makes sense.

# 13.2 EXERCISES

## Skill Development

**13-16.** A university recently completed a study to determine whether the number of classes students dropped during their college course work was related to whether the student lived on campus or commuted. The following data were collected:

|  | Where Lived | |
| --- | --- | --- |
| Courses Dropped | Commuted | On Campus |
| 0 | 70 | 90 |
| 1 | 60 | 50 |
| 2 | 45 | 28 |
| 3 | 42 | 47 |
| Over 3 | 36 | 48 |

Using an $\alpha = 0.05$ level, determine whether the number of courses dropped is independent of where a student lives.

**13-17.** The billing department of a national cable company accepts payment in one of four ways: in person at a local office, by mail, by credit card, or by electronic funds transfer from a bank account. The cable company randomly sampled 400 customers to determine if there is a relationship between the customer's age and the payment method used. The following sample results were obtained:

|  | Age of Customer | | | |
| --- | --- | --- | --- | --- |
| Payment Method | 20–30 | 31–40 | 41–50 | Over 50 |
| In Person | 8 | 12 | 11 | 13 |
| By Mail | 29 | 67 | 72 | 50 |
| By Credit Card | 26 | 19 | 5 | 7 |
| By Funds Transfer | 23 | 35 | 17 | 6 |

Based on the sample data, can the cable company conclude that there is a relationship between the age of the customer and the payment method used? Conduct the appropriate test at the $\alpha = 0.01$ level of significance.

**13-18.** A contingency analysis table has been constructed from data obtained in a phone survey of customers in a market area in which respondents were asked to indicate whether they owned a domestic or foreign car and whether they were a member of a union or not. The following contingency table is provided:

|        | Union | |
|--------|-------|-----|
| Car    | Yes   | No  |
| Domestic | 155 | 470 |
| Foreign  | 40  | 325 |

a.  Use the chi-square approach to test whether type of car owned (domestic or foreign) is independent of union membership. Test using an $\alpha = 0.05$ level.
b.  Calculate the $p$-value for this hypothesis test.

**13-19.** Use the following contingency table for this exercise:

|        | $C_1$ | $C_2$ |
|--------|-------|-------|
| $R_1$  | 51    | 207   |
| $R_2$  | 146   | 185   |
| $R_3$  | 240   | 157   |

a.  State the relevant null and alternative hypotheses.
b.  Calculate the expected values for each of the cells.
c.  Compute the chi-square test statistic for the hypothesis test.
d.  Determine the appropriate critical value and reach a decision for the hypothesis test. Use a significance level of 0.05.
e.  Obtain the $p$-value for this hypothesis test.

**13-20.** A manufacturer of sports drinks has randomly sampled 198 men and 202 women. Each sampled participant was asked to taste an unflavored version and a flavored version of a new sports drink currently in development. The participants' preferences are shown in the table.

|       | Flavored | Unflavored |
|-------|----------|------------|
| Men   | 101      | 97         |
| Women | 68       | 134        |

a.  State the relevant null and alternative hypotheses.
b.  Conduct the appropriate test and state a conclusion. Use a level of significance of 0.05.

**13-21.** A marketing research firm is conducting a study to determine if there is a relationship between an individual's age and the individual's preferred source of news. The research firm asked 1,000 individuals to list their preferred source for news: newspaper, radio and television, or the Internet. The following results were obtained:

| Preferred News Source | Age of Respondent | | | |
|-----------------------|-------|-------|-------|---------|
|                       | 20–30 | 31–40 | 41–50 | Over 50 |
| Newspaper             | 19    | 62    | 95    | 147     |
| Radio/TV              | 27    | 125   | 168   | 88      |
| Internet              | 104   | 113   | 37    | 15      |

At the 0.01 level of significance, can the marketing research firm conclude that there is a relationship between the age of the individual and the individual's preferred source for news?

**13-22.** A loan officer wished to determine if the marital status of loan applicants was independent of the approval of loans. The following table presents the results of her survey:

|          | Approved | Rejected |
|----------|----------|----------|
| Single   | 213      | 189      |
| Married  | 374      | 231      |
| Divorced | 358      | 252      |

a.  Conduct the appropriate hypothesis test that will provide an answer to the loan officer. Use a significance level of 0.01.
b.  Calculate the $p$-value for the hypothesis test in part a.

**13-23.** An instructor in a large accounting class is interested in determining whether the grades that students get are related to how close to the front of the room the students sit. He has categorized the room seating as "Front," "Middle," and "Back." The following data were collected over two sections with 400 total students. Based on the sample data, can he conclude that there is a dependency relationship between seating location and grade using a significance level equal to 0.05?

|        | A  | B   | C   | D  | F | Total |
|--------|----|-----|-----|----|---|-------|
| Front  | 18 | 55  | 30  | 3  | 0 | 106   |
| Middle | 7  | 42  | 95  | 11 | 1 | 156   |
| Back   | 3  | 15  | 104 | 14 | 2 | 138   |
| Total  | 28 | 112 | 229 | 28 | 3 | 400   |

## Business Applications

**13-24.** To determine if there is a difference between the investing preferences of mid-level managers working in the public and private sectors in an eastern city, a random sample of 320 public-sector employees and 380 private-sector employees was taken. Suppose the sampled participants were asked about their retirement investment decisions and classified as being either "aggressive," if they invested in only stocks or stock mutual funds, or "balanced," if they invested in some combination of stocks, bonds, cash, and other. The following results were found:

|         | Aggressive | Balanced |
|---------|------------|----------|
| Public  | 164        | 156      |
| Private | 236        | 144      |

a.  State the hypothesis of interest and conduct the appropriate hypothesis test to determine whether there is a relationship between employment sector and investing preference. Use a level of significance of 0.01.
b.  State the conclusion of the test conducted in part a.

**13-25.** The following table classifies a stock's price change as up, down, or no change for both today's and yesterday's prices. Price changes were examined for 100 days. A financial theory states that stock prices follow what is called a "random walk." This means, in part, that the price change today for a stock must be independent of yesterday's price change. Test the hypothesis that daily stock price changes for this stock are independent. Let $\alpha = 0.05$.

| | Price Change Previous Day | | |
|---|---|---|---|
| Price Change Today | Up | No Change | Down |
| Up | 14 | 16 | 12 |
| No Change | 6 | 8 | 6 |
| Down | 16 | 14 | 8 |

**13-26.** A local appliance retailer handles four washing machine models for a major manufacturer: standard, deluxe, superior, and XLT. The marketing manager has recently conducted a study on the purchasers of the washing machines. The study recorded the model of appliance purchased and the credit account balance of the customer at the time of purchase. The sample data are in the following table. Based on these data, is there evidence of a relationship between the account balance and the model of washer purchased? Use a significance level of 0.025. Conduct the test using a *p*-value approach.

| | Washer Model Purchased | | | |
|---|---|---|---|---|
| Credit Balance | Standard | Deluxe | Superior | XLT |
| Under $200 | 10 | 16 | 40 | 5 |
| $200–$800 | 8 | 12 | 24 | 15 |
| Over $800 | 16 | 12 | 16 | 30 |

**13-27.** A random sample of 980 heads of households was taken from the customer list for State Bank and Trust. Those sampled were asked to classify their own attitudes and their parents' attitudes toward borrowing money as follows:
A: Borrow only for real estate and car purchases
B: Borrow for short-term purchases such as appliances and furniture
C: Never borrow money
The following table indicates the responses from those in the study:

| | Respondent | | |
|---|---|---|---|
| Parent | A | B | C |
| A | 240 | 80 | 20 |
| B | 180 | 120 | 40 |
| C | 180 | 80 | 40 |

Test the hypothesis that the respondents' borrowing habits are independent from what they believe their parents' attitudes to be. Let $\alpha = 0.01$.

**13-28.** An online undergraduate business school uses a learning management software system (LMS) that allows instructors to know how many hours students are connected to the course website. When studying for the midterm exam, students have a choice of using two study guides. One study guide comes with the textbook and is professionally prepared by the publisher. The second study guide is prepared by the professor. Suppose the department chairperson is interested in determining whether a student's hours connected to the LMS (study time) is independent of the student's choice of study guide. The following data show the numbers of students organized by study time and selected study guide:

| | Study Guide Chosen | |
|---|---|---|
| Connect Time | Publisher's Guide | Professor's Guide |
| Under 6 Hours | 104 | 109 |
| 6–12 hours | 160 | 163 |
| 12–20 hours | 142 | 146 |
| Over 20 hours | 133 | 156 |

Determine whether the hours connected to the LMS is independent of the study guide the student chooses to use for the exam. Test at the alpha = 0.025 level using the *p*-value approach.

**13-29.** The Barclaycard Business Travel Survey is conducted annually. One question asked in two consecutive years was, "Have you considered reducing hours spent away from home to increase quality of life?" The following table represents the responses:

| | Year 1 | Year 2 | Total |
|---|---|---|---|
| Yes—have reduced | 400 | 384 | 784 |
| Yes—not been able to | 400 | 300 | 700 |
| No—not certain | 400 | 516 | 916 |
| Total | 1,200 | 1,200 | 2,400 |

a. Determine if the response to the survey question was independent of the year in which the question was asked. Use a significance level of 0.05.
b. Determine if there is a significant difference between the proportion of travelers who say they have reduced hours spent away from home between the two years.

## Computer Software Exercises

**13-30.** An article in *USA TODAY* reported on a study conducted by Daniel Vinson on the relationship between anger and propensity to be injured (*source:* Steve Sternberg, "Study: Angry people are a danger to themselves," http://usatoday30.usatoday.com, Feb. 5, 2006). Suppose the file titled **Angry** contains data on the emotions reported by a sample of patients just before they were injured.

a. Use the data in the file to construct a contingency table.

b. Determine if the type of emotion felt by patients just before they were injured is independent of the severity of that emotion. Use a contingency analysis and a significance level of 0.05.

**13-31.** According to the *Huffington Post*, approximately 10% of the population of Europe and North America is left-handed (*source:* Katie Kens, "11 Little-known facts about left-handers," www.huffingtonpost.com, Aug. 13, 2015). Suppose a survey of 501 randomly selected men and women at a midwestern university provided data on dominant hand and gender. The data are in a file called **Left-Handed**. Based on these data, determine if the "handedness" of an individual is independent of gender. Use a significance level of 0.01 and a *p*-value approach.

**13-32.** Suppose that a survey was mailed to a random sample of 400 guests of a national hotel chain. A total of 62 customers responded to the survey. Among the issues of interest was whether there is a relationship between the likelihood that customers will stay at the chain again and whether this was the customer's first stay at

the chain. The data are contained in the file called **Hotel Survey**. Using a significance level equal to 0.05, test to see whether these sample data imply a relationship between the two variables. Discuss the results.

**13-33.** ECCO (Electronic Controls Company) makes backup alarms that are used on such equipment as forklifts and delivery trucks. Suppose the quality manager performed a study involving a random sample of 110 warranty claims. One of the questions the manager wanted to answer was whether there is a relationship between the type of warranty complaint and the plant at which the alarm was made. The data are in the file called **ECCO**.

a. Calculate the expected values for the cells in this analysis. Suggest a way in which cells can be combined to assure that the expected value of each cell is at least 5 so that as many level combinations of the two variables as possible are retained.

b. Using a significance level of 0.01, conduct a relevant hypothesis test and provide an answer to the manager's question.

**13-34.** Referring to Exercise 13-33, can the quality manager conclude that the type of warranty problem is independent of the shift on which the alarm was manufactured? Test using a significance level of 0.05. Discuss your results.

# 13 Overview

## Summary

**13.1** | **Introduction to Goodness-of-Fit Tests** (pg. 498–510)

**OUTCOME 1** Utilize the chi-square goodness-of-fit test to determine whether data from a process fit a specified distribution.

- We can use the chi-square goodness-of-fit test to determine if a set of data comes from a specific hypothesized distribution.
- Recall that several of the procedures presented in Chapters 8–12 require that the sampled populations are normally distributed.
- In order to verify this requirement, the goodness-of-fit test determines if the observed set of values agrees with a set of data obtained from a specified probability distribution.
- The goodness-of-fit test is most often used to verify a normal distribution. However, it can also be used to detect other probability distributions.

**13.2** | **Introduction to Contingency Analysis** (pg. 510–520)

**OUTCOME 2** Set up a contingency analysis table and perform a chi-square test of independence.

- There are many business situations in which the level of data measurement for the variable of interest is either nominal or ordinal, not interval or ratio.
- Some situations involve multiple population proportions for which two-population procedures are not applicable.
- In each of these cases, the proportions relate to characteristic categories of the variable of interest.
- Situations that involve categorical data call for a new statistical tool known as *contingency analysis* to help analysts make decisions when multiple proportions are involved.
- Contingency analysis can be used when a level of data measurement is either nominal or ordinal and the values are determined by counting the number of occurrences in each category.

## Equations

**(13.1) Chi-Square Goodness-of-Fit Test Statistic** pg. 500

$$\chi^2 = \sum_{i=1}^{k} \frac{(o_i - e_i)^2}{e_i}$$

**(13.2) Chi-Square Contingency Test Statistic** pg. 512

$$\chi^2 = \sum_{i=1}^{r} \sum_{j=1}^{c} \frac{(o_{ij} - e_{ij})^2}{e_{ij}} \quad \text{with } df = (r-1)(c-1)$$

**(13.3) Expected Cell Frequencies** pg. 516

$$e_{ij} = \frac{(i\text{th row total})(j\text{th column total})}{\text{Total sample size}}$$

## Key Term

Contingency table pg. 511

## Chapter Exercises

### Conceptual Questions

**13-35.** Locate a journal article that uses either contingency analysis or a goodness-of-fit test. Discuss the article, paying particular attention to the reasoning behind using the particular statistical test.

**13-36.** Find a marketing research book (or borrow one from a friend). Does it discuss either of the tests considered in this chapter? If yes, outline the discussion. If no, determine where in the text such a discussion would be appropriate.

**13-37.** One of the topics in Chapter 10 was hypothesis testing for the difference between two population proportions. For the test to have validity, conditions were set on the sample sizes with respect to the sample proportions. A 2 × 2 contingency table may also be utilized to test the difference between proportions of two independent populations. This procedure has conditions placed on the expected value of each cell. Discuss the relationship between these two conditions.

**13-38.** A 2 × 2 contingency table and a hypothesis test of the difference between two population proportions can be used to analyze the same data set. However, besides all the similarities of the two methods, the hypothesis test of the difference between two proportions has two advantages. Identify these advantages.

### Business Applications

**13-39.** The College Bookstore has four checkout stands. The store manager noticed that registers 3 and 4 served fewer students than registers 1 and 2. She is not sure whether the layout of the store channels customers into these registers, whether the checkout clerks in these lines are simply slower than in the other two, or whether she was just seeing random differences. The manager kept a record of which stands the next 1,000 students chose for checkout. The students checked out of the four stands according to the following pattern:

| Stand 1 | Stand 2 | Stand 3 | Stand 4 |
|---------|---------|---------|---------|
| 338 | 275 | 201 | 186 |

Based on these data, can the store manager conclude that the proportions of students who use the four checkout stands are equal? (Use an $\alpha = 0.05$.)

**13-40.** A regional cancer treatment center has had success treating localized cancers with a linear accelerator. Whereas admissions for further treatment nationally average 2.1 per patient per year, the center's director thinks that readmissions with the new treatment are Poisson distributed, with a mean of 1.2 patients per year. He has collected the following data on a random sample of 300 patients:

| Readmissions Last Year | Patients |
|------------------------|----------|
| 0 | 139 |
| 1 | 87 |
| 2 | 48 |
| 3 | 14 |
| 4 | 8 |
| 5 | 1 |
| 6 | 1 |
| 7 | 0 |
| 8 | 2 |
| | 300 |

a. Adjust the data so that you can test the director's claim using a test statistic whose sampling distribution can be approximated by a chi-square distribution.

b. Assume the Type I error rate is to be controlled at 0.05. Do you agree with the director's claim? Why? Conduct a statistical procedure to support your opinion.

**13-41.** Cooper Manufacturing has a contract with the U.S. Air Force to produce a part for a new fighter plane being manufactured. The part is a bolt with specifications requiring that the length be normally distributed with a mean of 3.05 inches and a standard deviation of 0.015 inch. As part of the company's quality-control efforts, each day Cooper's engineers select a random sample of 100 bolts produced that day and carefully measure the bolts to determine whether they meet the specifications. The following data were collected yesterday:

| Length (inches) | Frequency |
|-----------------|-----------|
| Under 3.030 | 5 |
| 3.030 and under 3.035 | 16 |
| 3.035 and under 3.040 | 7 |
| 3.040 and under 3.050 | 20 |
| 3.050 and under 3.060 | 36 |
| 3.060 and under 3.065 | 8 |
| 3.065 and over | 8 |

Based on these sample data, what should Cooper's engineers conclude about the production output if they test using an $\alpha = 0.01$? Discuss.

**13-42.** Suppose the Cooper Company, discussed in Exercise 13-41, has a second contract with a private firm for which it makes fuses for an electronic instrument. The quality-control department at Cooper periodically selects a random sample of five fuses and tests each fuse to determine whether it is defective. Based on these findings, the production process is either shut down (if too many defectives are observed) or allowed to run. The quality-control department believes that the sampling process follows a binomial distribution, and it has been using the binomial distribution to compute the probabilities associated with the sampling outcomes.

The contract allows for at most 5% defectives. The head of quality control recently compiled a list of the sampling results for the past 300 days in which five randomly selected fuses were tested, with the following frequency distribution for the number of defectives observed. She is concerned that the binomial distribution with a sample size of 5 and a probability of defectives of 0.05 may not be appropriate.

| Number of Defectives | Frequency |
|---|---|
| 0 | 209 |
| 1 | 33 |
| 2 | 43 |
| 3 | 10 |
| 4 | 5 |
| 5 | 0 |

a. Calculate the expected values for the cells in this analysis. Suggest a way in which cells can be combined to assure that the expected value of each cell is at least 5.
b. Using a significance level of 0.10, what should the quality-control manager conclude based on these sample data? Discuss.

**13-43.** A study commissioned by the Missouri Traffic Safety Commission was conducted to determine if the proportion of adults who have been in an auto accident in the past three years is related to the age of the driver. Based on the following data, conduct the appropriate goodness-of-fit test using a significance level of 0.01:

| Age Group | 18–24 | 25–34 | 35–44 | 45–54 | 55–64 | 65 and over |
|---|---|---|---|---|---|---|
| Percentage | 11 | 17 | 19 | 20 | 14 | 19 |

**13-44.** A *Wall Street Journal* article discussed the drop in sports participation by youth in the United States (*source:* Ryan Wallerson, "Youth participation weakens in basketball, football, baseball, soccer," www.wsj.com, Jan. 31, 2014). Suppose a survey was conducted that compared activity levels to the levels found nine years earlier. A random selection of children (7- to 11-year-olds) indicating their favorite outdoor activity is given in the following table:

| | Bicycling | Swimming | Baseball | Fishing | Touch Football |
|---|---|---|---|---|---|
| Earlier | 68 | 60 | 29 | 25 | 16 |
| Present | 47 | 42 | 22 | 18 | 10 |

Construct a contingency analysis to determine if the type of preferred outdoor activity is dependent on the year in this survey. Use a significance level of 0.05 and the *p*-value approach.

## Computer Software Exercises

**13-45.** A 2015 study done by the Employee Benefit Research Institute (EBRI) found about 22% of workers said they and/or their spouses were very confident that they are saving enough for retirement. The file titled **Retirement** contains the total savings and investments. Use a contingency analysis and an alpha level of 0.05 to determine if the amount of total savings and investments is dependent on the age of the worker.

**13-46.** The manager at the Sacramento, California, airport recently conducted a study of passengers departing from the airport. A random sample of 100 passengers was selected. The data are in the file called **Airline Passengers**. An earlier study showed the following usage by airline:

| Delta | 20% |
|---|---|
| Horizon | 10% |
| Northwest | 10% |
| Skywest | 3% |
| Southwest | 25% |
| United | 32% |

a. If the manager wishes to determine whether the airline usage pattern has changed from that reported in the earlier study, state the appropriate null and alternative hypotheses.
b. Based on the sample data, what should the manager conclude? Test using a significance level of 0.01.

**13-47.** A pharmaceutical company claims that if a blood pressure drug is taken properly, the amount of blood pressure decrease will be normally distributed with a mean equal to 10 points on the diastolic reading and a standard deviation equal to 4.0. One hundred patients were administered the drug, and data were collected showing the reduction in blood pressure at the end of the test period. The data are in the file labeled **Blood Pressure**.

Using a goodness-of-fit test and a significance level equal to 0.05, what conclusion should be reached with respect to the distribution of diastolic blood pressure reduction? Discuss.

**13-48.** A survey conducted by a large brokerage company addressed the proportion of African Americans and White Americans who have money invested in the stock market. Suppose the file titled **Stockrace** contains data obtained in the surveys. The survey asked 500 African American and 500 White respondents if they personally had money invested in the stock market.

a. Create a contingency table using the data in the file.
b. Conduct a contingency analysis to determine if the proportion of African Americans differs from the proportion of White Americans who invest in stocks. Use a significance level of 0.05.

**13-49.** The state transportation department recently conducted a study of motorists in Idaho. Two main factors of interest were whether the vehicle was insured with liability insurance and whether the driver was wearing a seat belt. A random sample of 100 cars was stopped at various locations throughout the state. The data are in the file called **Liabins**. The investigators were interested in determining whether seat belt status is independent of insurance status.

Conduct the appropriate hypothesis test using a 0.05 level of significance and discuss your results.

## Case 13.1    National Oil Company—Part 2

Chad Williams sat back in his airline seat to enjoy the hour-long flight between Los Angeles and Oakland, California. (See Case 6.3.) The hour would give him time to reflect on his upcoming trip to Australia and the work he had been doing the past week in Los Angeles.

Chad is one man on a six-person crew employed by National Oil Company to literally walk the Earth searching for oil. His college degrees in geology and petroleum engineering landed him the job with National, but he never dreamed he would be doing the exciting work he now does. Chad and his crew spend several months in special locations around the world using highly sensitive electronic equipment for oil exploration.

The upcoming trip to Australia is one that Chad has been looking forward to since it was announced that his crew would be going there to search the Outback for oil. In preparation for the trip, the crew has been in Los Angeles at National's engineering research facility working on some new equipment that will be used in Australia.

Chad's thoughts centered on the problem he was having with a particular component part on the new equipment. The specifications called for 200 of the components, with each having a diameter of between 0.15 and 0.18 inch. The only available supplier of the component manufactures the components in New Jersey to specifications calling for normally distributed output, with a mean of 0.16 inch and a standard deviation of 0.02 inch.

Chad faces two problems. First, he is unsure that the supplier actually does produce parts with a mean of 0.16 inch and a standard deviation of 0.02 inch according to a normal distribution. Second, if the parts are made to specifications, he needs to determine how many components to purchase so that he receives enough acceptable components to make two oil exploration devices.

The supplier has sent Chad the following data for 330 randomly selected components. Chad believes that the supplier is honest and that he can rely on the data.

Chad needs to have a report ready for Monday indicating whether he believes the supplier delivers at its stated specifications and, if so, how many of the components National should order to have enough acceptable components to outfit two oil exploration devices.

| Diameter (Inch) | Frequency |
| --- | --- |
| Under 0.14 | 5 |
| 0.14 and under 0.15 | 70 |
| 0.15 and under 0.16 | 90 |
| 0.16 and under 0.17 | 105 |
| 0.17 and under 0.18 | 50 |
| Over 0.18 | 10 |
| Total | 330 |

### Required Tasks:

1. State the problems faced by Chad Williams.
2. Identify the statistical test Chad can use to determine whether the supplier's claim is true.
3. State the null and alternative hypotheses for the test to determine whether the supplier's claim is true.
4. Assuming that the supplier produces output whose diameter is normally distributed with a mean of 0.16 inch and a standard deviation of 0.02 inch, determine the expected frequencies that Chad would expect to see in a sample of 330 components.
5. Based on the observed and expected frequencies, calculate the appropriate test statistic.
6. Calculate the critical value of the test statistic. Select an alpha value.
7. State a conclusion. Do the sample data support the supplier's claim with respect to the specifications of the component parts?
8. Provide a short report that summarizes your analysis and conclusion.

## Case 13.2    Bentford Electronics—Part 1

On Saturday morning, Jennifer Bentford received a call at her home from the production supervisor at Bentford Electronics Plant 1. The supervisor indicated that she and the supervisors from Plants 2, 3, and 4 had agreed that something must be done to improve company morale and thereby increase the production output of their plants. Jennifer, president of Bentford Electronics, agreed to set up a

Monday morning meeting with the supervisors to see if they could arrive at a plan for accomplishing these objectives.

By Monday, each supervisor had compiled a list of several ideas, including a four-day work week and interplant competitions of various kinds. A second meeting was set for Wednesday to discuss the issue further.

Following the Wednesday afternoon meeting, Jennifer and her plant supervisors agreed to implement a weekly contest called the NBE Game of the Week. The plant that produced the most each week would be considered the NBE Game of the Week winner and would receive 10 points. The second-place plant would receive 7 points, and the third- and fourth-place plants would receive 3 points and 1 point, respectively. The contest would last 26 weeks. At the end of that period, a $200,000 bonus would be divided among the employees in the four plants proportional to the total points accumulated by each plant.

The announcement of the contest created a lot of excitement and enthusiasm at the four plants. No one complained about the rules because the four plants were designed and staffed to produce equally.

At the close of the contest, Jennifer called the supervisors into a meeting, at which time she asked for data to determine whether the contest had significantly improved productivity. She indicated that she had to know this before she could authorize a second contest. The supervisors, expecting this request, had put together the following data:

| Units Produced (4 Plants Combined) | Before-Contest Frequency | During-Contest Frequency |
|---|---|---|
| 0–2,500 | 11 | 0 |
| 2,501–8,000 | 23 | 20 |
| 8,001–15,000 | 56 | 83 |
| 15,001–20,000 | 15 | 52 |
| | 105 days | 155 days |

Jennifer examined the data and indicated that the contest looked to be a success, but she wanted to base her decision to continue the contest on more than just an observation of the data. "Surely there must be some way to statistically test the worth of this contest," Jennifer stated. "I have to see the results before I will authorize the second contest."

# 14 Introduction to Linear Regression and Correlation Analysis

## WHY YOU NEED TO KNOW

Many business situations will require you to consider the relationship between two or more variables. For example, a stock analyst might be interested in the relationship between stock prices and the dividends issued by a publicly traded company. A company might be interested in examining the relationship between product sales and the amount of money spent on advertising. In another case, consider a loan manager at a bank who is interested in determining the market value of a home before granting a real estate loan to a customer. She would begin by collecting data on a sample of comparable properties that have sold recently. In addition to the selling price, she would collect data on other factors, such as the size and

### Quick Prep

**Review** the methods for testing a null hypothesis using the *t*-distribution in Chapter 9.

**Review** confidence intervals discussed in Chapter 8.

**Make sure** you review the discussion about scatter plots in Chapter 2.

**Review** the concepts associated with selecting a simple random sample in Chapter 1.

**Review** the *F*-distribution and the approach for finding critical values from the *F*-table as discussed in Chapters 11 and 12.

age of the property. She might then analyze the relationship between the price and the other variables and use this relationship to determine an appraised price for the property in question.

We introduce simple linear regression and correlation analysis in this chapter. These techniques are two of the most often applied statistical procedures that business decision makers use for analyzing the relationship between two variables. In Chapter 15, we will extend the discussion to include three or more variables.

<table>
<tr><td>14.1</td><td>**Scatter Plots and Correlation**</td></tr>
</table>

**Scatter Plot**

A two-dimensional plot showing the values for the joint occurrence of two quantitative variables. The scatter plot may be used to graphically represent the relationship between two variables. It is also known as a scatter diagram.

Decision-making situations that call for understanding the relationship between two quantitative variables are aided by the use of **scatter plots**, or *scatter diagrams*. Figure 14.1 shows scatter plots that depict several potential relationships between values of a dependent variable, *y*, and an independent variable, *x*. A *dependent* (or *response*) *variable* is the variable whose variation we wish to explain (e.g., the value of a house). An *independent* (or *explanatory*) *variable* is a variable used to explain variation in the dependent variable (e.g., the size of the house). In Figure 14.1, (a) and (b) are examples of strong *linear* (or straight line) relationships between *x* and *y*. Note that the linear relationship can be either positive (as the *x* variable increases, the *y* variable also increases) or negative (as the *x* variable increases, the *y* variable decreases).

Figures 14.1 (c) and (d) illustrate situations in which the relationship between the *x* and *y* variables is nonlinear. Many possible nonlinear relationships can occur. The scatter plot is very useful for visually identifying the nature of the relationship.

Figures 14.1 (e) and (f) show examples in which there is no identifiable relationship between the two variables. This means that as *x* increases, *y* sometimes increases and sometimes decreases but with no particular pattern.

**Correlation Coefficient**

A quantitative measure of the strength of the linear relationship between two variables. The correlation ranges from −1.0 to +1.0. A correlation of ±1.0 indicates a perfect linear relationship, whereas a correlation of 0 indicates no linear relationship.

### The Correlation Coefficient

In addition to analyzing the relationship between two variables graphically, we can also measure the strength of the linear relationship between two variables using a measure called the **correlation coefficient**.

The correlation coefficient of two variables can be estimated from sample data using Equation 14.1 or the algebraic equivalent, Equation 14.2.

**FIGURE 14.1 Two-Variable Relationships**

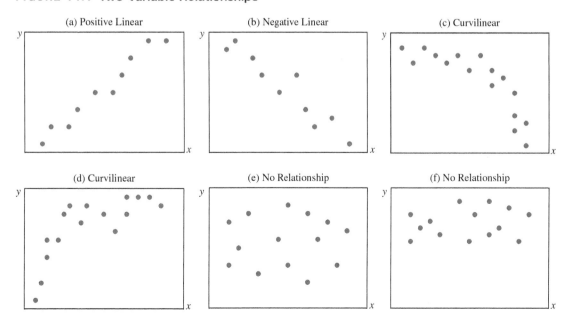

FIGURE 14.2 **Correlation between Two Variables**

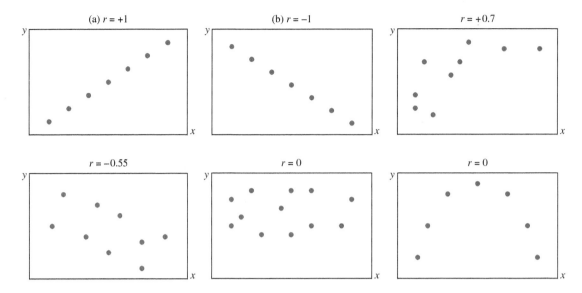

**Sample Correlation Coefficient**

OUTCOME **1**

$$r = \frac{\Sigma(x - \bar{x})(y - \bar{y})}{\sqrt{[\Sigma(x - x)^2][\Sigma(y - \bar{y})^2]}}$$ (14.1)

or the algebraic equivalent:

$$r = \frac{n\Sigma xy - \Sigma x \Sigma y}{\sqrt{[n(\Sigma x^2) - (\Sigma x)^2][n(\Sigma y^2) - (\Sigma y)^2]}}$$ (14.2)

where:

$r$ = Sample correlation coefficient

$n$ = Sample size

$x$ = Value of the independent variable

$y$ = Value of the dependent variable

The sample correlation coefficient computed using Equations 14.1 and 14.2 is called the *Pearson product moment correlation*. The sample correlation coefficient, $r$, can range from a perfect positive correlation, $+1.0$, to a perfect negative correlation, $-1.0$. A perfect correlation occurs if all points on the scatter plot fall on a straight line. If two variables have no linear relationship, the correlation between them is 0 and there is no linear relationship between the $x$ and $y$ variables. Consequently, the more the correlation differs from 0.0, the stronger the linear relationship between the two variables. The sign of the correlation coefficient indicates the direction of the relationship.

Figure 14.2 illustrates some examples of correlation between two variables. Once again, for the correlation coefficient to equal plus or minus 1.0, all the $(x, y)$ points must form a perfectly straight line. The more the points depart from a straight line, the weaker (closer to 0.0) the correlation is between the two variables.

**BUSINESS APPLICATION** **Testing for Significant Correlations**

Excel Tutorial

**Midwest Distribution Company** Consider the Midwest Distribution Company, which supplies soft drinks and snack foods to convenience stores in Michigan, Illinois, and Iowa. Although Midwest Distribution has been profitable, the director of marketing has been concerned about the rapid turnover in her sales force. In the course of exit interviews, she discovered a major concern with the compensation structure.

**FIGURE 14.3 Excel 2016 Scatter Plot of Sales vs. Years with Midwest Distribution**

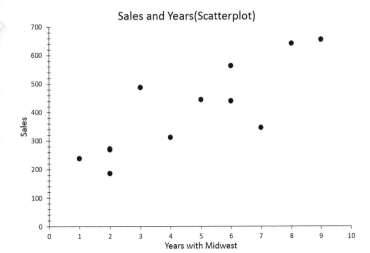

Midwest Distribution has a two-part wage structure: a base salary and a commission computed on monthly sales. Typically, about half of the total wages paid come from the base salary, which increases with longevity with the company. This portion of the wage structure is not an issue. The concern expressed by departing employees is that new employees tend to be given parts of the sales territory previously covered by existing employees and are assigned prime customers as a recruiting inducement. This is seen to adversely affect the commissions of the existing sales staff.

At issue, then, is the relationship between sales (on which commissions are paid) and number of years with the company. The data for a random sample of 12 sales representatives are in the file called **Midwest**. The first step is to develop a scatter plot of the data.

The scatter plot for the Midwest data is shown in Figure 14.3. Based on this plot, total sales and years with the company appear to be positively linearly related. However, the strength of this relationship is uncertain. That is, how close do the points come to being on a straight line? To answer this question, we need a quantitative measure of the strength of the linear relationship between the two variables. That measure is the correlation coefficient.

We can use Equation 14.1 to determine the correlation between sales and years with the company. Table 14.1 shows the manual calculations used to determine this correlation coefficient of 0.8325. However, because the calculations are rather tedious and long, we almost always use computer software such as Excel to perform the calculation. You can use either Excel's CORREL function or the Correlation tool in the Data Analysis tools, as shown in Figure 14.4. The $r = 0.8325$ indicates that there is a fairly strong positive correlation between these two variables for the sample data.

**Significance Test for the Correlation**    Although a correlation coefficient of 0.8325 seems quite large (relative to 0), you should remember that this value is based on a sample of 12 data points and is subject to sampling error. Therefore, we need to use a formal hypothesis-testing procedure to determine whether the linear relationship between sales and years with the company is statistically significant.

**FIGURE 14.4 Excel 2016 Correlation Output for Midwest Distribution**

| | A | B | C |
|---|---|---|---|
| 1 | | Sales | Years with Midwest |
| 2 | Sales | 1 | |
| 3 | Years with Midwest | 0.8325 | 1 |

**TABLE 14.1** Correlation Coefficient Calculations for the Midwest Distribution Example

| Sales | Years | | | | | |
|---|---|---|---|---|---|---|
| y | x | $x - \bar{x}$ | $y - \bar{y}$ | $(x - \bar{x})(y - \bar{y})$ | $(x - \bar{x})^2$ | $(y - \bar{y})^2$ |
| 487 | 3 | −1.58 | 82.42 | −130.22 | 2.50 | 6,793.06 |
| 445 | 5 | 0.42 | 40.42 | 16.98 | 0.18 | 1,633.78 |
| 272 | 2 | −2.58 | −132.58 | 342.06 | 6.66 | 17,577.46 |
| 641 | 8 | 3.42 | 236.42 | 808.56 | 11.70 | 55,894.42 |
| 187 | 2 | −2.58 | −217.58 | 561.36 | 6.66 | 47,341.06 |
| 440 | 6 | 1.42 | 35.42 | 50.30 | 2.02 | 1,254.58 |
| 346 | 7 | 2.42 | −58.58 | −141.76 | 5.86 | 3,431.62 |
| 238 | 1 | −3.58 | −166.58 | 596.36 | 12.82 | 27,748.90 |
| 312 | 4 | −0.58 | −92.58 | 53.70 | 0.34 | 8,571.06 |
| 269 | 2 | −2.58 | −135.58 | 349.80 | 6.66 | 18,381.94 |
| 655 | 9 | 4.42 | 250.42 | 1,106.86 | 19.54 | 62,710.18 |
| 563 | 6 | 1.42 | 158.42 | 224.96 | 2.02 | 25,096.90 |
| $\Sigma$ = 4,855 | $\Sigma$ = 55 | | | $\Sigma$ = 3,838.92 | $\Sigma$ = 76.92 | $\Sigma$ = 276,434.92 |

$$\bar{y} = \frac{\Sigma y}{n} = \frac{4,855}{12} = 404.58 \qquad \bar{x} = \frac{\Sigma x}{n} = \frac{55}{12} = 4.58$$

**Using Equation 14.1,**

$$r = \frac{\Sigma(x - \bar{x})(y - \bar{y})}{\sqrt{\Sigma(x - \bar{x})^2 \, \Sigma(y - \bar{y})^2}} = \frac{3,838.92}{\sqrt{(76.92)(276,434.92)}} = 0.8325$$

The null and alternative hypotheses to be tested are

$$H_0: \rho = 0 \quad \text{(no correlation)}$$
$$H_A: \rho \neq 0 \quad \text{(correlation exists)}$$

where the Greek symbol $\rho$ (rho) represents the population correlation coefficient.

We must test whether the sample data support or refute the null hypothesis. The test procedure utilizes the $t$-test statistic in Equation 14.3.

**OUTCOME 2**  **Test Statistic for Correlation**

$$t = \frac{r}{\sqrt{\dfrac{1 - r^2}{n - 2}}} \qquad df = n - 2 \tag{14.3}$$

where:

$$t = \text{Number of standard errors } r \text{ is from } 0$$
$$r = \text{Sample correlation coefficient}$$
$$n = \text{Sample size}$$

The degrees of freedom for this test are $n - 2$.

Figure 14.5 shows the hypothesis test for the Midwest Distribution example using an alpha level of 0.05. Recall that the sample correlation coefficient was $r = 0.8325$. Based on these sample data, we can conclude that there is a significant positive linear relationship in the population between years of experience and total sales for Midwest Distribution sales representatives. The implication is that the more years an employee has been with the company, the more sales that employee generates. This runs counter to the claims made by some of the departing employees. The manager will probably want to look further into the situation to see whether a problem might exist in certain regions.

**FIGURE 14.5 Correlation Significance Test for the Midwest Distribution Example**

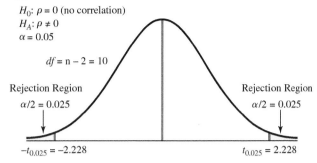

**Hypotheses:**

$H_0$: $\rho = 0$ (no correlation)
$H_A$: $\rho \neq 0$
$\alpha = 0.05$

$df = n - 2 = 10$

Rejection Region
$\alpha/2 = 0.025$

Rejection Region
$\alpha/2 = 0.025$

$-t_{0.025} = -2.228$

$t_{0.025} = 2.228$

The calculated $t$-value is

$$t = \frac{r}{\sqrt{\dfrac{1-r^2}{n-2}}} = \frac{0.8325}{\sqrt{\dfrac{1-0.6931}{10}}} = 4.752$$

**Decision Rule:**

If $t > t_{0.025} = 2.228$, reject $H_0$.
If $t < -t_{0.025} = -2.228$, reject $H_0$.
Otherwise, do not reject $H_0$.
Because $4.752 > 2.228$, reject $H_0$.

Based on the sample evidence, we conclude there is a significant positive linear relationship between years with the company and sales volume.

The $t$-test for determining whether the population correlation is significantly different from 0 requires the following assumptions:

**Assumptions**

1. The data are interval or ratio level.
2. The two variables ($y$ and $x$) are distributed as a *bivariate normal* distribution.

Although the formal mathematical representation is beyond the scope of this text, *two variables are bivariate normal if their joint distribution is normally distributed*. Although the $t$-test assumes a bivariate normal distribution, it is robust—that is, correct inferences can be reached even with slight departures from the normal-distribution assumption. (See Michael H. Kutner et al., *Applied Linear Statistical Models*, for further discussion of bivariate normal distributions.)

## Correlation Analysis

**Stock Portfolio Analysis**   A student intern at the investment firm of McMillan & Associates was given the assignment of determining whether there is a positive correlation between the number of individual stocks in a client's portfolio ($x$) and the annual rate of return ($y$) for the portfolio. The intern selected a simple random sample of 10 client portfolios and determined the number of individual company stocks and the annual rate of return earned by the client on his or her portfolio. To determine whether there is a statistically significant positive correlation between the two variables, the following steps can be employed:

**STEP 1  Specify the population parameter of interest.**

The intern wishes to determine whether the number of stocks in a client's portfolio is positively correlated with the rate of return earned by the client. The parameter of interest is, therefore, the population correlation, $\rho$.

**STEP 2  Formulate the appropriate null and alternative hypotheses.**

Because the intern was asked to determine whether a positive correlation exists between the variables of interest, the hypothesis test is one tailed, as follows:

$$H_0: \rho \le 0$$
$$H_A: \rho > 0$$

STEP 3    **Specify the level of significance.**
A significance level of 0.05 is chosen.

STEP 4    **Compute the correlation coefficient and the test statistic.**
Compute the sample correlation coefficient using Equation 14.1 or 14.2, or by using software such as Excel.

The following sample data were obtained:

| Number of Stocks | Rate of Return |
|:---:|:---:|
| 9 | 0.13 |
| 16 | 0.16 |
| 25 | 0.21 |
| 16 | 0.18 |
| 20 | 0.18 |
| 16 | 0.19 |
| 20 | 0.15 |
| 20 | 0.17 |
| 16 | 0.13 |
| 9 | 0.11 |

> The Excel 2016 function for computing the correlation coefficient is
>
> = **CORREL**(data range for Stocks, data range for Return)

Using Equation 14.1, we get

$$r = \frac{\Sigma (x - \bar{x})(y - \bar{y})}{\sqrt{[\Sigma (x - \bar{x})^2][\Sigma (y - \bar{y})^2]}} = 0.7796$$

We compute the *t*-test statistic using Equation 14.3:

$$t = \frac{r}{\sqrt{\dfrac{1 - r^2}{n - 2}}} = \frac{0.7796}{\sqrt{\dfrac{1 - 0.7796^2}{10 - 2}}} = 3.52$$

STEP 5    **Construct the rejection region and decision rule.**
For an alpha level equal to 0.05, the one-tailed, upper-tailed, critical value for $n - 2 = 10 - 2 = 8$ degrees of freedom is $t_{0.05} = 1.8595$. The decision rule is:

If $t > 1.8595$, reject the null hypothesis.

Otherwise, do not reject the null hypothesis.

STEP 6    **Reach a decision.**
Because

$$t = 3.52 > 1.8595, \text{ reject the null hypothesis.}$$

STEP 7    **Draw a conclusion.**
Because we reject the null hypothesis, the sample data do support the contention that there is a positive linear relationship between the number of individual stocks in a client's portfolio and the portfolio's rate of return.

**TRY EXERCISE 14-3** (pg. 533)

**Cause-and-Effect Interpretations**    Care must be used when interpreting the correlation results. For example, even though we found a significant linear relationship between years of experience and sales for the Midwest Distribution sales force, the correlation does not imply cause and effect. Although an increase in experience may, in fact, cause sales to change, simply because the two variables are correlated does not guarantee a cause-

and-effect relationship. Two seemingly unconnected variables may be highly correlated. For example, over a period of time, teachers' salaries in North Dakota might be highly correlated with the price of grapes in Spain. Yet we doubt that a change in grape prices will *cause* a corresponding change in salaries for teachers in North Dakota, or vice versa. When a correlation exists between two seemingly unrelated variables, the correlation is said to be a *spurious correlation*. You should take great care to avoid basing conclusions on spurious correlations. A statistically significant correlation alone does not prove that a cause-and-effect relationship exists.

# 14.1 EXERCISES

## Skill Development

**14-1.** An industry study was recently conducted in which the sample correlation between units sold and marketing expenses was 0.57. The sample size for the study was 15 companies. Based on the sample results, test to determine whether there is a significant positive correlation between these two variables. Use an $\alpha = 0.05$.

**14-2.** The following data for the dependent variable, $y$, and the independent variable, $x$, have been collected using simple random sampling:

| $x$ | $y$ |
|-----|-----|
| 10 | 120 |
| 14 | 130 |
| 16 | 170 |
| 12 | 150 |
| 20 | 200 |
| 18 | 180 |
| 16 | 190 |
| 14 | 150 |
| 16 | 160 |
| 18 | 200 |

a. Construct a scatter plot for these data. Based on the scatter plot, how would you describe the relationship between the two variables?

b. Compute the correlation coefficient.

**14-3.** A random sample of the following two variables was obtained:

| $x$ | 29 | 48 | 28 | 22 | 28 | 42 | 33 | 26 | 48 | 44 |
|-----|----|----|----|----|----|----|----|----|----|----|
| $y$ | 16 | 46 | 34 | 26 | 49 | 11 | 41 | 13 | 47 | 16 |

a. Calculate the correlation between these two variables.

b. Conduct a test of hypothesis to determine if there exists a correlation between the two variables in the population. Use a significance level of 0.10.

**14-4.** A random sample of two variables, $x$ and $y$, produced the following observations:

| $x$ | $y$ |
|-----|-----|
| 19 | 7 |
| 13 | 9 |
| 17 | 8 |
| 9 | 11 |
| 12 | 9 |
| 25 | 6 |
| 20 | 7 |
| 17 | 8 |

a. Develop a scatter plot for the two variables and describe what relationship, if any, exists.

b. Compute the correlation coefficient for these sample data.

c. Test to determine whether the population correlation coefficient is negative. Use a significance level of 0.05 for the hypothesis test.

**14-5.** You are given the following data for variables $x$ and $y$:

| $x$ | $y$ |
|-----|-----|
| 3.0 | 1.5 |
| 2.0 | 0.5 |
| 2.5 | 1.0 |
| 3.0 | 1.8 |
| 2.5 | 1.2 |
| 4.0 | 2.2 |
| 1.5 | 0.4 |
| 1.0 | 0.3 |
| 2.0 | 1.3 |
| 2.5 | 1.0 |

a. Plot these variables in scatter plot format. Based on this plot, what type of relationship appears to exist between the two variables?

b. Compute the correlation coefficient for these sample data. Indicate what the correlation coefficient measures.

c. Test to determine whether the population correlation coefficient is positive. Use the $\alpha = 0.01$ level to conduct the test. Be sure to state the null and alternative hypotheses and show the test and decision rule clearly.

**14-6.** For each of the following circumstances, perform the indicated hypothesis tests:

a. $H_A: \rho > 0$, $r = 0.53$, and $n = 30$ with $\alpha = 0.01$, using a test-statistic approach

b. $H_A: \rho \neq 0$, $r = -0.48$, and $n = 20$ with $\alpha = 0.05$, using a $p$-value approach

c. $H_A: \rho \neq 0$, $r = 0.39$, and $n = 45$ with $\alpha = 0.02$, using a test-statistic approach

d. $H_A: \rho < 0$, $r = 0.34$, and $n = 25$ with $\alpha = 0.05$, using a test-statistic approach

## Business Applications

**14-7.** The federal No Child Left Behind Act required periodic testing in standard subjects. A random sample of 50 junior high school students from a southern city was selected, and each student's scores on a standardized mathematics examination and a standardized English examination were recorded. School administrators were interested in the relationship between the two scores. Suppose the correlation coefficient for the two examination scores is 0.75.

a. Provide an explanation of the sample correlation coefficient in this context.

b. Using a level of significance of $\alpha = 0.01$, test to determine whether there is a positive linear relationship between mathematics scores and English scores for junior high school students in the city in the study.

**14-8.** A bank's chief financial officer is interested in whether there is a relationship between account balances and the number of times a card is used each month. A random sample of 50 accounts was selected. The account balance and the number of charges during the past month were the two variables recorded. The correlation coefficient for the two variables was $-0.23$.

a. Discuss what the $r = -0.23$ measures. Make sure to frame your discussion in terms of the two variables mentioned here.

b. Using an $\alpha = 0.10$ level, test to determine whether there is a significant linear relationship between account balance and the number of card uses during the past month. State the null and alternative hypotheses and show the decision rule.

c. Consider the decision you reached in part b. Describe the type of error you could have made in the context of this problem.

**14-9.** Many banks issue MasterCard credit cards to their customers. A main factor in determining whether a credit card will be profitable to the bank is the average monthly balance the customer maintains on the card that is subject to finance charges. Bank analysts wish to determine whether there is a relationship between the average monthly credit card balance and the income stated on the original credit card application form. The following sample data have been collected from existing credit card customers:

| Income | Credit Balance |
|---|---|
| $43,000 | $345 |
| $35,000 | $1,370 |
| $47,000 | $1,140 |
| $55,000 | $201 |
| $55,000 | $56 |
| $59,000 | $908 |
| $28,000 | $2,345 |
| $43,000 | $104 |
| $54,000 | $0 |
| $36,000 | $1,290 |
| $39,000 | $130 |
| $31,000 | $459 |
| $30,000 | $0 |
| $37,000 | $1,950 |
| $39,000 | $240 |

a. Indicate which variable is to be the independent variable and which is to be the dependent variable in the bank's analysis and indicate why.

b. Construct a scatter plot for these data and describe what, if any, relationship appears to exist between these two variables.

c. Calculate the correlation coefficient for these two variables and test to determine whether there is a significant correlation at the $\alpha = 0.05$ level.

**14-10.** Amazon.com has become one of the most successful online merchants. Two measures of its success are sales and net income/loss figures (all figures in $million). These values for the years 1995–2014 are as follows (*source:* Amazon Annual Reports 1995–2014):

| Year | Net Income/Loss | Sales |
|---|---|---|
| 1995 | −0.3 | 0.5 |
| 1996 | −5.7 | 15.7 |
| 1997 | −27.5 | 147.7 |
| 1998 | −124.5 | 609.8 |
| 1999 | −719.9 | 1,639.8 |
| 2000 | −1,411.2 | 2,761.9 |
| 2001 | −567.3 | 3,122.9 |
| 2002 | −149.1 | 3,933 |
| 2003 | 35.3 | 5,263.7 |
| 2004 | 588.5 | 6,921 |
| 2005 | 359 | 8,490 |
| 2006 | 190 | 10,711 |
| 2007 | 476 | 14,835 |
| 2008 | 645 | 19.17 |
| 2009 | 902 | 24.51 |
| 2010 | 1,152 | 34.2 |
| 2011 | 631 | 48.08 |
| 2012 | −39 | 61.09 |
| 2013 | 274 | 74.45 |
| 2014 | −241 | 88.99 |

a. Produce a scatter plot for Amazon's net income/loss and sales figures for the period 1995 to 2014. Does there appear to be a linear relationship between these two variables? Explain your response.

b. Calculate the correlation coefficient between Amazon's net income/loss and sales figures for the period 1995 to 2014.

c. Conduct a hypothesis test to determine if a positive correlation exists between Amazon's net income/loss and sales figures. Use a significance level of 0.05 and assume that these figures form a random sample.

14-11. A concert promoter has randomly sampled 100 concertgoers to determine whether there is a linear relationship between income level and the number of concerts attended per year. The correlation between these two variables was 0.68. Based on this calculation, is there a statistically significant correlation between income and number of concerts? Use an alpha = 0.05 level.

14-12. A regional retailer would like to determine if the variation in average monthly store sales can, in part, be explained by the size of the store measured in square feet. A random sample of 21 stores was selected and the store size and average monthly sales were computed. The results are shown here.

| Store Size (Sq. Ft) | Average Monthly Sales |
|---|---|
| 17,400 | $581,241.00 |
| 15,920 | $538,275.00 |
| 17,440 | $636,059.00 |
| 17,320 | $574,477.00 |
| 15,760 | $558,043.00 |
| 20,200 | $689,256.00 |
| 15,280 | $552,569.00 |
| 17,000 | $584,737.00 |
| 11,920 | $470,551.00 |
| 12,400 | $520,798.00 |
| 15,640 | $619,703.00 |
| 12,560 | $465,416.00 |
| 21,680 | $730,863.00 |
| 14,120 | $501,501.00 |
| 16,680 | $624,255.00 |
| 14,920 | $567,043.00 |
| 18,360 | $612,974.00 |
| 18,440 | $618,122.00 |
| 16,720 | $691,403.00 |
| 19,880 | $719,275.00 |
| 17,880 | $536,592.00 |

a. Construct a scatter plot using these data. What, if any relationship, appears to exist between the two variables?

b. Calculate the sample correlation coefficient between store size and average monthly sales.

c. Conduct a hypothesis test to determine if a positive correlation exists between store size and average monthly sales. Use a level of significance of 0.025.

## Computer Software Exercises

14-13. Goldie's Billiards, Inc., is a retailer of billiard supplies. It stands out among billiard suppliers because of the research it does to assure its products are top notch. One experiment was conducted to measure the speed a cue ball attains when it is struck by various weighted pool cues. The conjecture is that a light cue generates faster speeds while breaking the balls at the beginning of a game of pool. Goldie's research generated the data in the file titled **Breakcue**.

a. To determine if there is a negative relationship between the weight of the pool cue and the speed attained by the cue ball, calculate a correlation coefficient.

b. Conduct a test of hypothesis to determine if there is a negative relationship between the weight of the pool cue and the speed attained by the cue ball. Use a significance level of 0.025 and a $p$-value approach.

14-14. Customers who made online purchases last quarter from an Internet retailer were randomly sampled from the retailer's database. The dollar value of each customer's quarterly purchases along with the time the customer spent shopping the company's online catalog that quarter were recorded. The sample results are in the file **Online**.

a. Create a scatter plot of the variables Time ($x$) and Purchases ($y$). What relationship, if any, appears to exist between the two variables?

b. Compute the correlation coefficient for these sample data. What does the correlation coefficient measure?

c. Conduct a hypothesis test to determine if there is a positive relationship between time viewing the retailer's catalog and dollar amount purchased. Use a level of significance equal to 0.025. Provide a managerial explanation of your results.

14-15. A regional accreditation board for colleges and universities is interested in determining whether a relationship exists between student applicant verbal SAT scores and in-state tuition costs at the university. Data have been collected on a sample of colleges and universities and are in the file called **SAT Scores**.

a. Develop a scatter plot for these two variables and discuss what, if any, relationship you see between the two variables based on the scatter plot.

b. Compute the sample correlation coefficient.

c. Based on the correlation coefficient computed in part b, test to determine whether the population correlation coefficient is positive for these two variables. That is, can we expect schools that charge higher in-state tuition will attract students with higher average verbal SAT scores? Test using a 0.05 significance level.

**14-16.** A department store manager randomly sampled  customers in her store. She recorded the number of minutes each customer spent in the store and then asked the customers how many miles they had traveled to get to the store from their home. These data are in a file called **Shopping**.

a. Create a scatter plot using time spent in the store as the dependent variable. Discuss what the plot shows.
b. Test to determine whether there is significant correlation between the two variables. Use an alpha = 0.05.

**14.2**   **Simple Linear Regression Analysis**

In the Midwest Distribution application in Section 14.1, we determined that the relationship between years of experience and total sales is linear and statistically significant, based on the correlation analysis performed. Because hiring and training costs have been increasing, we would like to use this relationship to help formulate a more acceptable wage package for the sales force.

**Simple Linear Regression**
The method of regression analysis in which a single independent variable is used to explain the variation in the dependent variable.

The statistical method we will use to analyze the relationship between years of experience and total sales is *regression analysis*. When we have only two variables—a dependent variable, such as sales, and an independent variable, such as years with the company—the technique is referred to as *simple regression analysis*. When the relationship between the dependent variable and the independent variable is linear, the technique is **simple linear regression**.

### The Regression Model Assumptions

The objective of simple linear regression (which we shall call *regression analysis*) is to represent the relationship between values of $x$ and $y$ with a model of the form shown in Equation 14.4.

**Simple Linear Regression Model (Population Model)**

$$y = \beta_0 + \beta_1 x + \varepsilon \qquad (14.4)$$

where:

$$y = \text{Value of the dependent variable}$$
$$x = \text{Value of the independent variable}$$
$$\beta_0 = \text{Population's } y \text{ intercept}$$
$$\beta_1 = \text{Slope of the population regression line}$$
$$\varepsilon = \text{Random error term}$$

The simple linear regression population model described in Equation 14.4 has four assumptions:

**Assumptions**
1. Individual values of the error terms, $\varepsilon$, are statistically independent of one another, and these values represent a random sample from the population of possible $\varepsilon$ values at each level of $x$.
2. For a given value of $x$, there can exist many values of $y$ and therefore many values of $\varepsilon$. Further, the distribution of possible $\varepsilon$ values for any $x$ value is normal.
3. The distributions of possible $\varepsilon$ values have equal variances for all values of $x$.
4. The means of the dependent variable, $y$, for all specified values of the independent variable, $\mu_{y|x}$, can be connected by a straight line called the population regression model.

Figure 14.6 illustrates assumptions 2, 3, and 4. The regression model (straight line) connects the average of the $y$ values for each level of the independent variable, $x$. The actual

**FIGURE 14.6 Graphical Display of Linear Regression Assumptions**

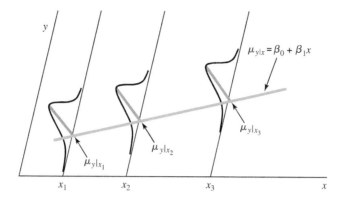

$y$ values for each level of $x$ are normally distributed around the mean of $y$. Finally, observe that the spread of possible $y$ values is the same regardless of the level of $x$. The population regression line is determined by two values, $\beta_0$ and $\beta_1$. These values are known as the population *regression coefficients*. $\beta_0$ identifies the $y$ intercept and $\beta_1$ the slope of the regression line. Under the regression assumptions, the coefficients define the true population model. For each observation, the actual value of the dependent variable, $y$, for any $x$ is the sum of two components:

$$y = \underset{\text{Linear component}}{\beta_0 + \beta_1 x} + \underset{\text{Random error component}}{\varepsilon}$$

The random error component, $\varepsilon$, may be positive, zero, or negative, depending on whether a single value of $y$ for a given $x$ falls above, on, or below the population regression line. Section 15.5 in Chapter 15 discusses how to check whether assumptions have been violated and the possible courses of action if the violations occur.

## Meaning of the Regression Coefficients

**Regression Slope Coefficient**
The average change in the dependent variable for a unit change in the independent variable. The slope coefficient may be positive, negative, or zero, depending on the relationship between the two variables.

Coefficient $\beta_1$, the **regression slope coefficient** of the population regression line, measures the average change in the value of the dependent variable, $y$, for each unit change in $x$. The regression slope can be positive, zero, or negative, depending on the relationship between $x$ and $y$. For example, a positive population slope of 12 ($\beta_1 = 12$) means that for a 1-unit increase in $x$, we can expect an average 12-unit increase in $y$. Correspondingly, if the population slope is negative 12 ($\beta_1 = -12$), we can expect an average decrease of 12 units in $y$ for a 1-unit increase in $x$.

The population's $y$ intercept, $\beta_0$, indicates the mean value of $y$ when $x$ is 0. However, this interpretation holds only if the population could have $x$ values equal to 0. When this cannot occur, $\beta_0$ does not have a meaningful interpretation in the regression model.

**BUSINESS APPLICATION**    **Simple Linear Regression Analysis**

**Midwest Distribution** (*continued*)    The Midwest Distribution marketing manager has data for a sample of 12 sales representatives. In Section 14.1, she established that a significant linear relationship exists between years of experience and total sales using correlation analysis. (Recall that the sample correlation between the two variables was $r = 0.8325$.) Now she would like to estimate the regression equation that defines the *true* linear relationship (that is, the population's linear relationship) between years of experience and sales. Figure 14.3 showed the scatter plot for two variables: years with the company and sales. We need to use the sample data to estimate $\beta_0$ and $\beta_1$, the true intercept and slope of the line representing the relationship between two variables. The *regression line* through the sample data is the best estimate of the population regression line. However, there

FIGURE 14.7 **Possible Regression Lines**

are an infinite number of possible regression lines for a set of points. For example, Figure 14.7 shows three of the possible different lines that pass through the Midwest Distribution data. Which line should be used to estimate the true regression model?

We must establish a criterion for selecting the best line. We use the **least squares criterion**. To understand the least squares criterion, you need to know about prediction error, or **residual**, which is the difference between the actual $y$ coordinate of an $(x, y)$ point and the predicted value of that $y$ coordinate produced by the regression line. Figure 14.8 shows how the prediction error is calculated for the employee who was with Midwest for four years $(x = 4)$ using one possible regression line (where $\hat{y}$ is the predicted sales value). The predicted sales value is

$$\hat{y} = 150 + 60(4) = 390$$

However, the actual sales $(y)$ for this employee is 312 (see Table 14.2). Thus, when $x = 4$, the difference between the observed value, $y = 312$, and the predicted value, $\hat{y} = 390$, is $312 - 390 = -78$. The residual (or prediction error) for this case when $x = 4$ is $-78$. Table 14.2 shows the calculated prediction errors and the sum of the squared errors for each of the three regression lines shown in Figure 14.7.[1] Of these three potential regression models, the line with the equation $\hat{y} = 150 + 60x$ has the smallest sum of squared errors. However, is there a better line than this? That is, would $\sum_{i=1}^{n}(y_i - \hat{y}_i)^2$ be smaller for some other line? One way to determine this is to calculate the sum of the squared errors for all other regression lines. However, because there are an infinite number of these lines, this approach is not feasible. Fortunately, through the use of calculus, we can derive equations to directly determine the slope and intercept estimates such that $\sum_{i=1}^{n}(y_i - \hat{y}_i)^2$ is minimized.[2] We do this by letting the estimated regression model be of the form shown in Equation 14.5.

**Least Squares Criterion**
The criterion for determining a regression line that minimizes the sum of squared prediction errors.

**Residual**
The difference between the actual value of the dependent variable and the value predicted by the regression model.

---
[1]The reason we are using the sum of the squared residuals is that the sum of the residuals will be zero for the best regression line (the positive values of the residuals will balance the negative values).
[2]The calculus derivation of the least squares equations is contained in the Kutner et al. reference shown at the end of the book.

**TABLE 14.2** Sum of Squared Errors for Three Linear
Equations for Midwest Distribution

From Figure 14.7(a):

$\hat{y} = 450 + 0x$

|  |  |  | Residual |  |
| --- | --- | --- | --- | --- |
| $x$ | $\hat{y}$ | $y$ | $y - \hat{y}$ | $(y - \hat{y})^2$ |
| 3 | 450 | 487 | 37 | 1,369 |
| 5 | 450 | 445 | −5 | 25 |
| 2 | 450 | 272 | −178 | 31,684 |
| 8 | 450 | 641 | 191 | 36,481 |
| 2 | 450 | 187 | −263 | 69,169 |
| 6 | 450 | 440 | −10 | 100 |
| 7 | 450 | 346 | −104 | 10,816 |
| 1 | 450 | 238 | −212 | 44,944 |
| 4 | 450 | 312 | −138 | 19,044 |
| 2 | 450 | 269 | −181 | 32,761 |
| 9 | 450 | 655 | 205 | 42,025 |
| 6 | 450 | 563 | 113 | 12,769 |
|  |  |  |  | $\Sigma = 301,187$ |

From Figure 14.7(b):

$\hat{y} = 250 + 40x$

|  |  |  | Residual |  |
| --- | --- | --- | --- | --- |
| $x$ | $\hat{y}$ | $y$ | $y - \hat{y}$ | $(y - \hat{y})^2$ |
| 3 | 370 | 487 | 117 | 13,689 |
| 5 | 450 | 445 | −5 | 25 |
| 2 | 330 | 272 | −58 | 3,364 |
| 8 | 570 | 641 | 71 | 5,041 |
| 2 | 330 | 187 | −143 | 20,449 |
| 6 | 490 | 440 | −50 | 2,500 |
| 7 | 530 | 346 | −184 | 33,856 |
| 1 | 290 | 238 | −52 | 2,704 |
| 4 | 410 | 312 | −98 | 9,604 |
| 2 | 330 | 269 | −61 | 3,721 |
| 9 | 610 | 655 | 45 | 2,025 |
| 6 | 490 | 563 | 73 | 5,329 |
|  |  |  |  | $\Sigma = 102,307$ |

From Figure 14.7(c):

$\hat{y} = 150 + 60x$

|  |  |  | Residual |  |
| --- | --- | --- | --- | --- |
| $x$ | $\hat{y}$ | $y$ | $y - \hat{y}$ | $(y - \hat{y})^2$ |
| 3 | 330 | 487 | 157 | 24,649 |
| 5 | 450 | 445 | −5 | 25 |
| 2 | 270 | 272 | 2 | 4 |
| 8 | 630 | 641 | 11 | 121 |
| 2 | 270 | 187 | −83 | 6,889 |
| 6 | 510 | 440 | −70 | 4,900 |
| 7 | 570 | 346 | −224 | 50,176 |
| 1 | 210 | 238 | 28 | 784 |
| 4 | 390 | 312 | −78 | 6,084 |
| 2 | 270 | 269 | −1 | 1 |
| 9 | 690 | 655 | −35 | 1,225 |
| 6 | 510 | 563 | 53 | 2,809 |
|  |  |  |  | $\Sigma = 97,667$ |

FIGURE 14.8 **Computation of Regression Error for the Midwest Distribution Example**

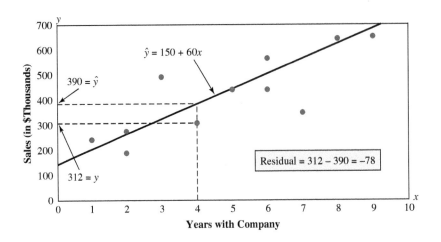

OUTCOME **3**

**Estimated Regression Model (Sample Model)**

$$\hat{y} = b_0 + b_1 x \qquad (14.5)$$

where:

$\hat{y}$ = Estimated, or predicted, $y$ value

$b_0$ = Unbiased estimate of the regression intercept, found using Equation 14.8

$b_1$ = Unbiased estimate of the regression slope, found using Equation 14.6 or 14.7

$x$ = Value of the independent variable

Equations 14.6 and 14.8 are referred to as the solutions to the *least squares equations* because they provide the slope and intercept that minimize the sum of squared errors. Equation 14.7 is the algebraic equivalent of Equation 14.6 and may be easier to use when the computation is performed using a calculator.

**Least Squares Equation**

Regression Slope Coefficient

$$b_1 = \frac{\Sigma (x_1 - \bar{x})(y_i - \bar{y})}{\Sigma (x_i - \bar{x})^2} \qquad (14.6)$$

or the algebraic equivalent:

$$b_1 = \frac{\Sigma xy - \dfrac{\Sigma x \Sigma y}{n}}{\Sigma x^2 - \dfrac{(\Sigma x)^2}{n}} \qquad (14.7)$$

Regression Intercept $\qquad b_0 = \bar{y} - b_1 \bar{x} \qquad (14.8)$

Table 14.3 shows the manual calculations, which are subject to rounding, for the least squares estimates for the Midwest Distribution example. However, you will almost always use a software package to perform these computations. (Figure 14.9 shows the Excel output.) In this case, the "best" regression line, given the least squares criterion, is $\hat{y} = 175.8288 + 49.9101x$. Figure 14.10 shows the predicted sales values along with the prediction errors and squared errors associated with this best simple linear regression line. Keep in mind that the prediction errors are also referred to as residuals. From Figure 14.10, the sum of the squared errors is 84,834.29. This is the smallest sum of squared residuals possible for this set of sample data. No other simple linear regression line through these 12 $(x, y)$ points will produce a smaller sum of squared errors. Equation 14.9 presents a formula that we can use to calculate the sum of squared errors manually.

**TABLE 14.3** Manual Calculations for Least Squares Regression Coefficients for the Midwest Distribution Example

| y | x | xy | $x^2$ | $y^2$ |
|---|---|---|---|---|
| 487 | 3 | 1,461 | 9 | 237,169 |
| 445 | 5 | 2,225 | 25 | 198,025 |
| 272 | 2 | 544 | 4 | 73,984 |
| 641 | 8 | 5,128 | 64 | 410,881 |
| 187 | 2 | 374 | 4 | 34,969 |
| 440 | 6 | 2,640 | 36 | 193,600 |
| 346 | 7 | 2,422 | 49 | 119,716 |
| 238 | 1 | 238 | 1 | 56,644 |
| 312 | 4 | 1,248 | 16 | 97,344 |
| 269 | 2 | 538 | 4 | 72,361 |
| 655 | 9 | 5,895 | 81 | 429,025 |
| 563 | 6 | 3,378 | 36 | 316,969 |
| $\Sigma y = 4{,}855$ | $\Sigma x = 55$ | $\Sigma xy = 26{,}091$ | $\Sigma x^2 = 329$ | $\Sigma y^2 = 2{,}240{,}687$ |

$$\bar{y} = \frac{\Sigma y}{n} = \frac{4{,}855}{12} = 404.58 \qquad \bar{x} = \frac{\Sigma x}{n} = \frac{55}{12} = 4.58$$

$$b_1 = \frac{\Sigma xy - \dfrac{\Sigma x \Sigma y}{n}}{\Sigma x^2 - \dfrac{(\Sigma x)^2}{n}} = \frac{26{,}091 - \dfrac{55(4{,}855)}{12}}{329 - \dfrac{(55)^2}{12}}$$

$$= 49.91$$

Then

$$b_0 = \bar{y} - b_1\bar{x} = 404.58 - 49.91(4.58) = 175.99$$

The least squares regression line is, therefore,

$$\hat{y} = 175.99 + 49.91x$$

There is a slight difference between the manual calculation and the computer result due to rounding.

**FIGURE 14.9** Excel 2016 Midwest Distribution Regression Results

Excel 2016 Instructions
1. Open file: **Midwest.xlsx**.
2. Select **Data** > **Data Analysis**.
3. Select **Regression**.
4. Define y (*Sales*) and x (*Years with Midwest*) variable data range.
5. Select **Labels**.
6. Select **Residuals**.
7. Select output location.

| | A | B | C | D | E | F | G |
|---|---|---|---|---|---|---|---|
| 1 | SUMMARY OUTPUT | | | | | | |
| 3 | *Regression Statistics* | | | | | | |
| 4 | Multiple R | 0.8325 | | | | | |
| 5 | R Square | 0.6931 | | | | | |
| 6 | Adjusted R Square | 0.6624 | | | | | |
| 7 | Standard Error | 92.1055 | | | | | |
| 8 | Observations | 12 | | | | | |
| 10 | ANOVA | | | | | | |
| 11 | | *df* | *SS* | *MS* | *F* | *Significance F* | |
| 12 | Regression | 1 | 191,600.62 | 191,600.62 | 22.59 | 0.0008 | |
| 13 | Residual | 10 | 84,834.29 | 8,483.43 | | | |
| 14 | Total | 11 | 276,434.92 | | | | |
| 16 | | *Coefficients* | *Standard Error* | *t Stat* | *P-value* | *Lower 95%* | *Upper 95%* |
| 17 | Intercept | 175.8288 | 54.9899 | 3.1975 | 0.0095 | 53.3037 | 298.3539 |
| 18 | Years with Midwest | 49.9101 | 10.5021 | 4.7524 | 0.0008 | 26.5100 | 73.3102 |
| 19 | | | | | | | |
| 20 | RESIDUAL OUTPUT | | | | | | |
| 21 | *Observation* | *Predicted* | *Residuals* | | | | |
| 22 | 1 | 325.56 | 161.44 | | | | |
| 23 | 2 | 425.38 | 19.62 | | | | |
| 24 | 3 | 275.65 | -3.65 | | | | |

SSE = 84,834.29

Estimated regression equation is:
$\hat{y} = 175.8288 + 49.9101x$

**FIGURE 14.10** Excel 2016 Residuals and Squared Residuals for the Midwest Distribution Example

| 22 | RESIDUAL OUTPUT | | | |
|---|---|---|---|---|
| 23 | Observation | Predicted Sales | Residuals | Residuals² |
| 24 | 1 | 325.56 | 161.44 | 26063.18 |
| 25 | 2 | 425.38 | 19.62 | 384.98 |
| 26 | 3 | 275.65 | -3.65 | 13.31 |
| 27 | 4 | 575.11 | 65.89 | 4341.57 |
| 28 | 5 | 275.65 | -88.65 | 7858.64 |
| 29 | 6 | 475.29 | -35.29 | 1245.33 |
| 30 | 7 | 525.20 | -179.20 | 32112.41 |
| 31 | 8 | 225.74 | 12.26 | 150.33 |
| 32 | 9 | 375.47 | -63.47 | 4028.33 |
| 33 | 10 | 275.65 | -6.65 | 44.21 |
| 34 | 11 | 625.02 | 29.98 | 898.83 |
| 35 | 12 | 475.29 | 87.71 | 7693.17 |
| 36 | | Sum = | 0.00 | 84834.29 |

Sum of residuals = 0.0
SSE = 84,834.29

**Sum of Squared Residuals (Errors)—Manual Calculation**

$$SSE = \sum y^2 - b_0 \sum y - b_1 \sum xy \tag{14.9}$$

Figure 14.11 shows the scatter plot of sales and years of experience and the least squares regression line for Midwest Distribution. This line is the *best fit* for these sample data. The regression line passes through the point corresponding to $(\bar{x}, \bar{y})$. This will always be the case.

## Least Squares Regression Properties

There are several important properties of least squares regression:

1. The sum of the residuals from the least squares regression line is 0 (Equation 14.10). The total underprediction by the regression model is exactly offset by the total overprediction (see Figure 14.10).

**Sum of Residuals**

$$\sum_{i=1}^{n} (y_i - \hat{y}_i) = 0 \tag{14.10}$$

2. The sum of the squared residuals is the minimum (Equation 14.11).

**Sum of Squared Residuals (Errors)**

$$SSE = \sum_{i=1}^{n} (y_i - \hat{y}_i)^2 \tag{14.11}$$

This property provided the basis for developing the equations for $b_0$ and $b_1$.

3. The simple regression line always passes through the mean of the $y$ variable, $\bar{y}$, and the mean of the $x$ variable, $\bar{x}$. So, to manually draw any simple linear regression line, all you need to do is to draw a line connecting the least squares $y$ intercept with the $(\bar{x}, \bar{y})$ point.
4. The least squares coefficients are unbiased estimates of $\beta_0$ and $\beta_1$. Thus, the expected values of $b_0$ and $b_1$ equal $\beta_0$ and $\beta_1$, respectively.

**FIGURE 14.11** Excel 2016 Least Squares Regression Line for Midwest Distribution

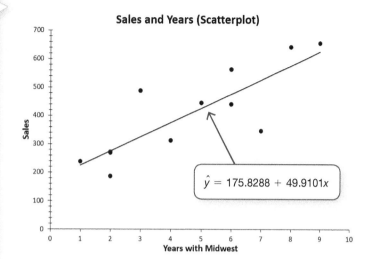

$$\hat{y} = 175.8288 + 49.9101x$$

---

**EXAMPLE 14-2** **Simple Linear Regression and Correlation**

**Foresight Investments**   The investment firm Foresight Investments wants to manage the pension fund of a major Chicago retailer. For their presentation to the retailer, the Foresight analysts want to use simple linear regression to model the relationship between profits and numbers of employees for 50 companies in the firm's portfolio. The data for the analysis are in the file **Employees**. This analysis can be done using the following steps:

Excel Tutorial

**STEP 1**  **Specify the independent and dependent variables.**
The objective in this example is to model the linear relationship between number of employees (the independent variable) and each company's profits in millions (the dependent variable).

**STEP 2**  **Develop a scatter plot to graphically display the relationship between the independent and dependent variables.**
Figure 14.12 shows the scatter plot, where the dependent variable, $y$, is company profits and the independent variable, $x$, is number of employees. There appears to be a slight positive linear relationship between the two variables.

**STEP 3**  **Calculate the correlation coefficient and the linear regression equation.**
Do either manually using Equations 14.1, 14.6 (or 14.7), and 14.8 or by using Excel. Figure 14.13 shows the Excel regression results. The sample correlation coefficient (called "Multiple R" in Excel) is

$$r = 0.3638$$

The regression equation is

$$\hat{y} = 2{,}556.88 + 0.0048x$$

The regression slope is estimated to be 0.0048, which means that for each additional employee, the average increase in company profit is 0.0048 million dollars, or \$4,800. The intercept can only be interpreted when a value equal to zero for the $x$ variable (employees) is plausible. Clearly, no company has zero employees, so the intercept in this case has no meaning other than it locates the height of the regression line for $x = 0$.

**Excel 2016 Instructions**

1. Open file: **Employees.xlsx**.
2. Move the *Profits* column to the right of the *Employees* column.
3. Select data to be used in the chart (Employees and Profits).
4. On the **Insert** tab, click **Insert Scatter (X,Y)** or **Bubble Chart**, and then click the **Scatter** option.
5. Use the **Design** tab of **Chart Tools** to add titles and remove the grid lines.
6. Use the **Design** tab of **Chart Tools** to move the chart to a new worksheet.

**FIGURE 14.12 Excel 2016 Scatter Plot for Foresight Investments**

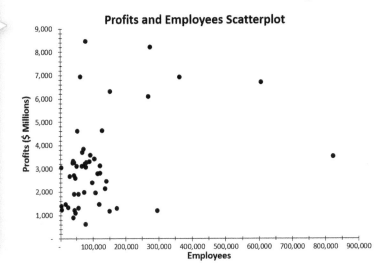

**FIGURE 14.13 Excel 2016 Regression Results for Foresight Investments**

**Excel 2016 Instructions**

1. Open file: **Employees.xlsx**.
2. Select **Data > Data Analysis**.
3. Select **Regression**.
4. Define *y* (*Profits*) and *x* (*Employees*) variable data range.
5. Select **Labels**.
6. Select **Residuals**.
7. Select output location.

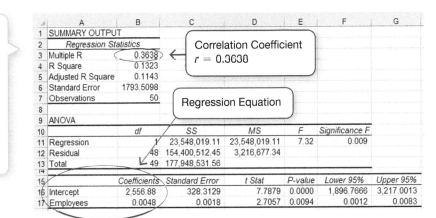

**TRY EXERCISE 14-17** (pg. 551)

**OUTCOME 4**

## Significance Tests in Regression Analysis

In Section 14.1, we pointed out that the correlation coefficient computed from sample data is a point estimate of the population correlation coefficient and is subject to sampling error. We also introduced a test of significance for the correlation coefficient. Likewise, the regression coefficients developed from a sample of data are also point estimates and are subject to sampling error. Therefore, we need a test procedure to determine whether the regression slope coefficient is statistically significant.

**Significance of the Slope Coefficient**   For a simple linear regression model (one independent variable), there are three equivalent statistical tests:

1. Test for significance of the correlation between $x$ and $y$
2. Test for significance of the regression slope coefficient
3. Test for significance of the coefficient of determination

Section 14.1 introduced the first of these tests. The second one deals specifically with the significance of the regression slope coefficient. The null and alternative hypotheses to be tested are

$$H_0: \beta_1 = 0$$
$$H_A: \beta_1 \neq 0$$

To test the significance of the simple linear regression slope coefficient, we are interested in determining whether the population regression slope coefficient is 0. A slope of 0 implies that there is no linear relationship between the $x$ and $y$ variables and that the $x$ variable, in its linear form, is of no use in explaining the variation in $y$. If the linear relationship is useful, then we should reject the hypothesis that the regression slope is 0. However, because the estimated regression slope coefficient, $b_1$, is calculated from sample data, it is subject to sampling error. Therefore, even though $b_1$ is not 0, we must determine whether its difference from 0 is greater than would generally be attributed to sampling error.

If we selected several samples from the same population and for each sample determined the least squares regression line, we would likely get regression lines with different slopes and different $y$ intercepts. This is analogous to getting different sample means from different samples when we attempt to estimate a population mean. Just as the distribution of possible sample means has a standard error, the possible regression slopes also have a standard error, which is estimated using Equation 14.12.

### Simple Regression Estimator for the Standard Error of the Slope

$$s_{b_1} = \frac{s_\varepsilon}{\sqrt{\Sigma (x - \bar{x})^2}}$$

(**14.12**)

where:

$s_{b_1}$ = Estimate of the standard error of the least squares slope

$s_\varepsilon = \sqrt{\dfrac{SSE}{n - 2}}$ = Sample standard error of the estimate (the measure of deviation of the actual $y$ values around the regression line)

Because the sample regression slope most likely does not equal the true population slope, we must test to determine whether the true slope could possibly be 0. A slope of 0 in the linear model means that the independent variable does not explain any variation in the dependent variable, nor is it useful in predicting the dependent variable. The null and alternative hypotheses to be tested are

$$H_0: \beta_1 = 0$$
$$H_A: \beta_1 \neq 0$$

To test the significance of a slope coefficient, we use the $t$-test value in Equation 14.13.

### Simple Linear Regression Test Statistic for Test of the Significance of the Slope

$$t = \frac{b_1 - \beta_1}{s_{b_1}} \qquad df = n - 2$$

(**14.13**)

where:

$b_1$ = Sample regression slope coefficient

$\beta_1$ = Hypothesized slope (usually $\beta_1 = 0$)

$s_{b_1}$ = Estimator of the standard error of the slope

---

**BUSINESS APPLICATION**

## Significance Test for the Regression Slope Coefficient

**Midwest Distribution** (*continued*)   For Midwest Distribution, the Excel 2016 regression output in Figure 14.14 shows $b_1 = 49.91$. The question is whether this value is different enough from 0 to have not been caused by sampling error. We find the answer by performing the statistical test shown in Figure 14.15.

**FIGURE 14.14 Excel 2016 Regression Results for Midwest Distribution**

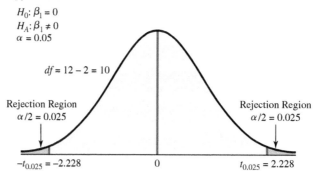

| | A | B | C | D | E | F | G |
|---|---|---|---|---|---|---|---|
| 1 | SUMMARY OUTPUT | | | | | | |
| 2 | Regression Statistics | | | | | | |
| 3 | Multiple R | 0.8325 | | | | | |
| 4 | R Square | 0.6931 | | | | | |
| 5 | Adjusted R Square | 0.6624 | | | | | |
| 6 | Standard Error | 92.1055 | | | | | |
| 7 | Observations | 12 | | | | | |
| 8 | | | | | | | |
| 9 | ANOVA | | | | | | |
| 10 | | df | SS | MS | F | Significance F | |
| 11 | Regression | 1 | 191,600.62 | 191,600.62 | 22.59 | 0.0008 | |
| 12 | Residual | 10 | 84,834.29 | 8,483.43 | | | |
| 13 | Total | 11 | 276,434.92 | | | | |
| 15 | | Coefficients | Standard Error | t Stat | P-value | Lower 95% | Upper 95% |
| 16 | Intercept | 175.83 | 54.99 | 3.20 | 0.0095 | 53.30 | 298.35 |
| 17 | Years with Midwest | 49.91 | 10.50 | 4.75 | 0.0008 | 26.51 | 73.31 |

The calculated *t*-statistic and *p*-value for testing whether the regression slope = 0.0

Standard error of the regression slope = 10.50

**FIGURE 14.15 Significance Test of the Regression Slope for Midwest Distribution**

**Hypotheses:**

$H_0: \beta_1 = 0$
$H_A: \beta_1 \neq 0$
$\alpha = 0.05$

$df = 12 - 2 = 10$

Rejection Region
$\alpha/2 = 0.025$

Rejection Region
$\alpha/2 = 0.025$

$-t_{0.025} = -2.228$    0    $t_{0.025} = 2.228$

The calculated *t* is

$$t = \frac{b_1 - \beta_1}{s_{b_1}} = \frac{49.91 - 0}{10.50} = 4.752$$

**Decision Rule:**

If $t > t_{0.025} = 2.228$, reject $H_0$.
If $t < -t_{0.025} = -2.228$, reject $H_0$.
Otherwise, do not reject $H_0$.

Because 4.752 > 2.228, we reject the null hypothesis and conclude that the true slope is not 0. Thus, the simple linear relationship that utilizes the independent variable, years with the company, is useful in explaining the variation in the dependent variable, sales volume.

The calculated *t*-value of 4.752 shown in Figures 14.14 and 14.15 exceeds the critical value, $t = 2.228$, from the *t*-distribution with 10 degrees of freedom and $\alpha/2 = 0.025$. This indicates that we should reject the hypothesis that the true regression slope is 0. Thus, years of experience can be used to help explain the variation in an individual representative's sales. This is not a coincidence. This test is always equivalent to the test for $\rho$ presented earlier.

The output shown in Figure 14.14 also contains the *p*-value for the calculated *t*-statistic. As with other situations involving two-tailed hypothesis tests, if the *p*-value is less than $\alpha/2$, the null hypothesis is rejected. In this case, because *p*-value = 0.0008 < 0.025, we reject the null hypothesis.

BUSINESS APPLICATION    ## The Coefficient of Determination, $R^2$

**Midwest Distribution** (*continued*)   Recall that the Midwest Distribution marketing manager was analyzing the relationship between the number of years an employee had been with the company (independent variable) and the sales generated by the employee (dependent variable). We note when looking at the sample data for 12 employees (see Table 14.3) that sales vary among employees. Regression analysis aims to determine the extent to which an independent variable can explain this variation. In this case, does the number of years with the company help explain the variation in sales from employee to employee?

We can use the *SST* (total sum of squares) to measure the variation in the dependent variable. *SST* is computed using Equation 14.14. For Midwest Distribution, the total sum of squares for sales is provided in the output generated by Excel, as shown in Figure 14.14. As you can see, the total sum of squares in sales that needs to be explained is 276,434.92. Note that the *SST* value is in squared units and has no particular meaning.

**Total Sum of Squares**

$$SST = \sum_{i=1}^{n} (y_i - \bar{y})^2 \tag{14.14}$$

where:

$$SST = \text{Total sum of squares}$$
$$n = \text{Sample size}$$
$$y_i = i\text{th value of the dependent variable}$$
$$\bar{y} = \text{Average value of the dependent variable}$$

The least squares regression line is computed so that the sum of squared residuals is minimized (recall the discussion of the least squares equations). The sum of squared residuals is also called the *sum of squares error* (*SSE*) and is defined by Equation 14.15.

*SSE* represents the amount of the total sum of squares in the dependent variable that *is not explained* by the least squares regression line. Excel refers to *SSE* as *sum of squares residual*. This value is shown in the regression output in Figure 14.14.

**Sum of Squares Error (Residual)**

$$SSE = \sum_{i=1}^{n} (y_i - \hat{y}_i)^2 \tag{14.15}$$

where:

$$n = \text{Sample size}$$
$$y_i = i\text{th value of the dependent variable}$$
$$\hat{y}_i = i\text{th predicted value of } y \text{ given the } i\text{th value of } x$$

$$SSE = \sum (y - \hat{y})^2 = 84{,}834.29$$

Thus, of the total sum of squares (*SST* = 276,434.92), the regression model leaves *SSE* = 84,834.29 unexplained. Then, the portion of the total sum of squares that *is explained* by the regression line is called the *sum of squares regression* (*SSR*) and is calculated by Equation 14.16.

**Sum of Squares Regression**

$$SSR = \sum_{i=1}^{n} (\hat{y}_i - \bar{y})^2 \qquad \textbf{(14.16)}$$

where:

$$\hat{y}_i = \text{Estimated value of } y \text{ for each value of } x$$
$$\bar{y} = \text{Average value of the } y \text{ variable}$$

The sum of squares regression $(SSR = 191,600.62)$ is provided in the regression output shown in Figure 14.14. You should also note that the following holds:

$$SST = SSR + SSE$$

For the Midwest Distribution example, in the Excel output we get

$$276,434.92 = 101,600.62 + 84,834.29$$

**Coefficient of Determination**
The portion of the total variation in the dependent variable that is explained by its relationship with the independent variable. The coefficient of determination is also called $R$-squared and is denoted as $R^2$.

We can use these calculations to compute an important measure in regression analysis called the **coefficient of determination**. The coefficient of determination is calculated using Equation 14.17.

**Coefficient of Determination, $R^2$**

$$R^2 = \frac{SSR}{SST} \qquad \textbf{(14.17)}$$

For the Midwest Distribution example, the proportion of variation in sales that can be explained by its linear relationship with the years of sales force experience is

$$R^2 = \frac{SSR}{SST} = \frac{191,600.62}{276,434.92} = 0.6931$$

This means that 69.31% of the variation in the sales data for this sample can be explained by the linear relationship between sales and years of experience. Notice that $R^2$ is part of the regression output in Figure 14.14.

$R^2$ can be a value between 0 and 1.0. If there is a perfect linear relationship between two variables, then the coefficient of determination, $R^2$, is 1.0. This corresponds to a situation in which the least squares regression line passes through each of the points in the scatter plot. Many decision makers use $R^2$ to indicate how well the linear regression line fits the $(x, y)$ data points. The better the fit, the closer $R^2$ will be to 1.0. $R^2$ will be close to 0 when there is a weak linear relationship.

Keep in mind that $R^2 = 0.6931$ is based on the random sample of size 12 and is subject to sampling error. Thus, just because $R^2 = 0.6931$ for the sample data does not mean that knowing the number of years an employee has worked for the company explains 69.31% of the variation in sales for the population of all employees with the company. Likewise, just because $R^2 > 0.0$ for the sample data does not mean that the population coefficient of determination, noted as $\rho^2$ (rho-squared), is greater than zero.

However, a statistical test exists for testing the following null and alternative hypotheses:

$$H_0: \rho^2 = 0$$
$$H_A: \rho^2 > 0$$

The test statistic is an $F$-test with the test statistic defined as shown in Equation 14.18.

**Test Statistic for Significance of the Coefficient of Determination**

$$F = \frac{\dfrac{SSR}{1}}{\dfrac{SSE}{(n-2)}} \qquad df = (D_1 = 1, D_2 = n-2) \tag{14.18}$$

where:

$$SSR = \text{Sum of squares regression}$$
$$SSE = \text{Sum of squares error}$$

For the Midwest Distribution example, the test statistic is computed using Equation 14.18 as follows:

$$F = \frac{\dfrac{191,600.62}{1}}{\dfrac{84,834.29}{12-2}} = 22.58$$

The critical value from the $F$-distribution table in Appendix H for $\alpha = 0.05$ and for 1 and 10 degrees of freedom is 4.965. This gives the following decision rule:

> If $F > 4.965$, reject the null hypothesis.
>
> Otherwise, do not reject the null hypothesis.

Because $F = 22.58 > 4.965$, we reject the null hypothesis and conclude that the population coefficient of determination $(\rho^2)$ is greater than zero. This means the independent variable explains a significant proportion of the variation in the dependent variable.

---

**HOW TO DO IT** (Example 14-3)

**Simple Linear Regression Analysis**

**1.** Define the independent $(x)$ and dependent $(y)$ variables and select a simple random sample of pairs of $(x, y)$ values.

**2.** Develop a scatter plot of $y$ and $x$. You are looking for a linear relationship between the two variables.

**3.** Compute the correlation coefficient for the sample data.

**4.** Calculate the least squares regression line for the sample data.

**5.** Conduct either of the following tests for determining whether the regression model is statistically significant:

    **a.** Test to determine whether the true regression slope is 0. The test statistic with $df = n - 2$ is

$$t = \frac{b_1 - \beta_1}{s_{b_1}} = \frac{b_1 - 0}{s_{b_1}}$$

    **b.** Test to see whether $\rho$ is significantly different from 0. The test statistic is

$$t = \frac{r}{\sqrt{\dfrac{1 - r^2}{n - 2}}}$$

**6.** Reach a decision.

**7.** Draw a conclusion.

---

**EXAMPLE 14-3**   **Simple Linear Regression Analysis**

 **Vantage Electronic Systems**   Consider the example involving Vantage Electronic Systems in Deerfield, Michigan, which started out supplying electronic equipment for the automobile industry but in recent years has ventured into other areas. One area is visibility sensors used by airports to provide takeoff and landing information and by transportation departments to detect low visibility on roadways during fog and snow. The recognized leader in the visibility sensor business is the SCR Company, which makes a sensor called the Scorpion. The research and development (R&D) department at Vantage has recently performed a test on its new unit by locating a Vantage sensor and a Scorpion sensor side by side. Various data, including visibility measurements, were collected at randomly selected points in time over a two-week period. These data are contained in a file called **Vantage**.

STEP 1   **Define the independent $(x)$ and dependent $(y)$ variables.**
The analysis included a simple linear regression using the Scorpion visibility measurement as the dependent variable, $y$, and the Vantage visibility measurement as the independent variable, $x$.

STEP 2   **Develop a scatter plot of $y$ and $x$.**
The scatter plot is shown in Figure 14.16. There does not appear to be a strong linear relationship.

STEP 3   **Compute the correlation coefficient for the sample data.**
We can use Equation 14.1 or Equation 14.2 for manual computation, or we can use Excel. The sample correlation coefficient is

$$r = 0.5778$$

STEP 4 **Calculate the least squares regression line for the sample data.**
Again, we can use Excel to do the computations.
The least squares regression equation is

$$\hat{y} = 0.586 + 3.017x$$

**FIGURE 14.16** Scatter Plot—Example 14-3

STEP 5 **Conduct either of the following tests to determine whether the regression model is statistically significant (or whether the population correlation is equal to 0).**
The null and alternative hypotheses to test the correlation coefficient are

$$H_0: \rho = 0$$
$$H_A: \rho \neq 0$$

The $t$-test statistic using Equation 14.3 is

$$t = \frac{r}{\sqrt{\dfrac{1 - r^2}{n - 2}}} = \frac{0.5778}{\sqrt{\dfrac{1 - 0.5778^2}{280 - 2}}} = 11.8$$

The $t = 11.8$ exceeds the critical $t$ for any reasonable level of $\alpha$ for 278 degrees of freedom, so we reject the null hypothesis and conclude that there is a statistically significant linear relationship between the visibility measures for the two visibility sensors.

Alternatively, the null and alternative hypotheses to test the regression slope coefficient are

$$H_0: \beta_1 = 0$$
$$H_A: \beta_1 \neq 0$$

The $t$-test statistic is

$$t = \frac{b_1 - \beta_1}{s_{b_1}} = \frac{3.017 - 0}{0.2557} = 11.8$$

STEP 6 **Reach a decision.**
The $t$-test statistic of 11.8 exceeds the critical $t$-value for any reasonable level of $\alpha$ for 278 degrees of freedom.

STEP 7 **Draw a conclusion.**
The population regression slope coefficient is not equal to 0. This means that knowing the Vantage visibility reading provides useful help in knowing what the Scorpion visibility reading will be.

Excel Tutorial

**TRY EXERCISE 14-17** (pg. 551)

## 14.2 EXERCISES

### Skill Development

**14-17.** You are given the following sample data for variables $y$ and $x$:

| $y$ | 140 | 120 | 80 | 100 | 130 | 90 | 110 | 120 | 130 | 130 | 100 |
|-----|-----|-----|----|-----|-----|----|-----|-----|-----|-----|-----|
| $x$ | 5 | 3 | 2 | 4 | 5 | 4 | 4 | 5 | 6 | 5 | 4 |

a. Develop a scatter plot for these data and describe what, if any, relationship exists.
b. (1) Compute the correlation coefficient. (2) Test to determine whether the correlation is significant at the significance level of 0.05. Conduct this hypothesis test using the $p$-value approach. (3) Compute the regression equation based on these sample data and interpret the regression coefficients.
c. Test the significance of the overall regression model using a significance level equal to 0.05.

**14-18.** You are given the following sample data for variables $x$ and $y$:

| $x$ (independent) | $y$ (dependent) |
|-------------------|-----------------|
| 1 | 16 |
| 7 | 50 |
| 3 | 22 |
| 8 | 59 |
| 11 | 63 |
| 5 | 46 |
| 4 | 43 |

a. Construct a scatter plot for these data and describe what, if any, relationship appears to exist.
b. Compute the regression equation based on these sample data and interpret the regression coefficients.
c. Based on the sample data, what percentage of the total variation in the dependent variable can be explained by the independent variable?
d. Test the significance of the overall regression model using a significance level of 0.01.
e. Test to determine whether the true regression slope coefficient is equal to 0. Use a significance level of 0.01.

**14-19.** The following data for the dependent variable, $y$, and the independent variable, $x$, have been collected using simple random sampling:

| $x$ | $y$ |
|-----|-----|
| 10 | 120 |
| 14 | 130 |
| 16 | 170 |
| 12 | 150 |
| 20 | 200 |

| $x$ | $y$ |
|-----|-----|
| 18 | 180 |
| 16 | 190 |
| 14 | 150 |
| 16 | 160 |
| 18 | 200 |

a. Develop a simple linear regression equation for these data.
b. Calculate the sum of squared residuals, the total sum of squares, and the coefficient of determination.
c. Calculate the standard error of the estimate.
d. Calculate the standard error for the regression slope.
e. Conduct the hypothesis test to determine whether the regression slope coefficient is equal to 0. Test using $\alpha = 0.02$.

**14-20.** Consider the following sample data for the variables $y$ and $x$:

| $x$ | 30.3 | 4.8 | 15.2 | 24.9 | 8.6 | 20.1 | 9.3 | 11.2 |
|-----|------|-----|------|------|-----|------|-----|------|
| $y$ | 14.6 | 27.9 | 17.6 | 15.3 | 19.8 | 13.2 | 25.6 | 19.4 |

a. Calculate the linear regression equation for these data.
b. Determine the predicted $y$ value when $x = 10$.
c. Estimate the change in the $y$ variable resulting from the increase in the $x$ variable of 10 units.
d. Conduct a hypothesis test to determine if an increase of 1 unit in the $x$ variable will result in the decrease of the average value of the $y$ variable. Use a significance of 0.025.

**14-21.** Examine the following sample data for the variables $y$ and $x$:

| $x$ | 1 | 2 | 3 | 4 | 5 |
|-----|---|---|---|---|---|
| $y$ | 4 | 2 | 5 | 8 | 9 |

a. Construct a scatter plot of these data. Describe the relationship between $x$ and $y$.
b. Calculate the sum of squares error for the following equations: (1) $\hat{y} = 0.8 + 1.60x$, (2) $\hat{y} = 1 + 1.50x$, and (3) $\hat{y} = 0.7 + 1.60x$.
c. Which of these equations provides the "best" fit of these data? Describe the criterion you used to determine "best" fit.
d. Determine the regression line that minimizes the sum of squares error.

### Business Applications

**14-22.** The Skelton Manufacturing Company recently did a study of its customers. A random sample of 50 customer accounts was pulled from the computer records. Two variables were observed:

$y$ = Total dollar volume of business this year

$x$ = Miles customer is from corporate headquarters

The following statistics were computed:

$$\hat{y} = 2{,}140.23 - 10.12x$$
$$s_{b_1} = 3.12$$

a. Interpret the regression slope coefficient.
b. Using a significance level of 0.01, test to determine whether it is true that the farther a customer is from the corporate headquarters, the smaller the total dollar volume of business.

**14-23.** A shipping company believes that the variation in the cost of a customer's shipment can be explained by differences in the weight of the package being shipped. To investigate whether this relationship is useful, a random sample of 20 customer shipments was selected, and the weight (in lb) and the cost (in dollars, rounded to the nearest dollar) for each shipment were recorded. The following results were obtained:

| Weight (x) | Cost (y) |
|------------|----------|
| 8 | 11 |
| 6 | 8 |
| 5 | 11 |
| 7 | 11 |
| 12 | 17 |
| 9 | 11 |
| 17 | 27 |
| 13 | 16 |
| 8 | 9 |
| 18 | 25 |
| 17 | 21 |
| 17 | 24 |
| 10 | 16 |
| 20 | 24 |
| 9 | 21 |
| 5 | 10 |
| 13 | 21 |
| 6 | 16 |
| 6 | 11 |
| 12 | 20 |

a. Construct a scatter plot for these data. What, if any, relationship appears to exist between the two variables?
b. Compute the linear regression model based on the sample data. Interpret the slope and intercept coefficients.
c. Test the significance of the overall regression model using a significance level equal to 0.05.
d. What percentage of the total variation in shipping cost can be explained by the regression model you developed in part b?

**14-24.** A company that makes designer sandals has tracked the production volume over the past six months. Because of the company's focus on quality, some sandals need to be reworked before they are fit to be shipped to customers. The number of pairs of sandals produced and the number of pairs that require rework are given in the table.

| Total Pairs Produced | 9,202 | 12,146 | 13,844 | 16,454 | 19,710 | 21,582 |
|----------------------|-------|--------|--------|--------|--------|--------|
| Pairs Reworked | 2,074 | 2,395 | 3,188 | 3,632 | 4,694 | 5,652 |

a. Conduct a simple linear regression analysis of these data in which the number of pairs reworked is predicted by the total pairs produced. Test the significance of the model using an $\alpha = 0.10$.
b. How much does the number of reworked pairs of sandals change when the total production increases by 100 units?
c. When the total production reaches 7,500 sandals, what would you expect the number of reworked pairs to be?

**14-25.** A regional retailer would like to determine if the variation in average monthly store sales can, in part, be explained by the size of the store measured in square feet. A random sample of 21 stores was selected and the store size and average monthly sales were computed. The following results are shown:

| Store Size (Sq. Ft) | Average Monthly Sales |
|---------------------|-----------------------|
| 17,400 | $581,241.00 |
| 15,920 | $538,275.00 |
| 17,440 | $636,059.00 |
| 17,320 | $574,477.00 |
| 15,760 | $558,043.00 |
| 20,200 | $689,256.00 |
| 15,280 | $552,569.00 |
| 17,000 | $584,737.00 |
| 11,920 | $470,551.00 |
| 12,400 | $520,798.00 |
| 15,640 | $619,703.00 |
| 12,560 | $465,416.00 |
| 21,680 | $730,863.00 |
| 14,120 | $501,501.00 |
| 16,680 | $624,255.00 |
| 14,920 | $567,043.00 |
| 18,360 | $612,974.00 |
| 18,440 | $618,122.00 |
| 16,720 | $691,403.00 |
| 19,880 | $719,275.00 |
| 17,880 | $536,592.00 |

a. Compute the simple linear regression model using the sample data to determine whether variation in average monthly sales can be explained by store size. Interpret the slope and intercept coefficients.
b. Test for the significance of the slope coefficient of the regression model. Use a level of significance of 0.05.

c. Based on the estimated regression model, what percentage of the total variation in average monthly sales can be explained by store size?

**14-26.** An engineering firm is interested in investigating whether the variability in the cost of small projects (defined as projects under $10 million) can be accounted for, in part, by differences in the number of direct engineering consulting hours billed. A sample of 24 small projects was randomly selected from small projects performed over the past two years. The number of engineering consulting hours billed for each project, along with the project's cost, was recorded. The results are shown in the table.

| Project | Billed Consulting Hours | Total Project Cost |
|---|---|---|
| 1 | 3,932 | $4,323,826 |
| 2 | 3,097 | $3,750,964 |
| 3 | 2,972 | $3,579,570 |
| 4 | 3,994 | $5,053,149 |
| 5 | 4,906 | $5,528,308 |
| 6 | 5,147 | $5,631,967 |
| 7 | 4,003 | $5,257,756 |
| 8 | 4,279 | $5,681,909 |
| 9 | 3,158 | $4,133,012 |
| 10 | 4,123 | $4,596,329 |
| 11 | 2,566 | $3,344,851 |
| 12 | 3,253 | $3,868,200 |
| 13 | 3,888 | $4,865,998 |
| 14 | 3,177 | $4,042,509 |
| 15 | 3,938 | $5,067,679 |
| 16 | 3,135 | $4,111,731 |
| 17 | 5,142 | $6,554,583 |
| 18 | 5,091 | $6,042,445 |
| 19 | 4,301 | $5,192,769 |
| 20 | 2,914 | $3,581,835 |
| 21 | 3,890 | $4,745,269 |
| 22 | 2,869 | $3,353,559 |
| 23 | 3,683 | $5,169,469 |
| 24 | 4,217 | $5,147,689 |

a. Develop a scatter plot based on the sample data. What type of relationship, if any, appears to exist between the two variables?

b. Calculate the sample correlation coefficient between billed consulting hours and total project costs.

c. Conduct a hypothesis test to determine if there is a significant positive correlation between billed consulting hours and total project costs. Use a level of significance of 0.01.

## Computer Software Exercises

**14-27.** The file **Online** contains a random sample of 48 customers who made purchases last quarter from an online retailer. The file contains information related to the time each customer spent viewing the online catalog and the dollar amount of purchases made. The

retailer would like to analyze the sample data to determine whether a relationship exists between the time spent viewing the online catalog and the dollar amount of purchases.

a. Compute the regression equation based on these sample data and interpret the regression coefficients.

b. Compute the coefficient of determination and interpret its meaning.

c. Test the significance of the overall regression model using a significance level of 0.01.

d. Test to determine whether the true regression slope coefficient is equal to 0. Use a significance level of 0.01 to conduct the hypothesis test.

**14-28.** The National Football League (NFL) commissioner's office staff wants you to developed a simple linear regression model with 2015 average home attendance used as the dependent variable and the total number of games won during the season as the independent variable. The staff is interested in determining whether games won could be used as a predictor for average attendance. Develop the simple linear regression model. The data are in the file called **2015 NFL** (*source:* http://espn.go.com/nfl/).

a. What percentage of the total variation in average home attendance is explained by knowing the number of home games the team won?

b. What is the standard error of the estimate for this regression model?

c. Using $\alpha = 0.05$, test to determine whether the regression slope coefficient is significantly different from 0.

d. After examining the regression analysis results, what should the NFL staff conclude about how the average attendance is related to the number of games the team won?

**14-29.** Sarin Employment helps midsized to large companies fill management and high-level technical positions. The file **Sarin** contains data for a sample of job placements that the company has facilitated. The data show the number of days it took to fill the job and the fee billed to the client.

a. Construct a scatter plot showing fees as the dependent variable. What is the apparent relationship between the two variables?

b. Calculate the correlation coefficient and test to determine whether there is a statistically significant linear relationship between the two variables. Use an alpha = 0.05.

c. Develop a regression model with fees charged as the dependent variable. What percentage of the variation is explained by the number of days required to fill the position?

**14-30.** A department store manager randomly sampled customers who shopped in her store. She recorded the number of minutes each customer spent in the store and then asked the customers how many miles they traveled to get to the store from their home. These data are in a file called **Shopping**.

a. Develop a regression model using the miles traveled as the independent variable to explain the variation in the number of minutes the customer spends in the store.

b. What percentage of the variation in the dependent variable is explained by the independent variable?

c. Using an alpha $= 0.05$ level, test whether the regression slope coefficient is statistically significant.

**14-31.** The file called **Vehicles** contains data for a sample of motor vehicles that are owned by an insurance company. Data are available for two variables: highway mileage and curb weight.

a. Develop a scatter plot for these data. Discuss what the plot implies about the relationship between the two variables. Assume that you wish to predict highway mileage by using vehicle curb weight.

b. Compute the correlation coefficient for the two variables and test to determine whether there is a linear relationship between the curb weight (independent variable) and the highway mileage (dependent variable) of automobiles.

c. Compute the linear regression equation based on the sample data. Provide an estimate of the average

highway mileage you would expect to obtain from a vehicle that weighs 4,012 pounds.

**14-32.** The Nevada Gaming Control Board's "Gaming Revenue Report" published on December 31, 2015, contains a wide array of data pertaining to gambling revenues in Nevada. The file titled **Nevada Revenues** contains data for seven of the most popular gambling games on both the number of units for each game statewide and the total winnings by the house (in $thousands).

a. Develop a scatter plot for the two variables showing number of units as the independent variable and winnings as the dependent variable. Calculate the correlation coefficient and discuss.

b. Develop a regression model that uses the number of units to explain the variation in winnings between the various games. Show the regression equation and indicate what percentage of the variation in the dependent variable is explained by the independent variable.

c. Referring to part b, test to determine whether the regression coefficient is statistically significant using an alpha $= 0.05$ level.

---

**OUTCOME 5**    **14.3**    **Uses for Regression Analysis**

Regression analysis is a statistical tool that is used for two main purposes: description and prediction. This section discusses these two applications.

## Regression Analysis for Description

**BUSINESS APPLICATION**    **Using Regression Analysis for Decision Making**

**Vehicle Mileage**    The U.S. Department of Energy website (www.fueleconomy.gov) contains data on 2015 automobile mileage ratings. Analysts collected data on a variety of variables for a sample of cars. Included among those data are the Environmental Protection Agency (EPA)'s combined highway and city miles per gallon and the estimated number of grams of $CO_2$ emitted per mile driven. The analysts are interested in the relationship between combined mileage ($x$) and grams of $CO_2$ emitted ($y$). The data are in the file **EPA Mileage 2015**.

The analysts can develop a simple linear regression model using Excel. The Excel output is shown in Figure 14.17. For these sample data, the coefficient of determination $R^2 = 0.9194$ indicates that knowing the combined miles per gallon of the car explains 91.94% of the variation in the grams of $CO_2$ emitted. The estimated regression equation is

$$\hat{y} = 703.39 - 13.64x$$

Before the analysts attempt to describe the relationship between miles per gallon and $CO_2$ emitted, they need to test whether there is a statistically significant linear relationship between the two variables. To do this, they can apply the $t$-test described in Section 14.2 to test the following null and alternative hypotheses:

$$H_0: \beta_1 = 0$$
$$H_A: \beta_1 \neq 0$$

at the significance level

$$\alpha = 0.05$$

**FIGURE 14.17  Excel 2016 Regression Results for the Car Mileage Study**

| | A | B | C | D | E | F | G |
|---|---|---|---|---|---|---|---|
| 1 | SUMMARY OUTPUT | | | | | | |
| 2 | *Regression Statistics* | | | | | | |
| 3 | Multiple R | 0.9589 | | | | | |
| 4 | R Square | 0.9194 | | | | | |
| 5 | Adjusted R Square | 0.9180 | | | | | |
| 6 | Standard Error | 16.2705 | | | | | |
| 7 | Observations | 58 | | | | | |
| 8 | | | | | | | |
| 9 | ANOVA | | | | | | |
| 10 | | *df* | *SS* | *MS* | *F* | *Significance F* | |
| 11 | Regression | 1 | 169115.67 | 169115.67 | 638.83 | 0.0000 | |
| 12 | Residual | 56 | 14824.81 | 264.73 | | | |
| 13 | Total | 57 | 183940.48 | | | | |
| 14 | | | | | | | |
| 15 | | *Coefficients* | *Standard Error* | *t Stat* | *P-value* | *Lower 95%* | *Upper 95%* |
| 16 | Intercept | 703.39 | 14.20 | 49.53 | 0.0000 | 674.94 | 731.84 |
| 17 | Combined MPG | -13.64 | 0.54 | -25.28 | 0.0000 | -14.72 | -12.56 |

The calculated *t*-statistic and the corresponding *p*-value are shown in Figure 14.17. Because

$$p\text{-value} = 0.0000 < 0.05$$

the null hypothesis, $H_0$, is rejected and the analysts can conclude that the population regression slope is not equal to 0.

The sample slope, $b_1$, equals $-13.64$. This means that for each 1-unit increase in miles per gallon, the $CO_2$ emission is estimated to decrease by an average of 13.64 grams. However, $b_1$ is subject to sampling error and is considered a *point estimate* for the true regression slope coefficient. From earlier discussions about point estimates in Chapters 8 and 10, we expect that $b_1 \neq \beta_1$. Therefore, to help describe the relationship between the independent variable and the dependent variable, we need to develop a *confidence interval estimate* for $\beta_1$. Equation 14.19 is used to do this.

**OUTCOME 6**   Confidence Interval Estimate for the Regression Slope, Simple Linear Regression

$$b_1 \pm t s_{b_1} \tag{14.19}$$

or equivalently,

$$b_1 \pm t \frac{s_\varepsilon}{\sqrt{\Sigma(x - \bar{x})^2}} \qquad df = n - 2$$

where:

$s_{b_1} = $ Standard error of the regression slope coefficient
$s_\varepsilon = $ Standard error of the estimate

The regression output shown in Figure 14.17 contains the 95% confidence interval estimate for the slope coefficient, which is

$$-14.72 \text{ ——— } -12.56$$

Thus, at the 95% confidence level, based on the sample data, the analysts can conclude that a 1-mile-per-gallon increase will result in a drop in $CO_2$ emissions by an average amount between approximately 12.56 and 14.72 grams.

There are many other situations in which the prime purpose of regression analysis is description. Economists use regression analysis for descriptive purposes as they search for a way of explaining the economy. Market researchers also use regression analysis, among other techniques, in an effort to describe the factors that influence the demand for products.

**EXAMPLE 14-4**

### Developing a Confidence Interval Estimate for the Regression Slope

**Home Prices** Home values are determined by a variety of factors. One factor is the size of the house (square feet). Groves & Associates Real Estate conducted a study aimed at estimating the average value of each additional square foot of space in a house. They collected a simple random sample of 319 homes sold by the agency within the past year. Here are the steps to compute a confidence interval estimate for the regression slope coefficient:

STEP 1 **Define the $y$ (dependent) and $x$ (independent) variables.**
The dependent variable is sales price, and the independent variable is square feet.

STEP 2 **Obtain the sample data.**
The study consists of sales prices and corresponding square feet for a random sample of 319 homes. The data are in a file called **First-City**.

STEP 3 **Compute the regression equation and the standard error of the slope coefficient.**
These computations can be performed using Excel to obtain these values.

| | Coefficients | Standard Error |
|---|---|---|
| Intercept ($b_0$) | 39,838.48 | 7,304.95 |
| Square Feet ($b_1$) | 75.70 | 3.78 |

The point estimate for the regression slope coefficient is $75.70. Thus, for a 1-square-foot increase in the size of a house, the house price increases by an average of $75.70. This is a point estimate and is subject to sampling error.

STEP 4 **Construct and interpret the confidence interval estimate for the regression slope using Equation 14.19.**
The confidence interval estimate is

$$b_1 \pm ts_{b_1}$$

where the degrees of freedom for the critical $t$ is $319 - 2 = 317$. The critical $t$ for a 95% confidence interval estimate is approximately 1.97, and the interval estimate is

$$\$75.70 \pm 1.97(\$3.78)$$
$$\$75.70 \pm \$7.45$$
$$\$68.25 \text{————} \$83.15$$

So, for a 1-square-foot increase in house size, at the 95% confidence level, the homes increase in price by an average of between $68.25 and $83.15.

**TRY EXERCISE 14-34** (pg. 560)

OUTCOME 5 ## Regression Analysis for Prediction

**BUSINESS APPLICATION**

### Predicting Hospital Costs Using Regression Analysis

**Excel Tutorial**

**Freedom Hospital** One of the main uses of regression analysis is *prediction*. You may need to predict the value of the dependent variable based on the value of the independent variable. Consider the administrator for Freedom Hospital, who has been asked by the hospital's board of directors to develop a model to predict the total charges for a geriatric patient. The file **Patients** contains the data that the administrator has collected.

**FIGURE 14.18** **Excel 2016 (XLSTAT) Prediction Interval for Freedom Hospital**

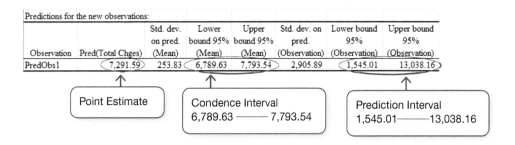

| | | Std. dev. on pred. (Mean) | Lower bound 95% (Mean) | Upper bound 95% (Mean) | Std. dev. on pred. (Observation) | Lower bound 95% (Observation) | Upper bound 95% (Observation) |
|---|---|---|---|---|---|---|---|
| Observation | Pred(Total Chges) | | | | | | |
| PredObs1 | 7,291.59 | 253.83 | 6,789.63 | 7,793.54 | 2,905.89 | 1,545.01 | 13,038.16 |

Point Estimate

Condence Interval
6,789.63 ——— 7,793.54

Prediction Interval
1,545.01 ———13,038.16

**Excel 2016 Instructions**

1. Open file: **Patients.xlsx**.
2. In an empty spreadsheet cell, type the *x*-variable value for which prediction is desired (5).
3. Click **Add-Ins > XLSTAT arrow**.
4. Point to **Modeling Data**; click **Linear regression**.
5. On the **General** tab, define the *y*-variable range (*Total Charges*).
6. Check Quantitative under X/Explanatory variables.
7. Define the *x*-variable range (*Length of Stay*).
8. On the **Options** tab, set the desired confidence interval (95% default).
9. On the **Prediction** tab, check **Prediction**, check **Quantitative**, click in the **Quantitative** box, and then click the spreadsheet cell from Step 2 containing the desired prediction value.
10. Click **OK**; on the **XLSTAT-Selection** dialog box, verify the following is listed:
    Y-Quantitative
    X-Quantitative
    Prediction/Quantitative
11. Click **Continue**.
12. Click **Summary statistics** arrow; scroll down and click **Prediction for the new observations**.

Although the Regression tool in Excel works well for generating the simple linear regression equation and other useful information, it does not provide predicted values for the dependent variable. However, the XLSTAT add-ins do provide predictions.

The administrator is attempting to construct a simple linear regression model, with total charges as the dependent (*y*) variable and length of stay as the independent (*x*) variable. Figure 14.18 shows the Excel (and XLSTAT) regression output. The administrator is interested in predicting total charges for a patient whose length of stay is 5 days. To do this using XLSTAT, as Step 7 of the Excel instructions shows, $x = 5$ is used. The point estimate for the predicted total charges is shown in Figure 14.18 as $7,291.59. The true charges will be either higher or lower than this amount. To account for this uncertainty, the administrator can develop a prediction interval, which is similar to the confidence interval estimates developed in Chapter 8.

**OUTCOME 6**    **Confidence Interval for the Average *y*, Given *x***    The administrator might like a 95% confidence interval for the *average*, or expected, value for charges of patients who stay in the hospital five days. The confidence interval for the expected value of a dependent variable, given a specific level of the independent variable, is determined by Equation 14.20. Observe that the specific value of *x* used to provide the prediction is denoted as $x_p$.

**Confidence Interval for $E(y)|x_p$**

$$\hat{y} \pm ts_\varepsilon \sqrt{\frac{1}{n} + \frac{(x_p - \bar{x})^2}{\Sigma(x - \bar{x})^2}}$$

(14.20)

where:

$\hat{y}$ = Point estimate of the dependent variable
$t$ = Critical value with $n - 2$ degrees of freedom
$n$ = Sample size
$x_p$ = Specific value of the independent variable
$\bar{x}$ = Mean of the independent variable observations in the sample
$s_\varepsilon$ = Estimate of the standard error of the estimate

Although the confidence interval estimate can be computed manually using Equation 14.20, using your computer is much easier. Excel with XLSTAT generates the confidence interval estimate for the average value of the dependent variable for a given value of the *x* variable, as shown in Figure 14.18. Given a five-day length of stay, the point estimate for the average total charges is $7,291.59, and at the 95% confidence level, the administrators believe the average total charges will be in the interval $6,789.63 to $7,793.54.

**Prediction Interval for a Particular *y*, Given *x*** The confidence interval just discussed is for the average value of *y* given $x_p$. The administrator might also be interested in predicting the total charges for a *particular* patient with a five-day stay, rather than the average of the charges for all patients staying five days. Developing this 95% prediction interval requires only a slight modification to Equation 14.20. This prediction interval is given by Equation 14.21.

**Prediction Interval for $y \mid x_p$**

$$\hat{y} \pm ts_\varepsilon \sqrt{1 + \frac{1}{n} + \frac{(x_p - \bar{x})^2}{\Sigma(x - \bar{x})^2}} \qquad \textbf{(14.21)}$$

As was the case with the confidence interval application discussed previously, the manual computations required to use Equation 14.21 can be onerous. We recommend using software such as Excel with XLSTAT to find the prediction interval. Figure 14.18 also shows the prediction interval for an individual value of the dependent variable (total charges) for a five-day length of stay.

Based on this regression model, at the 95% confidence level, the hospital administrators can predict the total charges for any specific patient with a length of stay of five days to be between $1,545.01 and $13,038.16. As you can see, this prediction has extremely poor precision. We doubt any hospital administrator will use a prediction interval that is so wide. Although the regression model explains a significant proportion of variation in the dependent variable, it is relatively imprecise for predictive purposes. To improve the precision, we might decrease the confidence level or increase the sample size and redevelop the model.

The prediction interval for a specific value of the dependent variable is wider (less precise) than the confidence interval for predicting the average value of the dependent variable. This will always be the case, as seen in Equations 14.20 and 14.21. From an intuitive viewpoint, we should expect to come closer to predicting an average value than a single value.

Note that, the term $(x_p - \bar{x})^2$ has a particular effect on the confidence interval determined by both Equations 14.20 and 14.21. The farther $x_p$ (the value of the independent variable used to predict *y*) is from $\bar{x}$, the greater $(x_p - \bar{x})^2$ becomes. Figure 14.19 shows two regression lines developed from two samples with the same set of *x* values. We have made both lines pass through the same $(\bar{x}, \bar{y})$ point; however, they have different slopes and intercepts. At $x_p = x_1$, the two regression lines give predictions of *y* that are close to each other. However, for $x_p = x_2$, the predictions of *y* are quite different. Thus, when $x_p$ is close to $\bar{x}$, the problems caused by variations in regression slopes are not as great as when $x_p$ is far from $\bar{x}$. Figure 14.20 shows the prediction intervals over the range of possible $x_p$ values. The band around the estimated regression line bends away from the regression line as $x_p$ moves in either direction from $\bar{x}$.

**OUTCOME** **7** ## Common Problems Using Regression Analysis

Regression is perhaps the most widely used statistical tool other than descriptive statistical techniques. Because it is so widely used, you need to be aware of the common problems encountered when the technique is employed.

One potential problem occurs when decision makers apply regression analysis for predictive purposes. The conclusions and inferences made from a regression line are statistically valid only over the range of the data contained in the sample used to develop the regression line. For instance, in the Midwest Distribution example, we analyzed the performance of

**FIGURE 14.19 Regression Lines Illustrating the Increase in Potential Variation in _y_ as _x_ₚ Moves Farther from _x̄_**

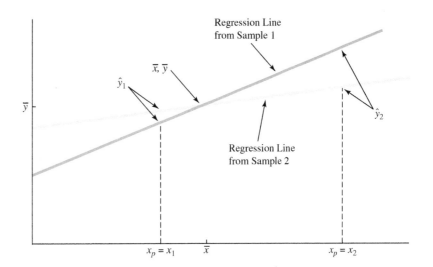

sales representatives with one to nine years of experience. Therefore, predicting sales levels for employees with one to nine years of experience is justified. However, if we were to try to predict the sales performance of someone with more than nine years of experience, the relationship between sales and experience might be different. Because no observations were taken for experience levels beyond the one-to-nine-year range, we have no information about what might happen outside that range. Figure 14.21 shows a case in which the true relationship between sales and experience reaches a peak value at about 20 years and then starts to decline. If a linear regression equation were used to predict sales based on experience levels beyond the relevant range of data, large prediction errors could occur.

A second important consideration, one that was discussed previously, involves correlation and causation. The fact that a significant linear relationship exists between two variables does not imply that one variable causes the other. Although there may be a cause-and-effect relationship, you should not infer that such a relationship is present based on only regression and/or correlation analysis. You should also recognize that a cause-and-effect relationship between two variables is not necessary for regression analysis to be an effective tool. What matters is that the regression model accurately reflects the relationship between the two variables and that the relationship remains stable.

Many users of regression analysis mistakenly believe that a high coefficient of determination $(R^2)$ guarantees that the regression model will be a good predictor. You should remember that $R^2$ is a measure of the variation in the dependent variable explained by the independent variable. Although the least squares criterion assures us that $R^2$ will be maximized (because the sum of squares error is a minimum) for the given set of sample data,

**FIGURE 14.20 Confidence Intervals for _y_|_x_ₚ and _E_(_y_)|_x_ₚ**

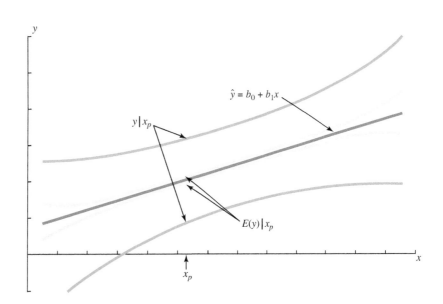

FIGURE 14.21 **Graph for a Sales Peak at 20 Years**

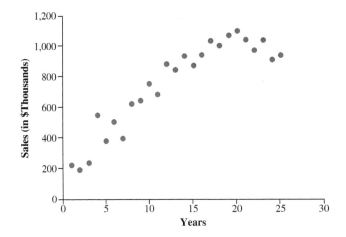

the value applies only to those data used to develop the model. Thus, $R^2$ measures the fit of the regression line to the sample data. There is no guarantee that there will be an equally good fit with new data. The only true test of a regression model's predictive ability is how well the model actually predicts.

Finally, we should mention that you might find a large $R^2$ with a large standard error. This can happen if the total sum of squares is large in comparison to the $SSE$. Then, even though $R^2$ is relatively large, so too is the estimate of the model's standard error. Thus, confidence and prediction intervals may simply be too wide for the model to be used in many situations. We will discuss this more fully in Chapter 15.

# 14.3 EXERCISES

## Skill Development

**14-33.** The following data have been collected by an accountant who is performing an audit of paper products at a large office supply company. The dependent variable, $y$, is the time (in minutes) the accountant takes to count the units. The independent variable, $x$, is the number of units on the computer inventory record.

| y | 23.1 | 100.5 | 242.9 | 56.4 | 178.7 | 10.5 | 94.2 | 200.4 | 44.2 | 128.7 | 180.5 |
|---|------|-------|-------|------|-------|------|------|-------|------|-------|-------|
| x | 24 | 120 | 228 | 56 | 190 | 13 | 85 | 190 | 32 | 120 | 230 |

a. Develop a scatter plot for these data.
b. Determine the regression equation representing the data. Is the model significant? Test using a significance level of 0.10 and the $p$-value approach.
c. Develop a 90% confidence interval estimate for the true regression slope and interpret this interval estimate. Based on this interval, could you conclude the accountant takes an additional minute to count each additional unit?

**14-34.** You are given the following sample data:

| x | y |
|----|---|
| 10 | 3 |
| 6 | 7 |
| 9 | 3 |

| x | y |
|---|---|
| 3 | 8 |
| 2 | 9 |
| 8 | 5 |
| 3 | 7 |

a. Develop a scatter plot for these data.
b. Determine the regression equation for the data.
c. Develop a 95% confidence interval estimate for the true regression slope and interpret this interval estimate.
d. Provide a 95% prediction interval estimate for a particular $y$, given $x_p = 7$.

**Exercises 14-35** and **14-36** refer to the following output for a simple linear regression model:

**Summary Output**

| Regression Statistics | |
|---|---|
| Multiple $R$ | 0.1027 |
| $R$-Square | 0.0105 |
| Adjusted $R$-Square | −0.0030 |
| Standard Error | 9.8909 |
| Observations | 75 |

ANOVA

|  | df | SS | MS | F | Significance F |
|---|---|---|---|---|---|
| Regression | 1 | 76.124 | 76.12 | 0.778 | 0.3806 |
| Residual | 73 | 7,141.582 | 97.83 | | |
| Total | 74 | 7,217.706 | | | |

|  | Coefficients | Standard Error | t-statistic |
|---|---|---|---|
| Intercept | 4.0133 | 3.878 | 1.035 |
| $x$ | 0.0943 | 0.107 | 0.882 |

|  | p-value | Lower 95% | Upper 95% |
|---|---|---|---|
| Intercept | 0.3041 | −3.715 | 11.742 |
| $x$ | 0.3806 | −0.119 | 0.307 |

**14-35.** Referring to the displayed regression model, what percentage of the variation in the $y$ variable is explained by the $x$ variable in the model?

**14-36.** Construct and interpret a 90% confidence interval estimate for the regression slope coefficient.

**14-37.** You are given the following summary statistics from a regression analysis:

$$\hat{y} = 200 + 150x$$
$$SSE = 25.25$$
$$SSX = \text{Sum of square } X = \Sigma(x - \bar{x})^2 = 99,645$$
$$n = 18$$
$$\bar{x} = 52.0$$

  a. Determine the point estimate for $y$ if $x_p = 48$ is used.

  b. Provide a 95% confidence interval estimate for the average $y$, given $x_p = 48$. Interpret this interval.

  c. Provide a 95% prediction interval estimate for a particular $y$, given $x_p = 48$. Interpret.

  d. Discuss the difference between the estimates provided in parts b and c.

**14-38.** The sales manager at City Real Estate Company is interested in describing the relationship between condo sales prices and the number of weeks the condo is on the market before it sells. He has collected a random sample of 17 low-end condos that have sold within the past three months. These data are as follows:

| Weeks on the Market | Selling Price |
|---|---|
| 23 | $76,500 |
| 48 | $102,000 |
| 9 | $53,000 |
| 26 | $84,200 |
| 20 | $73,000 |
| 40 | $125,000 |
| 51 | $109,000 |
| 18 | $60,000 |
| 25 | $87,000 |

| Weeks on the Market | Selling Price |
|---|---|
| 62 | $94,000 |
| 33 | $76,000 |
| 11 | $90,000 |
| 15 | $61,000 |
| 26 | $86,000 |
| 27 | $70,000 |
| 56 | $133,000 |
| 12 | $93,000 |

  a. Develop a simple linear regression model to explain the variation in selling price based on the number of weeks the condo is on the market.

  b. Test to determine whether the regression slope coefficient is significantly different from 0 using a significance level equal to 0.05.

  c. Construct and interpret a 95% confidence interval estimate for the regression slope coefficient.

**14-39.** A sample of ten yields the following data:

| $x$ | 10 | 8 | 11 | 7 | 10 | 11 | 6 | 7 | 15 | 9 |
|---|---|---|---|---|---|---|---|---|---|---|
| $y$ | 103 | 85 | 115 | 73 | 97 | 102 | 65 | 75 | 155 | 95 |

  a. Provide a 95% confidence interval for the average $y$ when $x_p = 9.4$.

  b. Provide a 95% confidence interval for the average $y$ when $x_p = 10$.

  c. Obtain the margin of errors for both part a and part b. Explain why the margin of error obtained in part b is larger than that in part a.

**14-40.** A regression analysis from a sample of 15 produced the following:

$$\Sigma(x_i - \bar{x})(y_i - \bar{y}) = 156.4$$
$$\Sigma(x_i - \bar{x})^2 = 173.5$$
$$\Sigma(y_i - \bar{y})^2 = 181.6$$
$$\Sigma(y_i - \hat{y})^2 = 40.621$$
$$\bar{x} = 13.4 \text{ and } \bar{y} = 56.4$$

  a. Produce the regression line.

  b. Determine if there is a linear relationship between the dependent and independent variables. Use a significance level of 0.05 and a $p$-value approach.

  c. Calculate a 90% confidence interval for the amount the dependent variable changes when the independent variable increases by 1 unit.

## Business Applications

**14-41.** The following data have been collected by an accountant who is performing an audit of account balances for a major city billing department. The population from which the data were collected is the accounts for which the customer had indicated the balance was incorrect. The dependent variable, $y$, is the actual account balance as verified by the accountant. The independent variable, $x$, is the computer account balance.

| y | 233 | 10 | 24 | 56 | 78 | 102 | 90 | 200 | 344 | 120 | 18 |
|---|-----|----|----|----|----|-----|----|-----|-----|-----|----|
| x | 245 | 12 | 22 | 56 | 90 | 103 | 85 | 190 | 320 | 120 | 23 |

a. Compute the least squares regression equation.
b. If the computer account balance was 100, what would you expect to be the actual account balance as verified by the accountant?
c. The computer balance for Timothy Jones is listed as 100 in the computer account record. Provide a 90% interval estimate for Mr. Jones's actual account balance.
d. Provide a 90% interval estimate for the average of all customers' actual account balances in which a computer account balance is the same as that of Mr. Jones (part c). Interpret.

**14-42.** American Outfitters sells and services exercise equipment such as treadmills, ellipticals, and stair climbers to gymnasiums and recreational centers. The company's management would like to determine if there is a relationship between the number of minutes required to complete a routine service call and the number of machines serviced. A random sample of 12 records revealed the following information concerning the number of machines serviced and the time (in minutes) to complete the routine service call:

| Number of Machines | Service Time (minutes) |
|--------------------|------------------------|
| 11 | 115 |
| 8 | 60 |
| 9 | 80 |
| 10 | 90 |
| 7 | 55 |
| 6 | 65 |
| 8 | 70 |
| 4 | 33 |
| 10 | 95 |
| 5 | 50 |
| 5 | 40 |
| 12 | 110 |

a. Estimate the least squares regression equation.
b. If a gymnasium had six machines, how many minutes should American Outfitters expect a routine service call to require?
c. Provide a 90% confidence interval for the average amount of time required to complete a routine service call when the number of machines being serviced is nine.
d. Provide a 90% prediction interval for the time required to complete a particular routine service call for a gymnasium that has seven machines.

**14-43.** A golf ball maker is interested in analyzing the relationship between the club head speed that a male player achieves when he uses a driver and the distance the ball travels. The sample data are listed here.

| Club Speed (mph) | Distance (yards) |
|------------------|------------------|
| 95 | 233 |
| 97 | 246 |
| 91 | 202 |
| 103 | 240 |
| 118 | 289 |
| 88 | 196 |
| 93 | 227 |
| 97 | 217 |
| 101 | 245 |
| 104 | 256 |
| 95 | 218 |

a. Develop a regression model using club speed as the independent variable and driving distance as the dependent variable. What percentage of the variation in the driving distance is explained by the club head speed?
b. Test using an alpha $= 0.05$ level to determine if the regression slope coefficient is statistically significant.
c. Use the regression model to describe the relationship between club speed and driving distance.

**14-44.** An engineering firm is interested in investigating whether the variability in the cost of small projects (defined as projects under $10 million) can be accounted for, in part, by differences in the number of direct engineering consulting hours billed. A random sample of 24 small projects was selected from small projects performed over the past two years. The number of engineering consulting hours billed for each project, along with the project's cost, was recorded. The results are shown here:

| Project | Billed Consulting Hours | Total Project Cost |
|---------|-------------------------|--------------------|
| 1 | 3,932 | $4,323,826 |
| 2 | 3,097 | $3,750,964 |
| 3 | 2,972 | $3,579,570 |
| 4 | 3,994 | $5,053,149 |
| 5 | 4,906 | $5,528,308 |
| 6 | 5,147 | $5,631,967 |
| 7 | 4,003 | $5,257,756 |
| 8 | 4,279 | $5,681,909 |
| 9 | 3,158 | $4,133,012 |
| 10 | 4,123 | $4,596,329 |
| 11 | 2,566 | $3,344,851 |
| 12 | 3,253 | $3,868,200 |
| 13 | 3,888 | $4,865,998 |

| Project | Billed Consulting Hours | Total Project Cost |
|---------|-------------------------|--------------------|
| 14 | 3,177 | $4,042,509 |
| 15 | 3,938 | $5,067,679 |
| 16 | 3,135 | $4,111,731 |
| 17 | 5,142 | $6,554,583 |
| 18 | 5,091 | $6,042,445 |
| 19 | 4,301 | $5,192,769 |
| 20 | 2,914 | $3,581,835 |
| 21 | 3,890 | $4,745,269 |
| 22 | 2,869 | $3,353,559 |
| 23 | 3,683 | $5,169,469 |
| 24 | 4,217 | $5,147,689 |

a. Develop a simple linear regression model to explain the variation in total project cost based on the number of billed consulting hours.
b. Construct and interpret a 95% confidence interval estimate for the regression slope coefficient.
c. Provide a 95% confidence interval for the average value of $y$ when $x_p = 3,500$.
d. Provide a 95% prediction interval for the total project cost for the a particular project with 3,500 billed consulting hours.

**14-45.** A regional retailer would like to determine if the variation in average monthly store sales can, in part, be explained by the size of the store measured in square feet. A random sample of 21 stores was selected and the store size and average monthly sales were computed. The following results are shown:

| Store Size (Sq. Ft) | Average Monthly Sales |
|---------------------|-----------------------|
| 17,400 | $581,241 |
| 15,920 | $538,275 |
| 17,440 | $636,059 |
| 17,320 | $574,477 |
| 15,760 | $558,043 |
| 20,200 | $689,256 |
| 15,280 | $552,569 |
| 17,000 | $584,737 |
| 11,920 | $470,551 |
| 12,400 | $520,798 |
| 15,640 | $619,703 |
| 12,560 | $465,416 |
| 21,680 | $730,863 |
| 14,120 | $501,501 |
| 16,680 | $624,255 |
| 14,920 | $567,043 |
| 18,360 | $612,974 |
| 18,440 | $618,122 |
| 16,720 | $691,403 |
| 19,880 | $719,275 |
| 17,880 | $536,592 |

a. Develop a simple linear regression model to explain the variation in average monthly sales based on the size of the store.
b. Construct and interpret a 95% confidence interval estimate for the regression slope coefficient.
c. Provide a 90% confidence interval for the average value of monthly sales when $x_p = 15,000$.
d. Provide a 90% prediction interval for a particular month's average sales when the store size is 15,000 square feet.

## Computer Software Exercises

**14-46.** The Billingsley Health Center offers a regimented weight loss program for its customers. The program director monitored 83 men. The file called **BHC** contains data on the weight they lost after six weeks on the program and the average daily calories each consumed per day.
a. Develop a regression model using calories as the independent variable and weight lost as the dependent variable. What percentage of the variation in weight lost is explained by the average calories consumed?
b. Test using an alpha $= 0.05$ level to determine if the regression slope coefficient is statistically significant.
c. Use the regression model to describe the relationship between calories consumed and weight lost.

**14-47.** A manufacturer produces washdown motors for the food service industry. The company manufactures the motors to order by modifying a base model to meet the specifications requested by the customer. The motors are produced in a batch environment, with the batch size equal to the number ordered. The manufacturer has recently sampled 50 customer orders. The motor manufacturer would like to determine if there is a relationship between the cost of producing the order and the order size so that it could estimate the cost of producing a particular size order. The sampled data are in the file **Washdown Motors**.
a. Use the sample data to estimate the least squares regression model.
b. Provide an interpretation of the regression coefficients.
c. Test the significance of the overall regression model using a significance level of 0.01.
d. The company has just received an order for 30 motors. Use the regression model developed in part a to estimate the cost of producing this particular order.
e. Referring to part d, what is the 90% confidence interval for an average cost of an order of 30 motors?

**14-48.** The Bureau of Labor Statistics (BLS) publishes data on many characteristics of the U.S. economy. The file called **BLS County Data** contains information on every county in the United States for which data are available (*source:* www.bls.gov/data/). Two of the variables are the 2014 county unemployment rate and the 2014 median household income.

a. Construct a regression equation with median income as the dependent variable and unemployment rate as the independent variable. Indicate what percentage of the variation in median income is explained by knowing the county's unemployment rate in 2014.

b. Write a short description of the relationship between these two variables. Use the regression model output to assist you with the discussion.

# 14 Overview

## Summary

14.1

### Scatter Plots and Correlation (pg. 527–536)

**OUTCOME 1** Calculate and interpret the correlation between two variables.

**OUTCOME 2** Determine whether the correlation is significant.

- A **scatter plot**, or scatter diagram, is a two-dimensional plot showing the values for the joint occurrence of two quantitative variables.
- The scatter plot may be used to graphically represent the relationship between two variables.
- The **correlation coefficient** is the numerical quantity that measures the strength of the linear relationship between two variables.
- The sample correlation coefficient, $r$, can range from a perfect positive correlation, $+1.0$, to a perfect negative correlation, $-1.0$.
- A test based on the $t$-distribution can determine whether the population correlation coefficient is significantly different from 0 and, therefore, whether a linear relationship exists between the dependent and independent variables.

14.2

### Simple Linear Regression Analysis (pg. 536–554)

**OUTCOME 3** Calculate the simple linear regression equation for a set of data and know the basic assumptions behind regression analysis.

**OUTCOME 4** Determine whether a regression model is significant.

- Regression analysis is the technique used to analyze the relationship between the dependent variable and the independent variable.
- This technique is called **simple linear regression** when the relationship between these two variables is linear.
- The population regression model is determined by three values known as the population regression coefficients: (1) the $y$ intercept, (2) the slope of the regression line, and (3) the random error term.
- The **least squares criterion** is used to determine the best estimate of the population regression line by choosing values for the $y$ intercept and slope that produce the smallest possible sum of squared prediction errors.
- Testing whether the population's slope coefficient is equal to zero is a way to determine if there is no linear relationship between the dependent and independent variables.
- The test for the simple linear regression is equivalent to the test that the correlation coefficient is significant.
- The **coefficient of determination** is a less involved mesure that indicates the goodness of fit of the regression equation to the data.
- Simple linear regression is one of the most often applied statistical tools for analyzing the relationship between two variables.

14.3

### Uses for Regression Analysis (pg. 554–564)

**OUTCOME 5** Recognize regression analysis applications for purposes of description and prediction.

**OUTCOME 6** Calculate and interpret confidence intervals for the regression analysis.

**OUTCOME 7** Recognize some potential problems if regression analysis is used incorrectly.

- Regression analysis is a statistical tool that is used for two main purposes: description and prediction.
- Description involves describing the plausible values the population slope coefficient may attain.
- To do this, we employ a confidence interval estimator of the population slope.
- A prediction interval is the estimator used to predict a particular $y$ for a given $x$.
- Any procedure in statistics is valid only if the assumptions it is built on are valid. This is particularly true in regression analysis. Therefore, before using a regression model for description or prediction, you should check to see if the assumptions associated with linear regression analysis are valid.
- Residual analysis is the procedure used to determine whether the assumptions associated with linear regression analysis are valid.

## Equations

**(14.1) Sample Correlation Coefficient** pg. 528

$$r = \frac{\Sigma (x - \bar{x})(y - \bar{y})}{\sqrt{[\Sigma (x - \bar{x})^2][\Sigma (y - \bar{y})^2]}}$$

**(14.2) or the algebraic equivalent** pg. 528

$$r = \frac{n\Sigma xy - \Sigma x \Sigma y}{\sqrt{[n(\Sigma x^2) - (\Sigma x)^2][n(\Sigma y^2) - (\Sigma y)^2]}}$$

**(14.3) Test Statistic for Correlation** pg. 530

$$t = \frac{r}{\sqrt{\dfrac{1 - r^2}{n - 2}}} \qquad df = n - 2$$

**(14.4) Simple Linear Regression Model (Population Model)** pg. 536

$$y = \beta_0 + \beta_1 x + \varepsilon$$

**(14.5) Estimated Regression Model (Sample Model)** pg. 540

$$\hat{y} = b_0 + b_1 x$$

**(14.6) Least Squares Equation** pg. 540
**Regression Slope Coefficient**

$$b_1 = \frac{\Sigma (x_i - \bar{x})(y_i - \bar{y})}{\Sigma (x_i - \bar{x})^2}$$

**(14.7) or the algebraic equivalent** pg. 540

$$b_1 = \frac{\Sigma xy - \dfrac{\Sigma x \Sigma y}{n}}{\Sigma x^2 - \dfrac{(\Sigma x)^2}{n}}$$

**(14.8) Regression Intercept** pg. 540

$$b_0 = \bar{y} - b_1 \bar{x}$$

**(14.9) Sum of Squared Residuals (Errors)—Manual Calculation** pg. 542

$$SSE = \Sigma y^2 - b_0 \Sigma y - b_1 \Sigma xy$$

**(14.10) Sum of Residuals** pg. 542

$$\sum_{i=1}^{n} (y_i - \hat{y}_i) = 0$$

**(14.11) Sum of Squared Residuals (Errors)** pg. 542

$$SSE = \sum_{i=1}^{n} (y_i - \hat{y}_i)^2$$

**(14.12) Simple Regression Estimator for the Standard Error of the Slope** pg. 545

$$s_{b_1} = \frac{s_\varepsilon}{\sqrt{\Sigma (x - \bar{x})^2}}$$

**(14.13) Simple Linear Regression Test Statistic for Test of the Significance of the Slope** pg. 545

$$t = \frac{b_1 - \beta_1}{s_{b_1}} \qquad df = n - 2$$

**(14.14) Total Sum of Squares** pg. 547

$$SST = \sum_{i=1}^{n} (y_i - \bar{y})^2$$

**(14.15) Sum of Squares Error (Residual)** pg. 547

$$SSE = \sum_{i=1}^{n} (y_i - \hat{y}_i)^2$$

**(14.16) Sum of Squares Regression** pg. 548

$$SSR = \sum_{i=1}^{n} (\hat{y}_i - \bar{y})^2$$

**(14.17) Coefficient of Determination, $R^2$** pg. 548

$$R^2 = \frac{SSR}{SST}$$

**(14.18) Test Statistic for Significance of the Coefficient of Determination** pg. 549

$$F = \frac{\dfrac{SSR}{1}}{\dfrac{SSE}{(n - 2)}} \qquad df = (D_1 = 1, D_2 = n - 2)$$

**(14.19) Confidence Interval Estimate for the Regression Slope, Simple Linear Regression** pg. 555

$$b_1 \pm t s_{b_1}$$

or equivalently,

$$b_1 \pm t \frac{s_\varepsilon}{\sqrt{\Sigma (x - \bar{x})^2}} \qquad df = n - 2$$

**(14.20) Confidence Interval for $E(y)|x_p$** pg. 557

$$\hat{y} \pm t s_\varepsilon \sqrt{\frac{1}{n} + \frac{(x_p - \bar{x})^2}{\Sigma (x - \bar{x})^2}}$$

**(14.21) Prediction Interval for $y|x_p$** pg. 558

$$\hat{y} \pm t s_\varepsilon \sqrt{1 + \frac{1}{n} + \frac{(x_p - \bar{x})^2}{\Sigma (x - \bar{x})^2}}$$

## Key Terms

Coefficient of determination pg. 547
Correlation coefficient pg. 527
Least squares criterion pg. 538

Regression slope coefficient pg. 537
Residual pg. 538
Scatter plot pg. 527

Simple linear regression pg. 536

## Chapter Exercises

### Conceptual Questions

**14-49.** A manager has computed the correlation between the number of hours of sick leave taken per year and the annual salary for the 200 employees in her company. The hours of sick leave range from 0 to 39, and the correlation coefficient for the two variables is $r = -0.19$. Suppose the manager has another employee who was not included in the analysis and this employee took 71 hours of sick leave during the year. Discuss whether the correlation computed from the other 200 employees applies to this new employee.

**14-50.** A statistics student was recently working on a class project that required him to compute a correlation coefficient for two variables. After careful work, he arrived at a correlation coefficient of 0.45. Interpret this correlation coefficient for the student who did the calculations.

**14-51.** Referring to Exercise 14-50, another student in the same class computed a regression equation relating the same two variables. The slope of the equation was found to be $-0.735$. After trying several times and always coming up with the same result, she felt that she must have been doing something wrong since the value was negative and she knew that this could not be right. Comment on this student's conclusion.

**14-52.** If we select a random sample of data for two variables and, after computing the correlation coefficient, conclude that the two variables may have zero correlation, can we say that there is no relationship between the two variables? Discuss.

**14-53.** Discuss why prediction intervals that attempt to predict a particular $y$ value are less precise than confidence intervals for predicting an average $y$ value.

**14-54.** Consider the two following scenarios:
  a. The number of new workers hired per week in your county has a high positive correlation with the average weekly temperature. Can you conclude that an increase in temperature causes an increase in the number of new hires? Discuss.
  b. Suppose the stock price and the common dividends declared for a certain company have a high positive correlation. Are you safe in concluding on the basis of the correlation coefficient that an increase in the common dividends declared causes an increase in the stock price? Present other reasons than the correlation coefficient that might lead you to conclude that an increase in common dividends declared causes an increase in the stock price.

**14-55.** Consider the following set of data:

| x | 48 | 27 | 34 | 24 | 49 | 29 | 39 | 38 | 46 | 32 |
|---|----|----|----|----|----|----|----|----|----|----|
| y | 47 | 23 | 31 | 20 | 50 | 48 | 47 | 47 | 42 | 47 |

  a. Calculate the correlation coefficient of these two variables.
  b. Multiply each value of the variable $x$ by 5 and add 10 to the resulting products. Now multiply each value of the variable $y$ by 3 and subtract 7 from the resulting products. Finally, calculate the correlation coefficient of the new $x$ and $y$ variables.
  c. Describe the principle that the example developed in parts a and b demonstrates.

**14-56.** Go to the library and locate an article in a journal related to your major (*Journal of Marketing*, *Journal of Finance*, etc.) that uses linear regression. Discuss the following:
  a. How the author chose the dependent and independent variables
  b. How the data were gathered
  c. What statistical tests the author used
  d. What conclusions the analysis allowed the author to draw

### Business Applications

**14-57.** The Olson Brothers Organic Milk Company recently studied a random sample of 30 of its distributors and found the correlation between sales and advertising dollars to be 0.67.
  a. Is there a significant linear relationship between sales and advertising? If so, is it fair to conclude that advertising causes sales to increase?
  b. If a regression model were developed using sales as the dependent variable and advertising as the independent variable, determine the proportion of the variation in sales that would be explained by its relationship to advertising. Discuss what this says about the usefulness of using advertising to predict sales.

**14-58.** At State University, a study was done to establish whether a relationship exists between students' graduating grade point average (GPA) and the SAT verbal score when the student originally entered the university. The sample data are reported as follows:

| GPA | 2.5 | 3.2 | 3.5 | 2.8 | 3.0 | 2.4 | 3.4 | 2.9 | 2.7 | 3.8 |
|-----|-----|-----|-----|-----|-----|-----|-----|-----|-----|-----|
| SAT | 640 | 700 | 550 | 540 | 620 | 490 | 710 | 600 | 505 | 710 |

a. Develop a scatter plot for these data and describe what, if any, relationship exists between the two variables, GPA and SAT score.

b. (1) Compute the correlation coefficient. (2) Does it appear that the success of students at State University is related to the SAT verbal scores of those students? Conduct a statistical procedure to answer this question. Use a significance level of 0.01.

c. (1) Compute the regression equation based on these sample data if you wish to predict the university GPA using the student SAT score. (2) Interpret the regression coefficients.

**14-59.** The Farmington City Council recently commissioned a study of park users in their community. Data were collected on the age of the person surveyed and the amount of hours he or she spent in the park in the past month. The data collected were as follows:

| Time in Park | Age | Time in Park | Age |
|---|---|---|---|
| 7.2 | 16 | 4.4 | 48 |
| 3.5 | 15 | 8.8 | 18 |
| 6.6 | 28 | 4.9 | 24 |
| 5.4 | 16 | 5.1 | 33 |
| 1.5 | 29 | 1.0 | 56 |
| 2.3 | 38 | | |

a. Draw a scatter plot for these data and discuss what, if any, relationship appears to be present between the two variables.

b. Compute the correlation coefficient between age and the amount of time spent in the park. Provide an explanation to the Farmington City Council of what the correlation measures.

c. Test to determine whether the amount of time spent in the park decreases with increases in the age of the park user. Use a significance level of 0.10. Use a $p$-value approach to conduct this hypothesis test.

**14-60.** Referring to Exercise 14-58, suppose a student has an SAT score of 400. What is her estimate GPA at State University? Discuss the ramifications of using the model developed in Exercise 14-58 to estimate this student's GPA.

**14-61.** A regional airline company recently performed a customer survey in which it asked a random sample of 100 passengers to indicate their income and the total cost of the airfares they purchased for pleasure trips during the past year. A regression model was developed for the purpose of determining whether income could be used as a variable to explain the variation in the total cost of airfare on airlines in a year. The following regression results were obtained:

$$\hat{y} = 0.25 + 0.0150x$$
$$s_\varepsilon = 721.44$$
$$R^2 = 0.65$$
$$s_{b_1} = 0.0000122$$

a. Determine whether income is a significant variable in explaining the variation in the total cost of airfare on airlines in a year by using a 90% confidence interval.

b. Assuming that no one who was surveyed had an income of 0 dollars, can the regression model intercept have a meaningful interpretation? Explain.

**14-62.** One of the advances that have helped to alleviate carpal tunnel syndrome is ergonomic keyboards. The ergonomic keyboards may also increase typing speed. Ten administrative assistants were chosen to type on both standard and ergonomic keyboards. The resulting typing speeds follow:

| Ergonomic | 69 | 80 | 60 | 71 | 73 | 64 | 63 | 70 | 63 | 74 |
|---|---|---|---|---|---|---|---|---|---|---|
| Standard | 70 | 68 | 54 | 56 | 58 | 64 | 62 | 51 | 64 | 53 |

a. Produce a scatter plot of the typing speeds of administrative assistants using ergonomic and standard keyboards. Does there appear to be a linear relationship between these two variables? Explain your response.

b. Calculate the correlation coefficient of the typing speeds of administrative assistants using ergonomic and standard keyboards.

c. Conduct a hypothesis test to determine if a positive correlation exists between the typing speeds of administrative assistants using ergonomic and standard keyboards. Use a significance level of 0.05.

**14-63.** A company is considering recruiting new employees from a particular college and plans to place a great deal of emphasis on the student's college GPA. However, the company is aware that not all schools have the same grading standards, so it is possible that a student at this school might have a lower or higher GPA than a student from another school, yet really be on par with the other student. To make this comparison between schools, the company has administered a test that uses a sample size of 400 students. With the results of the test, it has developed a regression model that it uses to predict student GPA. The following equation represents the model:

$$\hat{y} = 1.0 + 0.028x$$

The $R^2$ for this model is 0.88 and the standard error of the estimate is 0.20, based on the sample data used to develop the model. Note that the dependent variable is the GPA and the independent variable is test score, which can range from 0 to 100. For the sample data used to develop the model, the following values are known:

$$\bar{y} = 2.76$$
$$\bar{x} = 68$$
$$\Sigma(x - \bar{x})^2 = 148,885.73$$

a. Based on the information given in this problem, can you conclude that as the test score increases, the GPA will also increase, using a significance level of 0.05?

b. Suppose a student interviews with this company, takes the company test, and scores 80 correct. What is the 90% prediction interval estimate for this student's GPA? Interpret the interval.

c. Suppose the student in part b actually has a 2.90 GPA at this school. Based on this evidence, what might be concluded about this person's actual GPA compared with other students with the same GPA at other schools? Discuss the limitations you might place on this conclusion.

d. Suppose a second student with a 2.45 GPA took the test and scored 65 correct. What is the 90% prediction interval for this student's "real" GPA? Interpret.

## Computer Software Exercises

**14-64.** The board of directors for a small regional bank has decided to look into all aspects of revenues and costs. One service the bank offers is free checking, and the board is interested in whether the costs of this service are offset by revenues from interest earned on the deposits. One aspect in studying checking accounts is to determine whether changes in the average checking account balance can be explained by knowing the number of checks written per month. The sample data selected are contained in the data file named **Checking**.

a. Draw a scatter plot for these data.

b. Develop the least squares regression equation for these data.

c. Develop the 90% confidence interval estimate for the change in the average checking account balance when a person who formerly wrote 25 checks a month doubles the number of checks used.

d. Test to determine if an increase in the number of checks written by an individual can be used to predict the checking account balance of that individual. Use $\alpha = 0.05$. Comment on this result and the result of part c.

**14-65.** Geoff Black is the owner and president of Black Manufacturing. He has a passion for process improvement and has invested in his employees over the years to help them be more productive. Recently he hired a consulting firm to come into his plant and provide specialized training to employees who work on one of the production lines. Prior to the training, the consultant selected a sample of employees and measured the time in minutes that it took each employee to complete a specific task. Then, after the training, the employees were measured again on this same task and the times were recorded. The data are in the file called **Black**. Geoff is interested in seeing whether there is a linear relationship between the pre-training times and the post-training times.

a. (1) Develop a plot showing the relationship between the two variables. (2) Describe the relationship as being either linear or curvilinear.

b. (1) Develop a simple linear regression model with post-training time as the dependent variable.

(2) Write a short report describing the model and indicating the important measures.

**14-66.** Terry Downes owns a commercial cleaning company. He has conducted a survey of customers to determine how satisfied they are with the work performed. He devised a rating scale between 0 and 100, with 0 being poor and 100 being excellent service. He selected a random sample of 14 customers and asked them to rate the service. He also recorded the number of worker hours spent in the customer's facility. These are in the file named **Downes**.

a. (1) Draw a scatter plot showing these two variables, with the $y$ variable on the vertical axis and the $x$ variable on the horizontal axis. (2) Describe the relationship between these two variables.

b. (1) Develop a linear regression model to explain the variation in the service rating. (2) Write a short report describing the model and showing the results of pertinent hypothesis tests, using a significance level of 0.10.

**14-67.** Exercise 14-13 involved a file called **Breakcue**, which contains data on pool cue weights and the speed of pool balls struck with various cues. In that problem you were asked to calculate the correlation coefficient, and to test whether the two variables are significantly correlated. The next step is for you to develop a regression model pool ball speed as the dependent variable. Once the model is developed, determine the percentage of variation in ball speed that can be explained by knowing the cue weight. Finally, test to determine whether the regression slope coefficient is significantly different than zero testing at the alpha $= 0.05$.

**14-68.** Living donor kidney transplants are becoming more prevalent with the advent of kidney exchanges such as the National Kidney Registry (NKR). In order for a transplant to take place, there needs to be a good match between a willing donor and the recipient. Many factors must be considered in an effort to try to maximize the number of years the transplant will last. One factor of interest is the age of the donor. The following regression model was developed from a sample of living donor transplants that have been performed since 1987. The sample data are in a file called **Kidney**. The dependent variable is actual kidney life years (KLYs), which is the number of years from the time the transplant was performed until the kidney failed. (*Source:* OPTN/SRTR 2012 Annual Data Report, http://srtr.transplant.hrsa.gov.)

### SUMMARY OUTPUT

#### Regression Statistics

| | |
|---|---|
| Multiple R | 0.025856023 |
| R Square | 0.000668534 |
| Adjusted R Square | −0.007262668 |
| Standard Error | 8.951033668 |
| Observations | 128 |

ANOVA

| | df | SS | MS | F | Significance F |
|---|---|---|---|---|---|
| Regression | 1 | 6.753529839 | 6.75353 | 0.084292 | 0.772040921 |
| Residual | 126 | 10,095.24647 | 80.121 | | |
| Total | 127 | 10,102 | | | |

| | Coefficients | Standard Error | t-statistic | p-value | Lower 95% | Upper 95% |
|---|---|---|---|---|---|---|
| Intercept | 18.40339162 | 2.6083577 | 7.055547 | 1.02E-10 | 13.24152841 | 23.56525482 |
| Donor Age at Donation | 0.019323404 | 0.066556645 | 0.29033 | 0.772041 | −0.11239024 | 0.151037048 |

a. Based on these data, can we conclude that older donors are associated with a lower number of kidney life years? Test using an alpha = 0.10 level.
b. What percentage of the variation in KLYs is explained by donor age?
c. Based only this regression model, should transplant centers use donor age as a donor selection criterion? Explain.

**14-69.** The Grinfield Service Company's marketing director is interested in analyzing the relationship between her company's sales and the advertising dollars spent. In the course of her analysis, she selected a random sample of 20 weeks and recorded the sales for each week and the amount spent on advertising. These data are in the file called **Grinfield**.

a. Identify the independent and dependent variables.
b. Draw a scatter plot with the dependent variable on the vertical axis and the independent variable on the horizontal axis.
c. The marketing director wishes to know if increasing the amount spent on advertising increases sales. As a first attempt, use a statistical test that will provide the required information. Use a significance level of 0.025. On careful consideration, the marketing manager realizes that it takes a certain amount of time for the effect of advertising to register in terms

of increased sales. She therefore asks you to calculate a correlation coefficient for sales of the current week against amount of advertising spent in the previous week and to conduct a hypothesis test to determine if, under this model, increasing the amount spent on advertising increases sales. Again, use a significance level of 0.025.

**14-70.** Refer to the Grinfield Service Company discussed in Exercise 14-69.
a. Develop the least squares regression equation for these variables. Plot the regression line on the scatter plot.
b. Develop a 95% confidence interval estimate for the increase in sales resulting from increasing the advertising budget by $50. Interpret the interval.
c. Discuss whether it is appropriate to interpret the intercept value in this model. Under what conditions is it appropriate? Discuss.
d. Develop a 90% confidence interval for the mean sales amount achieved during all weeks in which advertising is $200 for the week.
e. Suppose you are asked to use this regression model to predict the weekly sales when advertising is set at $100. What would you reply to the request? Discuss.

## Case 14.1    A & A Industrial Products

Alex Court, the cost accountant for A & A Industrial Products, was puzzled by the repair cost analysis report he had just reviewed. This was the third consecutive report in which unscheduled plant repair costs were out of line with the repair cost budget allocated to each plant. A & A budgets for both scheduled maintenance and unscheduled repair costs for its plants' equipment, mostly large industrial machines. Budgets for scheduled maintenance activities are easy to estimate and are based on the equipment manufacturer's recommendations. The unscheduled repair costs, however, are harder to determine. Historically, A & A has estimated unscheduled maintenance using a formula based on the average

number of hours of operation between major equipment failures at a plant. Specifically, plants were given a budget of $65.00 per hour of operation between major failures. Alex had arrived at this amount by dividing aggregate historical repair costs by the total number of hours between failures. Then plant averages were used to estimate unscheduled repair cost. For example, if a plant averaged 450 hours of run time before a major repair occurred, the plant was allocated a repair budget of 450 × $65 = $29,250 per repair. If the plant was expected to be in operation 3,150 hours per year, the company would anticipate seven unscheduled repairs (3,150/450) annually and budget $204,750 for annual unscheduled repair costs.

Alex was becoming more and more convinced that this approach was not working. Not only was upper management upset about the variance between predicted and actual costs of repair, but plant managers believed that the model did not account for potential differences among the company's three plants when allocating dollars for unscheduled repairs. At the weekly management meeting, Alex was informed that he needed to analyze his cost projections further and produce a report that provided a more reliable method for predicting repair costs. On leaving the meeting, Alex had his assistant randomly pull 64 unscheduled repair reports. The data are in the file **A & A Costs**. The management team is anxiously waiting for Alex's analysis.

### Required Tasks:

1. Identify the major issue(s) of the case.
2. Analyze the overall cost allocation issues by developing a scatter plot of cost vs. hours of operation. Which variable, cost or hours of operation, should be the dependent variable? Explain why.
3. Fit a linear regression equation to the data.
4. Explain how the results of the linear regression equation could be used to develop a cost allocation formula. State any adjustments or modifications you have made to the regression output to develop a cost allocation formula that can be used to predict repair costs.
5. Sort the data by plant.
6. Fit a linear regression equation to each plant's data.
7. Explain how the results of the individual plant regression equations can help the manager determine whether a different linear regression equation could be used to develop a cost allocation formula for each plant. State any adjustments or modifications you have made to the regression output to develop a cost allocation formula.
8. Based on the individual plant regression equations, determine whether there is reason to believe there are differences among the repair costs of the company's three plants.
9. Summarize your analysis and findings in a report to the company's manager.

---

## Case 14.2    Sapphire Coffee—Part 1

Jennie Garcia could not believe that her career had moved so far so fast. When she left graduate school with a master's degree in anthropology, she intended to work at a local coffee shop until something else came along that was more related to her academic background. But after a few months, she came to enjoy the business, and in a little more than a year, she was promoted to store manager. When the company for which she worked continued to grow, Jennie was given oversight of a few stores.

Now, eight years after she started as a barista, Jennie was in charge of operations and planning for the company's southern region. As a part of her responsibilities, Jennie tracks store revenues and forecasts coffee demand. Historically, Sapphire Coffee based its demand forecast on the number of stores, believing that each store sold approximately the same amount of coffee. This approach seemed to work well when the company had shops of similar size and layout, but as the company grew, stores became more varied. Now, some stores had drive-thru windows, a feature that top management added to some stores believing that it would increase coffee sales for customers who wanted a cup of coffee on their way to work but were too rushed to park and enter the store to place an order.

Jennie noticed that weekly sales seemed to be more variable across stores in her region and was wondering what, if anything, might explain the differences. The company's financial vice president had also noticed the increased differences in sales across

stores and was wondering what might be happening. In an e-mail to Jennie, he stated that weekly store sales are expected to average $5.00 per square foot. Thus, a 1,000-square-foot store would have average weekly sales of $5,000. He asked that Jennie analyze the stores in her region to see if this rule of thumb was a reliable measure of a store's performance.

The vice president of finance was expecting the analysis to be completed by the weekend. Jennie decided to randomly select weekly sales records for 53 stores. The data are in the file **Sapphire Coffee-1**. A full analysis needs to be sent to the corporate office by Friday.

### Required Tasks:

1. Identify the major issue(s) of the case.
2. Develop a scatter plot of the variables store size vs. weekly sales. Identify the dependent variable. Briefly describe the relationship between the two variables.
3. Fit a linear regression equation to the data. Does the variable store size explain a significant amount of the variation in weekly sales?
4. Based on the estimated regression equation, does it appear that the $5.00 per square foot weekly sales expectation the company currently uses is a valid one?
5. Summarize your analysis and findings in a report to the company's vice president of finance.

---

## Case 14.3    Alamar Industries

While driving home in northern Kentucky at 8:00 P.M., Juan Alamar wondered whether his father had done him any favor by retiring early and letting him take control of the family machine tool–restoration business. When his father started the business of overhauling machine tools (both for resale and on a contract basis), American companies dominated the tool manufacturing

market. During the past 30 years, however, the original equipment industry had been devastated, first by competition from Germany and then from Japan. Although foreign competition had not yet invaded the overhaul segment of the business, Juan had heard about foreign companies establishing operations on the West Coast.

The foreign competitors were apparently stressing the high-quality service and operations that had been responsible for their great inroads into the original equipment market. Last week, Juan attended a daylong conference on total quality management that had discussed the advantages of competing for the Baldrige Award, the national quality award established in 1987. Presenters from past Baldrige winners, including Xerox, Federal Express, Cadillac, and Motorola, stressed the positive effects on their companies of winning and said similar effects would be possible for any company. This assertion of only positive effects was what Juan questioned. He was certain that the effect on his remaining free time would not be positive.

The Baldrige Award considers seven corporate dimensions of quality. Although the award is not based on a numerical score, an overall score is calculated. The maximum score is 1,000, with most recent winners scoring about 800. Juan did not doubt the award was good for the winners, but he wondered about the non-winners. In particular, he wondered about any relationship between attempting to improve quality according to the Baldrige dimensions and company profitability. Individual company scores are not released, but Juan was able to talk to one of the conference presenters, who shared some anonymous data, such as companies'

scores in the year they applied, their returns on investment (ROIs) in the year applied, and returns on investment in the year after application. Juan decided to commit the company to a total quality management process if the data provided evidence that the process would lead to increased profitability.

| Baldrige Score | ROI Application Year | ROI Next Year |
| --- | --- | --- |
| 470 | 11% | 13% |
| 520 | 10 | 11 |
| 660 | 14 | 15 |
| 540 | 12 | 12 |
| 600 | 15 | 16 |
| 710 | 16 | 16 |
| 580 | 11 | 12 |
| 600 | 12 | 13 |
| 740 | 16 | 16 |
| 610 | 11 | 14 |
| 570 | 12 | 13 |
| 660 | 17 | 19 |

## Case 14.4    Continental Trucking

Norm Painter is the newly hired cost analyst for Continental Trucking. Continental is a nationwide trucking firm, and until recently, most of its routes were driven under regulated rates. These rates were set to allow small trucking firms to earn an adequate profit, leaving little incentive to work to reduce costs by efficient management techniques. In fact, the greatest effort was made to try to influence regulatory agencies to grant rate increases.

A recent rash of deregulation has made the long-distance trucking industry more competitive. Norm has been hired to analyze Continental's whole expense structure. As part of this study, Norm is looking at truck repair costs. Because the trucks are involved in long hauls, they inevitably break down. In the past, little preventive maintenance was done, and if a truck broke down in the middle of a haul, either a replacement tractor was sent or an independent contractor finished the haul. The truck was then repaired at the nearest local shop. Norm is sure this procedure has led to more expense than if major repairs had been made before the trucks failed.

Norm thinks that some method should be found for determining when preventive maintenance is needed. He believes that fuel consumption is a good indicator of possible breakdowns; as trucks begin to run badly, they will consume more fuel. Unfortunately, the major determinants of fuel consumption are the weight of a truck and headwinds. Norm picks a sample of a single truck model

and gathers data relating fuel consumption to truck weight. All trucks in the sample are in good condition. He separates the data by direction of the haul, realizing that winds tend to blow predominantly out of the west.

Although he can rapidly gather future data on fuel consumption and haul weight, now that Norm has these data, he is not quite sure what to do with them.

| East-West Haul | | West-East Haul | |
| --- | --- | --- | --- |
| Miles/Gallon | Haul Weight | Miles/Gallon | Haul Weight |
| 4.1 | 41,000 lb | 4.3 | 40,000 lb |
| 4.7 | 36,000 | 4.5 | 37,000 |
| 3.9 | 37,000 | 4.8 | 36,000 |
| 4.3 | 38,000 | 5.2 | 38,000 |
| 4.8 | 32,000 | 5.0 | 35,000 |
| 5.1 | 37,000 | 4.7 | 42,000 |
| 4.3 | 46,000 | 4.9 | 37,000 |
| 4.6 | 35,000 | 4.5 | 36,000 |
| 5.0 | 37,000 | 5.2 | 42,000 |
| | | 4.8 | 41,000 |

# 15 Multiple Regression Analysis and Model Building

## WHY YOU NEED TO KNOW

Chapter 14 introduced linear regression and correlation analyses for examining the relationship between a dependent variable and a single independent variable. As you might expect, business problems are not limited to linear relationships involving only two variables. Many practical situations involve relationships among three or more variables. For example, a vice president of planning for an automobile manufacturer would be interested in the relationship between her company's automobile sales and the variables that influence those sales. Included in her analysis might be such independent or explanatory variables as automobile

### Quick Prep

Review the methods for testing a null hypothesis using the *t*-distribution in Chapter 9.

Review confidence intervals discussed in Chapter 8.

Make sure you review the discussion about scatter plots in Chapters 2 and 14.

Review the concepts associated with simple linear regression and correlation analysis presented in Chapter 14.

In Chapter 14, review the steps involved in using the *t*-distribution for testing the significance of a correlation coefficient and a regression coefficient.

price, competitors' sales, and advertising, as well as economic variables such as disposable personal income, the inflation rate, and the unemployment rate.

When multiple independent variables are included in an analysis simultaneously, the technique introduced in this chapter—multiple linear regression—is very useful. When a relationship between variables is nonlinear, we may be able to transform the independent variables in ways that allow us to use multiple linear regression analysis to model the nonlinear relationships. This chapter examines the general topic of model building by extending the concepts of simple linear regression analysis provided in Chapter 14.

## 15.1 | Introduction to Multiple Regression Analysis

Chapter 14 introduced the concept of simple linear regression analysis. The simple regression model is characterized by two variables: $y$, the *dependent variable*, and $x$, the *independent variable*. The single independent variable explains some variation in the dependent variable, but unless $x$ and $y$ are perfectly correlated, the proportion explained will be less than 100%. In multiple regression analysis, we add more independent variables to the regression model to account for some of the as yet unexplained variation in the dependent variable. Multiple regression is merely an extension of simple regression analysis; however, as we expand the model from one independent variable to two or more, there are some new considerations.

The general format of a *multiple regression model for the population* is given by Equation 15.1.

**Population Multiple Regression Model**

$$y = \beta_0 + \beta_1 x_1 + \beta_2 x_2 + \cdots + \beta_k x_k + \varepsilon \tag{15.1}$$

where:

$\beta_0 = $ Population's regression constant

$\beta_j = $ Population's regression coefficient for each variable, $x_j = 1, 2, \ldots, k$

$k = $ Number of independent variables

$\varepsilon = $ Model error

Four assumptions similar to those that apply to the simple linear regression model also apply to the multiple regression model.

**Assumptions**

1. Individual model errors, $\varepsilon$, are statistically independent of one another, and these values represent a random sample from the population of possible errors at each level of $x$.
2. For a given value of $x$, there can be many values of $y$, and therefore many possible values for $\varepsilon$. Further, the possible model errors for any level of $x$ are normally distributed.
3. The distributions of possible $\varepsilon$ values have equal variances at each level of $x$.
4. The means of the dependent variable, $y$, for all specified values of $x$ can be connected with a line called the population regression model.

Equation 15.1 represents the multiple regression model for the population. However, in most instances, you will be working with a random sample from the population. Given the preceding assumptions, the estimated multiple regression model, based on the sample data, is of the form shown in Equation 15.2.

**TABLE 15.1** Sample Data to Illustrate the Difference between Simple and Multiple Regression Models

| (A) One Independent Variable | | (B) Two Independent Variables | | |
|---|---|---|---|---|
| $y$ | $x_1$ | $y$ | $x_1$ | $x_2$ |
| 564.99 | 50 | 564.99 | 50 | 10 |
| 601.06 | 60 | 601.06 | 60 | 13 |
| 560.11 | 40 | 560.11 | 40 | 14 |
| 616.41 | 50 | 616.41 | 50 | 12 |
| 674.96 | 60 | 674.96 | 60 | 15 |
| 630.58 | 45 | 630.58 | 45 | 16 |
| 554.66 | 53 | 554.66 | 53 | 14 |

**Estimated Multiple Regression Model**

$$\hat{y} = b_0 + b_1 x_1 + b_2 x_2 + \cdots + b_k x_k \qquad (15.2)$$

This estimated model is an extension of an estimated simple regression model. The principal difference is that whereas the estimated simple regression model is the equation for a straight line in a two-dimensional space, the estimated multiple regression model forms a hyperplane (or response surface) through multidimensional space. Each regression coefficient represents a different slope. Therefore, using Equation 15.2, we can estimate a value of the dependent variable using values of two or more independent variables. The **regression hyperplane** represents the relationship between the dependent variable and $k$ independent variables.

**Regression Hyperplane**
The multiple regression equivalent of the simple regression line. The plane typically has a different slope for each independent variable.

For example, Table 15.1(Part A) shows sample data for a dependent variable, $y$, and one independent variable, $x_1$. Figure 15.1 shows a scatter plot and the regression line for the simple regression analysis for $y$ and $x_1$. The points are plotted in two-dimensional space, and the regression model is represented by a line through the points such that the sum of squares error $[SSE = \Sigma(y - \hat{y})^2]$ is minimized.

If we add variable $x_2$ to the model, as shown in Table 15.1(Part B), the resulting multiple regression equation becomes

$$\hat{y} = 307.71 + 2.85 x_1 + 10.94 x_2$$

For the time being, don't worry about how this equation was computed. We will discuss that shortly. Note, however, that the $(y, x_1, x_2)$ points form a three-dimensional space, as shown in Figure 15.2. The regression equation forms a slice (hyperplane) through the data such that $\Sigma(y - \hat{y})^2$ is minimized. This is the same *least squares criterion* that is used with simple linear regression.

**FIGURE 15.1 Simple Regression Line**

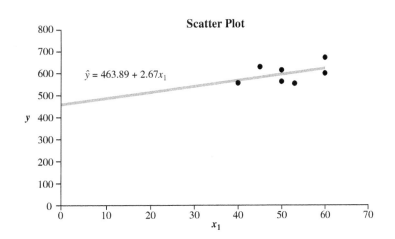

**FIGURE 15.2 Multiple Regression Hyperplane for Population**

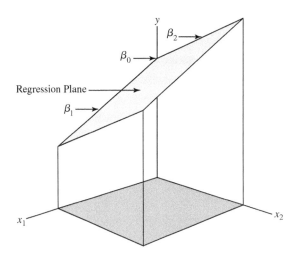

The mathematics for developing the least squares regression equation for simple linear regression involves differential calculus. The same is true for the multiple regression equation, but the mathematical derivation is beyond the scope of this text.[1]

Multiple regression analysis is usually performed with the aid of a computer and appropriate software. Excel contains a procedure for performing multiple regression, and the XLSTAT Excel add-in expands Excel's capabilities. However, other statistical software such as SPSS and SAS offer more extensive regression options. Each software package presents the output in a slightly different format; however, the same basic information will appear in all regression output.

OUTCOME **1**

## Basic Model-Building Concepts

**Model**

A representation of an actual system using either a physical or a mathematical portrayal.

An important activity in business decision making is referred to as **model** building. Analysts often use models to test changes in a system without actually having to change the real system. Models are also used to help describe a system or to predict the output of a system based on certain specified inputs. You are probably quite aware of physical models. Airlines use flight simulators to train pilots. Wind tunnels are used to determine the aerodynamics of automobile designs. Golf ball makers use a physical model of a golfer called Iron Mike that can be set to swing golf clubs in a very controlled manner to determine how far a golf ball will fly. Although physical models are very useful in business decision making, our emphasis in this chapter is on statistical models that are developed using multiple regression analysis.

Modeling is both an art and a science. Determining an appropriate model is a challenging task, but it can be made manageable by employing a model-building process consisting of the following three components: model specification, model building, and model diagnosis.

**Model Specification**    *Model specification*, or model identification, is the process of determining the dependent variable, deciding which independent variables should be included in the model, and obtaining the sample data for all variables. As with any statistical procedure, the larger the sample size the better, because the potential for extreme sampling error is reduced when the sample size is large. However, at a minimum, the sample size used to compute a regression model must be at least one greater than the number of independent variables.[2] If we are thinking of developing a regression model with five

---

[1]For a complete treatment of the matrix algebra approach for estimating multiple regression coefficients, consult *Applied Linear Statistical Models* by Kutner et al., listed at the end of the book in References.
[2]There are mathematical reasons for this sample-size requirement that are beyond the scope of this text. In essence, there is no unique solution to the regression coefficient in Equation 15.2 if the sample size is not at least one larger than the number of independent variables.

independent variables, the minimum sample size required is six. However, as a practical matter, the sample size should be at least four times the number of independent variables. Thus, if we had five independent variables ($k = 5$), we would want a sample size of at least 20.

**Model Building**   *Model building* is the process of actually constructing a mathematical equation in which some or all of the independent variables are used in an attempt to explain the variation in the dependent variable.

OUTCOME **2**

> **HOW TO DO IT**
>
> **Model Specification**
>
> **1.** Decide what question you want to ask. The question usually indicates the dependent variable. In Chapter 14, we discussed how simple linear regression analysis can be used to describe the relationship between a dependent and an independent variable.
>
> **2.** List the potential independent variables for your model. Here, your knowledge of the situation you are modeling guides you in identifying potential independent variables.
>
> **3.** Gather the sample data (observations) for all variables.

> **HOW TO DO IT**
>
> **Developing a Multiple Regression Model**
>
> **1.** Specify the model by determining the dependent variable and potential independent variables, and select the sample data.
>
> **2.** Formulate the model by computing the correlation coefficients for the dependent variable and each independent variable, and for each independent variable with all other independent variables. Also compute the multiple regression equation. Perform the computations using Excel.

**Correlation Coefficient**
A quantitative measure of the strength of the linear relationship between two variables. The correlation coefficient, $r$, ranges from $-1.0$ to $+1.0$.

---

**EXAMPLE 15-1**   **Developing a Multiple Regression Model**

 **First City Real Estate**   First City Real Estate executives wish to build a model to predict sales prices for residential property. Such a model will be valuable when they work with potential sellers who might list their homes with First City. This can be done using the following steps:

STEP **1**   **Model specification.**
The question being asked is How can the real estate firm determine the selling price for a house? Thus, the dependent variable is the sales price. This is what the managers want to be able to predict. The managers met in a brainstorming session to determine a list of possible independent (explanatory) variables. Some variables, such as "condition of the house," were eliminated because of lack of data. Others, such as "curb appeal" (the appeal of the house to people as they drive by), were eliminated because the values for these variables would be too subjective and difficult to quantify. From a wide list of possibilities, the managers selected the following variables as good candidates:

$$x_1 = \text{Home size (in square feet)}$$
$$x_2 = \text{Age of house}$$
$$x_3 = \text{Number of bedrooms}$$
$$x_4 = \text{Number of bathrooms}$$
$$x_5 = \text{Garage size (number of cars)}$$

The managers obtained data for a sample of 319 residential properties that had sold within the previous two months in an area served by two of First City's offices. For each house in the sample, they collected the sales price and values for each potential independent variable. The data are in the file **First City**.

STEP **2**   **Model building.**
We develop the regression model by including independent variables from among those for which we have complete data. There is no way to determine whether an independent variable will be a good predictor variable by analyzing the individual variable's descriptive statistics, such as the mean and standard deviation. Instead, we need to look at the correlation between each independent variable and the dependent variable, which is measured by the **correlation coefficient**.

**Excel 2016 Instructions**

1. Open file: **First City.xlsx**.
2. Use the Homes-Sample 1 Worksheet.
3. Select **Data > Data Analysis**.
4. Select **Correlation**.
5. Define **Input Range**—all rows and columns.
6. Click **Labels**.
7. Specify output location.
8. Click **OK**.

**FIGURE 15.3 Excel 2016 Results Showing First City Real Estate Correlation Matrix**

| | A | B | C | D | E | F | G |
|---|---|---|---|---|---|---|---|
| 1 | | Price | Sq. Feet | Age | Bedrooms | Bathrooms | Garage # |
| 2 | Price | 1 | | | | | |
| 3 | Sq. Feet | 0.7477 | 1 | | | | |
| 4 | Age | -0.4852 | -0.0729 | 1 | | | |
| 5 | Bedroom | 0.5401 | 0.7059 | -0.2024 | 1 | | |
| 6 | Bathroom | 0.6655 | 0.6293 | -0.3871 | 0.5996 | 1 | |
| 7 | Garage # | 0.6935 | 0.4163 | -0.4374 | 0.3120 | 0.4646 | 1 |

Correlation between age and square feet = $-0.0729$: Older homes tend to be smaller

**Correlation Matrix**

A table showing the pairwise correlations between all variables (dependent and independent).

When we have multiple independent variables and one dependent variable, we can look at the correlation between all pairs of variables by developing a **correlation matrix**. Each correlation is computed using one of the equations in Equation 15.3. The appropriate formula is determined by whether the correlation is being calculated for an independent variable and the dependent variable or for two independent variables.

**Correlation Coefficient**

$$r = \frac{\Sigma(x - \bar{x})(y - \bar{y})}{\sqrt{\Sigma(x - \bar{x})^2 \Sigma(y - \bar{y})^2}} \quad \text{or} \quad r = \frac{\Sigma(x_i - \bar{x}_i)(x_j - \bar{x}_j)}{\sqrt{\Sigma(x_i - \bar{x}_i)^2 \Sigma(x_j - \bar{x}_j)^2}} \quad \textbf{(15.3)}$$

One $x$ variable with $y$          One $x$ variable with another $x$

The actual calculations are done using Excel's correlation tool, and the result is shown in Figure 15.3. The output provides the correlation between $y$ and each $x$ variable and between each pair of independent variables. Recall that in Chapter 14, we used a $t$-test (see Equation 14.3) to test whether a correlation coefficient is statistically significant:

$$H_0: \rho = 0 \quad H_A: \rho \neq 0$$

We will conduct the test with a significance level of

$$\alpha = 0.05$$

Given degrees of freedom equal to

$$n - 2 = 319 - 2 = 317$$

we find the critical $t$ (see Appendix E) for a two-tailed test is approximately $\pm 1.967$.

Any correlation coefficient that generates a $t$-value greater than 1.967 or less than $-1.967$ is determined to be significant.

For now, we will focus on the correlations in the first column in Figure 15.3, which measures the strength of the linear relationship between each independent variable and the dependent variable, sales price. For example, the $t$-statistic for price and square feet is

The Excel 2016 function for the critical $t$-value is
= T.INV.2T(alpha,df)
= T.INV.2T(.05,317)

$$t = \frac{r}{\sqrt{\dfrac{1 - r^2}{n - 2}}} = \frac{0.7477}{\sqrt{\dfrac{1 - 0.7477^2}{319 - 2}}} = 20.048$$

Because

$$t = 20.048 > 1.967$$

we reject $H_0$ and conclude that the correlation between sales price and square feet is statistically significant.

Similar calculations for the other independent variables with price show that all variables are statistically correlated with price. This indicates that a significant linear relationship exists between each independent variable and sales price. Variable $x_1$, square feet, has the highest correlation at 0.7477. Variable $x_2$, age of the house, has the lowest correlation at $-0.4852$. The negative correlation implies that older homes tend to have lower sales prices.

FIGURE 15.4 **First City Real Estate Scatter Plots**

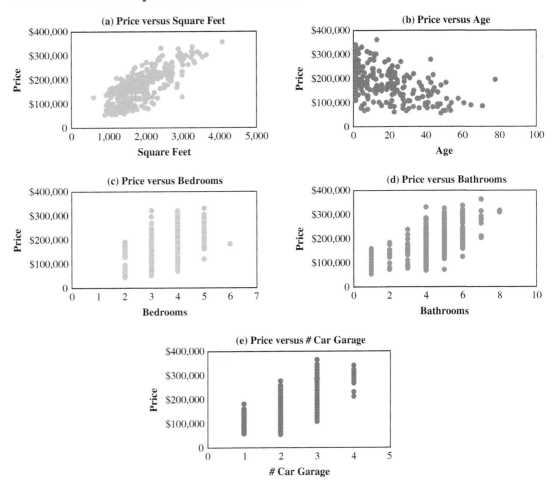

As we discussed in Chapter 14, it is always a good idea to develop scatter plots to visualize the relationship between two variables. Figure 15.4 shows the scatter plots for each independent variable and the dependent variable, sales price. In each case, the plots indicate a linear relationship between the independent variable and the dependent variable. Note that some of the independent variables (bedrooms, bathrooms, garage size) are quantitative but discrete. The scatter plots for these variables show points at each level of the independent variable rather than over a continuum of values.

**OUTCOME 3**    **Computing the Regression Equation**    First City's goal is to develop a regression model to predict the appropriate selling price for a home, using certain measurable characteristics. The first attempt at developing the model will be to run a multiple regression computer program using all available independent variables. The regression outputs from Excel are shown in Figure 15.5.

The estimate of the multiple regression model given in Figure 15.5 is

$$\hat{y} = 31{,}127.6 + 63.07(\text{sq. feet}) - 1{,}144.44(\text{age}) - 8{,}410.38(\text{bedrooms})$$
$$+ 3{,}521.95(\text{bathrooms}) + 28{,}203.54(\text{garage})$$

The coefficient for each independent variable represents an estimate of the average change in the dependent variable for a 1-unit change in the independent variable, holding all other independent variables constant. For example, for houses of the same age, with the same number of bedrooms, bathrooms, and garage size, a 1-square-foot increase in the size of the house is estimated to increase its price by an average of $63.07. Likewise, for houses with the same square footage, bedrooms, bathrooms, and garages, a 1-year increase in the age of the house is estimated to result in an average drop in sales price of $1,144.44. We interpret the other coefficients in the same way. Note, in each case, that we are interpreting the regression coefficient for one independent variable while holding the other variables constant. (Take

**FIGURE 15.5 Excel 2016 Multiple Regression Model Results for First City Real Estate**

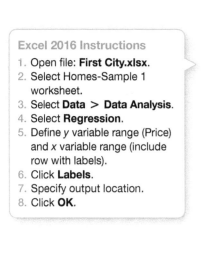

| | A | B | C | D | E | F | G |
|---|---|---|---|---|---|---|---|
| 1 | SUMMARY OUTPUT | | | | | | |
| 3 | *Regression Statistics* | | | | | | |
| 4 | Multiple R | 0.9034 | | | | | |
| 5 | R Square | 0.8161 | | | | | |
| 6 | Adjusted R Sq. | 0.8131 | | | | | |
| 7 | Standard Error | 27,350.25 | | | | | |
| 8 | Observations | 319 | | | | | |
| 10 | ANOVA | | | | | | |
| 11 | | *df* | *SS* | *MS* | *F* | *Significance F* | |
| 12 | Regression | 5 | 1038897292243 | 207779458449 | 277.77 | 0.0000 | |
| 13 | Residual | 313 | 234135351519 | 748036267 | | | |
| 14 | Total | 318 | 1273032643762 | | | | |
| 16 | | *Coefficients* | *Standard Error* | *t Stat* | *P-value* | *Lower 95%* | *Upper 95%* |
| 17 | Intercept | 31127.60 | 9539.67 | 3.26 | 0.0012 | 12357.62 | 49897.59 |
| 18 | Sq. Feet | 63.07 | 4.02 | 15.70 | 0.0000 | 55.16 | 70.97 |
| 19 | Age | -1144.44 | 112.78 | -10.15 | 0.0000 | -1366.34 | -922.53 |
| 20 | Bedrooms | -8410.38 | 3002.51 | -2.80 | 0.0054 | -14318.04 | -2502.72 |
| 21 | Bathrooms | 3521.95 | 1581.00 | 2.23 | 0.0266 | 411.23 | 6632.68 |
| 22 | Garage # | 28203.54 | 2858.69 | 9.87 | 0.0000 | 22578.86 | 33828.23 |

*Multiple coefficient of determination*
*Standard error of the estimate*
*Sums of squares*
*Regression coefficients*

note of the negative sign on the regression coefficient for $x_3$, number of bedrooms. This implies that, holding the other variables constant, each additional bedroom is associated with an average *reduction* in sales price of \$8,410.38. This counterintuitive result is a potential issue with our model that we will address later in this section.)

To estimate the value of a residential property, First City Real Estate brokers will substitute values for the independent variables into the regression equation. For example, suppose a broker is considering a house with the following characteristics:

$$x_1 = \text{Square feet} = 2,100$$
$$x_2 = \text{Age} = 15$$
$$x_3 = \text{Number of bedrooms} = 4$$
$$x_4 = \text{Number of bathrooms} = 3$$
$$x_5 = \text{Size of garage} = 2$$

The point estimate for the sales price is

$$\hat{y} = 31,127.6 + 63.07(2,100) - 1,144.44(15) - 8,410.38(4) + 3,521.95(3) + 28,203.54(2)$$
$$\hat{y} = \$179,739.41$$

**The Coefficient of Determination** You learned in Chapter 14 that the *coefficient of determination*, $R^2$, measures the proportion of variation in the dependent variable that can be explained by the dependent variable's relationship to a single independent variable. When there are multiple independent variables in a model, $R^2$ is called the **multiple coefficient of determination** and is used to determine the proportion of variation in the dependent variable that is explained by all the independent variables in the model. Equation 15.4 is used to compute $R^2$ for a multiple regression model.

**Multiple Coefficient of Determination ($R^2$)**
The proportion of the total variation of the dependent variable in a multiple regression model that is explained by its relationship to the independent variables. It is, as is the case in the simple linear model, called R-squared and is denoted as $R^2$.

**Multiple Coefficient of Determination ($R^2$)**

$$R^2 = \frac{\text{Sum of squares regression}}{\text{Total sum of squares}} = \frac{SSR}{SST} \tag{15.4}$$

As shown in Figure 15.5, $R^2 = 0.8161$. Both *SSR* and *SST* are also included in the Excel 2016 output. Therefore, you can also use Equation 15.4 to get $R^2$, as follows:

$$R^2 = \frac{SSR}{SST} = \frac{1.0389\text{E} + 12}{1.27303\text{E} + 12} = 0.8161$$

More than 81% of the variation in sales price can be explained by the linear relationship of the five independent variables in the regression model to the dependent variable. However, as we shall shortly see, not all independent variables are equally important to the model's ability to explain this variation.

**Model Diagnosis**    *Model diagnosis* is the process of analyzing the quality of the model you have constructed by determining how well a specified model fits the data you just gathered. You will examine output values such as *R*-squared and the standard error of the model. At this stage, you will also assess the extent to which the model's assumptions appear to be satisfied. (Section 15.5 is devoted to examining whether a model meets the regression analysis assumptions.) If the model is unacceptable in any of these areas, you will be forced to return to the model-specification step and begin again. However, you will be the final judge of whether the model provides acceptable results, and you will always face time and cost constraints.

You should use the simplest available model that will meet your needs. The objective of model building is to help you make better decisions. You do not need to feel that a sophisticated model is better if a simpler one will provide acceptable results.

Therefore, before First City actually uses this regression model to estimate the sales price of a house, several questions should be answered:

1. Is the overall model significant?
2. Are the individual variables significant?
3. Is the standard deviation of the model error too large to provide meaningful results?
4. Is multicollinearity a problem?
5. Have the regression analysis assumptions been satisfied?

We shall answer the first four questions in order. We will have to wait until Section 15.5 before we have the procedures to answer the fifth important question.

OUTCOME **4**    **Is the Model Significant?**    Because the regression model we constructed is based on a sample of data from a population of home sales and is subject to sampling error, we need to test the statistical significance of the overall regression model. The specific null and alternative hypotheses tested for First City Real Estate are

$$H_0: \beta_1 = \beta_2 = \beta_3 = \beta_4 = \beta_5 = 0$$
$$H_A: \text{At least one } \beta_i \neq 0$$

If the null hypothesis is true and all the slope coefficients are simultaneously equal to zero, the overall regression model is not useful for predictive or descriptive purposes.

The *F*-test is a method for testing whether the regression model explains a significant proportion of the variation in the dependent variable (and whether the overall model is significant). The *F*-test statistic for a multiple regression model is shown in Equation 15.5.

*F*-Test Statistic

$$F = \frac{\dfrac{SSR}{k}}{\dfrac{SSE}{n - k - 1}} \tag{15.5}$$

where:

$$SSR = \text{Sum of squares regression} = \sum (\hat{y} - \bar{y})^2$$
$$SSE = \text{Sum of squares error} = \sum (y - \hat{y})^2$$
$$n = \text{Sample size}$$
$$k = \text{Number of independent variables}$$
$$\text{Degrees of freedom} = D_1 = k \quad \text{and} \quad D_2 = (n - k - 1)$$

The ANOVA portion of the Excel 2016 output shown in Figure 15.5 contains values for *SSR*, *SSE*, and *F*. The general format of the ANOVA table in a regression analysis is as follows:

**ANOVA**

|  | df | SS | MS | F | Significance F |
|---|---|---|---|---|---|
| Regression | $k$ | SSR | $MSR = SSR/k$ | $MSR/MSE$ | computed $p$-value |
| Residual | $n - k - 1$ | SSE | $MSE = SSE/(n - k - 1)$ |  |  |
| Total | $n - 1$ | SST |  |  |  |

The ANOVA portion of the output from Figure 15.5 is as follows:

**ANOVA**

| Source | df | SS | MS | F | Significance F |
|---|---|---|---|---|---|
| Regression | 5 | 1.04E + 12 | 2.08E + 11 | 277.8 | 0.0000 |
| Residual | 313 | 2.34E + 11 | 7.48E + 08 |  |  |
| Total | 318 | 1.27E + 12 |  |  |  |

We can test the model's significance

$$H_0: \beta_1 = \beta_2 = \beta_3 = \beta_4 = \beta_5 = 0$$
$$H_A: \text{At least one } \beta_i \neq 0$$

by either comparing the calculated $F$-value, 277.8, with a critical value for a given alpha level

$$\alpha = 0.01$$

and $k = 5$ and $n - k - 1 = 313$ degrees of freedom using Excel's F.INV.RT function ($F_{0.01} = 3.076$) or comparing the $p$-value in the Excel 2016 output (Significance $F$) with a specified alpha level. Because

$$F = 277.8 > 3.076, \text{ reject } H_0$$

or because

$$p\text{-value} \approx 0.0 < 0.01, \text{ reject } H_0$$

> The Excel 2016 function for the critical $F$-value is
>
> = **F.INV.RT(alpha,k,n-k-1)**
> = **F.INV.RT(0.01,5,313)**

we should conclude that the regression model *does* explain a significant proportion of the variation in sales price. Thus, the overall model is statistically significant. This means we can conclude that at least one of the regression slope coefficients is not equal to zero.

Excel provides a measure called $R$-sq(adj), which is the **adjusted $R$-squared** value (see Figure 15.5). It is calculated by Equation 15.6.

**Adjusted *R*-Squared**

A measure of the percentage of explained variation in the dependent variable that takes into account the relationship between the sample size and the number of independent variables in the regression model.

> **Adjusted *R*-Squared**
> A measure of the percentage of explained variation in the dependent variable in a multiple regression model that takes into account the relationship between the sample size and the number of independent variables in the regression model.

$$R\text{-sq(adj)} = R_A^2 = 1 - (1 - R^2)\left(\frac{n - 1}{n - k - 1}\right) \tag{15.6}$$

where:

$$n = \text{Sample size}$$
$$k = \text{Number of independent variables}$$
$$R^2 = \text{Coefficient of determination}$$

Adding independent variables to the regression model will always increase $R^2$, even if these variables have no relationship to the dependent variable. Therefore, as we increase the number of independent variables (regardless of the quality of the variables), $R^2$ will increase. However, each additional variable results in the loss of one degree of freedom. This is viewed as part of the cost of adding the specified variable. The addition to $R^2$ may not justify the reduction in degrees of freedom. The $R_A^2$ value takes into account this cost and adjusts the $R_A^2$ value accordingly. $R_A^2$ will always be less than $R^2$. When we add a variable that does not contribute its fair share to the explanation of the variation in the dependent variable, the $R_A^2$ value may actually decline, even though $R^2$ will always increase. The adjusted $R$-squared is a particularly important measure when the number of independent variables is large relative to the sample size. It takes into account the relationship between sample size and number of variables. $R^2$ may appear artificially high if the number of variables is large compared with the sample size.

In this example, in which the sample size is quite large relative to the number of independent variables, the adjusted $R$-squared is 81.3%, only slightly less than $R^2 = 81.6\%$.

**OUTCOME 4**

**Are the Individual Variables Significant?**    We have concluded that the overall model is significant. This means that *at least* one independent variable explains a significant proportion of the variation in sales price. This does not mean that *all* the variables are significant, however. To determine which variables are significant, we test the following hypotheses:

$$H_0: \beta_j = 0$$
$$H_A: \beta_j \neq 0 \quad \text{for all } j$$

We can test the significance of each independent variable using significance level

$$\alpha = 0.05$$

and a $t$-test, as discussed in Chapter 14. The calculated $t$-values should be compared to the critical $t$-value with

$$n - k - 1 = 319 - 5 - 1 = 313$$

degrees of freedom, which is approximately

$$t \approx \pm 1.968$$

The Excel 2016 function for the critical $t$-value is

= T.INV.2T(alpha,n-k-1)
= T.INV.2T(0.05,313)

for $\alpha = 0.05$. The calculated $t$-value for each variable is provided on the Excel 2016 computer printout in Figure 15.5. Recall that the $t$-statistic is determined by dividing the regression coefficient by the estimate of the standard deviation of the regression coefficient, as shown in Equation 15.7.

### *t*-Test for Significance of Each Regression Coefficient

$$t = \frac{b_j - 0}{s_{b_j}} \qquad df = n - k - 1 \qquad \text{(15.7)}$$

where:

$b_j$ = Sample slope coefficient for the $j$th independent variable

$s_{b_j}$ = Estimate of the standard deviation for the $j$th sample slope coefficient

For example, the $t$-value for square feet shown in Figure 15.5 is 15.70. This was computed using Equation 15.7, as follows:

$$t = \frac{b_j - 0}{s_{b_j}} = \frac{63.1 - 0}{4.02} = 15.70$$

Because

$$t = 15.70 > 1.968, \text{ we reject } H_0$$

and conclude that, given the other independent variables in the model, the regression slope for square feet is not zero.

We can also look at the Excel 2016 output in Figure 15.5 and compare the $p$-value for each regression slope coefficient with alpha. If the $p$-value is less than alpha, we reject the null hypothesis and conclude that the independent variable is statistically significant in the model. Both the $t$-test and the $p$-value techniques will give the same results.

These $t$-tests are *conditional* tests. This means the null hypothesis is that *the value of each slope coefficient is 0, given that the other independent variables are already in the model*. Figure 15.6 shows the hypothesis tests for each independent variable using a 0.05 significance level. We conclude that all five independent variables in the model are significant. When a regression model is to be used for prediction, the model should contain no

**FIGURE 15.6 Significance Tests for Each Independent Variable in the First City Real Estate Example**

**Hypotheses:**

$H_0: \beta_j = 0$, given all other variables are already in the model
$H_A: \beta_j \neq 0$, given all other variables are already in the model
$\alpha = 0.05$

**Decision Rule:**
If $t < -1.968$ or
$t > 1.968$, reject $H_0$.
Otherwise, do not reject $H_0$.

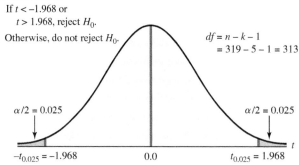

$df = n - k - 1$
$= 319 - 5 - 1 = 313$

$\alpha/2 = 0.025$    $\alpha/2 = 0.025$

$-t_{0.025} = -1.968$    $0.0$    $t_{0.025} = 1.968$    $t$

The test is:
For $\beta_1$: Calculated $t$ (from printout) = 15.70
    Because $15.70 > 1.968$, reject $H_0$.
For $\beta_2$: Calculated $t = -10.15$
    Because $-10.15 < -1.968$, reject $H_0$.
For $\beta_3$: Calculated $t = -2.80$
    Because $-2.80 < -1.968$, reject $H_0$.
For $\beta_4$: Calculated $t = 2.23$
    Because $2.23 > 1.968$, reject $H_0$.
For $\beta_5$: Calculated $t = 9.87$
    Because $9.87 > 1.968$, reject $H_0$.

insignificant variables. If insignificant variables are present, they should be dropped and a new regression equation obtained before the model is used for prediction purposes. We will have more to say about this later.

**Is the Standard Deviation of the Regression Model Too Large?** The purpose of developing the First City regression model is to be able to determine values of the dependent variable when corresponding values of the independent variables are known. We can get an indication of how good the regression model is by looking at the relationship between the measured values of the dependent variable and those values that would be predicted by the regression model. The standard deviation of the regression model (also called the *standard error of the estimate*) measures the dispersion of observed home sale values, $y$, around the values predicted by the regression model. The standard error of the estimate is shown in Figure 15.5 and can be computed using Equation 15.8.

**Standard Error of the Estimate**

$$s_\varepsilon = \sqrt{\frac{SSE}{n - k - 1}} = \sqrt{MSE} \tag{15.8}$$

where:

$$SSE = \text{Sum of squares error (residual)}$$
$$n = \text{Sample size}$$
$$k = \text{Number of independent variables}$$

Examining Equation 15.8 closely, we see that this standard error of the estimate is the square root of the mean square error of the residuals found in the analysis of variance table.

Sometimes, even though a model has a high $R^2$, the standard error of the estimate will be too large to provide adequate precision for confidence and prediction intervals. A rule of thumb that we have found useful is to examine the range $\pm 2s_\varepsilon$. Taking into account the mean value of the dependent variable, if this range is acceptable from a practical viewpoint, then we might consider the standard error of the estimate acceptable.[3]

In this First City Real Estate Company example, the standard error, shown in Figure 15.5, is \$27,350. Thus, the rough prediction range for the price of an individual home is

$$\pm 2(\$27{,}350) = \pm \$54{,}700$$

Given that the mean price of homes in this study is in the low \$200,000s, a potential error of \$54,700 high or low is probably not acceptable. Not many homeowners would be willing to have their appraisal value set by a model with a possible error this large. Even though the model is statistically significant, the company needs to take steps to reduce the standard deviation of the estimate. Subsequent sections of this chapter discuss some ways we can attempt to reduce it.

**OUTCOME 5**

**Is Multicollinearity a Problem?**    Even if the overall regression model is significant and each independent variable is significant, decision makers should still examine the regression model to determine whether it appears reasonable. This is referred to as checking for *face validity*. Specifically, you should check to see that the signs on the regression coefficients are consistent with the signs on the correlation coefficients between the independent variables and the dependent variable. Does any regression coefficient have an unexpected sign?

Before answering this question for the First City Real Estate example, we should review what the regression coefficients mean. First, the constant term, $b_0$, is the estimate of the model's y intercept. If the data used to develop the regression model contain values of $x_1, x_2, x_3, x_4$, and $x_5$ that are simultaneously 0 (such as would be the case for vacant land), then $b_0$ is the mean value of $y$, given that $x_1, x_2, x_3, x_4$, and $x_5$ all equal 0. Under these conditions $b_0$ would estimate the average value of a vacant lot. However, in the First City example, no vacant land was in the sample, so $b_0$ has no particular meaning.

The coefficient for square feet, $b_1$, estimates the average change in sales price corresponding to a change in house size of 1 square foot, holding the other independent variables constant. The value shown in Figure 15.5 for $b_1$ is 63.1. The coefficient is positive, indicating that an increase in square footage is associated with an increase in sales price. This relationship is expected. All other things being equal, bigger houses should sell for more money.

Likewise, the coefficient for $x_5$, the size of the garage, is positive, indicating that an increase in size is also associated with an increase in price. This is expected. The coefficient for $x_2$, the age of the house, is negative, indicating that an older house is worth less than a similar younger house. This also seems reasonable. Finally, variable $x_4$ for bathrooms has the expected positive sign. However, the coefficient for variable $x_3$, the number of bedrooms, is $-\$8{,}410.4$, meaning that if we hold the other variables constant but increase the number of bedrooms by one, the average price will *drop* by \$8,410.40. Does this seem reasonable?

From the correlation matrix shown in Figure 15.3, the correlation between variable $x_3$, the number of bedrooms, and $y$, the sales price, is $+0.540$. This indicates that without considering the other independent variables, the linear relationship between number of bedrooms and sales price is positive. But why does the regression coefficient for variable $x_3$ turn out to be negative in the model? The answer lies in what is called **multicollinearity**.

**Multicollinearity**

A high correlation between two independent variables such that the two variables contribute redundant information to the model. When highly correlated independent variables are included in the regression model, they can adversely affect the regression results.

Multicollinearity occurs when independent variables are correlated with each other and therefore overlap with respect to the information they provide in explaining the variation in the dependent variable. For example, $x_3$ and the other independent variables have the following correlations (see Figure 15.3):

$$r_{x_3, x_1} = \phantom{-}0.706$$
$$r_{x_3, x_2} = -0.202$$
$$r_{x_3, x_4} = \phantom{-}0.600$$
$$r_{x_3, x_5} = \phantom{-}0.312$$

---

[3]The actual confidence interval for prediction of a new observation requires the use of matrix algebra. However, when the sample size is large and dependent variable values near the means of the dependent variables are used, the rule of thumb given here is a close approximation. Refer to *Applied Linear Statistical Models* by Kutner et al. for further discussion.

All four correlations have $t$-values indicating a significant linear relationship. Refer to the correlation matrix in Figure 15.3 to see that other independent variables are also correlated with each other.

The problems caused by multicollinearity, and how to deal with them, continue to be of prime concern to statisticians. From a decision maker's viewpoint, you should be aware that multicollinearity can (and often does) exist and recognize the basic problems it can cause. The following are some of the most obvious problems and indications of severe multicollinearity:

1. The signs on the coefficients are unexpected and therefore potentially incorrect.
2. There is a sizable change in the values of the previously estimated coefficients when a new variable is added to the model.
3. A variable that was previously significant in the regression model becomes insignificant when a new independent variable is added.
4. The estimate of the standard deviation of the model error increases when a variable is added to the model.

Mathematical approaches exist for dealing with multicollinearity and reducing its impact. Although these procedures are beyond the scope of this text, one suggestion is to eliminate the variables that are the chief cause of the multicollinearity problems.

If the independent variables in a regression model are correlated and multicollinearity is present, another potential problem is that the $t$-tests for the significance of the individual independent variables may be misleading. That is, a $t$-test may indicate that the variable is not statistically significant when in fact it is.

One method of measuring multicollinearity is known as the **variance inflation factor** (*VIF*). Equation 15.9 is used to compute the *VIF* for each independent variable.

**Variance Inflation Factor**

A measure of how much the variance of an estimated regression coefficient increases if the independent variables are correlated. A *VIF* equal to 1.0 for a given independent variable indicates that this independent variable is not correlated with the remaining independent variables in the model. The greater the multicollinearity, the larger the *VIF*.

**Variance Inflation Factor**

$$VIF = \frac{1}{1 - R_j^2}$$

**(15.9)**

where:

$R_j^2 =$ Coefficient of determination when the $j$th independent variable is regressed against the remaining $k - 1$ independent variables

The XLSTAT add-in to Excel contains an option that provides *VIF* values.

Figure 15.7 on page 587 shows the Excel (XLSTAT) output of the *VIF*s for the First City Real Estate example. Generally, if $VIF < 5$ for a particular independent variable, then we do not consider multicollinearity a problem for that variable. $VIF \geq 5$ implies that the correlation between the independent variables is too extreme and should be dealt with by dropping variables from the model. As Figure 15.7 illustrates, the *VIF* values for all the independent variables are less than 5, so based on variance inflation factors, even though the sign on the variable bedrooms has switched from positive to negative, the other multicollinearity issues do not exist among these independent variables. If the negative sign on $x_3$, the number of bedrooms, is troubling, however, you could consider dropping the variable and creating a new regression model without bedrooms.

**Confidence Interval Estimation for Regression Coefficients** Previously, we showed how to determine whether the regression coefficients are statistically significant. This was necessary because the estimates of the regression coefficients are developed from sample data and are subject to sampling error. The issue of sampling error also comes into play when we interpret the slope coefficients.

Consider again the regression model for First City Real Estate shown in Figure 15.8. The regression coefficients shown are *point estimates* for the true regression coefficients. For example, the coefficient for the variable square feet is $b_1 = 63.1$. We interpret this to mean

**FIGURE 15.7 Excel 2016 (XLSTAT) Multiple Regression Model Results for First City Real Estate with Variance Inflation Factors**

| | A | B | C | D | E | F | G |
|---|---|---|---|---|---|---|---|
| 31 | Multicollinearity statistics: | | | | | | |
| 32 | | | | | | | |
| 33 | Statistic | Sq. Feet | Age | Bedrooms | Bathrooms | Garage # | |
| 34 | Tolerance | 0.3732 | 0.6765 | 0.4518 | 0.4565 | 0.6449 | |
| 35 | VIF | 2.6795 | 1.4783 | 2.2133 | 2.1906 | 1.5506 | |

**Excel 2016 Instructions**

1. Open XLSTAT and Enable Macros.
2. Open file: **First City.xlsx**.
3. Select Homes-Sample 1 worksheet.
4. On the **XLSTAT** tab, click **Modeling Data**, and then click **Linear Regression**.
5. Define the y/dependent variables (price) and the x/explanatory variables (quantitative), and check Variable labels.
6. On the **Outputs** tab, select **Descriptive Statistics**, **Correlation**, **Multicollinearity Statistics**, and **Analysis of Variance**; then click **OK**.

that, holding the other variables constant, for each increase in the size of a home by 1 square foot, the average price of a house is estimated to increase by $63.07. But like all point estimates, this is subject to sampling error. Chapter 14 introduced the concept of confidence interval estimates for the regression coefficients. That same concept applies in multiple regression models. We can use Equation 15.10 to develop the confidence interval estimate for the regression coefficients.

**Confidence Interval Estimate for the Regression Slope**

$$b_j \pm ts_{b_j} \qquad (15.10)$$

where:

$$b_j = \text{Point estimate for the regression coefficient for } x_j$$
$$t = \text{Critical } t\text{-value for the specified confidence level}$$
$$s_{b_j} = \text{Standard error of the } j\text{th regression coefficient}$$

**FIGURE 15.8 Excel 2016 Multiple Regression Model Results for First City Real Estate**

**Excel 2016 Instructions**

1. Open file: **First City.xlsx**.
2. Select Homes-Sample 1 worksheet.
3. Select **Data > Data Analysis**.
4. Select **Regression**.
5. Define the y-variable range (Price) and the x-variable range (include row with labels).
6. Click **Labels**.
7. Specify output location.
8. Click **OK**.

95% confidence interval estimates for the regression coefficients

| | A | B | C | D | E | F | G |
|---|---|---|---|---|---|---|---|
| 1 | SUMMARY OUTPUT | | | | | | |
| 2 | Regression Statistics | | | | | | |
| 3 | Multiple R | 0.9034 | | | | | |
| 4 | R Square | 0.8161 | | | | | |
| 5 | Adjusted R Square | 0.8131 | | | | | |
| 6 | Standard Error | 27,350.25 | | | | | |
| 7 | Observations | 319 | | | | | |
| 8 | | | | | | | |
| 9 | ANOVA | | | | | | |
| 10 | | df | SS | MS | F | Significance F | |
| 11 | Regression | 5 | 1.0389E+12 | 2.0778E+11 | 277.77 | 0.0000 | |
| 12 | Residual | 313 | 2.3414E+11 | 7.4804E+08 | | | |
| 13 | Total | 318 | 1.2730E+12 | | | | |
| 14 | | | | | | | |
| 15 | | Coefficients | Standard Error | t Stat | P-value | Lower 95% | Upper 95% |
| 16 | Intercept | 31127.60 | 9539.67 | 3.26 | 0.0012 | 12357.62 | 49897.59 |
| 17 | Sq. Feet | 63.07 | 4.02 | 15.70 | 0.0000 | 55.16 | 70.97 |
| 18 | Age | -1144.44 | 112.78 | -10.15 | 0.0000 | -1366.34 | -922.53 |
| 19 | Bedrooms | -8410.38 | 3002.51 | -2.80 | 0.0054 | -14318.04 | -2502.72 |
| 20 | Bathrooms | 3521.95 | 1581.00 | 2.23 | 0.0266 | 411.23 | 6632.68 |
| 21 | Garage # | 28203.54 | 2858.69 | 9.87 | 0.0000 | 22578.86 | 33828.22 |

The Excel output in Figure 15.8 provides the confidence interval estimate for each regression coefficient. For example, the 95% interval estimate for square feet is

$$\$55.16 \text{———} \$70.97$$

To manually calculate the confidence interval associated with the square feet variable, use Equation 15.10 as

> The Excel 2016 function for the critical $t$-value is
> $= T.INV.2T(alpha,n-k-1)$
> $= T.INV.2T(0.05,313)$

$$b_1 \pm ts_{b_1}$$
$$63.1 \pm 1.968(4.017)$$
$$63.1 \pm 7.90$$
$$\$55.16 \text{———} \$70.97$$

We interpret this interval as follows: If we hold the other variables constant and use a 95% confidence level, then a change in square feet by 1 foot is estimated to generate an average change in home price of between $55.16 and $70.97. Each of the other regression coefficients can be interpreted in the same manner.

## 15.1 EXERCISES

### Skill Development

**15-1.** The following output is associated with a multiple regression model with three independent variables:

|  | df | SS | MS | F | Significance F |
|---|---|---|---|---|---|
| Regression | 3 | 16,646.091 | 5,548.697 | 5.328 | 0.007 |
| Residual | 21 | 21,871.669 | 1,041.508 | | |
| Total | 24 | 38,517.760 | | | |

|  | Coefficients | Standard Error | t Stat | p-value |
|---|---|---|---|---|
| Intercept | 87.790 | 25.468 | 3.447 | 0.002 |
| $x_1$ | −0.970 | 0.586 | −1.656 | 0.113 |
| $x_2$ | 0.002 | 0.001 | 3.133 | 0.005 |
| $x_3$ | −8.723 | 7.495 | −1.164 | 0.258 |

|  | Lower 95% | Upper 95% | Lower 90% | Upper 90% |
|---|---|---|---|---|
| Intercept | 34.827 | 140.753 | 43.966 | 131.613 |
| $x_1$ | −2.189 | 0.248 | −1.979 | 0.038 |
| $x_2$ | 0.001 | 0.004 | 0.001 | 0.004 |
| $x_3$ | −24.311 | 6.864 | −21.621 | 4.174 |

a. What is the regression model associated with these data?
b. Is the model statistically significant?
c. How much of the variation in the dependent variable can be explained by the model?
d. Are all of the independent variables in the model significant? If not, which are not and how can you tell?

e. How much of a change in the dependent variable will be associated with a one-unit change in $x_2$? In $x_3$?
f. Do any of the 95% confidence interval estimates of the slope coefficients contain zero? If so, what does this indicate?

**15-2.** You are given the following estimated regression equation involving a dependent and two independent variables:

$$\hat{y} = 12.67 + 4.14x_1 + 8.72x_2$$

a. Interpret the values of the slope coefficients in the equation.
b. Estimate the value of the dependent variable when $x_1 = 4$ and $x_2 = 9$.

**15-3.** In working for a local retail store, you have developed the following estimated regression equation:

$$\hat{y} = 22,167 - 412x_1 + 818x_2 - 93x_3 - 71x_4$$

where:

$y =$ Weekly sales
$x_1 =$ Local unemployment rate
$x_2 =$ Weekly average high temperature
$x_3 =$ Number of activities in the local community
$x_4 =$ Average price of gasoline per gallon

a. Interpret the values of $b_1$, $b_2$, $b_3$, and $b_4$ in this estimated regression equation.
b. What is the estimated sales if the unemployment rate is 5.7%, the average high temperature is 61°, there are 14 activities, and the average price of gasoline is $3.39 per gallon?

**15-4.** The following correlation matrix is associated with the same data used to build the regression model in Exercise 15-1:

|       | $y$     | $x_1$   | $x_2$   | $x_3$ |
|-------|---------|---------|---------|-------|
| $y$   | 1       |         |         |       |
| $x_1$ | $-0.406$ | 1       |         |       |
| $x_2$ | 0.459   | 0.051   | 1       |       |
| $x_3$ | $-0.244$ | 0.504   | 0.272   | 1     |

Does this output indicate any potential multicollinearity problems with the analysis?

**15-5.** Consider the following set of data:

| $x_1$ | 29 | 48 | 28 | 22 | 28 | 42 | 33 | 26 | 48 | 44 |
|-------|----|----|----|----|----|----|----|----|----|----|
| $x_2$ | 15 | 37 | 24 | 32 | 47 | 13 | 43 | 12 | 58 | 19 |
| $y$   | 16 | 46 | 34 | 26 | 49 | 11 | 41 | 13 | 47 | 16 |

a. Obtain the estimated regression equation.
b. Develop the correlation matrix for this set of data. Select the independent variable whose correlation magnitude is the smallest with the dependent variable. Determine if its correlation with the dependent variable is significant.
c. Determine if the overall model is significant. Use a significance level of 0.05.
d. Calculate the variance inflation factor for each of the independent variables. Indicate if multicollinearity exists between the two independent variables.

**15-6.** Consider the following set of data:

| $x_2$ | 10  | 8  | 11  | 7  | 10 | 11  | 6  |
|-------|-----|----|-----|----|----|-----|----|
| $x_1$ | 50  | 45 | 37  | 32 | 44 | 51  | 42 |
| $y$   | 103 | 85 | 115 | 73 | 97 | 102 | 65 |

a. Obtain the estimated regression equation.
b. Examine the coefficient of determination and the adjusted coefficient of determination. Does it seem that either of the independent variables' addition to $R^2$ does not justify the reduction in degrees of freedom that results from its addition to the regression model? Support your assertions.
c. Conduct a hypothesis test to determine if the dependent variable increases when $x_2$ increases. Use a significance level of 0.025 and the $p$-value approach.
d. Construct a 95% confidence interval for the coefficient of $x_1$.

## Computer Software Exercises

**15-7.** An investment analyst collected data about 20 randomly chosen companies. The data obtained from publicly available sources consisted of the 52-week-high stock prices, the price-to-earnings (PE) ratio, and the market value of the company. These data are in the file titled **Investment**.
a. Produce a regression equation to predict the market value using the 52-week-high stock price and the PE ratio of the company.
b. Determine if the overall model is significant. Use a significance level of 0.05.
c. Suppose a particular company had a 52-week-high stock price of 31 and a PE ratio of 19. Estimate its market value for that time period. (*Note:* Its actual market value for that time period was $1,536.)

**15-8.** Paul Browning owns and operates a medium size company that sells sunglasses and beachwear to retailers in the United States and Canada. He is interested in analyzing the effect of several variables on the weekly sales for his business. He has collected data for a sample of 20 weeks, and the data are in a file called **Browning**.
a. Develop the correlation matrix for these five variables.
b. Select the two variables that are most highly correlated with sales and develop a regression equation that Paul could use to predict weekly sales.
c. Determine if the overall model is statistically significant. Test using an alpha $= 0.10$ level.
d. Test to determine whether the two regression coefficients are statistically significant. Use an alpha level of 0.10 and test using the $p$-value approach.

**15-9.** Refer to Exercise 15-8 and the data file **Browning**.
a. Produce a regression equation that will predict the sales as a function of the other four variables.
b. Determine if the overall model is significant. Use a significance level of 0.05.
c. Conduct a test of hypothesis to discover if any of the independent variables are insignificant. Test using an alpha $= 0.05$.

**15-10.** The National Association of Theatre Owners is the largest exhibition trade organization in the world, representing more than 26,000 movie screens in all 50 states and in more than 20 countries worldwide. Its membership includes the largest cinema chains and hundreds of independent theater owners. It publishes statistics concerning the movie sector of the economy. The file titled **Flicks** contains data on total U.S. box office grosses ($billion), total number of admissions (billion), average U.S. ticket price ($), and number of movie screens for the years 1987–2015 (*source:* National Association of Theatre Owners, www.natoonline.org/data, Feb. 2016).
a. Construct a regression equation that uses the other variables to predict total U.S. box office grosses.
b. Determine if the overall model is significant. Use a significance level of 0.05.
c. Determine the range of plausible values for the change in box office grosses if the average ticket price were increased by $1. Use a confidence level of 95%.

**15-11.** The athletic director of State University is interested in developing a multiple regression model that might be used to explain the variation in attendance at football games at his school. A sample of 16 games was selected from home games played during the past ten seasons. Data for the following factors were determined:

$y$ = Game attendance

$x_1$ = Team win/loss percentage to date

$x_2$ = Opponent win/loss percentage to date

$x_3$ = Games played this season

$x_4$ = Temperature at game time

The data collected are in the file called **Football**.

a. Produce scatter plots for each independent variable versus the dependent variable. Based on the scatter plots, produce a model that you believe represents the relationship between the dependent variable and the group of predictor variables represented in the scatter plots.

b. Based on the correlation matrix developed from these data, comment on whether you think a multiple regression model will be effectively developed from these data.

c. Use the sample data to estimate the multiple regression model that contains all four independent variables.

d. What percentage of the total variation in the dependent variable is explained by the four independent variables in the model?

e. Test to determine whether the overall model is statistically significant. Use $\alpha = 0.05$.

f. Which, if any, of the independent variables is statistically significant? Use a significance level of $\alpha = 0.08$ and the $p$-value approach to conduct these tests.

g. Estimate the standard deviation of the model error and discuss whether this regression model is acceptable as a means of predicting the football attendance at State University at any given game.

h. Define the term *multicollinearity* and indicate the potential problems that multicollinearity can cause for this model. Indicate what, if any, evidence there is of multicollinearity problems with this regression model. Use the variance inflation factor to assist you in this analysis.

i. Develop a 95% confidence interval estimate for each of the regression coefficients and interpret each estimate. Comment on whether the interpretation of the intercept is relevant in this situation.

---

**OUTCOME 6**    **15.2**    **Using Qualitative Independent Variables**

In Example 15-1 involving the First City Real Estate Company, the independent variables were quantitative and ratio level. However, in many situations you may wish to use a qualitative, lower-level variable as an explanatory variable.

If a variable is nominal and numerical codes are assigned to the categories, you already know not to perform mathematical calculations using those data. The results would be meaningless. Yet we may wish to use a variable such as marital status, gender, or geographical location as an independent variable in a regression model. Likewise, if the variable of interest is coded as an ordinal variable, such as education level or job performance ranking, computing means and variances is also inappropriate. Then how do we incorporate qualitative, nominal, and ordinal variables into a multiple regression analysis? The answer lies in using what are called **dummy (or indicator) variables**.

**Dummy Variable**
A variable that is assigned a value equal to either 0 or 1, depending on whether the observation possesses a given characteristic.

For instance, consider the variable gender, which can take on two possible values: male or female. Gender can be converted to a dummy variable as follows:

$$x_1 = 1 \text{ if female}$$
$$x_1 = 0 \text{ if male}$$

Thus, a data set consisting of males and females will have corresponding values for $x_1$ equal to 0s and 1s, respectively. Note that it makes no difference which gender is coded 1 and which is coded 0.

If a categorical variable has more than two mutually exclusive outcome possibilities, you must create multiple dummy variables. Consider the variable marital status, with the following possible outcomes:

never married        married        divorced        widowed

In this case, marital status has four values. To account for all the possibilities, you create three dummy variables, one less than the number of possible outcomes for the original variable. They could be coded as follows:

$$x_1 = 1 \text{ if never married, 0 if not}$$
$$x_2 = 1 \text{ if married, 0 if not}$$
$$x_3 = 1 \text{ if divorced, 0 if not}$$

Note that we don't need the fourth variable because we would know that a person is widowed if $x_1 = 0$, $x_2 = 0$, and $x_3 = 0$. If the person isn't single, married, or divorced, he or she must be widowed. *Always use one fewer dummy variable than categories.* The mathematical reason that the number of dummy variables must be one less than the number of possible responses is called the *dummy variable trap.* Perfect multicollinearity is introduced, and we cannot obtain the least squares regression estimates if the number of dummy variables equals the number of possible categories.

---

**EXAMPLE 15-2**    Incorporating Dummy Variables

**Salary Analysis**    To illustrate the effect of incorporating dummy variables into a regression model, consider the sample data displayed in the scatter plot in Figure 15.9. Suppose that the population from which the sample was selected consists of individuals between the ages of 24 and 60 who are working in the U.S. high-tech industry. Data for annual salary ($y$) and age ($x_1$) are available. The objective is to determine whether a model can be generated to explain the variation in annual salary. Even though age and annual salary are significantly correlated ($r = 0.686$) at the $\alpha = 0.05$ level, the coefficient of determination, $R^2$, is only 47%. Therefore, we would likely search for other independent variables that could help us to further explain the variation in annual salary.

Suppose we can determine which of the 16 people in the sample had a Master of Business Administration (MBA) degree. Figure 15.10 shows the scatter plot for these

**FIGURE 15.9 Salary Data—Scatter Plot**

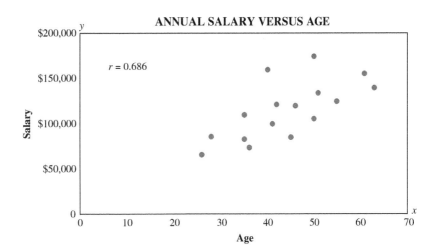

**FIGURE 15.10 Impact of a Dummy Variable**

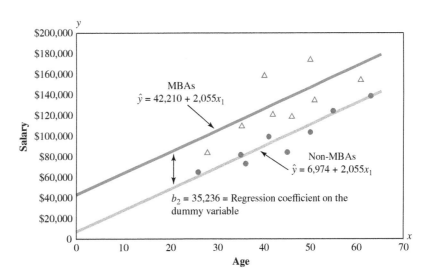

**TABLE 15.2 Salary Data Including MBA Variable**

| Salary ($) | Age | MBA |
|---|---|---|
| 65,000 | 26 | 0 |
| 85,000 | 28 | 1 |
| 74,000 | 36 | 0 |
| 83,000 | 35 | 0 |
| 110,000 | 35 | 1 |
| 160,000 | 40 | 1 |
| 100,000 | 41 | 0 |
| 122,000 | 42 | 1 |
| 85,000 | 45 | 0 |
| 120,000 | 46 | 1 |
| 105,000 | 50 | 0 |
| 135,000 | 51 | 1 |
| 125,000 | 55 | 0 |
| 175,000 | 50 | 1 |
| 156,000 | 61 | 1 |
| 140,000 | 63 | 0 |

same data, with the MBA data represented by triangles. To incorporate a qualitative variable into the analysis, use the following steps:

**STEP 1  Code the qualitative variable as a dummy variable.**
Create a new variable, $x_2$, which is a dummy variable coded as

$$x_2 = 1 \text{ if MBA, } 0 \text{ if not}$$

The data with the new variable are shown in Table 15.2.

**STEP 2  Develop a multiple regression model with the dummy variables incorporated as independent variables.**
The two-independent-variable population multiple regression model has the following form:

$$y = \beta_0 + \beta_1 x_1 + \beta_2 x_2 + \varepsilon$$

Using Excel, we get the regression equation

$$\hat{y} = 6,974 + 2,055x_1 + 35,236x_2$$

Because the dummy variable, $x_2$, has been coded 0 or 1 depending on MBA status, incorporating it into the regression model is like having two simple linear regression lines with the same slopes but different intercepts. For instance, when $x_2 = 0$, the regression equation is

$$\hat{y} = 6,974 + 2,055x_1 + 35,236(0)$$
$$= 6,974 + 2,055x_1$$

This line is shown in Figure 15.10. However, when $x_2 = 1$ (the individual has an MBA), the regression equation is

$$\hat{y} = 6,974 + 2,055x_1 + 35,236(1)$$
$$= 42,210 + 2,055x_1$$

This regression line is also shown in Figure 15.10. As you can see, incorporating the dummy variable affects the regression intercept. In this case, the intercept for individuals with an MBA degree is $35,236 higher than for those without an MBA. We interpret the regression coefficient on this dummy variable as follows: Based on these data, and holding age ($x_1$) constant, we estimate that individuals with an MBA degree make an average of $35,236 per year more in salary than their non-MBA counterparts.

**TRY EXERCISE 15-17** (pg. 595)

---

**BUSINESS APPLICATION**    **Regression Models Using Dummy Variables**

Excel Tutorial

**First City Real Estate** (*continued*)  The regression model developed in Example 15-1 for First City Real Estate showed potential because the overall model was statistically significant. Looking back at Figure 15.8, we see that the model explained nearly 82% of the variation in sales prices for the homes in the sample. All of the independent variables were significant. However, the standard error of the estimate was quite high at $27,350.

The managers have decided to try to improve the model. First, they have decided to add a new variable: area. However, at this point, the only area variable they have access to defines whether the home is in the foothills. Because this is a categorical variable with two possible outcomes (foothills or not foothills), a dummy variable can be created as follows:

$$x_6(\text{area}) = 1 \text{ if foothills, } 0 \text{ if not}$$

Of the 319 homes in the sample, 249 were homes in the foothills and 70 were not. Figure 15.11a shows the revised Excel multiple regression with the variable area added. This model

**FIGURE 15.11a** Excel 2016 Output—First City Real Estate Revised Regression Model

| | A | B | C | D | E | F | G |
|---|---|---|---|---|---|---|---|
| 1 | SUMMARY OUTPUT | | | | | | |
| 2 | *Regression Statistics* | | | | | | |
| 3 | Multiple R | 0.9506 | | | | | |
| 4 | R Square | 0.9036 | | | | | |
| 5 | Adjusted R Square | 0.9018 | | | | | |
| 6 | Standard Error | 19828.3910 | | | | | |
| 7 | Observations | 319 | | | | | |
| 9 | ANOVA | | | | | | |
| 10 | | *df* | *SS* | *MS* | *F* | *Significance F* | |
| 11 | Regression | 6 | 1.1504E+12 | 1.9173E+11 | 487.6514 | 3.1114E-155 | |
| 12 | Residual | 312 | 1.2267E+11 | 3.9317E+08 | | | |
| 13 | Total | 318 | 1.2730E+12 | | | | |
| 15 | | *Coefficients* | *Standard Error* | *t Stat* | *P-value* | *Lower 95%* | *Upper 95%* |
| 16 | Intercept | -6817.34 | 7273.96 | -0.94 | 0.349 | -21129.56 | 7494.88 |
| 17 | Sq. Feet | 63.33 | 2.91 | 21.75 | 0.000 | 57.60 | 69.06 |
| 18 | Age | -333.84 | 94.88 | -3.52 | 0.000 | -520.53 | -147.14 |
| 19 | Bedrooms | -8444.83 | 2176.76 | -3.88 | 0.000 | -12727.82 | -4161.84 |
| 20 | Bathrooms | -949.19 | 1176.55 | -0.81 | 0.420 | -3264.17 | 1365.78 |
| 21 | Garage # | 26246.43 | 2075.75 | 12.64 | 0.000 | 22162.19 | 30330.68 |
| 22 | Area | 62040.98 | 3684.61 | 16.84 | 0.000 | 54791.16 | 69290.81 |

Improved R-square, adjusted R-square, and standard error

Dummy variable coefficient

is an improvement over the original model because the adjusted *R*-squared has increased from 0.8131 to 0.9018 and the standard error of the estimate has decreased from $27,350 to $19,828. The conditional *t*-tests show that all of the regression models' slope coefficients, except the one for the variable bathrooms, differ significantly from 0.

The resulting regression model with the coefficients rounded to one decimal place is

$$\hat{y} = -6{,}817.3 + 63.3(\text{sq. feet}) - 333.8(\text{age}) - 8{,}444.8(\text{bedrooms}) - 949.2(\text{bathrooms}) + 26{,}246.4(\text{garage}) + 62{,}041(\text{area})$$

Because the variable bathrooms is not significant in the presence of the other variables, we can remove the variable and rerun the multiple regression. The resulting model shown in Figure 15.11b is

$$\text{Price} = -7{,}050.2 + 62.5(\text{sq. feet}) - 322(\text{age}) - 8{,}830(\text{bedrooms}) + 26{,}053.9(\text{garage}) + 61{,}370.1(\text{area})$$

Based on the sample data and this regression model, we estimate that a house with the same characteristics (square footage, age, bedrooms, and garage size) is worth an average of $61,370 more if it is located in the foothills (based on how the dummy variable was coded).

**FIGURE 15.11b** Excel 2016 Output—First City Real Estate Revised Regression Model (Bathrooms Removed)

| | A | B | C | D | E | F | G |
|---|---|---|---|---|---|---|---|
| 1 | SUMMARY OUTPUT | | | | | | |
| 2 | *Regression Statistics* | | | | | | |
| 3 | Multiple R | 0.9505 | | | | | |
| 4 | R Square | 0.9034 | | | | | |
| 5 | Adjusted R Square | 0.9019 | | | | | |
| 6 | Standard Error | 19817.33 | | | | | |
| 7 | Observations | 319 | | | | | |
| 9 | ANOVA | | | | | | |
| 10 | | *df* | *SS* | *MS* | *F* | *Significance F* | |
| 11 | Regression | 5 | 1.1501E+12 | 2.3002E+11 | 585.7049 | 1.6906E-156 | |
| 12 | Residual | 313 | 1.2292E+11 | 3.9273E+08 | | | |
| 13 | Total | 318 | 1.2730E+12 | | | | |
| 15 | | *Coefficients* | *Standard Error* | *t Stat* | *P-value* | *Lower 95%* | *Upper 95%* |
| 16 | Intercept | -7050.23 | 7264.18 | -0.97 | 0.333 | -21343.02 | 7242.55 |
| 17 | Sq. Feet | 62.49 | 2.72 | 22.99 | 0.000 | 57.14 | 67.84 |
| 18 | Age | -321.99 | 93.69 | -3.44 | 0.001 | -506.33 | -137.65 |
| 19 | Bedrooms | -8830.01 | 2122.57 | -4.16 | 0.000 | -13006.32 | -4653.69 |
| 20 | Garage # | 26053.86 | 2060.83 | 12.64 | 0.000 | 21999.03 | 30108.70 |
| 21 | Area | 61370.08 | 3587.54 | 17.11 | 0.000 | 54311.35 | 68428.82 |

There are still signals of multicollinearity problems. The coefficient on the independent variable bedrooms is negative, when we might expect homes with more bedrooms to sell for more. Also, the standard error of the estimate is still very large ($19,817) and does not provide the precision the managers need to set prices for homes. More work needs to be done before the model is complete.

---

**Possible Improvements to the First City Appraisal Model** Because the standard error of the estimate is still too high, we look to improve the model. We could start by identifying possible problems:

1. We may be missing useful independent variables.
2. We may have included independent variables that should not have been included.

There is no sure way of determining the correct model specification. However, a recommended approach is for the decision maker to try adding variables or removing variables from the model.

We begin by removing the bedrooms variable, which has an unexpected sign on the regression slope coefficient. (*Note:* If the regression model's sole purpose is for prediction, independent variables with unexpected signs do not automatically pose a problem and do not necessarily need to be deleted. However, insignificant variables should be deleted.) The resulting model is shown in Figure 15.12. Now, all the variables in the model have the expected signs. However, the standard error of the estimate has increased slightly.

Adding other explanatory variables might help. For instance, consider whether the house has central air conditioning, which might affect the sales price. If we can identify whether a house has air conditioning, we could add a dummy variable coded as follows:

$$\text{If air conditioning, } x_7 = 1$$

$$\text{If no air conditioning, } x_7 = 0$$

Other potential independent variables might include a more detailed location variable, a measure of the physical condition, or whether the house has one or two stories. Can you think of others?

The First City example illustrates that even though a regression model may pass the statistical tests of significance, it may not be functional. Good appraisal models can be developed using multiple regression analysis, provided more detail is available about such characteristics as finish quality, landscaping, location, neighborhood characteristics, and so forth. The cost and effort required to obtain these data can be relatively high.

Developing a multiple regression model is more of an art than a science. The real decisions revolve around how to select the best set of independent variables for the model.

---

**FIGURE 15.12 Excel 2016 Output for the First City Real Estate Revised Model**

**Excel 2016 Instructions**

1. Open file: **First City.xlsx**.
2. Select Homes-Sample 2 worksheet.
3. Select **Data > Data Analysis**.
4. Select **Regression**.
5. Define the *y*-variable range (Price) and the *x*-variable range (include row with labels).
6. Click **Labels**.
7. Specify output location.
8. Click **OK**.

|  | A | B | C | D | E | F | G |
|---|---|---|---|---|---|---|---|
| 1 | SUMMARY OUTPUT | | | | | | |
| 2 | *Regression Statistics* | | | | | | |
| 3 | Multiple R | 0.9477 | | | | | |
| 4 | R Square | 0.8981 | | | | | |
| 5 | Adjusted R Square | 0.8968 | | | | | |
| 6 | Standard Error | 20325.37 | | | | | |
| 7 | Observations | 319 | | | | | |
| 8 | | | | | | | |
| 9 | ANOVA | | | | | | |
| 10 | | *df* | *SS* | *MS* | *F* | *Significance F* | |
| 11 | Regression | 4 | 1.1433E+12 | 2.8583E+11 | 691.88 | ########## | |
| 12 | Residual | 314 | 1.2972E+11 | 4.1312E+08 | | | |
| 13 | Total | 318 | 1.2730E+12 | | | | |
| 14 | | | | | | | |
| 15 | | *Coefficients* | *Standard Error* | *t Stat* | *P-value* | *Lower 95%* | *Upper 95%* |
| 16 | Intercept | -25617.33 | 5878.26 | -4.36 | 0.0000 | -37183.09 | -14051.57 |
| 17 | Sq. Feet | 54.83 | 2.05 | 26.74 | 0.0000 | 50.80 | 58.87 |
| 18 | Age | -261.30 | 94.92 | -2.75 | 0.0063 | -448.05 | -74.54 |
| 19 | Garage # | 26753.30 | 2106.62 | 12.70 | 0.0000 | 22608.43 | 30898.17 |
| 20 | Area | 60578.04 | 3674.32 | 16.49 | 0.0000 | 53348.64 | 67807.45 |

All variables are statistically significant and have the expected signs on the coefficients.

## 15.2 EXERCISES

### Skill Development

**15-12.** Consider the following regression model:

$$y = \beta_0 + \beta_1 x_1 + \beta_2 x_2 + \varepsilon$$

where:

$$x_1 = \text{A quantitative variable}$$
$$x_2 = \begin{cases} 1 \text{ if } x_1 < 20 \\ 0 \text{ if } x_1 \geq 20 \end{cases}$$

The following estimated regression equation was obtained from a sample of 30 observations:

$$\hat{y} = 24.1 + 5.8x_1 + 7.9x_2$$

a. Provide the estimated regression equation for instances in which $x_1 < 20$.
b. Determine the value of $\hat{y}$ when $x_1 = 10$.
c. Provide the estimated regression equation for instances in which $x_1 > 20$.
d. Determine the value of $\hat{y}$ when $x_1 = 30$.

**15-13.** You are considering developing a regression equation relating a dependent variable to two independent variables. One of the variables can be measured on a ratio scale, but the other is a categorical variable with two possible levels.
a. Write a multiple regression equation relating the dependent variable to the independent variables.
b. Interpret the meaning of the coefficients in the regression equation.

**15-14.** You are considering developing a regression equation relating a dependent variable to two independent variables. One of the variables can be measured on a ratio scale, but the other is a categorical variable with four possible levels.
a. How many dummy variables do you need to represent the categorical variable?
b. Write a multiple regression equation relating the dependent variable to the independent variables.
c. Interpret the meaning of the coefficients in the regression equation.

**15-15.** A real estate agent wishes to estimate the monthly rental for apartments based on the size (square feet) and the location of the apartments. She chose the following model:

$$y = \beta_0 + \beta_1 x_1 + \beta_2 x_2 + \varepsilon$$

where:

$$x_1 = \text{Square footage of the apartment}$$
$$x_2 = \begin{cases} 1 \text{ if located in town center} \\ 0 \text{ if not located in town center} \end{cases}$$

This linear regression model was fitted to a sample of size 50 to produce the following regression equation:

$$\hat{y} = 145 + 1.2x_1 + 300x_2$$

a. Predict the average monthly rent for an apartment located in the town center that has 1,500 square feet.
b. Predict the average monthly rent for an apartment located in the suburbs that has 1,500 square feet.
c. Interpret $b_2$ in the context of this exercise.

### Business Applications

**15-16.** The Brentworth Utility Corporation is developing a multiple regression model that it plans to use to predict customers' utility usage. The analyst currently has three quantitative variables ($x_1$, $x_2$, and $x_3$) in the model, but she is dissatisfied with the $R$-squared and the estimate of the standard deviation of the model's error. Two variables she thinks might be useful are whether the house has a gas water heater or an electric water heater and whether the house was constructed after the 1974 energy crisis or before.

Provide the format for the model she should use to predict customers' utility usage. Specify the dummy variables to be used, the values these variables could assume, and what each value will represent.

**15-17.** A consumer association was interested in building a regression model to explain the variation in automobile highway mileage. At one stage of the analysis, the estimate of the model took the following form:

$$\hat{y} = 34.20 - 0.003x_1 + 4.56x_2$$

where:

$$x_1 = \text{Vehicle weight}$$
$$x_2 = \begin{cases} 1, \text{ if standard transmission} \\ 0, \text{ if automatic transmission} \end{cases}$$

a. Interpret the regression coefficient for variable $x_1$.
b. Interpret the regression coefficient for variable $x_2$.
c. Present an estimate of a model that would predict the average EPA mileage rating for an automobile with standard transmission as a function of the vehicle's weight.
d. Cadillac's STS-V with automatic transmission weighs approximately 3,973 pounds. Provide an estimate of the average highway mileage you would expect to obtain from this model.
e. Discuss the effect of a dummy variable being incorporated in a regression equation like this one. Use a graph if it is helpful.

**15-18.** A real estate agent wishes to determine the selling price of residences using the size (square feet) and whether the residence is a condominium or a single-family home. A sample of 20 residences was obtained with the following results:

| Price ($) | Type | Square Feet | Price ($) | Type | Square Feet |
|---|---|---|---|---|---|
| 199,700 | Family | 1,500 | 200,600 | Condo | 1,375 |
| 211,800 | Condo | 2,085 | 208,000 | Condo | 1,825 |
| 197,100 | Family | 1,450 | 210,500 | Family | 1,650 |
| 228,400 | Family | 1,836 | 233,300 | Family | 1,960 |
| 215,800 | Family | 1,730 | 187,200 | Condo | 1,360 |
| 190,900 | Condo | 1,726 | 185,200 | Condo | 1,200 |
| 312,200 | Family | 2,300 | 284,100 | Family | 2,000 |
| 313,600 | Condo | 1,650 | 207,200 | Family | 1,755 |
| 239,000 | Family | 1,950 | 258,200 | Family | 1,850 |
| 184,400 | Condo | 1,545 | 203,100 | Family | 1,630 |

a. Produce a regression equation to predict the selling price for residences using a model of the following form:

$$y_i = \beta_0 + \beta_1 x_1 + \beta_2 x_2 + \varepsilon$$

where:

$$x_1 = \text{Square footage} \quad \text{and} \quad x_2 = \begin{cases} 1 \text{ if a condo} \\ 0 \text{ if a single-family home} \end{cases}$$

b. Interpret the parameters $\beta_1$ and $\beta_2$ in the model given in part a.

c. Produce an equation that describes the relationship between the selling price and the square footage of (1) condominiums and (2) single-family homes.

d. Conduct a test of hypothesis to determine if the relationship between the selling price and the square footage is different between condominiums and single-family homes.

## Computer Software Exercises

**15-19.** For years, import cars have held an advantage over domestic automobiles according to data collected by J.D. Power and Associates, which generates a widely respected report on initial quality. The 2015 data are in the file **PP100**. Initial quality is measured by the number of problems per 100 vehicles (PP100). (*Source:* J.D. Power, "Korean brands lead industry in initial quality, while Japanese brands struggle to keep up with pace of improvement," www.jdpower.com, June 17, 2015.)

a. Produce a regression equation to predict PP100 for vehicles in the model

$$y_i = \beta_0 + \beta_1 x_1 + \beta_2 x_2 + \varepsilon$$

where $x_1 = 1$ if considered an Asian brand and 0 if not and $x_2 = 1$ if considered a European car and 0 if not.

b. Interpret the parameters $\beta_0$, $\beta_1$, and $\beta_2$ in the model given in part a.

c. Conduct a test of hypothesis to determine if the regression model developed in part a is statistically significant. Test using an alpha = 0.05.

**15-20.** The gross state product (GSP) is the sales of all goods and services originating in a state minus the costs of any goods or services used to generate the sales. The file **State GSP** contains the 2015 GSP for all 50 U.S. states along with several other variables, including gasoline prices, the 2015 population estimate, 2014 median household income, and the political party of the state's governor. (See the Description Tab on the data file for sources for these variables.)

a. Develop a correlation matrix for the variables in this data file. Identify the variable that has the highest correlation with the GSP variable.

b. Construct a simple linear regression model for explaining the variation in state GSP using the highest correlated independent variable. Test whether the regression coefficient for this variable is statistically significant using alpha = 0.05.

c. Construct a multiple linear regression model using state GSP as the dependent variable. What percent of the variation does the model explain?

d. Referring to the model developed in part c, indicate whether the overall model is significant and which, if any, of the independent variables, are significant. Use alpha = 0.05.

e. Construct a 95% confidence interval estimate for the regression coefficient on the dummy variable, governor political party, and interpret.

**15-21.** The Gilmore Accounting firm, in an effort to explain variation in client profitability, collected the data found in the file called **Gilmore**, where:

$y = $ Net profit earned from the client
$x_1 = $ Number of hours spent working with the client
$x_2 = $ Type of client:
    1, if manufacturing
    2, if service
    3, if governmental

a. Develop a scatter plot of each independent variable versus the client income variable. Comment on what, if any, relationship appears to exist in each case.

b. Run a simple linear regression analysis using only variable $x_1$ as the independent variable. Describe the resulting estimate fully.

c. Test to determine if the number of hours spent working with the client is useful in predicting client profitability.

**15-22.** Refer again to the data from the Gilmore Accounting firm found in the data file **Gilmore** (see Exercise 15-21).

a. Incorporate the client type into the regression analysis using dummy variables. Describe the resulting multiple regression estimate.

b. Test to determine if this model is useful in predicting the net profit earned from the client.

c. Test to determine if the number of hours spent working with the client is useful in this model in predicting the net profit earned from a client.

d. Considering the tests you have performed, construct a model and its estimate for predicting the net profit earned from the client.

e. Predict the average difference in profit if the client is governmental versus in manufacturing. Also state this in terms of a 95% confidence interval estimate.

**15-23.** There has been conjecture by high school administrators at a particular high school in Delaware about whether a relationship exists between the average math SAT score and the average verbal SAT score and the gender of the student taking the SAT examination. To help better understand the situation, suppose data have been collected on a sample of students in the school. Consider the following relationship:

$$y_i = \beta_0 + \beta_1 x_1 + \beta_2 x_2 + \varepsilon$$

where:

$$x_1 = \text{Average verbal SAT score} \quad \text{and} \quad x_2 = \begin{cases} 0 \text{ if female} \\ 1 \text{ if male} \end{cases}$$

a. Use the data provided in the file **MathSAT** to compute the linear regression equation to predict the average math SAT score using the gender and the average verbal SAT score of the students taking the SAT examination.

b. Interpret the parameters in the model.

c. Conduct a hypothesis test to determine if the gender of the student taking the SAT examination is a significant predictor of the student's average math SAT score for a given average verbal SAT score.

d. Predict the average math SAT score of female students who have an average verbal SAT score of 500.

**OUTCOME 7**    **15.3**    ## Working with Nonlinear Relationships

Section 14.1 in Chapter 14 showed that there are a variety of ways in which two variables can be related. Correlation and regression analysis techniques are tools for measuring and modeling linear relationships between variables. Many situations in business have a linear relationship between two variables, and regression equations that model that relationship are appropriate to use in these situations. However, there are also many instances in which the relationship between two variables is curvilinear rather than linear. For instance, demand for electricity has grown at an almost exponential rate relative to the population growth in some areas. Advertisers believe that a diminishing returns relationship occurs between sales and advertising if advertising is allowed to grow too large. These two situations are shown in Figures 15.13 and 15.14, respectively. They represent just two of the great many possible curvilinear relationships that exist between two variables.

As you will soon see, models with nonlinear relationships become more complicated than models showing only linear relationships. Although complicated models are sometimes

**FIGURE 15.13 Exponential Relationship of Increased Demand for Electricity versus Population Growth**

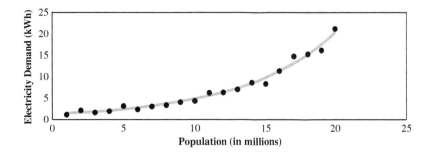

**FIGURE 15.14 Diminishing Returns Relationship of Advertising versus Sales**

necessary, decision makers should use them with caution for two reasons. First, people use decision aids they understand and don't use those they don't understand. So, the more complicated a model is, the less likely it is to be used. Second, the scientific principle of parsimony suggests using the simplest model possible that provides a reasonable fit of the data, because complex models typically do not reflect the underlying phenomena that produce the data in the first place.

This section provides a brief introduction to how linear regression analysis can be used in dealing with curvilinear relationships. To model such curvilinear relationships, we must incorporate terms into the multiple regression model that will create "curves" in the model we are building. Including terms whose independent variable has an exponent larger than 1 generates these curves. When a model possesses such terms, we refer to it as a *polynomial model*. The general equation for a polynomial with one independent variable is given in Equation 15.11.

**Polynomial Population Regression Model**

$$y = \beta_0 + \beta_1 x + \beta_2 x^2 + \cdots + \beta_p x^p + \varepsilon \tag{15.11}$$

where:

$\beta_0 =$ Population's regression constant
$\beta_j =$ Population's regression coefficient for variable $x^j$; $j = 1, 2, \ldots, p$
$p =$ Order (or degree) of the polynomial
$\varepsilon =$ Model error

The order, or degree, of the model is determined by the largest exponent of the independent variable in the model. For instance, the model

$$y = \beta_0 + \beta_1 x + \beta_2 x^2 + \varepsilon$$

is a second-order polynomial because the largest exponent in any term of the polynomial is 2. You will note that this model contains terms of all orders less than or equal to 2. A polynomial with this property is said to be a *complete* polynomial. Therefore, the previous model is referred to as a complete *second-order regression model*. A second-order model produces a parabola. The parabola opens either upward ($\beta_2 > 0$) or downward ($\beta_2 < 0$), as shown in Figure 15.15. You will notice that the models in Figures 15.13, 15.14, and 15.15 have a single curve.

As more curves appear in the data, the order of the polynomial must be increased. A general (complete) third-order polynomial is given by the equation

$$y = \beta_0 + \beta_1 x + \beta_2 x^2 + \beta_3 x^3 + \varepsilon$$

This produces a curvilinear model that reverses the direction of the initial curve to produce a second curve, as shown in Figure 15.16. Note that there are two curves in the third-order model. In general, a $p$th-order polynomial exhibits $p - 1$ curves.

**FIGURE 15.15 Second-Order Regression Models**

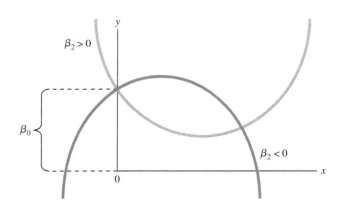

**FIGURE 15.16 Third-Order Regression Models**

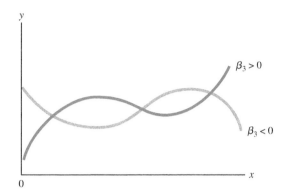

Although polynomials of all orders exist in the business sector, perhaps second-order polynomials are the most common. Sharp reversals in the curvature of a relationship between variables in the business environment usually point to some unexpected or perhaps severe changes that were not foreseen. The vast majority of organizations try to avoid such reverses. For this reason, and the fact that this is an introductory business statistics book, we will direct most of our attention to second-order polynomials.

The following examples illustrate two of the most common instances in which curvilinear relationships can be used in decision making. They should give you an idea of how to approach similar situations.

OUTCOME **7**    EXAMPLE 15-3    **Modeling Curvilinear Relationships**

**Ashley Investment Services**   Since the downturn in the stock market during the summer and fall of 2008, analysts at Ashley Investment Services have been extra busy analyzing new investment opportunities. The director of personnel has noticed an increased number of employees suffering from "burnout," in which physical and emotional fatigue hurt job performance. Although he cannot change the job's pressures, he has read that the more time a person spends socializing with coworkers away from the job, the higher the degree of burnout. With the help of the human resources lab at the local university, the personnel director has administered a questionnaire to company employees. A burnout index has been computed from the responses to the survey. Likewise, the survey responses are used to determine quantitative measures of socialization. Sample data from questionnaires are contained in the file **Ashley**. The following steps can be used to model the relationship between the socialization index and the burnout index for Ashley employees:

STEP 1    **Specify the model by determining the dependent and potential independent variables.**
The dependent variable is the burnout index. The company wishes to explain the variation in burnout level. One potential independent variable is the socialization index.

STEP 2    **Formulate the model.**
We begin by proposing that a linear relationship exists between the two variables. Figure 15.17 shows the linear regression analysis results using Excel. The correlation between the two variables is $r = 0.818$, which is statistically different from zero at any reasonable significance level. The estimate of the population linear regression model shown in Figure 15.17 is

$$\hat{y} = -66.164 + 9.589x$$

STEP 3    **Perform diagnostic checks on the model.**
The sample data and the regression line are plotted in Figure 15.18. The line appears to fit the data. However, a closer inspection reveals instances in which several consecutive points lie above or below the line. The points are not

**FIGURE 15.17 Excel 2016 Output of a Simple Linear Regression for Ashley Investment Services**

|   | A | B | C | D | E | F | G |
|---|---|---|---|---|---|---|---|
| 1 | SUMMARY OUTPUT | | | | | | |
| 2 | *Regression Statistics* | | | | | | |
| 3 | Multiple R | 0.818 | | | | | |
| 4 | R Square | 0.669 | | | | | |
| 5 | Adjusted R Square | 0.651 | | | | | |
| 6 | Standard Error | 159.992 | | | | | |
| 7 | Observations | 20 | | | | | |
| 9 | ANOVA | | | | | | |
| 10 | | *df* | *SS* | *MS* | *F* | *Significance F* | |
| 11 | Regression | 1 | 932504.166 | 932504.166 | 36.430 | 0.000 | |
| 12 | Residual | 18 | 460751.584 | 25597.310 | | | |
| 13 | Total | 19 | 1393255.750 | | | | |
| 15 | | *Coefficients* | *Standard Error* | *t Stat* | *P-value* | *Lower 95%* | *Upper 95%* |
| 16 | Intercept | -66.164 | 112.444 | -0.588 | 0.564 | -302.400 | 170.073 |
| 17 | Socialization Measure | 9.589 | 1.589 | 6.036 | 0.000 | 6.251 | 12.927 |

Regression coefficients

randomly dispersed around the regression line, as should be the case given the regression analysis assumptions.

We can use an *F*-test to test whether a regression model explains a significant amount of variation in the dependent variable.

$$H_0: \rho^2 = 0$$
$$H_A: \rho^2 > 0$$

From the ANOVA section of the output in Figure 15.17,

$$F = 36.43$$

which has *p*-value $\approx 0.000$.

Thus, we conclude that the simple linear model is statistically significant. However, we should also examine the data to determine if a better model can be developed based on a curvilinear relationship.

STEP 4 **Model the curvilinear relationship.**
Finding instances of nonrandom patterns in the residuals for a regression model indicates the possibility of using a curvilinear relationship rather than a linear one. One possible approach to modeling the curvilinear nature of the data in the Ashley Investment Services example is with the use of polynomials. From Figure 15.18, we can see that there appears to be one curve in the data. This suggests fitting the second-order polynomial

$$y = \beta_0 + \beta_1 x + \beta_2 x^2 + \varepsilon$$

**FIGURE 15.18 Plot of Regression Line for the Ashley Investment Services Example**

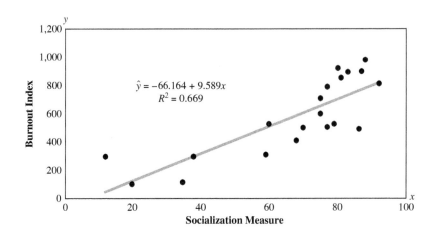

**FIGURE 15.19** Excel 2016 Output of a Second-Order Polynomial Fit for Ashley Investment Services

Excel 2016 Instructions

1. Open file: **Ashley.xlsx**.
2. Use Excel equation to create a new variable in column C (i.e., for the first data value in row 2, use =A2^2; then copy down to the last row).
3. Rearrange columns so both independent variables are together.
4. Select **Data > Data Analysis**.
5. Select **Regression**.
6. Define the y-variable range (Burnout) and the x-variable range (include Socialization Measure and the new variable—include row with labels).
7. Click **Labels**.
8. Specify output location.
9. Click **OK**.

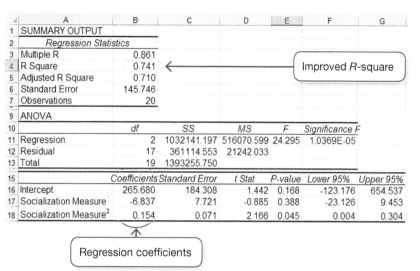

**FIGURE 15.20** Plot of Second-Order Polynomial Model for Ashley Investment Services

Before fitting the estimate for this population model, you will need to create the new independent variable by squaring the socialization measure variable. In Excel, use the formula option to create the new variable. Figure 15.19 shows the output after fitting this second-order polynomial model.

STEP 5 **Perform diagnostics on the revised curvilinear model.**
Notice that the second-order polynomial provides a model whose estimated regression equation has an $R^2$ of 0.741. This is higher than the $R^2$ of 0.669 for the linear model. Figure 15.20 shows the plot of the second-order polynomial model. Comparing Figure 15.20 with Figure 15.18, we can see that the polynomial model does appear to fit the sample data better than the linear model.

TRY EXERCISE 15-24 (pg. 608)

## Analyzing Interaction Effects

BUSINESS APPLICATION    Dealing with Interaction

**Ashley Investment Services** (*continued*)   Referring to Example 15-3 involving Ashley Investment Services, the director of personnel wondered if the effects of burnout differ between male and female workers. He therefore identified the gender of the previously surveyed employees (see file **Ashley-2**). A multiple scatter plot of the data appears as Figure 15.21.

Excel Tutorial

**FIGURE 15.21** **Excel 2016 Multiple Scatter Plot for Ashley Investment Services**

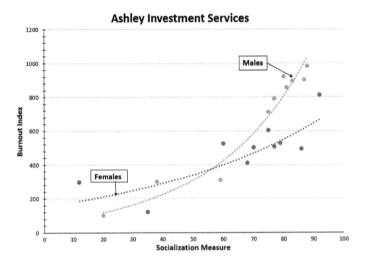

**Excel 2016 Instructions**

1. Open file: **Ashley-2.xlsx**.
2. Select Socialization Measure and Burnout Index columns.
3. Select **Insert** tab.
4. Select **Scatter > Scatter with only Markers**.
5. Select **Chart** and click the right mouse button—choose **Select Data**.
6. Click on **Add** on the **Legend Entry (Series)** section.
7. Enter **Series Name**—(Females)—for **Series X Values**, select data from Socialization column for row corresponding to females (rows 2–11). **For Series Y Values**, select data from the Burnout column corresponding to females (rows 2–11).
8. Repeat step 7 for males.
9. Select **Layout** tab to remove legend and to add chart and axis titles.
10. Select data points for males—right click and select **Add Trendline > Exponential**.
11. Repeat step 10 for females.
12. Click **Design** tab and click **Move Chart**.

The personnel director tried to determine the relationship between the burnout index and socialization measure for men and women. The graphical result is presented in Figure 15.21. Note that both relationships appear to be curvilinear, with similarly shaped curves. As we showed earlier, curvilinear shapes often can be modeled by the second-order polynomial

$$\hat{y} = b_0 + b_1 x_1 + b_2 x_1^2$$

However, the regression equations that estimate this second-order polynomial for men and women are not the same. The two equations seem to have different locations and different rates of curvature. Whether an employee is a man or woman seems to change the basic relationship between burnout index ($y$) and socialization measure ($x_1$). To represent this difference, the equation's coefficients $b_0$, $b_1$, and $b_2$ must be different for male and female employees. Thus, we could use two models, one for each gender. Alternatively, we could use one model for both male and female employees by incorporating a dummy independent variable with two levels, shown as

$$x_2 = 1 \text{ if male, } 0 \text{ if female}$$

As $x_2$ changes values from 0 to 1, it affects the values of the coefficients $b_0$, $b_1$, and $b_2$. Suppose the director fitted the second-order model for the female employees only. He obtained the following regression equation:

$$\hat{y} = 291.70 - 4.62 x_1 + 0.102 x_1^2$$

**Interaction**
The case in which one independent variable (such as $x_2$) affects the relationship between another independent variable ($x_1$) and the dependent variable ($y$).

The equation for only male employees was

$$\hat{y} = 149.59 - 4.40 x_1 + 0.160 x_1^2$$

To explain how a change in gender can cause this kind of change, we introduce **interaction**. In our example, gender ($x_2$) interacts with the relationship between socialization measure $(x_1)$ and burnout index $(y)$. The question is How do we obtain the interaction terms to model such a relationship? To answer this question, we first obtain the model for the basic relationship between the $x_1$ and the $y$ variables. The population model is

$$y = \beta_0 + \beta_1 x_1 + \beta_2 x_1^2 + \varepsilon$$

To obtain the interaction terms, we multiply the terms on the right-hand side of this model by the variable that is interacting with this relationship between $y$ and $x_1$. In this case, that interacting variable is $x_2$. Then the interaction terms are

$$\beta_3 x_2 + \beta_4 x_1 x_2 + \beta_5 x_1^2 x_2$$

Notice that we have changed the coefficient subscripts so we do not duplicate those in the original model. Then the interaction terms are added to the original model to produce the **composite model**:

$$y = \beta_0 + \beta_1 x_1 + \beta_2 x_1^2 + \beta_3 x_2 + \beta_4 x_1 x_2 + \beta_5 x_1^2 x_2 + \varepsilon$$

**Composite Model**
The model that contains both the basic terms and the interaction terms.

Note that the model for women is obtained by substituting $x_2 = 0$ into the composite model. This gives

$$y = \beta_0 + \beta_1 x_1 + \beta_2 x_1^2 + \beta_3(0) + \beta_4 x_1(0) + \beta_5 x_1^2(0) + \varepsilon$$
$$= \beta_0 + \beta_1 x_1 + \beta_2 x_1^2 + \varepsilon$$

Similarly, for men we substitute the value $x_2 = 1$. The model then becomes

$$y = \beta_0 + \beta_1 x_1 + \beta_2 x_1^2 + \beta_3(1) + \beta_4 x_1(1) + \beta_5 x_1^2(1) + \varepsilon$$
$$= (\beta_0 + \beta_3) + (\beta_1 + \beta_4)x_1 + (\beta_2 + \beta_5)x_1^2 + \varepsilon$$

This process illustrates how the coefficients are changed for different values of $x_2$ and, therefore, how $x_2$ is interacting with the relationship between $x_1$ and $y$. Once we know $\beta_3, \beta_4$, and $\beta_5$, we know the effect of the interaction of gender on the original relationship between the burnout index $(y)$ and the socialization measure $(x_1)$. To estimate the composite model, we need to create the required variables, as shown in Figure 15.22. Figure 15.23 shows the regression for the composite model. The estimate for the composite model is

$$\hat{y} = 291.706 - 4.615x_1 + 0.102x_1^2 - 142.113x_2 + 0.215x_1x_2 + 0.058x_1^2x_2$$

We obtain the model for females by substituting $x_2 = 0$, which gives

$$\hat{y} = 291.706 - 4.615x_1 + 0.102x_1^2 - 142.113(0) + 0.215x_1(0) + 0.058x_1^2(0)$$
$$\hat{y} = 291.706 - 4.615x_1 + 0.102x_1^2$$

**FIGURE 15.22 Excel 2016 Data Preparation for Estimating Interactive Effects for Second-Order Model for Ashley Investment Services**

**Excel 2016 Instructions**
1. Open file: **Ashley-2.xlsx**.
2. Use Excel formulas to create new variables in columns C, E, and F.

| | A | B | C | D | E | F |
|---|---|---|---|---|---|---|
| 1 | | Socialization | | | | |
| 2 | Burnout Index $y$ | Measure $x_1$ | Squared $x_1^2$ | Gender $x_2$ | $x_1x_2$ | $x_1^2x_2$ |
| 3 | 296 | 12 | 144 | 0 | 0 | 0 |
| 4 | 120 | 35 | 1225 | 0 | 0 | 0 |
| 5 | 525 | 60 | 3600 | 0 | 0 | 0 |
| 6 | 410 | 68 | 4624 | 0 | 0 | 0 |
| 7 | 501 | 70 | 4900 | 0 | 0 | 0 |
| 8 | 600 | 75 | 5625 | 0 | 0 | 0 |
| 9 | 506 | 77 | 5929 | 0 | 0 | 0 |
| 10 | 527 | 79 | 6241 | 0 | 0 | 0 |
| 11 | 493 | 86 | 7396 | 0 | 0 | 0 |
| 12 | 810 | 92 | 8464 | 0 | 0 | 0 |
| 13 | 100 | 20 | 400 | 1 | 20 | 400 |
| 14 | 300 | 38 | 1444 | 1 | 38 | 1444 |
| 15 | 310 | 59 | 3481 | 1 | 59 | 3481 |
| 16 | 709 | 75 | 5625 | 1 | 75 | 5625 |
| 17 | 791 | 77 | 5929 | 1 | 77 | 5929 |
| 18 | 920 | 80 | 6400 | 1 | 80 | 6400 |
| 19 | 855 | 81 | 6561 | 1 | 81 | 6561 |
| 20 | 892 | 83 | 6889 | 1 | 83 | 6889 |
| 21 | 900 | 87 | 7569 | 1 | 87 | 7569 |
| 22 | 980 | 88 | 7744 | 1 | 88 | 7744 |

Excel 2016 Instructions

1. Open file: **Ashley-2.xlsx**.
2. Use Excel equation to create new variables (see Figure 15.22 Excel 2016 Instructions).
3. Rearrange columns so all independent variables are together.
4. Select **Data > Data Analysis**.
5. Select **Regression**.
6. Define the y-variable range (Burnout) and the x-variable range (include Socialization Measure and the new variables—include row with labels).
7. Click **Labels**.
8. Specify output location.
9. Click **OK**.

**FIGURE 15.23 Excel 2016 Composite Model for Ashley Investment Services**

|  | A | B | C | D | E | F | G |
|---|---|---|---|---|---|---|---|
| 1 | SUMMARY OUTPUT | | | | | | |
| 2 | *Regression Statistics* | | | | | | |
| 3 | Multiple R | 0.953 | | | | | |
| 4 | R Square | 0.909 | | | | | |
| 5 | Adjusted R Square | 0.876 | | | | | |
| 6 | Standard Error | 95.363 | | | | | |
| 7 | Observations | 20 | | | | | |
| 9 | ANOVA | | | | | | |
| 10 | | *df* | *SS* | *MS* | *F* | *Significance F* | |
| 11 | Regression | 5 | 1265938.35 | 253187.67 | 27.84 | 0.0000 | |
| 12 | Residual | 14 | 127317.40 | 9094.10 | | | |
| 13 | Total | 19 | 1393255.75 | | | | |
| 15 | | *Coefficients* | *Standard Error* | *t Stat* | *P-value* | *Lower 95%* | *Upper 95%* |
| 16 | Intercept | 291.706 | 143.207 | 2.037 | 0.061 | -15.441 | 598.854 |
| 17 | Socialization Measure x1 | -4.615 | 6.060 | -0.762 | 0.459 | -17.612 | 8.382 |
| 18 | Socialization Squared x12 | 0.102 | 0.057 | 1.772 | 0.098 | -0.021 | 0.224 |
| 19 | Gender x2 | -142.113 | 267.112 | -0.532 | 0.603 | -715.011 | 430.785 |
| 20 | x1x2 | 0.215 | 11.153 | 0.019 | 0.985 | -23.707 | 24.136 |
| 21 | x12x2 | 0.058 | 0.101 | 0.569 | 0.578 | -0.160 | 0.275 |

Regression coefficients for the composite model

For males, we substitute $x_2 = 1$, giving

$$\hat{y} = 291.706 - 4.615x_1 + 0.102x_1^2 - 142.113(1) + 0.215x_1(1) + 0.058x_1^2(1)$$
$$\hat{y} = 149.593 - 4.40x_1 + 0.160x_1^2$$

Note that these equations for male and female employees are the same as those we found earlier when we generated two separate regression models, one for each gender.

In this example, we have looked at a case in which a dummy variable interacts with the relationship between another independent variable and the dependent variable. However, the interacting variable need not be a dummy variable. It can be any independent variable. Also, strictly speaking, interaction is not said to exist if the only effect of the interaction variable is to change the y intercept of the equation, relating another independent variable to the dependent variable. Therefore, when you examine a scatter plot to detect interaction, you are trying to determine if the relationships produced, when the interaction variable changes values, are parallel or not. If the relationships are parallel, that indicates that only the y intercept is being affected by the change of the interacting variable and that interaction does not exist. Figure 15.24 demonstrates this concept graphically.

**FIGURE 15.24 Graphical Evidence of Interaction**

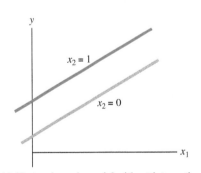

**(a) First-order polynomial without interaction**

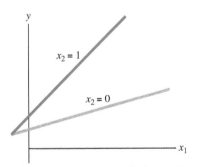

**(b) First-order polynomial with interaction**

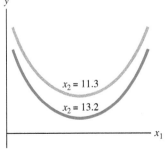

**(c) Second-order polynomial without interaction**

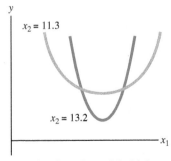

**(d) Second-order polynomial with interaction**

## Partial *F*-Test

So far we have presented the procedures required to test the significance of either one or all of the coefficients in a regression model. For instance, in Example 15-3, we used a hypothesis test to determine that a second-order model involving the socialization measure fit the sample data better than the linear model. Comparing the *R*-squared values was the mechanism used to establish this. We could have determined whether both the linear and quadratic components were useful in predicting the burnout index level by testing the hypothesis $H_0: \beta_1 = \beta_2 = 0$. However, more complex models occur. The interaction model involving Ashley Investment Services containing five predictor variables was

$$y_i = \beta_0 + \beta_1 x_{1i} + \beta_2 x_{1i}^2 + \beta_3 x_{2i} + \beta_4 x_{1i} x_{2i} + \beta_5 x_{1i}^2 x_{2i} + \varepsilon$$

Two of these predictor variables ($x_{1i} x_{2i}$ and $x_{1i}^2 x_{2i}$) in the model would indicate that interaction is evident in this regression model. If the two interaction variables were absent, the model would be

$$y_i = \beta_0 + \beta_1 x_{1i} + \beta_2 x_{1i}^2 + \beta_3 x_{2i} + \varepsilon$$

To determine whether there is statistical evidence of interaction, we must determine if the coefficients of the interaction terms are all equal to 0. If they are, there is no interaction. Otherwise, at least some interaction exists. For the Ashley Investment example, we test the hypotheses

$$H_0: \beta_4 = \beta_5 = 0$$
$$H_A: \text{Either } \beta_4 \text{ or } \beta_5 \neq 0$$

Earlier in this chapter, we introduced procedures for testing whether all of the coefficients of a model equaled 0. In that case, you use the analysis of variance *F*-test found in the ANOVA section of the Excel output. However, to test whether there is significant interaction, we must test more than one but fewer than all the regression coefficients. The method for doing this is the *partial F*-test. This test relies on the fact that if we are given a choice between two models, one model is a better fit if its sum of squares error (*SSE*) is significantly smaller than for the other model. Therefore, to determine if interaction exists in our model, we must obtain the *SSE* for the model with the interaction terms and for the model without the interaction terms. The model without the interaction terms is called the *reduced model*. The model containing the interaction terms is called the *complete model*. We will denote the respective sum of squares as $SSE_R$ and $SSE_C$.

The test is based on the idea that the *SSE* will be significantly reduced if not all of the regression coefficients being tested equal zero. Of course, if the *SSE* is significantly reduced, then $SSE_R - SSE_C$ must be significantly different from zero. To determine if this difference is significantly different from zero, we use the *partial F*-test statistic given by Equation 15.12.

**Partial *F*-Test Statistic**

$$F = \frac{(SSE_R - SSE_C)/(c - r)}{MSE_C}$$

(**15.12**)

where:

$$MSE_C = \text{Mean square error for the complete model}$$
$$= SSE_C/(n - c - 1)$$
$$r = \text{The number of coefficients in the reduced model}$$
$$c = \text{The number of coefficients in the complete model}$$
$$n = \text{Sample size}$$

The numerator of this test statistic is basically the average *SSE* per degree of freedom reduced by including the coefficients being tested in the model. This is compared to the average *SSE* per degree of freedom for the complete model. If these averages are significantly different, then the null hypothesis is rejected. This test statistic has an *F*-distribution whose numerator degrees of freedom ($D_1$) equals the number of parameters being tested ($c - r$) and whose denominator degrees of freedom equals the degrees of freedom for the complete model ($n - c - 1$).

We are now prepared to determine if the director's data indicate a significant interaction between gender and the relationship between the socialization measure and the burnout index. In order to conduct the test of hypothesis, the director produced regression equations for both models (Figures 15.25a and 15.25b). He obtained $SSE_C$ and $MSE_C$ from Figure 15.25a and $SSE_R$ from Figure 15.25b. He was then able to conduct the hypothesis test to determine if there was any interaction. Figure 15.25c displays this test.

Since the null hypothesis was rejected, we can conclude that interaction does exist in this model. Apparently, gender of the employee does affect the relationship between the burnout index and the socialization measure. The relationship between the burnout index and the socialization measure is different for men versus women.

You must be very careful with interpretations of regression coefficients when interaction exists. Notice that the equation that contains interaction terms is given by

$$\hat{y}_i = 291.706 - 4.615x_{1i} + 0.102x_{1i}^2 - 142.113x_{2i} + 0.215x_{1i}x_{2i} + 0.058x_{1i}^2x_{2i}$$

When interpreting the coefficient $b_1$, you may be tempted to say that the burnout index decreases by an average of 4.61 units for every unit the socialization measure ($x_{1i}$) increases, *holding all other predictor variables constant.* However, this is not true; three other components of this regression equation contain $x_{1i}$. When $x_{1i}$ increases by one unit, $x_{1i}^2$ will also increase. In addition, the interaction terms also contain $x_{1i}$, and therefore those terms will change as well. This being the case, every time the variable $x_2$ changes, the rate of change of the interaction terms is also affected. Perhaps you can see this more clearly if we rewrite the equation as

$$\hat{y}_i = (291.706 - 142.113x_{2i}) + (0.215x_{2i} - 4.615)x_{1i} + (0.102 + 0.058x_{2i})x_{1i}^2$$

In this form, you can see that the coefficients of $x_{1i}$ and $x_{1i}^2$ change whenever $x_2$ changes. Thus, the interpretation of any of these components depends on the value $x_2$ as well as $x_{1i}$. Whenever interaction or higher-order components are present, you should be very careful in your attempts to interpret the results of your regression analysis.

**Excel 2016 Instructions**

1. Open file: **Ashley-2.xlsx**.
2. Use Excel equation to create new variables (see Figure 15.22 Excel 2016 Instructions).
3. Rearrange columns so all independent variables are together.
4. Select **Data > Data Analysis**.
5. Select **Regression**.
6. Define the *y*-variable range (Burnout) and the *x*-variable range (include Socialization Measure, Gender and the new variables—include row with labels).
7. Click **Labels**.
8. Specify output location.
9. Click **OK**.

**FIGURE 15.25a Excel 2016 Sum of Squares for the Complete Model**

| | A | B | C | D | E | F | G |
|---|---|---|---|---|---|---|---|
| 1 | SUMMARY OUTPUT | | | | | | |
| 2 | *Regression Statistics* | | | | | | |
| 3 | Multiple R | 0.953 | | | | | |
| 4 | R Square | 0.909 | | | | | |
| 5 | Adjusted R Square | 0.876 | | | | | |
| 6 | Standard Error | 95.363 | | | | | |
| 7 | Observations | 20 | | | | | |
| 9 | ANOVA | | | | | | |
| 10 | | | *df* | *SS* | *MS* | *F* | *Significance F* |
| 11 | Regression | | 5 | 1265938.35 | 253187.67 | 27.84 | 0.0000 |
| 12 | Residual | | 14 | 127317.40 | 9094.10 | | |
| 13 | Total | | 19 | 1393255.75 | | | |
| 15 | | | *Coefficients* | *Standard Error* | *t Stat* | *P-value* | *Lower 95%* | *Upper 95%* |
| 16 | Intercept | | 291.706 | 143.207 | 2.037 | 0.061 | -15.441 | 598.854 |
| 17 | Socialization Measure x1 | | -4.615 | 6.060 | -0.762 | 0.459 | -17.612 | 8.382 |
| 18 | Socialization Squared x12 | | 0.102 | 0.057 | 1.772 | 0.098 | -0.021 | 0.224 |
| 19 | Gender x2 | | -142.113 | 267.112 | -0.532 | 0.603 | -715.011 | 430.785 |
| 20 | x1x2 | | 0.215 | 11.153 | 0.019 | 0.985 | -23.707 | 24.136 |
| 21 | x12x2 | | 0.058 | 0.101 | 0.569 | 0.578 | -0.160 | 0.275 |

$SSE_C$

$MSE_C$

**FIGURE 15.25b Excel 2016 Sum of Squares for the Reduced Model**

| | A | B | C | D | E | F | G |
|---|---|---|---|---|---|---|---|
| 1 | SUMMARY OUTPUT | | | | | | |
| 2 | *Regression Statistics* | | | | | | |
| 3 | Multiple R | 0.913 | | | | | |
| 4 | R Square | 0.834 | | | | | |
| 5 | Adjusted R Square | 0.802 | | | | | |
| 6 | Standard Error | 120.376 | | | | | |
| 7 | Observations | 20 | | | | | |
| 9 | ANOVA | | | | | | |
| 10 | | *df* | *SS* | *MS* | *F* | *Significance F* | |
| 11 | Regression | 3 | 1161410.57 | 387136.86 | 26.72 | 0.000 | |
| 12 | Residual | 16 | 231845.18 | 14490.32 | | | |
| 13 | Total | 19 | 1393255.75 | | | | |
| 15 | | Coefficients | Standard Error | t Stat | P-value | Lower 95% | Upper 95% |
| 16 | Intercept | 192.337 | 154.193 | 1.247 | 0.230 | -134.537 | 519.211 |
| 17 | Socialization Measure x1 | -6.571 | 6.378 | -1.030 | 0.318 | -20.091 | 6.949 |
| 18 | Socialization Squared x1$^2$ | 0.149 | 0.059 | 2.536 | 0.022 | 0.024 | 0.273 |
| 19 | Gender x2 | 161.318 | 54.010 | 2.987 | 0.009 | 46.822 | 275.814 |

$SSE_R$ (labeled pointing to cell C11 region)

---

**Excel 2016 Instructions**

1. Open file: **Ashley-2.xlsx**.
2. Use Excel equation to create new variable in column C (i.e., for first data value use = **A2^2**; then copy down to end last row of data).
3. Rearrange columns so all independent variables are together.
4. Select **Data > Data Analysis**.
5. Select **Regression**.
6. Define the *y*-variable range (Burnout) and the *x*-variable range (include Socialization Measure, Gender and the new variable — include row with labels).
7. Click **Labels**.
8. Specify output location.
9. Click **OK**.

---

**FIGURE 15.25c Partial *F* Hypothesis Test for Interaction**

$H_0: \beta_4 = \beta_5 = 0$
$H_A:$ At least one of the $\beta_i \neq 0$

$\alpha = 0.05$

**Test Statistic:**

$$F = \frac{(SSE_R - SSE_C)/(c - r)}{MSE_C} = \frac{(231,845.18 - 127,317.40)/(5 - 3)}{9,094.1} = 5.747$$

**Rejection Region:**

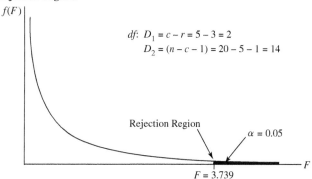

df: $D_1 = c - r = 5 - 3 = 2$
$D_2 = (n - c - 1) = 20 - 5 - 1 = 14$

Rejection Region

$\alpha = 0.05$

$F = 3.739$

Because partial-$F = 5.747 > 3.739$, we reject $H_0$.

## 15.3 EXERCISES

### Skill Development

**15-24.** Consider the following values for the dependent and independent variables:

| x | y |
|---|---|
| 5 | 10 |
| 15 | 15 |
| 40 | 25 |
| 50 | 44 |
| 60 | 79 |
| 80 | 112 |

a. Develop a scatter plot of the data. Does the plot suggest a linear or nonlinear relationship between the dependent and independent variables?

b. Develop an estimated linear regression equation for the data. Is the relationship significant? Test at an $\alpha = 0.05$ level.

c. Develop a regression equation of the form $\hat{y} = b_0 + b_1 x + b_2 x^2$. Does this equation provide a better fit to the data than the equation in part b?

**15-25.** Consider the following values for the dependent and independent variables:

| x | y |
|---|---|
| 6 | 5 |
| 9 | 20 |
| 14 | 28 |
| 18 | 30 |
| 22 | 33 |
| 27 | 35 |

a. Develop a scatter plot of the data. Does the plot suggest a linear or nonlinear relationship between the dependent and independent variables?

b. Develop an estimated linear regression equation for the data. Is the relationship significant? Test at an $\alpha = 0.05$ level.

c. Develop a regression equation of the form $\hat{y} = b_0 + b_1 \ln(x)$. Does this equation provide a better fit to the data than the equation in part b?

**15-26.** Examine the following data:

| x | 2 | 8 | 9 | 12 | 15 | 22 | 21 | 25 | 37 | 39 |
|---|---|---|---|----|----|----|----|----|----|----|
| y | 4 | 75 | 175 | 415 | 620 | 7,830 | 7,551 | 7,850 | 11,112 | 11,617 |

a. Construct a scatter plot of the data. Determine the order of the polynomial that is represented by the data.

b. Obtain an estimate of the model identified in part a.

c. Conduct a test of hypothesis to determine if a third-order, as opposed to a second-order, polynomial is a

better representation of the relationship between $y$ and $x$. Use a significance level of 0.05 and the $p$-value approach.

**15-27.** Examine the following two sets of data:

| When $x_2 = 1$ | | When $x_2 = 0$ | |
|---|---|---|---|
| $x_1$ | y | $x_1$ | y |
| 1 | 2 | 2 | 3 |
| 4 | 15 | 3 | 9 |
| 5 | 23 | 6 | 5 |
| 7 | 52 | 7 | 10 |
| 8 | 60 | 9 | 48 |
| 12 | 154 | 10 | 50 |
| 11 | 122 | 14 | 87 |
| 14 | 200 | 13 | 51 |
| 19 | 381 | 16 | 63 |
| 20 | 392 | 21 | 202 |

a. Produce a distinguishable scatter plot for each of the data sets on the same graph. Does it appear that there is interaction between $x_2$ and the relationship between $y$ and $x_1$? Support your assertions.

b. Consider the following model to represent the relationship among $y$, $x_1$, and $x_2$:

$$y_i = \beta_0 + \beta_1 x_1 + \beta_2 x_1^2 + \beta_3 x_1 x_2 + \beta_4 x_1^2 x_2 + \varepsilon$$

Produce the estimated regression equation for this model.

c. Conduct a test of hypothesis for each interaction term. Use a significance level of 0.05 and the $p$-value approach.

d. Based on the two hypothesis tests in part c, does it appear that there is interaction between $x_2$ and the relationship between $y$ and $x_1$? Support your assertions.

**15-28.** Consider the following data:

| x | 1 | 4 | 5 | 7 | 8 | 12 | 11 | 14 | 19 | 20 |
|---|---|---|---|---|---|----|----|----|----|----|
| y | 1 | 54 | 125 | 324 | 512 | 5,530 | 5,331 | 5,740 | 7,058 | 7,945 |

a. Construct a scatter plot of the data. Determine the order of the polynomial that is represented by this data.

b. Obtain an estimate of the model identified in part a.

c. Conduct a test of hypothesis to determine if a third-order, as opposed to a first-order, polynomial is a better representation of the relationship between $y$ and $x$. Use a significance level of 0.05 and the $p$-value approach.

**15-29.** A regression equation to be used to predict a dependent variable with four independent variables is developed from a sample of size 10. The resulting equation is

$$\hat{y} = 32.8 + 0.470x_1 + 0.554x_2 - 4.77x_3 + 0.929x_4$$

Two other equations are developed from the sample:

$$\hat{y} = 12.4 + 0.60x_1 + 1.60x_2$$

and

$$\hat{y} = 49.7 - 5.38x_3 + 1.35x_4$$

The respective sum of squares errors for the three equations are 201.72, 1,343, and 494.6.

a. Use the summary information to determine if the independent variables $x_3$ and $x_4$ belong in the complete regression model. Use a significance level of 0.05.

b. Repeat part a for the independent variables $x_1$ and $x_2$. Use the $p$-value approach and a significance level of 0.05.

## Computer Software Exercises

**15-30.** In a bit of good news for male students, based on 2014 data, American men have closed the gap with women on life span, according to the Population Reference Bureau (*source:* "Life expectancy at birth, by gender, 1970 and 2014," www.prb.org/DataFinder). Male life expectancy attained a record 76 years, and women's reached 81 years. The Population Reference Bureau provided the data given in the file titled **LifeExpectancy**.

a. Produce a scatter plot depicting the relationship between the life expectancies of women and men.

b. Determine the order of the polynomial that is represented on the scatter plot in part a. Produce the estimated regression equation that represents this relationship.

c. Determine if women's average life expectancy can be used in a second-order polynomial to predict the average life expectancy of men. Use a significance level of 0.05.

d. Use the estimated regression equation computed in part b to predict the average length of life of men when women's length of life equals 100. What does this tell you about the wisdom (or lack thereof) of extrapolation in regression models?

**15-31.** The Gilmore Accounting firm mentioned in Exercise 15-21, in an effort to explain variation in client profitability, collected the data in the file called **Gilmore**, where:

$y$ = Net profit earned from the client

$x_1$ = Number of hours spent working with the client

$x_2$ = Type of client:

    1, if manufacturing
    2, if service
    3, if governmental

Gilmore has asked if it needs the client type in addition to the number of hours spent working with the client to predict the net profit earned from the client. You are asked to provide this information.

a. Fit a model to the data that incorporates the number of hours spent working with the client and the type of client as independent variables. (*Hint:* Client type has three levels.)

b. Fit a second-order model to the data, again using dummy variables for client type. Does this model provide a better fit than the model in part a? Which model would you recommend be used?

**15-32.** McCullom's International Grains is constantly searching out areas in which to expand its market. Such markets present different challenges, since tastes in the international market are often different from domestic tastes. India is one country on which McCullom's has focused. Paddy is a grain used widely in India, but its characteristics are unknown to McCullom's. Charles Walters has been assigned to take charge of handling this grain. He has researched its various characteristics. During his research he came across an old article, "Determination of Biological Maturity and Effect of Harvesting and Drying Conditions on Milling Quality of Paddy" [*Journal of Agricultural Engineering Research* (1975), pp. 353–361], that examines the relationship between $y$, the yield (kg/ha) of paddy, as a function of $x$, the number of days after flowering at which harvesting took place. The data that appeared in the article are in a file called **Paddy**.

a. Construct a scatter plot of the yield (kg/ha) of paddy as a function of the number of days after flowering at which harvesting took place. Display at least two models that would explain the relationship you see in the scatter plot.

b. Conduct tests of hypotheses to determine if the models you selected are useful in predicting the yield of paddy.

c. Consider the model that includes the second-order term $x^2$. Would a simple linear regression model be preferable to the model containing the second-order term? Conduct a hypothesis test using the $p$-value approach to arrive at your answer.

d. Which model should Charles use to predict the yield of paddy? Explain your answer.

**15-33.** The National Association of Realtors Existing-Home Sales Series provides measurements on the residential real estate market. One of the measurements is the Housing Affordability Index (HAI). It is a measure of the financial ability of U.S. families to buy a house: 100 means that families earning the national median income have just the amount of money needed to qualify for a mortgage on a median-priced home, higher than 100 means they have more than enough, and lower than 100 means they have less than enough. The file titled **Index** contains the HAI and associated variables for the years 2013–2015 (*source:* "Housing Affordability Index," National Association of Realtors, www.realtor.org).

a. Construct a scatter plot relating the HAI to the median family income.

b. Obtain the estimated regression equation for a second-order polynomial model.

c. Discuss the model you developed in part b. Perform appropriate statistical tests using alpha = 0.05.

**15-34.** Badeaux Brothers Louisiana Treats ships packages of Louisiana coffee, cakes, and Cajun spices to customers around the United States. The cost to ship these products depends primarily on the weight of the package. Badeaux charges the customers for shipping and then ships the product itself. As a part of a study of whether it is economically feasible to continue to ship products itself, Badeaux sampled 20 recent shipments to determine what, if any, relationship exists between shipping costs and package weight. These data are in the file called **Badeaux**.

a. Develop a scatter plot of the data with the dependent variable, cost, on the vertical axis and the independent variable, weight, on the horizontal axis. Does there appear to be a relationship between the two variables? Is the relationship linear?

b. Compute the sample correlation coefficient between the two variables. Conduct a test, using a significance level of 0.05, to determine whether the population correlation coefficient is significantly different from zero.

c. Badeaux Brothers has been using a simple linear regression equation to predict the cost of shipping various items. Would you recommend it use a second-order polynomial model instead? Is the second-order polynomial model a significant improvement on the simple linear regression equation?

d. Badeaux Brothers has made a decision to stop shipping products if the shipping charges exceed $100. The company has asked you to determine the maximum weight for future shipments. Do this for both the first- and second-order models you have developed.

**15-35.** The National Association of Theatre Owners is the largest exhibition trade organization in the world, representing more than 26,000 movie screens in all 50 states and in more than 20 countries worldwide. Its membership includes the largest cinema chains and hundreds of independent theater owners. It publishes statistics concerning the movie sector of the economy. The file titled **Flicks** contains data on total U.S. box office grosses ($billion), total number of admissions (billion), average U.S. ticket price ($), and number of movie screens for the years 1987–2015 (*source:* National Association of Theatre Owners, www.natoonline.org, Feb. 2016). One concern is the effect the increasing ticket prices have on the number of individuals who go to the theaters to view movies.

a. Construct a scatter plot depicting the relationship between the total number of admissions and the U.S. ticket price. Test to determine whether the overall model is significant, and also test whether the independent variables have significant coefficients. Test at the alpha = 0.01 level.

b. Create a new variable, the square root of price, and run the model with price and the new variable as the independent variables. Obtain the estimated regression equation.

**15-36.** The Energy Information Administration (EIA), created by Congress in 1977, is a statistical agency of the U.S. Department of Energy. It provides data, forecasts, and analyses to promote sound policymaking and public understanding regarding energy and its interaction with the economy and the environment. One of the most important areas of analysis is petroleum. The file titled **Crude** contains data for the period 1995 through 2010 concerning the price, supply, and demand for fuel (*source:* "Petroleum & other liquids," U.S. Energy Information Administration, www.eia.gov). One concern was the increase in imported oil into the United States.

a. Examine the relationship between the price of gasoline and the annual amount of imported crude oil. Construct a scatter plot depicting this relationship.

b. Determine the order of the polynomial that would fit the data displayed in part a. Express "Imports" in millions of gallons, that is, $3,146,454/1,000,000 = 3.146454$. Produce an estimate of this polynomial.

c. Is a linear or quadratic model more appropriate for predicting the price of gasoline using the annual quantity of imported oil? Conduct the appropriate hypothesis test to substantiate your answer.

**15-37.** Recall from Exercise 15-7 that an investment analyst collected data obtained from publicly available sources on 20 randomly chosen companies. The data consisted of the 52-week-high stock prices, PE ratio, and market value of the company. These data are in the file titled **Investment**. The analyst wishes to produce a regression equation to predict the market value using the 52-week-high stock price and the PE ratio of the company. He creates a complete second-degree polynomial.

a. Construct an estimate of the regression equation using the indicated variables.

b. Determine if any of the quadratic terms are useful in predicting the average market value. Use a *p*-value approach with a significance level of 0.10.

c. Determine if any of the PE ratio terms are useful in predicting the average market value. Use a test statistic approach with a significance level of 0.05.

**OUTCOME** **8**    **15.4**    ## Stepwise Regression

One option in regression analysis is to bring all possible independent variables into the model in one step. This is what we have done in the previous sections. We use the term *full regression* to describe this approach. Another option for developing a regression model is called *stepwise regression*. Stepwise regression, as the name implies, develops the least squares regression equation in steps, either through *forward selection*, *backward elimination*, or *standard stepwise* regression.

### Forward Selection

The forward selection procedure begins (Step 1) by selecting a single independent variable from all those available. The independent variable selected at Step 1 is the variable that is most highly correlated with the dependent variable. A *t*-test is used to determine if this variable explains a significant amount of the variation in the dependent variable. At Step 1, if the variable is statistically significant, it is selected to be part of the final model used to predict the dependent variable. If it is not significant, the process is terminated. If no variables are found to be significant, the researcher will have to search for different independent variables than the ones already tested.

In the next step (Step 2), a second independent variable is selected based on its ability to explain the remaining unexplained variation in the dependent variable. Recall that the coefficient of determination $(R^2)$ measures the proportion of variation explained by all of the independent variables in the model. Thus, after we select the first variable (say, $x_1$), $R^2$ indicates the percentage of variation this variable explains. The forward selection routine then computes all possible two-variable regression models, with $x_1$ included, and determines the $R^2$ for each model. The coefficient of partial determination at Step 2 is the proportion of the as yet unexplained variation (after $x_1$ is in the model) that the additional variable explains. The independent variable that adds the most to $R^2$, given the variable(s) already in the model, is the one we select. Then, we conduct a *t*-test to determine if the newly added variable is significant. This process continues until either we have entered all available independent variables or the remaining independent variables do not add appreciably to $R^2$. For the forward selection procedure, the model begins with no variables. We enter variables one at a time, and after a variable is entered, it cannot be removed.

### Backward Elimination

Backward elimination is the reverse of the forward selection procedure. In the backward elimination procedure, all variables are forced into the model to begin the process. Then we remove the variables one insignificant variable at a time until no more insignificant variables are found. Once we have removed a variable from the model, it cannot be re-entered.

---

**EXAMPLE 15-4**    Applying Forward Selection Stepwise Regression Analysis

**B.T. Longmont Company**    The B.T. Longmont Company operates a large retail department store in Macon, Georgia. Like other department stores, Longmont incurs heavy losses due to shoplifting and employee pilferage. The store's security manager wants to develop a regression model to explain the monthly dollar loss. The following steps can be used when developing a multiple regression model using stepwise regression:

STEP 1    **Specify the model by determining the dependent variable and potential independent variables.**
The dependent variable ($y$) is the monthly dollar losses due to shoplifting and pilferage. The security manager has identified the following potential independent variables:

$x_1 =$ Average monthly temperature (degrees Fahrenheit)
$x_2 =$ Number of sales transactions
$x_3 =$ Dummy variable for holiday month (1 if holiday during month, 0 if not)
$x_4 =$ Number of persons on the store's monthly payroll

The data are in the file **Longmont**.

**Excel 2016 Instructions**
1. Open file: **Longmont.xlsx.**
2. Select **Data.**
3. Select **Data Analysis >** **Correlation.**
4. Specify data range (include labels).
5. Click **Labels.**
6. Specify output location.
7. Click **OK.**

**FIGURE 15.26 Excel 2016 Correlation Matrix Output for the B.T. Longmont Company**

| | A | B | C | D | E | F |
|---|---|---|---|---|---|---|
| 1 | | Average Temperature | Number of Sales Transactions | Holiday | Employees | Shoplifting Loss |
| 2 | Average Temperature | 1 | | | | |
| 3 | Number of Sales Transactions | -0.0241 | 1 | | | |
| 4 | Holiday | -0.1432 | 0.0626 | 1 | | |
| 5 | Employees | -0.0821 | 0.9185 | -0.1966 | 1 | |
| 6 | Shoplifting Loss | 0.2858 | 0.6307 | 0.1361 | 0.4132 | 1 |

STEP 2 **Formulate the regression model.**

The correlation matrix for the data is presented in Figure 15.26. The forward selection procedure will select the independent variable that is most highly correlated with the dependent variable. By examining the bottom row in the correlation matrix in Figure 15.26, we can see that the variable $x_2$, number of sales transactions, is most highly correlated ($r = 0.6307$) with dollars lost. Once we enter this variable into the model, the remaining independent variables are entered based on their ability to explain the remaining variation in the dependent variable.

Figure 15.27 shows the XLSTAT forward stepwise regression output. At Step 1 of the process, variable $x_2$, number of sales transactions, enters the model.

STEP 3 **Perform diagnostic checks on the model.**

Figure 15.27 reports the $R^2$ value 0.398. The standard error of the estimate is the square root of the mean square residual:

$$s_\varepsilon = \sqrt{MSE} = \sqrt{MS \text{ residual}} = \sqrt{128,163.96} = 358$$

STEP 4 **Continue to formulate and diagnose the model by adding other independent variables.**

The next variable we select is the one that can do the most to increase $R^2$. If we were doing this manually, we would try each variable to see which one yields the highest $R^2$, given that the transactions variable is already in the model. The XLSTAT add-in for Excel does this automatically. As shown in Figures 15.27 and 15.28, the variable selected in Step 2 of the process is $x_4$, employees. Using the ANOVA section, we can determine $R^2$ and $s_\varepsilon$ as before:

$$R^2 = \frac{SSR}{SST} = \frac{1,833,270.52}{3,192,631.53} = 0.5742$$

$$s_\varepsilon = \sqrt{MS \text{ residual}} = \sqrt{97,097.21} = 311.6$$

The model now explains 57.42% of the variation in the dependent variable. The $t$-values for both slope coefficients exceed $t = 2.145$ (the critical value from the $t$-distribution table with a one-tailed area equal to 0.025 and $n - k - 1 = 17 - 2 - 1 = 14$ degrees of freedom), so we conclude that both variables are significant in explaining the variation in the dependent variable, shoplifting loss.

The forward selection routine continues to enter variables as long as each additional variable explains a significant amount of the remaining variation in the dependent variable. Note that XLSTAT allows us to set the significance level in terms of a $p$-value. Then, as long as the calculated $p$-value for an incoming variable is less than our limit, the variable is allowed to enter the model.

In this example, with the $p$-value limit set at 0.05, neither of the two remaining independent variables would explain a significant amount of the

**FIGURE 15.27** Excel 2016 (XLSTAT) Forward Selection Results for the B.T. Longmont Company—Step 1

Regression of variable Shoplifting Loss:
Summary of the variables selection:

| Nbr. of variables | Variables | Variable IN/OUT | Status | MSE | R² |
|---|---|---|---|---|---|
| 1 | Number of Sales Transactions | Number of Sales Transactions | IN | 128163.9557 | 0.3978 |
| 2 | Number of Sales Transactions / Employees | Employees | IN | 97097.2147 | 0.5742 |

Excel 2016 (XLSTAT) Instructions

1. Open XLSTAT and Enable Macros.
2. Open file: **Longmont.xlsx**.
3. On the **XLSTAT** tab, click **Modeling Data**, and then click **Linear Regression**.
4. Define the y/dependent variables (shoplifting loss) and the x/explanatory variables (quantitative), and check Variable labels.
5. On the **Options** tab, click **Model Selection** and then select **Forward**.
6. On the **Outputs** tab, select **Descriptive Statistics**, **Correlation**, and **Analysis of Variance**, and then click **OK**.

**FIGURE 15.28** Excel 2016 (XLSTAT) Forward Selection Results for the B.T. Longmont Company—Step 2

Excel 2016 (XLSTAT) Instructions
1. See Figure 15.27.

Analysis of variance:

| Source | DF | Sum of squares | Mean squares | F | Pr > F |
|---|---|---|---|---|---|
| Model | 2 | 1833270.52 | 916635.26 | 9.4404 | 0.0025 |
| Error | 14 | 1359361.01 | 97097.21 | | |
| Corrected Total | 16 | 3192631.53 | | | |

Computed against model Y=Mean(Y)

Model parameters:

| Source | Value | Standard error | t | Pr > |t| | Lower bound (95%) | Upper bound (95%) |
|---|---|---|---|---|---|---|
| Intercept | 4600.8049 | 1010.5449 | 4.5528 | 0.0005 | 2433.4016 | 6768.2082 |
| Average Temperature | 0.0000 | 0.0000 | | | | |
| Number of Sales Transactions | 0.2034 | 0.0559 | 3.6422 | 0.0027 | 0.0836 | 0.3232 |
| Holiday | 0.0000 | 0.0000 | | | | |
| Employees | -21.5674 | 8.9559 | -2.4082 | 0.0304 | -40.7758 | -2.3589 |

Equation of the model:
Shoplifting Loss = 4601+0.20343*Number of Sales Transactions-21.56736*Employees

remaining variation in the dependent variable. The procedure is, therefore, terminated. The resulting regression equation provided by forward selection is

$$\hat{y} = 4{,}601 + 0.2034(\text{number of sales transactions}) - 21.5674(\text{employees})$$

Note that average temperature and the dummy variables for holiday did not enter the model. This implies that, given that the other variables are already included, knowing whether the month in question has a holiday or knowing its average temperature does not add significantly to the model's ability to explain the variation in the dependent variable.

The B.T. Longmont Company can now use this regression model to explain variation in shoplifting and pilferage losses based on knowing the number of sales transactions and the number of employees.

**TRY EXERCISE 15-39** (pg. 616)

## Standard Stepwise Regression

The standard stepwise procedure (sometimes referred to as *forward stepwise regression*—not to be confused with forward selection) combines attributes of both backward elimination and forward selection. The standard stepwise method serves one more important function. If two or more independent variables are correlated, a variable selected in an early step may become insignificant when other variables are added at later steps. The standard stepwise procedure will drop this insignificant variable from the model. Standard stepwise regression also offers a means of observing multicollinearity problems, because we can see how the regression model changes as each new variable is added to it. XLSTAT refers to the standard stepwise method as simply *Stepwise*.

The standard stepwise procedure is widely used in decision-making applications and is generally recognized as a useful regression method. However, you should exercise care when using this procedure because it is easy to rely too heavily on the automatic selection process. Remember that the order of variable selection is conditional, based on the variables already in the model. There is no guarantee that stepwise regression will lead you to the best set of independent variables from those available. Decision makers still must use common sense in applying regression analysis to make sure they have usable regression models.

## Best Subsets Regression

Another method for developing multiple regression models is called the *best subsets* method. As the name implies, the best subsets method works by selecting subsets from the chosen possible independent variables to form models. The user can then select the "best" model based on a variety of possible measures such as adjusted $R^2$. The XLSTAT add-in for Excel contains a procedure for performing best subsets regression called Best model.

| EXAMPLE 15-5 | Applying Best Subsets Regression |

**Winston Investment Advisors** Charles L. Winston, founder and CEO at Winston Investment Advisors in Burbank, California, is interested in developing a regression model to explain the variation in dividends paid per share by U.S. companies. Such a model would be useful in advising his clients. The following steps show how to develop such a model using the best subsets regression approach:

STEP 1 **Specify the model.**

Some publicly traded companies pay higher dividends than others. The CEO is interested in developing a multiple regression model to explain the variation in dividends per share paid to shareholders. The dependent variable will be dividends per share. The CEO met with other analysts in his firm to identify potential independent variables for which data would be readily available. The following list was selected:

$x_1$ = Return on equity (net income/equity)

$x_2$ = Earnings per share

$x_3$ = Current assets in millions of dollars

$x_4$ = Year-ending stock price

$x_5$ = Current ratio (current assets/current liabilities)

A random sample of 35 publicly traded U.S. companies was selected. For each company in the sample, the analysis obtained data from the companies' publicly available annual reports on the dividend per share paid last year and the year-ending data on the five independent variables. These data are in the file **Company Financials**.

STEP 2 **Formulate the regression model.**

The CEO is interested in developing the "best" regression model for explaining the variation in the dependent variable, dividends per share. The approach is to use best subsets, which requires that multiple regression models be developed, each containing a different mix of variables. The models tried will contain from one to five independent variables. The resulting models will be evaluated by comparing values for adjusted $R^2$. A model with a high adjusted $R^2$ is desirable.

**FIGURE 15.29 Excel 2016 (XLSTAT): Best Subsets Regression Output for Winston Investment Advisors**

**Regression of variable Dividends/Share:**

Summary of the variables selection:

| Nbr. of variables | Variables | MSE | R² | Adjusted R² | Mallows' Cp |
|---|---|---|---|---|---|
| 1 | Current Assets (Millions) | 0.2963 | 0.5081 | 0.4932 | 7.4227 |
| 2 | Earnings/ Share / Current Assets (Millions) | 0.2422 | 0.6102 | **0.5858** | 1.4493 |
| 3 | Earnings/ Share / Current Assets (Millions) / Year Ending Stock Price | 0.2426 | 0.6218 | 0.5852 | 2.5425 |
| 4 | ROE / Earnings/ Share / Current Assets (Millions) / Year Ending Stock Price | 0.2463 | 0.6284 | 0.5789 | 4.0281 |
| 5 | ROE / Earnings/ Share / Current Assets (Millions) / Year Ending Stock Price / Current Ratio | 0.2545 | 0.6288 | 0.5648 | 6.0000 |

The best model for the selected selection criterion is displayed in blue

*Note:* Mallow's $C_p$ is a measure of the precision of the regression coefficients and we look for $C_p$ to be less than the number of independent variables in the model.

**Excel 2016 (XLSTAT) Instructions**

1. Open XLSTAT and Enable Macros.
2. Open file: **Company Financials.xlsx**.
3. On the **XLSTAT** tab, click **Modeling Data**, and then click **Linear Regression**.
4. Define the *y*/dependent variables (dividends per share) and the *x*/explanatory variables (quantitative), and check Variable labels.
5. On the **Options** tab, check **Model Selection**, and then select **Best Model**.
6. On the **Outputs** tab, select **Descriptive Statistics**, **Correlation**, and **Analysis of Variance**, and then click **OK**.

The XLSTAT Excel add-in can be used to perform best subsets regression analysis. Figure 15.29 shows the output. Notice that several combinations of models with $k = 1$ to $k = 5$ independent variables are included. These models appear to be good candidates based on *R*-squared, adjusted *R*-squared, and *MSE*, values.

There is little difference in these models in terms of the statistics shown. We can examine any of them in more detail by looking at further XLSTAT output. For instance, the model that contains the variables earnings per share, current assets, and year-ending stock price is shown in Figure 15.30. Note that although this model is among the best with respect to adjusted *R*-squared, two of the three variables (earnings per share and year-ending stock price) have statistically insignificant regression coefficients. Figure 15.31 shows the regression model with two statistically significant variables remaining. The *R*-squared value is 0.61, the adjusted $R^2$ has increased, and the overall model is statistically significant.

**FIGURE 15.30 Excel 2016 (XLSTAT): One Potential Model for Winston Investment Advisors**

Analysis of variance:

| Source | DF | Sum of squares | Mean squares | F | Pr > F |
|---|---|---|---|---|---|
| Model | 3 | 12.3632 | 4.1211 | 16.9900 | < 0.0001 |
| Error | 31 | 7.5193 | 0.2426 | | |
| Corrected Total | 34 | 19.8825 | | | |

Computed against model Y=Mean(Y)

Model parameters:

| Source | Value | Standard error | t | Pr > \|t\| | Lower bound (95%) | Upper bound (95%) |
|---|---|---|---|---|---|---|
| Intercept | 0.2009 | 0.1696 | 1.1846 | 0.2452 | -0.1450 | 0.5469 |
| Earnings/Share | 0.0650 | 0.0386 | 1.6836 | 0.1023 | -0.0137 | 0.1438 |
| Current Assets (Millions) | 0.0000 | 0.0000 | 4.2494 | 0.0002 | 0.0000 | 0.0000 |
| Year Ending Stock Price | 0.0056 | 0.0057 | 0.9755 | 0.3369 | -0.0061 | 0.0173 |

**FIGURE 15.31** Excel 2016 (XLSTAT): A Final Model for Winston Investment Advisors

| Goodness of fit statistics: | |
|---|---|
| Observations | 35.0000 |
| Sum of weights | 35.0000 |
| DF | 32.0000 |
| R² | 0.6102 |
| Adjusted R² | 0.5858 |
| MSE | 0.2422 |
| RMSE | 0.4921 |
| DW | 1.9562 |

Analysis of variance:

| Source | DF | Sum of squares | Mean squares | F | Pr > F |
|---|---|---|---|---|---|
| Model | 2 | 12.1324 | 6.0662 | 25.0472 | < 0.0001 |
| Error | 32 | 7.7501 | 0.2422 | | |
| Corrected Total | 34 | 19.8825 | | | |

*Computed against model Y=Mean(Y)*

Model parameters:

| Source | Value | Standard error | t | Pr > \|t\| | Lower bound (95%) | Upper bound (95%) |
|---|---|---|---|---|---|---|
| Intercept | 0.3385 | 0.0941 | 3.5966 | 0.0011 | 0.1468 | 0.5303 |
| ROE | 0.0000 | 0.0000 | | | | |
| Earnings/Share | 0.0882 | 0.0305 | 2.8947 | 0.0068 | 0.0261 | 0.1502 |
| Current Assets (Millions) | 0.00000312 | 0.0000 | 4.7216 | < 0.0001 | 0.0000 | 0.0000 |
| Year Ending Stock Price | 0.0000 | 0.0000 | | | | |
| Current Ratio | 0.0000 | 0.0000 | | | | |

Equation of the model:
Dividends/Share = 0.33855+0.08816*Earnings/Share+0.00000312*Current Assets (Millions)

TRY EXERCISE 15-42 (pg. 617)

# 15.4 EXERCISES

## Skill Development

**15-38.** Suppose you have four potential independent variables, $x_1$, $x_2$, $x_3$, and $x_4$, from which you want to develop a multiple regression model. Using stepwise regression, $x_2$ and $x_4$ entered the model.
a. Why did only two variables enter the model? Discuss.
b. Suppose a full regression with only variables $x_2$ and $x_4$ had been run. Would the resulting model be different from the stepwise model that included only these two variables? Discuss.
c. Now, suppose a full regression model had been developed, with all four independent variables in the model. Which would have the higher $R^2$ value, the full regression model or the stepwise model? Discuss.

**15-39.** You are given the following data:

| y | $x_1$ | $x_2$ | $x_3$ | y | $x_1$ | $x_2$ | $x_3$ |
|---|---|---|---|---|---|---|---|
| 33 | 9 | 192 | 40 | 45 | 12 | 296 | 52 |
| 44 | 11 | 397 | 47 | 25 | 9 | 235 | 27 |
| 34 | 10 | 235 | 37 | 53 | 10 | 295 | 57 |
| 60 | 13 | 345 | 61 | 45 | 13 | 335 | 50 |
| 20 | 11 | 245 | 23 | 37 | 11 | 243 | 41 |
| 30 | 7 | 235 | 35 | 44 | 13 | 413 | 51 |

a. Determine the appropriate correlation matrix and use it to predict which variable will enter in the first step of a stepwise regression model.

b. Use standard stepwise regression to construct a model, entering all significant variables.
c. Construct a full regression model. What are the differences in the model? Which model explains the most variation in the dependent variable?

**15-40.** You are given the following data:

| y | $x_1$ | $x_2$ | $x_3$ | y | $x_1$ | $x_2$ | $x_3$ |
|---|---|---|---|---|---|---|---|
| 45 | 40 | 41 | 39 | 45 | 42 | 39 | 37 |
| 41 | 31 | 41 | 35 | 43 | 37 | 52 | 41 |
| 43 | 45 | 49 | 39 | 34 | 40 | 47 | 40 |
| 38 | 43 | 41 | 41 | 49 | 35 | 44 | 44 |
| 50 | 42 | 42 | 51 | 45 | 39 | 40 | 45 |
| 39 | 48 | 40 | 42 | 40 | 43 | 30 | 42 |
| 50 | 44 | 44 | 41 | 43 | 53 | 34 | 34 |

a. Determine the appropriate correlation matrix and use it to predict which variable will enter in the first step of a stepwise regression model.
b. Use standard stepwise regression to construct a model, entering all significant variables.
c. Construct a full regression model. What are the differences in the model? Which model explains the most variation in the dependent variable?

**15-41.** The following data represent a dependent variable and four independent variables:

| y | $x_1$ | $x_2$ | $x_3$ | $x_4$ | y | $x_1$ | $x_2$ | $x_3$ | $x_4$ |
|---|---|---|---|---|---|---|---|---|---|
| 61 | 37 | 13 | 10 | 21 | 68 | 35 | 17 | 3 | 33 |
| 69 | 35 | 19 | 9 | 23 | 48 | 12 | 8 | 8 | 30 |
| 37 | 25 | 7 | 11 | 32 | 65 | 37 | 11 | 17 | 19 |
| 24 | 23 | 14 | 7 | 31 | 66 | 34 | 15 | 2 | 33 |
| 22 | 23 | 6 | 7 | 18 | 45 | 30 | 9 | 24 | 31 |

a. Use the standard stepwise regression to produce an estimate of a multiple regression model to predict y. Use 0.15 as the alpha to enter and to remove.

b. Change the alpha to enter to 0.01. Repeat the standard stepwise procedure.

c. Change the alpha to remove to 0.35, leaving alpha to enter to be 0.15. Repeat the standard stepwise procedure.

d. Change the alpha to remove to 0.15, leaving alpha to enter to be 0.05. Repeat the standard stepwise procedure.

e. Change the alpha to remove and to enter to 0.35. Repeat the standard stepwise procedure.

f. Compare the estimated regression equations developed in parts a−e.

**15-42.** Consider the following data:

| y | $x_1$ | $x_2$ | $x_3$ | $x_4$ | y | $x_1$ | $x_2$ | $x_3$ | $x_4$ |
|---|---|---|---|---|---|---|---|---|---|
| 61 | 37 | 18 | 2 | 13 | 69 | 35 | 21 | 2 | 20 |
| 37 | 25 | 5 | 4 | 10 | 24 | 23 | 7 | 6 | 9 |
| 22 | 23 | 12 | 7 | 4 | 68 | 35 | 15 | 3 | 14 |
| 48 | 12 | 6 | 2 | 15 | 65 | 37 | 19 | 2 | 19 |
| 66 | 34 | 14 | 3 | 25 | 45 | 30 | 12 | 3 | 12 |

a. Use standard stepwise regression to produce an estimate of a multiple regression model to predict y.

b. Use forward selection stepwise regression to produce an estimate of a multiple regression model to predict y.

c. Use backward elimination stepwise regression to produce an estimate of a multiple regression model to predict y.

d. Use best subsets regression to produce an estimate of a multiple regression model to predict y.

## Computer Software Exercises

**15-43.** The U.S. Energy Information Administration compiles data on the U.S. energy industry. The file titled **Energy** contains data on the number of megawatt-hours of electricity sold in the United States to all customers, the average price per kilowatt-hour, the total number of customers, and the total volume of carbon dioxide emissions for the years 2005–2015 (*source:* www.eia .gov/electricity).

a. Produce a correlation matrix for the four variables. Predict which variable will be the first to enter the model if a forward stepwise approach is used to explain the variation in carbon dioxide emissions. Explain.

b. Use XLSTAT to run a forward stepwise model to maximize adjusted $R^2$ with carbon dioxide as the dependent variable. Discuss the resulting model and associated statistics.

c. Discuss what steps you would take to improve this model.

**15-44.** The White Cover Snowmobile Association gives out pamphlets, maps, and other tourist-related information to people who call a toll-free number and request the information. The association orders the packets of information from a document-printing company and likes to have enough available to meet the immediate need without having too many sitting around taking up space. The marketing manager decided to develop a multiple regression model to be used in predicting the number of calls during the coming week. A random sample of 12 weeks is selected, with the following variables:

$y$ = Number of calls

$x_1$ = Number of advertisements placed the previous week

$x_2$ = Number of calls received the previous week

$x_3$ = Number of airline tour bookings into Western cities for the current week

These data are in the file called **Winter Adventures**.

a. Develop the multiple regression model for predicting the number of calls received, using backward elimination stepwise regression.

b. At the final step of the analysis, how many variables are in the model?

c. Discuss why the variables were removed from the model in the order shown by the stepwise regression.

**15-45.** Refer to Exercise 15-44.

a. Develop the correlation matrix that includes all independent variables and the dependent variable. Predict the order that the variables will be selected into the model if forward selection stepwise regression is used.

b. Use forward selection stepwise regression to develop a model for predicting the number of calls that the company will receive. Write a report that describes what has taken place at each step of the regression process.

c. Compare the forward selection stepwise regression results in part b with the backward elimination results determined in Exercise 15-44. Which model would you choose? Explain your answer.

**15-46.** An investment analyst collected data obtained from publicly available sources on 20 randomly chosen companies. The data consisted of the 52-week-high stock prices, PE ratios, and market values of the companies. These data are in the file titled **Investment**. The analyst wishes to produce a regression equation to predict the market value using the 52-week-high stock price and the PE ratio of the company. He creates a complete second-degree polynomial.

a. Produce two scatter plots: (1) market value versus stock price and (2) market value versus PE ratio. Do the scatter plots support the analyst's decision to produce a second-order polynomial? Support your assertion with statistical reasoning.

b. Use forward selection stepwise regression to eliminate any unneeded components from the analyst's model.

c. Does forward selection stepwise regression support the analyst's decision to produce a second-order polynomial? Support your assertion with statistical reasoning.

**15-47.** A variety of sources suggest that individuals assess their health, at least in part, by estimating their percentage of body fat. A widely accepted measure of body fat uses an underwater weighing technique. There are, however, more convenient methods that use only a scale and a measuring tape. An article in the *Journal of Statistics Education* [Vol. 4, no. 1 (1996)] by Roger W. Johnson explored regression models to predict body fat. The file titled **Bodyfat** lists a portion of the data presented in the article.

a. Use best subsets stepwise regression to establish the relationship between body fat and the variables in the specified file.

b. Predict the body fat of an individual whose age is 21, weight is 170 pounds, height is 70 inches, chest circumference is 100 centimeters, abdomen is 90 centimeters, hip is 105 centimeters, and thigh is 60 centimeters around.

---

OUTCOME **9**  **15.5**  ## Determining the Aptness of the Model

In Section 15.1, we discussed these basic steps involved in building a multiple regression model:

1. Specify the model.
2. Build the model.
3. Perform diagnostic checks on the model.

The final step is the diagnostic step in which you examine the model to determine how well it performs. In Section 15.2, we discussed several statistics that you need to consider when performing the diagnostic step, including analyzing $R^2$, adjusted $R^2$, and the standard error of the estimate. In addition, we discussed the concept of multicollinearity and the possible impacts when multicollinearity is present. Section 15.3 introduced another diagnostic step that involves looking for potential curvilinear relationships between the independent variables and the dependent variable. We presented some basic data transformation techniques for dealing with curvilinear situations. However, a major part of the diagnostic process involves an analysis of how well the model fits the regression analysis assumptions.

The assumptions required to use multiple regression include the following:

**Assumptions**

1. Individual model errors, $\varepsilon$, are statistically independent of one another, and these values represent a random sample from the population of possible residuals at each level of $x$.
2. For a given value of $x$, there can be many values of $y$, and therefore many possible values for $\varepsilon$. Further, the possible $\varepsilon$ values for any level of $x$ are normally distributed.
3. The distributions of possible $\varepsilon$ values have equal variances at each level of $x$.
4. The means of the dependent variable, $y$, for all specified values of $x$ can be connected with a line called the population regression model.

The degree to which a regression model satisfies these assumptions is called *aptness*.

## Analysis of Residuals

**Residual**

The difference between the actual value of the dependent variable and the value predicted by the regression model.

The **residual** is computed using Equation 15.13.

**Residual**

$$e_i = y_i - \hat{y}_i \tag{15.13}$$

We can compute residual value for each observation in the data set and then learn a great deal about the aptness of the regression model by analyzing the residuals. The principal means of residual analysis is a study of residual plots. The following problems can be inferred through graphical analysis of residuals:

1. The underlying regression function is not linear.
2. The residuals do not have a constant variance.
3. The residuals are not independent.
4. The residual terms are not normally distributed.

We will address each of these in order. The regression options in Excel provide extensive residual analysis.

**Checking for Linearity** A plot of the residuals (on the vertical axis) versus the independent variable (on the horizontal axis) is useful for detecting whether a linear function is the appropriate regression function. Figure 15.32 illustrates two different residual plots. Figure 15.32a shows residuals that systematically depart from 0. When $x$ is small, the residuals are negative. When $x$ is in the midrange, the residuals are positive, and for large $x$ values, the residuals are negative again. This type of plot suggests that the relationship between $y$ and $x$ is nonlinear. Figure 15.32b shows a plot in which residuals do not show a systematic variation around 0, implying that the relationship between $x$ and $y$ is linear.

**FIGURE 15.32 Residual Plots Showing Linear and Nonlinear Patterns**

**(a) Nonlinear Pattern**

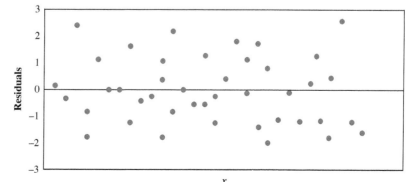

**(b) Linear Pattern**

If a linear model is appropriate, we expect the residuals to band around 0 with no systematic pattern displayed. If the residual plot shows a systematic pattern, it may be possible to transform the independent variable (refer to Section 15.3) so that the revised model will produce residual plots that will not systematically vary from 0.

**BUSINESS APPLICATION**   **Residual Analysis**

**First City Real Estate** (*continued*)   We have been using First City Real Estate to introduce multiple regression tools throughout this chapter. Remember, the managers wish to develop a multiple regression model for predicting the sales prices of homes in their market. Suppose that the most current model (**First City-3**) incorporates a transformation of the lot size variable as log of lot size. The output for this model is shown in Figure 15.33. Notice the model now has an $R^2$ value of 96.9%.

[ X ]
Excel Tutorial

**FIGURE 15.33** **Excel 2016 Output of First City Real Estate Appraisal Model**

**Excel 2016 Instructions**

1. Open file: **First City-3.xlsx**.
2. Select **Data > Data Analysis**.
3. Select **Regression**.
4. Define the *y*-variable range (Price) and the *x*-variable range (using only the *x* variables shown in this figure).
5. Click **Labels**.
6. Specify output location.
7. Click on desired **Residuals** plots and/or **Normal Probability Plot**.
8. Click **OK**.

| | A | B | C | D | E | F | G |
|---|---|---|---|---|---|---|---|
| 1 | SUMMARY OUTPUT | | | | | | |
| 2 | *Regression Statistics* | | | | | | |
| 3 | Multiple R | 0.984 | | | | | |
| 4 | R Square | 0.969 | | | | | |
| 5 | Adjusted R Square | 0.968 | | | | | |
| 6 | Standard Error | 11249.939 | | | | | |
| 7 | Observations | 319 | | | | | |
| 9 | ANOVA | | | | | | |
| 10 | | *df* | *SS* | *MS* | *F* | *Significance F* | |
| 11 | Regression | 4 | 1,233,292,449,609.03 | 308,323,112,402.26 | 2,436.160 | 0.0000 | |
| 12 | Residual | 314 | 39,740,194,152.73 | 126,561,127.87 | | | |
| 13 | Total | 318 | 1,273,032,643,761.75 | | | | |
| 15 | | *Coefficients* | *Standard Error* | *t Stat* | *P-value* | *Lower 95%* | *Upper 95%* |
| 16 | Intercept | -521919.558 | 10797.808 | -48.336 | 0.000 | -543164.759 | -500674.357 |
| 17 | Sq. Feet | 17.858 | 1.682 | 10.617 | 0.000 | 14.548 | 21.167 |
| 18 | Bedrooms | -2319.027 | 1173.993 | -1.975 | 0.049 | -4628.915 | -9.140 |
| 19 | Garage # | 6012.740 | 1299.306 | 4.628 | 0.000 | 3456.294 | 8569.187 |
| 20 | log Lot Size | 159050.177 | 3286.101 | 48.401 | 0.000 | 152584.618 | 165515.737 |

**FIGURE 15.34** **First City Real Estate Residual Plots versus the Independent Variables**

(a) Residuals versus Square Feet (Response Is Price)

(b) Residuals versus Garage (Response Is Price)

(c) Residuals versus Bedrooms (Response Is Price)

(d) Residuals versus Log Lot (Response Is Price)

There are currently four independent variables in the model: square feet, bedrooms, garage size, and the log of lot size. Excel provides procedures for automatically producing residual plots. Figure 15.34 shows the plots of the residuals against each of the independent variables. The transformed variable, log lot size, has a residual pattern that shows a systematic pattern (Figure 15.34d). The residuals are positive for small values of log lot size, negative for intermediate values of log lot size, and positive again for large values of log lot size. This pattern suggests that the curvature of the relationship between sales prices of homes and lot size is even more pronounced than the logarithm implies. Potentially, a second- or third-degree polynomial in the lot size should be pursued.

### Do the Residuals Have Equal Variances at All Levels of Each x Variable?

Residual plots also can be used to determine whether the residuals have a constant variance. Consider Figure 15.35, in which the residuals are plotted against an independent variable. The plot in Figure 15.35a shows an example in which as $x$ increases, the residuals become less variable. Figure 15.35b shows the opposite situation. When $x$ is small, the residuals are tightly packed around 0, but as $x$ increases, the residuals become more variable. Figure 15.35c shows an example in which the residuals exhibit a constant variance around the zero mean.

When we have developed a multiple regression model, we can analyze the equal variance assumption by plotting the residuals against the fitted $(\hat{y})$ values. When the residual plot is cone-shaped, as in Figure 15.36, it suggests that the assumption of equal variance has been violated. This is evident because the residuals are wider on one end than the other. That indicates that the standard error of the estimate is larger on one end than the other—that is, it is not constant.

Figure 15.37 shows the residuals plotted against the $\hat{y}$ values for First City Real Estate's appraisal model. We have drawn a band around the residuals that shows that the variance of the residuals stays quite constant over the range of the fitted values.

**FIGURE 15.35** **Residual Plots Showing Constant and Nonconstant Variances**

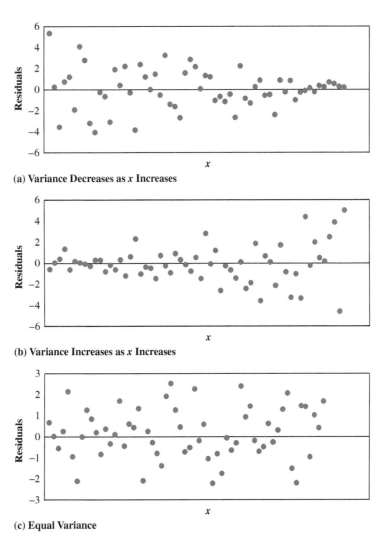

(a) Variance Decreases as $x$ Increases

(b) Variance Increases as $x$ Increases

(c) Equal Variance

**FIGURE 15.36** **Residual Plots against the Fitted ($\hat{y}$) Values**

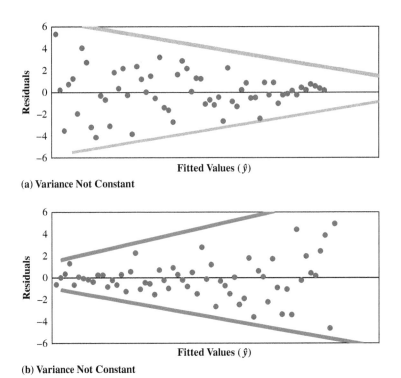

(a) Variance Not Constant

(b) Variance Not Constant

**FIGURE 15.37** **Excel 2016 Plot of Residuals versus Fitted Values for First City Real Estate**

**Residuals vs Fitted Values**
**(Response is Price)**

**Are the Residuals Independent?** If the data used to develop the regression model are measured over time, we can use a plot of the residuals against time to determine whether the residuals are correlated. Figure 15.38a shows an example in which the residual plot against time suggests independence. The residuals in Figure 15.38a appear to be randomly distributed around the mean of zero over time. However, in Figure 15.38b, the plot suggests that the residuals are not independent, because in the early time periods, the residuals are negative, and in later time periods, the residuals are positive. This, or any other nonrandom pattern in the residuals over time, indicates that the assumption of independent residuals has been violated. Generally, this means some variable associated with the passage of time has been omitted from the model. Often, time is used as a surrogate for other time-related variables in a regression model. Chapter 16 will discuss time-series data analysis and forecasting techniques in more detail and will address the issue of incorporating the time variable into the

**FIGURE 15.38 Plot of Residuals against Time**

**(a) Independent Residuals**

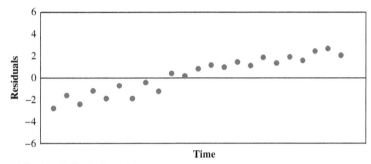

**(b) Residuals Not Independent**

model. In Chapter 16, we introduce a procedure called the Durbin-Watson test to determine whether residuals are correlated over time.

**Checking for Normally Distributed Error Terms**    The need for normally distributed model errors occurs when we want to test a hypothesis about the regression model. Small departures from normality do not cause serious problems. However, if the model errors depart dramatically from a normal distribution, there is cause for concern. Examining the sample residuals will allow us to detect such dramatic departures. One method for graphically analyzing the residuals is to form a frequency histogram of the residuals to determine whether the general shape is normal. We can then use the chi-square goodness-of-fit test presented in Chapter 13 to test whether the residuals fit a normal distribution.

Another method for determining normality is to calculate and plot the *standardized residuals*. In Chapter 3, you learned that we can standardize a random variable by subtracting its mean and dividing the result by its standard deviation. The mean of the residuals is zero. Therefore, dividing each residual by an estimate of its standard deviation gives the standardized residual. Although the proof is beyond the scope of this text, it can be shown that the standardized residual for any particular observation for a simple linear regression model is found using Equation 15.14.

**Standardized Residual for Linear Regression**

$$s_{e_i} = \frac{e_i}{s_\varepsilon \sqrt{1 + \frac{1}{n} + \frac{(x_i - \bar{x})^2}{\Sigma x^2 - \frac{(\Sigma x)^2}{n}}}} \qquad \textbf{(15.14)}$$

where:

$e_i$ = $i$th residual value

$s_\varepsilon$ = Standard error of the estimate

$x_i$ = Value of $x$ used to generate the predicted $y$ value for the $i$th observation

Computing the standardized residual for an observation in a multiple regression model is too complicated to be done by hand. However, the standardized residuals are generated from most statistical software, including Excel. The Excel tutorial illustrates the methods required to generate the standardized residuals and residual plots. Because other problems such as nonconstant variance and nonindependent residuals can result in residuals that seem to be abnormal, you should check these other factors before addressing the normality assumption.

Recall that for a normal distribution, approximately 68% of the values fall within $\pm$ 1 standard deviation of the mean, 95% fall within $\pm$ 2 standard deviations of the mean, and virtually all values fall within $\pm$ 3 standard deviations of the mean.

Figure 15.39 illustrates the histogram of the residuals for the First City Real Estate example. The distribution of residuals looks to be close to a normal distribution. Figure 15.40 shows the histogram for the standardized residuals, which has the same basic shape as the residual distribution in Figure 15.39.

## Corrective Actions

If, based on analyzing the residuals, you decide the model constructed is not appropriate, but you still want a regression-based model, some corrective action may be warranted. There are three approaches that may work: Transform some of the existing independent variables, remove some variables from the model, or start over in the development of the regression model.

Earlier in this chapter, we discussed a basic approach involved in variable transformation. In general, we use the transformations of the independent variables (such as raising $x$ to

**FIGURE 15.39 Excel 2016 Histogram of Residuals for First City Real Estate**

Excel 2016 Instructions
1. Open file: **First City-3.xlsx**.
2. Follow the instructions from Steps 2–5 on Figure 15.37.
3. Create a Histogram using the Residuals column.
4. Add appropriate titles.

**FIGURE 15.40 Excel 2016 Histogram of Standardized Residuals for First City Real Estate**

Excel 2016 Instructions
1. Open file: **First City-3.xlsx**.
2. Follow the instructions from Steps 3–5 on Figure 15.33.
3. Check **Residuals** option.
4. Check **Standardized Residuals** option.
5. Specify output location.
6. Click **OK**.
7. Create a Histogram using the Standardized Residuals column.
8. Add appropriate titles.

a power, taking the square root of $x$, or taking the log of $x$) to make the data better conform to a linear relationship. If the model suffers from nonlinearity and if the residuals have a non-constant variance, we may want to transform both the independent and dependent variables. In cases in which the normality assumption is not satisfied, transforming the dependent variable is often useful. In many instances, a log transformation works. In some instances, a transformation involving the product of two independent variables will help. A more detailed discussion is beyond the scope of this text. However, you can read more about this subject in the Kutner et al. reference listed at the end of the book.

The alternative of using a different regression model means that we respecify the model to include new independent variables or remove existing variables from the model. In most modeling situations, we are in a continual state of model respecification. We are always seeking to improve the regression model by finding new independent variables.

# 15.5 EXERCISES

## Skill Development

**15-48.** You are given the following values for dependent and independent variables:

| y | x |
|------|------|
| 25 | 36 |
| 400 | 81 |
| 784 | 196 |
| 900 | 324 |
| 1,089 | 484 |
| 1,225 | 729 |
| 2,025 | 1,089 |

a. Develop the linear regression model.
b. Test to determine whether the regression model is statistically significant using alpha = 0.05.
c. Calculate the residuals and standardized residuals.

**15-49.** Referring to Exercise 15-48, develop a plot of the residuals against the $x$ values. Discuss whether this plot implies that any of the regression assumptions have been violated.

**15-50.** Consider the following values for an independent and dependent variable:

| x | y |
|----|----|
| 6 | 5 |
| 9 | 20 |
| 14 | 28 |
| 18 | 30 |
| 22 | 33 |
| 27 | 35 |
| 33 | 45 |

a. Determine the estimated linear regression equation relating the dependent and independent variables.
b. Is the regression equation you found significant? Test at the $\alpha = 0.05$ level.
c. Determine both the residuals and standardized residuals. Is there anything about the residuals that

would lead you to question whether the assumptions necessary to use regression analysis are satisfied? Discuss.

**15-51.** Consider the following values for an independent and dependent variable:

| x | y |
|----|-----|
| 6 | 5 |
| 9 | 20 |
| 14 | 28 |
| 18 | 15 |
| 22 | 27 |
| 27 | 31 |
| 33 | 32 |
| 50 | 60 |
| 61 | 132 |
| 75 | 160 |

a. Determine the estimated linear regression equation relating the dependent and independent variables.
b. Is the regression equation you found significant? Test at the $\alpha = 0.05$ level.
c. Determine both the residuals and standardized residuals. Is there anything about the residuals that would lead you to question whether the assumptions necessary to use regression analysis are satisfied?

**15-52.** Examine the following data set:

| y | x |
|----|----|
| 25 | 10 |
| 35 | 10 |
| 14 | 10 |
| 45 | 20 |
| 52 | 20 |
| 41 | 20 |
| 65 | 30 |
| 63 | 30 |
| 68 | 30 |

a. Determine the estimated regression equation for this data set.
b. Calculate the residuals for this regression equation.
c. Produce the appropriate residual plot to determine if the linear function is the appropriate regression function for this data set.
d. Use a residual plot to determine if the residuals have a constant variance.
e. Produce a residual plot to determine if the residuals are independent. Assume the order of appearance is the time order of the data.

**15-53.** Examine the following data set:

| $y$ | $x_1$ | $x_2$ |
|-----|-----|-----|
| 25 | 5 | 25 |
| 35 | 5 | 5 |
| 14 | 5 | 5 |
| 45 | 25 | 40 |
| 52 | 25 | 5 |
| 41 | 25 | 25 |
| 65 | 30 | 30 |
| 63 | 30 | 30 |
| 68 | 30 | 25 |
| 75 | 40 | 30 |

a. Determine the estimated regression equation for this data set.
b. Calculate the residuals and the standardized residuals for this regression equation.
c. Produce the appropriate residual plot to determine if the linear function is the appropriate regression function for this data set.
d. Use a residual plot to determine if the residuals have a constant variance.
e. Produce the appropriate residual plot to determine if the residuals are independent.

## Computer Software Exercises

**15-54.** Refer to Exercise 15-8 in which Paul Browning is interested in analyzing the effects of several variables on the weekly sales for his business. He has collected data for a sample of 20 weeks on the following variables:

Sales ($000)
Average weekly high temperature
Advertising spending ($000)
Number of website hits
Number of customer orders received

The data are in a file called **Browning**.
a. Determine the estimated regression equation for this data set.
b. Calculate the residuals and the standardized residuals for this regression equation.
c. Produce the appropriate residual plot(s) to determine if the linear function is the appropriate regression function for this data set.

d. Use a residual plot to determine if the residuals have a constant variance.

**15-55.**  The White Cover Snowmobile Association promotes snowmobiling in both the Upper Midwest and the Rocky Mountain region. The industry has been affected in the West because of uncertainty associated with conflicting court rulings about the number of snowmobiles allowed in national parks. The association advertises in outdoor- and tourist-related publications and then sends out pamphlets, maps, and other regional related information to people who call a toll-free number and request the information. The association orders the packets from a document-printing company and likes to have enough available to meet the immediate need without having too many sitting around taking up space. The marketing manager decided to develop a multiple regression model to be used in predicting the number of calls that will be received in the coming week. A random sample of 12 weeks is selected, with the following variables:

$y$ = Number of calls
$x_1$ = Number of advertisements placed the previous week
$x_2$ = Number of calls received the previous week
$x_3$ = Number of airline tour bookings into western cities for the current week

The data are in the file called **Winter Adventures**.
a. Construct a multiple regression model using all three independent variables. Write a short report discussing the model.
b. Based on the appropriate residual plots, what can you conclude about the constant variance assumption? Discuss.
c. Based on the appropriate residual analysis, does it appear that the residuals are independent? Discuss.
d. Use an appropriate analysis of the residuals to determine whether the regression model meets the assumption of normally distributed error terms. Discuss.

**15-56.** The athletic director of State University is interested in developing a multiple regression model that might be used to explain the variation in attendance at football games at his school. A sample of 16 games was selected from home games played during the past 10 seasons. Data for the following factors were determined:

$y$ = Game attendance
$x_1$ = Team win/loss percentage to date
$x_2$ = Opponent win/loss percentage to date
$x_3$ = Games played this season
$x_4$ = Temperature at game time

The sample data are in the file called **Football**.
a. Build a multiple regression model using all four independent variables. Write a short report that outlines the characteristics of this model.
b. Develop a table of residuals for this model. What is the average residual value? Why do you suppose it came out to this value? Discuss.

c. Based on the appropriate residual plot, what can you conclude about the constant variance assumption? Discuss.

d. Based on the appropriate residual analysis, does it appear that the model errors are independent? Discuss.

e. Can you conclude, based on the appropriate method of analysis, that the model error terms are approximately normally distributed?

**15-57.** In Exercise 15-47, you were asked to use best subsets stepwise regression to establish the relationship between body fat and the independent variables weight, abdomen circumference, and thigh circumference based on data in the file **Bodyfat**. This is an extension of that exercise.

a. Construct an estimate of the regression equation using the same variables.

b. Produce the appropriate residual plots to determine if the linear function is the appropriate regression function for this data set.

c. Use a residual plot to determine if the residuals have a constant variance.

d. Produce the appropriate residual plot to determine if the residuals are independent. Assume the data were extracted in the order listed.

e. Construct a probability plot to determine if the error terms are normally distributed.

**15-58.** Refer to Exercises 15-10 and 15-35. The file titled **Flicks** contains data on total U.S. box office grosses ($billion), total number of admissions (billion), average U.S. ticket price ($), and number of movie screens for the years 1987–2015 (*source:* National Association of Theatre Owners, www.natoonline.org/data, Feb. 2016).

a. Construct a regression equation that uses the other variables to predict total U.S. box office grosses.

b. Produce the appropriate residual plots to determine if the linear function is the appropriate regression function for this data set.

c. Square each of the independent variables and add them to the model on which the regression equation in part a was built. Produce the new regression equation.

d. Use residual plots to determine if the quadratic model in part c alleviates the problem identified in part b.

# 15 Overview

## Summary

**OUTCOME 1** Understand the general concepts behind model building using multiple regression analysis.

**OUTCOME 2** Apply multiple regression analysis to business decision-making situations.

**OUTCOME 3** Analyze the computer output for a multiple regression model and interpret the regression results.

**OUTCOME 4** Test hypotheses about the significance of a multiple regression model and test the significance of the independent variables in the model.

**OUTCOME 5** Recognize potential problems when using multiple regression analysis and take steps to correct the problems.

- Multiple linear regression analysis examines the relationship between a dependent variable and more than one independent variable.
- The three steps in determining the appropriate relationship are **model specification**, **model building**, and **model diagnosis**.
- The purpose of the model is to explain variation in the dependent variable.
- Useful independent variables are those that are highly correlated with the dependent variable.
- $R^2$ is the coefficient of determination and indicates the percentage of variation explained by the regression model.
- **Multicollinearity** occurs when independent variables are highly correlated in multiple regression models.

**OUTCOME 6** Incorporate qualitative variables into a regression model by using dummy variables.

- Independent variables are not always quantitative and ratio level.
- Examples of qualitative independent variables include marital status, homeownership, employment, and type of car owned.
- Dummy variables are used to incorporate qualitative variables into multiple regression analysis.
- **Dummy variables** are numerical codes, 0 or 1, depending on whether the observation has the indicated characteristic.
- Be careful to ensure that you use one fewer dummy variable than categories to avoid the dummy variable trap.

**OUTCOME 7** Apply regression analysis to situations in which the relationship between the independent variable(s) and the dependent variable is nonlinear.

- Sometimes business situations involve a nonlinear relationship between the dependent and independent variables.
- Regression models with nonlinear relationships are more complicated to build and analyze.
- Follow this procedure for regression models with nonlinear relationships:
  - Plot the data to see the relationships between the dependent variable and independent variable. Exponential or second-or third-order polynomial relationships are commonly found.
  - Once the appropriate relationship is determined, modify the independent variable and use it in the model.

**15.4**  **Stepwise Regression** (pg. 611–618)

**OUTCOME 8** Understand the uses of stepwise regression.

- Stepwise regression develops the regression equation through either forward selection, backward elimination, or standard stepwise regression.
- **Forward selection** begins with the selection of a single independent variable that is most highly correlated with the dependent variable. More variables are added to the model as long as they reduce a significant amount of the remaining variation in the dependent variable.
- **Backward elimination** starts with all variables included in the model. Variables are removed one at a time until no more insignificant variables are found.
- **Standard stepwise** is similar to forward selection. However, if two or more variables are correlated, a variable selected in an early step may become insignificant when other variables are added at later steps. The standard stepwise procedure drops this insignificant variable from the model.

**15.5**  **Determining the Aptness of the Model** (pg. 618–627)

**OUTCOME 9** Analyze the extent to which a regression model satisfies the regression assumptions.

- Determining the aptness of a model relies on the analysis of residuals.
- A **residual** is the difference between the observed value of the dependent variable and the value predicted by the model. Residuals should be randomly scattered about the regression line with a normal distribution and constant variance.
- If the model constructed is not appropriate, corrective action should be taken, such as transforming independent variables, dropping or adding variables, or starting over with the model-building process.

## Equations

**(15.1) Population Multiple Regression Model** pg. 574

$$y = \beta_0 + \beta_1 x_1 + \beta_2 x_2 + \ldots + \beta_k x_k + \varepsilon$$

**(15.2) Estimated Multiple Regression Model** pg. 575

$$\hat{y} = b_0 + b_1 x_1 + b_2 x_2 + \ldots + b_k x_k$$

**(15.3) Correlation Coefficient** pg. 578

$$r = \frac{\Sigma (x - \bar{x})(y - \bar{y})}{\sqrt{\Sigma (x - \bar{x})^2 \Sigma (y - \bar{y})^2}} \quad \text{or} \quad r = \frac{\Sigma (x_i - \bar{x}_i)(x_j - \bar{x}_j)}{\sqrt{\Sigma (x_i - \bar{x}_i)^2 \Sigma (x_j - \bar{x}_j)^2}}$$

One $x$ variable with $y$      One $x$ variable with another $x$

**(15.4) Multiple Coefficient of Determination ($R^2$)** pg. 580

$$R^2 = \frac{\text{Sum of squares regression}}{\text{Total sum of squares}} = \frac{SSR}{SST}$$

**(15.5) F-Test Statistic** pg. 581

$$F = \frac{\dfrac{SSR}{k}}{\dfrac{SSE}{n - k - 1}}$$

**(15.6) Adjusted R-Squared** pg. 582

$$R\text{-sq(adj)} = R_A^2 = 1 - (1 - R^2)\left(\frac{n - 1}{n - k - 1}\right)$$

**(15.7) t-Test for Significance of Each Regression Coefficient** pg. 583

$$t = \frac{b_j - 0}{s_{b_j}} \qquad df = n - k - 1$$

**(15.8) Standard Error of the Estimate** pg. 584

$$s_\varepsilon = \sqrt{\frac{SSE}{n - k - 1}} = \sqrt{MSE}$$

**(15.9) Variance Inflation Factor** pg. 586

$$VIF = \frac{1}{1 - R_j^2}$$

**(15.10) Confidence Interval Estimate for the Regression Slope** pg. 587

$$b_j \pm t s_{b_j}$$

**(15.11) Polynomial Population Regression Model** pg. 598

$$y = \beta_0 + \beta_1 x + \beta_2 x^2 + \cdots + \beta_p x^p + \varepsilon$$

**(15.12) Partial F-Test Statistic** pg. 605

$$F = \frac{(SSE_R - SSE_C)/(c - r)}{MSE_C}$$

**(15.13) Residual** pg. 619

$$e_i = y_i - \hat{y}_i$$

**(15.14) Standardized Residual for Linear Regression** pg. 623

$$s_{e_i} = \frac{e_i}{s_\varepsilon \sqrt{1 + \dfrac{1}{n} + \dfrac{(x_i - \bar{x})^2}{\Sigma x^2 - \dfrac{(\Sigma x)^2}{n}}}}$$

## Key Terms

Adjusted *R*-squared pg. 582
Composite model pg. 603
Correlation coefficient pg. 577
Correlation matrix pg. 578
Dummy variable pg. 590

Interaction pg. 602
Model pg. 576
Multicollinearity pg. 585
Multiple coefficient of determination
($R^2$) pg. 580

Regression hyperplane pg. 575
Residual pg. 619
Variance inflation factor (*VIF*) pg. 586

## Chapter Exercises

### Conceptual Questions

**15-59.** Discuss the logic of how the first two variables are selected in a forward stepwise regression model.

**15-60.** Go to the library or use the Internet to find three articles that make use of a regression model with more than one independent variable. For each article, write a short summary covering the following points:

Purpose for using the model
How the variables in the model were selected
How the data in the model were selected
Any possible violations of the needed assumptions
The conclusions drawn from using the model

**15-61.** Discuss in your own terms the similarities and differences between simple linear regression analysis and multiple regression analysis.

**15-62.** Discuss what is meant by the least squares criterion as it pertains to multiple regression analysis. Is the least squares criterion any different for simple regression analysis? Discuss.

**15-63.** List the basic assumptions of regression analysis and discuss in your own terms what each means.

**15-64.** What does it mean if we have developed a multiple regression model and have concluded that the model is apt?

**15-65.** Consider the following model:

$$\hat{y} = 5 + 3x_1 + 5x_2$$

a. Provide an interpretation of the coefficient of $x_1$.
b. Is the interpretation provided in part a true regardless of the value of $x_2$? Explain.
c. Now consider the model
$\hat{y} = 5 + 3x_1 + 5x_2 + 4x_1x_2$. Let $x_2 = 1$. Give an interpretation of the coefficient of $x_1$ when $x_2 = 1$.
d. Repeat part c when $x_2 = 2$. Is the interpretation provided in part a true regardless of the value of $x_2$? Explain.
e. Considering your answers to parts c and d, what type of regression components has conditional interpretations?

### Computer Software Exercises

**15-66.** Amazon.com has become one of the most successful online merchants. Two measures of its success are sales and net income/loss figures. The data can be found in the file **Amazon** (*source:* Wikinvest, www.wikinvest.com).
a. Produce a scatter plot for Amazon's net income/loss and sales figures for the period 1995–2015.

Determine the order (or degree) of the polynomial that could be used to predict Amazon's net income/loss using sales figures for the period 1995–2015.
b. Produce the polynomial model indicated by these data.
c. Test to determine whether the overall model from part b is statistically significant. Use a significance level of 0.10.
d. Conduct a hypothesis test to determine if curvature exists in the model that predicts Amazon's net income/loss using sales figures from part b. Use a significance level of 0.02 and the test statistic approach.

**The following information applies to Exercises 15-67, 15-68, and 15-69.**
A publishing company in New York is attempting to develop a model to help predict textbook sales for books it is considering for future publication. The marketing department has collected data on several variables from a random sample of 15 books. These data are given in the file **Textbooks**.

**15-67.** Develop the correlation matrix showing the correlation between all possible pairs of variables. Test statistically to determine which independent variables are significantly correlated with the dependent variable, book sales. Use a significance level of 0.05.

**15-68.** Develop a multiple regression model containing all four independent variables. Show clearly the regression coefficients. Write a short report discussing the model. In your report make sure you cover the following issues:
a. How much of the total variation in book sales can be explained by these four independent variables? Can you conclude that the model is significant at the 0.05 level?
b. Develop a 95% confidence interval for each regression coefficient and interpret these confidence intervals.
c. Which of the independent variables can you conclude is significant in explaining the variation in book sales? Test using $\alpha = 0.05$.
d. How much of the variation in the dependent variable is explained by the independent variables? Is the model statistically significant at the $\alpha = 0.01$ level? Discuss.
e. How much, if at all, does adding one more page to the book impact the sales volume of the book?

Develop and interpret a 95% confidence interval estimate to answer this question.

f. Perform the appropriate analysis to determine the aptness of this regression model. Discuss your results and conclusions.

**15-69.** The publishing company recently came up with some additional data for the 15 books in the original sample. Two new variables, production expenditures $(x_5)$ and number of prepublication reviewers $(x_6)$, have been added. These additional data are as follows:

| Book | $x_5$ ($) | $x_6$ |
|------|-----------|-------|
| 1 | 38,000 | 5 |
| 2 | 86,000 | 8 |
| 3 | 59,000 | 3 |
| 4 | 80,000 | 9 |
| 5 | 29,500 | 3 |
| 6 | 31,000 | 3 |
| 7 | 40,000 | 5 |
| 8 | 69,000 | 4 |
| 9 | 51,000 | 4 |
| 10 | 34,000 | 6 |
| 11 | 20,000 | 2 |
| 12 | 80,000 | 5 |
| 13 | 60,000 | 5 |
| 14 | 87,000 | 8 |
| 15 | 29,000 | 3 |

Incorporating these additional data, calculate the correlation between each of these additional variables and the dependent variable, book sales.

a. Test the significance of the correlation coefficients, using $\alpha = 0.05$. Comment on your results.

b. Develop a multiple regression model that includes all six independent variables. Which, if any, variables would you recommend be retained if this model is going to be used to predict book sales for the publishing company? For any statistical tests you might perform, use a significance level of 0.05. Discuss your results.

c. Use the $F$-test approach to test the null hypothesis that all slope coefficients are 0. Test with a significance level of 0.05. What do these results mean? Discuss.

d. Do multicollinearity problems appear to be present in the model? Discuss the potential consequences of multicollinearity with respect to the regression model.

e. Discuss whether the standard error of the estimate is small enough to make this model useful for predicting the sales of textbooks.

f. Plot the residuals against the predicted value of $y$ and comment on what this plot means relative to the aptness of the model.

g. Compute the standardized residuals and form these into a frequency histogram. What does this indicate about the normality assumption?

h. Comment on the overall aptness of this model and indicate what might be done to improve the model.

**The following information applies to Exercises 15-70 through 15-79.**

The J. J. McCracken Company has authorized its marketing research department to make a study of customers who have been issued a McCracken charge card. The marketing research department hopes to identify the significant variables that explain the variation in customer purchases. Once these variables are determined, the department intends to try to attract new customers who are predicted to make a high volume of purchases.

Twenty-five customers were selected at random, and values for the following variables are recorded in the file called **McCracken**:

$$y = \text{Average monthly purchases (in dollars)}$$
$$\text{at McCracken}$$
$$x_1 = \text{Customer age}$$
$$x_2 = \text{Customer family income}$$
$$x_3 = \text{Family size}$$

**15-70.** A first step in regression analysis often involves developing a scatter plot of the data. Make the scatter plots of all the possible pairs of variables, and with a brief statement indicate what each plot says about the relationship between the two variables.

**15-71.** Compute the correlation matrix for these data. Develop the decision rule for testing the significance of each coefficient. Which, if any, correlations are not significant? Use $\alpha = 0.05$.

**15-72.** Use forward selection stepwise regression to develop the multiple regression model. The variable $x_2$, family income, was brought into the model. Discuss the reason for this.

**15-73.** Test the significance of the regression model at Step 1 of the process. Justify the significance level you have selected.

**15-74.** Develop a 95% confidence level for the slope coefficient for the family income variable at Step 1 of the model. Be sure to interpret this confidence interval.

**15-75.** Describe the regression model at Step 2 of the analysis. Be sure to discuss the effect of adding a new variable on the standard error of the estimate and on $R^2$.

**15-76.** Referring to Exercise 15-74, suppose the manager of McCracken's marketing department questions the appropriateness of adding a second variable. How would you respond to her question?

**15-77.** Looking carefully at the stepwise regression model, you can see that the value of the slope coefficient for variable $x_2$, family income, changes as a new variable is added to the regression model. Discuss why this change takes place.

**15-78.** Analyze the stepwise regression model. Write a report to the marketing manager pointing out the strengths and weaknesses of the model. Be sure to comment on the department's goal of being able to use the model to predict which customers will make a high volume of purchases from McCracken.

**15-79.** Plot the residuals against the predicted value of y and comment on what this plot means relative to the aptness of the model.

a. Compute the standardized residuals and form these in a frequency histogram. What does this indicate about the normality assumption?

b. Comment on the overall aptness of this model and indicate what might be done to improve the model.

**15-80.** An investment analyst collected data from publicly available sources on 20 randomly chosen companies. The data consisted of the 52-week-high stock prices, PE ratios, and market values of the companies. These data are in the file titled **Investment**. The analyst wishes to produce a regression equation to predict the market value using the 52-week-high stock price and the PE ratio of the company. He creates a complete second-degree polynomial.

a. Construct an estimate of the regression equation using the indicated variables.

b. Produce the appropriate residual plots to determine if the polynomial function is the appropriate regression function for this data set.

c. Use a residual plot to determine if the residuals have a constant variance.

d. Produce the appropriate residual plot to determine if the residuals are independent. Assume the data were extracted in the order listed.

e. Construct a probability plot to determine if the error terms are normally distributed.

**15-81.** Badeaux Brothers Louisiana Treats ships packages of Louisiana coffee, cakes, and Cajun spices to customers around the United States. The cost to ship these products depends primarily on the weight of the package. Badeaux charges the customers for shipping and then ships the product itself. As a part of a study of whether it is economically feasible to continue to ship products themselves, Badeaux sampled 20 recent shipments to determine what if any relationship exists between shipping costs and package weight. The data are contained in the file **Badeaux**.

a. Develop a scatter plot of the data with the dependent variable, cost, on the vertical axis and the independent variable, weight, on the horizontal axis. Does there appear to be a relationship between the two variables? Is the relationship linear?

b. Compute the sample correlation coefficient between the two variables. Conduct a test, using an alpha value of 0.05, to determine whether the population

correlation coefficient is significantly different from zero.

c. Determine the simple linear regression model for these data. Plot the simple linear regression model together with the data. Would a nonlinear model better fit the sample data?

d. Now develop a nonlinear model and plot the model against the data. Does the nonlinear model provide a better fit than the linear model developed in part c?

**15-82.** The State Tax Commission must download information files each morning. The time to download the files depends primarily on the size of the files. The commission has asked your computer consulting firm to determine what, if any, relationship exists between download time and size of files. The commission randomly selected a sample of days and provided the information contained in the file **Tax Commission**.

a. Develop a scatter plot of the data with the dependent variable, download time, on the vertical axis and the independent variable, size, on the horizontal axis. Does there appear to be a relationship between the two variables? Is the relationship linear?

b. Compute the sample correlation coefficient between the two variables. Conduct a test, using an alpha value of 0.05, to determine whether the population correlation coefficient is significantly different from zero.

c. Determine the simple linear regression model for these data. Plot the simple linear regression model together with the data. Would a nonlinear model better fit the sample data?

d. Now determine a nonlinear model and plot the model against the data. Does the nonlinear model provide a better fit than the linear model developed in part c?

**15-83.** Refer to the State Department of Transportation data set called **Liabins**. The department was interested in determining the level of compliance with the state's mandatory liability insurance law, as well as other things. *Assume the data were collected using a simple random sampling process.* Develop the best possible linear regression model using vehicle year as the dependent variable and any or all of the other variables as potential independent variables. Assume that your objective is to develop a predictive model. Write a report that discusses the steps you took to develop the final model. Include a correlation matrix and all appropriate statistical tests. Use an $\alpha = 0.05$. If you are using a nominal or ordinal variable, remember that you must make sure it is in the form of one or more dummy variables.

## Case 15.1    Dynamic Weighing, Inc.

In 2005, Stanley Ahlon and three financial partners formed Dynamic Weighing, Inc. The company was based on an idea Stanley had for developing a scale to weigh trucks in motion and thus eliminate the need for every truck to stop at weigh stations along highways. This dynamic scale would be placed in the highway approximately one-quarter mile from the regular weigh

station. The scale would have a minicomputer that would automatically record truck speeds, axle weights, and climate variables, including temperature, wind, and moisture. Stanley and his partners believed that state transportation departments in the United States would be the primary market for such a scale.

As with many technological advances, developing the dynamic scale has been difficult. When the scale finally proved accurate for trucks traveling 40 miles per hour, it would not perform for trucks traveling at higher speeds. However, eight months ago, Stanley announced that the dynamic scale was ready to be field-tested by the Nebraska State Department of Transportation under a grant from the federal government. Stanley explained to his financial partners, and to Nebraska transportation officials, that the dynamic weight would not exactly equal the static weight (truck weight on a static scale). However, he was sure a statistical relationship between dynamic weight and static weight could be determined, which would make the dynamic scale useful.

Nebraska officials, along with people from Dynamic Weighing, installed a dynamic scale on a major highway in Nebraska. Each month for six months, data were collected for a random sample of trucks weighed on both the dynamic scale and a static scale. Table 15.3 presents these data.

Once the data were collected, the next step was to determine whether, based on this test, the dynamic scale measurements could be used to predict static weights. A complete report will be submitted to the U.S. government and to Dynamic Weighing.

**TABLE 15.3** Test Data for the Dynamic Weighing Example

| Month | Front-Axle Static Weight (lb) | Front-Axle Dynamic Weight (lb) | Truck Speed (mph) | Temperature (°F) | Moisture (%) |
|---|---|---|---|---|---|
| January | 1,800 | 1,625 | 52 | 21 | 0.00 |
| | 1,311 | 1,904 | 71 | 17 | 0.15 |
| | 1,504 | 1,390 | 48 | 13 | 0.40 |
| | 1,388 | 1,402 | 50 | 19 | 0.10 |
| | 1,250 | 1,100 | 61 | 24 | 0.00 |
| February | 2,102 | 1,950 | 55 | 26 | 0.10 |
| | 1,410 | 1,475 | 58 | 32 | 0.20 |
| | 1,000 | 1,103 | 59 | 38 | 0.15 |
| | 1,430 | 1,387 | 43 | 24 | 0.00 |
| | 1,073 | 948 | 59 | 18 | 0.40 |
| March | 1,502 | 1,493 | 62 | 34 | 0.00 |
| | 1,721 | 1,902 | 67 | 36 | 0.00 |
| | 1,113 | 1,415 | 48 | 42 | 0.21 |
| | 978 | 983 | 59 | 29 | 0.32 |
| | 1,254 | 1,149 | 60 | 48 | 0.00 |
| April | 994 | 1,052 | 58 | 37 | 0.00 |
| | 1,127 | 999 | 52 | 34 | 0.21 |
| | 1,406 | 1,404 | 59 | 40 | 0.40 |
| | 875 | 900 | 47 | 48 | 0.00 |
| | 1,350 | 1,275 | 68 | 51 | 0.00 |
| May | 1,102 | 1,120 | 55 | 52 | 0.00 |
| | 1,240 | 1,253 | 57 | 57 | 0.00 |
| | 1,087 | 1,040 | 62 | 63 | 0.00 |
| | 993 | 1,102 | 59 | 62 | 0.10 |
| | 1,408 | 1,400 | 67 | 68 | 0.00 |
| June | 1,420 | 1,404 | 58 | 70 | 0.00 |
| | 1,808 | 1,790 | 54 | 71 | 0.00 |
| | 1,401 | 1,396 | 49 | 83 | 0.00 |
| | 933 | 1,004 | 62 | 88 | 0.40 |
| | 1,150 | 1,127 | 64 | 81 | 0.00 |

## Case 15.2    Glaser Machine Works

Glaser Machine Works has experienced a significant change in its business operations over the past 50 years. Glaser started business as a machine shop that produced specialty tools and products for the timber and lumber industry. This was a logical fit, given its location in the southern part of the United States. However, over the years, Glaser looked to expand its offerings beyond the lumber and timber industry. Initially, its small size coupled with its rural location made it difficult to attract the attention of large companies that could use its products. All of that began to change as Glaser developed the ability not only to fabricate parts and tools but also to assemble products for customers who needed special components in large quantities. Glaser's business really took off when first foreign and then domestic automakers began to build automobile plants in the southern United States. Glaser was able to provide quality parts quickly for firms that expected high quality and responsive delivery. Many of Glaser's customers operated with little inventory and required that suppliers be able to provide shipments with short lead times.

As part of its relationship with the automobile industry, Glaser was expected to buy into the lean-manufacturing and quality-improvement initiatives of its customers. Glaser had always prided itself on its quality, but as the number and variety of its products increased, along with ever higher expectations by its customers, Glaser knew that it would have to respond by ensuring its quality and operations were continually improving. Of recent concern was the performance of its manufacturing line 107B. This line produced a component part for a Japanese automobile company. The Japanese firm had initially been pleased with Glaser's performance, but lately the number of defects was approaching an unacceptable level. Managers of the 107B line knew the line and its workers had been asked to ramp up production to meet increased demand and that some workers were concerned with the amount of overtime being required. There was also concern about the second shift now being run at 107B. Glaser had initially run only one shift, but when demand for its product became so high that there was not sufficient capacity with one shift, additional workers were hired to operate a night shift.

Management was wondering if the new shift had been stretched beyond its capabilities. Glaser plant management asked Kristi Johnson, the assistant production supervisor for line 107B, to conduct an analysis of product defects for the line. Kristi randomly selected several days of output and counted the number of defective parts produced on the 107B line. This information, along with other data, is contained in the file **Glaser Machine Works**. Kristi promised to have a full report for the management team by the end of the month.

### Required Tasks:

1. Identify the primary issue of the case.
2. Identify a statistical model you might use to help analyze the case.
3. Develop a multiple regression model that you can use to help Kristi analyze the product defects for line 107B. Be sure to carefully specify the dependent variable and the independent variables.
4. Discuss how you will model the variables overtime hours, supervisor training, and shift.
5. Run the regression model you developed and interpret the results.
6. Which variables are significant?
7. Provide a short report that describes your analysis and explains in managerial terms the findings of your model. Be sure to explain which variables, if any, are significant explanatory variables. Provide a recommendation to management.

## Case 15.3    Hawlins Manufacturing

Ross Hawlins had done it all at Hawlins Manufacturing, a company founded by his grandfather 63 years ago. Among his many duties, Ross oversaw all the plant's operations, a task that had grown in responsibility given the company's rapid growth over the past three decades. When Ross's grandfather founded the company, there were only two manufacturing sites. Expansion and acquisition of competitors over the years had caused that number to grow to more than 50 manufacturing plants in 18 states.

Hawlins had a simple process that produced only two products, but the demand for these products was strong, and Ross had spent millions of dollars upgrading his facilities over the past decade. Consequently, most of the company's equipment was less than 10 years old on average. Hawlins's two products were made for local markets because prohibitive shipping costs prevented shipping them long distances. Product demand was sufficiently strong to support two manufacturing shifts (day and night) at every plant, and every plant had the capability to manufacture both products sold by Hawlins. Recently, the management team at Hawlins noticed that there were differences in output levels across the various plants. They were uncertain what, if anything, might explain these differences. Clearly, if some plants were more productive than others, there might be some meaningful insights that could be standardized across plants to boost overall productivity.

Ross asked Lisa Chandler, an industrial engineer at the company's headquarters, to conduct a study of the plant's productivity. Lisa randomly sampled 159 weeks of output from various plants together with the number of plant employees working that week, the plants' average age in years, the product mix produced that week (either product A or B), and whether the output was from the day or night shift. The sampled data are in the file **Hawlins Manufacturing**. The Hawlins management team is expecting a written report and a presentation by Lisa when it meets again next Tuesday.

### Required Tasks:

1. Identify the primary issue of the case.
2. Identify a statistical model you might use to help analyze the case.

3. Develop a multiple regression model for Lisa. Be sure to carefully specify the dependent variable and the independent variables.
4. Discuss how you can include the type of product (A or B) and the shift (day or night) in the regression model.
5. Run the regression model you developed and interpret the results.

6. Which variables are significant?
7. Provide a short report that describes your analysis and explains in management terms the findings of your model. Be sure to explain which variables, if any, are significant explanatory variables. Provide a recommendation to management.

## Case 15.4    Sapphire Coffee—Part 2

Jennie Garcia could not believe that her career had moved so far so fast. When she left graduate school with a master's degree in anthropology, she intended to work at a local coffee shop until something else came along that was more related to her academic background. But after a few months, she came to enjoy the business, and in a little over a year, she was promoted to store manager. When the company for whom she worked continued to grow, Jennie was given oversight of a few stores.

Now, eight years after she started as a barista, Jennie is in charge of operations and planning for the company's southern region. As a part of her responsibilities, Jennie tracks store revenues and forecasts coffee demand. Historically, Sapphire Coffee based its demand forecast on the number of stores, believing that each store sold approximately the same amount of coffee. This approach seemed to work well when the company had shops of similar size and layout, but as the company grew, stores became more varied. Now, some stores have drive-thru windows, a feature that top management added to some stores believing that it would increase coffee sales for customers who wanted a cup of coffee on their way to work but who were too rushed to park and enter the store.

Jennie noticed that weekly sales seemed to be more variable across stores in her region and was wondering what, if anything, might explain the differences. The company's financial vice president had also noticed the increased differences in sales across stores and was wondering what might be happening. In an e-mail to Jennie, he stated that weekly store sales are expected to average $5.00 per square foot. Thus, a 1,000-square-foot store would have average weekly sales of $5,000. He asked that Jennie analyze the stores in her region to see if this rule of thumb was a reliable measure of a store's performance.

Jennie had been in the business long enough to know that a store's size, although an important factor, was not the only thing that might influence sales. She had never been convinced of the efficacy of the drive-thru window, believing that it detracted from the coffee house experience that so many of Sapphire Coffee's customers had come to expect. The VP of finance was expecting the analysis to be completed by the weekend. Jennie decided to randomly select weekly sales records for 53 stores, along with each store's size, whether it was located close to a college, and whether it had a drive-thru window. The data are in the file **Sapphire Coffee-2**. A full analysis would need to be sent to the corporate office by Friday.

## Case 15.5    Wendell Motors

Wendell Motors manufactures and ships small electric motors and drives to a variety of industrial and commercial customers in and around St. Louis. Wendell is a small operation with a single manufacturing plant. Wendell's products are different from other motor and drive manufacturers because Wendell produces only small motors (25 horsepower or less) and because its products are used in a variety of industries and businesses that appreciate Wendell's quality and speed of delivery. Because it has only one plant, Wendell ships motors directly from the plant to its customers. Wendell's reputation for quality and speed of delivery allows it to maintain low inventories of motors and to ship make-to-order products directly.

As part of its ongoing commitment to lean manufacturing and continuous process improvement, Wendell carefully monitors the costs associated with both production and shipping. The manager of shipping for Wendell, Tyler Jenkins, regularly reports the shipping costs to Wendell's management team. Because few finished goods inventories are maintained, competitive delivery times often require that Wendell expedite shipments. This is almost always the case for those customers who operate their business around the clock every day of the week. Such customers might maintain their

own backup safety stock of a particular motor or drive, but circumstances often require that replacement products be rushed through production and then expedited to the customer.

Wendell's management team wondered if these special orders were too expensive to handle in this way and if it might be less expensive to produce and hold certain motors as finished goods inventory, enabling off-the-shelf delivery using less expensive modes of shipping. This might especially be true for orders that must be filled on a holiday, incurring an additional shipping charge. At the last meeting of the management team, Tyler was asked to analyze expedited shipping costs and to develop a model that could be used to estimate the cost of expediting a customer's order.

Donna Layton, an industrial engineer in the plant, was asked to prepare an inventory cost analysis to determine the expenses involved in holding additional finished goods inventory. Tyler began his analysis by randomly selecting 45 expedited shipping records. The sampled data can be found in the file **Wendell Motors**. The management team expects a full report in five days. Tyler knew he would need a model for explaining shipping costs for expedited orders and that he would also need to answer questions about what effect, if any, shipping on a holiday had on costs.

# 16 Analyzing and Forecasting Time-Series Data

## WHY YOU NEED TO KNOW

Every organization has a need to forecast the demand for the goods or services it provides. A retail clothing store must forecast the demand for the shirts it sells by shirt size. A concert promoter must forecast how many tickets will be sold before signing contracts with a performer and the concert venue. Apple relies on forecasts of the units of a new iPhone that will be demanded in order to determine how many to produce. Your state's elected officials must forecast tax revenues in order to establish a budget each year. For many organizations, the ability to forecast plays a major role in determining the general success of the organization.

---

**Quick Prep**

Review the steps used to develop a line chart discussed in Chapter 2.

Make sure you understand the steps necessary to construct and interpret linear and nonlinear regression models in Chapters 14 and 15.

Review the concepts and properties associated with means discussed in Chapter 3.

Forecasters won't have access to a crystal ball on which to rely for an accurate prediction of the future. Instead, they need tools to assist in preparing the forecast. This chapter provides an introduction to several basic forecasting techniques and illustrates how and when to apply them. We urge you to focus on the material and take with you the tools that will give you a competitive advantage over those who are not familiar with forecasting techniques.

## 16.1    Introduction to Forecasting and Time-Series Data

The concepts of forecasting and planning are often confused. Planning is the process of determining how to deal with the future. On the other hand, *forecasting* is the process of predicting what the future will be like. Forecasts are used as inputs for the planning process.

The two broad categories of forecasting techniques are qualitative and quantitative. *Qualitative forecasting* techniques are based on expert opinion and judgment. *Quantitative forecasting* techniques are based on statistical methods for analyzing quantitative historical data. This chapter focuses on quantitative forecasting techniques.

In general, analysts use quantitative forecasting techniques whenever the following conditions are true: There are historical data relating to the variable to be forecast, the historical data can be quantified, and we are willing to assume that the historical pattern will continue into the future.

## General Forecasting Issues

Forecasting is both an art and a science. The role of a forecaster is to model a real-world system. Determining the appropriate forecasting model is a challenging task, but it can be made manageable by employing the model-building process discussed in Chapter 15, which consisted of **model specification**, **model building**, and **model diagnosis**. As we will point out in later sections, guidelines exist for determining which forecasting techniques may be more appropriate than others in certain situations. However, you may have to specify (and try) several model forms for a given situation before deciding on one that is most acceptable. For example, if Walmart expects future sales patterns at a given store to look like the past sales patterns, then any model it develops should adequately fit the past data to have a reasonable chance of forecasting the future. As a forecaster, you will spend much time selecting a model's specification and estimating its parameters to reach an acceptable fit of the past data. You will need to determine how well a model fits past data, how well it performs in mock forecasting trials, and how well its assumptions appear to be satisfied. If the model is unacceptable in any of these areas, you will be forced to return to the model specification step and begin again.

An important consideration when you are developing a forecasting model is to use the simplest available model that will meet your forecasting needs. The objective of forecasting is to provide good forecasts. Do not think that a sophisticated approach is better if a simpler one will work just as well. As in football, in which some players specialize in defense and others in offense, forecasting techniques have been developed for special situations, which are generally dependent on the **forecasting horizon**. For the purpose of categorizing forecasting techniques in most business situations, the forecast horizon, or lead time, is typically divided into four categories:

1. Immediate term—less than one month
2. Short term—one to three months
3. Medium term—three months to two years
4. Long term—two years or more

As we introduce various forecasting techniques, we will indicate the forecasting horizon(s) for which each is typically best suited.

In addition to determining the desired forecasting horizon, the forecaster must determine the **forecasting period**. For instance, the forecasting period might be a day, a week, a month, a quarter, or a year. Thus, the forecasting horizon consists of one or more forecasting periods. If we want to use quantitative forecasting techniques, historical quantitative data must be

**Model Specification**
The process of selecting the forecasting technique to be used in a particular situation.

**Model Building**
The process of estimating the specified model's parameters to achieve an adequate fit of the historical data.

**Model Diagnosis**
The process of determining how well a model fits past data and how well the model's assumptions appear to be satisfied.

**Forecasting Horizon**
The number of future periods covered by a forecast. It is sometimes referred to as *forecast lead time*.

**Forecasting Period**
The unit of time for which forecasts are to be made.

**Forecasting Interval**
The frequency with which new
forecasts are prepared.

available for a similar period. If we want weekly forecasts, weekly historical data must be
available. The **forecasting interval** is generally the same length as the forecast period. That
is, if the forecast period is one week, then we will provide a new forecast each week.

## Components of a Time Series

All quantitative forecasting models use past measurements of the variable of interest to generate a forecast of the future. The past data, measured over time, are called *time-series data*.
The decision maker who plans to develop a quantitative forecasting model must analyze the
relevant time-series data.

OUTCOME **1**     **BUSINESS APPLICATION**     **Identifying Time-Series Components**

**Phillips Office Supply**   The owner of Phillips Office Supply needs to
forecast revenues for the copier service portion of the company in order to
make sure he has ample cash flows to operate the business. When forecasting this portion of the company's revenue for next year, he plans to consider
the historical pattern over the prior four years. The questions are whether
demand for copier services has generally increased or decreased and
whether there have been times during the year when demand was typically
higher than at other times. The forecasters can perform a time-series analysis of the historical sales.

Table 16.1 presents the time-series data for the revenue generated by this portion of the
company for the four-year period. An effective means for analyzing these data is to develop a
*time-series plot*, or line chart, as shown in Figure 16.1. By graphing the data, we can see the

**TABLE 16.1** Time-Series Data for Sales Revenues (Thousands of Dollars)

| Month | Billing Total | | | |
|---|---|---|---|---|
| | 2013 | 2014 | 2015 | 2016 |
| January | 170 | 390 | 500 | 750 |
| February | 200 | 350 | 470 | 700 |
| March | 190 | 300 | 510 | 680 |
| April | 220 | 320 | 480 | 710 |
| May | 180 | 310 | 530 | 710 |
| June | 230 | 350 | 500 | 660 |
| July | 220 | 380 | 540 | 630 |
| August | 260 | 420 | 580 | 670 |
| September | 300 | 460 | 630 | 700 |
| October | 330 | 500 | 690 | 720 |
| November | 370 | 540 | 770 | 850 |
| December | 390 | 560 | 760 | 880 |

**FIGURE 16.1 Time-Series
Plot for Phillips Office Supply
Copier Services Revenue**

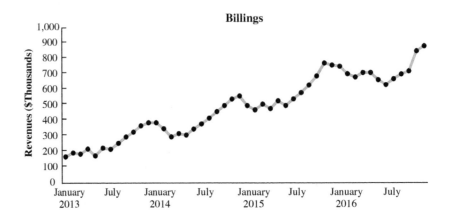

firm's revenues over the past four years. The time-series plot is an important tool in identifying the time-series components. All time-series data exhibit one or more of the following:

1. Trend component
2. Seasonal component
3. Cyclical component
4. Random component

**Trend Component**   A *trend* is the long-term increase or decrease in a variable being measured over time. Figure 16.1 shows that Phillips's revenues exhibited an upward trend over the four-year period. In other examples, the time series may exhibit a downward trend.

Trends can be classified as **linear** or nonlinear. We can observe a trend when a time series is measured in any time increment, such as years, quarters, months, or days. Figure 16.1 shows a good example of a positive linear trend. Time-series data that exhibit a linear trend tend to increase or decrease at a fairly constant rate. However, not all trends are linear. Many time series show a *nonlinear trend*. For instance, Figure 16.2 shows the total annual game attendance for the New York Yankees Major League Baseball team in the 14 years from 2001 to 2014 (*source:* "New York Yankee attendance data," www.baseball-almanac.com). Attendance was fairly flat between 2001 and 2003, increased dramatically between 2003 and 2006, slowed down again through 2008, and then declined sharply since 2008.

**Seasonal Component**   Another component that may be present in time-series data is the **seasonal component**. Many time series show a repeating pattern over time. For instance, Figure 16.1 shows a time series that exhibits a wavelike pattern. This pattern repeats throughout the time series. Copier services revenues reach an annual maximum around January and then decline to an annual minimum around April. This pattern repeats every 12 months. The shortest period of repetition for a pattern is known as its *recurrence period*. A seasonal component's recurrence period is at most one year. If the time series exhibits a repetitious pattern with a recurrence period longer than a year, the time series is said to exhibit a cyclical effect—a concept we will explore shortly.

In analyzing past sales data for a retail toy store, we would expect to see sales increase in the months leading into Christmas and then substantially decrease after Christmas. Automobile gasoline sales might show a seasonal increase during the summer months, when people drive more, and a decrease during the cold winter months. These predictable highs and lows at specific times during the year indicate seasonality in data.

To view seasonality in a time series, the data must be measured quarterly, monthly, weekly, or daily. Annual data do not show seasonal patterns of highs and lows. Figure 16.3 shows quarterly sales data for a hotel chain from summer 2010 through winter 2015. Notice that the data exhibit a definite seasonal pattern. The local maximums occur in the spring. The recurrence period of the component in the time series is, therefore, one year. The winter quarter tends to be low, whereas the following quarter (spring) is the high quarter each year.

We can also observe seasonality in time-series data measured over time periods shorter than a year. For example, the number of checks processed daily by a bank may show

**FIGURE 16.2** **New York Yankees Annual Attendance Showing a Nonlinear Trend**

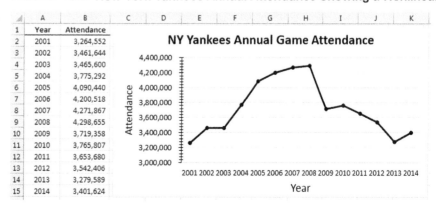

| | A | B |
|---|---|---|
| 1 | Year | Attendance |
| 2 | 2001 | 3,264,552 |
| 3 | 2002 | 3,461,644 |
| 4 | 2003 | 3,465,600 |
| 5 | 2004 | 3,775,292 |
| 6 | 2005 | 4,090,440 |
| 7 | 2006 | 4,200,518 |
| 8 | 2007 | 4,271,867 |
| 9 | 2008 | 4,298,655 |
| 10 | 2009 | 3,719,358 |
| 11 | 2010 | 3,765,807 |
| 12 | 2011 | 3,653,680 |
| 13 | 2012 | 3,542,406 |
| 14 | 2013 | 3,279,589 |
| 15 | 2014 | 3,401,624 |

**FIGURE 16.3 Hotel Sales by Quarter**

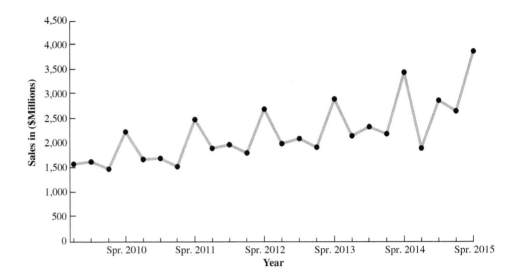

predictable highs and lows at certain times during a month. The pattern of customers arriving at the bank during any hour may be "seasonal" within a day, with more customers arriving near opening time, around the lunch hour, and near closing time.

**Cyclical Component** If you observe time-series data over a long enough time span, you may see sustained periods of high values followed by periods of lower values. If the recurrence period of these fluctuations is longer than a year, the data are said to contain a **cyclical component**.

**Cyclical Component**
A wavelike pattern within the time series that repeats throughout the time series and has a recurrence period of more than one year.

National economic measures such as the unemployment rate, gross national product, stock market indexes, and personal saving rates tend to cycle. The cycles vary in length and magnitude. That is, some cyclical time series may have longer runs of high and low values than others. Also, some time series may exhibit deeper troughs and higher crests than others. Figure 16.4 shows monthly housing starts in the United States between 1995 and 2016 (*source:* "Housing starts: total: new privately owned housing units started," Economic Research, Federal Reserve Bank of St. Louis, https://research.stlouisfed.org, Mar. 16, 2016). Note the definite cyclical pattern, with low periods in 1995, 1997, 2001, and 2009. Although the pattern resembles the shape of a seasonal component, the length of the recurrence period identifies this pattern as being the result of a cyclical component.

**FIGURE 16.4 Time-Series Plot of Housing Starts**

**Random Component**    Although not all time series possess a trend, seasonal, or cyclical component, virtually all time series have a **random component**. The random component is often referred to as "noise" in the data. A time series with no identifiable pattern is completely random and contains only noise. In addition to other components, each of the time series in Figures 16.1 through 16.4 contains random fluctuations.

In the following sections of this chapter, you will see how various forecasting techniques deal with the time-series components. An important first step in forecasting is to identify which components are present in the time series to be analyzed. As we have shown, constructing a time-series plot is the first step in this process.

**Random Component**
Changes in time-series data that are unpredictable and cannot be associated with a trend, seasonal, or cyclical component.

OUTCOME **2**

## Introduction to Index Numbers

When analyzing time-series data, decision makers must often compare one value measured at one point in time with other values measured at different points in time. For example, a real estate broker may wish to compare house prices in 2016 with house prices in previous years. A common procedure for making relative comparisons is to begin by determining a **base period index** to which all other data values can be fairly compared.

Equation 16.1 is used to make relative comparisons for data found in different periods by calculating a *simple index number*.

**Base Period Index**
The time-series value to which all other values in the time series are compared. The index number for the base period is defined as 100.

**Simple Index Number**

$$I_t = \frac{y_t}{y_0}(100) \tag{16.1}$$

where:

$I_t$ = Index number at time period $t$
$y_t$ = Value of the time series at time $t$
$y_0$ = Value of the time series at the index base period

---

**EXAMPLE 16-1**    Computing Simple Index Numbers

Cranston Supplies, Inc.    The managers at Cranston Supplies, Inc., a company that makes electronic control equipment for electric utilities, are considering the purchase of another controls manufacturer in North Carolina. The agents representing the company's current owners have touted their company's rapid sales growth over the past 10 years as a reason for their asking price. To gain a better understanding of the sales history, Cranston executives wish to convert the company's sales data to index numbers. The following steps can be used to do this:

STEP 1 **Obtain the time-series data.**
The company has sales data for each of the 10 years since 2006.

STEP 2 **Select a base period.**
Cranston managers have selected 2006 as the index base period. Sales in 2006 were $14.0 million.

STEP 3 **Compute the simple index numbers for each year using Equation 16.1.**
For instance, sales in 2007 were $15.2 million. Using Equation 16.1, the index for 2007 is

$$I_t = \frac{y_t}{y_0}(100)$$

$$I_{2007} = \frac{15.2}{14.0}(100) = 108.6$$

For the 10 years, we get:

| Year | Sales ($ millions) | Index |
|------|---------------------|-------|
| 2006 | 14.0 | 100.0 |
| 2007 | 15.2 | 108.6 |
| 2008 | 17.8 | 127.1 |
| 2009 | 21.4 | 152.9 |
| 2010 | 24.6 | 175.7 |
| 2011 | 30.5 | 217.9 |
| 2012 | 29.8 | 212.9 |
| 2013 | 32.4 | 231.4 |
| 2014 | 37.2 | 265.7 |
| 2015 | 39.1 | 279.3 |

**TRY EXERCISE 16-8** (pg. 643)

Referring to Example 16-1, we can use the index numbers to determine the percentage change from the base year for any year. For instance, sales in 2013 have an index of 231.4. This means that sales in 2013 were 131.4% above sales in the base year of 2006. Sales in 2015 were 179.3% higher than they were in 2006.

Note that although we can use the index number to compare values between any one time period and the base period and we can express the difference in percentage-change terms, we cannot compute period-to-period changes by subtracting the index numbers. For instance, in Example 16-1, when comparing sales for 2014 and 2015, we cannot say that the growth has been

$$279.3 - 265.7 = 13.6\%$$

To determine the actual percentage growth, we do the following:

$$\frac{279.3 - 265.7}{265.7}100 = 5.1\%$$

Thus, the sales growth rate between 2014 and 2015 was 5.1%, not 13.6%.

## Using Index Numbers to Deflate a Time Series

A common use of index numbers is to convert values measured at different times into more directly comparable values. For instance, if your wages increase, but at a rate less than inflation, you will in fact be earning less in "real terms." A company experiencing increasing sales at a rate of increase less than inflation is actually not increasing in "real terms."

**BUSINESS APPLICATION** Deflating Time-Series Values Using Index Values

**Wyman-Gorman Company** The Wyman-Gorman Company designs and produces forgings, primarily for internal combustion engines. Suppose that in 2005 the company experienced financial difficulty and declared bankruptcy. Table 16.2 shows sales in millions of dollars for the company for 1996 to 2005. Also shown is the PPI (Producer Price Index) for the same years. Finally, sales, adjusted to 1984 dollars, are shown. (*Source:* "Producer Price Index by industries: total manufacturing industries," Economic Research, Federal Reserve Bank of St. Louis, https://research.stlouisfed.org, Mar. 15, 2016.)

Equation 16.2 is used to determine the adjusted time-series values.

**Deflation Formula**

$$y_{adj_t} = \frac{y_t}{I_t}(100) \tag{16.2}$$

where:

$y_{adj_t}$ = Deflated time-series value at time $t$
$y_t$ = Actual value of the time series at time $t$
$I_t$ = Index (such as CPI or PPI) at time $t$

TABLE 16.2 Deflated Sales Data—Using Producer Price Index (PPI)

| Year | Sales ($ millions) | PPI (Base = 1984) | Sales ($ millions, adjusted to 1984 dollars) |
|------|--------------------|--------------------|----------------------------------------------|
| 1996 | 610.3 | 127.7 | 477.9 |
| 1997 | 473.1 | 127.6 | 370.8 |
| 1998 | 383.5 | 124.4 | 308.3 |
| 1999 | 425.5 | 125.5 | 339.0 |
| 2000 | 384.1 | 132.7 | 289.4 |
| 2001 | 341.1 | 134.2 | 254.2 |
| 2002 | 310.3 | 131.1 | 236.7 |
| 2003 | 271.6 | 138.1 | 196.7 |
| 2004 | 371.6 | 142.7 | 260.4 |
| 2005 | 390.2 | 157.4 | 247.9 |

For instance, in 1996 sales were $610.3 million. The PPI for that year was 127.7. The sales, adjusted to 1984 dollars, are

$$y_{adj_{1996}} = \frac{610.3}{127.7}(100) = \$477.9$$

# 16.1 EXERCISES

## Skill Development

**16-1.** What is meant by time-series data? Give an example.

**16-2.** Explain the difference between time-series data and cross-sectional data. Are these two types of data sets mutually exclusive? What do they have in common? How do they differ?

**16-3.** What are the differences between quantitative and qualitative forecasting techniques? Under what conditions is it appropriate to use a quantitative technique?

**16-4.** Provide an example of a business decision that requires (1) a short-term forecast, (2) a medium-term forecast, and (3) a long-term forecast.

**16-5.** What is meant by the trend component of a time series? How is a linear trend different from a nonlinear trend?

**16-6.** Must a seasonal component be associated with the seasons (fall, spring, summer, winter) of the year? Provide an example of a seasonal effect that is not associated with the seasons of the year.

**16-7.** A Greek entrepreneur followed the olive harvests. He noted that olives ripen in September. Each March he tried to determine if the upcoming olive harvest would be especially bountiful. If his analysis indicated it would, he entered into agreements with the owners of all the olive oil presses in the region. In exchange for a small deposit months ahead of the harvest, he obtained the right to lease the presses at market prices during the harvest. If he was correct about the harvest and demand for olive oil presses boomed, he could make a great deal of money. Identify the following quantities in the context of this scenario:

a. forecasting horizon
b. category that applies to the forecasting horizon identified in part a
c. forecasting period
d. forecasting interval

**16-8.** Consider the following median selling prices ($thousands) for homes in a community:

| Year | Price |
|------|-------|
| 1 | 320 |
| 2 | 334 |
| 3 | 329 |
| 4 | 344 |
| 5 | 358 |
| 6 | 347 |
| 7 | 383 |
| 8 | 404 |
| 9 | 397 |
| 10 | 411 |

a. Use year 1 as a base year and construct a simple index number to show how the median selling price has increased.
b. Determine the actual percentage growth in the median selling price between the base year and year 10.
c. Determine the actual percentage growth in the median selling price between the base year and year 5.
d. Determine the actual percentage growth in the median selling price between year 5 and year 10.

**16-9.** The following values represent advertising rates paid by a regional catalog retailer that advertises either on radio or in newspapers:

| Year | Radio Rates ($) | Newspaper Rates ($) |
|------|-----------------|---------------------|
| 1 | 300 | 400 |
| 2 | 310 | 420 |
| 3 | 330 | 460 |
| 4 | 346 | 520 |
| 5 | 362 | 580 |
| 6 | 380 | 640 |
| 7 | 496 | 660 |

Determine a relative index for each type of advertisement using year 1 as the base year.

## Business Applications

Exercises **16-10 and 16-11** refer to Gallup Construction and Paving, a company whose primary business is constructing homes in planned communities in the Upper Midwest. Suppose the company has kept a record of the relative costs of labor and materials in its market areas for the last 11 years. These data are as follows:

| Year | Hourly Wages ($) | Average Material Cost ($) |
|------|------------------|---------------------------|
| 2005 | 30.10 | 66,500 |
| 2006 | 30.50 | 68,900 |
| 2007 | 31.70 | 70,600 |
| 2008 | 32.50 | 70,900 |
| 2009 | 34.00 | 71,200 |

| Year | Hourly Wages ($) | Average Material Cost ($) |
|------|------------------|---------------------------|
| 2010 | 35.50 | 71,700 |
| 2011 | 35.10 | 72,500 |
| 2012 | 35.05 | 73,700 |
| 2013 | 34.90 | 73,400 |
| 2014 | 33.80 | 74,100 |
| 2015 | 34.20 | 74,000 |

**16-10.** Using 2005 as the base year, construct a separate index for each component in the construction of a house.

**16-11.** Plot both series of data and comment on the trend you see in both plots.

## Computer Software Exercises

**16-12.** The federal funds rate is the interest rate charged by banks when banks borrow "overnight" from each other. The file titled **The Fed** contains the federal funds rates for the period 1955–2015 (*source:* "Effective federal funds rate," Economic Research, Federal Reserve Bank of St. Louis, https://research.stlouisfed.org, Mar. 7, 2016). Construct a time-series plot for the federal funds rates for the period 1955–2015.

**16-13.** Refer back to Exercise 16-12 involving the federal funds rate and the data in the file called **The Fed**. Describe the time-series components that are present in the data.

**16-14.** Refer back to Exercises 16-12 and 16-13 involving the federal funds rate and the data in the file called **The Fed**. Indicate the recurrence period for any cyclical or seasonal component in the data.

---

| 16.2 | **Trend-Based Forecasting Techniques** |

As we discussed in Section 16.1, some time series exhibit an increasing or decreasing trend. Further, the trend may be linear or nonlinear. A plot of the data will be very helpful in identifying which, if any, of these trends exist.

### Developing a Trend-Based Forecasting Model

In this section, we introduce *trend-based forecasting techniques*. As the name implies, these techniques are used to identify the presence of a trend and to model that trend. Once the trend model has been defined, it is used to provide forecasts for future time periods.

**OUTCOMES 3, 4**   **BUSINESS APPLICATION**   Linear Trend Forecasting

**The Taft Ice Cream Company**   The Taft Ice Cream Company is a family-operated company that sells gourmet ice cream to resort areas, primarily on the North Carolina coast. Figure 16.5 displays the annual sales data for the 10-year period 2006–2015 and shows the time-series plot illustrating that sales have trended up in the 10-year period. These data are in a file called **Taft**.

Taft's owners are considering expanding their ice cream manufacturing facilities. As part of the bank's financing requirements, the managers are asked to supply a forecast of future

**FIGURE 16.5 Excel 2016 Output Showing Taft Ice Cream Sales Trend Line**

Excel 2016 Instructions

1. Open file: **Taft.xlsx**.
2. Select the data in the *Sales* column.
3. Select **Insert > Line or Area Chart**.
4. Click **Select Data**.
5. Under **Horizontal (categories) Axis Labels**, select data in *Year* column.
6. On the **Design** tab, enter desired title and horizontal and vertical axes labels.
7. Remove gridlines.

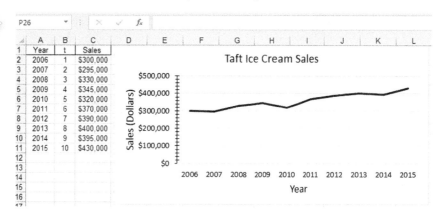

**Excel Tutorial**

sales. Recall from our earlier discussions that the forecasting process has three steps: (1) model specification, (2) model fitting, and (3) model diagnosis.

**STEP 1 Model specification.**

The time-series plot in Figure 16.5 indicates that sales have exhibited a linear growth pattern. A possible forecasting tool is a linear trend (straight-line) model.

**STEP 2 Model fitting.**

Because we have specified a linear trend model, the process of fitting can be accomplished using least squares regression analysis of a form described by Equation 16.3.

**Linear Trend Model**

$$y_t = \beta_0 + \beta_1 t + \varepsilon_t \tag{16.3}$$

where:

$$y_t = \text{Value of the trend at time } t$$
$$\beta_0 = y \text{ intercept of the trend line}$$
$$\beta_1 = \text{Slope of the trend line}$$
$$t = \text{Time period } (t = 1, 2, \dots )$$
$$\varepsilon_t = \text{Model error at time } t$$

We let the first period in the time series be $t = 1$, the second period be $t = 2$, and so forth. The values for time form the independent variable, with sales being the dependent variable. From Chapter 14, we know that the least squares regression equations for the slope and intercept are estimated by Equations 16.4 and 16.5. Here the sums are taken over the values of $t$ ($t = 1, 2, \dots$).

**Least Squares Equations Estimates**

$$b_1 = \frac{\Sigma t y_t - \dfrac{\Sigma t \Sigma y_t}{n}}{\Sigma t^2 - \dfrac{(\Sigma t)^2}{n}} \tag{16.4}$$

$$b_0 = \frac{\Sigma y_t}{n} - b_1 \frac{\Sigma t}{n} \tag{16.5}$$

where:

$$n = \text{Number of periods in the time series}$$
$$t = \text{Time period (independent variable)}$$
$$y_t = \text{Dependent variable at time } t$$

**FIGURE 16.6 Excel 2016 Output for Taft Ice Cream Trend Model**

| | A | B | C | D | E | F | G |
|---|---|---|---|---|---|---|---|
| 1 | SUMMARY OUTPUT | | | | | | |
| 2 | | | | | | | |
| 3 | Regression Statistics | | | | | | |
| 4 | Multiple R | 0.9551 | | | | | |
| 5 | R Square | 0.9123 | | | | | |
| 6 | Adjusted R Square | 0.9013 | | | | | |
| 7 | Standard Error | 14513.58 | | | | | |
| 8 | Observations | 10 | | | | | |
| 9 | | | | | | | |
| 10 | ANOVA | | | | | | |
| 11 | | df | SS | MS | F | Significance F | |
| 12 | Regression | 1 | 17,527,348,484.85 | 17,527,348,484.85 | 83.2084 | 0.0000 | |
| 13 | Residual | 8 | 1,685,151,515.15 | 210,643,939.39 | | | |
| 14 | Total | 9 | 19,212,500,000.00 | | | | |
| 15 | | | | | | | |
| 16 | | Coefficients | Standard Error | t Stat | P-value | Lower 95% | Upper 95% |
| 17 | Intercept | 277,333.33 | 9914.6611 | 27.9720 | 0.0000 | 254,470.08 | 300,196.58 |
| 18 | t | 14,575.76 | 1597.8923 | 9.1219 | 0.0000 | 10,891.01 | 18,260.50 |

Linear trend equation:
Sales = 277,333.33 + 14,575.76t

We can use the linear regression procedure in Excel to compute the least squares trend model. Figure 16.6 shows the Excel output for the Taft Ice Cream Company example. The least squares trend model is

$$\hat{y}_t = b_0 + b_1 t$$

$$\hat{y}_t = 277,333.33 + 14,575.76t$$

For a forecast, we use $F_t$ as the forecast value or predicted value at time period $t$. Thus,

$$F_t = 277,333.33 + 14,575.76t$$

STEP 3 **Model diagnosis.**

The linear trend regression output in Figure 16.6 offers some conclusions about the potential capabilities of our model. $R$-squared $= 0.9123$ shows that for these 10 years of data, the linear trend model explains more than 91% of the variation in sales. The $p$-value for the regression slope coefficient to four decimal places is 0.0000. This means that time ($t$) can be used to explain a significant portion of the variation in sales. Figure 16.7 shows the plot of the trend line through the data. You can see that the trend model fits the historical data quite closely. Although these results are a good sign, the model diagnosis step requires further analysis.

## Comparing the Forecast Values to the Actual Data

The slope of the trend line indicates that Taft Ice Cream Company has experienced an average increase in sales of $14,575.76 per year over the 10-year period. We can find the linear trend model's fitted sales values for periods $t = 1$ through $t = 10$ by substituting for $t$ in the following forecast equation:

$$F_t = 277,333.33 + 14,575.76t$$

For example, for $t = 1$, we get

$$F_t = 277,333.33 + 14,575.76(1)$$
$$= \$291,909.09$$

The Excel 2016 functions that can be used to calculate the linear trend model forecast are

= **FORECAST.LINEAR**(x,known _ys,known_xs).
= **FORECAST.LINEAR**(1,Sales,t) where Sales are the known_ys and t are the known_xs

Note that the actual sales figure, $y_1$, for period 1 was $300,000. The difference between the actual sales in time $t$ and the forecast values in time $t$, found using the trend model, is called the *forecast error* or the *residual*. Figure 16.8 shows the forecasts for periods 1 through 10 and the forecast errors at each period.

Computing the forecast error by comparing the trend-line values with actual past data is an important part of the model diagnosis step. The errors measure how closely the

**FIGURE 16.7 Excel 2016 Output for Taft Ice Cream Trend Line**

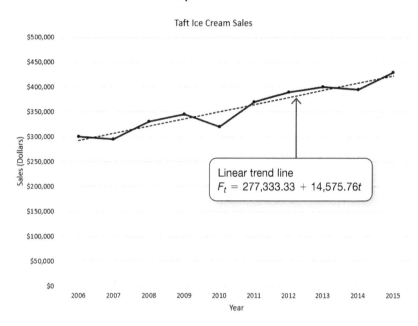

**FIGURE 16.8 Excel 2016 Residual Output for Taft Ice Cream**

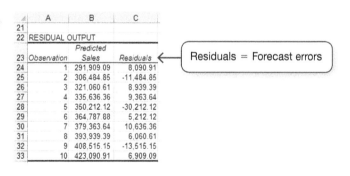

model fits the actual data at each point. A perfect fit would lead to residuals of 0 each time. We would like to see small residuals and an overall good fit. Two commonly used measures of fit are mean square residual, or *mean square error* (*MSE*), and *mean absolute deviation* (*MAD*). These measures are computed using Equations 16.6 and 16.7, respectively. *MAD* measures the average magnitude of the forecast errors. *MSE* is a measure of the variability in the forecast errors. The forecast error is the observed value, $y_t$, minus the predicted value, $F_t$.

**FIGURE 16.9 Excel 2016 *MSE* and *MAD* Computations for Taft Ice Cream**

| | A | B | C | D | E | F |
|---|---|---|---|---|---|---|
| 21 | | | | | | |
| 22 | RESIDUAL OUTPUT | | | | | |
| 23 | Observation | Predicted Sales | Residuals | Squared Residuals | | Absolute Value of Residuals |
| 24 | 1 | 291,909.09 | 8,090.91 | 65,462,809.92 | | 8,090.91 |
| 25 | 2 | 306,484.85 | -11,484.85 | 131,901,744.72 | | 11,484.85 |
| 26 | 3 | 321,060.61 | 8,939.39 | 79,912,764.00 | | 8,939.39 |
| 27 | 4 | 335,636.36 | 9,363.64 | 87,677,685.95 | | 9,363.64 |
| 28 | 5 | 350,212.12 | -30,212.12 | 912,772,268.14 | | 30,212.12 |
| 29 | 6 | 364,787.88 | 5,212.12 | 27,166,207.53 | | 5,212.12 |
| 30 | 7 | 379,363.64 | 10,636.36 | 113,132,231.40 | | 10,636.36 |
| 31 | 8 | 393,939.39 | 6,060.61 | 36,730,945.82 | | 6,060.61 |
| 32 | 9 | 408,515.15 | -13,515.15 | 182,659,320.48 | | 13,515.15 |
| 33 | 10 | 423,090.91 | 6,909.09 | 47,735,537.19 | | 6,909.09 |
| 34 | | | | | | |
| 35 | | | Sum = | 1,685,151,515.15 | | 110,424.24 |
| 36 | | | | | | |
| 37 | | | MSE = | 168,515,151.52 | MAD = | 11,042.42 |

**Excel 2016 Instructions**

1. Open file: **Taft.xlsx**.
2. Select **Data > Data Analysis**.
3. Select **Regression**.
4. Click **Residuals**.
5. Enter range for *y* variable (*Sales*).
6. Enter range for *x* variable ($t = 1,2,3,. . .$).
7. Click **Labels**.
8. Specify output location.
9. Create a new column of squared residuals (i.e., cell D25 use equation = C25^2).
10. Create a column of absolute values of the residuals (i.e., cell F25 use equation = ABS(C25)).
11. Use Equations 16.6 and 16.7 to calculate *MSE* and *MAD*.

**Mean Square Error**

$$MSE = \frac{\sum (y_t - F_t)^2}{n} \qquad (16.6)$$

**Mean Absolute Deviation**

$$MAD = \frac{\sum |y_t - F_t|}{n} \qquad (16.7)$$

where:

$$y_t = \text{Actual value at time } t$$
$$F_t = \text{Predicted value at time } t$$
$$n = \text{Number of time periods}$$

Figure 16.9 shows the *MSE* and *MAD* calculations using Excel for the Taft Ice Cream example. The *MAD* value of $11,042.42 indicates that the linear trend model has an average absolute error of $11,042.42 per period. The *MSE* (in squared units) equals 168,515,151.52. The square root of the *MSE* (often referred to as *RMSE*, root mean square error) is $12,981.34, and although it is not equal to the *MAD* value, it does provide similar information about the relationship between the forecast values and the actual values of the time series.[1]

These error measures are particularly helpful when we want to compare two or more forecasting techniques. We can compute the *MSE* and/or the *MAD* for each forecasting technique. The forecasting technique that gives the smallest *MSE* or *MAD* is generally considered to provide the best fit.

**Autocorrelation**   In addition to examining the fit of the forecasts to the actual time series, the model-diagnosis step should examine how a model meets the assumptions of the regression model. One regression assumption is that the error terms are uncorrelated, or independent.

---

[1] Technically this is the square root of the average squared distance between the forecasts and the observed data values. Algebraically, of course, this is not the same as the average forecast error, but it is comparable.

When regression is used with time-series data, the assumption of independence could be violated. That is, the error terms may be correlated over time. We call this *serial correlation*, or **autocorrelation**.

**Autocorrelation**

Correlation of the error terms (residuals) when the residuals at points in time are related.

For a time-series variable, the value of $y$ at time period $t$ is commonly related to the value of $y$ at previous time periods. If a relationship between $y_t$ and $y_{t-1}$ exists, we conclude that first-order autocorrelation exists. If $y_t$ is related to $y_{t-2}$, second-order autocorrelation exists, and so forth. If the time-series values are autocorrelated, the assumption that the error terms are independent is violated.

The autocorrelation can be positive or negative. For instance, when the values are first-order positively autocorrelated, we expect a positive residual to be followed by a positive residual in the next period, and we expect a negative residual to be followed by another negative residual. With negative first-order autocorrelation, we expect a positive residual to be followed by a negative residual, followed by a positive residual, and so on. The presence of autocorrelation can have adverse consequences on tests of statistical significance in a regression model. Thus, you need to be able to detect the presence of autocorrelation and take action to remove the problem. The *Durbin-Watson statistic*, shown in Equation 16.8, is used to test whether residuals are autocorrelated.

**Durbin-Watson Statistic**

$$d = \frac{\sum_{t=2}^{n}(e_t - e_{t-1})^2}{\sum_{t=1}^{n}e_t^2}$$

(16.8)

where:

$$d = \text{Durbin-Watson test statistic}$$
$$e_t = (y_t - \hat{y}_t) = \text{Residual at time } t$$
$$n = \text{Number of time periods in the time series}$$

Figure 16.10 shows the XLSTAT add-in for Excel output providing the Durbin-Watson statistic for the Taft Ice Cream data, as follows:

$$d = \frac{\sum_{t=2}^{n}(e_t - e_{t-1})^2}{\sum_{t=1}^{n}e_t^2} = 2.65$$

Examining Equation 16.8, we see that if successive values of the residual are close in value, the Durbin-Watson $d$ statistic will be small. This would describe residuals that are positively correlated.

Two Excel 2016 functions can be used to compute the Durbin-Watson statistic from the residual output:

=**SUMXMY2**($e_2$:$e_n$,$e_1$:$e_{n-1}$)/ **SUMSQ**($e_1$:$e_n$), where e refers to the residual.

Excel 2016 (XLSTAT) Instructions

1. Open **XLSTAT** and **Enable Macros**.
2. Open file: **Taft.xlsx**.
3. On the **XLSTAT** tab, click **Modeling data > Linear regression**.
4. Select Y/Dependent variable: (**Sales**) and X/Explanatory variables: Quantitative: **t**.
5. Select the **Variable labels** check box.
6. On the **Outputs** tab, select **Descriptive statistics**, **Correlations**, and **Analysis of variance**. Click **OK**.

**FIGURE 16.10** Excel 2016 (XLSTAT)—Durbin-Watson Statistic: Taft Ice Cream Company Example

| | A | B | C | D |
|---|---|---|---|---|
| 25 | | Goodness of fit statistics: | | |
| 26 | | Observations | 10.00 | |
| 27 | | Sum of weights | 10.00 | |
| 28 | | DF | 8.00 | |
| 29 | | R² | 0.91 | |
| 30 | | Adjusted R² | 0.90 | Durbin-Watson Statistic |
| 31 | | MSE | 210643939.39 | |
| 32 | | RMSE | 14513.58 | |
| 33 | | DW | 2.65 | |

The value of the Durbin-Watson statistic can range from 0 to 4. A value of 2 indicates no autocorrelation. However, like any other statistics computed from a sample, the Durbin-Watson $d$ is subject to sampling error. We may wish to conduct a formal hypothesis test to determine whether positive autocorrelation exists:

$$H_0: \rho = 0$$
$$H_A: \rho > 0$$

If the $d$ statistic is too small, we will reject the null hypothesis and conclude that positive autocorrelation exists. If the $d$ statistic is too large, we will not reject and will not be able to conclude that positive autocorrelation exists. Appendix N contains a table of one-tailed Durbin-Watson critical values for $\alpha = 0.05$ and $\alpha = 0.01$ levels. (*Note:* The critical values in Appendix N are for one-tailed tests with $\alpha = 0.05$ or 0.01. For a two-tailed test, the alpha is doubled.) The Durbin-Watson table provides two critical values: $d_L$ and $d_U$. In this test for positive autocorrelation, the decision rule is

> If $d < d_L$, *reject $H_0$* and conclude that positive autocorrelation exists.
>
> If $d > d_U$, do not reject $H_0$ and conclude that no positive autocorrelation exists.
>
> If $d_L < d < d_U$, the test is inconclusive.

The Durbin-Watson test is not reliable for sample sizes smaller than 15. Therefore, for the Taft Ice Cream Company application, we are unable to conduct the hypothesis test for autocorrelation. However, Example 16-2 shows a Durbin-Watson test carried out.

---

**EXAMPLE 16-2** | **Testing for Autocorrelation**

**Wilder Foods, Inc.** Wilder Foods supplied hamburger patties to fast-food companies such as Burger King and Jack in the Box and operated between 1997 and 2014, when it was sold to another meat supplier. During this time, revenues grew steadily. Suppose that the time-series plot in Figure 16.11 shows the trend in sales revenues for the company. The data are in a file called **Wilder Foods**.

Assume that the managers of the current company developed a linear trend regression model they hope to use to forecast revenue for the next two years to determine whether they can support adding another processing line to their Ohio factory. They are now interested in determining whether the linear model is subject to positive autocorrelation. To test for this, the following steps can be used:

STEP 1 **Specify the model.**
Based on a study of the line chart, the forecasting model is to be a simple linear trend regression model, with revenue as the dependent variable and time ($t$) as the independent variable.

**FIGURE 16.11 Time-Series Plot of Wilder Foods Revenue Data**

**FIGURE 16.12** **Wilder Foods Trend Line**

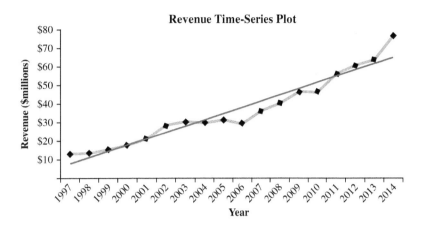

STEP 2  **Fit the model.**

Because we have specified a linear trend model, to accomplish the process of fitting we can use least squares regression analysis and Excel to estimate the slope and intercept for the model. Fitting the 18 data points with a least squares line, we find the following:

$$F_t = 5.0175 + 3.3014t$$

STEP 3  **Diagnose the model.**

The following values were also found:

$$R^2 = 0.935$$
$$F\text{-statistic} = 230.756$$
$$\text{Standard error} = 4.78$$

The large $F$-statistic indicates that the model explains a significant amount of variation in revenues over time. However, looking at a plot of the trend line shown in Figure 16.12, we see a pattern of actual revenue values first above and then below the trend line. This pattern indicates possible autocorrelation among the error terms. We will test for autocorrelation by calculating the Durbin-Watson $d$ statistic or by using the XLSTAT add-in for Excel. The Excel output is shown in Figure 16.13.

Figure 16.13 shows the Durbin-Watson $d$ statistic as

$$d = 0.6612$$

**FIGURE 16.13** **Excel 2016 (XLSTAT) Output—Durbin-Watson Statistic for Wilder Foods**

| | A | B | C | D | E |
|---|---|---|---|---|---|
| 22 | | | | | |
| 23 | | Regression of variable Revenue (millions $): | | | |
| 25 | | Goodness of fit statistics: | | | |
| 26 | | Observations | 18 | | |
| 27 | | Sum of weights | 18 | | |
| 28 | | DF | 16 | | |
| 29 | | R² | 0.9352 | | |
| 30 | | Adjusted R² | 0.9311 | Durbin- | |
| 31 | | MSE | 22.8838 | Watson | |
| 32 | | RMSE | 4.7837 | Statistic | |
| 33 | | DW | 0.6612 | | |

**Excel 2016 (XLSTAT) Instructions**

1. Open **XLSTAT** and **Enable Macros**.
2. Open file: **Wilder Foods.xlsx**.
3. On the **XLSTAT** tab, click **Modeling data > Linear regression**.
4. Select Y/Dependent variable Revenue (C1:C19):  and X/Explanatory variables: Quantitative: Time Period (A1:A19)
5. Select the **Variable labels check box**.
6. On the **Outputs** tab, select **Descriptive statistics**, **Correlations**, and **Analysis of variance**. Click **OK**.

The null and alternative hypotheses for testing for positive autocorrelation are

$$H_0: \rho = 0$$
$$H_A: \rho > 0$$

We next go to the Durbin-Watson table (Appendix N) for $\alpha = 0.05$, sample size 18, and number of independent variables, $P = 1$. The values from the table for $d_L$ and $d_U$ are

$$d_L = 1.16 \quad \text{and} \quad d_U = 1.39$$

The decision rule for testing whether we have positive autocorrelation is

If $d < 1.16$, reject $H_0$ and conclude that positive autocorrelation exists.
If $d > 1.39$, do not reject $H_0$ and conclude that no positive autocorrelation exists.
If $1.16 < d < 1.39$, the test is inconclusive.

Because $d = 0.661 < d_L = 1.16$, we must reject the null hypothesis and conclude that significant positive autocorrelation exists in the regression model. This means that the assumption of uncorrelated error terms has been violated in this case. Thus, the linear trend model is not the appropriate model to provide the annual revenue forecasts for the next two years.

**TRY EXERCISE 16-15** (pg. 664)

**True Forecasts and Split Samples**   Although a decision maker is interested in how well a forecasting technique can fit historical data, the real test comes with how well it forecasts future values. Recall in the Taft example starting on page 644, we had 10 years of historical data. If we wish to forecast ice cream sales for year 11 using the linear trend model, we substitute $t = 11$ into the forecast equation to produce a forecast as follows:

> The Excel 2016 function is
>
> **=FORECAST.LINEAR**(11, Sales, time periods)

$$F_{11} = 277,333.33 + 14,575.76(11) = \$437,666.69$$

This method of forecasting is called *trend projection*. To determine how well our trend model actually forecasts, we would have to wait until we know the actual sales amount for period 11.

A model's true forecasting ability is determined by how well it forecasts future values, not by how well it fits historical values. However, having to wait until after the forecasting period to know how effective a forecast is doesn't help us assess a model's effectiveness ahead of time. This problem can be partially overcome by using *split samples*, which involves dividing a time series into two groups. We put the first $(n_1)$ periods of historical data in the first group. These $(n_1)$ periods will be used to develop the forecasting model. The second group contains the remaining $(n_2)$ periods of historical data, which will be used to test the model's forecasting ability. These data are called the *holdout data*. Usually, three to five periods are held out, depending on the total number of periods in the time series.

In the Taft Ice Cream business application, we have only 10 years of historical data, so we will hold out the last three periods and use the first seven periods to develop the linear trend model. The computations are performed as before, using Excel or Equations 16.4 and 16.5. Because we are using a different data set to develop the linear equation, we get a slightly different trend line than when all 10 periods were used. The new trend line is

$$F_t = 277,142.85 + 14,642.85t$$

Now we use this model to provide forecasts for periods 8 through 10 by using trend projection. These forecasts are given in the table.

| Year | Actual | Forecast | Error |
|------|--------|----------|-------|
| $t$ | $y_t$ | $F_t$ | $(y_t - F_t)$ |
| 8 | 400,000 | 394,285.65 | 5,714.35 |
| 9 | 395,000 | 408,928.50 | −13,928.50 |
| 10 | 430,000 | 423,571.35 | 6,428.65 |

Then we can compute the *MSE* and the *MAD* values for periods 8 through 10:

$$MSE = \frac{\left[(5,714.35)^2 + (-13,928.50)^2 + (6,428.65)^2\right]}{3} = 89,328,149.67$$

and

$$MAD = \frac{(|5,714.35| + |-13,928.50| + |6,428.65|)}{3} = 8,690.50$$

These values may be compared with those produced using other forecasting techniques or evaluated against the forecaster's own standards. Smaller values are preferred. Other factors should also be considered. For instance, in some cases, the forecast values might tend to be higher (or lower) than the actual values. This may imply the linear trend model isn't the best model to use. Forecasting models that tend to over- or underforecast are said to contain *forecast bias*. Equation 16.9 is used as an estimator of the bias.

**Forecast Bias**

$$\text{Forecast bias} = \frac{\sum(y_t - F_t)}{n} \tag{16.9}$$

The forecast bias can be either positive or negative. A positive value indicates a tendency to underforecast. A negative value indicates a tendency to overforecast. The estimated bias taken from the forecasts for periods 8 through 10 in our example is

$$\text{Forecast bias} = \frac{(5,714.35) + (-13,928.50) + (6,428.65)}{3} = -595.17$$

This means that, on average, the model overforecasts sales by $595.17.

Suppose that on the basis of our bias estimate, we judge that the linear trend model does an acceptable job in forecasting. Then we will use all available data (periods 1 through 10) to develop a linear trend model (see Figure 16.6), and use a trend projection to forecast for future time periods by substituting appropriate values for *t* into the trend model

$$F_t = 277,333.33 + 14,575.76t$$

For the purpose of the bank loan application, the Taft Ice Cream Company needs to forecast sales for the next three years (periods 11 through 13). Assuming the linear trend model is acceptable, these forecasts are

$$F_{11} = 277,333.33 + 14,575.76(11) = \$437,666.69$$
$$F_{12} = 277,333.33 + 14,575.76(12) = \$452,242.45$$
$$F_{13} = 277,333.33 + 14,575.76(13) = \$466,818.21$$

## Nonlinear Trend Forecasting

You may encounter a time series that exhibits a nonlinear trend. Figure 16.2 showed an example of a nonlinear trend. When the historical data show a nonlinear trend, you should consider using a nonlinear trend forecasting model. A common method for dealing with nonlinear trends is to use a modification of the linear trend method. This extension calls for making a data transformation before applying the least squares regression analysis.

Excel Tutorial

| BUSINESS APPLICATION | Forecasting Nonlinear Trends |

**Harrison Equipment Company** Consider Harrison Equipment Company, which leases large construction equipment to contractors in the Southwest. The lease arrangements call for Harrison to perform all repairs and maintenance on this equipment. Figure 16.14 shows a line chart for the repair costs for a crawler tractor leased to a contractor in Phoenix for the past 20 quarters. The data are contained in the file **Harrison**.

**OUTCOME 4**

**STEP 1 Model specification.**

Harrison Equipment is interested in forecasting future repair costs for the crawler tractor. Recall that the first step in forecasting is model specification. Even though the plot in Figure 16.14 indicates a sharp upward nonlinear trend, the forecaster may start by specifying a linear trend model.

**STEP 2 Model fitting.**

As a part of the model-fitting step, the forecaster could use Excel's regression procedure to obtain the linear forecasting model shown in Figure 16.15. As shown, the linear trend model is

$$F_t = -1,022.69 + 570.94t$$

**STEP 3 Model diagnosis.**

The fit is pretty good with $R$-squared $= 0.821$ and a standard error of 1,618.45. But we need to look closer. Figure 16.15 shows a plot of the trend line compared with the actual data. A close inspection indicates the linear trend model may not be best for this case. Notice that the linear model underforecasts, then overforecasts, then underforecasts again. From this we might suspect positive autocorrelation.

We can establish the following null and alternative hypotheses:

$$H_0: \rho = 0$$
$$H_A: \rho > 0$$

**FIGURE 16.14 Excel 2016 Time-Series Plot for Harrison Equipment Repair Costs**

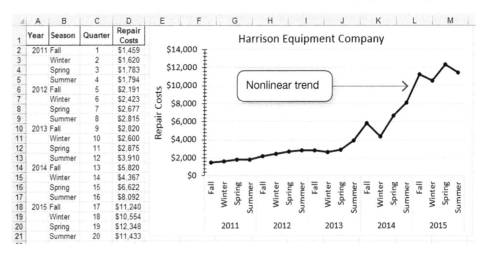

**Excel 2016 Instructions**

1. Open file: **Harrison.xlsx**.
2. Select data in the *Repair Costs* data column.
3. Click **Insert > Line or Area Chart**.
4. On the **Design** tab, click **Select Data**.
5. Under **Horizontal (categories) Axis Labels**, select data in *Year* and *Season* columns.
6. On the **Design** tab, enter desired title and vertical title.
7. Remove gridlines.

**FIGURE 16.15 Excel 2016 (XLSTAT) Output for the Harrison Equipment Company Linear Trend Model**

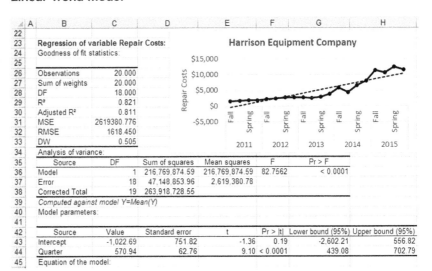

**Excel 2016 (XLSTAT) Instructions**

1. Open **XLSTAT** and **Enable Macros**.
2. Open file: **Harrison.xlsx**.
3. On the **XLSTAT** tab, click **Modeling data > Linear regression**.
4. Select Y/Dependent variable: Repair Costs and X/ Explanatory variables: Quantitative: Quarter.
5. Select the **Variable labels check box**.
6. On the **Outputs** tab, select **Descriptive statistics**, **Correlations**, and **Analysis of variance**. Click **OK**.
7. Paste chart output from Figure 16.14.
8. Right click on the line and select **Add Trendline**, select **Linear** and click **Close**.

We could use Equation 16.11 to manually compute the Durbin-Watson $d$ statistic, or more likely, we would use XLSTAT. The calculated Durbin-Watson is

$$d = 0.505$$

The $d_L$ critical value from the Durbin-Watson table in Appendix N for $\alpha = 0.05$ and a sample size of $n = 20$ and $p = 1$ independent variable is 1.20.

Because $d = 0.505 < d_L = 1.20$, we reject the null hypothesis.

We conclude that the error terms are significantly positively autocorrelated. We need to repeat the model-building process.

**Model Specification**   After examining Figure 16.15 and determining the results of the test for positive autocorrelation, we conclude that a nonlinear trend will likely provide a better fit for these data. To account for the nonlinear growth trend, which starts out slowly and then builds rapidly, we might consider transforming the time variable by squaring $t$ to form a model of the form

$$y = \beta_0 + \beta_1 t + \beta_2 t^2 + \varepsilon$$

This transformation is suggested because the growth in costs appears to be increasing at an increasing rate. Other nonlinear trends may require different types of transformations, such as taking a square root or natural log. Each situation must be analyzed separately. (See the reference by Kutner et al. for further discussion of transformations.)

**Model Fitting**   Figure 16.16 shows the Excel regression results, and Figure 16.17 on page 657 shows the revised time-series plot using the polynomial transformation. The resulting nonlinear trend regression model rounded to one decimal place is

$$F_t = 2{,}318.7 - 340.4t + 43.4t^2$$

**Model Diagnosis**   Visually, the transformed model now looks more appropriate. The fit is much better as the $R$-squared value is increased to 0.9466 and the standard error is reduced to 910.35. The null and alternative hypotheses for testing whether positive autocorrelation exists are

$$H_0: \rho = 0$$
$$H_A: \rho > 0$$

FIGURE 16.16 Excel 2016 (XLSTAT) Transformed Regression Model for Harrison Equipment

| | A | B | C | D | E | F | G | H |
|---|---|---|---|---|---|---|---|---|
| 25 | | Regression of variable Repair Costs: | | | | | | |
| 26 | | Goodness of fit statistics: | | | | | | |
| 27 | | Observations | 20.0000 | | | | | |
| 28 | | Sum of weights | 20.0000 | | | | | |
| 29 | | DF | 17.0000 | | | | | |
| 30 | | R² | 0.9466 | | | | | |
| 31 | | Adjusted R² | 0.9403 | | | | | |
| 32 | | MSE | 828,734.5238 | | | | | |
| 33 | | RMSE | 910.3486 | | | | | |
| 34 | | DW | 1.6301 | | | | | |
| 35 | | Analysis of variance: | | | | | | |
| 36 | | Source | DF | Sum of squares | Mean squares | F | Pr > F | |
| 37 | | Model | 2 | 249,830,241.65 | 124,915,120.82 | 150.73 | < 0.0001 | |
| 38 | | Error | 17 | 14,088,486.90 | 828,734.52 | | | |
| 39 | | Corrected Total | 19 | 263,918,728.55 | | | | |
| 40 | | Computed against model Y=Mean(Y) | | | | | | |
| 41 | | Model parameters: | | | | | | |
| 42 | | | | | | | | |
| 43 | | Source | Value | Standard error | t | Pr > \|t\| | Lower bound (95%) | Upper bound (95%) |
| 44 | | Intercept | 2,318.7307 | 677.2827 | 3.4236 | 0.0032 | 889.7892 | 3,747.6723 |
| 45 | | Quarter | -340.3603 | 148.5386 | -2.2914 | 0.0350 | -653.7493 | -26.9712 |
| 46 | | Quarter2 | 43.3951 | 6.8706 | 6.3161 | < 0.0001 | 28.8994 | 57.8908 |
| 47 | | Equation of the model: | | | | | | |
| 48 | | | | | | | | |
| 49 | | Repair Costs = 2319-340.36025*Quarter+43.39514*Quarter2 | | | | | | |

**Excel 2016 (XLSTAT) Instructions**

1. Open **XLSTAT** and **Enable Macros**.
2. Open file: **Harrison.xlsx**.
3. Insert a new column between columns C and D.
4. Create a new variable, $Quarter^2$, in the newly added column.
5. On the **XLSTAT** tab, click **Modeling data > Linear regression**.
6. Select Y/Dependent variable: Repair Costs and X/Explanatory variables: Quantitative: Quarter and $Quarter^2$.
7. Select the **Variable labels check box**.
8. On the **Outputs** tab, select **Descriptive statistics**, **Correlations**, and **Analysis of variance**. Click **OK**.

As seen in Figure 16.16, the calculated Durbin-Watson statistic is

$$d = 1.6301$$

The $d_L$ and $d_U$ critical values from the Durbin-Watson table in Appendix N for $\alpha = 0.05$ and a sample size of $n = 20$ and $p = 2$ independent variables are 1.10 and 1.54, respectively.

Because $d = 1.6301 > 1.54$, the Durbin-Watson test indicates that there is no positive autocorrelation. Given this result and the improvements to $R$-squared and the standard error of the estimate, the nonlinear model is judged superior to the original linear model.

We can obtain forecasts for periods 21 and 22, using this latest model (coefficients rounded to one decimal place), with the trend projection method.

For period $t = 21$:

$$F_{21} = 2,318.7 - 340.4(21) + 43.4(21^2) = \$14,310$$

For period $t = 22$:

$$F_{22} = 2,318.7 - 340.4(22) + 43.4(22^2) = \$15,836$$

Using transformations is often a very effective way to improve the fit of a time series. However, a forecaster should be careful not to get caught up in an exercise of "curve-fitting." One suggestion is that only explainable terms—terms that can be justified—be used for transforming data. For instance, in our example, we might well expect repair costs to increase at a faster rate as a tractor gets older and begins to wear out. Thus, the $t^2$ transformation seems to make sense.

**Some Words of Caution** The trend projection method relies on the future behaving in a manner similar to the past. In the previous example, if equipment repair costs continue to

**FIGURE 16.17** Excel 2016 Transformed Model for Harrison Equipment Company

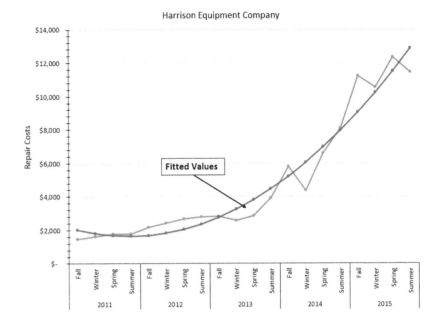

**Excel 2016 Instructions**

1. Open **XLSTAT** and **Enable Macros**.
2. Open file: **Harrison.xlsx**.
3. Insert a new column between columns C and D.
4. Create a new variable, *Quarter*$^2$ in the newly added column.
5. On the **XLSTAT** tab, click **Modeling data > Linear regression**.
6. Select Y/Dependent variable: Repair Costs and X/Explanatory variables: Quantitative: Quarter and Quarter$^2$.
7. Select the **Variable labels check box**.
8. On the **Outputs** tab, select **Descriptive statistics**, **Correlations**, **Analysis of variance**, and **Predictions and residuals**. Click **OK**.
9. Under **Prediction and residuals** on the **Regression** worksheet, select the data Repair Costs and **Pred (Repair Costs)**.
10. Click **Insert > Line or Area Chart > Line with Markers**.
11. Under **Horizontal (categories) Axis Labels**, select **Edit**, select data in *Year* and *Quarter* columns on the **Data** worksheet.
12. Click **OK** twice.
13. On the **Design** tab, enter desired title and vertical title.
14. Move the chart to a new sheet.
15. Remove markers from predicted values line, gridlines, and legends.

follow the pattern displayed over the past 20 quarters, these forecasts may prove acceptable. However, if the future pattern changes, there is no reason to believe these forecasts will be close to actual costs.

## Adjusting for Seasonality

In Section 16.1, we discussed seasonality in a time series. The seasonal component represents those changes (highs and lows) in the time series that occur at approximately the same time every period. If the forecasting model you are using does not already explicitly account for seasonality, you should adjust your forecast to take into account the seasonal component. The linear and nonlinear trend models discussed thus far do not automatically incorporate the seasonal component. Forecasts using these models should be adjusted as illustrated in the following application.

**Excel Tutorial**

**Seasonal Index**

A number used to quantify the effect of seasonality in time-series data.

**BUSINESS APPLICATION**   **Forecasting with Seasonal Data**

**Majestic Mountain Ski Resort**   Most businesses in the tourist industry know that sales are seasonal. For example, at the Majestic Mountain Ski Resort, business peaks at two times during the year: winter for skiing and summer for golf and tennis. These peaks can be identified in a time series if the sales data are measured on at least a quarterly basis. Figure 16.18 shows the quarterly sales data for the past four years in spreadsheet form. The line chart for these data is also shown. The data are in the file **Majestic Mountain**. The time-series plot clearly shows that the summer and winter quarters are the busy times. There has also been a slightly increasing linear trend in sales over the four years.

Majestic Mountain Resort wants to forecast sales for each quarter of the coming year, and it hopes to use a linear trend model. When the historical data show a trend and seasonality, we need to adjust the trend-based forecasting model to incorporate the seasonality. One method for doing this involves computing **seasonal indexes**.

For instance, when we have quarterly data, we can develop four seasonal indexes, one each for winter, spring, summer, and fall. A seasonal index below 1.00 indicates that the quarter has a value that is typically below the average value for the year. On the other hand, an index greater than 1.00 indicates that the quarter's value is typically higher than the yearly average.

**FIGURE 16.18  Excel 2016 Majestic Mountain Resort Quarterly Sales Data**

**Excel 2016 Instructions**

1. Open file: **Majestic Mountain.xlsx**.
2. Select data in the *Sales* column.
3. Click **Insert > Line or Area Chart > Line with Markers**.
4. Click **Select Data**.
5. Under **Horizontal (categories) Axis Labels**, select data in *Year* and *Season* columns.
6. On the **Design** tab, enter desired title and enter vertical title.
7. Remove the gridlines.

**Computing Seasonal Indexes**   Although there are several methods for computing the seasonal indexes, the procedure introduced here is the *ratio-to-moving-average method*. This method assumes that the actual time-series data can be represented as a product of the four time-series components—trend, seasonal, cyclical, and random—which produces the multiplicative model shown in Equation 16.10.

**Multiplicative Time-Series Model**

$$y_t = T_t \times S_t \times C_t \times I_t \qquad (16.10)$$

where:

$y_t$ = Value of the time series at time $t$

$T_t$ = Trend value at time $t$

$S_t$ = Seasonal value at time $t$

$C_t$ = Cyclical value at time $t$

$I_t$ = Irregular or random value at time $t$

**Moving Average**
The successive averages of $n$ consecutive values in a time series.

The first step in the ratio-to-moving-average method is to remove the seasonal and irregular components, $S_t$ and $I_t$, from the data, leaving the combined trend and cyclical components, $T_t$ and $C_t$. We do this by first computing successive four-period moving averages for the time series. A moving average is the average of $n$ consecutive values of a time series. Using the Majestic Mountain sales data in Figure 16.19, we find that the **moving average** using the first four quarters is

$$\frac{205 + 96 + 194 + 102}{4} = 149.25$$

This moving average is associated with the middle time period of the data values in the moving average. The middle period of the first four quarters is 2.5 (between quarter 2 and quarter 3).

We find the second moving average by dropping the value from period 1 and adding the value from period 5, as follows:

$$\frac{96 + 194 + 102 + 230}{4} = 155.50$$

This moving average is associated with time period 3.5, the middle period between quarters 3 and 4.

**FIGURE 16.19** **Excel 2016 Seasonal Index—Step 1: Majestic Mountain Resort Moving Average Values**

| | Season | Year | Quarter | Sales | 4-Period Moving Average | |
|---|---|---|---|---|---|---|
| 1 | Season | Year | Quarter | Sales | 4-Period Moving Average | |
| 2 | Winter | 2012 | 1 | 205 | | |
| 3 | Spring | | 2 | 96 | 149.25 | Each moving average corresponds to the midpoint between its cell and the following cell. |
| 4 | Summer | | 3 | 194 | 155.50 | |
| 5 | Fall | | 4 | 102 | 157.75 | |
| 6 | Winter | 2013 | 5 | 230 | 170.50 | |
| 7 | Spring | | 6 | 105 | 175.00 | |
| 8 | Summer | | 7 | 245 | 185.50 | |
| 9 | Fall | | 8 | 120 | 186.75 | |
| 10 | Winter | 2014 | 9 | 272 | 189.25 | |
| 11 | Spring | | 10 | 110 | 187.75 | |
| 12 | Summer | | 11 | 255 | 193.75 | |
| 13 | Fall | | 12 | 114 | 198.75 | |
| 14 | Winter | 2015 | 13 | 296 | 202.50 | |
| 15 | Spring | | 14 | 130 | 209.00 | |
| 16 | Summer | | 15 | 270 | | |
| 17 | Fall | | 16 | 140 | | |

**Excel 2016 Instructions**

1. Open file: **Majestic Mountain.xlsx**.
2. Create a new column labeled *4 Period Moving Average* in Column E.
3. On the **Data** tab, click **Data Analysis > Moving Average**.
4. For **Input Range** select **D1:D17**, check **Labels in First Row**, for **Interval** type 4, and for **Output Range** select **E2**.
5. Delete any #N/A's.
6. Select remaining 4-Period Moving Average data and drag up to move to cell **E3**.
7. Select cells **E3:E15**, click the arrow on the message to the left of the first cell, and click **Ignore Error**.

Figure 16.19 shows the moving averages for the Majestic Mountain sales data in Excel spreadsheet form. We selected 4 data values for the moving average because we have quarterly data; with monthly data, we would have used 12 data values.

The next step is to compute the *centered moving averages* by averaging each successive pair of moving averages. Centering the moving averages is necessary so that the resulting moving average will be associated with one of the data set's original time periods. In this example, Majestic Mountain is interested in quarterly sales data—that is, time periods 1, 2, 3, and so on. Therefore, the moving averages we have representing time periods 2.5, 3.5, and so forth are not of interest to Majestic Mountain.[2] Centering these averaged time-series values, however, produces moving averages for the (quarterly) time periods of interest. For example, we average the first two moving averages to produce the first centered moving average. We get

$$\frac{149.25 + 155.5}{2} = 152.38$$

This centered moving average is associated with quarter 3. The centered moving averages are shown in Figure 16.20. These values estimate the $T_t \times C_t$ value.

If the number of data values used for a moving average is odd, the moving average will be associated with the time period of the middle observation. In such cases, we would not have to center the moving average, as we did in Figure 16.20, because the moving averages would already be associated with one of the time periods from the original time series.

Next, we estimate the $S_t \times I_t$ value. We do this by dividing the actual sales value for each quarter by the corresponding centered moving average, as in Equation 16.11. As an example, we examine the third time period: summer of 2012. The sales value of 194 is divided by the centered moving average of 152.38 to produce 1.273. This value is called the *ratio-to-moving-average*. Figure 16.21 shows these values for the Majestic Mountain data.

**Ratio-to-Moving-Average**

$$S_t \times I_t = \frac{y_t}{T_t \times C_t} \tag{16.11}$$

**FIGURE 16.20** **Excel 2016 Seasonal Index—Step 2: Majestic Mountain Resort Centered Moving Averages**

**Excel 2016 Instructions**

1. Follow Figure 16.19 Instructions 1–3.
2. Create a new column labeled *Centered Moving Average* in Column F.
3. Create a 2-period moving average for the range E4:E15 using Output Range F3. Remove the check from Labels in First Row.
4. Delete any #N/A's.

| | A | B | C | D | E | F |
|---|---|---|---|---|---|---|
| | | | | | 4-Period Moving Average | Centered Moving Average |
| 1 | Season | Year | Quarter | Sales | | |
| 2 | Winter | 2012 | 1 | 205 | | |
| 3 | Spring | | 2 | 96 | 149.25 | |
| 4 | Summer | | 3 | 194 | 155.50 | 152.375 |
| 5 | Fall | | 4 | 102 | 157.75 | 156.625 |
| 6 | Winter | 2013 | 5 | 230 | 170.50 | 164.125 |
| 7 | Spring | | 6 | 105 | 175.00 | 172.750 |
| 8 | Summer | | 7 | 245 | 185.50 | 180.250 |
| 9 | Fall | | 8 | 120 | 186.75 | 186.125 |
| 10 | Winter | 2014 | 9 | 272 | 189.25 | 188.000 |
| 11 | Spring | | 10 | 110 | 187.75 | 188.500 |
| 12 | Summer | | 11 | 255 | 193.75 | 190.750 |
| 13 | Fall | | 12 | 114 | 198.75 | 196.250 |
| 14 | Winter | 2015 | 13 | 296 | 202.50 | 200.625 |
| 15 | Spring | | 14 | 130 | 209.00 | 205.750 |
| 16 | Summer | | 15 | 270 | | |
| 17 | Fall | | 16 | 140 | | |

---

[2] Excel's tabular format does not allow the uncentered moving averages to be displayed with their "interquarter" time periods. That is, 149.25 is associated with time period 2.5, 155.50 with time period 3.5, and so forth.

**FIGURE 16.21** Excel 2016 Seasonal Index—Step 3: Majestic Mountain Resort Ratio-to-Moving-Averages

| | Season | Year | Quarter | Sales | 4-Period Moving Average | Centered Moving Average | Ratio to Moving Average |
|---|---|---|---|---|---|---|---|
| 1 | Season | Year | Quarter | Sales | 4-Period Moving Average | Centered Moving Average | Ratio to Moving Average |
| 2 | Winter | 2012 | 1 | 205 | | | |
| 3 | Spring | | 2 | 96 | 149.25 | | |
| 4 | Summer | | 3 | 194 | 155.50 | 152.38 | 1.27 |
| 5 | Fall | | 4 | 102 | 157.75 | 156.63 | 0.65 |
| 6 | Winter | 2013 | 5 | 230 | 170.50 | 164.13 | 1.40 |
| 7 | Spring | | 6 | 105 | 175.00 | 172.75 | 0.61 |
| 8 | Summer | | 7 | 245 | 185.50 | 180.25 | 1.36 |
| 9 | Fall | | 8 | 120 | 186.75 | 186.13 | 0.64 |
| 10 | Winter | 2014 | 9 | 272 | 189.25 | 188.00 | 1.45 |
| 11 | Spring | | 10 | 110 | 187.75 | 188.50 | 0.58 |
| 12 | Summer | | 11 | 255 | 193.75 | 190.75 | 1.34 |
| 13 | Fall | | 12 | 114 | 198.75 | 196.25 | 0.58 |
| 14 | Winter | 2015 | 13 | 296 | 202.50 | 200.63 | 1.48 |
| 15 | Spring | | 14 | 130 | 209.00 | 205.75 | 0.63 |
| 16 | Summer | | 15 | 270 | | | |
| 17 | Fall | | 16 | 140 | | | |

**Excel 2016 Instructions**

1. Open file: **Majestic Mountain.xlsx**.
2. Follow Instructions 2–7 in Figure 16.19 and instructions 2–4 in Figure 16.20.
3. Create a new column of ratio-to-moving-averages using equation = **D4/F4** in column G beginning in cell G4.
4. Copy equation down to cell G15.

The final step in determining the seasonal indexes is to compute the mean ratio-to-moving-average value for each season. Each quarter's ratio-to-moving-average is averaged over the years to produce the *seasonal index* for that quarter. Figure 16.22 shows the seasonal indexes. The seasonal index for the winter quarter is 1.441. This indicates that sales for Majestic Mountain during the winter are 44.1% above the average for the year. Also, sales in the spring quarter are only 60.8% of the average for the year.

One important point about the seasonal indexes is that the sum of the indexes is equal to the number of seasonal indexes. That is, the average of all seasonal indexes equals 1.0. In the Majestic Mountain Resort example, we find

| Summer | | Fall | | Winter | | Spring |
|---|---|---|---|---|---|---|
| 1.323 | + | 0.626 | + | 1.441 | + | 0.608 = 3.998 (difference from 4 due to rounding) |

Likewise, in an example with monthly data instead of quarterly data, we would generate 12 seasonal indexes, one for each month. The sum of these indexes should be 12.

**FIGURE 16.22** Excel 2016 Seasonal Index—Step 4: Majestic Mountain Resort Mean Ratios

| | Summer | Fall | Winter | Spring | |
|---|---|---|---|---|---|
| 20 | Summer | Fall | Winter | Spring | |
| 21 | 1.273 | 0.651 | 1.401 | 0.608 | |
| 22 | 1.359 | 0.645 | 1.447 | 0.584 | |
| 23 | 1.337 | 0.581 | 1.475 | 0.632 | |
| 24 | | | | | |
| 25 | 3.969 | 1.877 | 4.324 | 1.823 | = Total |
| 26 | 1.323 | 0.626 | 1.441 | 0.608 | = Average |

Seasonal indexes:
Summer = 1.323
Fall = 0.626
Winter = 1.441
Spring = 0.608

**Excel 2016 Instructions**

1. Open file: **Majestic Mountain.xlsx**.
2. Complete Instructions 2–7 in Figure 16.19. Complete Instructions 2–4 in Figure 16.20. Complete Instructions 3 and 4 in Figure 16.21.
3. Rearrange the ratio-to-moving-average values below the data, organizing them by season of the year (summer, fall, etc.).
4. Compute the total and average for each season's column.

**The Need to Normalize the Indexes**   If the sum of the seasonal indexes does not equal the number of time periods in the recurrence period of the time series, we need to make an adjustment. In the Majestic Mountain Resort example, the sum of the four seasonal indexes may have been something other than 4 (the recurrence period). In such cases, we must adjust the seasonal indexes by multiplying each by the number of time periods in the recurrence period over the sum of the unadjusted seasonal indexes. For quarterly data such as the Majestic Mountain Resort example, we would multiply each seasonal index by 4/(Sum of the unadjusted seasonal indexes). Performing this multiplication will *normalize* the seasonal indexes. This adjustment is necessary if the seasonal adjustments are going to even out over the recurrence period.

**Deseasonalizing**   A strong seasonal component may partially mask a trend in the time-series data. Consequently, to identify the trend, we first remove the effect of the seasonal component. This is called *deseasonalizing* the time series.

Again, we assume that the multiplicative model shown in Equation 16.10 is appropriate:

$$y_t = T_t \times S_t \times C_t \times I_t$$

Deseasonalizing is accomplished by dividing $y_t$ by the appropriate seasonal index, $S_t$, as shown in Equation 16.12.

**Deseasonalization**

$$T_t \times C_t \times I_t = \frac{y_t}{S_t} \tag{16.12}$$

For time period 1, which is the winter quarter, the seasonal index is 1.441. The deseasonalized value for $y_1$ is

$$\frac{205}{1.441} = 142.26$$

Figure 16.23 presents the deseasonalized values and the graph of these deseasonalized sales data for the Majestic Mountain example. This shows that there has been a gentle upward trend over the four years.

**FIGURE 16.23 Excel 2016 Deseasonalized Time Series for Majestic Mountain Sales Data**

**Excel 2016 Instructions**

1. Open file: **Majestic Mountain.xlsx**.
2. Follow instructions for Figures 16.18 through 16.22.
3. Add a new column labeled *Seasonal Index* and enter the seasonal indexes calculated earlier for the appropriate season.
4. Create a new column to contain the deseasonalized values. Enter Equation 16.12. For example, in cell I2, use =**D2/1.441**. In cell I2, use =**D3/0.608**, etc.
5. On the **Design** tab, **Select Data** and under Legend Entries (Series), add the Deseasonalized data.

Once the data have been deseasonalized, the next step is to determine the trend based on the deseasonalized data. As in the previous examples of trend estimation, we can use Excel to develop the linear model for the deseasonalized data. The results are shown in Figure 16.24. The linear regression trend line equation is

$$F_t = 142.096 + 4.683t$$

We can use this trend line and the trend projection method to forecast sales for period $t = 17$:

$$F_{17} = 142.096 + 4.683(17) = 221.707 = \$221,707$$

**Seasonally Unadjusted Forecast**

A forecast made for seasonal data that does not include an adjustment for the seasonal component in the time series.

This is a **seasonally unadjusted forecast** because the time-series data used in developing the trend line were deseasonalized.

Now we need to adjust the forecast for period 17 to reflect the quarterly fluctuations. We do this by multiplying the unadjusted forecast values by the appropriate seasonal index. In this case, period 17 corresponds to the winter quarter. The winter quarter has a seasonal index of 1.441, indicating a high sales period. The adjusted forecast is

$$F_{17} = (221.707)(1.441) = 319.480, \text{ or } \$319,480$$

The seasonally adjusted forecasts for each quarter in 2015 are as follows:

| Quarter (2015) | $t$ | Unadjusted Forecast | Index | Adjusted Forecast |
|---|---|---|---|---|
| Winter | 17 | 221.707 | 1.441 | 319.480 = $319,480 |
| Spring | 18 | 226.390 | 0.608 | 137.645 = $137,645 |
| Summer | 19 | 231.073 | 1.323 | 305.710 = $305,710 |
| Fall | 20 | 235.756 | 0.626 | 147.583 = $147,583 |

We can use the seasonally adjusted trend model when a time series exhibits both a trend and seasonality. This process allows for a better identification of the trend and produces forecasts that are more sensitive to seasonality in the data.

**HOW TO DO IT**

**The Seasonal Adjustment Process: The Multiplicative Model**

**1.** Compute each moving average from the $k$ appropriate consecutive data values, where $k$ is the number of values in one period of the time series.

**2.** Compute the centered moving averages.

**3.** Isolate the seasonal component by computing the ratio-to-moving-average values.

**4.** Compute the seasonal indexes by averaging the ratio-to-moving-average values for comparable periods.

**5.** Normalize the seasonal indexes (if necessary).

**6.** Deseasonalize the time series by dividing the actual data by the appropriate seasonal index.

**7.** Use least squares regression to develop the trend line using the deseasonalized data.

**8.** Develop the unadjusted forecasts using trend projection.

**9.** Seasonally adjust the forecasts by multiplying the unadjusted forecasts by the appropriate seasonal index.

**FIGURE 16.24 Excel 2016 Regression Trend Line of Majestic Mountain Deseasonalized Data**

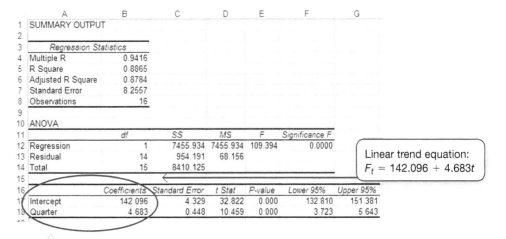

| | A | B | C | D | E | F | G |
|---|---|---|---|---|---|---|---|
| 1 | SUMMARY OUTPUT | | | | | | |
| 2 | | | | | | | |
| 3 | Regression Statistics | | | | | | |
| 4 | Multiple R | 0.9416 | | | | | |
| 5 | R Square | 0.8865 | | | | | |
| 6 | Adjusted R Square | 0.8784 | | | | | |
| 7 | Standard Error | 8.2557 | | | | | |
| 8 | Observations | 16 | | | | | |
| 9 | | | | | | | |
| 10 | ANOVA | | | | | | |
| 11 | | df | SS | MS | F | Significance F | |
| 12 | Regression | 1 | 7455.934 | 7455.934 | 109.394 | 0.0000 | |
| 13 | Residual | 14 | 954.191 | 68.156 | | | |
| 14 | Total | 15 | 8410.125 | | | | |
| 15 | | | | | | | |
| 16 | | Coefficients | Standard Error | t Stat | P-value | Lower 95% | Upper 95% |
| 17 | Intercept | 142.096 | 4.329 | 32.822 | 0.000 | 132.810 | 151.381 |
| 18 | Quarter | 4.683 | 0.448 | 10.459 | 0.000 | 3.723 | 5.643 |

Linear trend equation:
$F_t = 142.096 + 4.683t$

**Excel 2016 Instructions**

1. Open file: **Majestic Mountain.xlsx.**
2. Follow instructions for Figures 16.19 through 16.23.
3. Click **Data.**
4. Select **Data Analysis > Regression.**
5. Specify the *y*-variable range (column I, deseasonalized data) and the *x*-variable range (column C, the quarter).
6. Click **OK.**

## 16.2 EXERCISES

### Skill Development

Exercises **16-15 to 16-19** refer to Tran's Furniture Store, which has maintained monthly sales records for the past 48 months, with the following results:

| Month | Sales ($) | Month | Sales ($) |
|---|---|---|---|
| 1 (Jan.) | 23,500 | 25 (Jan.) | 31,000 |
| 2 | 21,700 | 26 | 30,400 |
| 3 | 18,750 | 27 | 29,800 |
| 4 | 22,000 | 28 | 32,500 |
| 5 | 23,000 | 29 | 34,500 |
| 6 | 26,200 | 30 | 33,800 |
| 7 | 27,300 | 31 | 34,200 |
| 8 | 29,300 | 32 | 36,700 |
| 9 | 31,200 | 33 | 39,700 |
| 10 | 34,200 | 34 | 42,400 |
| 11 | 39,500 | 35 | 43,600 |
| 12 | 43,400 | 36 | 47,400 |
| 13 (Jan.) | 23,500 | 37 (Jan.) | 32,400 |
| 14 | 23,400 | 38 | 35,600 |
| 15 | 21,400 | 39 | 31,200 |
| 16 | 24,200 | 40 | 34,600 |
| 17 | 26,900 | 41 | 36,800 |
| 18 | 29,700 | 42 | 35,700 |
| 19 | 31,100 | 43 | 37,500 |
| 20 | 32,400 | 44 | 40,000 |
| 21 | 34,500 | 45 | 43,200 |
| 22 | 35,700 | 46 | 46,700 |
| 23 | 42,000 | 47 | 50,100 |
| 24 | 42,600 | 48 | 52,100 |

**16-15.** Based on the Durbin-Watson statistic, is there evidence of autocorrelation in these data? Use a linear trend model.

**16-16.** Using the multiplicative model, estimate the $T \times C$ portion by computing a 12-month moving average and then the centered 12-month moving average.

**16-17.** Estimate the $S \times I$ portion of the multiplicative model by finding the ratio-to-moving-averages for the time-series data. Determine whether these ratio-to-moving-averages are stable from year to year.

**16-18.** Extract the irregular component by taking the normalized average of the ratio-to-moving-averages. Make a table that shows the normalized seasonal indexes. Interpret what the index for January means relative to the index for July.

**16-19.** Refer back to your work in the previous three exercises.
   a. Determine a seasonally adjusted linear trend forecasting model. Compare this model with an unadjusted linear trend model. Use both models to forecast Tran's sales for period 49.
   b. Which of the two models developed has the lower *MAD* and lower *MSE*?

**16-20.** Consider the following sales data, in millions of dollars:

| 2013 | 2015 |
|---|---|
| 1st quarter 152 | 1st quarter 217 |
| 2nd quarter 162 | 2nd quarter 209 |
| 3rd quarter 157 | 3rd quarter 202 |
| 4th quarter 167 | 4th quarter 221 |

| 2014 | 2016 |
|---|---|
| 1st quarter 182 | 1st quarter 236 |
| 2nd quarter 192 | 2nd quarter 242 |
| 3rd quarter 191 | 3rd quarter 231 |
| 4th quarter 197 | 4th quarter 224 |

Plot these data. Based on your visual observations, what time-series components are present in the data?

**16-21.** Refer to the sales data for the years 2013 through 2016 shown in Exercise 16-20 and determine the seasonal indexes for the 1st, 2nd, 3rd, and 4th quarters.

**16-22.** Refer to the sales data for the years 2013 through 2016 shown in Exercise 16-20 and fit the linear trend model to the data. Determine the *MAD* and *MSE* values. Comment on the adequacy of the linear model based on these measures of forecast error.

**16-23.** Refer to the sales data for the years 2013 through 2016 shown in Exercise 16-20 and provide a seasonally adjusted forecast using the linear trend model for each quarter of the year 2017. Also use the seasonal index values computed in Exercise 16-21 to provide seasonally adjusted forecasts for the four quarters of 2017.

**16-24.** Examine the following time series:

| $t$ | 1 | 2 | 3 | 4 | 5 | 6 | 7 | 8 | 9 | 10 |
|---|---|---|---|---|---|---|---|---|---|---|
| $y_t$ | 52 | 72 | 58 | 66 | 68 | 60 | 46 | 43 | 17 | 3 |

   a. Produce a scatter plot and indicate the appropriate forecasting model for this time series.
   b. Construct the equation for the forecasting model identified in part a.
   c. Produce forecasts for time periods 11, 12, 13, and 14.
   d. Obtain the forecast bias for the forecasts produced in part c if the actual time series values are $-35$, $-41$, $-79$, and $-100$ for periods 11 through 14, respectively.

**16-25.** Examine the following quarterly data:

| $t$ | 1 | 2 | 3 | 4 | 5 | 6 | 7 | 8 | 9 | 10 | 11 | 12 |
|---|---|---|---|---|---|---|---|---|---|---|---|---|
| $y_t$ | 2 | 12 | 23 | 20 | 18 | 32 | 48 | 41 | 35 | 52 | 79 | 63 |

   a. Compute the four-period moving averages for this set of data.
   b. Compute the centered moving averages from the moving averages of part a.

c. Compute the ratio-to-moving-averages values.
d. Calculate the seasonal indexes. Normalize them if necessary.
e. Deseasonalize the time series.
f. Produce the trend line using the deseasonalized data.
g. Produce seasonally adjusted forecasts for each of the time periods 13, 14, 15, and 16.

## Business Applications

**16-26.** A May 2015 *Wall Street Journal* article indicated that 2015 college graduates had the highest average student debt ever, at slightly more than $35,000 per student. The following data show the average student debt for the years 1993 through 2015 (*source:* Jeffrey Sparshott, "Congratulations, class of 2015. You're the most indebted ever (for now)," http://blogs.wsj.com, May 8, 2015).

| Year | Average Student Loan Debt (Rounded to nearest $000) |
|---|---|
| 1993 | $10,000 |
| 1994 | $11,000 |
| 1995 | $12,000 |
| 1996 | $13,000 |
| 1997 | $14,000 |
| 1998 | $15,000 |
| 1999 | $16,000 |
| 2000 | $17,000 |
| 2001 | $17,000 |
| 2002 | $17,000 |
| 2003 | $18,000 |
| 2004 | $18,000 |
| 2005 | $20,000 |
| 2006 | $21,000 |
| 2007 | $22,000 |
| 2008 | $24,000 |
| 2009 | $25,000 |
| 2010 | $26,000 |
| 2011 | $28,000 |
| 2012 | $30,000 |
| 2013 | $31,000 |
| 2014 | $33,000 |
| 2015 | $35,000 |

a. Produce a time-series plot of these data. Indicate the time-series components that exist in the data.
b. Conduct a test of hypothesis to determine if there is a linear trend in these data. Use a significance level of 0.10 and the *p*-value approach.
c. Use the model developed in part b to provide a forecast for the average student debt in 2017.

**16-27.** Suppose the owner of a convenience store has tracked the average price of a gallon of milk for the months between July 2013 and June 2015. The data are shown in the table.

| Month | Price ($) | Month | Price ($) | Month | Price ($) |
|---|---|---|---|---|---|
| 7/13 | 2.247 | 3/14 | 2.344 | 11/14 | 2.637 |
| 8/13 | 2.108 | 4/14 | 2.642 | 12/14 | 2.289 |
| 9/13 | 2.111 | 5/14 | 2.532 | 1/15 | 2.357 |
| 10/13 | 2.352 | 6/14 | 2.375 | 2/15 | 2.628 |
| 11/13 | 2.374 | 7/14 | 2.592 | 3/15 | 2.626 |
| 12/13 | 2.192 | 8/14 | 2.774 | 4/15 | 2.903 |
| 1/14 | 1.989 | 9/14 | 3.031 | 5/15 | 3.417 |
| 2/14 | 2.130 | 10/14 | 2.943 | 6/15 | 3.301 |

a. Produce a time-series plot of the average prices of milk. Identify any time-series components that exist in the data.
b. Identify the recurrence period of the time series. Determine the seasonal index for each month in the recurrence period.
c. Fit a linear trend model to the deseasonalized data.
d. Provide a seasonally adjusted forecast using the linear trend model for July 2015 and July 2019.

**16-28.** Suppose that a company was formed five years ago to provide meal delivery, physical therapy, and minor housekeeping services to senior citizens. The revenue data for the first five years of operation are shown here.

| Revenue ($10,000s) | | | | | |
|---|---|---|---|---|---|
| | 2011 | 2012 | 2013 | 2014 | 2015 |
| January | 23 | 67 | 72 | 76 | 81 |
| February | 34 | 63 | 64 | 75 | 72 |
| March | 45 | 65 | 64 | 77 | 71 |
| April | 48 | 71 | 77 | 81 | 83 |
| May | 46 | 75 | 79 | 86 | 85 |
| June | 49 | 70 | 72 | 75 | 77 |
| July | 60 | 72 | 71 | 80 | 79 |
| August | 65 | 75 | 77 | 82 | 84 |
| September | 67 | 80 | 79 | 86 | 91 |
| October | 60 | 78 | 78 | 87 | 86 |
| November | 71 | 89 | 87 | 91 | 94 |
| December | 76 | 94 | 92 | 96 | 99 |

a. Plot these data. Based on your visual observations, what time-series components are present in the data?
b. Determine the seasonal index for each month.
c. (1) Fit a linear trend model to the deseasonalized data for the years 2011–2015 and determine the *MAD* and *MSE* for forecasts for each of the months in 2016. (2) Conduct a test of hypothesis to determine if the linear trend model fits the existing data. (3) Comment on the adequacy of the linear trend model based on the measures of forecast error and the hypothesis test you conducted.
d. Use the seasonal index values computed in part b to provide seasonal adjusted forecasts for each month of the year 2016.

**16-29.** The manager of a financial investment office is evaluated based on the number of new clients generated each quarter. The following data reflect the number of new customers added during each quarter from 2012 to 2015.

| 2012 | 2013 |
|---|---|
| 1st quarter 218 | 1st quarter 250 |
| 2nd quarter 190 | 2nd quarter 220 |
| 3rd quarter 236 | 3rd quarter 265 |
| 4th quarter 218 | 4th quarter 241 |

| 2014 | 2015 |
|---|---|
| 1st quarter 244 | 1st quarter 229 |
| 2nd quarter 228 | 2nd quarter 221 |
| 3rd quarter 263 | 3rd quarter 248 |
| 4th quarter 240 | 4th quarter 231 |

a. Plot the time series and discuss the components that are present in the data.

b. Referring to part a, fit the linear trend model to the data for the years 2012 through 2014. Then use the resulting model to forecast the number of new brokerage customers for each quarter in the year 2015. Compute the MAD and MSE for these forecasts and discuss the results.

c. Using the data for the years 2012 through 2014, determine the seasonal indexes for each quarter.

d. Develop a seasonally unadjusted forecast for the four quarters of year 2015.

e. Using the seasonal indexes computed in part d, determine the seasonally adjusted forecast for each quarter for the year 2015. Compute the MAD and MSE for these forecasts.

f. Examine the values for the MAD and MSE in parts b and e. Which of the two forecasting techniques would you recommend the manager use to forecast the number of new clients generated each quarter? Support your choice by giving your rationale.

## Computer Software Exercises

**16-30.** Logan Pickens is a plan/build construction company specializing in resort area construction projects. Suppose that plan/build companies typically have a cash flow problem, since they tend to be paid in lump sums when projects are completed or hit milestones. The data file **Logan-Pickens** contains month-end cash balances for the past 16 months.

a. Plot the data as a time-series graph. Discuss what the graph implies about the relationship between cash balance and the time variable, month.

b. Fit a linear trend model to the data. Compute the coefficient of determination for this model and show the trend line on the time-series graph. Discuss the appropriateness of the linear trend model. What are the strengths and weaknesses of the model?

c. Referring to part b, compute the MAD and MSE for the 16 data points.

d. Use the $t^2$ transformation approach and recompute the linear model using the transformed time variable. Plot the new trend line against the transformed data. Discuss whether this model appears to provide a better fit than did the model without the transformation. Compare the coefficients of determination for the two models. Which model seems to be superior, using the coefficient of determination as the criterion?

e. Refer to part d. Compute the MAD and MSE for the 16 data values. Discuss how these compare to those that you computed in part c, prior to transformation. Do the measures of fit ($R^2$, MSE, or MAD) agree on the best model to use for forecasting purposes?

**16-31.** Refer to Exercise 16-30.

a. Use the linear trend model (without transformation) for the first 15 months and provide a cash balance forecast for month 16. Then make the $t^2$ transformation and develop a new linear trend forecasting model based on months 1 through 15. Forecast the cash balance for month 16. Now compare the accuracy of the forecasts with and without the transformation. Which of the two forecast models do you prefer? Explain your answer.

b. Provide a 95% prediction interval for the cash balance forecast for month 16 using the linear trend model both with and without the transformation. Which interval has the widest width? On this basis, which procedure would you choose?

**16-32.** The federal funds rate is the interest rate charged by banks when banks borrow "overnight" from each other. The file titled **The Fed** contains the federal funds rates for the period 1955 through 2015 (*source:* "Effective federal funds rate," Economic Research, Federal Reserve Bank of St. Louis, https://research.stlouisfed.org, Mar. 7, 2016).

a. Produce a scatter plot of the federal funds rates for the period 1955 through 2015. Identify any time-series components in the data.

b. Identify the recurrence period of the time series.

c. Fit a nonlinear trend model that uses coded years and coded years squared as predictors for the deseasonalized data.

d. Provide a forecast using the nonlinear trend model for 2016.

**16-33.** The United States Census Bureau collects and maintains data on a wide range of statistics associated with the U.S. economy. For example, the file **Retail Sales** contains the total retail sales in January for the years 2001 through 2016 (*source:* www.census.gov/retail).

a. Produce a time-series plot of these data. Indicate the time-series components in the data.

b. Conduct a test of hypothesis to determine if there is a linear trend in these data. Use a significance level of 0.10 and the *p*-value approach.

c. Provide forecasts for e-commerce retail sales for January 2017.

**16.3**    **Forecasting Using Smoothing Methods**

The trend-based forecasting technique introduced in the previous section is widely used and can be very effective in many situations. However, it has a disadvantage in that it gives as much weight to the earliest data in the time series as it does to the data that are close to the period for which the forecast is required. Also, this trend approach does not provide an opportunity for the model to "learn" or "adjust" to changes in the time series.

A class of forecasting techniques called *smoothing models* is widely used to overcome these problems and to provide forecasts in situations in which there is no pronounced trend in the data. These models attempt to "smooth out" the random or irregular component in the time series by an averaging process. In this section we introduce two frequently used smoothing-based forecasting techniques: single exponential smoothing and double exponential smoothing. Double exponential smoothing offers a modification to the single exponential smoothing model that specifically deals with trends.

OUTCOME **5**

## Exponential Smoothing

The trend-based forecasting methods discussed in Section 16.2 are used in many forecasting situations. As we showed, the least squares trend line is computed using all available historical data. Each observation is given equal input in establishing the trend line, thus allowing the trend line to reflect all the past data. If the future pattern looks like the past, the forecast should be reasonably accurate.

However, in many situations involving time-series data, the more recent the observation, the more indicative it is of possible future values. For example, this month's sales are probably a better indicator of next month's sales than are sales from 20 months ago. However, the regression analysis approach to trend-based forecasting does not take this fact into account. The data from 20 periods ago are given the same weight as data from the most current period in developing a forecasting model. This equal valuation can be a drawback to the trend-based forecasting approach.

**Exponential Smoothing**
A time-series and forecasting technique that produces an exponentially weighted moving average in which each forecast is dependent on all previous observed values.

With **exponential smoothing**, current observations can be weighted more heavily than older observations in determining the forecast. Therefore, if in recent periods the time-series values are much higher (or lower) than those in earlier periods, the forecast can be made to reflect this difference. The extent to which the forecast reflects the current data depends on the weights assigned by the decision maker.

We will introduce two classes of exponential smoothing models: single exponential smoothing and double exponential smoothing. Double smoothing is used when a time series exhibits a linear trend. Single smoothing is used when no linear trend is present in the time series. Both single and double exponential smoothing are appropriate for short-term forecasting and for time series that are not seasonal.

**Single Exponential Smoothing**    Just as its name implies, single exponential smoothing uses a single smoothing constant. Equations 16.13 and 16.14 represent two equivalent methods for forecasting using single exponential smoothing.

**Exponential Smoothing Model**

$$F_{t+1} = F_t + \alpha(y_t - F_t) \tag{16.13}$$

or

$$F_{t+1} = \alpha y_t + (1 - \alpha)F_t \tag{16.14}$$

where:

$F_{t+1}$ = Forecast value for period $t + 1$

$y_t$ = Actual value of the time series at time $t$

$F_t$ = Forecast value for period $t$

$\alpha$ = Alpha (smoothing constant $0 \leq \alpha \leq 1$)

The logic of the exponential smoothing model is that the forecast made for the next period will equal the forecast made for the current period, plus or minus some adjustment factor. The adjustment factor is determined by the difference between this period's forecast and the actual value $(y_t - F_t)$, multiplied by the smoothing constant, $\alpha$. The idea is that if we forecast low, we will adjust next period's forecast upward, by an amount determined by the smoothing constant.

---

**EXAMPLE 16-3**   **Developing a Single Exponential Smoothing Model**

**Dawson Graphic Design**   Consider the past 13 weeks of incoming sales calls from potential customers for Dawson Graphic Design located in Orlando, Florida. These data and their line graph are shown in Figure 16.25. The data showing the number of incoming calls from potential customers are in the file **Dawson**. Suppose the current time period is the end of week 13 and we wish to forecast the number of incoming calls for week 14 using a single exponential smoothing model. The following steps can be used:

**Excel Tutorial**

STEP 1   **Specify the model.**
Because the data do not exhibit a pronounced trend and because we are interested in a short-term forecast (one period ahead), we can use the single exponential smoothing model with a single smoothing constant.

STEP 2   **Fit the model.**
We start by selecting a value for $\alpha$, the smoothing constant, between 0.0 and 1.0. The closer $\alpha$ is to 0.0, the less influence the current observations have in determining the forecast. Small $\alpha$ values will result in greater smoothing of the time series. Likewise, when $\alpha$ is near 1.0, the current observations have greater impact in determining the forecast and less smoothing will occur. There is no firm rule for selecting the value for the smoothing constant. However, in general, if the time series is quite stable, a small $\alpha$ should be used to lessen the impact of random or irregular fluctuations. Because the time series shown in Figure 16.25 appears to be relatively stable, we will use $\alpha = 0.20$ in this example.

The forecast value for period $t = 14$ is found using Equation 16.14, as follows:

$$F_{14} = 0.20y_{13} + (1 - 0.20)F_{13}$$

**FIGURE 16.25** **Incoming Sales Calls Data and Line Graph for Dawson Graphic Design**

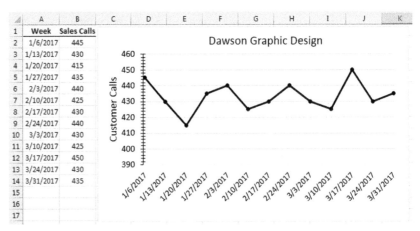

---

**Excel 2016 Instructions**

1. Open file: **Dawson.xlsx**.
2. Select data in the *Sales Calls* data column.
3. Click **Insert > Insert Line or Area Chart > Line with Markers**.
4. Click **Select Data**.
5. On the **Design** tab, enter desired title.
6. Remove gridlines.
7. Under **Horizontal (Categories) Axis Labels**, select data in *Week* column.

This demonstrates that the forecast for period 14 is a weighted average of the actual number of calls in period 13 and the forecast for period 13. Although we know the number of calls for period 13, we don't know the forecast for period 13. However, we can determine it by

$$F_{13} = 0.20y_{12} + (1 - 0.20)F_{12}$$

Again, this forecast is a weighted average of the actual number of calls in period 12 and the forecast calls for period 12. We continue in this manner until we get to

$$F_2 = 0.20y_1 + (1 - 0.20)F_1$$

This requires a forecast for period 1. Because we have no data before period 1 from which to develop a forecast, a rule often used is to assume that $F_1 = y_1$.[3]

### Forecast for period 1 = Actual value in period 1

Because setting the starting value is somewhat arbitrary, we need to obtain as much historical data as possible to "warm" the model and dampen out the effect of the starting value. In our example, we have 13 periods of data to warm the model before the forecast for period 14 is made. Note that when we use an exponential smoothing model, the effect of the initial forecast is reduced by $(1 - \alpha)$ in the forecast for period 2, then reduced again for period 3, and so on. After sufficient periods, any error due to the arbitrary initial forecast should be very small.

Figure 16.26 shows the results of using the single exponential smoothing equation and Excel for weeks 1 through 13. For week 1, $F_1 = y_1 = 445$. Then, for week 2, we get

$$F_2 = 0.20y_1 + (1 - 0.20)F_1$$
$$F_2 = (0.20)445 + (1 - 0.20)445.00 = 445.00$$

For week 3,

$$F_3 = 0.20y_2 + (1 - 0.20)F_2$$
$$F_3 = (0.20)430 + (1 - 0.20)455.00 = 442.00$$

At the end of week 2, after we see what actually happened to the number of calls in week 2, our forecast for week 3 is 442 calls. This is a 3-unit decrease

**FIGURE 16.26 Dawson Graphic Design Single Exponential Smoothing—Excel Spreadsheet**

Excel 2016 Instructions

1. Open file: **Dawson.xlsx**.
2. Create and label two new columns.
3. Enter the smoothing constant (alpha) in an empty cell (e.g., B17 = 0.20).
4. Enter initial starting value forecast for period 1 in cell C2 = 445.
5. Use text Equation 16.14 to create forecast of period t + 1 in cell D2.
6. Copy equation down to cell D12.
7. Create a formula in cell **C3** that refers to the forecast for period t + 1 in cell **D2**.
8. Copy the formula down through cell **C15**.

| | A | B | C | D |
|---|---|---|---|---|
| | Week | Sales Calls | Forecast for Period t | Forecast for period t+1 |
| 1 | | | | |
| 2 | 1/6/2017 | 445 | 445 | 445 |
| 3 | 1/13/2017 | 430 | 445 | 442 |
| 4 | 1/20/2017 | 415 | 442 | 436.60 |
| 5 | 1/27/2017 | 435 | 436.60 | 436.28 |
| 6 | 2/3/2017 | 440 | 436.28 | 437.02 |
| 7 | 2/10/2017 | 425 | 437.02 | 434.62 |
| 8 | 2/17/2017 | 430 | 434.62 | 433.70 |
| 9 | 2/24/2017 | 440 | 433.70 | 434.96 |
| 10 | 3/3/2017 | 430 | 434.96 | 433.97 |
| 11 | 3/10/2017 | 425 | 433.97 | 432.17 |
| 12 | 3/17/2017 | 450 | 432.17 | 435.74 |
| 13 | 3/24/2017 | 430 | 435.74 | 434.59 |
| 14 | 3/31/2017 | 435 | 434.59 | 434.67 |
| 15 | | | 434.67 | |
| 16 | | | | |
| 17 | Alpha = | 0.2 | | |

Forecast for period 14 = 434.67 or rounded to 435

Alpha = 0.2

---

[3] Another approach for establishing the starting value, $F_1$, is to use the mean value for some portion of the available data. Regardless of the method used, the quantity of available data should be large enough to dampen out the impact of the starting value.

over the forecast for week 2 of 445 calls. The actual number of calls in week 2 was 430, rather than 445. The number of calls for week 2 was 15 units lower than the forecast for that time period. Because the actual calls were less than the forecast, we must make an adjustment. The 3-unit adjustment is determined by multiplying the smoothing constant by the forecast error [$(0.20)(15) = 3$], as specified in Equation 16.13. The adjustment compensates for the forecast error in week 2.

Continuing for week 4 again using Equation 16.14, we have

$$F_4 = 0.20y_3 + (1 - 0.20)F_3$$
$$F_4 = (0.20)415 + (1 - 0.20)442.00 = 436.60$$

Recall that our forecast for week 3 was 442. However, the actual number of calls was lower than forecast at 415, and we overforecast by 27 calls. The adjustment for week 4 is then $0.20(27) = 5.40$, and the forecast for week 4 is $442 - 5.40 = 436.60$.

This process continues through the data until we are ready to forecast week 14, as shown in Figure 16.26:

$$F_{14} = 0.20y_{13} + (1 - 0.20)F_{13}$$
$$F_{14} = (0.20)435 + (1 - 0.20)434.59 = 434.67$$

Dawson Graphic Design managers would forecast 435 incoming calls for week 14. If we wished to forecast week 15 calls, we would either use the week 14 forecast or wait until we know the actual week 14 calls and then update the smoothing equations to get a new forecast for week 15.

STEP 3  **Diagnose the model.**

However, before we actually use the exponential smoothing forecast for decision-making purposes, we need to determine how successfully the model fits the historical data. Unlike the trend-based forecast, which uses least squares regression, there is no need to use split samples to test the forecasting ability of an exponential smoothing model, because the forecasts are "true forecasts." The forecast for a given period is made before considering the actual value for that period.

Figure 16.27 shows the *MAD* for the forecast model with $\alpha = 0.20$ and a plot of the forecast values versus the actual call values. This plot shows the smoothing that has occurred. Note that we don't include period 1 in the *MAD* calculation, since the forecast is arbitrarily set equal to the actual value.

Our next step would be to try different smoothing constants and find the *MAD* for each new $\alpha$. We would make the forecast for period 14 using the smoothing constant that generates the smallest *MAD*.

**TRY EXERCISE 16-34** (pg. 677)

A major advantage of the single exponential smoothing model is that it is easy to update. In Example 16-3, the forecast for week 15 using this model is found by simply plugging the actual data value for week 14, once it is known, into the smoothing formula:

$$F_{15} = \alpha y_{14} + (1 - \alpha)F_{14}$$

We do not need to go back and recompute the entire model, as would have been necessary with a trend-based regression model.

OUTCOME **5**  **Double Exponential Smoothing**  When the time series has an increasing or decreasing trend, we use a modification to the single exponential smoothing model to explicitly account for the trend. The resulting technique is called *double exponential smoothing*. The double exponential smoothing model is often referred to as *exponential smoothing with trend*. In double exponential smoothing, a second smoothing constant, beta ($\beta$), is included to account for the trend. Equations 16.15, 16.16, and 16.17 are needed to provide the forecasts.

**FIGURE 16.27** Excel 2016 Output for Dawson Graphic Design *MAD* Computation for Single Exponential Smoothing, $\alpha = 0.20$

| | Week | Sales Calls | Forecast for Period t | Forecast for period t+1 | Forecast Error | Absolute Forecast Error |
|---|---|---|---|---|---|---|
| 2 | 1/6/2017 | 445 | 445 | 445 | | |
| 3 | 1/13/2017 | 430 | 445 | 442 | -15 | 15 |
| 4 | 1/20/2017 | 415 | 442 | 436.60 | -27 | 27 |
| 5 | 1/27/2017 | 435 | 436.60 | 436.28 | -1.60 | 1.60 |
| 6 | 2/3/2017 | 440 | 436.28 | 437.02 | 3.72 | 3.72 |
| 7 | 2/10/2017 | 425 | 437.02 | 434.62 | -12.02 | 12.02 |
| 8 | 2/17/2017 | 430 | 434.62 | 433.70 | -4.62 | 4.62 |
| 9 | 2/24/2017 | 440 | 433.70 | 434.96 | 6.30 | 6.30 |
| 10 | 3/3/2017 | 430 | 434.96 | 433.97 | -4.96 | 4.96 |
| 11 | 3/10/2017 | 425 | 433.97 | 432.17 | -8.97 | 8.97 |
| 12 | 3/17/2017 | 450 | 432.17 | 435.74 | 17.83 | 17.83 |
| 13 | 3/24/2017 | 430 | 435.74 | 434.59 | -5.74 | 5.74 |
| 14 | 3/31/2017 | 435 | 434.59 | 434.67 | 0.41 | 0.41 |
| 15 | | | 434.67 | | | |
| 16 | | | | | Sum = | 108.16 |
| 17 | Alpha = | 0.2 | | | | |
| 18 | | | | | MAD = | 9.01 |

**Excel 2016 Instructions**

1. Open file: **Dawson.xlsx**.
2. Follow Figure 16.26 instructions.
3. Create a column of forecast errors using an Excel equation starting in E3.
4. Starting in cell F3, create a column of absolute forecast errors using Excel's **ABS** function.
5. Compute the *MAD* starting in F3 using the **AVERAGE** function for the column on absolute errors.
6. Select *Sales Calls* and *Forecast* data.
7. Click **Insert > Insert Line or Area Chart > Line with Markers**.
8. On the **Design** tab, enter desired title and vertical title.
9. Remove gridlines.
10. Double-click the **Vertical (Value) Axis** and change the minimum bound to **390** and the maximum bound to **460**.

**Double Exponential Smoothing Model**

$$C_t = \alpha y_t + (1 - \alpha)(C_{t-1} + T_{t-1}) \tag{16.15}$$

$$T_t = \beta(C_t - C_{t-1}) + (1 - \beta)T_{t-1} \tag{16.16}$$

$$F_{t+1} = C_t + T_t \tag{16.17}$$

where:

$$y_t = \text{Value of the time series at time } t$$
$$\alpha = \text{Constant-process smoothing constant}$$
$$\beta = \text{Trend-smoothing constant}$$
$$C_t = \text{Smoothed constant-process value for period } t$$
$$T_t = \text{Smoothed trend value for period } t$$
$$F_{t+1} = \text{Forecast value for period } t + 1$$
$$t = \text{Current time period}$$

Equation 16.15 is used to smooth the time-series data; Equation 16.16 is used to smooth the trend; and Equation 16.17 combines the two smoothed values to form the forecast for period $t + 1$.

Excel Tutorial

**EXAMPLE 16-4** **Double Exponential Smoothing**

**Billingsley Insurance Company** Billingsley Insurance Company has maintained data on the number of automobile claims filed at its Denver office over the past 12 months. These data, which are in the file **Billingsley**, are listed and graphed in Figure 16.28. The claims manager wants to forecast the number of claims for month 13. A double exponential smoothing model can be developed using the following steps:

STEP 1 **Specify the model.**

The time series contains a strong upward trend, so we select a double exponential smoothing model.

As with single exponential smoothing, we must select starting values. In the case of the double exponential smoothing model, we must select initial values for $C_0$, $T_0$, and the smoothing constants $\alpha$ and $\beta$. The choice of smoothing constant values ($\alpha$ and $\beta$) depends on the same issues as those discussed earlier for single exponential smoothing. That is, use larger smoothing constants when we want less smoothing and values closer to 0 when we want more smoothing. The larger the smoothing constant value, the more impact that current data will have on the forecast. Suppose we use $\alpha = 0.2$ and $\beta = 0.3$ in this example. There are several approaches for selecting starting values for $C_0$ and $T_0$. The method we use here is to fit the least squares trend to the historical data,

$$\hat{y}_t = b_0 + b_1 t$$

where the y intercept, $b_0$, is used as the starting value, $C_0$, and the slope, $b_1$, is used as the starting value for the trend, $T_0$. We can use the regression procedure in Excel to perform these calculations, giving

$$\hat{y}_t = 34.273 + 4.112t$$

So

$$C_0 = 34.273 \quad \text{and} \quad T_0 = 4.112$$

Keep in mind that these are arbitrary starting values, and as with single exponential smoothing, their effect will be dampened out as we proceed through the sample data to the current period. The more historical data we have, the less impact the starting values will have on the forecast.

**FIGURE 16.28** **Excel 2016 Billingsley Insurance Company Data and Time-Series Plot**

**Excel 2016 Instructions**

1. Open file: **Billingsley.xlsx**.
2. Select data in the *Claims* data column.
3. **Insert Line or Area Chart > Line with Markers**.
4. On the **Design** tab, enter desired title and vertical title.

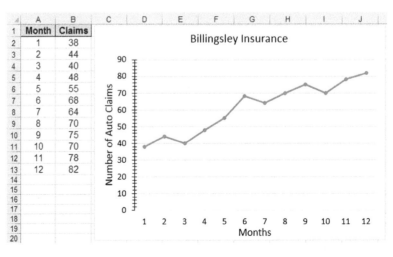

| Month | Claims |
|-------|--------|
| 1 | 38 |
| 2 | 44 |
| 3 | 40 |
| 4 | 48 |
| 5 | 55 |
| 6 | 68 |
| 7 | 64 |
| 8 | 70 |
| 9 | 75 |
| 10 | 70 |
| 11 | 78 |
| 12 | 82 |

STEP 2  **Fit the model.**

The forecast for period 1 made at the beginning of period 1 is

$$F_1 = C_0 + T_0$$
$$F_1 = 34.273 + 4.112 = 38.385$$

At the close of period 1, in which actual claims were 38, the smoothing equations are updated as follows:

$$C_1 = 0.20(38) + (1 - 0.20)(34.273 + 4.112) = 38.308$$
$$T_1 = 0.30(38.308 - 34.273) + (1 - 0.30)(4.112) = 4.089$$

Next, the forecast for period 2 is

$$F_2 = 38.308 + 4.089 = 42.397$$

We then repeat the process through period 12 to find the forecast for period 13.

STEP 3  **Diagnose the model.**

Figure 16.29 shows the results of the computations and the *MAD* value. The forecast for period 13 is

$$F_{13} = C_{12} + T_{12}$$
$$F_{13} = 83.867 + 3.908 = 87.776$$

Based on this double exponential smoothing model, the number of claims for period 13 is forecast to be about 88. However, before settling on this forecast, we should try different smoothing constants to determine whether we can find a smaller *MAD*.

**FIGURE 16.29 Excel 2016 Double Exponential Smoothing Spreadsheet for Billingsley Insurance**

| | A | B | C | D | E | F | G |
|---|---|---|---|---|---|---|---|
| 1 | Month | Claims | Constant | Trend | Forecast | Forecast Error | Absolute Error |
| 2 | $t$ | $y_t$ | $C_t$ | $T_t$ | $F_{t+1}=C_t+F_t$ | $Y_t-F_t$ | $|Y_t-F_t|$ |
| 3 | 1 | 38 | 38.308 | 4.089 | 38.385 | -0.385 | 0.385 |
| 4 | 2 | 44 | 42.717 | 4.185 | 42.397 | 1.603 | 1.603 |
| 5 | 3 | 40 | 45.522 | 3.771 | 46.902 | -6.902 | 6.902 |
| 6 | 4 | 48 | 49.034 | 3.693 | 49.293 | -1.293 | 1.293 |
| 7 | 5 | 55 | 53.182 | 3.830 | 52.728 | 2.272 | 2.272 |
| 8 | 6 | 68 | 59.209 | 4.489 | 57.012 | 10.988 | 10.988 |
| 9 | 7 | 64 | 63.759 | 4.507 | 63.698 | 0.302 | 0.302 |
| 10 | 8 | 70 | 68.613 | 4.611 | 68.266 | 1.734 | 1.734 |
| 11 | 9 | 75 | 73.579 | 4.718 | 73.224 | 1.776 | 1.776 |
| 12 | 10 | 70 | 76.637 | 4.220 | 78.297 | -8.297 | 8.297 |
| 13 | 11 | 78 | 80.286 | 4.048 | 80.857 | -2.857 | 2.857 |
| 14 | 12 | 82 | 83.867 | 3.908 | 84.334 | -2.334 | 2.334 |
| 15 | | | | | 87.776 | | |
| 16 | | | | | | | |
| 17 | Alpha = | 0.2 | | $C_0$ | 34.27 | Sum = | 40.745 |
| 18 | Beta = | 0.3 | | $T_0$ | 4.112 | MAD = | 3.395 |

Month 13 forecast

**Excel 2016 Instructions**

1. Open file: **Billingsley.xlsx**.
2. Create one new row with 5 new columns as shown in Figure 16.29.
3. Place smoothing constants (alpha and beta) in empty cells (B17 and B18).
4. Place starting values for the constant process and the trend in empty cells (D17 and D18).
5. Use text equations 16.15 and 16.16 to create values for $C_t$ and $T_t$ in columns C and D.
6. Use text Equation 16.17 to create forecasts in column E.
7. Calculate forecast errors in column F by subtracting column E value from column B value.
8. Calculate a column of absolute errors using Excel's **ABS** function.
9. Calculate the *MAD* by using Excel's **AVERAGE** function ( = **AVERAGE (G3:G14)**).

**TRY 16-39** (pg. 678)

As you can see, the computations required for double exponential smoothing are somewhat tedious and are ideally suited for your computer. We have used Excel formulas to develop our model. However, Excel 2016 has a built-in forecasting tool that you can use; it is illustrated in the following section.

## Forecasting with Excel 2016

The Excel 2016 Forecast Sheet option makes it easy for the decision maker to develop smoothing forecasts using historical time-series data. This feature can produce forecasts for data when there are no trend or seasonality effects, as well as in those instances where trend and/or seasonal effects exist. The Forecast Sheet option uses the historical data to produce a forecast and a forecast chart and to generate confidence intervals for the predicted values.

To create a forecast using the Excel 2016 Forecast Sheet option, we need two data series: (1) a series with date or time entries and (2) a corresponding series of data values. The date entries must have consistent time intervals (e.g., the first of every month, or every Monday of the week). It is acceptable to have up to 30% of the data points missing, or to have several data values with an identical time stamp. To create a forecast, use the following steps:

STEP 1 In a worksheet, enter the dates and the corresponding data values in two separate columns. As our example, we will use the number of weekly sales calls made by a pharmaceutical company's sales staff over a 13-week period between January 5, 2016, and March 29, 2016.

STEP 2 Select the two columns of entries.

STEP 3 On the **Data** tab in the Forecast group, click **Forecast Sheet**. The results are shown in Figure 16.30.

STEP 4 In the **Create Forecast Worksheet** box, select the line chart or the column chart option. These options determine the visual presentation for the forecast. Generally, line charts are used to describe time-series data, so that is the recommended option. Also, in the **Forecast End** box, select an end date. The date specifies the forecast horizon—that is, the length of time into the future for which forecasts are desired. The line chart is shown in Figure 16.31.

FIGURE 16.30 **Excel 2016 Forecast Sheet**

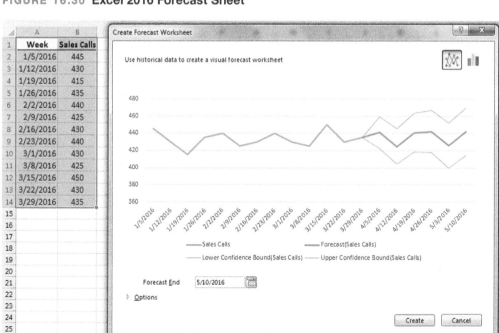

FIGURE 16.31 **Excel 2016 Line Chart**

STEP 5    To produce the forecast for the next several weeks in April 2016, click **Create**. A new worksheet will be generated that includes the historical and predicted values along with the chart that you selected in Step 4. This new worksheet will be positioned before the worksheet that contains the forecast time stamps and data. The results are shown in Figure 16.32.

Excel 2016's Forecast Sheet option also gives us the opportunity to change the settings we use to produce the forecasts. We can make changes by selecting the *Options* choice in the lower left-hand side of the worksheet, as shown in Figure 16.33.

The various options that a forecasting analyst might want to modify are briefly discussed below.

- Forecast Start—This is the date for the forecast to begin. If you choose a date before the end of the historical data, then only those points prior to the start date are used to generate the forecast. If you start your forecast before the last historical period, then you will have an opportunity to compare the predicted to the actual historical data. This will provide you with a graphical measure of the forecast accuracy. Caution should be

FIGURE 16.32 **Excel 2016 Forecast Output**

| | A | B | C | D | E | F | G |
|---|---|---|---|---|---|---|---|
| 1 | Week | Sales Calls | Forecast | Lower Confidence | Upper Confidence | Statistic | Value |
| 2 | 1/5/2016 | 445 | | | | Alpha | 0.50 |
| 3 | 1/12/2016 | 430 | | | | Beta | 0.00 |
| 4 | 1/19/2016 | 415 | | | | Gamma | 0.00 |
| 5 | 1/26/2016 | 435 | | | | MASE | 0.88 |
| 6 | 2/2/2016 | 440 | | | | SMAPE | 0.02 |
| 7 | 2/9/2016 | 425 | | | | MAE | 10.66 |
| 8 | 2/16/2016 | 430 | | | | RMSE | 11.70 |
| 9 | 2/23/2016 | 440 | | | | | |
| 10 | 3/1/2016 | 430 | | | | | |
| 11 | 3/8/2016 | 425 | | | | | |
| 12 | 3/15/2016 | 450 | | | | | |
| 13 | 3/22/2016 | 430 | | | | | |
| 14 | 3/29/2016 | 435 | 435 | 435.00 | 435.00 | | |
| 15 | 4/5/2016 | | 441.0012438 | 422.71 | 459.30 | | |
| 16 | 4/12/2016 | | 424.7460942 | 404.27 | 445.22 | | |
| 17 | 4/19/2016 | | 440.8067875 | 418.36 | 463.25 | | |
| 18 | 4/26/2016 | | 442.2305533 | 417.96 | 466.50 | | |
| 19 | 5/3/2016 | | 425.9754036 | 400.00 | 451.95 | | |
| 20 | 5/10/2016 | | 442.0360969 | 414.46 | 469.61 | | |

FIGURE 16.33 **Excel 2016 Forecasting (Optional Settings)**

exercised, however, because starting a forecast too early will not always produce as accurate a prediction as could be obtained using all the historical data.

- Confidence Interval—To show a confidence interval for your forecast predictions, check the Confidence Interval option.

- Timeline Range and Values Range—These are the ranges of values included in the analysis. These may be changed; however, the two ranges must match each other.

- Seasonality—This option captures the length of the seasonal pattern. It is automatically detected by the algorithm used to produce the forecasts. You can, however, override the automatic value by manually entering the seasonal period effect. *Note:* If there are less than two cycles of seasonality in the historical data, then Excel's algorithm cannot automatically identify a seasonal pattern. In this case the prediction becomes a linear trend model.

- Fill Missing Points Using—The forecasting model can work with up to 30% of the data missing. This feature identifies how you want Excel to handle missing observations. Excel uses a weighted average of neighboring values to interpolate for a missing value. To assign missing data a value of zero, select the **Zeros** option from the list.

- Duplicate Aggregates—For multiple data values having identical time stamps, Excel averages the values. To change this to a different calculation method, select the desired calculation from the list provided.

- Include Forecast Statistics—To include forecast statistics, check this box. Doing so will include measures of forecast accuracy (MAE and RMSE) and the forecasting smoothing coefficients (alpha for the level component, beta for the trend component, and gamma for the seasonal component) on the forecast sheet. The mean absolute error (denoted as *MAE*) is identical to the mean absolute deviation (*MAD*) discussed earlier. The root mean square error (denoted as *RMSE*) is the square root of the mean square error (*MSE*), also discussed earlier.

**FIGURE 16.34** Excel 2016 Billingsley Monthly Claims

| | A | B |
|---|---|---|
| 1 | **Month** | **Claims** |
| 2 | 5/15/2016 | 38 |
| 3 | 6/15/2016 | 44 |
| 4 | 7/15/2016 | 40 |
| 5 | 8/15/2016 | 48 |
| 6 | 9/15/2016 | 55 |
| 7 | 10/15/2016 | 68 |
| 8 | 11/15/2016 | 64 |
| 9 | 12/15/2016 | 70 |
| 10 | 1/15/2017 | 75 |
| 11 | 2/15/2017 | 70 |
| 12 | 3/15/2017 | 78 |
| 13 | 4/15/2017 | 82 |

**EXAMPLE 16-5** **Forecasting with Excel 2016**

**Billingsley Insurance Company** (*continued*)   In Example 16-4, we featured the Billingsley Insurance Company, which maintained data on the number of automobile claims filed at its Denver office over the past 12 months. The data, which are in the file **Billingsley-2**, are listed in Figure 16.34.

**FIGURE 16.35** Excel 2016 Billingsley Forecast Output

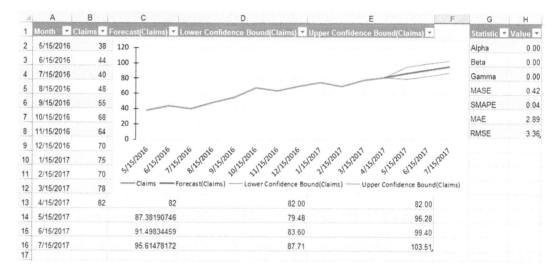

The claims manager wants to forecast the number of claims for May, June, and July using the Excel 2016 Forecast Sheet option. The Excel 2016 output displaying the forecast and forecast statistics is shown in Figure 16.35.

In this example, there are not enough data to identify a seasonal pattern, so Excel 2016 automatically selects a linear trend model, as indicated by the alpha and beta smoothing constant values of 0.00.

# 16.3 EXERCISES

## Skill Development

**16-34.** The following table represents two years of data:

| Year 1 | | Year 2 | |
|---|---|---|---|
| 1st quarter | 242 | 1st quarter | 272 |
| 2nd quarter | 252 | 2nd quarter | 267 |
| 3rd quarter | 257 | 3rd quarter | 276 |
| 4th quarter | 267 | 4th quarter | 281 |

a. Prepare a single exponential smoothing forecast for the first quarter of year 3 using an alpha value of 0.10. Let the initial forecast value for quarter 1 of year 1 be 250.

b. Prepare a single exponential smoothing forecast for the first quarter of year 3 using an alpha value of 0.25. Let the initial forecast value for quarter 1 of year 1 be 250.

c. Calculate the *MAD* value for the forecasts you generated in parts a and b. Which alpha value provides the smaller *MAD* value at the end of the 4th quarter in year 2?

**16-35.** The following data represent enrollment in a major at your university for the past six semesters. (*Note:* Semester 1 is the oldest data; semester 6 is the most recent data.)

| Semester | Enrollment |
|---|---|
| 1 | 87 |
| 2 | 110 |
| 3 | 123 |
| 4 | 127 |
| 5 | 145 |
| 6 | 160 |

a. Prepare a graph of enrollment for the six semesters.
b. Based on the graph you prepared in part a, does it appear that a trend is present in the enrollment figures?
c. Prepare a single exponential smoothing forecast for semester 7 using an alpha value of 0.35. Assume that the initial forecast for semester 1 is 90.
d. Prepare a double exponential smoothing forecast for semester 7 using an alpha value of 0.20 and a beta value of 0.25. Assume that the initial smoothed constant value for semester 1 is 80 and the initial smoothed trend value for semester 1 is 10.
e. Calculate the *MAD* values for the simple exponential smoothing model and the double exponential smoothing model at the end of semester 6. Which model appears to be doing the better job of forecasting course enrollment? Don't include period 1 in the calculation.

**16-36.** The following data represent the average number of employees in outlets of a large consumer electronics retailer:

| Year | 2006 | 2007 | 2008 | 2009 | 2010 | 2011 | 2012 | 2013 | 2014 | 2015 |
|------|------|------|------|------|------|------|------|------|------|------|
| Number | 20.6 | 17.3 | 18.6 | 21.5 | 23.2 | 19.9 | 18.7 | 15.6 | 19.7 | 20.4 |

a. Construct a time-series plot of this time series. Does it appear that a linear trend exists in the time series?
b. Calculate forecasts for each of the years in the time series. Use a smoothing constant of 0.25 and single exponential smoothing.
c. Calculate the *MAD* value for the forecasts you generated in part b.
d. Construct a single exponential smoothing forecast for 2016. Use a smoothing constant of 0.25.

**16-37.** A brokerage company is interested in forecasting the number of new accounts the office will obtain next month. It has collected the following data for the past 12 months:

| Month | Accounts |
|-------|----------|
| 1 | 19 |
| 2 | 20 |
| 3 | 21 |
| 4 | 25 |
| 5 | 26 |
| 6 | 24 |
| 7 | 24 |
| 8 | 21 |
| 9 | 27 |
| 10 | 30 |
| 11 | 24 |
| 12 | 30 |

a. Produce a time-series plot for these data. Specify the exponential forecasting model that should be used to obtain next month's forecast.

b. Assuming a double exponential smoothing model, fit the least squares trend to the historical data, to determine the smoothed constant-process value and the smoothed trend value for period 0.
c. Produce the forecasts for periods 1 through 12 using $\alpha = 0.15$, $\beta = 0.25$. Indicate the number of new accounts the company may expect to receive next month based on the forecast model.
d. Calculate the *MAD* for this model.

## Business Applications

**16-38.** With tax revenues declining in many states, school districts have been searching for ways to cut costs without affecting classroom academics. Suppose one district has been looking at the cost of extracurricular activities ranging from band trips to athletics. The district business manager has gathered the past six months' costs for these activities as shown here.

| Month | Expenditures ($) |
|-------|------------------|
| September | 23,586.41 |
| October | 23,539.22 |
| November | 23,442.06 |
| December | 23,988.71 |
| January | 23,727.13 |
| February | 23,799.69 |

Using this past history, prepare a single exponential smoothing forecast for March using an $\alpha$ value of 0.25.

**16-39.** A May 2015 *Wall Street Journal* article indicated that 2015 college graduates had the highest average student debt ever, at slightly more than $35,000 per student. The following data show the average student debt for the years 1993 through 2015 (*source:* Jeffrey Sparshott, "Congratulations, class of 2015. You're the most indebted ever (for now)," http://blogs.wsj.com, May 8, 2015).

| Year | Average Student Loan Debt (Rounded to nearest $000) |
|------|------------------------------------------------------|
| 1993 | $10,000 |
| 1994 | $11,000 |
| 1995 | $12,000 |
| 1996 | $13,000 |
| 1997 | $14,000 |
| 1998 | $15,000 |
| 1999 | $16,000 |
| 2000 | $17,000 |
| 2001 | $17,000 |
| 2002 | $17,000 |
| 2003 | $18,000 |
| 2004 | $18,000 |
| 2005 | $20,000 |

| Year | Average Student Loan Debt (Rounded to nearest $000) |
|------|------|
| 2006 | $21,000 |
| 2007 | $22,000 |
| 2008 | $24,000 |
| 2009 | $25,000 |
| 2010 | $26,000 |
| 2011 | $28,000 |
| 2012 | $30,000 |
| 2013 | $31,000 |
| 2014 | $33,000 |
| 2015 | $35,000 |

Use Excel 2016 to forecast the average debt for 2017 and 2018 college graduates. Discuss the model that Excel 2016 has selected as the best model.

**16-40.** The human resources manager for a medium-sized business is interested in predicting the dollar value of medical expenditures filed by employees of her company for the year 2017. From her company's database, she has collected the following information for the previous seven years:

| Year | Medical Claims |
|------|------|
| 2010 | $405,642.43 |
| 2011 | $407,180.60 |
| 2012 | $408,203.30 |
| 2013 | $410,088.03 |
| 2014 | $411,085.64 |
| 2015 | $412,200.39 |
| 2016 | $414,043.90 |

a. Prepare a graph of medical expenditures for the years 2010 through 2016. Which forecasting technique do you think is most appropriate for this time series: single exponential smoothing or double exponential smoothing? Why?

b. Use an $\alpha$ value of 0.25 and a $\beta$ value of 0.15 to produce a double exponential forecast for the medical claims data. Use linear trend analysis to obtain the starting values for $C_0$ and $T_0$.

c. Compute the MAD value for your model for the years 2010 through 2016. Also produce a graph of your forecast values.

**16-41.** An electrical utility in a city in a South American country needs to forecast the highest temperature in the city on any day during a month. The following data reflect the highest recorded temperature (in degrees Fahrenheit) on any day during each month between June 2015 and June 2016:

| Year | Month | High Temperature |
|------|------|------|
| 2015 | June | 99.1 |
| 2015 | July | 102.8 |

| Year | Month | High Temperature |
|------|------|------|
| 2015 | August | 103.5 |
| 2015 | September | 104.6 |
| 2015 | October | 94 |
| 2015 | November | 96.8 |
| 2015 | December | 97.3 |
| 2016 | January | 99.7 |
| 2016 | February | 99.6 |
| 2016 | March | 101.6 |
| 2016 | April | 107.5 |
| 2016 | May | 95.9 |
| 2016 | June | 101.3 |

a. Construct a time-series plot of these data. Does it appear that a linear trend exists in the time series?

b. Calculate forecasts for each of the months in the time series. Use a smoothing constant of 0.25.

c. Calculate the MAD for the forecasts you generated in part b.

d. Construct a single exponential smoothing forecast for July 2016. Use a smoothing constant of 0.25.

## Computer Software Exercises

**16-42.** Harrison Avenue Burgers is a fast-food business that has been open since 1992. Suppose that the owner has monitored the average price he has paid for a pound of hamburger on an annual basis since 1992. The data are in a file called **Hamburger**.

a. Produce a time-series plot for these data. Specify the exponential forecasting model that should be used to obtain next year's forecast.

b. Assuming a double exponential smoothing model, fit the least squares trend to the historical data to determine the smoothed constant-process value and the smoothed trend value for period 0.

c. Use $\alpha = 0.20$ and $\beta = 0.30$ to forecast the average yearly hamburger price for the year 2010.

d. Calculate the MAD for this model.

**16-43.** The file **Museum Visitors** contains the number of visitors to a city museum each day during the period January 3 to March 20, 2015.

a. Use the data and the Excel 2016 Forecast Sheet option to develop a forecast for the number of museum visitors on March 21, 2015.

b. Report the forecast value and its 95% confidence interval.

**16-44.** The sales manager at Grossmieller Importers in New York City needs to determine a monthly forecast for the number of men's golf shirts that will be sold so that he can order an appropriate number of packing boxes. Grossmieller ships shirts to retail stores throughout the United States and Canada. Shirts are packed six to a box. Data for the past 16 months are in the data file called **Grossmieller**.

a. Plot the sales data using a time-series plot. Based on the graph, what time-series components are present? Discuss.
b. (1) Use a single exponential smoothing model with $\alpha = 0.30$ to forecast sales for month 17. Assume that the initial forecast for period 1 is 36,000. (2) Compute the *MAD* for this model. (3) Graph the smoothing-model-fitted values on the time-series plot.
c. (1) Referring to part b, try different alpha levels to determine which smoothing constant value you would recommend. (2) Indicate why you have selected this value and then develop the forecast for month 17. (3) Compare this to the forecast you got using $\alpha = 0.30$ in part b.

**16-45.** Refer back to Exercise 16-44, in which the sales manager for Grossmieller Imports of New York City needs to forecast monthly sales.
a. Discuss why a double exponential smoothing model might be preferred over a single exponential smoothing model.
b. (1) Develop a double exponential smoothing model using $\alpha = 0.20$ and $\beta = 0.30$ as smoothing constants. To obtain the starting values, use the regression trend line approach discussed in this section. (2) Determine the forecast for month 17. (3) Also compute the *MAD* for this model. (4) Graph the fitted values on the time-series graph.
c. Compare the results for this double exponential smoothing model with the "best" single exponential smoothing model developed in part c of Exercise 16-44. Discuss which model is preferred.
d. Referring to part b, try different alpha and beta values in an attempt to determine an improved forecast model for monthly sales. For each model, show the forecast for period 17 and the *MAD*. Write a short report that compares the different models.
e. Referring to part d and to part c of Exercise 16-44, write a report to the Grossmieller sales manager that indicates your choice for the forecasting model, complete with your justification for the selection.

# 16 Overview

## Summary

**16.1**

### Introduction to Forecasting and Time-Series Data (pg. 637–644)

**OUTCOME 1** Identify the components present in a time series.

**OUTCOME 2** Understand and compute basic index numbers.

- Quantitative forecasting techniques rely on data gathered in the past to forecast what will happen in the future.
- **Time-series analysis** is a commonly used quantitative forecasting technique that involves looking for patterns in the past data that are likely to continue into the future. It examines four components: trend, seasonal, cyclical, and random.
  - A **trend** is the long-term increase or decrease in a variable being measured over time; it can be linear or nonlinear.
  - A **seasonal** component is present if the data show a repeating pattern over time.
  - A **cyclical** component exists when there are sustained periods of high values followed by periods of lower values and the recurrence period of these fluctuations is longer than a year.
  - The **random** component is often referred to as "noise" in the data.
- Not all time series possess a trend, seasonal, or cyclical component, but virtually all time series have a random component.
- When analyzing time-series data, we often compare one value measured at one point in time with other values measured at different points in time.
- One procedure for making relative comparisons is to determine a **base period index** to which all other data values can be fairly compared.

**16.2**

### Trend-Based Forecasting Techniques (pg. 644–666)

**OUTCOME 3** Apply the fundamental steps in developing and implementing forecasting models.

**OUTCOME 4** Apply trend-based forecasting models, including linear trend, nonlinear trend, and seasonally adjusted trend.

- Trend-based forecasting techniques begin by identifying and modeling that trend. Once we have defined the trend model, we use it to provide forecasts for future time periods.
- Regression analysis is often used to identify the trend component.
- The **mean square error (*MSE*)** or **mean absolute deviation (*MAD*)** is used to determine how well a trend fits the data.
- In general, the lower the *MSE* and *MAD*, the better the model fits the actual data.
- Using regression analysis to determine the trend carries some risk. One risk is that the error terms in the analysis are not independent.
- Related error terms indicate autocorrelation in the data. We use the **Durbin-Watson statistic** to test for this.
- The **ratio-to-moving-average method** is one option for computing seasonal indexes. Once we have determined the seasonal indexes, we use them to deseasonalize the data to allow for a better trend forecast. The indexes are then used to determine a seasonally adjusted forecast.
- Determining the trend and seasonal components in a time-series model allows us to better identify the cyclical and random components.

**16.3**

### Forecasting Using Smoothing Methods (pg. 667–680)

**OUTCOME 5** Use smoothing-based forecasting models, including single and double exponential smoothing.

- A disadvantage of trend-based forecasting is that it gives as much weight to the earliest data in the series as it does to the data that are close to the period for which the forecast is required. It does not allow the model to "learn" or "adjust" to changes in the time series.

- **Exponential smoothing** allows us to weight current observations more heavily than older observations when determining the forecast.
- **Single exponential smoothing** models are used when we see only random fluctuations in the data.
- **Double exponential smoothing** models are used if the data seem to combine random variations with a trend.

## Equations

**(16.1) Simple Index Number** pg. 641

$$I_t = \frac{y_t}{y_0}(100)$$

**(16.2) Deflation Formula** pg. 642

$$y_{adj_t} = \frac{y_t}{I_t}(100)$$

**(16.3) Linear Trend Model** pg. 645

$$y_t = \beta_0 + \beta_1 t + \varepsilon_t$$

**(16.4) Least Squares Equations Estimates** pg. 645

$$b_1 = \frac{\sum ty_t - \frac{\sum t \sum y_t}{n}}{\sum t^2 - \frac{(\sum t)^2}{n}}$$

**(16.5)**

$$b_0 = \frac{\sum y_t}{n} - b_1 \frac{\sum t}{n}$$

**(16.6) Mean Square Error** pg. 648

$$MSE = \frac{\sum (y_t - F_t)^2}{n}$$

**(16.7) Mean Absolute Deviation** pg. 648

$$MAD = \frac{\sum |y_t - F_t|}{n}$$

**(16.8) Durbin-Watson Statistic** pg. 649

$$d = \frac{\sum_{t=2}^{n} (e_t - e_{t-1})^2}{\sum_{t=1}^{n} e_t^2}$$

**(16.9) Forecast Bias** pg. 653

$$\text{Forecast bias} = \frac{\sum (y_t - F_t)}{n}$$

**(16.10) Multiplicative Time-Series Model** pg. 659

$$y_t = T_t \times S_t \times C_t \times I_t$$

**(16.11) Ratio-to-Moving-Average** pg. 660

$$S_t \times I_t = \frac{y_t}{T_t \times C_t}$$

**(16.12) Deseasonalization** pg. 662

$$T_t \times C_t \times I_t = \frac{y_t}{S_t}$$

**(16.13) Exponential Smoothing Model** pg. 667

$$F_{t+1} = F_t + \alpha(y_t - F_t)$$

or

**(16.14)**

$$F_{t+1} = \alpha y_t + (1 - \alpha)F_t$$

**(16.15) Double Exponential Smoothing Model** pg. 671

$$C_t = \alpha y_t + (1 - \alpha)(C_{t-1} + T_{t-1})$$

**(16.16)**

$$T_t = \beta(C_t - C_{t-1}) + (1 - \beta)T_{t-1}$$

**(16.17)** pg. 673

$$F_{t+1} = C_t + T_t$$

## Key Terms

Autocorrelation pg. 649
Base period index pg. 641
Cyclical component pg. 640
Exponential smoothing pg. 667
Forecasting horizon pg. 637
Forecasting interval pg. 638

Forecasting period pg. 637
Linear trend pg. 639
Model diagnosis pg. 637
Model building pg. 637
Model specification pg. 637
Moving average pg. 659

Random component pg. 641
Seasonal component pg. 639
Seasonal index pg. 658
Seasonally unadjusted forecast pg. 663

## Chapter Exercises

### Conceptual Questions

**16-46.** Go to the library or use the Internet to find data showing your state's population for the past 20 years. Plot these data and indicate which of the time-series components are present.

**16-47.** A time series exhibits the pattern stated below. Indicate the time-series component described.
  a. The pattern is wavelike with a recurrence period of nine months.
  b. The time series is steadily increasing.

c. The pattern is wavelike with a recurrence period of two years.

d. The pattern is unpredictable.

e. The pattern steadily decreases, with a wavelike shape that reoccurs every 10 years.

**16-48.** Identify the businesses in your community that you expect to have sales that exhibit a seasonal component. Discuss.

**16-49.** Discuss the difference between a cyclical component and a seasonal component. Which component is more predictable: seasonal or cyclical? Discuss and illustrate with examples.

**16-50.** In the simple linear regression model, confidence and prediction intervals are utilized to provide interval estimates for an average and a particular value, respectively, of the dependent variable. The linear trend model in time series is an application of simple linear regression. This being said, discuss whether a confidence or a prediction interval is the relevant interval estimate for a linear trend model's forecast.

## Business Applications

Exercises **16-51 through 16-54** refer to Malcar Autoparts Company, which has started producing replacement control microcomputers for automobiles. Sales data since 2000 are as follows:

| Year | Sales ($) |
| --- | --- |
| 2000 | 240,000 |
| 2001 | 218,000 |
| 2002 | 405,000 |
| 2003 | 587,000 |
| 2004 | 795,000 |
| 2005 | 762,000 |
| 2006 | 998,000 |
| 2007 | 1,217,000 |
| 2008 | 1,570,000 |
| 2009 | 1,947,000 |
| 2010 | 2,711,000 |
| 2011 | 3,104,000 |
| 2012 | 2,918,000 |
| 2013 | 4,606,000 |
| 2014 | 5,216,000 |
| 2015 | 5,010,000 |

**16-51.** a. Graph these data and indicate whether they appear to have a linear trend.

b. Develop a simple linear regression model with time as the independent variable. Using this regression model, describe the trend and the strength of the linear trend over the 16 years. Is the trend line statistically significant? Plot the trend line against the actual data.

c. Compute the *MAD* for this model.

d. Provide the Malcar Autoparts Company an estimate of its expected sales for the next 5 years.

**16-52.** Develop a single exponential smoothing model using $\alpha = 0.20$. Use as a starting value the average of the first 6 years' data. Determine the forecasted value for year 2016.

a. Compute the *MAD* for this model.

b. Plot the forecast values against the actual data.

c. Use the same starting value but try different smoothing constants (say, 0.05, 0.10, 0.25, and 0.30) in an effort to reduce the *MAD*.

d. Is it possible to answer part d of Exercise 16-51 using this forecasting technique? Explain your answer.

**16-53.** Develop a double exponential smoothing model using smoothing constants $\alpha = 0.20$ and $\beta = 0.40$. As starting values, use the least squares trend line slope and intercept values.

a. Compute the *MAD* for this model.

b. Plot the forecast values against the actual data.

c. Use the same starting values but try different smoothing constants $[$ say, $(\alpha, \beta) = (0.10, 0.50)$, $(0.30, 0.30)$, and $(0.40, 0.20)]$ in an effort to reduce the *MAD*.

**16-54.** Using whatever diagnostic tools you are familiar with, determine which of the three forecasting methods utilized to forecast sales for Malcar Autoparts Company in the previous three exercises provides superior forecasts. Explain the reasons for your choice.

**16-55.** Below are income and sales data for a company for the years 2003 through 2015.

| Year | Net Income/Loss | Sales |
| --- | --- | --- |
| 2003 | −0.3 | 0.5 |
| 2004 | −5.7 | 15.7 |
| 2005 | −27.5 | 147.7 |
| 2006 | −124.5 | 609.8 |
| 2007 | −719.9 | 1,639.8 |
| 2008 | −1,411.2 | 2,761.9 |
| 2009 | −567.3 | 3,122.9 |
| 2010 | −149.1 | 3,933 |
| 2011 | 35.3 | 5,263.7 |
| 2012 | 588.5 | 6,921 |
| 2013 | 359 | 8,490 |
| 2014 | 190 | 10,711 |
| 2015 | 476 | 14,835 |

a. Produce a time-series plot for these data. Specify the exponential forecasting model that should be used to obtain the following years' forecasts.

b. Assuming a double exponential smoothing model, fit the least squares trend to the historical data to determine the smoothed constant-process value and the smoothed trend value for period 0.

c. Produce the forecasts for periods 1 through 13 using $\alpha = 0.10$ and $\beta = 0.20$. Indicate the sales the company should expect for 2016 based on the forecast model.

d. Calculate the *MAD* for this model.

**16-56.** A pizza outlet has tracked the total number of super deluxe pizzas sold in each of the past six years. The data are as follows:

| Year | 2011 | 2012 | 2013 | 2014 | 2015 | 2016 |
|------|------|------|------|------|------|------|
| Sales | 2,074 | 2,395 | 3,138 | 3,632 | 4,694 | 5,652 |

a. Produce a time-series plot of these data. Indicate the time-series components in the data.
b. Provide a forecast for the number of super deluxe pizzas that this outlet will sell in 2017.

**16-57.** Loans and leases data for the years 2005 through 2015 for Glen Lake Bank are shown below:

| Year | Loans & Leases ($ millions) |
|------|------|
| 2005 | 511 |
| 2006 | 616 |
| 2007 | 718 |
| 2008 | 909 |
| 2009 | 1,135 |
| 2010 | 1,371 |
| 2011 | 1,677 |
| 2012 | 1,871 |
| 2013 | 2,021 |
| 2014 | 1,904 |
| 2015 | 2,354 |

a. Produce a time-series plot of these data.
b. Fit a linear trend line to these data.
c. Conduct a hypothesis test of the significance of the linear trend model developed in part b. Use a level of significance of 0.05.
d. For the years 2005 through 2015, what is the predicted value each year for loans and leases based on the linear trend model estimated in part b?
e. Calculate the MAD for the linear trend forecasts developed in part d.
f. Use the linear trend model to forecast loans and leases for Glen Lake Bank for 2016 and 2017.

**16-58.** Tom and Hank's is a regional convenience store with locations in several small towns along the Mississippi River. The quarterly revenues for the store beginning with the first quarter of 2005 and ending with the second quarter of 2011 are listed here.

| Period | Sales | Period | Sales |
|--------|-------|--------|-------|
| 1Q05 | $ 304,072 | 2Q08 | $ 414,776 |
| 2Q05 | $ 241,146 | 3Q08 | $ 409,164 |
| 3Q05 | $ 251,719 | 4Q08 | $ 397,537 |
| 4Q05 | $ 247,792 | 1Q09 | $ 548,111 |
| 1Q06 | $ 370,434 | 2Q09 | $ 412,788 |
| 2Q06 | $ 291,448 | 3Q09 | $ 417,408 |
| 3Q06 | $ 297,308 | 4Q09 | $ 406,495 |
| 4Q06 | $ 286,892 | 1Q10 | $ 586,479 |
| 1Q07 | $ 415,718 | 2Q10 | $ 468,013 |
| 2Q07 | $ 325,157 | 3Q10 | $ 480,706 |
| 3Q07 | $ 336,537 | 4Q10 | $ 466,087 |
| 4Q07 | $ 387,424 | 1Q11 | $ 667,478 |
| 1Q08 | $ 546,057 | 2Q11 | $ 522,337 |

a. Produce a time-series plot of the store's revenues by quarter. What time-series components appear to be present in the data?
b. Determine a seasonal index for each quarter.
c. Use the seasonal indexes from part b to deseasonalize the data.
d. Graph and fit a linear trend line to the deseasonalized data.
e. Use the linear trend line and the seasonal indexes to produce a seasonally adjusted forecast for each quarter beginning with quarter 1 of 2005 and ending with quarter 2 of 2011.
f. Calculate a forecast error for each period.
g. Calculate the MAD and the MSE for the forecasts you developed in part e.
h. Using the seasonal indexes and the estimated trend line, produce forecasts for quarters 3 and 4 of 2011.

**16-59.** Suppose the median weekly earnings for recent college graduates in a large city by quarter for the years 2008 to 2015 are as follows:

| Quarter | Median Weekly Earnings | Quarter | Median Weekly Earnings |
|---------|------|---------|------|
| Qtr 1 2008 | 816 | Qtr 1 2012 | 947.2 |
| Qtr 2 2008 | 816 | Qtr 2 2012 | 961.6 |
| Qtr 3 2008 | 835.2 | Qtr 3 2012 | 963.2 |
| Qtr 4 2008 | 836.8 | Qtr 4 2012 | 964.8 |
| Qtr 1 2009 | 852.8 | Qtr 1 2013 | 976 |
| Qtr 2 2009 | 865.6 | Qtr 2 2013 | 972.8 |
| Qtr 3 2009 | 864 | Qtr 3 2013 | 984 |
| Qtr 4 2009 | 889.6 | Qtr 4 2013 | 982.4 |
| Qtr 1 2010 | 908.8 | Qtr 1 2014 | 998.4 |
| Qtr 2 2010 | 894.4 | Qtr 2 2014 | 990.4 |
| Qtr 3 2010 | 912 | Qtr 3 2014 | 1006.4 |
| Qtr 4 2010 | 921.6 | Qtr 4 2014 | 1020.8 |
| Qtr 1 2011 | 929.6 | Qtr 1 2015 | 1033.6 |
| Qtr 2 2011 | 931.2 | Qtr 2 2015 | 1016 |
| Qtr 3 2011 | 936 | Qtr 3 2015 | 1046.4 |
| Qtr 4 2011 | 940.8 | Qtr 4 2015 | 1038.4 |

a. Prepare a line chart of the median earnings by quarter.
b. Estimate a linear trend line for this set of data.
c. Produce a forecast for each quarter using the linear trend line you estimated in part b.
d. Calculate the error for each period.
e. Calculate the Durbin-Watson statistic and test for positive autocorrelation. Use a level of significance of 0.05.

## Computer Software Exercises

**16-60.** Freddie Mac is an organization that was formed by the U.S. Congress in 1970 to help maintain mortgage interest rates at a fairly stable level. The data in the file **Mortgage Rates** show the 30-year fixed mortgage interest rates on a monthly basis from January 2002 through December 2015 (*source:* www.freddiemac.

com). Use Excel 2016 to construct a model for forecasting the monthly interest rates for 2016. Describe the model using the output from the Excel 2016 forecasting tools.

**16-61.** The data file **Armstead** contains the monthly number of calls received by Armstead & Associates, a company that was formed to help individuals locate long-term health care assistance for elderly individuals.

a. Plot these data in a time-series graph. Based on the graph, what time-series components are in the data?

b. Develop the seasonal indexes for each month. Describe what the seasonal index for August means.

c. Fit a linear trend model to the deseasonalized data for months 1 through 48 and determine the *MAD*. Comment on the adequacy of the linear trend model based on these measures of forecast error.

d. Provide a seasonally unadjusted forecast using the linear trend model for each month of the year.

e. Use the seasonal index values computed in part b to provide seasonal adjusted forecasts for months 49 through 52.

**16-62.** Referring to Exercise 16-61, the managers at Armstead & Associates need to forecast monthly call volumes in order to have sufficient capacity. Develop a single exponential smoothing model using $\alpha = 0.30$. Use as a starting value the average of the first six months' data.

a. Compute the *MAD* for this model.

b. Plot the forecast values against the actual data.

c. Use the same starting value but try different smoothing constants (say, 0.10, 0.20, 0.40, and 0.50) in an effort to reduce the *MAD*.

d. Reflect on the type of time series for which the single exponential smoothing model is designed to provide forecasts. Does it surprise you that the *MAD* for this method is relatively large for these data? Explain your reasoning.

**16-63.** The file **Phone Expenses** contains the monthly telephone expenses for a small business.

a. Produce a time-series graph of the data. Can you identify a trend or seasonal component?

b. Using the Excel 2016 Forecast Sheet option, produce a forecast for January 2016.

c. Report the smoothing parameter and the forecast error measurements produced by Excel's forecast.

d. Use the Excel 2016 forecast function, FORECAST. ETS, to produce a forecast for January 4, 2016. Assume there is no seasonality in the data.

e. Use the Excel 2016 forecast function, FORECAST. ETS.CONFINT, to compute a 95% confidence interval for the January 4, 2016, forecast for phone expenses.

**16-64.** The file **Quarterly Shipments** lists the number of units shipped each quarter for the years 2011–2015 by Bourff Manufacturing.

a. Produce a time-series graph of the data. Are there noticeable trend and seasonal components?

b. Using the Excel 2016 Forecast Sheet option, produce a forecast for each quarter of 2016.

c. Report the smoothing parameters and error measurements produced by Excel's Forecast Sheet option.

d. Report the 90% confidence interval estimates for the forecast generated.

e. Briefly comment on the forecasting model's performance.

## Case 16.1     Park Falls Chamber of Commerce

Before starting a career as a university professor in 2008, Masao Sugiyama was president of the Chamber of Commerce in Park Falls. He was also the owner of the only full-service hardware store in this small farming town. Being president of the Chamber of Commerce had been considered largely a ceremonial post because business conditions had not changed in Park Falls for as long as anyone could remember. In early 2006, however, Masao read an article in a business periodical that made him think he needed to take a more active interest in the business conditions of his town.

The article concerned Walmart, the largest retailer in the United States. Walmart had been expanding primarily by locating in small towns and avoiding large suburban areas. The Park Falls merchants had not dealt with either Lowes or Home Depot because these companies had located primarily in large urban centers, but a supplier had recently mentioned that both Lowes and Home Depot were considering locating stores in smaller towns.

Masao realized that he should know more about all three retailers. He asked the son of a friend to locate the sales data, which are in a file called **Park Falls**.

Suppose that Masao was interested in what all these data tell him. How much faster had Walmart grown than the other two firms? Was there any evidence that Walmart's growth had leveled off? Did Lowes seem to be rebounding, based on sales? Were seasonal fluctuations an issue in these sales figures? Was there any evidence that one firm was more affected by the cyclical component than the others? Now flash forward to the current year. Locate similar sales data for these three companies since 2005 and answer the previous questions based on these new data.

## Case 16.2     The St. Louis Companies

An irritated Roger Hatton finds himself sitting in the St. Louis airport after hearing that his flight to Chicago has been delayed—and, if the storm in Chicago continues, possibly canceled. Because he must get to Chicago if at all possible, Roger is stuck at the airport. He decides he might as well try to get some work done, so he opens his laptop computer and calls up the **Claimnum** file.

Roger was recently assigned as an analyst in the worker's compensation section of the St. Louis Companies, one of the biggest issuers of worker's compensation insurance in the country. Until this year, the revenues and claim costs for all parts of the company were grouped together to determine any yearly profit or loss. Therefore, no one really knew if an individual department was profitable. Now, however, the new president is looking at each part of the company as a profit center. The clear implication is that money-losing departments may not have a future unless they develop a clear plan to become profitable.

When Roger asked the accounting department for a listing, by client, of all policy payments and claims filed and paid, he was told that the information is available but he may have to wait two or three months to get it. He was able to determine, however, that the department has been keeping track of the clients who file frequent (at least one a month) claims and the total number of firms that purchase worker's compensation insurance. Using the data from this report, Roger divides the number of clients filing frequent claims by the corresponding number of clients.

Staring at these figures, Roger feels there should be some way to use them to project what the next several years may hold if the company doesn't change its underwriting policies.

## Case 16.3     Wagner Machine Works

Mary Lindsey has recently agreed to leave her upper-level management job at a major paper manufacturing firm and return to her hometown to take over the family machine-products business. The U.S. machine-products industry had a strong position of world dominance until recently, when it was devastated by foreign competition, particularly from Germany and Japan. Among the many problems facing the American industry is that it is made up of many small firms that must compete with foreign industrial giants.

Wagner Machine Works, the company Mary is taking over, is one of the few survivors in its part of the state, but it, too, faces increasing competitive pressure. Mary's father let the business slide as he approached retirement, and Mary sees the need for an immediate modernization of their plant. She has arranged for a loan from the local bank, but now she must forecast sales for the next three years to ensure that the company has enough cash flow to repay the debt. Surprisingly, Mary finds that her father has no forecasting system in place, and she cannot afford the time or money to install a system like that used at her previous company.

Wagner Machine Works' quarterly sales (in millions of dollars) for the past 15 years are shown in the table.

While looking at these data, Mary wonders whether they can be used to forecast sales for the next three years. She wonders how much, if any, confidence she can have in a forecast made with these data. She also wonders if the recent increase in sales is due to growing business or just to inflationary price increases in the national economy.

### Required Tasks:

1. Identify the central issue in the case.
2. Plot the quarterly sales for the past 15 years for Wagner Machine Works.

| | Quarter | | | |
|---|---|---|---|---|
| **Year** | **1** | **2** | **3** | **4** |
| 2001 | 10,490 | 11,130 | 10,005 | 11,058 |
| 2002 | 11,424 | 12,550 | 10,900 | 12,335 |
| 2003 | 12,835 | 13,100 | 11,660 | 13,767 |
| 2004 | 13,877 | 14,100 | 12,780 | 14,738 |
| 2005 | 14,798 | 15,210 | 13,785 | 16,218 |
| 2006 | 16,720 | 17,167 | 14,785 | 17,725 |
| 2007 | 18,348 | 18,951 | 16,554 | 19,889 |
| 2008 | 20,317 | 21,395 | 19,445 | 22,816 |
| 2009 | 23,335 | 24,179 | 22,548 | 25,029 |
| 2010 | 25,729 | 27,778 | 23,391 | 27,360 |
| 2011 | 28,886 | 30,125 | 26,049 | 30,300 |
| 2012 | 30,212 | 33,702 | 27,907 | 31,096 |
| 2013 | 31,715 | 35,720 | 28,554 | 34,326 |
| 2014 | 35,533 | 39,447 | 30,046 | 37,587 |
| 2015 | 39,093 | 44,650 | 30,046 | 37,587 |

3. Identify any patterns that are evident in the quarterly sales data.
4. If you have identified a seasonal pattern, estimate quarterly seasonal factors.
5. Deseasonalize the data using the quarterly seasonal factors developed.
6. Run a regression model on the deseasonalized data using the time period as the independent variable.
7. Develop a seasonally adjusted forecast for the next three years.
8. Prepare a report that includes graphs and analysis.

# 17 Introduction to Nonparametric Statistics

## WHY YOU NEED TO KNOW

In previous chapters, we introduced a wide variety of statistical techniques that are used extensively to analyze data and aid in the decision-making process. However, some of these estimation and hypothesis-testing techniques may not be appropriate in certain situations if the assumptions of those procedures are not satisfied. For instance, housing prices are particularly important when a company considers potential locations for a new manufacturing plant because the company would like affordable housing to be available for employees who transfer to the new location. A company that is in the midst of relocation has taken a sample of real estate listings from the four cities in contention for the new plant and would like to make a

### Quick Prep

**Review** the concepts associated with hypothesis testing for a single population mean using the *t*-distribution in Chapter 9.

**Make sure** you are familiar with the steps involved in testing hypotheses for the difference between two population means discussed in Chapter 10.

**Review** the concepts and assumptions associated with analysis of variance in Chapter 12.

statistically valid comparison of home prices based on this sample information. The analysis of variance (ANOVA) *F*-test introduced in Chapter 12 seems appropriate. However, this test is based on the assumptions that all populations are normally distributed and have equal variances. Unfortunately, housing prices are generally not normally distributed. Most cities have home prices that are highly right-skewed, with most prices clustered around the median and a few very expensive houses that pull the mean value up. A class of statistical techniques known as nonparametric statistics is available for situations such as the housing price case.

## 17.1 The Wilcoxon Signed Rank Test for One Population Median

Up to this point, the text has presented a wide array of statistical tools for describing data and for drawing inferences about a population based on sample information from that population. These tools are widely used in decision-making situations. However, you will also encounter decision situations with major departures from the required assumptions. For example, many populations, such as family income levels and house prices, are highly skewed. In other instances, the level of data measurement is too low (ordinal or nominal) to warrant use of the techniques presented earlier. In such cases, the alternative is to employ a *nonparametric statistical procedure* that has been developed to meet specific inferential needs. Such procedures have fewer restrictive assumptions concerning data level and underlying probability distributions. There are a great many nonparametric procedures that cover a wide range of applications. The purpose of this chapter is to introduce the concept of nonparametric statistics and illustrate some of the more frequently used methods.

OUTCOME **1**

### The Wilcoxon Signed Rank Test—Single Population

Chapter 9 introduced examples that involved testing hypotheses about a single population mean. Recall that if the data are interval or ratio level and the population is normally distributed, we can use a *t*-test to test whether a population mean has a specified value. However the *t*-test is not appropriate in cases in which the data level is ordinal or populations are not believed to be approximately normally distributed. To overcome data limitation issues, we can use a nonparametric statistical technique known as the *Wilcoxon signed rank test*. This test makes no highly restrictive assumption about the shape of the population distribution.

The Wilcoxon test is used to test hypotheses about a population median rather than a population mean. The basic logic of the Wilcoxon test is straightforward. Because the median is the midpoint in a population, allowing for sampling error, we expect approximately half the data values in a random sample to be below the hypothesized median and about half to be above it. We reject the hypothesized median if the actual data distribution shows too large a departure from this expectation.

---

**BUSINESS APPLICATION** Applying the Wilcoxon Signed Rank Test

**Student Loans** In 2015, *USA Today* reported that the total outstanding debt for student loans in the United States had reached $1.2 trillion (*source:* Nicholas Rayfield, "National student loan debt reaches a bonkers $1.2 trillion," http://college.usatoday.com, Apr. 8, 2015). Suppose that a university president in the Southwest has stated that the median student loan balance for graduates of her university exceeds $35,000. She also stated that the distribution of student loan balances is right skewed with some students amassing very high loans, which is why she used the median as a measure of the center rather than the mean.

The student body vice president at this university is interested in testing whether the president's assertion is right and wants to apply an appropriate statistical test to do so. The *t*-test from Chapter 9, which requires that the population be normally distributed, is not appropriate. Besides, that test is used for hypotheses about population means, not medians. Thus, the VP will need to use a test that can accommodate a nonnormal distribution and that can be used in a test of the median of a population. The Wilcoxon signed rank test can be used to test whether the population median exceeds $35,000. The student body leader has selected a simple random sample of $n = 10$ graduates and collected data on the loan balance for each student.

As with all tests, we start by stating the appropriate null and alternative hypotheses. The null and alternative hypotheses for the one-tailed test involving a population median are

$$H_0: \widetilde{\mu} \leq \$35,000$$
$$H_A: \widetilde{\mu} > \$35,000$$

We will conduct the test using

$$\alpha = 0.05$$

For small samples, we test the hypothesis using a $W$-test statistic determined by the following steps:

STEP 1  **Collect the sample data.**

STEP 2  **Compute $d_i$, the deviation between each value and the hypothesized median.**

STEP 3  **Convert the $d_i$ values to absolute differences.**

STEP 4  **Determine the rank for each $d_i$ value, eliminating any zero $d_i$ values.**
The lowest $d_i$ value receives a rank of 1. If observations are tied, assign the average rank of the tied observations to each of the tied values.

STEP 5  **For any data value greater than the hypothesized median, place the rank in an $R+$ column. For data values less than the hypothesized median, place the rank in an $R-$ column.**

STEP 6  **The test statistic W is the sum of the ranks in the $R+$ column.**
For a lower-tailed test, use the sum in the $R-$ column. For an "equal to" hypothesis, use either sum.

Table 17.1 shows the results for a random sample of 10 loan balances and the calculations for $W$. We test the hypothesis by comparing the calculated $W$-value with the critical values for the Wilcoxon signed rank test that are shown in Appendix O. Both upper and lower critical values are shown, corresponding to $n = 5$ to $n = 20$ for various levels of alpha. Note that $n$ equals the number of nonzero $d_i$ values. In this example, we have $n = 9$ nonzero $d_i$ values. The lower critical value for $n = 9$ and a one-tailed $\alpha = 0.05$ is 8. The corresponding upper-tailed critical value is 37.

Because this is an upper-tailed test, we are interested in only the upper critical value $W_{0.05}$. Therefore, the decision rule is

$$\text{If } W > 37, \text{ reject } H_0.$$

Because $W = 29.5 < 37$, we do not reject the null hypothesis and are unable to conclude that the median loan balance for graduates from this university exceeds $35,000 as the president suggested in her speech.

**TABLE 17.1  Wilcoxon Ranking Table for the Student Loans Example**

| Loan Balance ($) | $d_i = x_i - \$35,000$ | $\lvert d_i \rvert$ | Rank | $R+$ | $R-$ |
|---|---|---|---|---|---|
| 36,400 | 1,400 | 1,400 | 2 | 2 | |
| 38,500 | 3,500 | 3,500 | 3 | 3 | |
| 27,000 | −8,000 | 8,000 | 8 | | 8 |
| 35,000 | 0 | 0 | | | |
| 29,000 | −6,000 | 6,000 | 6.5 | | 6.5 |
| 40,000 | 5,000 | 5,000 | 5 | 5 | |
| 52,000 | 17,000 | 17,000 | 9 | 9 | |
| 34,000 | −1,000 | 1,000 | 1 | | 1 |
| 38,900 | 3,900 | 3,900 | 4 | 4 | |
| 41,000 | 6,000 | 6,000 | 6.5 | 6.5 | |
| | | | Total $= W = 29.5$ | | 15.5 |

The student loan balance example illustrates how the Wilcoxon signed rank test is used when the sample sizes are small. The $W$-test statistic approaches a normal distribution as $n$ increases. Therefore, for sample sizes >20, the Wilcoxon test can be approximated using the normal distribution where the test statistic is a $z$-value as shown in Equation 17.1.

**Large-Sample Wilcoxon Signed Rank Test Statistic**

$$z = \frac{W - \dfrac{n(n+1)}{4}}{\sqrt{\dfrac{n(n+1)(2n+1)}{24}}} \qquad \text{(17.1)}$$

where:

$$W = \text{Sum of the } R+ \text{ ranks}$$
$$n = \text{Number of nonzero } d_i \text{ values}$$

---

**EXAMPLE 17-1** | **Wilcoxon Signed Rank Test, One Sample, $n > 20$**

**Executive Salaries** Suppose testimony at a civil trial about executive salaries indicated that the median salary in 2015 for C-level executives (CEO, CFO, CIO, etc.) in the United States was less than $276,200. A party to the suit being tried decided to test this assertion. A random sample of 25 C-level executives was selected. Since we would expect that executive salaries are highly right-skewed, and because we are testing a population median, a $t$-test is not appropriate. Instead we can conduct a large-sample Wilcoxon signed rank test using the following steps:

STEP 1 **Specify the null and alternative hypotheses.**
In this case, the null and alternative hypotheses are

$$H_0: \tilde{\mu} \geq \$276,200$$
$$H_A: \tilde{\mu} < \$276,200 \text{ (claim)}$$

STEP 2 **Determine the significance level for the test.**
The test will be conducted using

$$\alpha = 0.01$$

STEP 3 **Collect the sample data and compute the $W$-test statistic.**
Using the steps outlined on page 689, we manually compute the $W$-test statistic as shown in Table 17.2.

STEP 4 **Compute the $z$-test statistic.**
The $z$-test statistic using the sum of the positive ranks is

$$z = \frac{W - \dfrac{n(n+1)}{4}}{\sqrt{\dfrac{n(n+1)(2n+1)}{24}}} = \frac{70 - \dfrac{25(25+1)}{4}}{\sqrt{\dfrac{25(25+1)(2(25)+1)}{24}}} = -2.49$$

STEP 5 **Reach a decision.**
The critical value for a one-tailed lower tail test for alpha = 0.01 from the standard normal distribution is −2.33.
Because $z = -2.49 < -2.33$, we reject the null hypothesis.

STEP 6 **Draw a conclusion.**
Thus, based on the sample data, those conducting the investigation should conclude the median executive salary is less than $276,200.

**TABLE 17.2    Wilcoxon Ranking Table for Executive Salaries Example**

| Salary = $x_i$ ($) | $d_i$ | $|d_i|$ | Rank | R+ | R− |
|---|---|---|---|---|---|
| 273,000 | −3,200 | 3,200 | 1 | | 1 |
| 269,900 | −6,300 | 6,300 | 2 | | 2 |
| 263,500 | −12,700 | 12,700 | 3 | | 3 |
| 260,600 | −15,600 | 15,600 | 4 | | 4 |
| 259,200 | −17,000 | 17,000 | 5 | | 5 |
| 257,200 | −19,000 | 19,000 | 6 | | 6 |
| 256,500 | −19,700 | 19,700 | 7 | | 7 |
| 255,400 | −20,800 | 20,800 | 8 | | 8 |
| 255,200 | −21,000 | 21,000 | 9 | | 9 |
| 297,750 | 21,550 | 21,550 | 10 | 10 | |
| 254,200 | −22,000 | 22,000 | 11 | | 11 |
| 300,750 | 24,550 | 24,550 | 12 | 12 | |
| 249,500 | −26,700 | 26,700 | 13 | | 13 |
| 303,000 | 26,800 | 26,800 | 14 | 14 | |
| 304,900 | 28,700 | 28,700 | 15 | 15 | |
| 245,900 | −30,300 | 30,300 | 16 | | 16 |
| 243,500 | −32,700 | 32,700 | 17 | | 17 |
| 237,650 | −38,550 | 38,550 | 18 | | 18 |
| 316,250 | 40,050 | 40,050 | 19 | 19 | |
| 234,500 | −41,700 | 41,700 | 20 | | 20 |
| 228,900 | −47,300 | 47,300 | 21 | | 21 |
| 217,000 | −59,200 | 59,200 | 22 | | 22 |
| 212,400 | −63,800 | 63,800 | 23 | | 23 |
| 204,500 | −71,700 | 71,700 | 24 | | 24 |
| 202,600 | −73,600 | 73,600 | 25 | | 25 |
| | | | | 70 | 255 |

**TRY EXERCISE 17-1** (pg. 691)

## 17.1 EXERCISES

### Skill Development

**17-1.** Consider the following set of observations:

9.0   15.6   21.1   11.1   13.5   9.2   13.6   15.8   12.5   18.7   18.9

You should not assume these data come from a normal distribution. Test the hypothesis that the median of these data is greater than or equal to 14.

**17-2.** Consider the following set of observations:

10.21   13.65   12.30   9.51   11.32   12.77   6.16   8.55   11.78   12.32

You should not assume these data come from a normal distribution. Test the hypothesis that these data come from a distribution with a median less than or equal to 10.

**17-3.** Consider the following set of observations:

3.1   4.8   2.3   5.6   2.8   2.9   4.4

You should not assume these data come from a normal distribution. Test the hypothesis that these data come from a distribution with a median equal to 4. Use $\alpha = 0.10$.

### Business Applications

**17-4.** Sigman Corporation makes batteries that are used in highway signals in rural areas. The company managers claim that the median life of a battery exceeds 4,000 hours. To test this claim, they have selected a random sample of $n = 12$ batteries and have traced their life spans between installation and failure. They obtained the following data:

| 1,973 | 4,838 | 3,805 | 4,494 | 4,738 | 5,249 |
| 4,459 | 4,098 | 4,722 | 5,894 | 3,322 | 4,800 |

a. State the appropriate null and alternative hypotheses.
b. Assuming that the test is to be conducted using a 0.05 level of significance, what conclusion should the managers reach based on these sample data? Be sure to examine the required normality assumption.

**17-5.** A cable television customer call center has a goal that the median time for each completed call should not exceed four minutes. The operations manager does not want to incorrectly conclude that the goal isn't being satisfied unless sample data justify that conclusion. A sample of 12 calls was selected, and the following times (in seconds) were recorded:

194  278  302  140  245  234  268  208  102  190  220  255

a. Construct the appropriate null and alternative hypotheses.
b. Based on the sample data, what should the operations manager conclude? Test at the 0.05 significance level.

**17-6.** The human resources manager at an insurance company has stated that during their initial 6-month period of employment, new sales personnel spent a median of 119 hours per month in training. A random sample of 20 new salespersons was selected. The numbers of hours they spent in training in a randomly chosen month are listed here:

| 163 | 103 | 112 | 96 | 134 |
| 147 | 102 | 95 | 134 | 126 |
| 189 | 126 | 135 | 114 | 129 |
| 142 | 111 | 103 | 89 | 115 |

Do the data support the manager's claim? Conduct the appropriate hypothesis test with a significance level of 0.05.

**17-7.** A company that packages salted peanuts has stated that each package weighs 11 ounces. However, the company has received complaints that the packages are underweight. To assess this situation, suppose the company has conducted a statistical study that concluded the average weight of packages is indeed 11 ounces. A consumer organization, while acknowledging the finding that the mean weight is 11 ounces, claims that more than 50% of the peanut packages weigh less than 11 ounces and that a few heavy packages pull the mean up, thereby cheating a majority of customers. A sample of 20 peanut packages was selected. The data obtained follow:

| 10.9 | 11.7 | 10.5 | 11.8 | 10.2 |
| 11.5 | 10.8 | 11.2 | 11.8 | 10.7 |
| 10.6 | 10.9 | 11.6 | 11.2 | 11.0 |
| 10.7 | 10.8 | 10.5 | 11.3 | 10.1 |

Test the consumer organization's claim at a significance level of 0.05.

**17-8.** Fillmore, Inc.'s quality-control division is constantly monitoring various parameters related to its products. One investigation addressed the life of incandescent light bulbs (in hours). Initially, the division was satisfied with examining the average length of life. However, a recent sample taken from the production floor gave them pause for thought. The data follow:

| 1,100 | 1,140 | 1,550 | 1,210 | 1,280 | 840 | 1,620 | 1,500 |
| 1,460 | 1,940 | 2,080 | 1,350 | 1,150 | 730 | 2,410 | 1,060 |
| 1,150 | 1,260 | 1,760 | 1,250 | 1,500 | 1,560 | 1,210 | 1,440 |
| 1,770 | 1,270 | 1,210 | 1,230 | 1,230 | 2,100 | 1,630 | 500 |

Their initial efforts indicated that the average length of life of the light bulbs was 1,440 hours.

a. Construct a box and whisker plot of these data. On this basis, draw a conclusion concerning the distribution of the population from which this sample was drawn.
b. Conduct a hypothesis test to determine if the median length of life of the light bulbs is longer than the average length of life. Use $\alpha = 0.05$.

**17-9.** Suppose an engine oil manufacturer wants to verify the viscosity of its premium 30-weight oil. A simple random sample of specimens taken from automobiles running at normal temperatures was obtained. The viscosities observed were as follows:

| 25 | 24 | 21 | 35 | 25 |
| 25 | 35 | 38 | 32 | 36 |
| 35 | 29 | 30 | 27 | 28 |
| 27 | 31 | 32 | 30 | 30 |

Determine if the median viscosity at normal running temperatures is equal to 30 as advertised for the company's premium 30-weight oil. (Use $\alpha = 0.05$.)

## Computer Software Exercises

**17-10.** The marketing manager of a cell phone company is working on an advertising plan that is to be directed toward a young audience. In conjunction with the plan, the manager states that the median age of the company's customers is younger than 40. Before actually completing the advertising plan, the manager decided to randomly sample customers. Among the questions asked in a survey of 50 customers was the customers' ages. The data are in the file **Cell Phone Survey**.

a. Examine the sample data. Is the variable being measured a discrete or a continuous variable? Does it seem feasible that these data could have come from a normal distribution?

b. The marketing manager must support his statement concerning customer age in an upcoming board meeting. Using a significance level of 0.10, provide support for the marketing manager.

17-11. The Wilson Company uses a great deal of water in the process of making industrial milling equipment. To determine whether the company is complying with federal clean water laws, sample measures are taken

every so often. One requirement is that the median pH level must be lower than 7.4. A sample of 95 pH measures has been taken. The data for these measures are in the file **Wilson Water**.

a. Carefully examine the data. Use an appropriate procedure to determine if the data could have been sampled from a normal distribution. (*Hint:* Review the goodness-of-fit test in Chapter 13.)

b. Based on the sample data of pH level, what should the company conclude about its current status on meeting federal requirements? Test the hypothesis at the 0.05 level.

---

### 17.2  Nonparametric Tests for Two Population Medians

One of the assumptions for the *t*-test involving two population means is that the two populations are normally distributed. Another assumption is that the data are interval or ratio level. In this section we introduce two nonparametric techniques that do not require such stringent assumptions and data requirements: the *Mann–Whitney U-test*[1] and the *Wilcoxon matched-pairs signed rank test*. Both tests can be used with ordinal (ranked) data, and neither requires that the populations be normally distributed. We use the Mann–Whitney *U*-test when the samples are independent, whereas the Wilcoxon matched-pairs signed rank test is used when the design has paired samples.

## The Mann–Whitney *U*-Test

OUTCOME **2**

**BUSINESS APPLICATION**    Testing Two Population Medians

**Baramusa County Highway District**   The workforce of the Baramusa County Highway District (BCHD) is made up of the rural and urban divisions. A few months ago, several rural division supervisors began claiming that the urban division employees waste gravel from the county gravel pit. The supervisors claimed the urban division uses more gravel per mile of road maintenance than the rural division. Suppose in response to these claims, the BCHD materials manager decides to conduct a test by selecting a random sample from the district's job-cost records of jobs performed by the urban (U) division and another sample of jobs performed by the rural (R) division. The yards of gravel per mile for each job are recorded.

Even though the data are ratio level, the manager is not willing to make the normality assumptions necessary to employ the two-sample *t*-test (discussed in Chapter 10). However, the Mann–Whitney *U*-test will allow him to compare the gravel use of the two divisions.

The Mann–Whitney *U*-test is one of the most commonly used nonparametric tests to compare samples from two populations in those cases when the following assumptions are satisfied:

**Assumptions**

1. The two samples are independent and random.
2. The value measured is a continuous variable.
3. The measurement scale used is at least ordinal.
4. If they differ, the distributions of the two populations will differ only with respect to central location.

---

[1] An equivalent test to the Mann–Whitney *U*-test is the Wilcoxon rank-sum test.

The fourth point is instrumental in setting the null and alternative hypotheses. We are interested in determining whether two populations have the same or different medians. The test can be performed using the following steps:

STEP 1    **State the appropriate null and alternative hypotheses.**
In this situation, the variable of interest is cubic yards of gravel used. This is a ratio-level variable. However, the populations are suspected to be skewed, so the materials manager has decided to test the following hypotheses, stated in terms of the population medians:

$$H_0\colon \widetilde{\mu}_U \leq \widetilde{\mu}_R \; (\text{Median urban gravel use is less than or equal to median rural use.})$$

$$H_A\colon \widetilde{\mu}_U > \widetilde{\mu}_R \; (\text{Urban median exceeds rural median.})$$

STEP 2    **Specify the desired level of significance.**
The decision makers have determined that the test will be conducted using

$$\alpha = 0.05$$

STEP 3    **Select the sample data and compute the appropriate test statistic.**
Computing the test statistic manually requires several steps:

1. Combine the raw data from the two samples into one set of numbers, keeping track of the sample from which each value came.
2. Rank the numbers in this combined set from low to high. Note that we expect no ties to occur because the values are considered to have come from continuous distributions. However, in actual situations ties sometimes occur. When they do, we give tied observations the average of the rank positions for which they are tied. For instance, if the lowest four data points were each 460, each of the four 460s would receive a rank of $(1 + 2 + 3 + 4)/4 = 10/4 = 2.5$.[2]
3. Separate the two samples, listing each observation with the rank it has been assigned. This leads to the rankings shown in Table 17.3.
   The logic of the Mann–Whitney $U$-test is based on the idea that if the sum of the rankings of one sample differs greatly from the sum of the rankings of the second sample, we should conclude that there is a difference in the population medians.

**TABLE 17.3** Ranking of Yards of Gravel per Mile for the Baramusa County Highway District Example

| Urban ($n_1 = 12$) | | Rural ($n_2 = 12$) | |
|---|---|---|---|
| Yards of Gravel | Rank | Yards of Gravel | Rank |
| 460 | 2 | 600 | 6 |
| 830 | 16 | 652 | 9 |
| 720 | 12 | 603 | 7 |
| 930 | 20 | 594 | 5 |
| 500 | 4 | 1,402 | 23 |
| 620 | 8 | 1,111 | 21 |
| 703 | 11 | 902 | 18 |
| 407 | 1 | 700 | 10 |
| 1,521 | 24 | 827 | 15 |
| 900 | 17 | 490 | 3 |
| 750 | 13 | 904 | 19 |
| 800 | 14 | 1,400 | 22 |
| | $\Sigma R_1 = 142$ | | $\Sigma R_2 = 158$ |

---

[2] Noether provides an adjustment when ties occur. He, however, points out that using the adjustment has little effect unless a large proportion of the observations are tied or there are ties of considerable extent. See the References at the end of the book.

4. Calculate a $U$-value for each sample, as shown in Equations 17.2 and 17.3.

**$U$-Statistics**

$$U_1 = n_1 n_2 + \frac{n_1(n_1 + 1)}{2} - \sum R_1 \qquad \textbf{(17.2)}$$

$$U_2 = n_1 n_2 + \frac{n_2(n_2 + 1)}{2} - \sum R_2 \qquad \textbf{(17.3)}$$

where:

$n_1$ and $n_2$ = Sample sizes from Populations 1 and 2

$\Sigma R_1$ and $\Sigma R_2$ = Sum of ranks for samples 1 and 2

For our example using the ranks in Table 17.3,

$$U_1 = 12(12) + \frac{12(13)}{2} - 142$$
$$= 80$$

$$U_2 = 12(12) + \frac{12(13)}{2} - 158$$
$$= 64$$

Note that $U_1 + U_2 = n_1 n_2$. This is always the case, and it provides a good check on the correctness of the rankings in Table 17.3.

5. Select the $U$-value to be the test statistic.

The Mann–Whitney $U$ tables in Appendices K and L give the lower tail of the $U$-distribution. For one-tailed tests such as our Baramusa County example, we look at the alternative hypothesis to determine whether $U_1$ or $U_2$ should be selected as the test statistic. Recall that

$$H_A: \widetilde{\mu}_U > \widetilde{\mu}_R$$

If the alternative hypothesis indicates that Population 1 will have a higher median, then we select $U_1$ as the test statistic. If Population 2 is expected to have a higher median, then we should select $U_2$ as the test statistic. The reason is that the population with the larger median should have the larger sum of ranked values, thus producing the smaller $U$-value. It is very important to note that this selection must be made in terms of the alternative hypothesis and not on the basis of the $U$-values obtained from the samples.

Now, we select the $U$-value that the alternative hypothesis indicates should be the smaller and call this $U$. Because Population 1 (urban) should have the smaller $U$-value (larger median) if the alternative hypothesis is true, the sample data give

$$U = 80$$

This is actually larger than the $U$-value for the rural population, but we still use it as the test statistic because the alternative hypothesis indicates that $\widetilde{\mu}_U > \widetilde{\mu}_R$.[3]

STEP 4 **Determine the critical value for the Mann–Whitney $U$-test.**
For sample sizes less than 9, use the Mann–Whitney $U$ table in Appendix K for the appropriate sample size. For sample sizes from 9 to 20, as in this example, the null hypothesis can be tested by comparing $U$ with the appropriate critical value given in the Mann–Whitney $U$ table in Appendix L. We begin by locating the part of the table associated with the desired significance level. In this case, we have a one-tailed test with

$$\alpha = 0.05$$

---

[3] For a two-tailed test, you should select the smaller $U$-value as your test statistic. This will force you toward the lower tail. If the $U$-value is smaller than the critical value in the Mann–Whitney $U$ table, you will reject the null hypothesis.

Go across the top of the Mann–Whitney $U$ table to locate the value corresponding to the sample size from Population 2 (rural) and down the left side of the table to the sample size from Population 1 (urban).

In the Baramusa County example, both sample sizes are 12, so to use the Mann–Whitney table in Appendix L for a one-tailed test at $\alpha = 0.05$, go across the top of the table to

$$n_2 = 12$$

and down the left-hand side to

$$n_1 = 12$$

The intersection of these column and row values gives a critical value of

$$U_{0.05} = 42$$

We can now form the decision rule as follows:

If $U < 42$, reject $H_0$.

Otherwise, do not reject $H_0$.

STEP 5  **Reach a decision.**
Now because

$$U = 80 > 42$$

we do not reject the null hypothesis.

STEP 6  **Draw a conclusion.**
Therefore, based on the sample data, there is not sufficient evidence to conclude that the median yards of gravel per mile used by the urban division is greater than that for the rural division.

## Mann–Whitney *U*-Test—Large Samples

In a situation with sample sizes in excess of 20, we cannot use the previous approaches to the Mann–Whitney $U$-test because of table limitations. However, the $U$-statistic approaches a normal distribution as the sample sizes increase, and the Mann–Whitney $U$-test can be conducted using a normal approximation approach, where the mean and standard deviation for the $U$-statistic are as given in Equations 17.4 and 17.5, respectively.

**Mean and Standard Deviation for *U*-Statistic**

$$\mu = \frac{n_1 n_2}{2} \tag{17.4}$$

$$\sigma = \sqrt{\frac{(n_1)(n_2)(n_1 + n_2 + 1)}{12}} \tag{17.5}$$

where:

$n_1$ and $n_2$ = Sample sizes from Populations 1 and 2

Equations 17.4 and 17.5 are used to form the $Z$-test statistic in Equation 17.6.

**Mann–Whitney *U*-Test Statistic**

$$z = \frac{U - \dfrac{n_1 n_2}{2}}{\sqrt{\dfrac{(n_1)(n_2)(n_1 + n_2 + 1)}{12}}} \tag{17.6}$$

**Large-Sample Test of Two Population Medians**

**Tax Preparation Software**   Over the past few years, two tax preparation software products have emerged as the leaders—TurboTax and H&R Block. These software options have most likely pulled business away from accounting firms and tax services businesses. Suppose a claim has been made that the median household income for families that use software to do their own taxes is lower than the median income for those who use a professional tax service.

The claim was made in terms of the median (as opposed to the mean) income because data such as household income are notorious for having large outliers and skewed distributions. In such cases, the median, which is not sensitive to outliers, is a preferable measure of the center of the data. Both the use of the median and the potential skewness of the income distribution are reasons to use a nonparametric procedure such as the Mann–Whitney test.

We can use the Mann–Whitney $U$-test to test the assertion about median incomes for families using software versus professional tax preparation services. The correct procedure is as follows:

STEP **1**   **Specify the null and alternative hypotheses.**
Given the statement that the median household income for software customers (S) is lower than the median for those customers who use professional services (P) the null and alternative hypotheses to be tested are

$$H_0: \widetilde{\mu}_S \geq \widetilde{\mu}_P$$

$$H_A: \widetilde{\mu}_S < \widetilde{\mu}_P \text{ (claim)}$$

STEP **2**   **Specify the desired level of significance.**
The test is to be conducted using

$$\alpha = 0.05$$

STEP **3**   **Select the random sample and compute the test statistic.**

To test this hypothesis, take a survey in the market area for a total of 548 households in this case (144 who use professional services and 404 who use tax software). The results of the survey are in the file **Tax Software**. Because of the sample size, we can use the large-sample approach to the Mann–Whitney $U$-test. To compute the test statistic shown in Equation 17.6, use the following steps:

1. Convert the income data to ranks. The sample data and ranks are in a file called **Tax Software-Ranks**. Note that when data are tied in value, they share the same average rank. For example, if four values are tied for the fifth position, each one is assigned the average of rankings 5, 6, 7, and 8, or $(5 + 6 + 7 + 8)/4 = 6.5$.
2. Next, compute the $U$-value. The sum of the ranks for the families using a professional service is

$$\Sigma R_1 = 41,204$$

and the sum of the ranks for the families using software is

$$\Sigma R_2 = 109,222$$

3. Based on sample sizes of

$$n_1 = 144$$

tax professional users and

$$n_2 = 404$$

software users, compute the $U$-values using Equations 17.2 and 17.3:

$$U_1 = 144(404) + \frac{144(145)}{2} - 41,204 = 27,412$$

$$U_2 = 144(404) + \frac{404(405)}{2} - 109,222 = 30,746$$

Because the alternative hypothesis predicts that families that use a tax professional will have a higher median, we select $U_1$ to be $U$. Thus,

$$U = 27,412$$

4. Now substitute appropriate values into Equations 17.4 and 17.5:

$$\mu = \frac{n_1 n_2}{2} = \frac{144(404)}{2} = 29,088$$

and

$$\sigma = \sqrt{\frac{(n_1)(n_2)(n_1 + n_2 + 1)}{12}} = \sqrt{\frac{(144)(404)(144 + 404 + 1)}{12}} = 1,631.43$$

5. Compute the test statistic using Equation 17.6:

$$z = \frac{U - \frac{n_1 n_2}{2}}{\sqrt{\frac{(n_1)(n_2)(n_1 + n_2 + 1)}{12}}} = \frac{27,412 - 29,088}{\sqrt{\frac{(144)(404)(144 + 404 + 1)}{12}}}$$

$$= \frac{-1,676}{1,631.43} = -1.027$$

STEP 4 **Determine the critical value for the test.**
Based on a one-tailed test with $\alpha = 0.05$, the critical value from the standard normal distribution table is

$$z_{0.05} = -1.645$$

STEP 5 **Reach a decision.**
Since $z = -1.027 > -1.645$, we cannot reject the null hypothesis.

STEP 6 **Draw a conclusion.**
The claim that families that use a tax professional to prepare their income taxes have a higher median income than families that use a tax software package is not supported by the sample data.

---

OUTCOME 3    **The Wilcoxon Matched-Pairs Signed Rank Test**    The Mann–Whitney $U$-test is a very useful nonparametric technique. However, as discussed in the Baramusa County Highway District example, its use is limited to those situations in which the samples from the two populations are independent. As we discussed in Chapter 10, you will encounter decision situations in which the samples are paired and, therefore, are not independent.

The *Wilcoxon matched-pairs signed rank test* has been developed for situations in which we have related samples and are unwilling or unable (due to data-level limitations) to use the paired-sample *t*-test. It is useful when the two related samples have a measurement scale that allows us to determine not only whether the pairs of observations differ but also the magnitude of any difference. We can use the Wilcoxon matched-pairs test in those cases in which the following assumptions are satisfied:

**Assumptions**

1. The differences are measured on a continuous variable.
2. The measurement scale used is at least interval.
3. The distribution of the population differences is symmetric about their median.

**EXAMPLE 17-2** | **Small-Sample Wilcoxon Test**

**Financial Software Associates**　Financial Software Associates develops and markets financial planning software. To differentiate its products from the other packages on the market, Financial Software has built many macros into its software. According to the company, once a user learns the macro keystrokes, complicated financial computations become much easier to perform.

As part of its product-development testing program, software engineers at Financial Software have selected a focus group of seven people who frequently use spreadsheet packages. Each person is given complicated financial and accounting data and is asked to prepare a detailed analysis. The software tracks the amount of time each person takes to complete the task. Once the analysis is complete, these same seven individuals are given a training course in Financial Software add-ons. After the training course, they are given a similar set of data and are asked to do the same analysis. Again, the systems software determines the time needed to complete the analysis.

You should recognize that the samples in this application are not independent because the same subjects are used in both cases. If the software engineers performing the analysis are unwilling to make the normal distribution assumption required of the paired-sample $t$-test, they can use the Wilcoxon matched-pairs signed rank test. This test can be conducted using the following steps:

STEP 1　**Specify the appropriate null and alternative hypotheses.**
The null and alternative hypotheses being tested are

$$H_0: \widetilde{\mu}_b \leq \widetilde{\mu}_a$$
$$H_A: \widetilde{\mu}_b > \widetilde{\mu}_a \text{ (Median time will be less } after \text{ the training.)}$$

STEP 2　**Specify the desired level of significance.**
The test will be conducted using

$$\alpha = 0.025$$

STEP 3　**Collect the sample data and compute the test statistic.**
The data are shown in Table 17.4.

First, we convert the data in Table 17.4 to differences. The column of differences, $d$, gives the "before minus after" differences. The next column is the rank of the $d$ values from low to high. Note that the ranks are determined without considering the sign on the $d$ value. However, once the rank is determined, the original sign on the $d$ value is attached to the rank. For example, $d = 13$ is given a rank of 7, whereas $d = -4$ has a rank of $-3$.

The final column is titled "Ranks with Smallest Expected Sum." To determine the values in this column, we take the absolute values of either the positive or the negative ranks, depending on which group has the smallest expected sum of absolute-valued ranks. We look to the alternative hypothesis, which is

$$H_A: \widetilde{\mu}_b > \widetilde{\mu}_a$$

**TABLE 17.4** Financial Software Associates Ranked Data

| Subject | Before Training | After Training | d | Rank of d | Ranks with Smallest Expected Sum |
|---------|-----------------|----------------|------|-----------|----------------------------------|
| 1 | 24 | 11 | 13 | 7 | |
| 2 | 20 | 18 | 2 | 1 | |
| 3 | 19 | 23 | −4 | −3 | 3 |
| 4 | 20 | 15 | 5 | 4 | |
| 5 | 13 | 16 | −3 | −2 | 2 |
| 6 | 28 | 22 | 6 | 5 | |
| 7 | 15 | 8 | 7 | 6 | |
| | | | | | ── |
| | | | | | $T = 5$ |

Because the before median is predicted to exceed the after median, we would expect the positive differences to exceed the negative differences. Therefore, the negative ranks should have the smaller sum, and therefore should be used in the final column, as shown in Table 17.4. In this case, the test statistic, $T$, is equal to the sum of the absolute values of these negative ranks. Thus, $T = 5$.

STEP 4 **Determine the critical value.**
To determine whether $T$ is small enough to reject the null hypothesis, we consult the Wilcoxon table of critical $T$-values in Appendix M. If the calculated $T$ is less than or equal to the critical $T$ from the table, we reject the null hypothesis. For instance, with $\alpha = 0.025$ for our one-tailed test and $n = 7$, we get a critical value of

$$T_{0.025} = 2$$

STEP 5 **Reach a decision.**
Because

$$T = 5 > 2, \text{do not reject } H_0.$$

STEP 6 **Draw a conclusion.**
Based on these sample data, Financial Software Associates does not have a statistical basis for stating that its product will reduce the median time required to perform complicated financial analyses.

**TRY EXAMPLE 17-20** (pg. 702)

**Ties in the Data**   If the two measurements of an observed data pair have the same values and, therefore, a $d$ value of 0, we drop that case from the analysis and the sample size is reduced accordingly. Note that this procedure favors rejecting the null hypothesis because we are eliminating cases in which the two sample points have exactly the same values.

If two or more $d$ values have the same absolute values, we assign the same average rank to each one using the same approach as with the Mann–Whitney $U$-test. For example, if we have two $d$ values that tie for ranks 4 and 5, we average them as $(4 + 5)/2 = 4.5$ and assign both a rank of 4.5. Studies have shown that this method of assigning ranks to ties has little effect on the Wilcoxon test results. For a more complete discussion of the effect of ties on the Wilcoxon matched-pairs signed rank test, please see the text by Marascuilo and McSweeney in the reference section at the end of this book.

**Large-Sample Wilcoxon Test**   If the sample size (number of matched pairs) exceeds 25, we cannot use the Wilcoxon table of critical $T$-values in Appendix M. However, it can be shown that for large samples, the distribution of $T$-values is approximately normal, with a mean and standard deviation given by Equations 17.7 and 17.8, respectively.

**Wilcoxon Mean and Standard Deviation**

$$\mu = \frac{n(n + 1)}{4} \tag{17.7}$$

$$\sigma = \sqrt{\frac{n(n + 1)(2n + 1)}{24}} \tag{17.8}$$

where:

$$n = \text{Number of paired values}$$

The *Wilcoxon test statistic* is given by Equation 17.9.

### Wilcoxon Test Statistic

$$z = \frac{T - \dfrac{n(n+1)}{4}}{\sqrt{\dfrac{n(n+1)(2n+1)}{24}}}$$    (17.9)

Then, the $z$-value is compared to the critical value from the standard normal table in the usual manner.

## 17.2 EXERCISES

### Skill Development

**17-12.** For each of the following tests, determine which of the two $U$ statistics ($U_1$ or $U_2$) you would choose, the appropriate test statistic, and the rejection region for the Mann–Whitney test:

a. $H_A: \tilde{\mu}_1 < \tilde{\mu}_2, \alpha = 0.05, n_1 = 5$, and $n_2 = 10$
b. $H_A: \tilde{\mu}_1 > \tilde{\mu}_2, \alpha = 0.05, n_1 = 15$, and $n_2 = 12$
c. $H_A: \tilde{\mu}_1 \neq \tilde{\mu}_2, \alpha = 0.10, n_1 = 12$, and $n_2 = 17$
d. $H_A: \tilde{\mu}_1 < \tilde{\mu}_2, \alpha = 0.05, n_1 = 22$, and $n_2 = 25$
e. $H_A: \tilde{\mu}_1 \neq \tilde{\mu}_2, \alpha = 0.10, n_1 = 44$, and $n_2 = 15$

**17-13.** The following sample data have been collected from two independent samples from two populations. Test the claim that the second population median will exceed the median of the first population.

| Sample 1 | | Sample 2 | |
|---|---|---|---|
| 12 | 11 | 9 | 20 |
| 21 | 14 | 18 | 7 |
| 15 | 12 | 16 | 12 |
| 10 | 8 | 17 | 19 |

a. State the appropriate null and alternative hypotheses.
b. If you are unwilling to assume that the two populations are normally distributed, based on the sample data, what should you conclude about the null hypothesis? Test using $\alpha = 0.05$.

**17-14.** The following sample data have been collected from independent samples from two populations. The claim is that the first population median will be larger than the median of the second population.

| Sample 1 | | Sample 2 | |
|---|---|---|---|
| 4.4 | 2.6 | 3.7 | 4.2 |
| 2.7 | 2.4 | 3.5 | 5.2 |
| 1.0 | 2.0 | 4.0 | 4.4 |
| 3.5 | 2.8 | 4.9 | 4.3 |
| 2.8 | | 3.1 | |

a. State the appropriate null and alternative hypotheses.
b. Using the Mann–Whitney $U$-test, based on the sample data, what should you conclude about the null hypothesis? Test using $\alpha = 0.05$.

**17-15.** The following sample data have been collected from two independent random samples from two populations. Test the claim that the first population median will exceed the median of the second population.

| Sample 1 | | Sample 2 | |
|---|---|---|---|
| 50 | 43 | 38 | 31 |
| 47 | 46 | 44 | 38 |
| 44 | 72 | 38 | 39 |
| 48 | 40 | 37 | 54 |
| 40 | 55 | 43 | 41 |
| 36 | 38 | 44 | 40 |

a. State the appropriate null and alternative hypotheses.
b. Using the Mann–Whitney $U$-test, based on the sample data, what should you conclude about the null hypothesis? Test using a significance level of 0.01.

**17-16.** Determine the rejection region for the Mann–Whitney $U$-test in each of the following cases:

a. $H_A: \tilde{\mu}_1 < \tilde{\mu}_2, \alpha = 0.05, n_1 = 3$, and $n_2 = 15$
b. $H_A: \tilde{\mu}_1 \neq \tilde{\mu}_2, \alpha = 0.10, n_1 = 5$, and $n_2 = 20$
c. $H_A: \tilde{\mu}_1 < \tilde{\mu}_2, \alpha = 0.025, n_1 = 9$, and $n_2 = 12$
d. $H_A: \tilde{\mu}_1 \neq \tilde{\mu}_2, \alpha = 0.10, n_1 = 124$, and $n_2 = 25$

**17-17.** The following sample data have been collected from independent samples from two populations. Do the populations have different medians? Test at a significance level of 0.05. Use the Mann–Whitney $U$-test.

| Sample 1 | | Sample 2 | |
|---|---|---|---|
| 550 | 489 | 594 | 538 |
| 483 | 480 | 542 | 505 |
| 379 | 433 | 447 | 486 |

| Sample 1 | | Sample 2 | |
|---|---|---|---|
| 438 | 436 | 466 | 425 |
| 398 | 540 | 560 | 497 |
| 582 | 415 | 447 | 511 |
| 528 | 532 | 526 | 576 |
| 502 | 412 | 446 | 558 |
| 352 | 572 | 573 | 500 |
| 488 | 579 | 542 | 467 |
| 400 | 556 | 473 | 556 |
| 451 | 383 | 418 | 383 |
| 571 | 515 | 511 | 515 |
| 382 | 501 | 510 | 501 |
| 588 | 353 | 577 | 353 |
| 465 | 369 | 585 | |
| 492 | 475 | 436 | |
| 384 | 470 | 461 | |
| 563 | 595 | 545 | |
| 506 | 361 | 441 | |

**17-18.** For each of the following tests, determine which of the two sums of absolute ranks (negative or positive) you would choose, the appropriate test statistic, and the rejection region for the Wilcoxon matched-pairs signed rank test:
a. $H_A: \tilde{\mu}_1 < \tilde{\mu}_2, \alpha = 0.025, n = 15$
b. $H_A: \tilde{\mu}_1 > \tilde{\mu}_2, \alpha = 0.01, n = 12$
c. $H_A: \tilde{\mu}_1 \neq \tilde{\mu}_2, \alpha = 0.05, n = 9$
d. $H_A: \tilde{\mu}_1 < \tilde{\mu}_2, \alpha = 0.05, n = 26$
e. $H_A: \tilde{\mu}_1 \neq \tilde{\mu}_2, \alpha = 0.10, n = 44$

**17-19.** You are given two paired samples with the following information:

| Item | Sample 1 | Sample 2 |
|---|---|---|
| 1 | 3.4 | 2.8 |
| 2 | 2.5 | 3.0 |
| 3 | 7.0 | 5.5 |
| 4 | 5.9 | 6.7 |
| 5 | 4.0 | 3.5 |
| 6 | 5.0 | 5.0 |
| 7 | 6.2 | 7.5 |
| 8 | 5.3 | 4.2 |

a. Based on these paired samples, test at the $\alpha = 0.05$ level whether the true median paired difference is 0.
b. Answer part a assuming the data given here were sampled from normal distributions with equal variances.

**17-20.** You are given two paired samples with the following information:

| Item | Sample 1 | Sample 2 |
|---|---|---|
| 1 | 19.6 | 21.3 |
| 2 | 22.1 | 17.4 |
| 3 | 19.5 | 19.0 |
| 4 | 20.0 | 21.2 |

| Item | Sample 1 | Sample 2 |
|---|---|---|
| 5 | 21.5 | 20.1 |
| 6 | 20.2 | 23.5 |
| 7 | 17.9 | 18.9 |
| 8 | 23.0 | 22.4 |
| 9 | 12.5 | 14.3 |
| 10 | 19.0 | 17.8 |

Based on these paired samples, test at the $\alpha = 0.05$ level whether the true median paired difference is 0.

**17-21.** You are given two paired samples with the following information:

| Item | Sample 1 | Sample 2 |
|---|---|---|
| 1 | 1,004 | 1,045 |
| 2 | 1,245 | 1,145 |
| 3 | 1,360 | 1,400 |
| 4 | 1,150 | 1,000 |
| 5 | 1,300 | 1,350 |
| 6 | 1,450 | 1,350 |
| 7 | 900 | 1,140 |

Based on these paired samples, test at the $\alpha = 0.05$ level whether the true median paired difference is 0.

**17-22.** From a recent study we have collected the following data from two independent random samples:

| Sample 1 | Sample 2 |
|---|---|
| 405 | 300 |
| 450 | 340 |
| 290 | 400 |
| 370 | 250 |
| 345 | 270 |
| 460 | 410 |
| 425 | 435 |
| 275 | 390 |
| 380 | 225 |
| 330 | 210 |
| 500 | 395 |
| 215 | 315 |

Suppose we do not wish to assume normal distributions. Use the appropriate nonparametric test to determine whether the populations have equal medians. Test at $\alpha = 0.05$.

**17-23.** You are given two paired samples with the following information:

| Item | Sample 1 | Sample 2 |
|---|---|---|
| 1 | 234 | 245 |
| 2 | 221 | 224 |
| 3 | 196 | 194 |
| 4 | 245 | 267 |
| 5 | 234 | 230 |
| 6 | 204 | 198 |

Based on these paired samples, test at the $\alpha = 0.05$ level whether the true median paired difference is 0.

**17-24.** Consider the following data for two paired samples:

| Case | Sample 1 | Sample 2 |
|------|----------|----------|
| 1 | 258 | 304 |
| 2 | 197 | 190 |
| 3 | 400 | 500 |
| 4 | 350 | 340 |
| 5 | 237 | 250 |
| 6 | 400 | 358 |
| 7 | 370 | 390 |
| 8 | 130 | 100 |

a. Test the following null and alternative hypotheses at an $\alpha = 0.05$ level:

$H_0$: There is no difference between the two population distributions.

$H_A$: There is a difference between the two populations.

b. Answer part a as if the samples were independent samples from normal distributions with equal variances.

## Business Applications

**17-25.** National Reading Academy claims that graduates of its program have a higher median reading speed per minute than people who do not take the course. Suppose an independent agency conducted a study to determine whether this claim was justified. Researchers from the agency selected a random sample of people who had taken the speed reading course and another random sample of people who had not taken the course. The agency was unwilling to make the assumption that the populations were normally distributed. Therefore, a nonparametric test was needed. The following summary data were observed:

| With Course | Without Course |
|-------------|----------------|
| $n = 7$ | $n = 5$ |
| Sum of ranks $= 42$ | Sum of ranks $= 36$ |

Assuming that higher ranks imply more words per minute being read, what should the testing agency conclude based on the sample data? Test at an $\alpha = 0.05$ level.

**17-26.** The makers of a new mobile theater system that is vision correctable so there is no need to wear glasses while using the device have done a marketing research study in which they asked two independently selected groups to rate the mobile theater system on a scale of 1 to 100, with 100 being perfect satisfaction. The first group consisted of college students. The second group consisted of working professionals between the ages of 30 and 40 years. The company hoped to be able to say

that the product would receive the same median ranking from each group. The following summary data were recorded:

| College Students | Business Professionals |
|------------------|------------------------|
| $n = 10$ | $n = 8$ |
| Sum of ranks $= 92$ | Sum of ranks $= 79$ |

Based on these data, what should the company conclude? Test at the $\alpha = 0.02$ level.

**17-27.** Property taxes are based on assessed values of property. In most states, the law requires that assessed values be "at or near" market value of the property. In one northwestern county, a tax protest group claimed that assessed values are higher than market values. To address this claim, the county tax assessor, together with representatives from the protest group, has selected 15 properties at random that have sold within the past six months. Both parties agree that the sales price was the market value at the time of the sale.

The assessor then listed the assessed values and the sales values side by side, as shown.

| House | Assessed Value ($) | Market Value ($) |
|-------|--------------------|--------------------|
| 1 | 302,000 | 198,000 |
| 2 | 176,000 | 182,400 |
| 3 | 149,000 | 154,300 |
| 4 | 198,500 | 198,500 |
| 5 | 214,000 | 218,000 |
| 6 | 235,000 | 230,000 |
| 7 | 305,000 | 298,900 |
| 8 | 187,500 | 190,000 |
| 9 | 150,000 | 149,800 |
| 10 | 223,000 | 222,000 |
| 11 | 178,500 | 180,000 |
| 12 | 245,000 | 250,900 |
| 13 | 167,000 | 165,200 |
| 14 | 219,000 | 220,700 |
| 15 | 334,000 | 320,000 |

a. Assuming that the population of assessed values and the population of market values have the same distribution shape and that they may differ only with respect to medians, state the appropriate null and alternative hypotheses.

b. Test the hypotheses using an $\alpha = 0.01$ level.

c. Discuss why one would not assume that the samples were obtained from normal distributions for this problem. What characteristic about the market values of houses would lead you to conclude that these data were not normally distributed?

**17-28.** Suppose a state tax commission conducted a study to determine whether there is a difference in median deductions taken for charitable contributions depending on whether a tax return is filed as a single or a joint return. A random sample from each category was selected, with the following results:

| Single | Joint |
|---|---|
| $n = 6$ | $n = 8$ |
| Sum of ranks $= 43$ | Sum of ranks $= 62$ |

Based on these data, what should the tax commission conclude? Use an $\alpha = 0.05$ level.

**17-29.** A cattle feedlot operator has collected data for 40 matched pairs of cattle showing weight gain on two different feed supplements. His purpose in collecting the data is to determine whether there is a difference in the median weight gains for the two supplements. He has no preconceived idea about which supplement might produce higher weight gain. He wishes to test using an $\alpha = 0.05$ level.

Assuming that the $T$-value for these data is 480, what should the operator conclude concerning which supplement produces higher weight gain? Use the large-sample Wilcoxon matched-pairs signed rank test normal approximation. Conduct the test using a $p$-value approach.

**17-30.** Radio advertisements have been stressing the virtues of an online audio program to help children learn to read. To test whether this program can cause a quick improvement in reading ability, suppose 10 children were given a nationally recognized reading test that measures reading ability. The same 10 children were then given the tapes to listen to for 4 hours spaced over a 2-day period. The children then were tested again. The test scores were as follows:

| Child | Before | After |
|---|---|---|
| 1 | 60 | 63 |
| 2 | 40 | 38 |
| 3 | 78 | 77 |
| 4 | 53 | 50 |
| 5 | 67 | 74 |
| 6 | 88 | 96 |
| 7 | 77 | 80 |
| 8 | 60 | 70 |
| 9 | 64 | 65 |
| 10 | 75 | 75 |

If higher scores are better, use the Wilcoxon matched-pairs signed rank test to test whether this tape program produces quick improvement in reading ability. Use an $\alpha = 0.025$.

**17-31.** The Montgomery Athletic Shoe Company has developed a new shoe-sole material it thinks provides superior wear compared with the old material the company has been using for its running shoes. The company selected 10 cross-country runners and supplied each runner with a pair of shoes. Each pair had one sole made of the old material and the other made of the new material. The shoes were monitored until the soles wore out. The following lifetimes (in hours) were recorded for each material:

| Runner | Old Material | New Material |
|---|---|---|
| 1 | 45.5 | 47.0 |
| 2 | 50.0 | 51.0 |
| 3 | 43.0 | 42.0 |
| 4 | 45.5 | 46.0 |
| 5 | 58.5 | 58.0 |
| 6 | 49.0 | 50.5 |
| 7 | 29.5 | 39.0 |
| 8 | 52.0 | 53.0 |
| 9 | 48.0 | 48.0 |
| 10 | 57.5 | 61.0 |

a. If the populations from which these samples were taken could be considered to have normal distributions, determine if the soles made of the new material have a longer mean lifetime than those made from the old material. Use a significance level of 0.025.

b. Suppose you were not willing to consider that the populations have normal distributions. Make the determination requested in part a.

c. Given only the information in this problem, which of the two procedures indicated in parts a and b would you choose to use? Give reasons for your answer.

## Computer Software Exercises

**17-32.** There has been a debate over whether children who are placed in child-care facilities while their parents work suffer as a result. A *New York Times* article suggests that this might not be the case and, in fact, children might actually do better when their mothers work (*source:* Claire Cain Miller, "Mounting evidence of advantages for children of working mothers," www.nytimes.com, May 15, 2015). To investigate this premise, suppose a nonprofit organization called Child Care America conducted a small study in which children were observed playing in neutral settings (not at home or at a day-care center). Over 20 hours of observation, 15 children who did not go to day care and 21 children who had spent much time in day care were observed. The variable of interest was the total minutes of play in which each child was actively interacting with other students. Child Care America leaders hoped to show that the children who had been in day care would have a higher median time in interactive situations than the stay-at-home children. The file **Children** contains the results of the study.

a. Conduct a hypothesis test to determine if the hopes of the Child Care America leaders can be substantiated. Use a significance level of 0.05, and write a short statement that describes the results of the test.

b. Based on the outcome of the hypothesis test, which statistical error might have been committed?

**17-33.** An intern for the California Transportation Department conducted a study on a stretch of interstate highway south of San Francisco to determine whether the mean speed for California vehicles exceeded the mean speed for out-of-state vehicles. A total of 140 California cars were included in the study, and 75 out-of-state cars were included. Radar was used to measure the speed. The file **Speed-Test** contains the data.

a. Past studies have indicated that the speeds at which both out-of-state and California drivers drive have normal distributions. Use a *p*-value approach and an alpha = 0.10 to conduct the relevant hypothesis test. Discuss the results of this test in a short written statement.

b. Describe, in the context of this problem, what a Type I error would be.

**17-34.** When a new home-use digital blood pressure monitor is developed it needs to be tested for accuracy. One such test is to have a physician first determine the blood pressure of a randomly sampled group of participants. Then each participant is measured using the home test monitor. If the mean blood pressures are the same for the two measurements the home test monitor is considered to be accurate.

In one such test, 15 people were randomly selected to be in the sample. The blood pressure readings for these people using both methods are in the file **BloodPressure**.

a. Based on the sample data and a significance level equal to 0.05, what conclusion should the engineers reach regarding the latest blood pressure monitor? Discuss your answer in a short written statement.

b. Conduct the test as a paired *t*-test.

c. Discuss whether the two-sample test conducted in part a or the paired-difference test conducted in part b is more appropriate to analyze the data presented in this problem.

**17-35.** J.A. Hersh and Company provides court transcription services. The company is considering two keyboard styles for its staff to use in court. One factor that will influence its decision is the ease of use. Suppose that to test the keyboards, nine court transcribers were selected and asked to transcribe a 75-minute court hearing using each keyboard. The transcribers then rated the systems on a scale of 0 to 100. The resulting ratings are in the file **Hersh**.

a. Which measurement level describes the data collected for this analysis?

b. (1) Could a normal distribution describe the population distribution from which these data were sampled? (2) Which measure of central tendency would be appropriate to describe the center of the populations from which these data were sampled?

c. Choose the appropriate hypothesis-testing procedure to determine if there is a difference in the measures of central tendency you selected in part b between these two keyboards. Use a significance level of 0.01.

d. Which keyboard would you recommend the Hersh Corporation adopt? Support your answer with statistical reasoning.

OUTCOME **4** | 17.3 | **Kruskal–Wallis One-Way Analysis of Variance**

Section 17.2 showed that the Mann–Whitney *U*-test is a useful nonparametric procedure for determining whether two independent samples are from populations with the same median. However, as discussed in Chapter 12, many decisions involve comparing more than two populations. Chapter 12 introduced one-way analysis of variance and showed how, if the assumptions of normally distributed populations with equal variances are satisfied, the *F*-distribution can be used to test the hypothesis of equal population means. However, what if decision makers are not willing to assume normally distributed populations? In that case, they can turn to a nonparametric procedure to compare the populations. *Kruskal–Wallis one-way analysis of variance* is the nonparametric counterpart to the one-way ANOVA procedure. It is applicable any time the variables in question satisfy the following conditions:

**Assumptions** {
1. They have a continuous distribution.
2. The data are at least ordinal.
3. The samples are independent.
4. The samples come from populations whose only possible difference is that at least one may have a different central location than the others.

Using the Kruskal–Wallis One-Way ANOVA Test

**Gibson & Schmidt Travel, Inc.** Gibson & Schmidt Travel, Inc., is considering outsourcing its information systems activities, including general accounting, client reservations, and billing. The company's information systems manager has reduced the possible suppliers to three, each using different computer systems. One critical factor in the decision is downtime. When the system goes down, online applications stop and normal activities are interrupted. The information systems manager received from each supplier a list of firms using its service. From these lists, the manager selected random samples of nine users of each service. Suppose, through a series of phone calls, she found the number of hours of downtime in the previous month for each service. At issue is whether the three systems have the same or different median numbers of downtimes. If the manager is unwilling to make the assumptions of normality and equal variances required for the one-way ANOVA technique introduced in Chapter 12, she can apply the Kruskal–Wallis nonparametric test using the following steps:

STEP 1 **Specify the appropriate null and alternative hypotheses to be tested.**
In this application the information systems manager is interested in determining whether a difference exists between median downtimes for the three systems. Thus, the null and alternative hypotheses are

$$H_0: \tilde{\mu}_A = \tilde{\mu}_B = \tilde{\mu}_C$$
$$H_A: \text{Not all population medians are equal.}$$

STEP 2 **Specify the desired level of significance for the test.**
The test will be conducted using a significance level equal to

$$\alpha = 0.10$$

STEP 3 **Collect the sample data and compute the test statistic.**
The data represent a random sample of downtimes from each service. The samples are independent. To use the Kruskal–Wallis ANOVA, first replace each downtime measurement by its *relative ranking* within all groups combined. The smallest downtime is given a rank of 1, the next smallest a rank of 2, and so forth, until all downtimes for the three services have been replaced by their relative rankings. Table 17.5 shows the sample data and the rankings for the 27 observations. Notice that the rankings are summed for each service. The Kruskal–Wallis test will determine whether these sums are so different that it is not likely that they came from populations with equal medians.

If the samples actually do come from populations with equal medians (that is, the three services have the same per-month median downtime), then the $H$-statistic, calculated by Equation 17.10, will be approximately distributed as a chi-square

**TABLE 17.5** Sample Data and Rankings of System Downtimes for the Gibson & Schmidt Travel Example

| Service A | | Service B | | Service C | |
|---|---|---|---|---|---|
| Data | Ranking | Data | Ranking | Data | Ranking |
| 4.0 | 11 | 6.9 | 19 | 0.5 | 1 |
| 3.7 | 10 | 11.3 | 23 | 1.4 | 4 |
| 5.1 | 15 | 21.7 | 27 | 1.0 | 2 |
| 2.0 | 6 | 9.2 | 20 | 1.7 | 5 |
| 4.6 | 12 | 6.5 | 17 | 3.6 | 9 |
| 9.3 | 21 | 4.9 | 14 | 5.2 | 16 |
| 2.7 | 8 | 12.2 | 25 | 1.3 | 3 |
| 2.5 | 7 | 11.7 | 24 | 6.8 | 18 |
| 4.8 | 13 | 10.5 | 22 | 14.1 | 26 |
| Sum of ranks = 103 | | Sum of ranks = 191 | | Sum of ranks = 84 | |

variable with $k - 1$ degrees of freedom, where $k$ equals the number of populations (systems in this application) under study.

**H-Statistic**

$$H = \frac{12}{N(N + 1)} \sum_{i=1}^{k} \frac{R_i^2}{n_i} - 3(N + 1) \tag{17.10}$$

where:

$N$ = Sum of the sample sizes from all populations
$k$ = Number of populations
$R_i$ = Sum of ranks in the sample from the $i$th population
$n_i$ = Size of the sample from the $i$th population

Using Equation 17.10, the $H$-statistic is

$$H = \frac{12}{N(N + 1)} \sum_{i=1}^{k} \frac{R_i^2}{n_i} - 3(N + 1)$$

$$= \frac{12}{27(27 + 1)} \left[ \frac{103^2}{9} + \frac{191^2}{9} + \frac{84^2}{9} \right] - 3(27 + 1) = 11.50$$

STEP 4 **Determine the critical value from the chi-square distribution.**
If $H$ is larger than $\chi^2$ from the chi-square distribution with $k - 1$ degrees of freedom in Appendix G, the hypothesis of equal medians should be rejected. The critical value for $\alpha = 0.10$ and

$$k - 1 = 3 - 1 = 2$$

degrees of freedom is

$$\chi^2_{0.10} = 4.6052$$

STEP 5 **Reach a decision.**
Since $H = 11.50 > 4.6052$, reject the null hypothesis based on these sample data.

STEP 6 **Draw a conclusion.**
The Kruskal–Wallis one-way ANOVA shows that the information systems manager should conclude, based on the sample data, that the three services *do not* have equal median downtimes. From this analysis, the supplier with System B would most likely be eliminated from consideration unless other factors such as price or service support offset the apparent longer downtimes.

**EXAMPLE 17-3**  **Using Kruskal–Wallis One-Way ANOVA**

Excel Tutorial

**Amalgamated Sugar**   Amalgamated Sugar is one of many manufacturing companies that have implemented an effort called total productive maintenance (TPM). The objective of TPM is to increase the overall operating effectiveness of the company's equipment. One component of the TPM process attempts to reduce unplanned machine downtime. The first step is to gain an understanding of the current downtime situation. To do this, suppose Amalgamated Sugar has collected a sample of 20 days for each of the three shifts (day, swing, and graveyard). The variable of interest is the minutes of unplanned downtime per shift per day. The minutes are tabulated by summing the downtime minutes for all equipment in the plant. The Kruskal–Wallis test can be performed using the following steps:

STEP **1** **State the appropriate null and alternative hypotheses.**
The Kruskal–Wallis one-way ANOVA procedure can test whether the medians are equal, as follows:

$$H_0: \tilde{\mu}_1 = \tilde{\mu}_2 = \tilde{\mu}_3$$
$$H_A: \text{Not all population medians are equal.}$$

STEP **2** **Specify the desired significance level for the test.**
The test will be conducted using an

$$\alpha = 0.05$$

STEP **3** **Collect the sample data and compute the test statistic.**
The sample data are in the **Amalgamated** file.
Excel (with the XLSTAT add-in) can be used to perform the Kruskal–Wallis nonparametric ANOVA test. Figure 17.1 illustrates the Excel XLSTAT output for these sample data. The calculated $H$-statistic is

$$H = 0.1859$$

> The Excel 2016 function for the chi-square critical value is
>
> $= \text{CHISQ.INV.RT}(\text{alpha,df})$
> $= \text{CHISQ.INV.RT}(0.05,2)$

STEP **4** **Determine the critical value from the chi-square distribution.**
The critical value for $\alpha = 0.05$ and $k - 1 = 2$ degrees of freedom is

$$\chi^2_{0.05} = 5.9915$$

**FIGURE 17.1** **Excel 2016 (XLSTAT) Kruskal–Wallis ANOVA Output for Amalgamated Sugar**

p-value: Asymptotic p-value
Summary statistics ▼
Summary statistics:

| Variable | Observations | Obs. with missing data | Obs. without missing data | Minimum | Maximum | Mean | Std. deviation |
|---|---|---|---|---|---|---|---|
| Day | 20 | 0 | 20 | 87.00 | 902.00 | 426.05 | 248.96 |
| Swing | 20 | 0 | 20 | 98.00 | 845.00 | 445.65 | 212.53 |
| Graveyard | 20 | 0 | 20 | 78.00 | 921.00 | 415.50 | 201.72 |

Kruskal-Wallis test:

| | |
|---|---|
| K (Observed value) | 0.1859 |
| K (Critical value) | 5.9915 |
| DF | 2 |
| p-value (Two-tailed) | 0.9112 |
| alpha | 0.05 |

An approximation has been used to compute the p-value.

Test interpretation:
H0: The samples come from the same population.
Ha: The samples do not come from the same population.

As the computed p-value is greater than the significance level alpha=0.05, one cannot reject the null hypothesis H0.

The risk to reject the null hypothesis H0 while it is true is 91.12%.

Ties have been detected in the data and the appropriate corrections have been applied.

> Test statistic
> p-value
> Conclusion

**Excel 2016 (XLSTAT) Instructions**

1. Open XLSTAT and Enable Macros.
2. Open File: **Amalgamated.xlsx.**
3. On the **XLSTAT** tab, click **Nonparametric Tests**, and then click **Comparison of k samples (Kruskal-Wallis, Friedman, ...).**
4. Under **Samples**, highlight the data (**B1:D21**).
5. Select the **Column labels** and **Kruskal-Wallis test** checkboxes.
6. Under the **Options** tab, if necessary, change the **significance level** to 5 and check **Asymptotic p-value** and click **OK.**
7. In the XLSTAT selection dialog box, click **Samples** and then click **Continue.**

STEP 5  **Reach a decision.**

Because

$$H = 0.1859 < 5.9915$$

we do not reject the null hypothesis.

The XLSTAT output provides the $p$-value associated with the $H$-statistic. The $p$-value of 0.9112 far exceeds an alpha of 0.05.

STEP 6  **Draw a conclusion.**

Based on the sample data, the three shifts do not appear to differ with respect to median equipment downtime. The company can now begin to work on steps that will reduce the downtime across the three shifts.

**TRY EXERCISE 17-37** (pg. 710)

## Limitations and Other Considerations

The Kruskal–Wallis one-way ANOVA does *not* require the assumption of normality and is, therefore, often used instead of the ANOVA technique discussed in Chapter 12. However, the Kruskal–Wallis test as discussed here applies only if the sample size of each population is at least 5, the samples are independently selected, and each population has the same distribution except for a possible difference in central location.

When ranking observations, we sometimes encounter ties. When ties occur, each observation is given the average rank for which it is tied. The $H$-statistic is influenced by ties and should be corrected by dividing Equation 17.10 by Equation 17.11.

**Correction for Tied Rankings—Kruskal–Wallis Test**

$$1 - \frac{\sum_{i=1}^{g} (t_i^3 - t_i)}{N^3 - N} \tag{17.11}$$

where:

$g$ = Number of different groups of ties

$t_i$ = Number of tied observations in the $i$th tied group of scores

$N$ = Total number of observations

The correct formula for calculating the Kruskal–Wallis $H$ statistic when ties are present is Equation 17.12.

Correcting for ties increases $H$, which makes rejecting the null hypothesis more likely than if the correction is not used. A rule of thumb is that if no more than 25% of the observations are involved in ties, the correction factor is not required. The XLSTAT add-in to Excel for performing the Kruskal–Wallis test automatically provides the adjusted $H$-statistic.

**$H$-Statistic Corrected for Tied Rankings**

$$H = \frac{\dfrac{12}{N(N+1)} \sum_{i=1}^{k} \dfrac{R_i^2}{n_i} - 3(N+1)}{1 - \dfrac{\sum_{i=1}^{g} (t_i^3 - t_i)}{N^3 - N}} \tag{17.12}$$

## 17.3 EXERCISES

### Skill Development

**17-36.** You are given the following sample data:

| Group 1 | Group 2 | Group 3 |
|---------|---------|---------|
| 21 | 17 | 29 |
| 25 | 15 | 38 |
| 36 | 34 | 28 |
| 35 | 22 | 27 |
| 33 | 16 | 14 |
| 23 | 19 | 26 |
| 31 | 30 | 39 |
| 32 | 20 | 36 |

a. State the appropriate null and alternative hypotheses to test whether there is a difference in the medians of the three populations.
b. Based on the sample data and a significance level of 0.05, what conclusion should you reach about the medians of the three populations if you are not willing to make the assumption that the populations are normally distributed?
c. Test the hypothesis stated in part a, assuming that the populations are normally distributed with equal variances.
d. Which of the procedures described in parts b and c would you select to analyze the data? Explain your reasoning.

**17-37.** You are given the following sample data:

| Group 1 | Group 2 | Group 3 |
|---------|---------|---------|
| 10 | 8 | 13 |
| 9 | 6 | 12 |
| 11 | 8 | 12 |
| 12 | 9 | 11 |
| 13 | 10 | 13 |
| 12 | 10 | 15 |

a. State the appropriate null and alternative hypotheses for determining whether a difference exists in the median value for the three populations.
b. Based on the sample data, use the Kruskal–Wallis ANOVA procedure to test the null hypothesis using $\alpha = 0.05$. What conclusion should you reach?

**17-38.** You are given the following data:

| Group 1 | Group 2 | Group 3 | Group 4 |
|---------|---------|---------|---------|
| 20 | 28 | 17 | 21 |
| 27 | 26 | 15 | 23 |
| 26 | 21 | 18 | 19 |
| 22 | 29 | 20 | 17 |
| 25 | 30 | 14 | 20 |
| 30 | 25 | | |
| 23 | | | |

a. State the appropriate null and alternative hypotheses for determining whether a difference exists in the median value for the four populations.
b. Based on the sample data, use the Kruskal–Wallis one-way ANOVA procedure to test the null hypothesis. What conclusion should you reach using a significance level of 0.10? Discuss.
c. Determine the $H$-value adjusted for ties.
d. Given the results in part b, is it necessary to use the $H$-value adjusted for ties? If it is, conduct the hypothesis test using this adjusted value of $H$. If not, explain why not.

**17-39.** A study was conducted in which samples were selected independently from four populations. The sample size from each population was 20. The data were converted to ranks. The sum of the ranks for the data from each sample is as follows:

| | Sample 1 | Sample 2 | Sample 3 | Sample 4 |
|---|---------|---------|---------|---------|
| Sum of ranks | 640 | 780 | 460 | 1,360 |

a. State the appropriate null and alternative hypotheses if we wish to determine whether the populations have equal medians.
b. Use the information in this exercise to perform a Kruskal–Wallis one-way ANOVA.

### Business Applications

**17-40.** The U.S. Beef Producers Association is just one organization that conducts various studies, the results of which are often used in advertising campaigns. Assume that one study involved a quality perception test. Three grades of beef were involved: choice, standard, and economy. A random sample of people was provided pieces of choice-grade beefsteak and asked to rate its quality on a scale of 1 to 100. A second sample of people was given pieces of standard-grade beefsteak, and a third sample was given pieces of economy-grade beefsteak, with instructions to rate the beef on the 100-point scale. The following data were obtained:

| Choice | Standard | Economy |
|--------|----------|---------|
| 78 | 67 | 65 |
| 87 | 80 | 62 |
| 90 | 78 | 70 |
| 87 | 80 | 66 |
| 89 | 67 | 70 |
| 90 | 70 | 73 |

a. What measurement level do these data possess? Would it be appropriate to assume that such data could be obtained from a normal distribution? Explain your answers.

b. Based on the sample data, what conclusions should be reached concerning the median quality perception scores for the three grades of beef? Test using an $\alpha = 0.01$.

**17-41.** Suppose a study was conducted by the sports department of a television station in which the objective was to determine whether a difference exists between median annual salaries of National Basketball Association (NBA) players, National Football League (NFL) players, and Major League Baseball (MLB) players. The analyst in charge of the study believes that the normal distribution assumption is violated in this study. Thus, she thinks a nonparametric test is in order. The following summary data have been collected:

| NBA | NFL | MLB |
|---|---|---|
| $n = 20$ | $n = 30$ | $n = 40$ |
| $\Sigma R_i = 1,655$ | $\Sigma R_i = 1,100$ | $\Sigma R_i = 1,340$ |

a. Why would the sports department address the median as the parameter of interest in this analysis, as opposed to the mean? Explain your answer.
b. What characteristics of the salaries of professional athletes suggest that such data are not normally distributed? Explain.
c. Based on these data, what can be concluded about the median salaries for the three sports? Test at an $\alpha = 0.05$. Assume no ties.

**17-42.** Referring to Exercise 17-41, assume there were 40 ties at eight different salary levels. The following shows how many scores were ties at each salary level:

| Level | t |
|---|---|
| 1 | 2 |
| 2 | 3 |
| 3 | 2 |
| 4 | 4 |
| 5 | 8 |
| 6 | 10 |
| 7 | 6 |
| | 5 |

a. Given the results in the previous exercise, is it necessary to use the $H$-value adjusted for ties?
b. If your answer to part a is yes, conduct the test of hypothesis using this adjusted value of $H$. If it is not, explain why not.

**17-43.** Suppose as part of your job you are responsible for installing emergency lighting in state office buildings. Bids have been received from four manufacturers of battery-operated emergency lights. The costs are about equal, so the decision will be based on the length of time the lights last before failing. A sample of four lights from each manufacturer has been tested, with the

following values (time in hours) recorded for each manufacturer:

| Type A | Type B | Type C | Type D |
|---|---|---|---|
| 1,024 | 1,270 | 1,121 | 923 |
| 1,121 | 1,325 | 1,201 | 983 |
| 1,250 | 1,426 | 1,190 | 1,087 |
| 1,022 | 1,322 | 1,122 | 1,121 |

Using $\alpha = 0.01$, what conclusion for the four manufacturers should you reach about the median length of time the lights last before failing? Explain.

## Computer Software Exercises

**17-44.** As purchasing agent for your company, you have primary responsibility for securing high-quality raw materials at the best possible prices. One particular material your company uses a great deal of is aluminum. After careful study, you have been able to reduce the prospective vendors to three. It is unclear whether these three vendors produce aluminum that is equally durable.

To compare durability, the recommended procedure is to put pressure on aluminum until it cracks. The vendor whose aluminum requires the highest median pressure will be judged to provide the most durable product. To carry out this test, 14 pieces from each vendor have been selected. These data are in the file **Aluminum**. (The data are pounds per square inch pressure.) Using $\alpha = 0.05$, what should the company conclude about whether there is a difference in the median strengths of the three vendors' aluminum?

**17-45.** A large metropolitan police force is considering changing from full-size to mid-size cars. The police force sampled cars from each of three manufacturers. The number sampled represents the number that the manufacturer was able to provide for the test. Each car was driven for 5,000 miles, and the operating cost per mile was computed. The operating costs, in cents per mile, for the 12 cars are provided in the file **Police**. Perform the appropriate ANOVA test on these data. Assume a significance level of 0.05. State the appropriate null and alternative hypotheses. Do the experimental data provide evidence that the median operating costs per mile for the three types of police cars are different?

**17-46.** A nationwide moving company is considering five different types of nylon tie-down straps. The purchasing department randomly selected straps from each company and determined their breaking strengths in pounds. The sample data are contained in the file **Nylon**. Based on your analysis, with a Type I error rate of 0.05, can you conclude that a difference exists among the median breaking strengths of the types of nylon ropes?

# 17 Overview

## Summary

**17.1**

### The Wilcoxon Signed Rank Test for One Population Median (pg. 688–693)

**OUTCOME 1** Use the Wilcoxon signed rank test for a population median.

- The **Wilcoxon signed rank test** is used to test a population in situations where the $t$-test is not appropriate, such as if the data are ordinal or the populations are not normally distributed.
- This test makes no restrictive assumption about the shape of the population distribution.
- The Wilcoxon test is used to test hypotheses about a population median rather than a population mean. This follows the logic that the median is the midpoint in a population, so we expect approximately half the data values in a random sample to lie below the hypothesized median and about half to lie above it.
- We reject the hypothesized median if the actual data distribution shows too large a departure from a 50-50 split.

**17.2**

### Nonparametric Tests for Two Population Medians (pg. 693–705)

**OUTCOME 2** Recognize the situations for which the Mann–Whitney $U$-test for the difference between two population medians applies and apply it in a decision-making context.

**OUTCOME 3** Be able to apply the Wilcoxon matched-pairs signed rank test for related samples.

- Two nonparametric techniques do not require the distribution and data-level assumptions that the $t$-test requires to find the difference between two population medians: the Mann–Whitney $U$-test and the Wilcoxon matched-pairs signed rank test.
- Both tests can be used with ordinal data, and neither requires a normal distribution.
- We use the **Mann–Whitney $U$-test** when the samples are independent.
- We use the **Wilcoxon matched-pairs signed rank test** when the design has paired samples.

**17.3**

### Kruskal–Wallis One-Way Analysis of Variance (pg. 705–711)

**OUTCOME 4** Perform nonparametric analysis of variance using the Kruskal–Wallis one-way ANOVA.

- Frequently decision makers must decide among three or more alternatives. The **Kruskal–Wallis one-way analysis of variance** is the nonparametric counterpart to the one-way ANOVA procedure presented in Chapter 12.
- The Kruskal–Wallis one-way analysis of variance has the following assumptions:
  - The distributions are continuous.
  - The data are at least ordinal.
  - The samples are independent.
  - The samples come from populations whose only possible difference is that one might have a different central location than the others.
- Refer to Figure 17.2 for assistance in determining which nonparametric test to use in different situations.

**FIGURE 17.2 Nonparametric Tests Introduced in Chapter 17**

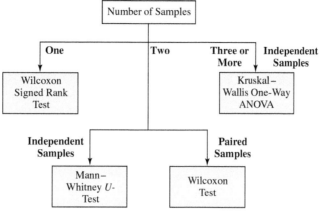

Other Commonly Used Nonparametric Tests:
Friedman Test: randomized block ANOVA
Sign Test: test for randomness
Runs Test: test for randomness
Spearman Rank Correlation: measure of the linear relationship between two variables

## Equations

**(17.1) Large-Sample Wilcoxon Signed Rank Test Statistic** pg. 690

$$z = \frac{W - \dfrac{n(n+1)}{4}}{\sqrt{\dfrac{n(n+1)(2n+1)}{24}}}$$

**(17.2) $U$-Statistics** pg. 695

$$U_1 = n_1 n_2 + \frac{n_1(n_1+1)}{2} - \sum R_1$$

**(17.3)**
$$U_2 = n_1 n_2 + \frac{n_2(n_2+1)}{2} - \sum R_2$$

**(17.4) Mean and Standard Deviation for $U$-Statistic** pg. 696

$$\mu = \frac{n_1 n_2}{2}$$

**(17.5)**
$$\sigma = \sqrt{\frac{(n_1)(n_2)(n_1+n_2+1)}{12}}$$

**(17.6) Mann–Whitney $U$-Test Statistic** pg. 696

$$z = \frac{U - \dfrac{n_1 n_2}{2}}{\sqrt{\dfrac{(n_1)(n_2)(n_1+n_2+1)}{12}}}$$

**(17.7) Wilcoxon Mean and Standard Deviation** pg. 700

$$\mu = \frac{n(n+1)}{4}$$

**(17.8)**
$$\sigma = \sqrt{\frac{n(n+1)(2n+1)}{24}}$$

**(17.9) Wilcoxon Test Statistic** pg. 701

$$z = \frac{T - \dfrac{n(n+1)}{4}}{\sqrt{\dfrac{n(n+1)(2n+1)}{24}}}$$

**(17.10) $H$-Statistic** pg. 707

$$H = \frac{12}{N(N+1)} \sum_{i=1}^{k} \frac{R_i^2}{n_i} - 3(N+1)$$

**(17.11) Correction for Tied Rankings—Kruskal–Wallis Test** pg. 709

$$1 - \frac{\sum\limits_{i=1}^{g}(t_i^3 - t_i)}{N^3 - N}$$

**(17.12) $H$-Statistic Corrected for Tied Rankings** pg. 709

$$H = \frac{\dfrac{12}{N(N+1)} \sum\limits_{i=1}^{k} \dfrac{R_i^2}{n_i} - 3(N+1)}{1 - \dfrac{\sum\limits_{i=1}^{g}(t_i^3 - t_i)}{N^3 - N}}$$

## Chapter Exercises

### Conceptual Questions

**17-47.** Find an organization you think would be interested in data that violate the measurement scale or known distribution assumptions necessary to use the statistical tools found in Chapters 10–12 (retail stores are a good candidate). Determine to what extent this organization considers these problems and whether it uses any of the techniques discussed in this chapter.

**17-48.** Discuss the data conditions that would lead you to use the Kruskal–Wallis test as opposed to the ANOVA procedure introduced in Chapter 12. Present an example illustrating these conditions.

**17-49.** In the library, find two journal articles that use one of the nonparametric tests discussed in this chapter. Prepare a brief outline of the articles, paying particular attention to the reasons given for using the particular test.

**17-50.** As an example of how the sampling distribution for the Mann–Whitney $U$-test is derived, consider two samples with sizes $n_1 = 2$ and $n_2 = 3$. The distribution is obtained under the assumption that the two variables, say $x$ and $y$, are identically distributed. Under this assumption, each measurement is equally likely to obtain one of the ranks between 1 and $n_1 + n_2$.

  a. List all the possible sets of two ranks that could be obtained from five ranks. Calculate the Mann–Whitney $U$-value for each of these sets of two ranks.

  b. The number of ways in which we may choose $n_1$ ranks from $n_1 + n_2$ is given by $(n_1 + n_2)!/n_1!n_2!$. Calculate this value for $n_1 = 2$ and $n_2 = 3$. Now calculate the probability of any one of the possible Mann–Whitney $U$-values.

  c. List all the possible Mann–Whitney $U$-values you obtained in part a. Then using part b, calculate the probability that each of these $U$-values occurs, thereby producing the sampling distribution for the Mann–Whitney $U$-statistic when $n_1 = 2$ and $n_2 = 3$.

**17-51.** Let us examine how to obtain the sampling distribution of the Wilcoxon test statistic. Consider the sampling distributions of the positive ranks from a sample size of 4. The ranks to be considered are, therefore, 1, 2, 3, and 4. Under the null hypothesis, the differences to be ranked are distributed symmetrically about zero. Thus, each difference is just as likely to be positively as negatively ranked.

  a. For a sample size of four, there are $2^4 = 16$ possible sets of signs associated with the four ranks. List the 16 possible sets of ranks that could be positive—that is, (none), (1), (2), ..., (1, 2, 3, 4). Each of these sets of positive ranks (under the null hypothesis) has the same probability of occurring.

  b. Calculate the sum of the ranks of each set specified in part a.

  c. Using parts a and b, produce the sampling distribution for the Wilcoxon test statistic when $n = 4$.

### Business Applications

**17-52.** Most students who attend Upper Mountain Community College buy their textbooks online from one of two different booksellers because the college does not have a bookstore. The following data represent sample amounts that students spend on books per term:

| Company 1 ($) | Company 2 ($) |
|---|---|
| 246 | 300 |
| 211 | 305 |
| 235 | 308 |
| 270 | 325 |
| 411 | 340 |
| 310 | 295 |
| 450 | 320 |
| 502 | 330 |
| 311 | 240 |
| 200 | 360 |

  a. Do these data indicate a difference in mean textbook prices for the two companies? Apply the Mann–Whitney $U$-test with a significance level of 0.10.

  b. Apply the $t$-test to determine whether the data indicate a difference between the mean amounts spent on books at the two companies. Use a significance level of 0.10. Indicate what assumptions must be made to apply the $t$-test.

**17-53.** The Hunter Family Corporation owns roadside diners in numerous locations across the country. For the past few months, the company has undertaken a new advertising study. Initially, company executives selected 22 of its retail outlets that were similar with respect to sales volume, profitability, location, climate, economic status of customers, and experience of store management. Each of the outlets was randomly assigned one of two advertising plans promoting a new sandwich product. The accompanying data represent the number of new sandwiches sold during the specific test period at each retail outlet.

Hunter executives want you to determine which of the two advertising plans leads to the highest average sales levels for the new product. They are not willing to make the assumptions necessary for you to use the $t$-test. They do not wish to have an error rate higher than 0.05.

| Advertising Plan 1 ($) | Advertising Plan 2 ($) |
|---|---|
| 1,711 | 2,100 |
| 1,915 | 2,210 |
| 1,905 | 1,950 |
| 2,153 | 3,004 |
| 1,504 | 2,725 |
| 1,195 | 2,619 |
| 2,103 | 2,483 |

| Advertising Plan 1 ($) | Advertising Plan 2 ($) |
|---|---|
| 1,601 | 2,520 |
| 1,580 | 1,904 |
| 1,475 | 1,875 |
| 1,588 | 1,943 |

**17-54.** The Miltmore Corporation performs consulting services for companies that think they have image problems. Recently, a beer producer approached Miltmore concerned that the company's image, relative to its two closest competitors, had diminished. Suppose Miltmore conducted an image study in which a random sample of 8 people was asked to rate the beer company's image. Five people were asked to rate Competitor A's image, and 10 people were asked to rate Competitor B's image. The image ratings were made on a 100-point scale, with 100 being the best possible rating. Here are the results of the sampling:

| Beer Company | Competitor A | Competitor B |
|---|---|---|
| 40 | 95 | 50 |
| 60 | 53 | 80 |
| 70 | 55 | 82 |
| 40 | 92 | 87 |
| 55 | 90 | 93 |
| 90 | | 51 |
| 20 | | 63 |
| 20 | | 72 |
| | | 96 |
| | | 88 |

a. Based on these sample results, should the company conclude there is an image difference among the three competitors? Use a significance level equal to 0.05.

b. Why might the decision maker wish to use parametric ANOVA rather than the corresponding nonparametric test? Discuss.

**17-55.** The Style-Rite Company developed a new material and is in the process of test-marketing jackets made from the material. As part of this test-marketing effort, 10 people were each supplied with a jacket made from the original material and were asked to wear it for two months, washing it at least twice during that time. A second group of 10 people was each given a jacket made from the new material and asked to wear it for two months with the same washing requirements.

Following the two-month trial period, the individuals were asked to rate the jackets on a scale of 0 to 100, with 0 being the worst performance rating and 100 being the best. The ratings for each material are shown as follows:

| Original Material | New Material |
|---|---|
| 76 | 55 |
| 34 | 90 |
| 70 | 72 |
| 23 | 17 |
| 45 | 56 |
| 80 | 69 |
| 10 | 91 |
| 46 | 95 |
| 67 | 86 |
| 75 | 74 |

The company expects that, on the average, the performance ratings will be superior for the new material.

a. Examine the data given. What characteristics of these data sets would lead you to reject the assumption that the data came from populations that had normal distributions and equal variances?

b. Do the sample data support this belief at a significance level of 0.05? Discuss.

**17-56.** Suppose a study was conducted by a federal government agency to determine attitudes regarding the agency's policies in western U.S. states. One part of the study asked respondents to rate the performance of the agency on its responsiveness to environmental issues. The following responses were obtained for a sample of 12 urban residents and 10 rural residents. The ratings are on a 1 to 100 scale, with 100 being perfect.

| Urban | Rural |
|---|---|
| 76 | 55 |
| 90 | 80 |
| 86 | 94 |
| 60 | 40 |
| 43 | 85 |
| 96 | 92 |
| 50 | 77 |
| 20 | 68 |
| 30 | 35 |
| 82 | 59 |
| 75 | |
| 84 | |

a. Based on the sample data, should the agency conclude that there is no difference between the urban and rural residents with respect to median environmental rating? Test using a significance level of 0.02.

b. Perform the appropriate parametric statistical test and indicate the assumptions necessary to use this test that were not required by the Mann–Whitney tests. Use a significance level of 0.02. (Refer to Chapter 10 if needed.)

**17-57.** The manager of credit card operations for a small regional bank has determined that last year's median credit card balance was $1,989.32. A sample of 18 customer balances this year revealed the following figures, in dollars:

| Sample 1 | Sample 2 | Sample 3 | Sample 4 | Sample 5 | Sample 6 |
|---|---|---|---|---|---|
| 1,827.85 | 1,992.75 | 2,012.35 | 1,955.64 | 2,023.19 | 1,998.52 |

| Sample 7 | Sample 8 | Sample 9 | Sample 10 | Sample 11 | Sample 12 |
|---|---|---|---|---|---|
| 2,003.75 | 1,752.55 | 1,865.32 | 2,013.13 | 2,225.35 | 2,100.35 |

| Sample 13 | Sample 14 | Sample 15 | Sample 16 | Sample 17 | Sample 18 |
|---|---|---|---|---|---|
| 2,002.02 | 1,850.37 | 1,995.35 | 2,001.18 | 2,252.54 | 2,035.75 |

Based on the 18 customer balances sampled, is there enough evidence to allow you to conclude the median balance has changed? Test at the 0.05 level of significance.

**17-58.** During the production of a textbook, there are many steps between when the author begins preparing the manuscript and when the book is finally printed and bound. Tremendous effort is made to minimize the number of errors of any type in the text. One type of error that is especially difficult to eliminate is the typographical error that can creep in when the book is typeset. The Prolythic Type Company does contract work for many publishers. As part of its quality control efforts, it charts the number of corrected errors per page in its manuscripts. In one particularly difficult to typeset book, the following data were observed for a sample of 15 pages (in sequence):

| Page | 1 | 2 | 3 | 4 | 5 | 6 | 7 | 8 | 9 | 10 | 11 | 12 | 13 | 14 | 15 |
|---|---|---|---|---|---|---|---|---|---|---|---|---|---|---|---|
| Errors | 2 | 4 | 1 | 0 | 6 | 7 | 4 | 2 | 9 | 4 | 3 | 6 | 2 | 4 | 2 |

Is there sufficient evidence to conclude the median number of errors per page is greater than 6?

**17-59.** A Vermont company is monitoring a process that fills maple syrup bottles. When the process is filling correctly, the median average fill in an 8-ounce bottle of syrup is 8.03 ounces. The last 15 bottles sampled revealed the following levels of fill:

| | | | | | | | |
|---|---|---|---|---|---|---|---|
| 7.95 | 8.02 | 8.07 | 8.06 | 8.05 | 8.04 | 7.97 | 8.01 |
| 8.04 | 8.05 | 8.08 | 8.11 | 7.99 | 8.00 | 8.02 | |

a. Formulate the null and alternative hypotheses needed in this situation.
b. Do the sample values support the null or alternative hypothesis?

## Computer Software Exercises

**17-60.** Suppose a car manufacturer is experimenting with three new methods of pollution control. The testing lab must determine whether the three methods produce equal pollution reductions. To do this, readings from a calibrated carbon monoxide meter are taken from groups of engines randomly assigned to one of the three control

methods. The file **Pollution** contains data from such a study. Determine whether the three pollution-control methods produce equal results.

**17-61.** A business statistics instructor at State University has been experimenting with her testing procedure. Assume that this term, she has taken the approach of giving two tests over each section of material. The first test is a problem-oriented exam, in which students have to set up and solve applications problems. The exam is worth 50 points. The second test, given a day later, is a multiple-choice test covering the concepts introduced in the section of the text covered by the exam. This exam is also worth 50 points.

In one class of 15 students, the observed test scores over the first section of material in the course are contained in the file **State University**.

a. If the instructor is unwilling to make the assumptions for the paired-sample $t$-test, what should she conclude based on these data about the distribution of scores for the two tests if she tests at a significance level of 0.05?
b. In the context of this problem, define a Type II error.

**17-62.** Consider a situation in which brands of tires are being tested for tread wear. To control for vehicle and driver variation, one tire of each brand is put on the front wheels of 10 cars. The cars are driven under normal driving conditions for a total of 15,000 miles. The tread wear is then measured using a very sophisticated instrument. The data that were observed are in the file **Tread Wear**. (Note that the larger the number, the less wear in the tread.)

a. What would be the possible objection in this case for employing the paired-sample $t$-test? Discuss.
b. Assuming that the decision makers in this situation are not willing to make the assumptions required to perform the paired-sample $t$-test, what decision should they reach using the appropriate nonparametric test if a significance level of 0.05 is used? Discuss.

**17-63.** High Fuel Company markets a gasoline additive for automobiles that it claims will increase a car's miles per gallon (mpg) performance. In an effort to determine whether High Fuel's claim is valid, a consumer testing agency randomly selected eight makes of automobiles. Each car's tank was filled with gasoline and driven around a track until empty. Then the car's tank was refilled with gasoline and the additive, and the car was driven until the gas tank was empty again. The miles per gallon were measured for each car with and without the additive. The results are reported in the file **High Fuel**.

The testing agency is unwilling to accept the assumption that the underlying probability distribution is normally distributed, but it would still like to perform a statistical test to determine the validity of High Fuel's claim.

a. What statistical test would you recommend the testing agency use in this case? Why?
b. Conduct the test that you believe to be appropriate. Use a significance level of 0.025.

c. State your conclusions based on the test you have just conducted. Do the test's findings support High Fuel's claim?

**17-64.** A company assembles remote controls for television sets. The company's design engineers have developed a revised design that they think will make it faster to assemble the controls. To test whether the new design leads to faster assembly, 14 assembly workers were randomly selected and each worker was asked to assemble a control using the current design and then assemble a control using the revised design. The times in seconds to assemble the controls are shown in the file **Remote Control**. The company's engineers are unable

to assume that the assembly times are normally distributed, but they would like to test whether assembly times are lower with the revised design.

a. What statistical test do you recommend the company use? Why?
b. State the null and alternative hypotheses of interest to the company.
c. At the 0.025 level of significance, is there any evidence to support the engineers' belief that the revised design reduces assembly time?
d. How might the company's management use the results of the statistical test?

## Case 17.1  Bentford Electronics—Part 2

On Saturday morning, Jennifer Bentford received a call at her home from the production supervisor at Bentford Electronics Plant 1. The supervisor indicated that she and the supervisors from Plants 2, 3, and 4 had agreed that something must be done to improve company morale and, thereby, increase the production output of their plants. Jennifer, President of Bentford Electronics, agreed to set up a Monday morning meeting with the supervisors to see if they could arrive at a plan for accomplishing these objectives.

By Monday each supervisor had compiled a list of several ideas, including a 4-day work week and interplant competition of various kinds.

After listening to the discussion for some time, Jennifer asked if anyone knew whether there was a difference in average daily output for the four plants. When she heard no positive response, she told the supervisors to select a random sample of daily production reports from each plant and test whether there was a difference. They were to meet again on Wednesday afternoon with test results.

By Wednesday morning, the supervisors had collected the following data on units produced:

| Plant 1 | Plant 2 | Plant 3 | Plant 4 |
|---|---|---|---|
| 4,306 | 1,853 | 2,700 | 1,704 |
| 2,852 | 1,948 | 2,705 | 2,320 |
| 1,900 | 2,702 | 2,721 | 4,150 |
| 4,711 | 4,110 | 2,900 | 3,300 |
| 2,933 | 3,950 | 2,650 | 3,200 |
| 3,627 | 2,300 | 2,480 | 2,975 |

The supervisors had little trouble collecting the data, but they were at a loss about how to determine whether there was a difference in the outputs of the four plants. Jerry Gibson, the company's research analyst, told the supervisors there were statistical procedures that could be used to test hypotheses regarding multiple samples if the daily output was distributed in a bell shape (normal distribution) at each plant. The supervisors expressed dismay because no one thought his or her output was normally distributed. Jerry indicated that there were techniques that did not require the normality assumption, but he did not know what they were.

The meeting with Jennifer was scheduled to begin in 3 hours, so he needed some statistical-analysis help immediately.

# 18 Introducing Business Analytics

**Business Analytics**

The application of well-known statistical tools, combined with powerful software, to transform large and complex data sets into information that is useful for making decisions.

## WHY YOU NEED TO KNOW

If you read current business publications such as *Fortune*, *Bloomberg Businessweek*, *Barron's*, or *The Wall Street Journal*, you come across stories about how companies are using **business analytics** to extract useful information from *big data*.

It is fast becoming the case that to be competitive in the job market, business graduates need to have a basic familiarity with business analytics. Earlier chapters in this text have provided you with a set of statistical tools that you must have in today's business world. However, you also need to build upon these tools by adding a basic understanding of business analytics. Your first employer will likely be a data-driven company. These companies know that

### Quick Prep

Review the discussion associated with data sources, data collection and data classification in Chapter 1.

Review the discussion on constructing charts and graphs in Chapter 2.

Review the material on calculating numerical measures of the center and variation in Chapter 3.

data are an asset and a key competitive advantage if they can glean hidden information from the data to answer important questions about their customers, their own operations, and their competitors. The more you know about business analytics, the more value you will bring to your organization.

There are scores of examples of real companies successfully using business analytics in different ways to be more competitive. For example, Nicole Laskowski's article "Ten analytics success stories in a nutshell" profiles companies such as Coca-Cola, CVS Health, and Lockheed Martin (*source:* www.searchcio.techtarget.com, Apr. 2015). Project managers for Coca-Cola's Minute Maid and Simply Orange used business analytics tools to determine that oranges come in more than 600 different flavors. They then used analytics to develop an algorithm that allows the company to produce orange juice that has consistent texture and taste regardless of which orange supplier they are using. Managers at CVS Health employed business analytics tools to segment customers who use their call center into six groups based on behavior characteristics. In addition, CVS Health analyzed data associated with their call center agents to effectively match agents to customers in a more optimal way, resulting in faster call times and improved customer service. Finally, decision makers at Lockheed Martin used business analytics to analyze data the company had collected but was not using. They were able to analyze data measurements from the aerospace company's many past programs and projects to determine predictive indicators that now are helping the company assess the potential success of new programs.

These are just a small sampling of the benefits that organizations are achieving by using business analytics. The emphasis on this relatively new area of study will continue to grow, so the more you can do to increase your business analytics understanding and experience, the more competitive you will be in the marketplace—both as you seek your first professional position and as you move through your career. This chapter provides a brief introduction to the subject of business analytics and offers a taste for how one commercially available business analytics software product, *Microsoft Power BI Desktop*, uses many of the statistical techniques introduced in earlier chapters to transform data into information.

---

**OUTCOME 1**   **18.1**   ## What Is Business Analytics?

**Business Intelligence**
The process of collecting data from inside or outside an organization that are thought to relate to or influence the performance of the organization.

We have defined business analytics as the application of statistical tools, combined with powerful software, to transform large and complex data sets into information that is useful for making decisions. But before business analytics can be put into practice, an organization must have data to analyze. In Chapter 1, we introduced a number of concepts associated with data collection and different types and levels of data. These concepts and methodology form the basis of what has become known as **business intelligence**.

If you are involved in business intelligence within an organization, your data acquisition efforts will potentially cover a wide range of areas, including internal operations, competitors, customers, vendor interactions, and regional, national, and international economic data. Although a key component of business intelligence is gathering data that are relevant to the decision makers in your organization, it is also a good idea to include extra potentially irrelevant data. Then business analytics can be used to identify the meaningful data and to provide effective ways of communicating the information in the data so that the decision makers can reach the proper conclusions. Thus, although the field of business intelligence is critically important to the field of business analytics, the two are not identical.

A common misconception is that business analytics can provide answers to unknown questions. Some people believe that business analytics tools and software can magically scan a vast array of data and kick out information that no one in the organization had thought about needing. That is not how it works! Instead, business analytics offers powerful ways of digging through data to examine relationships to a far greater extent than was possible in the past. The process starts with questions posed by decision makers—such as, Which customers are profitable? What is causing us to miss our on-time delivery targets? Why are our inventories so high? How can we increase customer satisfaction? Each of these broad questions will be followed by more detailed questions that allow you to drill down into the data. Usually the

**Descriptive Analytics**
Tools, techniques, and methodologies used to examine data for the purpose of looking back to the past to answer questions about what *has* happened.

**Predictive Analytics**
Statistical models and forecasting methods applied to data to answer questions about what *could* happen.

early output from the business analytics effort generates more questions and the process can continue until an "answer" is hopefully found.

Business analytics is typically divided into two categories: **descriptive analytics** and **predictive analytics**.

Some business analytics experts add a third category they refer to as **prescriptive analytics**, which involves the use of optimization tools and simulation models to answer the question, What should we do now? In this text, we won't address this third category, but we do provide an introductory discussion of descriptive and predictive analytics. After that, in Section 18.2, we provide an introduction to the Microsoft Power BI Desktop software.

## Descriptive Analytics

Descriptive analytics is actually based on the topics covered in Chapters 2 and 3 in this text. Chapter 2 introduced you to a variety of graphical tools, such as bar charts, histograms, pie charts, scatter plots, and trend charts. These tools are useful for helping us visualize data, but our coverage was quite basic with examples and applications that didn't dive too deeply into the available data. Chapter 3 added a discussion of numerical measures of the center and spread in the data using such measures as the mean, median, and standard deviation. But again, the examples tended to focus on one variable at a time and were based on data that were collected sometime in the past.

Using descriptive analytics along with powerful software such as Microsoft Power BI Desktop, analysts can combine the visual and numerical tools to examine relationships in the data that lay far below the surface level discussed in Chapters 2 and 3. In fact, in many instances descriptive analytics can combine historical data and current data collected in "real time" so that the output reflects not only what happened in the past but also what is happening right now. For example, a retail grocery store manager might be able to monitor her inventory of perishable produce to see what is, and is not, selling at any particular point in time. Thus, rather than basing decisions on "old" data, the manager might use real-time business analytics to see that inventories of lettuce and carrots are moving lower together while starches like corn and potatoes are not selling. She might use this information to create an expedited order for more lettuce and carrots and to mark down the price on potatoes and corn to spur sales. Or, the descriptive analytics output might show that when beef sales are up, potatoes and red wine sales also are up, so her decision might be to lower beef prices in order to drive potato and wine sales.

**Key Performance Indicator**
A measure of how well an organization is achieving its most important business objectives. High-level KPIs measure the overall performance of the organization; low-level KPIs measure the success of an organization's subunits (departments, centers, etc.).

An important outgrowth of descriptive analytics has been the expanded use of **dashboards**, which are real-time graphical and numerical displays of **key performance indicators** (KPIs) for an organization. Figures 18.1 and 18.2 show two examples of dashboards.

The examples in Figures 18.1 and 18.2 illustrate that dashboards can express data in many different ways. In Figure 18.1, the dashboard shows financial metrics using numerical measures as well as graphs and charts. The dashboard in Figure 18.2 uses three different

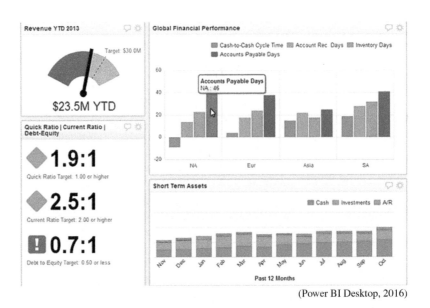

(Power BI Desktop, 2016)

**FIGURE 18.2 Dashboard of Quality Metrics**

(*Source:* Microsoft Power BI dashboard examples, https://app.powerbi.com)

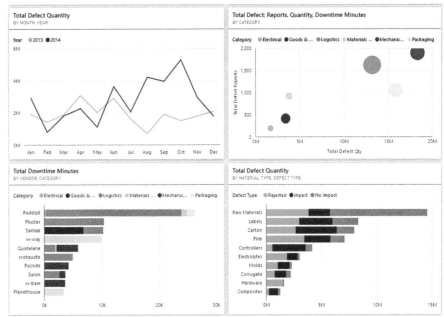

(Power BI Desktop, 2016)

types of charts to display the organization's quality performance. It would be ideal for these dashboards to be updated in real time as new data become available. Further-more, decision makers would have these charts on their desktops to monitor on a regular basis, and they could click on one of the dashboard displays to open up other related displays to drill down into the data behind the high-level metrics.

**BUSINESS APPLICATION**    Drilling Down into the Data

**West Branch Manufacturing**    Although using dashboards to provide information on key measures for an organization is an important component of business analytics, decision makers may also wish to explore their data in other ways. For example, the managers at West Branch Manufacturing wish to examine the safety data for their company. As a first pass, the managers learn that last year the company experienced 2,540 safety-related incidents. To put this into context, suppose the managers ask to see the safety incident trend for the past five years shown in Figure 18.3. The trend is downward, which is positive for the company. But not all safety incidents are equivalent. Figure 18.4 breaks down the recent year's safety incidents by type.

**FIGURE 18.3 Safety Incident Trend for West Branch Manufacturing**

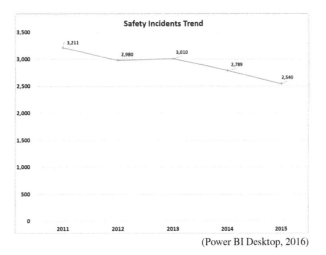

(Power BI Desktop, 2016)

**FIGURE 18.4** **Safety Incident Breakdown for West Branch Manufacturing**

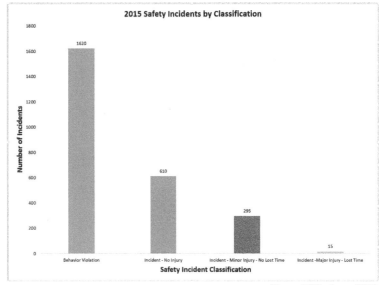

(Power BI Desktop, 2016)

Figure 18.4 shows that the vast majority of safety incidents were behavior violations. Because the managers are aware that poor safety behaviors can lead to incidents that result in injuries and lost work time, they might dig deeper into the behavior violation data. The results are shown in Figure 18.5. Three types of violations account for most of the behavior issues: lack of protection for ear (31%), eye (30%), and hand (26%). While these three classes of poor safety behaviors can lead to serious safety incidents, the lock out–tag out violation has the potential to be most serious. This behavior violation occurs when a machine operator opens a machine for maintenance or setup but fails to lock the machine so that it can't inadvertently start up with the employee inside. Figure 18.6 identifies the plant and shift for the lock out–tag out incidents. The Minnesota plant had 47 of these incidents, with 30 of these occurring on the "swing" shift that runs from 4:00 P.M. to midnight. The North Carolina plant also had a high number of these issues on the "swing" shift (30 out of 38).

This application illustrates how a manager can drill down into the data to find information that will help in making decisions. The West Branch managers started out at the highest level just knowing the total number of safety incidents, but through a series of queries they

**FIGURE 18.5** **Safety Behavior Breakdown for West Branch Manufacturing**

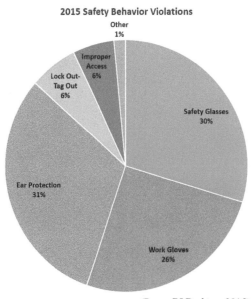

(Power BI Desktop, 2016)

**FIGURE 18.6  Lock Out–Tag Out Behavior Violation Breakdown for West Branch Manufacturing**

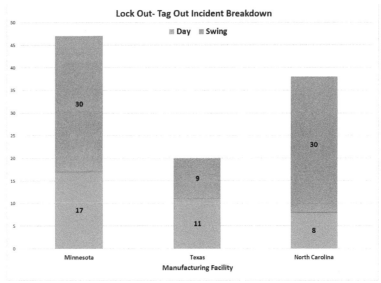

(Power BI Desktop, 2016)

were able to learn more about one particular type of safety incident. This should give you a feel for how managers apply descriptive analytics. Section 18.2 will take this a step further by showing how Microsoft's Power BI Desktop software is used to create dashboards and perform descriptive analytics.

## Predictive Analytics

As we indicated earlier, business decision makers use descriptive analytics to examine current and past data with the objective of providing information about what has happened in the past or what is currently happening. They often use predictive analytics when they want to know about the future.

Predictive analytics may identify patterns and relationships in the data that allow decision makers to anticipate a behavior or action. For example, a lending company might use predictive analytics to forecast whether a mortgage borrower will eventually pay off the loan or default on the mortgage. A company in the business of supplying backup data storage for clients might use predictive analytics to anticipate when system failures will occur in their networks and server farms. These analytics tools tend to identify and focus on specific variables that are "leading" indicators of the future.

But predictive analytics is also used when decision makers want to understand the current situation and then use this knowledge to help determine a future result. In this regard, predictive analytics seeks to define causal patterns (or relationships) between potential independent indicator variables and a dependent variable of interest. For example, through an analysis of large quantities of customer purchases, a grocery store chain may determine that a customer who buys beer also tends to buy chips and other snack products. Or, as another example, a business that analyzes great volumes of purchase invoices may determine that invoices with certain vendor addresses or dollar amounts tend to be fraudulent.

A 2010 white paper published by Oracle called "Predictive analytics: Bringing the tools to the data" presents a nice way of organizing the predictive analytics tools shown in Table 18.1. Don't panic if you are not familiar with all the algorithms listed in the table. While the topics covered in Chapters 1–17 of this text have provided an introduction to many useful and important statistical tools, the predictive analytics tools go far beyond in sophistication and are outside the scope of this text. A good predictive analyst must have a large bag of tools to be able to tackle many different potential applications.

Not only do these predictive analytics tools require a fairly advanced mathematics and statistics background, but special software is also needed. One such software system called R is a powerful aid in predictive analytics. Thomas Miller has written a book that is a particularly good introduction to both predictive analytics and the R software.[1] But many companies

---

[1] Thomas W. Miller, *Modeling Techniques in Predictive Analytics: Business Problems and Solutions with R*, Pearson Education, Inc., 2014.

**TABLE 18.1  Predictive Analytics Tools and Techniques**

| Function | Algorithm | Application |
|---|---|---|
| Classification | Logistic Regression<br>Decision Trees<br>Naïve Bayes<br>Support Vector Machine | Response modeling<br>Recommending "next likely product"<br>Employee retention<br>Credit default modeling |
| Regression | Multiple Regression<br>Support Vector Machine | Credit scoring<br>Customer profitability modeling |
| Anomaly detetection | One Class SVM (Support Vector Machine) | Claims fraud<br>Network intrusion |
| Attribute importance | Minimum Description Length (MDL) | Surgery preparation, triage<br>Net promoter score |
| Association rules | Apriori | Market basket analysis<br>Link analysis |
| Clustering | Hierarchial K-Means<br>Hierarchial O-Cluster | Customer segmentation<br>Gene and Protein analysis |
| Feature Extraction | Nonnegative Matrix Factorization | Text analysis, search |

(*Source:* Frank Buytendijk and Lucie Trepanier, Oracle Corporation, "Predictive analytics: Bringing the tools to the data," Sept. 2010)

use other software systems such as JMP, SAS, and IBM Predictive Analytics. For a summary of free predictive analytics software, see www.predictiveanalyticstoday.com.

Regardless of the software your organization uses for predictive analytics, you should know that the range of applications is very wide, including such diverse areas as health care, crime prevention, tax collections, and public welfare fraud detection. In health care, predictive analytics is used to tie symptoms and treatments to long-term patient outcomes. Insurance companies use predictive analytics to help identify fraudulent claims before payments are paid and to recover previously paid claims that are deemed fraudulent after the fact. The IRS uses predictive analytics to identify tax returns that are either underreporting income or overstating deductions. Departments of Health and Welfare are able to save taxpayer funds by determining when applications for such things as unemployment benefits, low-income housing, or food stamps are based on fraudulent data.

Predictive analytics is a growing field and can be of potential value to organizations where the outcomes of repetitive decisions may have significant financial impact, where there are lots of available data, and where predictive models can be developed to automate these repetitive decisions. Regardless of your professional role in a business, you will no doubt play a part in the predictive analytics effort in your company. We encourage you to look for opportunities to expand your knowledge of the subject.

## 18.1 EXERCISES

### Conceptual Questions

**18-1.** Using the material presented in this chapter and other sources of information, discuss the difference between *business intelligence* and *business analytics*.

**18-2.** What is the purpose of a dashboard in business analytics?

**18-3.** Locate an example of a dashboard and describe what it shows. Then discuss how a manager might use it.

**18-4.** What is meant by the terms *descriptive analytics* and *predictive analytics*? Under what conditions might each be used?

**18-5.** Based on your own independent research, discuss one or more examples of successful business analytics applications. Be sure to indicate how business analytics was used to improve decision making within the organization.

## Business Applications

**Exercises 18-6 through 18-12** ask you to drill down into financial data for companies in the data file called **Public Companies** [*source:* Bloomberg (2014), *Bloomberg Professional*. [Online]. Available at: Subscription Service (Accessed: Sept. 8, 2014)].

**18-6.** Use Excel to determine the number of companies and the mean revenue for the companies listed in the data file **Public Companies**.

**18-7.** Referring to Exercise 18-6, calculate the mean revenue broken down by the stock exchange where the firm's shares are traded. Display the results in graphical form.

**18-8.** Referring to Exercises 18-6 and 18-7, determine the number of companies traded on the New York Stock Exchange (NYSE) that have revenues of $20 billion or higher.

**18-9.** Referring to Exercises 18-6 through 18-8, determine the number of companies traded on the NYSE that have revenues higher than $20 billion by state of incorporation. Prepare a graph to display these data.

**18-10.** Referring to Exercises 18-6 through 18-9, of those NYSE companies that have revenues of $20 billion or higher and that are incorporated in Delaware, calculate the mean and standard deviation of revenues per employee.

**18-11.** Referring to Exercise 18-10, calculate the mean number of board members for Delaware-incorporated companies that are traded on the NYSE and that have revenues of $20 billion or higher and have revenues per employee in excess of $1 million. Create a graph that displays the distribution of the number of board members for firms that meet these requirements.

**18-12.** Referring to Exercises 18-10 and 18-11, calculate the mean percentage of female board members for Delaware-incorporated companies that have ten or more board members, that are traded on the NYSE with revenues of $20 billion or higher, and that have revenues per employee in excess of $1 million.

---

## 18.2 Data Visualization Using Microsoft Power BI Desktop

In this section, we introduce the fundamentals of data visualization using the Microsoft Power BI Desktop software. This Power BI software may be available on your university's computers or it can be downloaded free of charge.[2] Although the Official Microsoft Blog indicates that there are more than 500,000 unique users of Power BI at over 45,000 companies, the Capterra website lists many additional data visualization (descriptive analytics) software solutions.[3,4] Each software has its own capabilities and advantages, but all are used to provide a decision maker with a powerful way of gleaning information from large and complex data sets.

The objective of this section is to provide a sense of what data visualization software is capable of and in particular to introduce the fundamentals of Power BI. Power BI allows you to work with many different data formats such as Excel, CSV, and Text as well as Access, Oracle, MySQL, and Sybase databases. It allows you to work with multiple data files that can be linked together. Power BI Desktop is one of several Power BI platforms developed by Microsoft that enable users to create, share, and use information embedded in the data available to their organization.

**BUSINESS APPLICATION** Using Microsoft Power BI

**Bordon's Plumbing Supply** Archie Bordon founded Bordon's Plumbing Supply in 1993 to provide high-quality plumbing fixtures and parts to stores and contractors in the Midwest. After Archie's son, Mark, joined the company in 2007 as Vice President, the company has grown substantially. Bordon's now has sales territories throughout the United States, Canada, Australia, and several European countries. Bordon's Plumbing Supply will serve as an example for introducing some of the potential uses of Microsoft Power BI to transform the company's data into useful information.

The data that Bordon's Plumbing has available are stored in three Excel files. The first is **Bordon's SalesData**, which contains detailed information about the sales orders for the previous month. The next file is **Bordon's ProductData**, which contains information about Bordon's product catalog. The last file is called **Bordon's Territories**, and it contains data

---

[2] You can access and download Microsoft Power BI Desktop at https://powerbi.microsoft.com/en-us/desktop/. The free version has a limit of 1GB data capacity.

[3] See James Phillips, "Over 500,000 unique users from 45,000 companies across 185 countries helped shape the new Power BI," http://blogs.microsoft.com, July 10, 2015.

[4] See http://www.capterra.com/.

that indicate where Bordon's has sales offices. We will use these three files to demonstrate how to use Power BI.

### Working with Power BI    Power BI has five main components:

- Visualizations
- Data sets
- Reports
- Dashboards
- Tiles

We will first provide a brief look at what each of these components entails, and then we will show some of the key Power BI features.

**Visualizations**    As the name implies, a **visualization** is a pictorial display of the data using charts, graphs, tables, and other visual tools. Power BI contains many visualization options. Figure 18.7 shows some visualizations from the Bordon's Plumbing Supply sales data.

Visualizations vary in complexity—from a single value or table of values, like the product inventory list in Figure 18.7, to something more complex like the map depicting total order quantity by sales territory. In the end, you should select the visual methods that help us best understand the data.

**Data Sets**    A **data set** can be as simple as a single Excel file or a collection of files from a variety of sources. For example, in the Bordon's Plumbing application, the data set consists of three Excel files: **Bordon's SalesData**, **Bordon's Territories**, and **Bordon's ProductData**. Figure 18.8 shows a portion of each of these three files. Once the data files are loaded into Power BI, they are generally referred to as **tables**.

If we are working with large data sets that have many variables, we may want to eliminate some variables (columns) or cases (rows) that are not of interest. This can make it easier to get the desired information from the analysis. This can be done prior to working with Power BI or from within the software. For example, in the Bordon's Plumbing example, if the owners are

**FIGURE 18.7 Bordon's Plumbing Supply Visualization Examples**

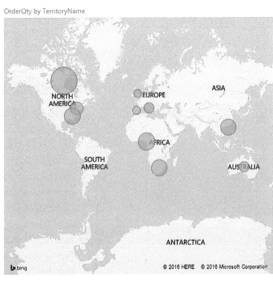

(Power BI Desktop, 2016)

**FIGURE 18.8 Data Set for Bordon's Plumbing Supply Example**

(a) Sales

| CustomerID | ProductID | OrderQty | UnitPrice | SalesPersonID | TerritoryID | Customer Name |
|---|---|---|---|---|---|---|
| 1160 | 10786 | 5 | 451.9589 | 82 | 16 | Alamo Pricing |
| 1160 | 10784 | 5 | 451.9589 | 82 | 16 | Alamo Pricing |
| 1160 | 10788 | 5 | 451.9589 | 82 | 16 | Alamo Pricing |
| 1160 | 10784 | 5 | 451.9589 | 82 | 16 | Alamo Pricing |
| 1214 | 10789 | 5 | 451.9589 | 82 | 16 | Bartsow Stores |
| 1214 | 10787 | 5 | 451.9589 | 82 | 16 | Bartsow Stores |
| 1214 | 10784 | 5 | 451.9589 | 82 | 16 | Bartsow Stores |
| 1269 | 10788 | 5 | 451.9589 | 82 | 16 | Dragon Fly Products |
| 1269 | 10784 | 5 | 451.9589 | 82 | 16 | Dragon Fly Products |
| 1269 | 10788 | 5 | 451.9589 | 82 | 16 | Dragon Fly Products |
| 1017 | 10783 | 5 | 451.9589 | 82 | 16 | Granger Products |
| 1017 | 10788 | 5 | 451.9589 | 82 | 16 | Granger Products |
| 1017 | 10786 | 5 | 451.9589 | 82 | 16 | Granger Products |
| 1017 | 10788 | 5 | 451.9589 | 82 | 16 | Granger Products |
| 1017 | 10782 | 5 | 451.9589 | 82 | 16 | Granger Products |
| 1017 | 10788 | 5 | 451.9589 | 82 | 16 | Granger Products |

(b) Territories

| TerritoryID | TerritoryName | TerritoryGroup |
|---|---|---|
| 11 | Southeast US | North America |
| 12 | Northwest US | North America |
| 13 | Northeast US | North America |
| 14 | Central US | North America |
| 15 | Southwest US | North America |
| 16 | Canada | North America |
| 17 | Italy | Europe |
| 18 | Australia | Pacific |
| 19 | Spain | Europe |
| 20 | United Kingdom | Europe |

(c) Products

| ProductID | Name | ProductNumber | Color | StandardCost | ListPrice | ProductSubcategoryID | ProductCategory | Inventory |
|---|---|---|---|---|---|---|---|---|
| 10727 | Y33 Shower Drain | HL-U509-R | Stainless | 13.0863 | 34.99 | 151 | Accessories | 5460 |
| 10728 | X 55 Shower Drain | HL-U509 | Powder Blue | 13.0863 | 34.99 | 151 | Accessories | 867 |
| 10729 | Shower Drain Seal - Y033 | SO-B909-M | White | 3.3963 | 9.5 | 143 | Accessories | 5211 |
| 10730 | Shower Drain Seal - W980 | SO-B909-L | White | 3.3963 | 9.5 | 143 | Accessories | 4316 |
| 10731 | X 5988 Shower Drain | HL-U509-B | Blue | 13.0863 | 34.99 | 151 | Accessories | 1603 |
| 10732 | X 6988 Shower Drain Seal | CA-1098 | Multi | 6.9223 | 8.99 | 139 | Accessories | 4217 |
| 10733 | 400 CR Toilet Fill Valve | LJ-0192-S | Multi | 38.4923 | 49.99 | 141 | Accessories | 5066 |
| 10734 | 405 CR Toilet Fill Valve | LJ-0192-M | Multi | 38.4923 | 49.99 | 141 | Accessories | 4046 |

(Power BI Desktop, 2016)

not interested in analyzing sales from outside the United States, they might eliminate all non–United States data. We can also create new variables to analyze. For instance, Bordon's Plumbing managers may wish to create a new variable in the Product table that is the ratio ListPrice/StandardCost. We show how to use formulas to add variables later in this section. After we have the data set the way we want it, we are ready to create Power BI reports.

**Reports** A Power BI **report** is a set of visual items that are related to each other. A report can consist of a single page or multiple pages. For example, the owners of Bordon's Plumbing might want to see a report that shows a bar chart of total order quantities from all the sales territories on one page followed by other pages that look at the sales in more detail, with each page representing a territory. Figures 18.9 and 18.10 show two pages for this report.

Figure 18.9 is page 1 of a multiple-page report. It shows that total monthly sales from all sales territories are $26.06 million; it also shows how the sales volume is broken down among

**FIGURE 18.9 Bordon's Plumbing Supply Report— Page 1**

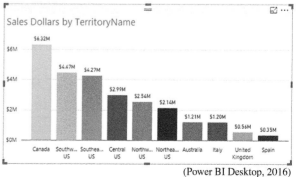

(Power BI Desktop, 2016)

FIGURE 18.10 Bordon's
Plumbing Supply Report—
Page 2

FIGURE 18.10 Bordon's
Plumbing Supply Report—
Page 2

Australia Sales

$1.21M

Australia Sales By Product Catagory

Australia Sales by Customer

Australia Sales

| Color | OrderQty | Sales Dollars |
|---|---|---|
| Blue | 57 | $3,003.13 |
| Chrome | 132 | $217,429.64 |
| Multi | 140 | $7,452.27 |
| Powder Blue | 439 | $352,805.76 |
| Stainless | 669 | $630,561.26 |
| White | 13 | $496.60 |
| **Total** | **1450** | **$1,211,748.66** |

(Power BI Desktop, 2016)

the sales territories. Figure 18.10 is a secondary page with only Australia sales data shown. It shows $1.21 million total sales last month from Australian customers and also breaks down this sales figure by product category, by customer, and by product color.

**Dashboards** In Section 18.1, we indicated that dashboards are real-time graphical and numerical displays of key performance indicators. A dashboard is one or more pages of important information and measures that are commonly shared with others in the organization. There is no specific way to create a dashboard, but three commonly accepted practices are:

- The most important information is displayed as the biggest visual.
- The most important information is in the upper corner.
- Appropriate charts and graphs convey the information effectively.

Figure 18.11 shows a sample dashboard. Product Inventory is the key factor being displayed. The dashboard shows that currently the company has 265,658 units of product in inventory. The total dollar value is $202.94 million. There are four products with inventory levels higher than 6,000 units. Finally, bathtubs account for almost half of the total inventory value at $91.67 million. These values will be updated in real time as new data become available.

**Tiles** A single graph, chart, or table is referred to as a **tile** in Power BI. Four tiles are shown in the dashboard in Figure 18.11. We can move these tiles to different areas on the page, and we can size them to the height and width we deem most appropriate.

We now illustrate how to use Power BI to create visualizations, reports, and dashboards like those shown in the Bordon's Plumbing Supply application.

FIGURE 18.11 Bordon's
Plumbing Supply Dashboard
Example

Total Inventory -Units On Hand

265,658

Inventory (Qty > 6,000 Units)

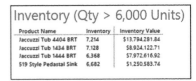

Total Inventory Value

$202.94M

Inventory Value By Product Catagory

(Power BI Desktop, 2016)

## Using Microsoft Power BI Desktop

As you read through the remainder of this section it may be helpful for you to have Power BI Desktop open on your computer so that you can replicate the examples provided.

**The Opening Power BI Screen**   Assuming that we have already downloaded Power BI onto our computer, when we open the software the opening screen will be the one shown in Figure 18.12. We have circled the area on this screen that we will discuss first. To create a new Power BI application, click on **Get Data**. To use a previously created application, click on one of the **Recent Sources** listed.

**OUTCOME** **3**   **Getting the Data**   For our purposes here, we are assuming that we want to create a new Power BI application. Thus, we click on **Get Data**, which opens the screen shown in Figure 18.13.

**FIGURE 18.12**  **Power BI Opening Screen**

(Power BI Desktop, 2016)

**FIGURE 18.13**  **Power BI Get Data Screen**

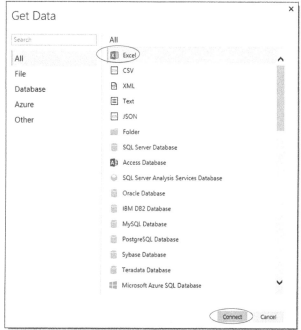

(Power BI Desktop, 2016)

Figure 18.13 shows that Power BI can import a wide range of data including Excel workbooks and CSV files, and a range of databases including Access, SQL, and Sybase. The data for the Bordon's Plumbing Supply application are in Excel format, so we select that file type and then click on the **Connect** button. Three Excel files are required for this application (**Bordon's ProductData**, **Bordon's SalesData**, and **Bordon's Territories**). These files are available on the website for this text where you have accessed other Excel files. Once we locate those files, we select **Bordon's ProductData** and then click on **Open**. We will see the screen shown in Figure 18.14.

Clicking on the **Load** button will bring the Excel file into Power BI, as shown in Figure 18.15. Note the three buttons in the upper left-hand corner of the Power BI "canvas." These

**FIGURE 18.14 Data Screen for the Bordon's ProductData Excel File**

(Power BI Desktop, 2016)

**FIGURE 18.15 Bordon's ProductData Load Screen**

(Power BI Desktop, 2016)

buttons are used to display different views within Power BI: Visualization, Data, and Relationships. In this case, we want to see the data, so we click on the middle button. We are able to see the column headings, which are descriptive of the data in each column.

We now load the other two Excel files using the **Get Data** option on the Power BI **File** tab. These screens are shown in Figures 18.16 and 18.17. As these three files are loaded into Power BI, the file names and **fields** (variables) are listed on the right-hand side, as shown in Figure 18.18. The fields are particularly important because we will be working with them as we create Power BI visualizations for these data. Once a file is loaded into Power BI, it is referred to as a **table**.

**Relationships among the Data Tables** One of the major strengths of business analytics software such as Microsoft Power BI is its capability to pull information from multiple data tables if those tables are related. As we have just seen, the Bordon's Plumbing Supply application involves three data files (tables). Once these tables are loaded, Power BI attempts to **autodetect** relationships among them by checking the variable names (Fields) to see if there is a match. In Figure 18.18, we see that the **ProductData** table has a variable called ProductID. The **SalesData** table also has a variable called ProductID. Thus, these two tables

**FIGURE 18.16 Bordon's ProductData Load Screen**

| ProductID | Name | ProductNumber | Color | StandardCost | ListPrice | ProductSubcategoryID | ProductCategory | Inventory |
|---|---|---|---|---|---|---|---|---|
| 10727 | Y33 Shower Drain | HL-U509-R | Stainless | 13.0863 | 34.99 | 151 | Accessories | 5460 |
| 10728 | X 55 Shower Drain | HL-U509 | Powder Blue | 13.0863 | 34.99 | 151 | Accessories | 867 |
| 10729 | Shower Drain Seal - Y033 | SO-B909-M | White | 3.3963 | 9.5 | 143 | Accessories | 5211 |
| 10730 | Shower Drain Seal - W980 | SO-B909-L | White | 3.3963 | 9.5 | 143 | Accessories | 4316 |
| 10731 | X 5988 Shower Drain | HL-U509-B | Blue | 13.0863 | 34.99 | 151 | Accessories | 1603 |
| 10732 | X 6988 Shower Drain Seal | CA-1098 | Multi | 6.9223 | 8.99 | 139 | Accessories | 4217 |
| 10733 | 400 CR Toilet Fill Valve | LJ-0192-S | Multi | 38.4923 | 49.99 | 141 | Accessories | 5066 |
| 10734 | 405 CR Toilet Fill Valve | LJ-0192-M | Multi | 38.4923 | 49.99 | 141 | Accessories | 4046 |
| 10735 | 420 CR Toilet Fill Valve | LJ-0192-J | Multi | 38.4923 | 49.99 | 141 | Accessories | 5551 |

(Power BI Desktop, 2016)

**FIGURE 18.17 Bordon's Territories Load Screen**

| TerritoryID | TerritoryName | TerritoryGroup |
|---|---|---|
| 11 | Southeast US | North America |
| 12 | Northwest US | North America |
| 13 | Northeast US | North America |
| 14 | Central US | North America |
| 15 | Southwest US | North America |
| 16 | Canada | North America |
| 17 | Italy | Europe |
| 18 | Australia | Pacific |
| 19 | Spain | Europe |
| 20 | United Kingdom | Europe |

(Power BI Desktop, 2016)

**FIGURE 18.18 Bordon's Data Tables and Fields**

(Power BI Desktop, 2016)

**FIGURE 18.19 Data File Relationships for Bordon's Plumbing Example**

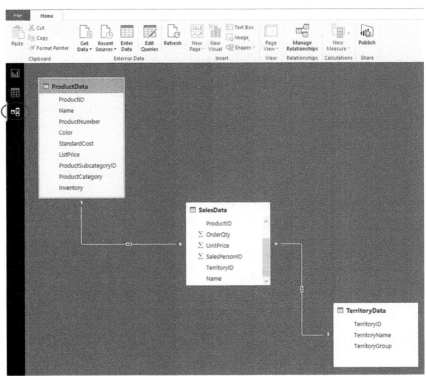

(Power BI Desktop, 2016)

are related based on this common variable. The **SalesData** table and the **TerritoryData** table are related based on the common variable TerritoryID. Power BI shows these relationships visually in Figure 18.19. If the variable names are not close enough for Power BI to establish the relationships, the **Manage Relationships** tool under the **Home** tab can be used to link files using variables that have different names for the same information.

The most common relationship is called **Many to One**. In Figure 18.19, the relationship between SalesData and ProductData is a Many to One relationship, since the sales data contain numerous sales records, many of which may have the same value for ProductID, while each row in the ProductData file has a unique value for the ProductID variable. For a more complete discussion of Power BI relationships, access the Power BI tutorials located at https://powerbi.microsoft.com.

**OUTCOME 4**

**Creating a Visualization** Now that we know that our three tables are related, we can pull information from two or more of them simultaneously to create visualization tiles, reports, and dashboards. To do this, we need to start with a blank "canvas" in the visualization view, as shown in Figure 18.20. On the upper right-hand side of the screen is an array of available Power BI visualization tools.

Refer back to Figure 18.7 where we showed three visualization examples for Bordon's Plumbing Supply. The column chart in the upper left-hand corner shows the total number of units ordered in each of the company's sales territories. To create this visualization, we need to define the variables to be used from the appropriate data files. Figure 18.21 and the following steps show how this is done.

STEP 1 Look to the visualization tool icons on the upper left and select the "clustered column chart." Hover the cursor over these icons to see the name of the tool.

STEP 2 Select the data variable from the appropriate file and drag it to the **Value** location. In this example, the OrderQTY variable contains the data we wish to graph.

STEP 3 Select the variable to be used to group records for the OrderQTY variable and drag it to the **AXIS** location. This will be TerritoryName.

You may want to resize the chart, add titles, sort the data, modify colors, or make other changes to make the visualization more appropriate for conveying the information.

**FIGURE 18.20** Blank Canvas

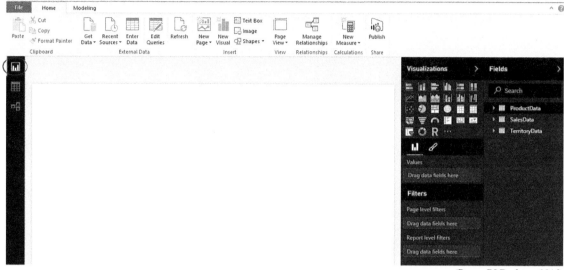

(Power BI Desktop, 2016)

STEP 4 Resize the chart by selecting it and dragging from a side or from a corner.

STEP 5 Sort the columns by selecting the chart and then clicking on the three dots in the upper right-hand corner of the chart. Select **SORT BY** and then pick the variable (OrderQTY).

The results of Steps 4 and 5 are illustrated in Figure 18.22.

STEP 6 Add a title to the graph by clicking the pencil icon that is below the visualization icons on the right side of the screen. (See Figure 18.23.) Make sure the graph is selected. Select and expand the **Title** options. Enter the desired title, and then pick the desired font size and font color.

STEP 7 Modify the colors on the columns by selecting and expanding the **Data Colors**/. You might also want to add annotations and data values to the graph to provide information for the user.

**FIGURE 18.21** Creating a Bar Chart—Bordon's Plumbing Supply Example

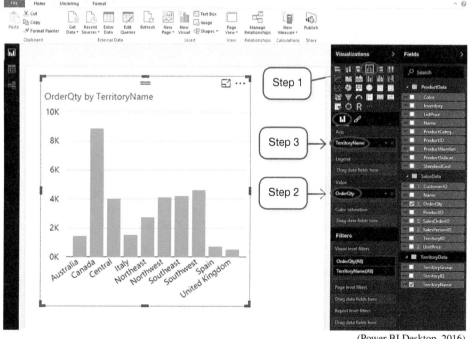

(Power BI Desktop, 2016)

FIGURE 18.22 **Resizing and Sorting**

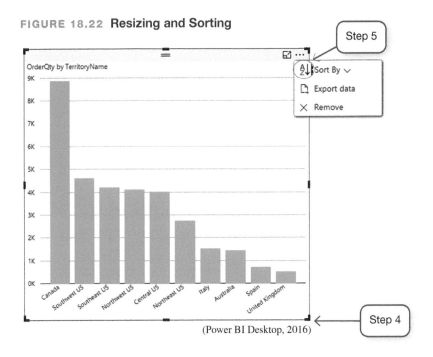

(Power BI Desktop, 2016)

FIGURE 18.23 **Titles and Colors**

(Power BI Desktop, 2016)

STEP 8  Add data values by selecting and expanding **Data Labels**. Make choices for how the data labels are displayed.

STEP 9  Add annotations by using the **Text Box** tool. We want to call out the "United States Territories," which are all the same color.

Figure 18.24 shows the annotation and data labels.

At this point, we suggest that you open Microsoft Power BI Desktop and the three Bordon's Plumbing Supply data tables and replicate the steps we have just shown. Then try using different visualization tools and some of the graph modification features that we have shown. You might also look at Figure 18.9 and create the other visualizations that are shown there.

**FIGURE 18.24 Data Labels and Annotations**

(Power BI Desktop, 2016)

**OUTCOME 5**

**Creating New Variables**  In the process of analyzing data, we often need to create new variables (new columns) from existing variables in the data set. For example, the managers at Bordon's Plumbing Supply might be interested in creating a new variable called ProfitMargin that is the difference between a product's list price and its standard cost. Both of these variables are in the **ProductData** table. To create the new variable in Power BI, the managers can use the **DAX** (Data Analysis Expressions) programming language. The DAX methodology is very similar to how equations are created in Excel and includes constants, operators, and functions that allow users to create new variables from the data they already have available.

**STEP 1**  To create the new variable, make sure the appropriate table is selected and the data are visible (see Figure 18.15). Then click the down arrow on the **New Measure** tool on the Home tab and select **New Column**. This creates a new blank column.

**STEP 2**  In the section below the tools and above the data, use the DAX language to write the equation for the new variable. To do this, start by typing the name of the new variable followed by the = sign (ProfitMargin =). Next, click on variable name, ListPrice, which will result in the table name and variable name being automatically inserted, giving ProfitMargin = ProductData[ListPrice]. Next, insert a − sign and click on the variable name, StandardCost. This will result in the completed equation:

ProfitMargin = ProductData[ListPrice] − ProductData[StandardCost]

After the Enter key is pressed, values for the new variable, ProfitMargin, will be generated for each product, as shown in Figure 18.25. Note that the **Format** tool on the **Modeling** tab is used to display the data in dollars and cents.

Bordon's Plumbing managers might also wish to create a new column in the **SalesData** table that is the total dollar value (Sales Dollars) for each customer order. This variable is computed by multiplying the OrderQty variable by ListPrice in the **ProductData** table. Following Steps 1 and 2 above, the DAX equation for creating this variable is

Sales Dollars = SalesData[OrderQty]*RELATED(ProductData[ListPrice])

The term **RELATED** is used to access information from the two related tables, SalesData and ProductData. The SalesData table has multiple customers who have ordered various quantities (OrderQTY) with the same ProductID. The ProductData table provides the

FIGURE 18.25 **New Variable Created**

(Power BI Desktop, 2016)

ListPrice for each individual ProductID. The RELATED command indicates that there is a Many to One relationship between the two tables.

**Creating a Measure** One of the most useful and powerful features of Power BI is its capability to develop **measures**. Measures are a feature that allows you to perform calculations (sums, average, etc.) on your data when you interact with the visuals that have been created. We can illustrate the creation of measures using the new variables that were just created. Bordon's Plumbing managers might want to create a visualization that shows the average OrderValue and average ProfitMargin.

STEP 1 In many cases, measures are created automatically when we drag a variable onto the canvas. To illustrate, we can drag the new Sales Dollars variable onto a blank canvas as shown in Figure 18.26. (For some variables, such as text variables, that do not automatically create a measure, we need to drag the variable to the **Value** space in order to begin developing the measure.) What we have at this point is a default column chart that shows the sum of the individual sales dollars associated with all orders during the month. Power BI defaults to the sum for quantitative variables rather than displaying the individual values. The data behind this sum come from the **SalesData** table, which contains 4,980 individual sales transactions for the month. Many data sets that you will encounter are much larger than this, so displaying the raw data values is not practical.

STEP 2 The managers are interested in the average sales dollars per order. The average can be calculated by changing the **Value** display as shown in Figure 18.27. Note that the column chart has now changed from displaying the sum of all order values to the average of all order values.

FIGURE 18.26 **Creating the Sales Dollars Measure—Step 1**

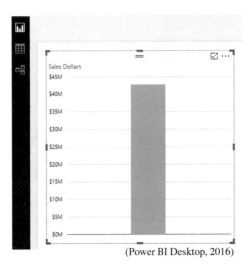

(Power BI Desktop, 2016)

**FIGURE 18.27  Calculating an Average Measure**

(Power BI Desktop, 2016)

**FIGURE 18.28  Changing the Measure's Context—Average Order Value by Sales Territory**

(Power BI Desktop, 2016)

We can modify a measure's *context* if we wish to display the measure in a different way. For instance, the managers may want to see Sales Dollars displayed by TerritoryName.

STEP 3  To change the context of the measure so that it is displayed by sales territory, drag the variable TerritoryName onto the column chart, giving the result shown in Figure 18.28.

We can create new measures from the original variables in our data set or from variables that we created earlier. Suppose the Bordon's Plumbing managers want to track profits, where the profit on each sale is the difference between the sales dollars for an order and the standard cost associated with the order. Previously we created the Sales Dollars variable. Now we need to create a variable in the **SalesData** table called Order Cost. The DAX command to do this is

Order Cost = SalesData[OrderQty] * RELATED(ProductData[StandardCost])

We calculate the Total Profit measure with the following DAX command:

Total Profit = SUM(SalesData[Sales Dollars]) − SUM(SalesData[Order Cost])

**FIGURE 18.29** **New Measure—Profit**

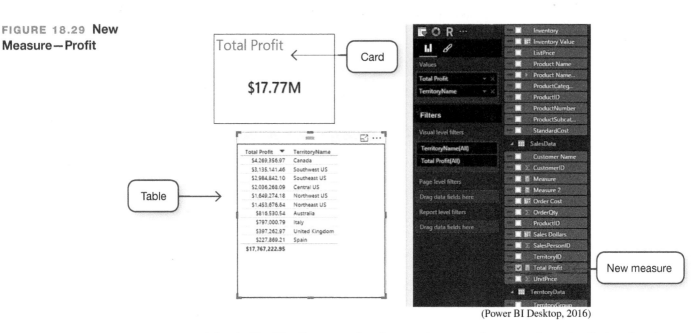

(Power BI Desktop, 2016)

After the Total Profit measure is reformatted as general currency, it can be displayed as shown in Figure 18.29. Notice that we have used the **Card** visualization tool to show the total profit and the **Table** tool to display the profits by sales region. To create the table, we first select the new measure, Total Profit, and the **Table** tool. Then we drag the Territory Name variable to the table.

The DAX language is very powerful, and the examples presented here have only scratched the surface. To learn more about DAX, go to the Guided Learning option under the Learning tab at https://powerbi.microsoft.com/en-us/.

**Creating Multiple-Page Reports** In Power BI, a report consists of one or more pages that offer multiple perspectives on the data using a variety of visualizations. Each page in a report may have a single graph, chart, or table, or it may have a variety of different visualizations. A given data set can have multiple reports attached to it, but only one report may be open at a time. The examples we have provided so far are actually reports. Once you save your work as a Power BI Desktop file, it is considered to be a report.

Suppose we start with the visualizations in Figure 18.30 and think of this as the first page of our report for Bordon's Plumbing Supply. According to the data, the company earned $17.77 million in profit this month. Most of the profit ($15.3 million) was generated in North America. The territory with the highest profit this month was Canada ($4.3 million).

**FIGURE 18.30** **Bordon's Plumbing Supply—Report Page 1**

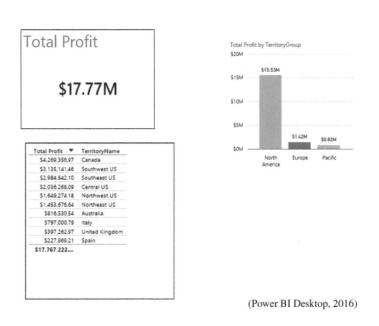

(Power BI Desktop, 2016)

**FIGURE 18.31** Bordon's Plumbing Supply—Report Page 2—Sliced by Territory Group

(Power BI Desktop, 2016)

Suppose we want to look at these data in more detail. To add a second page to the report, we just click on the **+** tab at the bottom left of the screen. Subsequent pages of the report will allow the managers to examine each of the three territory groups in more detail. To do this, we can make use of the Power BI **filters**. In particular, we could use **Page Level Filters** to create three additional pages, one for each of the three territory groups. An alternative approach is to make use of the **Slicer** tool. To do this, we select the Slicer tool from the Power BI icons and drag the TerritoryGroup variable into the Slicer space. Then we create the visualizations we want to display. Figure 18.31 is Page 2 of the report, which shows the Total Profit in one card and a table with the total units sold and total sales dollars for the territory group selected. In Figure 18.31, we have selected North America.

This section has introduced you to the basics of Power BI Desktop. As you begin to use this software, you will see that it has many more features that allow you to create a wide range of visualizations to help you transform your data into useful information. You can learn much more about Power BI by going to the Learning Tab on the Microsoft website, https://powerbi.microsoft.com/en-us/. Regardless of which business analytics software you use, the key is to use it to transform the data into information that will help you make better business decisions.

## 18.2 EXERCISES

### Computer Software Exercises

**18-13.** Refer to the Bordon's Plumbing Supply application in this section, open Microsoft Power BI Desktop, and use **Get Data** to open three files: **Bordon's SalesData**, **Bordon's Territories**, and **Bordon's ProductData**. Use the appropriate button in Power BI to illustrate the links between these three files.

**18-14.** Referring to Exercise 18-13, open a blank canvas and create a "card" that shows the number of sales transactions last month.

**18-15.** Referring to Exercises 18-13 and 18-14, use Power BI to construct a horizontal bar chart that shows the total number of units ordered by territory group. Add labels to the bars.

**18-16.** Referring to the Bordon's Plumbing Supply application, change the format for the Standard Cost variable in the ProductData table to one showing the values in dollars and cents. Then develop a "card" display that shows the mean standard cost.

**18-17.** Referring to Exercise 18-16, develop a table that shows average Standard Cost, maximum and minimum Standard Cost, and standard deviation for Standard Cost.

**18-18.** Referring to Exercises 18-13 through 18-17, create a new column called Sales Margin in the SalesData table that is the difference between ListPrice and UnitPrice. Note that ListPrice is found in the ProductData table, so you will need to use the RELATED command. Once you have created the new column, set the column, format as General Currency with two decimal places, and create a clustered column chart that shows the average Sales Margin by Territory Group for the orders that Bordon's received this month. Add a title and data labels to the chart.

**18-19.** Referring to Exercise 18-18, create a map that displays the average Sales Margin per order displayed by Territory Name.

**18-20.** Referring to Exercises 18-13 through 18-19, create a new column called Sales Dollars (Sales Dollars = Order QTY * List Price) and then construct a table that shows the following fields for any sales dollars amount exceeding $50,000:

> Customer Name
> OrderQty
> Product Name
> Sales Dollars (Sort from high to low)

**18-21.** Referring to Exercise 18-20, construct a pie chart that shows total sales dollars broken down by TerritoryGroup. Add value labels and discuss.

**Exercises 18-22 through 18-30** relate to data that are collected and maintained by the Art Form Corporation. Suppose that Art Form maintained employee records starting in 1987 and the Excel file called **Employees** contains information about every employee hired by the company between 1987 and 2013, when Art Form was sold to a European company. A second Excel file called **Job Classification** contains the descriptions of the different job classes used by Art Form. A third file called **Employee Race** contains the race codes and descriptions.

**18-22.** After opening Power BI, open each of the three files. Change the names of the tables to the following: Race Codes, Job Codes, and Master Employee Data. Show how the three files are related.

**18-23.** Create a clustered column chart that shows the number of employees hired each year. Make sure the chart has a title. In what year were the most employees hired? How many were hired that year?

**18-24.** Referring to Exercise 18-23, create a stacked column chart that illustrates how many employees of each gender (male and female) were hired in each year. In what year were the most females hired? How many females were hired that year?

**18-25.** Create a table that illustrates the mean and standard deviation for the number of days each employee has been with the company by gender.

**18-26.** Referring to Exercise 18-25, create the same table but only for employees who are no longer working for the company. Discuss.

**18-27.** Create a chart that shows the average age when the employee was hired broken down by gender for those employees who stayed with the company less than five years.

**18-28.** Create a pie chart that breaks down the number of employees hired in 2010 by race classification. What are the two races that were most frequently hired in 2010?

**18-29.** Create a card display that shows the mean last job performance rating for all emplyees who have worked for the company since 1987. Also create a card display that shows how many employees had a performance rating of 100.

**18-30.** Prepare a report that describes the Art Form employee data. Select two or three themes and prepare a multiple-page Power BI visualization report using a variety of visual tools. Prepare a Word document that discusses the visualizations you created.

# 18 Overview

## Summary

**OUTCOME 1** Distinguish between business intelligence and business analytics.

**OUTCOME 2** Know the difference between descriptive analytics and predictive analytics.

- The process of collecting data from inside or outside an organization that are thought to relate to or influence the performance of the organization is called **business intelligence**.
- The use of tools, techniques, and methodologies to examine data for the purpose of looking back to the past to answer questions about what *has* happened is called **descriptive analytics**.
- The application of statistical models and forecasting methods to data to answer questions about what *could* happen is called **predictive analytics**.
- A measure of how well an organization is achieving its most important business objectives is called a **key performance indicator (KPI)**.

**OUTCOME 3** Be able to load data files into Power BI and establish relationship links between files.

**OUTCOME 4** Create dashboards and reports with appropriate visualizations using Power BI.

**OUTCOME 5** Use Power BI to create new variables and new measures.

- The five main Power BI components are visualizations, data sets, reports, dashboards, and tiles.
- A **visualization** is a pictorial display of data using charts, graphs, tables, and other visual tools.
- A **data set** is a collection of one or more files that contain data of interest.
- A set of visual items that are related to each other consisting of one or more pages in Power BI is called a **report**.
- One or more pages that contain visualizations of data fields and measures that are typically updated in real time are referred to as **dashboards**.
- A **tile** is a single graph, chart, or table in Microsoft Power BI.

## Key Terms

## Case 18.1 — New York City Taxi Trips

New York City Taxi data were made available to the public through the efforts of Christopher Whong, Civic Technologist & Developer at the NYC Department of City Planning. Whong used the Freedom of Information Act to obtain individual taxi trip data for New York City for the year 2013. These data are available at www.andresmh.com/nyctaxitrips/. **New York City Taxi.PBIX** is a Microsoft Power BI report file that contains a portion of the NYC taxi data for 2013. This file contains data for 14,776,615 individual trips for the month of January 2013. In addition to the taxi medallion and hack license numbers, the variables included are the system used by the taxi to process credit card payments (CMT = Creative Mobile Technologies and VTS = Verifone), passenger pickup date and time, payment type (CSH = cash, CRD = credit card), fare amount, surcharge amount, Metropolitan Transportation Authority (MTA) tax, tip amount, toll charge, total fare, and day of week code.

## Required Tasks:

1. Open Microsoft Power BI and click on **Open Other Reports**. Locate the file **New York City Taxi.PBIX** in the same web location as other files used in this text.

2. Assume you are working for an agency that oversees taxi services in New York City. After examining the data fields, develop a plan for analyzing these data. Identify a number of questions that you believe can be answered from these data. Make sure that you think beyond the surface information and that your plan requires you to drill down into the data.

3. Construct a Microsoft Power BI report consisting of multiple pages and a variety of appropriate visualizations to answer the questions you posed in part 2. Create at least one new variable and one measure in your analysis.

# Appendix Tables

| | | | | | | | | |
|---|---|---|---|---|---|---|---|---|
| 1511 | 4745 | 8716 | 2793 | 9142 | 4958 | 5245 | 8312 | 8925 |
| 6249 | 7073 | 0460 | 0819 | 0729 | 6806 | 2713 | 6595 | 5149 |
| 2587 | 4800 | 3455 | 7565 | 1196 | 7768 | 6137 | 4941 | 0488 |
| 0168 | 1379 | 7838 | 7487 | 7420 | 5285 | 8045 | 6679 | 1361 |
| 9664 | 9021 | 4990 | 5570 | 4697 | 7939 | 5842 | 5353 | 7503 |
| 1384 | 4981 | 2708 | 6437 | 2298 | 6230 | 7443 | 9425 | 5384 |
| 6390 | 8953 | 4292 | 7372 | 7197 | 2121 | 6538 | 2093 | 7629 |
| 6944 | 8134 | 0704 | 8500 | 6996 | 3492 | 4397 | 8802 | 3253 |
| 3610 | 3119 | 7442 | 6218 | 7623 | 0546 | 8394 | 3286 | 4463 |
| 9865 | 0028 | 1783 | 9029 | 2858 | 8737 | 7023 | 0444 | 8575 |
| 7044 | 6712 | 7530 | 0018 | 0945 | 8803 | 4467 | 0979 | 1342 |
| 9304 | 4857 | 5476 | 8386 | 1540 | 5760 | 9815 | 7191 | 3291 |
| 1717 | 8278 | 0072 | 2636 | 3217 | 1693 | 6081 | 1330 | 3458 |
| 2461 | 3598 | 5173 | 9666 | 6165 | 7438 | 6805 | 2357 | 6994 |
| 8240 | 9856 | 0075 | 7599 | 8468 | 7653 | 6272 | 0573 | 4344 |
| 1697 | 6805 | 1386 | 2340 | 6694 | 9786 | 0536 | 6423 | 1083 |
| 4695 | 2251 | 8962 | 5638 | 9459 | 5578 | 0676 | 2276 | 4724 |
| 3056 | 8558 | 3020 | 7509 | 5105 | 4283 | 5390 | 5715 | 8405 |
| 6887 | 9035 | 8520 | 6571 | 3233 | 7175 | 2859 | 1615 | 3349 |
| 1267 | 8824 | 5588 | 2821 | 1247 | 0967 | 4355 | 1385 | 0727 |
| 4369 | 9267 | 9377 | 8205 | 6479 | 7002 | 0649 | 4731 | 7086 |
| 2888 | 0333 | 5347 | 4849 | 5526 | 2975 | 5295 | 5071 | 6011 |
| 9893 | 7251 | 6243 | 4617 | 9256 | 4039 | 4800 | 9393 | 3263 |
| 8927 | 3977 | 6054 | 5979 | 8566 | 8120 | 2566 | 4449 | 2414 |
| 2676 | 7064 | 2198 | 3234 | 3796 | 5506 | 4462 | 5121 | 9052 |
| 0775 | 7316 | 2249 | 5606 | 9411 | 3818 | 5268 | 7652 | 6098 |
| 3828 | 9178 | 3726 | 0743 | 4075 | 3560 | 9542 | 3922 | 7688 |
| 3281 | 3419 | 6660 | 7968 | 1238 | 2246 | 2164 | 4567 | 1801 |
| 0328 | 7471 | 5352 | 2019 | 5842 | 1665 | 5939 | 6337 | 9102 |
| 8406 | 1826 | 8437 | 3078 | 9068 | 1425 | 1232 | 0573 | 7751 |
| 7076 | 8418 | 6778 | 1292 | 2019 | 3506 | 7474 | 0141 | 6544 |
| 0446 | 8641 | 3249 | 5431 | 4068 | 6045 | 1939 | 5626 | 1867 |
| 3719 | 9712 | 7472 | 1517 | 8850 | 6862 | 6990 | 5475 | 6227 |
| 5648 | 0563 | 6346 | 1981 | 9512 | 0659 | 5694 | 6668 | 2563 |
| 3694 | 8582 | 3434 | 4052 | 8392 | 3883 | 5126 | 0477 | 4034 |
| 3554 | 9876 | 4249 | 9473 | 9085 | 6594 | 2434 | 9453 | 8883 |
| 4934 | 8446 | 4646 | 2054 | 1136 | 1023 | 6295 | 6483 | 9915 |
| 7835 | 1506 | 0019 | 5011 | 0563 | 4450 | 1466 | 6334 | 2606 |
| 1098 | 2113 | 8287 | 3487 | 8250 | 2269 | 1876 | 3684 | 8856 |
| 1186 | 2685 | 7225 | 8311 | 3835 | 8059 | 9163 | 2539 | 6487 |
| 4618 | 1522 | 0627 | 0448 | 0669 | 4086 | 4083 | 0881 | 4270 |
| 5529 | 4173 | 5711 | 7419 | 2535 | 5876 | 8435 | 2564 | 3031 |
| 0754 | 5808 | 8458 | 2218 | 9180 | 6213 | 5280 | 4753 | 0696 |
| 5865 | 0806 | 2070 | 7986 | 4800 | 3076 | 2866 | 0515 | 7417 |
| 6168 | 8963 | 0235 | 1514 | 7875 | 2176 | 3095 | 1171 | 7892 |
| 7479 | 4144 | 6697 | 2255 | 5465 | 7233 | 4981 | 3553 | 8144 |
| 4608 | 6576 | 9422 | 4198 | 2578 | 1701 | 4764 | 7460 | 3509 |
| 0654 | 2483 | 6001 | 4486 | 4941 | 1500 | 3502 | 9693 | 1956 |
| 3000 | 9694 | 6616 | 5599 | 7759 | 1581 | 9896 | 2312 | 8140 |
| 2686 | 3675 | 5760 | 2918 | 0185 | 7364 | 9985 | 5930 | 9869 |
| 4713 | 4121 | 5144 | 5164 | 8104 | 0403 | 4984 | 3877 | 8772 |
| 9281 | 6522 | 7916 | 8941 | 6710 | 1670 | 1399 | 5961 | 4714 |
| 5736 | 9419 | 5022 | 6955 | 3356 | 5732 | 1042 | 0527 | 7441 |
| 2383 | 0408 | 2821 | 7313 | 5781 | 6951 | 7181 | 0608 | 2864 |
| 8740 | 8038 | 7284 | 6054 | 2246 | 1674 | 9984 | 0355 | 0775 |

## Cumulative Binomial Distribution Table

$$P(x \le X) = \sum_{i=0}^{X} \frac{n!}{i!(n-i)!} p^i (1-p)^{n-i}$$

### n = 1

| x | p = 0.01 | p = 0.02 | p = 0.03 | p = 0.04 | p = 0.05 | p = 0.06 | p = 0.07 | p = 0.08 | p = 0.09 |
|---|---|---|---|---|---|---|---|---|---|
| 0 | 0.9900 | 0.9800 | 0.9700 | 0.9600 | 0.9500 | 0.9400 | 0.9300 | 0.9200 | 0.9100 |
| 1 | 1.0000 | 1.0000 | 1.0000 | 1.0000 | 1.0000 | 1.0000 | 1.0000 | 1.0000 | 1.0000 |

| x | p = 0.10 | p = 0.15 | p = 0.20 | p = 0.25 | p = 0.30 | p = 0.35 | p = 0.40 | p = 0.45 | p = 0.50 |
|---|---|---|---|---|---|---|---|---|---|
| 0 | 0.9000 | 0.8500 | 0.8000 | 0.7500 | 0.7000 | 0.6500 | 0.6000 | 0.5500 | 0.5000 |
| 1 | 1.0000 | 1.0000 | 1.0000 | 1.0000 | 1.0000 | 1.0000 | 1.0000 | 1.0000 | 1.0000 |

| x | p = 0.55 | p = 0.60 | p = 0.65 | p = 0.70 | p = 0.75 | p = 0.80 | p = 0.85 | p = 0.90 | p = 0.91 |
|---|---|---|---|---|---|---|---|---|---|
| 0 | 0.4500 | 0.4000 | 0.3500 | 0.3000 | 0.2500 | 0.2000 | 0.1500 | 0.1000 | 0.0900 |
| 1 | 1.0000 | 1.0000 | 1.0000 | 1.0000 | 1.0000 | 1.0000 | 1.0000 | 1.0000 | 1.0000 |

| x | p = 0.92 | p = 0.93 | p = 0.94 | p = 0.95 | p = 0.96 | p = 0.97 | p = 0.98 | p = 0.99 | p = 1.00 |
|---|---|---|---|---|---|---|---|---|---|
| 0 | 0.0800 | 0.0700 | 0.0600 | 0.0500 | 0.0400 | 0.0300 | 0.0200 | 0.0100 | 0.0000 |
| 1 | 1.0000 | 1.0000 | 1.0000 | 1.0000 | 1.0000 | 1.0000 | 1.0000 | 1.0000 | 1.0000 |

### n = 2

| x | p = 0.01 | p = 0.02 | p = 0.03 | p = 0.04 | p = 0.05 | p = 0.06 | p = 0.07 | p = 0.08 | p = 0.09 |
|---|---|---|---|---|---|---|---|---|---|
| 0 | 0.9801 | 0.9604 | 0.9409 | 0.9216 | 0.9025 | 0.8836 | 0.8649 | 0.8464 | 0.8281 |
| 1 | 0.9999 | 0.9996 | 0.9991 | 0.9984 | 0.9975 | 0.9964 | 0.9951 | 0.9936 | 0.9919 |
| 2 | 1.0000 | 1.0000 | 1.0000 | 1.0000 | 1.0000 | 1.0000 | 1.0000 | 1.0000 | 1.0000 |

| x | p = 0.10 | p = 0.15 | p = 0.20 | p = 0.25 | p = 0.30 | p = 0.35 | p = 0.40 | p = 0.45 | p = 0.50 |
|---|---|---|---|---|---|---|---|---|---|
| 0 | 0.8100 | 0.7225 | 0.6400 | 0.5625 | 0.4900 | 0.4225 | 0.3600 | 0.3025 | 0.2500 |
| 1 | 0.9900 | 0.9775 | 0.9600 | 0.9375 | 0.9100 | 0.8775 | 0.8400 | 0.7975 | 0.7500 |
| 2 | 1.0000 | 1.0000 | 1.0000 | 1.0000 | 1.0000 | 1.0000 | 1.0000 | 1.0000 | 1.0000 |

| x | p = 0.55 | p = 0.60 | p = 0.65 | p = 0.70 | p = 0.75 | p = 0.80 | p = 0.85 | p = 0.90 | p = 0.91 |
|---|---|---|---|---|---|---|---|---|---|
| 0 | 0.2025 | 0.1600 | 0.1225 | 0.0900 | 0.0625 | 0.0400 | 0.0225 | 0.0100 | 0.0081 |
| 1 | 0.6975 | 0.6400 | 0.5775 | 0.5100 | 0.4375 | 0.3600 | 0.2775 | 0.1900 | 0.1719 |
| 2 | 1.0000 | 1.0000 | 1.0000 | 1.0000 | 1.0000 | 1.0000 | 1.0000 | 1.0000 | 1.0000 |

| x | p = 0.92 | p = 0.93 | p = 0.94 | p = 0.95 | p = 0.96 | p = 0.97 | p = 0.98 | p = 0.99 | p = 1.00 |
|---|---|---|---|---|---|---|---|---|---|
| 0 | 0.0064 | 0.0049 | 0.0036 | 0.0025 | 0.0016 | 0.0009 | 0.0004 | 0.0001 | 0.0000 |
| 1 | 0.1536 | 0.1351 | 0.1164 | 0.0975 | 0.0784 | 0.0591 | 0.0396 | 0.0199 | 0.0000 |
| 2 | 1.0000 | 1.0000 | 1.0000 | 1.0000 | 1.0000 | 1.0000 | 1.0000 | 1.0000 | 1.0000 |

### n = 3

| x | p = 0.01 | p = 0.02 | p = 0.03 | p = 0.04 | p = 0.05 | p = 0.06 | p = 0.07 | p = 0.08 | p = 0.09 |
|---|---|---|---|---|---|---|---|---|---|
| 0 | 0.9703 | 0.9412 | 0.9127 | 0.8847 | 0.8574 | 0.8306 | 0.8044 | 0.7787 | 0.7536 |
| 1 | 0.9997 | 0.9988 | 0.9974 | 0.9953 | 0.9928 | 0.9896 | 0.9860 | 0.9818 | 0.9772 |
| 2 | 1.0000 | 1.0000 | 1.0000 | 0.9999 | 0.9999 | 0.9998 | 0.9997 | 0.9995 | 0.9993 |
| 3 | 1.0000 | 1.0000 | 1.0000 | 1.0000 | 1.0000 | 1.0000 | 1.0000 | 1.0000 | 1.0000 |

| x | p = 0.10 | p = 0.15 | p = 0.20 | p = 0.25 | p = 0.30 | p = 0.35 | p = 0.40 | p = 0.45 | p = 0.50 |
|---|---|---|---|---|---|---|---|---|---|
| 0 | 0.7290 | 0.6141 | 0.5120 | 0.4219 | 0.3430 | 0.2746 | 0.2160 | 0.1664 | 0.1250 |
| 1 | 0.9720 | 0.9393 | 0.8960 | 0.8438 | 0.7840 | 0.7183 | 0.6480 | 0.5748 | 0.5000 |
| 2 | 0.9990 | 0.9966 | 0.9920 | 0.9844 | 0.9730 | 0.9571 | 0.9360 | 0.9089 | 0.8750 |
| 3 | 1.0000 | 1.0000 | 1.0000 | 1.0000 | 1.0000 | 1.0000 | 1.0000 | 1.0000 | 1.0000 |

| x | p = 0.55 | p = 0.60 | p = 0.65 | p = 0.70 | p = 0.75 | p = 0.80 | p = 0.85 | p = 0.90 | p = 0.91 |
|---|---|---|---|---|---|---|---|---|---|
| 0 | 0.0911 | 0.0640 | 0.0429 | 0.0270 | 0.0156 | 0.0080 | 0.0034 | 0.0010 | 0.0007 |
| 1 | 0.4253 | 0.3520 | 0.2818 | 0.2160 | 0.1563 | 0.1040 | 0.0608 | 0.0280 | 0.0228 |
| 2 | 0.8336 | 0.7840 | 0.7254 | 0.6570 | 0.5781 | 0.4880 | 0.3859 | 0.2710 | 0.2464 |
| 3 | 1.0000 | 1.0000 | 1.0000 | 1.0000 | 1.0000 | 1.0000 | 1.0000 | 1.0000 | 1.0000 |

| x | p = 0.92 | p = 0.93 | p = 0.94 | p = 0.95 | p = 0.96 | p = 0.97 | p = 0.98 | p = 0.99 | p = 1.00 |
|---|---|---|---|---|---|---|---|---|---|
| 0 | 0.0005 | 0.0003 | 0.0002 | 0.0001 | 0.0001 | 0.0000 | 0.0000 | 0.0000 | 0.0000 |
| 1 | 0.0182 | 0.0140 | 0.0104 | 0.0073 | 0.0047 | 0.0026 | 0.0012 | 0.0003 | 0.0000 |
| 2 | 0.2213 | 0.1956 | 0.1694 | 0.1426 | 0.1153 | 0.0873 | 0.0588 | 0.0297 | 0.0000 |
| 3 | 1.0000 | 1.0000 | 1.0000 | 1.0000 | 1.0000 | 1.0000 | 1.0000 | 1.0000 | 1.0000 |

*(continued)*

### $n = 4$

| x | $p = 0.01$ | $p = 0.02$ | $p = 0.03$ | $p = 0.04$ | $p = 0.05$ | $p = 0.06$ | $p = 0.07$ | $p = 0.08$ | $p = 0.09$ |
|---|---|---|---|---|---|---|---|---|---|
| 0 | 0.9606 | 0.9224 | 0.8853 | 0.8493 | 0.8145 | 0.7807 | 0.7481 | 0.7164 | 0.6857 |
| 1 | 0.9994 | 0.9977 | 0.9948 | 0.9909 | 0.9860 | 0.9801 | 0.9733 | 0.9656 | 0.9570 |
| 2 | 1.0000 | 1.0000 | 0.9999 | 0.9998 | 0.9995 | 0.9992 | 0.9987 | 0.9981 | 0.9973 |
| 3 | 1.0000 | 1.0000 | 1.0000 | 1.0000 | 1.0000 | 1.0000 | 1.0000 | 1.0000 | 0.9999 |
| 4 | 1.0000 | 1.0000 | 1.0000 | 1.0000 | 1.0000 | 1.0000 | 1.0000 | 1.0000 | 1.0000 |

| x | $p = 0.10$ | $p = 0.15$ | $p = 0.20$ | $p = 0.25$ | $p = 0.30$ | $p = 0.35$ | $p = 0.40$ | $p = 0.45$ | $p = 0.50$ |
|---|---|---|---|---|---|---|---|---|---|
| 0 | 0.6561 | 0.5220 | 0.4096 | 0.3164 | 0.2401 | 0.1785 | 0.1296 | 0.0915 | 0.0625 |
| 1 | 0.9477 | 0.8905 | 0.8192 | 0.7383 | 0.6517 | 0.5630 | 0.4752 | 0.3910 | 0.3125 |
| 2 | 0.9963 | 0.9880 | 0.9728 | 0.9492 | 0.9163 | 0.8735 | 0.8208 | 0.7585 | 0.6875 |
| 3 | 0.9999 | 0.9995 | 0.9984 | 0.9961 | 0.9919 | 0.9850 | 0.9744 | 0.9590 | 0.9375 |
| 4 | 1.0000 | 1.0000 | 1.0000 | 1.0000 | 1.0000 | 1.0000 | 1.0000 | 1.0000 | 1.0000 |

| x | $p = 0.55$ | $p = 0.60$ | $p = 0.65$ | $p = 0.70$ | $p = 0.75$ | $p = 0.80$ | $p = 0.85$ | $p = 0.90$ | $p = 0.91$ |
|---|---|---|---|---|---|---|---|---|---|
| 0 | 0.0410 | 0.0256 | 0.0150 | 0.0081 | 0.0039 | 0.0016 | 0.0005 | 0.0001 | 0.0001 |
| 1 | 0.2415 | 0.1792 | 0.1265 | 0.0837 | 0.0508 | 0.0272 | 0.0120 | 0.0037 | 0.0027 |
| 2 | 0.6090 | 0.5248 | 0.4370 | 0.3483 | 0.2617 | 0.1808 | 0.1095 | 0.0523 | 0.0430 |
| 3 | 0.9085 | 0.8704 | 0.8215 | 0.7599 | 0.6836 | 0.5904 | 0.4780 | 0.3439 | 0.3143 |
| 4 | 1.0000 | 1.0000 | 1.0000 | 1.0000 | 1.0000 | 1.0000 | 1.0000 | 1.0000 | 1.0000 |

| x | $p = 0.92$ | $p = 0.93$ | $p = 0.94$ | $p = 0.95$ | $p = 0.96$ | $p = 0.97$ | $p = 0.98$ | $p = 0.99$ | $p = 1.00$ |
|---|---|---|---|---|---|---|---|---|---|
| 0 | 0.0000 | 0.0000 | 0.0000 | 0.0000 | 0.0000 | 0.0000 | 0.0000 | 0.0000 | 0.0000 |
| 1 | 0.0019 | 0.0013 | 0.0008 | 0.0005 | 0.0002 | 0.0001 | 0.0000 | 0.0000 | 0.0000 |
| 2 | 0.0344 | 0.0267 | 0.0199 | 0.0140 | 0.0091 | 0.0052 | 0.0023 | 0.0006 | 0.0000 |
| 3 | 0.2836 | 0.2519 | 0.2193 | 0.1855 | 0.1507 | 0.1147 | 0.0776 | 0.0394 | 0.0000 |
| 4 | 1.0000 | 1.0000 | 1.0000 | 1.0000 | 1.0000 | 1.0000 | 1.0000 | 1.0000 | 1.0000 |

### $n = 5$

| x | $p = 0.01$ | $p = 0.02$ | $p = 0.03$ | $p = 0.04$ | $p = 0.05$ | $p = 0.06$ | $p = 0.07$ | $p = 0.08$ | $p = 0.09$ |
|---|---|---|---|---|---|---|---|---|---|
| 0 | 0.9510 | 0.9039 | 0.8587 | 0.8154 | 0.7738 | 0.7339 | 0.6957 | 0.6591 | 0.6240 |
| 1 | 0.9990 | 0.9962 | 0.9915 | 0.9852 | 0.9774 | 0.9681 | 0.9575 | 0.9456 | 0.9326 |
| 2 | 1.0000 | 0.9999 | 0.9997 | 0.9994 | 0.9988 | 0.9980 | 0.9969 | 0.9955 | 0.9937 |
| 3 | 1.0000 | 1.0000 | 1.0000 | 1.0000 | 1.0000 | 0.9999 | 0.9999 | 0.9998 | 0.9997 |
| 4 | 1.0000 | 1.0000 | 1.0000 | 1.0000 | 1.0000 | 1.0000 | 1.0000 | 1.0000 | 1.0000 |
| 5 | 1.0000 | 1.0000 | 1.0000 | 1.0000 | 1.0000 | 1.0000 | 1.0000 | 1.0000 | 1.0000 |

| x | $p = 0.10$ | $p = 0.15$ | $p = 0.20$ | $p = 0.25$ | $p = 0.30$ | $p = 0.35$ | $p = 0.40$ | $p = 0.45$ | $p = 0.50$ |
|---|---|---|---|---|---|---|---|---|---|
| 0 | 0.5905 | 0.4437 | 0.3277 | 0.2373 | 0.1681 | 0.1160 | 0.0778 | 0.0503 | 0.0313 |
| 1 | 0.9185 | 0.8352 | 0.7373 | 0.6328 | 0.5282 | 0.4284 | 0.3370 | 0.2562 | 0.1875 |
| 2 | 0.9914 | 0.9734 | 0.9421 | 0.8965 | 0.8369 | 0.7648 | 0.6826 | 0.5931 | 0.5000 |
| 3 | 0.9995 | 0.9978 | 0.9933 | 0.9844 | 0.9692 | 0.9460 | 0.9130 | 0.8688 | 0.8125 |
| 4 | 1.0000 | 0.9999 | 0.9997 | 0.9990 | 0.9976 | 0.9947 | 0.9898 | 0.9815 | 0.9688 |
| 5 | 1.0000 | 1.0000 | 1.0000 | 1.0000 | 1.0000 | 1.0000 | 1.0000 | 1.0000 | 1.0000 |

| x | $p = 0.55$ | $p = 0.60$ | $p = 0.65$ | $p = 0.70$ | $p = 0.75$ | $p = 0.80$ | $p = 0.85$ | $p = 0.90$ | $p = 0.91$ |
|---|---|---|---|---|---|---|---|---|---|
| 0 | 0.0185 | 0.0102 | 0.0053 | 0.0024 | 0.0010 | 0.0003 | 0.0001 | 0.0000 | 0.0000 |
| 1 | 0.1312 | 0.0870 | 0.0540 | 0.0308 | 0.0156 | 0.0067 | 0.0022 | 0.0005 | 0.0003 |
| 2 | 0.4069 | 0.3174 | 0.2352 | 0.1631 | 0.1035 | 0.0579 | 0.0266 | 0.0086 | 0.0063 |
| 3 | 0.7438 | 0.6630 | 0.5716 | 0.4718 | 0.3672 | 0.2627 | 0.1648 | 0.0815 | 0.0674 |
| 4 | 0.9497 | 0.9222 | 0.8840 | 0.8319 | 0.7627 | 0.6723 | 0.5563 | 0.4095 | 0.3760 |
| 5 | 1.0000 | 1.0000 | 1.0000 | 1.0000 | 1.0000 | 1.0000 | 1.0000 | 1.0000 | 1.0000 |

| x | $p = 0.92$ | $p = 0.93$ | $p = 0.94$ | $p = 0.95$ | $p = 0.96$ | $p = 0.97$ | $p = 0.98$ | $p = 0.99$ | $p = 1.00$ |
|---|---|---|---|---|---|---|---|---|---|
| 0 | 0.0000 | 0.0000 | 0.0000 | 0.0000 | 0.0000 | 0.0000 | 0.0000 | 0.0000 | 0.0000 |
| 1 | 0.0002 | 0.0001 | 0.0001 | 0.0000 | 0.0000 | 0.0000 | 0.0000 | 0.0000 | 0.0000 |
| 2 | 0.0045 | 0.0031 | 0.0020 | 0.0012 | 0.0006 | 0.0003 | 0.0001 | 0.0000 | 0.0000 |
| 3 | 0.0544 | 0.0425 | 0.0319 | 0.0226 | 0.0148 | 0.0085 | 0.0038 | 0.0010 | 0.0000 |
| 4 | 0.3409 | 0.3043 | 0.2661 | 0.2262 | 0.1846 | 0.1413 | 0.0961 | 0.0490 | 0.0000 |
| 5 | 1.0000 | 1.0000 | 1.0000 | 1.0000 | 1.0000 | 1.0000 | 1.0000 | 1.0000 | 1.0000 |

### $n = 6$

| x | $p = 0.01$ | $p = 0.02$ | $p = 0.03$ | $p = 0.04$ | $p = 0.05$ | $p = 0.06$ | $p = 0.07$ | $p = 0.08$ | $p = 0.09$ |
|---|---|---|---|---|---|---|---|---|---|
| 0 | 0.9415 | 0.8858 | 0.8330 | 0.7828 | 0.7351 | 0.6899 | 0.6470 | 0.6064 | 0.5679 |

| x | | | | | | | | | |
|---|---|---|---|---|---|---|---|---|---|
| 1 | 0.9985 | 0.9943 | 0.9875 | 0.9784 | 0.9672 | 0.9541 | 0.9392 | 0.9227 | 0.9048 |
| 2 | 1.0000 | 0.9998 | 0.9995 | 0.9988 | 0.9978 | 0.9962 | 0.9942 | 0.9915 | 0.9882 |
| 3 | 1.0000 | 1.0000 | 1.0000 | 1.0000 | 0.9999 | 0.9998 | 0.9997 | 0.9995 | 0.9992 |
| 4 | 1.0000 | 1.0000 | 1.0000 | 1.0000 | 1.0000 | 1.0000 | 1.0000 | 1.0000 | 1.0000 |
| 5 | 1.0000 | 1.0000 | 1.0000 | 1.0000 | 1.0000 | 1.0000 | 1.0000 | 1.0000 | 1.0000 |
| 6 | 1.0000 | 1.0000 | 1.0000 | 1.0000 | 1.0000 | 1.0000 | 1.0000 | 1.0000 | 1.0000 |

| x | $p = 0.10$ | $p = 0.15$ | $p = 0.20$ | $p = 0.25$ | $p = 0.30$ | $p = 0.35$ | $p = 0.40$ | $p = 0.45$ | $p = 0.50$ |
|---|---|---|---|---|---|---|---|---|---|
| 0 | 0.5314 | 0.3771 | 0.2621 | 0.1780 | 0.1176 | 0.0754 | 0.0467 | 0.0277 | 0.0156 |
| 1 | 0.8857 | 0.7765 | 0.6554 | 0.5339 | 0.4202 | 0.3191 | 0.2333 | 0.1636 | 0.1094 |
| 2 | 0.9842 | 0.9527 | 0.9011 | 0.8306 | 0.7443 | 0.6471 | 0.5443 | 0.4415 | 0.3438 |
| 3 | 0.9987 | 0.9941 | 0.9830 | 0.9624 | 0.9295 | 0.8826 | 0.8208 | 0.7447 | 0.6563 |
| 4 | 0.9999 | 0.9996 | 0.9984 | 0.9954 | 0.9891 | 0.9777 | 0.9590 | 0.9308 | 0.8906 |
| 5 | 1.0000 | 1.0000 | 0.9999 | 0.9998 | 0.9993 | 0.9982 | 0.9959 | 0.9917 | 0.9844 |
| 6 | 1.0000 | 1.0000 | 1.0000 | 1.0000 | 1.0000 | 1.0000 | 1.0000 | 1.0000 | 1.0000 |

| x | $p = 0.55$ | $p = 0.60$ | $p = 0.65$ | $p = 0.70$ | $p = 0.75$ | $p = 0.80$ | $p = 0.85$ | $p = 0.90$ | $p = 0.91$ |
|---|---|---|---|---|---|---|---|---|---|
| 0 | 0.0083 | 0.0041 | 0.0018 | 0.0007 | 0.0002 | 0.0001 | 0.0000 | 0.0000 | 0.0000 |
| 1 | 0.0692 | 0.0410 | 0.0223 | 0.0109 | 0.0046 | 0.0016 | 0.0004 | 0.0001 | 0.0000 |
| 2 | 0.2553 | 0.1792 | 0.1174 | 0.0705 | 0.0376 | 0.0170 | 0.0059 | 0.0013 | 0.0008 |
| 3 | 0.5585 | 0.4557 | 0.3529 | 0.2557 | 0.1694 | 0.0989 | 0.0473 | 0.0159 | 0.0118 |
| 4 | 0.8364 | 0.7667 | 0.6809 | 0.5798 | 0.4661 | 0.3446 | 0.2235 | 0.1143 | 0.0952 |
| 5 | 0.9723 | 0.9533 | 0.9246 | 0.8824 | 0.8220 | 0.7379 | 0.6229 | 0.4686 | 0.4321 |
| 6 | 1.0000 | 1.0000 | 1.0000 | 1.0000 | 1.0000 | 1.0000 | 1.0000 | 1.0000 | 1.0000 |

| x | $p = 0.92$ | $p = 0.93$ | $p = 0.94$ | $p = 0.95$ | $p = 0.96$ | $p = 0.97$ | $p = 0.98$ | $p = 0.99$ | $p = 1.00$ |
|---|---|---|---|---|---|---|---|---|---|
| 0 | 0.0000 | 0.0000 | 0.0000 | 0.0000 | 0.0000 | 0.0000 | 0.0000 | 0.0000 | 0.0000 |
| 1 | 0.0000 | 0.0000 | 0.0000 | 0.0000 | 0.0000 | 0.0000 | 0.0000 | 0.0000 | 0.0000 |
| 2 | 0.0005 | 0.0003 | 0.0002 | 0.0001 | 0.0000 | 0.0000 | 0.0000 | 0.0000 | 0.0000 |
| 3 | 0.0085 | 0.0058 | 0.0038 | 0.0022 | 0.0012 | 0.0005 | 0.0002 | 0.0000 | 0.0000 |
| 4 | 0.0773 | 0.0608 | 0.0459 | 0.0328 | 0.0216 | 0.0125 | 0.0057 | 0.0015 | 0.0000 |
| 5 | 0.3936 | 0.3530 | 0.3101 | 0.2649 | 0.2172 | 0.1670 | 0.1142 | 0.0585 | 0.0000 |
| 6 | 1.0000 | 1.0000 | 1.0000 | 1.0000 | 1.0000 | 1.0000 | 1.0000 | 1.0000 | 1.0000 |

$$n = 7$$

| x | $p = 0.01$ | $p = 0.02$ | $p = 0.03$ | $p = 0.04$ | $p = 0.05$ | $p = 0.06$ | $p = 0.07$ | $p = 0.08$ | $p = 0.09$ |
|---|---|---|---|---|---|---|---|---|---|
| 0 | 0.9321 | 0.8681 | 0.8080 | 0.7514 | 0.6983 | 0.6485 | 0.6017 | 0.5578 | 0.5168 |
| 1 | 0.9980 | 0.9921 | 0.9829 | 0.9706 | 0.9556 | 0.9382 | 0.9187 | 0.8974 | 0.8745 |
| 2 | 1.0000 | 0.9997 | 0.9991 | 0.9980 | 0.9962 | 0.9937 | 0.9903 | 0.9860 | 0.9807 |
| 3 | 1.0000 | 1.0000 | 1.0000 | 0.9999 | 0.9998 | 0.9996 | 0.9993 | 0.9988 | 0.9982 |
| 4 | 1.0000 | 1.0000 | 1.0000 | 1.0000 | 1.0000 | 1.0000 | 1.0000 | 0.9999 | 0.9999 |
| 5 | 1.0000 | 1.0000 | 1.0000 | 1.0000 | 1.0000 | 1.0000 | 1.0000 | 1.0000 | 1.0000 |
| 6 | 1.0000 | 1.0000 | 1.0000 | 1.0000 | 1.0000 | 1.0000 | 1.0000 | 1.0000 | 1.0000 |
| 7 | 1.0000 | 1.0000 | 1.0000 | 1.0000 | 1.0000 | 1.0000 | 1.0000 | 1.0000 | 1.0000 |

| x | $p = 0.10$ | $p = 0.15$ | $p = 0.20$ | $p = 0.25$ | $p = 0.30$ | $p = 0.35$ | $p = 0.40$ | $p = 0.45$ | $p = 0.50$ |
|---|---|---|---|---|---|---|---|---|---|
| 0 | 0.4783 | 0.3206 | 0.2097 | 0.1335 | 0.0824 | 0.0490 | 0.0280 | 0.0152 | 0.0078 |
| 1 | 0.8503 | 0.7166 | 0.5767 | 0.4449 | 0.3294 | 0.2338 | 0.1586 | 0.1024 | 0.0625 |
| 2 | 0.9743 | 0.9262 | 0.8520 | 0.7564 | 0.6471 | 0.5323 | 0.4199 | 0.3164 | 0.2266 |
| 3 | 0.9973 | 0.9879 | 0.9667 | 0.9294 | 0.8740 | 0.8002 | 0.7102 | 0.6083 | 0.5000 |
| 4 | 0.9998 | 0.9988 | 0.9953 | 0.9871 | 0.9712 | 0.9444 | 0.9037 | 0.8471 | 0.7734 |
| 5 | 1.0000 | 0.9999 | 0.9996 | 0.9987 | 0.9962 | 0.9910 | 0.9812 | 0.9643 | 0.9375 |
| 6 | 1.0000 | 1.0000 | 1.0000 | 0.9999 | 0.9998 | 0.9994 | 0.9984 | 0.9963 | 0.9922 |
| 7 | 1.0000 | 1.0000 | 1.0000 | 1.0000 | 1.0000 | 1.0000 | 1.0000 | 1.0000 | 1.0000 |

| x | $p = 0.55$ | $p = 0.60$ | $p = 0.65$ | $p = 0.70$ | $p = 0.75$ | $p = 0.80$ | $p = 0.85$ | $p = 0.90$ | $p = 0.91$ |
|---|---|---|---|---|---|---|---|---|---|
| 0 | 0.0037 | 0.0016 | 0.0006 | 0.0002 | 0.0001 | 0.0000 | 0.0000 | 0.0000 | 0.0000 |
| 1 | 0.0357 | 0.0188 | 0.0090 | 0.0038 | 0.0013 | 0.0004 | 0.0001 | 0.0000 | 0.0000 |
| 2 | 0.1529 | 0.0963 | 0.0556 | 0.0288 | 0.0129 | 0.0047 | 0.0012 | 0.0002 | 0.0001 |
| 3 | 0.3917 | 0.2898 | 0.1998 | 0.1260 | 0.0706 | 0.0333 | 0.0121 | 0.0027 | 0.0018 |
| 4 | 0.6836 | 0.5801 | 0.4677 | 0.3529 | 0.2436 | 0.1480 | 0.0738 | 0.0257 | 0.0193 |
| 5 | 0.8976 | 0.8414 | 0.7662 | 0.6706 | 0.5551 | 0.4233 | 0.2834 | 0.1497 | 0.1255 |
| 6 | 0.9848 | 0.9720 | 0.9510 | 0.9176 | 0.8665 | 0.7903 | 0.6794 | 0.5217 | 0.4832 |
| 7 | 1.0000 | 1.0000 | 1.0000 | 1.0000 | 1.0000 | 1.0000 | 1.0000 | 1.0000 | 1.0000 |

| x | $p = 0.92$ | $p = 0.93$ | $p = 0.94$ | $p = 0.95$ | $p = 0.96$ | $p = 0.97$ | $p = 0.98$ | $p = 0.99$ | $p = 1.00$ |
|---|---|---|---|---|---|---|---|---|---|
| 0 | 0.0000 | 0.0000 | 0.0000 | 0.0000 | 0.0000 | 0.0000 | 0.0000 | 0.0000 | 0.0000 |

*(continued)*

| | | | | | | | | | |
|---|---|---|---|---|---|---|---|---|---|
| 1 | 0.0000 | 0.0000 | 0.0000 | 0.0000 | 0.0000 | 0.0000 | 0.0000 | 0.0000 | 0.0000 |
| 2 | 0.0001 | 0.0000 | 0.0000 | 0.0000 | 0.0000 | 0.0000 | 0.0000 | 0.0000 | 0.0000 |
| 3 | 0.0012 | 0.0007 | 0.0004 | 0.0002 | 0.0001 | 0.0000 | 0.0000 | 0.0000 | 0.0000 |
| 4 | 0.0140 | 0.0097 | 0.0063 | 0.0038 | 0.0020 | 0.0009 | 0.0003 | 0.0000 | 0.0000 |
| 5 | 0.1026 | 0.0813 | 0.0618 | 0.0444 | 0.0294 | 0.0171 | 0.0079 | 0.0020 | 0.0000 |
| 6 | 0.4422 | 0.3983 | 0.3515 | 0.3017 | 0.2486 | 0.1920 | 0.1319 | 0.0679 | 0.0000 |
| 7 | 1.0000 | 1.0000 | 1.0000 | 1.0000 | 1.0000 | 1.0000 | 1.0000 | 1.0000 | 1.0000 |

## $n = 8$

| $x$ | $p = 0.01$ | $p = 0.02$ | $p = 0.03$ | $p = 0.04$ | $p = 0.05$ | $p = 0.06$ | $p = 0.07$ | $p = 0.08$ | $p = 0.09$ |
|---|---|---|---|---|---|---|---|---|---|
| 0 | 0.9227 | 0.8508 | 0.7837 | 0.7214 | 0.6634 | 0.6096 | 0.5596 | 0.5132 | 0.4703 |
| 1 | 0.9973 | 0.9897 | 0.9777 | 0.9619 | 0.9428 | 0.9208 | 0.8965 | 0.8702 | 0.8423 |
| 2 | 0.9999 | 0.9996 | 0.9987 | 0.9969 | 0.9942 | 0.9904 | 0.9853 | 0.9789 | 0.9711 |
| 3 | 1.0000 | 1.0000 | 0.9999 | 0.9998 | 0.9996 | 0.9993 | 0.9987 | 0.9978 | 0.9966 |
| 4 | 1.0000 | 1.0000 | 1.0000 | 1.0000 | 1.0000 | 1.0000 | 0.9999 | 0.9999 | 0.9997 |
| 5 | 1.0000 | 1.0000 | 1.0000 | 1.0000 | 1.0000 | 1.0000 | 1.0000 | 1.0000 | 1.0000 |
| 6 | 1.0000 | 1.0000 | 1.0000 | 1.0000 | 1.0000 | 1.0000 | 1.0000 | 1.0000 | 1.0000 |
| 7 | 1.0000 | 1.0000 | 1.0000 | 1.0000 | 1.0000 | 1.0000 | 1.0000 | 1.0000 | 1.0000 |
| 8 | 1.0000 | 1.0000 | 1.0000 | 1.0000 | 1.0000 | 1.0000 | 1.0000 | 1.0000 | 1.0000 |

| $x$ | $p = 0.10$ | $p = 0.15$ | $p = 0.20$ | $p = 0.25$ | $p = 0.30$ | $p = 0.35$ | $p = 0.40$ | $p = 0.45$ | $p = 0.50$ |
|---|---|---|---|---|---|---|---|---|---|
| 0 | 0.4305 | 0.2725 | 0.1678 | 0.1001 | 0.0576 | 0.0319 | 0.0168 | 0.0084 | 0.0039 |
| 1 | 0.8131 | 0.6572 | 0.5033 | 0.3671 | 0.2553 | 0.1691 | 0.1064 | 0.0632 | 0.0352 |
| 2 | 0.9619 | 0.8948 | 0.7969 | 0.6785 | 0.5518 | 0.4278 | 0.3154 | 0.2201 | 0.1445 |
| 3 | 0.9950 | 0.9786 | 0.9437 | 0.8862 | 0.8059 | 0.7064 | 0.5941 | 0.4770 | 0.3633 |
| 4 | 0.9996 | 0.9971 | 0.9896 | 0.9727 | 0.9420 | 0.8939 | 0.8263 | 0.7396 | 0.6367 |
| 5 | 1.0000 | 0.9998 | 0.9988 | 0.9958 | 0.9887 | 0.9747 | 0.9502 | 0.9115 | 0.8555 |
| 6 | 1.0000 | 1.0000 | 0.9999 | 0.9996 | 0.9987 | 0.9964 | 0.9915 | 0.9819 | 0.9648 |
| 7 | 1.0000 | 1.0000 | 1.0000 | 1.0000 | 0.9999 | 0.9998 | 0.9993 | 0.9983 | 0.9961 |
| 8 | 1.0000 | 1.0000 | 1.0000 | 1.0000 | 1.0000 | 1.0000 | 1.0000 | 1.0000 | 1.0000 |

| $x$ | $p = 0.55$ | $p = 0.60$ | $p = 0.65$ | $p = 0.70$ | $p = 0.75$ | $p = 0.80$ | $p = 0.85$ | $p = 0.90$ | $p = 0.91$ |
|---|---|---|---|---|---|---|---|---|---|
| 0 | 0.0017 | 0.0007 | 0.0002 | 0.0001 | 0.0000 | 0.0000 | 0.0000 | 0.0000 | 0.0000 |
| 1 | 0.0181 | 0.0085 | 0.0036 | 0.0013 | 0.0004 | 0.0001 | 0.0000 | 0.0000 | 0.0000 |
| 2 | 0.0885 | 0.0498 | 0.0253 | 0.0113 | 0.0042 | 0.0012 | 0.0002 | 0.0000 | 0.0000 |
| 3 | 0.2604 | 0.1737 | 0.1061 | 0.0580 | 0.0273 | 0.0104 | 0.0029 | 0.0004 | 0.0003 |
| 4 | 0.5230 | 0.4059 | 0.2936 | 0.1941 | 0.1138 | 0.0563 | 0.0214 | 0.0050 | 0.0034 |
| 5 | 0.7799 | 0.6846 | 0.5722 | 0.4482 | 0.3215 | 0.2031 | 0.1052 | 0.0381 | 0.0289 |
| 6 | 0.9368 | 0.8936 | 0.8309 | 0.7447 | 0.6329 | 0.4967 | 0.3428 | 0.1869 | 0.1577 |
| 7 | 0.9916 | 0.9832 | 0.9681 | 0.9424 | 0.8999 | 0.8322 | 0.7275 | 0.5695 | 0.5297 |
| 8 | 1.0000 | 1.0000 | 1.0000 | 1.0000 | 1.0000 | 1.0000 | 1.0000 | 1.0000 | 1.0000 |

| $x$ | $p = 0.92$ | $p = 0.93$ | $p = 0.94$ | $p = 0.95$ | $p = 0.96$ | $p = 0.97$ | $p = 0.98$ | $p = 0.99$ | $p = 1.00$ |
|---|---|---|---|---|---|---|---|---|---|
| 0 | 0.0000 | 0.0000 | 0.0000 | 0.0000 | 0.0000 | 0.0000 | 0.0000 | 0.0000 | 0.0000 |
| 1 | 0.0000 | 0.0000 | 0.0000 | 0.0000 | 0.0000 | 0.0000 | 0.0000 | 0.0000 | 0.0000 |
| 2 | 0.0000 | 0.0000 | 0.0000 | 0.0000 | 0.0000 | 0.0000 | 0.0000 | 0.0000 | 0.0000 |
| 3 | 0.0001 | 0.0001 | 0.0000 | 0.0000 | 0.0000 | 0.0000 | 0.0000 | 0.0000 | 0.0000 |
| 4 | 0.0022 | 0.0013 | 0.0007 | 0.0004 | 0.0002 | 0.0001 | 0.0000 | 0.0000 | 0.0000 |
| 5 | 0.0211 | 0.0147 | 0.0096 | 0.0058 | 0.0031 | 0.0013 | 0.0004 | 0.0001 | 0.0000 |
| 6 | 0.1298 | 0.1035 | 0.0792 | 0.0572 | 0.0381 | 0.0223 | 0.0103 | 0.0027 | 0.0000 |
| 7 | 0.4868 | 0.4404 | 0.3904 | 0.3366 | 0.2786 | 0.2163 | 0.1492 | 0.0773 | 0.0000 |
| 8 | 1.0000 | 1.0000 | 1.0000 | 1.0000 | 1.0000 | 1.0000 | 1.0000 | 1.0000 | 1.0000 |

## $n = 9$

| $x$ | $p = 0.01$ | $p = 0.02$ | $p = 0.03$ | $p = 0.04$ | $p = 0.05$ | $p = 0.06$ | $p = 0.07$ | $p = 0.08$ | $p = 0.09$ |
|---|---|---|---|---|---|---|---|---|---|
| 0 | 0.9135 | 0.8337 | 0.7602 | 0.6925 | 0.6302 | 0.5730 | 0.5204 | 0.4722 | 0.4279 |
| 1 | 0.9966 | 0.9869 | 0.9718 | 0.9522 | 0.9288 | 0.9022 | 0.8729 | 0.8417 | 0.8088 |
| 2 | 0.9999 | 0.9994 | 0.9980 | 0.9955 | 0.9916 | 0.9862 | 0.9791 | 0.9702 | 0.9595 |
| 3 | 1.0000 | 1.0000 | 0.9999 | 0.9997 | 0.9994 | 0.9987 | 0.9977 | 0.9963 | 0.9943 |
| 4 | 1.0000 | 1.0000 | 1.0000 | 1.0000 | 1.0000 | 0.9999 | 0.9998 | 0.9997 | 0.9995 |
| 5 | 1.0000 | 1.0000 | 1.0000 | 1.0000 | 1.0000 | 1.0000 | 1.0000 | 1.0000 | 1.0000 |
| 6 | 1.0000 | 1.0000 | 1.0000 | 1.0000 | 1.0000 | 1.0000 | 1.0000 | 1.0000 | 1.0000 |
| 7 | 1.0000 | 1.0000 | 1.0000 | 1.0000 | 1.0000 | 1.0000 | 1.0000 | 1.0000 | 1.0000 |
| 8 | 1.0000 | 1.0000 | 1.0000 | 1.0000 | 1.0000 | 1.0000 | 1.0000 | 1.0000 | 1.0000 |
| 9 | 1.0000 | 1.0000 | 1.0000 | 1.0000 | 1.0000 | 1.0000 | 1.0000 | 1.0000 | 1.0000 |

| x | p = 0.10 | p = 0.15 | p = 0.20 | p = 0.25 | p = 0.30 | p = 0.35 | p = 0.40 | p = 0.45 | p = 0.50 |
|---|---|---|---|---|---|---|---|---|---|
| 0 | 0.3874 | 0.2316 | 0.1342 | 0.0751 | 0.0404 | 0.0207 | 0.0101 | 0.0046 | 0.0020 |
| 1 | 0.7748 | 0.5995 | 0.4362 | 0.3003 | 0.1960 | 0.1211 | 0.0705 | 0.0385 | 0.0195 |
| 2 | 0.9470 | 0.8591 | 0.7382 | 0.6007 | 0.4628 | 0.3373 | 0.2318 | 0.1495 | 0.0898 |
| 3 | 0.9917 | 0.9661 | 0.9144 | 0.8343 | 0.7297 | 0.6089 | 0.4826 | 0.3614 | 0.2539 |
| 4 | 0.9991 | 0.9944 | 0.9804 | 0.9511 | 0.9012 | 0.8283 | 0.7334 | 0.6214 | 0.5000 |
| 5 | 0.9999 | 0.9994 | 0.9969 | 0.9900 | 0.9747 | 0.9464 | 0.9006 | 0.8342 | 0.7461 |
| 6 | 1.0000 | 1.0000 | 0.9997 | 0.9987 | 0.9957 | 0.9888 | 0.9750 | 0.9502 | 0.9102 |
| 7 | 1.0000 | 1.0000 | 1.0000 | 0.9999 | 0.9996 | 0.9986 | 0.9962 | 0.9909 | 0.9805 |
| 8 | 1.0000 | 1.0000 | 1.0000 | 1.0000 | 1.0000 | 0.9999 | 0.9997 | 0.9992 | 0.9980 |
| 9 | 1.0000 | 1.0000 | 1.0000 | 1.0000 | 1.0000 | 1.0000 | 1.0000 | 1.0000 | 1.0000 |

| x | p = 0.55 | p = 0.60 | p = 0.65 | p = 0.70 | p = 0.75 | p = 0.80 | p = 0.85 | p = 0.90 | p = 0.91 |
|---|---|---|---|---|---|---|---|---|---|
| 0 | 0.0008 | 0.0003 | 0.0001 | 0.0000 | 0.0000 | 0.0000 | 0.0000 | 0.0000 | 0.0000 |
| 1 | 0.0091 | 0.0038 | 0.0014 | 0.0004 | 0.0001 | 0.0000 | 0.0000 | 0.0000 | 0.0000 |
| 2 | 0.0498 | 0.0250 | 0.0112 | 0.0043 | 0.0013 | 0.0003 | 0.0000 | 0.0000 | 0.0000 |
| 3 | 0.1658 | 0.0994 | 0.0536 | 0.0253 | 0.0100 | 0.0031 | 0.0006 | 0.0001 | 0.0000 |
| 4 | 0.3786 | 0.2666 | 0.1717 | 0.0988 | 0.0489 | 0.0196 | 0.0056 | 0.0009 | 0.0005 |
| 5 | 0.6386 | 0.5174 | 0.3911 | 0.2703 | 0.1657 | 0.0856 | 0.0339 | 0.0083 | 0.0057 |
| 6 | 0.8505 | 0.7682 | 0.6627 | 0.5372 | 0.3993 | 0.2618 | 0.1409 | 0.0530 | 0.0405 |
| 7 | 0.9615 | 0.9295 | 0.8789 | 0.8040 | 0.6997 | 0.5638 | 0.4005 | 0.2252 | 0.1912 |
| 8 | 0.9954 | 0.9899 | 0.9793 | 0.9596 | 0.9249 | 0.8658 | 0.7684 | 0.6126 | 0.5721 |
| 9 | 1.0000 | 1.0000 | 1.0000 | 1.0000 | 1.0000 | 1.0000 | 1.0000 | 1.0000 | 1.0000 |

| x | p = 0.92 | p = 0.93 | p = 0.94 | p = 0.95 | p = 0.96 | p = 0.97 | p = 0.98 | p = 0.99 | p = 1.00 |
|---|---|---|---|---|---|---|---|---|---|
| 0 | 0.0000 | 0.0000 | 0.0000 | 0.0000 | 0.0000 | 0.0000 | 0.0000 | 0.0000 | 0.0000 |
| 1 | 0.0000 | 0.0000 | 0.0000 | 0.0000 | 0.0000 | 0.0000 | 0.0000 | 0.0000 | 0.0000 |
| 2 | 0.0000 | 0.0000 | 0.0000 | 0.0000 | 0.0000 | 0.0000 | 0.0000 | 0.0000 | 0.0000 |
| 3 | 0.0000 | 0.0000 | 0.0000 | 0.0000 | 0.0000 | 0.0000 | 0.0000 | 0.0000 | 0.0000 |
| 4 | 0.0003 | 0.0002 | 0.0001 | 0.0000 | 0.0000 | 0.0000 | 0.0000 | 0.0000 | 0.0000 |
| 5 | 0.0037 | 0.0023 | 0.0013 | 0.0006 | 0.0003 | 0.0001 | 0.0000 | 0.0000 | 0.0000 |
| 6 | 0.0298 | 0.0209 | 0.0138 | 0.0084 | 0.0045 | 0.0020 | 0.0006 | 0.0001 | 0.0000 |
| 7 | 0.1583 | 0.1271 | 0.0978 | 0.0712 | 0.0478 | 0.0282 | 0.0131 | 0.0034 | 0.0000 |
| 8 | 0.5278 | 0.4796 | 0.4270 | 0.3698 | 0.3075 | 0.2398 | 0.1663 | 0.0865 | 0.0000 |
| 9 | 1.0000 | 1.0000 | 1.0000 | 1.0000 | 1.0000 | 1.0000 | 1.0000 | 1.0000 | 1.0000 |

$n = 10$

| x | p = 0.01 | p = 0.02 | p = 0.03 | p = 0.04 | p = 0.05 | p = 0.06 | p = 0.07 | p = 0.08 | p = 0.09 |
|---|---|---|---|---|---|---|---|---|---|
| 0 | 0.9044 | 0.8171 | 0.7374 | 0.6648 | 0.5987 | 0.5386 | 0.4840 | 0.4344 | 0.3894 |
| 1 | 0.9957 | 0.9838 | 0.9655 | 0.9418 | 0.9139 | 0.8824 | 0.8483 | 0.8121 | 0.7746 |
| 2 | 0.9999 | 0.9991 | 0.9972 | 0.9938 | 0.9885 | 0.9812 | 0.9717 | 0.9599 | 0.9460 |
| 3 | 1.0000 | 1.0000 | 0.9999 | 0.9996 | 0.9990 | 0.9980 | 0.9964 | 0.9942 | 0.9912 |
| 4 | 1.0000 | 1.0000 | 1.0000 | 1.0000 | 0.9999 | 0.9998 | 0.9997 | 0.9994 | 0.9990 |
| 5 | 1.0000 | 1.0000 | 1.0000 | 1.0000 | 1.0000 | 1.0000 | 1.0000 | 1.0000 | 0.9999 |
| 6 | 1.0000 | 1.0000 | 1.0000 | 1.0000 | 1.0000 | 1.0000 | 1.0000 | 1.0000 | 1.0000 |
| 7 | 1.0000 | 1.0000 | 1.0000 | 1.0000 | 1.0000 | 1.0000 | 1.0000 | 1.0000 | 1.0000 |
| 8 | 1.0000 | 1.0000 | 1.0000 | 1.0000 | 1.0000 | 1.0000 | 1.0000 | 1.0000 | 1.0000 |
| 9 | 1.0000 | 1.0000 | 1.0000 | 1.0000 | 1.0000 | 1.0000 | 1.0000 | 1.0000 | 1.0000 |
| 10 | 1.0000 | 1.0000 | 1.0000 | 1.0000 | 1.0000 | 1.0000 | 1.0000 | 1.0000 | 1.0000 |

| x | p = 0.10 | p = 0.15 | p = 0.20 | p = 0.25 | p = 0.30 | p = 0.35 | p = 0.40 | p = 0.45 | p = 0.50 |
|---|---|---|---|---|---|---|---|---|---|
| 0 | 0.3487 | 0.1969 | 0.1074 | 0.0563 | 0.0282 | 0.0135 | 0.0060 | 0.0025 | 0.0010 |
| 1 | 0.7361 | 0.5443 | 0.3758 | 0.2440 | 0.1493 | 0.0860 | 0.0464 | 0.0233 | 0.0107 |
| 2 | 0.9298 | 0.8202 | 0.6778 | 0.5256 | 0.3828 | 0.2616 | 0.1673 | 0.0996 | 0.0547 |
| 3 | 0.9872 | 0.9500 | 0.8791 | 0.7759 | 0.6496 | 0.5138 | 0.3823 | 0.2660 | 0.1719 |
| 4 | 0.9984 | 0.9901 | 0.9672 | 0.9219 | 0.8497 | 0.7515 | 0.6331 | 0.5044 | 0.3770 |
| 5 | 0.9999 | 0.9986 | 0.9936 | 0.9803 | 0.9527 | 0.9051 | 0.8338 | 0.7384 | 0.6230 |
| 6 | 1.0000 | 0.9999 | 0.9991 | 0.9965 | 0.9894 | 0.9740 | 0.9452 | 0.8980 | 0.8281 |
| 7 | 1.0000 | 1.0000 | 0.9999 | 0.9996 | 0.9984 | 0.9952 | 0.9877 | 0.9726 | 0.9453 |
| 8 | 1.0000 | 1.0000 | 1.0000 | 1.0000 | 0.9999 | 0.9995 | 0.9983 | 0.9955 | 0.9893 |
| 9 | 1.0000 | 1.0000 | 1.0000 | 1.0000 | 1.0000 | 1.0000 | 0.9999 | 0.9997 | 0.9990 |
| 10 | 1.0000 | 1.0000 | 1.0000 | 1.0000 | 1.0000 | 1.0000 | 1.0000 | 1.0000 | 1.0000 |

| x | p = 0.55 | p = 0.60 | p = 0.65 | p = 0.70 | p = 0.75 | p = 0.80 | p = 0.85 | p = 0.90 | p = 0.91 |
|---|---|---|---|---|---|---|---|---|---|
| 0 | 0.0003 | 0.0001 | 0.0000 | 0.0000 | 0.0000 | 0.0000 | 0.0000 | 0.0000 | 0.0000 |
| 1 | 0.0045 | 0.0017 | 0.0005 | 0.0001 | 0.0000 | 0.0000 | 0.0000 | 0.0000 | 0.0000 |

*(continued)*

| | | | | | | | | | |
|---|---|---|---|---|---|---|---|---|---|
| 2 | 0.0274 | 0.0123 | 0.0048 | 0.0016 | 0.0004 | 0.0001 | 0.0000 | 0.0000 | 0.0000 |
| 3 | 0.1020 | 0.0548 | 0.0260 | 0.0106 | 0.0035 | 0.0009 | 0.0001 | 0.0000 | 0.0000 |
| 4 | 0.2616 | 0.1662 | 0.0949 | 0.0473 | 0.0197 | 0.0064 | 0.0014 | 0.0001 | 0.0001 |
| 5 | 0.4956 | 0.3669 | 0.2485 | 0.1503 | 0.0781 | 0.0328 | 0.0099 | 0.0016 | 0.0010 |
| 6 | 0.7340 | 0.6177 | 0.4862 | 0.3504 | 0.2241 | 0.1209 | 0.0500 | 0.0128 | 0.0088 |
| 7 | 0.9004 | 0.8327 | 0.7384 | 0.6172 | 0.4744 | 0.3222 | 0.1798 | 0.0702 | 0.0540 |
| 8 | 0.9767 | 0.9536 | 0.9140 | 0.8507 | 0.7560 | 0.6242 | 0.4557 | 0.2639 | 0.2254 |
| 9 | 0.9975 | 0.9940 | 0.9865 | 0.9718 | 0.9437 | 0.8926 | 0.8031 | 0.6513 | 0.6106 |
| 10 | 1.0000 | 1.0000 | 1.0000 | 1.0000 | 1.0000 | 1.0000 | 1.0000 | 1.0000 | 1.0000 |

| $x$ | $p = 0.92$ | $p = 0.93$ | $p = 0.94$ | $p = 0.95$ | $p = 0.96$ | $p = 0.97$ | $p = 0.98$ | $p = 0.99$ | $p = 1.00$ |
|---|---|---|---|---|---|---|---|---|---|
| 0 | 0.0000 | 0.0000 | 0.0000 | 0.0000 | 0.0000 | 0.0000 | 0.0000 | 0.0000 | 0.0000 |
| 1 | 0.0000 | 0.0000 | 0.0000 | 0.0000 | 0.0000 | 0.0000 | 0.0000 | 0.0000 | 0.0000 |
| 2 | 0.0000 | 0.0000 | 0.0000 | 0.0000 | 0.0000 | 0.0000 | 0.0000 | 0.0000 | 0.0000 |
| 3 | 0.0000 | 0.0000 | 0.0000 | 0.0000 | 0.0000 | 0.0000 | 0.0000 | 0.0000 | 0.0000 |
| 4 | 0.0000 | 0.0000 | 0.0000 | 0.0000 | 0.0000 | 0.0000 | 0.0000 | 0.0000 | 0.0000 |
| 5 | 0.0006 | 0.0003 | 0.0002 | 0.0001 | 0.0000 | 0.0000 | 0.0000 | 0.0000 | 0.0000 |
| 6 | 0.0058 | 0.0036 | 0.0020 | 0.0010 | 0.0004 | 0.0001 | 0.0000 | 0.0000 | 0.0000 |
| 7 | 0.0401 | 0.0283 | 0.0188 | 0.0115 | 0.0062 | 0.0028 | 0.0009 | 0.0001 | 0.0000 |
| 8 | 0.1879 | 0.1517 | 0.1176 | 0.0861 | 0.0582 | 0.0345 | 0.0162 | 0.0043 | 0.0000 |
| 9 | 0.5656 | 0.5160 | 0.4614 | 0.4013 | 0.3352 | 0.2626 | 0.1829 | 0.0956 | 0.0000 |
| 10 | 1.0000 | 1.0000 | 1.0000 | 1.0000 | 1.0000 | 1.0000 | 1.0000 | 1.0000 | 1.0000 |

$$n = 11$$

| $x$ | $p = 0.01$ | $p = 0.02$ | $p = 0.03$ | $p = 0.04$ | $p = 0.05$ | $p = 0.06$ | $p = 0.07$ | $p = 0.08$ | $p = 0.09$ |
|---|---|---|---|---|---|---|---|---|---|
| 0 | 0.8953 | 0.8007 | 0.7153 | 0.6382 | 0.5688 | 0.5063 | 0.4501 | 0.3996 | 0.3544 |
| 1 | 0.9948 | 0.9805 | 0.9587 | 0.9308 | 0.8981 | 0.8618 | 0.8228 | 0.7819 | 0.7399 |
| 2 | 0.9998 | 0.9988 | 0.9963 | 0.9917 | 0.9848 | 0.9752 | 0.9630 | 0.9481 | 0.9305 |
| 3 | 1.0000 | 1.0000 | 0.9998 | 0.9993 | 0.9984 | 0.9970 | 0.9947 | 0.9915 | 0.9871 |
| 4 | 1.0000 | 1.0000 | 1.0000 | 1.0000 | 0.9999 | 0.9997 | 0.9995 | 0.9990 | 0.9983 |
| 5 | 1.0000 | 1.0000 | 1.0000 | 1.0000 | 1.0000 | 1.0000 | 1.0000 | 0.9999 | 0.9998 |
| 6 | 1.0000 | 1.0000 | 1.0000 | 1.0000 | 1.0000 | 1.0000 | 1.0000 | 1.0000 | 1.0000 |
| 7 | 1.0000 | 1.0000 | 1.0000 | 1.0000 | 1.0000 | 1.0000 | 1.0000 | 1.0000 | 1.0000 |
| 8 | 1.0000 | 1.0000 | 1.0000 | 1.0000 | 1.0000 | 1.0000 | 1.0000 | 1.0000 | 1.0000 |
| 9 | 1.0000 | 1.0000 | 1.0000 | 1.0000 | 1.0000 | 1.0000 | 1.0000 | 1.0000 | 1.0000 |
| 10 | 1.0000 | 1.0000 | 1.0000 | 1.0000 | 1.0000 | 1.0000 | 1.0000 | 1.0000 | 1.0000 |
| 11 | 1.0000 | 1.0000 | 1.0000 | 1.0000 | 1.0000 | 1.0000 | 1.0000 | 1.0000 | 1.0000 |

| $x$ | $p = 0.10$ | $p = 0.15$ | $p = 0.20$ | $p = 0.25$ | $p = 0.30$ | $p = 0.35$ | $p = 0.40$ | $p = 0.45$ | $p = 0.50$ |
|---|---|---|---|---|---|---|---|---|---|
| 0 | 0.3138 | 0.1673 | 0.0859 | 0.0422 | 0.0198 | 0.0088 | 0.0036 | 0.0014 | 0.0005 |
| 1 | 0.6974 | 0.4922 | 0.3221 | 0.1971 | 0.1130 | 0.0606 | 0.0302 | 0.0139 | 0.0059 |
| 2 | 0.9104 | 0.7788 | 0.6174 | 0.4552 | 0.3127 | 0.2001 | 0.1189 | 0.0652 | 0.0327 |
| 3 | 0.9815 | 0.9306 | 0.8389 | 0.7133 | 0.5696 | 0.4256 | 0.2963 | 0.1911 | 0.1133 |
| 4 | 0.9972 | 0.9841 | 0.9496 | 0.8854 | 0.7897 | 0.6683 | 0.5328 | 0.3971 | 0.2744 |
| 5 | 0.9997 | 0.9973 | 0.9883 | 0.9657 | 0.9218 | 0.8513 | 0.7535 | 0.6331 | 0.5000 |
| 6 | 1.0000 | 0.9997 | 0.9980 | 0.9924 | 0.9784 | 0.9499 | 0.9006 | 0.8262 | 0.7256 |
| 7 | 1.0000 | 1.0000 | 0.9998 | 0.9988 | 0.9957 | 0.9878 | 0.9707 | 0.9390 | 0.8867 |
| 8 | 1.0000 | 1.0000 | 1.0000 | 0.9999 | 0.9994 | 0.9980 | 0.9941 | 0.9852 | 0.9673 |
| 9 | 1.0000 | 1.0000 | 1.0000 | 1.0000 | 1.0000 | 0.9998 | 0.9993 | 0.9978 | 0.9941 |
| 10 | 1.0000 | 1.0000 | 1.0000 | 1.0000 | 1.0000 | 1.0000 | 1.0000 | 0.9998 | 0.9995 |
| 11 | 1.0000 | 1.0000 | 1.0000 | 1.0000 | 1.0000 | 1.0000 | 1.0000 | 1.0000 | 1.0000 |

| $x$ | $p = 0.55$ | $p = 0.60$ | $p = 0.65$ | $p = 0.70$ | $p = 0.75$ | $p = 0.80$ | $p = 0.85$ | $p = 0.90$ | $p = 0.91$ |
|---|---|---|---|---|---|---|---|---|---|
| 0 | 0.0002 | 0.0000 | 0.0000 | 0.0000 | 0.0000 | 0.0000 | 0.0000 | 0.0000 | 0.0000 |
| 1 | 0.0022 | 0.0007 | 0.0002 | 0.0000 | 0.0000 | 0.0000 | 0.0000 | 0.0000 | 0.0000 |
| 2 | 0.0148 | 0.0059 | 0.0020 | 0.0006 | 0.0001 | 0.0000 | 0.0000 | 0.0000 | 0.0000 |
| 3 | 0.0610 | 0.0293 | 0.0122 | 0.0043 | 0.0012 | 0.0002 | 0.0000 | 0.0000 | 0.0000 |
| 4 | 0.1738 | 0.0994 | 0.0501 | 0.0216 | 0.0076 | 0.0020 | 0.0003 | 0.0000 | 0.0000 |
| 5 | 0.3669 | 0.2465 | 0.1487 | 0.0782 | 0.0343 | 0.0117 | 0.0027 | 0.0003 | 0.0002 |
| 6 | 0.6029 | 0.4672 | 0.3317 | 0.2103 | 0.1146 | 0.0504 | 0.0159 | 0.0028 | 0.0017 |
| 7 | 0.8089 | 0.7037 | 0.5744 | 0.4304 | 0.2867 | 0.1611 | 0.0694 | 0.0185 | 0.0129 |
| 8 | 0.9348 | 0.8811 | 0.7999 | 0.6873 | 0.5448 | 0.3826 | 0.2212 | 0.0896 | 0.0695 |
| 9 | 0.9861 | 0.9698 | 0.9394 | 0.8870 | 0.8029 | 0.6779 | 0.5078 | 0.3026 | 0.2601 |
| 10 | 0.9986 | 0.9964 | 0.9912 | 0.9802 | 0.9578 | 0.9141 | 0.8327 | 0.6862 | 0.6456 |
| 11 | 1.0000 | 1.0000 | 1.0000 | 1.0000 | 1.0000 | 1.0000 | 1.0000 | 1.0000 | 1.0000 |

| x | p = 0.92 | p = 0.93 | p = 0.94 | p = 0.95 | p = 0.96 | p = 0.97 | p = 0.98 | p = 0.99 | p = 1.00 |
|---|---|---|---|---|---|---|---|---|---|
| 0 | 0.0000 | 0.0000 | 0.0000 | 0.0000 | 0.0000 | 0.0000 | 0.0000 | 0.0000 | 0.0000 |
| 1 | 0.0000 | 0.0000 | 0.0000 | 0.0000 | 0.0000 | 0.0000 | 0.0000 | 0.0000 | 0.0000 |
| 2 | 0.0000 | 0.0000 | 0.0000 | 0.0000 | 0.0000 | 0.0000 | 0.0000 | 0.0000 | 0.0000 |
| 3 | 0.0000 | 0.0000 | 0.0000 | 0.0000 | 0.0000 | 0.0000 | 0.0000 | 0.0000 | 0.0000 |
| 4 | 0.0000 | 0.0000 | 0.0000 | 0.0000 | 0.0000 | 0.0000 | 0.0000 | 0.0000 | 0.0000 |
| 5 | 0.0001 | 0.0000 | 0.0000 | 0.0000 | 0.0000 | 0.0000 | 0.0000 | 0.0000 | 0.0000 |
| 6 | 0.0010 | 0.0005 | 0.0003 | 0.0001 | 0.0000 | 0.0000 | 0.0000 | 0.0000 | 0.0000 |
| 7 | 0.0085 | 0.0053 | 0.0030 | 0.0016 | 0.0007 | 0.0002 | 0.0000 | 0.0000 | 0.0000 |
| 8 | 0.0519 | 0.0370 | 0.0248 | 0.0152 | 0.0083 | 0.0037 | 0.0012 | 0.0002 | 0.0000 |
| 9 | 0.2181 | 0.1772 | 0.1382 | 0.1019 | 0.0692 | 0.0413 | 0.0195 | 0.0052 | 0.0000 |
| 10 | 0.6004 | 0.5499 | 0.4937 | 0.4312 | 0.3618 | 0.2847 | 0.1993 | 0.1047 | 0.0000 |
| 11 | 1.0000 | 1.0000 | 1.0000 | 1.0000 | 1.0000 | 1.0000 | 1.0000 | 1.0000 | 1.0000 |

*n* = 12

| x | p = 0.01 | p = 0.02 | p = 0.03 | p = 0.04 | p = 0.05 | p = 0.06 | p = 0.07 | p = 0.08 | p = 0.09 |
|---|---|---|---|---|---|---|---|---|---|
| 0 | 0.8864 | 0.7847 | 0.6938 | 0.6127 | 0.5404 | 0.4759 | 0.4186 | 0.3677 | 0.3225 |
| 1 | 0.9938 | 0.9769 | 0.9514 | 0.9191 | 0.8816 | 0.8405 | 0.7967 | 0.7513 | 0.7052 |
| 2 | 0.9998 | 0.9985 | 0.9952 | 0.9893 | 0.9804 | 0.9684 | 0.9532 | 0.9348 | 0.9134 |
| 3 | 1.0000 | 0.9999 | 0.9997 | 0.9990 | 0.9978 | 0.9957 | 0.9925 | 0.9880 | 0.9820 |
| 4 | 1.0000 | 1.0000 | 1.0000 | 0.9999 | 0.9998 | 0.9996 | 0.9991 | 0.9984 | 0.9973 |
| 5 | 1.0000 | 1.0000 | 1.0000 | 1.0000 | 1.0000 | 1.0000 | 0.9999 | 0.9998 | 0.9997 |
| 6 | 1.0000 | 1.0000 | 1.0000 | 1.0000 | 1.0000 | 1.0000 | 1.0000 | 1.0000 | 1.0000 |
| 7 | 1.0000 | 1.0000 | 1.0000 | 1.0000 | 1.0000 | 1.0000 | 1.0000 | 1.0000 | 1.0000 |
| 8 | 1.0000 | 1.0000 | 1.0000 | 1.0000 | 1.0000 | 1.0000 | 1.0000 | 1.0000 | 1.0000 |
| 9 | 1.0000 | 1.0000 | 1.0000 | 1.0000 | 1.0000 | 1.0000 | 1.0000 | 1.0000 | 1.0000 |
| 10 | 1.0000 | 1.0000 | 1.0000 | 1.0000 | 1.0000 | 1.0000 | 1.0000 | 1.0000 | 1.0000 |
| 11 | 1.0000 | 1.0000 | 1.0000 | 1.0000 | 1.0000 | 1.0000 | 1.0000 | 1.0000 | 1.0000 |
| 12 | 1.0000 | 1.0000 | 1.0000 | 1.0000 | 1.0000 | 1.0000 | 1.0000 | 1.0000 | 1.0000 |

| x | p = 0.10 | p = 0.15 | p = 0.20 | p = 0.25 | p = 0.30 | p = 0.35 | p = 0.40 | p = 0.45 | p = 0.50 |
|---|---|---|---|---|---|---|---|---|---|
| 0 | 0.2824 | 0.1422 | 0.0687 | 0.0317 | 0.0138 | 0.0057 | 0.0022 | 0.0008 | 0.0002 |
| 1 | 0.6590 | 0.4435 | 0.2749 | 0.1584 | 0.0850 | 0.0424 | 0.0196 | 0.0083 | 0.0032 |
| 2 | 0.8891 | 0.7358 | 0.5583 | 0.3907 | 0.2528 | 0.1513 | 0.0834 | 0.0421 | 0.0193 |
| 3 | 0.9744 | 0.9078 | 0.7946 | 0.6488 | 0.4925 | 0.3467 | 0.2253 | 0.1345 | 0.0730 |
| 4 | 0.9957 | 0.9761 | 0.9274 | 0.8424 | 0.7237 | 0.5833 | 0.4382 | 0.3044 | 0.1938 |
| 5 | 0.9995 | 0.9954 | 0.9806 | 0.9456 | 0.8822 | 0.7873 | 0.6652 | 0.5269 | 0.3872 |
| 6 | 0.9999 | 0.9993 | 0.9961 | 0.9857 | 0.9614 | 0.9154 | 0.8418 | 0.7393 | 0.6128 |
| 7 | 1.0000 | 0.9999 | 0.9994 | 0.9972 | 0.9905 | 0.9745 | 0.9427 | 0.8883 | 0.8062 |
| 8 | 1.0000 | 1.0000 | 0.9999 | 0.9996 | 0.9983 | 0.9944 | 0.9847 | 0.9644 | 0.9270 |
| 9 | 1.0000 | 1.0000 | 1.0000 | 1.0000 | 0.9998 | 0.9992 | 0.9972 | 0.9921 | 0.9807 |
| 10 | 1.0000 | 1.0000 | 1.0000 | 1.0000 | 1.0000 | 0.9999 | 0.9997 | 0.9989 | 0.9968 |
| 11 | 1.0000 | 1.0000 | 1.0000 | 1.0000 | 1.0000 | 1.0000 | 1.0000 | 0.9999 | 0.9998 |
| 12 | 1.0000 | 1.0000 | 1.0000 | 1.0000 | 1.0000 | 1.0000 | 1.0000 | 1.0000 | 1.0000 |

| x | p = 0.55 | p = 0.60 | p = 0.65 | p = 0.70 | p = 0.75 | p = 0.80 | p = 0.85 | p = 0.90 | p = 0.91 |
|---|---|---|---|---|---|---|---|---|---|
| 0 | 0.0001 | 0.0000 | 0.0000 | 0.0000 | 0.0000 | 0.0000 | 0.0000 | 0.0000 | 0.0000 |
| 1 | 0.0011 | 0.0003 | 0.0001 | 0.0000 | 0.0000 | 0.0000 | 0.0000 | 0.0000 | 0.0000 |
| 2 | 0.0079 | 0.0028 | 0.0008 | 0.0002 | 0.0000 | 0.0000 | 0.0000 | 0.0000 | 0.0000 |
| 3 | 0.0356 | 0.0153 | 0.0056 | 0.0017 | 0.0004 | 0.0001 | 0.0000 | 0.0000 | 0.0000 |
| 4 | 0.1117 | 0.0573 | 0.0255 | 0.0095 | 0.0028 | 0.0006 | 0.0001 | 0.0000 | 0.0000 |
| 5 | 0.2607 | 0.1582 | 0.0846 | 0.0386 | 0.0143 | 0.0039 | 0.0007 | 0.0001 | 0.0000 |
| 6 | 0.4731 | 0.3348 | 0.2127 | 0.1178 | 0.0544 | 0.0194 | 0.0046 | 0.0005 | 0.0003 |
| 7 | 0.6956 | 0.5618 | 0.4167 | 0.2763 | 0.1576 | 0.0726 | 0.0239 | 0.0043 | 0.0027 |
| 8 | 0.8655 | 0.7747 | 0.6533 | 0.5075 | 0.3512 | 0.2054 | 0.0922 | 0.0256 | 0.0180 |
| 9 | 0.9579 | 0.9166 | 0.8487 | 0.7472 | 0.6093 | 0.4417 | 0.2642 | 0.1109 | 0.0866 |
| 10 | 0.9917 | 0.9804 | 0.9576 | 0.9150 | 0.8416 | 0.7251 | 0.5565 | 0.3410 | 0.2948 |
| 11 | 0.9992 | 0.9978 | 0.9943 | 0.9862 | 0.9683 | 0.9313 | 0.8578 | 0.7176 | 0.6775 |
| 12 | 1.0000 | 1.0000 | 1.0000 | 1.0000 | 1.0000 | 1.0000 | 1.0000 | 1.0000 | 1.0000 |

| x | p = 0.92 | p = 0.93 | p = 0.94 | p = 0.95 | p = 0.96 | p = 0.97 | p = 0.98 | p = 0.99 | p = 1.00 |
|---|---|---|---|---|---|---|---|---|---|
| 0 | 0.0000 | 0.0000 | 0.0000 | 0.0000 | 0.0000 | 0.0000 | 0.0000 | 0.0000 | 0.0000 |
| 1 | 0.0000 | 0.0000 | 0.0000 | 0.0000 | 0.0000 | 0.0000 | 0.0000 | 0.0000 | 0.0000 |
| 2 | 0.0000 | 0.0000 | 0.0000 | 0.0000 | 0.0000 | 0.0000 | 0.0000 | 0.0000 | 0.0000 |

*(continued)*

| | | | | | | | | | |
|---|---|---|---|---|---|---|---|---|---|
| **3** | 0.0000 | 0.0000 | 0.0000 | 0.0000 | 0.0000 | 0.0000 | 0.0000 | 0.0000 | 0.0000 |
| **4** | 0.0000 | 0.0000 | 0.0000 | 0.0000 | 0.0000 | 0.0000 | 0.0000 | 0.0000 | 0.0000 |
| **5** | 0.0000 | 0.0000 | 0.0000 | 0.0000 | 0.0000 | 0.0000 | 0.0000 | 0.0000 | 0.0000 |
| **6** | 0.0002 | 0.0001 | 0.0000 | 0.0000 | 0.0000 | 0.0000 | 0.0000 | 0.0000 | 0.0000 |
| **7** | 0.0016 | 0.0009 | 0.0004 | 0.0002 | 0.0001 | 0.0000 | 0.0000 | 0.0000 | 0.0000 |
| **8** | 0.0120 | 0.0075 | 0.0043 | 0.0022 | 0.0010 | 0.0003 | 0.0001 | 0.0000 | 0.0000 |
| **9** | 0.0652 | 0.0468 | 0.0316 | 0.0196 | 0.0107 | 0.0048 | 0.0015 | 0.0002 | 0.0000 |
| **10** | 0.2487 | 0.2033 | 0.1595 | 0.1184 | 0.0809 | 0.0486 | 0.0231 | 0.0062 | 0.0000 |
| **11** | 0.6323 | 0.5814 | 0.5241 | 0.4596 | 0.3873 | 0.3062 | 0.2153 | 0.1136 | 0.0000 |
| **12** | 1.0000 | 1.0000 | 1.0000 | 1.0000 | 1.0000 | 1.0000 | 1.0000 | 1.0000 | 1.0000 |

$$n = 13$$

| $x$ | $p = 0.01$ | $p = 0.02$ | $p = 0.03$ | $p = 0.04$ | $p = 0.05$ | $p = 0.06$ | $p = 0.07$ | $p = 0.08$ | $p = 0.09$ |
|---|---|---|---|---|---|---|---|---|---|
| **0** | 0.8775 | 0.7690 | 0.6730 | 0.5882 | 0.5133 | 0.4474 | 0.3893 | 0.3383 | 0.2935 |
| **1** | 0.9928 | 0.9730 | 0.9436 | 0.9068 | 0.8646 | 0.8186 | 0.7702 | 0.7206 | 0.6707 |
| **2** | 0.9997 | 0.9980 | 0.9938 | 0.9865 | 0.9755 | 0.9608 | 0.9422 | 0.9201 | 0.8946 |
| **3** | 1.0000 | 0.9999 | 0.9995 | 0.9986 | 0.9969 | 0.9940 | 0.9897 | 0.9837 | 0.9758 |
| **4** | 1.0000 | 1.0000 | 1.0000 | 0.9999 | 0.9997 | 0.9993 | 0.9987 | 0.9976 | 0.9959 |
| **5** | 1.0000 | 1.0000 | 1.0000 | 1.0000 | 1.0000 | 0.9999 | 0.9999 | 0.9997 | 0.9995 |
| **6** | 1.0000 | 1.0000 | 1.0000 | 1.0000 | 1.0000 | 1.0000 | 1.0000 | 1.0000 | 0.9999 |
| **7** | 1.0000 | 1.0000 | 1.0000 | 1.0000 | 1.0000 | 1.0000 | 1.0000 | 1.0000 | 1.0000 |
| **8** | 1.0000 | 1.0000 | 1.0000 | 1.0000 | 1.0000 | 1.0000 | 1.0000 | 1.0000 | 1.0000 |
| **9** | 1.0000 | 1.0000 | 1.0000 | 1.0000 | 1.0000 | 1.0000 | 1.0000 | 1.0000 | 1.0000 |
| **10** | 1.0000 | 1.0000 | 1.0000 | 1.0000 | 1.0000 | 1.0000 | 1.0000 | 1.0000 | 1.0000 |
| **11** | 1.0000 | 1.0000 | 1.0000 | 1.0000 | 1.0000 | 1.0000 | 1.0000 | 1.0000 | 1.0000 |
| **12** | 1.0000 | 1.0000 | 1.0000 | 1.0000 | 1.0000 | 1.0000 | 1.0000 | 1.0000 | 1.0000 |
| **13** | 1.0000 | 1.0000 | 1.0000 | 1.0000 | 1.0000 | 1.0000 | 1.0000 | 1.0000 | 1.0000 |

| $x$ | $p = 0.10$ | $p = 0.15$ | $p = 0.20$ | $p = 0.25$ | $p = 0.30$ | $p = 0.35$ | $p = 0.40$ | $p = 0.45$ | $p = 0.50$ |
|---|---|---|---|---|---|---|---|---|---|
| **0** | 0.2542 | 0.1209 | 0.0550 | 0.0238 | 0.0097 | 0.0037 | 0.0013 | 0.0004 | 0.0001 |
| **1** | 0.6213 | 0.3983 | 0.2336 | 0.1267 | 0.0637 | 0.0296 | 0.0126 | 0.0049 | 0.0017 |
| **2** | 0.8661 | 0.6920 | 0.5017 | 0.3326 | 0.2025 | 0.1132 | 0.0579 | 0.0269 | 0.0112 |
| **3** | 0.9658 | 0.8820 | 0.7473 | 0.5843 | 0.4206 | 0.2783 | 0.1686 | 0.0929 | 0.0461 |
| **4** | 0.9935 | 0.9658 | 0.9009 | 0.7940 | 0.6543 | 0.5005 | 0.3530 | 0.2279 | 0.1334 |
| **5** | 0.9991 | 0.9925 | 0.9700 | 0.9198 | 0.8346 | 0.7159 | 0.5744 | 0.4268 | 0.2905 |
| **6** | 0.9999 | 0.9987 | 0.9930 | 0.9757 | 0.9376 | 0.8705 | 0.7712 | 0.6437 | 0.5000 |
| **7** | 1.0000 | 0.9998 | 0.9988 | 0.9944 | 0.9818 | 0.9538 | 0.9023 | 0.8212 | 0.7095 |
| **8** | 1.0000 | 1.0000 | 0.9998 | 0.9990 | 0.9960 | 0.9874 | 0.9679 | 0.9302 | 0.8666 |
| **9** | 1.0000 | 1.0000 | 1.0000 | 0.9999 | 0.9993 | 0.9975 | 0.9922 | 0.9797 | 0.9539 |
| **10** | 1.0000 | 1.0000 | 1.0000 | 1.0000 | 0.9999 | 0.9997 | 0.9987 | 0.9959 | 0.9888 |
| **11** | 1.0000 | 1.0000 | 1.0000 | 1.0000 | 1.0000 | 1.0000 | 0.9999 | 0.9995 | 0.9983 |
| **12** | 1.0000 | 1.0000 | 1.0000 | 1.0000 | 1.0000 | 1.0000 | 1.0000 | 1.0000 | 0.9999 |
| **13** | 1.0000 | 1.0000 | 1.0000 | 1.0000 | 1.0000 | 1.0000 | 1.0000 | 1.0000 | 1.0000 |

| $x$ | $p = 0.55$ | $p = 0.60$ | $p = 0.65$ | $p = 0.70$ | $p = 0.75$ | $p = 0.80$ | $p = 0.85$ | $p = 0.90$ | $p = 0.91$ |
|---|---|---|---|---|---|---|---|---|---|
| **0** | 0.0000 | 0.0000 | 0.0000 | 0.0000 | 0.0000 | 0.0000 | 0.0000 | 0.0000 | 0.0000 |
| **1** | 0.0005 | 0.0001 | 0.0000 | 0.0000 | 0.0000 | 0.0000 | 0.0000 | 0.0000 | 0.0000 |
| **2** | 0.0041 | 0.0013 | 0.0003 | 0.0001 | 0.0001 | 0.0000 | 0.0000 | 0.0000 | 0.0000 |
| **3** | 0.0203 | 0.0078 | 0.0025 | 0.0007 | 0.0001 | 0.0000 | 0.0000 | 0.0000 | 0.0000 |
| **4** | 0.0698 | 0.0321 | 0.0126 | 0.0040 | 0.0010 | 0.0002 | 0.0000 | 0.0000 | 0.0000 |
| **5** | 0.1788 | 0.0977 | 0.0462 | 0.0182 | 0.0056 | 0.0012 | 0.0002 | 0.0000 | 0.0000 |
| **6** | 0.3563 | 0.2288 | 0.1295 | 0.0624 | 0.0243 | 0.0070 | 0.0013 | 0.0001 | 0.0001 |
| **7** | 0.5732 | 0.4256 | 0.2841 | 0.1654 | 0.0802 | 0.0300 | 0.0075 | 0.0009 | 0.0005 |
| **8** | 0.7721 | 0.6470 | 0.4995 | 0.3457 | 0.2060 | 0.0991 | 0.0342 | 0.0065 | 0.0041 |
| **9** | 0.9071 | 0.8314 | 0.7217 | 0.5794 | 0.4157 | 0.2527 | 0.1180 | 0.0342 | 0.0242 |
| **10** | 0.9731 | 0.9421 | 0.8868 | 0.7975 | 0.6674 | 0.4983 | 0.3080 | 0.1339 | 0.1054 |
| **11** | 0.9951 | 0.9874 | 0.9704 | 0.9363 | 0.8733 | 0.7664 | 0.6017 | 0.3787 | 0.3293 |
| **12** | 0.9996 | 0.9987 | 0.9963 | 0.9903 | 0.9762 | 0.9450 | 0.8791 | 0.7458 | 0.7065 |
| **13** | 1.0000 | 1.0000 | 1.0000 | 1.0000 | 1.0000 | 1.0000 | 1.0000 | 1.0000 | 1.0000 |

| $x$ | $p = 0.92$ | $p = 0.93$ | $p = 0.94$ | $p = 0.95$ | $p = 0.96$ | $p = 0.97$ | $p = 0.98$ | $p = 0.99$ | $p = 1.00$ |
|---|---|---|---|---|---|---|---|---|---|
| **0** | 0.0000 | 0.0000 | 0.0000 | 0.0000 | 0.0000 | 0.0000 | 0.0000 | 0.0000 | 0.0000 |
| **1** | 0.0000 | 0.0000 | 0.0000 | 0.0000 | 0.0000 | 0.0000 | 0.0000 | 0.0000 | 0.0000 |
| **2** | 0.0000 | 0.0000 | 0.0000 | 0.0000 | 0.0000 | 0.0000 | 0.0000 | 0.0000 | 0.0000 |
| **3** | 0.0000 | 0.0000 | 0.0000 | 0.0000 | 0.0000 | 0.0000 | 0.0000 | 0.0000 | 0.0000 |

| | | | | | | | | | |
|---|---|---|---|---|---|---|---|---|---|
| 4 | 0.0000 | 0.0000 | 0.0000 | 0.0000 | 0.0000 | 0.0000 | 0.0000 | 0.0000 | 0.0000 |
| 5 | 0.0000 | 0.0000 | 0.0000 | 0.0000 | 0.0000 | 0.0000 | 0.0000 | 0.0000 | 0.0000 |
| 6 | 0.0000 | 0.0000 | 0.0000 | 0.0000 | 0.0000 | 0.0000 | 0.0000 | 0.0000 | 0.0000 |
| 7 | 0.0003 | 0.0001 | 0.0001 | 0.0000 | 0.0000 | 0.0000 | 0.0000 | 0.0000 | 0.0000 |
| 8 | 0.0024 | 0.0013 | 0.0007 | 0.0003 | 0.0001 | 0.0000 | 0.0000 | 0.0000 | 0.0000 |
| 9 | 0.0163 | 0.0103 | 0.0060 | 0.0031 | 0.0014 | 0.0005 | 0.0001 | 0.0000 | 0.0000 |
| 10 | 0.0799 | 0.0578 | 0.0392 | 0.0245 | 0.0135 | 0.0062 | 0.0020 | 0.0003 | 0.0000 |
| 11 | 0.2794 | 0.2298 | 0.1814 | 0.1354 | 0.0932 | 0.0564 | 0.0270 | 0.0072 | 0.0000 |
| 12 | 0.6617 | 0.6107 | 0.5526 | 0.4867 | 0.4118 | 0.3270 | 0.2310 | 0.1225 | 0.0000 |
| 13 | 1.0000 | 1.0000 | 1.0000 | 1.0000 | 1.0000 | 1.0000 | 1.0000 | 1.0000 | 1.0000 |

$n = 14$

| $x$ | $p = 0.01$ | $p = 0.02$ | $p = 0.03$ | $p = 0.04$ | $p = 0.05$ | $p = 0.06$ | $p = 0.07$ | $p = 0.08$ | $p = 0.09$ |
|---|---|---|---|---|---|---|---|---|---|
| 0 | 0.8687 | 0.7536 | 0.6528 | 0.5647 | 0.4877 | 0.4205 | 0.3620 | 0.3112 | 0.2670 |
| 1 | 0.9916 | 0.9690 | 0.9355 | 0.8941 | 0.8470 | 0.7963 | 0.7436 | 0.6900 | 0.6368 |
| 2 | 0.9997 | 0.9975 | 0.9923 | 0.9833 | 0.9699 | 0.9522 | 0.9302 | 0.9042 | 0.8745 |
| 3 | 1.0000 | 0.9999 | 0.9994 | 0.9981 | 0.9958 | 0.9920 | 0.9864 | 0.9786 | 0.9685 |
| 4 | 1.0000 | 1.0000 | 1.0000 | 0.9998 | 0.9996 | 0.9990 | 0.9980 | 0.9965 | 0.9941 |
| 5 | 1.0000 | 1.0000 | 1.0000 | 1.0000 | 1.0000 | 0.9999 | 0.9998 | 0.9996 | 0.9992 |
| 6 | 1.0000 | 1.0000 | 1.0000 | 1.0000 | 1.0000 | 1.0000 | 1.0000 | 1.0000 | 0.9999 |
| 7 | 1.0000 | 1.0000 | 1.0000 | 1.0000 | 1.0000 | 1.0000 | 1.0000 | 1.0000 | 1.0000 |
| 8 | 1.0000 | 1.0000 | 1.0000 | 1.0000 | 1.0000 | 1.0000 | 1.0000 | 1.0000 | 1.0000 |
| 9 | 1.0000 | 1.0000 | 1.0000 | 1.0000 | 1.0000 | 1.0000 | 1.0000 | 1.0000 | 1.0000 |
| 10 | 1.0000 | 1.0000 | 1.0000 | 1.0000 | 1.0000 | 1.0000 | 1.0000 | 1.0000 | 1.0000 |
| 11 | 1.0000 | 1.0000 | 1.0000 | 1.0000 | 1.0000 | 1.0000 | 1.0000 | 1.0000 | 1.0000 |
| 12 | 1.0000 | 1.0000 | 1.0000 | 1.0000 | 1.0000 | 1.0000 | 1.0000 | 1.0000 | 1.0000 |
| 13 | 1.0000 | 1.0000 | 1.0000 | 1.0000 | 1.0000 | 1.0000 | 1.0000 | 1.0000 | 1.0000 |
| 14 | 1.0000 | 1.0000 | 1.0000 | 1.0000 | 1.0000 | 1.0000 | 1.0000 | 1.0000 | 1.0000 |

| $x$ | $p = 0.10$ | $p = 0.15$ | $p = 0.20$ | $p = 0.25$ | $p = 0.30$ | $p = 0.35$ | $p = 0.40$ | $p = 0.45$ | $p = 0.50$ |
|---|---|---|---|---|---|---|---|---|---|
| 0 | 0.2288 | 0.1028 | 0.0440 | 0.0178 | 0.0068 | 0.0024 | 0.0008 | 0.0002 | 0.0001 |
| 1 | 0.5846 | 0.3567 | 0.1979 | 0.1010 | 0.0475 | 0.0205 | 0.0081 | 0.0029 | 0.0009 |
| 2 | 0.8416 | 0.6479 | 0.4481 | 0.2811 | 0.1608 | 0.0839 | 0.0398 | 0.0170 | 0.0065 |
| 3 | 0.9559 | 0.8535 | 0.6982 | 0.5213 | 0.3552 | 0.2205 | 0.1243 | 0.0632 | 0.0287 |
| 4 | 0.9908 | 0.9533 | 0.8702 | 0.7415 | 0.5842 | 0.4227 | 0.2793 | 0.1672 | 0.0898 |
| 5 | 0.9985 | 0.9885 | 0.9561 | 0.8883 | 0.7805 | 0.6405 | 0.4859 | 0.3373 | 0.2120 |
| 6 | 0.9998 | 0.9978 | 0.9884 | 0.9617 | 0.9067 | 0.8164 | 0.6925 | 0.5461 | 0.3953 |
| 7 | 1.0000 | 0.9997 | 0.9976 | 0.9897 | 0.9685 | 0.9247 | 0.8499 | 0.7414 | 0.6047 |
| 8 | 1.0000 | 1.0000 | 0.9996 | 0.9978 | 0.9917 | 0.9757 | 0.9417 | 0.8811 | 0.7880 |
| 9 | 1.0000 | 1.0000 | 1.0000 | 0.9997 | 0.9983 | 0.9940 | 0.9825 | 0.9574 | 0.9102 |
| 10 | 1.0000 | 1.0000 | 1.0000 | 1.0000 | 0.9998 | 0.9989 | 0.9961 | 0.9886 | 0.9713 |
| 11 | 1.0000 | 1.0000 | 1.0000 | 1.0000 | 1.0000 | 0.9999 | 0.9994 | 0.9978 | 0.9935 |
| 12 | 1.0000 | 1.0000 | 1.0000 | 1.0000 | 1.0000 | 1.0000 | 0.9999 | 0.9997 | 0.9991 |
| 13 | 1.0000 | 1.0000 | 1.0000 | 1.0000 | 1.0000 | 1.0000 | 1.0000 | 1.0000 | 0.9999 |
| 14 | 1.0000 | 1.0000 | 1.0000 | 1.0000 | 1.0000 | 1.0000 | 1.0000 | 1.0000 | 1.0000 |

| $x$ | $p = 0.55$ | $p = 0.60$ | $p = 0.65$ | $p = 0.70$ | $p = 0.75$ | $p = 0.80$ | $p = 0.85$ | $p = 0.90$ | $p = 0.91$ |
|---|---|---|---|---|---|---|---|---|---|
| 0 | 0.0000 | 0.0000 | 0.0000 | 0.0000 | 0.0000 | 0.0000 | 0.0000 | 0.0000 | 0.0000 |
| 1 | 0.0003 | 0.0001 | 0.0000 | 0.0000 | 0.0000 | 0.0000 | 0.0000 | 0.0000 | 0.0000 |
| 2 | 0.0022 | 0.0006 | 0.0001 | 0.0000 | 0.0000 | 0.0000 | 0.0000 | 0.0000 | 0.0000 |
| 3 | 0.0114 | 0.0039 | 0.0011 | 0.0002 | 0.0000 | 0.0000 | 0.0000 | 0.0000 | 0.0000 |
| 4 | 0.0426 | 0.0175 | 0.0060 | 0.0017 | 0.0003 | 0.0000 | 0.0000 | 0.0000 | 0.0000 |
| 5 | 0.1189 | 0.0583 | 0.0243 | 0.0083 | 0.0022 | 0.0004 | 0.0000 | 0.0000 | 0.0000 |
| 6 | 0.2586 | 0.1501 | 0.0753 | 0.0315 | 0.0103 | 0.0024 | 0.0003 | 0.0000 | 0.0000 |
| 7 | 0.4539 | 0.3075 | 0.1836 | 0.0933 | 0.0383 | 0.0116 | 0.0022 | 0.0002 | 0.0001 |
| 8 | 0.6627 | 0.5141 | 0.3595 | 0.2195 | 0.1117 | 0.0439 | 0.0115 | 0.0015 | 0.0008 |
| 9 | 0.8328 | 0.7207 | 0.5773 | 0.4158 | 0.2585 | 0.1298 | 0.0467 | 0.0092 | 0.0059 |
| 10 | 0.9368 | 0.8757 | 0.7795 | 0.6448 | 0.4787 | 0.3018 | 0.1465 | 0.0441 | 0.0315 |
| 11 | 0.9830 | 0.9602 | 0.9161 | 0.8392 | 0.7189 | 0.5519 | 0.3521 | 0.1584 | 0.1255 |
| 12 | 0.9971 | 0.9919 | 0.9795 | 0.9525 | 0.8990 | 0.8021 | 0.6433 | 0.4154 | 0.3632 |
| 13 | 0.9998 | 0.9992 | 0.9976 | 0.9932 | 0.9822 | 0.9560 | 0.8972 | 0.7712 | 0.7330 |
| 14 | 1.0000 | 1.0000 | 1.0000 | 1.0000 | 1.0000 | 1.0000 | 1.0000 | 1.0000 | 1.0000 |

| $x$ | $p = 0.92$ | $p = 0.93$ | $p = 0.94$ | $p = 0.95$ | $p = 0.96$ | $p = 0.97$ | $p = 0.98$ | $p = 0.99$ | $p = 1.00$ |
|---|---|---|---|---|---|---|---|---|---|
| 0 | 0.0000 | 0.0000 | 0.0000 | 0.0000 | 0.0000 | 0.0000 | 0.0000 | 0.0000 | 0.0000 |

*(continued)*

| | | | | | | | | | |
|---|---|---|---|---|---|---|---|---|---|
| 1 | 0.0000 | 0.0000 | 0.0000 | 0.0000 | 0.0000 | 0.0000 | 0.0000 | 0.0000 | 0.0000 |
| 2 | 0.0000 | 0.0000 | 0.0000 | 0.0000 | 0.0000 | 0.0000 | 0.0000 | 0.0000 | 0.0000 |
| 3 | 0.0000 | 0.0000 | 0.0000 | 0.0000 | 0.0000 | 0.0000 | 0.0000 | 0.0000 | 0.0000 |
| 4 | 0.0000 | 0.0000 | 0.0000 | 0.0000 | 0.0000 | 0.0000 | 0.0000 | 0.0000 | 0.0000 |
| 5 | 0.0000 | 0.0000 | 0.0000 | 0.0000 | 0.0000 | 0.0000 | 0.0000 | 0.0000 | 0.0000 |
| 6 | 0.0000 | 0.0000 | 0.0000 | 0.0000 | 0.0000 | 0.0000 | 0.0000 | 0.0000 | 0.0000 |
| 7 | 0.0000 | 0.0000 | 0.0000 | 0.0000 | 0.0000 | 0.0000 | 0.0000 | 0.0000 | 0.0000 |
| 8 | 0.0004 | 0.0002 | 0.0001 | 0.0000 | 0.0000 | 0.0000 | 0.0000 | 0.0000 | 0.0000 |
| 9 | 0.0035 | 0.0020 | 0.0010 | 0.0004 | 0.0002 | 0.0000 | 0.0000 | 0.0000 | 0.0000 |
| 10 | 0.0214 | 0.0136 | 0.0080 | 0.0042 | 0.0019 | 0.0006 | 0.0001 | 0.0000 | 0.0000 |
| 11 | 0.0958 | 0.0698 | 0.0478 | 0.0301 | 0.0167 | 0.0077 | 0.0025 | 0.0003 | 0.0000 |
| 12 | 0.3100 | 0.2564 | 0.2037 | 0.1530 | 0.1059 | 0.0645 | 0.0310 | 0.0084 | 0.0000 |
| 13 | 0.6888 | 0.6380 | 0.5795 | 0.5123 | 0.4353 | 0.3472 | 0.2464 | 0.1313 | 0.0000 |
| 14 | 1.0000 | 1.0000 | 1.0000 | 1.0000 | 1.0000 | 1.0000 | 1.0000 | 1.0000 | 1.0000 |

$n = 15$

| x | p = 0.01 | p = 0.02 | p = 0.03 | p = 0.04 | p = 0.05 | p = 0.06 | p = 0.07 | p = 0.08 | p = 0.09 |
|---|---|---|---|---|---|---|---|---|---|
| 0 | 0.8601 | 0.7386 | 0.6333 | 0.5421 | 0.4633 | 0.3953 | 0.3367 | 0.2863 | 0.2430 |
| 1 | 0.9904 | 0.9647 | 0.9270 | 0.8809 | 0.8290 | 0.7738 | 0.7168 | 0.6597 | 0.6035 |
| 2 | 0.9996 | 0.9970 | 0.9906 | 0.9797 | 0.9638 | 0.9429 | 0.9171 | 0.8870 | 0.8531 |
| 3 | 1.0000 | 0.9998 | 0.9992 | 0.9976 | 0.9945 | 0.9896 | 0.9825 | 0.9727 | 0.9601 |
| 4 | 1.0000 | 1.0000 | 0.9999 | 0.9998 | 0.9994 | 0.9986 | 0.9972 | 0.9950 | 0.9918 |
| 5 | 1.0000 | 1.0000 | 1.0000 | 1.0000 | 0.9999 | 0.9999 | 0.9997 | 0.9993 | 0.9987 |
| 6 | 1.0000 | 1.0000 | 1.0000 | 1.0000 | 1.0000 | 1.0000 | 1.0000 | 0.9999 | 0.9998 |
| 7 | 1.0000 | 1.0000 | 1.0000 | 1.0000 | 1.0000 | 1.0000 | 1.0000 | 1.0000 | 1.0000 |
| 8 | 1.0000 | 1.0000 | 1.0000 | 1.0000 | 1.0000 | 1.0000 | 1.0000 | 1.0000 | 1.0000 |
| 9 | 1.0000 | 1.0000 | 1.0000 | 1.0000 | 1.0000 | 1.0000 | 1.0000 | 1.0000 | 1.0000 |
| 10 | 1.0000 | 1.0000 | 1.0000 | 1.0000 | 1.0000 | 1.0000 | 1.0000 | 1.0000 | 1.0000 |
| 11 | 1.0000 | 1.0000 | 1.0000 | 1.0000 | 1.0000 | 1.0000 | 1.0000 | 1.0000 | 1.0000 |
| 12 | 1.0000 | 1.0000 | 1.0000 | 1.0000 | 1.0000 | 1.0000 | 1.0000 | 1.0000 | 1.0000 |
| 13 | 1.0000 | 1.0000 | 1.0000 | 1.0000 | 1.0000 | 1.0000 | 1.0000 | 1.0000 | 1.0000 |
| 14 | 1.0000 | 1.0000 | 1.0000 | 1.0000 | 1.0000 | 1.0000 | 1.0000 | 1.0000 | 1.0000 |
| 15 | 1.0000 | 1.0000 | 1.0000 | 1.0000 | 1.0000 | 1.0000 | 1.0000 | 1.0000 | 1.0000 |

| x | p = 0.10 | p = 0.15 | p = 0.20 | p = 0.25 | p = 0.30 | p = 0.35 | p = 0.40 | p = 0.45 | p = 0.50 |
|---|---|---|---|---|---|---|---|---|---|
| 0 | 0.2059 | 0.0874 | 0.0352 | 0.0134 | 0.0047 | 0.0016 | 0.0005 | 0.0001 | 0.0000 |
| 1 | 0.5490 | 0.3186 | 0.1671 | 0.0802 | 0.0353 | 0.0142 | 0.0052 | 0.0017 | 0.0005 |
| 2 | 0.8159 | 0.6042 | 0.3980 | 0.2361 | 0.1268 | 0.0617 | 0.0271 | 0.0107 | 0.0037 |
| 3 | 0.9444 | 0.8227 | 0.6482 | 0.4613 | 0.2969 | 0.1727 | 0.0905 | 0.0424 | 0.0176 |
| 4 | 0.9873 | 0.9383 | 0.8358 | 0.6865 | 0.5155 | 0.3519 | 0.2173 | 0.1204 | 0.0592 |
| 5 | 0.9978 | 0.9832 | 0.9389 | 0.8516 | 0.7216 | 0.5643 | 0.4032 | 0.2608 | 0.1509 |
| 6 | 0.9997 | 0.9964 | 0.9819 | 0.9434 | 0.8689 | 0.7548 | 0.6098 | 0.4522 | 0.3036 |
| 7 | 1.0000 | 0.9994 | 0.9958 | 0.9827 | 0.9500 | 0.8868 | 0.7869 | 0.6535 | 0.5000 |
| 8 | 1.0000 | 0.9999 | 0.9992 | 0.9958 | 0.9848 | 0.9578 | 0.9050 | 0.8182 | 0.6964 |
| 9 | 1.0000 | 1.0000 | 0.9999 | 0.9992 | 0.9963 | 0.9876 | 0.9662 | 0.9231 | 0.8491 |
| 10 | 1.0000 | 1.0000 | 1.0000 | 0.9999 | 0.9993 | 0.9972 | 0.9907 | 0.9745 | 0.9408 |
| 11 | 1.0000 | 1.0000 | 1.0000 | 1.0000 | 0.9999 | 0.9995 | 0.9981 | 0.9937 | 0.9824 |
| 12 | 1.0000 | 1.0000 | 1.0000 | 1.0000 | 1.0000 | 0.9999 | 0.9997 | 0.9989 | 0.9963 |
| 13 | 1.0000 | 1.0000 | 1.0000 | 1.0000 | 1.0000 | 1.0000 | 1.0000 | 0.9999 | 0.9995 |
| 14 | 1.0000 | 1.0000 | 1.0000 | 1.0000 | 1.0000 | 1.0000 | 1.0000 | 1.0000 | 1.0000 |
| 15 | 1.0000 | 1.0000 | 1.0000 | 1.0000 | 1.0000 | 1.0000 | 1.0000 | 1.0000 | 1.0000 |

| x | p = 0.55 | p = 0.60 | p = 0.65 | p = 0.70 | p = 0.75 | p = 0.80 | p = 0.85 | p = 0.90 | p = 0.91 |
|---|---|---|---|---|---|---|---|---|---|
| 0 | 0.0000 | 0.0000 | 0.0000 | 0.0000 | 0.0000 | 0.0000 | 0.0000 | 0.0000 | 0.0000 |
| 1 | 0.0001 | 0.0000 | 0.0000 | 0.0000 | 0.0000 | 0.0000 | 0.0000 | 0.0000 | 0.0000 |
| 2 | 0.0011 | 0.0003 | 0.0001 | 0.0000 | 0.0000 | 0.0000 | 0.0000 | 0.0000 | 0.0000 |
| 3 | 0.0063 | 0.0019 | 0.0005 | 0.0001 | 0.0000 | 0.0000 | 0.0000 | 0.0000 | 0.0000 |
| 4 | 0.0255 | 0.0093 | 0.0028 | 0.0007 | 0.0001 | 0.0000 | 0.0000 | 0.0000 | 0.0000 |
| 5 | 0.0769 | 0.0338 | 0.0124 | 0.0037 | 0.0008 | 0.0001 | 0.0000 | 0.0000 | 0.0000 |
| 6 | 0.1818 | 0.0950 | 0.0422 | 0.0152 | 0.0042 | 0.0008 | 0.0001 | 0.0000 | 0.0000 |
| 7 | 0.3465 | 0.2131 | 0.1132 | 0.0500 | 0.0173 | 0.0042 | 0.0006 | 0.0000 | 0.0000 |
| 8 | 0.5478 | 0.3902 | 0.2452 | 0.1311 | 0.0566 | 0.0181 | 0.0036 | 0.0003 | 0.0002 |
| 9 | 0.7392 | 0.5968 | 0.4357 | 0.2784 | 0.1484 | 0.0611 | 0.0168 | 0.0022 | 0.0013 |
| 10 | 0.8796 | 0.7827 | 0.6481 | 0.4845 | 0.3135 | 0.1642 | 0.0617 | 0.0127 | 0.0082 |

| | | | | | | | | | |
|---|---|---|---|---|---|---|---|---|---|
| 11 | 0.9576 | 0.9095 | 0.8273 | 0.7031 | 0.5387 | 0.3518 | 0.1773 | 0.0556 | 0.0399 |
| 12 | 0.9893 | 0.9729 | 0.9383 | 0.8732 | 0.7639 | 0.6020 | 0.3958 | 0.1841 | 0.1469 |
| 13 | 0.9983 | 0.9948 | 0.9858 | 0.9647 | 0.9198 | 0.8329 | 0.6814 | 0.4510 | 0.3965 |
| 14 | 0.9999 | 0.9995 | 0.9984 | 0.9953 | 0.9866 | 0.9648 | 0.9126 | 0.7941 | 0.7570 |
| 15 | 1.0000 | 1.0000 | 1.0000 | 1.0000 | 1.0000 | 1.0000 | 1.0000 | 1.0000 | 1.0000 |

| $x$ | $p = 0.92$ | $p = 0.93$ | $p = 0.94$ | $p = 0.95$ | $p = 0.96$ | $p = 0.97$ | $p = 0.98$ | $p = 0.99$ | $p = 1.00$ |
|---|---|---|---|---|---|---|---|---|---|
| 0 | 0.0000 | 0.0000 | 0.0000 | 0.0000 | 0.0000 | 0.0000 | 0.0000 | 0.0000 | 0.0000 |
| 1 | 0.0000 | 0.0000 | 0.0000 | 0.0000 | 0.0000 | 0.0000 | 0.0000 | 0.0000 | 0.0000 |
| 2 | 0.0000 | 0.0000 | 0.0000 | 0.0000 | 0.0000 | 0.0000 | 0.0000 | 0.0000 | 0.0000 |
| 3 | 0.0000 | 0.0000 | 0.0000 | 0.0000 | 0.0000 | 0.0000 | 0.0000 | 0.0000 | 0.0000 |
| 4 | 0.0000 | 0.0000 | 0.0000 | 0.0000 | 0.0000 | 0.0000 | 0.0000 | 0.0000 | 0.0000 |
| 5 | 0.0000 | 0.0000 | 0.0000 | 0.0000 | 0.0000 | 0.0000 | 0.0000 | 0.0000 | 0.0000 |
| 6 | 0.0000 | 0.0000 | 0.0000 | 0.0000 | 0.0000 | 0.0000 | 0.0000 | 0.0000 | 0.0000 |
| 7 | 0.0000 | 0.0000 | 0.0000 | 0.0000 | 0.0000 | 0.0000 | 0.0000 | 0.0000 | 0.0000 |
| 8 | 0.0001 | 0.0000 | 0.0000 | 0.0000 | 0.0000 | 0.0000 | 0.0000 | 0.0000 | 0.0000 |
| 9 | 0.0007 | 0.0003 | 0.0001 | 0.0001 | 0.0000 | 0.0000 | 0.0000 | 0.0000 | 0.0000 |
| 10 | 0.0050 | 0.0028 | 0.0014 | 0.0006 | 0.0002 | 0.0001 | 0.0000 | 0.0000 | 0.0000 |
| 11 | 0.0273 | 0.0175 | 0.0104 | 0.0055 | 0.0024 | 0.0008 | 0.0002 | 0.0000 | 0.0000 |
| 12 | 0.1130 | 0.0829 | 0.0571 | 0.0362 | 0.0203 | 0.0094 | 0.0030 | 0.0004 | 0.0000 |
| 13 | 0.3403 | 0.2832 | 0.2262 | 0.1710 | 0.1191 | 0.0730 | 0.0353 | 0.0096 | 0.0000 |
| 14 | 0.7137 | 0.6633 | 0.6047 | 0.5367 | 0.4579 | 0.3667 | 0.2614 | 0.1399 | 0.0000 |
| 15 | 1.0000 | 1.0000 | 1.0000 | 1.0000 | 1.0000 | 1.0000 | 1.0000 | 1.0000 | 1.0000 |

$n = 20$

| $x$ | $p = 0.01$ | $p = 0.02$ | $p = 0.03$ | $p = 0.04$ | $p = 0.05$ | $p = 0.06$ | $p = 0.07$ | $p = 0.08$ | $p = 0.09$ |
|---|---|---|---|---|---|---|---|---|---|
| 0 | 0.8179 | 0.6676 | 0.5438 | 0.4420 | 0.3585 | 0.2901 | 0.2342 | 0.1887 | 0.1516 |
| 1 | 0.9831 | 0.9401 | 0.8802 | 0.8103 | 0.7358 | 0.6605 | 0.5869 | 0.5169 | 0.4516 |
| 2 | 0.9990 | 0.9929 | 0.9790 | 0.9561 | 0.9245 | 0.8850 | 0.8390 | 0.7879 | 0.7334 |
| 3 | 1.0000 | 0.9994 | 0.9973 | 0.9926 | 0.9841 | 0.9710 | 0.9529 | 0.9294 | 0.9007 |
| 4 | 1.0000 | 1.0000 | 0.9997 | 0.9990 | 0.9974 | 0.9944 | 0.9893 | 0.9817 | 0.9710 |
| 5 | 1.0000 | 1.0000 | 1.0000 | 0.9999 | 0.9997 | 0.9991 | 0.9981 | 0.9962 | 0.9932 |
| 6 | 1.0000 | 1.0000 | 1.0000 | 1.0000 | 1.0000 | 0.9999 | 0.9997 | 0.9994 | 0.9987 |
| 7 | 1.0000 | 1.0000 | 1.0000 | 1.0000 | 1.0000 | 1.0000 | 1.0000 | 0.9999 | 0.9998 |
| 8 | 1.0000 | 1.0000 | 1.0000 | 1.0000 | 1.0000 | 1.0000 | 1.0000 | 1.0000 | 1.0000 |
| 9 | 1.0000 | 1.0000 | 1.0000 | 1.0000 | 1.0000 | 1.0000 | 1.0000 | 1.0000 | 1.0000 |
| 10 | 1.0000 | 1.0000 | 1.0000 | 1.0000 | 1.0000 | 1.0000 | 1.0000 | 1.0000 | 1.0000 |
| 11 | 1.0000 | 1.0000 | 1.0000 | 1.0000 | 1.0000 | 1.0000 | 1.0000 | 1.0000 | 1.0000 |
| 12 | 1.0000 | 1.0000 | 1.0000 | 1.0000 | 1.0000 | 1.0000 | 1.0000 | 1.0000 | 1.0000 |
| 13 | 1.0000 | 1.0000 | 1.0000 | 1.0000 | 1.0000 | 1.0000 | 1.0000 | 1.0000 | 1.0000 |
| 14 | 1.0000 | 1.0000 | 1.0000 | 1.0000 | 1.0000 | 1.0000 | 1.0000 | 1.0000 | 1.0000 |
| 15 | 1.0000 | 1.0000 | 1.0000 | 1.0000 | 1.0000 | 1.0000 | 1.0000 | 1.0000 | 1.0000 |
| 16 | 1.0000 | 1.0000 | 1.0000 | 1.0000 | 1.0000 | 1.0000 | 1.0000 | 1.0000 | 1.0000 |
| 17 | 1.0000 | 1.0000 | 1.0000 | 1.0000 | 1.0000 | 1.0000 | 1.0000 | 1.0000 | 1.0000 |
| 18 | 1.0000 | 1.0000 | 1.0000 | 1.0000 | 1.0000 | 1.0000 | 1.0000 | 1.0000 | 1.0000 |
| 19 | 1.0000 | 1.0000 | 1.0000 | 1.0000 | 1.0000 | 1.0000 | 1.0000 | 1.0000 | 1.0000 |
| 20 | 1.0000 | 1.0000 | 1.0000 | 1.0000 | 1.0000 | 1.0000 | 1.0000 | 1.0000 | 1.0000 |

| $x$ | $p = 0.10$ | $p = 0.15$ | $p = 0.20$ | $p = 0.25$ | $p = 0.30$ | $p = 0.35$ | $p = 0.40$ | $p = 0.45$ | $p = 0.50$ |
|---|---|---|---|---|---|---|---|---|---|
| 0 | 0.1216 | 0.0388 | 0.0115 | 0.0032 | 0.0008 | 0.0002 | 0.0000 | 0.0000 | 0.0000 |
| 1 | 0.3917 | 0.1756 | 0.0692 | 0.0243 | 0.0076 | 0.0021 | 0.0005 | 0.0001 | 0.0000 |
| 2 | 0.6769 | 0.4049 | 0.2061 | 0.0913 | 0.0355 | 0.0121 | 0.0036 | 0.0009 | 0.0002 |
| 3 | 0.8670 | 0.6477 | 0.4114 | 0.2252 | 0.1071 | 0.0444 | 0.0160 | 0.0049 | 0.0013 |
| 4 | 0.9568 | 0.8298 | 0.6296 | 0.4148 | 0.2375 | 0.1182 | 0.0510 | 0.0189 | 0.0059 |
| 5 | 0.9887 | 0.9327 | 0.8042 | 0.6172 | 0.4164 | 0.2454 | 0.1256 | 0.0553 | 0.0207 |
| 6 | 0.9976 | 0.9781 | 0.9133 | 0.7858 | 0.6080 | 0.4166 | 0.2500 | 0.1299 | 0.0577 |
| 7 | 0.9996 | 0.9941 | 0.9679 | 0.8982 | 0.7723 | 0.6010 | 0.4159 | 0.2520 | 0.1316 |
| 8 | 0.9999 | 0.9987 | 0.9900 | 0.9591 | 0.8867 | 0.7624 | 0.5956 | 0.4143 | 0.2517 |
| 9 | 1.0000 | 0.9998 | 0.9974 | 0.9861 | 0.9520 | 0.8782 | 0.7553 | 0.5914 | 0.4119 |
| 10 | 1.0000 | 1.0000 | 0.9994 | 0.9961 | 0.9829 | 0.9468 | 0.8725 | 0.7507 | 0.5881 |
| 11 | 1.0000 | 1.0000 | 0.9999 | 0.9991 | 0.9949 | 0.9804 | 0.9435 | 0.8692 | 0.7483 |
| 12 | 1.0000 | 1.0000 | 1.0000 | 0.9998 | 0.9987 | 0.9940 | 0.9790 | 0.9420 | 0.8684 |
| 13 | 1.0000 | 1.0000 | 1.0000 | 1.0000 | 0.9997 | 0.9985 | 0.9935 | 0.9786 | 0.9423 |
| 14 | 1.0000 | 1.0000 | 1.0000 | 1.0000 | 1.0000 | 0.9997 | 0.9984 | 0.9936 | 0.9793 |

*(continued)*

| 15 | 1.0000 | 1.0000 | 1.0000 | 1.0000 | 1.0000 | 1.0000 | 0.9997 | 0.9985 | 0.9941 |
| 16 | 1.0000 | 1.0000 | 1.0000 | 1.0000 | 1.0000 | 1.0000 | 1.0000 | 0.9997 | 0.9987 |
| 17 | 1.0000 | 1.0000 | 1.0000 | 1.0000 | 1.0000 | 1.0000 | 1.0000 | 1.0000 | 0.9998 |
| 18 | 1.0000 | 1.0000 | 1.0000 | 1.0000 | 1.0000 | 1.0000 | 1.0000 | 1.0000 | 1.0000 |
| 19 | 1.0000 | 1.0000 | 1.0000 | 1.0000 | 1.0000 | 1.0000 | 1.0000 | 1.0000 | 1.0000 |
| 20 | 1.0000 | 1.0000 | 1.0000 | 1.0000 | 1.0000 | 1.0000 | 1.0000 | 1.0000 | 1.0000 |

| $x$ | $p = 0.55$ | $p = 0.60$ | $p = 0.65$ | $p = 0.70$ | $p = 0.75$ | $p = 0.80$ | $p = 0.85$ | $p = 0.90$ | $p = 0.91$ |
|---|---|---|---|---|---|---|---|---|---|
| 0 | 0.0000 | 0.0000 | 0.0000 | 0.0000 | 0.0000 | 0.0000 | 0.0000 | 0.0000 | 0.0000 |
| 1 | 0.0000 | 0.0000 | 0.0000 | 0.0000 | 0.0000 | 0.0000 | 0.0000 | 0.0000 | 0.0000 |
| 2 | 0.0000 | 0.0000 | 0.0000 | 0.0000 | 0.0000 | 0.0000 | 0.0000 | 0.0000 | 0.0000 |
| 3 | 0.0003 | 0.0000 | 0.0000 | 0.0000 | 0.0000 | 0.0000 | 0.0000 | 0.0000 | 0.0000 |
| 4 | 0.0015 | 0.0003 | 0.0000 | 0.0000 | 0.0000 | 0.0000 | 0.0000 | 0.0000 | 0.0000 |
| 5 | 0.0064 | 0.0016 | 0.0003 | 0.0000 | 0.0000 | 0.0000 | 0.0000 | 0.0000 | 0.0000 |
| 6 | 0.0214 | 0.0065 | 0.0015 | 0.0003 | 0.0000 | 0.0000 | 0.0000 | 0.0000 | 0.0000 |
| 7 | 0.0580 | 0.0210 | 0.0060 | 0.0013 | 0.0002 | 0.0000 | 0.0000 | 0.0000 | 0.0000 |
| 8 | 0.1308 | 0.0565 | 0.0196 | 0.0051 | 0.0009 | 0.0001 | 0.0000 | 0.0000 | 0.0000 |
| 9 | 0.2493 | 0.1275 | 0.0532 | 0.0171 | 0.0039 | 0.0006 | 0.0000 | 0.0000 | 0.0000 |
| 10 | 0.4086 | 0.2447 | 0.1218 | 0.0480 | 0.0139 | 0.0026 | 0.0002 | 0.0000 | 0.0000 |
| 11 | 0.5857 | 0.4044 | 0.2376 | 0.1133 | 0.0409 | 0.0100 | 0.0013 | 0.0001 | 0.0000 |
| 12 | 0.7480 | 0.5841 | 0.3990 | 0.2277 | 0.1018 | 0.0321 | 0.0059 | 0.0004 | 0.0002 |
| 13 | 0.8701 | 0.7500 | 0.5834 | 0.3920 | 0.2142 | 0.0867 | 0.0219 | 0.0024 | 0.0013 |
| 14 | 0.9447 | 0.8744 | 0.7546 | 0.5836 | 0.3828 | 0.1958 | 0.0673 | 0.0113 | 0.0068 |
| 15 | 0.9811 | 0.9490 | 0.8818 | 0.7625 | 0.5852 | 0.3704 | 0.1702 | 0.0432 | 0.0290 |
| 16 | 0.9951 | 0.9840 | 0.9556 | 0.8929 | 0.7748 | 0.5886 | 0.3523 | 0.1330 | 0.0993 |
| 17 | 0.9991 | 0.9964 | 0.9879 | 0.9645 | 0.9087 | 0.7939 | 0.5951 | 0.3231 | 0.2666 |
| 18 | 0.9999 | 0.9995 | 0.9979 | 0.9924 | 0.9757 | 0.9308 | 0.8244 | 0.6083 | 0.5484 |
| 19 | 1.0000 | 1.0000 | 0.9998 | 0.9992 | 0.9968 | 0.9885 | 0.9612 | 0.8784 | 0.8484 |
| 20 | 1.0000 | 1.0000 | 1.0000 | 1.0000 | 1.0000 | 1.0000 | 1.0000 | 1.0000 | 1.0000 |

| $x$ | $p = 0.92$ | $p = 0.93$ | $p = 0.94$ | $p = 0.95$ | $p = 0.96$ | $p = 0.97$ | $p = 0.98$ | $p = 0.99$ | $p = 1.00$ |
|---|---|---|---|---|---|---|---|---|---|
| 0 | 0.0000 | 0.0000 | 0.0000 | 0.0000 | 0.0000 | 0.0000 | 0.0000 | 0.0000 | 0.0000 |
| 1 | 0.0000 | 0.0000 | 0.0000 | 0.0000 | 0.0000 | 0.0000 | 0.0000 | 0.0000 | 0.0000 |
| 2 | 0.0000 | 0.0000 | 0.0000 | 0.0000 | 0.0000 | 0.0000 | 0.0000 | 0.0000 | 0.0000 |
| 3 | 0.0000 | 0.0000 | 0.0000 | 0.0000 | 0.0000 | 0.0000 | 0.0000 | 0.0000 | 0.0000 |
| 4 | 0.0000 | 0.0000 | 0.0000 | 0.0000 | 0.0000 | 0.0000 | 0.0000 | 0.0000 | 0.0000 |
| 5 | 0.0000 | 0.0000 | 0.0000 | 0.0000 | 0.0000 | 0.0000 | 0.0000 | 0.0000 | 0.0000 |
| 6 | 0.0000 | 0.0000 | 0.0000 | 0.0000 | 0.0000 | 0.0000 | 0.0000 | 0.0000 | 0.0000 |
| 7 | 0.0000 | 0.0000 | 0.0000 | 0.0000 | 0.0000 | 0.0000 | 0.0000 | 0.0000 | 0.0000 |
| 8 | 0.0000 | 0.0000 | 0.0000 | 0.0000 | 0.0000 | 0.0000 | 0.0000 | 0.0000 | 0.0000 |
| 9 | 0.0000 | 0.0000 | 0.0000 | 0.0000 | 0.0000 | 0.0000 | 0.0000 | 0.0000 | 0.0000 |
| 10 | 0.0000 | 0.0000 | 0.0000 | 0.0000 | 0.0000 | 0.0000 | 0.0000 | 0.0000 | 0.0000 |
| 11 | 0.0000 | 0.0000 | 0.0000 | 0.0000 | 0.0000 | 0.0000 | 0.0000 | 0.0000 | 0.0000 |
| 12 | 0.0001 | 0.0000 | 0.0000 | 0.0000 | 0.0000 | 0.0000 | 0.0000 | 0.0000 | 0.0000 |
| 13 | 0.0006 | 0.0003 | 0.0001 | 0.0000 | 0.0000 | 0.0000 | 0.0000 | 0.0000 | 0.0000 |
| 14 | 0.0038 | 0.0019 | 0.0009 | 0.0003 | 0.0001 | 0.0000 | 0.0000 | 0.0000 | 0.0000 |
| 15 | 0.0183 | 0.0107 | 0.0056 | 0.0026 | 0.0010 | 0.0003 | 0.0000 | 0.0000 | 0.0000 |
| 16 | 0.0706 | 0.0471 | 0.0290 | 0.0159 | 0.0074 | 0.0027 | 0.0006 | 0.0000 | 0.0000 |
| 17 | 0.2121 | 0.1610 | 0.1150 | 0.0755 | 0.0439 | 0.0210 | 0.0071 | 0.0010 | 0.0000 |
| 18 | 0.4831 | 0.4131 | 0.3395 | 0.2642 | 0.1897 | 0.1198 | 0.0599 | 0.0169 | 0.0000 |
| 19 | 0.8113 | 0.7658 | 0.7099 | 0.6415 | 0.5580 | 0.4562 | 0.3324 | 0.1821 | 0.0000 |
| 20 | 1.0000 | 1.0000 | 1.0000 | 1.0000 | 1.0000 | 1.0000 | 1.0000 | 1.0000 | 1.0000 |

$n = 25$

| $x$ | $p = 0.01$ | $p = 0.02$ | $p = 0.03$ | $p = 0.04$ | $p = 0.05$ | $p = 0.06$ | $p = 0.07$ | $p = 0.08$ | $p = 0.09$ |
|---|---|---|---|---|---|---|---|---|---|
| 0 | 0.7778 | 0.6035 | 0.4670 | 0.3604 | 0.2774 | 0.2129 | 0.1630 | 0.1244 | 0.0946 |
| 1 | 0.9742 | 0.9114 | 0.8280 | 0.7358 | 0.6424 | 0.5527 | 0.4696 | 0.3947 | 0.3286 |
| 2 | 0.9980 | 0.9868 | 0.9620 | 0.9235 | 0.8729 | 0.8129 | 0.7466 | 0.6768 | 0.6063 |
| 3 | 0.9999 | 0.9986 | 0.9938 | 0.9835 | 0.9659 | 0.9402 | 0.9064 | 0.8649 | 0.8169 |
| 4 | 1.0000 | 0.9999 | 0.9992 | 0.9972 | 0.9928 | 0.9850 | 0.9726 | 0.9549 | 0.9314 |
| 5 | 1.0000 | 1.0000 | 0.9999 | 0.9996 | 0.9988 | 0.9969 | 0.9935 | 0.9877 | 0.9790 |
| 6 | 1.0000 | 1.0000 | 1.0000 | 1.0000 | 0.9998 | 0.9995 | 0.9987 | 0.9972 | 0.9946 |
| 7 | 1.0000 | 1.0000 | 1.0000 | 1.0000 | 1.0000 | 0.9999 | 0.9998 | 0.9995 | 0.9989 |
| 8 | 1.0000 | 1.0000 | 1.0000 | 1.0000 | 1.0000 | 1.0000 | 1.0000 | 0.9999 | 0.9998 |

| 9 | 1.0000 | 1.0000 | 1.0000 | 1.0000 | 1.0000 | 1.0000 | 1.0000 | 1.0000 | 1.0000 |
| 10 | 1.0000 | 1.0000 | 1.0000 | 1.0000 | 1.0000 | 1.0000 | 1.0000 | 1.0000 | 1.0000 |
| 11 | 1.0000 | 1.0000 | 1.0000 | 1.0000 | 1.0000 | 1.0000 | 1.0000 | 1.0000 | 1.0000 |
| 12 | 1.0000 | 1.0000 | 1.0000 | 1.0000 | 1.0000 | 1.0000 | 1.0000 | 1.0000 | 1.0000 |
| 13 | 1.0000 | 1.0000 | 1.0000 | 1.0000 | 1.0000 | 1.0000 | 1.0000 | 1.0000 | 1.0000 |
| 14 | 1.0000 | 1.0000 | 1.0000 | 1.0000 | 1.0000 | 1.0000 | 1.0000 | 1.0000 | 1.0000 |
| 15 | 1.0000 | 1.0000 | 1.0000 | 1.0000 | 1.0000 | 1.0000 | 1.0000 | 1.0000 | 1.0000 |
| 16 | 1.0000 | 1.0000 | 1.0000 | 1.0000 | 1.0000 | 1.0000 | 1.0000 | 1.0000 | 1.0000 |
| 17 | 1.0000 | 1.0000 | 1.0000 | 1.0000 | 1.0000 | 1.0000 | 1.0000 | 1.0000 | 1.0000 |
| 18 | 1.0000 | 1.0000 | 1.0000 | 1.0000 | 1.0000 | 1.0000 | 1.0000 | 1.0000 | 1.0000 |
| 19 | 1.0000 | 1.0000 | 1.0000 | 1.0000 | 1.0000 | 1.0000 | 1.0000 | 1.0000 | 1.0000 |
| 20 | 1.0000 | 1.0000 | 1.0000 | 1.0000 | 1.0000 | 1.0000 | 1.0000 | 1.0000 | 1.0000 |
| 21 | 1.0000 | 1.0000 | 1.0000 | 1.0000 | 1.0000 | 1.0000 | 1.0000 | 1.0000 | 1.0000 |
| 22 | 1.0000 | 1.0000 | 1.0000 | 1.0000 | 1.0000 | 1.0000 | 1.0000 | 1.0000 | 1.0000 |
| 23 | 1.0000 | 1.0000 | 1.0000 | 1.0000 | 1.0000 | 1.0000 | 1.0000 | 1.0000 | 1.0000 |
| 24 | 1.0000 | 1.0000 | 1.0000 | 1.0000 | 1.0000 | 1.0000 | 1.0000 | 1.0000 | 1.0000 |
| 25 | 1.0000 | 1.0000 | 1.0000 | 1.0000 | 1.0000 | 1.0000 | 1.0000 | 1.0000 | 1.0000 |

| $x$ | $p = 0.10$ | $p = 0.15$ | $p = 0.20$ | $p = 0.25$ | $p = 0.30$ | $p = 0.35$ | $p = 0.40$ | $p = 0.45$ | $p = 0.50$ |
|---|---|---|---|---|---|---|---|---|---|
| 0 | 0.0718 | 0.0172 | 0.0038 | 0.0008 | 0.0001 | 0.0000 | 0.0000 | 0.0000 | 0.0000 |
| 1 | 0.2712 | 0.0931 | 0.0274 | 0.0070 | 0.0016 | 0.0003 | 0.0001 | 0.0000 | 0.0000 |
| 2 | 0.5371 | 0.2537 | 0.0982 | 0.0321 | 0.0090 | 0.0021 | 0.0004 | 0.0001 | 0.0000 |
| 3 | 0.7636 | 0.4711 | 0.2340 | 0.0962 | 0.0332 | 0.0097 | 0.0024 | 0.0005 | 0.0001 |
| 4 | 0.9020 | 0.6821 | 0.4207 | 0.2137 | 0.0905 | 0.0320 | 0.0095 | 0.0023 | 0.0005 |
| 5 | 0.9666 | 0.8385 | 0.6167 | 0.3783 | 0.1935 | 0.0826 | 0.0294 | 0.0086 | 0.0020 |
| 6 | 0.9905 | 0.9305 | 0.7800 | 0.5611 | 0.3407 | 0.1734 | 0.0736 | 0.0258 | 0.0073 |
| 7 | 0.9977 | 0.9745 | 0.8909 | 0.7265 | 0.5118 | 0.3061 | 0.1536 | 0.0639 | 0.0216 |
| 8 | 0.9995 | 0.9920 | 0.9532 | 0.8506 | 0.6769 | 0.4668 | 0.2735 | 0.1340 | 0.0539 |
| 9 | 0.9999 | 0.9979 | 0.9827 | 0.9287 | 0.8106 | 0.6303 | 0.4246 | 0.2424 | 0.1148 |
| 10 | 1.0000 | 0.9995 | 0.9944 | 0.9703 | 0.9022 | 0.7712 | 0.5858 | 0.3843 | 0.2122 |
| 11 | 1.0000 | 0.9999 | 0.9985 | 0.9893 | 0.9558 | 0.8746 | 0.7323 | 0.5426 | 0.3450 |
| 12 | 1.0000 | 1.0000 | 0.9996 | 0.9966 | 0.9825 | 0.9396 | 0.8462 | 0.6937 | 0.5000 |
| 13 | 1.0000 | 1.0000 | 0.9999 | 0.9991 | 0.9940 | 0.9745 | 0.9222 | 0.8173 | 0.6550 |
| 14 | 1.0000 | 1.0000 | 1.0000 | 0.9998 | 0.9982 | 0.9907 | 0.9656 | 0.9040 | 0.7878 |
| 15 | 1.0000 | 1.0000 | 1.0000 | 1.0000 | 0.9995 | 0.9971 | 0.9868 | 0.9560 | 0.8852 |
| 16 | 1.0000 | 1.0000 | 1.0000 | 1.0000 | 0.9999 | 0.9992 | 0.9957 | 0.9826 | 0.9461 |
| 17 | 1.0000 | 1.0000 | 1.0000 | 1.0000 | 1.0000 | 0.9998 | 0.9988 | 0.9942 | 0.9784 |
| 18 | 1.0000 | 1.0000 | 1.0000 | 1.0000 | 1.0000 | 1.0000 | 0.9997 | 0.9984 | 0.9927 |
| 19 | 1.0000 | 1.0000 | 1.0000 | 1.0000 | 1.0000 | 1.0000 | 0.9999 | 0.9996 | 0.9980 |
| 20 | 1.0000 | 1.0000 | 1.0000 | 1.0000 | 1.0000 | 1.0000 | 1.0000 | 0.9999 | 0.9995 |
| 21 | 1.0000 | 1.0000 | 1.0000 | 1.0000 | 1.0000 | 1.0000 | 1.0000 | 1.0000 | 0.9999 |
| 22 | 1.0000 | 1.0000 | 1.0000 | 1.0000 | 1.0000 | 1.0000 | 1.0000 | 1.0000 | 1.0000 |
| 23 | 1.0000 | 1.0000 | 1.0000 | 1.0000 | 1.0000 | 1.0000 | 1.0000 | 1.0000 | 1.0000 |
| 24 | 1.0000 | 1.0000 | 1.0000 | 1.0000 | 1.0000 | 1.0000 | 1.0000 | 1.0000 | 1.0000 |
| 25 | 1.0000 | 1.0000 | 1.0000 | 1.0000 | 1.0000 | 1.0000 | 1.0000 | 1.0000 | 1.0000 |

| $x$ | $p = 0.55$ | $p = 0.60$ | $p = 0.65$ | $p = 0.70$ | $p = 0.75$ | $p = 0.80$ | $p = 0.85$ | $p = 0.90$ | $p = 0.91$ |
|---|---|---|---|---|---|---|---|---|---|
| 0 | 0.0000 | 0.0000 | 0.0000 | 0.0000 | 0.0000 | 0.0000 | 0.0000 | 0.0000 | 0.0000 |
| 1 | 0.0000 | 0.0000 | 0.0000 | 0.0000 | 0.0000 | 0.0000 | 0.0000 | 0.0000 | 0.0000 |
| 2 | 0.0000 | 0.0000 | 0.0000 | 0.0000 | 0.0000 | 0.0000 | 0.0000 | 0.0000 | 0.0000 |
| 3 | 0.0000 | 0.0000 | 0.0000 | 0.0000 | 0.0000 | 0.0000 | 0.0000 | 0.0000 | 0.0000 |
| 4 | 0.0001 | 0.0000 | 0.0000 | 0.0000 | 0.0000 | 0.0000 | 0.0000 | 0.0000 | 0.0000 |
| 5 | 0.0004 | 0.0001 | 0.0000 | 0.0000 | 0.0000 | 0.0000 | 0.0000 | 0.0000 | 0.0000 |
| 6 | 0.0016 | 0.0003 | 0.0000 | 0.0000 | 0.0000 | 0.0000 | 0.0000 | 0.0000 | 0.0000 |
| 7 | 0.0058 | 0.0012 | 0.0002 | 0.0000 | 0.0000 | 0.0000 | 0.0000 | 0.0000 | 0.0000 |
| 8 | 0.0174 | 0.0043 | 0.0008 | 0.0001 | 0.0000 | 0.0000 | 0.0000 | 0.0000 | 0.0000 |
| 9 | 0.0440 | 0.0132 | 0.0029 | 0.0005 | 0.0000 | 0.0000 | 0.0000 | 0.0000 | 0.0000 |
| 10 | 0.0960 | 0.0344 | 0.0093 | 0.0018 | 0.0002 | 0.0000 | 0.0000 | 0.0000 | 0.0000 |
| 11 | 0.1827 | 0.0778 | 0.0255 | 0.0060 | 0.0009 | 0.0001 | 0.0000 | 0.0000 | 0.0000 |
| 12 | 0.3063 | 0.1538 | 0.0604 | 0.0175 | 0.0034 | 0.0004 | 0.0000 | 0.0000 | 0.0000 |
| 13 | 0.4574 | 0.2677 | 0.1254 | 0.0442 | 0.0107 | 0.0015 | 0.0001 | 0.0000 | 0.0000 |
| 14 | 0.6157 | 0.4142 | 0.2288 | 0.0978 | 0.0297 | 0.0056 | 0.0005 | 0.0000 | 0.0000 |
| 15 | 0.7576 | 0.5754 | 0.3697 | 0.1894 | 0.0713 | 0.0173 | 0.0021 | 0.0001 | 0.0000 |
| 16 | 0.8660 | 0.7265 | 0.5332 | 0.3231 | 0.1494 | 0.0468 | 0.0080 | 0.0005 | 0.0002 |
| 17 | 0.9361 | 0.8464 | 0.6939 | 0.4882 | 0.2735 | 0.1091 | 0.0255 | 0.0023 | 0.0011 |

*(continued)*

| | | | | | | | | | |
|----|--------|--------|--------|--------|--------|--------|--------|--------|--------|
| 18 | 0.9742 | 0.9264 | 0.8266 | 0.6593 | 0.4389 | 0.2200 | 0.0695 | 0.0095 | 0.0054 |
| 19 | 0.9914 | 0.9706 | 0.9174 | 0.8065 | 0.6217 | 0.3833 | 0.1615 | 0.0334 | 0.0210 |
| 20 | 0.9977 | 0.9905 | 0.9680 | 0.9095 | 0.7863 | 0.5793 | 0.3179 | 0.0980 | 0.0686 |
| 21 | 0.9995 | 0.9976 | 0.9903 | 0.9668 | 0.9038 | 0.7660 | 0.5289 | 0.2364 | 0.1831 |
| 22 | 0.9999 | 0.9996 | 0.9979 | 0.9910 | 0.9679 | 0.9018 | 0.7463 | 0.4629 | 0.3937 |
| 23 | 1.0000 | 0.9999 | 0.9997 | 0.9984 | 0.9930 | 0.9726 | 0.9069 | 0.7288 | 0.6714 |
| 24 | 1.0000 | 1.0000 | 1.0000 | 0.9999 | 0.9992 | 0.9962 | 0.9828 | 0.9282 | 0.9054 |
| 25 | 1.0000 | 1.0000 | 1.0000 | 1.0000 | 1.0000 | 1.0000 | 1.0000 | 1.0000 | 1.0000 |

| $x$ | $p = 0.92$ | $p = 0.93$ | $p = 0.94$ | $p = 0.95$ | $p = 0.96$ | $p = 0.97$ | $p = 0.98$ | $p = 0.99$ | $p = 1.00$ |
|-----|--------|--------|--------|--------|--------|--------|--------|--------|--------|
| 0 | 0.0000 | 0.0000 | 0.0000 | 0.0000 | 0.0000 | 0.0000 | 0.0000 | 0.0000 | 0.0000 |
| 1 | 0.0000 | 0.0000 | 0.0000 | 0.0000 | 0.0000 | 0.0000 | 0.0000 | 0.0000 | 0.0000 |
| 2 | 0.0000 | 0.0000 | 0.0000 | 0.0000 | 0.0000 | 0.0000 | 0.0000 | 0.0000 | 0.0000 |
| 3 | 0.0000 | 0.0000 | 0.0000 | 0.0000 | 0.0000 | 0.0000 | 0.0000 | 0.0000 | 0.0000 |
| 4 | 0.0000 | 0.0000 | 0.0000 | 0.0000 | 0.0000 | 0.0000 | 0.0000 | 0.0000 | 0.0000 |
| 5 | 0.0000 | 0.0000 | 0.0000 | 0.0000 | 0.0000 | 0.0000 | 0.0000 | 0.0000 | 0.0000 |
| 6 | 0.0000 | 0.0000 | 0.0000 | 0.0000 | 0.0000 | 0.0000 | 0.0000 | 0.0000 | 0.0000 |
| 7 | 0.0000 | 0.0000 | 0.0000 | 0.0000 | 0.0000 | 0.0000 | 0.0000 | 0.0000 | 0.0000 |
| 8 | 0.0000 | 0.0000 | 0.0000 | 0.0000 | 0.0000 | 0.0000 | 0.0000 | 0.0000 | 0.0000 |
| 9 | 0.0000 | 0.0000 | 0.0000 | 0.0000 | 0.0000 | 0.0000 | 0.0000 | 0.0000 | 0.0000 |
| 10 | 0.0000 | 0.0000 | 0.0000 | 0.0000 | 0.0000 | 0.0000 | 0.0000 | 0.0000 | 0.0000 |
| 11 | 0.0000 | 0.0000 | 0.0000 | 0.0000 | 0.0000 | 0.0000 | 0.0000 | 0.0000 | 0.0000 |
| 12 | 0.0000 | 0.0000 | 0.0000 | 0.0000 | 0.0000 | 0.0000 | 0.0000 | 0.0000 | 0.0000 |
| 13 | 0.0000 | 0.0000 | 0.0000 | 0.0000 | 0.0000 | 0.0000 | 0.0000 | 0.0000 | 0.0000 |
| 14 | 0.0000 | 0.0000 | 0.0000 | 0.0000 | 0.0000 | 0.0000 | 0.0000 | 0.0000 | 0.0000 |
| 15 | 0.0000 | 0.0000 | 0.0000 | 0.0000 | 0.0000 | 0.0000 | 0.0000 | 0.0000 | 0.0000 |
| 16 | 0.0001 | 0.0000 | 0.0000 | 0.0000 | 0.0000 | 0.0000 | 0.0000 | 0.0000 | 0.0000 |
| 17 | 0.0005 | 0.0002 | 0.0001 | 0.0000 | 0.0000 | 0.0000 | 0.0000 | 0.0000 | 0.0000 |
| 18 | 0.0028 | 0.0013 | 0.0005 | 0.0002 | 0.0000 | 0.0000 | 0.0000 | 0.0000 | 0.0000 |
| 19 | 0.0123 | 0.0065 | 0.0031 | 0.0012 | 0.0004 | 0.0001 | 0.0000 | 0.0000 | 0.0000 |
| 20 | 0.0451 | 0.0274 | 0.0150 | 0.0072 | 0.0028 | 0.0008 | 0.0001 | 0.0000 | 0.0000 |
| 21 | 0.1351 | 0.0936 | 0.0598 | 0.0341 | 0.0165 | 0.0062 | 0.0014 | 0.0001 | 0.0000 |
| 22 | 0.3232 | 0.2534 | 0.1871 | 0.1271 | 0.0765 | 0.0380 | 0.0132 | 0.0020 | 0.0000 |
| 23 | 0.6053 | 0.5304 | 0.4473 | 0.3576 | 0.2642 | 0.1720 | 0.0886 | 0.0258 | 0.0000 |
| 24 | 0.8756 | 0.8370 | 0.7871 | 0.7226 | 0.6396 | 0.5330 | 0.3965 | 0.2222 | 0.0000 |
| 25 | 1.0000 | 1.0000 | 1.0000 | 1.0000 | 1.0000 | 1.0000 | 1.0000 | 1.0000 | 1.0000 |

$$P(x \leq X) = \sum_{i=0}^{X} \frac{(\lambda t)^i e^{-\lambda t}}{i!}$$

$\lambda t$

| x | 0.005 | 0.01 | 0.02 | 0.03 | 0.04 | 0.05 | 0.06 | 0.07 | 0.08 | 0.09 |
|---|-------|------|------|------|------|------|------|------|------|------|
| 0 | 0.9950 | 0.9900 | 0.9802 | 0.9704 | 0.9608 | 0.9512 | 0.9418 | 0.9324 | 0.9231 | 0.9139 |
| 1 | 1.0000 | 1.0000 | 0.9998 | 0.9996 | 0.9992 | 0.9988 | 0.9983 | 0.9977 | 0.9970 | 0.9962 |
| 2 | 1.0000 | 1.0000 | 1.0000 | 1.0000 | 1.0000 | 1.0000 | 1.0000 | 0.9999 | 0.9999 | 0.9999 |
| 3 | 1.0000 | 1.0000 | 1.0000 | 1.0000 | 1.0000 | 1.0000 | 1.0000 | 1.0000 | 1.0000 | 1.0000 |

$\lambda t$

| x | 0.10 | 0.20 | 0.30 | 0.40 | 0.50 | 0.60 | 0.70 | 0.80 | 0.90 | 1.00 |
|---|------|------|------|------|------|------|------|------|------|------|
| 0 | 0.9048 | 0.8187 | 0.7408 | 0.6703 | 0.6065 | 0.5488 | 0.4966 | 0.4493 | 0.4066 | 0.3679 |
| 1 | 0.9953 | 0.9825 | 0.9631 | 0.9384 | 0.9098 | 0.8781 | 0.8442 | 0.8088 | 0.7725 | 0.7358 |
| 2 | 0.9998 | 0.9989 | 0.9964 | 0.9921 | 0.9856 | 0.9769 | 0.9659 | 0.9526 | 0.9371 | 0.9197 |
| 3 | 1.0000 | 0.9999 | 0.9997 | 0.9992 | 0.9982 | 0.9966 | 0.9942 | 0.9909 | 0.9865 | 0.9810 |
| 4 | 1.0000 | 1.0000 | 1.0000 | 0.9999 | 0.9998 | 0.9996 | 0.9992 | 0.9986 | 0.9977 | 0.9963 |
| 5 | 1.0000 | 1.0000 | 1.0000 | 1.0000 | 1.0000 | 1.0000 | 0.9999 | 0.9998 | 0.9997 | 0.9994 |
| 6 | 1.0000 | 1.0000 | 1.0000 | 1.0000 | 1.0000 | 1.0000 | 1.0000 | 1.0000 | 1.0000 | 0.9999 |
| 7 | 1.0000 | 1.0000 | 1.0000 | 1.0000 | 1.0000 | 1.0000 | 1.0000 | 1.0000 | 1.0000 | 1.0000 |

$\lambda t$

| x | 1.10 | 1.20 | 1.30 | 1.40 | 1.50 | 1.60 | 1.70 | 1.80 | 1.90 | 2.00 |
|---|------|------|------|------|------|------|------|------|------|------|
| 0 | 0.3329 | 0.3012 | 0.2725 | 0.2466 | 0.2231 | 0.2019 | 0.1827 | 0.1653 | 0.1496 | 0.1353 |
| 1 | 0.6990 | 0.6626 | 0.6268 | 0.5918 | 0.5578 | 0.5249 | 0.4932 | 0.4628 | 0.4337 | 0.4060 |
| 2 | 0.9004 | 0.8795 | 0.8571 | 0.8335 | 0.8088 | 0.7834 | 0.7572 | 0.7306 | 0.7037 | 0.6767 |
| 3 | 0.9743 | 0.9662 | 0.9569 | 0.9463 | 0.9344 | 0.9212 | 0.9068 | 0.8913 | 0.8747 | 0.8571 |
| 4 | 0.9946 | 0.9923 | 0.9893 | 0.9857 | 0.9814 | 0.9763 | 0.9704 | 0.9636 | 0.9559 | 0.9473 |
| 5 | 0.9990 | 0.9985 | 0.9978 | 0.9968 | 0.9955 | 0.9940 | 0.9920 | 0.9896 | 0.9868 | 0.9834 |
| 6 | 0.9999 | 0.9997 | 0.9996 | 0.9994 | 0.9991 | 0.9987 | 0.9981 | 0.9974 | 0.9966 | 0.9955 |
| 7 | 1.0000 | 1.0000 | 0.9999 | 0.9999 | 0.9998 | 0.9997 | 0.9996 | 0.9994 | 0.9992 | 0.9989 |
| 8 | 1.0000 | 1.0000 | 1.0000 | 1.0000 | 1.0000 | 1.0000 | 0.9999 | 0.9999 | 0.9998 | 0.9998 |
| 9 | 1.0000 | 1.0000 | 1.0000 | 1.0000 | 1.0000 | 1.0000 | 1.0000 | 1.0000 | 1.0000 | 1.0000 |

$\lambda t$

| x | 2.10 | 2.20 | 2.30 | 2.40 | 2.50 | 2.60 | 2.70 | 2.80 | 2.90 | 3.00 |
|---|------|------|------|------|------|------|------|------|------|------|
| 0 | 0.1225 | 0.1108 | 0.1003 | 0.0907 | 0.0821 | 0.0743 | 0.0672 | 0.0608 | 0.0550 | 0.0498 |
| 1 | 0.3796 | 0.3546 | 0.3309 | 0.3084 | 0.2873 | 0.2674 | 0.2487 | 0.2311 | 0.2146 | 0.1991 |
| 2 | 0.6496 | 0.6227 | 0.5960 | 0.5697 | 0.5438 | 0.5184 | 0.4936 | 0.4695 | 0.4460 | 0.4232 |
| 3 | 0.8386 | 0.8194 | 0.7993 | 0.7787 | 0.7576 | 0.7360 | 0.7141 | 0.6919 | 0.6696 | 0.6472 |
| 4 | 0.9379 | 0.9275 | 0.9162 | 0.9041 | 0.8912 | 0.8774 | 0.8629 | 0.8477 | 0.8318 | 0.8153 |
| 5 | 0.9796 | 0.9751 | 0.9700 | 0.9643 | 0.9580 | 0.9510 | 0.9433 | 0.9349 | 0.9258 | 0.9161 |
| 6 | 0.9941 | 0.9925 | 0.9906 | 0.9884 | 0.9858 | 0.9828 | 0.9794 | 0.9756 | 0.9713 | 0.9665 |
| 7 | 0.9985 | 0.9980 | 0.9974 | 0.9967 | 0.9958 | 0.9947 | 0.9934 | 0.9919 | 0.9901 | 0.9881 |
| 8 | 0.9997 | 0.9995 | 0.9994 | 0.9991 | 0.9989 | 0.9985 | 0.9981 | 0.9976 | 0.9969 | 0.9962 |
| 9 | 0.9999 | 0.9999 | 0.9999 | 0.9998 | 0.9997 | 0.9996 | 0.9995 | 0.9993 | 0.9991 | 0.9989 |
| 10 | 1.0000 | 1.0000 | 1.0000 | 1.0000 | 0.9999 | 0.9999 | 0.9999 | 0.9998 | 0.9998 | 0.9997 |
| 11 | 1.0000 | 1.0000 | 1.0000 | 1.0000 | 1.0000 | 1.0000 | 1.0000 | 1.0000 | 0.9999 | 0.9999 |
| 12 | 1.0000 | 1.0000 | 1.0000 | 1.0000 | 1.0000 | 1.0000 | 1.0000 | 1.0000 | 1.0000 | 1.0000 |

*(continued)*

$\lambda t$

| $x$ | 3.10 | 3.20 | 3.30 | 3.40 | 3.50 | 3.60 | 3.70 | 3.80 | 3.90 | 4.00 |
|---|---|---|---|---|---|---|---|---|---|---|
| 0 | 0.0450 | 0.0408 | 0.0369 | 0.0334 | 0.0302 | 0.0273 | 0.0247 | 0.0224 | 0.0202 | 0.0183 |
| 1 | 0.1847 | 0.1712 | 0.1586 | 0.1468 | 0.1359 | 0.1257 | 0.1162 | 0.1074 | 0.0992 | 0.0916 |
| 2 | 0.4012 | 0.3799 | 0.3594 | 0.3397 | 0.3208 | 0.3027 | 0.2854 | 0.2689 | 0.2531 | 0.2381 |
| 3 | 0.6248 | 0.6025 | 0.5803 | 0.5584 | 0.5366 | 0.5152 | 0.4942 | 0.4735 | 0.4532 | 0.4335 |
| 4 | 0.7982 | 0.7806 | 0.7626 | 0.7442 | 0.7254 | 0.7064 | 0.6872 | 0.6678 | 0.6484 | 0.6288 |
| 5 | 0.9057 | 0.8946 | 0.8829 | 0.8705 | 0.8576 | 0.8441 | 0.8301 | 0.8156 | 0.8006 | 0.7851 |
| 6 | 0.9612 | 0.9554 | 0.9490 | 0.9421 | 0.9347 | 0.9267 | 0.9182 | 0.9091 | 0.8995 | 0.8893 |
| 7 | 0.9858 | 0.9832 | 0.9802 | 0.9769 | 0.9733 | 0.9692 | 0.9648 | 0.9599 | 0.9546 | 0.9489 |
| 8 | 0.9953 | 0.9943 | 0.9931 | 0.9917 | 0.9901 | 0.9883 | 0.9863 | 0.9840 | 0.9815 | 0.9786 |
| 9 | 0.9986 | 0.9982 | 0.9978 | 0.9973 | 0.9967 | 0.9960 | 0.9952 | 0.9942 | 0.9931 | 0.9919 |
| 10 | 0.9996 | 0.9995 | 0.9994 | 0.9992 | 0.9990 | 0.9987 | 0.9984 | 0.9981 | 0.9977 | 0.9972 |
| 11 | 0.9999 | 0.9999 | 0.9998 | 0.9998 | 0.9997 | 0.9996 | 0.9995 | 0.9994 | 0.9993 | 0.9991 |
| 12 | 1.0000 | 1.0000 | 1.0000 | 0.9999 | 0.9999 | 0.9999 | 0.9999 | 0.9998 | 0.9998 | 0.9997 |
| 13 | 1.0000 | 1.0000 | 1.0000 | 1.0000 | 1.0000 | 1.0000 | 1.0000 | 1.0000 | 0.9999 | 0.9999 |
| 14 | 1.0000 | 1.0000 | 1.0000 | 1.0000 | 1.0000 | 1.0000 | 1.0000 | 1.0000 | 1.0000 | 1.0000 |

$\lambda t$

| $x$ | 4.10 | 4.20 | 4.30 | 4.40 | 4.50 | 4.60 | 4.70 | 4.80 | 4.90 | 5.00 |
|---|---|---|---|---|---|---|---|---|---|---|
| 0 | 0.0166 | 0.0150 | 0.0136 | 0.0123 | 0.0111 | 0.0101 | 0.0091 | 0.0082 | 0.0074 | 0.0067 |
| 1 | 0.0845 | 0.0780 | 0.0719 | 0.0663 | 0.0611 | 0.0563 | 0.0518 | 0.0477 | 0.0439 | 0.0404 |
| 2 | 0.2238 | 0.2102 | 0.1974 | 0.1851 | 0.1736 | 0.1626 | 0.1523 | 0.1425 | 0.1333 | 0.1247 |
| 3 | 0.4142 | 0.3954 | 0.3772 | 0.3594 | 0.3423 | 0.3257 | 0.3097 | 0.2942 | 0.2793 | 0.2650 |
| 4 | 0.6093 | 0.5898 | 0.5704 | 0.5512 | 0.5321 | 0.5132 | 0.4946 | 0.4763 | 0.4582 | 0.4405 |
| 5 | 0.7693 | 0.7531 | 0.7367 | 0.7199 | 0.7029 | 0.6858 | 0.6684 | 0.6510 | 0.6335 | 0.6160 |
| 6 | 0.8786 | 0.8675 | 0.8558 | 0.8436 | 0.8311 | 0.8180 | 0.8046 | 0.7908 | 0.7767 | 0.7622 |
| 7 | 0.9427 | 0.9361 | 0.9290 | 0.9214 | 0.9134 | 0.9049 | 0.8960 | 0.8867 | 0.8769 | 0.8666 |
| 8 | 0.9755 | 0.9721 | 0.9683 | 0.9642 | 0.9597 | 0.9549 | 0.9497 | 0.9442 | 0.9382 | 0.9319 |
| 9 | 0.9905 | 0.9889 | 0.9871 | 0.9851 | 0.9829 | 0.9805 | 0.9778 | 0.9749 | 0.9717 | 0.9682 |
| 10 | 0.9966 | 0.9959 | 0.9952 | 0.9943 | 0.9933 | 0.9922 | 0.9910 | 0.9896 | 0.9880 | 0.9863 |
| 11 | 0.9989 | 0.9986 | 0.9983 | 0.9980 | 0.9976 | 0.9971 | 0.9966 | 0.9960 | 0.9953 | 0.9945 |
| 12 | 0.9997 | 0.9996 | 0.9995 | 0.9993 | 0.9992 | 0.9990 | 0.9988 | 0.9986 | 0.9983 | 0.9980 |
| 13 | 0.9999 | 0.9999 | 0.9998 | 0.9998 | 0.9997 | 0.9997 | 0.9996 | 0.9995 | 0.9994 | 0.9993 |
| 14 | 1.0000 | 1.0000 | 1.0000 | 0.9999 | 0.9999 | 0.9999 | 0.9999 | 0.9999 | 0.9998 | 0.9998 |
| 15 | 1.0000 | 1.0000 | 1.0000 | 1.0000 | 1.0000 | 1.0000 | 1.0000 | 1.0000 | 0.9999 | 0.9999 |
| 16 | 1.0000 | 1.0000 | 1.0000 | 1.0000 | 1.0000 | 1.0000 | 1.0000 | 1.0000 | 1.0000 | 1.0000 |

$\lambda t$

| $x$ | 5.10 | 5.20 | 5.30 | 5.40 | 5.50 | 5.60 | 5.70 | 5.80 | 5.90 | 6.00 |
|---|---|---|---|---|---|---|---|---|---|---|
| 0 | 0.0061 | 0.0055 | 0.0050 | 0.0045 | 0.0041 | 0.0037 | 0.0033 | 0.0030 | 0.0027 | 0.0025 |
| 1 | 0.0372 | 0.0342 | 0.0314 | 0.0289 | 0.0266 | 0.0244 | 0.0224 | 0.0206 | 0.0189 | 0.0174 |
| 2 | 0.1165 | 0.1088 | 0.1016 | 0.0948 | 0.0884 | 0.0824 | 0.0768 | 0.0715 | 0.0666 | 0.0620 |
| 3 | 0.2513 | 0.2381 | 0.2254 | 0.2133 | 0.2017 | 0.1906 | 0.1800 | 0.1700 | 0.1604 | 0.1512 |
| 4 | 0.4231 | 0.4061 | 0.3895 | 0.3733 | 0.3575 | 0.3422 | 0.3272 | 0.3127 | 0.2987 | 0.2851 |
| 5 | 0.5984 | 0.5809 | 0.5635 | 0.5461 | 0.5289 | 0.5119 | 0.4950 | 0.4783 | 0.4619 | 0.4457 |
| 6 | 0.7474 | 0.7324 | 0.7171 | 0.7017 | 0.6860 | 0.6703 | 0.6544 | 0.6384 | 0.6224 | 0.6063 |
| 7 | 0.8560 | 0.8449 | 0.8335 | 0.8217 | 0.8095 | 0.7970 | 0.7841 | 0.7710 | 0.7576 | 0.7440 |
| 8 | 0.9252 | 0.9181 | 0.9106 | 0.9027 | 0.8944 | 0.8857 | 0.8766 | 0.8672 | 0.8574 | 0.8472 |
| 9 | 0.9644 | 0.9603 | 0.9559 | 0.9512 | 0.9462 | 0.9409 | 0.9352 | 0.9292 | 0.9228 | 0.9161 |
| 10 | 0.9844 | 0.9823 | 0.9800 | 0.9775 | 0.9747 | 0.9718 | 0.9686 | 0.9651 | 0.9614 | 0.9574 |
| 11 | 0.9937 | 0.9927 | 0.9916 | 0.9904 | 0.9890 | 0.9875 | 0.9859 | 0.9841 | 0.9821 | 0.9799 |
| 12 | 0.9976 | 0.9972 | 0.9967 | 0.9962 | 0.9955 | 0.9949 | 0.9941 | 0.9932 | 0.9922 | 0.9912 |
| 13 | 0.9992 | 0.9990 | 0.9988 | 0.9986 | 0.9983 | 0.9980 | 0.9977 | 0.9973 | 0.9969 | 0.9964 |
| 14 | 0.9997 | 0.9997 | 0.9996 | 0.9995 | 0.9994 | 0.9993 | 0.9991 | 0.9990 | 0.9988 | 0.9986 |
| 15 | 0.9999 | 0.9999 | 0.9999 | 0.9998 | 0.9998 | 0.9998 | 0.9997 | 0.9996 | 0.9996 | 0.9995 |
| 16 | 1.0000 | 1.0000 | 1.0000 | 0.9999 | 0.9999 | 0.9999 | 0.9999 | 0.9999 | 0.9999 | 0.9998 |
| 17 | 1.0000 | 1.0000 | 1.0000 | 1.0000 | 1.0000 | 1.0000 | 1.0000 | 1.0000 | 1.0000 | 0.9999 |
| 18 | 1.0000 | 1.0000 | 1.0000 | 1.0000 | 1.0000 | 1.0000 | 1.0000 | 1.0000 | 1.0000 | 1.0000 |

| | | | | | $\lambda t$ | | | | | |
|---|---|---|---|---|---|---|---|---|---|---|
| $x$ | **6.10** | **6.20** | **6.30** | **6.40** | **6.50** | **6.60** | **6.70** | **6.80** | **6.90** | **7.00** |
| 0 | 0.0022 | 0.0020 | 0.0018 | 0.0017 | 0.0015 | 0.0014 | 0.0012 | 0.0011 | 0.0010 | 0.0009 |
| 1 | 0.0159 | 0.0146 | 0.0134 | 0.0123 | 0.0113 | 0.0103 | 0.0095 | 0.0087 | 0.0080 | 0.0073 |
| 2 | 0.0577 | 0.0536 | 0.0498 | 0.0463 | 0.0430 | 0.0400 | 0.0371 | 0.0344 | 0.0320 | 0.0296 |
| 3 | 0.1425 | 0.1342 | 0.1264 | 0.1189 | 0.1118 | 0.1052 | 0.0988 | 0.0928 | 0.0871 | 0.0818 |
| 4 | 0.2719 | 0.2592 | 0.2469 | 0.2351 | 0.2237 | 0.2127 | 0.2022 | 0.1920 | 0.1823 | 0.1730 |
| 5 | 0.4298 | 0.4141 | 0.3988 | 0.3837 | 0.3690 | 0.3547 | 0.3406 | 0.3270 | 0.3137 | 0.3007 |
| 6 | 0.5902 | 0.5742 | 0.5582 | 0.5423 | 0.5265 | 0.5108 | 0.4953 | 0.4799 | 0.4647 | 0.4497 |
| 7 | 0.7301 | 0.7160 | 0.7017 | 0.6873 | 0.6728 | 0.6581 | 0.6433 | 0.6285 | 0.6136 | 0.5987 |
| 8 | 0.8367 | 0.8259 | 0.8148 | 0.8033 | 0.7916 | 0.7796 | 0.7673 | 0.7548 | 0.7420 | 0.7291 |
| 9 | 0.9090 | 0.9016 | 0.8939 | 0.8858 | 0.8774 | 0.8686 | 0.8596 | 0.8502 | 0.8405 | 0.8305 |
| 10 | 0.9531 | 0.9486 | 0.9437 | 0.9386 | 0.9332 | 0.9274 | 0.9214 | 0.9151 | 0.9084 | 0.9015 |
| 11 | 0.9776 | 0.9750 | 0.9723 | 0.9693 | 0.9661 | 0.9627 | 0.9591 | 0.9552 | 0.9510 | 0.9467 |
| 12 | 0.9900 | 0.9887 | 0.9873 | 0.9857 | 0.9840 | 0.9821 | 0.9801 | 0.9779 | 0.9755 | 0.9730 |
| 13 | 0.9958 | 0.9952 | 0.9945 | 0.9937 | 0.9929 | 0.9920 | 0.9909 | 0.9898 | 0.9885 | 0.9872 |
| 14 | 0.9984 | 0.9981 | 0.9978 | 0.9974 | 0.9970 | 0.9966 | 0.9961 | 0.9956 | 0.9950 | 0.9943 |
| 15 | 0.9994 | 0.9993 | 0.9992 | 0.9990 | 0.9988 | 0.9986 | 0.9984 | 0.9982 | 0.9979 | 0.9976 |
| 16 | 0.9998 | 0.9997 | 0.9997 | 0.9996 | 0.9996 | 0.9995 | 0.9994 | 0.9993 | 0.9992 | 0.9990 |
| 17 | 0.9999 | 0.9999 | 0.9999 | 0.9999 | 0.9998 | 0.9998 | 0.9998 | 0.9997 | 0.9997 | 0.9996 |
| 18 | 1.0000 | 1.0000 | 1.0000 | 1.0000 | 0.9999 | 0.9999 | 0.9999 | 0.9999 | 0.9999 | 0.9999 |
| 19 | 1.0000 | 1.0000 | 1.0000 | 1.0000 | 1.0000 | 1.0000 | 1.0000 | 1.0000 | 1.0000 | 1.0000 |
| 20 | 1.0000 | 1.0000 | 1.0000 | 1.0000 | 1.0000 | 1.0000 | 1.0000 | 1.0000 | 1.0000 | 1.0000 |

| | | | | | $\lambda t$ | | | | | |
|---|---|---|---|---|---|---|---|---|---|---|
| $x$ | **7.10** | **7.20** | **7.30** | **7.40** | **7.50** | **7.60** | **7.70** | **7.80** | **7.90** | **8.00** |
| 0 | 0.0008 | 0.0007 | 0.0007 | 0.0006 | 0.0006 | 0.0005 | 0.0005 | 0.0004 | 0.0004 | 0.0003 |
| 1 | 0.0067 | 0.0061 | 0.0056 | 0.0051 | 0.0047 | 0.0043 | 0.0039 | 0.0036 | 0.0033 | 0.0030 |
| 2 | 0.0275 | 0.0255 | 0.0236 | 0.0219 | 0.0203 | 0.0188 | 0.0174 | 0.0161 | 0.0149 | 0.0138 |
| 3 | 0.0767 | 0.0719 | 0.0674 | 0.0632 | 0.0591 | 0.0554 | 0.0518 | 0.0485 | 0.0453 | 0.0424 |
| 4 | 0.1641 | 0.1555 | 0.1473 | 0.1395 | 0.1321 | 0.1249 | 0.1181 | 0.1117 | 0.1055 | 0.0996 |
| 5 | 0.2881 | 0.2759 | 0.2640 | 0.2526 | 0.2414 | 0.2307 | 0.2203 | 0.2103 | 0.2006 | 0.1912 |
| 6 | 0.4349 | 0.4204 | 0.4060 | 0.3920 | 0.3782 | 0.3646 | 0.3514 | 0.3384 | 0.3257 | 0.3134 |
| 7 | 0.5838 | 0.5689 | 0.5541 | 0.5393 | 0.5246 | 0.5100 | 0.4956 | 0.4812 | 0.4670 | 0.4530 |
| 8 | 0.7160 | 0.7027 | 0.6892 | 0.6757 | 0.6620 | 0.6482 | 0.6343 | 0.6204 | 0.6065 | 0.5925 |
| 9 | 0.8202 | 0.8096 | 0.7988 | 0.7877 | 0.7764 | 0.7649 | 0.7531 | 0.7411 | 0.7290 | 0.7166 |
| 10 | 0.8942 | 0.8867 | 0.8788 | 0.8707 | 0.8622 | 0.8535 | 0.8445 | 0.8352 | 0.8257 | 0.8159 |
| 11 | 0.9420 | 0.9371 | 0.9319 | 0.9265 | 0.9208 | 0.9148 | 0.9085 | 0.9020 | 0.8952 | 0.8881 |
| 12 | 0.9703 | 0.9673 | 0.9642 | 0.9609 | 0.9573 | 0.9536 | 0.9496 | 0.9454 | 0.9409 | 0.9362 |
| 13 | 0.9857 | 0.9841 | 0.9824 | 0.9805 | 0.9784 | 0.9762 | 0.9739 | 0.9714 | 0.9687 | 0.9658 |
| 14 | 0.9935 | 0.9927 | 0.9918 | 0.9908 | 0.9897 | 0.9886 | 0.9873 | 0.9859 | 0.9844 | 0.9827 |
| 15 | 0.9972 | 0.9969 | 0.9964 | 0.9959 | 0.9954 | 0.9948 | 0.9941 | 0.9934 | 0.9926 | 0.9918 |
| 16 | 0.9989 | 0.9987 | 0.9985 | 0.9983 | 0.9980 | 0.9978 | 0.9974 | 0.9971 | 0.9967 | 0.9963 |
| 17 | 0.9996 | 0.9995 | 0.9994 | 0.9993 | 0.9992 | 0.9991 | 0.9989 | 0.9988 | 0.9986 | 0.9984 |
| 18 | 0.9998 | 0.9998 | 0.9998 | 0.9997 | 0.9997 | 0.9996 | 0.9996 | 0.9995 | 0.9994 | 0.9993 |
| 19 | 0.9999 | 0.9999 | 0.9999 | 0.9999 | 0.9999 | 0.9999 | 0.9998 | 0.9998 | 0.9998 | 0.9997 |
| 20 | 1.0000 | 1.0000 | 1.0000 | 1.0000 | 1.0000 | 1.0000 | 0.9999 | 0.9999 | 0.9999 | 0.9999 |
| 21 | 1.0000 | 1.0000 | 1.0000 | 1.0000 | 1.0000 | 1.0000 | 1.0000 | 1.0000 | 1.0000 | 1.0000 |

| | | | | | $\lambda t$ | | | | | |
|---|---|---|---|---|---|---|---|---|---|---|
| $x$ | **8.10** | **8.20** | **8.30** | **8.40** | **8.50** | **8.60** | **8.70** | **8.80** | **8.90** | **9.00** |
| 0 | 0.0003 | 0.0003 | 0.0002 | 0.0002 | 0.0002 | 0.0002 | 0.0002 | 0.0002 | 0.0001 | 0.0001 |
| 1 | 0.0028 | 0.0025 | 0.0023 | 0.0021 | 0.0019 | 0.0018 | 0.0016 | 0.0015 | 0.0014 | 0.0012 |
| 2 | 0.0127 | 0.0118 | 0.0109 | 0.0100 | 0.0093 | 0.0086 | 0.0079 | 0.0073 | 0.0068 | 0.0062 |
| 3 | 0.0396 | 0.0370 | 0.0346 | 0.0323 | 0.0301 | 0.0281 | 0.0262 | 0.0244 | 0.0228 | 0.0212 |
| 4 | 0.0940 | 0.0887 | 0.0837 | 0.0789 | 0.0744 | 0.0701 | 0.0660 | 0.0621 | 0.0584 | 0.0550 |
| 5 | 0.1822 | 0.1736 | 0.1653 | 0.1573 | 0.1496 | 0.1422 | 0.1352 | 0.1284 | 0.1219 | 0.1157 |
| 6 | 0.3013 | 0.2896 | 0.2781 | 0.2670 | 0.2562 | 0.2457 | 0.2355 | 0.2256 | 0.2160 | 0.2068 |
| 7 | 0.4391 | 0.4254 | 0.4119 | 0.3987 | 0.3856 | 0.3728 | 0.3602 | 0.3478 | 0.3357 | 0.3239 |
| 8 | 0.5786 | 0.5647 | 0.5507 | 0.5369 | 0.5231 | 0.5094 | 0.4958 | 0.4823 | 0.4689 | 0.4557 |

*(continued)*

| 9 | 0.7041 | 0.6915 | 0.6788 | 0.6659 | 0.6530 | 0.6400 | 0.6269 | 0.6137 | 0.6006 | 0.5874 |
| 10 | 0.8058 | 0.7955 | 0.7850 | 0.7743 | 0.7634 | 0.7522 | 0.7409 | 0.7294 | 0.7178 | 0.7060 |
| 11 | 0.8807 | 0.8731 | 0.8652 | 0.8571 | 0.8487 | 0.8400 | 0.8311 | 0.8220 | 0.8126 | 0.8030 |
| 12 | 0.9313 | 0.9261 | 0.9207 | 0.9150 | 0.9091 | 0.9029 | 0.8965 | 0.8898 | 0.8829 | 0.8758 |
| 13 | 0.9628 | 0.9595 | 0.9561 | 0.9524 | 0.9486 | 0.9445 | 0.9403 | 0.9358 | 0.9311 | 0.9261 |
| 14 | 0.9810 | 0.9791 | 0.9771 | 0.9749 | 0.9726 | 0.9701 | 0.9675 | 0.9647 | 0.9617 | 0.9585 |
| 15 | 0.9908 | 0.9898 | 0.9887 | 0.9875 | 0.9862 | 0.9848 | 0.9832 | 0.9816 | 0.9798 | 0.9780 |
| 16 | 0.9958 | 0.9953 | 0.9947 | 0.9941 | 0.9934 | 0.9926 | 0.9918 | 0.9909 | 0.9899 | 0.9889 |
| 17 | 0.9982 | 0.9979 | 0.9977 | 0.9973 | 0.9970 | 0.9966 | 0.9962 | 0.9957 | 0.9952 | 0.9947 |
| 18 | 0.9992 | 0.9991 | 0.9990 | 0.9989 | 0.9987 | 0.9985 | 0.9983 | 0.9981 | 0.9978 | 0.9976 |
| 19 | 0.9997 | 0.9997 | 0.9996 | 0.9995 | 0.9995 | 0.9994 | 0.9993 | 0.9992 | 0.9991 | 0.9989 |
| 20 | 0.9999 | 0.9999 | 0.9998 | 0.9998 | 0.9998 | 0.9998 | 0.9997 | 0.9997 | 0.9996 | 0.9996 |
| 21 | 1.0000 | 1.0000 | 0.9999 | 0.9999 | 0.9999 | 0.9999 | 0.9999 | 0.9999 | 0.9998 | 0.9998 |
| 22 | 1.0000 | 1.0000 | 1.0000 | 1.0000 | 1.0000 | 1.0000 | 1.0000 | 1.0000 | 0.9999 | 0.9999 |
| 23 | 1.0000 | 1.0000 | 1.0000 | 1.0000 | 1.0000 | 1.0000 | 1.0000 | 1.0000 | 1.0000 | 1.0000 |

$\lambda t$

| x | 9.10 | 9.20 | 9.30 | 9.40 | 9.50 | 9.60 | 9.70 | 9.80 | 9.90 | 10.00 |
|---|---|---|---|---|---|---|---|---|---|---|
| 0 | 0.0001 | 0.0001 | 0.0001 | 0.0001 | 0.0001 | 0.0001 | 0.0001 | 0.0001 | 0.0001 | 0.0000 |
| 1 | 0.0011 | 0.0010 | 0.0009 | 0.0009 | 0.0008 | 0.0007 | 0.0007 | 0.0006 | 0.0005 | 0.0005 |
| 2 | 0.0058 | 0.0053 | 0.0049 | 0.0045 | 0.0042 | 0.0038 | 0.0035 | 0.0033 | 0.0030 | 0.0028 |
| 3 | 0.0198 | 0.0184 | 0.0172 | 0.0160 | 0.0149 | 0.0138 | 0.0129 | 0.0120 | 0.0111 | 0.0103 |
| 4 | 0.0517 | 0.0486 | 0.0456 | 0.0429 | 0.0403 | 0.0378 | 0.0355 | 0.0333 | 0.0312 | 0.0293 |
| 5 | 0.1098 | 0.1041 | 0.0986 | 0.0935 | 0.0885 | 0.0838 | 0.0793 | 0.0750 | 0.0710 | 0.0671 |
| 6 | 0.1978 | 0.1892 | 0.1808 | 0.1727 | 0.1649 | 0.1574 | 0.1502 | 0.1433 | 0.1366 | 0.1301 |
| 7 | 0.3123 | 0.3010 | 0.2900 | 0.2792 | 0.2687 | 0.2584 | 0.2485 | 0.2388 | 0.2294 | 0.2202 |
| 8 | 0.4426 | 0.4296 | 0.4168 | 0.4042 | 0.3918 | 0.3796 | 0.3676 | 0.3558 | 0.3442 | 0.3328 |
| 9 | 0.5742 | 0.5611 | 0.5479 | 0.5349 | 0.5218 | 0.5089 | 0.4960 | 0.4832 | 0.4705 | 0.4579 |
| 10 | 0.6941 | 0.6820 | 0.6699 | 0.6576 | 0.6453 | 0.6329 | 0.6205 | 0.6080 | 0.5955 | 0.5830 |
| 11 | 0.7932 | 0.7832 | 0.7730 | 0.7626 | 0.7520 | 0.7412 | 0.7303 | 0.7193 | 0.7081 | 0.6968 |
| 12 | 0.8684 | 0.8607 | 0.8529 | 0.8448 | 0.8364 | 0.8279 | 0.8191 | 0.8101 | 0.8009 | 0.7916 |
| 13 | 0.9210 | 0.9156 | 0.9100 | 0.9042 | 0.8981 | 0.8919 | 0.8853 | 0.8786 | 0.8716 | 0.8645 |
| 14 | 0.9552 | 0.9517 | 0.9480 | 0.9441 | 0.9400 | 0.9357 | 0.9312 | 0.9265 | 0.9216 | 0.9165 |
| 15 | 0.9760 | 0.9738 | 0.9715 | 0.9691 | 0.9665 | 0.9638 | 0.9609 | 0.9579 | 0.9546 | 0.9513 |
| 16 | 0.9878 | 0.9865 | 0.9852 | 0.9838 | 0.9823 | 0.9806 | 0.9789 | 0.9770 | 0.9751 | 0.9730 |
| 17 | 0.9941 | 0.9934 | 0.9927 | 0.9919 | 0.9911 | 0.9902 | 0.9892 | 0.9881 | 0.9870 | 0.9857 |
| 18 | 0.9973 | 0.9969 | 0.9966 | 0.9962 | 0.9957 | 0.9952 | 0.9947 | 0.9941 | 0.9935 | 0.9928 |
| 19 | 0.9988 | 0.9986 | 0.9985 | 0.9983 | 0.9980 | 0.9978 | 0.9975 | 0.9972 | 0.9969 | 0.9965 |
| 20 | 0.9995 | 0.9994 | 0.9993 | 0.9992 | 0.9991 | 0.9990 | 0.9989 | 0.9987 | 0.9986 | 0.9984 |
| 21 | 0.9998 | 0.9998 | 0.9997 | 0.9997 | 0.9996 | 0.9996 | 0.9995 | 0.9995 | 0.9994 | 0.9993 |
| 22 | 0.9999 | 0.9999 | 0.9999 | 0.9999 | 0.9999 | 0.9998 | 0.9998 | 0.9998 | 0.9997 | 0.9997 |
| 23 | 1.0000 | 1.0000 | 1.0000 | 1.0000 | 0.9999 | 0.9999 | 0.9999 | 0.9999 | 0.9999 | 0.9999 |
| 24 | 1.0000 | 1.0000 | 1.0000 | 1.0000 | 1.0000 | 1.0000 | 1.0000 | 1.0000 | 1.0000 | 1.0000 |

$\lambda t$

| x | 11.00 | 12.00 | 13.00 | 14.00 | 15.00 | 16.00 | 17.00 | 18.00 | 19.00 | 20.00 |
|---|---|---|---|---|---|---|---|---|---|---|
| 0 | 0.0000 | 0.0000 | 0.0000 | 0.0000 | 0.0000 | 0.0000 | 0.0000 | 0.0000 | 0.0000 | 0.0000 |
| 1 | 0.0002 | 0.0001 | 0.0000 | 0.0000 | 0.0000 | 0.0000 | 0.0000 | 0.0000 | 0.0000 | 0.0000 |
| 2 | 0.0012 | 0.0005 | 0.0002 | 0.0001 | 0.0000 | 0.0000 | 0.0000 | 0.0000 | 0.0000 | 0.0000 |
| 3 | 0.0049 | 0.0023 | 0.0011 | 0.0005 | 0.0002 | 0.0001 | 0.0000 | 0.0000 | 0.0000 | 0.0000 |
| 4 | 0.0151 | 0.0076 | 0.0037 | 0.0018 | 0.0009 | 0.0004 | 0.0002 | 0.0001 | 0.0000 | 0.0000 |
| 5 | 0.0375 | 0.0203 | 0.0107 | 0.0055 | 0.0028 | 0.0014 | 0.0007 | 0.0003 | 0.0002 | 0.0001 |
| 6 | 0.0786 | 0.0458 | 0.0259 | 0.0142 | 0.0076 | 0.0040 | 0.0021 | 0.0010 | 0.0005 | 0.0003 |
| 7 | 0.1432 | 0.0895 | 0.0540 | 0.0316 | 0.0180 | 0.0100 | 0.0054 | 0.0029 | 0.0015 | 0.0008 |
| 8 | 0.2320 | 0.1550 | 0.0998 | 0.0621 | 0.0374 | 0.0220 | 0.0126 | 0.0071 | 0.0039 | 0.0021 |
| 9 | 0.3405 | 0.2424 | 0.1658 | 0.1094 | 0.0699 | 0.0433 | 0.0261 | 0.0154 | 0.0089 | 0.0050 |
| 10 | 0.4599 | 0.3472 | 0.2517 | 0.1757 | 0.1185 | 0.0774 | 0.0491 | 0.0304 | 0.0183 | 0.0108 |
| 11 | 0.5793 | 0.4616 | 0.3532 | 0.2600 | 0.1848 | 0.1270 | 0.0847 | 0.0549 | 0.0347 | 0.0214 |
| 12 | 0.6887 | 0.5760 | 0.4631 | 0.3585 | 0.2676 | 0.1931 | 0.1350 | 0.0917 | 0.0606 | 0.0390 |
| 13 | 0.7813 | 0.6815 | 0.5730 | 0.4644 | 0.3632 | 0.2745 | 0.2009 | 0.1426 | 0.0984 | 0.0661 |
| 14 | 0.8540 | 0.7720 | 0.6751 | 0.5704 | 0.4657 | 0.3675 | 0.2808 | 0.2081 | 0.1497 | 0.1049 |

| | | | | | | | | | | |
|---|---|---|---|---|---|---|---|---|---|---|
| **15** | 0.9074 | 0.8444 | 0.7636 | 0.6694 | 0.5681 | 0.4667 | 0.3715 | 0.2867 | 0.2148 | 0.1565 |
| **16** | 0.9441 | 0.8987 | 0.8355 | 0.7559 | 0.6641 | 0.5660 | 0.4677 | 0.3751 | 0.2920 | 0.2211 |
| **17** | 0.9678 | 0.9370 | 0.8905 | 0.8272 | 0.7489 | 0.6593 | 0.5640 | 0.4686 | 0.3784 | 0.2970 |
| **18** | 0.9823 | 0.9626 | 0.9302 | 0.8826 | 0.8195 | 0.7423 | 0.6550 | 0.5622 | 0.4695 | 0.3814 |
| **19** | 0.9907 | 0.9787 | 0.9573 | 0.9235 | 0.8752 | 0.8122 | 0.7363 | 0.6509 | 0.5606 | 0.4703 |
| **20** | 0.9953 | 0.9884 | 0.9750 | 0.9521 | 0.9170 | 0.8682 | 0.8055 | 0.7307 | 0.6472 | 0.5591 |
| **21** | 0.9977 | 0.9939 | 0.9859 | 0.9712 | 0.9469 | 0.9108 | 0.8615 | 0.7991 | 0.7255 | 0.6437 |
| **22** | 0.9990 | 0.9970 | 0.9924 | 0.9833 | 0.9673 | 0.9418 | 0.9047 | 0.8551 | 0.7931 | 0.7206 |
| **23** | 0.9995 | 0.9985 | 0.9960 | 0.9907 | 0.9805 | 0.9633 | 0.9367 | 0.8989 | 0.8490 | 0.7875 |
| **24** | 0.9998 | 0.9993 | 0.9980 | 0.9950 | 0.9888 | 0.9777 | 0.9594 | 0.9317 | 0.8933 | 0.8432 |
| **25** | 0.9999 | 0.9997 | 0.9990 | 0.9974 | 0.9938 | 0.9869 | 0.9748 | 0.9554 | 0.9269 | 0.8878 |
| **26** | 1.0000 | 0.9999 | 0.9995 | 0.9987 | 0.9967 | 0.9925 | 0.9848 | 0.9718 | 0.9514 | 0.9221 |
| **27** | 1.0000 | 0.9999 | 0.9998 | 0.9994 | 0.9983 | 0.9959 | 0.9912 | 0.9827 | 0.9687 | 0.9475 |
| **28** | 1.0000 | 1.0000 | 0.9999 | 0.9997 | 0.9991 | 0.9978 | 0.9950 | 0.9897 | 0.9805 | 0.9657 |
| **29** | 1.0000 | 1.0000 | 1.0000 | 0.9999 | 0.9996 | 0.9989 | 0.9973 | 0.9941 | 0.9882 | 0.9782 |
| **30** | 1.0000 | 1.0000 | 1.0000 | 0.9999 | 0.9998 | 0.9994 | 0.9986 | 0.9967 | 0.9930 | 0.9865 |
| **31** | 1.0000 | 1.0000 | 1.0000 | 1.0000 | 0.9999 | 0.9997 | 0.9993 | 0.9982 | 0.9960 | 0.9919 |
| **32** | 1.0000 | 1.0000 | 1.0000 | 1.0000 | 1.0000 | 0.9999 | 0.9996 | 0.9990 | 0.9978 | 0.9953 |
| **33** | 1.0000 | 1.0000 | 1.0000 | 1.0000 | 1.0000 | 0.9999 | 0.9998 | 0.9995 | 0.9988 | 0.9973 |
| **34** | 1.0000 | 1.0000 | 1.0000 | 1.0000 | 1.0000 | 1.0000 | 0.9999 | 0.9998 | 0.9994 | 0.9985 |
| **35** | 1.0000 | 1.0000 | 1.0000 | 1.0000 | 1.0000 | 1.0000 | 1.0000 | 0.9999 | 0.9997 | 0.9992 |
| **36** | 1.0000 | 1.0000 | 1.0000 | 1.0000 | 1.0000 | 1.0000 | 1.0000 | 0.9999 | 0.9998 | 0.9996 |
| **37** | 1.0000 | 1.0000 | 1.0000 | 1.0000 | 1.0000 | 1.0000 | 1.0000 | 1.0000 | 0.9999 | 0.9998 |
| **38** | 1.0000 | 1.0000 | 1.0000 | 1.0000 | 1.0000 | 1.0000 | 1.0000 | 1.0000 | 1.0000 | 0.9999 |
| **39** | 1.0000 | 1.0000 | 1.0000 | 1.0000 | 1.0000 | 1.0000 | 1.0000 | 1.0000 | 1.0000 | 0.9999 |
| **40** | 1.0000 | 1.0000 | 1.0000 | 1.0000 | 1.0000 | 1.0000 | 1.0000 | 1.0000 | 1.0000 | 1.0000 |

## Standard Normal Distribution Table

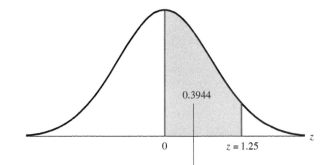

| z | 0 | 0.01 | 0.02 | 0.03 | 0.04 | 0.05 | 0.06 | 0.07 | 0.08 | 0.09 |
|---|---|------|------|------|------|------|------|------|------|------|
| 0.0 | 0.0000 | 0.0040 | 0.0080 | 0.0120 | 0.0160 | 0.0199 | 0.0239 | 0.0279 | 0.0319 | 0.0359 |
| 0.1 | 0.0398 | 0.0438 | 0.0478 | 0.0517 | 0.0557 | 0.0596 | 0.0636 | 0.0675 | 0.0714 | 0.0753 |
| 0.2 | 0.0793 | 0.0832 | 0.0871 | 0.0910 | 0.0948 | 0.0987 | 0.1026 | 0.1064 | 0.1103 | 0.1141 |
| 0.3 | 0.1179 | 0.1217 | 0.1255 | 0.1293 | 0.1331 | 0.1368 | 0.1406 | 0.1443 | 0.1480 | 0.1517 |
| 0.4 | 0.1554 | 0.1591 | 0.1628 | 0.1664 | 0.1700 | 0.1736 | 0.1772 | 0.1808 | 0.1844 | 0.1879 |
| 0.5 | 0.1915 | 0.1950 | 0.1985 | 0.2019 | 0.2054 | 0.2088 | 0.2123 | 0.2157 | 0.2190 | 0.2224 |
| 0.6 | 0.2257 | 0.2291 | 0.2324 | 0.2357 | 0.2389 | 0.2422 | 0.2454 | 0.2486 | 0.2517 | 0.2549 |
| 0.7 | 0.2580 | 0.2611 | 0.2642 | 0.2673 | 0.2704 | 0.2734 | 0.2764 | 0.2794 | 0.2823 | 0.2852 |
| 0.8 | 0.2881 | 0.2910 | 0.2939 | 0.2967 | 0.2995 | 0.3023 | 0.3051 | 0.3078 | 0.3106 | 0.3133 |
| 0.9 | 0.3159 | 0.3186 | 0.3212 | 0.3238 | 0.3264 | 0.3289 | 0.3315 | 0.3340 | 0.3365 | 0.3389 |
| 1.0 | 0.3413 | 0.3438 | 0.3461 | 0.3485 | 0.3508 | 0.3531 | 0.3554 | 0.3577 | 0.3599 | 0.3621 |
| 1.1 | 0.3643 | 0.3665 | 0.3686 | 0.3708 | 0.3729 | 0.3749 | 0.3770 | 0.3790 | 0.3810 | 0.3830 |
| 1.2 | 0.3849 | 0.3869 | 0.3888 | 0.3907 | 0.3925 | 0.3944 | 0.3962 | 0.3980 | 0.3997 | 0.4015 |
| 1.3 | 0.4032 | 0.4049 | 0.4066 | 0.4082 | 0.4099 | 0.4115 | 0.4131 | 0.4147 | 0.4162 | 0.4177 |
| 1.4 | 0.4192 | 0.4207 | 0.4222 | 0.4236 | 0.4251 | 0.4265 | 0.4279 | 0.4292 | 0.4306 | 0.4319 |
| 1.5 | 0.4332 | 0.4345 | 0.4357 | 0.4370 | 0.4382 | 0.4394 | 0.4406 | 0.4418 | 0.4429 | 0.4441 |
| 1.6 | 0.4452 | 0.4463 | 0.4474 | 0.4484 | 0.4495 | 0.4505 | 0.4515 | 0.4525 | 0.4535 | 0.4545 |
| 1.7 | 0.4554 | 0.4564 | 0.4573 | 0.4582 | 0.4591 | 0.4599 | 0.4608 | 0.4616 | 0.4625 | 0.4633 |
| 1.8 | 0.4641 | 0.4649 | 0.4656 | 0.4664 | 0.4671 | 0.4678 | 0.4686 | 0.4693 | 0.4699 | 0.4706 |
| 1.9 | 0.4713 | 0.4719 | 0.4726 | 0.4732 | 0.4738 | 0.4744 | 0.4750 | 0.4756 | 0.4761 | 0.4767 |
| 2.0 | 0.4772 | 0.4778 | 0.4783 | 0.4788 | 0.4793 | 0.4798 | 0.4803 | 0.4808 | 0.4812 | 0.4817 |
| 2.1 | 0.4821 | 0.4826 | 0.4830 | 0.4834 | 0.4838 | 0.4842 | 0.4846 | 0.4850 | 0.4854 | 0.4857 |
| 2.2 | 0.4861 | 0.4864 | 0.4868 | 0.4871 | 0.4875 | 0.4878 | 0.4881 | 0.4884 | 0.4887 | 0.4890 |
| 2.3 | 0.4893 | 0.4896 | 0.4898 | 0.4901 | 0.4904 | 0.4906 | 0.4909 | 0.4911 | 0.4913 | 0.4916 |
| 2.4 | 0.4918 | 0.4920 | 0.4922 | 0.4925 | 0.4927 | 0.4929 | 0.4931 | 0.4932 | 0.4934 | 0.4936 |
| 2.5 | 0.4938 | 0.4940 | 0.4941 | 0.4943 | 0.4945 | 0.4946 | 0.4948 | 0.4949 | 0.4951 | 0.4952 |
| 2.6 | 0.4953 | 0.4955 | 0.4956 | 0.4957 | 0.4959 | 0.4960 | 0.4961 | 0.4962 | 0.4963 | 0.4964 |
| 2.7 | 0.4965 | 0.4966 | 0.4967 | 0.4968 | 0.4969 | 0.4970 | 0.4971 | 0.4972 | 0.4973 | 0.4974 |
| 2.8 | 0.4974 | 0.4975 | 0.4976 | 0.4977 | 0.4977 | 0.4978 | 0.4979 | 0.4979 | 0.4980 | 0.4981 |
| 2.9 | 0.4981 | 0.4982 | 0.4982 | 0.4983 | 0.4984 | 0.4984 | 0.4985 | 0.4985 | 0.4986 | 0.4986 |
| 3.0 | 0.4987 | 0.4987 | 0.4987 | 0.4988 | 0.4988 | 0.4989 | 0.4989 | 0.4989 | 0.4990 | 0.4990 |

## Exponential Distribution Table

**Values of $e^{-\lambda a}$**

| $\lambda a$ | $e^{-\lambda a}$ | $\lambda a$ | $e^{-\lambda a}$ | $\lambda a$ | $e^{-\lambda a}$ | $\lambda a$ | $e^{-\lambda a}$ | $\lambda a$ | $e^{-\lambda a}$ |
|---|---|---|---|---|---|---|---|---|---|
| 0.00 | 1.0000 | 2.05 | 0.1287 | 4.05 | 0.0174 | 6.05 | 0.0024 | 8.05 | 0.0003 |
| 0.05 | 0.9512 | 2.10 | 0.1225 | 4.10 | 0.0166 | 6.10 | 0.0022 | 8.10 | 0.0003 |
| 0.10 | 0.9048 | 2.15 | 0.1165 | 4.15 | 0.0158 | 6.15 | 0.0021 | 8.15 | 0.0003 |
| 0.15 | 0.8607 | 2.20 | 0.1108 | 4.20 | 0.0150 | 6.20 | 0.0020 | 8.20 | 0.0003 |
| 0.20 | 0.8187 | 2.25 | 0.1054 | 4.25 | 0.0143 | 6.25 | 0.0019 | 8.25 | 0.0003 |
| 0.25 | 0.7788 | 2.30 | 0.1003 | 4.30 | 0.0136 | 6.30 | 0.0018 | 8.30 | 0.0002 |
| 0.30 | 0.7408 | 2.35 | 0.0954 | 4.35 | 0.0129 | 6.35 | 0.0017 | 8.35 | 0.0002 |
| 0.35 | 0.7047 | 2.40 | 0.0907 | 4.40 | 0.0123 | 6.40 | 0.0017 | 8.40 | 0.0002 |
| 0.40 | 0.6703 | 2.45 | 0.0863 | 4.45 | 0.0117 | 6.45 | 0.0016 | 8.45 | 0.0002 |
| 0.45 | 0.6376 | 2.50 | 0.0821 | 4.50 | 0.0111 | 6.50 | 0.0015 | 8.50 | 0.0002 |
| 0.50 | 0.6065 | 2.55 | 0.0781 | 4.55 | 0.0106 | 6.55 | 0.0014 | 8.55 | 0.0002 |
| 0.55 | 0.5769 | 2.60 | 0.0743 | 4.60 | 0.0101 | 6.60 | 0.0014 | 8.60 | 0.0002 |
| 0.60 | 0.5488 | 2.65 | 0.0707 | 4.65 | 0.0096 | 6.65 | 0.0013 | 8.65 | 0.0002 |
| 0.65 | 0.5220 | 2.70 | 0.0672 | 4.70 | 0.0091 | 6.70 | 0.0012 | 8.70 | 0.0002 |
| 0.70 | 0.4966 | 2.75 | 0.0639 | 4.75 | 0.0087 | 6.75 | 0.0012 | 8.75 | 0.0002 |
| 0.75 | 0.4724 | 2.80 | 0.0608 | 4.80 | 0.0082 | 6.80 | 0.0011 | 8.80 | 0.0002 |
| 0.80 | 0.4493 | 2.85 | 0.0578 | 4.85 | 0.0078 | 6.85 | 0.0011 | 8.85 | 0.0001 |
| 0.85 | 0.4274 | 2.90 | 0.0550 | 4.90 | 0.0074 | 6.90 | 0.0010 | 8.90 | 0.0001 |
| 0.90 | 0.4066 | 2.95 | 0.0523 | 4.95 | 0.0071 | 6.95 | 0.0010 | 8.95 | 0.0001 |
| 0.95 | 0.3867 | 3.00 | 0.0498 | 5.00 | 0.0067 | 7.00 | 0.0009 | 9.00 | 0.0001 |
| 1.00 | 0.3679 | 3.05 | 0.0474 | 5.05 | 0.0064 | 7.05 | 0.0009 | 9.05 | 0.0001 |
| 1.05 | 0.3499 | 3.10 | 0.0450 | 5.10 | 0.0061 | 7.10 | 0.0008 | 9.10 | 0.0001 |
| 1.10 | 0.3329 | 3.15 | 0.0429 | 5.15 | 0.0058 | 7.15 | 0.0008 | 9.15 | 0.0001 |
| 1.15 | 0.3166 | 3.20 | 0.0408 | 5.20 | 0.0055 | 7.20 | 0.0007 | 9.20 | 0.0001 |
| 1.20 | 0.3012 | 3.25 | 0.0388 | 5.25 | 0.0052 | 7.25 | 0.0007 | 9.25 | 0.0001 |
| 1.25 | 0.2865 | 3.30 | 0.0369 | 5.30 | 0.0050 | 7.30 | 0.0007 | 9.30 | 0.0001 |
| 1.30 | 0.2725 | 3.35 | 0.0351 | 5.35 | 0.0047 | 7.35 | 0.0006 | 9.35 | 0.0001 |
| 1.35 | 0.2592 | 3.40 | 0.0334 | 5.40 | 0.0045 | 7.40 | 0.0006 | 9.40 | 0.0001 |
| 1.40 | 0.2466 | 3.45 | 0.0317 | 5.45 | 0.0043 | 7.45 | 0.0006 | 9.45 | 0.0001 |
| 1.45 | 0.2346 | 3.50 | 0.0302 | 5.50 | 0.0041 | 7.50 | 0.0006 | 9.50 | 0.0001 |
| 1.50 | 0.2231 | 3.55 | 0.0287 | 5.55 | 0.0039 | 7.55 | 0.0005 | 9.55 | 0.0001 |
| 1.55 | 0.2122 | 3.60 | 0.0273 | 5.60 | 0.0037 | 7.60 | 0.0005 | 9.60 | 0.0001 |
| 1.60 | 0.2019 | 3.65 | 0.0260 | 5.65 | 0.0035 | 7.65 | 0.0005 | 9.65 | 0.0001 |
| 1.65 | 0.1920 | 3.70 | 0.0247 | 5.70 | 0.0033 | 7.70 | 0.0005 | 9.70 | 0.0001 |
| 1.70 | 0.1827 | 3.75 | 0.0235 | 5.75 | 0.0032 | 7.75 | 0.0004 | 9.75 | 0.0001 |
| 1.75 | 0.1738 | 3.80 | 0.0224 | 5.80 | 0.0030 | 7.80 | 0.0004 | 9.80 | 0.0001 |
| 1.80 | 0.1653 | 3.85 | 0.0213 | 5.85 | 0.0029 | 7.85 | 0.0004 | 9.85 | 0.0001 |
| 1.85 | 0.1572 | 3.90 | 0.0202 | 5.90 | 0.0027 | 7.90 | 0.0004 | 9.90 | 0.0001 |
| 1.90 | 0.1496 | 3.95 | 0.0193 | 5.95 | 0.0026 | 7.95 | 0.0004 | 9.95 | 0.0000 |
| 1.95 | 0.1423 | 4.00 | 0.0183 | 6.00 | 0.0025 | 8.00 | 0.0003 | 10.00 | 0.0000 |
| 2.00 | 0.1353 | | | | | | | | |

## Values of *t* for Selected Probabilities

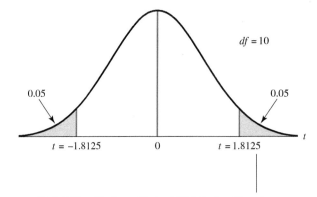

$df = 10$

0.05      0.05

$t = -1.8125$     0     $t = 1.8125$

Probabilites (Or Areas Under *t*-Distribution Curve)

| Conf. Level<br>One Tail<br>Two Tails | 0.1<br>0.45<br>0.9 | 0.3<br>0.35<br>0.7 | 0.5<br>0.25<br>0.5 | 0.7<br>0.15<br>0.3 | 0.8<br>0.1<br>0.2 | 0.9<br>0.05<br>0.1 | 0.95<br>0.025<br>0.05 | 0.98<br>0.01<br>0.02 | 0.99<br>0.005<br>0.01 |
|---|---|---|---|---|---|---|---|---|---|
| **d. f.** | | | | | *Values of t* | | | | |
| 1 | 0.1584 | 0.5095 | 1.0000 | 1.9626 | 3.0777 | 6.3138 | 12.7062 | 31.8205 | 63.6567 |
| 2 | 0.1421 | 0.4447 | 0.8165 | 1.3862 | 1.8856 | 2.9200 | 4.3027 | 6.9646 | 9.9248 |
| 3 | 0.1366 | 0.4242 | 0.7649 | 1.2498 | 1.6377 | 2.3534 | 3.1824 | 4.5407 | 5.8409 |
| 4 | 0.1338 | 0.4142 | 0.7407 | 1.1896 | 1.5332 | 2.1318 | 2.7764 | 3.7469 | 4.6041 |
| 5 | 0.1322 | 0.4082 | 0.7267 | 1.1558 | 1.4759 | 2.0150 | 2.5706 | 3.3649 | 4.0321 |
| 6 | 0.1311 | 0.4043 | 0.7176 | 1.1342 | 1.4398 | 1.9432 | 2.4469 | 3.1427 | 3.7074 |
| 7 | 0.1303 | 0.4015 | 0.7111 | 1.1192 | 1.4149 | 1.8946 | 2.3646 | 2.9980 | 3.4995 |
| 8 | 0.1297 | 0.3995 | 0.7064 | 1.1081 | 1.3968 | 1.8595 | 2.3060 | 2.8965 | 3.3554 |
| 9 | 0.1293 | 0.3979 | 0.7027 | 1.0997 | 1.3830 | 1.8331 | 2.2622 | 2.8214 | 3.2498 |
| 10 | 0.1289 | 0.3966 | 0.6998 | 1.0931 | 1.3722 | 1.8125◄ | 2.2281 | 2.7638 | 3.1693 |
| 11 | 0.1286 | 0.3956 | 0.6974 | 1.0877 | 1.3634 | 1.7959 | 2.2010 | 2.7181 | 3.1058 |
| 12 | 0.1283 | 0.3947 | 0.6955 | 1.0832 | 1.3562 | 1.7823 | 2.1788 | 2.6810 | 3.0545 |
| 13 | 0.1281 | 0.3940 | 0.6938 | 1.0795 | 1.3502 | 1.7709 | 2.1604 | 2.6503 | 3.0123 |
| 14 | 0.1280 | 0.3933 | 0.6924 | 1.0763 | 1.3450 | 1.7613 | 2.1448 | 2.6245 | 2.9768 |
| 15 | 0.1278 | 0.3928 | 0.6912 | 1.0735 | 1.3406 | 1.7531 | 2.1314 | 2.6025 | 2.9467 |
| 16 | 0.1277 | 0.3923 | 0.6901 | 1.0711 | 1.3368 | 1.7459 | 2.1199 | 2.5835 | 2.9208 |
| 17 | 0.1276 | 0.3919 | 0.6892 | 1.0690 | 1.3334 | 1.7396 | 2.1098 | 2.5669 | 2.8982 |
| 18 | 0.1274 | 0.3915 | 0.6884 | 1.0672 | 1.3304 | 1.7341 | 2.1009 | 2.5524 | 2.8784 |
| 19 | 0.1274 | 0.3912 | 0.6876 | 1.0655 | 1.3277 | 1.7291 | 2.0930 | 2.5395 | 2.8609 |
| 20 | 0.1273 | 0.3909 | 0.6870 | 1.0640 | 1.3253 | 1.7247 | 2.0860 | 2.5280 | 2.8453 |
| 21 | 0.1272 | 0.3906 | 0.6864 | 1.0627 | 1.3232 | 1.7207 | 2.0796 | 2.5176 | 2.8314 |
| 22 | 0.1271 | 0.3904 | 0.6858 | 1.0614 | 1.3212 | 1.7171 | 2.0739 | 2.5083 | 2.8188 |
| 23 | 0.1271 | 0.3902 | 0.6853 | 1.0603 | 1.3195 | 1.7139 | 2.0687 | 2.4999 | 2.8073 |
| 24 | 0.1270 | 0.3900 | 0.6848 | 1.0593 | 1.3178 | 1.7109 | 2.0639 | 2.4922 | 2.7969 |
| 25 | 0.1269 | 0.3898 | 0.6844 | 1.0584 | 1.3163 | 1.7081 | 2.0595 | 2.4851 | 2.7874 |
| 26 | 0.1269 | 0.3896 | 0.6840 | 1.0575 | 1.3150 | 1.7056 | 2.0555 | 2.4786 | 2.7787 |
| 27 | 0.1268 | 0.3894 | 0.6837 | 1.0567 | 1.3137 | 1.7033 | 2.0518 | 2.4727 | 2.7707 |
| 28 | 0.1268 | 0.3893 | 0.6834 | 1.0560 | 1.3125 | 1.7011 | 2.0484 | 2.4671 | 2.7633 |
| 29 | 0.1268 | 0.3892 | 0.6830 | 1.0553 | 1.3114 | 1.6991 | 2.0452 | 2.4620 | 2.7564 |
| 30 | 0.1267 | 0.3890 | 0.6828 | 1.0547 | 1.3104 | 1.6973 | 2.0423 | 2.4573 | 2.7500 |
| 40 | 0.1265 | 0.3881 | 0.6807 | 1.0500 | 1.3031 | 1.6839 | 2.0211 | 2.4233 | 2.7045 |
| 50 | 0.1263 | 0.3875 | 0.6794 | 1.0473 | 1.2987 | 1.6759 | 2.0086 | 2.4033 | 2.6778 |
| 60 | 0.1262 | 0.3872 | 0.6786 | 1.0455 | 1.2958 | 1.6706 | 2.0003 | 2.3901 | 2.6603 |
| 70 | 0.1261 | 0.3869 | 0.6780 | 1.0442 | 1.2938 | 1.6669 | 1.9944 | 2.3808 | 2.6479 |
| 80 | 0.1261 | 0.3867 | 0.6776 | 1.0432 | 1.2922 | 1.6641 | 1.9901 | 2.3739 | 2.6387 |
| 90 | 0.1260 | 0.3866 | 0.6772 | 1.0424 | 1.2910 | 1.6620 | 1.9867 | 2.3685 | 2.6316 |
| 100 | 0.1260 | 0.3864 | 0.6770 | 1.0418 | 1.2901 | 1.6602 | 1.9840 | 2.3642 | 2.6259 |
| 250 | 0.1258 | 0.3858 | 0.6755 | 1.0386 | 1.2849 | 1.6510 | 1.9695 | 2.3414 | 2.5956 |
| 500 | 0.1257 | 0.3855 | 0.6750 | 1.0375 | 1.2832 | 1.6479 | 1.9647 | 2.3338 | 2.5857 |
| ∞ | 0.1257 | 0.3853 | 0.6745 | 1.0364 | 1.2816 | 1.6449 | 1.9600 | 2.3263 | 2.5758 |

# APPENDIX G

## Values of $\chi^2$ for Selected Probabilities

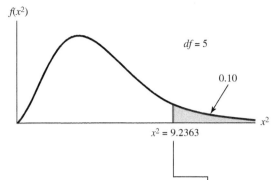

$f(x^2)$

$df = 5$

0.10

$x^2$

$x^2 = 9.2363$

PROBABILITIES (OR AREAS UNDER CHI-SQUARE DISTRIBUTION CURVE ABOVE GIVEN CHI-SQUARE VALUES)

| df | 0.995 | 0.99 | 0.975 | 0.95 | 0.90 | 0.10 | 0.05 | 0.025 | 0.01 | 0.005 |
|----|-------|------|-------|------|------|------|------|-------|------|-------|
| | | | | | | Values of Chi-Squared | | | | |
| 1 | 0.0000 | 0.0002 | 0.0010 | 0.0039 | 0.0158 | 2.7055 | 3.8415 | 5.0239 | 6.6349 | 7.8794 |
| 2 | 0.0100 | 0.0201 | 0.0506 | 0.1026 | 0.2107 | 4.6052 | 5.9915 | 7.3778 | 9.2104 | 10.5965 |
| 3 | 0.0717 | 0.1148 | 0.2158 | 0.3518 | 0.5844 | 6.2514 | 7.8147 | 9.3484 | 11.3449 | 12.8381 |
| 4 | 0.2070 | 0.2971 | 0.4844 | 0.7107 | 1.0636 | 7.7794 | 9.4877 | 11.1433 | 13.2767 | 14.8602 |
| 5 | 0.4118 | 0.5543 | 0.8312 | 1.1455 | 1.6103 | 9.2363 | 11.0705 | 12.8325 | 15.0863 | 16.7496 |
| 6 | 0.6757 | 0.8721 | 1.2373 | 1.6354 | 2.2041 | 10.6446 | 12.5916 | 14.4494 | 16.8119 | 18.5475 |
| 7 | 0.9893 | 1.2390 | 1.6899 | 2.1673 | 2.8331 | 12.0170 | 14.0671 | 16.0128 | 18.4753 | 20.2777 |
| 8 | 1.3444 | 1.6465 | 2.1797 | 2.7326 | 3.4895 | 13.3616 | 15.5073 | 17.5345 | 20.0902 | 21.9549 |
| 9 | 1.7349 | 2.0879 | 2.7004 | 3.3251 | 4.1682 | 14.6837 | 16.9190 | 19.0228 | 21.6660 | 23.5893 |
| 10 | 2.1558 | 2.5582 | 3.2470 | 3.9403 | 4.8652 | 15.9872 | 18.3070 | 20.4832 | 23.2093 | 25.1881 |
| 11 | 2.6032 | 3.0535 | 3.8157 | 4.5748 | 5.5778 | 17.2750 | 19.6752 | 21.9200 | 24.7250 | 26.7569 |
| 12 | 3.0738 | 3.5706 | 4.4038 | 5.2260 | 6.3038 | 18.5493 | 21.0261 | 23.3367 | 26.2170 | 28.2997 |
| 13 | 3.5650 | 4.1069 | 5.0087 | 5.8919 | 7.0415 | 19.8119 | 22.3620 | 24.7356 | 27.6882 | 29.8193 |
| 14 | 4.0747 | 4.6604 | 5.6287 | 6.5706 | 7.7895 | 21.0641 | 23.6848 | 26.1189 | 29.1412 | 31.3194 |
| 15 | 4.6009 | 5.2294 | 6.2621 | 7.2609 | 8.5468 | 22.3071 | 24.9958 | 27.4884 | 30.5780 | 32.8015 |
| 16 | 5.1422 | 5.8122 | 6.9077 | 7.9616 | 9.3122 | 23.5418 | 26.2962 | 28.8453 | 31.9999 | 34.2671 |
| 17 | 5.6973 | 6.4077 | 7.5642 | 8.6718 | 10.0852 | 24.7690 | 27.5871 | 30.1910 | 33.4087 | 35.7184 |
| 18 | 6.2648 | 7.0149 | 8.2307 | 9.3904 | 10.8649 | 25.9894 | 28.8693 | 31.5264 | 34.8052 | 37.1564 |
| 19 | 6.8439 | 7.6327 | 8.9065 | 10.1170 | 11.6509 | 27.2036 | 30.1435 | 32.8523 | 36.1908 | 38.5821 |
| 20 | 7.4338 | 8.2604 | 9.5908 | 10.8508 | 12.4426 | 28.4120 | 31.4104 | 34.1696 | 37.5663 | 39.9969 |
| 21 | 8.0336 | 8.8972 | 10.2829 | 11.5913 | 13.2396 | 29.6151 | 32.6706 | 35.4789 | 38.9322 | 41.4009 |
| 22 | 8.6427 | 9.5425 | 10.9823 | 12.3380 | 14.0415 | 30.8133 | 33.9245 | 36.7807 | 40.2894 | 42.7957 |
| 23 | 9.2604 | 10.1957 | 11.6885 | 13.0905 | 14.8480 | 32.0069 | 35.1725 | 38.0756 | 41.6383 | 44.1814 |
| 24 | 9.8862 | 10.8563 | 12.4011 | 13.8484 | 15.6587 | 33.1962 | 36.4150 | 39.3641 | 42.9798 | 45.5584 |
| 25 | 10.5196 | 11.5240 | 13.1197 | 14.6114 | 16.4734 | 34.3816 | 37.6525 | 40.6465 | 44.3140 | 46.9280 |
| 26 | 11.1602 | 12.1982 | 13.8439 | 15.3792 | 17.2919 | 35.5632 | 38.8851 | 41.9231 | 45.6416 | 48.2898 |
| 27 | 11.8077 | 12.8785 | 14.5734 | 16.1514 | 18.1139 | 36.7412 | 40.1133 | 43.1945 | 46.9628 | 49.6450 |
| 28 | 12.4613 | 13.5647 | 15.3079 | 16.9279 | 18.9392 | 37.9159 | 41.3372 | 44.4608 | 48.2782 | 50.9936 |
| 29 | 13.1211 | 14.2564 | 16.0471 | 17.7084 | 19.7677 | 39.0875 | 42.5569 | 45.7223 | 49.5878 | 52.3355 |
| 30 | 13.7867 | 14.9535 | 16.7908 | 18.4927 | 20.5992 | 40.2560 | 43.7730 | 46.9792 | 50.8922 | 53.6719 |

**F-Distribution Table: Upper 5% Probability (or 5% Area) under F-Distribution Curve**

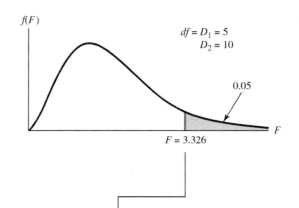

$df = D_1 = 5$
$D_2 = 10$

$0.05$

$F = 3.326$

**DENOMINATOR**

$df = D_2$                                                    NUMERATOR $df = D_1$

| | 1 | 2 | 3 | 4 | 5 | 6 | 7 | 8 | 9 | 10 |
|---|---|---|---|---|---|---|---|---|---|---|
| 1 | 161.446 | 199.499 | 215.707 | 224.583 | 230.160 | 233.988 | 236.767 | 238.884 | 240.543 | 241.882 |
| 2 | 18.513 | 19.000 | 19.164 | 19.247 | 19.296 | 19.329 | 19.353 | 19.371 | 19.385 | 19.396 |
| 3 | 10.128 | 9.552 | 9.277 | 9.117 | 9.013 | 8.941 | 8.887 | 8.845 | 8.812 | 8.785 |
| 4 | 7.709 | 6.944 | 6.591 | 6.388 | 6.256 | 6.163 | 6.094 | 6.041 | 5.999 | 5.964 |
| 5 | 6.608 | 5.786 | 5.409 | 5.192 | 5.050 | 4.950 | 4.876 | 4.818 | 4.772 | 4.735 |
| 6 | 5.987 | 5.143 | 4.757 | 4.534 | 4.387 | 4.284 | 4.207 | 4.147 | 4.099 | 4.060 |
| 7 | 5.591 | 4.737 | 4.347 | 4.120 | 3.972 | 3.866 | 3.787 | 3.726 | 3.677 | 3.637 |
| 8 | 5.318 | 4.459 | 4.066 | 3.838 | 3.688 | 3.581 | 3.500 | 3.438 | 3.388 | 3.347 |
| 9 | 5.117 | 4.256 | 3.863 | 3.633 | 3.482 | 3.374 | 3.293 | 3.230 | 3.179 | 3.137 |
| 10 | 4.965 | 4.103 | 3.708 | 3.478 | 3.326 | 3.217 | 3.135 | 3.072 | 3.020 | 2.978 |
| 11 | 4.844 | 3.982 | 3.587 | 3.357 | 3.204 | 3.095 | 3.012 | 2.948 | 2.896 | 2.854 |
| 12 | 4.747 | 3.885 | 3.490 | 3.259 | 3.106 | 2.996 | 2.913 | 2.849 | 2.796 | 2.753 |
| 13 | 4.667 | 3.806 | 3.411 | 3.179 | 3.025 | 2.915 | 2.832 | 2.767 | 2.714 | 2.671 |
| 14 | 4.600 | 3.739 | 3.344 | 3.112 | 2.958 | 2.848 | 2.764 | 2.699 | 2.646 | 2.602 |
| 15 | 4.543 | 3.682 | 3.287 | 3.056 | 2.901 | 2.790 | 2.707 | 2.641 | 2.588 | 2.544 |
| 16 | 4.494 | 3.634 | 3.239 | 3.007 | 2.852 | 2.741 | 2.657 | 2.591 | 2.538 | 2.494 |
| 17 | 4.451 | 3.592 | 3.197 | 2.965 | 2.810 | 2.699 | 2.614 | 2.548 | 2.494 | 2.450 |
| 18 | 4.414 | 3.555 | 3.160 | 2.928 | 2.773 | 2.661 | 2.577 | 2.510 | 2.456 | 2.412 |
| 19 | 4.381 | 3.522 | 3.127 | 2.895 | 2.740 | 2.628 | 2.544 | 2.477 | 2.423 | 2.378 |
| 20 | 4.351 | 3.493 | 3.098 | 2.866 | 2.711 | 2.599 | 2.514 | 2.447 | 2.393 | 2.348 |
| 24 | 4.260 | 3.403 | 3.009 | 2.776 | 2.621 | 2.508 | 2.423 | 2.355 | 2.300 | 2.255 |
| 30 | 4.171 | 3.316 | 2.922 | 2.690 | 2.534 | 2.421 | 2.334 | 2.266 | 2.211 | 2.165 |
| 40 | 4.085 | 3.232 | 2.839 | 2.606 | 2.449 | 2.336 | 2.249 | 2.180 | 2.124 | 2.077 |
| 50 | 4.034 | 3.183 | 2.790 | 2.557 | 2.400 | 2.286 | 2.199 | 2.130 | 2.073 | 2.026 |
| 100 | 3.936 | 3.087 | 2.696 | 2.463 | 2.305 | 2.191 | 2.103 | 2.032 | 1.975 | 1.927 |
| 200 | 3.888 | 3.041 | 2.650 | 2.417 | 2.259 | 2.144 | 2.056 | 1.985 | 1.927 | 1.878 |
| 300 | 3.873 | 3.026 | 2.635 | 2.402 | 2.244 | 2.129 | 2.040 | 1.969 | 1.911 | 1.862 |

**DENOMINATOR**

$df = D_2$                                                    NUMERATOR $df = D_1$

| | 11 | 12 | 13 | 14 | 15 | 16 | 17 | 18 | 19 | 20 |
|---|---|---|---|---|---|---|---|---|---|---|
| 1 | 242.981 | 243.905 | 244.690 | 245.363 | 245.949 | 246.466 | 246.917 | 247.324 | 247.688 | 248.016 |
| 2 | 19.405 | 19.412 | 19.419 | 19.424 | 19.429 | 19.433 | 19.437 | 19.440 | 19.443 | 19.446 |
| 3 | 8.763 | 8.745 | 8.729 | 8.715 | 8.703 | 8.692 | 8.683 | 8.675 | 8.667 | 8.660 |
| 4 | 5.936 | 5.912 | 5.891 | 5.873 | 5.858 | 5.844 | 5.832 | 5.821 | 5.811 | 5.803 |
| 5 | 4.704 | 4.678 | 4.655 | 4.636 | 4.619 | 4.604 | 4.590 | 4.579 | 4.568 | 4.558 |
| 6 | 4.027 | 4.000 | 3.976 | 3.956 | 3.938 | 3.922 | 3.908 | 3.896 | 3.884 | 3.874 |
| 7 | 3.603 | 3.575 | 3.550 | 3.529 | 3.511 | 3.494 | 3.480 | 3.467 | 3.455 | 3.445 |
| 8 | 3.313 | 3.284 | 3.259 | 3.237 | 3.218 | 3.202 | 3.187 | 3.173 | 3.161 | 3.150 |
| 9 | 3.102 | 3.073 | 3.048 | 3.025 | 3.006 | 2.989 | 2.974 | 2.960 | 2.948 | 2.936 |
| 10 | 2.943 | 2.913 | 2.887 | 2.865 | 2.845 | 2.828 | 2.812 | 2.798 | 2.785 | 2.774 |
| 11 | 2.818 | 2.788 | 2.761 | 2.739 | 2.719 | 2.701 | 2.685 | 2.671 | 2.658 | 2.646 |
| 12 | 2.717 | 2.687 | 2.660 | 2.637 | 2.617 | 2.599 | 2.583 | 2.568 | 2.555 | 2.544 |
| 13 | 2.635 | 2.604 | 2.577 | 2.554 | 2.533 | 2.515 | 2.499 | 2.484 | 2.471 | 2.459 |
| 14 | 2.565 | 2.534 | 2.507 | 2.484 | 2.463 | 2.445 | 2.428 | 2.413 | 2.400 | 2.388 |

| | | | | | | | | | | |
|---|---|---|---|---|---|---|---|---|---|---|
| 15 | 2.507 | 2.475 | 2.448 | 2.424 | 2.403 | 2.385 | 2.368 | 2.353 | 2.340 | 2.328 |
| 16 | 2.456 | 2.425 | 2.397 | 2.373 | 2.352 | 2.333 | 2.317 | 2.302 | 2.288 | 2.276 |
| 17 | 2.413 | 2.381 | 2.353 | 2.329 | 2.308 | 2.289 | 2.272 | 2.257 | 2.243 | 2.230 |
| 18 | 2.374 | 2.342 | 2.314 | 2.290 | 2.269 | 2.250 | 2.233 | 2.217 | 2.203 | 2.191 |
| 19 | 2.340 | 2.308 | 2.280 | 2.256 | 2.234 | 2.215 | 2.198 | 2.182 | 2.168 | 2.155 |
| 20 | 2.310 | 2.278 | 2.250 | 2.225 | 2.203 | 2.184 | 2.167 | 2.151 | 2.137 | 2.124 |
| 24 | 2.216 | 2.183 | 2.155 | 2.130 | 2.108 | 2.088 | 2.070 | 2.054 | 2.040 | 2.027 |
| 30 | 2.126 | 2.092 | 2.063 | 2.037 | 2.015 | 1.995 | 1.976 | 1.960 | 1.945 | 1.932 |
| 40 | 2.038 | 2.003 | 1.974 | 1.948 | 1.924 | 1.904 | 1.885 | 1.868 | 1.853 | 1.839 |
| 50 | 1.986 | 1.952 | 1.921 | 1.895 | 1.871 | 1.850 | 1.831 | 1.814 | 1.798 | 1.784 |
| 100 | 1.886 | 1.850 | 1.819 | 1.792 | 1.768 | 1.746 | 1.726 | 1.708 | 1.691 | 1.676 |
| 200 | 1.837 | 1.801 | 1.769 | 1.742 | 1.717 | 1.694 | 1.674 | 1.656 | 1.639 | 1.623 |
| 300 | 1.821 | 1.785 | 1.753 | 1.725 | 1.700 | 1.677 | 1.657 | 1.638 | 1.621 | 1.606 |

**DENOMINATOR**

$df = D_2$        **NUMERATOR** $df = D_1$

| | 24 | 30 | 40 | 50 | 100 | 200 | 300 |
|---|---|---|---|---|---|---|---|
| 1 | 249.052 | 250.096 | 251.144 | 251.774 | 253.043 | 253.676 | 253.887 |
| 2 | 19.454 | 19.463 | 19.471 | 19.476 | 19.486 | 19.491 | 19.492 |
| 3 | 8.638 | 8.617 | 8.594 | 8.581 | 8.554 | 8.540 | 8.536 |
| 4 | 5.774 | 5.746 | 5.717 | 5.699 | 5.664 | 5.646 | 5.640 |
| 5 | 4.527 | 4.496 | 4.464 | 4.444 | 4.405 | 4.385 | 4.378 |
| 6 | 3.841 | 3.808 | 3.774 | 3.754 | 3.712 | 3.690 | 3.683 |
| 7 | 3.410 | 3.376 | 3.340 | 3.319 | 3.275 | 3.252 | 3.245 |
| 8 | 3.115 | 3.079 | 3.043 | 3.020 | 2.975 | 2.951 | 2.943 |
| 9 | 2.900 | 2.864 | 2.826 | 2.803 | 2.756 | 2.731 | 2.723 |
| 10 | 2.737 | 2.700 | 2.661 | 2.637 | 2.588 | 2.563 | 2.555 |
| 11 | 2.609 | 2.570 | 2.531 | 2.507 | 2.457 | 2.431 | 2.422 |
| 12 | 2.505 | 2.466 | 2.426 | 2.401 | 2.350 | 2.323 | 2.314 |
| 13 | 2.420 | 2.380 | 2.339 | 2.314 | 2.261 | 2.234 | 2.225 |
| 14 | 2.349 | 2.308 | 2.266 | 2.241 | 2.187 | 2.159 | 2.150 |
| 15 | 2.288 | 2.247 | 2.204 | 2.178 | 2.123 | 2.095 | 2.085 |
| 16 | 2.235 | 2.194 | 2.151 | 2.124 | 2.068 | 2.039 | 2.030 |
| 17 | 2.190 | 2.148 | 2.104 | 2.077 | 2.020 | 1.991 | 1.981 |
| 18 | 2.150 | 2.107 | 2.063 | 2.035 | 1.978 | 1.948 | 1.938 |
| 19 | 2.114 | 2.071 | 2.026 | 1.999 | 1.940 | 1.910 | 1.899 |
| 20 | 2.082 | 2.039 | 1.994 | 1.966 | 1.907 | 1.875 | 1.865 |
| 24 | 1.984 | 1.939 | 1.892 | 1.863 | 1.800 | 1.768 | 1.756 |
| 30 | 1.887 | 1.841 | 1.792 | 1.761 | 1.695 | 1.660 | 1.647 |
| 40 | 1.793 | 1.744 | 1.693 | 1.660 | 1.589 | 1.551 | 1.537 |
| 50 | 1.737 | 1.687 | 1.634 | 1.599 | 1.525 | 1.484 | 1.469 |
| 100 | 1.627 | 1.573 | 1.515 | 1.477 | 1.392 | 1.342 | 1.323 |
| 200 | 1.572 | 1.516 | 1.455 | 1.415 | 1.321 | 1.263 | 1.240 |
| 300 | 1.554 | 1.497 | 1.435 | 1.393 | 1.296 | 1.234 | 1.210 |

**F-Distribution Table: Upper 2.5% Probability (or 2.5% Area) under F-Distribution Curve**

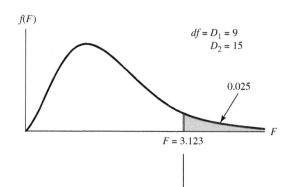

**DENOMINATOR**

$df = D_2$                                NUMERATOR $df = D_1$

| | 1 | 2 | 3 | 4 | 5 | 6 | 7 | 8 | 9 | 10 | 11 |
|---|---|---|---|---|---|---|---|---|---|---|---|
| 1 | 647.793 | 799.482 | 864.151 | 899.599 | 921.835 | 937.114 | 948.203 | 956.643 | 963.279 | 968.634 | 973.028 |
| 2 | 38.506 | 39.000 | 39.166 | 39.248 | 39.298 | 39.331 | 39.356 | 39.373 | 39.387 | 39.398 | 39.407 |
| 3 | 17.443 | 16.044 | 15.439 | 15.101 | 14.885 | 14.735 | 14.624 | 14.540 | 14.473 | 14.419 | 14.374 |
| 4 | 12.218 | 10.649 | 9.979 | 9.604 | 9.364 | 9.197 | 9.074 | 8.980 | 8.905 | 8.844 | 8.794 |
| 5 | 10.007 | 8.434 | 7.764 | 7.388 | 7.146 | 6.978 | 6.853 | 6.757 | 6.681 | 6.619 | 6.568 |
| 6 | 8.813 | 7.260 | 6.599 | 6.227 | 5.988 | 5.820 | 5.695 | 5.600 | 5.523 | 5.461 | 5.410 |
| 7 | 8.073 | 6.542 | 5.890 | 5.523 | 5.285 | 5.119 | 4.995 | 4.899 | 4.823 | 4.761 | 4.709 |
| 8 | 7.571 | 6.059 | 5.416 | 5.053 | 4.817 | 4.652 | 4.529 | 4.433 | 4.357 | 4.295 | 4.243 |
| 9 | 7.209 | 5.715 | 5.078 | 4.718 | 4.484 | 4.320 | 4.197 | 4.102 | 4.026 | 3.964 | 3.912 |
| 10 | 6.937 | 5.456 | 4.826 | 4.468 | 4.236 | 4.072 | 3.950 | 3.855 | 3.779 | 3.717 | 3.665 |
| 11 | 6.724 | 5.256 | 4.630 | 4.275 | 4.044 | 3.881 | 3.759 | 3.664 | 3.588 | 3.526 | 3.474 |
| 12 | 6.554 | 5.096 | 4.474 | 4.121 | 3.891 | 3.728 | 3.607 | 3.512 | 3.436 | 3.374 | 3.321 |
| 13 | 6.414 | 4.965 | 4.347 | 3.996 | 3.767 | 3.604 | 3.483 | 3.388 | 3.312 | 3.250 | 3.197 |
| 14 | 6.298 | 4.857 | 4.242 | 3.892 | 3.663 | 3.501 | 3.380 | 3.285 | 3.209 | 3.147 | 3.095 |
| 15 | 6.200 | 4.765 | 4.153 | 3.804 | 3.576 | 3.415 | 3.293 | 3.199 | 3.123 | 3.060 | 3.008 |
| 16 | 6.115 | 4.687 | 4.077 | 3.729 | 3.502 | 3.341 | 3.219 | 3.125 | 3.049 | 2.986 | 2.934 |
| 17 | 6.042 | 4.619 | 4.011 | 3.665 | 3.438 | 3.277 | 3.156 | 3.061 | 2.985 | 2.922 | 2.870 |
| 18 | 5.978 | 4.560 | 3.954 | 3.608 | 3.382 | 3.221 | 3.100 | 3.005 | 2.929 | 2.866 | 2.814 |
| 19 | 5.922 | 4.508 | 3.903 | 3.559 | 3.333 | 3.172 | 3.051 | 2.956 | 2.880 | 2.817 | 2.765 |
| 20 | 5.871 | 4.461 | 3.859 | 3.515 | 3.289 | 3.128 | 3.007 | 2.913 | 2.837 | 2.774 | 2.721 |
| 24 | 5.717 | 4.319 | 3.721 | 3.379 | 3.155 | 2.995 | 2.874 | 2.779 | 2.703 | 2.640 | 2.586 |
| 30 | 5.568 | 4.182 | 3.589 | 3.250 | 3.026 | 2.867 | 2.746 | 2.651 | 2.575 | 2.511 | 2.458 |
| 40 | 5.424 | 4.051 | 3.463 | 3.126 | 2.904 | 2.744 | 2.624 | 2.529 | 2.452 | 2.388 | 2.334 |
| 50 | 5.340 | 3.975 | 3.390 | 3.054 | 2.833 | 2.674 | 2.553 | 2.458 | 2.381 | 2.317 | 2.263 |
| 100 | 5.179 | 3.828 | 3.250 | 2.917 | 2.696 | 2.537 | 2.417 | 2.321 | 2.244 | 2.179 | 2.124 |
| 200 | 5.100 | 3.758 | 3.182 | 2.850 | 2.630 | 2.472 | 2.351 | 2.256 | 2.178 | 2.113 | 2.058 |
| 300 | 5.075 | 3.735 | 3.160 | 2.829 | 2.609 | 2.451 | 2.330 | 2.234 | 2.156 | 2.091 | 2.036 |

**DENOMINATOR**

$df = D_2$                                NUMERATOR $df = D_1$

| | 12 | 13 | 14 | 15 | 16 | 17 | 18 | 19 | 20 | 24 | 30 |
|---|---|---|---|---|---|---|---|---|---|---|---|
| 1 | 976.725 | 979.839 | 982.545 | 984.874 | 986.911 | 988.715 | 990.345 | 991.800 | 993.081 | 997.272 | 1001.405 |
| 2 | 39.415 | 39.421 | 39.427 | 39.431 | 39.436 | 39.439 | 39.442 | 39.446 | 39.448 | 39.457 | 39.465 |
| 3 | 14.337 | 14.305 | 14.277 | 14.253 | 14.232 | 14.213 | 14.196 | 14.181 | 14.167 | 14.124 | 14.081 |
| 4 | 8.751 | 8.715 | 8.684 | 8.657 | 8.633 | 8.611 | 8.592 | 8.575 | 8.560 | 8.511 | 8.461 |
| 5 | 6.525 | 6.488 | 6.456 | 6.428 | 6.403 | 6.381 | 6.362 | 6.344 | 6.329 | 6.278 | 6.227 |
| 6 | 5.366 | 5.329 | 5.297 | 5.269 | 5.244 | 5.222 | 5.202 | 5.184 | 5.168 | 5.117 | 5.065 |
| 7 | 4.666 | 4.628 | 4.596 | 4.568 | 4.543 | 4.521 | 4.501 | 4.483 | 4.467 | 4.415 | 4.362 |
| 8 | 4.200 | 4.162 | 4.130 | 4.101 | 4.076 | 4.054 | 4.034 | 4.016 | 3.999 | 3.947 | 3.894 |
| 9 | 3.868 | 3.831 | 3.798 | 3.769 | 3.744 | 3.722 | 3.701 | 3.683 | 3.667 | 3.614 | 3.560 |
| 10 | 3.621 | 3.583 | 3.550 | 3.522 | 3.496 | 3.474 | 3.453 | 3.435 | 3.419 | 3.365 | 3.311 |
| 11 | 3.430 | 3.392 | 3.359 | 3.330 | 3.304 | 3.282 | 3.261 | 3.243 | 3.226 | 3.173 | 3.118 |
| 12 | 3.277 | 3.239 | 3.206 | 3.177 | 3.152 | 3.129 | 3.108 | 3.090 | 3.073 | 3.019 | 2.963 |
| 13 | 3.153 | 3.115 | 3.082 | 3.053 | 3.027 | 3.004 | 2.983 | 2.965 | 2.948 | 2.893 | 2.837 |

| 14 | 3.050 | 3.012 | 2.979 | 2.949 | 2.923 | 2.900 | 2.879 | 2.861 | 2.844 | 2.789 | 2.732 |
| 15 | 2.963 | 2.925 | 2.891 | 2.862 | 2.836 | 2.813 | 2.792 | 2.773 | 2.756 | 2.701 | 2.644 |
| 16 | 2.889 | 2.851 | 2.817 | 2.788 | 2.761 | 2.738 | 2.717 | 2.698 | 2.681 | 2.625 | 2.568 |
| 17 | 2.825 | 2.786 | 2.753 | 2.723 | 2.697 | 2.673 | 2.652 | 2.633 | 2.616 | 2.560 | 2.502 |
| 18 | 2.769 | 2.730 | 2.696 | 2.667 | 2.640 | 2.617 | 2.596 | 2.576 | 2.559 | 2.503 | 2.445 |
| 19 | 2.720 | 2.681 | 2.647 | 2.617 | 2.591 | 2.567 | 2.546 | 2.526 | 2.509 | 2.452 | 2.394 |
| 20 | 2.676 | 2.637 | 2.603 | 2.573 | 2.547 | 2.523 | 2.501 | 2.482 | 2.464 | 2.408 | 2.349 |
| 24 | 2.541 | 2.502 | 2.468 | 2.437 | 2.411 | 2.386 | 2.365 | 2.345 | 2.327 | 2.269 | 2.209 |
| 30 | 2.412 | 2.372 | 2.338 | 2.307 | 2.280 | 2.255 | 2.233 | 2.213 | 2.195 | 2.136 | 2.074 |
| 40 | 2.288 | 2.248 | 2.213 | 2.182 | 2.154 | 2.129 | 2.107 | 2.086 | 2.068 | 2.007 | 1.943 |
| 50 | 2.216 | 2.176 | 2.140 | 2.109 | 2.081 | 2.056 | 2.033 | 2.012 | 1.993 | 1.931 | 1.866 |
| 100 | 2.077 | 2.036 | 2.000 | 1.968 | 1.939 | 1.913 | 1.890 | 1.868 | 1.849 | 1.784 | 1.715 |
| 200 | 2.010 | 1.969 | 1.932 | 1.900 | 1.870 | 1.844 | 1.820 | 1.798 | 1.778 | 1.712 | 1.640 |
| 300 | 1.988 | 1.947 | 1.910 | 1.877 | 1.848 | 1.821 | 1.797 | 1.775 | 1.755 | 1.688 | 1.616 |

**DENOMINATOR**

$df = D_2$　　　　　　　　　　　　　　　**NUMERATOR** $df = D_1$

| | **40** | **50** | **100** | **200** | **300** |
|---|---|---|---|---|---|
| 1 | 1005.596 | 1008.098 | 1013.163 | 1015.724 | 1016.539 |
| 2 | 39.473 | 39.478 | 39.488 | 39.493 | 39.495 |
| 3 | 14.036 | 14.010 | 13.956 | 13.929 | 13.920 |
| 4 | 8.411 | 8.381 | 8.319 | 8.288 | 8.278 |
| 5 | 6.175 | 6.144 | 6.080 | 6.048 | 6.037 |
| 6 | 5.012 | 4.980 | 4.915 | 4.882 | 4.871 |
| 7 | 4.309 | 4.276 | 4.210 | 4.176 | 4.165 |
| 8 | 3.840 | 3.807 | 3.739 | 3.705 | 3.693 |
| 9 | 3.505 | 3.472 | 3.403 | 3.368 | 3.357 |
| 10 | 3.255 | 3.221 | 3.152 | 3.116 | 3.104 |
| 11 | 3.061 | 3.027 | 2.956 | 2.920 | 2.908 |
| 12 | 2.906 | 2.871 | 2.800 | 2.763 | 2.750 |
| 13 | 2.780 | 2.744 | 2.671 | 2.634 | 2.621 |
| 14 | 2.674 | 2.638 | 2.565 | 2.526 | 2.513 |
| 15 | 2.585 | 2.549 | 2.474 | 2.435 | 2.422 |
| 16 | 2.509 | 2.472 | 2.396 | 2.357 | 2.343 |
| 17 | 2.442 | 2.405 | 2.329 | 2.289 | 2.275 |
| 18 | 2.384 | 2.347 | 2.269 | 2.229 | 2.215 |
| 19 | 2.333 | 2.295 | 2.217 | 2.176 | 2.162 |
| 20 | 2.287 | 2.249 | 2.170 | 2.128 | 2.114 |
| 24 | 2.146 | 2.107 | 2.024 | 1.981 | 1.966 |
| 30 | 2.009 | 1.968 | 1.882 | 1.835 | 1.819 |
| 40 | 1.875 | 1.832 | 1.741 | 1.691 | 1.673 |
| 50 | 1.796 | 1.752 | 1.656 | 1.603 | 1.584 |
| 100 | 1.640 | 1.592 | 1.483 | 1.420 | 1.397 |
| 200 | 1.562 | 1.511 | 1.393 | 1.320 | 1.293 |
| 300 | 1.536 | 1.484 | 1.361 | 1.285 | 1.255 |

## F-Distribution Table: Upper 1% Probability (or 1% Area) under F-Distribution Curve

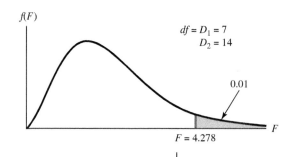

**DENOMINATOR**

$df = D_2$          **NUMERATOR** $df = D_1$

| | 1 | 2 | 3 | 4 | 5 | 6 | 7 | 8 | 9 | 10 | 11 |
|---|---|---|---|---|---|---|---|---|---|---|---|
| 1 | 4052.185 | 4999.340 | 5403.534 | 5624.257 | 5763.955 | 5858.950 | 5928.334 | 5980.954 | 6022.397 | 6055.925 | 6083.399 |
| 2 | 98.502 | 99.000 | 99.164 | 99.251 | 99.302 | 99.331 | 99.357 | 99.375 | 99.390 | 99.397 | 99.408 |
| 3 | 34.116 | 30.816 | 29.457 | 28.710 | 28.237 | 27.911 | 27.671 | 27.489 | 27.345 | 27.228 | 27.132 |
| 4 | 21.198 | 18.000 | 16.694 | 15.977 | 15.522 | 15.207 | 14.976 | 14.799 | 14.659 | 14.546 | 14.452 |
| 5 | 16.258 | 13.274 | 12.060 | 11.392 | 10.967 | 10.672 | 10.456 | 10.289 | 10.158 | 10.051 | 9.963 |
| 6 | 13.745 | 10.925 | 9.780 | 9.148 | 8.746 | 8.466 | 8.260 | 8.102 | 7.976 | 7.874 | 7.790 |
| 7 | 12.246 | 9.547 | 8.451 | 7.847 | 7.460 | 7.191 | 6.993 | 6.840 | 6.719 | 6.620 | 6.538 |
| 8 | 11.259 | 8.649 | 7.591 | 7.006 | 6.632 | 6.371 | 6.178 | 6.029 | 5.911 | 5.814 | 5.734 |
| 9 | 10.562 | 8.022 | 6.992 | 6.422 | 6.057 | 5.802 | 5.613 | 5.467 | 5.351 | 5.257 | 5.178 |
| 10 | 10.044 | 7.559 | 6.552 | 5.994 | 5.636 | 5.386 | 5.200 | 5.057 | 4.942 | 4.849 | 4.772 |
| 11 | 9.646 | 7.206 | 6.217 | 5.668 | 5.316 | 5.069 | 4.886 | 4.744 | 4.632 | 4.539 | 4.462 |
| 12 | 9.330 | 6.927 | 5.953 | 5.412 | 5.064 | 4.821 | 4.640 | 4.499 | 4.388 | 4.296 | 4.220 |
| 13 | 9.074 | 6.701 | 5.739 | 5.205 | 4.862 | 4.620 | 4.441 | 4.302 | 4.191 | 4.100 | 4.025 |
| 14 | 8.862 | 6.515 | 5.564 | 5.035 | 4.695 | 4.456 | 4.278 | 4.140 | 4.030 | 3.939 | 3.864 |
| 15 | 8.683 | 6.359 | 5.417 | 4.893 | 4.556 | 4.318 | 4.142 | 4.004 | 3.895 | 3.805 | 3.730 |
| 16 | 8.531 | 6.226 | 5.292 | 4.773 | 4.437 | 4.202 | 4.026 | 3.890 | 3.780 | 3.691 | 3.616 |
| 17 | 8.400 | 6.112 | 5.185 | 4.669 | 4.336 | 4.101 | 3.927 | 3.791 | 3.682 | 3.593 | 3.518 |
| 18 | 8.285 | 6.013 | 5.092 | 4.579 | 4.248 | 4.015 | 3.841 | 3.705 | 3.597 | 3.508 | 3.434 |
| 19 | 8.185 | 5.926 | 5.010 | 4.500 | 4.171 | 3.939 | 3.765 | 3.631 | 3.523 | 3.434 | 3.360 |
| 20 | 8.096 | 5.849 | 4.938 | 4.431 | 4.103 | 3.871 | 3.699 | 3.564 | 3.457 | 3.368 | 3.294 |
| 24 | 7.823 | 5.614 | 4.718 | 4.218 | 3.895 | 3.667 | 3.496 | 3.363 | 3.256 | 3.168 | 3.094 |
| 30 | 7.562 | 5.390 | 4.510 | 4.018 | 3.699 | 3.473 | 3.305 | 3.173 | 3.067 | 2.979 | 2.906 |
| 40 | 7.314 | 5.178 | 4.313 | 3.828 | 3.514 | 3.291 | 3.124 | 2.993 | 2.888 | 2.801 | 2.727 |
| 50 | 7.171 | 5.057 | 4.199 | 3.720 | 3.408 | 3.186 | 3.020 | 2.890 | 2.785 | 2.698 | 2.625 |
| 100 | 6.895 | 4.824 | 3.984 | 3.513 | 3.206 | 2.988 | 2.823 | 2.694 | 2.590 | 2.503 | 2.430 |
| 200 | 6.763 | 4.713 | 3.881 | 3.414 | 3.110 | 2.893 | 2.730 | 2.601 | 2.497 | 2.411 | 2.338 |
| 300 | 6.720 | 4.677 | 3.848 | 3.382 | 3.079 | 2.862 | 2.699 | 2.571 | 2.467 | 2.380 | 2.307 |

**DENOMINATOR**

$df = D_2$          **NUMERATOR** $df = D_1$

| | 12 | 13 | 14 | 15 | 16 | 17 | 18 | 19 | 20 | 24 | 30 |
|---|---|---|---|---|---|---|---|---|---|---|---|
| 1 | 6106.682 | 6125.774 | 6143.004 | 6156.974 | 6170.012 | 6181.188 | 6191.432 | 6200.746 | 6208.662 | 6234.273 | 6260.350 |
| 2 | 99.419 | 99.422 | 99.426 | 99.433 | 99.437 | 99.441 | 99.444 | 99.448 | 99.448 | 99.455 | 99.466 |
| 3 | 27.052 | 26.983 | 26.924 | 26.872 | 26.826 | 26.786 | 26.751 | 26.719 | 26.690 | 26.597 | 26.504 |
| 4 | 14.374 | 14.306 | 14.249 | 14.198 | 14.154 | 14.114 | 14.079 | 14.048 | 14.019 | 13.929 | 13.838 |
| 5 | 9.888 | 9.825 | 9.770 | 9.722 | 9.680 | 9.643 | 9.609 | 9.580 | 9.553 | 9.466 | 9.379 |
| 6 | 7.718 | 7.657 | 7.605 | 7.559 | 7.519 | 7.483 | 7.451 | 7.422 | 7.396 | 7.313 | 7.229 |
| 7 | 6.469 | 6.410 | 6.359 | 6.314 | 6.275 | 6.240 | 6.209 | 6.181 | 6.155 | 6.074 | 5.992 |
| 8 | 5.667 | 5.609 | 5.559 | 5.515 | 5.477 | 5.442 | 5.412 | 5.384 | 5.359 | 5.279 | 5.198 |
| 9 | 5.111 | 5.055 | 5.005 | 4.962 | 4.924 | 4.890 | 4.860 | 4.833 | 4.808 | 4.729 | 4.649 |
| 10 | 4.706 | 4.650 | 4.601 | 4.558 | 4.520 | 4.487 | 4.457 | 4.430 | 4.405 | 4.327 | 4.247 |
| 11 | 4.397 | 4.342 | 4.293 | 4.251 | 4.213 | 4.180 | 4.150 | 4.123 | 4.099 | 4.021 | 3.941 |
| 12 | 4.155 | 4.100 | 4.052 | 4.010 | 3.972 | 3.939 | 3.910 | 3.883 | 3.858 | 3.780 | 3.701 |
| 13 | 3.960 | 3.905 | 3.857 | 3.815 | 3.778 | 3.745 | 3.716 | 3.689 | 3.665 | 3.587 | 3.507 |

| 14 | 3.800 | 3.745 | 3.698 | 3.656 | 3.619 | 3.586 | 3.556 | 3.529 | 3.505 | 3.427 | 3.348 |
| 15 | 3.666 | 3.612 | 3.564 | 3.522 | 3.485 | 3.452 | 3.423 | 3.396 | 3.372 | 3.294 | 3.214 |
| 16 | 3.553 | 3.498 | 3.451 | 3.409 | 3.372 | 3.339 | 3.310 | 3.283 | 3.259 | 3.181 | 3.101 |
| 17 | 3.455 | 3.401 | 3.353 | 3.312 | 3.275 | 3.242 | 3.212 | 3.186 | 3.162 | 3.083 | 3.003 |
| 18 | 3.371 | 3.316 | 3.269 | 3.227 | 3.190 | 3.158 | 3.128 | 3.101 | 3.077 | 2.999 | 2.919 |
| 19 | 3.297 | 3.242 | 3.195 | 3.153 | 3.116 | 3.084 | 3.054 | 3.027 | 3.003 | 2.925 | 2.844 |
| 20 | 3.231 | 3.177 | 3.130 | 3.088 | 3.051 | 3.018 | 2.989 | 2.962 | 2.938 | 2.859 | 2.778 |
| 24 | 3.032 | 2.977 | 2.930 | 2.889 | 2.852 | 2.819 | 2.789 | 2.762 | 2.738 | 2.659 | 2.577 |
| 30 | 2.843 | 2.789 | 2.742 | 2.700 | 2.663 | 2.630 | 2.600 | 2.573 | 2.549 | 2.469 | 2.386 |
| 40 | 2.665 | 2.611 | 2.563 | 2.522 | 2.484 | 2.451 | 2.421 | 2.394 | 2.369 | 2.288 | 2.203 |
| 50 | 2.563 | 2.508 | 2.461 | 2.419 | 2.382 | 2.348 | 2.318 | 2.290 | 2.265 | 2.183 | 2.098 |
| 100 | 2.368 | 2.313 | 2.265 | 2.223 | 2.185 | 2.151 | 2.120 | 2.092 | 2.067 | 1.983 | 1.893 |
| 200 | 2.275 | 2.220 | 2.172 | 2.129 | 2.091 | 2.057 | 2.026 | 1.997 | 1.971 | 1.886 | 1.794 |
| 300 | 2.244 | 2.190 | 2.142 | 2.099 | 2.061 | 2.026 | 1.995 | 1.966 | 1.940 | 1.854 | 1.761 |

**DENOMINATOR**

| $df = D_2$ | | | NUMERATOR $df = D_1$ | | |
|---|---|---|---|---|---|
| | **40** | **50** | **100** | **200** | **300** |
| 1 | 6286.427 | 6302.260 | 6333.925 | 6349.757 | 6355.345 |
| 2 | 99.477 | 99.477 | 99.491 | 99.491 | 99.499 |
| 3 | 26.411 | 26.354 | 26.241 | 26.183 | 26.163 |
| 4 | 13.745 | 13.690 | 13.577 | 13.520 | 13.501 |
| 5 | 9.291 | 9.238 | 9.130 | 9.075 | 9.057 |
| 6 | 7.143 | 7.091 | 6.987 | 6.934 | 6.916 |
| 7 | 5.908 | 5.858 | 5.755 | 5.702 | 5.685 |
| 8 | 5.116 | 5.065 | 4.963 | 4.911 | 4.894 |
| 9 | 4.567 | 4.517 | 4.415 | 4.363 | 4.346 |
| 10 | 4.165 | 4.115 | 4.014 | 3.962 | 3.944 |
| 11 | 3.860 | 3.810 | 3.708 | 3.656 | 3.638 |
| 12 | 3.619 | 3.569 | 3.467 | 3.414 | 3.397 |
| 13 | 3.425 | 3.375 | 3.272 | 3.219 | 3.202 |
| 14 | 3.266 | 3.215 | 3.112 | 3.059 | 3.040 |
| 15 | 3.132 | 3.081 | 2.977 | 2.923 | 2.905 |
| 16 | 3.018 | 2.967 | 2.863 | 2.808 | 2.790 |
| 17 | 2.920 | 2.869 | 2.764 | 2.709 | 2.691 |
| 18 | 2.835 | 2.784 | 2.678 | 2.623 | 2.604 |
| 19 | 2.761 | 2.709 | 2.602 | 2.547 | 2.528 |
| 20 | 2.695 | 2.643 | 2.535 | 2.479 | 2.460 |
| 24 | 2.492 | 2.440 | 2.329 | 2.271 | 2.251 |
| 30 | 2.299 | 2.245 | 2.131 | 2.070 | 2.049 |
| 40 | 2.114 | 2.058 | 1.938 | 1.874 | 1.851 |
| 50 | 2.007 | 1.949 | 1.825 | 1.757 | 1.733 |
| 100 | 1.797 | 1.735 | 1.598 | 1.518 | 1.490 |
| 200 | 1.694 | 1.629 | 1.481 | 1.391 | 1.357 |
| 300 | 1.660 | 1.594 | 1.441 | 1.346 | 1.309 |

**p = 0.95**

| $D_2$ \ $D_1$ | 2 | 3 | 4 | 5 | 6 | 7 | 8 | 9 | 10 |
|---|---|---|---|---|---|---|---|---|---|
| 1 | 17.97 | 26.98 | 32.82 | 37.08 | 40.41 | 43.12 | 45.40 | 47.36 | 49.07 |
| 2 | 6.08 | 8.33 | 9.80 | 10.88 | 11.74 | 12.44 | 13.03 | 13.54 | 13.99 |
| 3 | 4.50 | 5.91 | 6.82 | 7.50 | 8.04 | 8.48 | 8.85 | 9.18 | 9.46 |
| 4 | 3.93 | 5.04 | 5.76 | 6.29 | 6.71 | 7.05 | 7.35 | 7.60 | 7.83 |
| 5 | 3.64 | 4.60 | 5.22 | 5.67 | 6.03 | 6.33 | 6.58 | 6.80 | 6.99 |
| 6 | 3.46 | 4.34 | 4.90 | 5.30 | 5.63 | 5.90 | 6.12 | 6.32 | 6.49 |
| 7 | 3.34 | 4.16 | 4.68 | 5.06 | 5.36 | 5.61 | 5.82 | 6.00 | 6.16 |
| 8 | 3.26 | 4.04 | 4.53 | 4.89 | 5.17 | 5.40 | 5.60 | 5.77 | 5.92 |
| 9 | 3.20 | 3.95 | 4.41 | 4.76 | 5.02 | 5.24 | 5.43 | 5.59 | 5.74 |
| 10 | 3.15 | 3.88 | 4.33 | 4.65 | 4.91 | 5.12 | 5.30 | 5.46 | 5.60 |
| 11 | 3.11 | 3.82 | 4.26 | 4.57 | 4.82 | 5.03 | 5.20 | 5.35 | 5.49 |
| 12 | 3.08 | 3.77 | 4.20 | 4.51 | 4.75 | 4.95 | 5.12 | 5.27 | 5.39 |
| 13 | 3.06 | 3.73 | 4.15 | 4.45 | 4.69 | 4.88 | 5.05 | 5.19 | 5.32 |
| 14 | 3.03 | 3.70 | 4.11 | 4.41 | 4.64 | 4.83 | 4.99 | 5.13 | 5.25 |
| 15 | 3.01 | 3.67 | 4.08 | 4.37 | 4.59 | 4.78 | 4.94 | 5.08 | 5.20 |
| 16 | 3.00 | 3.65 | 4.05 | 4.33 | 4.56 | 4.74 | 4.90 | 5.03 | 5.15 |
| 17 | 2.98 | 3.63 | 4.02 | 4.30 | 4.52 | 4.70 | 4.86 | 4.99 | 5.11 |
| 18 | 2.97 | 3.61 | 4.00 | 4.28 | 4.49 | 4.67 | 4.82 | 4.96 | 5.07 |
| 19 | 2.96 | 3.59 | 3.98 | 4.25 | 4.47 | 4.65 | 4.79 | 4.92 | 5.04 |
| 20 | 2.95 | 3.58 | 3.96 | 4.23 | 4.45 | 4.62 | 4.77 | 4.90 | 5.01 |
| 24 | 2.92 | 3.53 | 3.90 | 4.17 | 4.37 | 4.54 | 4.68 | 4.81 | 4.92 |
| 30 | 2.89 | 3.49 | 3.85 | 4.10 | 4.30 | 4.46 | 4.60 | 4.72 | 4.82 |
| 40 | 2.86 | 3.44 | 3.79 | 4.04 | 4.23 | 4.39 | 4.52 | 4.63 | 4.73 |
| 60 | 2.83 | 3.40 | 3.74 | 3.98 | 4.16 | 4.31 | 4.44 | 4.55 | 4.65 |
| 120 | 2.80 | 3.36 | 3.68 | 3.92 | 4.10 | 4.24 | 4.36 | 4.47 | 4.56 |
| ∞ | 2.77 | 3.31 | 3.63 | 3.86 | 4.03 | 4.17 | 4.29 | 4.39 | 4.47 |

| $D_2$ \ $D_1$ | 11 | 12 | 13 | 14 | 15 | 16 | 17 | 18 | 19 | 20 |
|---|---|---|---|---|---|---|---|---|---|---|
| 1 | 50.59 | 51.96 | 53.20 | 54.33 | 55.36 | 56.32 | 57.22 | 58.04 | 58.83 | 59.56 |
| 2 | 14.39 | 14.75 | 15.08 | 15.38 | 15.65 | 15.91 | 16.14 | 16.37 | 16.57 | 16.77 |
| 3 | 9.72 | 9.95 | 10.15 | 10.35 | 10.52 | 10.69 | 10.84 | 10.98 | 11.11 | 11.24 |
| 4 | 8.03 | 8.21 | 8.37 | 8.52 | 8.66 | 8.79 | 8.91 | 9.03 | 9.13 | 9.23 |
| 5 | 7.17 | 7.32 | 7.47 | 7.60 | 7.72 | 7.83 | 7.93 | 8.03 | 8.12 | 8.21 |
| 6 | 6.65 | 6.79 | 6.92 | 7.03 | 7.14 | 7.24 | 7.34 | 7.43 | 7.51 | 7.59 |
| 7 | 6.30 | 6.43 | 6.55 | 6.66 | 6.76 | 6.85 | 6.94 | 7.02 | 7.10 | 7.17 |
| 8 | 6.05 | 6.18 | 6.29 | 6.39 | 6.48 | 6.57 | 6.65 | 6.73 | 6.80 | 6.87 |
| 9 | 5.87 | 5.98 | 6.09 | 6.19 | 6.28 | 6.36 | 6.44 | 6.51 | 6.58 | 6.64 |
| 10 | 5.72 | 5.83 | 5.93 | 6.03 | 6.11 | 6.19 | 6.27 | 6.34 | 6.40 | 6.47 |
| 11 | 5.61 | 5.71 | 5.81 | 5.90 | 5.98 | 6.06 | 6.13 | 6.20 | 6.27 | 6.33 |
| 12 | 5.51 | 5.61 | 5.71 | 5.80 | 5.88 | 5.95 | 6.02 | 6.09 | 6.15 | 6.21 |
| 13 | 5.43 | 5.53 | 5.63 | 5.71 | 5.79 | 5.86 | 5.93 | 5.99 | 6.05 | 6.11 |
| 14 | 5.36 | 5.46 | 5.55 | 5.64 | 5.71 | 5.79 | 5.85 | 5.91 | 5.97 | 6.03 |
| 15 | 5.31 | 5.40 | 5.49 | 5.57 | 5.65 | 5.72 | 5.78 | 5.85 | 5.90 | 5.96 |
| 16 | 5.26 | 5.35 | 5.44 | 5.52 | 5.59 | 5.66 | 5.73 | 5.79 | 5.84 | 5.90 |
| 17 | 5.21 | 5.31 | 5.39 | 5.47 | 5.54 | 5.61 | 5.67 | 5.73 | 5.79 | 5.84 |
| 18 | 5.17 | 5.27 | 5.35 | 5.43 | 5.50 | 5.57 | 5.63 | 5.69 | 5.74 | 5.79 |
| 19 | 5.14 | 5.23 | 5.31 | 5.39 | 5.46 | 5.53 | 5.59 | 5.65 | 5.70 | 5.75 |
| 20 | 5.11 | 5.20 | 5.28 | 5.36 | 5.43 | 5.49 | 5.55 | 5.61 | 5.66 | 5.71 |
| 24 | 5.01 | 5.10 | 5.18 | 5.25 | 5.32 | 5.38 | 5.44 | 5.49 | 5.55 | 5.59 |
| 30 | 4.92 | 5.00 | 5.08 | 5.15 | 5.21 | 5.27 | 5.33 | 5.38 | 5.43 | 5.47 |
| 40 | 4.82 | 4.90 | 4.98 | 5.04 | 5.11 | 5.16 | 5.22 | 5.27 | 5.31 | 5.36 |
| 60 | 4.73 | 4.81 | 4.88 | 4.94 | 5.00 | 5.06 | 5.11 | 5.15 | 5.20 | 5.24 |
| 120 | 4.64 | 4.71 | 4.78 | 4.84 | 4.90 | 4.95 | 5.00 | 5.04 | 5.09 | 5.13 |
| ∞ | 4.55 | 4.62 | 4.68 | 4.74 | 4.80 | 4.85 | 4.89 | 4.93 | 4.97 | 5.01 |

*Note:* $D_1 = K$ populations and $D_2 = N - K$.

$p = 0.99$

| $D_2$ \ $D_1$ | 2 | 3 | 4 | 5 | 6 | 7 | 8 | 9 | 10 |
|---|---|---|---|---|---|---|---|---|---|
| 1 | 90.03 | 135.0 | 164.3 | 185.6 | 202.2 | 215.8 | 227.2 | 237.0 | 245.6 |
| 2 | 14.04 | 19.02 | 22.29 | 24.72 | 26.63 | 28.20 | 29.53 | 30.68 | 31.69 |
| 3 | 8.26 | 10.62 | 12.17 | 13.33 | 14.24 | 15.00 | 15.64 | 16.20 | 16.69 |
| 4 | 6.51 | 8.12 | 9.17 | 9.96 | 10.58 | 11.10 | 11.55 | 11.93 | 12.27 |
| 5 | 5.70 | 6.98 | 7.80 | 8.42 | 8.91 | 9.32 | 9.67 | 9.97 | 10.24 |
| 6 | 5.24 | 6.33 | 7.03 | 7.56 | 7.97 | 8.32 | 8.61 | 8.87 | 9.10 |
| 7 | 4.95 | 5.92 | 6.54 | 7.01 | 7.37 | 7.68 | 7.94 | 8.17 | 8.37 |
| 8 | 4.75 | 5.64 | 6.20 | 6.62 | 6.96 | 7.24 | 7.47 | 7.68 | 7.86 |
| 9 | 4.60 | 5.43 | 5.96 | 6.35 | 6.66 | 6.91 | 7.13 | 7.33 | 7.49 |
| 10 | 4.48 | 5.27 | 5.77 | 6.14 | 6.43 | 6.67 | 6.87 | 7.05 | 7.21 |
| 11 | 4.39 | 5.15 | 5.62 | 5.97 | 6.25 | 6.48 | 6.67 | 6.84 | 6.99 |
| 12 | 4.32 | 5.05 | 5.50 | 5.84 | 6.10 | 6.32 | 6.51 | 6.67 | 6.81 |
| 13 | 4.26 | 4.96 | 5.40 | 5.73 | 5.98 | 6.19 | 6.37 | 6.53 | 6.67 |
| 14 | 4.21 | 4.89 | 5.32 | 5.63 | 5.88 | 6.08 | 6.26 | 6.41 | 6.54 |
| 15 | 4.17 | 4.84 | 5.25 | 5.56 | 5.80 | 5.99 | 6.16 | 6.31 | 6.44 |
| 16 | 4.13 | 4.79 | 5.19 | 5.49 | 5.72 | 5.92 | 6.08 | 6.22 | 6.35 |
| 17 | 4.10 | 4.74 | 5.14 | 5.43 | 5.66 | 5.85 | 6.01 | 6.15 | 6.27 |
| 18 | 4.07 | 4.70 | 5.09 | 5.38 | 5.60 | 5.79 | 5.94 | 6.08 | 6.20 |
| 19 | 4.05 | 4.67 | 5.05 | 5.33 | 5.55 | 5.73 | 5.89 | 6.02 | 6.14 |
| 20 | 4.02 | 4.64 | 5.02 | 5.29 | 5.51 | 5.69 | 5.84 | 5.97 | 6.09 |
| 24 | 3.96 | 4.55 | 4.91 | 5.17 | 5.37 | 5.54 | 5.69 | 5.81 | 5.92 |
| 30 | 3.89 | 4.45 | 4.80 | 5.05 | 5.24 | 5.40 | 5.54 | 5.65 | 5.76 |
| 40 | 3.82 | 4.37 | 4.70 | 4.93 | 5.11 | 5.26 | 5.39 | 5.50 | 5.60 |
| 60 | 3.76 | 4.28 | 4.59 | 4.82 | 4.99 | 5.13 | 5.25 | 5.36 | 5.45 |
| 120 | 3.70 | 4.20 | 4.50 | 4.71 | 4.87 | 5.01 | 5.12 | 5.21 | 5.30 |
| ∞ | 3.64 | 4.12 | 4.40 | 4.60 | 4.76 | 4.88 | 4.99 | 5.08 | 5.16 |

| $D_2$ \ $D_1$ | 11 | 12 | 13 | 14 | 15 | 16 | 17 | 18 | 19 | 20 |
|---|---|---|---|---|---|---|---|---|---|---|
| 1 | 253.2 | 260.0 | 266.2 | 271.8 | 277.0 | 281.8 | 286.3 | 290.4 | 294.3 | 298.0 |
| 2 | 32.59 | 33.40 | 34.13 | 34.81 | 35.43 | 36.00 | 36.53 | 37.03 | 37.50 | 37.95 |
| 3 | 17.13 | 17.53 | 17.89 | 18.22 | 18.52 | 18.81 | 19.07 | 19.32 | 19.55 | 19.77 |
| 4 | 12.57 | 12.84 | 13.09 | 13.32 | 13.53 | 13.73 | 13.91 | 14.08 | 14.24 | 14.40 |
| 5 | 10.48 | 10.70 | 10.89 | 11.08 | 11.24 | 11.40 | 11.55 | 11.68 | 11.81 | 11.93 |
| 6 | 9.30 | 9.48 | 9.65 | 9.81 | 9.95 | 10.08 | 10.21 | 10.32 | 10.43 | 10.54 |
| 7 | 8.55 | 8.71 | 8.86 | 9.00 | 9.12 | 9.24 | 9.35 | 9.46 | 9.55 | 9.65 |
| 8 | 8.03 | 8.18 | 8.31 | 8.44 | 8.55 | 8.66 | 8.76 | 8.85 | 8.94 | 9.03 |
| 9 | 7.65 | 7.78 | 7.91 | 8.03 | 8.13 | 8.23 | 8.33 | 8.41 | 8.49 | 8.57 |
| 10 | 7.36 | 7.49 | 7.60 | 7.71 | 7.81 | 7.91 | 7.99 | 8.08 | 8.15 | 8.23 |
| 11 | 7.13 | 7.25 | 7.36 | 7.46 | 7.56 | 7.65 | 7.73 | 7.81 | 7.88 | 7.95 |
| 12 | 6.94 | 7.06 | 7.17 | 7.26 | 7.36 | 7.44 | 7.52 | 7.59 | 7.66 | 7.73 |
| 13 | 6.79 | 6.90 | 7.01 | 7.10 | 7.19 | 7.27 | 7.35 | 7.42 | 7.48 | 7.55 |
| 14 | 6.66 | 6.77 | 6.87 | 6.96 | 7.05 | 7.13 | 7.20 | 7.27 | 7.33 | 7.39 |
| 15 | 6.55 | 6.66 | 6.76 | 6.84 | 6.93 | 7.00 | 7.07 | 7.14 | 7.20 | 7.26 |
| 16 | 6.46 | 6.56 | 6.66 | 6.74 | 6.82 | 6.90 | 6.97 | 7.03 | 7.09 | 7.15 |
| 17 | 6.38 | 6.48 | 6.57 | 6.66 | 6.73 | 6.81 | 6.87 | 6.94 | 7.00 | 7.05 |
| 18 | 6.31 | 6.41 | 6.50 | 6.58 | 6.65 | 6.73 | 6.79 | 6.85 | 6.91 | 6.97 |
| 19 | 6.25 | 6.34 | 6.43 | 6.51 | 6.58 | 6.65 | 6.72 | 6.78 | 6.84 | 6.89 |
| 20 | 6.19 | 6.28 | 6.37 | 6.45 | 6.52 | 6.59 | 6.65 | 6.71 | 6.77 | 6.82 |
| 24 | 6.02 | 6.11 | 6.19 | 6.26 | 6.33 | 6.39 | 6.45 | 6.51 | 6.56 | 6.61 |
| 30 | 5.85 | 5.93 | 6.01 | 6.08 | 6.14 | 6.20 | 6.26 | 6.31 | 6.36 | 6.41 |
| 40 | 5.69 | 5.76 | 5.83 | 5.90 | 5.96 | 6.02 | 6.07 | 6.12 | 6.16 | 6.21 |
| 60 | 5.53 | 5.60 | 5.67 | 5.73 | 5.78 | 5.84 | 5.89 | 5.93 | 5.97 | 6.01 |
| 120 | 5.37 | 5.44 | 5.50 | 5.56 | 5.61 | 5.66 | 5.71 | 5.75 | 5.79 | 5.83 |
| ∞ | 5.23 | 5.28 | 5.35 | 5.40 | 5.45 | 5.49 | 5.54 | 5.57 | 5.61 | 5.65 |

*Source:* Extracted from H. L. Harter and D. S. Clemm, "The Probability Integrals of the Range and of the Studentized Range—Probability Integral, Percentage Points, and Moments of the Range," *Wright Air Development Technical Report 58-484*, Vol. 1, 1959.

## Critical Values of *r* in the Runs Test

a. Lower Tail: Too Few Runs

| $n_1$ \ $n_2$ | 2 | 3 | 4 | 5 | 6 | 7 | 8 | 9 | 10 | 11 | 12 | 13 | 14 | 15 | 16 | 17 | 18 | 19 | 20 |
|---|---|---|---|---|---|---|---|---|---|---|---|---|---|---|---|---|---|---|---|
| 2 | | | | | | | | | | | 2 | 2 | 2 | 2 | 2 | 2 | 2 | 2 | 2 |
| 3 | | | | 2 | 2 | 2 | 2 | 2 | 2 | 2 | 2 | 2 | 2 | 3 | 3 | 3 | 3 | 3 | 3 |
| 4 | | | 2 | 2 | 2 | 3 | 3 | 3 | 3 | 3 | 3 | 3 | 3 | 3 | 4 | 4 | 4 | 4 | 4 |
| 5 | | 2 | 2 | 3 | 3 | 3 | 3 | 3 | 4 | 4 | 4 | 4 | 4 | 4 | 4 | 4 | 5 | 5 | 5 |
| 6 | 2 | 2 | 3 | 3 | 3 | 3 | 4 | 4 | 4 | 4 | 5 | 5 | 5 | 5 | 5 | 5 | 6 | 6 | |
| 7 | 2 | 2 | 3 | 3 | 3 | 4 | 4 | 5 | 5 | 5 | 5 | 5 | 6 | 6 | 6 | 6 | 6 | 6 | |
| 8 | 2 | 3 | 3 | 3 | 4 | 4 | 5 | 5 | 5 | 6 | 6 | 6 | 6 | 6 | 7 | 7 | 7 | 7 | |
| 9 | 2 | 3 | 3 | 4 | 4 | 5 | 5 | 5 | 6 | 6 | 6 | 7 | 7 | 7 | 7 | 8 | 8 | 8 | |
| 10 | 2 | 3 | 3 | 4 | 5 | 5 | 5 | 6 | 6 | 7 | 7 | 7 | 7 | 8 | 8 | 8 | 8 | 9 | |
| 11 | 2 | 3 | 4 | 4 | 5 | 5 | 6 | 6 | 7 | 7 | 7 | 8 | 8 | 8 | 9 | 9 | 9 | 9 | |
| 12 | 2 | 2 | 3 | 4 | 4 | 5 | 6 | 6 | 7 | 7 | 7 | 8 | 8 | 8 | 9 | 9 | 9 | 10 | 10 |
| 13 | 2 | 2 | 3 | 4 | 5 | 5 | 6 | 6 | 7 | 7 | 8 | 8 | 9 | 9 | 9 | 10 | 10 | 10 | 10 |
| 14 | 2 | 2 | 3 | 4 | 5 | 5 | 6 | 7 | 7 | 8 | 8 | 9 | 9 | 9 | 10 | 10 | 10 | 11 | 11 |
| 15 | 2 | 3 | 3 | 4 | 5 | 6 | 6 | 7 | 7 | 8 | 8 | 9 | 9 | 10 | 10 | 11 | 11 | 11 | 12 |
| 16 | 2 | 3 | 4 | 4 | 5 | 6 | 6 | 7 | 8 | 8 | 9 | 9 | 10 | 10 | 11 | 11 | 11 | 12 | 12 |
| 17 | 2 | 3 | 4 | 4 | 5 | 6 | 7 | 7 | 8 | 9 | 9 | 10 | 10 | 11 | 11 | 11 | 12 | 12 | 13 |
| 18 | 2 | 3 | 4 | 5 | 5 | 6 | 7 | 8 | 8 | 9 | 9 | 10 | 10 | 11 | 11 | 12 | 12 | 13 | 13 |
| 19 | 2 | 3 | 4 | 5 | 6 | 6 | 7 | 8 | 8 | 9 | 10 | 10 | 11 | 11 | 12 | 12 | 13 | 13 | 13 |
| 20 | 2 | 3 | 4 | 5 | 6 | 6 | 7 | 8 | 9 | 9 | 10 | 10 | 11 | 12 | 12 | 13 | 13 | 13 | 14 |

b. Upper Tail: Too Many Runs

| $n_1$ \ $n_2$ | 2 | 3 | 4 | 5 | 6 | 7 | 8 | 9 | 10 | 11 | 12 | 13 | 14 | 15 | 16 | 17 | 18 | 19 | 20 |
|---|---|---|---|---|---|---|---|---|---|---|---|---|---|---|---|---|---|---|---|
| 2 | | | | | | | | | | | | | | | | | | | |
| 3 | | | | | | | | | | | | | | | | | | | |
| 4 | | | | 9 | 9 | | | | | | | | | | | | | | |
| 5 | | | 9 | 10 | 10 | 11 | 11 | | | | | | | | | | | | |
| 6 | | | 9 | 10 | 11 | 12 | 12 | 13 | 13 | 13 | 13 | | | | | | | | |
| 7 | | | | 11 | 12 | 13 | 13 | 14 | 14 | 14 | 14 | 15 | 15 | 15 | | | | | |
| 8 | | | | 11 | 12 | 13 | 14 | 14 | 15 | 15 | 16 | 16 | 16 | 16 | 17 | 17 | 17 | 17 | 17 |
| 9 | | | | | 13 | 14 | 14 | 15 | 16 | 16 | 16 | 17 | 17 | 18 | 18 | 18 | 18 | 18 | 18 |
| 10 | | | | | 13 | 14 | 15 | 16 | 16 | 17 | 17 | 18 | 18 | 18 | 19 | 19 | 19 | 20 | 20 |
| 11 | | | | | 13 | 14 | 15 | 16 | 17 | 17 | 18 | 19 | 19 | 19 | 20 | 20 | 20 | 21 | 21 |
| 12 | | | | | 13 | 14 | 16 | 16 | 17 | 18 | 19 | 19 | 20 | 20 | 21 | 21 | 21 | 22 | 22 |
| 13 | | | | | | 15 | 16 | 17 | 18 | 19 | 19 | 20 | 20 | 21 | 21 | 22 | 22 | 23 | 23 |
| 14 | | | | | | 15 | 16 | 17 | 18 | 19 | 20 | 20 | 21 | 22 | 22 | 23 | 23 | 23 | 24 |
| 15 | | | | | | 15 | 16 | 18 | 18 | 19 | 20 | 21 | 22 | 22 | 23 | 23 | 24 | 24 | 25 |
| 16 | | | | | | | 17 | 18 | 19 | 20 | 21 | 21 | 22 | 23 | 23 | 24 | 25 | 25 | 25 |
| 17 | | | | | | | 17 | 18 | 19 | 20 | 21 | 22 | 23 | 23 | 24 | 25 | 25 | 26 | 26 |
| 18 | | | | | | | 17 | 18 | 19 | 20 | 21 | 22 | 23 | 24 | 25 | 25 | 26 | 26 | 27 |
| 19 | | | | | | | 17 | 18 | 20 | 21 | 22 | 23 | 23 | 24 | 25 | 26 | 26 | 27 | 27 |
| 20 | | | | | | | 17 | 18 | 20 | 21 | 22 | 23 | 24 | 25 | 25 | 26 | 27 | 27 | 28 |

*Source:* Adapted from Frieda S. Swed and C. Eisenhart, "Tables for testing randomness of grouping in a sequence of alternatives," *Ann. Math. Statist.* 14 (1943): 83–86, with the permission of the publisher.

### $n_2 = 3$

| $U$ \ $n_1$ | 1 | 2 | 3 |
|---|---|---|---|
| 0 | .250 | .100 | .050 |
| 1 | .500 | .200 | .100 |
| 2 | .750 | .400 | .200 |
| 3 | | .600 | .350 |
| 4 | | | .500 |
| 5 | | | .650 |

### $n_2 = 4$

| $U$ \ $n_1$ | 1 | 2 | 3 | 4 |
|---|---|---|---|---|
| 0 | .200 | .067 | .028 | .014 |
| 1 | .400 | .133 | .057 | .029 |
| 2 | .600 | .267 | .114 | .057 |
| 3 | | .400 | .200 | .100 |
| 4 | | .600 | .314 | .171 |
| 5 | | | .429 | .243 |
| 6 | | | .571 | .343 |
| 7 | | | | .443 |
| 8 | | | | .557 |

### $n_2 = 5$

| $U$ \ $n_1$ | 1 | 2 | 3 | 4 | 5 |
|---|---|---|---|---|---|
| 0 | .167 | .047 | .018 | .008 | .004 |
| 1 | .333 | .095 | .036 | .016 | .008 |
| 2 | .500 | .190 | .071 | .032 | .016 |
| 3 | .667 | .286 | .125 | .056 | .028 |
| 4 | | .429 | .196 | .095 | .048 |
| 5 | | .571 | .286 | .143 | .075 |
| 6 | | | .393 | .206 | .111 |
| 7 | | | .500 | .278 | .155 |
| 8 | | | .607 | .365 | .210 |
| 9 | | | | .452 | .274 |
| 10 | | | | .548 | .345 |
| 11 | | | | | .421 |
| 12 | | | | | .500 |
| 13 | | | | | .579 |

### $n_2 = 6$

| $U$ \ $n_1$ | 1 | 2 | 3 | 4 | 5 | 6 |
|---|---|---|---|---|---|---|
| 0 | .143 | .036 | .012 | .005 | .002 | .001 |
| 1 | .286 | .071 | .024 | .010 | .004 | .002 |
| 2 | .428 | .143 | .048 | .019 | .009 | .004 |
| 3 | .571 | .214 | .083 | .033 | .015 | .008 |
| 4 | | .321 | .131 | .057 | .026 | .013 |
| 5 | | .429 | .190 | .086 | .041 | .021 |
| 6 | | .571 | .274 | .129 | .063 | .032 |
| 7 | | | .357 | .176 | .089 | .047 |
| 8 | | | .452 | .238 | .123 | .066 |
| 9 | | | .548 | .305 | .165 | .090 |
| 10 | | | | .381 | .214 | .120 |
| 11 | | | | .457 | .268 | .155 |
| 12 | | | | .545 | .331 | .197 |
| 13 | | | | | .396 | .242 |
| 14 | | | | | .465 | .294 |
| 15 | | | | | .535 | .350 |
| 16 | | | | | | .409 |
| 17 | | | | | | .469 |
| 18 | | | | | | .531 |

### $n_2 = 7$

| $U$ \ $n_1$ | 1 | 2 | 3 | 4 | 5 | 6 | 7 |
|---|---|---|---|---|---|---|---|
| 0 | .125 | .028 | .008 | .003 | .001 | .001 | .000 |
| 1 | .250 | .056 | .017 | .006 | .003 | .001 | .001 |
| 2 | .375 | .111 | .033 | .012 | .005 | .002 | .001 |
| 3 | .500 | .167 | .058 | .021 | .009 | .004 | .002 |
| 4 | .625 | .250 | .092 | .036 | .015 | .007 | .003 |
| 5 | | .333 | .133 | .055 | .024 | .011 | .006 |
| 6 | | .444 | .192 | .082 | .037 | .017 | .009 |
| 7 | | .556 | .258 | .115 | .053 | .026 | .013 |
| 8 | | | .333 | .158 | .074 | .037 | .019 |
| 9 | | | .417 | .206 | .101 | .051 | .027 |
| 10 | | | .500 | .264 | .134 | .069 | .036 |
| 11 | | | .583 | .324 | .172 | .090 | .049 |
| 12 | | | | .394 | .216 | .117 | .064 |
| 13 | | | | .464 | .265 | .147 | .082 |
| 14 | | | | .538 | .319 | .183 | .104 |
| 15 | | | | | .378 | .223 | .130 |
| 16 | | | | | .438 | .267 | .159 |
| 17 | | | | | .500 | .314 | .191 |
| 18 | | | | | .562 | .365 | .228 |
| 19 | | | | | | .418 | .267 |
| 20 | | | | | | .473 | .310 |
| 21 | | | | | | .527 | .355 |
| 22 | | | | | | | .402 |
| 23 | | | | | | | .451 |
| 24 | | | | | | | .500 |
| 25 | | | | | | | .549 |

$$n_2 = 8$$

| U \ n_1 | 1 | 2 | 3 | 4 | 5 | 6 | 7 | 8 | t | Normal |
|---|---|---|---|---|---|---|---|---|---|---|
| 0 | .111 | .022 | .006 | .002 | .001 | .000 | .000 | .000 | 3.308 | .001 |
| 1 | .222 | .044 | .012 | .004 | .002 | .001 | .000 | .000 | 3.203 | .001 |
| 2 | .333 | .089 | .024 | .008 | .003 | .001 | .001 | .000 | 3.098 | .001 |
| 3 | .444 | .133 | .042 | .014 | .005 | .002 | .001 | .001 | 2.993 | .001 |
| 4 | .556 | .200 | .067 | .024 | .009 | .004 | .002 | .001 | 2.888 | .002 |
| 5 | | .267 | .097 | .036 | .015 | .006 | .003 | .001 | 2.783 | .003 |
| 6 | | .356 | .139 | .055 | .023 | .010 | .005 | .002 | 2.678 | .004 |
| 7 | | .444 | .188 | .077 | .033 | .015 | .007 | .003 | 2.573 | .005 |
| 8 | | .556 | .248 | .107 | .047 | .021 | .010 | .005 | 2.468 | .007 |
| 9 | | | .315 | .141 | .064 | .030 | .014 | .007 | 2.363 | .009 |
| 10 | | | .387 | .184 | .085 | .041 | .020 | .010 | 2.258 | .012 |
| 11 | | | .461 | .230 | .111 | .054 | .027 | .014 | 2.153 | .016 |
| 12 | | | .539 | .285 | .142 | .071 | .036 | .019 | 2.048 | .020 |
| 13 | | | | .341 | .177 | .091 | .047 | .025 | 1.943 | .026 |
| 14 | | | | .404 | .217 | .114 | .060 | .032 | 1.838 | .033 |
| 15 | | | | .467 | .262 | .141 | .076 | .041 | 1.733 | .041 |
| 16 | | | | .533 | .311 | .172 | .095 | .052 | 1.628 | .052 |
| 17 | | | | | .362 | .207 | .116 | .065 | 1.523 | .064 |
| 18 | | | | | .416 | .245 | .140 | .080 | 1.418 | .078 |
| 19 | | | | | .472 | .286 | .168 | .097 | 1.313 | .094 |
| 20 | | | | | .528 | .331 | .198 | .117 | 1.208 | .113 |
| 21 | | | | | | .377 | .232 | .139 | 1.102 | .135 |
| 22 | | | | | | .426 | .268 | .164 | .998 | .159 |
| 23 | | | | | | .475 | .306 | .191 | .893 | .185 |
| 24 | | | | | | .525 | .347 | .221 | .788 | .215 |
| 25 | | | | | | | .389 | .253 | .683 | .247 |
| 26 | | | | | | | .433 | .287 | .578 | .282 |
| 27 | | | | | | | .478 | .323 | .473 | .318 |
| 28 | | | | | | | .522 | .360 | .368 | .356 |
| 29 | | | | | | | | .399 | .263 | .396 |
| 30 | | | | | | | | .439 | .158 | .437 |
| 31 | | | | | | | | .480 | .052 | .481 |
| 32 | | | | | | | | .520 | | |

*Source:* Reproduced from H. B. Mann and D. R. Whitney, "On a test of whether one of two random variables is stochastically larger than the other," *Ann. Math. Statist.* 18 (1947): 52–54, with the permission of the publisher.

## Mann-Whitney $U$ Test Critical Values ($9 \leq n \leq 20$)

Critical Values of $U$ for a One-Tailed Test at $\alpha = 0.001$ or for a Two-Tailed Test at $\alpha = 0.002$

| $n_1$ \ $n_2$ | 9 | 10 | 11 | 12 | 13 | 14 | 15 | 16 | 17 | 18 | 19 | 20 |
|---|---|---|---|---|---|---|---|---|---|---|---|---|
| 1 | | | | | | | | | | | | |
| 2 | | | | | | | | | | | | |
| 3 | | | | | | | | | 0 | 0 | 0 | 0 |
| 4 | | 0 | 0 | 0 | 1 | 1 | 1 | 2 | 2 | 3 | 3 | 3 |
| 5 | 1 | 1 | 2 | 2 | 3 | 3 | 4 | 5 | 5 | 6 | 7 | 7 |
| 6 | 2 | 3 | 4 | 4 | 5 | 6 | 7 | 8 | 9 | 10 | 11 | 12 |
| 7 | 3 | 5 | 6 | 7 | 8 | 9 | 10 | 11 | 13 | 14 | 15 | 16 |
| 8 | 5 | 6 | 8 | 9 | 11 | 12 | 14 | 15 | 17 | 18 | 20 | 21 |
| 9 | 7 | 8 | 10 | 12 | 14 | 15 | 17 | 19 | 21 | 23 | 25 | 26 |
| 10 | 8 | 10 | 12 | 14 | 17 | 19 | 21 | 23 | 25 | 27 | 29 | 32 |
| 11 | 10 | 12 | 15 | 17 | 20 | 22 | 24 | 27 | 29 | 32 | 34 | 37 |
| 12 | 12 | 14 | 17 | 20 | 23 | 25 | 28 | 31 | 34 | 37 | 40 | 42 |
| 13 | 14 | 17 | 20 | 23 | 26 | 29 | 32 | 35 | 38 | 42 | 45 | 48 |
| 14 | 15 | 19 | 22 | 25 | 29 | 32 | 36 | 39 | 43 | 46 | 50 | 54 |
| 15 | 17 | 21 | 24 | 28 | 32 | 36 | 40 | 43 | 47 | 51 | 55 | 59 |
| 16 | 19 | 23 | 27 | 31 | 35 | 39 | 43 | 48 | 52 | 56 | 60 | 65 |
| 17 | 21 | 25 | 29 | 34 | 38 | 43 | 47 | 52 | 57 | 61 | 66 | 70 |
| 18 | 23 | 27 | 32 | 37 | 42 | 46 | 51 | 56 | 61 | 66 | 71 | 76 |
| 19 | 25 | 29 | 34 | 40 | 45 | 50 | 55 | 60 | 66 | 71 | 77 | 82 |
| 20 | 26 | 32 | 37 | 42 | 48 | 54 | 59 | 65 | 70 | 76 | 82 | 88 |

Critical Values of $U$ for a One-Tailed Test at $\alpha = 0.01$ or for a Two-Tailed Test at $\alpha = 0.02$

| $n_1$ \ $n_2$ | 9 | 10 | 11 | 12 | 13 | 14 | 15 | 16 | 17 | 18 | 19 | 20 |
|---|---|---|---|---|---|---|---|---|---|---|---|---|
| 1 | | | | | | | | | | | | |
| 2 | | | | | 0 | 0 | 0 | 0 | 0 | 0 | 1 | 1 |
| 3 | 1 | 1 | 1 | 2 | 2 | 2 | 3 | 3 | 4 | 4 | 4 | 5 |
| 4 | 3 | 3 | 4 | 5 | 5 | 6 | 7 | 7 | 8 | 9 | 9 | 10 |
| 5 | 5 | 6 | 7 | 8 | 9 | 10 | 11 | 12 | 13 | 14 | 15 | 16 |
| 6 | 7 | 8 | 9 | 11 | 12 | 13 | 15 | 16 | 18 | 19 | 20 | 22 |
| 7 | 9 | 11 | 12 | 14 | 16 | 17 | 19 | 21 | 23 | 24 | 26 | 28 |
| 8 | 11 | 13 | 15 | 17 | 20 | 22 | 24 | 26 | 28 | 30 | 32 | 34 |
| 9 | 14 | 16 | 18 | 21 | 23 | 26 | 28 | 31 | 33 | 36 | 38 | 40 |
| 10 | 16 | 19 | 22 | 24 | 27 | 30 | 33 | 36 | 38 | 41 | 44 | 47 |
| 11 | 18 | 22 | 25 | 28 | 31 | 34 | 37 | 41 | 44 | 47 | 50 | 53 |
| 12 | 21 | 24 | 28 | 31 | 35 | 38 | 42 | 46 | 49 | 53 | 56 | 60 |
| 13 | 23 | 27 | 31 | 35 | 39 | 43 | 47 | 51 | 55 | 59 | 63 | 67 |
| 14 | 26 | 30 | 34 | 38 | 43 | 47 | 51 | 56 | 60 | 65 | 69 | 73 |
| 15 | 28 | 33 | 37 | 42 | 47 | 51 | 56 | 61 | 66 | 70 | 75 | 80 |
| 16 | 31 | 36 | 41 | 46 | 51 | 56 | 61 | 66 | 71 | 76 | 82 | 87 |
| 17 | 33 | 38 | 44 | 49 | 55 | 60 | 66 | 71 | 77 | 82 | 88 | 93 |
| 18 | 36 | 41 | 47 | 53 | 59 | 65 | 70 | 76 | 82 | 88 | 94 | 100 |
| 19 | 38 | 44 | 50 | 56 | 63 | 69 | 75 | 82 | 88 | 94 | 101 | 107 |
| 20 | 40 | 47 | 53 | 60 | 67 | 73 | 80 | 87 | 93 | 100 | 107 | 114 |

Critical Values of $U$ for a
One-Tailed Test at $\alpha = 0.025$
or for a Two-Tailed
Test at $\alpha = 0.05$

| $n_1$ \ $n_2$ | 9 | 10 | 11 | 12 | 13 | 14 | 15 | 16 | 17 | 18 | 19 | 20 |
|---|---|---|---|---|---|---|---|---|---|---|---|---|
| 1 | | | | | | | | | | | | |
| 2 | 0 | 0 | 0 | 1 | 1 | 1 | 1 | 1 | 2 | 2 | 2 | 2 |
| 3 | 2 | 3 | 3 | 4 | 4 | 5 | 5 | 6 | 6 | 7 | 7 | 8 |
| 4 | 4 | 5 | 6 | 7 | 8 | 9 | 10 | 11 | 11 | 12 | 13 | 13 |
| 5 | 7 | 8 | 9 | 11 | 12 | 13 | 14 | 15 | 17 | 18 | 19 | 20 |
| 6 | 10 | 11 | 13 | 14 | 16 | 17 | 19 | 21 | 22 | 24 | 25 | 27 |
| 7 | 12 | 14 | 16 | 18 | 20 | 22 | 24 | 26 | 28 | 30 | 32 | 34 |
| 8 | 15 | 17 | 19 | 22 | 24 | 26 | 29 | 31 | 34 | 36 | 38 | 41 |
| 9 | 17 | 20 | 23 | 26 | 28 | 31 | 34 | 37 | 39 | 42 | 45 | 48 |
| 10 | 20 | 23 | 26 | 29 | 33 | 36 | 39 | 42 | 45 | 48 | 52 | 55 |
| 11 | 23 | 26 | 30 | 33 | 37 | 40 | 44 | 47 | 51 | 55 | 58 | 62 |
| 12 | 26 | 29 | 33 | 37 | 41 | 45 | 49 | 53 | 57 | 61 | 65 | 69 |
| 13 | 28 | 33 | 37 | 41 | 45 | 50 | 54 | 59 | 63 | 67 | 72 | 76 |
| 14 | 31 | 36 | 40 | 45 | 50 | 55 | 59 | 64 | 67 | 74 | 78 | 83 |
| 15 | 34 | 39 | 44 | 49 | 54 | 59 | 64 | 70 | 75 | 80 | 85 | 90 |
| 16 | 37 | 42 | 47 | 53 | 59 | 64 | 70 | 75 | 81 | 86 | 92 | 98 |
| 17 | 39 | 45 | 51 | 57 | 63 | 67 | 75 | 81 | 87 | 93 | 99 | 105 |
| 18 | 42 | 48 | 55 | 61 | 67 | 74 | 80 | 86 | 93 | 99 | 106 | 112 |
| 19 | 45 | 52 | 58 | 65 | 72 | 78 | 85 | 92 | 99 | 106 | 113 | 119 |
| 20 | 48 | 55 | 62 | 69 | 76 | 83 | 90 | 98 | 105 | 112 | 119 | 127 |

Critical Values of $U$ for a
One-Tailed Test at $\alpha = 0.05$
or for a Two-Tailed
Test at $\alpha = 0.10$

| $n_1$ \ $n_2$ | 9 | 10 | 11 | 12 | 13 | 14 | 15 | 16 | 17 | 18 | 19 | 20 |
|---|---|---|---|---|---|---|---|---|---|---|---|---|
| 1 | | | | | | | | | | | 0 | 0 |
| 2 | 1 | 1 | 1 | 2 | 2 | 2 | 3 | 3 | 3 | 4 | 4 | 4 |
| 3 | 3 | 4 | 5 | 5 | 6 | 7 | 7 | 8 | 9 | 9 | 10 | 11 |
| 4 | 6 | 7 | 8 | 9 | 10 | 11 | 12 | 14 | 15 | 16 | 17 | 18 |
| 5 | 9 | 11 | 12 | 13 | 15 | 16 | 18 | 19 | 20 | 22 | 23 | 25 |
| 6 | 12 | 14 | 16 | 17 | 19 | 21 | 23 | 25 | 26 | 28 | 30 | 32 |
| 7 | 15 | 17 | 19 | 21 | 24 | 26 | 28 | 30 | 33 | 35 | 37 | 39 |
| 8 | 18 | 20 | 23 | 26 | 28 | 31 | 33 | 36 | 39 | 41 | 44 | 47 |
| 9 | 21 | 24 | 27 | 30 | 33 | 36 | 39 | 42 | 45 | 48 | 51 | 54 |
| 10 | 24 | 27 | 31 | 34 | 37 | 41 | 44 | 48 | 51 | 55 | 58 | 62 |
| 11 | 27 | 31 | 34 | 38 | 42 | 46 | 50 | 54 | 57 | 61 | 65 | 69 |
| 12 | 30 | 34 | 38 | 42 | 47 | 51 | 55 | 60 | 64 | 68 | 72 | 77 |
| 13 | 33 | 37 | 42 | 47 | 51 | 56 | 61 | 65 | 70 | 75 | 80 | 84 |
| 14 | 36 | 41 | 46 | 51 | 56 | 61 | 66 | 71 | 77 | 82 | 87 | 92 |
| 15 | 39 | 44 | 50 | 55 | 61 | 66 | 72 | 77 | 83 | 88 | 94 | 100 |
| 16 | 42 | 48 | 54 | 60 | 65 | 71 | 77 | 83 | 89 | 95 | 101 | 107 |
| 17 | 45 | 51 | 57 | 64 | 70 | 77 | 83 | 89 | 96 | 102 | 109 | 115 |
| 18 | 48 | 55 | 61 | 68 | 75 | 82 | 88 | 95 | 102 | 109 | 116 | 123 |
| 19 | 51 | 58 | 65 | 72 | 80 | 87 | 94 | 101 | 109 | 116 | 123 | 130 |
| 20 | 54 | 62 | 69 | 77 | 84 | 92 | 100 | 107 | 115 | 123 | 130 | 138 |

*Source:* Adapted and abridged from Tables 1, 3, 5, and 7 of D. Auble, "Extended tables for the Mann-Whitney statistic," *Bulletin of the Institute of Educational Research at Indiana University* 1, No. 2 (1953), with the permission of the publisher.

**Critical Values of *T* in the Wilcoxon Matched-Pairs Signed-Ranks Test (*n* ≤ 25)**

| *n* | LEVEL OF SIGNIFICANCE FOR ONE-TAILED TEST | | |
| | 0.025 | 0.01 | 0.005 |
| | LEVEL OF SIGNIFICANCE FOR TWO-TAILED TEST | | |
| | 0.05 | 0.02 | 0.01 |
|---|---|---|---|
| 6 | 0 | — | — |
| 7 | 2 | 0 | — |
| 8 | 4 | 2 | 0 |
| 9 | 6 | 3 | 2 |
| 10 | 8 | 5 | 3 |
| 11 | 11 | 7 | 5 |
| 12 | 14 | 10 | 7 |
| 13 | 17 | 13 | 10 |
| 14 | 21 | 16 | 13 |
| 15 | 25 | 20 | 16 |
| 16 | 30 | 24 | 20 |
| 17 | 35 | 28 | 23 |
| 18 | 40 | 33 | 28 |
| 19 | 46 | 38 | 32 |
| 20 | 52 | 43 | 38 |
| 21 | 59 | 49 | 43 |
| 22 | 66 | 56 | 49 |
| 23 | 73 | 62 | 55 |
| 24 | 81 | 69 | 61 |
| 25 | 89 | 77 | 68 |

*Source:* Adapted from Table 1 of F. Wilcoxon, *Some Rapid Approximate Statistical Procedures* (New York: American Cyanamid Company, 1949), 13.

APPENDIX N

Critical Values $d_L$
and $d_U$ of the Durbin-
Watson Statistic $D$
(Critical Values Are
One-Sided)

$\alpha = .05$

| $n$ | $P = 1$ $d_L$ | $d_U$ | $P = 2$ $d_L$ | $d_U$ | $P = 3$ $d_L$ | $d_U$ | $P = 4$ $d_L$ | $d_U$ | $P = 5$ $d_L$ | $d_U$ |
|---|---|---|---|---|---|---|---|---|---|---|
| 15 | 1.08 | 1.36 | .95 | 1.54 | .82 | 1.75 | .69 | 1.97 | .56 | 2.21 |
| 16 | 1.10 | 1.37 | .98 | 1.54 | .86 | 1.73 | .74 | 1.93 | .62 | 2.15 |
| 17 | 1.13 | 1.38 | 1.02 | 1.54 | .90 | 1.71 | .78 | 1.90 | .67 | 2.10 |
| 18 | 1.16 | 1.39 | 1.05 | 1.53 | .93 | 1.69 | .82 | 1.87 | .71 | 2.06 |
| 19 | 1.18 | 1.40 | 1.08 | 1.53 | .97 | 1.68 | .86 | 1.85 | .75 | 2.02 |
| 20 | 1.20 | 1.41 | 1.10 | 1.54 | 1.00 | 1.68 | .90 | 1.83 | .79 | 1.99 |
| 21 | 1.22 | 1.42 | 1.13 | 1.54 | 1.03 | 1.67 | .93 | 1.81 | .83 | 1.96 |
| 22 | 1.24 | 1.43 | 1.15 | 1.54 | 1.05 | 1.66 | .96 | 1.80 | .86 | 1.94 |
| 23 | 1.26 | 1.44 | 1.17 | 1.54 | 1.08 | 1.66 | .99 | 1.79 | .90 | 1.92 |
| 24 | 1.27 | 1.45 | 1.19 | 1.55 | 1.10 | 1.66 | 1.01 | 1.78 | .93 | 1.90 |
| 25 | 1.29 | 1.45 | 1.21 | 1.55 | 1.12 | 1.66 | 1.04 | 1.77 | .95 | 1.89 |
| 26 | 1.30 | 1.46 | 1.22 | 1.55 | 1.14 | 1.65 | 1.06 | 1.76 | .98 | 1.88 |
| 27 | 1.32 | 1.47 | 1.24 | 1.56 | 1.16 | 1.65 | 1.08 | 1.76 | 1.01 | 1.86 |
| 28 | 1.33 | 1.48 | 1.26 | 1.56 | 1.18 | 1.65 | 1.10 | 1.75 | 1.03 | 1.85 |
| 29 | 1.34 | 1.48 | 1.27 | 1.56 | 1.20 | 1.65 | 1.12 | 1.74 | 1.05 | 1.84 |
| 30 | 1.35 | 1.49 | 1.28 | 1.57 | 1.21 | 1.65 | 1.14 | 1.74 | 1.07 | 1.83 |
| 31 | 1.36 | 1.50 | 1.30 | 1.57 | 1.23 | 1.65 | 1.16 | 1.74 | 1.09 | 1.83 |
| 32 | 1.37 | 1.50 | 1.31 | 1.57 | 1.24 | 1.65 | 1.18 | 1.73 | 1.11 | 1.82 |
| 33 | 1.38 | 1.51 | 1.32 | 1.58 | 1.26 | 1.65 | 1.19 | 1.73 | 1.13 | 1.81 |
| 34 | 1.39 | 1.51 | 1.33 | 1.58 | 1.27 | 1.65 | 1.21 | 1.73 | 1.15 | 1.81 |
| 35 | 1.40 | 1.52 | 1.34 | 1.58 | 1.28 | 1.65 | 1.22 | 1.73 | 1.16 | 1.80 |
| 36 | 1.41 | 1.52 | 1.35 | 1.59 | 1.29 | 1.65 | 1.24 | 1.73 | 1.18 | 1.80 |
| 37 | 1.42 | 1.53 | 1.36 | 1.59 | 1.31 | 1.66 | 1.25 | 1.72 | 1.19 | 1.80 |
| 38 | 1.43 | 1.54 | 1.37 | 1.59 | 1.32 | 1.66 | 1.26 | 1.72 | 1.21 | 1.79 |
| 39 | 1.43 | 1.54 | 1.38 | 1.60 | 1.33 | 1.66 | 1.27 | 1.72 | 1.22 | 1.79 |
| 40 | 1.44 | 1.54 | 1.39 | 1.60 | 1.34 | 1.66 | 1.29 | 1.72 | 1.23 | 1.79 |
| 45 | 1.48 | 1.57 | 1.43 | 1.62 | 1.38 | 1.67 | 1.34 | 1.72 | 1.29 | 1.78 |
| 50 | 1.50 | 1.59 | 1.46 | 1.63 | 1.42 | 1.67 | 1.38 | 1.72 | 1.34 | 1.77 |
| 55 | 1.53 | 1.60 | 1.49 | 1.64 | 1.45 | 1.68 | 1.41 | 1.72 | 1.38 | 1.77 |
| 60 | 1.55 | 1.62 | 1.51 | 1.65 | 1.48 | 1.69 | 1.44 | 1.73 | 1.41 | 1.77 |
| 65 | 1.57 | 1.63 | 1.54 | 1.66 | 1.50 | 1.70 | 1.47 | 1.73 | 1.44 | 1.77 |
| 70 | 1.58 | 1.64 | 1.55 | 1.67 | 1.52 | 1.70 | 1.49 | 1.74 | 1.46 | 1.77 |
| 75 | 1.60 | 1.65 | 1.57 | 1.68 | 1.54 | 1.71 | 1.51 | 1.74 | 1.49 | 1.77 |
| 80 | 1.61 | 1.66 | 1.59 | 1.69 | 1.56 | 1.72 | 1.53 | 1.74 | 1.51 | 1.77 |
| 85 | 1.62 | 1.67 | 1.60 | 1.70 | 1.57 | 1.72 | 1.55 | 1.75 | 1.52 | 1.77 |
| 90 | 1.63 | 1.68 | 1.61 | 1.70 | 1.59 | 1.73 | 1.57 | 1.75 | 1.54 | 1.78 |
| 95 | 1.64 | 1.69 | 1.62 | 1.71 | 1.60 | 1.73 | 1.58 | 1.75 | 1.56 | 1.78 |
| 100 | 1.65 | 1.69 | 1.63 | 1.72 | 1.61 | 1.74 | 1.59 | 1.76 | 1.57 | 1.78 |

$n =$ Number of observations; $P =$ Number of independent variables.
*Source*: Computed from TSP 4.5 based on R. W. Farebrother, "A Remark on Algorithms AS106, AS153, and AS155: The Distribution of a Linear Combination of Chi-Square Random Variables," *Journal of the Royal Statistical Society*, Series C (Applied Statistics), 1984, 29, pp. 323–333.

$$\alpha = .01$$

| n | P = 1 | | P = 2 | | P = 3 | | P = 4 | | P = 5 | |
|---|---|---|---|---|---|---|---|---|---|---|
| | $d_L$ | $d_U$ | $d_L$ | $d_U$ | $d_L$ | $d_U$ | $d_L$ | $d_U$ | $d_L$ | $d_U$ |
| 15 | .81 | 1.07 | .70 | 1.25 | .59 | 1.46 | .49 | 1.70 | .39 | 1.96 |
| 16 | .84 | 1.09 | .74 | 1.25 | .63 | 1.44 | .53 | 1.66 | .44 | 1.90 |
| 17 | .87 | 1.10 | .77 | 1.25 | .67 | 1.43 | .57 | 1.63 | .48 | 1.85 |
| 18 | .90 | 1.12 | .80 | 1.26 | .71 | 1.42 | .61 | 1.60 | .52 | 1.80 |
| 19 | .93 | 1.13 | .83 | 1.26 | .74 | 1.41 | .65 | 1.58 | .56 | 1.77 |
| 20 | .95 | 1.15 | .86 | 1.27 | .77 | 1.41 | .68 | 1.57 | .60 | 1.74 |
| 21 | .97 | 1.16 | .89 | 1.27 | .80 | 1.41 | .72 | 1.55 | .63 | 1.71 |
| 22 | 1.00 | 1.17 | .91 | 1.28 | .83 | 1.40 | .75 | 1.54 | .66 | 1.69 |
| 23 | 1.02 | 1.19 | .94 | 1.29 | .86 | 1.40 | .77 | 1.53 | .70 | 1.67 |
| 24 | 1.04 | 1.20 | .96 | 1.30 | .88 | 1.41 | .80 | 1.53 | .72 | 1.66 |
| 25 | 1.05 | 1.21 | .98 | 1.30 | .90 | 1.41 | .83 | 1.52 | .75 | 1.65 |
| 26 | 1.07 | 1.22 | 1.00 | 1.31 | .93 | 1.41 | .85 | 1.52 | .78 | 1.64 |
| 27 | 1.09 | 1.23 | 1.02 | 1.32 | .95 | 1.41 | .88 | 1.51 | .81 | 1.63 |
| 28 | 1.10 | 1.24 | 1.04 | 1.32 | .97 | 1.41 | .90 | 1.51 | .83 | 1.62 |
| 29 | 1.12 | 1.25 | 1.05 | 1.33 | .99 | 1.42 | .92 | 1.51 | .85 | 1.61 |
| 30 | 1.13 | 1.26 | 1.07 | 1.34 | 1.01 | 1.42 | .94 | 1.51 | .88 | 1.61 |
| 31 | 1.15 | 1.27 | 1.08 | 1.34 | 1.02 | 1.42 | .96 | 1.51 | .90 | 1.60 |
| 32 | 1.16 | 1.28 | 1.10 | 1.35 | 1.04 | 1.43 | .98 | 1.51 | .92 | 1.60 |
| 33 | 1.17 | 1.29 | 1.11 | 1.36 | 1.05 | 1.43 | 1.00 | 1.51 | .94 | 1.59 |
| 34 | 1.18 | 1.30 | 1.13 | 1.36 | 1.07 | 1.43 | 1.01 | 1.51 | .95 | 1.59 |
| 35 | 1.19 | 1.31 | 1.14 | 1.37 | 1.08 | 1.44 | 1.03 | 1.51 | .97 | 1.59 |
| 36 | 1.21 | 1.32 | 1.15 | 1.38 | 1.10 | 1.44 | 1.04 | 1.51 | .99 | 1.59 |
| 37 | 1.22 | 1.32 | 1.16 | 1.38 | 1.11 | 1.45 | 1.06 | 1.51 | 1.00 | 1.59 |
| 38 | 1.23 | 1.33 | 1.18 | 1.39 | 1.12 | 1.45 | 1.07 | 1.52 | 1.02 | 1.58 |
| 39 | 1.24 | 1.34 | 1.19 | 1.39 | 1.14 | 1.45 | 1.09 | 1.52 | 1.03 | 1.58 |
| 40 | 1.25 | 1.34 | 1.20 | 1.40 | 1.15 | 1.46 | 1.10 | 1.52 | 1.05 | 1.58 |
| 45 | 1.29 | 1.38 | 1.24 | 1.42 | 1.20 | 1.48 | 1.16 | 1.53 | 1.11 | 1.58 |
| 50 | 1.32 | 1.40 | 1.28 | 1.45 | 1.24 | 1.49 | 1.20 | 1.54 | 1.16 | 1.59 |
| 55 | 1.36 | 1.43 | 1.32 | 1.47 | 1.28 | 1.51 | 1.25 | 1.55 | 1.21 | 1.59 |
| 60 | 1.38 | 1.45 | 1.35 | 1.48 | 1.32 | 1.52 | 1.28 | 1.56 | 1.25 | 1.60 |
| 65 | 1.41 | 1.47 | 1.38 | 1.50 | 1.35 | 1.53 | 1.31 | 1.57 | 1.28 | 1.61 |
| 70 | 1.43 | 1.49 | 1.40 | 1.52 | 1.37 | 1.55 | 1.34 | 1.58 | 1.31 | 1.61 |
| 75 | 1.45 | 1.50 | 1.42 | 1.53 | 1.39 | 1.56 | 1.37 | 1.59 | 1.34 | 1.62 |
| 80 | 1.47 | 1.52 | 1.44 | 1.54 | 1.42 | 1.57 | 1.39 | 1.60 | 1.36 | 1.62 |
| 85 | 1.48 | 1.53 | 1.46 | 1.55 | 1.43 | 1.58 | 1.41 | 1.60 | 1.39 | 1.63 |
| 90 | 1.50 | 1.54 | 1.47 | 1.56 | 1.45 | 1.59 | 1.43 | 1.61 | 1.41 | 1.64 |
| 95 | 1.51 | 1.55 | 1.49 | 1.57 | 1.47 | 1.60 | 1.45 | 1.62 | 1.42 | 1.64 |
| 100 | 1.52 | 1.56 | 1.50 | 1.58 | 1.48 | 1.60 | 1.46 | 1.63 | 1.44 | 1.65 |

$n$ = Number of observations; $P$ = Number of independent variables.
*Source:* Computed from TSP 4.5 based on R. W. Farebrother, "A Remark on Algorithms AS106, AS153, and AS155: The Distribution of a Linear Combination of Chi-Square Random Variables," *Journal of the Royal Statistical Society*, Series C (Applied Statistics), 1984, 29, pp. 323–333.

**Lower and Upper Critical Values *W* of Wilcoxon Signed-Ranks Test**

| | One-Tailed: $\alpha = .05$ | $\alpha = .025$ | $\alpha = .01$ | $\alpha = .005$ |
|---|---|---|---|---|
| | Two-Tailed: $\alpha = .10$ | $\alpha = .05$ | $\alpha = .02$ | $\alpha = .01$ |
| *n* | (Lower, Upper) | | | |
| 5 | 0,15 | —,— | —,— | —,— |
| 6 | 2,19 | 0,21 | —,— | —,— |
| 7 | 3,25 | 2,26 | 0,28 | —,— |
| 8 | 5,31 | 3,33 | 1,35 | 0,36 |
| 9 | 8,37 | 5,40 | 3,42 | 1,44 |
| 10 | 10,45 | 8,47 | 5,50 | 3,52 |
| 11 | 13,53 | 10,56 | 7,59 | 5,61 |
| 12 | 17,61 | 13,65 | 10,68 | 7,71 |
| 13 | 21,70 | 17,74 | 12,79 | 10,81 |
| 14 | 25,80 | 21,84 | 16,89 | 13,92 |
| 15 | 30,90 | 25,95 | 19,101 | 16,104 |
| 16 | 35,101 | 29,107 | 23,113 | 19,117 |
| 17 | 41,112 | 34,119 | 27,126 | 23,130 |
| 18 | 47,124 | 40,131 | 32,139 | 27,144 |
| 19 | 53,137 | 46,144 | 37,153 | 32,158 |
| 20 | 60,150 | 52,158 | 43,167 | 37,173 |

*Source:* Adapted from Table 2 of F. Wilcoxon and R. A. Wilcox, *Some Rapid Approximate Statistical Procedures* (Pearl River, NY: Lederle Laboratories, 1964), with permission of the American Cyanamid Company.

| Number of Observations in Subgroup | $d_2$ | $d_3$ | $D_3$ | $D_4$ | $A_2$ |
|---|---|---|---|---|---|
| 2 | 1.128 | 0.853 | 0 | 3.267 | 1.880 |
| 3 | 1.693 | 0.888 | 0 | 2.575 | 1.023 |
| 4 | 2.059 | 0.880 | 0 | 2.282 | 0.729 |
| 5 | 2.326 | 0.864 | 0 | 2.114 | 0.577 |
| 6 | 2.534 | 0.848 | 0 | 2.004 | 0.483 |
| 7 | 2.704 | 0.833 | 0.076 | 1.924 | 0.419 |
| 8 | 2.847 | 0.820 | 0.136 | 1.864 | 0.373 |
| 9 | 2.970 | 0.808 | 0.184 | 1.816 | 0.337 |
| 10 | 3.078 | 0.797 | 0.223 | 1.777 | 0.308 |
| 11 | 3.173 | 0.787 | 0.256 | 1.744 | 0.285 |
| 12 | 3.258 | 0.778 | 0.283 | 1.717 | 0.266 |
| 13 | 3.336 | 0.770 | 0.307 | 1.693 | 0.249 |
| 14 | 3.407 | 0.763 | 0.328 | 1.672 | 0.235 |
| 15 | 3.472 | 0.756 | 0.347 | 1.653 | 0.223 |
| 16 | 3.532 | 0.750 | 0.363 | 1.637 | 0.212 |
| 17 | 3.588 | 0.744 | 0.378 | 1.622 | 0.203 |
| 18 | 3.640 | 0.739 | 0.391 | 1.609 | 0.194 |
| 19 | 3.689 | 0.733 | 0.404 | 1.596 | 0.187 |
| 20 | 3.735 | 0.729 | 0.415 | 1.585 | 0.180 |
| 21 | 3.778 | 0.724 | 0.425 | 1.575 | 0.173 |
| 22 | 3.819 | 0.720 | 0.435 | 1.565 | 0.167 |
| 23 | 3.858 | 0.716 | 0.443 | 1.557 | 0.162 |
| 24 | 3.895 | 0.712 | 0.452 | 1.548 | 0.157 |
| 25 | 3.931 | 0.708 | 0.459 | 1.541 | 0.153 |

*Source:* Reprinted from ASTM-STP 15D by kind permission of the American Society for Testing and Materials.

# Answers to Selected Odd-Numbered Exercises

This section contains summary answers to most of the odd-numbered Exercises in the text. The *Student Solutions Manual* contains fully developed solutions to all odd-numbered Exercises and shows clearly how each answer is determined.

## CHAPTER 1

**1-1.** Descriptive; use charts, graphs, tables, and numerical measures.

**1-3.** A bar chart is used whenever you want to display data that have already been categorized, while a histogram is used to display data over a range of values for the factor under consideration.

**1-5.** The company could use statistical inference to determine if its parts last longer.

**1-7.** The appropriate chart is a histogram where the horizontal axis contains the number of missed days and the area of the bars represents the number of employees who missed each number of days.

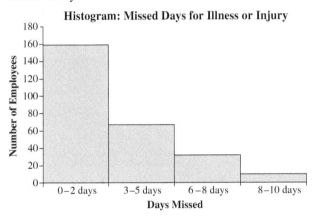

**1-11.** An experiment would not be appropriate. Use a written survey or telephone survey.

**1-13.** An experiment is any process that generates data as its outcome. The plan for performing the experiment in which the variable of interest is defined is referred to as an experimental design.

**1-15. a.** observation of bike riders
**b.** a telephone survey to gas stations in the state
**c.** a written survey of passengers

**1-17.** These data could have been collected through a survey. Employees of the USDA could provide periodic reports of fire ant activity in their region.

**1-25. a.** Because the population is spread over a large geographical area, a cluster random sample could be selected to reduce travel costs.
**b.** A stratified random sample would probably be used to keep the sample size as small as possible.
**c.** Most likely a convenience sample would be used, since doing a statistical sample would be too difficult.

**1-27.** Whenever a descriptive numerical measure such as an average is calculated from the entire population, it is a parameter. The corresponding measure calculated from a subset of the population—that is to say, a sample—is a statistic.

**1-29.** From a numbered list of all customers who own a certificate of deposit, the bank would need to randomly determine a starting point between 1 and $k$, where $k = 25,000/1,000 = 25$. Once this value is determined, the bank would select that numbered customer as the first sampled customer and then select every 25th customer until 100 customers are sampled.

**1-31.** Values computed from a sample are always considered statistics. For a value, such as an average, to be considered a parameter, it must be computed from all items in the population.

**1-35.** These values are statistics. A poll looks at a sample of eligible voters rather than all eligible voters.

**1-37.** This is a statistical sample. With this method, every employee has an equal chance of being selected.

**1-43.** Qualitative data are categories or numerical values that represent categories. Quantitative data are data that are purely numerical.

**1-45.** Nominal data involve placing observations in separate categories according to some measurable characteristic. Ordinal data also involve placing observations in separate categories, but the categories can be rank-ordered.

**1-47. a.** The data are cross-sectional. The data are collected from 2,300 customers at approximately the same point in time.
**b.** This is a ratio-level, quantitative variable. The data represent a measurement of time.
**c.** The data are ordinal, with a numerical value representing customers' rating of level of service.

**1-49. a.** cross-sectional
**b.** time-series
**c.** cross-sectional
**d.** cross-sectional
**e.** time-series

**1-53.** interval or ratio data

## CHAPTER 2

**2-3. a.** $2^k \geq n$ or $2^{10} = 1,024 \geq 1,000$. Thus, use $k = 10$ classes.

**b.** $w = \dfrac{\text{High} - \text{Low}}{\text{Classes}} = \dfrac{2,900 - 300}{10} = \dfrac{2,600}{10} = 260$ (round to 300)

**2-7. a.** The data do not require grouping. The frequency distribution is:

| $x$ | Frequency |
|-----|-----------|
| 0 | 0 |
| 1 | 0 |
| 2 | 1 |
| 3 | 1 |
| 4 | 10 |
| 5 | 15 |
| 6 | 13 |
| 7 | 13 |
| 8 | 5 |
| 9 | 1 |
| 10 | 1 |
| | 60 |

**b.** The following histrogram could be developed:

**c.** The relative frequency distribution shows the fraction of values that fall at each value of $x$.

| $x$ | Frequency | Relative Frequency |
|---|---|---|
| 0 | 0 | 0.00 |
| 1 | 0 | 0.00 |
| 2 | 1 | 0.02 |
| 3 | 1 | 0.02 |
| 4 | 10 | 0.17 |
| 5 | 15 | 0.25 |
| 6 | 13 | 0.22 |
| 7 | 13 | 0.22 |
| 8 | 5 | 0.08 |
| 9 | 1 | 0.02 |
| 10 | 1 | 0.02 |
| | 60 | |

**d.** The relative frequency histogram is shown below.

**e.** The two histograms look exactly alike because the same data are being graphed. The bars represent either the frequency or the relative frequency.

**2-9. a.** Because the number of possible values for the variable is relatively small, there is no need to group the data into classes. The resulting frequency distribution is

| $x$ | Frequency |
|---|---|
| 0 | 0 |
| 1 | 2 |
| 2 | 3 |
| 3 | 3 |
| 4 | 10 |
| 5 | 16 |

| $x$ | Frequency |
|---|---|
| 6 | 19 |
| 7 | 7 |
| 8 | 9 |
| 9 | 6 |
| 10 | 2 |
| 11 | 2 |
| 12 | 1 |
| Total = | 80 |

This frequency distribution shows the manager that most customer receipts have four to eight line items.

**b.** A histogram is a graph of a frequency distribution for a quantitative variable. The resulting histogram is shown here.

**2-13. a.** There are $n = 100$ values in the data. Using the $2^k \geq n$ guideline, we would need at least $k = 7$ classes.

**b.** Using $k = 7$ classes, we determine the class width as follows:

$$w = \frac{\text{High} - \text{Low}}{\text{Classes}} = \frac{\$376,644 - \$87,429}{7} = \frac{\$289,215}{7} = \$41,316.43.$$

Rounded up to the nearest $1,000, the class width is $42,000.

**c.** The grouped frequency distribution with seven classes and a class width of $42,000 will depend on the starting point for the first class.

**d.** A possible histogram for the frequency distribution in part c is shown here.

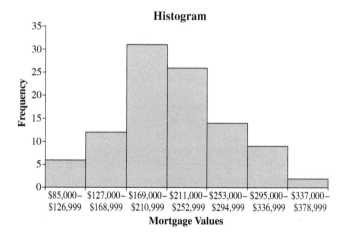

Interpretation should involve a discussion of the range of values, considering where the major classes are located.

**2-17. a.**

**b.** The tread life of at least 50% of the tires is 60,000 miles or more. The top 10% had a tread life greater than 66,000, and the longest tread tread life is 74,000 miles. Additional information will vary.

**c.**

Students may say that the 12 classes give better information because they can see more detail about the number of miles the tires can go.

**2-19.**

**2-23. a.** The pie chart is shown as follows:

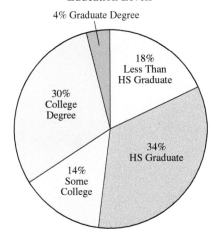

**b.** The horizontal bar chart is shown as follows:

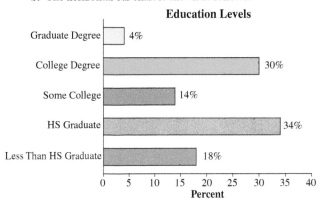

**2-25. a.** The bar chart:

**b.** The pie chart:

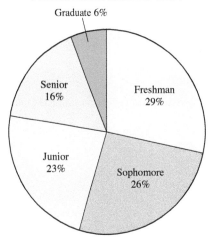

**c.** A case can be made for either a bar chart or a pie chart. Pie charts are especially good at showing how the total is divided into parts. The bar chart is better to draw attention to specific results.

**2-31. a.** A bar chart would be more appropriate.

**b.**

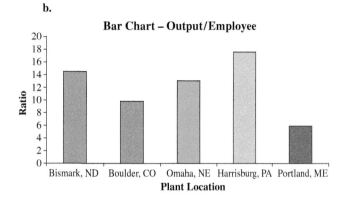

**Bar Chart – Output/Employee**

**2-33.**

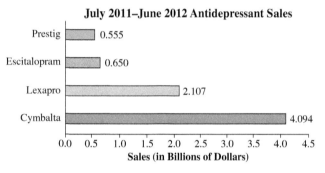

**July 2011–June 2012 Antidepressant Sales**

**2-37. a.** Stem and leaf diagram for the number of days to collect payment:

| Stem unit: | 10 |
| --- | --- |
| 2 | 2 4 8 9 |
| 3 | 0 1 2 3 3 4 5 5 5 6 6 7 8 8 9 |
| 4 | 1 3 5 7 8 |
| 5 | 5 6 6 |
| 6 | 0 5 6 |

**b.** Most payments are collected within the range of 30–39 days.

**2-41. a.** A bar chart is an appropriate graph because there are two categories, males and females. A pie chart could also be used to display the data.

**b.**

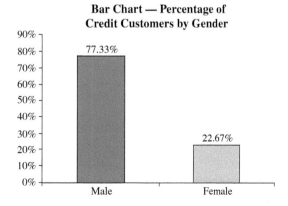

**Bar Chart — Percentage of Credit Customers by Gender**

This shows that a clear majority of credit card holders are males (77.33%).

**2-43. a.**

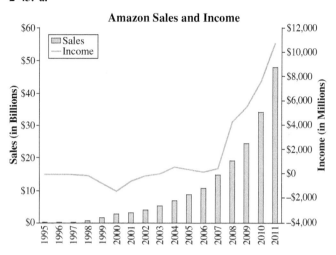

**Amazon Sales and Income**

**b.** There appears to be a linear relationship between the sales volume and the years in business.

**c.** In the time period between 2000 and 2001, Amazon experienced a decrease in its losses. Prior to this time, each year produced increased losses.

**2-45. a.**

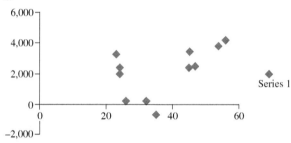

Based on the scatter diagram, there appears to be a curvilinear relationship between the dependent and independent variables.

**b.**

**Profit Line Chart — Deleted Extreme Data Point**

Having removed the three extreme data points, we see that the relationship between the dependent and independent variables seems to be linear and positive.

**2-47.**

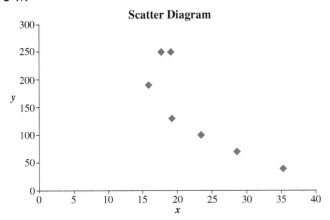

Scatter Diagram

There is a negative linear relationship between the two variables.

**2-49. a.**

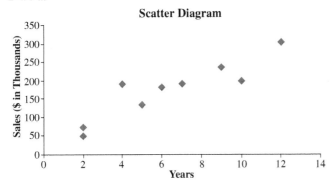

Scatter Diagram

**b.** The relationship appears to be linear and positive.
**c.** $30k to $35k per year

**2-51.**

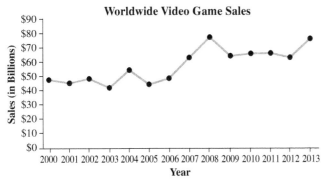

Worldwide Video Game Sales

The line chart illustrates that over the 14-year period between 2000 and 2013, video game sales have grown quite steadily from just under $50 billion to over $70 billion.

**2-53.**

S&P Oil Companies Combined Net Income

**2-55.**

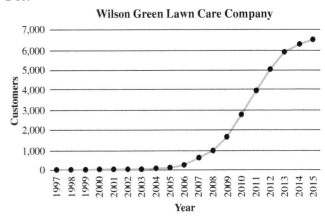

Wilson Green Lawn Care Company

Since 2006, there has been a very steep growth in the number of customers.

**2-59. a.**

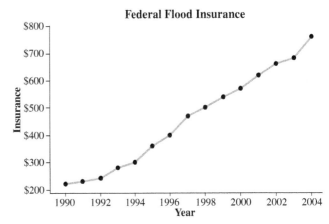

Federal Flood Insurance

**b.** The relationship appears to be linear and positive.
**c.** The average equals the sum of the yearly growth divided by the number of years which is $540/14 = $38.57 per year.

**2-61. a.**

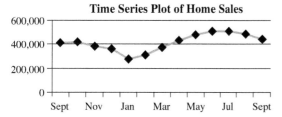

**Time Series Plot of Home Sales**

**b.**

**Sales vs Median Price**

Note: You must convert the sales data into tens of thousands, which is obtained by (sales × 100); for example, 6.76 million becomes 6.76(100) = 676 (tens of thousands). This is done to produce comparable *y* axes for the two line plots.

**c.** It appears from the line plots that the monthly sales have been fluctuating greatly during this period, dipping in January, heading back up during the summer months, and then declining again. The median sales price has shown a steady minor decline during the period.

**2-63.** Bar charts are visual representations of frequency distributions constructed from nominal or ordinal data.

**2-67. a.** Using the $2^k \geq n$ guideline:

$$2^k \geq 48 = 2^6 \geq 48$$

To determine the class width, $(17.5 - 0.3)/6 = 2.87$, which is rounded up to 3 to make it easier.

| Classes | Frequency |
|---|---|
| 0.1–3 | 27 |
| 3.1–6.0 | 9 |
| 6.1–9.0 | 6 |
| 9.1–12 | 4 |
| 12.1–15.0 | 0 |
| 15.1–18.0 | 2 |

**b.** Stem and leaf diagram. Stem unit: 1.

| | |
|---|---|
| 0 | 3 4 5 7 |
| 1 | 0 0 0 0 4 5 5 9 |
| 2 | 0 0 0 0 0 4 5 5 5 7 |
| 3 | 0 0 0 0 0 2 5 5 5 5 6 |
| 4 | 0 0 0 |
| 5 | |
| 6 | 4 5 |
| 7 | 5 |
| 8 | 3 |
| 9 | 0 0 2 |
| 10 | |

| | |
|---|---|
| 11 | 0 |
| 12 | 0 0 |
| 13 | |
| 14 | |
| 15 | |
| 16 | 0 |
| 17 | 5 |

**c.**

**Pie Chart—Miles**

- ≥ 6: 25%
- < 3: 46%
- 3 to < 6: 29%

**d.**

**Bar Chart—Proportion of Employees**

**2-69. a.**

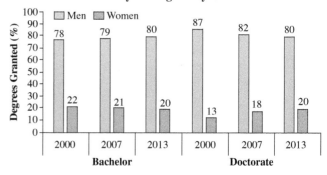

**Physics Degrees by Gender**

| | Men | Women |
|---|---|---|
| Bachelor 2000 | 78 | 22 |
| Bachelor 2007 | 79 | 21 |
| Bachelor 2013 | 80 | 20 |
| Doctorate 2000 | 87 | 13 |
| Doctorate 2007 | 82 | 18 |
| Doctorate 2013 | 80 | 20 |

**b.** There has been a slight decline in the percentage of physics bachelor degrees granted to women and an increase in the percentage of doctorate degrees granted to women.

**2-71. a.** Based on the following table, the total percent of classes that hold at least 120 seconds (2 minutes) is $0.0311 + 0.0244 + 0.0171 + 0.0301 = 0.1029$.

| Classes (in seconds) | Number | Relative Frequency |
|---|---|---|
| $< 15$ | 456 | 0.0899 |
| $15 < 30$ | 718 | 0.1415 |
| $30 < 45$ | 891 | 0.1756 |
| $45 < 60$ | 823 | 0.1622 |
| $60 < 75$ | 610 | 0.1202 |
| $75 < 90$ | 449 | 0.0885 |
| $90 < 105$ | 385 | 0.0759 |
| $105 < 120$ | 221 | 0.0435 |
| $120 < 150$ | 158 | 0.0311 |
| $150 < 180$ | 124 | 0.0244 |
| $180 < 240$ | 87 | 0.0171 |
| $\geq 240$ | 153 | 0.0301 |

Ogive

Note: For this exercise the class widths are not equal.

**b.** The number of people who have to wait 120 seconds (2 minutes) or longer is

$$158 + 124 + 87 + 153 = 522 \times \$30 = \$15,660 \text{ per month}$$

**2-75. a.** Using the $2^k \geq n$ guideline, we have

$$2^k \geq 100 \text{ so } 2^7 = 128.$$

Determine the width: $(310,495 = 70,464)/7 = 34,290$. Round to 35,000.

| Classes | Frequency |
|---|---|
| 70,000–104,999 | 43 |
| 105,000–139,999 | 34 |
| 140,000–174,999 | 13 |
| 175,000–209,999 | 5 |
| 210,000–244,999 | 2 |
| 245,000–279,999 | 1 |
| 280,000–314,999 | 2 |

**b.**

| Classes | Frequency | Relative Frequency | Cumulative Relative Frequency |
|---|---|---|---|
| 70,000–104,999 | 43 | 0.43 | 0.43 |
| 105,000–139,999 | 34 | 0.34 | 0.77 |
| 140,000–174,999 | 13 | 0.13 | 0.9 |
| 175,000–209,999 | 5 | 0.05 | 0.95 |
| 210,000–244,999 | 2 | 0.02 | 0.97 |
| 245,000–279,999 | 1 | 0.01 | 0.98 |
| 280,000–314,999 | 2 | 0.02 | 1 |

**2-77. a.**

Histogram–Diesel Prices

## CHAPTER 3

When applicable, the first few exercises in each section will be done following the appropriate step-by-step procedures outlined in the corresponding sections of the chapter. Later exercises will provide key points and the answers to the questions, but all answers can be arrived at using the appropriate steps.

**3-1.** $Q_1 = 4,423$; Median $= 5,002$; $Q_3 = 5,381$; $\bar{x} = 4956.60$

**3-3.** $Q_1 = \dfrac{13.5 + 13.6}{2} = 13.55$

$Q_3 = \dfrac{15.5 + 15.9}{2} = 15.7$

**3-7. a.** $(31.2 + 32.2)/2 = 31.7$

**b.** 25th percentile $= (12.1 + 13)/2 = 12.55$
75th percentile $= (26.7 + 31.2)/2 = 28.95$

**c.** $(20.8 + 22.8)/2 = 21.8$

**3-9. a.** Mean $= 19$; Median $= (19 + 19)/2 = 19$; Mode $= 19$

**b.** Symmetric; Mean = Median = Mode

**c.**

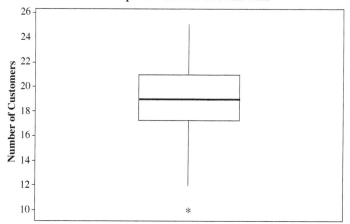

Boxplot of Number of Customers

**3-11. a.** 11,213.48

**b.** Use weighted average.

**3-13. a.** Mean = 114.21; Median = 107.50; Mode = 156

**b.** skewed right

**c.**

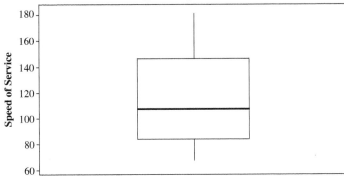

**Boxplot of Speed of Service**

**3-15. a.** 562.99

**b.** 551.685

**3-17. a.** FDIC commercial banks mean assets =

$$\frac{\sum x_i}{n} = \frac{14,727}{5,410} = \$2.72 \text{ billion}$$

FDIC savings institutions mean assets =

$$\frac{\sum x_i}{n} = \frac{1,074}{860} = \$1.25 \text{ billion}$$

**b.** The mean deposits for commercial banks are about $1.50 billion greater than the mean for savings institutions.

**c.** Because these data include all banks and savings entities, the means would be considered parameters.

**3-19. a.** Mean = 0.33

**b.** Median = 0.31

**c.** Mode = 0.24

**d.** 80th percentile = .40 minute

**3-21. a.** Mean = 108.13; Median = 107.9; symmetric or slightly right skewed based on mean > median

**b.** First quartile = 106.8; Third quartile = 110.4

**3-25. a.** Range = 8 − 0 = 8

**b.** 3.99

**c.** 1.998

**3-27. a.** 16.87

**b.** 4.11

**3-29.** Standard deviation = 2.8

**3-31. a.** Standard deviation = 7.21; IQR = 24 − 12 = 12

**b.** Range = Largest − Smallest = 30 − 6 = 24; Standard deviation = 6.96; IQR = 12

**c.** $\sigma^2$ is smaller than $s^2$ by a factor of $(N-1)/N$. $\sigma$ is smaller than $s$ by a factor of $\sqrt{(N-1)/N}$. The range and IQR are not affected.

**3-33. a.** The variance is 815.79 and the standard deviation is 28.56.

**b.** Interquartile range overcomes the susceptibility of the range to being highly influenced by extreme values.

**3-35. a.** Range = 33 − 21 = 12

$$\bar{x} = \frac{\sum_{i=1}^{n} x_i}{n} = 261/10 = 26.1$$

$$s^2 = \frac{\sum_{i=1}^{n} (x - \bar{x})^2}{n-1} = 148.9/(10-1) = 16.5444$$

$$s = \sqrt{s^2} = \sqrt{16.5444} = 4.0675$$

Interquartile range = 28 − 23 = 5

**b.** Ages are lower at this college than for the U.S. colleges and universities as a group.

**3-37. a.** The range is 113.0, the IQR is 62.25, the variance is 1,217.14, and the standard deviation is 34.89.

**b.** No

**c.** Adding a constant to all the data values leaves the variance unchanged.

**3-39. a.** Mean = −1.2

Standard deviation = 0.940

IQR = $Q_3 - Q_1$ = 0.9 − (−1.6) = 0.7

**b.** Change is negative over this period of time.

**c.** The standard deviation measures the average price fluctuation from month to month for the price index.

**3-41. a.** Men spent an average of $397, whereas women spent an average of $378 for their phones. The standard deviation for men was nearly twice that for women.

**b.** Business users spent an average of $446.67 on their phones, whereas home users spent an average of $385.74 The variation in phone costs for the two groups was about equal.

**3-43. a.** The population mean is

$$\mu = \frac{\sum x}{N} = \$178,465$$

**b.** The population median is
$\tilde{\mu} = \$173,000$

**c.** The range is
R = High − Low
R = $361,100 − $54,100
= $307,000

**d.** The population standard deviation is

$$\sigma = \sqrt{\frac{\sum (x-\mu)^2}{N}} = \$63,172$$

**3-47. a.** at least 75% in the range 2,600 to 3,400; $\mu \pm 2(\sigma)$

**b.** The range 2,400 to 3,600 should contain at least 89% of the data values.

**c.** less than 11%

**3-49. a.** The mean = 1,487/14 = 106.21
The standard deviation = 25.008

**b.** $CV$ = 23.55%

**c.** The range from 31.19 to 181.24 should contain at least 89% of the data values.

**3-51.** For Distribution A: $CV = \frac{\sigma}{\mu}(100) = \frac{100}{500}(100) = 20\%$

For Distribution B: $CV = \frac{\sigma}{\mu}(100) = \frac{4.0}{10.0}(100) = 40\%$

**3-53. a.** $z = \frac{800 - \bar{x}}{s} = \frac{800 - 1,000}{250} = -0.80$

**b.** $z = 0.80$

**c.** $z = 0.00$

**3-55. a.** $\bar{x} = \frac{1,530}{30} = 51$; Variance = 510.55; Standard deviation = 22.60

**b.** $51 \pm 22.60$, $51 \pm 2(22.60)$, $51 \pm 3(22.60)$, i.e., (28.4, 73.6), (5.8, 96.2), and (−16.8, 118.8). There are (19/30)100% = 63.3% of the data within (28.4, 73.6), (30/30)100% = 100% of the data within (5.8, 96.2), (30/30)100% = 100% of the data within (−16.8, 118.8).

**c.** bell-shaped population

**3-57. a.**

|                    | Drug A | Drug B |
|--------------------|--------|--------|
| Mean               | 234.75 | 270.92 |
| Standard Deviation | 13.92  | 19.90  |

   **b.** Based on the sample means of the time each drug is effective, Drug B appears to be effective longer than Drug A.

   **c.** Based on the standard deviation of effect time, Drug B exhibits a higher variability in effect time than Drug A.

   **d.** Drug A, $CV = 5.93\%$; Drug B, $CV = 7.35\%$. Drug B has a higher coefficient of variation and the greater relative spread.

**3-59.** Existing supplier: $CV = \dfrac{0.078}{3.75}(100) = 2.08\%$

New supplier: $CV = \dfrac{0.135}{18.029}(100) = 0.75\%$

**3-61.** Anyone scoring below 61.86 (rounded to 62) will be rejected without an interview.
Anyone scoring higher than 91.98 (rounded to 92) will be sent directly to the company.

**3-63.** $CV = \dfrac{3{,}083.45}{11{,}144.48}(100) = 27.67\%$

At least 75% of CPA firms will compute a tax owed between $4,977.58 and $17,311.38.

**3-65. a.** Mean $= 17.8\%$; Median $= 16.3\%$

   **b.** Standard deviation $= 9.08\%$

   **c.** Coefficient of variation $= 9.08/17.8(100) = 51\%$

   **d.** The 90th percentile is 31.12%. Since this employee has a percent body fat of 29, he or she is not at or above the 90th percentile.

**3-73.** The mode is a useful measure of location for a set of data if the data set is large and involves nominal or ordinal data.

**3-75. a.** To determine how many standard deviations the data points are from the mean, calculate the standardized sample data:

$z = \dfrac{x - \bar{x}}{s}$; $z = (19 - 28)/9 = -1$. The Empirical Rule indicates that 68% of the data are within 1 standard deviation from the mean. This area is half of that. So the proportion of players between 19 and 28 is $0.68/2 = 0.34$.

   **b.** Calculate the standardized sample data: $z = \dfrac{x - \bar{x}}{s}$; $z = (37 - 28)/9 = 1$. This is the same area as in part a except it is on the right-hand side of the mean. Thus, the proportion of players between 28 and 37 is $0.68/2 = 0.34$.

   **c.** If 68% of the players are between 19 and 37, then $100 - 68 = 32\%$ are outside of this interval, half of whom are older than 37. Thus, the proportion of players older than 37 is $0.32/2 = 0.16$.

**3-77. a.** 364.42

   **b.** Variance $= 16{,}662.63$; Standard deviation $= 129.08$

**3-81. a.** Comparing only the mean bushels/acre, you would say that Seed type C produces the greatest average yield per acre.

   **b.** $CV$ of Seed type A $= 25/88 = 0.2841$ or 28.41%
$CV$ of Seed type B $= 15/56 = 0.2679$ or 26.79%
$CV$ of Seed type C $= 16/100 = 0.1600$ or 16%
Seed type C shows the least relative variability.

   **c.** Seed type A: 68% between 63 and 113
95% between 38 and 138
approximately 100% between 13 and 163

Seed type B: 68% between 41 and 71
95% between 26 and 86
approximately 100% between 11 and 101

Seed type C: 68% between 84 and 116
95% between 68 and 132
approximately 100% between 52 and 148

   **d.** Seed type A

   **e.** Seed type C

**3-87. a.** Mean $= 54.00$
Standard deviation $= 3.813$

   **b.** $\bar{x} \pm 1s = 54 \pm (3.813) = (50.187, 57.813)$
$\bar{x} \pm 2s = (46.374, 61.626)$
$\bar{x} \pm 3s = (42.561, 65.439)$

   **c.** The Empirical Rule indicates that 95% of the data are contained within $\bar{x} \pm 2s$. This would mean that each tail has $(1 - 0.95)/2 = 0.025$ of the data. Therefore, the costume should be priced at $46.37.

**3-89. a.** Mean $= 0.1303$; Median $= 0.09$; Standard deviation $= 0.262$

   **b.** It means that the closing price for GE stock is an average of approximately 13 cents ($0.1303) higher than the opening price.

## CHAPTER 4

**4-1.** independent events

**4-3.** V, V   V, C   V, S   C, V   C, C   C, S
S, V   S, C   S, S

**4-5. a.** subjective probability based on expert opinion

   **b.** relative frequency based on previous customer return history

   **c.** classical assessment

**4-7.** $1/3 = 0.333333$

**4-9. a.** $P(\text{Brown}) = \#\text{Brown}/\text{Total} = 310/982 = 0.3157$

   **b.** $P(\text{YZ-99}) = \#\text{YZ-99}/\text{Total} = 375/982 = 0.3819$

   **c.** $P(\text{YZ-99 and Brown}) = 205/982 = 0.2088$

   **d.** not mutually exclusive, since their joint probability is 0.1324

**4-11. a.** No

   **b.** 8 outcomes

   **d.** $E_1 = 3/8$; $E_2 = 7/8$; $E_3 = 4/8$

   **e.** 0.375, 0.875, 0.50

**4-15.**

| Type of Ad     | Occurrences |
|----------------|-------------|
| Help-wanted ad | 204         |
| Real estate ad | 520         |
| Other ad       | 306         |
| Total          | 1,030       |

   **a.** 0.1981

   **b.** relative frequency

   **c.** yes

**4-17. a.** relative frequency assessment method

   **b.** $P(\#1) = \dfrac{4{,}100}{5{,}900} = 0.69$

**4-19. a.** $3{,}122/21{,}768 = 0.1434$

   **b.** relative frequency assessment method

**4-21. a.** $P(\text{Caesarean}) = \dfrac{22}{50} = 0.44$

   **b.** New births may not exactly match the 50 in this study.

**4-23.** The following joint frequency table (developed using Excel's pivot table feature) summarizes the data.

|           | Electrical | Mechanical | Total |
| --------- | ---------- | ---------- | ----- |
| Lincoln   | 28         | 39         | 67    |
| Tyler     | 64         | 69         | 133   |
| Total     | 92         | 108        | 200   |

  **a.** $133/200 = 0.665$
  **b.** $108/200 = 0.54$
  **c.** $28/200 = 0.14$

**4-25. a.** $\dfrac{43}{100} = 0.43$

  **b.** $\dfrac{5 + 6 + 6}{100} = 0.17$

  **c.** For Pepsi, Probability $= \dfrac{5 + 6 + 6}{12 + 12 + 11} = \dfrac{17}{35} = 0.486$

  For Coke, Probability $= \dfrac{6 + 6 + 6}{12 + 12 + 11} = \dfrac{18}{35} = 0.514$

  **d.** For Pepsi, Probability $= \dfrac{7 + 6 + 8 + 5}{19 + 16 + 14 + 16} = \dfrac{26}{65}$
  $= 0.4$

  For Coke, Probability $= \dfrac{12 + 10 + 6 + 11}{19 + 16 + 14 + 16} = \dfrac{39}{65}$
  $= 0.6$

**4-27. a.** $(0.9)(1 - 0.5) = 0.45$
  **b.** $(0.6)(0.8) = 0.48$

**4-29.** $P(\text{Senior 1 and senior 2}) = \left(\dfrac{5}{10}\right)\left(\dfrac{4}{9}\right) = \dfrac{20}{90} = 0.22$

**4-31. a.** $P(E_1 \text{ and } B) = P(E_1 | B)P(B) = 0.25(0.30) = 0.075$
  **b.** $P(E_1 \text{ or } B) = P(E_1) + P(B) - P(E_1 \text{ and } B)$
  $\qquad = 0.35 + 0.30 - 0.075 = 0.575$
  **c.** $P(E_1 \text{ and } E_2 \text{ and } E_3) = P(E_1)P(E_2)P(E_3)$
  $\qquad = (0.35)(0.15)(0.40) = 0.021$

**4-33. a.** $P(B) = \dfrac{\text{Number of drives from } B}{\text{Total drives}} = \dfrac{195}{700} = 0.2786$

  **b.** $P(\text{Defect}) = \dfrac{\text{Number of defective drives}}{\text{Total drives}} = \dfrac{50}{700} = 0.0714$

  **c.** $P(\text{Defect} | B) = \dfrac{P(\text{Defect and } B)}{P(B)} = \dfrac{0.0214}{0.2786} = 0.0769$

  $P(\text{Defect} | B) = \dfrac{\text{Number of defective drives from } B}{\text{Number of drives from } B}$
  $\qquad = \dfrac{15}{195} = 0.0769$

**4-35. a.** $0.316$
  **b.** $0.518$
  **c.** $0.39$

**4-37.** They cannot get to 99.9% on color copies.

**4-39.** $P(\text{Free gas}) = 0.00015 + 0.00585 + 0.0005 = 0.0065$

**4-41. a.** $P(\text{NFL}) = 105/200 = 0.5250$
  **b.** $P(\text{College degree and NBA}) = 40/200 = 0.20$
  **c.** $10/50 = 0.20$
  **d.** The two events are not independent.

**4-43.** $P(\text{Line 1} | \text{Defective}) = (0.05)(0.4)/0.0725 = 0.2759$
  $P(\text{Line 2} | \text{Defective}) = (0.10)(0.35)/0.0725 = 0.4828$
  $P(\text{Line 3} | \text{Defective}) = (0.07)(0.25)/0.0725 = 0.2413$
  The unsealed cans probably came from Line 2.

**4-45.** $P(\text{Supplier A} | \text{Defective}) = (0.15)(0.3)/0.115 = 0.3913$
  $P(\text{Supplier B} | \text{Defective}) = (0.10)(0.7)/0.115 = 0.6087$
  Supplier B is more likely to have supplied the defective parts.

**4-47. a.** $0.29$
  **b.** $0.82$
**4-49. a.** $0.76$
  **b.** $0.988$
  **c.** $0.00012$
**4-51. a.** Probability first sampled device is defective $= 3/50$
  $\qquad = 0.06$
  **b.** $(47/50)(3/49) = 0.0576$
  **c.** $(47/50)(46/49)(45/48) = 0.8273$
**4-53. a.** $0.1856$
  **b.** $0.50$
  **c.** $0.0323$
  **d.** If they are independent, 22/62, which equals 0.3548, should equal (52/62) (27/62), which equals (0.8387)(0.4355) = 0.3653, since they are not equal they are not independent.
**4-55. a.** $0.0644$
  **b.** $0.4888$
**4-57. a.** $0.50$
  **b.** $0.755$
  **c.** $0.269$
**4-63. a.** $0.80; 0.40; 0.20; 0.60$
  **b.** $A | \overline{B}$ and $\overline{A} | \overline{B}$ are complements.
**4-65. a.** $0.0156$
  **b.** $0.1563$
  **c.** $0.125, 0.50$
**4-67. a.** relative frequency of occurrence
  **b.** $P(\text{both you and friend are audited}) =$
  $(0.0086)(0.0086) = 0.000074$
  **c.** $P(\text{Harry and Sarah are not audited}) =$
  $0.8378 \cdot 0.9947 = 0.8334$
**4-69. a.** relative frequency assessment approach
  **b.** $0.028$
  **c.** $0.349$
  **d.** yes
**4-71.** Clerk 1 is most likely responsible for the boxes that raised the complaints.
**4-73. a.** $18/50 = 0.36$
  **b.** $4/50 = 0.08$
  **c.** $3/20 = 0.15$
  **d.** $10/18 = 0.556$
  **e.** There are a higher proportion of females whose functional background is marketing and a higher proportion of males whose functional background is operations.
**4-75. a.** $265/500 = 0.53$
  **b.** $200/500 = 0.40$
  **c.** $105/235 = 0.4468$
  **d.** $P(\text{Yes}) = 265/500 = 0.53$
  $P(\text{No}) = 235/500 = 0.47$
  Is $P(\text{Yes}) = P(\text{Yes|Male})$?
  $P(\text{Yes|Male}) = 195/300 = 0.65$
  Since $0.65 \neq 0.53$, responses are not independent of gender.
**4-77. a.** $P(\text{Mountain bike but no road bike}) = 0.59$
  **b.** $P(\text{Road bike but no mountain bike}) = 0.19$
  **c.** $P(\text{Neither road bike nor mountain bike}) = 0.16$
  **d.** $P(\text{Mountain bike given road bike}) = 0.06/0.25 = 0.24$
**4-79. a.** $P(\text{Food and no coffee}) = 0.10$
  **b.** $P(\text{Coffee and no food}) = 0.40$
  **c.** $P(\text{No coffee and no food}) = 0.10$
**4-81. a.** $P(\text{No train}) = 0.39$
  **b.** $P(\text{Neither train nor bus}) = 0.11$
  **c.** $P(\text{Train|Bus}) = 0.06/0.24 = 0.25$

**4-83. a.** $P(\text{Salt Lake}) = 24/110 = 0.2182$
    **b.** $P(\text{Wiring}) = 23/110 = 0.2091$
    **c.** $P(\text{Salt Lake and wiring}) = 8/110 = 0.0727$
    **d.** $P(\text{Day shift and Salt Lake and cracked lens}) = 8/110 = 0.0727$
    **e.** The most likely profile is the largest number; that is, the Boise day shift for cracked lens.

# CHAPTER 5

**5-1. a.** discrete random variable
    **b.** The possible values are $x = \{0, 1, 2, 3, 4, 5, 6\}$.
**5-3. a.** number of children under 22 living in a household
    **b.** discrete
**5-5.** 3.7 days
**5-7. a.** 130
    **b.** 412.50
    **c.** $\sqrt{412.50} = 20.31$
**5-9. a.** 15.75
    **b.** 20.75
    **c.** 78.75
    **d.** increases the expected value by an amount equal to the constant added
    **e.** the expected value being multiplied by that same constant
**5-11. a.** 3.51
    **b.** 1.2845
**5-13. a.** 2.87 days
    **b.** $\sigma = \sqrt{1.4931} = 1.22$
    **c.** 1.65 days to 4.09 days
**5-15. a.** \$58,300
    **b.** \$57,480; Then, comparing this \$57,480 expected cost with the expected cost of \$58,300 found in part a, the best decision based on cost alone is to switch to the new supplier.
**5-17. a.** Small firm profits = \$135,000
    Mid-sized profits = \$155,000
    Large firm profits = \$160,000
    **b.** Small firm: $\sigma = \$30,000$
    Mid-sized firm: $\sigma = \$90,000$
    Large firm: $\sigma = \$156,604.60$
    **c.** The large firm has the largest expected profit.

**5-21. a.**

| $x$ | $P(x)$ |
|-----|--------|
| 14 | 0.008 |
| 15 | 0.024 |
| 16 | 0.064 |
| 17 | 0.048 |
| 18 | 0.184 |
| 19 | 0.216 |
| 20 | 0.240 |
| 21 | 0.128 |
| 22 | 0.072 |
| 23 | 0.016 |

    **b.** 19.168; $\sigma = \sqrt{\sigma^2} = \sqrt{3.1638} = 1.7787$
    **c.** The probability of lasting more than 19 months is 0.456 so the quality control department is incorrect.
**5-23.** 0.2668
**5-25. a.** $P(x = 5) = 0.0746$
    **b.** $P(x \geq 7) = 0.2143$
    **c.** 4
    **d.** $\sigma = \sqrt{npq} = \sqrt{20(0.20)(0.80)} = 1.7889$
**5-27.** 0.1030
**5-29. a.** 0.1442
    **b.** 0.8002

    **c.** $E(x) = np = (7)(0.65) = 4.55$
    **d.** $1 - 0.9510 = 0.0490$
**5-31. a.** 0.0688
    **b.** 0.0031
    **c.** 0.1467
    **d.** 0.8470
    **e.** 0.9987
**5-33. a.** 3.2
    **b.** 1.386
    **c.** 0.4060
    **d.** 0.9334
**5-35. a.** $n = 10; p = 0.38; q = 1 - p = 1 - 0.38 = 0.62$
    Step 2: Use Equation 5.4 to find the expected value:
    $\mu_x = E(x) = np = 10(0.38) = 3.8$
    **b.** $n = 10; p = 0.62; q = 1 - p = 1 - 0.62 = 0.38;$
    $p(x = 10) = 0.0084$
    **c.** 0.2017
    **d.** It is unlikely that the employees reflect the California trend.
**5-37. a.** 0.5580
    **b.** 0.8784
    **c.** An increase in sample size would be required.
**5-39. a.** Expected number $= 8(0.37) = 2.96$
    **b.** 1.3656
    **c.** $P(x \leq 2) = 0.3811$
**5-41. a.** 0.3179
    **b.** 0.2174
    **c.** 0.25374
**5-43. a.** 0.3666
    **b.** 0.0003
**5-45. a.** 0.372
    **b.** The estimate of 12 may be too high.
**5-49. a.** 18 corporations
    **b.** Binomial with $n = 35$ and $p = 0.63$;
    $P(x \leq 18) = 0.1087$
**5-51. a.** 0.0498
    **b.** 0.1512
**5-53.** 0.175
**5-55. a.** 0.4242
    **b.** 0.4242
    **c.** 0.4696
**5-57. a.** $P(x = 3) = 0.5$
    **b.** $P(x = 5) = 0$
    **c.** 0.1667
    **d.** Since $0.6667 > 0.25$, then $x' = 2$.
**5-59.** $P(x \geq 10) = 1 - 0.8305 = 0.1695$
**5-61.** 0.0015
**5-63. a.** $P(x = 4) = 0.4696$
    **b.** $P(x = 3) = 0.2167$
    **c.** 0.9680
**5-65. a.** 0.0355
    **b.** 0.0218
    **c.** 0.0709
**5-67. a.** 0.0632
    **b.** 120 spicy hot dogs
**5-69. a.** 0.02759
    **b.** 0.0000026
    **c.** 0.00008
**5-71. a.** 8
    **b.** $\lambda t = 1(3) = 3$
    **c.** 0.0119
    **d.** It is very unlikely. Therefore, we believe the goal has not been met.

**5-75. a.** This means the trials are dependent.
   **b.** does not imply that the trials are independent

**5-77. a.**

| x | P(x) | xP(x) |
|---|------|-------|
| 0 | 0.56 | 0.00 |
| 1 | 0.21 | 0.21 |
| 2 | 0.13 | 0.26 |
| 3 | 0.07 | 0.21 |
| 4 | 0.03 | 0.12 |
|   |      | 0.80 |

   **b.** Standard deviation = 1.0954; Variance = 1.20

**5-79.** 0.0020

**5-81.** 0.6244

**5-83. a.** 2.0
   **b.** 1.4142
   **c.** because outcomes are equally likely

**5-85. a.** $E(x) = 750; E(y) = 100$
   **b.** $StDev(x) = 844.0972; StDev(y) = 717.635$
   **c.** $CV(x) = 844.0972/750 = 1.1255$
   $CV(y) = 717.635/100 = 7.1764$

**5-87. a.** 0.3501
   **b.** 0.3250

**5-89. a.** 0.01178
   **b.** 0.42067
   **c.** 0.9999

**5-91. a.** 0.00758
   **b.** 0.98224
   **c.** Expected value = $np$. If $p = 0.008$, then 125 parts need to be sampled.

**5-93. a.** 0.08879
   **b.** 0.0926

## CHAPTER 6

**6-1. a.** $\dfrac{225 - 200}{20} = \dfrac{25}{20} = 1.25$

   **b.** $\dfrac{190 - 200}{20} = \dfrac{-10}{20} = -0.50$

   **c.** $\dfrac{240 - 200}{20} = \dfrac{40}{20} = 2.00$

**6-3. a.** 0.4901
   **b.** 0.6826
   **c.** 0.0279

**6-5. a.** 0.4750
   **b.** 0.05
   **c.** 0.0904
   **d.** 0.9759
   **e.** 0.8413

**6-7. a.** 0.9270
   **b.** 0.6678
   **c.** 0.9270
   **d.** 0.8413
   **e.** 0.3707

**6-9. a.** $x = 1.29(0.50) + 5.5 = 6.145$
   **b.** $\mu = 6.145 - (1.65)(0.50) = 5.32$

**6-11. a.** 0.0027
   **b.** 0.2033
   **c.** 0.1085

**6-13. a.** 0.0668
   **b.** 0.228
   **c.** 0.7745

**6-15. a.** 0.3446
   **b.** 0.673
   **c.** 51.30
   **d.** 0.9732

**6-17. a.** 0.1762
   **b.** 0.3446
   **c.** 0.4401
   **d.** 0.0548

**6-19.** The mean and standard deviation of the random variable are 15,000 and 1,250, respectively.
   **a.** 0.0548
   **b.** 0.0228
   **c.** $\mu = 15,912$ (approximately)

**6-21.** about $3,367.35

**6-23. a.** 0.1949
   **b.** 0.9544
   **c.** Median = Mean = $1,200

**6-25.** $P(x < 1.0) = 0.5000 - 0.4761 = 0.0239$

**6-27. a.** $P(0.747 \le x \le 0.753) = 0.6915 - 0.1587 = 0.5328$

   **b.** $\sigma = \dfrac{0.753 - 0.75}{2.33} = 0.001$

**6-31. a.** Yes. It appears that this distribution is approximately normal but slightly right-skewed.
   **b.** 23.88%

**6-33. a.** 0.75

   **b.** $Q_1 = \dfrac{4 + 0.25}{0.0625} = 8; Q_2 = \dfrac{4 + 0.50}{0.0625} = 12;$

   $Q_3 = \dfrac{4 + 0.75}{0.0625} = 16$

   **c.** 14.43
   **d.** 0.92

**6-35. a.** 0.9179
   **b.** 0.0498
   **c.** 0.0323
   **d.** 0.9502
   **e.** $1 - 0.0498 = 0.9502$

**6-37. a.** 0.3935
   **b.** 0.2865

**6-39. a.** 0.7143
   **b.** 0.1429
   **c.** 0.0204

**6-41. a.** 0.4084; yes
   **b.** 40,840

**6-43. a.** 0.2992
   **b.** 0.5397

**6-45. a.** 0.0183
   **b.** 0.3679

**6-47. a.** 0.0498
   **b.** 0.4493
   **c.** approximately, $\lambda = 0.08917$

**6-49. a.** Mean = 3.87; Standard deviation = $2.60
   **b.** 0.4606

**6-55. a.** 0.1353
   **b.** 0.1353

**6-57. a.** 0.486583
   **b.** 0.452934
   **c.** 0.18888

**6-59.** 0.5507

**6-61.** Machine 1: 0.4236
   Machine 2: 0.4772

**6-63.** 0.142778

**6-65.** They need an additional $11,500 - 500 = 11,000$ parking spaces.

**6-67.** **a.** 0.3406
  **b.** 0.5580

**6-71.** **a.** 0.3085
  **b.** Prob $(x \geq 82) =$ Prob $(z \leq (82 - 78)/12) =$ Prob $(z \geq 0.33) = 0.5 - 0.1293 = 0.3707$
  **c.** $x = 93.36$ minutes

**6-73.** **a.** 0.1587
  **b.** 0.0912
  **c.** 0.5328
  **d.** 26.13 minutes

**6-75.** **a.** 0.1711
  **b.** The distribution is nearly normally distributed.
  **c.** 0.1401

**6-77.** **a.** This histogram definitely has the shape of an exponential distribution, positively skewed.
  **b.** Mean $= 20,000$; Standard deviation $= 22,535$
  **c.** 13,862.9 hours
  **d.** $1 - 0.7135 = 0.2865$

**6-79.** **a.** The income appears to be right-skewed.
  **b.** Mean $= 32,801.09$; Standard deviation $= 6,230.097$
  **c.** 0.123
  **d.** $23,580.60

## CHAPTER 7

**7-1.** $-18.50$

**7-3.** $\bar{x} - \mu = 10.17 - 11.38 = -1.21$

**7-5.** **a.** $-4.125$
  **b.** $-13.458$ to $11.042$
  **c.** $-9.208$ to $9.208$

**7-9.** $\bar{x} - \mu = 6.8 - 6.0 = 0.80$

**7-11.** **a.** $\mu = \dfrac{\sum x}{N} = \dfrac{864}{20} = 43.20$ days
  **b.** $\bar{x} = \dfrac{\sum x}{n} = \dfrac{206}{5} = 41.20$ days;
    Sampling error $= 41.20 - 43.20 = -2$
  **c.** $-28.4$ days to $40.4$ days

**7-13.** **a.** $3,445.30
  **b.** $-$29.70

**7-15.** **a.** 1,432.08
  **b.** 87.12
  **c.** $-175.937$ to $178.634$

**7-17.** **a.** Mean of sales $= 2,764.83$
  **b.** Mean of sample $= 2,797.16$
  **c.** $32.33 million
  **d.** Smallest $= -$170.47$; Largest $= $218.41$

**7-21.** **a.** Sampling error $= \bar{x} - \mu = $15.84 - $20.00 = -$4.16$
  **b.** Random and nonrandom samples can produce sampling error, and the error is computed the same way.

**7-23.** $P(\bar{x} > 2,100) = 0.5000 - 0.3907 = 0.1093$

**7-25.** $\sigma_{\bar{x}} = \dfrac{\sigma}{\sqrt{n}} = \dfrac{40}{\sqrt{25}} = 8$

**7-27.** 0.0016

**7-29.** **a.** 0.3936
  **b.** 0.2119
  **c.** 0.1423
  **d.** 0.0918

**7-31.** **a.** 0.8413
  **b.** 0.1587

**7-33.** **a.** 0.3830
  **b.** 0.9544
  **c.** 0.9736

**7-35.** $P(\bar{x} \geq 4.2) = 0.5000 - 0.4998 = 0.0002$

**7-37.** $P(\bar{x} \leq 33.5) = 0.5000 - 0.2019 = 0.2981$

**7-39.** **a.** **Descriptive Statistics: Video Price**

| Variable | Mean | StDev |
|---|---|---|
| Video price | 58.53 | 2.53 |

  **b.** $P(\bar{x} \leq 58.52) = 0.00$
  **c.** $P(\bar{x} \leq 58.52) = 0.0094$
  **d.** Yes; it appears that the average retail price has declined.

**7-43.** $10,977,744

**7-45.** **a.** 0.8621
  **b.** 0.0146
  **c.** 0.8475
  **d.** 0.7422

**7-47.** **a.** $\bar{p} - p = 0.65 - 0.70 = -0.05$
  **b.** 0.1379

**7-49.** **a.** 0.8221
  **b.** 0.6165

**7-51.** **a.** 0.9015
  **b.** 0.0049

**7-53.** **a.** $p = \dfrac{x}{n} = \dfrac{27}{60} = 0.45$
  **b.** $P(p \geq 0.45) = 0.5000 - 0.2852 = 0.2148$

**7-55.** $P(p \geq 0.09) = 0.5000 - 0.4838 = 0.0162$

**7-57.** **a.** 0.0793
  **b.** 0.5000
  **c.** 0.0000

**7-59.** **a.** 0.0749
  **b.** Yes, 0.0359 is a small probability.
  **c.** Yes, the probability of such a result is essentially 0.

**7-61.** **a.** 58.86%
  **b.** 0.9476
  **c.** Yes, probability is essentially 0.

**7-63.** **a.** 131 over $100,000 and 65 of $100,000 or less
  **b.** 0.6684
  **c.** 0.2946

**7-67.** Sample averages are less variable than the population.

**7-69.** **a.** 405.55
  **b.** 159.87

**7-71.** A sample size of 1 is sufficient.

**7-73.** **a.** 0.2643
  **b.** 0.3752
  **c.** 0.0951

**7-75.** **a.** 0.0294
  **b.** 0.1190

**7-77.** **a.** Mean $= 4.5$; Standard Deviation $= 0.8660$
  **b.** 0.9783

**7-79.** **a.** $P(x > 16.10) = 0.5000 - 0.1554 = 0.3446$
  **b.** $P(\bar{x} > 16.10) = 0.5000 - 0.4177 = 0.0823$

**7-81.** Note that because of the small population, the finite correction factor is used.
  **a.** 0.1112
  **b.** Either the mean or the standard deviation or both may have changed.
  **c.** 0.2483

**7-83.** 0.02275

**7-85.** **a.** $n = 25$: 0.96; $n = 50$: 0.98; $n = 100$: 0.95; $n = 200$: 0.965

**b.** $n = 25$: Mean $= 0.97$; Standard deviation $= 0.0341$
$n = 50$: Mean $= 0.97$; Standard deviation $= 0.0241$
$n = 100$: Mean $= 0.97$; Standard deviation $= 0.0171$
$n = 200$: Mean $= 0.97$; Standard deviation $= 0.0121$

**c.** $n = 25$: $-0.01$; $n = 50$: $0.01$; $n = 100$: $-0.02$;
$n = 200$: $-0.005$

**d.** $0.0493$

**7-87. a.** $0.79$

**b.** Sampling error $= \bar{p} - p = 0.79 - 0.74 = 0.05$

**c.** $0.8926$

**d.** $0.74 \pm 0.02 = (0.72, 0.76)$; $P(0.72 \le \bar{p} \le 0.76)$
$= 0.9794 - 0.0206 = 0.9588$

**7-89. a.** $0.216$

**b.** $0.3275$

# CHAPTER 8

**8-1.** $15.86$ ———— $20.94$

**8-3.** $293.18$ ———— $306.82$

**8-5.** $1,180.10$ ———— $1,219.90$

**8-7. a.** $1.69$ ———— $4.31$

**b.** $1.38$ ———— $4.62$

**8-9.** $97.62$ ———— $106.38$

**8-11. a.** $(11.6028, 15.1972)$

**b.** $(29.9590, 32.4410)$

**c.** $(2.9098, 6.0902)$

**d.** $(18.3192, 25.0808)$

**8-13. a.** $\$13.945$ ———— $\$14.515$

**b.** There is no reason to believe that this is out of line.

**8-15. a.** $(4,780.25, 5,219.75)$

**b.** $219.75$

**c.** $109.875$; $n = 716$

**8-17. a.** $\$5.29$ ———— $\$13.07$

**b.** These sample data do not dispute the study.

**8-19. a.** $\bar{x} = 129.5$, $s = 9.204$

**b.** $(122.92, 136.08)$

**8-21. a.** $163.5026$ ———— $171.5374$

**b.** Increasing the sample size or decreasing the level of confidence can reduce the margin of error.

**8-23. a.** $6.5368$

**b.** $6.3881$ ———— $6.6855$

**8-25. a.** $256.01$, $80.68$ (calculated using Excel's STDEV.S function)

**b.** $242.01$ ———— $270.01$

**c.** $\pm 14.00$ seconds

**8-27.** $189$

**8-29.** $918$

**8-31.** $3,684.21$

**8-33. a.** $n = 62$

**b.** $n = 5,726$

**c.** $n = 2$

**d.** $n = 306$

**e.** $n = 35$

**8-35.** $249$

**8-37. a.** $883$ less the 20 items in the pilot sample $= 863$ more

**b. 1.** Reduce the confidence level to something less than 95%.

**2.** Increase the margin of error beyond 0.25 pound.

**3.** Use combination of decreasing the confidence level and increasing the margin of error.

**8-39.** $325 - 50 = 275$

**8-41. a.** $246$

**b.** $6,147$

**c.** $\$0.44$ to $\$0.51$

**8-43. a.** $60$

**b.** No, sample size is almost four times as large; $n = 239$

**8-45. a.** $292$

**b.** $165$ additional households must be sampled.

**8-47. a.** $1,599$

**b.** Reduce the confidence level (lowers the $z$-value) or increase the margin of error or some combination of the two

**8-49.** $1,698 - 75 = 1,623$

**8-51.** $0.224$ ———— $0.336$

**8-53. a.** The sampling distribution can be approximated by a normal distribution.

**b.** $(0.286, 0.414)$

**c.** $0.286$ to $0.414$

**d.** $0.064$

**8-55. a.** $\bar{p} = 0.175$

**b.** $(0.057, 0.293)$

**c.** $n = 888$

**8-57. a.** $0.324$ ———— $0.436$

**b.** $9,604$

**8-59.** $\bar{p} = 345/1,000 = 0.345$

**a.** between $0.3155$ and $0.3745$

**b.** $179.20$ ———— $212.716$

**c.** $\bar{p} = (280/690) = 0.4058$; between $0.3692$ and $0.4424$

**d.** $1,225$

**8-61.** $0.895$ ———— $0.925$

**8-63. a.** $0.6617$ ———— $0.7183$

**b.** $2,401$

**c.** $0.969$

**8-65. a.** $0.1167$

**b.** $(0.0736, 0.1598)$

**8-67. a.** Use Excel to count the number of Miami respondents who plan to retire after age 65. The number is 47; therefore, the best point estimate for the population proportion is calculated from the sample information as $x/n = 47/90 = 0.5222$.

**b.** A 99% confidence level is desired. The critical value from the standard normal distribution table is $z = 2.575$. Compute the confidence interval using Equation 8.10:

$$\bar{p} \pm z\sqrt{\frac{\bar{p}(1 - \bar{p})}{n}} = 0.5222 \pm 2.575\sqrt{\frac{0.5222(1 - 0.5222)}{90}}$$

$$= 0.5222 \pm 0.1356 = 0.3866 \text{————} 0.6578$$

**c.** The sample size could be increased. This would reduce the margin of error while leaving the confidence level unchanged.

**8-69. a.** males $= 0.3513$ ———— $0.6087$
females $= 0.1133$ ———— $0.3267$

**b.** males $>$ females

**8-75.** $0.2564 \le p \le 0.4936$

**8-77. a.** $0.7265$ ———— $0.7935$

**b.** $25,427.50$ ———— $27,772.50$

**8-79. a.** $22$ more

**b.** $\$770$

**8-81. a.** $5.21$

**b.** $390$

**c.** $2.00$ work hours

**8-83. a.** $0.7003$ ———— $0.7741$

**b.** $32,279.4674$ ———— $33,322.7227$

**8-85. a.** $40.178$ to $40.882$

**b.** $n = 393$

# CHAPTER 9

**9-1. a.** $z = 1.96$

**b.** $t = -1.6991$

**c.** $t = \pm 2.4033$

**d.** $z = \pm 1.645$

**9-3. a.** $z_\alpha = 1.645$

**b.** $t_{\alpha/2} = \pm 2.5083$

**c.** $z_{\alpha/2} = \pm 2.575$

**d.** $-t_\alpha = -1.5332$

**e.** invalid

**9-5.** **a.** Reject the null hypothesis if the calculated value of the test statistic, $z$, is greater than 2.575 or less than $-2.575$. Otherwise, do not reject.

**b.** $z = -3.111$

**c.** Reject the null hypothesis.

**9-7.** **a.** Reject the null hypothesis if the calculated value of the test statistic, $t$, is less than the critical value of $-2.0639$. Otherwise, do not reject.

**b.** $t = -1.875$

**c.** Do not reject the null hypothesis.

**9-9.** **a.** Reject the null hypothesis if the calculated value of the test statistic, $t$, is greater than 1.3277. Otherwise, do not reject.

**b.** $t = 0.78$

**c.** Do not reject the null hypothesis.

**9-11.** **a.** Type I error

**b.** Type II error

**c.** Type I error

**d.** no error

**e.** Type II error

**f.** no error

**9-13.** **a.** $H_0: \mu \geq 30{,}000$
$H_A: \mu < 30{,}000$

**b.** $29{,}588.75

**c.** Do not reject.

**d.** Type II

**9-15.** **a.** $H_0: \mu \geq 3{,}600$
$H_A: \mu < 3{,}600$

**b.** Since $t = -0.85 > -1.8331$, the null hypothesis is not rejected.

**9-17.** **a.** $H_0: \mu \geq 55$
$H_A: \mu < 55$

**b.** Because $t = -0.93 > -2.4620$, the null hypothesis is not rejected.

**9-19.** $H_0: \mu \geq \$3{,}971$
$H_A: \mu < \$3{,}971$
Yes, because $t = -31.16$ is much less than the critical $t$-value of $-2.0736$, reject $H_0$.

**9-21.** **a.** Since $t = -2.567 < -2.014$, reject the null hypothesis based on these sample data.

**b.** no statistical error

**9-23.** **a.** $z = 1.96$

**b.** $z = -1.645$

**c.** $z = \pm 2.33$

**d.** $z = \pm 1.645$

**9-25.** Since $-2.17 > -2.33$, don't reject.

**9-27.** **a.** Reject the null hypothesis if the calculated value of the test statistic, $z$, is less than the critical value of the test statistic $z = -1.96$. Otherwise, do not reject.

**b.** $z = -2.0785$

**c.** reject

**9-29.** **a.** $p$-value = 0.05

**b.** $p$-value = 0.5892

**c.** $p$-value = 0.1902

**d.** $p$-value = 0.0292

**9-31.** Yes, because $z = -3.145$ is less than $-2.055$, reject $H_0$.

**9-33.** Since $z = 0.73 < 1.645$, do not reject the null hypothesis.

**9-35.** Since $0.00087 < 0.025$, reject the null hypothesis.

**9-37.** **a.** $H_0: p \leq 0.40$
$H_A: p > 0.40$

**b.** Since $z = 1.43 < 1.645$, do not reject the null hypothesis.

**9-39.** **a.** $H_0: p \leq 0.10$
$H_A: p > 0.10$

**b.** Since $p$-value = 0.1736 is greater than 0.05, don't reject.

**9-41.** **a.** Since $z = 2.85 > 1.96$, reject $H_0$.

**b.** $p$-value = $0.5 - 0.4978 = 0.0044$

**9-43.** Yes, because $z = 2.36$ is greater than 1.645, reject $H_0$.

**9-45.** **a.** $H_0: p \geq 0.50$
$H_A: p < 0.50$

**b.** Since $z = -6.08 < -2.05$, reject the null hypothesis.

**9-47.** **a.** $H_0: p \geq 0.95$
$H_A: p < 0.95$

**b.** Since $z = -4.85 < -1.645$, reject the null hypothesis.

**9-49.** **a.** 0.80

**b.** 0.20

**c.** As the sample size increases, the power increases, and beta decreases.

**d.** Since $\bar{x} = 1.23$, then $1.0398 < 1.23 < 1.3062$; do not reject $H_0$.

**9-51.** 0.8888

**9-53.** 0.3228

**9-55.** **a.** 0.0084

**b.** 0.2236

**c.** 0.9160

**9-57.** **a.** 0.1685

**b.** 0.1469

**c.** 0.1190

**9-59.** **a.** $H_0: \mu \leq 400$
$H_A: \mu > 400$

**b.** $\beta = 0.5000 - 0.5000 = 0.00$

**9-61.** **a.** $H_0: \mu \geq 15$
$H_A: \mu < 15$

**b.** 0.0606

**9-63.** **a.** $H_0: \mu \geq \$53{,}300$
$H_A: \mu < \$53{,}300$

**b.** $\beta = 0.5000 - 0.4525 = 0.0475$

**9-65.** **a.** $H_0: \mu \geq \$8{,}968$
$H_A: \mu < \$8{,}968$

**b.** $= P(z > 1.24) = 0.5 + 0.3925 = 0.8925$

**9-67.** **a.** Since $t = -3.97 < -1.6991$, reject $H_0$.

**b.** 0.3557

**9-77.** **a.** If $\alpha$ is decreased, the rejection region is smaller, making it easier to accept $H_0$, so $\beta$ is increased.

**b.** If $n$ is increased, the test statistic is also increased, making it harder to accept $H_0$, so $\beta$ is decreased.

**c.** If $n$ is increased, the test statistic is also increased, making it harder to accept $H_0$, so $\beta$ is decreased and power is increased.

**d.** If $\alpha$ is decreased, the rejection region is smaller, making it easier to accept $H_0$, so $\beta$ is increased and power is decreased.

**9-79.** **a.** $z = \dfrac{\bar{x} - \mu}{\dfrac{\sigma}{\sqrt{n}}}$

**b.** $t = \dfrac{\bar{x} - \mu}{\dfrac{s}{\sqrt{n}}}$

**c.** $z = \dfrac{\bar{p} - p}{\sqrt{\dfrac{p(1 - p)}{n}}}$ but should not be used since $n(1 - p) < 5$.

**d.** A hypothesis test using a small ($< 30$) sample from a skewed distribution cannot be conducted using procedures in this text.

**9-81.** **a.** $H_0: \mu \leq 4{,}000$
$H_A: \mu > 4{,}000$

**b.** Since $t = 1.2668 < 1.7959$, do not reject.

**9-83.** **a.** $H_0$: $p \leq 0.50$
$H_A$: $p > 0.50$
**b.** Since $z = 5.889 > 1.645$, reject the null hypothesis. Since $z = 5.889$, the $p$-value is approximately zero.
**9-85.** **a.** Since $z = -1.5275 > -1.645$, do not reject.
**b.** Type II error means "accepting" a false null hypothesis
**9-87.** **a.** $H_0$: $p = 0.30$
$H_A$: $p \neq 0.30$
**b.** Since $p$-value $= 0.1902 > 0.05$, do not reject the null hypothesis.
**9-89.** **a.** $H_0$: $\mu \leq 3$
$H_A$: $\mu > 3$
**b.** If $z > 2.33$, reject $H_0$. Otherwise, do not reject $H_0$. Since $z = 1.98 < 2.33$, do not reject $H_0$.
**c.** Since $z = 1.98 < 2.33$, do not reject $H_0$.
**9-91.** Since $z = 2.04 > 1.645 = z_\alpha$, reject $H_0$.
**9-93.** **a.** $H_0$: $\mu \geq 40$
$H_A$: $\mu < 40$
**b.** Since $t = -0.8052 > -1.2991$, do not reject $H_0$.
**c.** Type II error. There is very little chance of ever being able to determine if you had made the error.
**d.** $\bar{x}_\alpha = 37.6123$
**e.** $p$-value $= 0.2122$
**9-95.** **a.** $H_0$: $\mu \leq 7.4$
$H_A$: $\mu > 7.4$
**b.** Since $t = 10.1335 > 1.6612$, reject $H_0$. Conclude that the average pH level is higher than 7.4.
**9-97.** **a.** $H_0$: $\mu \leq \$5$
$H_A$: $\mu > \$5$
**b.** No. Since $t = -2.2758 < 1.6766$, do not reject $H_0$.
**c.** The consumer group would be more concerned with a Type I error. The company would be more concerned with a Type II error.
**9-99.** **a.** Since $z = -1.1711 > -1.645$, do not reject and conclude that the speaker is correct and that "knock-offs" is greater than or equal to 30%.
**b.** A Type II error may have been committed.
**c.** In this case, we have $0.2687(294) > 5$ and $(1 - 0.2687)(294) > 5$, so we are okay.
**d.** Since $z = -0.827 > -1.645$, do not reject $H_0$ and conclude that the "knock-offs" are not less than 30% in the players.

**CHAPTER 10**
**10-1.** $-6.54 \leq (\mu_1 - \mu_2) \leq 0.54$
**10-3.** $-13.34 \leq (\mu_1 - \mu_2) \leq -6.2$
**10-5.** $-10.40 \leq (\mu_1 - \mu_2) \leq 61.08$
**10-7.** $-19.47 \leq (\mu_1 - \mu_2) \leq -7.93$
**10-9.** **a.** $-0.05$
**b.** $-0.0974 \leq (\mu_1 - \mu_2) \leq -0.0026$
**c.** The managers can conclude the two lines do not fill bags with equal average amounts.
**10-11.** **a.** 2.1783
$-0.1043$————2.7043
**b.** No difference in the setup time for the two additives.
**10-13.** $-\$35.07$ ———— $-\$45.13$
**10-15.** $-0.10 \leq (\mu_1 - \mu_2) \leq 0.30$
**10-17.** **a.** 2.35%
**b.** $-1,527.32$————4,926.82
**c.** It is plausible there is no difference.
**d.** 3,227
**10-19.** Since $t = -4.80 < -2.1199$, reject the null hypothesis.
**10-21.** Since $z = 5.26 > 1.645$, reject the null hypothesis.

**10-23.** **a.** $H_0$: $\mu_1 = \mu_2$
$H_A$: $\mu_1 \neq \mu_2$
If $t > 1.9698$ or $t < -1.9698$, reject $H_0$.
**b.** Since $5.652 > 1.9698$, reject $H_0$.
**10-25.** $t = 0.896$; do not reject the null hypothesis.
**10-27.** **a.** $t = 0.9785 > -1.677$; do not reject.
**b.** Type II error
**10-29.** $t = -4.30 < -1.6510$; reject the null hypothesis.
**10-31.** $p$-value $= P(t \leq 5.36) \cong 1.00$; the null hypothesis is not rejected.
**10-33.** $H_0$: $\mu_1 \leq \mu_2$
$H_A$: $\mu_1 > \mu_2$
Since $3.44 > 1.677$, reject the null hypothesis.
**10-35.** **a.** $-13.76$————$-1.00$; contains $-10$
**b.** $p$-value is greater than 0.025; do not reject the null hypothesis.
**c.** You should be able to reach the same conclusion using either a confidence interval estimate or a hypothesis test.
**10-37.** $-674.41 \leq \mu \leq -191.87$
**10-39.** **a.** $H_0$: $\mu_d \geq 0$
$H_A$: $\mu_d < 0$
**b.** Since $-3.64 < -1.3968$, reject $H_0$.
**c.** $-2.1998$ ——— $-0.7122$
**10-41.** **a.** (7.6232, 13.4434)
**b.** $t = 0.37 < 1.459$
**10-43.** **a.** The samples were matched pairs.
**b.** $0.005 < p$-value $< 0.01$
**10-45.** **a.** $t = -7.35 < -1.98$; reject $H_0$.
**b.** $-100.563$ ——— $-57.86$
**10-47.** Because $t = 6.11 > 2.5083$, reject the null hypothesis.
**10-49.** $t = 1.068 < 2.136$; do not reject.
**10-51.** **a.** The sampling distribution cannot be approximated with a normal distribution.
**b.** The sampling distribution cannot be approximated with a normal distribution.
**c.** The sampling distribution can be approximated with a normal distribution.
**d.** The sampling distribution cannot be approximated with a normal distribution.
**10-53.** $p$-value is greater than $\alpha$; do not reject the null hypothesis.
**10-55.** **a.** Since $z = 2.538 > 1.96$, reject $H_0$.
**b.** Since $z = 2.538 > -1.645$, fail to reject $H_0$.
**c.** Since $z = 2.538 > 1.96$, reject $H_0$.
**d.** Since $z = 1.987 < 2.33$, fail to reject $H_0$.
**10-57.** $z = -0.4111 > -2.575$; do not reject.
**10-59.** Since the test statistic $-1.809 > -1.96$, do not reject the null hypothesis.
**10-61.** **a.** The sampling distribution can be approximated with a normal distribution.
**b.** $p$-value $= 0.039 < 0.05$; reject.
**10-65.** **a.** Because $t = -2.35 < -2.1448$, reject the null hypothesis.
**b.** Because $t = -2.01 > -2.3646$, do not reject the null hypothesis.
**c.** There are three reasons for the difference in the outcomes of the two hypothesis tests: (1) the degrees of freedom is smaller in the paired samples' case, (2) the sample size is smaller for the paired samples' case, and (3) the standard deviation of the paired samples' case is larger.
**10-67.** Because $t = -1.529 > t = -3.2498$, do not reject the null hypothesis.
**10-69.** $2.35 \leq (\mu_1 - \mu_2) \leq 6.73$
**10-71.** **a.** Because $t = 1.89 > 1.6479$, reject the null hypothesis.
**b.** The required assumptions are independent samples, normal distributions, and equal population variances.

**10-73.** $z = 1.7025 < 2.33$; do not reject.
**10-75. a.** Yes, because $n_1\bar{p}_1 = 75(0.29) = 21.75 > 5$,
  $n_1(1 - \bar{p}_1) = 75(0.71) = 53.25 > 5$;
  $n_2\bar{p}_2 = 125(0.045) = 5.625 > 5$,
  $n_2(1 - \bar{p}_2) = 125(0.955) = 119.375 > 5$
  **b.** Since $z = 2.887 > 1.645$, reject $H_0$.
**10-77. a.** Since $t = -1.7832 > -2.345$, do not reject the null hypothesis.
  **b.** $p$-value $= 0.038$ results in significance for all alpha levels $> 0.038$
  **c.** Since $t = -1.8168 > -2.345$, do not reject. $p$-value $= 0.0354$
**10-79. a.** Since $0.5828 < 1.96$, do not reject $H_0$.

## CHAPTER 11

**11-1.** $74{,}953.7 \le \sigma^2 \le 276{,}472.2$
**11-3.** Since $\chi^2 = 12.39 > 10.1170$, do not reject the null hypothesis.
**11-5. a.** Because $\chi^2 = 17.82 < \chi^2_{0.05} = 19.6752$ and because $\chi^2 = 17.82 > \chi^2_{0.95} = 4.5748$, do not reject the null hypothesis.
  **b.** Because $\chi^2 = 12.96 < \chi^2_{0.025} = 31.5264$ and because $\chi^2 = 12.96 > \chi^2_{0.975} = 8.2307$, do not reject the null hypothesis.
**11-7. a.** Since $p$-value $< \alpha$, reject $H_0$.
  **b.** Since the test statistic $= 1.591 <$ the $\chi^2$ critical value $= 1.6899$, reject the null hypothesis.
  **c.** $p$-value $= 0.12$; do not reject.
**11-9.** $22.72 \le \sigma^2 \le 81.90$
**11-11.** $353.38 \le \sigma^2 \le 1{,}745.18$
**11-13. a.** $H_0: \mu \le 10$
  $H_A: \mu > 10$
  **b.** Since $1.2247 < 1.383$, do not reject $H_0$.
  **c.** Since $3.75 < 14.6837$, do not reject $H_0$.
**11-15.** Because $\chi^2 = 4.18 <$ the $\chi^2$ critical value $= 13.8484$, do not reject the null hypothesis.
**11-17.** $H_0: \sigma^2 \le 0.000278$
  $H_A: \sigma^2 > 0.000278$
  Since $p$-value $= 0.004 < 0.01$, reject the null hypothesis.
**11-19. a.** If the calculated $F > 2.278$, reject $H_0$; otherwise, do not reject $H_0$.
  **b.** Since $1.0985 < 2.278$, do not reject $H_0$.
**11-21. a.** $F = 3.619$
  **b.** $F = 3.106$
  **c.** $F = 3.051$
**11-23.** Since $F = 0.867 < 6.388 = F_{0.05}$, do not reject $H_0$.
**11-25.** Yes. Since $3.4807 > 1.984$, reject $H_0$.
**11-27.** $H_0: \sigma^2_A \le \sigma^2_T$
  $H_A: \sigma^2_A > \sigma^2_T$
  If the calculated $F > 2.534$, reject $H_0$; otherwise, do not reject $H_0$.
  $F = 0.202^2/0.14^2 = 2.0818$
  Since $2.0818 < 2.534$, do not reject $H_0$. There is not sufficient evidence to conclude that the Trenton plant produced parts that are less variable than the Altanta plant.
**11-29.** $H_0: \sigma^2_1 \le \sigma^2_2$
  $H_A: \sigma^2_1 > \sigma^2_2$
  $\alpha = 0.05$
  The critical value obtained from the $F$-distribution with Excel's F.INV.RT(0.05,19,19) function is 2.1683. Reject $H_0$ if $F > 2.1683$.
  The test statistic is $F = \dfrac{s^2_1}{s^2_2} = \dfrac{(2{,}636)^2}{(1{,}513)^2} = 3.035$.
  Since $F = 3.035 > 2.1683 = F_{0.05}$, reject $H_0$.

Based on these sample data, there is sufficient evidence to conclude that male executives exhibit greater variability in miles flown than do female executives.
**11-31.** Since $1.4833 < 3.027$, do not reject $H_0$.
**11-33.** Since $F = 33.9 > F$-critical $= 1.60$, reject the null hypothesis.
**11-39.** $0.753 \le \sigma^2 \le 2.819$
**11-41.** No. Since $1.2129 < 2.231$, do not reject.
**11-43. a.** Since $\chi^2 = 37.24 > \chi^2_U = 30.1435$, reject $H_0$.
  **b.** $P(x \ge 3) = 0.0230 + 0.0026 + 0.0002 = 0.0258$
**11-45.** Since $1.0757 < 1.5126$, do not reject the null hypothesis.
**11-47.** Since $91.848 > 69.2299$, do not reject the null hypothesis.
**11-49. a.** Since $\chi^2 = 103.66 < 169.278$, do not reject the null hypothesis.
  **b.** Type II error
**11-51. a.** Since $33.8887 < 36.8182$, reject the null hypothesis.
  **b.** Since $F = 1.3777 < 1.4464$, do not reject the null hypothesis.

## CHAPTER 12

**12-1. a.** $SSW = SST - SSB = 9{,}271{,}678{,}090 - 2{,}949{,}085{,}157$
  $= 6{,}322{,}592{,}933$
  $MSB = 2{,}949{,}085{,}157/2 = 1{,}474{,}542{,}579$
  $MSW = 6{,}322{,}592{,}933/72 = 87{,}813{,}791$
  $F = 1{,}474{,}542{,}579/87{,}813{,}791 = 16.79$
  **b.** Because the $F$-test statistic $= 16.79 > F_\alpha = 2.3778$, we reject.
**12-3. a.** $H_0: \mu_1 = \mu_2 = \mu_3$
  $H_A:$ Not all $\mu_j$ are equal
  **b.** Because $F = 9.84 >$ critical $F = 3.35$, we reject. Because $p$-value $= 0.0006 < \alpha = 0.05$, we reject.
  **c.** Critical range $= 6.0$; $\mu_1 > \mu_2$ and $\mu_3 > \mu_2$
**12-5. a.** $df_B = 3$
  **b.** $F = 11.1309$
  **c.** $H_0: \mu_1 = \mu_2 = \mu_3 = \mu_4$
  $H_A:$ At least two population means are different
  **d.** Since $11.1309 > 2.9467$, reject $H_0$.
**12-7. a.** Because $F = 5.03 > 3.885$, reject $H_0$.
  **b.** Population 1 and 2 means differ; no other differences.
**12-9. a.** Since $F = 7.131 > F_{\alpha=0.01} = 5.488$, reject.
  **b.** $CR = 1{,}226.88$; Venetti mean is greater than Edison mean.
**12-11. a.** Since $F = 1{,}459.78 > 3.885$, reject $H_0$.
  **b.** Critical range $= 9.36$
**12-13. a.** Since $10.48326 > 5.9525$, reject $H_0$ and conclude that at least two population means are different.
  **b.** Critical range $= 222.02$; eliminate Types D and A.
**12-15. a.** Since $0.01 < 0.05$, reject $H_0$ and conclude that at least two population means are different.
  **b.**

| | Absolute Differences | Critical Range | Significant? |
|---|---|---|---|
| Mini 1 − Mini 2 | 0.633 | 1.264 | No |
| Mini 1 − Mini 3 | 1.175 | 1.161 | Yes |
| Mini 2 − Mini 3 | 1.808 | 1.322 | Yes |

Student reports will vary, but they should recommend either 1 or 2, since there is no statistically significant difference between them.
  **c.** $0.978$——$2.678$ cents per mile; $\$293.40$——$\$803.40$ annual savings

**12-17.** **a.** $p$-value $= 0.000 < \alpha = 0.05$
 **b.** Average lifetimes differ between Delphi and Exide and also between Delphi and Johnson. There is not enough evidence to indicate that the average lifetimes for Exide and Johnson differ.

**12-19.** **a.** $H_0: \mu_1 = \mu_2 = \mu_3 = \mu_4$
  $H_A$: At least two population means are different
  $H_0: \mu_{b1} = \mu_{b2} = \mu_{b3} = \mu_{b4} = \mu_{b5} = \mu_{b6} = \mu_{b7} = \mu_{b8}$
  $H_A$: Not all block means are equal
 **b.** $F_{\text{Blocks}} = 46.87669$; $F_{\text{Groups}} = 13.8906$
 **c.** Since $46.876 > 2.487$, reject $H_0$.
 **d.** Since $p$-value $0.0000326 < 0.05$, reject.
 **e.** $LSD = 5.48$

**12-21.** Because $F = 14.3 > $ Critical $F = 3.326$, reject and conclude that blocking is effective.
Because $F = 0.1515 < $ Critical $F = 4.103$, do not reject.

**12-23.** **a.** Because $F = 32.12 > F_{\alpha=0.01} = 9.78$, reject the null hypothesis.
 **b.** Because $F = 1.673 < F_{\alpha=0.01} = 10.925$, do not reject the null hypothesis.

**12-25.** **a.** Because $F = 22.32 > F_{\alpha=0.05} = 6.944$, reject the null hypothesis.
 **b.** Because $F = 14.185 > F_{\alpha=0.05} = 6.944$, reject the null hypothesis.
 **c.** $LSD = 8.957$

**12-27.** **a.** Yes, $p$-value $= 0.000 < \alpha = 0.05$
 **b.** Yes, $p$-value $= 0.004 < \alpha = 0.05$
 **c.** $LSD = 1.55$; $\mu_1 < \mu_3$ and $\mu_2 < \mu_3$

**12-29.** **a.** No, $p$-value $= 0.628 > 0.05$
 **b.** Yes, $p$-value $= 0.000 < \alpha = 0.05$
 **c.** $LSD = 372.304$; $\mu_1 > \mu_2 > \mu_3$

**12-31.** **a.** $p$-value $= 0.854 > \alpha = 0.05$. Therefore, fail to reject $H_0$.
 **b.** Since $F = 47.10 > F_{0.05} = 5.143$, reject $H_0$.
 **c.** $p$-value $= 0.039 < \alpha = 0.05$. Therefore, reject $H_0$.

**12-33.** **a.** Since $0.4617 < 3.8853$, do not reject $H_0$.
 **b.** Since $2.3766 < 3.8853$, do not reject $H_0$.
 **c.** Since $5.7532 > 4.7472$, reject $H_0$.

**12-35.** **a.** Since $F = 39.63 > F_{0.05} = 3.633$, reject $H_0$.
 **b.** Since $F = 2.90 < F_{0.05} = 9.552$, fail to reject $H_0$.
 **c.** Since $F = 3.49 < F_{0.05} = 9.552$, fail to reject $H_0$.
  Since $F = 57.73 > F_{0.05} = 9.552$, reject $H_0$.

**12-37.** **a.** Because $F = 1.016 < F_{\alpha=0.05} = 2.728$, do not reject.
 **b.** Because $F = 1.157 < F_{\alpha=0.05} = 3.354$, do not reject.
 **c.** Because $F = 102.213 > F_{\alpha=0.05} = 3.354$, reject.

**12-39.** **a.** Since $p$-value $= 0.0570 > 0.01$, do not reject.
 **b.** Since $2.9945 < 6.0129$, do not reject.
 **c.** Since $p$-value $= 0.4829 > 0.1$, do not reject.

**12-41.** **a.** $\alpha = 0.025 < p$-value $= 0.849$. Therefore, fail to reject.
 **b.** Since $F = 15.61 > F_{0.025} = 3.353$, reject.
 **c.** Since $-2.0595 < t = -0.69 < 2.0595$, fail to reject.

**12-49.** **a.** $\alpha = 0.05 < p$-value $= 0.797$. Therefore, fail to reject.
 **b.** Since $F = 25.55 > F_{0.05} = 3.855$, reject.
 **c.** Since $F = 0.82 < F_{0.05} = 3.940$, do not reject.

**12-51.** **a.** Since $F = 3.752 > 2.657$, reject.
 **b.** Critical range $= 9.5099$; $\mu_I > \mu_B$, $\mu_I > \mu_{IF}$, and $\mu_I > \mu_G$

**12-53.** **a.** Since $0.59153 < 2.16795$, do not reject $H_0$, and conclude that there is no interaction.
 **b.** Since $0.74628 < 3.0648$, do not reject $H_0$.

**12-55.** **a.** Since $p$-value $= $ essentially zero $< 0.05$, reject $H_0$.
 **b.** Tukey-Kramer critical range $= 1,504.72$
 **c.** $n = 1.96^2(6,368,664)/500^2 = 97.8634$, or 98

**12-57.** **a.** Calculated $F = 7.4381 > $ Critical $F = 2.7249$, so reject the null hypothesis.
 **b.** Tukey-Kramer critical range $= 8.478$

## CHAPTER 13

**13-1.** Because $\chi^2 = 6.4607 < 13.2767$, do not reject.

**13-3.** Because $\chi^2 = 227.81 > 18.3070$, reject.

**13-5.** Because the calculated value of $616.987 > 15.0863$, reject.

**13-7.** Because the calculated value of $50.019 > $ critical value of $12.8325$, we reject $H_O$.

**13-9.** **a.** Since chi-square statistic $= 3.379443 < 11.3449$, do not reject.
 **b.** Based on the test, we have no reason to conclude that the company is not meeting its product specification.

**13-11.** Chi-square value $= 0.3647$; do not reject.

**13-13.** $H_0$: The arrival distribution is Poisson distributed.
 $H_A$: The arrival distribution is not Poisson distributed.
 Since chi-square $= 3.6549 < 7.3778$, do not reject.

**13-15.** Because the calculated value of $1.97433$ is less than the critical value of $14.0671$, do not reject.

**13-17.** $\chi^2 = 50.315 > 21.666$; reject.

**13-19.** **a.** $H_0$: The row and column variables are independent.
  $H_A$: The row and column variables are not independent.
 **b.** The expected cell values for all cells are

| | $C_1$ | $C_2$ | Total |
|---|---|---|---|
| $R_1$ | 114.35 | 143.65 | 258 |
| $R_2$ | 146.70 | 184.30 | 331 |
| $R_3$ | 175.95 | 221.05 | 397 |
| Sum | 437 | 549 | 986 |

 **c.** The test statistic is $104.905$.
 **d.** $\chi^2 = 104.905 > 5.9915$
 **e.** Since $\chi^2_{0.005} = 10.5965 < \chi^2 = 104.905$, then $p$-value $< 0.005$.

**13-21.** $\chi^2 = 300.531 > 16.8119$; reject.

**13-23.** $\chi^2 = 86.9 > 12.5916$; reject.

**13-25.** Since $1.2987 < 9.4877$, do not reject.

**13-27.** $\chi^2 = 27.4510 > 13.2767$; reject.

**13-29.** **a.** $\chi^2 = 29.302 > 5.9915$; reject.
 **b.** Since $-1.96 < z = 0.523 < 1.96$, do not reject.

**13-31.** $p$-value $= 0.000 < \alpha = 0.05$

**13-33.** **a.** One option is to combine Salt Lake City and Toronto into one group and wiring and sound into one group.
 **b.** $H_0$: Type of warranty problem and manufacturing plant are independent.
  $H_A$: Type of warranty problem and manufacturing plant are not independent.
  $6.7582 < 9.2104$; do not reject.

**13-39.** Since $59.464 > 7.81472$, reject.

**13-41.** $25.2975 > 16.8118$; reject.

**13-43.** $\chi^2 = 3.68 > 15.0863$; do not reject.

**13-45.** $\chi^2 = 50.91 > 5.9915$; reject.

**13-47.** $\chi^2 = 125.1905 > 9.4877$; reject $H_0$ and conclude that blood pressure loss is not normally distributed with $\mu = 10$ and $\sigma = 4$.

**13-49.** $\chi^2 = 1.155 < 7.8147$; do not reject.

## CHAPTER 14

**14-1.** $H_0: \rho \le 0.0$
 $H_A: \rho > 0.0$
 $\alpha = 0.05$, $t = 2.50$, $2.50 > 1.7709$; reject.

**14-3.** **a.** $r = 0.206$

**b.** $H_0: \rho = 0$

$H_A: \rho \neq 0$

$\alpha = 0.10, -1.8595 < t = 0.59 < 1.8595$; do not reject $H_0$.

**14-5.** **a.** a positive linear relationship

**b.** $r = 0.9239$

**c.** $H_0: \rho \leq 0$

$H_A: \rho > 0$

$df = 10 - 2 = 8$

Since $6.8295 > 2.8965$, reject $H_0$.

**14-7.** **a.** fairly strong positive correlation

**b.** $H_0: \rho \leq 0$

$H_A: \rho > 0$

$\alpha = 0.01, t = 7.856 > 2.4066$; reject the null hypothesis.

**14-9.** **a.** The dependent variable is the average credit card balance. The independent variable is the income variable.

**b.** does not appear to be a strong relationship

**c.** $H_0: \rho = 0.0$

$H_A: \rho \neq 0.0$

$\alpha = 0.05, t = -1.56 > -2.1604$; do not reject.

**14-11.** Because $9.18 > t = 1.9844$, reject $H_0$.

**14-13.** **a.** $-0.305$

**b.** Since $p$-value $= 0.0228 < \alpha = 0.025$ reject $H_0$.

**14-15.** **a.** a positive linear relationship

**b.** 0.51

**c.** $t = 15.87 > 1.647$; reject the null hypothesis.

**14-17.** **a.** a positive linear relationship

**b.** $r = 0.7540$

$H_0: p$-value $= 0.0073 < 0.05$; reject $H_0$.

$\hat{y} = 57.836 + 13.060x$

**c.** $t = 3.44$; reject.

**14-19.** **a.** $\hat{y} = 36.7 + 8.33x$

**b.** $SE = 1,467$

$SST = 7,050$

$$R^2 = \frac{SSR}{SST} = 1 - \frac{SSE}{SST} = 1 - \frac{1,467}{7,050} = 0.79$$

**c.** $S_e = \sqrt{\dfrac{SSE}{n-2}} = \sqrt{\dfrac{1,467}{8}} = 13.54$

**d.** $s_{b_1} = \dfrac{s_e}{\sqrt{\Sigma x^2 - \dfrac{(\Sigma x)^2}{n}}} = \dfrac{13.54}{\sqrt{2,452 - \dfrac{154^2}{10}}} = 1.51$

**e.** $t = 5.52 > 2.8965$; reject.

**14-21.** **b.** $\Sigma(y - \hat{y})^2 = 7.60$

$\Sigma(y - \hat{y})^2 = 7.75$

$\Sigma(y - \hat{y})^2 = 7.65$

**c.** the first equation

**d.** $\hat{y} = 0.8 + 1.60x$

**14-23.** **a.** a positive linear relationship

**b.** $\hat{y} = 4.5 + 1.101x$

**c.** $t = 7.7171 > 2.1009$; reject.

**14-25.** **a.** $\hat{y} = 171,205.83 + 25.316x$

**b.** $t = 7.08 > 2.0930$; reject.

**c.** $R^2 = 0.725$

**14-27.** **a.** $\hat{y} = 95.7391 + 0.8649x$

**b.** $R^2 = 0.5722$

**c.** $t = 8.095 > 2.68$; reject.

**d.** $t = 8.0983 > 2.68$

**14-29.** **a.**

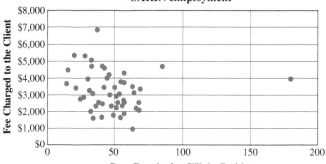

**SARIN Employment**

(y-axis: Fee Charged to the Client, $0 to $8,000; x-axis: Days Required to Fill the Position, 0 to 200)

**b.** Because $t = -0.787 > -2.01$, do not reject $H_0$ and conclude that the correlation is not significant.

**c.** 1.27% of the variation

**14-31.** **a.** a negative linear relationship

**b.** Since $-5.8061 < -2.0484$, reject $H_0$ and conclude that there is a correlation between curb weight and highway mileage.

**c.** $\hat{y} = 46.3298 + (-0.006)(4,012) = 22.2578$

**14-33.** **b.** Since $p$-value $= 0.0000 < \alpha < 0.10$, reject.

**c.** 0.7925——1.057314.34

**14-35.** $R^2 = 0.0105$

**14-37.** **a.** $\hat{y} = 200 + 150(48) = 7,400$

**b.** 7,399.37——7,400.63

**c.** 7,397.26——7,402.74

**14-39.** **a.** (93.15, 99.87)

**b.** (98.84, 105.76)

**c.** The margin of error for part a is 3.3687 and for part b is 3.4647.

**14-41.** **a.** $\hat{y} = -4.7933 + 1.0488x$

**b.** $\hat{y} = -4.7933 + 1.0488(100) = 100.09$

**c.** 82.205——117.975

**d.** 94.88——105.30

**14-43.** **a.** $R^2 = 0.872$, so 87.2% explained

**b.** Since $t = 7.814 >$ critical $t = 2.2622$, reject $H_0$ and conclude that club speed is significant in explaining distance variation.

**c.** For each additional 1 mph of club speed, the average driving distance increases by 3.006 yard.

**14-45.** **a.** $\hat{y} = 171,205.828 + 25.316x$

**b.** $17.829 < \beta_1 < 32.803$

**c.** 532,907——568,985

**d.** 478,607——623,286

**14-47.** **a.** $\hat{y} = 252.846 + 250.247x$

**b.** A $b_0$ value of 252.846 would be the fixed cost of production, even if order size was 0; $b_1$ of 250.247 means that for every one unit increase in order size, total production cost will increase by $250.247 on average.

**c.** $t = 28.61 > 2.6822$; reject $H_0$.

**d.** $\hat{y} = 252.846 + 250.247(30) = \$7,760.26$

**e.** \$7,601.12——\$7,919.41

**14-57.** **a.** Since $4.7757 > 2.0484$, reject.

**b.** $R^2 = (0.67)^2 = 0.4489$

**14-59.** **a.** a weak negative linear relationship

**b.** $-0.59974$

**c.** Since $p$-value $= 0.02557 <$ alpha $= 0.10$ reject $H_0$.

**14-61.** **a.** 0.01498——0.01502. Income is a significant variable, since the interval does not contain 0.

**14-63. a.** Since $53.8462 > 1.649$, reject $H_0$.
  **b.** $3.24 \pm 1.649(0.2)$; $2.91$——$3.57$
  **c.** GPS lower than expected
  **d.** $\hat{y} = 1.0 + 0.028(65) = 2.82$
     $2.491$——$3.149$
**14-65. a.** a positive linear relationship
  **b.** $\hat{y} = -19.3726 + 2.9026x$
**14-67.** 9.3% of the variation.
  The true population regression coefficient could be 0 because $p$-value $= 0.052 > \alpha/2 = 0.025$.
**14-69. a.** The independent variable is advertising and the dependent variable is sales.
  **c.** Since $-0.4173 < 2.1098$, do not reject.

## CHAPTER 15

**15-1. a.** $\hat{y} = 87.79 - 0.9705x_1 + 0.0023x_2 - 8.7233x_3$
  **b.** $F = 5.3276 > F_{0.05} = 3.0725$; also, $p$-value $= 0.00689 <$ any reasonable alpha. Therefore, reject $H_0: \beta_1 = \beta_2 = \beta_3 = 0$.
  **c.** $R^2 = 0.432$
  **d.** $x_1$ ($p$-value $= 0.1126 > \alpha = 0.05$; fail to reject $H_0: \beta_1 = 0$) and $x_3$ ($p$-value $= 0.258 > \alpha = 0.05$; fail to reject $H_0: \beta_3 = 0$) are not significant.
  **e.** $b_2 = 0.002$; $\hat{y}$ increases 0.002 for each one-unit increase of $x_2$.
     $b_3 = -8.7233$; $\hat{y}$ decreases 8.7233 for each one-unit increase of $x_3$.
  **f.** The confidence intervals for $\beta_1$ and $\beta_3$ contain 0.
**15-3. a.** $b_1 = -412$; $b_2 = 818$; $b_3 = -93$; $b_4 = -71$
  **b.** $\hat{y} = 22{,}167 - 412(5.7) + 818(61) - 93(14) - 71(3.39) = 68{,}173.91$
**15-5. a.** $y_i = 5.05 - 0.051x_1 + 0.888x_2$

  **b.**

|       | $y_i$ | $x_1$ |
|-------|-------|-------|
| $x_1$ | 0.206 |       |
| $x_2$ | 0.919 | 0.257 |

  $H_0: \rho = 0$
  $H_A: \rho \neq 0$
  $\alpha = 0.05$, $t = 0.5954$, $-2.306 < t = 0.5954 < 2.306$; fail to reject $H_0$.
  **c.** $H_0: \beta_1 = \beta_2 = 0$
     $H_A:$ At least one $\beta_i \neq 0$
     $\alpha = 0.05$, $F = 19.07$
     Since $F = 19.07 > 4.737$, reject $H_0$.

  **d.**

| Predictor | Coef    | SE Coef | T     | P     | VIF |
|-----------|---------|---------|-------|-------|-----|
| Constant  | 5.045   | 8.698   | 0.58  | 0.580 |     |
| $x_1$     | -0.0513 | 0.2413  | -0.21 | 0.838 | 1.1 |
| $x_2$     | 0.8880  | 0.1475  | 6.02  | 0.001 | 1.1 |

**15-7. a.** $\hat{y} = -977.1 + 11.20(\text{52-week high}) + 117.72(\text{PE})$
  **b.** $H_0: \beta_1 = \beta_2 = 0$
     $H_A:$ At least one $\beta_i \neq 0$
     $\alpha = 0.05$, $F = 39.80 > 3.592$; reject.
  **c.** $\hat{y} = 1{,}607$
**15-9. a.** $\hat{y} = 139.684 + 0.060x_1 - 1.170x_2 + 0.108x_3 - 0.007x_4$
  **b.** Since Significance $F = 0.00001135 < 0.05$, reject $H_0$. The overall model is significant.
  **c.** The number of websites has $p$-value $= 0.000$, which is less than alpha $= 0.05$, so it is significant. None of the other three variables is significant.

**15-11. a.** There is a positive linear relationship between team win/loss percentage and game attendance, opponent win/loss percentage and game attendance, and games played and game attendance. There is a negative linear relationship between temperature and game attendance.
  **b.** There is a significant relationship between game attendance and team win/loss percentage and games played.
  **c.** Attendance $= 14{,}122.24 + 63.15(\text{win/loss\%}) + 0.10$ (opponent win/loss) $+ 31.51(\text{games played}) - 55.46(\text{temperature})$
  **d.** $R^2 = 0.7753$, so 77.53% is explained.
  **e.** $H_0: \beta_1 = \beta_2 = \beta_3 = \beta_4 = 0$
     $H_A:$ At least one $\beta_i$ does not equal 0.
     Significance $F = 0.00143$; reject $H_0$.
  **f.** For team win/loss %, $p$-value $= 0.0014 < 0.08$.
     For opponent win/loss %, $p$-value $= 0.4953 > 0.08$.
     For games played, $p$-value $= 0.8621 > 0.08$.
     For temperature, $p$-value $= 0.3909 > 0.08$.
  **g.** 1,184.125; interval of $\pm 2(1{,}184.125)$

  **h.**

|                            | VIF   |
|----------------------------|-------|
| Team win/loss percentage   | 1.509 |
| Temperature                | 1.570 |
| Games played               | 1.964 |
| Opponent win/loss percentage | 1.314 |

  Multicollinearity is not a problem, since no VIF is greater than 5.
  **i.** Team Win/Loss %   $30.27$——$96.03$
     Opp. Win/Loss %   $21.41$——$41.60$
     Games Played   $358.35$——$421.37$
     Teamperature   $-192.13$——$81.21$
**15-15. a.** $x_2 = 1$, $\hat{y} = 145 + 1.2(1{,}500) + 300(1) = 2{,}245$
  **b.** $x_2 = 0$, $\hat{y} = 145 + 1.2(1{,}500) + 300(0) = 1{,}945$
  **c.** $b_2$ indicates the average premium paid for living in the city's town center.
**15-17. a.** decrease by 0.003
  **b.** increase by 4.56, holding the weight constant
  **c.** $\hat{y} = 34.2 - 0.003x_1 + 4.56(1) = 38.76 - 0.003x_1$
  **d.** $\hat{y} = 34.2 - 0.003(3{,}973) + 4.56(0) = 22.28$
  **e.** Incorporating the dummy variable essentially gives two regression equations with the same slope but different intercepts.
**15-19. a.** $y = 115.33 - 2.13(\text{Asian}) - 1.43(\text{European})$
  **b.** $\beta_0 = $ Average PP100 of U.S. vehicles $= 115.33$;
     $\beta_1 = -2.13$, implying that Asian vehicles have fewer defects reported than U.S. cars;
     $\beta_2 = -1.43$, the difference in the average PP100 between European and U.S. vehicles. European vehicles have fewer defects reported than do U.S. cars.
  **c.** Since Significance $F = 0.95288 > 0.05$, the overall model is not statistically significant. Neither of the dummy variables individually has a $p$-value $< 0.05$, so neither is statistically significant.
**15-21. a.** There appears to be a weak positive linear relationship between hours and net profit. There appears to be a weak negative linear relationship between client type and net profit.
  **b.** $\hat{y} = -1{,}012.0542 + 69.1471x_1$
  **c.** $p$-value $= 0.0531$; $R^2$ is only 0.3549.
**15-23. a.** $\hat{y} = 390 - 0.263x_1 - 37.0x_2$
     $H_0: \beta_1 = 0$
     $H_A: \beta_1 \neq 0$
  **b.** $\alpha = 0.05$. Since $t = -20.45 < -1.9921$, reject $H_0$.
  **c.** $\hat{y} = 390 - 37.0x_1 + 0.263x_2 = 390 - 37.0(1) + 0.263(500) = 484.5 \approx 485$

**15-25. a.** A linear line is possible; nonlinear is more likely.
    **b.** $\hat{y} = 4.937 + 1.2643x$; $p$-value $= 0.015 < \alpha = 0.05$, reject.
    **c.** $\hat{y} = -25.155 + 18.983 \ln x$ R-Sq for nonlinear model is higher than R-Sq for linear model.

**15-27. b.** two quadratic models; interaction between $x_2$ and quadratic relationship between $y$ and $x_2$;
$$\hat{y}_i = 4.710 - 2.121x_1 + 512x_1^2 + 1.397x_1x_2 + 0.529x_1^2x_2$$
    **c.** $\beta_3 x_1 x_2$ and $\beta_4 x_1^2 x_2$. So you must conduct two hypothesis tests:
      **i.** $H_0: \beta_3 = 0$
        $H_A: \beta_3 \neq 0$
        $\alpha = 0.05$, $p$-value $= 0.496$; fail to reject $H_0$.
      **ii.** For $\beta_4 = 0$, $p$-value $= 0.001$.
    **d.** There is an interaction between $x_2$ and the quadratic relationship between $x_1$ and $y$.

**15-29. a.** The complete model is $y_i = \beta_0 + \beta_1x_1 + \beta_2x_2 + \beta_3x_3 + \beta_4x_4 + \varepsilon_i$. The reduced model is $y_i = \beta_0 + \beta_1x_1 + \beta_2x_2 + \varepsilon_i$, $H_0: \beta_3 = \beta_4 = 0$, $H_A$: At least one $\beta_i \neq 0$. $SSE_C = 201.72$. So $MSE_C = SSE_C/(n - c - 1) = 201.72/(10 - 4 - 1) = 40.344$ and $SSE_R = 1,343$. $\alpha = 0.05$, $F = 14.144$; $14.144 > 5.786$; reject $H_0$.
    **b.** The complete model is $y_i = \beta_0 + \beta_1x_1 + \beta_2x_2 + \beta_3x_3 + \beta_4x_4 + \varepsilon_i$. The reduced model is $y_i = \beta_0 + \beta_1x_3 + \beta_2x_4 + \varepsilon_i$. $SSE_C = 201.72$. So $MSE_C = SSE_C/(n - c - 1) = 201.72/(10 - 4 - 1) = 40.344$ and $SSE_R = 494.6$. $H_0: \beta_1 = \beta_2 = 0$; $H_A$: At least one $\beta_i \neq 0$; $\alpha = 0.05$, $F = 3.63$. The numerator degrees of freedom are $c - r = 4 - 2 = 2$ and the denominator degrees of freedom are $n - c - 1 = 10 - 4 - 1 = 5$. $p$-value $= P(F \geq 3.630) = 0.1062$; fail to reject.

**15-31. a.** two dummy variables:
    $x_2 = 1$ if manufacturing, 0 otherwise
    $x_3 = 1$ if service, 0 otherwise
    Net profit $= -586.256 + 22.862x_1 + 2,302.267x_2 + 1,869.813x_3$
    **b.** Net profit $= 5,828.692 - 334.406x_1 + 4.557139_1$ sq $+ 2,694.801x_2 + 1287.495x_3$

**15-33. a.**

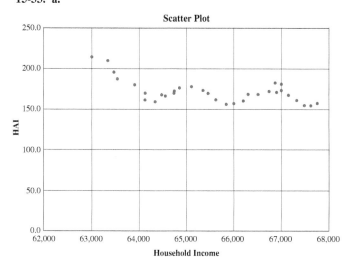

Scatter Plot

    **b.** $\hat{y} = 15422.57 - 0.460x + 0.0000035x^2$
    **c.** 52.2%, Significance $F = 0.00002227 < 0.05$. The scatterplot in part a indicates that a higher-order model might be appropriate.

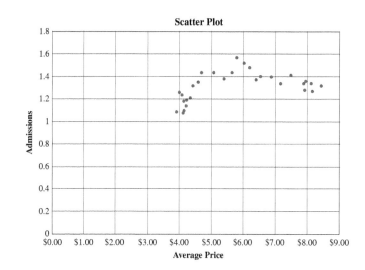

Scatter Plot

**15-35. a.** There appear to be four "curves," which suggests a fifth-order polynomial.
    **b.** $y = -6.571 - 1.291(\text{price}) + 6.443$ (square root of price) Overall $p$-value $= 0.000000167$. The individual coefficients have $p$-values $= 0.000$, which indicates that both variables are statistically significant.

**15-37. a.** $y = 2,857 - 26.36(\text{52-week high}) - 80.63(\text{PE}) + 0.1151(\text{52-week high})^2 + 2.3133(\text{PE})^2 + 0.5421$ (52-week-high x PE)
    **b.** $0.067 < 0.10$; reject $H_0$.
    **c.** Since $F = 38.107 > 3.344$, reject $H_0$. There is sufficient evidence to conclude that the PE ratio is useful in predicting the average market value.

**15-39. a.** $x_3$ is the most highly correlated with the dependent variable and should enter the model first.
    **b.** $\hat{y} = -3.097 + 0.973x_3$
    **c.** The difference between the two models is that the full model has two additional independent variables. Viewing the $p$-values and the adjusted $R^2$, however, it appears that $x_1$ and $x_2$ are not significant additions to the model. Based on the adjusted $R^2$, it is apparent that the equation developed used Standard Selection explains the most variation in the dependent variable adjusting for the number of independent variables in the model.

**15-41. a.** $\hat{y} = 6.36 + 1.52x_1$
    **b.** Here, no variables enter the model.
    **c.** $\hat{y} = 6.36 + 1.52x_1$
    **d.** $\hat{y} = 6.36 + 1.52x_1$
    **e.** $\hat{y} = 2.546 + 1.01x_1 + 1.6x_2$

**15-43. a.** The variable, number of customers, has the highest correlation with $CO_2$ emissions at $-0.828$. It would be the first variable to enter the model.
    **b.** $\hat{y} = 1,9270,266 - 0.1455(\text{customers}) + 416,725(\text{average price})$
    Significance $F = 0.0005$. Both independent variables are significant because of low $p$-values. $R^2$ is 0.851.

**15-45. a.** Calls received, ads placed, and then airline bookings
    **b.** $y = -18.3279 + 1.0679(\text{calls received previous week})$

**15-47. a.** $\hat{y} = -59.0 - 0.277(\text{weight}) + 0.950(\text{abdomen}) + 0.651(\text{thigh})$
    **b.** $\hat{y} = -59.00 - 0.277(170) + 0.950(90) + 0.651(60) = 18.47$

**15-49.**

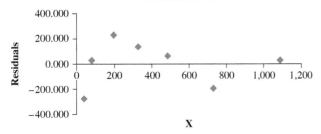

**X Residual Plot**

**15-51. a.** $\hat{y} = -16.02 + 2.1277x$
  **b.** $p$-value $= 0.000 < 0.05$
**15-55. a.** Calls $= -269.838 + 4.953$(ads previous week) $+$ $0.834$(calls previous week) $+ 0.089$(airline bookings). The overall model is not significant, and none of the independent variables are significant.
  **b.** The assumption of constant variance has not been violated.
  **c.** It is inappropriate to test for randomness using a plot of the residuals over time, since the weeks were randomly selected and are not in sequential, time-series order.
  **d.** Model meets the assumption of normally distributed error terms.
**15-57. a.** $\hat{y} = -59.00 - 0.277x_2 + 0.950x_5 + 0.65x_7$
  **b.** The residual plot supports the choice of the linear model.
  **c.** The residuals do have constant variances.
  **d.** The linear model appears to be sufficient.
  **e.** The error terms are normally distributed.
**15-63. a.** The relationship between the dependent variable and each independent variable is linear.
  **b.** The residuals are independent.
  **c.** The variances of the residuals are constant over the range of the independent variables.
  **d.** The residuals are normally distributed.
**15-65. a.** The average $y$ increases by three units, holding $x_2$ constant.
  **b.** Yes.
  **c.** The coefficient of $x_1$ indicates that the average $y$ increases by 7 units when $x_2 = 1$.
  **d.** The coefficient of $x_1$ indicates that the average $y$ increases by 11 units when $x_2 = 1$.
  **e.** Those coefficients affected by the interaction terms have conditional interpretations.
**15-69. a.** The critical $t$ for all pairs would be 2.1604; correlated pairs:
    Volumes sold ($y$) $-$ Production expenditures
    Volumes sold ($y$) $-$ Number of reviewers
    Volumes sold ($y$) $-$ Pages
    Volumes sold ($y$) $-$ Advertising budget
  **b.** All $p$-values $> 0.05$
  **c.** Critical $F = 3.581$; since $F = 9.1258 > 3.581$, conclude that the overall model is significant.
  **e.** $\pm 2(24,165.9419) = \pm 48,331.8$
  **f.** Constant variance assumption is satisfied.
  **g.** The residuals appear to be approximately normally distributed.
  **h.** The model satisfies the normal distribution assumption.
**15-71.** The $t$-critical for all pairs would be $\pm 2.0687$; correlated pairs are:
  For family size and age
  For purchase volume and age
  For purchase volume and family income

**15-73.** Significance $F = 0.0210 = p$-value.
**15-75.** Age entered the model.
    The $R^2$ at Step 1 was 0.2108, and the standard error at Step 1 was 36.3553. The $R^2$ at Step 2 is 0.3955, and the standard error at Step 2 is 32.5313.
**15-77.** Other variables that enter into the model partially overlap with the other included variables in their ability to explain the variation in the dependent variable.
**15-79. a.** normal distribution of the residuals
  **b.** The selected independent variables are not highly correlated with the dependent variable.
**15-83.** Only the drivers' gender (male or female), years of education, and the dummy variable associated with not being required to wear a seat belt are significant in the model. However, the available data explain only 14.959% of the variation in the year of the vehicle.

## CHAPTER 16

**16-3.** Generally, we can use quantitative forecasting techniques whenever we have historical data related to the variable of interest and we believe that the historical patterns will continue into the future.
**16-7. a.** 6 months
  **b.** a medium-term forecast
  **c.** a month
  **d.** 12 months
**16-9.**

| Year | Radio Advertising | Index | Newspaper Ad | Index |
|------|-------------------|--------|--------------|-------|
| 1 | 300 | 100.00 | 400 | 100 |
| 2 | 310 | 103.33 | 420 | 105 |
| 3 | 330 | 110.00 | 460 | 115 |
| 4 | 346 | 115.33 | 520 | 130 |
| 5 | 362 | 120.67 | 580 | 145 |
| 6 | 380 | 126.67 | 640 | 160 |
| 7 | 496 | 165.33 | 660 | 165 |

**16-11.**

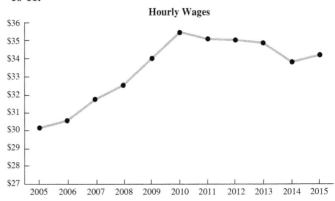

**Hourly Wages**

**Average Material Cost**

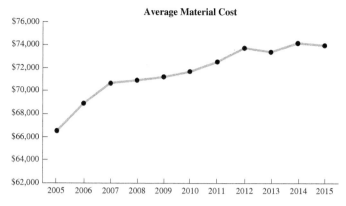

Both graphs show an upward trend with random components. The average material cost time series exhibits a cyclical component with a recurrence period of five years. The hourly wages time series may be indicating a cyclical component with a recurrence period of nine years. However, there are not enough data to determine if this pattern will repeat.

**16-13.** Two distinct areas exhibit a linear trend. In the time period 1955–1980 the trend is generally positive. In the time period 1981–2015 the trend is generally negative. A cyclical component is evidenced by the wave form, which reoccurs approximately every five years. Finally, random components are evident throughout the time series, indicated by the random deviations from a "smooth" wave form.

**16-15.** The following output is found using XLSTAT:

| Durbin-Watson Calculations | |
|---|---|
| Sum of Squared Difference of Residuals | 1057045997 |
| Sum of Squared Residuals | 1405827136 |
| Durbin-Watson Statistic | 0.751903253 |

Reject the hypothesis and conclude there is evidence of autocorrelation.

**16-19. a.** Unadjusted: $23,424.557 + 420.894(49) = 44,048.363$
Adjusted: $(25,249.568 + 339.974(49))0.849 = 35,580.142$
**b.** Both the *MAD* and *MSE* are lower for the seasonally adjusted model.

**16-21.**

| Quarter | Seasonal Index |
|---|---|
| 1 | 1.035013 |
| 2 | 1.020898 |
| 3 | 0.959934 |
| 4 | 0.984154 |

**16-23.**

| Quarter | Period | Seasonally Unadjusted Forecast | Seasonal Index | Seasonally Adjusted Forecast |
|---|---|---|---|---|
| Quarter 1 2017 | 17 | 250.15 | 1.035 | 258.91 |
| Quarter 2 2017 | 18 | 265.17 | 1.0209 | 261.52 |
| Quarter 3 2017 | 19 | 262.18 | 0.9599 | 251.68 |
| Quarter 4 2017 | 20 | 268.2 | 0.9842 | 263.95 |

**16-27. b.** The seasonal indexes are:

| Month | Index |
|---|---|
| 1 | 1.02349 |
| 2 | 1.07969 |
| 3 | 1.16502 |
| 4 | 1.12147 |
| 5 | 0.98695 |
| 6 | 0.83324 |
| 7 | 0.86807 |
| 8 | 0.91287 |
| 9 | 0.97699 |
| 10 | 1.07311 |
| 11 | 1.01382 |
| 12 | 0.94529 |

**c.** Forecast $= 1.98 + 0.0459(\text{month})$
**d.** $F_{25} = 1.98 + 0.0459(25) = 3.12$. Adjusted
$F_{25} = (1.02349)(3.12) = 3.19$
$F_{73} = 1.98 + 0.0458589(73) = 5.32$. Adjusted
$F_{73} = (1.02349)(5.32) = 5.44$

**16-29. a.** seasonal component to the data
**b.** $MSE = 976.34$, $MAD = 29.887$
**c.**

| Quarter | Index |
|---|---|
| 1 | 1.0084 |
| 2 | 0.9289 |
| 3 | 1.0811 |
| 4 | 0.9816 |

**d.**

| 2015 | Period | Forecast |
|---|---|---|
| Qtr 1 | 13 | 256.5620033 |
| Qtr 2 | 14 | 260.0884382 |
| Qtr 3 | 15 | 263.614873 |
| Qtr 4 | 16 | 267.1413079 |

**e.**

| Period | Adjusted Forecast |
|---|---|
| 13 | 258.7171 |
| 14 | 241.5964 |
| 15 | 284.9941 |
| 16 | 262.2259 |

$MSE = 912.7345$, $MAD = 29.7984$
**f.** The adjusted model has a lower *MSE* and *MAD*.

**16-31. a.** Forecast without transformation $= 36.0952 + 10.8714(16) = 210.0376$
Forecast with transformation $= 65.2986 + 0.6988(16)^2 = 244.1914$
Actual cash balance for month 16 was 305. The transformed model had a smaller error than the model without the transformation.
**b.** Model without transformation:

| For Individual Response y | |
|---|---|
| Interval Half Width | 48.27804189 |
| Prediction Interval Lower Limit | 161.7600534 |
| Prediction Interval Upper Limit | 258.3161371 |

Model with transformation:

| For Individual Response y | |
| --- | --- |
| Interval Half Width | 23.89550188 |
| Prediction Interval Lower Limit | 220.29634337 |
| Prediction Interval Upper Limit | 268.08734713 |

The model without the transformation has the wider interval.

**16-33. a.**

**January Total Retail Sales**

**b.** The *p*-value is essentially 0.000. Therefore, reject the null hypothesis. Enough evidence exists to indicate that there is a linear trend in these data.

**c.** $y = -14,782,800.17 + 8,174.06(\text{year})$
Forecast for 2017 is $-14,782,800.17 + 8,174.06(\text{year}) =$ $359,899.33.

**16-35. b.** A trend is present.

**c.** Forecast $= 136.78$, $MAD = 23.278$

**d.** Forecast $= 163.69$, $MAD = 7.655$

**e.** The double exponential smoothing forecast has a lower *MAD*.

**16-37. a.** The time series contains a strong upward trend, so a double exponential smoothing model is selected.

**b.** The equation is $\hat{y}_t = 19.364 + 0.7517t$. Since $C_0 = b_0$, $C_0 = 19.364$; $T_0 = b_1 = 0.7517$.

**c. Forecasts**

| Period | Forecast | Lower | Upper |
| --- | --- | --- | --- |
| 13 | 29.1052 | 23.9872 | 34.2231 |

**d.**

| Accuracy Measures | |
| --- | --- |
| *MAPE* | 8.58150 |
| *MAD* | 2.08901 |
| *MSE* | 6.48044 |

**16-39.** The forecast for 2017 is $38,260 and the forecast for 2018 is $40,522. The model selected by the Excel 2016 forecast method is double exponential smoothing with an alpha constant of 0.50 and a beta constant of 0.50.

**16-41. a.**

**High Temperatures**

There does not appear to be any trend component in this time series.

**b.** The forecast is calculated using the starting values $F_1 = F_2 = 99.1$. Then $F_3 = 0.25y_2 + (1 - 0.25)F_2 = 0.25(102.8) + 0.75(99.1) = 100.025$.

**c.** $MAD = 3.2652$

**d.** $F_{14} = 0.25y_{13} + (1 - 0.25)F_{13} = 0.25(101.3) + 0.75(100.22) = 100.49$

**16-43. b.** The forecast for March 21, 2015, is 571.364 with a 95% confidence interval of 492.84 to 649.89.

**16-45. a.** The double exponential smoothing model incorporates the trend effect.

**b.** From regression output, Initial constant $= 28,848$; Initial trend $= 2,488.96$. Forecast for period $17 = 72,450.17$; $MAD = 5,836.06$.

**c.** The *MAD* produced by the double exponential smoothing model at the end of month 16 is lower than the *MAD* produced by the single exponential smoothing model.

**d. and e.** Of the combinations considered, the minimum *MAD* at the end of month 16 occurs when alpha $= 0.05$ and beta $= 0.05$. The forecast for month 17 with alpha $= 0.05$ and beta $= 0.05$ is 71,128.45.

**16-47. a.** a seasonal component

**b.** The pattern is linear with a positive slope.

**c.** a cyclical component

**d.** a random component

**e.** a cyclical component

**16-51. a.** There appears to be an upward linear trend.

**b.** Forecast $= -682,238,010.3 + 342,385.3(\text{year})$ Since $F = 123.9719 > 4.6001$, conclude that there is a significant relationship.

**c.** $MAD = 461,216.7279$

**d.**

| Year | Forecast |
| --- | --- |
| 2016 | 4,929,275.00 |
| 2017 | 5,271,660.29 |
| 2018 | 5,614,045.59 |
| 2019 | 5,956,430.88 |
| 2020 | 6,298,816.18 |

**16-55. a.** The time series contains a strong upward trend, so a double exponential smoothing model is selected.

**b.** Since $C_0 = b_0$, $C_0 = -2,229.9$; $T_0 = b_1 = 1.12$.

**c.** Forecast(2008) $= 740.073$

**d.** $MAD = 89.975$

**16-57. b.** $y = 240.16 + 188.56(\text{time period})$

**c.** $t = 17.85 > 2.2281$; reject.

**e.** $MAD = 80.91$ or 81

**f.** Forecast for 2016 $= 2503$ (rounded) Forecast for 2017 $= 2691$ (rounded)

**16-59. b.** The estimated linear trend line equation is $y = 823.12 + 7.032x$.

**e.** Since $1.057 < 1.37$, reject $H_0$ and conclude that positive autocorrelation exists.

**16-61. a.** There appears to be a slight linear trend as well as a seasonal trend.

**d.**

| Month | Period | Forecast |
|---|---|---|
| January | 49 | 41,907.4368 |
| February | 50 | 42,247.4050 |
| March | 51 | 42,587.3731 |
| April | 52 | 42,927.3412 |
| May | 53 | 43,267.3072 |
| June | 54 | 43,607.2753 |
| July | 55 | 43,947.2434 |
| August | 56 | 44,287.2115 |
| September | 57 | 44,627.1796 |
| October | 58 | 44,967.1477 |
| November | 59 | 45,307.1157 |
| December | 60 | 45,647.0837 |

**e.** Deseasonalizing the data has increased $R^2$ and decreased the *MAD*.

| Month | Period | Unadjusted Forecast | Seasonal Index | Adjusted Forecast |
|---|---|---|---|---|
| January | 49 | 41,907.4368 | 0.8488 | 35,572.333 |
| February | 50 | 42,247.4050 | 0.8627 | 36,445.963 |
| March | 51 | 42,587.3731 | 0.7889 | 33,595.260 |
| April | 52 | 42,927.3412 | 0.8662 | 37,182.217 |

**16-63. a.** There does not appear to be a trend or seasonal component to these data.
**b.** The forecast for January 2016 is 12,350.
**c.** The alpha smoothing constant is 0.13, which indicates a single exponential smoothing model. The *MASE* is 0.56.
**d.** FORECAST.ETS returns a forecast for January 2016 of 12,350.
**e.** FORECAST.ETS.CONFINT gives a margin of error of 534, so the lower bound is 11,817 and the upper bound is 12,884.
**f.** FORECAST.ETS.STAT gives a *MASE* of 0.559. The *MASE* metric has a Statistic_Type of 4 in the FORE-CAST.ETS.STAT formula. Student comments will vary on the accuracy of the forecast, but a *MASE* of 0.56 indicates a fairly accurate forecast.

## CHAPTER 17

**17-1.** The hypotheses are $H_0: \widetilde{\mu} \geq 14$
$H_A: \widetilde{\mu} < 14$
$W = 36$
$n = 11, \alpha = 0.05$; reject if $W < 13$.
**17-3.** The hypotheses are $H_0: \widetilde{\mu} = 4$
$H_A: \widetilde{\mu} \neq 4$
$W = 9, W = 19$:
Critical values for $n = 7$, assuming $\alpha = 0.1$, are 3 and 25. Do not reject.
**17-5. a.** The hypotheses are $H_0: \widetilde{\mu} \leq 4$
$H_A: \widetilde{\mu} > 4$
**b.** Using the Wilcoxon signed rank test, $W = 26$: Upper-tailed test and $n = 12, \alpha = 0.05$, reject if $W > 61$. Since $W = 26$, do not reject.
**17-7.** $H_0: \widetilde{\mu} \geq 11$
$H_A: \widetilde{\mu} < 11$
Using the Wilcoxon signed rank test, $W = 92$: Reject if $W < 53$.
**17-9.** $H_0: \widetilde{\mu} = 30$
$H_A: \widetilde{\mu} \neq 30$

Using the Wilcoxon signed rank test, $W = 71.5$, $W = 81.5$: Because some of the differences are 0, $n = 17$. The upper and lower values for the Wilcoxon test are 34 and 119 for $\alpha = 0.05$. Do not reject.

**17-11. a.** Using data classes one standard deviation wide, with the data mean of 7.6306 and a standard deviation of 0.2218:

| $e$ | $o$ | $(o - e)^2/e$ |
|---|---|---|
| 14.9440 | 21 | 2.45417 |
| 32.4278 | 31 | 0.06287 |
| 32.4278 | 27 | 0.90851 |
| 14.9440 | 16 | 0.07462 Sum = 3.5002 |

Testing at the $\alpha = 0.05$ level, $\chi_\alpha^2 = 5.9915$.
**b.** Since we concluded the data come from a normal distribution, we test the following:
$H_0: \mu \geq 7.4$
$H_A: \mu < 7.4$
Decision rule: If $z < -1.645$, reject $H_0$; otherwise, do not reject. $Z = 10.13$
**17-13. a.** Putting the claim in the alternative hypothesis:
$H_0: \widetilde{\mu}_1 - \widetilde{\mu}_2 \geq 0$
$H_A: \widetilde{\mu}_1 - \widetilde{\mu}_2 < 0$
**b.** Test using the Mann–Whitney $U$-test.
$U_1 = 40, U_2 = 24$
Use $U_2$ as the test statistic. For $n_1 = 8$ and $n_2 = 8$ and $U = 24$, $p$-value $= 0.221$.
**17-15. a.** $H_0: \widetilde{\mu}_1 - \widetilde{\mu}_2 \leq 0$
$H_A: \widetilde{\mu}_1 - \widetilde{\mu}_2 > 0$
**b.** Since the alternative hypothesis indicates Population 1 should have the larger median, $U_1 = 40$.
$n_1 = 12$ and $n_2 = 12$. Reject if $U \leq 31$.
**17-17.** $H_0: \widetilde{\mu}_1 - \widetilde{\mu}_2 = 0$
$H_A: \widetilde{\mu}_1 - \widetilde{\mu}_2 \neq 0$
Mann-Whitney $U$-test

| $C_1$ | $N = 40$, | Median = 481.50 |
|---|---|---|
| $C_2$ | $N = 35$, | Median = 505.00 |

Point estimate for Sample 1 − Sample 2 is −25.00.
95.1% *CI* for Sample 1 − Sample 2 is (−62.00,9.00).
$W = 1,384.0$
Test of Sample 1 = Sample 2 vs. Sample 1 not = Sample 2 is significant at 0.1502.
**17-19. a.** $H_0: \widetilde{\mu}_1 - \widetilde{\mu}_2 = 0$
$H_A: \widetilde{\mu}_1 - \widetilde{\mu}_2 \neq 0$
With $n = 8$, reject if $T \leq 4$.
Since $T = 11.5$, do not reject the null hypothesis.
**b.** Use the paired-sample $t$-test; $p$-value $= 0.699$.
**17-21.** $H_0: \widetilde{\mu}_1 - \widetilde{\mu}_2 = 0$
$H_A: \widetilde{\mu}_1 - \widetilde{\mu}_2 \neq 0$
With $n = 7$, reject if $T \leq 2, T = 13$.
**17-23.** $H_0: \widetilde{\mu}_2 = \widetilde{\mu}_1$
$H_A: \widetilde{\mu}_2 \neq \widetilde{\mu}_1$
If $T \leq 0$, reject $H_0$; $T = 8$.
**17-25.** $H_0: \widetilde{\mu}_W - \widetilde{\mu}_{WO} \leq 0$
$H_A: \widetilde{\mu}_W - \widetilde{\mu}_{WO} > 0$
$U_1 = (7)(5) + (7)(7 + 1)/2 - 42 = 21$
$U_2 = (7)(5) + (5)(5 + 1)/2 - 36 = 14$
$U$-test $= 21$
Since 21 is not in the table, you cannot determine the exact $p$-value, but you know that the $p$-value will be greater than 0.562.
**17-27. a.** $H_0: \widetilde{\mu}_1 \leq \widetilde{\mu}_2$
$H_A: \widetilde{\mu}_1 > \widetilde{\mu}_2$
**b.** Since $T = 51$ is greater than 16, do not reject $H_0$.
**c.** Housing values are typically skewed.

**17-29.** $H_0: \widetilde{\mu}_1 = \widetilde{\mu}_2$
$H_A: \widetilde{\mu}_1 \neq \widetilde{\mu}_2$
$\mu = 40(40 + 1)/4 = 410$
$\sigma = \sqrt{40(40 + 1)(80 + 1)/24} = 74.3976$
$z = (480 - 410)/74.3976 = 0.94$
$p$-value $= (0.5 - 0.3264)2 = (0.1736)(2) = 0.3472$.
Do not reject $H_0$.

**17-31. a.** a paired-sample $t$-test
  $H_0: \mu_d \geq 0$
  $H_A: \mu_d < 0$
  $t = (-1.7)/(3.011091/\sqrt{10}) = -1.785$
  Since $-1.785 > t$ critical $= -2.2622$; do not reject $H_0$.
  **b.** $H_0: \widetilde{\mu}_O \geq \widetilde{\mu}_N$
  $H_A: \widetilde{\mu}_O < \widetilde{\mu}_N$
  $T = 5.5$. Since $5.5 < 6$, reject $H_0$ and conclude that the medians are not the same.
  **c.** Because you cannot assume the underlying populations are normal, you must use the technique from part b.

**17-33. a.** $H_0: \widetilde{\pi}_o \geq \widetilde{\mu}_C$
  $H_A: \widetilde{\pi}_o < \widetilde{\mu}_C$
  $U_1 = 4,297, U_2 = 6,203, \mu = 5,250, \sigma = 434.7413$
  $z = -2.19$
  $p$-value $= 0.0143$
  **b.** A Type I error is rejecting a true null hypothesis.

**17-35. a.** The data are ordinal.
  **b.** The median would be the best measure.
  **c.** $H_0: \widetilde{\mu}_1 = \widetilde{\mu}_2$
  $H_A: \widetilde{\mu}_1 \neq \widetilde{\mu}_2$
  Using $\alpha = 0.01$, if $T \leq 2$, reject $H_0$. Since $12.5 > 2$, do not reject $H_0$.
  **d.** The decision could be made based on some other factor, such as cost.

**17-37. a.** $H_0: \widetilde{\mu}_1 = \widetilde{\mu}_2 = \widetilde{\mu}_3$
  $H_A:$ Not all population medians are equal.
  **b.** $H = 10.98$. Since, with $\alpha = 0.05$, $\chi^2_\alpha = 5.9915$, and $H = 10.98$, reject.

**17-39. a.** $H_0: \widetilde{\mu}_1 = \widetilde{\mu}_2 = \widetilde{\mu}_3 = \widetilde{\mu}_4$
  $H_A:$ Not all population medians are equal.
  **b.** Use Equation 17.10.
  Selecting $\alpha = 0.05$, $\chi^2_\alpha = 7.8147$, since $H = 42.11$, reject the null hypothesis of equal medians.

**17-41. a.** Salaries in general are usually thought to be skewed.
  **b.** The top-salaried players get extremely high salaries compared to the other players.
  **c.** $H_0: \widetilde{\mu}_1 = \widetilde{\mu}_2 = \widetilde{\mu}_3$
  $H_A:$ Not all population medians are equal.
  $H = 52.531$. If $\alpha = 0.05$, $\chi^2_\alpha = 5.9915$. Reject.

**17-43.** $H_0: \widetilde{\mu}_1 = \widetilde{\mu}_2 = \widetilde{\mu}_3 = \widetilde{\mu}_4$
  $H_A:$ Not all population medians are equal.
  Using XLSTAT, $H$-test statistic $= 11.13971$. Adjusting for ties, the test statistic is 11.21, which is smaller than the critical value (11.34488). Do not reject.

**17-45.** $H_0: \widetilde{\mu}_1 = \widetilde{\mu}_2 = \widetilde{\mu}_3$
  $H_A:$ Not all population medians are equal.
  $H = 13.9818$, testing at $\alpha = 0.05$, $\chi^2_\alpha = 5.9915$
  Since $13.9818 > 5.9915$, reject $H_0$.

**17-53.** $H_0: \widetilde{\mu}_1 - \widetilde{\mu}_2 = 0$
  $H_A: \widetilde{\mu}_1 - \widetilde{\mu}_2 \neq 0$
  $U_1 = 107, U_2 = 14$; $U$-test statistic $= 14$ with $\alpha = 0.05, U_\alpha = 30$.
  Since $14 < 30$, reject $H_0$.

**17-55. a.** a nonparametric test
  **b.** $H_0: \widetilde{\mu}_O - \widetilde{\mu}_N \geq 0$
  $H_A: \widetilde{\mu}_O - \widetilde{\mu}_N < 0$
  $U_1 - 71, U_2 = 29$; $U$-test statistic $= 29$

  If $\alpha = 0.05, U_\alpha = 27$.
  Since $29 > 27$, do not reject $H_0$.

**17-57.** $H_0: \widetilde{\mu} = 1,989.32$
  $H_A: \widetilde{\mu} \neq 1,989.32$
  Find $W = 103, W = 68$.
  With $\alpha = 0.05$, reject if $W \leq 40$ or $W > 131$.

**17-59. a.** $H_0: \widetilde{\mu} = 8.03$
  $H_A: \widetilde{\mu} \neq 8.03$
  **b.** $W = 62.5, W = 57.5$
  This is a two-tailed test with $n = 15$.
  If $\alpha = 0.05$, reject if $W \leq 25$ or $W > 95$. There is not sufficient evidence to reject the null hyporhesis.

**17-61.** $H_0: \widetilde{\mu}_1 = \widetilde{\mu}_2$
  $H_A: \widetilde{\mu}_1 \neq \widetilde{\mu}_2$
  Constructing the paired difference table, $T = 44.5$. With $\alpha = 0.05$, reject if $T \leq 21$ or if $T \geq 84$.
  **b.** A Type II error would be accepting that the population medians are equal when actually they are not equal.

**17-63. a.** They should use the Wilcoxon matched-pairs signed rank test.
  **b.** $H_0: \widetilde{\mu}_{w/oA} \geq \widetilde{\mu}_A$
  $H_A: \widetilde{\mu}_{w/oA} < \widetilde{\mu}_A$
  $T = 6$
  Using $\alpha = 0.025, T_\alpha = 4$.
  **c.** Do not reject $H_0$.

## CHAPTER 18

**18-7.**

**Revenue by Stock Exchange**

**18-9.**

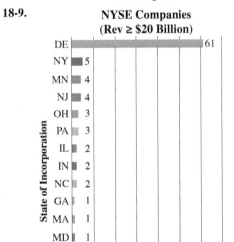

**NYSE Companies (Rev ≥ $20 Billion)**

The state of Delaware dominates; 61 of the 93 NYSE companies have revenues of $20 billion or higher.

**18-11.**

**Board Size Distribution**

The mean number of board members for companies that have the required characteristics is 10.69.

**18-13.**

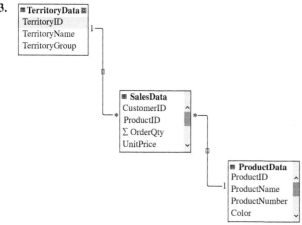

**18-17.**

| Average of StandardCost | Max of StandardCost | Min of StandardCost | Standard deviation of StandardCost |
|---|---|---|---|
| $826.06 | $2,171.29 | $3.40 | $640.18 |

**18-19.** **Average Sales Margin by Territory Group**

**18-21.** **Sales Dollars by Territory Group**

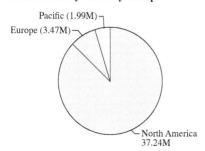

North American customers account for $37.24 million in total sales, which is about five times the total for the European and Pacific groups combined.

**18-23.**

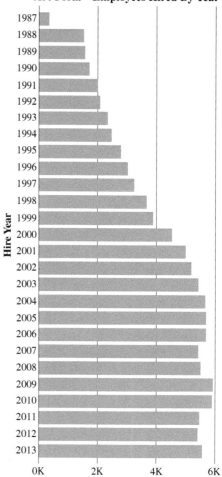

**Art Form – Employees Hired By Year**

In 2009, Art Form hired the most employees ever, 5,989.

**18-25.**

| Employee Gender | Average of Days with Company | Standard Deviation of Days with Company |
|---|---|---|
| F | 3,358 | 2,314 |
| M | 3,343 | 2,285 |
| **Total** | **3,349** | **2,297** |

**18-27.**

**Average Age by Gender (Tenure < 5 Years)**

**18-29.**

**46.96**

Average of Last Job Performance Rating (0-100)

# References

## CHAPTER 1

Berenson, Mark L., and David M. Levine, *Basic Business Statistics: Concepts and Applications*, 13th ed. (New York: Pearson Education, 2015).

Cryer, Jonathan D., and Robert B. Miller, *Statistics for Business: Data Analysis and Modeling*, 2nd ed. (Belmont, CA: Duxbury Press, 1994).

DeVeaux, Richard D., Paul F. Velleman, and David E. Bock, *Stats Data and Models*, 4th ed. (New York: Pearson Education, 2016).

Fowler, Floyd J., *Survey Research Methods*, 4th ed. (Thousand Oaks, CA: Sage Publications, 2009).

Hildebrand, David, and R. Lyman Ott, *Statistical Thinking for Managers*, 4th ed. (Belmont, CA: Duxbury Press, 1998).

John, J. A., D. Whitiker, and D. G. Johnson, *Statistical Thinking in Business*, 2nd ed. (Boca Raton, FL: CRC Press, 2006).

*Microsoft Excel 2016* (Redmond, WA: Microsoft Corp., 2016).

Scheaffer, Richard L., William Mendenhall, R. Lyman Ott, and Kenneth G. Gerow, *Elementary Survey Sampling*, 7th ed. (Independence, KY: Brooks/Cole, 2012).

Siegel, Andrew F., *Practical Business Statistics*, 7th ed. (Amsterdam: Elsevier, 2016).

## CHAPTER 2

Berenson, Mark L., and David M. Levine, *Basic Business Statistics: Concepts and Applications*, 13th ed. (New York: Pearson Education, 2015).

Cleveland, William S., "Graphs in Scientific Publications," *The American Statistician* 38 (November 1984), pp. 261–269.

Cleveland, William S., and R. McGill, "Graphical Perception: Theory, Experimentation, and Application to the Development of Graphical Methods," *Journal of the American Statistical Association* 79 (September 1984), pp. 531–554.

Cryer, Jonathan D., and Robert B. Miller, *Statistics for Business: Data Analysis and Modeling*, 2nd ed. (Belmont, CA: Duxbury Press, 1994).

DeVeaux, Richard D., Paul F. Velleman, and David E. Bock, *Stats Data and Models*, 4th ed. (New York: Pearson Education, 2016).

Few, Stephen, *Now You See It: Simple Visualization Techniques for Quantitative Analysis* (Oakland, CA: Analytics Press, 2009).

Few, Stephen, *Show Me the Numbers: Designing Tables and Graphs to Enlighten*, 2nd ed. (Burlingame, CA: Analytics Press, 2012).

*Microsoft Excel 2016* (Redmond, WA: Microsoft Corp., 2016).

Siegel, Andrew F., *Practical Business Statistics*, 7th ed. (Amsterdam: Elsevier, 2016).

Tufte, Edward R., *Envisioning Information* (Cheshire, CT: Graphics Press, 1990).

Tufte, Edward R., *The Visual Display of Quantitative Information*, 2nd ed. (Cheshire, CT: Graphics Press, 2001).

Tukey, John W., *Exploratory Data Analysis* (New York: Pearson Education, 1977).

## CHAPTER 3

Berenson, Mark L., and David M. Levine, *Basic Business Statistics: Concepts and Applications*, 13th ed. (New York: Pearson Education, 2015).

DeVeaux, Richard D., Paul F. Velleman, and David E. Bock, *Stats Data and Models*, 4th ed. (New York: Pearson Education, 2016).

Few, Stephen, *Now You See It: Simple Visualization Techniques for Quantitative Analysis* (Oakland, CA: Analytics Press, 2009).

Few, Stephen, *Show Me the Numbers: Designing Tables and Graphs to Enlighten*, 2nd ed. (Burlingame, CA: Analytics Press, 2012).

*Microsoft Excel 2016* (Redmond, WA: Microsoft Corp., 2016).

Siegel, Andrew F., *Practical Business Statistics*, 7th ed. (Amsterdam: Elsevier, 2016).

Tukey, John W., *Exploratory Data Analysis* (New York: Pearson Education, 1977).

## CHAPTER 4

Blyth, C. R., "Subjective vs. Objective Methods in Statistics," *The American Statistician* 26 (June 1972), pp. 20–22.

DeVeaux, Richard D., Paul F. Velleman, and David E. Bock, *Stats Data and Models*, 4th ed. (New York: Pearson Education, 2016).

Hogg, R. V., Elliot A. Tanis, and Dale Zimmerman, *Probability and Statistical Inference*, 9th ed. (New York: Pearson Education, 2015).

Larsen, Richard J., and Morris L. Marx, *An Introduction to Mathematical Statistics and Its Applications*, 5th ed. (New York: Pearson Education, 2012).

*Microsoft Excel 2016* (Redmond, WA: Microsoft Corp., 2016).

Mlodinow, Leonard, *The Drunkard's Walk: How Randomness Rules Our Lives* (New York: Vintage Books, 2008).

Raiffa, H., *Decision Analysis: Introductory Lectures on Choices Under Uncertainty* (New York: McGraw Hill, 1997).

Siegel, Andrew F., *Practical Business Statistics*, 7th ed. (Amsterdam: Elsevier, 2016).

## CHAPTER 5

DeVeaux, Richard D., Paul F. Velleman, and David E. Bock, *Stats Data and Models*, 4th ed. (New York: Pearson Education, 2016).

Hogg, R. V., Elliot A. Tanis, and Dale Zimmerman, *Probability and Statistical Inference*, 9th ed. (New York: Pearson Education, 2015).

Larsen, Richard J., and Morris L. Marx, *An Introduction to Mathematical Statistics and Its Applications*, 5th ed. (New York: Pearson Education, 2012).

*Microsoft Excel 2016* (Redmond, WA: Microsoft Corp., 2016).

Siegel, Andrew F., *Practical Business Statistics*, 7th ed. (Amsterdam: Elsevier, 2016).

## CHAPTER 6

Albright, Christian S., Wayne L. Winston, and Christopher Zappe, *Data Analysis for Managers with Microsoft Excel*, 2nd ed. (Independence, KY: Brooks/Cole, 2004).

DeVeaux, Richard D., Paul F. Velleman, and David E. Bock, *Stats Data and Models*, 4th ed. (New York: Pearson Education, 2016).

Hogg, R. V., Elliot A. Tanis, and Dale Zimmerman, *Probability and Statistical Inference*, 9th ed. (New York: Pearson Education, 2015).

Larsen, Richard J., and Morris L. Marx, *An Introduction to Mathematical Statistics and Its Applications*, 5th ed. (New York: Pearson Education, 2012).

*Microsoft Excel 2016* (Redmond, WA: Microsoft Corp., 2016).

Siegel, Andrew F., *Practical Business Statistics*, 7th ed. (Amsterdam: Elsevier, 2016).

**CHAPTER 7**

Berenson, Mark L., and David M. Levine, *Basic Business Statistics: Concepts and Applications*, 13th ed. (New York: Pearson Education, 2015).

Cochran, William G., *Sampling Techniques*, 3rd ed. (New York: Wiley, 1977).

DeVeaux, Richard D., Paul F. Velleman, and David E. Bock, *Stats Data and Models*, 4th ed. (New York: Pearson Education, 2016).

Hogg, R. V., Elliot A. Tanis, and Dale Zimmerman, *Probability and Statistical Inference*, 9th ed. (New York: Pearson Education, 2015).

Johnson, Richard A., and Dean W. Wichern, *Business Statistics: Decision Making with Data* (New York: Wiley, 1997).

Larsen, Richard J., and Morris L. Marx, *An Introduction to Mathematical Statistics and Its Applications*, 5th ed. (New York: Pearson Education, 2012).

*Microsoft Excel 2016* (Redmond, WA: Microsoft Corp., 2016).

**CHAPTER 8**

Berenson, Mark L., and David M. Levine, *Basic Business Statistics: Concepts and Applications*, 13th ed. (New York: Pearson Education, 2015).

DeVeaux, Richard D., Paul F. Velleman, and David E. Bock, *Stats Data and Models*, 4th ed. (New York: Pearson Education, 2016).

Hogg, R. V., Elliot A. Tanis, and Dale Zimmerman, *Probability and Statistical Inference*, 9th ed. (New York: Pearson Education, 2015).

Larsen, Richard J., and Morris L. Marx, *An Introduction to Mathematical Statistics and Its Applications*, 5th ed. (New York: Pearson Education, 2012).

*Microsoft Excel 2016* (Redmond, WA: Microsoft Corp., 2016).

Siegel, Andrew F., *Practical Business Statistics*, 7th ed. (Amsterdam: Elsevier, 2016).

**CHAPTER 9**

Berenson, Mark L., and David M. Levine, *Basic Business Statistics: Concepts and Applications*, 13th ed. (New York: Pearson Education, 2015).

Brown, L., et al., "Interval Estimation for a Binomial Proportion," *Statistical Science*, 2001, pp. 101–133.

DeVeaux, Richard D., Paul F. Velleman, and David E. Bock, *Stats Data and Models*, 4th ed. (New York: Pearson Education, 2016).

Hogg, R. V., Elliot A. Tanis, and Dale Zimmerman, *Probability and Statistical Inference*, 9th ed. (New York: Pearson Education, 2015).

Larsen, Richard J., and Morris L. Marx, *An Introduction to Mathematical Statistics and Its Applications,* 5th ed. (New York: Pearson Education, 2012).

*Microsoft Excel 2016* (Redmond, WA: Microsoft Corp., 2016).

Siegel, Andrew F., *Practical Business Statistics*, 7th ed. (Amsterdam: Elsevier, 2016).

**CHAPTER 10**

Berenson, Mark L., and David M. Levine, *Basic Business Statistics: Concepts and Applications*, 13th ed. (New York: Pearson Education, 2015).

Cryer, Jonathan D., and Robert B. Miller, *Statistics for Business: Data Analysis and Modeling*, 2nd ed. (Belmont, CA: Duxbury Press, 1994).

DeVeaux, Richard D., Paul F. Velleman, and David E. Bock, *Stats Data and Models*, 4th ed. (New York: Pearson Education, 2016).

Johnson, Richard A., and Dean W. Wichern, *Business Statistics: Decision Making with Data* (New York: Wiley, 1997).

Larsen, Richard J., Morris L. Marx, and Bruce Cooil, *Statistics for Applied Problem Solving and Decision Making* (Nashville, TN: South-Western, 1997).

*Microsoft Excel 2016* (Redmond, WA: Microsoft Corp., 2016).

Siegel, Andrew F., *Practical Business Statistics*, 7th ed. (Amsterdam: Elsevier, 2016).

**CHAPTER 11**

Berenson, Mark L., and David M. Levine, *Basic Business Statistics: Concepts and Applications*, 13th ed. (New York: Pearson Education, 2015).

Cryer, Jonathan D., and Robert B. Miller, *Statistics for Business: Data Analysis and Modeling*, 2nd ed. (Belmont, CA: Duxbury Press, 1994).

DeVeaux, Richard D., Paul F. Velleman, and David E. Bock, *Stats Data and Models*, 4th ed. (New York: Pearson Education, 2016).

Duncan, Acheson J., *Quality Control and Industrial Statistics*, 5th ed. (Burr Ridge, IL: Irwin, 1986).

Johnson, Richard A., and Dean W. Wichern, *Business Statistics: Decision Making with Data* (New York: Wiley, 1997).

Larsen, Richard J., Morris L. Marx, and Bruce Cooil, *Statistics for Applied Problem Solving and Decision Making* (Pacific Grove, CA: Duxbury Press, 1997).

Markowski, Carol, and Edmund Markowski, "Conditions for the Effectiveness of a Preliminary Test of Variance," *The American Statistician* 4 (November 1990), pp. 322–326.

*Microsoft Excel 2016* (Redmond, WA: Microsoft Corp., 2016).

Siegel, Andrew F., *Practical Business Statistics*, 7th ed. (Amsterdam: Elsevier, 2016).

**CHAPTER 12**

Berenson, Mark L., and David M. Levine, *Basic Business Statistics: Concepts and Applications*, 13th ed. (New York: Pearson Education, 2015).

Bowerman, Bruce L., and Richard T. O'Connell, *Linear Statistical Models: An Applied Approach*, 2nd ed. (Belmont, CA: Duxbury Press, 1990).

Cox, D. R., *Planning of Experiments* (New York: John Wiley & Sons, 1992).

Cryer, Jonathan D., and Robert B. Miller, *Statistics for Business: Data Analysis and Modeling*, 2nd ed. (Belmont, CA: Duxbury Press, 1994).

DeVeaux, Richard D., Paul F. Velleman, and David E. Bock, *Stats Data and Models*, 4th ed. (New York: Pearson Education, 2016).

Kutner, Michael H., Christopher J. Nachtsheim, John Neter, and William Li, *Applied Linear Statistical Models*, 5th ed. (New York: McGraw-Hill Irwin, 2005).

*Microsoft Excel 2016* (Redmond, WA: Microsoft Corp., 2016).

Montgomery, D. C., *Design and Analysis of Experiments*, 8th ed. (New York: John Wiley & Sons, 2013).

Searle, S. R., and R. F. Fawcett, "Expected Mean Squares in Variance Component Models Having Finite Populations," *Biometrics* 26 (1970), pp. 243–254.

## CHAPTER 13

Berenson, Mark L., and David M. Levine, *Basic Business Statistics: Concepts and Applications*, 13th ed. (New York: Pearson Education, 2015).

Conover, W. J., *Practical Nonparametric Statistics*, 3rd ed. (New York: Wiley, 1999).

DeVeaux, Richard D., Paul F. Velleman, and David E. Bock, *Stats Data and Models*, 4th ed. (New York: Pearson Education, 2016).

Higgins, James J., *Introduction to Modern Nonparametric Statistics*, 1st ed. (Pacific Grove, CA: Duxbury, 2004).

Marascuilo, Leonard, and M. McSweeney, *Nonparametric and Distribution Free Methods for the Social Sciences* (Monterey, CA: Brooks/Cole, 1977).

*Microsoft Excel 2016* (Redmond, WA: Microsoft Corp., 2016).

## CHAPTER 14

Berenson, Mark L., and David M. Levine, *Basic Business Statistics: Concepts and Applications*, 13th ed. (New York: Pearson Education, 2015).

Cryer, Jonathan D., and Robert B. Miller, *Statistics for Business: Data Analysis and Modeling*, 2nd ed. (Belmont, CA: Duxbury Press, 1994).

DeVeaux, Richard D., Paul F. Velleman, and David E. Bock, *Stats Data and Models*, 4th ed. (New York: Pearson Education, 2016).

Dielman, Terry E., *Applied Regression Analysis—A Second Course in Business and Economic Statistics*, 4th ed. (Belmont, CA: Duxbury Press, 2005).

Draper, Norman R., and Harry Smith, *Applied Regression Analysis*, 3rd ed. (New York: John Wiley & Sons, 1998).

Frees, Edward W., *Data Analysis Using Regression Models: The Business Perspective* (Upper Saddle River, NJ: Prentice Hall, 1996).

Kleinbaum, David G., Lawrence L. Kupper, Azhar Nizam, and Keith E. Muller, *Applied Regression Analysis and Multivariable Methods*, 4th ed. (Independence, KY: Thomson Brooks/Cole, 2008).

Kutner, Michael H., Christopher J. Nachtsheim, John Neter, and William Li, *Applied Linear Statistical Models*, 5th ed. (New York: McGraw-Hill Irwin, 2005).

*Microsoft Excel 2016* (Redmond, WA: Microsoft Corp., 2016).

## CHAPTER 15

Berenson, Mark L., and David M. Levine, *Basic Business Statistics: Concepts and Applications*, 13th ed. (New York: Pearson Education, 2015).

Bowerman, Bruce L., and Richard T. O'Connell, *Linear Statistical Models: An Applied Approach*, 2nd ed. (Belmont, CA: Duxbury Press, 1990).

Cryer, Jonathan D., and Robert B. Miller, *Statistics for Business: Data Analysis and Modeling*, 2nd ed. (Belmont, CA: Duxbury Press, 1994).

Demmert, Henry, and Marshall Medoff, "Game-Specific Factors and Major League Baseball Attendance: An Econometric Study," *Santa Clara Business Review* (1977), pp. 49–56.

DeVeaux, Richard D., Paul F. Velleman, and David E. Bock, *Stats Data and Models*, 4th ed. (New York: Pearson Education, 2016).

Dielman, Terry E., *Applied Regression Analysis—A Second Course in Business and Economic Statistics*, 4th ed. (Belmont, CA: Duxbury Press, 2005).

Draper, Norman R., and Harry Smith, *Applied Regression Analysis*, 3rd ed. (New York: John Wiley & Sons, 1998).

Frees, Edward W., *Data Analysis Using Regression Models: The Business Perspective* (Upper Saddle River, NJ: Prentice Hall, 1996).

Gloudemans, Robert J., and Dennis Miller, "Multiple Regression Analysis Applied to Residential Properties," *Decision Sciences* 7 (April 1976), pp. 294–304.

Kleinbaum, David G., Lawrence L. Kupper, Azhar Nizam, and Keith E. Muller, *Applied Regression Analysis and Multivariable Methods*, 4th ed. (Independence, KY: Thomson Brooks/Cole, 2008).

Kutner, Michael H., Christopher J. Nachtsheim, John Neter, and William Li, *Applied Linear Statistical Models*, 5th ed. (New York: McGraw-Hill Irwin, 2005).

*Microsoft Excel 2016* (Redmond, WA: Microsoft Corp., 2016).

## CHAPTER 16

Armstrong, J. Scott, "Forecasting by Extrapolation: Conclusions from 25 Years of Research," *Interfaces* 14, no. 6 (1984).

Bails, Dale G., and Larry C. Peppers, *Business Fluctuations: Forecasting Techniques and Applications*, 2nd ed. (Englewood Cliffs, NJ: Prentice Hall, 1992).

Berenson, Mark L., and David M. Levine, *Basic Business Statistics: Concepts and Applications*, 13th ed. (New York: Pearson Education, 2015).

Bowerman, Bruce L., and Richard T. O'Connell, *Linear Statistical Models: An Applied Approach*, 2nd ed. (Belmont, CA: Duxbury Press, 1990).

Brandon, Charles, R. Fritz, and J. Xander, "Econometric Forecasts: Evaluation and Revision," *Applied Economics* 15, no. 2 (1983).

Cryer, Jonathan D., and Robert B. Miller, *Statistics for Business: Data Analysis and Modeling*, 2nd ed. (Belmont, CA: Duxbury Press, 1994).

DeVeaux, Richard D., Paul F. Velleman, and David E. Bock, *Stats Data and Models*, 4th ed. (New York: Pearson Education, 2016).

Frees, Edward W., *Data Analysis Using Regression Models: The Business Perspective* (Upper Saddle River, NJ: Prentice Hall, 1996).

Granger, C. W. G., *Forecasting in Business and Economics*, 2nd ed. (Bringley, United Kingdom: Emerald, 2007).

Kutner, Michael H., Christopher J. Nachtsheim, John Neter, and William Li, *Applied Linear Statistical Models*, 5th ed. (New York: McGraw-Hill Irwin, 2005).

Makridakis, Spyros, Steven C. Wheelwright, and Rob J. Hyndman, *Forecasting: Methods and Applications*, 3rd ed. (New York: John Wiley & Sons, 1998).

McLaughlin, Robert L., "Forecasting Models: Sophisticated or Naive?" *Journal of Forecasting* 2, no. 3 (1983).

*Microsoft Excel 2016* (Redmond, WA: Microsoft Corp., 2016).

Montgomery, Douglas C., and Lynwood A. Johnson, *Forecasting and Time Series Analysis*, 2nd ed. (New York: McGraw-Hill, 1990).

The Ombudsman, "Research on Forecasting—A Quarter-Century Review, 1960–1984," *Interfaces* 16, no. 1 (1986).

Willis, R. E., *A Guide to Forecasting for Planners & Managers* (Englewood Cliffs, NJ: Prentice Hall, 1987).

Wonnacott, T. H., and R. J. Wonnacott, *Econometrics*, 2nd ed. (New York: John Wiley & Sons, 1979).

## CHAPTER 17

Berenson, Mark L., and David M. Levine, *Basic Business Statistics: Concepts and Applications*, 13th ed. (New York: Pearson Education, 2015).

Conover, W. J., *Practical Nonparametric Statistics*, 3rd ed. (New York: Wiley, 1999).

DeVeaux, Richard D., Paul F. Velleman, and David E. Bock, *Stats Data and Models*, 4th ed. (New York: Pearson Education, 2016).

Dunn, O. J., "Multiple Comparisons Using Rank Sums," *Technometrics* 6 (1964), pp. 241–252.

Marascuilo, Leonard, and M. McSweeney, *Nonparametric and Distribution Free Methods for the Social Sciences* (Monterey, CA: Brooks/Cole, 1977).

*Microsoft Excel 2016* (Redmond, WA: Microsoft Corp., 2016).

Noether, G. E., *Elements of Nonparametric Statistics* (New York: John Wiley & Sons, 1967).

## CHAPTER 18

Collie, Rob, and Avichal Singh, *Power Pivot and Power BI: The Excel User's Guide to DAX, Power Query, Power BI and Power Pivot in Excel 2010–2016*, 2nd ed. (Merritt Island, FL: Holy Macro! Books, 2016).

Lachev, Teo, *Applied Microsoft Power BI: Bring Your Data to Life* Atlanta, GA: (Prologika Press, 2016).

LeBlanc, Patrick, Jessica M. Moss, Dejan Sarka, and Dustin Ryan, *Applied Microsoft Business Intelligence* (Indianapolis, IN: John Wiley & Sons, 2015).

# Glossary

**Adjusted *R*-squared** A measure of the percentage of explained variation in the dependent variable in a multiple regression model that takes into account the relationship between the sample size and the number of independent variables in the regression model.

**Aggregate Price Index** An index that is used to measure the rate of change from a base period for a group of two or more items.

**All-Inclusive Classes** A set of classes that contains all the possible data values.

**Alternative Hypothesis** The hypothesis that includes all population values not included in the null hypothesis. The alternative hypothesis will be selected only if there is strong enough sample evidence to support it. The alternative hypothesis is deemed to be true if the null hypothesis is rejected.

**Arithmetic Average or Mean** The sum of all values divided by the number of values.

**Autocorrelation** Correlation of the error terms (residuals) occurs when the residuals at points in time are related.

**Balanced Design** An experiment has a balanced design if the factor levels have equal sample sizes.

**Bar Chart** A graphical representation of a categorical data set in which a rectangle or bar is drawn over each category or class. The length or height of each bar represents the frequency or percentage of observations or some other measure associated with the category. The bars may be vertical or horizontal. The bars may all be the same color or they may be different colors depicting different categories. Additionally, multiple variables can be graphed on the same bar chart.

**Base Period Index** The time-series value to which all other values in the time series are compared. The index number for the base period is defined as 100.

**Between-Sample Variation** Dispersion among the factor sample means.

**Bias** An effect that alters a statistical result by systematically distorting it; different from a random error, which may distort on any one occasion but balances out on the average.

**Binomial Probability Distribution Characteristics** A distribution that gives the probability of *x* successes in *n* trials in a process that meets the following conditions:

1. A trial has only two possible outcomes: a success or a failure.
2. There is a fixed number, *n*, of identical trials.
3. The trials of the experiment are independent of each other. This means that if one outcome is a success, this does not influence the chance of another outcome being a success.
4. The process must be consistent in generating successes and failures. That is, the probability, *p*, associated with a success remains constant from trial to trial.
5. If *p* represents the probability of a success, then $1 - p = q$ is the probability of a failure.

**Box and Whisker Plot** A graph that is composed of two parts: a box and the whiskers. The box has a width that ranges from the first quartile $(Q_1)$ to the third quartile $(Q_3)$. A vertical line through the box is placed at the median. Limits are located at a value that is 1.5 times the difference between $Q_1$ and $Q_3$ below $Q_1$ and above $Q_3$. The whiskers extend to the left to the lowest value within the limits and to the right to the highest value within the limits.

**Business Analytics** The application of well-known statistical tools, combined with powerful software, to transform large and complex data into useful information for decision-making purposes.

**Business Intelligence** The application of tools and technologies for gathering, storing, retrieving, and analyzing data that businesses collect and use. Also, the process of collecting data from inside or outside an organization that are thought to in some way relate to, or influence, the performance of the organization.

**Business Statistics** A collection of procedures and techniques that are used to convert data into meaningful information in a business environment.

**Census** An enumeration of the entire set of measurements taken from the whole population.

**Central Limit Theorem** For simple random samples of *n* observations taken from a population with mean $\mu$ and standard deviation $\sigma$, regardless of the population's distribution, provided the sample size is sufficiently large, the distribution of the sample means, $\bar{x}$, will be approximately normal with a mean equal to the population mean $(\mu_{\bar{x}} = \mu_x)$ and a standard deviation equal to the population standard deviation divided by the square root of the sample size $(\sigma_{\bar{x}} = \sigma/\sqrt{n})$. The larger the sample size, the better the approximation to the normal distribution.

**Class Boundaries** The upper and lower values of each class.

**Class Width** The distance between the lowest possible value and the highest possible value for a frequency class.

**Classical Probability Assessment** The method of determining probability based on the ratio of the number of ways an outcome or event of interest can occur to the number of ways *any* outcome or event can occur when the individual outcomes are equally likely.

**Closed-End Questions** Questions that require the respondent to select from a short list of defined choices.

**Cluster Sampling** A method by which the population is divided into groups, or clusters, that are each intended to be mini-populations. A simple random sample of *m* clusters is selected. The items chosen from a cluster can be selected using any probability sampling technique.

**Coefficient of Determination** The portion of the total variation in the dependent variable that is explained by its relationship with the independent variable. The coefficient of determination is also called *R*-squared and is denoted as $R^2$.

**Coefficient of Partial Determination** The measure of the marginal contribution of each independent variable, given that other independent variables are in the model.

**Coefficient of Variation** The ratio of the standard deviation to the mean expressed as a percentage. The coefficient of variation is used to measure variation relative to the mean.

**Complement** The complement of an event *E* is the collection of all possible outcomes not contained in event *E*.

**Completely Randomized Design** An experiment is completely randomized if it consists of the independent random selection of observations representing each level of one factor.

**Composite Model** The model that contains both the basic terms and the interaction terms.

**Conditional Probability** The probability that an event will occur given that some other event has already happened.

**Confidence Interval** An interval developed from sample values such that if all possible intervals of a given width were constructed, a percentage of these intervals, known as the confidence level, would include the true population parameter.

**Confidence Level** The percentage of all possible confidence intervals that will contain the true population parameter.

**Consistent Estimator** An unbiased estimator is said to be a consistent estimator if the difference between the estimator and the parameter tends to become smaller as the sample size becomes larger.

**Contingency Table** A table used to classify sample observations according to two or more identifiable characteristics. It is also called a *cross-tabulation table*.

**Continuous Data** Data whose possible values are uncountable and that may assume any value in an interval.

**Continuous Random Variables** Random variables that can assume an uncountably infinite number of values.

**Convenience Sampling** A sampling technique that selects the items from the population based on accessibility and ease of selection.

**Correlation Coefficient** A quantitative measure of the strength of the linear relationship between two variables. The correlation ranges from $-1.0$ to $+1.0$. A correlation of $\pm 1.0$ indicates a perfect linear relationship, whereas a correlation of 0 indicates no linear relationship.

**Correlation Matrix** A table showing the pairwise correlations between all variables (dependent and independent).

**Critical Value** The value corresponding to a significance level that determines those test statistics that lead to rejecting the null hypothesis and those that lead to a decision not to reject the null hypothesis.

**Cross-Sectional Data** A set of data values observed at a fixed point in time.

**Cumulative Frequency Distribution** A summary of a set of data that displays the number of observations with values less than or equal to the upper limit of each of its classes.

**Cumulative Relative Frequency Distribution** A summary of a set of data that displays the proportion of observations with values less than or equal to the upper limit of each of its classes.

**Cyclical Component** A wavelike pattern within the time series that repeats itself throughout the time series and has a recurrence period of more than one year.

**Data Array** Data that have been arranged in numerical order.

**Data Mining** The application of statistical techniques and algorithms to the analysis of large data sets.

**Degrees of Freedom** The number of independent data values available to estimate the population's standard deviation. If $k$ parameters must be estimated before the population's standard deviation can be calculated from a sample of size $n$, the degrees of freedom are equal to $n - k$.

**Demographic Questions** Questions relating to the respondents' characteristics, backgrounds, and attributes.

**Dependent Events** Two events are dependent if the occurrence of one event influences the probability of the other event occurring.

**Dependent Variable** A variable whose values are thought to be a function of, or dependent on, the values of another variable called the *independent variable*. On a scatter plot, the dependent variable is placed on the $y$ axis and is often called the response variable.

**Descriptive Analytics** Tools, techniques, and methodologies used to examine data for the purpose of looking back to the past to answer questions about what has happened.

**Discrete Data** Data that can take on a countable number of possible values.

**Discrete Random Variable** A random variable that can assume only a finite number of values or an infinite sequence of values, such as $[0, 1, 2, \ldots]$.

**Dummy Variable** A variable that is assigned a value equal to either 0 or 1, depending on whether the observation possesses a given characteristic.

**Empirical Rule** If the data distribution is bell-shaped, then the interval

$\mu \pm 1\sigma$ contains approximately 68% of the values

$\mu \pm 2\sigma$ contains approximately 95% of the values

$\mu \pm 3\sigma$ contains virtually all of the data values

**Equal-Width Classes** The distance between the lowest possible value and the highest possible value in each class is equal for all classes.

**Event** A collection of experimental outcomes.

**Expected Value** The mean of a probability distribution. The average value when the experiment that generates values for the random variable is repeated over the long run.

**Experiment** A process that produces a single outcome whose result cannot be predicted with certainty.

**Experimental Design** A plan for performing an experiment in which the variable of interest is defined. One or more factors are identified to be manipulated, changed, or observed so that the impact (or influence) on the variable of interest can be measured or observed.

**Experiment-Wide Error Rate** The proportion of experiments in which at least one of the set of confidence intervals constructed does not contain the true value of the population parameter being estimated.

**Exponential Smoothing** A time-series and forecasting technique that produces an exponentially weighted moving average in which each smoothing calculation or forecast is dependent on all previous observed values.

**External Validity** A characteristic of an experiment whose results can be generalized beyond the test environment so that the outcomes can be replicated when the experiment is repeated.

**Factor** A quantity under examination in an experiment as a possible cause of variation in the response variable.

**Forecasting Horizon** The number of future periods covered by a forecast. It is sometimes referred to as *forecast lead time*.

**Forecasting Interval** The frequency with which new forecasts are prepared.

**Forecasting Period** The unit of time for which forecasts are to be made.

**Frequency Distribution** A summary of a set of data that displays the number of observations in each of the distribution's distinct categories or classes.

**Frequency Histogram** A graph of a frequency distribution with the horizontal axis showing the classes, the vertical axis showing the frequency count, and (for equal class widths) the rectangles having a height equal to the frequency in each class.

**Hypergeometric Distribution** The hypergeometric distribution is formed by the ratio of the number of ways an event of interest can occur over the total number of ways any event can occur.

**Independent Events** Two events are independent if the occurrence of one event in no way influences the probability of the occurrence of the other event.

**Independent Samples** Samples selected from two or more populations in such a way that the occurrence of values in one sample has no influence on the probability of the occurrence of values in the other sample(s).

**Independent Variable** A variable whose values are thought to influence the values of the *dependent variable*. The independent variable, or explanatory variable, is often within the direct control of the decision maker. On a scatter plot, the independent variable, or explanatory variable, is graphed on the x axis.

**Interaction** The case in which one independent variable (such as $x_2$) affects the relationship between another independent variable ($x_1$) and a dependent variable ($y$).

**Internal Validity** A characteristic of an experiment in which data are collected in such a way as to eliminate the effects of variables within the experimental environment that are not of interest to the researcher.

**Interquartile Range** A measure of variation that is determined by computing the difference between the third and first quartiles.

**Key Performance Indicator** A measure of how well an organization is achieving its most important business objectives. High-level KPIs measure the overall performance of the organization, while low-level KPIs measure the success of an organization's subunits (departments, centers, etc.).

**Least Squares Criterion** The criterion for determining a regression line that minimizes the sum of squared prediction errors.

**Left-Skewed Data** A data distribution is left-skewed if the mean for the data is smaller than the median.

**Levels** The categories, measurements, or strata of a factor of interest in the current experiment.

**Line Chart** A two-dimensional chart showing time on the horizontal axis and the variable of interest on the vertical axis.

**Linear Trend** A long-term increase or decrease in a time series in which the rate of change is relatively constant.

**Margin of Error** The amount that is added to and subtracted from the point estimate to determine the endpoints of the confidence interval. Also, a measure of how close we expect the point estimate to be to the population parameter with the specified level of confidence.

**Mean** A numerical measure of the center of a set of quantitative measures computed by dividing the sum of the values by the number of values in the data.

**Median** The center value that divides a data array into two halves. We use $\tilde{\mu}$ to denote the population median and $M_d$ to denote the sample median.

**Mode** The value in a data set that occurs most frequently.

**Model** A representation of an actual system using either a physical or a mathematical portrayal.

**Model Diagnosis** The process of determining how well a model fits past data and how well the model's assumptions appear to be satisfied.

**Model Fitting** The process of estimating the specified model's parameters to achieve an adequate fit of the historical data.

**Model Specification** The process of selecting the forecasting technique to be used in a particular situation.

**Moving Average** The successive averages of $n$ consecutive values in a time series.

**Multicollinearity** A high correlation between two independent variables such that the two variables contribute redundant information to the model. When highly correlated independent variables are included in the regression model, they can adversely affect the regression results.

**Multiple Coefficient of Determination** The proportion of the total variation of the dependent variable in a multiple regression model that is explained by its relationship to the independent variables. It is, as is the case in the simple linear model, called $R$-squared and is denoted as $R^2$.

**Mutually Exclusive Classes** Classes that do not overlap so that a data value can be placed in only one class.

**Mutually Exclusive Events** Two events are mutually exclusive if the occurrence of one event precludes the occurrence of the other event.

**Nonstatistical Sampling Techniques** Those methods of selecting samples using convenience, judgment, or other nonchance processes.

**Normal Distribution** A bell-shaped distribution that has the following properties:

1. It is *unimodal*; that is, the normal distribution peaks at a single value.
2. It is *symmetric*; this means that the two areas under the curve between the mean and any two points equidistant on either side of the mean are identical. One side of the distribution is the mirror image of the other side.
3. The mean, median, and mode are equal.
4. The normal approaches the horizontal axis on either side of the mean toward plus and minus infinity ($\infty$). In more formal terms, the normal distribution is *asymptotic* to the x axis.
5. The amount of variation in the random variable determines the height and spread of the normal distribution.

**Null Hypothesis** The statement about the population parameter that will be assumed to be true during the conduct of the hypothesis test. The null hypothesis will be rejected only if the sample data provide substantial contradictory evidence.

**Ogive** The graphical representation of the cumulative relative frequency. A line is connected to points plotted above the upper limit of each class at a height corresponding to the cumulative relative frequency.

**One-Tailed Test** A hypothesis test in which the entire rejection region is located in one tail of the sampling distribution. In a one-tailed test, the entire alpha level is located in one tail of the distribution.

**One-Way Analysis of Variance** An analysis of variance design in which independent samples are obtained from two or more levels of a single factor for the purpose of testing whether the levels have equal means.

**Open-End Questions** Questions that allow respondents the freedom to respond with any value, words, or statements of their own choosing.

**Paired Samples** Samples that are selected in such a way that values in one sample are matched with the values in the second sample for the purpose of controlling for extraneous factors. Another term for paired samples is *dependent samples*.

**Parameter** A measure computed from the entire population. As long as the population does not change, the value of the parameter will not change.

**Pareto Principle** 80% of the problems come from 20% of the causes.

**Percentiles** The $p$th percentile in a data array is a value that divides the data set into two parts. The lower segment contains at least $p\%$ and the upper segment contains at least $(100 - p)\%$ of the data. The 50th percentile is the median.

**Pie Chart** A graph in the shape of a circle. The circle is divided into "slices" corresponding to the categories or classes to be displayed. The size of each slice is proportional to the magnitude of the displayed variable associated with each category or class.

**Pilot Sample** A sample taken from the population of interest of a size smaller than the anticipated sample size that is used to provide an estimate for the population standard deviation.

**Point Estimate** A single statistic, determined from a sample, that is used to estimate the corresponding population parameter.

**Population** The set of all objects or individuals of interest or the measurements obtained from all objects or individuals of interest.

**Population Mean** The average for all values in the population computed by dividing the sum of all values by the population size.

**Population Proportion** The fraction of values in a population that have a specific attribute.

**Power** The probability that the hypothesis test will correctly reject the null hypothesis when the hypothesis is false.

**Power Curve** A graph showing the probability that the hypothesis test will correctly reject a false null hypothesis for a range of possible "true" values for the population parameter.

**Predictive Analytics** Statistical models and forecasting methods applied to data to answer questions about what could happen.

**Probability** The chance that a particular event will occur. The probability value will be in the range 0 to 1. A value of 0 means the event will not occur. A probability of 1 means the event will occur. Anything between 0 and 1 reflects the uncertainty of the event occurring. The definition given is for a countable number of events.

*p*-**Value** The probability (assuming the null hypothesis is true) of obtaining a test statistic at least as extreme as the test statistic we calculated from the sample. The *p*-value is also known as the *observed significance level*.

**Qualitative Data** Data whose measurement scale is inherently categorical.

**Quantitative Data** Measurements whose values are inherently numerical.

**Quartiles** Quartiles in a data array are those values that divide the data set into four equal-sized groups. The median corresponds to the second quartile.

**Random Component** Changes in time-series data that are unpredictable and cannot be associated with a trend, seasonal, or cyclical component.

**Random Variable** A variable that takes on different numerical values based on chance.

**Range** A measure of variation that is computed by finding the difference between the maximum and minimum values in a data set.

**Regression Hyperplane** The multiple regression equivalent of the simple regression line. The plane typically has a different slope for each independent variable.

**Regression Slope Coefficient** The average change in the dependent variable for a unit change in the independent variable. The slope coefficient may be positive or negative, depending on the relationship between the two variables.

**Relative Frequency** The proportion of total observations that are in a given category. Relative frequency is computed by dividing the frequency in a category by the total number of observations. The relative frequencies can be converted to percentages by multiplying by 100.

**Relative Frequency Assessment** The method that defines probability as the number of times an event occurs divided by the total number of times an experiment is performed in a large number of trials.

**Research Hypothesis** The hypothesis the decision maker attempts to demonstrate to be true. Because this is the hypothesis deemed to be the most important to the decision maker, it will be declared true only if the sample data strongly indicate that it is true.

**Residual** The difference between the actual value of the dependent variable and the value predicted by the regression model.

**Right-Skewed Data** A data distribution is right-skewed if the mean for the data is larger than the median.

**Sample** A subset of the population.

**Sample Mean** The average for all values in the sample computed by dividing the sum of all sample values by the sample size.

**Sample Proportion** The fraction of items in a sample that have the attribute of interest.

**Sample Space** The collection of all outcomes that can result from a selection, decision, or experiment.

**Sampling Distribution** The distribution of all possible values of a statistic for a given sample size that has been randomly selected from a population.

**Sampling Error** The difference between a measure computed from a sample (a statistic) and the corresponding measure computed from the population (a parameter).

**Scatter Diagram, or Scatter Plot** A two-dimensional graph of plotted points in which the vertical axis represents values of one quantitative variable and the horizontal axis represents values of the other quantitative variable. Each plotted point has coordinates whose values are obtained from the respective variables.

**Seasonal Component** A wavelike pattern that is repeated throughout a time series and has a recurrence period of at most one year.

**Seasonal Index** A number used to quantify the effect of seasonality in time-series data.

**Seasonally Unadjusted Forecast** A forecast made for seasonal data that does not include an adjustment for the seasonal component in the time series.

**Significance Level** The maximum allowable probability of committing a Type I statistical error. The probability is denoted by the symbol $\alpha$.

**Simple Linear Regression** The method of regression analysis in which a single independent variable is used to predict the dependent variable.

**Simple Random Sample** A sample selected in such a manner that each possible sample of a given size has an equal chance of being selected.

**Simple Random Sampling** A method of selecting items from a population such that every possible sample of a specified size has an equal chance of being selected.

**Skewed Data** Data sets that are not symmetric. For skewed data, the mean is larger or smaller than the median.

**Standard Deviation** The positive square root of the variance.

**Standard Error** A value that measures the spread of the sample means around the population mean. The standard error is reduced when the sample size is increased.

**Standard Normal Distribution** A normal distribution that has a mean = 0.0 and a standard deviation = 1.0. The horizontal axis is scaled in *z*-values that measure the number of standard deviations a point is from the mean. Values above the mean have positive *z*-values. Values below the mean have negative *z*-values.

**Standardized Data Values** The number of standard deviations a value is from the mean. Standardized data values are sometimes referred to as *z*-scores.

**Statistic** A measure computed from a sample that has been selected from a population. The value of the statistic will depend on which sample is selected.

**Statistical Inference Procedures** Procedures that allow a decision maker to reach a conclusion about a set of data based on a subset of that data.

**Statistical Sampling Techniques** Those sampling methods that use selection techniques based on chance selection.

**Stratified Random Sampling** A statistical sampling method in which the population is divided into subgroups called *strata* so that each population item belongs to only one stratum. The objective is to form strata such that the population values of interest within each stratum are as much alike as possible. Sample items are selected from each stratum using the simple random sampling method.

**Structured Interview** An interview in which the questions are scripted.

**Student's** *t*-**Distributions** A family of distributions that is bell-shaped and symmetric like the standard normal distribution but with greater area in the tails. Each distribution in the *t*-family is defined by its degrees of freedom. As the degrees of freedom increase, the $t$-distribution approaches the normal distribution.

**Subjective Probability Assessment** The method that defines the probability of an event as reflecting a decision maker's state of mind regarding the chances that the particular event will occur.

**Symmetric Data** Data sets whose values are evenly spread around the center. For symmetric data, the mean and median are equal.

**Systematic Random Sampling** A statistical sampling technique that involves selecting every $k$th item in the population after a randomly selected starting point between 1 and $k$. The value of $k$ is determined as the ratio of the population size over the desired sample size.

**Tchebysheff's Theorem** Regardless of how data are distributed, *at least* $(1 - 1/k^2)$ of the values will fall within $k$ standard deviations of the mean. For example:

At least $\left(1 - \dfrac{1}{1^2}\right) = 0 = 0\%$ of the values will fall within

$k = 1$ standard deviation of the mean.

At least $\left(1 - \dfrac{1}{2^2}\right) = \dfrac{3}{4} = 75\%$ of the values will lie within $k = 2$ standard deviations of the mean.

At least $\left(1 - \dfrac{1}{3^2}\right) = \dfrac{8}{9} = 89\%$ of the values will lie within $k = 3$ standard deviations of the mean.

**Test Statistic** A function of the sampled observations that provides a basis for testing a statistical hypothesis.

**Time-Series Data** A set of consecutive data values observed at successive points in time.

**Total Quality Management** A journey to excellence in which everyone in the organization is focused on continuous process improvement directed toward increased customer satisfaction.

**Total Variation** The aggregate dispersion of the individual data values across the various factor levels.

**Two-Tailed Test** A hypothesis test in which the entire rejection region is split into the two tails of the sampling distribution. In a two-tailed test, the alpha level is split evenly between the two tails.

**Type I Error** *Rejecting* the null hypothesis when it is, in fact, true.

**Type II Error** *Failing to reject* the null hypothesis when it is, in fact, false.

**Unbiased Estimator** A characteristic of certain statistics in which the average of all possible values of the sample statistic equals a parameter, no matter the value of the parameter.

**Unstructured Interview** An interview that begins with one or more broadly stated questions, with further questions being based on the responses.

**Variance** The average of the squared distances of the data values from the mean.

**Variance Inflation Factor** A measure of how much the variance of an estimated regression coefficient increases if the independent variables are correlated. A *VIF* equal to 1.0 for a given independent variable indicates that this independent variable is not correlated with the remaining independent variables in the model. The greater the multicollinearity, the larger the *VIF*.

**Variation** A set of data exhibits variation if all the data are not the same value.

**Weighted Mean** The mean value of data values that have been weighted according to their relative importance.

**Within-Sample Variation** The dispersion that exists among the data values within a particular factor level.

# Index

# Credits

## Photographs

**FRONTMATTER**

p. vii, David F. Groebner; p. vii, Patrick W. Shannon; p. vii, Phillip C. Fry

**CHAPTER 1**

p. 1, Adike/Shutterstock; p. 14, Besjunior/Fotolia; p. 16, Fstockfoto/Shutterstock; p. 21, Joe Gough/Shutterstock

**CHAPTER 2**

p. 28, Monty Rakusen/Cultura/Getty Images; p. 33, Elena Elisseeva/Shutterstock; p. 40, ArenaCreative/Fotolia;
p. 42, ArenaCreative/Fotolia

**CHAPTER 3**

p. 73, HappyAlex/Fotolia; p. 76, Pavel Losevsky/Fotolia; p. 82, Beth Van Trees/Shutterstock; p. 86, Kurhan/Fotolia; p. 87, Jayzynism/Fotolia; p. 99, Svitlana Kataieva/Shutterstock; p. 108, JJAVA/Fotolia

**CHAPTER 4**

p. 128, Shots Studio/Shutterstock; p. 130, ExQuisine/Fotolia; p. 134, Elnur/Shutterstock; p. 143, Wavebreakmedia/Shutterstock; p. 157, Severija/Fotolia

**CHAPTER 5**

p. 172, PL.TH/Fotolia; p. 175, Oleksiy Mark/Fotolia; p. 181, Antonio Diaz/Shutterstock; p. 184, Szasz Fabian Ilka Erika/Shutterstock; p. 185, David Lee/Shutterstock; p. 186, Fernando Blanco Calzada/Shutterstock; p. 188, Fernando Blanco Calzada/Shutterstock; p. 188, Fernando Blanco Calzada/Shutterstock; p. 197, Discovod/Fotolia

**CHAPTER 6**

p. 212, Amble Design/Shutterstock; p. 215, Michael jung/Fotolia; p. 216, Claudio Zacc/Fotolia; p. 217, Atoss/Fotolia; p. 220, Pixel Embargo/Fotolia; p. 228, Robert Kneschke/Fotolia

**CHAPTER 7**

p. 239, BlueOrange Studio/Fotolia; p. 242, Nuxusseven/Fotolia; p. 249, Gunnar Pippel/Shutterstock; p. 254, Wuttichok/Fotolia; p. 264, Ra2 studio/Fotolia

**CHAPTER 8**

p. 277, Photodisc/Getty Images; p. 278, Avatar_023/Shutterstock; p. 282, Avatar_023/Shutterstock; p. 283, Avatar_023/Shutterstock; p. 286, Avatar_023/Shutterstock; p. 296, Marima/Fotolia; p. 298, Sandor Jackal/Fotolia; p. 303, Elzeva/Fotolia; p. 304, Elzeva/Fotolia

**CHAPTER 9**

p. 316, Denis Pepin/Shutterstock; p. 321, Peter Galbraith/Fotolia; p. 323, Peter Galbraith/Fotolia; p. 326, Minerva Studio/Fotolia; p. 327, Joyfull/Shutterstock; p. 329, Tund/Shutterstock; p. 340, Mike Flippo/Shutterstock

**CHAPTER 10**

p. 363, Anya Berkut/Fotolia; p. 366, Nick Barounis/Shutterstock; p. 368, Stocklight/Shutterstock; p. 370, Aastock/Shutterstock; p. 378, Tatiana Popova/Shutterstock; p. 379, TokyoRucx/Fotolia; p. 389, Sculpies/Shutterstock

**CHAPTER 11**

p. 410, Boggy/Fotolia; p. 412, WavebreakmediaMicro/Fotolia; p. 413, WavebreakmediaMicro/Fotolia; p. 415, Jens Ickler/123RF; p. 421, Brian Enright/Fotolia; p. 423, Touch Pix of EuToch/Shutterstock; p. 425, WavebreakmediaMicro/Fotolia

**CHAPTER 12**

p. 434, EpicStockMedia/Fotolia; p. 435, Volff/Fotolia; p. 439, Volff/Fotolia; p. 443, Tatiana Belova/Fotolia

**CHAPTER 13**

p. 497, Dcdebs/iStock/Getty Images; p. 498, Gwimages/Fotolia; p. 501, M.studio/Fotolia; p. 505, Arek_Malang/Shutterstock; p. 514, Jim Arbogast/Getty Images; p. 515, Nejron Photo/Shutterstock

**CHAPTER 14**

p. 526, Zjk/Fotolia; p. 528, Chad McDermott/Shutterstock; p. 531, Mdd/Shutterstock; p. 537, Chad McDermott/Shutterstock; p. 543, Nan/Fotolia; p. 545, Chad McDermott/Shutterstock; p. 547, Chad McDermott/Shutterstock; p. 549, Denis Dryashkin/Shutterstock; p. 556, Amble Design/Shutterstock

**CHAPTER 15**

p. 573, Hongqi Zhang/123RF; p. 577, Sai Yeung Chan/Shutterstock; p. 591, Pavel L Photo and Video/Shutterstock; p. 592, Sai Yeung Chan/Shutterstock; p. 614, Wavebreakmedia/Shutterstock; p. 620, Sai Yeung Chan/Shutterstock

**CHAPTER 16**

p. 636, Andres Rodriguez/Fotolia; p. 638, Tiler84/Fotolia; p. 644, PSD photography/Shutterstock; p. 650, Bill/Fotolia; p. 672, Deno/Fotolia; p. 677, Deno/Fotolia

**CHAPTER 17**

p. 687, Sergey Nivens/Fotolia; p. 688, Jorge Salcedo/Shutterstock; p. 690, Ilin Sergey/Shutterstock; p. 693, Philip Lange/Shutterstock; p. 697, John Kwan/Fotolia; p. 707, Harris Shiffman/Shutterstock

**CHAPTER 18**

p. 718, Rawpixel.com/Shutterstock; p. 721, Kzenon/Shutterstock

## Figures

**CHAPTERS 1–18**

Screenshots of Microsoft Excel, Microsoft® Excel, Microsoft Corporation. Reprinted with permission.

## CHAPTER 1

p. 16, Fig. 1-7, Microsoft® Excel, Microsoft Corporation. Reprinted with permission.

## CHAPTER 2

p. 40, Fig. 2-4, Microsoft® Excel, Microsoft Corporation. Reprinted with permission; p. 63, Fig. 2-16, Microsoft® Excel, Microsoft Corporation. Reprinted with permission.

## CHAPTER 5

p. 187, Fig. 5-3, Microsoft® Excel, Microsoft Corporation. Reprinted with permission; p. 197, Fig. 5-7, Microsoft® Excel, Microsoft Corporation. Reprinted with permission.

## CHAPTER 9

p. 334, Fig. 9-7, Microsoft® Excel, Microsoft Corporation. Reprinted with permission.

## CHAPTER 10

p. 380, Fig. 10-4, Microsoft® Excel, Microsoft Corporation. Reprinted with permission; p. 380, Fig. 10-5, Microsoft® Excel, Microsoft Corporation. Reprinted with permission; p. 381, Fig. 10-6, Microsoft® Excel, Microsoft Corporation. Reprinted with permission.

## CHAPTER 16

p. 646, Fig. 16-6, Microsoft® Excel, Microsoft Corporation. Reprinted with permission; p. 647, Fig. 16-8, Microsoft® Excel, Microsoft Corporation. Reprinted with permission; p. 648, Fig. 16-9, Microsoft® Excel, Microsoft Corporation. Reprinted with permission; p. 649, Fig. 16-10, Microsoft® Excel, Microsoft Corporation. Reprinted with permission; p. 651, Fig. 16-13, Microsoft® Excel, Microsoft Corporation. Reprinted with permission; p. 654, Fig. 16-14, Microsoft® Excel, Microsoft Corporation. Reprinted with permission; p. 655, Fig. 16-15, Microsoft® Excel, Microsoft Corporation. Reprinted with permission; p. 656, Fig. 16-16, Microsoft® Excel, Microsoft Corporation. Reprinted with permission; p. 658, Fig. 16-18, Microsoft® Excel, Microsoft Corporation. Reprinted with permission; p. 659, Fig. 16-19, Microsoft® Excel, Microsoft Corporation. Reprinted with permission; p. 660, Fig. 16-20, Microsoft® Excel, Microsoft Corporation. Reprinted with permission; p. 661, Fig. 16-21, Microsoft® Excel, Microsoft Corporation. Reprinted with permission; p. 661, Fig. 16-22, Microsoft® Excel, Microsoft Corporation. Reprinted with permission; p. 662, Fig. 16-23, Microsoft® Excel, Microsoft Corporation. Reprinted with permission; p. 663, Fig. 16-24, Microsoft® Excel, Microsoft Corporation. Reprinted with permission; p. 668, Fig. 16-25, Microsoft® Excel, Microsoft Corporation. Reprinted with permission; p. 669, Fig. 16-26, Microsoft® Excel, Microsoft Corporation. Reprinted with permission; p. 671, Fig. 16-27, Microsoft® Excel, Microsoft Corporation. Reprinted with permission; p. 672, Fig. 16-28, Microsoft® Excel, Microsoft Corporation. Reprinted with permission; p. 673, Fig. 16-29, Microsoft® Excel, Microsoft Corporation. Reprinted with permission; p. 677, Fig. 16-35, Microsoft® Excel, Microsoft Corporation. Reprinted with permission.

# Values of *t* for Selected Probabilities

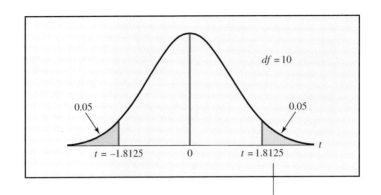

$df = 10$

0.05     0.05

$t = -1.8125$    0    $t = 1.8125$

## Probabilites (Or Areas Under *t*-Distribution Curve)

| Conf. Level | 0.1 | 0.3 | 0.5 | 0.7 | 0.8 | 0.9 | 0.95 | 0.98 | 0.99 |
|---|---|---|---|---|---|---|---|---|---|
| One Tail | 0.45 | 0.35 | 0.25 | 0.15 | 0.1 | 0.05 | 0.025 | 0.01 | 0.005 |
| Two Tails | 0.9 | 0.7 | 0.5 | 0.3 | 0.2 | 0.1 | 0.05 | 0.02 | 0.01 |
| d. f. | | | | | | Values of t | | | |
| 1 | 0.1584 | 0.5095 | 1.0000 | 1.9626 | 3.0777 | 6.3138 | 12.7062 | 31.8205 | 63.6567 |
| 2 | 0.1421 | 0.4447 | 0.8165 | 1.3862 | 1.8856 | 2.9200 | 4.3027 | 6.9646 | 9.9248 |
| 3 | 0.1366 | 0.4242 | 0.7649 | 1.2498 | 1.6377 | 2.3534 | 3.1824 | 4.5407 | 5.8409 |
| 4 | 0.1338 | 0.4142 | 0.7407 | 1.1896 | 1.5332 | 2.1318 | 2.7764 | 3.7469 | 4.6041 |
| 5 | 0.1322 | 0.4082 | 0.7267 | 1.1558 | 1.4759 | 2.0150 | 2.5706 | 3.3649 | 4.0321 |
| 6 | 0.1311 | 0.4043 | 0.7176 | 1.1342 | 1.4398 | 1.9432 | 2.4469 | 3.1427 | 3.7074 |
| 7 | 0.1303 | 0.4015 | 0.7111 | 1.1192 | 1.4149 | 1.8946 | 2.3646 | 2.9980 | 3.4995 |
| 8 | 0.1297 | 0.3995 | 0.7064 | 1.1081 | 1.3968 | 1.8595 | 2.3060 | 2.8965 | 3.3554 |
| 9 | 0.1293 | 0.3979 | 0.7027 | 1.0997 | 1.3830 | 1.8331 | 2.2622 | 2.8214 | 3.2498 |
| 10 | 0.1289 | 0.3966 | 0.6998 | 1.0931 | 1.3722 | 1.8125 | 2.2281 | 2.7638 | 3.1693 |
| 11 | 0.1286 | 0.3956 | 0.6974 | 1.0877 | 1.3634 | 1.7959 | 2.2010 | 2.7181 | 3.1058 |
| 12 | 0.1283 | 0.3947 | 0.6955 | 1.0832 | 1.3562 | 1.7823 | 2.1788 | 2.6810 | 3.0545 |
| 13 | 0.1281 | 0.3940 | 0.6938 | 1.0795 | 1.3502 | 1.7709 | 2.1604 | 2.6503 | 3.0123 |
| 14 | 0.1280 | 0.3933 | 0.6924 | 1.0763 | 1.3450 | 1.7613 | 2.1448 | 2.6245 | 2.9768 |
| 15 | 0.1278 | 0.3928 | 0.6912 | 1.0735 | 1.3406 | 1.7531 | 2.1314 | 2.6025 | 2.9467 |
| 16 | 0.1277 | 0.3923 | 0.6901 | 1.0711 | 1.3368 | 1.7459 | 2.1199 | 2.5835 | 2.9208 |
| 17 | 0.1276 | 0.3919 | 0.6892 | 1.0690 | 1.3334 | 1.7396 | 2.1098 | 2.5669 | 2.8982 |
| 18 | 0.1274 | 0.3915 | 0.6884 | 1.0672 | 1.3304 | 1.7341 | 2.1009 | 2.5524 | 2.8784 |
| 19 | 0.1274 | 0.3912 | 0.6876 | 1.0655 | 1.3277 | 1.7291 | 2.0930 | 2.5395 | 2.8609 |
| 20 | 0.1273 | 0.3909 | 0.6870 | 1.0640 | 1.3253 | 1.7247 | 2.0860 | 2.5280 | 2.8453 |
| 21 | 0.1272 | 0.3906 | 0.6864 | 1.0627 | 1.3232 | 1.7207 | 2.0796 | 2.5176 | 2.8314 |
| 22 | 0.1271 | 0.3904 | 0.6858 | 1.0614 | 1.3212 | 1.7171 | 2.0739 | 2.5083 | 2.8188 |
| 23 | 0.1271 | 0.3902 | 0.6853 | 1.0603 | 1.3195 | 1.7139 | 2.0687 | 2.4999 | 2.8073 |
| 24 | 0.1270 | 0.3900 | 0.6848 | 1.0593 | 1.3178 | 1.7109 | 2.0639 | 2.4922 | 2.7969 |
| 25 | 0.1269 | 0.3898 | 0.6844 | 1.0584 | 1.3163 | 1.7081 | 2.0595 | 2.4851 | 2.7874 |
| 26 | 0.1269 | 0.3896 | 0.6840 | 1.0575 | 1.3150 | 1.7056 | 2.0555 | 2.4786 | 2.7787 |
| 27 | 0.1268 | 0.3894 | 0.6837 | 1.0567 | 1.3137 | 1.7033 | 2.0518 | 2.4727 | 2.7707 |
| 28 | 0.1268 | 0.3893 | 0.6834 | 1.0560 | 1.3125 | 1.7011 | 2.0484 | 2.4671 | 2.7633 |
| 29 | 0.1268 | 0.3892 | 0.6830 | 1.0553 | 1.3114 | 1.6991 | 2.0452 | 2.4620 | 2.7564 |
| 30 | 0.1267 | 0.3890 | 0.6828 | 1.0547 | 1.3104 | 1.6973 | 2.0423 | 2.4573 | 2.7500 |
| 40 | 0.1265 | 0.3881 | 0.6807 | 1.0500 | 1.3031 | 1.6839 | 2.0211 | 2.4233 | 2.7045 |
| 50 | 0.1263 | 0.3875 | 0.6794 | 1.0473 | 1.2987 | 1.6759 | 2.0086 | 2.4033 | 2.6778 |
| 60 | 0.1262 | 0.3872 | 0.6786 | 1.0455 | 1.2958 | 1.6706 | 2.0003 | 2.3901 | 2.6603 |
| 70 | 0.1261 | 0.3869 | 0.6780 | 1.0442 | 1.2938 | 1.6669 | 1.9944 | 2.3808 | 2.6479 |
| 80 | 0.1261 | 0.3867 | 0.6776 | 1.0432 | 1.2922 | 1.6641 | 1.9901 | 2.3739 | 2.6387 |
| 90 | 0.1260 | 0.3866 | 0.6772 | 1.0424 | 1.2910 | 1.6620 | 1.9867 | 2.3685 | 2.6316 |
| 100 | 0.1260 | 0.3864 | 0.6770 | 1.0418 | 1.2901 | 1.6602 | 1.9840 | 2.3642 | 2.6259 |
| 250 | 0.1258 | 0.3858 | 0.6755 | 1.0386 | 1.2849 | 1.6510 | 1.9695 | 2.3414 | 2.5956 |
| 500 | 0.1257 | 0.3855 | 0.6750 | 1.0375 | 1.2832 | 1.6479 | 1.9647 | 2.3338 | 2.5857 |
| ∞ | 0.1257 | 0.3853 | 0.6745 | 1.0364 | 1.2816 | 1.6449 | 1.9600 | 2.3263 | 2.5758 |